T0269420

Book of Abstracts of the 70th Annual Meeting of the European Federation of Animal Science

EAAP
European Federation of Animal Science

The European Federation of Animal Science wishes to express its appreciation to the
Ministero delle Politiche Agricole Alimentari e Forestali (Italy) and the
Associazione Italiana Allevatori (Italy)
for their valuable support of its activities.

Book of Abstracts of the 70th Annual Meeting of the European Federation of Animal Science

Ghent, Belgium, 26th-30th August, 2019

EAAP Scientific Committee:

E. Strandberg
G. Savoini
H.A.M. Spoolder
H. Sauerwein
M. Lee
J.F. Hocquette
J. Conington
E.F. Knol
A.S. Santos
T. Veldkamp
I. Halachmi
G. Pollott

OASES
Online Academic Submission and Evaluation System

EAN: 9789086863396
e-EAN: 9789086868902
ISBN: 978-90-8686-339-6
e-ISBN: 978-90-8686-890-2
DOI: 10.3920/978-90-8686-890-2

ISSN 1382-6077

First published, 2019

© **Wageningen Academic Publishers**
The Netherlands, 2019

Wageningen Academic
P u b l i s h e r s

This work is subject to copyright. All rights are reserved, whether the whole or part of the material is concerned. Nothing from this publication may be translated, reproduced, stored in a computerised system or published in any form or in any manner, including electronic, mechanical, reprographic or photographic, without prior written permission from the publisher:
Wageningen Academic Publishers
P.O. Box 220
6700 AE Wageningen
The Netherlands
www.WageningenAcademic.com
copyright@WageningenAcademic.com

The individual contributions in this publication and any liabilities arising from them remain the responsibility of the authors.

The designations employed and the presentation of material in this publication do not imply the expression of any opinion whatsoever on the part of the European Federation of Animal Science concerning the legal status of any country, territory, city or area or of its authorities, or concerning the delimitation of its frontiers or boundaries.

The publisher is not responsible for possible damages, which could be a result of content derived from this publication.

Welcome

On behalf of the Belgian Organizing Committee, we are delighted to welcome you at the 70th Annual Meeting of EAAP in the historic and lively city of Ghent from the 26th to the 30th of August 2019. It is since 1977 that an EAAP meeting was organized in Belgium (Brussels). We are proud to say that the very first EAAP meeting was held in Ghent in 1950. Today, for the special 70th edition EAAP is back in Belgium.

The main theme of the 2019 meeting is '**Animal farming for a healthy world**' and the conference will deal with topics such as resource efficiency, animal welfare, diversification, agro-ecology, climate change and product quality. The EAAP commissions have designed a very attractive program with 74 scientific sessions and 2 poster sessions covering various disciplines such as animal genetics, animal health and welfare, animal nutrition, animal physiology, livestock farming systems, cattle, horse, pig, sheep and goat production, precision livestock farming and insects.

Enjoy this EAAP platform for scientists, policy makers and industry experts to meet and acquire new knowledge and to exchange experiences about the latest research results from many areas of animal science. This annual meeting is a unique opportunity for the application of new ideas in practice through many parallel sessions, a plenary meeting, poster presentations, and discussions about scientific achievements in livestock production. All these activities make the EAAP meeting one of the largest animal science meetings in the world. This year we welcome more than 1,250 participants from more than 65 countries worldwide.

Besides this scientific part of the Conference, we have prepared an attractive and interesting program of social events, which will start with the Welcome Ceremony and will include a Belgian Evening, a guided tour in Ghent, and a Conference Dinner, as well as technical tours and an accompanying persons program so everyone can enjoy a taste of the beauty and delicacies Belgium has to offer.

We hope that the 70th Annual Meeting of the EAAP in Ghent will be a great opportunity to meet and socialize with colleagues and friends. We wish you a pleasant stay in our city!

Sam De Campeneere
Chairman of the Belgian Organizing Committee

National Organisers of the 70th EAAP Annual Meeting

The 70th EAAP annual meeting is organized by ILVO (Flanders Research Institute for Agriculture, Fisheries and Food) under the patronage of the Department of Agriculture and Fisheries (DAF) from the Flemish Government. The organization of this event is also in close collaboration with other Belgian institutes, including Ghent University (UGent), the University of Leuven (KUL), the University of Antwerp (UA), the University of Liège (ULG), the University of Louvain (UCL) and the Walloon Research Centre for Agriculture (CRA-W).

Belgian Steering Committee

President:	*Sam De Campeneere* (ILVO)
Executive secretaries:	*Bart Sonck* (ILVO), *Sam Millet* (ILVO), *Johan De Boever* (ILVO)
Members:	*Hanne Geenen* (DAF), *Ivan Ryckaert* (DAF),
	Jürgen Vangeyte (ILVO), *Stefaan De Smet* (UGent),
	Nadia Everaert (ULG)

Belgian Scientific Committee

President:	*Johan De Boever* (ILVO)
Vice-president:	*Sam De Campeneere* (ILVO)
Executive secretaries:	*Sam Millet* (ILVO), *Bart Sonck* (ILVO)
Animal Genetics:	*Nadine Buys* (KUL), *Stefaan De Smet* (UGent),
	Nicolas Gengler (ULG)
Animal Health and Welfare:	*Ben Aernouts* (KUL), *Dominiek Maes* (UGent),
	Frank Tuyttens (ILVO)
Animal Nutrition:	*Yves Beckers* (ULG), *Eric Froidmont* (CRA-W), *Veerle Fievez* (UGent),
	Yvan Larondelle (UCL)
Animal Physiology:	*Nadia Everaert* (ULG), *Jo Leroy* (UA), *Geert Janssens* (UGent)
Livestock Farming Systems:	*Jérôme Bindelle* (ULG), *Didier Stilmant* (CRA-W),
	Daan Delbare (ILVO)
Cattle Production:	*Frédéric Dehareng* (CRA-W), *Geert Opsomer* (UGent),
	Leen Vandaele (ILVO)
Horse Production:	*Steven Janssens* (KUL), *Charlotte Sandersen* (ULG),
	Myriam Hesta (UGent)
Pig Production:	*Joris Michiels* (UGent), *Marijke Aluwé* (ILVO),
	Jeroen Dewulf (UGent)
Sheep and Goat:	*Pierre Rondia* (CRA-W), *Bert Driessen* (KUL)
Insects:	*Mik Van Der Borgt* (KUL), *Veerle Van linden* (ILVO)
Precision Livestock Farming:	*Jurgen Vangeyte* (ILVO), *Tomas Norton* (KUL),
	Hélène Soyeurt (ULG)

European Federation of Animal Science (EAAP)

President:	M. Gauly
Secretary General:	A. Rosati
Address:	Via G. Tomassetti 3, A/I
	I-00161 Rome, Italy
Phone/Fax:	+39 06 4420 2639
E-mail:	eaap@eaap.org
Web:	www.eaap.org

Council Members

President	Matthias Gauly (Italy)
Secretary General	Andrea Rosati
Vice-Presidents	Jeanne Bormann (Luxembourg)
	Johannes Sölkner (Austria)
Members	John Carty (Ireland)
	Isabel Casasus Peyo (Spain)
	Horia Grosu (Romania)
	Georgia Hadjipavlou (Cyprus)
	Mogens Vestergaards (Denmark)
	Lotta Rydhmer (Sweden)
	Bruno Ronchi (Italy)
	Maurer Veronika (Switzerland)
Auditor	Zdravko Barac (Croatia)
	Gerry Greally (Ireland)
Alternate Auditor	Andreas Hofer (Switzerland)
FAO Representative	Badi Besbes

European Federation of Animal Science has close established links with the sister organisations American Dairy Science Association, American Society of Animal Science, Canadian Society of Animal Science and Asociación Latinoamericana de Producción Animal.

Friends of EAAP

By creating the 'Friends of EAAP', EAAP offers the opportunity to industries to receive services from EAAP in change of a fixed sponsoring amount of support every year.
- The group of supporting industries are layered in three categories: 'silver', 'gold' and 'diamond' level.
- It is offered an important discount (one year free of charge) if the sponsoring industry will agree for a four years period.
- EAAP will offer the service to create a scientific network (with Research Institutes and Scientists) around Europe.
- Creation of a permanent Board of Industries within EAAP with the objective to inform, influence the scientific and organizational actions of EAAP, like proposing choices of the scientific sessions and invited speakers and to propose industry representatives for the Study Commissions.
- Organization of targeted workshops, proposed by industries.
- EAAP can represent and facilitate activities of the supporting industries toward international legislative and regulatory organizations.
- EAAP can facilitate the supporting industries to enter in consortia dealing with internationally supported research projects.

Furthermore EAAP offers, depending to the level of support (details on our website: www.eaap.org):
- Free entrances to the EAAP annual meeting and Gala dinner invitation.
- Free registration to journal *animal*.
- Inclusion of industry advertisement in the EAAP Newsletter, in the banner of the EAAP website, in the Book of Abstract and in the Programme Booklet of the EAAP annual meeting.
- Inclusion of industry leaflets in the annual meeting package.
- Presence of industry advertisements on the slides between presentations at selected standard sessions.
- Presence of industry logos and advertisements on the slides between presentations at the Plenary Sessions.
- Public Recognition by the EAAP President at the Plenary Opening Session of the annual meeting.
- Discounted stands at the EAAP annual meeting.
- Invitation to meetings (at every annual meeting) to discuss joint strategy EAAP/Industries with the EAAP President, Vice-President for Scientific affair, Secretary General and other selected members of the Council and of the Scientific Committee.

Contact and further information

If the industry you represent is interested to become 'Friend of EAAP' or want to have further information please contact jean-marc.perez0000@orange.fr or EAAP secretariat (eaap@eaap.org, phone : +39 06 44202639).

The Association

EAAP (The European Federation of Animal Science) organises every year an international meeting which attracts between 900 and 1,500 people. The main aims of EAAP are to promote, by means of active co-operation between its members and other relevant international and national organisations, the advancement of scientific research, sustainable development and systems of production; experimentation, application and extension; to improve the technical and economic conditions of the livestock sector; to promote the welfare of farm animals and the conservation of the rural environment; to control and optimise the use of natural resources in general and animal genetic resources in particular; to encourage the involvement of young scientists and technicians. More information on the organisation and its activities can be found at www.eaap.org

What is the YoungEAAP?

YoungEAAP is a group of young scientists organized under the EAAP umbrella. It aims to create a platform where scientists during their early career get the opportunity to meet and share their experiences, expectations and aspirations. This is done through activities at the Annual EAAP Meetings and social media. The large constituency and diversity of the EAAP member countries, commissions and delegates create a very important platform to stay up-to-date, close the gap between our training and the future employer expectations, while fine-tuning our skills and providing young scientists applied and industry-relevant research ideas.

Committee Members at a glace

- Christian Lambertz (President)
- Anamarija Smetko (Secretary)
- Torun Wallgren (Secretary)

YoungEAAP promotes Young and Early Career Scientists to:
- Stay up-to-date (i.e. EAAP activities, social media);
- Close the gap between our training and the future employer expectations;
- Fine-tune our skills through EAAP meetings, expand the special young scientists' sessions, and/ or start online webinars/trainings with industry and academic leaders;
- Meet to network and share our graduate school or early employment experiences;
- Develop research ideas, projects and proposals.

Who can be a Member of YoungEAAP?

All individual members of EAAP can join the YoungEAAP if they meet one of the following criteria: Researchers under 35 years of age OR within 10 years after PhD-graduation

Just request your membership form (christian.lambertz@fibl.org) and become member of this network!!!

71st Annual Meeting of the European Federation of Animal Science

Porto, Portugal,
August 31st to September 4th 2020

Organizing Committees

The 71st EAAP annual meeting is organized by the Portuguese Association of Animal Science (APEZ) under the patronage of the Ministry of agriculture, rural development and fisheries of Portugal.

The organization of this event is also in close collaboration with other Portuguese institutions, including National Institute of agrarian research (INIAV), University of Trás-os-Montes and Alto Douro (UTAD), University of Lisbon (ISA and FMV), University of Évora (UE), University of Porto (UP-ICBAS), University of Minho (UM), and the National Association of Politechnych institutes (ANP).

Portuguese Steering Committee

President:	Ana Sofia Santos (APEZ-UTAD)
Executive Secretaries:	Telma Pinto (APEZ), Elisabete Gomes Mena (UTAD-APEZ), Mariana Almeida (APEZ-UTAD), Pedro Vaz (APEZ)
Members:	André Almeida (APEZ-ISA), Ana Geraldo (APEZ), Divanildo Monteiro (UTAD-APEZ), Luis Ferreira (APEZ-UTAD)

Conference website: www.eaap2020.org

Commission on Animal Genetics

Erling Strandberg	President	Swedish University of Agricultural Sciences
	Sweden	erling.strandberg@slu.se
Eileen Wall	Vice-President	Reader in Integrative Animal Sciences
	United Kingdom	eileen.wall@sruc.ac.uk
Filippo Miglior	Vice-President	Ontario Genomics
	Canada	fmiglior@ontariogenomics.ca
Han Mulder	Vice-President	Wageningen University
	The Netherlands	han.mulder@wur.nl
Alessio Cecchinato	Secretary	Padova University
	Italy	alessio.cecchinato@unipd.it
Tessa Brinker	Industry rep.	EFFAB
	The Netherlands	tessa.brinker@effab.info
Marcin Pszczoła	Young Club	Poznan University of Life Sciences
	Poland	marcin.pszczola@gmail.com

Commission on Animal Nutrition

Giovanni Savoini	President	University of Milan
	Italy	giovanni.savoini@unimi.it
Eleni Tsiplakou	Vice-President	Agricultural University of Athens
	Greece	eltsiplakou@aua.gr
Geert Bruggeman	Secretary	Nuscience Group
	Belgium	geert.bruggeman@nusciencegroup.com
Roselinde Goselink	Secretary	Wageningen Livestock Research
	The Netherlands	roselinde.goselink@wur.nl
Daniele Bonvicini	Industry Rep	Prosol S.p.a
	Italy	d.bonvicini@prosol-spa.it
Susanne Kreuzer-Redmer	Young Club	Vetmed Vienna
	Austria	susanne.kreuzer-redmer@vetmeduni.ac.at

Commission on Health and Welfare

Hans Spoolder	President	ASG-WUR
	The Netherlands	hans.spoolder@wur.nl
Gurbuz Das	Vice-President	Leibniz Institute for Farm Animal Biology
	Germany	gdas@fbn-dummerstorf.de
Evangelia Sossidou	Vice-President	Hellenic Agricultural Organization
	Greece	sossidou.arig@nagref.gr
Stefano Messori	Secretary	SIRCAH
	France	stefano.messori@yahoo.it
Mariana Dantas	Young Club	University of Tras-os-Montes and Alto Douro
	Portugal	mdantas@utad.pt

Commission on Animal Physiology

Helga Sauerwein	President	University of Bonn
	Germany	sauerwein@uni-bonn.de
Isabelle Louveau	Vice president	INRA
	France	isabelle.louveau@inra.fr
Chris Knight	Secretary	University of Copenhagen
	Denmark	chkn@sund.ku.dk
Arnulf Troescher	Industry rep.	BASF
	Germany	arnulf.troescher@basf.com
Akos Kenez	Young Club	City University of Hong Kong"
	Hong Kong	akos.kenez@cityu.edu.hk

Commission on Livestock Farming Systems

Michael Lee	Acting-President	University of Bristol
	United Kingdom	michael.lee@rothamsted.ac.uk
Monika Zehetmeier	Acting Vice-President	Institute Agricultural Economics and Farm Management
	Germany	monika.zehetmeier@lfl.bayern.de
Raimon Ripoll Bosch	Secretary	Wageningen University
	The Netherlands	raimon.ripollbosch@wur.nl
Thomas Turini	Industry rep.	Terre-ECOS
	France	t.turini@terre-ecos.com

Commission on Cattle Production

Jean François Hocquette	President	INRA
	France	jean-francois.hocquette@inra.fr
Sven König	Vice-President	Justus-Liebig Universität Giessen
	Germany	sven.koenig@agrar.uni-giessen.de
Andreas Foskolos	Vice-President	Aberystwyth University
	United Kingdom	anf20@aber.ac.uk
Birgit Fürst-Waltl	Secretary	University of Natural Resources and Life Sciences
	Austria	birgit.fuerst-waltl@boku.ac.at
Massimo De Marchi	Secretary	Padova University
	Italy	massimo.demarchi@unipd.it
Ray Keatinge	Industry rep.	Agriculture & Horticulture Development Board
	United Kingdom	ray.keatinge@ahdb.org.uk
Akke Kok	Young Club	Wageningen Livestock Research
	The Netherlands	akke.kok@wur.nl

Commission on Sheep and Goat Production

Joanne Conington	President	SAC
	United Kingdom	joanne.conington@sac.ac.uk
Nóirín McHugh	Vice President	Teagasc
	Ireland	noirin.mchugh@teagasc.ie
Ouranios Tzamaloukas	Secretary	Cyprus University of Technology
	Cyprus	ouranios.tzamaloukas@cut.ac.cy
Vasco Augusto Piläo Cadavez	Secretary	CIMO - Mountain Research Centre
	Portugal	vcadavez@ipb.pt
John Yates	Industry rep.	Texel
	United Kingdom	johnyates@texel.co.uk

Commission on Pig Production

Egbert Knol	President	TOPIGS
	The Netherlands	egbert.knol@topigs.com
Sam Millet	Vice President	ILVO
	Belgium	sam.millet@ilvo.vlaanderen.be
Katja Nilsson	Secretary	Swedish University of Agricultural Science
	Sweden	katja.nilsson@slu.se
Paolo Trevisi	Secretary	Bologna University
	Italy	paolo.trevisi@unibo.it
Charlotte Grimberg	Young Club	University of Kiel
	Germany	cgrimberg@tierzucht.uni-kiel.de

Commission on Horse Production

Ana Sofia Santos	President	CECAV - UTAD - EUVG
	Portugal	assantos@utad.pt
Rhys Evans	Vice president	Norwegian University College of
	Norway	Agriculture and Rural Development
		rhys@hlb.no
Katharina Stock	Vice president	VIT-Vereinigte Informationssysteme
	Germany	Tierhaltung w.V.
		friederike.katharina.stock@vit.de
Klemen Potočnik	Vice president	University of Ljubljana
	Slovenija	klemen.potocnik@bf.uni-lj.si
Isabel Cervantes Navarro	Secretary	Complutense University of Madrid
	Spain	icervantes@vet.ucm.es
Melissa Cox	Industry rep.	CAG GmbH – Center for Animal Genetics
	Germany	melissa.cox@centerforanimalgenetics.de

Commission on Insects

Teun Veldkamp	President	Wageningen Livestock Research
	The Netherlands	teun.veldkamp@wur.nl
Michelle Epstein	Vice president	Medical University of Vienna
	Austria	michelle.epstein@meduniwien.ac.at
Jørgen Eilenberg	Secretary	University of Copenhagen
	Denmark	jei@plen.ku.dk
Alessandro Agazzi	Secretary	University of Milan
	Italy	alessandro.agazzi@unimi.it
Marian Peters	Industry rep.	IIC (International Insect Centre)
	The Netherlands	marianpeters@ngn.co.nl
Alexis Angot	Industry rep.	Ynsect
	France	aan@ynsect.com
Roel Boersma	Industry rep.	Protix
	The Netherlands	roel.boersma@protix.eu

Commission on Precision Livestock Farming (PLF)

Ilan Halachmi	President	Agriculture research organization (ARO)
	Israel	halachmi@volcani.agri.gov.il
Jarissa Maselyne	Vice president	ILVO
	Belgium	jarissa.maselyne@ilvo.vlaanderen.be
Matti Pastell	Vice president	Natural Resources Institute Finland (Luke)
	Finland	matti.pastell@luke.fi
Shelly Druyan	Secretary	ARO, The Volcani Center
	Israel	shelly.druyan@mail.huji.ac.il
Radovan Kasarda	Secretary	Slovak University of Agriculture in Nitra
	Slovakia	radovan.kasarda@uniag.sk
Malcolm Mitchell	Industry rep.	SRUC
	United Kingdom	malcolm.mitchell@sruc.ac.uk
Ines Adriaens	Young Club	KU Leuven
	Belgium	ines.adriaens@kuleuven.be

Sponsors

Nuscience & Agrifirm, Better Together

With more than 3.000 dedicated employees driven to excel every day, Royal Agrifirm Group contributes to a responsible food chain for future generations. Founded over 120 years ago, we are now a leading agricultural cooperative with an international network of subsidiaries in 16 countries within Europe, South America and Asia and a worldwide distribution network.

 With our multidisciplinary & Multi cultural team

 | | | | |
--- | --- | --- | --- | --- | ---
Nutrition | Physiology | Digitalization | Immunology | (Bio)technology | Chemistry

 With our global facilities

 In Vitro

> Cell & Tissue Models > Simulation Models
> Chromatographic Analysis > Microbiology
> Audio Visual Processing > Biomarker Studies

 In Vivo

> +15 Poultry Trial Centers
> +20 Pigs Trial Centers
> +35 Cattle Trial Centers

 With our global network

+35 partners
in Europe

+20 partners
Worldwide

 Want to be part of this?

> Discover our vacancies at:
WWW.NUSCIENCE.BE or send your curriculum vitae at JOBS@NUSCIENCEGROUP.COM.

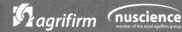

WE **CREATE**, **SELECT** AND **DELIVER** FEED ADDITIVES

Orffa develops, customizes, sources and offers a complete range of feed additives and concepts for the animal nutrition markets.

The challenges faced by the feed industry are ever-growing and become more and more complex. This requiresdedicated specialists who are able to engineer specific, applicable feed solutions. It is our mission to help the feed industry in finding such solutions.

Check our complete story and vacancies on our website www.orffa.com

Engineering your feed solutions

www.orffa.com

 ORFFA

Log 3 reduction in
Enterobacteriaceae?

Come visit us at EAAP, August 26–30!

EASTMAN

AN-9245   6/19

an
INSECT
TECHNOLOGY
GROUP
company

HAVE YOU EVER THOUGHT ABOUT INSECTS?
Circular Organics™ creates sustainable insect products
for agriculture and industry.

Cutting-edge insect technology
developed at our European R&D-centre.

Insect protein, lipids and chitin produced
from EU-approved agricultural side streams.

Natural fertilisers and soil improvement
products based on insect frass.

Kempen
Insect
Cluster

Kempen Insect Cluster is proudly supported by

AGENTSCHAP
INNOVEREN &
ONDERNEMEN

EFRO
EUROPEES FONDS
VOOR REGIONALE
ONTWIKKELING

www.circularorganics.com
info@circularorganics.com

There is no end to what we can achieve together.

www.avevebiochem.com

AVEVE
Biochem
Why animal nutrition is about people.
Expert in feed additives & specialties - Member of Arvesta

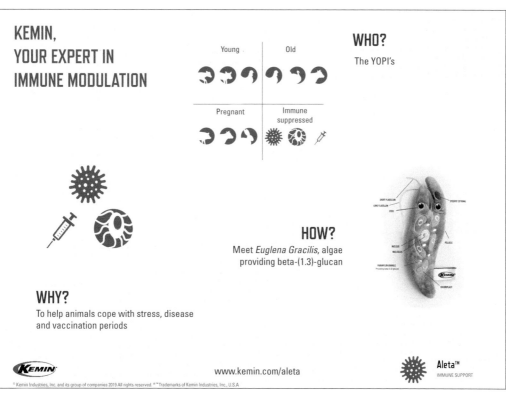

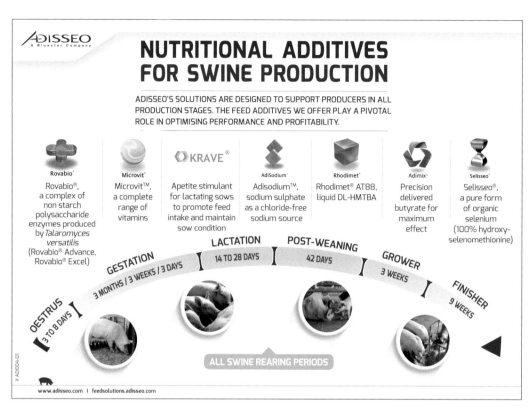

Accelerating animal genomic breakthroughs together.

A history of progress. A future of promise.

We are dedicated to advancing the future of agrigenomics. Our microarray and next-generation sequencing (NGS) technologies are helping researchers and breeders develop healthier and more productive livestock to address the mounting challenges of feeding the global population.

Visit our booth to learn more.

www.illumina.com/agrigenomics

© 2019 Illumina, Inc. All rights reserved.

illumına®

Thank you

to the 70th EAAP Annual Congress Sponsors and Friends

Platinum sponsor

Silver sponsors

EASTMAN

Bronze sponsors

ICEROBOTIC

Other sponsors

App sponsor

Session sponsors

session 6 and 28   session 27   session 37 and 57   training activities

Exhibitors

Acknowledgements

Scientific programme EAAP 2019

Monday 26 August 8.30 – 12.30	Monday 26 August 14.00 – 17.00	Tuesday 27 August 8.30 – 12.30
ANIMAL GENETIC RESOURCES SYMPOSIUM		
Session 01 What to conserve? Chair: Berg / Fernández	**Session 11** Raising awareness on the importance of animal genetic resources Chair: Leroy	
Session 02 Novelties in genomics research and their impact on genetic selection Chair: Pszczola	**Session 12** Novelties in genomics research and their impact on genetic selection Chair: Mulder	
Session 03 PLF possibilities for sheep, goats, poultry and horses Chair: Adriaens / Conington	**Session 13** Towards a climate smart European livestock farming Chair: Peyraud	
Session 04 Innovative dairy research and extension (Young Train) Chair: Lambertz	**Session 14** Differentiation of consumers oriented milk & meat products Chair: Penasa	
Session 05 Metabolomics and further OMICs techniques applied to livestock physiology Chair: Kenez	**Session 15** Innovations in sheep and goat breeding Chair: Tzamaloukas / McHugh	**Session 22** **Plenary session** and **Leroy Lecture** Chair: Gauly
Session 06 Insects in a circular economy Chair: Eilenberg / Veldkamp	**Session 16** Producing insects on different feeding substrates Chair: Eilenberg / Veldkamp	
Session 07 Dietary functional components: effects on animal performance, health and environment Chair: Tsiplakou / Fradinho	**Session 17** Alternative feed ingredients: former food, by-products, and new materials Chair: Pinotti	
WAGENINGEN ACADEMIC PUBLISHERS EARLY CAREER COMPETITION		
Session 08 Innovative approaches to pig production and pig research Chair: Trevisi	**Session 18** Chair: Wallgren	
Session 09 Challenges of livestock farming systems in relation to society Chair: De Olde	**Session 19** Heat stress and other environmental factors affecting performance: a physiology perspective Chair: Louveau	
Session 10 Tail biting and feather pecking Chair: Boyle / Sossidou	**Session 20** Parasites Chair: Das / Sotiraki	
	Session 21 Horse performance and welfare - Ethics Chair: Cox	
11.30 – 12.30 **Commission meeting** • Physiology	17.00 – 18.00 **Commission meeting** • Health and Welfare **Poster session 1** (session 1–32 and 44)	

Tuesday 27 August 14.00 – 18.00	Wednesday 28 August 8.30 – 12.30	Wednesday 28 August 14.00 – 17.00
Session 23 Crossbreeding Chair: Wall **Session 24** Epigenetics Chair: Toro **Session 25** PLF for animal health and welfare Chair: Mitchell / Ingrand **Session 26** Microbiome-host interactions and gut health Chair: Ghaffari / Beaumont **Session 27** Animal farming for a healthy world Chair: Amon / Metges / Zentek **Session 28** Standardisation of research methods and parameters Chair: Agazzi / Veldkamp **Session 29** Algae as animal feed Chair: Goselink **Session 30** Neonatal survival in pigs Chair: Nilsson **Session 31** How to address tradeoffs and synergies in livestock farming systems? Chair: Zehetmeier **Session 32** New sheep & goat projects day symposium Chair: Cadavez	**Session 33** Burning issues in biodiversity 1: what are the benefits from animal gene banks? Chair: Mäki-Tanila **Session 34** Gene editing: can we afford (not) to use precision technologies in livestock breeding? Chair: Granados Chapatte **Session 35** IOT session, first results from the project www.iof2020.eu (2017-2020), Chair: Maselyne / Halachmi / Pastell **Session 36** Beef farming and products towards the future Chair: Hocquette	**Session 43** Burning issues in biodiversity 2: fitter livestock farms from better gene banks Chair: Tixier-Boichard **Session 44** Free communications in animal genetics Chair: Berghof **Session 45** Alarm management, individual feed efficiency, data quality and data ownership, decision support systems Chair: Pastell / Kasarda / Maselyne
	STRATEGIES REDUCING ANTIMICROBIAL NEED (PART I)	**SUSTAINABLE PIG PRODUCTION SYSTEMS (PART I)**
	Session 37 Chair: Messori	**Session 46** Chair: Rauw
	Session 38 Insects in animal feed: beyond the protein concept Chair: Veldkamp / Eilenberg **Session 39** Societal concerns, with focus on environment and antimicrobial resistance, that can be addressed through animal nutrition: state of science and specialty feed ingredients Chair: Fink-Gremmels **Session 40** Various topics in pig production Chair: Millet **Session 41** Resilient livestock farming systems in the context of climate and market uncertainties Chair: Lee **Session 42** Equine production and equine products/use Chair: De Palo	**Session 47** Match making for research, industry, policy makers Chair: Peters / Veldkamp **Session 48** Implementing innovative solutions in animal nutrition: tools and success stories from the field to tackle environmental issues and to reduce the need of antibiotic use in animal farming Chair: Bruggeman **Session 49** Health in poultry and free communications Chair: Stadnicka **Session 50** YoungEAAP: should I stay or should I go? – pros and cons of alternative career options Chair: Lambertz **Session 51** Progress in making aquaculture more sustainable Chair: Sorgeloos **Session 52** Horses as agents of European culture, from the past to the future / The co-evolution of humans and horses Chair: Santos
	11.30 – 12.30 **Commission meeting** • Insect	17.00 – 18.00 **Poster session 2** (session 33–73 minus 44)

Thursday 29 August 8.30 – 11.30	Thursday 29 August 14.00 – 18.00
Session 53 Fibre from camelids and sheep Chair: Allain	**Session 64** Camelids as emerging food producing species in our changing climate Chair: Nagy
Session 54 Free communications Inbreeding; GxE Chair: Karapandža	**Session 65** Free communications Genomic prediction and GWAS Chair: Fernandez Martin
Session 55 Sensing cutting edge technologies in milk and livestock Chair: Mitchell / Druyan / Adriaens	**Session 66** Impact of new precision phenotyping technologies on animal breeding (in collaboration with GplusE) Chair: Cecchinato
Session 56 Metabolic diseases in dairy cows: strategies for their reduction Chair: Keatinge / Van Knegsel	**Session 67** Impact of environmental challenges and consumer demands on cattle response traits and farming systems Chair: König
STRATEGIES REDUCING ANTIMICROBIAL NEED (PART II)	**SUSTAINABLE PIG PRODUCTION SYSTEMS (PART II)**
Session 57 Chair: Kemp	**Session 68** Chair: Stefanski
	**INVALUABLE: LESSONS AND RESULTS FROM THE DANISH INSECT VALUE CHAIN**
Session 58	**Session 69**
	Chair: Heckmann
Session 59 Free Communications animal nutrition: dairy and beef cattle Chair: Savoini	**Session 70** Feed efficiency and enteric methane emission of cattle Chair: De Haas / Gransworthy
Session 60 Sow + gilt nutrition and management Chair: Bee	**Session 71** Limits in production growth - on level of cow and farm and industry (ADSA-EAAP seminar) Chair: Sauerwein / De Vries
Session 61 Livestock farming systems free communications Chair: Lee	**Session 72** Agroecological approaches in livestock farming systems Chair: Dumont
Session 62 Producing sheep & goats with reduced veterinary inputs Chair: Yates	**Session 73** Free communications animal nutrition Chair: Savoini
Session 63 Equine breeding systems Chair: Cervantes	

11.30 – 12.30

Commission meeting

- Cattle
- Genetics
- Horse

- LFS
- Nutrition
- Pig

- PLF
- Sheep and Goat

Scientific programme

Session 01. What to conserve?

Date: Monday 26 August 2019; 8.30 – 12.30
Chair: Berg / Fernández

Theatre Session 01

Benchmarking inbreeding coefficient estimators with a whole-genome sequenced Holstein pedigree   105
S.W. Alemu, N.K. Kadri, C. Charlier and T. Druet

Whole genome resequencing detects signature of selection in 23 European pig breeds and wild boars   105
S. Bovo, G. Schiavo, A. Ribani, M. Muñoz, V.J. Utzeri, M. Gallo, J. Riquet, G. Usai, R. Charneca,
J.P. Araujo, R. Quintanilla, V. Razmaite, M. Candek-Potokar, A. Fernandez, C. Ovilo, L. Fontanesi and
Treasure Consortium

Selection sweeps through variation in linkage disequilibrium in horses   106
N. Moravčíková, R. Kasarda, M. Halo and J. Candrák

Effect of population bottlenecks on patterns of deleterious variation in small captive populations   106
C. Bortoluzzi, M. Bosse, M.F.L. Derk, R.P.M.A. Crooijmans, M.A.M. Groenen and H.-J. Megens

Optimal management of gene and allelic diversity in subdivided populations   107
E. López-Cortegano, R. Pouso, A. Pérez-Figueroa, J. Fernández and A. Caballero

Assessment of the risk status of local breeds in Poland – preliminary results   107
G.M. Polak and J. Krupinski

Comparative genome analyses of two local cattle breeds reared in the Parmigiano Reggiano region   108
G. Schiavo, S. Bovo, A. Ribani, H. Kazemi, V.J. Utzeri, F. Bertolini, S. Dall'Olio and L. Fontanesi

Optimizing the creation of base populations for breeding programs using allelic information   108
D. Bersabé, B. Villanueva, M.A. Toro and J. Fernández

Genomic characterisation of a subsample of the Mangalica population in Germany   109
S. Addo, J. Schäler and D. Hinrichs

Genome-wide diversity and chromosomal inbreeding in German White-headed Mutton Sheep   109
S. Addo, D. Hinrichs and G. Thaller

Genetic characterization of a small closed island population of Norwegian coastal goat   110
P. Berg, L.F. Groeneveld, D.I. Våge and L. Grøva

Genetic diversity in eight chicken lines from the Norwegian live poultry gene bank   110
C. Brekke, L.F. Groeneveld, T.H.E. Meuwissen, N. Sæther, S. Weigend and P. Berg

Conservation genomic analyses of two Croatian autochthonous sheep breeds   111
M. Špehar, M. Ferenčaković, I. Curik, N. Karapandža, Z. Barać, T. Sinković and V. Cubric-Curik

Origin of native Danish cattle breeds   111
A.A. Schönherz, V.H. Nielsen and B. Guldbrandtsen

Poster Session 01

Session 02. Novelties in genomics research and their impact on genetic selection

Date: Monday 26 August 2019; 8.30 – 12.30
Chair: Pszczola

Theatre Session 02

Poster Session 02

Session 03. PLF possibilities for sheep, goats, poultry and horses

Date: Monday 26 August 2019; 8.30 – 12.30
Chair: Adriaens / Conington

Theatre Session 03

Poster Session 03

Session 04. Innovative dairy research and extension (Young Train)

Date: Monday 26 August 2019; 8.30 – 12.30
Chair: Lambertz

Theatre Session 04

Session 05. Metabolomics and further OMICs techniques applied to livestock physiology

Date: Monday 26 August 2019; 8.30 – 11.30
Chair: Kenez

Theatre Session 05

Poster Session 05

Session 06. Insects in a circular economy

Date: Monday 26 August 2019; 8.30 – 12.30
Chair: Eilenberg / Veldkamp

Theatre Session 06

Poster Session 06

Session 07. Dietary functional components: effects on animal performance, health and environment

Date: Monday 26 August 2019; 8.30 – 12.30
Chair: Tsiplakou / Fradinho

Theatre Session 07

Poster Session 07

Session 08. Innovative approaches to pig production and pig research (Wageningen Academic Publishers early career competition) - part 1

Date: Monday 26 August 2019; 8.30 – 12.30
Chair: Trevisi

Theatre Session 08

Session 09. Challenges of livestock farming systems in relation to society

Date: Monday 26 August 2019; 8.30 – 12.30
Chair: De Olde

Theatre Session 09

Poster Session 09

Net contribution of dairy production to food supply: impact of intensification and grassland share | 200
C. Bottheu-Noirfalise, D. Stilmant, E. Reding, B. Wyzen, J. Wester, M. Mathot, S. Hennart and
V. Decruyenaere

Session 10. Tail biting and feather pecking

Date: Monday 26 August 2019; 8.30 – 12.30
Chair: Boyle / Sossidou

Theatre Session 10

Poster Session 10

Session 11. Raising awareness on the importance of animal genetic resources

Date: Monday 26 August 2019; 14.00 – 17.00
Chair: Leroy

Theatre Session 11

Session 12. Novelties in genomics research and their impact on genetic selection

Date: Monday 26 August 2019; 14.00 – 17.00
Chair: Mulder

Theatre Session 12

Session 13. Towards a climate smart European livestock farming

Date: Monday 26 August 2019; 14.00 – 17.00
Chair: Peyraud

Theatre Session 13

Session 14. Differentiation of consumers oriented milk & meat products (e.g. A2A2 milk, pasture based milk & meat / hay milk / …)

Date: Monday 26 August 2019; 14.00 – 17.00
Chair: Penasa

Theatre Session 14

Poster Session 14

Session 15. Innovations in sheep and goat breeding

Date: Monday 26 August 2019; 14.00 – 17.00
Chair: Tzamaloukas / McHugh

Theatre Session 15

Poster Session 15

Session 16. Producing insects on different feeding substrates

Date: Monday 26 August 2019; 14.00 – 17.00
Chair: Eilenberg / Veldkamp

Theatre Session 16

Poster Session 16

Session 17. Alternative feed ingredients: former food, by-products, and new materials

Date: Monday 26 August 2019; 14.00 – 17.00
Chair: Pinotti

Theatre Session 17

Poster Session 17

Session 18. Innovative approaches to pig production and pig research (Wageningen Academic Publishers early career competition) - part 2

Date: Monday 26 August 2019; 14.00 – 17.00
Chair: Wallgren

Theatre Session 18

Session 19. Heat stress and other environmental factors affecting performance: a physiology perspective

Date: Monday 26 August 2019; 14.00 – 17.00
Chair: Louveau

Theatre Session 19

Poster Session 19

Session 20. Parasites

Date: Monday 26 August 2019; 14.00 – 17.00
Chair: Das / Sotiraki

Theatre Session 20

Session 21. Horse performance and welfare - Ethics

Date: Monday 26 August 2019; 14.00 – 17.00
Chair: Cox

Theatre Session 21

Poster Session 21

Session 22. Plenary Session and Leroy lecture

Date: Tuesday 27 August 2019; 8.30 – 12.30
Chair: Gauly

Theatre Session 22

Session 23. Crossbreeding

Date: Tuesday 27 August 2019; 14.00 – 18.00
Chair: Wall

Theatre Session 23

Poster Session 23

Session 24. Epigenetics

Date: Tuesday 27 August 2019; 14.00 – 18.00
Chair: Toro

Theatre Session 24

Poster Session 24

Session 25. PLF for animal health and welfare

Date: Tuesday 27 August 2019; 14.00 – 18.00
Chair: Mitchell / Ingrand

Theatre Session 25

Poster Session 25

Session 26. Microbiome-host interactions and gut health

Date: Tuesday 27 August 2019; 14.00 – 18.00
Chair: Ghaffari / Beaumont

Theatre Session 26

Session 27. Animal farming for a healthy world

Date: Tuesday 27 August 2019; 14.00 – 18.00
Chair: Amon / Metges / Zentek

Theatre Session 27

Session 28. Standardisation of research methods and parameters

Date: Tuesday 27 August 2019; 14.00 – 18.00
Chair: Agazzi / Veldkamp

Theatre Session 28

`invited` Resource conversion by black soldier fly larvae: towards standardisation of methods and reporting	324
G. Bosch, D.G.A.B. Oonincx, H.R. Jordan, J. Zhang, J.J.A. Van Loon, A. Van Huis and J.K. Tomberlin	
`invited` What synchronization efforts are worth the investment in mass insect rearing facilities	325
N. De Craene	
`invited` Development and evaluation of real-time PCR targets for the authentification of insect flours	325
F. Debode, A. Marien, D. Akhatova, C. Kohl, H. Sedefoglu, L. Mariscal-Diaz, A. Gérard, O. Fumière,	
J. Maljean, J. Hulin, F. Francis and G. Berben	
`invited` An optimised test for recording hygienic behaviour in the honeybee	326
E. Facchini, P. Bijma, G. Pagnacco, R. Rizzi and E.W. Brascamp	

Poster Session 28

Investigating genetic and phenotypic variability of honeybee queens' individual traits	326
E. Facchini, F. Turri, F. Pizzi, D. Laurino, M. Porporato, R. Rizzi and G. Pagnacco	
A cross-laboratory study on analytical variability of amino acid content in three insect species	327
D.G.A.B. Oonincx, G. Bosch, M. Van Der Borght, R. Smets, L. Gasco, A.J. Fascetti, Z. Yu, V. Johnson,	
J.K. Tomberlin and M.D. Finke	

Session 29. Algae as animal feed

Date: Tuesday 27 August 2019; 14.00 – 18.00
Chair: Goselink

Theatre Session 29

`invited` Current uses and challenges of microalgae in animal feeding	327
J.A.M. Prates	
Microalgae are suitable protein feeds for lactating dairy cows	328
M. Lamminen, A. Halmemies-Beauchet-Filleau, T. Kokkonen, S. Jaakkola and A. Vanhatalo	
The effects of *Schizochytrium* sp. lipid extract or fish oil on lamb meat fatty acids	328
A. Francisco, S.P. Alves, J. Santos-Silva and R.J.B. Bessa	
Dietary algal β-(1,3)-glucan and its role in modulating the immune system in pigs and poultry	329
N. Smeets, R. Spaepen and V. Van Hamme	
Algae as animal feed – poster pitches	329
R.M.A. Goselink	
`invited` Seaweed as a potential feed resource for farmed animals	330
M. Øverland, L.T. Mydland and A. Skrede	
Polysaccharides in seaweeds: necessary steps in order to use seaweeds in animal feed	330
W. Muizelaar, J. Krooneman and G. Van Duinkerken	

Session 30. Neonatal survival in pigs

Date: Tuesday 27 August 2019; 14.00 – 18.00
Chair: Nilsson

Theatre Session 30

Poster Session 30

Session 31. How to address tradeoffs and synergies in livestock farming systems?

Date: Tuesday 27 August 2019; 14.00 – 18.00
Chair: Zehetmeier

Theatre Session 31

Poster Session 31

Session 32. New sheep & goat projects day symposium

Date: Tuesday 27 August 2019; 14.00 – 18.00
Chair: Cadavez

Theatre Session 32

Poster Session 32

Session 33. Burning issues in biodiversity 1: what are the benefits from animal gene banks?

Date: Wednesday 28 August 2019; 8.30 – 12.30
Chair: Mäki-Tanila

Theatre Session 33

Poster Session 33

Session 35. IOT session, first results from the project www.iof2020.eu (2017-2020),

Date: Wednesday 28 August 2019; 8.30 – 12.30
Chair: Maselyne / Halachmi / Pastell

Theatre Session 35

Poster Session 35

Session 36. Beef farming and products towards the future

Date: Wednesday 28 August 2019; 8.30 – 12.30
Chair: Hocquette

Theatre Session 36

Poster Session 36

Session 37. Strategies reducing antimicrobial need

Date: Wednesday 28 August 2019; 8.30 – 12.30
Chair: Messori

Theatre Session 37

Poster Session 37

Session 38. Insects in animal feed: beyond the protein concept

Date: Wednesday 28 August 2019; 8.30 – 11.30
Chair: Veldkamp / Eilenberg

Theatre Session 38

Poster Session 38

Session 39. Societal concerns, with focus on environment and antimicrobial resistance, that can be addressed through animal nutrition: state of science and specialty feed ingredients

Date: Wednesday 28 August 2019; 8.30 – 12.30
Chair: Fink-Gremmels

Theatre Session 39

Poster Session 39

Session 40. Various topics in pig production

Date: Wednesday 28 August 2019; 8.30 – 12.30
Chair: Millet

Theatre Session 40

Poster Session 40

Session 41. Resilient livestock farming systems in the context of climate and market uncertainties

Date: Wednesday 28 August 2019; 8.30 – 12.30
Chair: Lee

Theatre Session 41

Poster Session 41

Session 42. Equine production and equine products/use

Date: Wednesday 28 August 2019; 8.30 – 12.30
Chair: De Palo

Theatre Session 42

Poster Session 42

Session 43. Burning issues in biodiversity 2: fitter livestock farms from better gene banks

Date: Wednesday 28 August 2019; 14.00 – 17.00
Chair: Tixier-Boichard

Theatre Session 43

Session 44. Free communications in animal genetics

Date: Wednesday 28 August 2019; 14.00 – 17.00
Chair: Berghof

Theatre Session 44

Poster Session 44

Session 45. Alarm management, individual feed efficiency, data quality and data ownership, decision support systems

Date: Wednesday 28 August 2019; 14.00 – 17.00
Chair: Pastell / Kasarda / Maselyne

Theatre Session 45

Session 46. Sustainable pig production systems

Date: Wednesday 28 August 2019; 14.00 – 17.00
Chair: Rauw

Theatre Session 46

Session 47. Match making for research, industry, policy makers

Date: Wednesday 28 August 2019; 14.00 – 17.00
Chair: Peters / Veldkamp

Theatre Session 47

Session 48. Implementing innovative solutions in animal nutrition: tools and success stories from the field to tackle environmental issues and to reduce the need of antibiotic use in animal farming

Date: Wednesday 28 August 2019; 14.00 – 17.00
Chair: Bruggeman

Theatre Session 48

Poster Session 48

Session 49. Health in poultry and free communications

Date: Wednesday 28 August 2019; 14.00 – 17.00
Chair: Stadnicka

Theatre Session 49

Poster Session 49

Session 50. YoungEAAP: should I stay or should I go? – pros and cons of alternative career options

Date: Wednesday 28 August 2019; 14.00 – 17.00
Chair: Lambertz

Theatre Session 50

Session 51. Progress in making aquaculture more sustainable

Date: Wednesday 28 August 2019; 14.00 – 17.00
Chair: Sorgeloos

Theatre Session 51

Poster Session 51

Session 52. Horses as agents of European culture, from the past to the future / The co-evolution of humans and horses

Date: Wednesday 28 August 2019; 14.00 – 17.00
Chair: Santos

Theatre Session 52

Session 53. Fibre from camelids and sheep

Date: Thursday 29 August 2019; 8.30 – 11.30
Chair: Allain

Theatre Session 53

Poster Session 53

Session 54. Free communications Inbreeding; GxE

Date: Thursday 29 August 2019; 8.30 – 11.30
Chair: Karapandža

Theatre Session 54

Poster Session 54

Session 55. Sensing cutting edge technologies in milk and livestock

Date: Thursday 29 August 2019; 8.30 – 11.30
Chair: Mitchell / Druyan / Adriaens

Theatre Session 55

Poster Session 55

Session 56. Metabolic diseases in dairy cows: strategies for their reduction

Date: Thursday 29 August 2019; 8.30 – 11.30
Chair: Keatinge / Van Knegsel

Theatre Session 56

Poster Session 56

Session 57. Strategies reducing antimicrobial need

Date: Thursday 29 August 2019; 8.30 – 12.30
Chair: Kemp

Theatre Session 57

Session 58. inValuable: lessons and results from the Danish insect value chain (part I)

Date: Thursday 29 August 2019; 8.30 – 12.30
Chair: Heckmann

Theatre Session 58

Session 59. Free Communications animal nutrition: dairy and beef cattle

Date: Thursday 29 August 2019; 8.30 – 11.30
Chair: Savoini

Theatre Session 59

Poster Session 59

Session 60. Sow + gilt nutrition and management

Date: Thursday 29 August 2019; 8.30 – 11.30
Chair: Bee

Theatre Session 60

Session 61. Livestock farming systems free communications

Date: Thursday 29 August 2019; 8.30 – 11.30
Chair: Lee

Theatre Session 61

Session 62. Producing sheep & goats with reduced veterinary inputs

Date: Thursday 29 August 2019; 8.30 – 11.30
Chair: Yates

Session 63. Equine breeding systems

Date: Thursday 29 August 2019; 8.30 – 11.30
Chair: Cervantes

Theatre Session 63

Poster Session 63

Session 64. Camelids as emerging food producing species in our changing climate

Date: Thursday 29 August 2019; 14.00 – 18.00
Chair: Nagy

Theatre Session 64

Poster Session 64

Session 65. Free communications Genomic prediction and GWAS

Date: Thursday 29 August 2019; 14.00 – 18.00
Chair: Fernandez Martin

Theatre Session 65

Poster Session 65

EMABG 2.0 – European master in animal breeding and genetics   600
G. Mészáros, J. Sölkner, T. Heams, I. Laissy, H. Simianer, E. Strandberg, A.M. Johansson, H. Komen,
D. Lont, G. Klementsdal and P. Berg

Prediction of genomic breeding values in Montana Tropical Composite cattle using single-step GBLUP   601
L. Grigoletto, L.F. Brito, H.R. Oliveira, J.P. Eler, F. Baldi and J.B. Ferraz

Session 66. Impact of new precision phenotyping technologies on animal breeding (in collaboration with GplusE)

Date: Thursday 29 August 2019; 14.00 – 18.00
Chair: Cecchinato

Theatre Session 66

invited The use of novel phenotyping technologies in animal breeding   601
H. Bovenhuis

Precision phenotyping using Fourier-transform infrared spectroscopy for expensive-to-measure traits   602
F. Tiezzi, H. Toledo-Alvarado, G. Bittante and A. Cecchinato

Towards a genomic evaluation of cheese-making traits including candidate SNP in Montbéliarde cows   602
M.P. Sanchez, T. Tribout, V. Wolf, M. El Jabri, N. Gaudillière, S. Fritz, C. Laithier, A. Delacroix-Buchet,
M. Brochard and D. Boichard

Feed efficiency: the next step in animal breeding   603
J.J. Bouwmeester-Vosman, P. Van Goor, E.C. Verduijn, C. Van Der Linde and G. De Jong

Genetic parameters for environmental traits in Australian dairy cattle   603
C.M. Richardson, T.T.T. Nguyen, M. Abdelsayed, B.G. Cocks, L. Marett, B. Wales and J.E. Pryce

Development of a NIRS method to assess the digestive ability in growing pigs   604
E. Labussière, P. Ganier, A. Condé, E. Janvier and J. Van Milgen

Keeping up with a healthy milk fatty acid profile require selection   604
M. Kargo, L. Hein, N.A. Poulsen and A.J. Buitenhuis

New phenotypes from milk MIR spectra: challenges to obtain reliable predictions   605
C. Grelet, P. Dardenne, H. Soyeurt, A. Vanlierde, J.A. Fernandez, N. Gengler and F. Dehareng

Zootechnical parameters added to the milk MIR spectra as predictive value to estimate CH_4 emissions   605
A. Vanlierde, F. Dehareng, N. Gengler, E. Froidmont, S. McParland, M. Kreuzer, M. Bell, P. Lund,
C. Martin, B. Kuhla and H. Soyeurt

Potential use of MIR spectra in the prediction of hoof disorders in Holstein Friesians   606
A. Mineur, E.C. Verduijn, H.M. Knijn, G. De Jong, H. Soyeurt, C. Grelet and N. Gengler

Validation of the prediction of body weight from dairy cow characteristics and milk MIR spectra   606
H. Soyeurt, E. Froidmont, I. Dufrasne, Z. Wang, N. Gengler, F. Dehareng, GplusE Consortium and
C. Grelet

Chromosome regions affecting MIR spectra indirect predictors of cheese-making properties in cattle   607
R.R. Mota, H. Hammami, S. Vanderick, S. Naderi, N. Gengler and GplusE Consortium

Using milk MIR spectra to identify candidate genes associated with climate-smart traits in cattle   607
H. Hammami, R.R. Mota, S. Vanderick, S. Naderi, C. Bastin, N. Gengler and GplusE Consortium

Poster Session 66

Session 67. Impact of environmental challenges and consumer demands on cattle response traits and farming systems

Date: Thursday 29 August 2019; 14.00 – 18.00
Chair: König

Theatre Session 67

Session 68. Sustainable pig production systems

Date: Thursday 29 August 2019; 14.00 – 18.00
Chair: Stefanski

Poster Session 68

Session 69. inValuable: lessons and results from the Danish insect value chain (part II)

Date: Thursday 29 August 2019; 14.00 – 18.00
Chair: Heckmann

Theatre Session 69

Session 70. Feed efficiency and enteric methane emission of cattle

Date: Thursday 29 August 2019; 14.00 – 18.00
Chair: De Haas / Gransworthy

Theatre Session 70

Poster Session 70

Session 71. Limits in production growth - on level of cow and farm and industry (ADSA-EAAP seminar)

Date: Thursday 29 August 2019; 14.00 – 18.00
Chair: Sauerwein / De Vries

Theatre Session 71

Poster Session 71

Session 72. Agroecological approaches in livestock farming systems

Date: Thursday 29 August 2019; 14.00 – 18.00
Chair: Dumont

Theatre Session 72

Poster Session 72

Grazing frequency using goats to control *Salsola ibérica* scrub in rangelands of northern Mexico 660
E. Ruiz-Martinez, O. Angel-García, V. Contreras-Villarreal, C.A. Meza Herrera, F.G. Véliz-Deras,
J.L. Morales-Cruz and L.R. Gaytán-Alemán

Agroecological evaluation of four production systems in very dry tropical forest in Colombia 661
A. Conde-Pulgarin, C.P. Alvarez-Ochoa, R. Frias-Navarro and D.E. Rodriguez-Robayo

Session 73. Free communications animal nutrition

Date: Thursday 29 August 2019; 14.00 – 18.00
Chair: Savoini

Theatre Session 73

Dietary zinc:copper ratios and net postprandial portal appearance of these minerals in pigs 661
D. Bueno Dalto, I. Audet and J.J. Matte

Nutritional evaluation of feeding lambs different supplementary dietary levels of *Morenga* dry leaves 662
A.Y.M.A.N. Hassan

A general method to relate feed intake and body mass across individuals and species 662
J.A.N. Filipe, M. Piles, W.M. Rauw and I. Kyriazakis

A meta-analysis assessing the bacterial inoculation of corn silage over 37-years period 663
C.H.S. Rabelo, A. Bernardi, C.J. Härter, A.W.L. Silva and R.A. Reis

invited The effect of hemp and Camelina cake feeding on the fatty acid composition of duck muscles 663
R. Juodka, V. Juskiene, R. Juska, R. Leikus, R. Nainiene and D. Stankeviciene

Defining a management goal that maximizes the intake rate in palisadegrass 664
G.F. Silva Neto, J. Bindelle, J. Rossetto, P.A. Nunes, Í.M. Monteiro, R. Becker, J.V. Savian, G.D. Farias,
R.T. Schons, E.A. Laca and P.C.F. Carvalho

Introducing *Lupinus* spp. seeds in Churra da Terra Quente lambs diets and its effect on growth 664
M. Almeida, A. Nunes, S. Garcia-Santos, C. Guedes, S. Silva and L. Ferreira

The effect of introduction of *Lupinus* spp. seeds in diets on lamb eating behaviour 665
M. Almeida, A. Nunes, S. Garcia-Santos, C. Guedes, L. Ferreira, G. Stilwell and S. Silva

Productivity and feeding behaviour of finishing pigs fed with a calming herbal extract 665
M. Fornós, E. Jiménez-Moreno, J. Gasa, V. Rodríguez-Estévez and D. Carrión

Reliability of genomic predictions for feed efficiency traits based on different pig lines 666
A. Aliakbari, E. Delpuech, Y. Labrune, J. Riquet and H. Gilbert

The effect of protein and salt level on performance and faecal consistency in weanling piglets 666
S. Millet, B. Ampe and M.D. Tokach

Semen and sperm quality parameters of Potchefstroom koekoek cockerels fed dietary Moringa oleifera l 667
A. Sebola

Session 01 — Theatre 1

Benchmarking inbreeding coefficient estimators with a whole-genome sequenced Holstein pedigree

S.W. Alemu, N.K. Kadri, C. Charlier and T. Druet
University of Liège, GIGA-R, Avenue de l'Hôpital 11 (B34), 4000 Liège, Belgium; tom.druet@uliege.be

Estimation of the inbreeding coefficient F is essential to study inbreeding depression (ID) and for the management of populations under conservation or included in breeding programs. Several methods using genetic markers to estimate the realised F have been proposed in the literature. These have often been compared either without knowledge of the true inbreeding coefficients or based on simulated datasets that inevitably rely on some assumptions (eventually including the arbitrary definition of a base population). Here we used whole-genome sequence data of a large Holstein cattle pedigree to compare different estimators of F under different scenarios. We compared approaches relying on the pedigree, on the genomic relationship matrix (GRM), on the correlation between the uniting gametes (UG), on the excess of homozygosity, on maximum likelihood estimators and on the identification of runs-of-homozygosity (ROH). ID results either from increased homozygosity at rare and young deleterious alleles or from reduced heterozygosity at frequent alleles presenting heterozygote advantage. Therefore, in our first scenario we computed the homozygous load (the number of homozygous genotypes) at low-frequency alleles. To enrich our dataset in young alleles, we selected derived and breed-specific alleles. In our second scenario, we estimated the heterozygosity at frequent alleles. We used the SNPs functional annotation to determine whether the properties of the estimators remain similar for different classes of variants. UG and GRM-based methods had the highest correlations (0.84 to 0.91) with rare alleles whereas ROH-based estimators presented low correlations (~0.23). When using only derived and breed-specific alleles, the differences were smaller (0.77 for UG and 0.52 for ROH-based). The UG had also stronger correlations (-0.96) with the heterozygosity at common alleles than ROH-based estimators (-0.60). Similar trends were observed for predicted of F (using genotypes of parents to predict F in their sequenced offspring). Finally, for locus-specific comparisons (e.g. using variants in a 1 Mb window) UG and ROH-based estimators presented similar correlations with the homozygous load or with the heterozygosity.

Session 01 — Theatre 2

Whole genome resequencing detects signature of selection in 23 European pig breeds and wild boars

S. Bovo[1], G. Schiavo[1], A. Ribani[1], M. Muñoz[2], V.J. Utzeri[1], M. Gallo[3], J. Riquet[4], G. Usai[5], R. Charneca[6], J.P. Araujo[7], R. Quintanilla[8], V. Razmaite[9], M. Candek-Potokar[10], A. Fernandez[2], C. Ovilo[2], L. Fontanesi[1] and Treasure Consortium[10]
[1]University of Bologna, Dept. of Agricultural and Food Sciences, Viale Fanin 46, 40127 Bologna, Italy, [2]INIA, Crta. de la Coruña, km. 7,5, 28040 Madrid, Spain, [3]ANAS, Via Nizza 53, 00198 Roma, Italy, [4]INRA, GenPhySE, 24 chemin de Borde-Rouge Auzeville Tolosane, 31326 Castanet Tolosan, France, [5]AGRIS SARDEGNA, Loc. Bonassai, 07100 Sassari, Italy, [6]Universidade de Évora, ICAAM, Apartado 94, 7000-803 Évora, Portugal, [7]IPCV, Praça General Barbosa, 4900 Viana do Castelo, Portugal, [8]IRTA, Animal Breeding and Genetics, Torre Marimon s/n, 08140 Caldes de Montbui (Barcelona), Spain, [9]Lithuanian University of Health Sciences, Animal Science Institute, R. Žebenkos 12, 82317 Baisogala, Lithuania, [10]KIS, Hacquetova ulica 17, 1000 Ljubljana, Slovenia; luca.fontanesi@unibo.it

Natural and artificial directional selection in cosmopolitan and autochthonous pig breeds and wild boars have shaped their genomes determining selective sweeps as final adaptation to different environmental conditions and production systems. In this study we analysed whole-genome re-sequencing data from 19 European local pig breeds, 3 cosmopolitan Italian breeds and Italian wild boars. A total of 23 DNA pools, each one obtained by using equimolar DNA from 30-35 individuals, were sequenced with an Illumina HiSeq machine, which produced more than 18.4 billion of paired-end reads with a sequencing depth of ~42X. More than 30 million of variants (18% not included in dbSNP yet) were annotated with VEP. Fixation index (Fst) and pooled heterozygosity (Hp) were used as statistics for sweep analyses interrogating 100 kb genome windows. On the whole, more than 400 outlier genome windows distributed along the 18 porcine autosomes and scaffolds were identified. Some of these genome regions harboured well known major genes affecting body shape/ size (e.g. *NR6A1*, *PLAG1*, *LCORL* and *CASP10*), coat colour (e.g. *KIT*, *MC1R*) and growth/fatness (e.g. *MC4R*). These results provide a first global variability analysis of European *Sus scrofa*. The authors thank ANAS for providing samples. This study has been partially funded by European Union's H2020 RIA program (grant agreement No 634476).

Selection sweeps through variation in linkage disequilibrium in horses
N. Moravčíková, R. Kasarda, M. Halo and J. Candrák
Slovak University of Agriculture in Nitra, Tr. A. Hlinku 2, 94976 Nitra, Slovak Republic; nina.moravcikova1@gmail.com

The objective of this study was to assess the impact of selection on the genome of Lipizzan and Norik of Muran breeds through analysis of selection footprints distribution and to describe the most important genomic regions reflecting the selective breeding for traits of interest. The genomic DNA in total of 71 animals (36 Lipizzan, 25 Norik of Muran) were genotyped by using GGP Equine70k chip in commercial lab. The quality control of genotyping data were prepared to exclude all of animals and SNP markers with more than 10% of missing data and loci with unknown chromosomal position or SNPs located on sex chromosomes. After SNP pruning the final database stored information for 62,730 markers. Two approaches were used to identify the genome-wide selection footprints; variation in linkage disequilibrium (LD) patterns between and within breeds and scan for the integrated haplotype homozygosity (iHS) score for each breed. The differences in genome-wide LD patterns between populations were quantified over sliding windows of 50 SNPs. The genomic regions under intensive selection pressure were defined by the top 0.01, 0.1, and 1 percentile of signals. Based on applied approaches, overall 25 regions located within 17 autosomes were detected across both breeds. The longest region located on chromosome 10 (25,185,746 to 30,023,465) contained 170 protein-coding genes, whereas the shortest one located on chromosome 3 (19,756,981 to 19,801,190) included two genes. In total 415 protein-coding genes were identified within detected selection sweeps. The results showed that the signals were strong mainly in genomic regions containing genes responsible for immunity related traits (e. g. toll like receptors), growth and muscle mass, body weight and reproduction.

Effect of population bottlenecks on patterns of deleterious variation in small captive populations
C. Bortoluzzi, M. Bosse, M.F.L. Derk, R.P.M.A. Crooijmans, M.A.M. Groenen and H.-J. Megens
Wageningen University and research, Animal Breeding and Genomics, Droevendaalsesteeg 1, 6708 PB Wageningen, the Netherlands; richard.crooijmans@wur.nl

Predictions about the consequences of a small population size on genetic variation are fundamental to population genetics. As small populations are more affected by genetic drift, purifying selection acting against deleterious alleles is predicted to be less efficient, therefore increasing the risk of inbreeding depression. However, the extent to which small populations are subjected to genetic drift considerably vary depending on the nature and time frame in which bottleneck occurs. Domesticated species are an excellent model to investigate the consequences of population bottlenecks on genetic and deleterious variation, because their history is dominated by known bottlenecks and intense artificial selection. Here, we use whole-genome sequencing data from 97 chickens representing 39 traditional fancy breeds to directly examine the impact of past and recent population bottlenecks on patterns of deleterious variation in small populations. Overall, we find that populations recently established from larger populations have a higher proportion of deleterious variants relative to populations that have been kept at small population sizes for a long period of time. We also observe that most deleterious variants in recently bottlenecked populations are found in long tracts of homozygous genotypes, suggesting that deleterious variants have increased in frequency because of inbreeding. Our results indicate that the timing and nature of population bottlenecks can substantially shape the deleterious variation landscape in small populations.

Optimal management of gene and allelic diversity in subdivided populations

E. López-Cortegano[1], R. Pouso[1], A. Pérez-Figueroa[2], J. Fernández[3] and A. Caballero[1]
[1]Universidad de Vigo, Departamento de Bioquímica, Genética e Inmunología, Facultad de Biología, 36310 Vigo, Spain, [2]Centro de Investigaciones Biomédicas (CINBIO), Universidad de Vigo, 36310 Vigo, Spain, [3]Instituto de Investigación y Tecnología Agraria y Alimentaria (INIA), Departamento de Mejora Genética, Crta. de la Coruña, km. 7.5, 28040 Madrid, Spain; e.lopez@uvigo.es

Genetic diversity is generally assessed by means of neutral molecular markers, and it is usually quantified by the expected heterozygosity under Hardy-Weinberg equilibrium, while allelic diversity is measured by the number of alleles per locus. These two measures of diversity are complementary because, whereas the former is directly related to genetic variance for quantitative traits and, therefore, to the short-term response to selection and adaptation, the latter is more sensitive to population bottlenecks and relates more to the long-term capacity of populations to adapt to changing environments. In the context of structured populations undergoing conservation programs, it is critical to decide the optimum management strategy in order to preserve as much of both diversity measures as possible. Here we first present a new release of the software Metapop for the analysis and management of diversity in subdivided populations, and illustrate its use with cattle data. This new update includes computation of allelic diversity measures, as well as a simulation mode to forecast the consequences of taking different management strategies over time. We examine through computer simulations the consequences of choosing a strategy based either on heterozygosity or allelic diversity in the context of different demographic histories for a structured population. Our results indicate that maximisation of allelic diversity can maintain large levels of both heterozygosity and allelic richness, and thus it should be the recommended strategy in conservation programs for structured populations.

Assessment of the risk status of local breeds in Poland – preliminary results

G.M. Polak and J. Krupinski
National Research Institute of Animal Production, Horse Breeding, Sarego, 2 Str., 31-047 Cracow, Poland; grazyna.polak@izoo.krakow.pl

The risk assessment is the basis for an early warning system against the loss of local and indigenous breeds of domestic animals. It is also one of the main factors determining the implementation of genetic resources conservation programmes. Currently in Poland, the assessment of status risk of local breeds has been consistent with the guidelines of the Common agriculture Policy and Rural Development Programs, setting thresholds for particular species. The aim of this work was to develop a new method, considering the specifics of Polish conditions. The presented preliminary studies take into account 4 breeds: two horses, one sheep and one cattle. Analysis was carried out in the Institute of Animal Production used the method developed, taking into account the FAO guidelines from 2013 and research of other countries. The model is based on: number of females, effective population size and 5 additional features: geographical concentration, existence of branded products, *ex situ* protection, origin testing, cooperation of breeders. The results show that out of the 4 breeds surveyed, 3 were at risk and 1 required constant monitoring.

Comparative genome analyses of two local cattle breeds reared in the Parmigiano Reggiano region
G. Schiavo[1], S. Bovo[1], A. Ribani[1], H. Kazemi[1], V.J. Utzeri[1], F. Bertolini[2], S. Dall'Olio[1] and L. Fontanesi[1]
[1]University of Bologna, Department of Agricultural and Food Sciences, Viale G. Fanin 46, 40127 Bologna, Italy, [2]Technical University of Denmark, National Institute of Aquatic Resources, Kemitorvet, Building 202, room 6118, 2800 Kgs. Lyngby, Denmark; luca.fontanesi@unibo.it

Reggiana and Modenese are autochthonous cattle breeds reared mainly in the provinces of Reggio Emilia and Modena (from which they took their names) located in Emilia Romagna region (North of Italy). This is the Parmigiano-Reggiano cheese production area. Reggiana and Modenese are characterized by a typical solid red and solid white coat colour, respectively. Since the mid of the 1990s, Reggiana conservation program is linked to a new brand of Parmigiano-Reggiano cheese made only of Reggiana milk that reverted the decreasing trend of its population. Conservation and management programs of small populations should be focused on maintaining genetic diversity by controlling inbreeding and, at the same time, on preserving genetic uniformity for breed-specific traits (i.e. coat colour) and increasing frequencies of favourable alleles (related to cheese-making properties for these breeds). In this study we genotyped 654 Reggiana and 109 Modenese cattle (about 25 and 10% of all animals of the two breeds, respectively) with the GeneSeek Genomic Profiler Bovine 150K (GGP-HD) and with several single nucleotide polymorphisms (SNPs) in coat colour, milk protein and disease associated genes. SNPs were used with their coordinates on the ARS-UCD1.2 bovine genome assembly. Quality filtering, statistics and runs of homozygosity (ROH) were computed with PLINK version 1.9. For each animal, the ROH derived genomic inbreeding coefficient (FROH) was calculated. Averaging results over all animals, Reggiana breed had a lower FROH than Modenese breed (0.064 ± 0.036 vs 0.084 ± 0.045). MC1R recessive allele was almost fixed in Reggiana breed. CSN3 allele B was more frequent in Modenese (0.70) than Reggiana (0.56). These results will be useful to analyse the effect of the past conservation programs and design new breeding strategies for a sustainable conservation of these two cattle genetic resources. We thank ANABORARE for the support. This work has been funded by the PSRN 'Dual Breeding' project.

Optimizing the creation of base populations for breeding programs using allelic information
D. Bersabé[1], B. Villanueva[1], M.A. Toro[2] and J. Fernández[1]
[1]INIA, Departamento de Mejora Genética Animal, Crta. de la Coruña Km. 7,5, 28040 Madrid, Spain, [2]Universidad Politécnica de Madrid, Departamento de Producción Agraria, Ciudad Universitaria, 28040 Madrid, Spain; bersabe.diego@inia.es

One of the critical elements to design a successful breeding program relies in the proper choice of the initial founders. Ideally, this base population should be set up in a way that captures most of the genetic variation contained in the available strains. In previous studies, it has been shown that the optimal number of candidates to be sampled from each strain that maximizes gene diversity in the base population can be computed as long as the number of molecular markers is high enough. However, an alternative optimization based on the maximization of the allelic diversity in the base population could also be considered, since it has been hypothesized that long-term response for selection might be more related to the number of alleles initially available. The aim of this study is to explore the consequences of using allelic information for optimizing the creation of base populations from where breeding programs start. Using computer simulations, we evaluate different scenarios that differ in the type of strains, the level of available information and the type of measure used to characterize the amount of genetic variation and perform the maximization (either expected heterozygosity or allelic diversity at SNP haplotypes of different sizes). Results show that strategies aimed at maximizing the expected heterozygosity are all equally effective, regardless of the size of the haplotype. Moreover, maximizing allelic diversity in terms of the number of effective alleles roughly produces the same results as those achieved for expected heterozygosity. However, maximizing the number of total alleles leads to reduced levels of expected heterozygosity, as well as to an allelic diversity that is not much greater than that obtained by maximizing the expected heterozygosity itself. This last effect is especially remarkable when the size of the haplotype is small. Further studies are required to assess the impact of these strategies on trait performance and genetic diversity levels when the base population undergoes a subsequent breeding program.

Genomic characterisation of a subsample of the Mangalica population in Germany
S. Addo[1,2], J. Schäler[2] and D. Hinrichs[2]
[1]*Kiel University, Institute of Animal Breeding and Husbandry, Olshausenstraße 40, 24098 Kiel, Germany, [2]University Kassel, Department of Animal Breeding, Nordbahnhofstr. 1a, 37213 Witzenhausen, Germany; dhinrichs@agrar.uni-kassel.de*

In Germany no official Herd book for the Mangalica pig breed exist and therefore there is more or less no pedigree information available. However, using genomic information generated with SNP chips could solve this problem. The aim of this study was to investigate the potential of SNP information in the German Mangalica population. In total 57 individuals were genotyped for 62,543 SNPs using a 70k SNP chip. After data processing, the number of SNPs reduced to 34,066 on 48 individuals. The analysed data set contains genomic information from 10 Blond Mangalica (BM), 20 Red Mangalica (RM), and 18 Swallow-Belly-Mangalica (SBM). The population structure of the data were analysed using multidimensional scaling (MDS) and a model based clustering algorithm (ADMIXTURE). The MDS showed two different clusters for the BM and the SBM but not for the RM. For the RM, only five individuals cluster together. The reason for this could be that most of the RM in the data set are crossbreed animals and not pure breed RM. The ADMIXTURE analysis showed that the use of three different clusters was optimal for the data of this study. Furthermore, the ADMIXTURE analysis confirmed the MDS results, especially with respect to RM, because again five individuals were not influenced by another cluster. The observed heterozygosity was 0.35, 0.30, and 0.33 for BM, RM, and SBM, respectively. Chromosomal inbreeding (F_ROH) was estimated using different approaches. On the one hand, F_ROH was estimated relative to the length of the chromosome and on the other hand, F_ROH was estimated relative to the length of the autosome. Finally, F_ROH calculations were done for data sets including only RM, BM, or SBM. The results show different pattern for chromosomal inbreeding in the BM, RM, and SBM. However, due to the small data sets, this could also be an artefact of the data. Overall, this study shows that SNP information has the potential for practical implementation in the Mangalica population in Germany. However to confirm the results of this study, further research is needed with larger data sets and additional breeds.

Genome-wide diversity and chromosomal inbreeding in German White-headed Mutton Sheep
S. Addo[1,2], D. Hinrichs[1] and G. Thaller[2]
[1]*University of Kassel, Department of Animal Breeding, Nordbahnhofstr. 1a, 37213 Witzenhausen, Germany, [2]Christian-Albrechts-University, Institute of Animal Breeding and Husbandry, Olshausenstr. 40, 24098 Kiel, Germany; saddo@tierzucht.uni-kiel.de*

The German White-headed Mutton sheep (GWM) is a monitoring population. Other breeds, e.g. Texel (TXL) and French Berrichone du Cher (BDC) were used to improve lamb quality in GWM. The aim of the study was to analyse genetic diversity in GWM and its historical relatedness to other breeds. Our data consisted of 40,753 quality filtered SNPs on 46 GWM and publicly available genotype data from 209 individuals belonging to 9 breeds including BDC, German Texel (GTX), Border Leicester (BRL), New Zealand Romney (ROM), Suffolk (SUF), East Friesian White (EFW), Mouton Charolais (CHL) and Estremadura and Andalusian Spanish Merino sheep (SME, SMA). Genomic inbreeding (F_{ROH}) was calculated based on runs of homozygosity (ROH) and compared with pedigree inbreeding (F_{PED}). For 31 GWM individuals with pedigree, F_{ROH} (0.040) was higher than F_{PED} (0.029). The correlation between F_{PED} and F_{ROH} was 0.82 and the observed heterozygosity ranged between 0.32 and 0.41. F_{ROH} was also calculated for each chromosome: by dividing the sum of ROH (SOR) by the length of chromosome or by dividing SOR by the length of autosome. For both approaches, F_{ROH} was highest in chromosomes 1, 2 and 3. The summation of chromosomal F_{ROH} within an individual reproduced the individual's total F_{ROH} when chromosomal inbreeding was calculated relative to the total length of autosomal genome. In a multidimensional scaling analysis, BDC clustered together with GWM. German Texel samples were placed closed to the joint cluster while samples of other breeds were further apart. This was confirmed by a model based admixture analysis, which showed varying levels of admixture of GWM with ancestry from BDC and GTX. Our analyses highlight the historical use of other breeds in the creation of the modern GWM population.

Genetic characterization of a small closed island population of Norwegian coastal goat

P. Berg[1], L.F. Groeneveld[2], D.I. Våge[1] and L. Grøva[3]
[1]NMBU, Arboretveien 6, 1432 Ås, Norway, [2]NordGen, Raveien 9, 1430 Ås, Norway, [3]NIBIO, Gunnars vei 6, 6630 Tingvoll,
Norway; linn.groeneveld@nordgen.org

The Norwegian Coastal goat is defined as a national and endangered breed in Norway. Currently, the Coastal goat population is subdivided, with the largest population fragment found on the mainland in the county of Selje on Norway's west coast. Two smaller population fragments are found on the islands of Skorpa and Sandsøy. The Skorpa population can be considered feral, since it has been unmanaged since the 1950s, when permanent residence on the island was given up. The census population size on Skorpa has been estimated to be around 100 individuals. The future of this population is challenged by the fact that it is unmanaged, but none the less classified as a livestock breed. A characterization project was initiated in 2014, to ensure that the future management of the Skorpa population is evidence-based. We present data on genetic diversity in the Skorpa coastal goat population and its relationship to the mainland coastal goat population, as well as the Norwegian milk goat population. Analyses were based on 50k SNPchip data (Illumina Beadchip) from 52 Norwegian milk goats and a total of 44 Coastal goat individuals, of which seven were from Skorpa and 37 from the mainland population fragment. After quality control for call rates, missing data, minor allele frequency and deviation from Hardy Weinberg, 45,772 autosomal SNPs were available for analysis. The three populations clearly separate in principal component analysis and clustering analyses based on genetic distances, with the Skorpa and mainland goat populations being significantly different from one another. The Skorpa island population is significantly more inbred than the mainland coastal goat population, as evident from both fewer segregating markers and longer runs of homozygosity. However, heterozygosity at segregating loci was as large in the Skorpa population as in the two other populations. Overall, genetic diversity in the island population was low and estimates of effective population size indicate that this population has an unsustainable effective size. Means of introducing genetic diversity into the population should be considered.

Genetic diversity in eight chicken lines from the Norwegian live poultry gene bank

C. Brekke[1], L.F. Groeneveld[2], T.H.E. Meuwissen[1], N. Sæther[3], S. Weigend[4] and P. Berg[1]
[1]NMBU, Arboretveien 6, 1432 Ås, Norway, [2]NordGen, Raveien 9, 1430 Ås, Norway, [3]NIBIO, Raveien 9, 1430 Ås, Norway,
[4]FLI, Höltystr. 10, 31535 Neustadt, Germany; cathrine.brekke@nmbu.no

Since 1995 there has been no active breeding of commercial egg layers in Norway. The last active Norbrid lines, as well as other endangered breeds are conserved in the Norwegian live poultry gene bank. Our aim was to evaluate genetic diversity within and between the Norwegian lines, as well as their genetic uniqueness in an international context. Included in the study are seven lines as well as the national Norwegian landrace, Jærhøns. The eight lines were genotyped with the Affymetrix® Axiom® Chicken Genotyping Array. Genomic relationships within and between the eight lines were estimated from 468,973 SNP markers. The observed heterozygosity ranged from 0.30 in Jærhøns to 0.46 in the Barred Plymouth Rock lines. The white egg layers were generally more inbred than the brown layers. Model-based ancestry analysis, as implemented in Admixture, confirmed that the eight lines are distinct populations. Comparative analyses were carried out with 72 international populations of different origins including two commercial lines. The last active Norwegian lines are more closely related to the current commercial lines than to the rest of the international set. Based on principal component analysis, the Norbrid brown egg layers are clearly separated from all others whereas the Norbrid white egglayers form a separate cluster together with two other Leghorn lines and the commercial white egglayer. To assess the conservation value in an international context, the Norwegian lines were ranked according to their relative contributions to the genetic diversity in the international set. Of the Norwegian lines, the brown egg layer Norbrid 7 adds most genetic diversity to the international set of 72 lines, whereas Jærhøns adds the least. The analysis showed that the Norwegian genebank lines are of conservation value both from a national and international perspective, as they all add genetic diversity to the global set. Moreover, the Norbrid lines that are closely related to the current commercial lines could be of interest for starting new breeding programs or as additional genetic variation for current commercial breeding populations.

Conservation genomic analyses of two Croatian autochthonous sheep breeds

M. Špehar[1], M. Ferenčaković[2], I. Curik[2], N. Karapandža[1], Z. Barać[1], T. Sinković[1] and V. Cubric-Curik[2]
[1]*Ministry of Agriculture, Vukovarska 78, 10000 Zagreb, Croatia, [2]University of Zagreb, Faculty of Agriculture, Svetošimunska cesta 25, 10000 Zagreb, Croatia; marija.spehar@mps.hr*

Istrian (IS) and Pag (PS) sheep breeds are Croatian autochthonous sheep breeds. We analysed genotypes of 175 animals (96 IS, and 75 PS) using Illumina OvineSNP50 BeadChip array and estimated genomic inbreeding levels (F_{ROH}) and linkage disequilibrium effective population size (Ne_{LD}). We also defined the genomic position of IS and PS concerning overall 671 animals of 22 sheep breeds (Italian, Spanish and some Merino). For this purpose, we used publicly available data from Dryad digital repository. The inbreeding level based on runs of homozygosity (F_{ROH}) was higher in IS compared to PS for F_{ROH}>2 Mb, 0.08 vs 0.04, F_{ROH}>8 Mb, 0.04 vs 0.03, and F_{ROH}>16 Mb, 0.02 vs 0.01 respectively. Software NeEstimator was used to estimate Ne from the parental generation of the sample based on the linkage disequilibrium (LD) method. Estimated Ne_{LD} for IS was 101 (95% CI=81-130), while higher Ne_{LD} of 222 (95% CI=161-343) was obtained for PS. Principal Components Analysis of the overall dataset showed separation of both, IS and PS from other breeds. However, PS was placed near to the cluster containing some Italian breeds (Massese, Appenninica, Sardinian White, Comisana, Leccese, and Laticauda) while IS was slightly remote. To our knowledge, this is the first genomic analysis of IS and PS providing valuable insights into their conservation status.

Origin of native Danish cattle breeds

A.A. Schönherz[1], V.H. Nielsen[2] and B. Guldbrandtsen[1]
[1]*Aarhus University, Department of Molecular Biology and Genetics, Center for Quantitative Genetics and Genomics, Blichers Allé 20, 8830 Tjele, Denmark, [2]Aarhus University, Danish Centre for Food and Agriculture, Blichers Allé 20, 8830 Tjele, Denmark; anna.schonherz@mbg.au.dk*

Domestication, breed formation and several centuries of selection as well as limited long distance exchange in traditional livestock breeding systems have resulted in divergent cattle breeds. Increasing global trade in farm animal genetics has caused dramatic changes in livestock breeding systems. Expansion of international breeding programs has resulted in extensive breed replacement and drastic population declines of native cattle breeds. Consequently, many native breeds experience increased rates of inbreeding and loss of genetic variation. In Denmark, four native cattle breeds are recognized by the national animal genetic resource conservation program: Red Danish Dairy Breed anno 1970 (RDM-70), the Black-and-White Danish Dairy Cattle anno 1965 (SDM-65), the Danish Shorthorn Cattle (SHO), and the Jutland Cattle (JYS). To minimize the loss of genetic resources, sustainable *in vivo* conservation programs are needed. Hence, this study was conducted to investigate the genetic diversity, breed origin, population structure, and migration dynamics of the native Danish cattle breeds in a Northern European context, providing information important for the establishment of sustainable *in vivo* conservation programs in these breeds. Data presented here will especially focus on findings related to RDM-70. RDM-70 is genetically distinct from other Northern European Red Dairy Cattle breeds with low levels of breed admixture. Average molecular inbreeding rates are as high as 22%. Overrepresentation of shorter ROH elements, however, indicates low genetic variation in the RDM-70 founder population rather than recent loss as a main cause. This was further supported by maintenance of genetic variation present prior to conservation initiation (prior 1970). Angler cattle showed closest genetic relatedness to RDM-70, confirming historical records on breed formation. In contrast, modern Danish Red Dairy Cattle (mRDC-DK), descending from RDM-70, were highly admixed and grouped within the phylogenetic clade of Alpine breeds distant from RDM-70 and other Northern European Red Dairy Cattle, indicating intensive crossing with Brown Swiss.

Cytogenetic investigations of Romanian cattle breeds

I. Nicolae
Research and Development Institute for Bovine Balotesti, Cattle Genetics, Bucuresti-Ploiesti, km 21, Ilfov, 077015 Balotesti, Romania; ioana_nicolae2002@yahoo.com

In cattle, the artificial insemination (AI) is widely used in Romanian farms. As such, genetic inherited abnormalities found in bulls used for semen collection can become dispersed and have the potential to cause significant economical loses due to reproductive failure and animal mortality. Considering these aspects, it is of outmost importance to include and maintain breeding animals (both AI bulls, and cows with reproductive disorders) under cytogenetic control. During the last 40 years, different kinds of abnormalities (e.g. 1/29 translocation, tandem translocation, ring-chromosome, xx/xy chimerism, chromosomal instabilities) were identified in the frame of the cytogenetic screening program implemented in Romania, at the Bovine RD Institute, mainly concerning proven bulls belonging to the Romanian Spotted, Romanian Black Spotted and Brown breeds. Cytogenetic investigation of more than 3,500 A.I. bulls has had both technical and economical positive effects for breeders and AI bull stations. On the one hand, the removal of genetic abnormalities carriers from reproduction avoided the dissemination of chromosomal rearrangements in offsprings. On the other hand, the expenses corresponding to the complete selection procedure of the carrier AI bulls were significantly reduced.

Effective population size and genomic inbreeding of Slovak Spotted Cattle

A. Trakovická, K. Lehocká, R. Kasarda, O. Kadlečík and N. Moravčíková
Slovak University of Agriculture in Nitra, Tr. A. Hlinku 2, 94976 Nitra, Slovak Republic; anna.trakovicka@uniag.sk

The aim of this study was to estimate the effective population size as well as to analyse the level of historical and recent inbreeding in Slovak Spotted cattle as one of the most important autochthonous livestock species in Slovakia. Overall 85 animals representing the nucleus of Slovak Spotted cattle were genotyped by using two platform; ICBF International Dairy and Beef and Illumina BovineSNP BedChip v2. Two different methods were used to calculate genomic inbreeding: genomic relationship approach (F_{GOF}) and scan for the distribution of runs of homozygosity segments across the genome (F_{ROH}). The results showed that ROH segments greater than 4 Mb cover in average 2.05% of the genome, while ROH segments >16 Mb achieved 0.43% that signalized recent inbreeding in analysed population. All of the F_{GOF} estimates were negative, which could be explained as possible result of the using the current population as the base population to estimation of allelic frequencies. Historical and recent effective population size was estimated based on the relationship between the extent of linkage disequilibrium (LD) and effective population size. Current N_e was analysed as the linear regression of estimates obtained for the last 50 generations. The highest values of LD were achieved at the shortest distance (20 kb bin=0.098), in the contrary, lowest were reached at the longest distance (50,000 kb bin=0.04). The average r^2 value ranged from 0.089 (total population) to 0.115 (bulls) for markers in 50 to 100 kb bin. The effective population size estimate for current generation at level 38.69 (decrease 7.64 animal per generation) clearly pointed out to the endangered status of Slovak Spotted cattle.

Breeding policies to optimise effective population size in captive populations

J.P. Gutiérrez[1], I. Cervantes[1], E. Moreno[2] and F. Goyache[3]
[1]Dpto. Producción Animal, Universidad Complutense de Madrid, 28040, Spain, [2]Estación Experimental de Zonas Áridas, CSIC, 04120, Almería, Spain, [3]Área de Genética y Reproducción Animal, SERIDA-Deva, 33394, Spain; icervantes@vet.ucm.es

Preservation of genetic variability has traditionally been faced via the proposal of breeding strategies aiming at the increase of the effective population (Ne). Protection provided by zoological reserves and parks may be increasingly necessary for short- and long-term survival. There is consensus on the fact that minimising coancestry in a cohort outperforms other breeding strategies. This research compares via computer simulations the performance of different breeding schemes aiming at limiting the loss of genetic diversity in small captive populations, assessed via fluctuations of Ne at both the short- and long-term. Nine mating strategies have been assayed: minimum coancestry (F), minimum increase in coancestry (dF), minimum weighted coancestry (mFm; accounting for the actual representation of each individual), minimum weighted increase in coancestry (dmdFdm), minimum offspring coancestry (C), minimum offspring increase in coancestry (dC), minimum offspring coancestry free (Cf; like strategy C, but females have unbalanced or no offspring), minimum offspring increase in coancestry free (dCf) and mixing F and C information of the offspring (mix). Twenty replicates were performed during 20 discrete generations. Methods were tested under scenarios simulated starting from the actual pedigree of Gazella cuvieri an endangered Sahelo-Saharan gazelle with a relative big population kept in captivity. All strategies provided an increase of Ne in the first generation. That increase was higher in scenarios where not all males were forced to produce offspring. This was due to the previous unbalance of pedigree depth and the differential contributions among parents. Ne decreased in all scenarios in the second generation. After third generation, scenarios forcing the use of all males increased Ne. The mix and the dCf strategies gave higher Ne values than the other strategies in the long term. However, dCf gave the worst result in the first generation in cases of poor pedigree depth. The practical use of the methodology depends on the planned horizon that might vary according to generation interval.

Role of the genetic resource in the settlement project in Mediterranean pastoral sheep breeding

T. Verdoux[1], M.O. Nozières-Petit[2] and A. Lauvie[2]
[1]M2 Student Montpellier SupAgro, 2 place Viala, 34060 Montpellier Cedex 01, France, [2]INRA UMR SELMET, 2 place Viala, 34060 Montpellier Cedex 01, France; anne.lauvie@inra.fr

Conducted in the frame of the ARIMNET PERFORM project, the goal of this study is to understand the role of the genetic resource in the creation and implementation of the installation project of sheep farmers (for meat production). The case studied is the one of a Mediterranean area concerned by sheep farming with pastoral component, in south east France. Comprehensive interviews have been conducted among 19 sheep farmers who have been settled for the last ten years. A first result of the study was to show that a large part of these farmers don't raise a single breed but several of them. This can either be a will to have crossbreeds (cross between conformation and rusticity judged interesting) or, for most of them, a temporary state toward the 'ideal' flock constituted progressively (for instance by absorption breeding of the minor breed). Concerning the breed choice, motivations are very diverse. However, farmers agree, no matter which breed they raise, on rusticity and territory adaptability capacity importance when they choose the breed. Other zootechnical criteria are also mentioned (conformation, maternal qualities, unseasoning, etc.) but also aesthetic or socio-cultural criteria (tradition, etc.). The accessibility of the resource when constituting the flock is also of importance (availability of breeding animals, economic accessibility). The role of the social dimension around the breed is also underlined. Network constituted around the breed is important for genetic resource collective management, but it can also have an impact on choices and breeders' individual practices (informal exchanges on technical aspects). Coherence between different parts of the system (commercialization, reproduction, alimentation) is very important in those pastoral systems, and the genetic resources choice is made in relation with those other dimensions of the system.

Molecular diversity of sheep breeds from the United Kingdom

E. Somenzi[1], M. Barbato[1], J.L. Edmunds[2], P. Orozco-Terwengel[2] and P. Ajmone-Marsan[1]
[1]Università Cattolica del Sacro Cuore, Dipartimento di Scienze Animali, della Nutrizione e degli Alimenti, via Emilia Parmense, 84, 29122 Piacenza, Italy, [2]Cardiff University, School of Biosciences, The Sir Martin Evans Building, Museum Avenue, Cardiff, CF10 3AX, United Kingdom; paolo.ajmone@unicatt.it

Sheep farming is the most important agricultural activity in the United Kingdom, with ~33 million sheep and 90 breeds, of which more than half are local breeds. The current changes in market demands and the consequent plummeting of lamb meat price are responsible for a reduced interest of sheep breeders on the less profitable local breeds. Hence, local breeds are experiencing a decrease in population size, increase of inbreeding, and the overall loss of genetic diversity. To evaluate the current status of British local breeds, we investigated the genetic diversity and population structure of 14 local sheep breeds chosen among the most representative. We included one commercial breed from UK and two from mainland Europe, all of them belonging to the Texel group. We analysed 515 individuals genotyped using the Ovine SNP50 BeadChip. We evaluated the molecular diversity through observed and expected heterozygosity. The population structure was also investigated through Admixture and PCA analyses. Our results highlighted high levels of inbreeding and low values of observed heterozygosity for many of the local breeds analysed, such as for South Wales Welsh Mountain. Only a few local breeds had low inbreeding values, and moderately high observed heterozygosity, probably due to the increased awareness of authorities on biodiversity, and the consequent implementation of conservation policies. We identified signs of admixture in some breeds, likely indicating the occurrence of crossbreeding events aimed at improving animal production. Despite a few positive exceptions, results underline that many UK local sheep breeds possess low levels of molecular diversity, along with unknown or partially incomplete history and partial admixture. High levels of inbreeding combined with low levels of heterozygosity may increase the risk of extinction of many UK breeds, especially those having a reduced census size. In this context, our results underline the importance of monitoring genetic diversity across generations to design and implement educated conservation plans.

The Facciuta della Valnerina goat as an economic opportunity in the mountain areas of central Italy

S.G. Giovannini, S.C. Ceccobelli and F.M.S. Sarti
Università degli Studi di Perugia, Dipartimento di Scienze Agrarie, Alimentari e Ambientali, Borgo XX giugno, 74, 06121, Perugia, Italy; samira.giovannini@gmail.com

The project originated by the consideration of a historical presence of a traditional goat named 'Facciuta della Valnerina' in the Umbria and Marche Appennine mountain territory. The recovering action of this goat is developed on the rule of biodiversity in that specific rural area, which has been deeply damaged by earthquake in the recent years. 'Facciuta' is a term that stands for goat with specifics morphological traits, as a black tan coat and two white strips on the face. The project has multiple goals to achieve: in the first place there's a genetic target which is about saving an old genetic type, at the same time it has got an anthropological and historical meaning on recovering traditional activities of wild breeding strongly bonded with the territory. At this purpose it has to be highlighted that the environment in this area is very harsh and so the goat breeding is one of the few possible economic solutions. Besides, must be considered that the specialized high-productive breeds cannot properly be reared there, so only native populations, as Facciuta is, can give some economic advantage. The 'Facciuta della Valnerina' goat indeed is well adapted to wild breeding, the strong adaptability of this genetic type at that rural areas could be useful to promote free range farming. The typical Facciuta products, namely cheese and kid meat, could also lead income to small farmers with low labour costs. The first step of the project was aimed to monitoring the number of goats with those specific characteristics in the Umbria and Marche's areas. Moreover, an identification sheet was created, in which for each goat were recorded the ID, the morphological description and the measures of seven biometric traits to evaluate the body size; if available, the genealogy of the animals was detected. The milk production and the organoleptic qualities of milk and cheese were also observed. The project could be seen as a way to reveal the potential rule that even unproductive areas can have; the aim is to give an economic value to the less-favoured area, as the Valnerina is, through the connection between the history and the tradition of a territory with the conservation of an ancient breed.

A conservation plan for the Danish Jutland horse

H.M. Nielsen[1] and M. Kargo[1,2]
[1]*Aarhus University, Center for Quantitative Genetics and Genomics, Department of Molecular Biology and Genetics, Blicher alle 20, 8830 Tjele, Denmark,* [2]*SEGES, Agro Food Park 15, 8200 Aarhus N, Denmark; morten.kargo@mbg.au.dk*

The Jutland breed is an original Danish horse breed, which previous was used as a draught horse in the farming industry. The aim of this study was to present a conservation plan for the Jutland horse. In year 2016, the Jutland horse breeding organization initiated a conservation project supported by the Danish Committee on farm animal genetic ressources. Using optimum contribution by the EVA software, we identified 12 colts for future breeding. With economic support, the breeding organization raised the 12 colts until March 2019 where they were scored for conformation. In March 2019, only 11 stallions out of the 12 stallions from the conservation project were available for mating. The project also included 88 breeding mares for which their owner agreed to participate in the project and thus to use one of the 11 stallions for breeding their mare in 2019. SEGES, the National Committee of Horses provided the data for the study. The pedigree contained 4,757 individuals and the average relationship between males and females was 0.184. In order to ensure ownership of the project we first assigned 4 matings to all stallions in the project. Next, 44 matings were distributed to the stallions using EVA where each male was allowed to mate a maximum of 8 mares. Out of the 11 stallions, 7 were selected using optimum contribution. Finally, we optimized the matings between the 88 mares and the 11 stallions while minimizing inbreeding in the next generation. The average inbreeding was 0.071 and the rate of inbreeding was 0.0025. In a third step, we also ran EVA with all approved stallions and all alive mares under 15 years. This provided input to the use of stallions from the project and other approved stallions, which were not part of the conservation project. This study demonstrates how to carry out a conservation plan for an endangered horse breed.

Does the Walloon Piétrain pig breed require preservation measures?

H. Wilmot[1], S. Vanderick[1], R. Reis Mota[1], P. Mayeres[2] and N. Gengler[1]
[1]*ULiège-GxABT, Passage des Déportés, 2, 5030 Gembloux, Belgium,* [2]*Walloon Breeding Association, Rue des Champs-Elysées, 4, 5590 Ciney, Belgium; helene.wilmot@student.uliege.be*

Piétrain pigs are used worldwide as terminal sires, and they are not expected to require preservation measures. However, the number of pure Piétrain breeders is dwindling everywhere and also in Wallonia (southern Belgium), the region where the breed is originated from. As a first step, the objective of this study was to assess the genetic diversity in Walloon Piétrain pig populations by using pedigree information. A total of 199 boars, whose breeders could be identified and which passed through performances testing using crossbred progeny at the performances recording station during the last ten years, were used for pedigree extraction. Kinship coefficients were determined and a classical multi-dimensional scaling (MDS) was performed on those boars for herd comparison purposes. In addition, breeders who have stopped their activity were identified in order to check genetic diversity loss overtime. Four groups were identified: a first cluster of Walloon Brabant breeders; a second core cluster, which suggests high levels of inbreeding, composed mainly by Hainaut breeders; a third cluster with great diversity, represented by a breeder whose animals may had German boars influence; and the last cluster formed by two breeders. The breeders from all four clusters are in ongoing pig breeding activities. However, due to their aging as well as the lack of new breeders, conservation measures establishment may be urgent in order to preserve genetic resources. As Walloon Piétrain breeders tend to keep very different phenotypes, complementary principal components analysis will be performed by using pseudo-phenotypes, by using deregressed estimated breeding values (EBV), to be compared with previous MDS results. Finally, single nucleotide polymorphisms (SNP) markers from different European Piétrain pig populations will be used to determine genomic relationships to further verify if Walloon Piétrain population(s) have intrinsic particularities, which would justify conservation measures.

The Suino Nero Cinghiato recovering on rural farms

S.G. Giovannini and F.M.S. Sarti
Università degli Studi di Perugia, Dipartimento di Scienze Agrarie, Alimentari e Ambientali, Borgo XX giugno, 74, 06121, Perugia, Italy; samira.giovannini@gmail.com

The Suino Nero Cinghiato project is focused on biodiversity recovering, but also has an economic target. The project was inspired by the presence of a painted belted pig in lots of frescoes in many churches in Appennine area. That pig had specific morphological traits: elongated face, black skin and black bristles and a white stripe running all around its trunk. The project has a multiple relevance: together with the genetic target in save old genotypes, have also an anthropological meaning in recovering traditional activities of processing pig meat. The Suino Nero Cinghiato is well adapted to wild breeding; it is in fact able to graze freely mainly feeding on herbs and acorns; due to its feeding habits the Suino Nero Cinghiato meat is evenly fat-veined, usually well separated by the lean parts; and it has an extraordinary taste and flavour. The project started with a breeding band made by one boar and three sows. From these animals, through selection on the morphological traits, was obtained a small group of heads who looked like the pig painted in the frescoes. Thanks to the received funds the project is grown up, the number of animals is increased and the fresh or processed meat selling could be now possible. The biggest marketing problem is the socio-economic structure of the involved farm companies. All of them are small and, so, able to produce only rather low quantities of product; specific market strategies are therefore necessary. In this situation, ten Umbrian companies have consequently chosen to found a contract network to increase marketing competitiveness, maintaining their autonomy and independence. In this way an overvalue for their production is achieved together with a long-term enhancement of that rural areas, deeply damaged by earthquake in the recent years. The result of this complex and interactive processes is very innovative and is based on the synergic actions of the contract network founding and of the scientific research. This project can be looked as a methodological way to give an economic aid to a rural area, through the connection of historical proof and local culture testimonies, with the scientific genetic recovering of an ancient breed and its special traditional products.

Conservation of the Dark Bee through an extensive organic beekeeping in the Swiss Alps

S.V. Garibay
Research Institute of Organic Agriculture FiBL, Department of International Cooperation, Ackerstrasse 113, Postfach 219, 5070 Frick, Switzerland; salvador.garibay@fibl.org

In this project, the conservation of the Dark Bee focus on the adaptation of bees to different environments and its conservation through selection of colonies that fulfil characteristics as vitality, adaptation behaviour to environment, and morphological characteristics. This project is located in the Alpine zone from central Switzerland, at three different levels of altitude namely 400 m, 1000 m and 1,500 m over sea level. Such location are different on climate condition (400 m warm vs 1,500 m cold), development of the colony (400 m fast vs 1,500 m slow), and flowering season (400 m early and 1,500 m late) etc. In each location, 12 bee colonies will be kept, starting the spring 2018 with artificial swarms. All swarms will be under an extensive organic beekeeping management and certified as organic. The natural wax comb production and the swarm process will be an integral part of its management. Varroa will be controlled also using the allowed organic acids. During the year 2018, the bee colonies (mother colonies) are maintained until the winter 2018-2019 and the selection will be carry out during early summer 2019 according specific traits, this before the colonies will swarm. Firs results are expected during the month of June 2019.

Sustainable breeding and conservation actions of *Apis mellifera* in Italy

G. Minozzi[1], G. Gandini[1] and G. Pagnacco[2]
[1]Università degli Studi di Milano, DIMEVET, Via Celoria 10, 20133, Italy, [2]CNR-IBBA Milano, Via Bassini, 20133, Italy; giulietta.minozzi@unimi.it

Since the end of the last glaciation, numerous subspecies of *Apis mellifera* appeared in Europe. In Italy the *Ligustica* variety of *A. mellifera* adapted to the mild climate and to the rich floristic biodiversity of the territory. Thanks to its favourable characteristics, this variety has spread since 1800 worldwide. With the arrival of the *Varroa destructor* in the 80s and with the increasingly dramatic use of pesticides in agriculture, the Italian bee is increasingly dependent on human interventions for its survival, both for the control of parasites and for the threat posed by pesticides. More recently, the effects of globalization have favoured in the Italian territory other subspecies of *A. mellifera* more performing in terms of productivity and resistance to pathogens. In this context, there is an urgent need to define strategies for the conservation of local ecotypes and for their valorisation on a productive scale. The control of matings becomes therefore a major point, and in this regard the BEENOMIX project has contributed to implement a specific selection breeding scheme applicable by any breeder able to carry out a selective activity. The BEENOMIX project developed and implemented a selection model based on productivity in terms of kg of honey and hygienic behaviour with a relative emphasis of 1 and 0.4, respectively. Two partially isolated populations are considered. The selection takes place between 108 families per year, from which the 6 best families will be identified and from which the new 108 families of the next cycle will be produced. In addition, from the best family every year a group of sister drone-producing queens (DPQ) is raised. The drones produced by the DPQ will fertilize the 108 virgin queens (VQ) in an isolated mating site. The mating site is chosen on different criteria, in particular to be isolated from the presence of other drones. The research was funded by the BEENOMIX project funded by the Lombardy Region.

Analysis of the consequences of genetic rescue and its dependence on purging

N. Pérez-Pereira[1], A. Caballero[1] and A. García-Dorado[2]
[1]Universidad de Vigo, Departamento de Bioquímica, Genética e Inmunología, Facultad de Biología, 36310 Vigo, Pontevedra, Spain, [2]Universidad Complutense, Departamento de Genética, Facultad de Biología, 28040 Madrid, Spain; noeperez@uvigo.es

Inbreeding depression, the reduction of fitness observed in populations with high inbreeding, has been a main concern in conservation programs. However, effective genetic purging of the recessive deleterious alleles expressed under inbreeding has been confirmed both theoretical and experimentally, especially under slow inbreeding. Therefore, this genetic purging needs to be taken into account to elaborate recommendations of management strategies as it becomes a possible determinant of its success. Particularly, the consequences of genetic purging on genetic rescue programs has not been explored to date. The term genetic rescue is used to refer to the introduction of new genetic variation from another population in order to reduce inbreeding depression and increase genetic diversity. However, despite its potential, the practice of this type of management is still controversial due to the associated risks. One of them is caused by the large inbreeding load that large populations tend to harbour, which can be potentially introduced by migrants and promote an increase of inbreeding depression later on. If the threatened population has reached the current situation by slow progressive reductions of size, the efficiency of genetic purging after the rescue may be lower than the one occurred prior to the rescue, leading to a negative effect on fitness in the medium term. We aim to explore by computer simulations the consequences of genetic purging as a determinant of the circumstances in which the rescue should or should not be recommended. The results indicate that when inbreeding has progressed slowly in the past allowing efficient purging and genetic rescue is performed under conditions in which purging is not effective, initial hybrid vigour is followed by an increased decline in fitness in the medium-long term.

Heritability of milk production traits in Red Polish cows included in the genetic resources program

P. Topolski, K. Żukowski and B. Szymik
National Research Institute of Animal Production, Department of Genetics and Cattle Breeding, Krakowska 1, 32-083 Balice near Cracow, Poland; piotr.topolski@izoo.krakow.pl

Heritability of milk production traits in the conservative population of Red Polish cows included in the genetic resources programme (GRC) were estimated. Data for the study were drawn from the SYMLEK system (basic population) belonging to Polish Federation of Cattle Breeders and Dairy Farmers and included the results of the evaluation (expressed as standardized values for 305 days of lactation) of five milk production traits: milk, fat and protein yields (kg) and fat and protein contents (%). The optimization of the dataset structure and the combination of fixed and random effects included in the observation models was carried out using proprietary software written in the R language, using packages in which the ML method and the ANOVA and GLM procedures were implemented. In total, 1,622 Red Polish cows records with the values of phenotypic evaluation of milk production traits registered for the first lactation were used in the analyses. Only cows records within the range ±2.50 SD for milk yield in the basic population and ±2.70 SD for the other traits in the basic population were included. The (co)variance components of milk production traits were estimated assuming an animal model using REML. The multi-traits model contained herd-year-season, milk production level-age at calving as fixed effects; additive genetic, sire-daughter interaction, permanent environmental and temporary environmental effects were considered as random effects. The estimated coefficients of heritability of milk production traits were generally characterized by low to moderate level in the conservative population of Red Polish cows. The highest heritability was reported for the protein content $h^2=0.39$, and the lowest for the fat yield $h^2=0.18$. The estimated coefficient of heritability for milk yield was 0.26.

Determining breed purity in local Dutch cattle with a DNA test to increase number of breeding animals

B. Hulsegge, J.J. Windig, S.J. Hiemstra and A. Schurink
Wageningen University & Research, Centre for Genetic Resources, the Netherlands, P.O. Box 338, 6700 AH Wageningen, the Netherlands; jack.windig@wur.nl

Breed registries have been established to maintain purity of cattle breeds and document ancestry of animals. The number of registered purebred breeding females of many local breeds is so low that they are 'at risk'. Including unregistered animals belonging to the same breed will improve the situation, but due to missing and incomplete pedigree data this is generally not possible. However, breed composition of unregistered individuals can be determined with genomic data. We set up a DNA test to determine breed purity for animals with unknown or incomplete pedigree. A reference population was build containing 572 purebred animals with known pedigree and breed of origin including 6 local Dutch breeds and Holstein Friesian. A limited number of SNPs (n=133) from the commercially available Bovine 50k/HD SNP panel was selected, using the combination of Principal Component Analysis and Random Forest, to discriminate amongst the breeds. STRUCTURE analysis using this panel confirmed the presence of seven breeds. An independent test population (n=147) containing both purebred and crossbred animals was used for validation. Using STRUCTURE and pre-specifying the reference population resulted in 88.9% correctly classified individuals. One pair of breeds split recently while another pair recently mixed, preventing complete separation by a genetic test. For these breeds a combination of genetic testing and breed identification on the basis of phenotypic characteristics is used. The test has now been implemented in practice and herd books are increasing their breeding population by registering purebred animals without pedigrees.

Neural networks and deep learning in genome-wide prediction

P. Waldmann

Swedish University of Agricultural Sciences, Department of Animal Breeding and Genetics, P.O. Box 7023, 75007 Uppsala, Sweden; patrik.waldmann@slu.se

Genome-wide prediction (GWP) has become the state-of-the-art method in artificial selection. A typical study includes high dimensional data with thousands to millions of genomic markers (SNPs) recorded in a number of individuals that usually is on the order of some hundreds to a few thousand. Different machine learning approaches have successfully been used in GWP, but the use of neural networks (NN) and deep learning (DL) is still scarce. In this talk, I will briefly review the theory behind NN and DL, and show how these models easily can be implemented in popular DL software such as mxnet and Keras. It is also shown that recent developments effectively can be used to regularize DL models. Models for the traditional Multilayer perceptron (MLP) and the Convolutional neural network (CNN) are presented. It is demonstrated, on both simulated and real data, that a proper implementation of the CNN can achieve a considerable improvement in prediction accuracy compared with the MLP and conventional GWP methods (i.e. GBLUP).

Deep learning – an alternative for genomic prediction?

T. Pook[1], S. Herzog[2], J. Heise[3] and H. Simianer[1]

[1]University of Goettingen, Animal Sciences, Albrecht-Thaer-Weg 3, 37075 Goettingen, Germany, [2]Max Planck Institute for Dynamics and Self-Organization, Am Faßberg 17, 37077 Goettingen, Germany, [3]IT Solutions for Animal Production (vit), Heinrich-Schroeder-Weg 1, 27283 Verden, Germany; torsten.pook@uni-goettingen.de

With increasing computational power, machine learning and especially deep learning have risen in popularity over the last few years. Applications range from image and speech recognition to new ways for designing intelligent agents. As for genetics, potential applications are numerous and their potential needs to be evaluated. In this work, we assess the potential usage of artificial neural networks (ANN) for genomic prediction. Model performances were evaluated on a real dataset of a cattle population (n=10,501) for multiple traits ranging from highly heritable ones like milk yield to traits with low heritability like non-return-rate. Secondly, we evaluated the performance on simulated datasets with up to 100,000 animals and known underlying effect structures. We considered a wide range of traits, ranging from simple ones caused by a few additive effects to traits caused by complex epistatic interactions. As the design space of ANN is wide, we compared a variety of different number of layers as well as other parameter settings like dropout rate, activation/target functions or number of epochs and discuss strategies to automatically fit those. Overall, networks using local convolutional layers performed best in our tests, leading to slight improvements in terms of prediction accuracy compared to commonly used methods like GBLUP and BayesR. Still, there are problems that need to be solved before any practical application. ANN definitely have some potential when non-additive genetic effects are to be accounted for as e.g. in prediction of phenotypes or crossbreds. However, most of the breeding programs in animal breeding still exclusively use additive breeding values. Secondly, an ANN can be somewhat of a black box and prediction outside the training set needs to be critically evaluated. Another major potential in the usage of ANNs is the possibility to include additional types of input variables (e.g. gene expression data, environmental data) which are difficult to account for and have limited effects with traditional methods.

Generalised sparse-group lasso for whole-genome regression and genomic selection

J. Klosa[1], N.R. Simon[2], V. Liebscher[3] and D. Wittenburg[1]
[1]Leibniz Institute for Farm Animal Biology (FBN), Institute of Genetics and Biometry, Wilhelm-Stahl-Allee 2, 18196 Dummerstorf, Germany, [2]University of Washington, Department of Biostatistics, Health Sciences Building F-650, Box 357232, 98195 Seattle, WA, USA, [3]University of Greifswald, Institute of Mathematics and Computer Science, Walther-Rathenau-Straße 47, 17489 Greifswald, Germany; klosa@fbn-dummerstorf.de

Whole-genome regression methods are typically used to uncover links between performance traits and genomic markers (e.g. single nucleotide polymorphisms; SNPs). The number of such markers likely exceeds the number of observations, which results in multicollinearity between predictor variables. Thus, the precision of marker-effect estimates may be altered. Possible mathematical treatment involves selection, shrinkage, or a combination of both. In this study, a generalisation of the sparse-group lasso is developed. The objective is, to conserve the properties of selection and shrinkage of the sparse-group lasso, to maintain its ability to create a solution that is sparse among and within groups, and to increase the precision of marker-effect estimation. Finally, this will result in a higher accuracy of the prediction of breeding values, which is an important property in genomic selection. For the statistical evaluation of the new method, simulated scenarios were analysed which resemble a dairy cattle population. Grouping of adjacent markers was based on linkage disequilibrium. Results were compared to GBLUP and common lasso variants, such as lasso, group lasso, and sparse-group lasso. The generalised sparse-group lasso outperformed all other methods if the number of quantitative trait loci (QTL) was large and if the QTL occurrence in groups was high. In such scenarios, the accuracy of predicted breeding values was increased up to 16%. The new method seems to be particularly promising for genomic selection because it helps to precisely identify individuals with the highest breeding values (i.e. selection candidates).

Genome-based discovery of trait networks in dairy cattle

S. Pegolo[1], M. Momen[2], G. Morota[2], G.J.M. Rosa[3,4], D. Gianola[4,5], G. Bittante[1] and A. Cecchinato[1]
[1]University of Padua, DAFNAE, Viale dell'università 16, 35020 Legnaro, Italy, [2]Virginia Tech, APSC, Litton-Reaves Hall, 175 W Campus Dr, 24061 Blacksburg VA, USA, [3]University of Wisconsin, Department of Biostatistics and Medical Informatics, 600 Highland Ave, 53792 Madison WI, USA, [4]University of Wisconsin, Department of Animal Sciences, 1675 Observatory Dr, 53706 Madison WI, USA, [5]University of Wisconsin, Department of Dairy Science, 1675 Observatory Dr, 53706 Madison WI, USA; sara.pegolo@unipd.it

Structural equation modelling (SEM) is a powerful tool for modelling phenotypic networks. We propose SEM-GWAS as a complementary approach to multi-trait GWAS (MTM-GWAS) methods for inferring networks among traits, making possible to partition single nucleotide polymorphisms (SNP) effects on each trait into direct and indirect. Our specific aim was to illustrate the application of SEM-GWAS in dairy cattle by using phenotypic traits related to udder health, i.e. milk yield (MY), somatic cell score (SCS), lactose (%, LACT), pH and casein (CN, % of total milk N), in a cohort of 1,158 Italian Brown Swiss cows, as a case example. Animals were genotyped with the Illumina BovineSNP50 Bead Chip v.2. A Bayesian multi-trait genomic best linear unbiased prediction model was fitted to the five traits to obtain posterior means of model residuals, used as input for inferring putative causal networks and directions among traits via the Hill-Climbing algorithm. The inferred structure led to a set of structural equations, which were used to estimate SEM parameters. We found negative path coefficients for MY→LACT and LACT→SCS, and positive for SCS→pH and LACT→CN. Joint use of MTM-GWAS and SEM-GWAS enabled to identify six significant SNPs for SCS, and their effects were decomposed into direct and indirect, i.e. mediated by the up-stream phenotypes in the network (MY and LACT). Pathway enrichment analyses confirmed an overrepresentation (false discovery rate <0.05) of pathways consistent with the traits biology (e.g. organic anion transmembrane transporter activity for pH and MY or Wnt signalling for SCS). In summary, SEM-GWAS may offer new insights on relationships among phenotypes and on the path of SNP effects, producing useful information for selective breeding of cows and for management decisions.

Parameter estimation in the ancestral regression with missing parental or grandparental genotypes

R.J.C. Cantet and N.S. Forneris

Facultad de Agronomía, Unibversidad de Buenos Aires, INPA-CONICET, Producción Animal, Av. San Martín 4453, C.A.B.A. Buenos Aires., 1417, Argentina; rcantet@agro.uba.ar

The ancestral regression (AR) is a genetic evaluation model that generalizes the Gaussian animal model and reduces the variance of the Mendelian residuals by introducing two linear parameters (β_S and β_D) per genotyped animal. Originally AR required that all four grandparents are genotyped. The goal of this research was to provide with a Bayesian algorithm for estimating β_S and β_D when one or more of the four grandparents plus any parent are not genotyped. The estimating algorithm is based on the property of the Markov AR process from which the covariance structure is written as $\Sigma = (I - B)^{-1} D (I - B')^{-1}$, where D is diagonal and contains the standardized variances of Mendelian errors, $\text{Var}(\varphi)$. In addition, each row of the triangular matrix B contains β_S and β_D, with positive sign in the columns of both grandsires and negative signs in the columns pertaining to both granddams, plus the two 0.5 values at the parental columns. The Cholesky root free decomposition of Σ^{-1} follows a Wishart density of small dimension, i.e. it includes the BV of all four grandparents, both parents and the individual X. However, due to the restrictions on the positive definitiveness of Σ, a more efficient approach is to sample individual elements of this 7×7 matrix. Sufficient statistics are the differences between the covariance between the grandsire and X, minus the covariance between the granddam and X. The $\text{Var}(\varphi)$ is sampled from a truncated chi-square density and the betas are sampled from a truncated normal in the interval -0.25 to 0.25. It was observed in simulated data that if either grandparent is not genotyped, the estimate of any beta goes to zero. The missing information may be supplied by ancestors and half-sibs of the grandparent, but the standard error of estimation is higher than when the complete information on grandparents and parents is employed.

Using Monte Carlo method to include polygenic effects into SNP-BLUP reliability estimation

H. Ben Zaabza, E. Mäntysaari and I. Strandén

Natural Resources Institute Finland (Luke), Jokioinen, Finland, Production systems, Myllytie 1 Jokioinen, Finland, 31600, Finland; hafedh.benzaabza@luke.fi

Breeding values estimated using SNP marker information by SNP-BLUP is computationally scalable for even large numbers of genotyped animals. However, typically genetic variation cannot be completely captured by SNP markers because of incomplete linkage disequilibrium between the loci that affect the trait and the SNP markers. The marker-based SNP-BLUP model can be augmented by residual polygenic (RPG) effects which gives a more accurate model. The inclusion of the RPG effects to SNP-BLUP leads to mixed model equations (MME) having a size of the number of genotyped animals plus all their ancestors. Consequently, the calculation of model reliabilities requiring elements of the inverse of MME coefficient matrix will become computationally challenging. We present a Monte Carlo (MC) sampling based method to estimate model reliabilities in SNP-BLUP having RPG effects. We compare the calculated reliabilities in the Finnish Red dairy cattle with different proportions of RPG effect and test it with different MC sample sizes. There were 15,905 genotyped animals with 46,914 SNPs, and the pedigree had 231,186 animals. Correlations of model reliabilities between exact and MC based approach were 0.9999 (0.9999), 0.9980 (0.9996), 0.9844 (0.9970) for RPG proportion of 0.01, 0.2 and 0.5, respectively, when number of MC samples was 1000 (5,000). MC sampling of the RPG effects for the reliability computations work the better the smaller the RPG proportion and the more MC samples are used.

Planning experimental designs for genomic evaluations: sample size and statistical power calculation

D. Wittenburg and M. Doschoris

Leibniz Institute for Farm Animal Biology, Institute of Genetics and Biometry, Wilhelm-Stahl-Allee 2, 18196 Dummerstorf, Germany; wittenburg@fbn-dummerstorf.de

Single nucleotide polymorphisms (SNPs) with significant impact on a trait can be identified with genome-wide association studies. The high linkage disequilibrium (LD) among SNPs makes it difficult to identify causative variants correctly. In contrast to a single-locus model, where single SNPs are investigated successively, a multi-locus model is the better choice because it is able to identify causative variants simultaneously, to state their positions more precisely and to account for existing dependencies. Sample size has not only a crucial impact on the precision of parameter estimates, it also ensures that the desired level of statistical power can be reached. Although it is possible to theoretically determine the required sample size for discovering or confirming a new variant in a single-locus model at a given power, it is not clear what sample size is needed in a multi-locus approach. We target this issue by employing the theoretical instead of observed dependence between SNPs (matrix K), which can be set up for any given population structure; the entries of K are functions of paternal and maternal LD. Initially, we base our power investigations on the commonly applied SNP-BLUP approach because it requires only one shrinkage parameter. We determine the t-test statistic for locally testing non-zero effects of SNPs and derive its distribution under the alternative hypothesis. The non-centrality parameter is decisive: it depends on assumptions of sample size, effect size and extent of shrinkage. It is also studied whether power depends on the number and location of true variants on the chromosome, either at one end or in the middle, and their locus-specific heritabilities. Assuming one half-sib family, preliminary results show that the required sample size is reduced by roughly 10-20% compared to calculations based on a single-locus model with adjusted type-I error rate. We will extend our investigations by allowing for multiple half-sib families and LD-group specific shrinkage and work out criteria for an optimal breeding design.

Use of genomic information to improve accuracy of prediction from group records

T.T. Chu[1,2], J.W.M. Bastiaansen[2], P. Berg[1,3] and H. Komen[2]

[1]Aarhus University, Center for Quantitative Genetics and Genomics, Blichers Allé 20, 8830 Tjele, Denmark, [2]Wageningen University & Research, Animal Breeding and Genomics, 6709 PG Wageningen, the Netherlands, [3]Norwegian University of Life Sciences, Department of Animal and Aquacultural Sciences, 1432 Ås, Norway; chu.thinh@mbg.au.dk

The accuracy of prediction from group records depends on the group structure. When information on genotypes is available before phenotyping, utilization of the information to form groups may improve the accuracy of prediction from group records. This study proposed two grouping methods based on genomic information including unsupervised clustering implemented in the STRUCTURE software and supervised clustering based on genomic relationship scripted in R. Estimation of variance components and prediction of GEBVs from group records were carried out with GBLUP models. Genetic variances estimated from individual records and from group records were consistent. Using genomic information to make groups led to a higher genomic relationship between group members than a random grouping of paternal half-sibs and random grouping of full-sibs. The genomic relationship between members of groups that were formed from the supervised clustering method depended on group sizes and family sizes. The grouping methods based on genomic information resulted in higher accuracy of GEBV prediction (1.2-11.7%) compared to the grouping methods based on pedigree information. In addition, grouping methods based on genomic information led to lower coancestry coefficients between top-ranking GEBV animals. Of the two proposed methods, supervised clustering based on the genomic relationship was superior in term of computation requirements, applicability and accuracy of GEBV prediction. In summary, the use of genotyping information for grouping gives additional accuracy of selection when group records are used in genomic selection breeding programs.

Genomic selection without own phenotypes exploits new mutational variance less than BLUP selection

H.A. Mulder[1], S.H. Lee[2,3], S. Clark[3], B.J. Hayes[4] and J.H.J. Van Der Werf[3]
[1]Wageningen University & Research Animal Breeding and Genomics, P.O. Box 338, 6700 AH Wageningen, the Netherlands, [2]Centre for Population Health Research, School of Health Sciences and Sansom Institute of Health Research, University of South Australia, Adelaide, Australia, [3]School of Environmental and Rural Science, University of New England, Armidale, 2351, NSW, Australia, [4]Centre for Animal Science, Queensland Alliance for Agriculture and Food, The University of Queensland, St Lucia, 4067, Queensland, Australia; han.mulder@wur.nl

De novo mutations (DNM) create new genetic variance and are an important driver for long-term selection response. We hypothesized that genomic selection without own phenotypes exploits new mutational variance less than pedigree-based BLUP selection. The objectives were to compare GBLUP selection without own phenotype (GBLUP), GBLUP selection with own phenotype (GBLUP_OP) and pedigree-based BLUP selection with own phenotype for 20 generations using Monte Carlo simulation. GBLUP resulted in the lowest genetic variance (24.7% of total genetic variance) and response due to DNM after 20 generations (9.4% of total response). GBLUP_OP had the highest genetic variance due to new DNM (47.8% of total genetic variance) and selection response due to DNM (30% of total response), while BLUP was in between with 27.1% of genetic variance due to DNM and 21.9% of response due to DNM. However, GBLUP_OP caused a rapid decline in total genetic variance, because the genetic variance due to old QTL rapidly eroded. GBLUP had very little selection pressure on DNM and as a consequence, GBLUP had the highest number of DNM with negative effects that were still segregating 10 generations after a DNM arose. Due to the decline in genetic variance for all selection strategies, selection limits are foreseen and these selection limits are lower for GBLUP (11.2 phenotypic SD from the base population mean) and GBLUP_OP (10.7 phenotypic SD) than for BLUP selection (11.8 phenotypic SD). It can be concluded that genomic selection without own phenotypes exploits new mutational variance less than BLUP selection with own phenotypes, it increases the probability of DNM with negative effects and it lowers selection limits. Sustainable breeding strategies should consider optimal ways to exploit DNM.

Genomic accuracy for indirect predictions based on SNP effects from single-step GBLUP

D. Lourenco[1], I. Aguilar[2], A. Legarra[3], S. Miller[4], S. Tsuruta[1] and I. Misztal[1]
[1]University of Georgia, Athens, GA, 30602, USA, [2]INIA, Las Brujas, 90200, Uruguay, [3]INRA, Castanet Tolosan, 31326, France, [4]Angus Genetics Inc., St. Joseph, MO, 64506, USA; danilino@uga.edu

Indirect predictions (IP) can be obtained for young animals that are not yet included or will never make up to an official genomic evaluation. Additionally, if lots of animals are genotyped but they do not contribute to the evaluation, having IP for them would reduce computing cost. This is beneficial if accuracy levels are maintained. If single-step GBLUP (ssGBLUP) is the official evaluation method, SNP effects can be obtained by backsolving genomic EBV (GEBV). Subsequently, IP are obtained by the sum of SNP effects weighted by gene content. When IP are released, a measure of accuracy that reflects the standard error of IP is needed. Our objective was to implement formulas to calculate accuracy for IP and to apply the method in a cattle population. The American Angus data used in this study consisted of 35k records for post-weaning gain and pedigree information for 202k animals, of which 60k were genotyped for 38.4k SNP (after quality control). From the genotyped animals, 2k were in the validation set. A complete dataset had phenotypes and pedigree up to 2013, genotypes up to 2014, and was used to calculate the benchmark accuracy by the inverse of the LHS. This accuracy is referred to as GEBV accuracy. A reduced dataset had phenotypes, pedigree and genotypes up to 2013, and was used to obtain the prediction error covariance (PEC) for SNP effects. This PEC was used to calculate IP accuracy. The process of obtaining IP accuracy involved several steps: factorization and inversion of the LHS of ssGBLUP mixed model equations; extraction of the inverse of the LHS for genotyped animals; calculation of SNP effects from GEBV; computation of PEC for SNP effects; calculation of IP; computation of accuracy of IP as a function of PEC for SNP effect and SNP content. Correlations between GEBV and IP accuracies were over 0.997, indicating the possibility of getting a good measure of IP accuracy based on ssGBLUP, without directly switching to a SNP-BLUP model. Obtaining IP accuracies from large-scale ssGBLUP evaluations, which use the algorithm for proven and young, will be the subject of future investigations.

Genomic optimal contribution selection in small cattle populations

J. Obsteter[1], J. Jenko[2], J.M. Hickey[3] and G. Gorjanc[3]
[1]Agricultural Institute of Slovenia, Hacquetova ulica 17, 1000, Ljubljana, Slovenia, [2]Geno Breeding and A.I. Association, Storhamargata 44, 2317 Hamar, Norway, [3]The Roslin Institute and Royal (Dick) School of Veterinary Studies, University of Edinburgh, Easter Bush, Midlothian, EH259RG, United Kingdom; jana.obsteter@kis.si

Sustainability is increasingly important in animal breeding. While breeding programs must deliver genetic gain to meet the agricultural needs and maintain competitiveness, they also have to preserve genetic variability to ensure future potential. The implementation of genomic selection in cattle breeding provides new challenges and opportunities in addressing short- and long-term goals, particularly in small cattle populations. Optimum contribution selection (OCS) is a method for maximizing genetic gain at an acceptable loss in genetic diversity, that is, it maximizes the efficiency of converting standing genetic diversity into genetic gain. It achieves this by optimizing genetic contributions of existing and new individuals to subsequent generations. With genomic selection, the OCS method can optimize the contributions of tested bulls while maximizing the conversion efficiency. Our aim was to quantify the impact of optimizing bull contributions on genetic gain and conversion efficiency compared to conventional and genomic truncation selection in the breeding program of a small cattle population using simulation. We compared three truncation selection scenarios with genomic and progeny testing and long or rapid turnover of bulls and five OCS scenarios with a different balance between short- and long-term goals. We show that OCS achieves comparable genetic gain at a higher conversion efficiency than truncation selection using progeny or genomic testing. As expected, using genomic testing about doubled genetic gain compared to progeny testing due to shorter generation interval. When we coupled genomic selection with rapid turnover of bulls we further increased genetic gain (140%) compared to the baseline progeny testing, yet maintained the conversion efficiency of the baseline progeny testing. OCS delivered higher conversion efficiency through the use of more bulls over a shorter period. Our results inform small breeding programs on how to optimize the use of genomically and progeny tested bulls while ensuring long-term genetic improvement.

Novel approaches to incorporate recessive lethal genes in genomic prediction of calf survival

G. Gebreyesus, G. Sahana, A.C. Sørensen and G. Su
Aarhus University, Molecular Biology and Genetics, Blichers Allé 20, P.O. Box 50, 8830, Tjele, Denmark; grum.gebreyesus@mbg.au.dk

Calf and young stock mortality present significant economic loss for dairy farmers in addition to the serious animal welfare issues in the dairy industry. The genetic component of death in calf and young stock during the rearing period can be attributed to polygenic effects and lethal genes. Genomic prediction for calf and young stock survival might thus benefit from incorporation of information on known recessive lethal genes. Implementation of genomic prediction for survival traits with underlying genetic architecture characterized by mixed recessive gene and polygenic inheritance is not straightforward. In this study, we propose a novel approach where genomic breeding values (GEBVs) are estimated from models considering only polygenic effects and then are transformed into risk probabilities, and the risk probability from lethal genes is computed separately from carrier status and population recessive allele frequencies. The approach was tested using simulated data. Genotypic data was simulated based on the Danish Holstein LD structure and include 40k markers, 1,880 QTL with polygenic effects and 20 lethal genes with recessive allele frequencies between 0.04 and 0.05. Phenotypes (binary traits) were generated normally distributed underlying QTL polygenic effect and recessive lethal genes with penetrance ranging from 60 to 100%. Linear (GBLUP), threshold and logit models were used to analyse the data with or without considering the genotypic information on lethal genes. Accuracy of predicted risk probabilities were computed as correlation between predicted and observed calf mortality of validation sires. The results indicate that our novel approach can greatly increase the accuracy of selection for survival traits compared with the predictions without distinguishing polygenic and lethal gene effects.

Optimization for Hamiltonian Monte Carlo

A. Arakawa, M. Nishio, M. Taniguchi and S. Mikawa
Institute of Livestock and Grassland Science, NARO, 2 Ikenodai Tsukuba, Ibaraki, 3050901, Japan; aisaku@affrc.go.jp

A Hamiltonian Monte Carlo (HMC) method has been more and more popular in Bayesian statisticians. The HMC method needs to be set two parameters for a leapfrog integration; a step size (ε) and a number of discretizing times (L), which deeply affect to estimate marginal posterior distributions. The user-friendly software, STAN, automatically decides these two parameters of the leapfrog. However, STAN needs high computing demand, which leads to spending a long time of computing to obtain accurate results of marginal distributions. The aim of this study was to investigate properties of the leapfrog integration for a normal and an inverse chi-square distributions, which are common in a general animal model, and then we applied the HMC method optimized the leapfrog integration to the animal model. First of all, we investigated factors affecting a trajectory of the leapfrog integrations for the normal and the inverse chi-square distributions. After finding optimal settings of ε and L for the leapfrog integrations, the optimized HMC method was applied to a simulated data generated under an infinitesimal animal model. We compared estimates and effective sample size (ESS) of variance components by the HMC with ones by Gibbs sampling (GS). The data was simulated by QMSim. A phenotypic variance was set to be 1.0, and heritability was assumed to be 0.50. The population size was 1000 animals with equal numbers of each sex. To obtain marginal posterior distributions with both methods, a total of 11,000 samples were simulated, the first 1000 of which was discarded as burn-in iterations. All results were obtained by the R language. Our results of the leapfrog integrations showed that the trajectories for the normal and the inverse of chi-square distributions were strongly influenced by a scale of variance and the degree of belief, respectively. We calibrated the HMC method to have twenty times per one rotation for the leapfrog trajectory. The HMC and the GS methods were not biased, and in the case of L=7 in this setting, the ESS by the calibrated HMC method was quite larger than ones by the GS. However, our optimized HMC method needs true values for variance.

Using sequence data to refine QTL mapping in French dairy goats

E. Talouarn, P. Bardou, G. Tosser-Klopp, R. Rupp and C. Robert-Granié
INRA, 24, Chemin de Borde Rouge, 31326 Castanet-Tolosan, France; estelle.talouarn@inra.fr

Goats were domesticated 10,500 years ago with the aim of supplying milk, meat and fibres. Since then, breeds have specialized and adapted to their local environment developing specific genetic profiles. The VarGoats project is an international resequencing program, which aims at covering at best the genetic diversity of the Capra species. To date the sequence data 829 *Capra hircus* of various breeds and geographical origins. Variant calling led to the identification of a total of 110,193,942 variants on the 29 autosomal chromosomes. Quality checks were applied to sequence variants using various indicators: quality, depth, minor allele frequency and position. For French Alpine and Saanen breeds the concordance with 50k genotypes (CaprineSNP50 BeadChip) was checked. Mean concordance rate was 98.22% and ranged from 94.00 to 99.96% among individuals. Imputation was tested on the 23,338,436 filtered variants using FImpute software. Pedigree was provided and imputation was performed in a within-breed leave-one-out scenario. Imputation quality was checked for 4 individuals on all chromosomes and for every sequenced animal on chromosome 29. Mean concordance rate for chromosome 29 ranged from 73.49 to 85.71% and from 72.72 to 79.27% in Alpine and Saanen breeds respectively. On all chromosomes, average correlation between true and imputed sequence were 0.77 and 0.76 in Alpine and Saanen respectively. Imputation was then applied to a population of 1,129 French Alpine and Saanen males with at least 10 daughters phenotypes. Association analysis was performed on production and functional traits using polygenic models with the GCTA software. Using genotype data imputed to sequence level allowed to improve the significance of QTL previously identified with 50k SNP-chip only. In Saanen breed, 1.6 Mb region on chromosome 19 was identified as significantly linked to 4 conformation traits, 3 semen production traits, somatic cell count, milk yield and protein and fat contents. These researches provided insights on how to implement a quality check, imputation that will ensure the quality of subsequent analyses, and on new SNPs to be added to the 50k SNP-chip.

Bias, slope and accuracies of genetic evaluation of milk yield in Manech dairy sheep

F.L. Macedo[1,2] and A. Legarra[2]
[1]FVet-UdelaR, Lasplaces 1550, 11600, Uruguay, [2]INRA, UMR 1388 GenPhySE, 31326 Castanet Tolosan, France; fernando.macedo@inra.fr

The recently presented method Linear Regression (LR), was proposed to benchmark models in genetic selection schemes. The principle of the method is to estimate biases and accuracies comparing estimated breeding values (EBVs) obtained from a partial data set (with old records, EBVp) and from a whole data set (with the addition of new records, EBVw). In this study, we estimate μ_{wp} (the bias or difference between average EBVp and TBV, with an expected value of 0), b_{wp} (the slope of regression of TBV on EBVp, with an expected value of 1) and ρ_{pw} (inversely proportional to increase in accuracy from 'partial' to 'whole') for Manech Tete Rousse dairy sheep. Pedigree contains 540,986 individuals born from 1950 to 2017 and the phenotypic data file includes milk production of 1,842,295 lactations. No marker data was used. The model used to estimate breeding values includes five fixed effects as well as the permanent effect for milk production. The heritability used was 0.30. The model takes account of heterogeneous residual variances and genetic groups. The comparisons involved parent average EBV of rams born from 2005 to 2015 (as EBVp) and results from first daughters (as EBVw). Finally, we obtained statistic from 11 pairs of EBVp-EBVw. The mean (mean ± standard error) of μ_{wp}, expressed in genetic standard deviation of trait, was 0.26±0.03 and the mean of b_{wp} was 0.80±0.09, indicating a larger bias. On other hand, the mean of ρ_{pw} was 0.51±0.04; in other words, accuracy doubles with first progeny performance. As conclusion, Manech Tete Rousse parent average EBVs are over dispersed for unknown reasons. Work financed by the ARDI project of Poctefa program (European Union FEDER funds), Metaprogram GENSEL (INRA) and La Region Occitanie.

Metafounder approaches for single-step genomic evaluations of Red Dairy cattle

A.A. Kudinov[1,2], E.A. Mäntysaari[2], G.P. Aamand[3], P. Uimari[1] and I. Strandén[2]
[1]University of Helsinki, Helsinki, 00014, Finland, [2]Natural Resources Institute Finland (Luke), Jokioinen, 31600, Finland, [3]Nordic Cattle Genetic Evaluation, Aarhus, 8200, Denmark; andrei.kudinov@luke.fi

Single-step genomic evaluation (ssGBLUP) utilizing simultaneously genomic and pedigree data can be a potentially powerful approach for breeding value estimation. However, conflicting marker and pedigree information still limit its use in practice. Typically, most of the genotyped animals have been born within the last decade, while pedigree usually dates back at least two decades from the oldest genotyped animals. This poses a time distance or base population problem between the founders of pedigree and genotyped pool affecting ssGBLUP evaluations. Unification of the base population between the two data sources has been proposed by the metafounder (MF) approach where the pedigree is augmented by related pseudo-individuals or MF groups that present initial pool of gametes and are born before the base animals. The MF groups are assumed to have a covariance structure denoted Γ. The MF groups help to define the same base for the genomic and pedigree data, and correct inbreeding in the pedigree relationship matrix (A) to form a MF group augmented relationship matrix A^{Γ}. In practice, defining MF groups for real pedigree can be complicated by multi breed structure of population. We tested two MF grouping strategies for Finnish Red dairy cattle (RDC) data which had pedigree included 19,757 genotyped (by 46,914 SNP markers) animals and their 69,728 relatives. The MF were based on breed and birth years within RDC breed. The MF4 strategy described RDC base animals by three founders, and MF6 by four RDC, one HOL, and one other breed founder. The two Γ matrices were created by $8Cov(P)$, where P was estimated by the marker allele frequencies across loci in 4 and 6 MF groups. Average (off)diagonal elements of Γ matrix were 0.67 (0.71) 0.59 (0.57) for MF4 and MF6 founders, respectively. Inbreeding trend in the A_{22}^{Γ} matrix of the genotyped animals was notably closer to that in the genomic relationship matrix G, compare to the original A_{22} relationship matrix. Correlation 0.63 of diagonals of original A_{22} and G matrix was increased up to 0.74 by using either 4 or 6 MF groups.

Effects of changing annotation and SNP density on genomic prediction accuracy in Brown Swiss cattle

F.R. Seefried[1], H. Pausch[2], I.M. Häfliger[3], C. Drögemüller[3] and M. Spengeler[1]
[1]Qualitas AG, Chamerstrasse 56, 6300 Zug, Switzerland, [2]Animal Genomics ETH Zürich, Eschikon 27, 8315 lindau, Switzerland, [3]Institute of Genetics Vetsuisse Faculty, Bremgartenstr.109a, 3001 Bern, Switzerland; franz.seefried@qualitasag.ch

Increasing genomic prediction accuracy is a major goal and a challenge under the framework of genomic selection. In a typical routine genomic evaluation scheme reference animals are genotyped with high-density (HD) arrays while young candidates are first genotyped using low-density (LD) arrays and subsequently imputed to higher density. In Switzerland different versions of HD and LD chips have been used since genomic selection has been implemented. The current HD array comprises approximately 150,000 SNP (150k), the LD array contains ~47,000 SNP (47k). However, since previous versions of HD arrays contained ~50,000 SNPs (50k) the genomic evaluations are based on 50k markers so far for both sub-populations: Brown Swiss and Original Braunvieh. The aim of the presented project was to evaluate the impact of updating genomic evaluation systems for BSW together with an update in the genome annotation due to new bovine reference sequence. Once, the existing 50k system was compared with a 50k system using new bovine genome assembly ARS-UCD1.2. Additionally two 150k-array scenarios using ARS-UCD1.2 annotation were evaluated: a standard scenario and a scenario where sequence-based genotypes from a total number 242 sequenced BSW animals and 70 putative QTL regions were added. QTL regions were identified during preliminary GWAS runs using imputed WGS data. SNP effects were estimated using the BayesC algorithm and de-regressed EBVs from a common joint conventional evaluation system for both sub-populations. Genomic validation data set consisted 5% youngest animals with reliable EBVs. Genomic prediction accuracy increased for 150k-scenarios across all trait – population combinations. This increase was much less obvious for the 50k ARS-UCD1.2 dataset. The increased genomic prediction accuracy was more pronounced in Original Braunvieh than in the Brown Swiss sub-population. Finally, adding sequence variants to SNP-array data outperformed standard array data systems in terms of genomic reliabilities however number of sequenced animals was relatively small.

Improved ssGTBLUP for single-step genomic evaluations

M. Koivula[1], I. Strandén[1], G.P. Aamand[2] and E.A. Mäntysaari[1]
[1]Natural Resources Institute Finland, Animal genetics, Myllytie 1, 31600 Jokioinen, Finland, [2]NAV Nordic Cattle Genetic Evaluation, Agro Food Park 15, 8200 Aarhus N, Denmark; minna.koivula@luke.fi

The increasing amount of genomic information has caused problems in the single-step evaluations. At first, the limiting factor was the number of genotyped animals, because the model required inverses of genomic and pedigree relationship matrices (G and A_{22}) of genotyped animals. In ssGTBLUP, these problems are avoided by expressing G^{-1} matrix through product of two rectangular matrices, and $(A_{22})^{-1}$ matrix through sparse matrix blocks. However, it seems that T convergence of the iterative preconditioned conjugate gradient (PCG) method deteriorates as the number of genotyped animals increases. Also, unknown genetic groups are problematic if they are in the pedigree but not accounted in A_{22} or G. Earlier it has been shown that ssGTBLUP has the same convergence properties in PCG iteration as the original single-step MME. The aim of this study is to test the ssGTBLUP with more advanced options. These were the inclusion of residual polygenic effect (RPG) and QP transformations to the T and $(A_{22})^{-1}$ matrix to better take account the genetic groups. We tested the new ssGTBLUP options with Nordic Holstein production data, including ca. 7.3 million cows in the data and 10 million animals in the pedigree. The genomic information was from ca. 127,000 animals. The G matrix for ssGBLUP and T matrix for ssGTBLUP were calculated including 30% of RPG effect, and with or without accounting genetic groups. Our results showed that without QP transformation the convergence of ssGTBLUP was better (less iteration rounds) than the original ssGBLUP, and the iteration was also faster, each round taking about 13.5 sec with ssGTBLUP and 63.5 sec with ssGBLUP. When QP transformation was implemented, the convergence of both ssGTBLUP and ssGBLUP improved considerably and the total amount of iterations decreased about 45% both in ssGTBLUP and ssGBLUP. The ssGTBLUP can further be made faster by using eigen decomposition approach. Thus, ssGTBLUP seems to offer a computationally reasonable approach for solving genomic breeding values using the single-step method.

Incorporating epistasis effects into the genomic prediction in water buffaloes (*Bubalus bubalis*)

S. Fernandes Lázaro[1], D.F.C. Diercles Francisco Cardoso[1], A.Vieira Nascimento[1], D.C. Becker Scalez[1], D.J. De Abreu Santos[2] and H. Tonhati[1]
[1]*Universidade Estadual Paulista (Unesp), College of Agricultural and Veterinary, Animal Science, Via de acesso Professor Paulo Donato Castellane, 14884-900, Jaboticabal-SP, Brazil, [2]University of Maryland, Animal Science, College Park, 20742, Maryland, USA; sirlenelazaro@yahoo.com.br*

This study evaluated the potential of epistasis models through of extensions of the additive model considering epistatic effects in the matrix equivalent to the genomic relationship matrix to benefit genomic prediction in water buffaloes. All the 422 samples used in this study were genotyped using the Axiom® Buffalo Genotyping Array, 90k. We assessed the predictive ability of different epistasis models on milk buffalo traits for three production traits and one reproductive trait (milk yield, fat and protein yield and age at first calving), comparing the results with those of exclusively additive models. We illustrate how the choice of marker coding ((0,1,2), (2,1,0), (-1,0,1) and (CE – represents markers which have the same configuration between two individuals)) implicitly specifies the models. All the four versions of the EGBLUP models provided similar results, but their predictive abilities were higher (0.642-0.749) for age at first calving (AFC), despite the complexity and low heritability (0.10). Therefore, the methods we used here could be a feasible approach for the incorporation of epistatic effects into the genomic prediction, and to contribute to higher prediction accuracy for traits of low to moderate heritability and can be used to select variables within EGBLUP models, which can then be used for genome-assisted prediction.

The performance of estimating genetic parameters by Hamiltonian Monte Carlo and No-U-Turn Samper

M.N. Nishio, A.A. Arakawa, K.I. Ishii and O.S. Sasaki
Institute of Livestock and Grassland Science, NARO, 2 Ikenodai Tsukuba, Ibaraki, 3053901, Japan; mtnishio@affrc.go.jp

Hamiltonian Monte Carlo (HMC) is a powerful Markov chain Monte Carlo (MCMC) algorithm that avoids the random walk behaviour and sensitivity to correlated parameters by taking a series of steps informed by first-order gradients of log posteriors. These features allow it to converge to high-dimensional target distributions more quickly than simpler methods such as random walk Metropolis or Gibbs sampling (GS). However, the performance of HMC is highly sensitive to two hyperparameters: a step size and a number of steps. No-U-Turn Sampler (NUTS) extends HMC by automating tuning these two parameters and has been widely used in many field because NUTS algorithm is packaged into the open-source and probabilistic programming language Stan. This study investigated the performance of estimating genetic parameters by HMC and NUTS in both simulated and real pig data. In simulated data, the population consisted of 1000 animals with equal number of each sex and the heritabilities of the simulated phenotypes were set to 0.2 and 0.5. In real phenotype analysis, we used 991 records of backfat thickness (BF) and loin eye muscle area (LEA), and 1,521 pedigree records in Japanese Duroc purebred pigs. For all data, univariate linear mixed model was used to estimate heritabilities. GS, HMC and NUTS were coded in R. A total of 10,000 iterations were run for all MCMC algorithms with first 1000 iterations discarded as warmup or burn-in. We estimated effective sample size (ESS) which provided the number of effectively independent draws from the posterior distribution obtained by the MCMC. In simulated data, the heritability estimates were close to the setting values for the all methods. For BF and LEA, the heritability estimates were almost the same among GS and NUTS whereas HMC could not estimate heritability because of no convergence. For both simulated and real pig data, the ESS estimates by NUTS were larger than those by GS. These results indicated that NUTS was computationally efficient approach in the field of animal breeding and HMC might require hands-on tuning of hyperparameters according to a trait and a population structure.

Small breed thinking big: the use of genomics in the MRY breed

B.L. Hollema[1], W.M. Stoop[1], G. De Jong[1] and F. Reinhardt[2]
[1]Coop. CRV, Animal Evaluation Unit, P.O. Box 454, 6800 AL Arnhem, the Netherlands, [2]VIT, Heideweg 1, 27283 Verden, Germany; baukje.hollema@crv4all.com

For small breeds it can be challenging to implement an effective genomic evaluation system. For the Meuse-Rhine-Yssel (MRY) breed, breeding company CRV has implemented a genomic evaluation system, with the goal to improve genetic progress in this small population. The MRY breed is a local Dutch red and white dual purpose breed, with approximately 7,500 purebred animals registered. Interest in the breed is rising, especially in the cross-breeding segment due to its good fertility, good longevity and high milk protein. The equivalent of the MRY breed in Germany is the Doppelnutzung (DN) breed, which has been promoted in recent years by a common MRY-DN ranking in both countries. To evaluate the added reliability of genomic evaluation, in 2017, 550 bulls were genotyped, of which 480 MRY from the Dutch-Flemish population and 70 DN from the German population. A validation study was performed in 2017 using approximately 400 bulls as training population and another 100 as validation bulls. Added reliability was assessed assuming a conventional parent average reliability of 30%. For production traits added reliability averaged 15.9%, for conformation traits added reliability averaged 9.6%, and for functional traits added reliability tended to be low. In 2018 2,000 cows were genotyped and added to the reference population, thereby increasing the reliability. Despite a relatively small reference population an increase in reliability was realized for production and conformation traits using genomic evaluation.

Genomic tools for use in the New Zealand deer industry

K.G. Dodds, S.J. Rowe, A.S. Hess, J.C. McEwan, T.P. Bilton, R. Brauning, A.J. Chappell and S.M. Clarke
AgResearch, Bioinformatics and Statistics, Invermay Agricultural Centre, Private Bag 50034, Mosgiel 9053, New Zealand; ken.dodds@agresearch.co.nz

The New Zealand deer industry has been using DNA-marker testing since the 1990s. This has been primarily for parentage assignment, as deer behaviour prevents manual recording of pedigree at birth. DNA markers have also been used to provide information about breed. The primary breeds are wapiti or elk (*Cervus canadensis*) and red deer (*Cervus elaphus*), which are regarded as distinct species, but there is also interest in estimating the components of differing European origin (Eastern or Western) in red deer. Since 2017 genotyping-by-sequencing (GBS) has been used as the marker system. This provides at least 70,000 SNPs for analysis. For cost reasons, sequencing is performed on many individuals at once, so that the genotypes are supported by few reads. Therefore, it is important to account for the partial information provided by these low depth reads in any analysis. Additionally, as the SNPs are untargeted, they include many at low minor allele frequency. The high genetic diversity (between wapiti and red deer) presents some additional challenges. We show how recently developed methods for low-depth GBS data are being used in the New Zealand deer industry for parentage, breed prediction and gender assignment, and consider how this GBS resource can be used for gene discovery and genomic selection.

The effects of homozygosity on health in a commercial turkey (*Meleagris gallopavo*) population

S. Adams[1], B.J. Wood[1,2], C. Maltecca[3] and C.F. Baes[1,4]
[1]*University of Guelph, Centre for Genetic Improvement of Livestock, Department of Animal Biosciences, 50 Stone Road East, N1G 2W1, Guelph, Canada,* [2]*Hybrid Turkeys, 650 Riverbend Drive, Suite C, N2K 3S2, Kitchener, Canada,* [3]*North Carolina State University, Department of Animal Science, 120 W Broughton Drive, 27607, Raleigh, USA,* [4]*University of Bern, Institute of Genetics, Vetsuisse Faculty, Bremgartenstr. 109a Postfach, 3001 Bern, Switzerland; sadams05@uoguelph.ca*

Inbreeding depression as a consequence of strong directional selection has become a growing concern in livestock production. It can negatively affect animal fitness and health, and this can have animal welfare implications. Inbreeding coefficients are commonly estimated from pedigree data, however genomic data can be used to better capture Mendelian sampling variance, resulting in more accurate inbreeding estimates. Continuous stretches of homozygous genotypes, or runs of homozygosity (ROH), can be used to characterize genomic data. These ROH are correlated with other measures of inbreeding, and can be used for estimation of autozygosity at the genomic level. Pedigree records (n=773,414), phenotypic records (n=5,297), and genotypic records (n=5,297), obtained using a dense SNP array (56,450 SNP), were collected from one commercial line between 2010 and 2018. Runs of homozygosity were detected and inbreeding coefficients based on ROH were calculated. Inbreeding coefficients calculated from ROH were higher than those calculated using pedigree for the same animals (F_{PED}=0.12; F_{ROH}=0.24). Taking into consideration that higher levels of homozygosity may result in reduced performance, these genomic measures of inbreeding provide information that can be used to investigate the biological implications of homozygosity. Identification of unfavourable haplotypes can allow for better management of these regions across time by minimizing the frequency of ROH associated with unfavourable effects in the next generation. Further investigation of the identified haplotypes and better understanding of their effects will aid in understanding the negative implications of homozygosity on traits that affect the fitness, health, and welfare of turkey populations.

Indirect predictions based on SNP effects from GBLUP with increasing number of genotyped animals

A. Garcia[1], Y. Masuda[1], S. Miller[2], I. Misztal[1] and D. Lourenco[1]
[1]*University of Georgia, Animal and Dairy Science, 425, River road, 30602, USA,* [2]*Angus Genetics Inc, 3201 Frederick Avenue, 64506, USA; andre.garcia@uga.edu*

With the increasing number of genotyped animals, the algorithm for proven and young (APY) can be used to compute the inverse of the genomic relationship matrix (G-1apy) needed for genomic BLUP (GBLUP) and single-step GBLUP (ssGBLUP). This algorithm also allows the use of all genotyped animals for the calculation of SNP effects from genomic EBV (GEBV). Obtaining SNP effects from GEBV is of interest because they can be used to calculate indirect predictions (IP), which are useful as interim evaluations for young genotyped animals or as genomic prediction for non-registered animals that are not included in official evaluations. The objective of the study was to evaluate the quality of IP obtained based on SNP effects from GBLUP, when the number of genotyped animals increase. The data and genotypes were provided by the American Angus Association and contained phenotypes for birth weight, weaning weight, and post-weaning gain. Phenotypes and genotypes were divided in 3 scenarios based on year of birth: genotyped animals born up to 2013 (n=114,937), 2014 (n=183,847) and 2015 (n=280,506). A 3-trait animal model was fit and GBLUP with APY was used to calculate GEBV and SNP effects. To calculate G-1apy, 19,021 animals were used as core, and those were randomly sampled from animals born up to 2013. In this way, the core animals remained the same for all scenarios, whereas the number of non-core animals increased as more genotyped animals were added. An alternative scenario had updated core animals for each year bracket. The SNP effects were calculated based on G-1apy and G-1 only for core animals (G-1core). IP were computed for all animals in each scenario by multiplying the matrix of centred genotypes by the SNP effects. To access the quality of IP, the correlation between the IP and GEBV was calculated for all animals. Correlations between GEBV and IP were greater than 0.99 for all traits in all scenarios. Despite the increase of non-core animals in APY, GEBV were successfully retrieved from SNP effects using IP. When SNP effects are calculated based on G-1core instead of G-1apy, updating the core animals as the number of genotyped animals increase seems to be the best choice.

Selection signature of tail fatness in Barbarine sheep using high density SNP markers

I. Baazaoui[1], S. Bedhiaf[2], K. Dodds[3], R. Brauning[3], R. Anderson[3], T. Van Stijn[3] and J. McEwan[3]
[1]Faculty of Sciences of Bizerte; University of Carthage, Zarzouna Bizerte, 7021, Tunisia, [2]INRAT, Rue Hédi Karray, 2049, Tunisia, [3]AgResearch Ltd, Invermay Agricultural Centre. Private Bag 50034. Mosgiel, 9053, New Zealand; imenbaazaoui18@gmail.com

Fat-tailed sheep is an important class of domestic sheep breeds, it represent about 25% of the world's sheep population. They are predominant in North Africa, reared in pastoral, transhumant and low input systems. The aim of this study is to perform a genome wide association related to tail fatness in Barbarine breed. A genotyping-by-sequencing data was used to provide dense genome-wide scan for 64,925 SNPs for a total of 105 Barbarine samples and 65 local thin tailed breeds. A genome wide association mapping was performed using a general linear model with correction for population structure using TASSEL.5 software. The strongest signal was observed on OAR 13 and OAR20 harbouring BMP2 and VEGFA genes, respectively. The VEGFA gene has a potential role in the control of energy metabolism and adipose tissue function and BMP2 gene has been identified in strong signals of fat tail selection in Mediterranean fat tailed breeds. The identification of genes with a role in the fat-tail phenotype contributes to the understanding of the physiology of fat deposition as well as the mechanisms of adaptation to better maintain future breeding options.

GBS data reveals the role of *KIT* gene in resistance to skin photosensitization in a black coat sheep

I. Baazaoui[1], S. Bedhiaf[2], K. Dodds[3], R. Brauning[3], R. Anderson[3], T. Van Stijn[3] and J. McEwan[3]
[1]Faculty of Sciences of Bizerte; University of Carthage, Jarzouna Bizerte, 7021, Tunisia, [2]INRAT, Rue Hédi Karray Ariana, 2049, Tunisia, [3]AgResearch Ltd, Invermay Agricultural Centre. Private Bag 50034. Mosgiel, 9053, New Zealand; imenbaazaoui18@gmail.com

Previous studies demonstrated a strong association between the pigmentation trait and resistance to photosensitization by environmental toxins in domestic animals. A local Tunisian 'Noir de Thibar' sheep breed, was recently selected to fix the black colour coat in order to avoid photosensitization following consumption of a wild grass (*Hypericum perforatum*), by white faced animals which caused major economic loss to sheep farmers. The aim of this study is to perform a genome wide association for resistance to *H. perforatum* photosensitivity. A genotyping-by-sequencing (GBS) data was used to provide medium genome-wide scan for 48,489 SNPs for a total of 38 Noire de Thibar and 85 Barbarine sheep samples that dispose black and white coat colour, respectively. A genome wide association mapping was performed using a general linear model (GLM) with correction for population structure using *TASSEL.5* software. The strongest selection signal identified two SNPs (rs399680744 and rs410179910) located on chromosome 6 and have a genomic position at 70,008,012 and 69,981,487 pb respectively. Gene annotation of the region ±200 kb from the position of significant SNPs harbour a candidate gene: *KIT* gene, implicated in coat pigmentation in different domestic animals which confirm the strong association between coat colour and the resistance to photosensitization by *H. perforatum* grass. This result brings a new insight into the genetic basis of adaptation to environmental constraints.

Assessing nanopore sequencing for detection of large structural variation in the Goettingen Minipig

J. Geibel[1,2], S. Hansen[3], A. Abdelwahed[3], S. Böhlken-Fascher[3], A.R. Sharifi[1,2], C.-P. Czerny[3], H. Simianer[1,2] and C. Reimer[1,2]
[1]*University of Goettingen, Center for Integrated Breeding Research, Albrecht-Thaer-Weg 3, 37075 Göttingen, Germany,* [2]*University of Goettingen, Animal Breeding and Genetics Group, Albrecht-Thaer-Weg 3, 37075 Göttingen, Germany,* [3]*University of Goettingen, Microbiology and Animal Hygiene, Burckhardtweg 2, 37077 Göttingen, Germany; johannes.geibel@uni-goettingen.de*

Recent developments in long-read sequencing technologies, especially Oxford Nanopore Technologies' nanopore sequencing, are said to come with substantial quality increase. To assess the usability of nanopore sequencing for detection and genotyping of structural variation in mammalian genomes, we processed a trio of the Göttingen Minipig at low coverage. The complete trio was sequenced applying a rapid barcoding sequencing protocol and the sire was additionally sequenced applying a $1D^2$ protocol, which is expected to increase the read quality. The barcoded approach led to 5.6 Gb in 25 h, which can still be increased in future runs, while reassigning the barcoded reads to the original samples failed for approximately 50% of the reads. An unsatisfying yield of the $1D^2$ approach was even reduced by a very small proportion of $1D^2$ reads, which could be restored from the 1D reads. The drawback of low base quality scores around 7.9 (83% accuracy), which could be slightly increased to 9.5 (88% accuracy) for the restored $1D^2$ reads, is compensated by very high read lengths (read length N50 of approximately 10 kb; longest read of 505 kb). Low coverages between 0.21 X and 0.72 X prohibited calling and genotyping of structural variations with sufficient confidence. Despite this drawback, we could identify 321 mendelian inherited variants larger than 50 bp between sire and offspring and 330 between dam and offspring, while requesting a minimum of five supporting reads in the total population. The results promise a flexible use of the technology combined with small initial investment costs and the unique possibility of directly observing structural variation from sequencing reads. However, a strong increase in coverage and by this a per-animal cost increase is necessary for reliable results.

Potential of precision livestock farming in small ruminant farming systems

C. Morgan-Davies[1], J.M. Gautier[2], E. González-García[3], I. Halachmi[4], G. Caja[5], L. Grøva[6], G. Molle[7], F. Kenyon[8], G. Lagriffoul[2], S. Schmoelzl[9], H. Wishart[1], A. Waterhouse[1] and D. McCracken[1]
[1]*SRUC, Hill & Mountain Research Centre, Kirkton, Crianlarich, FK20 8RU, United Kingdom,* [2]*Institut de l'Elevage, BP 42118, 31321 Castonet Tolosan Cedex, France,* [3]*INRA, SELMET, 2 Place Pierre Viala, 34060 Montpellier, France,* [4]*ARO, The Volcani Centre, Rishon LeTsiyon, 7505101, Israel,* [5]*Universitat Autònoma de Barcelona, 08193 Bellaterra, Barcelona, Spain,* [6]*NIBIO, Gunnars vei 6, 6630 Tingvoll, Norway,* [7]*Agris Sardegna, 07040, Olmedo, Italy,* [8]*Moredun Research Institute, Pentlands Science Park, Penicuik, United Kingdom,* [9]*CSIRO, Armidale, NSW 2350, Australia; claire.morgan-davies@sruc.ac.uk*

The benefit of precision livestock farming (PLF) is well recognised in the more intensive livestock sectors, such as dairy, pig and poultry. However, PLF has not been applied as widely in species where animals are considered to have a lower individual value or with less economic interest, as is the case in small ruminants (SR), or in extensive management systems. This is despite the very significant production, welfare, and labour efficiency advantages that can be achieved by applying PLF in these contexts. Despite their crucial role for the rural economy, society, and environment and their importance in ecosystem services such as biodiversity and maintaining cultural heritage, SR systems face issues such as challenging climatic and topographical conditions, lack of labour and low profitability that could be alleviated by introducing PLF technologies. Research on PLF for SR systems has been recently emerging, but perhaps lacks a joint up approach. This paper present an overview of current research and potential for future applications in several countries of PLF for SR systems on various themes, highlighting the wealth of potentially available solutions and prototypes. The topics presented cover feed and water intake, health, disease and parasite control and monitoring, fertility and reproduction management, grazing and predation control, animal locations and management monitoring, lambing and mismothering issues, as well as lactation monitoring. Issues relating to acceptability, economic relevance, technology readiness level (TRL) and industry engagement will also be discussed.

Virtual fencing for goats

S. Eftang[1,2] and K.E. Bøe[2]
[1]*Nofence AS, Evjevegen 8, 6631 Batnfjordsøra, Norway,* [2]*Norwegian University of Life Sciences, Animal and Aquacultural Sciences, P.O. Box: 5003, 1432 Ås, Norway; silje@nofence.no*

Virtual fencing is new precision livestock farming technology facilitating extensive production by giving livestock flexible access to pastures. It also contributes to better monitoring of grazing animals. The Nofence virtual fencing system establishes specific pasture borders defined by GPS-coordinates conveyed to animals via signals from a neck collar. When an animal crosses the virtual border, the collar starts playing a warning sound. A mild electric shock, with a stored energy of 0.1 Joule, is delivered by the collar if the animal does not respond by turning around upon hearing the sound. If an animal ignores three successive warning sounds and electric shocks, it is registered as escaped and the system is switched off until the animal has returned to the pasture. During autumn 2017, we performed a field study to collect data on the number of warning sounds and electric shocks given under three different scenarios, relating this to the welfare of goats fitted with Nofence collars. Ten commercial farms with a total of 92 goats (Boer and Kashmir breed) participated in the study. The goats were divided among three studies: 1. Experienced goats on their regular rangeland (10 groups, n=92), 2. Experienced goats being moved to new pastures or given a new, extended area (4 groups, n=45), and 3. Goats learning the system for the first time (6 groups, n=53). Data from collars were recorded for 7 days in study 1, while for study 2 and 3 the data were recorded for 5 days. Some of the goats were involved in all three studies. Goats in study 1 received a mean (range) of 10.6 (0.3-148.4) warning sounds and 0.4 (0-4.1) electric shocks /animal per day. Goats in study 2 received a mean of 4.8 (0-33.0) warnings sounds and 0.5 (0-3.4) electric shocks /animal per day, while in study 3, the means were, respectively, 5.9 (0.6-22.4) and 1.9 (0-7.6) /animal per day. We concluded that goats, within few days, started to associate the warning sound with the electric shock and hence learned the system, which in turn resulted in the animals getting relatively few electric shocks. There were large individual differences observed regarding the number of warning sounds.

Automatic monitoring of sheep pasture behaviour using motion sensor data

J. Maxa and S. Thurner
Bavarian State Research Centre for Agriculture, Institute of Agricultural Engineering and Animal Husbandry, Voettingerstr. 36, 85354 Freising, Germany; jan.maxa@lfl.bayern.de

Motion sensors are nowadays commonly applied in order to monitor animal behaviour and health status in indoor housing systems, predominantly for cattle. In contrast such sensors are not yet implemented for livestock and especially sheep on wide-range pastures. Nevertheless, information gained from such techniques can help to improve pasture management and welfare of grazing animals as well as to reduce daily workload of the farmer. The main objective of this study was therefore to develop a classification algorithm for detection of pasture behaviour of sheep based on data from tracking collars with motion sensors. Data needed for the development of the classification algorithm were obtained from a four day observation trial conducted on two paddocks located in Bavaria, Germany. In total five ewes of German Blackheaded Mutton Sheep were fitted with a tracking collar prototype collecting three-axis accelerometer and magnetometer data at 10 Hz. Four active (grazing, walking, running and other active activities) and two inactive (lying and standing, partially with ruminating) behavioural activities were recorded based on video observations for a maximum of eight hours per day. Motion sensor data from tracking collars were further aggregated into intervals of 10 seconds. For each interval the mean values with standard deviation were calculated for each axis corresponding to information about position of the neck and activity of the observed animal. Mixture models were applied to data of the training dataset in order to calculate thresholds among behaviours for a classification algorithm. The new classification algorithm used one axis of the magnetometer and enabled to distinguish among active and inactive behaviours of sheep. The algorithm was then applied to the test dataset resulting in a sensitivity and specificity of 96.1 and 94.3%, respectively. Furthermore, a possible development of a more complex classification algorithm via sensor fusion enabling to predict behaviours within the similar activity level will be investigated. Such information could give in future knowledge about the health status of sheep on wide-range pastures as well.

Weight based precision livestock farming approach for targeting anthelmintic use in grazing lambs

F. Kenyon[1], A. Hayward[1], D. McBean[1], E. Morgan[2], L. Andrews[1], A. Morrison[1] and L. Melville[1]
[1]Moredun Research Institute, Pentlands Science Park, EH26 0PZ, United Kingdom, [2]Queens University Belfast, 97 Lisburn Road, BT9 7BL, United Kingdom; fiona.kenyon@moredun.ac.uk

Gastrointestinal nematodes are a major cause of lost production in grazing livestock, control of which is complicated by the development of anthelmintic resistance. A performance-based targeted selective treatment (TST) approach, where individual lambs are treated when they do not meet pre-determined weight targets, maintains growth rates and drench efficacy, whilst reducing drench usage. However, this approach may not be suitable for all farms due to the frequency of weighing required. This study aimed to explore a less intensive approach by determining when anthelmintic treatment was required by monitoring weight gain in a proportion of lambs (sentinels). Four field trials were conducted, each using 100 co-grazing Texel-cross lambs of ~ 3 months old. In each year lambs were nominally split into 3 groups, where wormer treatment varied; TST (n=20): each lamb treated only if failing to reach a predetermined weight target using the Happy Factor model, strategic (n=20): group treated at weaning and 6 weeks post-weaning and mob (n=60): whole group treated when a proportion of lambs failed to reach their target weight. In studies 1 and 2, proportions of the mob used to determine treatment (sentinels) were 45 and 20%, respectively, with sentinels identified at the beginning of the study. In years 3 and 4 the sentinel lambs selected for monitoring were the first 20% weighed. We recorded lamb weight gain, faecal egg count (FEC), number of wormer treatments administered and the order in which animals were weighed. We assessed three measures of performance (body weight, weight gain, and total FEC) in the three groups (TST, 'strategic' and 'mob') at the end of each trial using t-tests (body weight) and Mann-Whitney tests (weight gain, FEC). In all trials there were no significant differences in body weight or weight gain between groups, however FEC varied between groups in some years. The number of drenches was the same in each group, but the timing differed. The study suggests that monitoring weight gain in a random 20% of lambs can determine when anthelmintic treatment is required for the group.

Water intake monitoring for health monitoring and evaluation of forage quality in sheep and goats

I. Halachmi[1], A. Godo[1], Y. Laper[1], V. Bloch[1], H. Levit[1], L. Rosenfeld[1], E. Vilenski[1] and T. Glasser[2]
[1]PLF Lab., Institute of Agricultural Engineering, Agricultural Research Organization (A.R.O.) – The Volcani Centre., P.O. Box 15159, HaMaccabim Road 68, Rishon LeZion, Israel, [2]Ramat Hanadiv, Nature Park, Zichron-Ya'akov, 3095302, Israel; halachmi@volcani.agri.gov.il

The development of an electromechanical system identified animal body weighed and water intake will be presented. This study is to prove that this system can indicate shifts in forage quality by data derived from water intake in real-time. Twenty two goats were divided to two groups of eleven goats each. One group has received alfalfa hay (high quality) while the other group has received wheat hay (low quality) every morning during nineteen days. Results clearly indicate a significant difference of water consumption, as measured by the automatic water trough among the two groups, regardless of the experimental period. Nonetheless, there were differences in total water consumption between experimental periods. These results enable the proof of concept that an automatic real-time watering system can evaluate feed quality shifts.

On field assessment of ultra-high frequency technology for sheep electronic identification

S. Duroy[1], J.M. Gautier[1], P.G. Grisot[1], F. Demarquet[2], D. Gautier[3], A. Hardy[4] and J.L. Przewozny[5]
[1]Institut de l'Elevage, Monvoisin, 35652 Le Rheu, France, [2]EPLEFPA Carmejane, Route d'Espinousse, 04510 Le Chaffaut Saint-Jurson, France, [3]CIIRPO, Le Mourier, 87800 Saint-Priest-Ligoure, France, [4]EPLEFPA Saint-Affrique, Route de Bournac, 12400 Saint-Affrique, France, [5]COOP de France Nouvelle-Aquitaine, Boulevard des arcades, 87000 Limoges, France; sebastien.duroy@idele.fr

The electronic identification of sheep and goats, mandatory in the European Union since 2010, is based on a low-frequency (LF) technology which enables to read animals one by one and up to 0,8 m at best with stationary readers. In France, after several years, few farmers are using radio frequency identification (RFID) for farm management and many stakeholders of the industry still have difficulties to get animal identification numbers in good conditions. Collecting animal identification numbers from batches of moving animals identified by LF requires to make animals going through individual corridors. Consequently, the flow of animals goes down and the waste of time can be difficult to accept by the stakeholders. Unlike the low-frequency technology, ultra-high frequency (UHF) RFID can read quickly simultaneously many tags with a longer read range. As UHF waves reflect on metals and cannot get through liquids and organic tissues, some have considered it was not suitable for live animals. Nowadays, UHF is more and more large scale used in many sectors of activity as logistic, medical, retails, industry... The technology is improving, and UHF tags can be used now close to liquids goods. As some trials of UHF identification on animals were emerging in different countries (deer in New-Zealand, pigs in Denmark, cattle in Scotland and USA...), the French Livestock Institute suggested in 2016 to the French sheep industry to carry out a field assessment of the UHF technology. The goal was to perform field trials and to get a practical experience about UHF: What is the reading rate on moving animals with UHF? Have metals and organic tissues an impact on readings? Do wave reflections generate unexpected readings? How UHF technology may solve problems which are still open with LF? This paper deals with these topics and presents the learnings got in different trials performed in sheep farms and collection centres.

Inferring social network centrality from behavioural tests in sheep: a novel method for precision livestock farming

R. Degrande, F. Bocquier and J.-B. Menassol
Montpellier SupAgro, MPRS, 2 place Pierre Viala, UMR SELMET bat.22, 34060, France, Metropolitan; rachel.degrande@gmail.com

In pasture-based systems activity sensors combined with GPS tracking are promising management tools for precision livestock farming. However, their performances are dependent upon the number and choice of the individuals to be equipped. We aim to propose simple field tests to rationalize the choice of the most representative individuals as a compromise between performances and costs. Thus, we start from the assumption that central individuals in sheep social networks have a larger impact on collective processes. The structure of the social network within a group of 55 animals derived from spatial associations automatically recorded every 5 minutes for 15 days thanks to collar-mounted radiofrequency sensors. Hierarchical rank and individual trend to lead, to initiate a collective movement and to explore a novel landscape were scored for each animal through field observations and tests for 7 days. Clustering individual properties and network analysis were performed using the R software (igraph package). Our results indicate that leadership behaviour depends on the motivation, when there is a competition for resource access then individuals with higher social statuses maintain leadership (Pearson, r=0.5297, P<0.05); otherwise for collective movements at pasture, leadership is provided by individuals that tend express initiation (Pearson, r=0.3976, P<0.05) this latter trait being correlated with exploratory traits (Pearson, r=0.5868, P<0.05). We found no significant correlation between individual behavioural scores and eigenvector centrality. This is certainly because the network is measured during the test period, thus a comparison between in-test and before-test period networks will conclude about the results. This first study needs to be assessed by long term monitoring as well as a dynamic network approach and establishes the basis of a novel on-farm method to rationalize the usage of embarked digital tools.

On-farm automated tracking of group-housed poultry

E.D. Ellen[1], M. Van Der Sluis[1,2], B. De Klerk[3], Y. De Haas[1], T. Hijink[1] and T.B. Rodenburg[1,2]
[1]Wageningen University & Research, Droevendaalsesteeg 1, 6708 PB Wageningen, the Netherlands, [2]Utrecht University, Yalelaan 2, 3584 CL Utrecht, the Netherlands, [3]Cobb Vantress, Koorstraat 2, 5831 GH Boxmeer, the Netherlands; esther.ellen@wur.nl

Production animals, including poultry, are increasingly kept in groups, making identification and tracking of individual animals challenging. Often video analyses are used, but these are time-consuming and prone to human error. Therefore, automated systems for monitoring individual animals are desired. One possible approach is the implementation of an ultra-wideband (UWB) system to track individual animals. A previous study implemented an UWB system for laying hens in a research setting. Hens were fitted with an UWB tag in a small backpack and, using stationary sensing beacons, the location of each bird could be determined. Compared to video tracking of individual birds, the UWB system was shown to be capable of detecting the bird's location with 85% accuracy. In the current study, the system was applied on a broiler farm to assess its on-farm applicability. At approximately 14 days of age, broilers were fitted with an UWB tag. Twelve birds were tracked with the UWB system on nineteen consecutive days, for one hour each day. Distances moved according to the UWB system were compared to those found on video and a moderately strong correlation between the UWB system and video tracking was found (Repeated measures correlation, r=0.71, P<0.0005). Furthermore, individual levels of activity were assessed using this setup. 137 birds from different genetic crosses were tracked near-continuously for seventeen consecutive days. First results indicate that for all crosses, the UWB system was found to be capable of detecting a decrease in activity over time. Overall, the UWB system appears well-suited for tracking of broilers. However, the UWB tags are relatively large. Therefore, tracking and monitoring of young broilers is not possible. Current work is focussing on implementing a passive radio frequency identification system, with smaller tags. Possibly, this system can track individual activity of broiler chickens throughout their life.

Automatic individual-broiler temperature measuring by thermal imaging camera

N. Barchilon[1,2], V. Bloch[2], I. Halachmi[2] and S. Druyan[1]
[1]Animal Science, Agricultural Research Organization (A.R.O.) – The Volcani Centre, P.O. Box 15159, HaMaccabim Road 68, Rishon LeZion, Israel, [2]PLF Lab., Institute of Agricultural Engineering, Agricultural Research Organization (A.R.O.) – The Volcani Centre, P.O. Box 15159, HaMaccabim Road 68, Rishon LeZion, Israel; halachmi@volcani.agri.gov.il

The heat stress of broilers in commercial broiler-houses significantly decreases their profitability. However, the today climate control systems use sensors measuring the temperature around the broilers, which can be different from the actual body temperature of the broilers. In this research, a method for measuring the body temperature of an individual broiler was designed and validated. The method is based on a low-cost IR camera calibrated by a thermistor, algorithm for the IR image processing and the Lasso regression model predicting the actual body temperature. A prototype utilizing this method was built and tested in a research broiler-house during 21 days for 15 broilers. The predicted body temperature was compared with the actual body temperature measured by implanted temperature loggers. The standard error of the method was 0.25 °C.

Genetic of the 3D morphology of jumping horses using geometric morphometrics

A. Ricard[1,2], P. Pourcelot[3], B. Dumont Saint Priest[1], N. Crevier-Denoix[3] and S. Danvy[1]
[1]Institut Français du Cheval et de l'Equitation, Pole développement, Innovation et Recherche, 61310 Exmes, France,
[2]Institut National de la Recherche Agronomique, Génétique Animale et Biologie Intégrative, 78350 Jouy en Josas, France,
[3]Institut National de la Recherche Agronomique, Ecole Nationale Vétérinaire d'Alfort, Unité 957, 94700 Maison Alfort,
France; anne.ricard@inra.fr

Morphological data were recorded on 2,097 jumping horses aged 4 and 5 with 3D technology using 3 cameras. For each horse, the anatomical landmarks of the main joints, including the head, have been identified, leading to a 25 point file describing the morphology of the horse with their coordinates in 3 dimensions. The study was made by dissociating the shapes of the general size through a Procrusted type analysis. After creation of new Procrustes coordinates, a principal component analysis reveals the predominant forms in playing simultaneously on all points. The Geomorph package under R was used. The heritability of these components has been estimated after correcting the influence of the person who made the marks on the images of the horses, the place and the date of measure, the age and sex of the horse, and the angle of the anterior and posterior canons with the vertical on the chosen image. The genealogies over 6 generations were used (18,029 horses). ASREML software was used. 10 components are needed to explain 80% of the variance. The first two components are the orientation of the limbs in the frontal plane: the first distinguishes the horses whose posteriors marks are close to the midline of the hoof point. The second is similar for the forelegs with a width at shoulders following the same movement. The third component is mainly explained on the sagittal plane and distinguishes big and short horses of long and small for the same general size. The heritability of the overall size is 0.21. The two first components are not heritable (0.09 and 0.06). Only components 3, 6 (length of the neckline) and 8 (high neck tie, long shoulder, short withers and hollow back line) are really heritable (respectively 0.27, 0.26 and 0.21). Genetic correlation with overall size were not null, denoting a strong allometry. Multiple phenotypic regression with performances in jumping competition do not exceed a r^2 of 2%.

Using a microchip to detect temperature variations before parturition in mares

J. Auclair-Ronzaud[1], L. Wimel[1] and P. Chavatte-Palmer[2]
[1]IFCE, Experimentation research center of Chamberet, Domaine de la Valade, 19370 Chamberet, France, [2]INRA, UMR BDR, Université Paris Saclay, 78350 Jouy-en-Josas, France; juliette.auclair-ronzaud@ifce.fr

In mammals, body temperature is an endogenous process regulated by thermoregulation, and following a circadian rhythm. In a previous study, the use of an identification microchip implanted in the neckline was determined as a new and non-invasive way to measure body temperature. Temperature variations depend on numerous parameters such as sex or activity. However, if the temperature variations are well determined under normal conditions, the impact of farming events on these variations has yet to be explored. One of the most important events is breeding and especially foaling. Predicting its occurrence is a great challenge for the horse-breeding sector. This study was conducted on 18 mares in 2018. They were kept in individual boxes of 20 square meters and fed twice a day (forage and concentrates). They had access to a shared paddock three times a week between 10am and 4 pm. Temperature was measured day and night every 2 hours for, at least, 12 days before parturition. After foaling mares' temperature was monitored for two days. A significant difference between days before parturition and parturition day was detected for 79% of the mares (P<0.001) with an average drop of 0.8 °C. In each case, individuals are their own control to detect a difference between days before foaling and the parturition day. Temperature started dropping one to three hours before foaling, depending on the individuals. As we were able to identify changes before parturition, our goal is to explore the possible use of this tool as a new predictor of foaling.

Effects of feeding systems on thermoregulation and semen quality of Merino and SA Mutton Merino rams

A.M. Du Preez, E.C. Webb and W.A. Van Niekerk
University of Pretoria, Animal Sciences, Pretoria, Hatfield, 0002, South Africa; amelia.may.dp@gmail.com

This study investigated the effects of feeding systems during the growing phase (5-12 mnth old) on scrotal thermoregulation capacity and semen quality of Merino (M) and SA Mutton Merino (SAM) rams. The effects of extensive feeding (C), extensive followed by intensive feeding (TR1) and intensive feeding (TR2) were evaluated in M and SAM rams. Growth, size, scrotal measurements, semen quality and post mortem gonadal measurements were recorded. Data was analysed by GLM ANOVA procedures, using IBM SPSS V 23. Differences between treatment means for breeds, feeding systems and interactions were tested by Bonferroni's multiple range test at $P<0.05$. Growth, size and gonadal measurements differed between breeds and feeding systems ($P<0.05$). Measurements to assess testicular thermoregulation were Pampiniform venous plexus circumference (PPC) and scanned scrotal neck fat (SSNF). PPC:SSNF gave an estimate of effective PPC (EffecPPC) and hence thermoregulation capacity. M and SAM had better PPC ($P<0.05$) in C than TR1. M had more SSNF in TR1 and SAM in C ($P<0.05$). M and SAM had less SSNF and smaller PPC in TR2 ($P<0.05$). PPC:SSNF of SAM and M in TR2 was better than SAM and M in C ($P<0.05$). EffecPPC:Testes mass (TM) of SAM in TR1 and C were better than TR2 ($P<0.05$). Merino in C had smaller EffecPPC:TM than M in TR1 or TR2. SAM and M in C had better semen quality than in TR1 and TR2. SAM in TR2 had correlations ($P<0.05$) between PPC:TM and percentage normal sperm (PNS)(r=-0.68), EffecPPC:TM and PNS (r=-0.70), PPC:TM and dag A defects (dAd)(r=0.84), EffecPPC:TM and dag D defects (dDd)(r=0.70). M in C had correlations between EffecPPC:TM and dAd (r=096), EffecPPC:TM and progressive motility (PM)(r=-0.85), confirming the importance of testes size. In TR1, PPC was important in M based on correlations of PPC:SSNF and dAd (r=-0.68), PPC:TM and dAd (r=-0.98), PPC:TM and dDd (r=-0.85), PPC:TM and PM (r=0.93), PPC:TM and PNS (r=0.82), EffecPPC:TM and dAd (r=-0.70) and between EffecPPC:TM and PNS (r=0.72). PPC:SSNF and EffecPPC:TM possibly misleading – requires further research. The current research suggest that SSNF does not adversely affect semen quality in young rams, and that size of testes and PPC is more important.

Artificial insemination without hormonal synchronization of ewes using an automated oestrus detector

N. Debus, G. Besche, A. Lurette, J.-D. Guyonneau, A. Tesniere, J.-B. Menassol and F. Bocquier
SELMET, Univ Montpellier, CIRAD, INRA, Montpellier SupAgro, 2 place Viala, 34060 Montpellier Cedex 1, France; jean-baptiste.menassol@supagro.fr

The achievement of hormone-free artificial insemination (AI) in small ruminants is a challenging goal in terms of technology, methodology, animal welfare and environmental impacts. This goal is getting closer thanks to an automated oestrus detector (Alpha-D® or Alpha). This original digital tool consists in a RFID reader worn by the male that is triggered at the time of mounting to read a transponder placed on the rump of the females. When triggered the detector stores the ID number of the ewe together with the date and time of each mount. Thus to consolidate hormone-free AI programs we conducted experiments over three consecutive years using the combination of Alpha and ram effect in two farms and using two different breeds, Mérinos d'Arles (meat breed) and Lacaune (dairy breed). All inseminations were performed intra-cervically, with fresh extended semen collected in the morning, on ewes soon after being detected in oestrus by the Alpha. Mérinos ewes were inseminated over 3 days centred on the 1st or 2nd peak of oestrus after the introduction of the rams depending on the year. Following the classification established by the Alpha 73, 51 and 77 ewes out of 399, 343 and 354 were inseminated in 2013, 2014 and 2015 respectively. The mean success rates of these AI averaged 37% in 2013, 65% in 2014 and 64% in 2015. In the Lacaune ewes, 99 and 118 ewes out of 259 and 224 were inseminated in 2014 and 2015 over 6 days centred on the 1st and 2nd peak of oestrus after ram effect. The success rates of these AI averaged 68% in 2014 and 74% in 2015. Our results revealed that ewes inseminated between 0 and 35 h after the beginning of the detected-oestrus had higher fertility rates (65±5% in Mérinos ewes and 75±5% in Lacaune ewes) than those inseminated after 35 h (35±5% in Mérinos ewes and 42±5% in Lacaune ewes; $P<0.05$). These first experiments show that in these conditions hormone-free AI programs are feasible. Now we have to develop a set of services around an integrated digital tool that AI centres could adopt along with farmers.

Can we compare plate meters with height sticks for measure grass height?

T.J. Jousset
IFCE, DIR, Domaine de la Valade, 19370 Chamberet, France; tristan.jousset@ifce.fr

Different grassmeters exist today, we can find the plate meter, which measure the height of grass compressed, or others like rulers. All these tools allow monitoring grass height when horses come in and come out of the pasture. It represents also a way to evaluate grass stocks or grass growth. The aim of this study is to find out a correlation between the different measurement tools and equations to switch from one to another. We had three different tools, first we used a height stick, which is a graduated ruler, and then the Grasshopper® (Irish) and the Jenquip® (New Zealand) which are plate meters. These two grassmeters have an identical sized circular metal plate with a diameter of 35.5 cm. The Grasshopper exerts 31.25% more pressure on the grass than the Jenquip, with applied pressures which are, respectively, of 6.3 and 4.8 kg/m². Regarding the precision of measurement, the height stick measures centimetre by centimetre, whereas the Jenquip has a precision of 0.5 cm and the Grasshopper has 0.1 cm precision. Measurements were recorded on permanent pasture. For each point, 1 measurement per tool was performed. In 2018, 297 measurement points were taken at the experimental station of Chamberet. On average, grass heights measured with the height stick are 10.1±5.7 cm, those measured with the Grasshopper are 7.7±4.0 cm and those measured with the Jenquip are 7.6±4.0 cm. Analysis provide following equations: Height[Grasshopper] = 0.99 × Height[Jenquip] (n=297; r²=0.87) Height[Grasshopper] = 1.29 + 0.63 × Height[Stick] (n=297; r²=0.78) We notice an almost linear relationship between grass heights measured with the Grasshopper and the Jenquip, with an origin and a slope near to the line X=Y. The comparison between the height stick and the Grasshopper shows a linear relationship, but with a different origin and slope from the line X=Y. This study shows that measurements made with the Grasshopper are identical to those of made with the Jenquip. However, for the height stick, it is necessary to convert measurements before any comparison of heights. Those equations allow results comparison between places and operators.

Using herbage disappearance method to estimate forage intake

G. Ortega, Y. Lopez, J.M. Garrido, D. Custodio and P. Chilibroste
Agronomy Faculty, Animal Science Department, Av Gral Eugenio Garzon 809, 12900, Uruguay; gortegaconforte@gmail.com

Four farmlets under a rotation of annual and perennial pastures were grazed with one of four 2×2 factorial arrangements of treatments: two stocking rates 1.5 (MSR) or 2.0 (HSR) milking cows per hectare and two pasture managements: A (4 cm residual sward height), and B (6 cm residual sward height for autumn and winter, and 9 cm for spring). Treatments grazed during 2017 and 2018 from March till December. Sward mass of each individual plot was assessed weekly through the double sample technique. Every fifteen days in each paddock five scales with three repetitions were cut in 0.51×0.30 m quadrant size. Relation between height and biomass was made for each pasture for pre and post grazing condition. Indirect estimates of herbage mass were done daily before and after grazing using a ruler. Mass cover was estimated (kg DM/ha) for each grazing session with a linear regression $Y_{(kg\ DM/ha)} = \beta_1{}_{(h\ cm)} + \beta_0$ for each type of pasture pre and post grazing y = 103.15x + 237.6 R²=0.44, n=274; y = 124.6x – 202.5 R²=0.3, n=70 for Medicago sativa, y = 98.8x – 264.33 R²=0.38 n=158; y = 103.5x – 359.1 R²=0.23 n=26, for *Avena byzanthina* with *Lolium multiflorum*, and y = 103,15x + 269 R²=0.51, n=1,127; y = 85.6x + 522.87 R²=0.05; n=266 for *Dactylis glomerata* pre and post grazing, respectively. Estimations were contrasted with estimation derived from a unique equation (UE) for all type of pastures involved = 103x + 269,32 R²=0,41 n=1,559; y = 75,157x + 609,49 R²=0,10 n=427 pre and post grazing respectively. Pasture dry matter intake per cow per day (PDMI) was estimated based on pasture disappearance before and after grazing corrected divided by the number of cows grazing. Differences were declared significant when P<0.05 (Tuckey). The PDMI estimated with UE equations was not different between 1.5B, 2.0B, and 2.0A treatments (14.4 kg DM/cow/day P>0.05) while 1.5A exhibited a lower value (11.9 kg DM/cow/day, P<0.05). Same result was obtained with the use of DE equations (13.8, 13.7, 13.4 kg DM/cow/day for 1.5B, 2.0B, 2.0A, (P>0.05) respectively, and 11.5 kg DM/cow/day for 1.5A (P<0.05). We concluded that it is possible to estimate PDMI trough herbage disappearance using either a UE or using specific regressions for each forage involved in the rotation.

Post-calving leukocyte immune-related genes are enhanced in Simmental compared with Holstein cows

V. Lopreiato[1], A. Minuti[2], D. Britti[1], C. Perri[1], F. Piccioli Cappelli[2], J.J. Loor[3] and E. Trevisi[2]
[1]Magna Graecia University, Interdepartmental Services Centre of Veterinary for Human and Animal Health, Viale Europa, 88100 Catanzaro, Italy, [2]Università Cattolica del Sacro Cuore, Department of Animal Sciences, Food and Nutrition, Via Emilia Parmense, 29122 Piacenza, Italy, [3]University of Illinois, Department of Animal Sciences and Division of Nutritional Sciences, W. Gregory Drive, 61801 Urbana, USA; vlopreiato@gmail.com

Simmental is a cattle breed selected for meat and milk production. However, in the last decades breeding programs started to address genetic selection of this breed to increase milk production for intensive production systems, but still maintaining the dual-purpose. The aim of this study was to investigate changes in the expression of genes involved in leukocytes function of cows highly specialized for milk production, Holstein, and cows selected for meat and milk production. The experimental design of this study involved Simmental (SM; 6 primiparous and 7 multiparous) and Holstein (HO; 6 primiparous and 6 multiparous) cows enrolled in two different farms. Blood was collected on d 3 after calving in PAXgene tubes (Preanalytix) to measure mRNA expression of 33 genes. Normalized data were subjected to MIXED model of SAS. Compared with HO, SM dairy cows had greater (P≤0.05) transcript abundance of *IL1B*, *TNF*, *IL1R*, *TNFRSF1A*, *CX3CR1*, *ITGB2*, *CD44*, *LGALS8*, and *IDO1* genes. In addition, SM cows tended to have higher gene expression of *CD16* (P=0.10), *MYD88* (P=0.07), *RPL13A* (P=0.10), *MPO* (P=0.07), and *ALOX5* (P=0.10) compared with HO. In contrast, compared with SM, HO cows had greater (P<0.05) abundance of *TLR2*, *MMP9*, *LTF*, and *S100A8* genes. Gene expression profiling shown that leukocytes of SM cows have been characterized by an increased function of chemotaxis, cell-cell interaction, pathogen-recognition pathway, and inflammatory mediators, that in turn might be beneficial for mounting a better response to the homeorhetic adaptation of the new lactation. In conclusion, these results unreval a breed-specific immune response between SM and HO cows in the first days post-calving under intensive dairy management, providing important biological insights into the immune-function differentiation among cattle breeds and functional information to be used in the breeding programs.

Health traits and lactation survival in relation to conformation traits in German Holstein cows

A. Rolfes and H.H. Swalve
Martin-Luther University Halle-Wittenberg, Institute of Agricultural and Nutritional Sciences, Theodor-Lieser-Str. 11, 06120 Halle, Germany; anke.rolfes@landw.uni-halle.de

For many breeders and breeding organizations of dairy cattle, conformation traits still play an important role. The objectives of the present study were (1) to examine relationships between health traits and conformation traits, and (2) to study differences in conformation traits comparing cows culled within the first lactation to cows reaching four or more lactations. Data originated from large contract herds in North-Eastern Germany. In these herds, all cows are scored for conformation and health traits for all sorts of diseases are recorded under a supervised scheme. For conformation, data consists of the usual linear recording on a 1 to 9 scale. Data set I comprised 57,829 first lactation records with information on health and conformation for years of first calving from 2008 to 2015. For data set II, conformation data was truncated to include years of first calving from 2008 to 2013 in order to ensure that all cows had the chance to reach a fourth calving till 2018 which was the cut-off point for survival information. Data set II contained 8,645 cows culled within the first lactation and 26,141 cows that finished at least three lactations. A linear mixed model was used for data analysis. When comparing groups of affected cows vs unaffected cows for various health traits (data set I), the traits dairy form, rear legs side view, hock quality, rear legs rear view, fore udder attachment, udder depth, BCS and especially locomotion showed significant differences in conformation scores for several diseases within the first lactation of 0.20 to 0.56 (P<0.0001). The comparison of survival to differing numbers of parity, i.e. contrasting cows culled within first lactation against cows with more than three lactations (data set II), revealed significant differences of at least 0.10 in average conformation scores (P<0.0001) for the traits hock quality, rear udder width, suspensory ligament, fore udder attachment, udder depth, body condition score (BCS) and locomotion. In conclusion, some conformation traits are very valuable indicators of health and longevity while others do not exhibit any relationships at all.

Neural Network models as a new method to diagnose ketosis in dairy cows

E.A. Bauer, E. Ptak and W. Jagusiak
University of Agriculture in Krakow, Department of Genetics and Animal Breeding, Al.A. Mickiewicza 24/28, 31-120, Poland; e.bauer@ur.krakow.pl

Clinical ketosis is a common disease in dairy cattle, especially in the days after calving, and it is often undiagnosed. The procedure currently used for detection of cows at risk of ketosis is not effective enough. Many cows are diagnosed too late, causing serious economic losses. The aim of the study was to estimate relationship between milk composition and β-hydroxybutyric acid level in blood an indicators of ketosis with using one of the artificial neural networks (ANN) model multi layer perceptron (MLP). An ANN is a powerful computational technique for correlating data by using a large number of simple processing elements. The ANN simulator implemented in the *STATISTICA*13® package was used in the study. A classical feedforward ANN between 8 to 15 hidden layers, eight input layers and one output layer, with a backpropagation learning algorithm and sigmoid transfer functions, was used in this investigation. A neural model was a essentially a regression model for establishing correlations between a vector of independent variables (inputs) and a vector of dependent variables (outputs). The study was repeated with 17,800 architectures for MLP type. The analyses included data set on population of 1,085 polish Holstein-Friesian cows made available by the Polish Cattle Breeders and Milk Producers. The set of milk data composition as: milk yield(kg), protein(%), fat(%), fat-to-protein ratio(%), lactose(%), somatic cell count(cell/ml) and content of urea(mmol/l), β-hydroxybutyric acid (BHBA) ≥1.2 mmol/l, and acetone was used as input variables. Our results show that the benefit of milk composition information to predictive ability of the prediction model is high. The ANN models shows a Pearson correlation coefficient of >0.965 respectively obtained in the data learned sets. By reducing the number of variables, preliminary results were obtained: at high parameters of sensitivity analysis for ACE and BHBA. Ketosis can be accurately modelled by using only these two milk composition components. The results and conclusions of this study provide new grounds for improvement of ANN to diagnosis ketosis. The Research was financed by the National Science Center, Poland no.2017/25/N/NZ5/00793.

The bovine colostrum miRNome and its implications for the neonate

I. Van Hese[1,2], K. Goossens[2], G. Opsomer[1] and L. Vandaele[2]
[1]*Faculty of veterinary medicine, Ghent University, Reproduction, Obstretrics and Herd Health, Salisburylaan 133, 9820 Merelbeke, Belgium,* [2]*Flanders Research Institute for Agriculture, Fisheries and Food (ILVO), animal science unit, Scheldeweg 68, 9090 Melle, Belgium; ilke.vanhese@ilvo.vlaanderen.be*

Calves are born without circulating antibodies because the bovine placenta prevents maternal antibodies from crossing over to foetal circulation. Instead, transfer of passive immunity takes place through the ingestion of colostrum, which is the first milk a cow produces and contains a large portion of immunoglobulins G (IgG). Antibodies can only be absorbed in the small intestines within 24 hours after birth. Serum IgG concentrations lower than 10 g/l at an age of 1 week (also referred to as failure of passive transfer (FPT)) are associated with higher risk of illness or death at a young age. Colostrum provides many other important components like growth factors, hormones, vitamins and minerals. Recently, the presence of microRNAs (miRNAs) in bovine colostrum and milk was confirmed. MiRNAs are short, non-coding RNAs that regulate gene expression at a post-transcriptional stage. Colostrum is in particular rich in immune- and development-related miRNAs. This study investigates the miRNome of bovine colostrum in order to find a link with the immune and health status of the calf. Samples of first colostrum were taken at the ILVO research barn from 108 dairy (n=76) and beef (n=32) cows immediately after parturition. Calves received a total of 6 litre of colostrum in the first 24 h of life. Blood samples were taken from calves at the age of 3 days and serum IgG was measured using electrophoresis. Serum IgG concentrations (mean ± SD) averaged 20.28±8.68 g/l. The prevalence of FPT (serum IgG <10 g/l) was 10.3%. RNA was extracted from colostrum with the miRNeasy mini kit (QIAGEN) and Small RNA sequencing is currently performed to determine miRNA expression in colostrum. After processing the sequencing results, differential expression analyses will be performed between calves with high levels of serum IgG and calves with low levels of serum IgG. Analyses are in process and results will be available for the conference. Ilke Van Hese is an SB PhD fellow at FWO, Research Foundation – Flanders, project number 1S20218N.

Not all inbreeding is depressing

H.P. Doekes[1,2], R.F. Veerkamp[2], P. Bijma[2], S.J. Hiemstra[1], G. De Jong[3] and J.J. Windig[1,2]
[1]Centre for Genetic Resources the Netherlands, Wageningen University & Research, P.O. Box 16, 6700 AA Wageningen, the Netherlands, [2]Animal Breeding and Genomics, Wageningen University & Research, P.O. Box 338, 6700 AH Wageningen, the Netherlands, [3]Cooperation CRV, Wassenaarweg 20, 6843 NW Arnhem, the Netherlands; harmen.doekes@wur.nl

Inbreeding decreases animal performance (inbreeding depression), but not all inbreeding is expected to be equally harmful. Inbreeding on recent ancestors is expected to be more harmful than inbreeding on more ancient ancestors, because of purging. Purging is the removal of deleterious recessive alleles over time by selection. We investigated the effects of recent and ancient inbreeding on yield, fertility and udder health of 38,792 Holstein Friesian first-parity cows, using linear mixed models. Pedigree data were used to compute traditional inbreeding (F_{PED}) and genotype data were used to identify regions of homozygosity (ROH) and compute ROH-based inbreeding (F_{ROH}). Inbreeding depression was apparent, e.g. a 1% increase in F_{ROH} was associated with a decrease in 305-d milk yield of 36.3 kg (SE=2.4), an increase in calving interval of 0.48 days (SE=0.15) and an increase in mean somatic cell score in day 150 to 400 of 0.86 units (SE=0.28). Distinguishing recent from ancient inbreeding gave mixed results. For example, only very long ROHs (which indicate recent inbreeding) significantly increased calving interval, whereas both long and short ROH decreased protein yield. When F_{PED} was split into new and ancestral components, based on whether alleles were *identical by descent* for the first time or not, there was clear evidence of purging. For example, a 1% increase in new inbreeding was associated with a 2.2 kg (SE=0.4) decrease in protein yield, compared to a 0.9 kg (SE=0.8) increase for ancestral inbreeding. The mixed results may be partly due to difficulties in estimating ancient inbreeding. Distant ancestors are less well registered, and short ROHs may be less reliable than long ROHs. Furthermore, purging may have acted on some, but not all alleles. To conclude, there is purging in the population, but purging effects are only partly reflected by the difference in inbreeding depression due to recent and ancient inbreeding.

Genome-wide association study for methane concentration emitted by dairy cows

M. Sypniewski, T. Strabel and M. Pszczola
Poznan University of Life Sciences, Genetics and Animal Breeding, Wolynska 33, 60-637, Poznan, Poland;
mateusz.sypniewski@up.poznan.pl

Methane (CH_4) emission has a potential impact on the economic and environmental aspects of the livestock industry. The amount of methane emitted from livestock can be quantified in several ways, many of which are indirect. The aim of this study is to investigate the genetic background of a direct CH_4 measurement of methane concentration. The studied trait was the average daily concentrations of CH_4 measured in particles per million (ppm). We aimed to assess the variability of a proposed trait by estimating heritability and conducting genome-wide association studies (GWAS) using records of genotyped and ungenotyped animals. CH_4 concentrations were performed on 495 Polish Holstein – Friesian cows at two commercial farms over a period of 20 months. CH_4 concentration was measured using Fourier transform infrared (FTIR) analyser installed in the Automated Milking System (AMS). Overall 335 of 495 animals available were genotyped with 50k SNP chip. The model used in this study included the fixed effect of herd-year-week, general lactation curve fitted with up to 4th degree Legendre polynomials and cow's lactation number, random effect of the animal and random permanent environment effect. All analyses were performed in BLUPF90 family software Heritability estimates were acquired with the REML procedure using a single-step approach. GWAS analyses were performed as post analyses of solutions obtained from single-step BLUP analyses. To identify statistically significant SNPs, we used a sliding window of 20 adjacent SNPs and calculated the variance explained. Estimated heritability reached 0.08 and the highest amount of variance explained by 20 adjacent SNPs reached 0.4% while none of the analysed SNPs exceeded the 10-5 threshold for statistical significance. Results of this study indicate that methane is a highly polygenic trait.

On the genomic regions affecting milk lactose content in dairy cattle

A. Costa[1], H. Schwarzenbacher[2], G. Mészáros[3], B. Fuerst-Waltl[3], C. Fuerst[2], J. Sölkner[3] and M. Penasa[1]
[1]University of Padova, Department of Agronomy, Food, Natural Resources, Animals and Environment, Viale dell'Università 16, 35020 Legnaro (PD), Italy, [2]ZuchtData EDV-Dienstleistungen GmbH, Dresdner Strasse 89/19, 1200 Vienna, Austria, [3]University of Natural Resources and Life Sciences Vienna (BOKU), Department of Sustainable Agricultural Systems, Division Livestock Sciences, Gregor Mendel-Strasse 33, 1180 Vienna, Austria; angela.costa.1@phd.unipd.it

Lactose is the major carbohydrate of bovine milk and its concentration may be used as an indicator of udder health in cows. In fact, negative and moderate phenotypic and genetic correlations of lactose content with somatic cells and mastitis have been reported in literature. Besides, lactose content is more heritable than other milk traits, including somatic cells. Despite this, only few studies have explored the genes affecting lactose in cattle so far. Therefore, a genome-wide association study for lactose content was carried out using Austrian Fleckvieh bulls' de-regressed proofs, obtained in MiX99 software basing on 16 million test-day records. Following the conventional quality control, 40,964 SNPs (54k chip) in 2,854 bulls (estimated daughter contribution ≥10) were available for the association analysis in GEMMA software. After false discovery rate correction, 22 SNPs spread across the genome were found as significant. The highest P-value was found within the gene *MFSD6* (BTA2, 5.757 Mb), known for its role in macrophages reception and intra/extra cellular transport facilitation. A variant at the end of BTA1 was within *SH3BP5*, a gene involved in white cells development and activity. One significant SNP was found within *EFNA1* (BTA3), reported to regulate inflammatory response in udder of cows; moreover, variants were around 58 Mb on BTA20, region known to affect SCS/mastitis. Other signals were within and/or nearby genes related to transmembrane molecular transport (e.g. *KCNK1, TFRC, RASGEF1C, LOC515333* and *SLC35F3*). The four most significant SNPs (P<5.05×10^{-07}) explained 5% of the genetic variance; these results thus support that lactose content is a polygenic trait with a complex genetic architecture and affected by several genomic regions with small effect size.

Genetic correlations between energy status indicator traits and fertility in Nordic Red Dairy cows

T. Mehtiö, P. Mäntysaari and M.H. Lidauer
Natural Resources Institute Finland (Luke), Myllytie 1, 31600 Jokioinen, Finland; terhi.mehtio@luke.fi

Negative energy balance in early lactation may predispose cows to health and fertility problems, and therefore should be considered in dairy cattle breeding programs. However, measuring energy status (ES) is challenging, and thus easily measured indicator traits are of interest. In this study we estimated genetic variation in ES indicator traits and their genetic correlation with fertility in early lactation. The six indicator traits considered were: plasma non-esterified fatty acid (NEFA) concentration (mmol/l) either predicted by multiple linear regression including milk fat:protein (FPR), fatty acids (FA) C10, C14, C18:1 cis-9, C14*C18:1 cis-9, and days in milk ($NEFA_{FA}$) or predicted directly from milk mid-infrared spectra ($NEFA_{MIR}$), milk FA C18:1 cis-9 (g/100 ml milk), FPR, beta-hydroxybutyrate (BHB, mmol/l milk), and acetone (mmol/l milk). Interval from calving to first insemination (ICF) was considered as the fertility trait. The data consisted of 16,294 primiparous Nordic Red Dairy Cattle cows with first test-day record and NEFA predictions from early lactation (week 2 to 6) during 2015 to 2018. From these cows, 13,852 cows had ICF observations, and 10,677 cows had BHB and acetone observations. The fitted multivariate animal model included the fixed effects of herd×year, year×month, and calving age as well as linear regressions on days in milk for ES indicator traits and linear regression on heterozygosity for ICF, and the random additive genetic and residual effects. The pedigree was pruned to four generations and had 59,712 animals. The heritability estimates were 0.13, 0.16, 0.10, 0.05, 0.12, 0.16, and 0.04 for $NEFA_{FA}$, $NEFA_{MIR}$, C18:1 cis-9, FPR, BHB (log-transformed), acetone (log-transformed), and ICF, respectively. Strong genetic correlations were found between $NEFA_{FA}$ and C18:1 cis-9 (0.94), $NEFA_{FA}$ and $NEFA_{MIR}$ (0.91), and BHB and acetone (0.96). The genetic correlations with ICF were 0.63, 0.55, 0.55, 0.47, 0.58 and 0.46 for $NEFA_{FA}$, $NEFA_{MIR}$, C18:1 cis-9, FPR, BHB and acetone, respectively. In conclusion, $NEFA_{FA}$ had the highest genetic correlation with ICF compared to other ES indicator traits, and is capable to determine the genetic variation of ES in early lactation.

Combining rapeseed meal with brewers' grains, lowers methane emission intensity in dairy cows

D. Van Wesemael[1,2], L. Vandaele[2], S. De Campeneere[2], V. Fievez[1] and N. Peiren[2]
[1]Ghent University, Laboratory for Animal Nutrition and Animal Product Quality, Coupure Links 653, 9000 Gent, Belgium, [2]Flanders Research Institute for Agriculture, Fisheries and Food, Animal Sciences Unit, Scheldeweg 68, 9090 Melle, Belgium; dorien.vanwesemael@ilvo.vlaanderen.be

As dietary fat supplementation has proven to reduce enteric methane (CH_4) emissions, we assessed, in a previous *in vivo* trial, the effect of brewers' grains (BG), a fat-rich by-product, on CH_4 emissions and milk production (MP) of dairy cattle. The trial only showed an effect on CH_4 emissions and MP when BG was used in combination with rapeseed meal (RSM) and not when combined with soybean meal (SBM), although fat intake was increased in both diets. The present *in vivo* trial, with 36 high-productive cows (35.5±4.8 kg milk/d and 97±39 DIM), investigated the effect of RSM alone or combined with BG. The control group (n=12) received the whole trial the reference diet (ref), without BG or RSM. Both other groups (n=12 both) switched diets twice: after 3 weeks on ref they switched to 8 weeks on their treatment diets, being BG and RSM (group dBG-RSM) and RSM alone (group dRSM), and finally to 4 weeks washout period on the reference diet (assigned as WOdBG-RSM and WOdRSM, respectively). Methane emissions were measured with a GreenFeed unit. Data analysis was based on a linear mixed model with treatment (ref, dBG-RSM, dRSM, WOdBG-RSM, WOdRSM) as fixed effect and week and cow as random effect. Also a random slope for week within cow was included. MP was significantly higher for dBG-RSM (33.9 kg/d) than for ref, dRSM and WOdBG-RSM (31.7, 31.4 and 31.3 kg/d; all P<0.01). CH_4 production was significantly lower for dBG-RSM and dRSM (486 and 492 g/d; both P<0.01) compared with ref (537 g/d), and was increased again in the final weeks for WOdBG-RSM and WOdRSM (515 g/d; P=0.04 and 506 g/d; P=0.84). CH_4 emission intensity (g CH_4/kg of milk) was significantly lower for dBG-RSM (14.3 g/kg) compared to ref (16.6 g/kg; P<0.01), dRSM (15.8 g/kg; P=0.03) and WOdBG-RSM (16.4 g/kg; P<0.01). We can conclude that in dairy cattle the combination of BG and RSM increases MP and decreases CH_4 emissions, and that RSM alone decreases CH_4 production, but not CH_4 emission intensity.

Estimation of protein autonomy of livestock farming in Wallonia with a focus on dairy farming

C. Battheu-Noirfalise[1], S. Hennart[1], M. Hoffmann[2], V. Decruyenaere[1] and D. Stilmant[1]
[1]Walloon Agricultural Research Centre (CRAW), 100 rue du Serpont, 6800 Libramont, Belgium, [2]Lycée Technique Agricole, 72 avenue Salentiny, 9080 Ettelbruck, Luxembourg; c.battheu@cra.wallonie.be

High levels of production of current livestock farming systems imply high densities of energy and protein in the diets. In Wallonia, high protein-density components are complicated to get at local scale. At this time, a significant part of the necessary protein complementation is based on imported cakes. These importations, these moves of primary material remain difficult to define. How could this be considered in the definition of protein autonomy of livestock farming system? A methodology has been used, in the AUTOPROT project, to reach this target at the Walloon scale. Autonomy is defined as the ratio between the resource locally available and the Walloon herd needs. The needs where calculated as the sum of the amount of animal of each livestock category multiplied by its theoretical needs in Digestible Crude Protein (DCP). The needs of the dairy sector were specified by elaborating standard diet for each agronomic region for both conventional and organic sectors. The protein production was evaluated as the sum of each production multiplied by its theoretical content in DCP. Different types of autonomy where defined: total protein autonomy (all crops), fodder protein autonomy (all fodders resources), protein autonomy taking into account the nutrition needs of each livestock type and this for ruminant sand monogastrics. Even if total protein autonomy reaches 96%, fodder autonomy only reaches 82% and autonomy taking into account animal physiology reaches 71%, showing that the way we consider protein autonomy is crucial. Based on physiological needs, for ruminants, the estimation of autonomy is 78% while it is only of 19% for monogastrics. Dairy cows in production are responsible of most of the imported protein-rich cakes (31%). Beef herd only account for 19%. Total ruminant herd valorises 89% of the imported cakes while monogastrics, in minority in Walloon area, only use 11% of these resources. These preliminary results suggest that research should focus on the ruminant sector, and in particular the dairy sector, to diminish protein dependence of Wallonia.

Successful economic management differs between intensive and extensive dairy farms
A.-C. Dalcq[1], Y. Beckers[1], B. Wyzen[2], E. Reding[2], P. Delhez[1,3] and H. Soyeurt[1]
[1]ULiège-GXABT, Passage des Déportés 2, 5030 Gembloux, Belgium, [2]AWE, Rue des Champs Elysées 4, 5590 Ciney, Belgium, [3]FNRS, Egmont 5, 1000 Bruxelles, Belgium; anne-catherine.dalcq@uliege.be

Intensification is a great economic, environmental and social issue and dairy producers will take different directions regarding it in the future, impacting a lot of farm components. The objective of the study was to highlight the management success factors of extensive and intensive dairy system based on field data. A total of 2,554 accountings collected from 476 producers of the Southern part of Belgium between 2007 and 2014 were used. Intensification was defined as the maximisation of the most limiting factor, the agricultural area. Therefore, milk production per hectare of forage area was used to divide the dataset in 4 classes with 7,500; 9,500 and 11,500 litres per hectare as benchmarks. In the most intensive (INT) and the most extensive (EXT) classes, the correlation between economic results variables (ERv: gross margin per cow (GMc) and income per cow) on the one hand and management variables (MANv) on the other hand was calculated. GMc showed different relationships with MANv depending on the intensification classes. The percentage of multiparous cows in the herd was more correlated with GMc for EXT than INT (r_{EXT}=0.22 vs r_{INT}=0.11), the number of calvings per cow was correlated with GMc for EXT but not for INT (r_{EXT}=0.16). Considering feed, GMc was correlated with concentrates consumption and the percentage of area for meadow, corn silage and forage silage only for EXT (r_{EXT}=-0.14, r_{EXT}=0.29, r_{EXT}=-0.29, r_{EXT}=-0.09). This shows an interest for extensive producers in accentuating their system. Regarding fixed costs, buildings value was negatively correlated with GMc for EXT but positively correlated for INT, suggesting the interest in investing in the living space for cows in intensive system but not in extensive one (r_{EXT}=-0.13, r_{INT}=0.10). The number of cows per worker was negatively correlated for INT (r_{INT}=-0.15) but not correlated for EXT. The factors highlighted in this study could allow advising producers regarding their management practises depending on the intensification level of their dairy system.

What drive the environmental performance of dairy farms?
T.T.S. Siqueira and D. Galliano
INRA-UMR 1248 AGIR, 24, chemin de Borderouge, CS 52627 Castanet Tolosan, France; tiago.siqueira@purpan.fr

Understanding what drive the farms' environmental performance can help the animal sector to deal with global socio-ecological issues. We analysed the data of 47,211 dairy farms from the 2010 French Agricultural Census using ordered probit models to study the correlation between farm's internal and external factors and farms' environmental performance. The environmental performances are approached by an individual score based on the adoption of nine agri-environmental practices. First, we tested the role of farm's internal factors related to the characteristics of the farmer, farm's structure, and governance. Second, we tested the role of external factors related with commercial and regulatory followed by spatial features. The results show that internal factors like contracting agri-environmental insurance, the share of family working force, the size and the corporate legal status of the farm are negatively correlated with the environmental performance. The age of the farmer, the share of land in property, the use of specialized technical software and farm diversification are positively correlated with the score. The knowing of the successor, the educational level and the gender of the farmer is not correlated with the score. In terms of external factors, the statistical analysis highlights the strong positive correlation of positioning on alternative markets, short circuits, organic products, or quality markets with the environmental score. As the literature commonly suggests, our results also show that environmental regulations are positively correlated with the environmental performance. The results also show the central role of the spatial environment of the farm and, more specifically, the environmental score of neighbouring farms as a major driver of the environmental performance. Finally, polices to promote locally farmer's experience exchange, to supporting diversification, high quality products and short circuits can bust the environmental performance of dairy farms.

Suckling of dairy calves by their dams: consequences on animal performances, behaviour and welfare

A.N. Nicolao[1,2], B.M. Martin[1], M.B. Bouchon[3], E.S. Sturaro[2] and D.P. Pomiès[1]
[1]INRA, UMR 1213 Herbivores, Centre de Theix, 63122 Saint-Genès-Champanelle, France, [2]University of Padova, DAFNAE, Viale dell'Università 16, 35020 Legnaro, Italy, [3]INRA, UE 1414 Herbipôle, La Borie, 15190 Marcenat, France; alessandra.nicolao@phd.unipd.it

In most European dairy farms, calves are separated from their mothers immediately or within few hours after birth. This early separation is increasingly questioning the society about animal welfare and some farmers are keeping calves with their mothers until weaning. The consequences of this practice have been studied mainly from the animal welfare point of view. This study aimed at studying the impact of this practice on milk yield, milk composition, growth of calves and animal behaviour and welfare. A classic rearing system ('Control') was compared to a suckling rearing system ('Mother') involving 14 cows each with their calves monitored for 15 weeks after calving. In the 'Control' group calves were separated from their mothers immediately after birth and fed with an automatic milk feeder until weaning at 12 weeks. In the 'Mother' group, mother-calf contact was allowed from birth to weaning for 9 h between the morning and evening milkings. During the first 8 weeks of lactation, the milked milk loss in the 'Mother' group was 11.3 kg/d. Milk fat content was lower in 'Mother' group than the 'Control' group (-9.3 g/kg) while milk protein content was higher (+0.9 g/kg). Milk somatic cells count and calves' growth were not significantly different between the two groups. Weaning distress affected more both cows and calves from the 'Mother' group than calves from the 'Control' group that were only affected by the change of diet. In the 'Mother' group, cows and calves' vocalisations were maximum on the day after the weaning and stopped 7 d later. In this experiment, we found that allowing dairy calves to suckle their mother was responsible for an important impairment of cow's performances. Nevertheless, all calves, including males, were reared with their mother. In practice, only the replacement female calves are reared until weaning, which suggest the feasibility of this practice in on-farm conditions.

Mitigation strategies improve performance and reduce inflammation in aflatoxin challenged dairy cows

R.T. Pate and F.C. Cardoso
University of Illinois, Urbana, 61801, USA; rtpate2@illinois.edu

Mitigation strategies are vital in minimizing health and economic risks associated with dairy cattle exposure to aflatoxin (AF). The objectives of this study were to evaluate lactation performance and inflammatory markers during an AF challenge when two mitigation strategies were implemented: (1) trace mineral injection; and (2) dietary supplementation of aluminosilicate clay. In the first experiment, 58 Holstein cows [BW (734±60 kg); DIM (191±93)] were assigned to 1 of 3 treatments in a randomized complete block design: saline injection and no AF challenge (NEG), saline injection and AF challenge (POS), and trace mineral injection [15 mg/ml Cu, 5 mg/ml Se, 60 mg/ml Zn, and 10 mg/ml Mn (Multimin 90, Multimin North America, Fort Collins, CO)] and AF challenge (MM). MUN and BUN were greater for cows in POS (14.3 mg/dl and 16.5 mg/dl, respectively) than cows in MM (13.3 mg/dl and 15.8 mg/dl, respectively; P=0.03 and 0.04, respectively). Cows in MM tended to have higher plasma glutathione peroxidase activity (30.2 nmol/min/ml) than cows in POS (24.2 nmol/min/ml; P=0.10). An upregulation of liver GPX1 was observed for cows in POS (1.07±0.05) compared to cows in MM (0.95±0.05; P=0.01). In the second experiment, 16 Holstein cows [BW (758±76 kg); DIM (157±43)] were assigned to 1 of 4 treatments in a replicated 4×4 Latin square design: no adsorbent and no AF challenge (CON), no adsorbent and AF challenge (POS), 113 g of aluminosilicate clay top-dressed on the TMR (adsorbent; PMI Nutritional Additives, Arden Hills, MN) with AF challenge (F4), and 227 g of adsorbent with AF challenge (F8). Milk AFM1 concentration decreased as concentration of adsorbent in the diet increased (P=0.02). NFKB1 expression was greater in liver of cows in POS (0.78±0.04) than cows in CON (0.70±0.04; P=0.04). MTOR expression was greater in liver of cows in CON (1.19±0.06) than cows in POS (0.96±0.06; P=0.04). When compared with cows in CON, cows in POS had greater odds ratio for hepatocyte inflammation (OR=5.14; 95%CI=0.97-27.33; P=0.05). In conclusion, AF exposure reduced function and increased inflammation of the liver. Additionally, mitigation strategies aided performance and induced a positive antioxidant response during AF challenge.

Intensive feeding increases liver ceramide concentrations in Holstein fattening bulls

A. Kenez[1], S. Baessler[2], C. Koch[3], T. Scheu[3] and K. Huber[2]
[1]City University of Hong Kong, Department of Infectious Diseases and Public Health, Hong Kong, Hong Kong, [2]University of Hohenheim, Institute of Animal Science, Stuttgart, Germany, [3]Hofgut Neumühle, Educational and Research Centre for Animal Husbandry, Münchweiler an der Alsenz, Germany; akos.kenez@cityu.edu.hk

Sphingolipids are ubiquitous elements of cell membranes. However, several sphingolipids, including ceramides, have been identified as signalling factors modulating metabolic stress response in insulin resistance and inflammation. These lipid mediators might be involved in driving metabolic health status under intensive growth in young bulls. We hypothesized that a carbohydrate-rich intensive nutritional regimen triggers alteration in liver sphingolipid metabolic pathways in fattening bulls. Holstein bulls intended for beef production were randomly assigned to an intensive (n=15) or a control nutritional regimen (n=15). Diets were based on corn- and grass-silage and the intensive group received additionally 6 kg/day/animal concentrate feed for the last 8 months of the fattening period. Animals were slaughtered and liver samples were subjected to a targeted liquid chromatography – tandem mass spectrometry analysis in The Metabolomics Innovation Centre, Canada. Data of the quantified 77 sphingolipids were analysed by multivariate statistical methods. Principal component analysis revealed a clear separation between liver sphingolipid concentrations according to dietary treatments. Significantly different lipids included 5 ceramides, 4 ceramide-phosphates, and 6 sphingomyelins that had greater concentrations in the intensively fed group and 4 further sphingolipids that had greater concentrations in the restrictively fed group (P<0.05 and fold-change >1.2). Higher ceramide production was previously associated with insulin resistance in Holstein cows. Ceramide-phosphates were shown to activate eicosanoid production in humans, which indicates a possible role in modulating inflammation and immune response. In conclusion, the intensive nutritional regimen induced significant alterations in liver ceramide profiles in young Holstein bulls, compared to controls, with an indication for a higher activity of metabolic pathways associated with insulin resistance and inflammation due to intensive feeding.

Serum concentrations of 17 steroids in transition cows with divergent body condition score

K. Schuh[1,2], S. Häussler[2], G. Dusel[1], C. Koch[3], C. Prehn[4], J. Adamski[4], H. Sadri[5] and H. Sauerwein[2]
[1]University of Applied Sciences Bingen, Department of Life Sciences and Engineering, Animal Nutrition and Hygiene Unit, 55411 Bingen, Germany, [2]University of Bonn, Institute of Animal Science, Physiology Unit, 53115 Bonn, Germany, [3]Hofgut Neumühle, Educational and Research Centre for Animal Husbandry, 1, 67728 Münchweiler an der Alenz, Germany, [4]Helmholtz Zentrum München, Genome Analysis Center, Research Unit Molecular Endocrinology and Metabolism, 85764 Neuherberg, Germany, [5]University of Tabriz, Department of Clinical Science, Faculty of Veterinary Medicine, 5166616471 Tabriz, Iran; sauerwein@uni-bonn.de

Steroids are lipophilic molecules which accumulate in adipose tissue (AT). With increasing lipid mobilization in dairy cows peripartum, steroid release from AT into the circulation likely changes as well. Our aim was to test if the periparturient patterns of steroid hormones in serum might differ between over-conditioned and normal cows. Multiparous Holstein cows were classified 15 weeks ante partum based on their previous BCS and backfat thickness (BFT) as either normal (NBCS, n=18, BCS<3.5 or BFT<1.2 cm) or over-conditioned (HBCS, n=18, BCS>3.75 or BFT>1.4 cm). Until dry-off, HBCS were fed on 7.2 NEL MJ/kg DM and NBCS on 6.8 NEL MJ/kg DM; thereafter all cows received the same diets; differences in BCS and BFT were maintained. A targeted metabolomic approach (AbsoluteIDQTM Stero17 Kit, Biocrates) was applied in serum samples from d -49, 3, 21, 84 relative to calving. Data were analysed by a linear mixed model with repeated measurements (fixed effects: group, time, and parity and their interactions; random: cow). The concentration of 15 out of 17 steroids changed with time, mostly decreasing towards lactation. Group differences were limited to testosterone (P=0.017) with HBCS having 48% greater concentrations than NBCS on d 84. For aldosterone, a group × time interaction was observed (P=0.033) due to higher values in HBCS cows in week 1. Parity was affecting the concentrations of dehydroepiandrosterone, etiocholanolone, dihydrotestosterone, and oestradiol with greater concentrations in younger animals. Elevated testosterone, considered as predictor of increased AT accumulation in women, might have similar associations in dairy cows including compromised fertility.

Milk metabolomics of heat-stressed goats with and without intramammary lipopolysaccharide challenge

S. Love, A. Contreras-Jodar, N. Mehaba, X. Such, G. Caja and A.A.K. Salama
Universitat Autonoma de Barcelona, Group of Research in Ruminants (G2R), Edifici V, Campus de la UAB, 08193 Bellaterra, Spain; ahmed.salama@uab.es

The aim of the current study was to evaluate the effect of both heat stress and simulated intramammary infection on milk metabolomics in dairy goats. Eight multiparous Murciano-Granadina dairy goats (2.2±0.1 l/d; 100±5 DIM, 42±2 kg BW) were maintained under 2 environmental conditions varying in temperature, relative humidity (RH) and temperature humidity index (THI): (1) 4 goats under thermoneutral (TN; 15 to 20 °C, RH=50±5%, THI=59 to 65); and (2) 4 goats under heat stress conditions (HS; 35 °C from 09.00 to 21.00 and 28 °C from 21.00 to 09.00, RH=45±5%, THI=75 to 83). Adaptation of 11 d to the experimental treatments was allowed. On d 12 each animal had one udder half infused with 10 µg *Escherichia coli* lipopolysaccharide (LPS) and the other udder half as the control with 0.9% saline (CON). Rectal temperature, respiratory rate, feed intake and water consumption were daily recorded. Additionally, milk samples (0, 4, 6, 12, and 24 h after LPS administration) were collected and analysed by ^{1}H NMR spectroscopy operating at 600 MHz. Productive data were analysed by the Mixed Procedure for repeated measures of SAS. Metabolomic data were processed by the R program (pls package), in which the principal component analysis and partial least square–discriminant analysis were used to identify possible metabolite markers in milk. Compared to TN goats, HS goats had greater (P<0.001) rectal temperature (39.91 vs 38.64 °C), respiratory rate (146 vs 39 breaths/min), water consumption (10.1 vs 6.0 l/d), but lower (P<0.05) DM intake (1.87 vs 2.59 kg/d). Citrate increased ($\log_2$ fold change=1.79; P=0.010) in milk of HS goats. On the other hand, the LPS challenge resulted in an increment (P<0.05) in milk N-acetyl carbohydrates, choline, beta-hydroxybutyrate, and L-lactate. The extent of increment of these metabolites by LPS injection varied between TN and HS conditions. In conclusion, high ambient temperatures and udder inflammation dramatically affected the metabolomic profile of milk. Response of milk metabolomic profile to intramammary LPS challenge varied between thermoneutral and heat stress conditions.

Getting prepared for metabolomics studies: deliberate selection of samples from large animal studies

R. Riosa[1], M.H. Ghaffari[1], M. Iwersen[2], D. Suess[2], M. Drillich[2], C. Parys[3] and H. Sauerwein[1]
[1]Institute of Animal Sciences, Physiology and Hygiene Unit, Katzenburgweg 7-9, 53115 Bonn, Germany, [2]University of Veterinary Medicine Vienna, University Clinic for Ruminants, Veterinärplatz 1, 1210 Vienna, Austria, [3]Evonik Nutrition & Care GmbH, Rodenbacher Chaussee 4, 63457 Hanau-Wolfgang, Germany; rriosa@uni-bonn.de

Large samples sizes in animal trials increase the statistical power, decrease false discovery rates and improve reproducibility. When aiming to decipher physiological mechanisms underlying potential treatment effects by means of metabolomics, however, selecting meaningful sample subsets is desirable for a cost-efficient use of the complex analyses. Based on a trial comprising 1,864 cows fed the same diet with or without a methionine (M, Mepron®) supplement, we herein aimed to identify samples comprising the most informational value based on performance and health data, focusing on endometritis. After a first screening of the data base according to duration of M supplementation, availability of production data and blood samples, we further screened the remaining 1,036 cows for the occurrence of clinical endometritis (CE, assessed by veterinary diagnostics protocols) to create 4 subgroups: control healthy (CH; n=77), control with CE (CCE; n=36), M healthy (MH; n=57) and M with CE (MCE; n=43). Then we compared these groups for verifying that they differed in performance using the MIXED procedure of SAS. CH yielded more milk than CE cows, also when corrected for energy content (P<0.05). Contents of milk fat, fat yield and the milk fat to protein ratio were affected by treatment and disease (P<0.05). Somatic cells counts were lower in MCE compared to CCE but were higher in MH compared to CH (P<0.05). Body condition score and back fat thickness were higher in CH compared to CE (P<0.05), and in MCE compared to MH (P=0.065). Calving to conception intervals were shorter in healthy animal (P<0.05). These results suggest that the selection was meaningful as statistical interactions were found between treatment and disease. Future 'omics' analyses on these selected cows will permit us to obtain a complete picture of the mechanisms involved and to identify markers which can then be targeted on all cows from the starting database.

Identifying the role of rumen microbiome on bovine milk fatty acid profile using -omics approaches

M.D. Auffret[1], I. Cabeza-Luna[1,2], M. Mora-Ortiz[2], R.J. Dewhurst[1], D. Humphries[2], M. Watson[3], R. Roehe[1] and S. Stergiadis[2]
[1]Scotland's Rural College (SRUC), Roslin Institute Building, Edinburgh, EH25 9RG, United Kingdom, [2]University of Reading, Centre for Dairy Research, Reading, Reading, RG6 6AR, United Kingdom, [3]University of Edinburgh, Roslin Institute Building, Edinburgh, EH25 9RG, Roslin Institute Building, Edinburgh, EH25 9RG, United Kingdom; marc.auffret@sruc.ac.uk

A number of studies focussed on reducing milk saturated FA (SFA) due to their associations with increased risk in cardiovascular diseases and increasing nutritionally-desirable monounsaturated FA (MUFA). However, the role of the rumen microbiome on milk FA profile has not been investigated. The aim of this research was to identify rumen microbial mechanisms explaining the significant variability in milk SFA and MUFA between cows with extreme milk FA profile by using a combination of metagenomics, metabolomics and modelling data analyses. Total DNA was extracted from rumen fluid from 16 lactating Holstein cows (8 low-SFA, averaging 63.7±1.7% total FA vs 8 high-SFA averaging 70.3±1.7% total FA) receiving the same diet (63:37 forage:concentrate). Microbial genera and gene abundances were based on Kraken and KEGG genes databases, respectively. Milk FA and rumen metabolite profiles were analysed using gas chromatography and 700 MHz NMR spectroscopy, respectively. Partial least square (PLS) analysis was used to identify genera and their genes explaining high variability in SFA and MUFA and orthogonal PLS discriminant analysis to determine the genes which were highly correlated with rumen metabolites. The lactic acid bacteria (LAB) *Lactobacillus*, *Leuconostoc* and *Weissella* were more abundant (P<0.05) in high-SFA cows and carrying genes involved in lactate synthesis (K03778) and acidic pH and stress resistances (K00383, K06148). Both LAB and their genes showed a positive correlation with butyrate and propionate and negative correlation with xanthine and hypoxanthine, reinforcing the idea of a more acidic rumen pH. This study provides evidence that high-SFA cows are more sensitive to high concentrate diets (such the one used in this study), leading to increased presence of LAB and a modified pattern of released microbial metabolites (e.g. butyrate) which are associated with mammary synthesis of SFA and reduction of MUFA.

Differential expression of bovine mRNA isoforms with functional consequence on mastitis host defence

V. Asselstine[1], J.F. Medrano[2], F. Miglior[1], A. Islas-Trejo[2] and A. Cánovas[1]
[1]University of Guelph, Centre for Genetic Improvement of Livestock, Department of Animal Biosciences, 50 Stone Road E, N1G 2W1, Guelph, Canada, [2]University of California-Davis, Department of Animal Science, 1 Shields Ave, 95616, Davis CA, USA; vasselst@uoguelph.ca

Bovine immune response associated with mastitis is a very complex biological process. Some isoforms have been previously identified and associated with mastitis resistance in dairy cows. However, using RNA-Sequencing to study the somatic cell transcriptome to identify specific mRNA isoforms with functional relevance associated with mastitis is a novel topic. Milk somatic cells fraction was isolated from 12 milk samples of six Holstein cows to identify differentially expressed (DE) mRNA isoforms, that could impact the cow's immune response to intramammary infections. RNA-Sequencing was performed on RNA samples from milk somatic cells from each cow from two separate quarters; one classified as healthy (n=6), one as mastitic (n=6). These sequences were then mapped to the bovine reference genome using the CLC genomics workbench. Lastly, transcript levels were quantified in reads per kilo base per million mapped reads (RPKM). Differential transcript expression analysis was performed between healthy and mastitic groups, identifying 516 DE transcripts (FDR<0.05 and FC>+/-2) which corresponded to 502 annotated genes. Among them, 435 genes had only one mRNA isoform, while 67 genes showed two or more mRNA isoforms, with at least one of them being DE (80 mRNA isoforms DE, corresponding to 67 genes). Functional analysis including gene ontology, gene network and metabolic pathway analysis was performed using the list of 80 DE mRNA isoforms (FDR<0.05 and FC>+/-2; 67 genes) which identified six metabolic pathways significantly enriched and involved with immune system processes (P<0.05). Among them, three of the genes with at least one DE mRNA isoform identified, *MyD88* (BTA22), *IL1RAP* (BTA1) and *IRAK2* (BTA22), demonstrated their involvement with immune system processes and regulation. In conclusion, we identified potential functional candidate genes with at least one mRNA isoform being DE that could lead to differences in the immune response of Holstein dairy cows.

Metabolomics of high and low feed efficient dairy cows reveal novel physiological mechanisms

X. Wang and H.N. Kadarmideen
Technical University of Denmark, Department of Applied Mathematics and Computer Science, Richard Peterson Plads, Building 324, 2800 Kongens Lyngby, Denmark; hajak@dtu.dk

Feed efficiency is an important trait in cattle production. Gas chromatography mass spectrometry (GC-MS) system is widely used High throughput metabolite profiling. In this study, we report on GC-MS metabolic profiling of cows and their impact on Residual Feed Intake (RFI) with an aim to generate a better understanding of metabolic mechanisms belonging to high and low RFI Jerseys and Holsteins and to identify potential predictive metabolic biomarkers for RFIs. GC-MS was used in this study for the annotation and metabolites were classified as among amino acid, tricarboxylic acid and fatty acid groups. They were analysed to reveal their associations with low and high RFI in two breeds by repeated measures ANOVA. Liver transcriptomics (RNAseq) revealed differentially expressed genes (DEGs) between high and low RFI groups in each breed. Integrative analysis of metabolomics with transcriptomics was performed to explore interactions between functionally related metabolites and DEGs in the created metabolite networks. Comparison between low and high RFI groups showed significant ($P<0.001$) differences in the amino acid group in both Jersey and Holstein cows and in the fatty acid group in Holstein cows. The significant breed effect ($P<0.05$) was found for Isoleucine, Leucine, Octadecanoic acid, Ornithine, Pentadecanoic acid and Valine, whereas significant RFI effect ($P<0.05$) only existed for Leucine. Metabolite set enrichment analysis (MSEA) for 34 identified metabolites revealed important pathways: Aminoacyl-tRNA biosynthesis, Alanine, aspartate and glutamate metabolism and Citrate cycle (TCA cycle). Finally, four genes (ELOVL6, LYZ, CTH and HACL1) were associated with seven metabolites using gene-metabolite interaction network analyses. Our study provided valuable knowledge on metabolic pathways involved in low- and high-RFIs cows and provided potential metabolic biomarkers for understanding novel biochemical mechanisms, for example, over-expressed ELOVL6 consistent with up-regulated Palmitoleic acid in the network could be applied in further selection programs.

Machine learning approach reveals a metabolic signature of over-conditioned cows

M. Hosseini Ghaffari
University of Bonn, Institute for Animal Science, Physiology & Hygiene, Katzenburgweg 7-9, 53115, Germany; morteza1@uni-bonn.de

Our objective was to obtain a metabolic signature associated with over-conditioning in dairy cows. We used an animal model with a targeted difference in body condition score (BCS) at dry-off. Multiparous Holstein cows were allocated to two groups that were fed differently to reach either high (HBCS: 7.2 NEL MJ/kg dry matter (DM) or normal BCS (NBCS: 6.8 NEL MJ/kg DM) at dry-off. The allocation was also based on differences in their BCS and BFT in the previous and the ongoing lactation; thereafter all cows received the same diets whereby the differences in BCS and BFT were maintained. A targeted metabolomic approach was applied in plasma samples from d -49, +3, +21, +84 relative to calving. We used various machine learning (ML) algorithms (PLS-DA, Random Forest, Information-Gain-Ranking) to identify the most important plasma metabolite features contributing most to the group differences. In early lactation, 12 metabolites (acetylcarnitine, hexadecanoyl-carnitine, hydroxyhexadecenoyl-carnitine, octadecanoyl-carnitine, octadecenoyl-carnitine, hydroxybutyryl-carnitine, glycine, leucine, phosphatidylcholine-diacyl-C40:3, trans-4-hydroxyproline, carnosine, and creatinine) were the most significant variables for differentiating the two groups identified by the ML algorithms. Among the affected pathways, branched-chain amino acids (leucine and valine, before calving) and mitochondrial β-oxidation (after calving) were the most enriched. Overall, HBCS cows were accumulating more long-chain acylcarnitines in plasma pointing to mitochondrial overload and impairment of ß-oxidation rate being associated with over-conditioning in early lactation.

A follow-up on blood metabolic clusters in dairy cows

L. Foldager[1,2] and K.L. Ingvartsen[2]
[1]Aarhus University, Bioinformatics Research Centre, C.F. Møllers Allé 8, 8000 Aarhus, Denmark, [2]Aarhus University, Dept. of Animal Science, Blichers Allé 20, 8830 Tjele, Denmark; leslie@anis.au.dk

Prediction of blood metabolic clusters by milk biomarkers were presented in two recent publications from the GplusE project, http://gpluse.eu. These clusters differentiating the metabolic profile of the cows were calculated by k-means clustering on basis of plasma glucose, BHB and NEFA and serum IGF-1. The objective is to follow-up on characteristics of these clusters in a large Danish dataset. 321 cows with 610 lactations (1st to 5th) were used: 108 Danish Red, 130 Danish Holstein and 83 Jersey. Cows were randomised to a low or normal energy density TMR, which they remained on throughout the experiment; see [3] for further description. Plasma glucose, BHB, NEFA and IGF-1 were measured days -10, 25, 74 and 256. Clusters were determined by k-means clustering of standardised measures within breed, parity (1, 2 and 3+) and sampling day. The standardisations were within breed and parity. Boxplots of standardised measures by breed, parity, day and cluster were examined and interpreted. In early lactation, one cluster differed from the two others by showing strong signs of physiological imbalance (PI; low glucose, high NEFA and high BHB). However, these differences were obliterated before 10 weeks. PI in primiparous cows is not due to higher EKM and differences among clusters in multiparous cows appear small. There were no apparent association between clusters and fat/lactose proportions but in early lactation, PI-cluster cows had higher citrate and fat/protein proportion of the milk. These cows also showed the most marked signs of negative energy balance and lowest plasma IGF-1 levels. In mid and late lactation, connections to energy balance and IGF-1 level seems to be weak. PI is not due to milk yield but more likely difficulties in adapting to lactation. Particularly heifers and parity3+ have a group with significant PI but it is obliterated before 10 weeks. At least the imbalanced heifers have higher citrate and fat/protein content but no difference in fat/lactose.

High-quality hay feeding in early lactating cows: energy partitioning and milk fatty acid profile

R. Khiaosa-Ard[1], M.-T. Kleefisch[1], Q. Zebeli[1] and F. Klevenhusen[2]
[1]University of Veterinary Medicine Vienna, Veterinärplatz 1, 1210 Vienna, Austria, [2]The German Federal Institute for Risk Assessment (BfR), Max-Dohrn-Str. 8-10, 10589 Berlin, Germany; ratchaneewan.khiaosa-ard@vetmeduni.ac.at

High quality (HQ) hay is rich in energy due to the enrichment of water-soluble carbohydrates and protein compared to traditional fibre-rich hay used in dairy cow feeding. Feeding such HQ hay could reduce or eliminate the dependency on cereal grains depending on cow productivity level. This could result in less competition over human-edible food sources and for the cow better rumen health. Using 24 Simmental cows, 4 different diets were compared from calving to 28 d afterward. The diets were Control (60% fibre-rich hay + 40% concentrate), 60HQ (60% HQ hay + 40% concentrate), 75HQ (75% HQ hay + 25% concentrate) and 100HQ (100% HQ hay). Performance data were recorded daily and milk constituents, milk fatty acid profile, and blood metabolites of weekly samples were analysed. Both 60HQ and 75HQ promoted energy intake compared to the other two groups (about +30 MJ/d, net energy lactation, P<0.05). 75HQ resulted in the best energy balance, whereas 60HQ partitioned energy towards milk production (+6 kg compared with the others, P<0.05) and secretion of milk de novo fatty acids (+145 g/d compared to 75HQ, P<0.05). Cows in Control and 100HQ showed more severe negative energy balances, elevated serum non-esterified fatty acids (NEFA) and betahydroxybutyrate (BHBA), and lowered proportion of de novo fatty acids at the expense of 18:1 n-9 in the milk fat. The ratio of milk de novo fatty acids to 18:1 n-9 was closely associated with the levels of NEFA (R^2=0.791) and BHBA (R^2=0.813) in blood, with ratios below 1 suggesting intensive body fat mobilization and development of subclinical ketosis. All HQ groups increased the level of conjugated linoleic acids in milk fat compared to Control (P<0.05). The level of 18:3 n-3 in milk fat increased with increasing amount of the HQ hay in the diet. In conclusion, feeding HQ hay can reduce concentrate feeds in the diet of early lactating cows. The proportion of HQ hay affects the energy partitioning of the cows. Certain groups of milk fatty acids are closely related to blood variables in early lactating cows.

Untargeted metabolomics analysis to discover biomarkers predictive of metritis in dairy cows

M.G. Colazo[1], M. Zhu[2], E. Dervishi[3], L. Li[2] and G. Plastow[3]
[1]Alberta Agriculture and Forestry, Livestock Systems Section, 303, 7000 – 113 Street NW, T6H 5T6, Edmonton, Canada, [2]University of Alberta, Chemistry, 11227 Saskatchewan Drive, T6G 2G2, Edmonton, Canada, [3]University of Alberta, Agricultural, Food & Nutritional Science, 8215, 112 Street, T6G 2C8, Edmonton, Canada; marcos.colazo@gov.ab.ca

The purpose of this study was to identify serum biomarkers predictive of metritis in lactating dairy cows. Blood samples were collected 6.9±4.9 d before calving from 1,096 cows in 11 dairy herds in Alberta, Canada. Serum samples from 29 cows subsequently diagnosed with metritis were chosen for an initial untargeted metabolomics analysis. Samples from 56 control cows from the same herd (accounting for parity, body condition score and time of calving) with no diagnosis of disease were used to search for differences in metabolic signatures. We applied 4-channel isotope labelling (dansyl and DmPA reactions) to analyse four different subgroups of metabolites (amine/phenol, ketone/aldehyde, carboxylic acid and hydroxyl submetabolome). A receiver operating characteristic (ROC) curve of each significant metabolite was generated to determine the sensitivity and specificity for differentiating metritis from control. Preliminary results showed that 5,464 peak pairs (metabolites) were commonly detected from the four submetabolomes. The volcano plot showed that there were 1,886 significantly changed peak pairs between metritis and control samples. In samples from cows with metritis, the number of peak pairs down- and up-regulated (FC>1.5, q<0.05; P<0.05) were 17 and 91, 123 and 98, 756 and 594, and 108 and 99 in the amine/phenol, ketone/aldehyde, carboxylic acid and hydroxyl submetabolome, respectively. Several significantly changed metabolites were found to have 98 to 100% sensitivity and specificity. These metabolites are being validated as potential biomarkers predictive of metritis in lactating dairy cows.

Network analysis uncovers the interplay among miRNA, mRNA and iron concentration in Nelore muscle

W.J.S. Diniz[1], P. Banerjee[2], G. Mazzoni[2], L.L. Coutinho[3], A.S.M. Cesar[3], C.F. Gromboni[4], A.R.A. Nogueira[5], H.N. Kadarmideen[2] and L.C.A. Regitano[5]
[1]Federal University of São Carlos, Genetics, Washington Luís Highway, 3565-905, São Carlos, São Paulo, Brazil, [2]Technical University of Denmark, Anker Engelunds Vej 1 Bygning, 2800, Kgs. Lyngby, Denmark, [3]University of São Paulo, Avenida Pádua Dias, 11, 13418-900, Piracicaba, Brazil, [4]Bahia Federal Institute of Education, Science and Technology, campus Ilhéus, Rodovia Ilhéus-Itabuna, km 13, S/N – Alto Mirante, 45603-255, Ilhéus, Bahia, Brazil, [5]Embrapa Pecuária Sudeste, Rod. Washington Luiz, Km 234 – Fazenda Canchim, 13560-970, São Carlos, SP, Brazil; hajak@dtu.dk

Beef is a source of nutrients, mainly iron (Fe), which is essential and plays an important role in many biological processes. Although studies have pointed out Fe modulating gene expression and miRNA biogenesis, this interplay is still unclear. To uncover the pathways and regulatory networks underlying Fe metabolism in Nelore cattle muscle (n=50), we performed a miRNA and mRNA co-expression and regulatory network analyses based on Weighted Gene Co-expression Network method. Next, by fitting a linear model, we identified eight modules associated with Fe concentration (P<0.05). By integrating the associated miRNA – mRNA modules and intersecting with TargetScan prediction, we identified 25 hub miRNAs and 2,677 pairs inversely correlated (P<0.05). Among the pairs, we found the transferrin receptor gene (TFRC) targeted by miR-103, -107, -15a, and 29e. The interplay between transferrin (TF) and TFRC control the level of free Fe in the biological fluids. We also identified the transcription factors YY1 and SP1, both targets of miR-29 family, which bind the human TF and modulate its expression level. From the pathway analysis, the gene targets were underlying the AMPK, insulin, mTOR, and thyroid hormone signalling pathways, which are interrelated and nutrient modulated. The associations identified here among miRNAs with Fe and their potential gene targets support the hypotheses of an intricate interplay among them, which can affect animal health and production.

Comparison of postpartum nutritional and metabolic profiles in dairy cows with or without metritis

E. Dervishi[1], G. Plastow[1], B. Hoff[2] and M.G. Colazo[3]
[1]University of Alberta, Agricultural, Food & Nutritional Science, 8215, 112 Street, T6G 2C8, Edmonton, Canada, [2] University of Guelph, Animal Health Laboratory, 419 Gordon Street, N1G 2W1, Guelph, Canada, [3]Alberta Agriculture and Forestry, Livestock Systems Section, 7000, 113 Street NW, T6H 5T6, Canada; marcos.colazo@gov.ab.ca

Metabolic profile testing can be an effective tool for the diagnosis of nutritional and metabolic diseases in dairy cows on a herd basis. The purpose was to compare postpartum nutritional and metabolic profiles of dairy cows with or without metritis. Blood samples were collected 8.0±3.9 d postpartum from 1,029 cows in 11 dairy herds in Alberta, Canada. Serum samples from 29 cows diagnosed with metritis within 21 d postpartum and from 27 control cows from the same herd (accounting for parity, body condition score (BCS) and time of calving) with no diagnosis of disease were used to search for differences in the nutritional and metabolic profile between both groups. Circulating total protein, albumin, globulin, urea, Ca, P, Mg, Na, Cl, β-hydroxybutyrate (BHB), non-esterified fatty acids (NEFA), glucose, cholesterol, haptoglobin, gamma-glutamyl transferase (GGT), aspartate aminotransferase (AST), and glutamate dehydrogenase (GLDH) were quantified. Statistical analyses were performed using a general linear model. Farm, BCS, day of sample collection and calving season were not significant, therefore, were not included in the final model. Cows with metritis had greater serum concentration of haptoglobin (0.77±0.12 vs 0.19±0.12 g/l) and lower concentration of Ca (2.27±0.02 vs 2.31±0.02 mmol/l), urea (3.43±0.19 vs 4.08±0.19 mmol/l) and albumin (31.46±0.65 vs 33.94±0.65 g/l) when compared with healthy cows (P<0.05). In addition, circulating total protein tended to be lower in cows with metritis when compared to healthy cows (64.48±1.31 vs 66.70±1.31 g/l; P=0.06). Haptoglobin is an acute phase protein, which binds to haemoglobin and inhibits bacterial proliferation. An increase in circulating haptoglobin in cows with metritis suggests that an inflammatory response was already taking place in these animals. In addition, lower concentration of urea, albumin and total protein suggests an impairment of protein utilization in cows diagnosed with metritis.

Identification of miRNA related with fatness traits in pig

K. Ropka-Molik, G. Żak, K. Pawlina-Tyszko, K. Piórkowska and M. Tyra
National Research Institute of Animal Production, Ul. Sarego 2, 31-047 Kraków, Poland; grzegorz.zak@izoo.krakow.pl

Processes related to the growth and development of adipose tissue are conditioned by both genetic and environmental factors. The exact molecular mechanisms associated with the intensity of lipid metabolism affecting the body's fatness has not been fully understood. The aim of the study was to identify the expression profile of miRNAs in fat tissue in the pig of different breeds and to identify miRNAs potentially associated with the fatness of carcasses. All analysed pigs representing 3 breeds (Pietrain, Hampshire, Large White; 8 pigs per breed) and in each breed the pigs with different fatness phenotypes were selected. The sequencing of miRNA was performed on the HiScanSQ system with the 81 single-end cycles. The analysis of the miRNAome profile of adipose tissue of the three pig breeds showed the differential expression of panel of miRNAs, nine of which were identified in all three analysed breeds. Among them were miRNAs also identified by other authors as strongly related with fatness traits: let-7a-5p, miR-143-3p, miR-143-3p, miR-24-3p and miR-378a-3p. The significant molecular pathways deregulated by detected miRNAs were Hippo signalling pathways, ECM-receptor interaction and PI3K/AKT/mTOR pathway. The present study allows to identify miRNA characteristic for adipose tissue of pigs with different fatness. Obtained results pinpointed miRNAs, which can be a main modulator of subcutaneous fat development (let-7a, miR-143, miR-24, miR-378). Supported by project no. 01-18-03-21.

Primary bovine hepatocyte lipolytic genes may respond differently to liver X receptor α perturbation than murine

S.J. Erb and H.M. White
University of Wisconsin-Madison, Dairy Science, 1675 Observatory Dr, 53706, USA; serb2@wisc.edu

The lipase patatin-like phospholipase domain-containing protein 3 (PNPLA3) may play a vital role in hepatic lipid accumulation and is responsive to fatty acid regulation through sterol regulatory element-binding protein 1c (SREBP1c) mediation in humans. Although preliminary research has indicated that PNPLA3 is associated with fatty liver in dairy cows, it is not yet known if PNPLA3 is regulated by SREBP1c. The objective of this study was to determine if PNPLA3 protein abundance is regulated by SREBP1c in bovine with an agonist, 22(R)-hydroxycholesterol (AGO), or antagonist, simvastatin (ANT), of liver X receptor (LXR) alpha used previously in murine models. Primary bovine hepatocytes were isolated from bull calves (n=3; <7 d of age) and cultured in monolayers. After 24 h, cells were treated in triplicate with 0 or 100 μM of either the AGO or ANT. After 24 h of incubation, cells were harvested in TRIzol for subsequent RNA and protein isolation. Gene expression of LXRα, LXRβ, carbohydrate (Ch) REBP1, and *SREBP1c* were determined by quantitative real-time PCR and expressed relative to reference genes. Protein abundance of SREBP1c and PNPLA3 were determined through Western blot analysis, normalized to total lane protein, and expressed relative to each respective 0 μM treatment. All data were analysed (PROC MIXED; SAS, 9.4) with fixed effect of treatment and random effect of calf. Treatment with AGO tended to increase LXRα (0.58 vs 0.77±0.17 arbitrary units (AU); P=0.07) and ChREBP1 (0.54 vs 0.65±0.25 AU; P=0.09) gene expression. There was no effect of AGO (P>0.1) on expression of LXRβ or *SREBP1c*. Treatment with ANT did not alter (P>0.1) LXRα, LXRβ, ChREBP1, or *SREBP1c* gene expression. Treatment with neither AGO nor ANT altered protein abundance (P>0.1). In contrast to murine models, primary bovine hepatocytes did not respond as predicted to 100 μM of this ANT. While AGO tended to increase LXRα as expected, there was no impact on expression of *SREBP1c* gene or abundance of SREBP1c or PNPLA3 protein. Further research on the regulation of bovine *SREBP1c* and subsequent protein is needed to understand the potential in regulating bovine PNPLA3.

Impact of melatonin on some hormonal and biochemical parameters in anoestrous buffalo cows

K.A. El Battawy
National Research Centre, Animal Reproduction and A.I Dept., University of Veterinary Medicine, Vienna, Austria, Josef Bauman gasse, 1021 Vienna, Austria; ekhairi@gmx.at

Ten anoestrous buffalo cows were used and were assigned on two groups (five cows in each group). The first group served as control while the other one was used as experimental group. Each animal in the experimental group received melatonin as implant at the base of the ear at the beginning of July 1. The purpose of this investigation was to assess the effect of melatonin implants on some hormonal and biochemical parameters during summer. Blood samples were collected from cows in both control and experimental groups before and after administration of melatonin implants to assess the concentrations of progesterone, prolactin, triglycerides, total cholesterol, insulin, glucose, free fatty acids, alanine aminotransferase and aspartate aminotransferase. Results elaborated that melatonin administration induced a significant (P<0.05) increase in the levels of progesterone, triglycerides and total cholesterol. Furthermore, melatonin caused slight increase in the concentrations of glucose and insulin. On the other hand, melatonin induced a significant (P<0.05) decrease in the levels of prolactin, and free fatty acids. Regarding the activity of liver enzymes, melatonin didn't cause any influence. In conclusion, melatonin plays an important role in regaining cyclicity in anoestrous buffaloes through minimizing prolactin concentration and elevating the levels of progesterone, triglycerides and total cholesterol.

Insect production in a life cycle perspective: the way to environmentally efficient production

S. Bruun and J. Eilenberg
University of Copenhagen, Department of Plant and Environmental Sciences, Thorvaldsensvej 40, 1871 Frederiksberg C, Denmark; jei@plen.ku.dk

Insects has been heralded as sustainable alternative to the mammals in livestock production systems because they may have more efficient feed conversion than mammals. In addition, insects is supposed to lead to fewer emissions of greenhouse gases. However, before we can conclude the insect production is more sustainable there are many aspects to consider. First, we should stress that insects in production are a very diverse group, so evaluations should be done for each insect species of relevance. In general, insect production systems based on waste streams are much more efficient than systems based on feed that could otherwise have been used directly as animal feed or even human food. Here insects that are able to feed on cellulose that is not digestible by non-ruminants may be of special interest. There are even insects able to digest wood that could potentially be used. Another aspect to consider is what the insects are used for. Direct human consumption is better from an environmental point of view than systems where the insects are used as animal feed. Even when they are used for food insects can play different roles ranging from a protein source to a snack. While it may be true that some insect are producing very low amounts of methane in their guts compared to ruminants, this may in fact not always be the case as we know form termites. In fact, it is possible that insects that digest cellulose has a tendency to produce methane just like ruminants. As feed production constitute a major part of the impacts associated with insect production, the advantage of insect production to a large extend stems from the fact that they have low feed conversion ratios. Mammals based production systems has been developed and optimized for many centuries and the feed conversion ratios have improved considerable in this process. Similar optimization of the insect production systems are likely to take place if insect production becomes widespread. Exactly how much the efficiency can be improved in this way will play an important role in the sustainability of the developed systems.

Environmental impacts of biomass conversion by insects: life cycle assessment perspective

S. Smetana[1], E. Schmitt[2], V. Heinz[1] and A. Mathys[3]
[1]DIL e.V., Professor-von-Klitzing-Straße 7, 49610, Quakenbrueck, Germany, [2]Protix, Industriestraat 3, 5107 NC Dongen, the Netherlands, [3]ETH Zurich, Institute of Food, Nutrition and Health, Laboratory of Sustainable Food Processing, Schmelzbergstrasse 9, 8092, Zurich, Switzerland; s.smetana@dil-ev.de

The lack of protein sources in several parts of the world is triggering the search for locally produced and sustainable alternatives. Insect production is recognized as a potential solution. This study is referring to a few life cycle assessments (LCA) of biomass conversions by insects for various purposes. Comparison is performed based on type of modelling (attributional, consequential), models applied (IMPACT 2002+, ReCiPe, separate impact categories), scope of application (waste treatment, feed and food, bio-materials, biofuels) for *Hermetia illucens* (black soldier fly) and different scales of production. The aim of the current study was the assessment of environmental impacts of insect-based intermediate products (usable for feed and food) with reliance on systemized multi-season dataset of *H. illucens* production and processing. This was done by considering a highly productive pilot plant with insight on future upscaling scenarios applying attributional (identification of the optimal production and allocation between products) and consequential LCA approaches for the definition of more sustainable options. Systematized approach allowed for justified identification of positive and negative scenarios of insect application for various purposes and in comparison to different benchmarks. It has been confirmed that in most scenarios, insect biomass is more environmentally beneficial than other animal-derived sources. Insect biomass has a potential to be close in environmental impact to plant-based materials only in certain scenarios. Application of insect biomass for the production of bio-materials of higher value is more beneficial than benchmark products (antimicrobial components, bio-lubricants, structural components) if insects are produced on waste materials. The efficiency of waste materials application for insect feeding highly depends on the type of waste. The study identifies strategies for more sustainable use of insect biomass transformation technologies from multi-dimensional perspective.

Infact: from insect to surfactant

G.R. Verheyen, I. Noyens, T. Ooms, S. Vreysen and S. Van Miert
Thomas More University of Applied Sciences, RADIUS, Kleinhoefstraat 4, 2440 Geel, Belgium;
geert.verheyen@thomasmore.be

Black soldier flies (BSF) can be cultivated on organic waste streams, thereby providing a useful system for converting waste into biomass. After harvesting the larvae, bio-based chemicals like chitin, proteins and fats, can be extracted and implemented in technical applications. BSF-fat contains about 50% of lauric acid and the fatty acid (FA) profile resembles the profile of coconut oil (CO) and palm kernel oil (PKO). Because BSF can be reared massively on small surfaces, they may provide a durable alternative to CO and PKO, which have a large impact on land-use and loss of biodiversity. In the INFACT-project, FA derived from the BSF and CO are coupled to amino acids (glycine and lysine) to generate amino acid-based surfactants. The surfactants are made by chemical procedures, but a more green, thermal amidation process is also investigated. This type of surfactant (currently produced commercially using CO) is non-toxic, mild and biodegradable and can be used in cosmetics and plant protection applications. After hydrolysis of the fat and conversion of the FA into acid chlorides, glycine-FA surfactants are synthesised with the Schotten-Baumann reaction at a yield of approximately 65% and >95% purity. Yields for the thermal amidation protocol are more variable (average 55%). Purity is variable but always >70%. Additional purification steps have good impact on the final purity. Lysine-FA surfactants have been synthesised with the Schotten-Baumann reaction, but purity is <50%. The mixture is complex, containing double and single bound surfactants and a lot of free FA. Using a TBTU coupling reagent. Better purities (>90% of double bound surfactants) are obtained. The glycine-FA surfactant has been implemented in several cosmetic formulations, including a shower foam, shower gel, hand cream and a w/o emulsion. Preliminary data indicate that in-house manufactured surfactants do not yet perform as well as commercial ones and improved purification of the surfactants is necessary.

Effects of feeding substrate on greenhouse gas emissions during black soldier fly larval development

C. Sandrock[1], C. Walter[1], J. Wohlfahrt[1], M. Krauss[1], S.L. Amelchanka[2], J. Berard[2], F. Leiber[1] and M. Kreuzer[2]
[1]Research Institute of Organic Agriculture, FiBL, Livestock Sciences, Ackerstrasse 113, 5070 Frick, Switzerland, [2]AgroVet Strickhof, Eschikon 27, 8315 Lindau, Switzerland; christoph.sandrock@fibl.org

Black soldier fly larvae (BSFL) based protein is considered a valuable alternative diet component of fish, poultry and pig. It may provide a twofold sustainability increase through more efficient nutrient recycling from agri-food waste and mitigating over-exploitation of limited marine and terrestrial resources. Yet, data on environmental impact criteria of BSFL rearing, e.g. greenhouse gas (GHG) emissions, imperative for overall life cycle assessments, are still scarce. Seven days old BSFL were incubated in open circuit respiration chambers at 28.5 °C and 50% relative humidity for 14 days during which four feeding events took place. The same pre-consumer food waste types were mixed in different (mirrored) proportions to obtain two feeding substrates of equal dry matter but high fibre (A) versus well digestible nutrient (B) contents. CO_2, CH_4 and N_2O were measured continuously for three chamber replicates each containing triplicated boxes of 10,000 BSFL. Before each feeding and at harvest larvae were temporarily separated from residue material to assess phase-specific biomass gain. Larvae of both feeding treatments survived equally and entered the prepupal stage, yet BSFL biomass production was higher for feed B. Similar for both feeding substrates, feeding events had major impact on all gas emission profiles. Likewise, BSFL developmental phases displayed characteristic responses, with strongly skewed biomass versus emission profiles towards the 6th larval instar. Overall, CO_2 emissions were similar between feeding treatments, CH_4 production was higher for feed A while N_2O production was higher for feed B. CH_4 and N_2O production during BSFL rearing likely correspond to particular contexts of type of feed, substrate depth, excess nutrients and associated microbiota. Total GHG emissions were high and comparable to conventional livestock biomass production. Comprehensive sustainability assessments of BSFL feed protein production would thus need to consider feeding substrates and rearing regimes, as well as duration of larval development until harvest.

The environmental potential of the conversion of organic resources by black soldier fly larvae

H.H.E. Van Zanten[1], A. Zamprogna[1], M. Veenenbos[2], N.P. Meijer[3], H.J. Van Der Fels-Klerx[3], J.J.A. Van Loon[2] and G. Bosch[4]
[1]Wageningen University & Research, Animal Production Systems group, De Elst 1, 6708 WD Wageningen, the Netherlands, [2]Wageningen University & Research, Laboratory of Entomology, Droevendaalsesteeg 1, 6708 PB Wageningen, the Netherlands, [3]Wageningen University & Research, RIKILT, Akkermaalsbos 2, 6708 WB Wageningen, the Netherlands, [4]Wageningen University & Research, Animal Nutrition Group, De Elst 1, 6708 WD, Wageningen, the Netherlands; hannah.vanzanten@wur.nl

The livestock sector is in need for sustainable feed sources, because of the increased demand for animal-source food and the already high environmental costs associated with especially the production of feed. Insect-based feed is often suggested as a sustainable alternative as insects can grow on a wide variety of low-value residual organic resources and generally have high protein and lipid contents. Although insects as feed show potential, their potential depends on the characteristics of the biomass resources used to feed the insects. The aim of this study was to assess the environmental opportunities of black soldier fly (BSF) larvae reared on different biomass sources using literature data on the efficiency of BSF larvae in converting such biomass resources into animal feed. The current EU legislative framework was used to classify the various biomass resources for rearing insects. Our results showed that resources within the legal groups that are, at the moment, not allowed in EU as animal feed (e.g. food waste or manure), have in general a lower environmental impact than the ones that are currently allowed (e.g. sorghum or wheat). Furthermore, BSF larvae reared on a resource that contains food and feed products generally have higher environmental impacts than conventional feed protein sources (fishmeal and soybean meal). While BSF larvae reared on a resource containing residual streams do offer potential to replace conventional feed protein sources and, thereby, to lower the environmental impact of food production. Using insects as feed, therefore, has potential to lower the environmental impact but a careful examination of the resource is needed as well as the assessments of potential food safety risks are required to relax EU legislation to bring promising residual streams into the food chain via BSF larvae.

Insect rearing and the law

J. Jacobs
Circular Organics, Slachthuisstraat 120/6, 2300 Turnhout, Belgium; johan.jacobs@circularorganics.com

Insect farming on an industrial scale, and the use of insect products in the markets of food, animal feed and technical applications, is rapidly growing into a fully-fledged business. Insects can have applications in multiple sectors, and an important economic and sustainability aspect of breeding insects is their ability to grow on relatively low value side streams. A range of side streams are forbidden for use as insect feed for reasons of food safety, even when the insects are reared specifically as a waste treatment agent, or as a source of raw materials for use outside the food chain. It's therefor safe to say that the existing legislation was not written down with insect bioconversion in mind. It contains numerous stipulations that make perfect sense in traditional livestock farming but are actually counterproductive in the development of the insect industry. This also hampers insect bioconversion in playing a role in the circular economy. We will provide an analysis of the legal framework in the European Union for insect rearing for different purposes, and the many aspects this covers. What pieces of legislation apply now 'by default' to insect farming, and how do we create a comprehensive 'insect rearing' legal framework? Elements such as the definition of insects as 'farm animals'; insect welfare; feedstock rules; 'slaughtering' of insects; processing and placing on the market of insect products; 'organic' insect products; and insect frass will be discussed. Insects can be the bridge from the organic waste sector to the feed and chemical industry. For this to happen, rather than adapting minor clauses in a specific Regulation, the relevant legislative packages must be reviewed, and integrated in a comprehensive Regulation on insect breeding and rearing for all purposes.

EU guidance document on best hygiene practices for the insect sector

C. Derrien

IPIFF, Secretariat, Avenue Adolphe Lacomblé 59, BTE 8, 1030 Brussels, Belgium; christophe.derrien@ipiff.org

With the insects as food and feed (for food-producing animals) market growing at an accelerating rate in regulations, science and technology several factors need to be simultaneously addressed. An important factor being Hygiene in production of insects-based products. The overarching objective of the IPIFF's Guide on Good Hygiene Practices is to help insect producers for food and/or animal feed purposes (hereafter referred to as 'insect producers') to achieve a high level of consumer protection and animal health through the production of safe products. To this end, the Guide provides guidance to insect producers to effectively apply EU food and feed safety legislation, while providing an incentive for them to develop a robust food and feed safety management system. In addition, this Guide specifies requirements enabling to: (1) ensure that insect producers conform to their stated food and feed safety policy and demonstrate their commitments in this regard; and (2) help the insect producers to effectively communicate food and feed safety issues to the regulatory authorities, and when needed, to their suppliers, customers and relevant interested parties (i.e. consumers) in the food and feed chains. To achieve the above objectives, the Guide has drawn on the skills and expertise of companies directly involved in the production of insects, either for human consumption or for animal feed purposes. Furthermore, consultations also involve several European representative organizations of the food and feed business sectors and other interested parties. The resulting outcome on the implementation of the Guide will provide Food safety and Hygiene control in the hands of every insect and based product producer.

Chemical food safety related to using supermarket returns for rearing *Hermetica Illucens*

H.J. Van Der Fels-Klerx[1], M.M. Nijkamp[1], E. Schmitt[2] and J.J. Van Loon[3]

[1]Wageningen Food Safety Research, Toxicology & Agrochains, Akkermaalsbos 2, 6708 WB Wageningen, the Netherlands, [2]Protix BV, Industrielaan 3, 5107 NC Dongen, the Netherlands, [3]Wageningen University, Entomology, P.O. Box 16, 6700 AA Wageningen, the Netherlands; ine.vanderfels@wur.nl

Insects are considered an alternative source of proteins for feed and food production in Europe. One of the insect species with great potential for large scale production is *Hermetica illucens* (black soldier fly; BSF). To become economically feasible and to contribute to circular economy, BSF rearing should be performed on substrates with little or no alternative use for feed and food production. In Europe, plant-derived materials and supermarket returns without meat and fish are currently legally allowed to be used. Supermarket returns are a group of foods that were produced for human consumption, but are no longer intended for human consumption. The aim of this study was to investigate the chemical safety related to the use of supermarket returns as substrate for rearing BSF. A small scale experiment (in triplicate) and a large scale experiment (in duplicate) were done with rearing BSF larvae on four different types of supermarket returns and a control (5 treatment groups) at, respectively, Wageningen University and at Protix. BSF larvae were reared from day 4 since hatching till day 10. After harvest they were frozen and sent to RIKILT (Wageningen University and Research) for chemical analyses. A range of possible contaminants were analysed in both the larvae samples and the substrates, including four heavy metals, dioxins and PCBs, PAHs and mineral oils. Survival rates and dry matter increase were not significantly different between the five treatment groups. Concentrations of the chemical compounds in the larvae were generally low. However, the bio-accumulation factor (the ratio of contaminant concentration in larvae to its concentration in substrate) was high for some of the contaminants considered. More detailed results are available at the start of the conference and will be presented.

Impact of natural contaminated substrates on insects: bioaccumulation and excretion

G. Leni[1], M. Cirlini[1], A. Caligiani[1], T. Tedeschi[1], J. Jacobs[2], N. Gianotten[3], J. Roels[4], L. Bastiaens[5] and S. Sforza[1]
[1]University of Parma, Food and Drug Department, Parco Area delle Scienze, 27/A, 43124 Parma, Italy, [2]Millibeter, Slachthuisstraat 120/6, 2300 Turnhout, Belgium, [3]Protifarm, Harderwijkerweg 141, 3852 Ermelo, the Netherlands, [4]Innovatiesteunpunt, Diestsevest 40, 3000 Leuven, Belgium, [5]VITO, Boeretang 200, 2400 Mol, Belgium; stefano.sforza@unipr.it

Insects are considered as suitable alternative feed for livestock production and their use is nowadays regulated in the EU (EC Regulation No 893/2017). Insects have the ability to grow on a different spectrum of substrates, which could be naturally contaminated by anthropogenic (agrochemicals) or naturally occurring (mycotoxins) contaminants. Very little it is known about their potential bioaccumulation in insect organism, but some studies suggest the potential detoxification effects of insects respect to mycotoxins. In the present study we investigated the ability of *Alphitobius diaperinus* and *Hermetia illucens* to retent, bioaccumulate or excrete mycotoxins and pesticides present as contaminants in the substrates used for their growth. The presence of contaminants was investigated on the different sidestreams used for feeding insects: rice bran (RB), wheat middling (WM), apple pomace, corn distillation residues (DDGS), olive pomace and rapeseed meal. LC-MS/MS and LC-UV/FLD methods were applied for the detection of common mycotoxins found in crops and/or vegetables. Trichothecene B, nivalenol, fumonisins (FB1 and FB2) and traces of Zearalenone were detected in some feeds for *A. Diaperinus* and *H. illucens* growth. The occurrence of pesticides was evaluated in a large screening of more than 100 of compounds (including phytochemicals, pyrethroids, organochlorine and organophosphate pesticides) with a GC-MS approach, showing the presence in some substrates of piperonyl butoxide, deltamethrin and pirimiphos methyl, cypermethrin and chlorpyrifos methyl. The presence of the identified contaminants was investigated also in larvae and rearing residue in order to evaluate their bioaccumulation. This study received funding from BBI Joint Undertaking in the framework of the European Union's Horizon 2020 research and innovation program, under grant agreement No 720715.

Molecular tools for insect detection in animal feed

E. Daniso, F. Tulli, G. Cardinaletti, R. Cerri and E. Tibaldi
University of Udine, Department of Agricolture, Food, Environment and Animal Science, via Sondrio 2A, 33100, udine, Italy; daniso.enrico@spes.uniud.it

Recently, European Commission (Reg. 2017/893/EC) allowed the use of processed insects in aquafeed and among the class Insect, only seven species have been allowed. DNA detection and identification of animal materials by means of PCR is now considered the official analytical method, thanks to its high sensitivity, to: (1) determine the species origin of processed animal proteins (PAPs); (2) give evidence of the absence of not allowed ingredients; (3) confirm the presence of authorised ingredients in animals feed. Among the detection/identification of PAPs, insect species identification may represent a new challenge and PCR methods the solution. In the present study, 7 specific primers against a conservative mitochondrial gene have been designed and validated for cross reactivity with different biological matrices, giving no cross reactivity. In detail, the most promising species to be used in feed according to availability and market price, *Tenebrio molitor* (TM) and *Hermetia illucens* (HI), were tested for amplification efficiency and sensitivity by six serial tenfold dilutions of pure HI and TM DNA in three replicates. The limit of detection (LOD5) defined, as the lowest dilution level for which three replicates still gave a positive result, was equal to 0.1μg for both the target species. HI DNA at 10 ng/μl in purity gave a reproducible signal at Ct=21.51 and TM DNA at 10 ng/μl gave a signal at Ct=20.24. TM and HI were detected also in practical feed with graded insect inclusion and the qPCR analysis output resulted in Ct values correlated to the inclusion level of the target species. The efficiency of PCR system for HI detection was described by a linear regression (y = -0.1665x + 39.879 R^2=0.977, n=10) obtained by plotting Ct-values against log DNA concentrations within the range of 0.03 to 0.5 g/g. Similar results were observed for *T. molitor*. The selected primers sequence revealed good efficiency and affinity in qPCR method and are promising candidates to be used in a portable biosensor for field analysis based on a fluorescence detection technique as a faster and easier detection system useful at industrial level. Research funded by Ager – SUSHIN project cod.2016-0112.

Bacterial populations in Dutch cricket and morio worm culture: a pilot study

O. Haenen[1,2], A. Borghuis[2], B. Weller[2], J. Van Eijk[2], E. Van Gelderen[1], E. Weerman[2], L. Bonte[2], B. De Ruiter[3], L. Dingboom[2], R. Petie[1], M. Cans[3] and P. De Cocq[3]

[1]Wageningen Bioveterinary Research, Bacteriology & Epidemiology, P.O. Box 65, 8200 AB Lelystad, the Netherlands, [2]HAS University of Applied Sciences, Applied Biology, P.O. Box 90108, 5200 MA 's-Hertogenbosch, the Netherlands, [3]Protifarm, Harderwijkerweg 141B, 3852 AB Ermelo, the Netherlands; o.haenen@has.nl

Insect culture gows, for protein for feed and food. Insects can be cultured on various left over streams. The Dutch Council on Animal Affairs warned for health risks for insects and man, and urged to support veterinary healthy, contact- and food-safe insect culture. The lectureship INVIS (Healthy, sustainable and safe insect culture for fish feed and food) of HAS cooperates with fish disease experts and epidemiologists of WBVR. So far few bacteria are identified as pathogens for cultured insects. The aims of this pilot were bacteriology of crickets (*Gryllodes sigillatus* and *Acheta domesticus*) and morio worms (*Zophobas morio*) and their substrates and housing, during the production cycle, in the insect farm Protifarm. Isolated bacteria were identified by biochemical methods, and by MALDI-TOF (matrix assisted laser desorption/ionisation time-of-flight). Various species of bacteria were detected, mostly commensals to humans and animals. Regarding insect pathogenic bacteria, in the morio worms, *Bacillus pumilus, Enterobacter cloacae, E. kobei, Klebsiella pneumoniae* were detected. In the house cricket, *A. domesticus*, and *Lysinibacillus sphaericus* was detected in its drinking water. Some of the commensal bacteria may turn zoonotic in rare cases, only when humans are strongly immunocompromised. It was concluded, that no alarming results were found, given the fact, that standard hygiene measures are practiced to prevent for infections. This pilot is a basis for further (inter)national research of the bacterioflora at insect farms, to support a healthy insect production chain, for healthy aquaculture and chicken feed, food, for education, and for science.

GBS-based genomic prediction for the improvement of natural enemies in biocontrol

S. Xia[1], B.A. Pannebakker[2], M.A.M. Groenen[1], B.J. Zwaan[2] and P. Bijma[1]

[1]Wageningen University & Research, Animal Breeding and Genomics, P.O. Box 338, 6700 AH Wageningen, the Netherlands, [2]Wageningen University& Research, Laboratory of Genetics, Droevendaalsesteeg 1, 6708 PB Wageningen, the Netherlands; shuwen.xia@wur.nl

Biocontrol is a strategy to reduce the population density of pests, such as insects, weeds and diseases, by using the natural enemies of the individuals causing those pests. Biocontrol is an appealing strategy in agriculture, because it holds the promise of controlling pests without the need for pesticides. Genetic selection of biocontrol populations may offer a solution to further improve the performance of natural enemies in practical biocontrol, in particular using the method of Genomic Prediction (GP). However, to our knowledge, the utility of GP has not yet been demonstrated for populations of natural enemies. Here we demonstrate proof-of-principle for the use of GP in a natural enemy population. We applied GP based on genotyping-by-sequencing (GBS) SNP data, using the parasitoid wasp *Nasiona vitripennis* as a model organism. A total of 1,230 individuals from two generations (G0 and G3) with genotypes for 8,639 SNPs were included in the analysis for wing aspect ratio (the ratio of wing length to width). Genomic best linear unbiased prediction (GBLUP) was applied to predict genomic breeding values (GEBVs). To assess the accuracy of GP, different cross-validation strategies were carried out: (1) across-generations validation; (2) 5-fold cross-validation within generations and combined dataset of G0 and G3. Accuracy was computed as the correlation between GEBVs and the observed phenotypes of individuals in the validation group divided by the square root of estimated heritability of validation group. The accuracy varied from 0.49 to 0.59 in across-generations validation, while higher variation in accuracy was observed in 5-fold cross-validation scenarios, ranging from 0.49 to 0.81. To conclude, our results indicate the potential of GP to predict breeding values in natural enemies.

How (not) to trigger pupation in *Tenebrio molitor*

D. Deruytter, C.L. Coudron, J. Claeys and S. Teerlinck
Inagro, Insect and aquaculture, Ieperseweg 87, 8800, Belgium; david.deruytter@inagro.be

A mealworm production system requires a continuous replenishment of adults to compensate for the loss of deceased ones. Part of the production must therefore be used to maintain reproduction. Due to biological variation, not all mealworms grow equally fast or pupate at the same age or body weight. all developmental stages (mealworm, pupae and beetle) will co-exist at a certain point in a tray at the same time. To avoid this and cannibalism, separation of pupae from mealworms is needed. This however requires manual labour before all mealworms have undergone metamorphosis. Furthermore, each time the pupae are handled there is a risk of damaging them resulting in death or deformed beetles. Triggering simultaneous pupation in mealworms would greatly enhance the efficiency of a mealworm production system both in time and number of successful pupations. Different physiological conditions were evaluated on mealworms with an average weight >90 mg in pilot scale conditions. Standard rearing conditions for the mealworms are: *ad libitum* wheat bran, a daily supply of a fresh vegetable mixture, a climate controlled environment (27 °C, 70% relative humidity) and complete darkness except for occasional lighting when technicians are working in the production hall. The hypothesis is that by changing these conditions mealworms experience a form of stress that triggers pupation, for example by mimicking springtime. Starvation, altering rearing density, changing the environmental temperature (31, 14, 5 and 1 °C), different light regimes (long days with light:dark = 20:4 and short days with light:dark = 4:20) and combinations of the variables mentioned were assessed. Pupation speed was evaluated as time needed for 80% of the mealworms to pupate (number of days passing to achieve 90% pupation, starting at 10%) and compared to pupation speed at standard conditions. Initial results indicate that cooling the larvae for 1 to 7 days does not alter pupation speed and may result in a higher mortality. In contrast to what was expected, long days decrease pupation speed and short days have little influence on the pupation.

Insects as food – a pilot study for industrial production

K. Wendin, V. Olsson, S. Forsberg, J. Gerberich, K. Birch, J. Berg, M. Langton, F. Davidsson, S. Stuffe, P. Andersson, S. Rask, F. Cedergaardh and I. Jönsson
Kristianstad University, Food and Meal Science, Elmetorpsvägen 15, 291 88 Kristianstad, Sweden; karin.wendin@hkr.se

Despite the many papers reporting on disgust factors of eating insects in Western cultures, the interest of insects as food is increasing, not least because they are nutritious, sustainable and tasty! The time has come to take the next step by making insects available not only as delicious restaurant food, but also for industrial production of foods and meals based on insects. The sensory attributes are of greatest importance to increase understanding of insects as a main ingredient in production and shelf life. By the use of factorial designs with mealworms as main ingredient, the aim was to evaluate the sensory impact of additions such as salt, oil/water and antioxidant agent. Also the impact of particle size of the mealworms was evaluated. Cooked fresh mealworms cut or ground into different particle sizes, oil, water, salt and rosemary were blended according to a factorial design. The resulting products were evaluated by descriptive sensory analysis in addition to instrumental measurements of viscosity and colour. Nutritional contents were calculated. Results showed that particle size of the mealworms had a great impact, i.e. an increased particle size increased the yellowness and the perceived coarseness. Further, both viscosity and crispiness increased. An increased particle size also meant a decreased odour, probably due to decreased exposure of particle surface. Increased salt content did, as expected, increase saltiness. It also increased the nutty flavour, probably due to the polarity of Sodium Chloride. Different ratios of oil/water did not seem to impact the sensory properties. With reference to the anti-oxidative effects of carnosic acid and carnosol, addition of rosemary had a significant impact on shelf life in terms of decreased rancidity and colour changes. All samples were high in protein content. All factors, but especially particle size of the mealworm fraction, influenced the sensory attributes.

Identification of volatile compounds from *Allomyrina dichotoma* larva according to extraction methods

S.J. Bae, Y.J. Kim and D.J. Shin
CJ Cheiljedang, Food R&D Center, 55, Gnunggro-ro 12beon-gil, Yeongtong-gu, Siwonsi, Gyeonggi-do 16495, Korea, South; sujin.bae1@cj.net

The purpose of this study was to provide better method to investigate the characteristics of volatile flavour compounds and aroma-active compounds of an edible insect, *Allomyrina dichotoma* larva by comparing two different extraction methods of solid-phase microextraction (SPME) and liquid-liquid continuous extraction/solvent-assisted flavour evaporation (LLCE/SAFE). The volatiles from *A. dichotoma* larva were initially isolated by SPME and LLCE/SAFE and then afterwards they were analysed by gas chromatography-mass spectrometry-olfactometry (GC-MS-O). As a result, 82 volatile components and 19 aroma-active compounds were identified by SPME while LLCE/SAFE detected 72 and 9 of volatile components and aroma-active compounds respectively. Isobutyl methyl ketone, acetoin, and 2-heptanol were believed to play an important role in earth and mushroom flavour for SPME method. Hexanol, 3-octanol, methional and methionol were considered the major aroma-active components in *A. dichotoma* larva for LLCE/SAFE method. It appeared that SPME extraction is more suitable method than LLCE/SAFE for the *A. dichotoma* larva because of its good extraction capacities and this study would be useful for study on reduction of the off-flavour in the edible insect and possibility of use as food materials.

Designing a cage for *Hermetia illucens*: summarizing two years work

D. Deruytter, C.L. Coudron, J. Claeys and S. Teerlinck
Inagro, Insect and aquaculture, Ieperseweg 87, 8800, Belgium; david.deruytter@inagro.be

Rearing insects on an industrial scale requires a constant and reliable source of eggs or newborn larvae. These can either be bought from a hatchery or, what is currently most common, bred on site. In any case, the space, labour and energy needed to breed them should be minimized. It is known that black soldier fly do need fairly specific climatic conditions to reproduce and that without these conditions no eggs or only infertile eggs are deposited. Although the basic conditions are roughly known (27 °C, high humidity and (sun)light), many aspects of black soldier fly breeding are still unknown or not optimized. At the experimental insect pilot plant of Inagro and within the Entomospeed project, black soldier fly has been bred for several years. The acquired knowledge was used to design a new black soldier fly cage for our pilot facility. The initial requirements were: (1) maximize the number of eggs per cubic meter; (2) both batch and continuous breeding should be possible; (3) a single person has to be able to handle the cage; (4) easy cleaning and hygienization; (5) durable design that can last for years; (6) versatile enough to be usable in all conditions. After two years, the result is a durable aluminium cage of 120 by 80 by 140 cm with a plastic removable bottom plate and mech on the sides. Two drawers are installed in the bottom plate for pupae and attractant. The drawer design ensures that the flies cannot reach the attractant and that the pupae can be added without entering the cage (minimizing possibility of escape). Wheels ensure that they can be easily transported to and from the climate room (e.g. to the wash station). Further experiments were performed to assess optimal climate conditions, light source (intensity and wave length), fly density, cage colour, interior design, etc. A proposal is made to scale up the cage to meet the industrial demand.

Artificial breeding of black soldier fly

B. Hoc, R. Caparros Megido and F. Francis
Gembloux Agro-Bio- Tech, Entomologie Fonctionnelle et Evolutive, 2 Passage des Déportés, 5030 Gembloux, Belgium;
bertrand.hoc@doct.uliege.be

Insects present a new source of protein and lipid produced locally. Compared to conventional animal production, insects require little space, do not consume much water, have a high conversion rate especially for organic waste and emit only small quantities of greenhouse gases. Several species have been selected as breeding models in Occident as mealworm (*Tenebrio molitor*) and house cricket (*Acheta domestica*) for food or black soldier fly (BSF) larvae (*Hermetia illucens*) for feed. The BSF larvae are voracious consumers of decomposing organic matter and polyphagous diet as the macronutrient quality (mainly lipids and proteins) of these larvae make them excellent candidates for various applications such as waste and organic material management, incorporation in animal feed or as an alternative energy source. The development of its rearing requires specific methods and facilities that meet the requirements of the different stages of the species life cycle (reproduction, magnification and broodstock selection). Nevertheless, rearing development in temperate regions requires artificial processes to continuously produce high quality eggs and larvae. This study aims at evaluating the influence of broodstock selection, density and light duration on *H. illucens* reproduction. A laying device to collect eggs has been developed and aims at centralize oviposition and simplify eggs collection. Two broodstock populations with opposite sex-ratio were selected and their respective eggs productions have been evaluated for five breeding densities. Finally, four light durations (2, 6, 12 and 18 h of daily light) were simulated to evaluate the impact of light on reproduction. The laying device effectively centralized oviposition and the eggs can be quantified by separating them from the substrate. Broodstock selection of differentiated sex-ratio populations led to obtained a higher spawning effort for female dominant and the oviposition rates do not increase significantly beyond 6 hours of illumination.

Fractionation of insect biomass – BBI-InDIRECT project

L. Soetemans[1], G. Leni[2], Q. Simons[1], S. Van Roy[1], A. Caligiani[2], S. Sforza[2] and L. Bastiaens[1]
[1]VITO NV (Flemish Insititute for Technonolgical Research), Sustainable Chemistry, Boeretang 200, 2400 Mol, Belgium,
[2]University of Parma, Department of Food and Drug, Via Universita 12, 43100 Parma, Italy; leen.bastiaens@vito.be

Biomass of insect larvae are rich in lipids, proteins and chitine, and can as such be considered as a feedstock for these compounds. The larvae can be used as a whole, or after fractionation. Within the BBI-H2020 InDIRECT project (www.bbi-indirect.eu), two fractionation approaches have been studied. A first biorefinery approach is based on a mechanical separation of the larvae generating (1) a juice fraction with a low chitin content (0.5%) and rich in lipids and proteins, and (2) a pulp fraction containing nearly all chitin besides some lipids and proteins. Further chemical supported fractionation of the juice resulted in a protein enriched fraction (up to 60% protein) and a lipid enriched fraction (up to 85%). In conclusion, depending on the envisioned application and the fractionation process can be tailored. The pulp fraction was further fractionated aiming at purified chitin and covalorisation of a protein rich fraction. Chitin was converted to the more water soluble chitosan and chitosan oligomers. The whole fractionation approach was applied on larvae from the black soldier fly (provided by Millibeter, B) as well as on the lesser mealworm (provided by Protifarm, NL), indicating that the biorefinery approach that was developed is rather generic and not insect species specific. The approach was developed at lab scale and translated to a pilot scale process to generate fractions for application tests related to feed and chemistry. The second approach is based on enzymatic-assisted hydrolysis of the whole larvae biomass. The proteolytic activity of enzymes allowed to extract and solubilize in a single fraction up to 80% of total proteins, in form of peptides. The peptide-fraction is more digestible, rich in bioactive peptides and free amino acids which make it suitable as feed for aquaculture. The remaining pellet, rich in chitin, represents an economical value for the chemistry sector. This project received funding from the Bio Based Industries Joint Undertaking under the European Union's Horizon 2020 research and innovation programme under grant agreement No 720715.

Effects of selected insecticides on black soldier fly (*Hermetia illucens*) larvae

N.P. Meijer[1], T.C. Rijk[1], J.J.A. Van Loon[2] and H.J. Van Der Fels-Klerx[1]
[1]*RIKILT Wageningen University & Research, Wageningen Campus P.O. Box 230, 6700 AE Wageningen, the Netherlands,* [2]*Wageningen University, Department of Plant Sciences, Laboratory of Entomology, Wageningen Campus P.O. Box 16, 6700 AA Wageningen, the Netherlands; ine.vanderfels@wur.nl*

The effects of chronic exposure of bees to persistent insecticides, particularly neonicotinoids, have come under increased scrutiny due to their association with sub-lethal effects and colony collapse disorder. Effects of insecticides residues (lethal/sub-lethal) on those insect species considered as mini-livestock, as well as accumulation in larvae, are largely unknown. We hypothesized that since insecticides specifically target insects, active substances present in the substrate (even at levels at or below the legal limit) may still affect insects. We selected nine insecticides, each with a different mode of action. These were added to the feed of seven-day old black soldier fly larvae (*Hermetia illucens*) at a concentration equal to the European Union legal limit in maize (Regulation (EC) No 396/2005) or for feed (Directive 2002/32/EC) (1×ML, study phase 1). In phase 2, this was done at a level ten times higher or lower, depending on the results of phase 1. The larvae were harvested at day fourteen. We find that two of the nine insecticides significantly affect the mortality and growth of the larvae in phase 1, which suggests that insect feed with concentrations of these insecticides at, or slightly below, the EU legal limit may be unsafe for reared black soldier fly larvae. In phase 2, the insecticides which affected growth and mortality in phase 1 appeared to have no significant effects at the lower concentration. Only one of the insecticides spiked to a higher concentration than in phase 1 appears to slightly affect the growth and mortality of the larvae. The results suggest bioaccumulation for one of the selected insecticides; but it appears that the remaining eight are partly converted into different metabolites. We recommend BSF farmers to take into account the possible contamination of substrates with insecticides for insect performance and safety of the resulting insects for feed and food.

New source of protein: edible insect to fight malnutrition in south of Madagascar

A. Ravelomanana[1] and B.L. Fisher[2]
[1]*Madagascar Biodiversity Center, Natural Capital, Parc Botanique et Zoologique de Tsimbazaza, Antananarivo, BP 6257, Madagascar,* [2]*California Academy of Sciences, Entomology, 55 Music Concourse Drive, San Francisco, CA 94118, USACalifornia, USA; ravelomanana.njaka@gmail.com*

Each year, severe drought occurs in the South of Madagascar and the rate of malnutrition remains high up to 70% in some localities. The government supported by different NGOs strengthen sustainable access to fortified foods. Moringa becomes very famous and largely used at a national level for nutritional supplementation. However, insect traditional food in many different regions of Madagascar less known. Insects are indeed a rich source of fat, iron, zinc, as well as protein and vitamin. It represents a new source of protein, cheap and it can be used as fortified food. Our goal is to promote insects as food by determining which insect species can be farm for large scale production and establish a protocol for best insect rearing. That goals allow to improve the availability of insect in the market and improve the quality of fortified food to tackle the problem of food security in the South. We started by farming four insect species. Criteria used to select the insect species were: native species- short life cycle – can be scaled for production. Indoor farming was the facility used for this farming and among the main variable established for our protocol were the access to the water, humidity, and temperature. Among the four insect reared, *Gryllus madagascariensis* (cricket) was the best species for rearing. Chicken feed (content 20% of protein), humidity at 65% and mean temperature 31 °C were the optimal parameters to get high output around 70% from juvenile to adult cricket. There is a positive correlation between temperature and reduction of time for hatching. One of the important outcomes is sexual maturity with three weeks less in our facility compare to thirteen weeks in the natural habitat. This is a very important step to save time. The nutrition value of this species is well established and the incorporation of edible insect flours on rice or maize has been demonstrated to be successfully accepted in tropical countries.

Diversity of wild and edible insects for forest sustainability in Madagascar

A. Ravelomanana[1] and B.L. Fisher[2]
[1]Madagascar Biodiversity Center, Natural Capital, Parc Botanique et Zoologique de Tsimbazaza, BP 6257 Antananarivo 101, Madagascar, [2]California Academy of Sciences, Entomology, 55 Music Concourse Drive, San Francisco, CA 94118, USA; ravelomanana.njaka@gmail.com

As part of Asia, eating insects is part of Malagasy culture. Insects are highly nutritious and rich in proteins. Every year, new insect species are described from Madagascar but less is known about the diversity of edible insects. A review of the practices of entomophagy is studied to evaluate the role of this traditional diet. Our objective is the inventory of edible insect near the major forest under treat (eastern) and in the south where the rate of people malnourished is higher. The data was collected by questionnaire, oral interview on edible insect consumption, consumers and marketers. Twenty localities belong to ten regions were visited. This inventory allowed to identify 49 edible insect species. The diversity of edible insect belongs to the order Orthoptera (30 sp.) was quite higher, followed by Coleoptera (6 sp.), Hemiptera (3 sp.), Hymenoptera (4 sp.) and Lepidoptera (1 sp.). Among these edible insect species, the consumption of Orthoptera and Coleoptera were the highest based on the oral interview. Among those edible insect species, *Locusta migratoria* (Orthoptera) and *Zanna madagascariens* (Hemiptera) are largely consumed by local people. These two species can be farmed locally or produced at large scale. This farming can resolve the problem of malnutrition and reduce the demand for bushmeat for people leaving near the forest. Taste appears to be the main motivation while cost and nutrition value was not considered by consumers. Roasting or boiling was the main preparation for consumption. Our survey highlighted also that the acceptance of using insects as food is high only at young age. People need to be educated to increase consumption and on its nutritional value. To reach the objective of the contribution of insects on forest management, collaborative efforts from local people, scientists, insect farmers, and consumers are needed. Another aspect as overharvesting of forest insects and social aspects should undertake to better estimate the contribution of edible insect on forest management.

Plant polyphenols: effects on rumen microbiota composition and methane emission

M. Mele[1] and A. Buccioni[2]
[1]University of Pisa, Dipartimento Scienze Agrarie, Alimentari. Agro-ambientali, via del Borghetto, 80, 56124 Pisa, Italy, [2]University of Florence, Dipartimento di Scienze e Tecnologie Agrarie, Alimentari, Ambientali e Forestali, via delle Cascine, 5, 50100 Firenze, Italy; marcello.mele@unipi.it

Plant polyphenols (PP) entry in the ruminant diets being present in both forage and concentrate ingredients. Polyphenols interact with rumen microbiota, affecting carbohydrate fermentation, protein degradation, and lipid metabolism. A relevant connection exists between fibre fermentation and methane production in the rumen. Methane production is also usually reduced when tannins are included in the diet of ruminants, probably as a consequence of the inhibition of fibre digestion. In fact, the production of acetic acid in the rumen is associated with increased H_2 production, which, in turn, is the substrate for methane production. Indeed, the microbiome of high-methane producing animals is different from that of low-methane-producing animals. Among PP, the effect of tannins on methane production has been largely investigated. Although it is widely recognized that methane declined when dietary tannins increased, the number of experiments where the effect of dietary PP (both tannic and nontannic) on methane production was measured along with the composition of rumen microbiota is still scarce. The major part of the experimental evidences is from *in vitro* trials. However, data from *in vivo* experiments are in general consistent with the *in vitro* data, although the effects of tannins on methane production had a minor extent. Data from literature are quite consistent about a general depressive effect of polyphenols on gram-positive fibrolytic bacteria and ciliate protozoa, resulting in a reduction of volatile fatty acid production. Considering that archaea are in association with protozoa, the reduced protozoa population on the presence of tannins could partly contribute to explain results about decreased methane production. Unlike condensed tannins, some evidence suggests that hydrolysable tannins may reduce methane emission by directly interacting with rumen microbiota without affecting fibre digestion. Flavonoids have been suggested to reduce ruminal methanogenesis by acting as H_2 sinks or by inhibiting methanogens.

Effect of feed additives on rumen bacteria of beef calves in transition to a high-concentrate diet

S. Yuste[1], Z. Amanzougarene[1], G. De La Fuente[2], M. Fondevila[1] and A. De Vega[1]
[1]Universidad de Zaragoza-IA2, M. Servet, 177, 50013 Zaragoza, Spain; [2]Universidad de Lleida, Dept. Ciència Animal, Av. Alcalde Rovira Roure, 191, 25198 Lleida, Spain; syuste@unizar.es

Changes in rumen bacterial community during the adaptation of 7-month-old beef calves (212 ± 27.0 kg) from a milk/grass diet to a high-energy diet consisting of a cereal-based concentrate plus wheat straw, both given *ad libitum* (diet C), were studied. Eighteen rumen-cannulated Limousine crossbred male calves were randomly assigned to three diets (C; C plus 20 g/kg of a 65:35 chestnut and quebracho tannin extract, T; and C plus 6 g/kg medium-chain fatty acid (MCFA) mixture, M) over a 28-d period. Rumen fluid was collected on days 0, 7, 14 and 28 before the morning feeding. Microbial DNA from rumen fluid was extracted with a commercial kit and V1-V3 regions of the 16S rRNA gene were sequenced using the Ion Torrent Platform. Permutational ANOVA and partial least squares discriminant analyses were performed to assess any changes in rumen bacterial community induced by the feed additives and/or days of sampling. Results indicated that the addition of tannins or MCFA did not modify ($P=0.98$) the general structure of the bacterial community. By contrast, it was affected by sampling day ($P<0.001$), indicating that dietary shift altered the microbiota. Similarly, diet did not affect the relative abundance of the main phyla and genera ($P<0.05$) but there was an effect of sampling day. There was a decrease ($P<0.01$) of *Firmicutes*, *Bacteroidetes*, *Fibrobacteres* and *Tenericutes*, and an increase of *Proteobacteria* and *Actinobacteria* on d28 compared to d0. Moreover, there was an increase of the relative abundance of *Prevotella*, *Ruminococcus* and *Succinivibrio*, and a decrease of *Butyrivibrio* and *Fibrobacter* ($P<0.05$). Diversity indexes (Shannon and richness) declined over sampling days ($P<0.05$), indicating a possible selection of bacteria more tolerant to the new environmental conditions. Transition to a high-concentrate diet modified the general structure of the bacterial community and reduced diversity. Tannins and MCFA, at the doses used in the present experiment, did not exert any effect on ruminal microbiota.

Effect of legumes on forage composition, steer performance, enteric methane emission and economics

B.M. Kelln[1], J.J. McKinnon[1], G.B. Penner[1], T.A. McAllister[2], S.N. Acharya[2], B. Biligetu[3], K. Larson[1] and H.A. Lardner[1]
[1]University of Saskatchewan, Animal and Poultry Science, 52 Campus Drive, Saskatoon, Saskatchewan, S7N 5B4, Canada, [2]Agriculture and Agri-Food Canada, Lethbridge Research Center, Lethbridge, Alberta, T1J 4B1, Canada, [3]University of Saskatchewan, Department of Plant Sciences, 52 Campus Drive, Saskatoon, Saskatchewan, S7N 5B4, Canada; bmk371@mail.usask.ca

A 3-yr study was conducted to determine the effects of sod-seeded non-bloat legumes into mixed pasture, on forage botanical composition and quality, steer performance, methane emissions and system economics. The study site was a 24-ha meadow bromegrass-alfalfa (*Bromus riparius-Medicago sativa* L.) pasture subdivided into fifteen, 2-ha paddocks. Each paddock was randomly assigned to 1 of 3 replicated treatments, sod-seeded: (1) sainfoin (*Onobrychis viciaefolia*) (SAIN) (n=6); (2) cicer milkvetch (*Astragalus cicer*) (CMV) (n=6); and no sod-seeded legume (CON) (n=3) in a completely randomized design. Each yr, 45 steers (326 ± 26.3 kg) were randomly allocated to treatment paddocks along with, 15 ruminally cannulated cows to determine effect of legumes on rumen fermentation and methane emission using SF_6 measurement technique, over 6-d sampling. Sainfoin composition declined (10 to 2%) while CMV increased (11 to 14%) in stand over the 3-yr trial. Crude protein content was greater ($P<0.01$) in SAIN and CMV (14.8%), compared to CON forage (12.6%) and neutral detergent fibre was lower ($P<0.01$) in legume treatments (64.16, 59.52, 59.04%; CON, SAIN, CMV). Numerically, steer average daily gain was 20% greater in sod-seeded paddocks compared to CON paddocks. Total short chain fatty acid concentration did not differ ($P=0.29$) (102.3, 104.9, and 113.8 mM) for CON, SAIN and CMV, respectively. Acetic to propionic ratio was lower ($P<0.05$) for CMV (4.18 mM) compared to CON and SAIN (4.47 and 4.46 mM), respectively. Enteric methane levels (l/d) of grazing animals did not differ ($P=0.68$) between treatments. Economic return on investment, estimated by crop production, beef production and grazing costs, increased ($P<0.05$) by 57 $/ha from grazing CMV and SAIN paddocks compared to CON paddocks. Study results suggest that sod-seeding legumes into mixed pasture can be a valuable pasture rejuvenation strategy.

Sugar for dairy cows – simulating fodder beet supplementation to reduce environmental pollution

A. Fleming, R.H. Bryant and P. Gregorini
Lincoln University, Agriculture and life sciences, P.O. Box 85084, 7647, Lincoln, Christchurch, New Zealand;
anita.fleming@lincoln.ac.nz

New Zealand pastoral diary producers are under increasing scrutiny to reduce environmental pollutants: urinary nitrate excretion (UN) and enteric methane (CH_4) emissions from livestock. While dietary manipulations are an affordable option, they often swap one pollutant for another. Supplementation of herbage with fodder beet, (*Beta vulgaris* M., FB) a sugar dense and protein deficient bulb, is reported to reduce UN excretion however, the effect on CH_4 production is unknown. The objective of this research was to estimate the potential of FB to maintain or increase milk yield while reducing environmental impacts. MINDY, a dynamic and mechanistic model of the grazing ruminant was used to simulate 90 diets using a factorial arrangement of FB allowance, herbage allowance and time of allocation. The FB allocations were 0, 2, 4 or 7 kg DM/cow/day, and herbage allocations (HA) were 18, 24 or 48 kg DM/cow/day above ground level. In the simulations all combinations were offered either in the morning or afternoon, or split across two equal meals. MINDY was initialised as a four year old Frisian × Jersey dairy cow at 120 days of lactation and 535 kg liveweight. Treatments (diets) were screened using Pareto front analysis to select diets that were the best compromise between low UN and CH_4 per kg milk yield. Both FB and HA incrementally improved daily dry matter intake and milk production. Pareto front analysis 'frontier' contained five diets which were all 7FB. Four of the treatments were 18HA, and one was 24HA split across two meals. The 7FB and 18HA of these diets, reduced intake of N and thereby N excreted in urine. Fodder beet increased enteric CH_4 production curvilinearly thus, 2FB and 4FB treatments caused pollution swapping of UN for CH_4 and limited the potential for FB to improve pollution. Although the results suggest pollution swapping can be avoided by supplementing herbage with FB, the level required may be detrimental to animal health and does not provide a practical approach to improving the environmental impacts of dairy production.

Effect of feeding *Candida utilis* yeast on intestinal development and health in post-weaning piglets

I.M. Håkenåsen[1], M. Øverland[1], A.Y.M. Sundaram[2], R. Ånestad[1] and L.T. Mydland[1]
[1]Norwegian University of Life Sciences, Department of Animal and Aquacultural Sciences, Arboretveien 6, 1433 Ås, Norway, [2]Oslo University Hospital, Department of Medical Genetics, Kirkeveien 166, 0450 Oslo, Norway; ingrid.marie.hakenasen@nmbu.no

Weaning is a critical phase for piglets as they are experiencing major changes, including abrupt transition in diet while their digestive system is still developing. *Candida utilis* yeast produced on sugars from Norwegian spruce by-products is a promising novel protein source with bioactive compounds known to improve intestinal health. The objective of this study was to examine the effect of feeding a diet with *C. utilis* yeast on development of gastrointestinal tract (GIT) function and health in post-weaning (PW) piglets. Sixty-four crossbred piglets were weaned at 28 days of age and fed a commercial-like control diet, or an experimental diet containing 14.6% *C. utilis* yeast, substituting 40% of the protein in the control diet. Piglets were sampled at day 0, 2, 4, 7 and 14 PW to determine digestive enzyme activities in jejunal digesta and gene expression using RNA sequencing of jejunal and ileal tissue. Ileal digesta was collected for determination of apparent ileal digestibility (AID) of major nutrients and minerals. Feeding yeast did not affect growth performance of the piglets, but AID of N and P were higher in piglets fed the yeast diet. Overall, enzyme activities were not affected by diet, but trypsin activity increased by time PW, whereas lipase activity decreased from day 2 to day 7 PW. Control piglets had greater number of significantly differentially expressed (DE) genes in jejunum on day 7 compared to day 0 than piglets fed the yeast diet (2,846 vs 96). In ileum, most DE genes were observed in control piglets at day 7 compared to day 2. Few genes were DE between dietary treatments, but on day 7, 447 jejunal genes were DE between the control piglets and the piglets fed the yeast diet. A more detailed discussion will be presented. The results indicate that *C. utilis* yeast is a high quality protein source with positive effects on GIT development in PW piglets.

Total or partial replacement of copper sulphate by copper bis-glycinate on performance of weaned pigs

R. Davin[1], Y.C. Link[2], S.-C. Wall[2] and F. Molist[1]
[1]Schothorst Feed Research, Meerkoetenweg 26, 8218 NA Lelystad, the Netherlands, [2]Phytobiotics Futterzusatzstoffe GmbH, Wallufer Str. 10a, 65343 Eltville, Germany; rdavin@schothorst.nl

The goal of this trial was to determine the effect of copper bis-glycinate, as total or partial replacement of copper sulphate, on the growth performance of weaned pigs. A total of 384 twenty-six day-old piglets (male:female ratio 1:1) with an average BW of 7.59 kg were weaned and allocated (6 piglets/pen) to 4 dietary treatments at random: a treatment with 120 mg/kg Cu as $CuSO_4$ (CuSO-120), two treatments with Cu bis-glycinate (Plexomin® Cu) at two levels, 60 or 120 mg/kg (Plexo-60 or Plexo-120), and a treatment with 120 mg/kg Cu containing 60 mg/kg Cu from each Cu product (Combo). Each dietary treatment was provided to 16 pen-replicates. Experimental diets were provided in two phases, pre-starter: d 0 to 14, and starter: d 14 to 28. Pigs and feed were weighed, and faecal consistency scored on a pen basis at the start and end of each phase, and FCR calculated. Data were analysed with a one-way ANOVA and significant differences were estimated with a Fisher's test. Significance was declared at P<0.05. All treatments had a similar ADG, ADFI and FCR during the starter phase (P>0.05), the ADG0-14d being 210, 207, 220 and 218 g/d for CuSO-120, Plexo-60, Plexo-120 and Combo, respectively. Dietary treatment affected ADG and ADFI during the grower phase (P<0.05), but not FCR (P=0.23). Plexo-120 and Combo had a higher ADG14-28d than CuSO-120 (553 and 565 vs 526 g/d, respectively), whereas Plexo-60 (529 g/d) was not significantly different to neither CuSO-120 nor Plexo-120. ADFI14-28d was higher for Combo than in CuSO-120 and Plexo-60 (781 vs 741 and 734 g/d) and Plexo-120 had an intermediate value (764 g/d). Considering the overall period (0-28d), Plexo-60 had a similar ADG and ADFI than CuSO-120, whereas Plexo-120 and Combo had greater ADG and ADFI than CuSO-120. ADG0-28d was 362, 363, 381 and 388 g/d for CuSO-120, Plexo-60, Plexo-120 and Combo, respectively. Faecal score was not affected by any treatment along the course of the study. In conclusion, copper bis-glycinate can improve the piglets' performance compared to copper sulphate when fed at an equivalent copper level, or maintain it when fed at 60 mg/kg.

Hydroxychloride trace minerals improve growth performance and carcass in pigs: a meta-analysis

S. Van Kuijk, M. Jacobs, C. Smits and Y. Han
Trouw Nutrition, R&D, Stationsstraat 77, 3811 MH Amersfoort, the Netherlands; sandra.van.kuijk@trouwnutrition.com

In Europe only low amounts of trace minerals are allowed in diets for pigs in the grower/finisher phase, because of environmental reasons. Therefore, it is important to select a good source of trace minerals with a high bioavailability and low reactivity in feed and animals. Hydroxychloride trace minerals (HTM; IntelliBond; Trouw Nutrition, Netherlands) can be a good alternative for inorganic trace minerals (ITM). HTM have homogeneous particle sizes and a low surface reactivity. Several studies were conducted and combined into a meta-analysis to compare the effect of HTM to that ITM in grower/finisher pigs. In total 6 studies were performed throughout Europe, in which the pigs were fed maximum allowed Zn (80 ppm) and Cu (15 ppm) levels from sulphates or HTM. The studies started when the pigs were around 9-10 weeks of age and lasted until slaughter. The body weight (BW), average daily gain (ADG), feed intake (FI), and gain:feed (G:F) were measured at the start and end of each of 3 feeding phases. The carcass yield, back fat thickness and lean meat % were measured at commercial slaughterhouses. Raw data from each study was combined into a meta-analysis. The BW, ADG, FI, G:F and back fat thickness were analysed using the MIXED model of SAS. Analysis of BW, ADG, FI and G:F per feeding phase included time as repeated measure. The carcass yield and lean meat % were analysed using the GLIMMIX procedure of SAS. In all models, the variations within and between study were taken into account. In the overall study periods G:F tended (P=0.0925) to be improved by 1.3% in HTM pigs. In the analysis per feeding phase a 3.9% improvement in ADG (P=0.0625) and G:F (P=0.0031) was observed in the last feeding phase before slaughter. In the slaughterhouse, the lean meat % was 0.6% higher (P=0.0009) in HTM fed pigs compared to ITM fed pigs. None of the other parameters were significantly different between the mineral sources. The results of the 6 studies suggest that HTM can improve growth performance compared to ITM. As a result, a higher lean meat % was obtained in HTM fed pigs.

Influence of zinc source and level on performance and tissue mineral content in fattening pigs

S. Villagómez-Estrada[1], D. Solà-Oriol[1], S. Van Kuijk[2], D. Melo-Durán[1] and J.F. Pérez[1]
[1]Universitat Autònoma de Barcelona, Animal Nutrition and Welfare Service, Animal and Food Science department, Barcelona, 08193, Spain, [2]Trouw Nutrition, Research and Development, Stationsstraat 77, AG Amersfoort, Utrecht, 3800, the Netherlands; sandra.villagomez@uab.cat

Growing pig diets traditionally include Zn (150-200 ppm) that exceed animals requirements. However, due to its negative effects on environment, the maximum permitted level in the European Union is 120 ppm. Consequently, the reduction of the dietary supply of Zn, through the evaluation of alternative sources, is one of the possible approaches to prevent this public health risk. The present study evaluated the effect of Zn as sulphate or hydroxychloride at added levels of 20 ppm (nutritional) or 80 ppm (high), on the productive performance, mineral tissue concentration and mineral apparent total digestibility of fattening pigs. A total of 444 commercial 63 days old growing pigs (18.7±0.20 kg) were distributed into 4 treatments in a randomized complete block design and 2×2 factorial arrangement. Each treatment had 9 replicates, consisted in 3 replicates with 11 and 6 replicates with 15 pigs. The study data were analysed taking into account the main factors and their interaction with ANOVA using GLM procedure of the statistical package SAS. Productive performance was not affected by a two-way interaction of source by level during the trial. Feeding pigs with higher doses produced a significant decrease in average daily feed intake (1.69 vs 1.76 kg/d; P=0.037) compared to those fed the nutritional level. No differences in pig performance were observed between sulphate and hydroxychloride sources. High inclusion levels increased the Zn concentration in the liver and metacarpal bone. Nevertheless, the apparent total tract digestibility was higher for pigs fed hydroxychloride at high inclusion than those fed sulphate at the same level (42.6 vs 32.8%; P=0.014). In conclusion, fed diets with sulphate or hydroxychloride promoted similar effects in pig performance. The lack of significant differences in performance parameters between two dietary levels shows that diets with levels close to those recommended by NRC (2012) could not reduce the productive performance.

Effect of dietary guanidinoacetic acid supplementation on growth performance in nursery piglets

J. Morales[1], M. Rademacher[2], U. Braun[3], M. Rodríguez[1] and A. Manso[1]
[1]PigCHAMP Pro Europa, 40006, Segovia, Spain, [2]Evonik Nutrition & Care GmbH, 63457, Hanau-Wolfgang, Germany, [3]AlzChem Trostberg GmbH, 83308, Trostberg, Germany; joaquin.morales@pigchamp-pro.com

Guanidinoacetic acid (GAA) is an amino acid (AA) derivate from arginine and glycine. It is a naturally occurring substance in animals and the immediate metabolic precursor for creatine in the body. However, because of the rapid growth in early life, these AA are in high demand easily becoming deficient in pigs. The objective of the study was to evaluate the efficacy of dietary CreAMINO® (96% GAA) supplementation in nursery piglets. At 28 days of age, 336 piglets were weaned and allotted to 56 pens (6 piglets/pen) according to body weight (BW) and sex, in order to produce 4 similar groups, each of 14 pens. Four different doses of CreAMINO were compared: control, 0.06, 0.09 and 0.12%. The 42 d nursery phase was divided in 2 feeding periods: 14 d of prestarter (8-12 kg BW) and 28 d of starter (12-26 kg BW). Performance was evaluated at days 0, 14 and 42. At day 42, blood from 12 piglets (6 male, 6 female) per treatment was sampled in the morning from the jugular vein for GAA, creatine and creatinine analysis. Statistical analysis was performed as a completely randomized design by GLM of SAS using treatment as fixed effect and initial BW as covariate. Supplementation of 0.12% CreAMINO significantly (P<0.05) increased average daily gain (ADG) in the pre- starter phase (d 1-14) as well as from d 1 to 28 and the overall period up to day 42. Feed conversion ratio (FCR) was significantly (P<0.05) improved at 0.12% CreAMINO supplementation from d 1-14 and from d 1-28. There were no differences (P>0.05) in average daily feed intake (ADFI) in any of the studied phases. However, a trend in BW at day 42 and ADFI in global nursery phase was observed (P=0.054 and P=0.090, respectively), which was highest in the 0.12% supplemented CreAMINO group. Numerical elevations were observed in GAA, creatine and creatinine serum concentration compared to the untreated control group. From this experiment we conclude that feed supplementation of up to 0.12% CreAMINO positively affects FCR, ADG and BW in the nursery period compared with the control group whilst lower supplemental levels scored in between.

Dietary hydroxy methionine supply in pigs: associated changes in muscle biological processes
F. Gondret[1], N. Le Floc'h[1], D.I. Batonon Alavo[2], Y. Mercier[2], M.H. Perruchot[1] and B. Lebret[1]
[1]PEGASE, INRA, Agrocampus Ouest, 35590 Saint Gilles, France; [2]DISSEO France, SAS 03600 Commentry, France;
florence.gondret@inra.fr

Methionine (Met) is an essential amino acid (AA) in pigs and the third limiting AA in cereal based diets. Met supplementation improves animal's performance and protein gain. Besides, Met is considered as an important factor in the control of oxidative status with potential benefits on health and meat quality. This study aimed at investigating the effects of dietary Met provided in excess relative to growth requirements on muscle metabolism in finishing pigs. During the last 14 days before slaughter, crossbred pigs were individually fed diets (13.6% protein, 5.8% fat, 10.8 MJ/kg Net energy, 0.73% Lys) supplemented (Met++) or not (CONT) with OH-Met (n=15 pigs/diet). Dietary Met content was 0.22 and 1.1% in CONT and Met++ diets, respectively. Growth rate, carcass lean meat content and Longissimus muscle (LM) weight at slaughter did not differ between the two groups. In LM, the free AA profile was modified (P<0.05) towards greater Lys, Pro, 3-methylHis and α-aminobutyric acid concentrations and lower Gln content in the Met++ pigs compared with CONT pigs, whereas Met and taurine contents did not significantly differ between the two groups. Glutathione content, the major non-enzymatic intracellular antioxidant synthetized from Met and Cys, was increased (P<0.001) in LM of Met++ pigs, but the ratio of oxidized to reduced glutathione forms did not differ between the two groups. Hexokinase and lactate dehydrogenase, two enzymes involved in glycolysis, displayed similar activities in the two groups. On the opposite, the activity of citrate synthase involved in oxidative metabolism was lower (P=0.02) in LM of Met++ pigs compared with CONT pigs. The expression of genes involved in protein synthesis, protein catabolism, energy metabolism, and autophagy, a process tuned by oxidation/redox signalling, was studied. Altogether, the results suggest an extensive utilization of the Met in excess for glutathione synthesis. A short-term extra dietary Met supply in finishing pigs altered muscle AA and energy metabolisms, which could in turn modify post-mortem biochemical processes of muscle transformation into meat and affect pork quality traits.

Can a concentrate diet enriched by vitamin E keep the vitamin E level high in calves after weaning?
S. Lashkari[1], M. Vestergaard[1], C.B. Hansen[2], K. Krogh[3], P. Theilgaard[4] and S.K. Jensen[1]
[1]Aarhus university, Animal science, Blichers Allé 20 Postbox 50, 8830 Tjele, Denmark, [2]Slagtekalverådgivning, Herningvej 23, 7300 Jelling, Denmark, [3]LVK, Sallvej 145, 8450 Hammel, Denmark, [4]Vilofoss, Ulsnæs 34, 6300 Gråsten, Denmark;
mogens.vestergaard@anis.au.dk

After weaning, calves often have plasma vitamin E levels lower than the recommended level (2 mg/l). Therefore, the present study was conducted to evaluate the effect of vitamin E supplementation on plasma vitamin E level and growth performance of calves around weaning. The experiment was conducted at three Danish commercial farms producing rosé veal calves. Calves were purchased from private dairy farms at 3-4 weeks of age and approximately 60 kg live weight. The growth experiment included a total of 307 calves. The standard concentrate was fed unsupplemented (CON) or supplemented with 488±24 (average ± SEM) mg RRR-α-tocopherol (TRT)/kg concentrate. Energy concentration of TRT concentrate was 5% lower than CON concentrate. In farm A 22 (10 CON and 12 TRT), in farm B 20 (10 CON and 10 TRT) and in farm C 15 (6 CON and 9 TRT) calves were used for blood sampling. Plasma vitamin E content was analysed at start, at weaning, and 2 to 3 weeks after weaning. Plasma vitamin E content was 2.12±0.43 (LSMeans ± SEM) and 2.38±0.42 (mg/l) before weaning in CON and TRT calves, respectively, with no significant difference. At weaning, plasma vitamin E content in TRT calves (5.09±0.58) was higher (P=0.04) than for CON calves (3.35±0.55, mg/l). Similarly, after weaning, plasma vitamin E content of TRT calves (3.71±0.27) was higher (P<0.0001) than CON calves (1.27±0.29 mg/l). Average daily gain (ADG) until weaning was not affected by vitamin E supplementation, but ADG from weaning to end of experiment was lower in TRT compared to CON calves in two of three herds (herd B: 731 vs 871 g/d; P<0.05 and herd C: 886 vs 1,005 g/d; P=0.10). However, ADG from start to end of experiment was not significantly affected by vitamin E supplement, despite being numerically 7 to 8% lower in two of three herds. In conclusion, vitamin E supplementation in the alcohol form increased the plasma vitamin E level well above the recommended level to keep an optimum immune function of calves after weaning time.

Performance of heat-stressed dairy goats supplemented with rumen-protected methionine
N. Mehaba, W. Coloma, A. Salama, E. Albanell, G. Caja and X. Such
Universitat Autonoma de Barcelona, Group of Research in Ruminants (G2R), Edifici V, Campus UAB, 08193 Bellaterra, Spain; ahmed.salama@uab.es

The objective of the current study was to evaluate the response of heat-stressed dairy goats to feed supplementation with rumen-protected methionine. Eight multiparous Murciano-Granadina dairy goats (89.6±0.3 DIM, 2.40±0.06 l/d and 43.3±1.5 kg BW) maintained in individual pens were used in 4×4 Latin square (4 periods; 21 d each). Factors were: (1) TN (15 to 20 °C); (2) HS (12 h/d at 35±0.5 °C and 12 h/d at 27.9±0.07 °C); (3) control diet (CON); and (4) diet supplemented with 2.6 g/d rumen-protected methionine (MET). Feed intake, water consumption, rectal temperature, respiratory rate, and milk yield were daily recorded. Additionally, milk samples were collected weekly for composition. Data were analysed by the Mixed Procedure of SAS for repeated measurements. Compared to TN, HS goats had greater (P<0.01) rectal temperature (39.61 vs 38.60 °C), respiratory rate (131 vs 32 breaths/min) and water consumption (6.53 vs 4.6 l/d). Although heat stress reduced (P<0.05) DM intake by 13%, milk yield was similar between TN and HS goats. However, HS goats experienced reduced (P<0.05) milk fat and protein contents by 15 and 10%, respectively. The MET did not affect DM intake, water consumption or milk yield in both TN and HS goats. However, an interaction between ambient temperature and methionine supplementation was detected (P<0.05), where MET supplementation increased milk fat content by 4.6% in HS, but not in TN goats. Consequently, fat-corrected milk was greater (P<0.05) in MET compared to CON goats under HS conditions by 14%. In conclusion, heat stress reduced milk fat and protein contents, and supplementation with rumen-protected methionine improved milk fat content under heat stress conditions.

Effect of supplementing sweet potato vines on goat feed intake, growth parameters and nematodes
F.N. Fon[1], C.F. Luthuli[1,2] and N.W. Kunene[1]
[1]University of Zululand, Agriculture, 1 Main Road, KwDlangezwa, 3886, South Africa, [2]KwaZulu-Natal Department of Agriculture and Rural Development, Grass and Soil Science Research, 16 Old Ntubatuba Road, 3880, South Africa; fonf@unizulu.ac.za

Majority of smallholder farmers in Zululand rural communities South Africa owns goats but forage shortages especially in winter calls for alternative crop residues with feed potential not used and are available in this area that can reduce cost and improve production. Sweet potatoes vines are one of those residues that are least exploited with 1990 cultivar used by most farmers. This study measured the effects of supplementing goats feed with sweet potato vines from 1990 cultivar, on feed intake, growth attributes, and anthelminthic activities (haemonchus). Nguni goats (32 yearling male) with a similar body weight of 21.84 kg were randomly allocated to four treatments with eight goats per treatment. Four levels of fresh sweet potato vine were included in diets (hay *ad libitum*) as follows: T1 (0%), T2 (1.5 kg), T3 (2.0 kg) and T4 (3.0 kg). The feeding trial lasted for 8 weeks. Anthelminthic potential was measured by using FAMACHA to determine anaemic levels and faecal egg count (FEC) for infestation. Data was analysed using the general linear models (GLM) procedure of statistical package of social sciences (SPSS, 2015). The results revealed that feed intake increased (P<0.04) with increasing levels of SPV supplementation. Goats fed T3 had the highest (P<0.05) final weight (FW), total weight gain (TWG) and average daily gain of 26.05 kg, 4.18 kg and 74.56 g/day, respectively. The egg per gram was reduced (P<0.05) in all groups of goats fed sweet potato vines (T4 was highest with 78.74%), but no reduction in control group. Supplementing the goats' diet with sweet potato vines has the potential to improve digestibility, production as well as animal health. It also one of the first study measuring the anthelminthic potential of sweet potato vines. Therefore, more research is required in a larger scale to establish the right supplementation quantity by smallholder farmers to favour both performance and health as well as determine the active compounds (nematode inhibitors).

Influence of Arabic gum extract on reproduction, oxidative stress and immunity of doe rabbits

I.T. El-Ratel[1] and S. El Naggar[2]
[1] Faculty of Agriculture, Damietta University, Poultry Production Department, 34517, Egypt, [2]National Research Centre, Animal Production Department, El Buhouth St., Dokki, Giza, Egypt; soddernaggar 75@gmail.com

This study investigates the effect of daily oral administration with Arabic gum extract levels (50, 100 and 150 mg/kg LBW), on *in vivo* and *in vitro* reproductive performance, oxidative stress and immunity of NZW doe rabbits. Nili-parous NZW does (n=100) were assigned to four groups (25 doe/group). Does in the 1st, 2nd, 3rd and 4th groups were daily received oral 0.5 ml saline solution (G1) as a control group, 50 (G2), 100 (G3) and 150 (G4) mg/kg LBW from Arabic gum extract for 60 days as a treatment period, respectively. At end of treatment, does were naturally mated with 20 untreated fertile NZW bucks (5 bucks/group). Conception and kindling rates and litter size (total and live) of does were calculated. The ovulatory response, embryo quality and embryo production *in vitro* were determined. Serum concentration of biochemical, antioxidant capacity and immunity markers, and hepatic and renal histology of does at the end of the experiment were determined. Results showed that does in G3 showed significantly (P<0.05) better reproductive performance (conception and viability rats and litter size), antioxidant status (total antioxidant capacity, glutathione peroxidase, glutathione S-transferase, superoxide dismutase and catalase), lipid peroxidation (malondialdehyde), immune response (IgA, IgG and IgM), ovulatory response (CLs number and ovulation rate), embryo quality, and blastocysts production as compared to other groups. In conclusion, orally administrated with Arabic gum extract (100 mg/kg LBW) as a natural antioxidant, had a positive impact on *in vivo* and *in vitro* reproductive performance, antioxidative status and immunity without adverse effects on liver and kidney functions of doe rabbits.

Effect of probiotic on ruminant performance

H.M. Khattab[1], H.M. Gado[1], H.M. Metwally[1], A.F. El-Barbary[2] and S. Elmashed[3]
[1]Agriculture college, Ain shams university, Egypt, Animal production, Hadaeq Shubra, Shubra El-Khima, Cairo, 11351, Egypt, [2]private farming, animal breeding, el obour, 12681, Egypt, [3]Animal production Research Institute, Animal production, Nadi Elsaid, Dokki, Giza, 112411, Egypt; sh.elmashed@hotmail.com

This work aimed to study the effect of probiotic treatments on 200 multiparous Holstein dairy cows after calving milk production and composition and animal health. All cows in two groups were almost similar (n=100) in milk production, days in milk and number of lactation seasons. Both groups were fed the same TMR but the treatment group was supplemented by probiotic ZAD (10 ml/cow/day). ZAD was added and mixed to the TMR at the time of feeding once per day. Representative fresh samples of TMR were collected weekly and stored at -20 °C until. Milk yield of each cow was recorded after each milking (3 times/day) during 12 weeks of experiment period. Milk samples were collected biweekly and analysed immediately for fat, protein and lactose. Blood samples were collected from Jugular vein. Results obtained indicated that treated cows ate insignificantly more dry matter. Control group consumed 15.5 (kg/day) and treatment group consumed 15.8 (kg/day). Average daily milk yield was significantly higher in treatment group (41.7 l/day) compared to control group (39.57 l/day). There was significant increase in daily milk fat yield. Similar results were obtained for protein milk percentage and yield. Milk lactose percentage was 4.79% for cows fed diets without supplementation and 4.83% for cows fed diets with supplementation with probiotic ZAD. Serum total protein increased significantly from 11.52 (g/dl) for control group up to 11.85(g/dl) for treated group. Blood urea concentration in treated group was 34.77 (mg/dl) significantly higher than control 33.91 (mg/dl). Alkaline phosphates increased significantly from 21.105 U/l in control group up to 26.92 U/l in treated group. Cholesterol concentration was reduced significantly due to treatment from 240.98 mg/dl in control cows to 190.13 mg/dl in treated cows..T3 concentration increased significantly as a response to treatment by ZAD. It was concluded that ZAD supplementation improved productive performance of dairy cows with no harmful effect on their health.

Effect of probiotics treatment on agricultural by products used in ruminant nutrition

H.M. Metwally[1], H.M. Gado[1], A.A. El-Giziry[2], A.M.I.R.A. Abd Elmaqsoud[2] and S. Elmashed[2]
[1]*Agriculture college, Ain shams university, Animal production, Hadaeq Shubra, 11351, P.O. Box 68, Shubra El-Khima, Cairo, 11351, Egypt,* [2]*Animal production Research Institute, Nadi Elsaid, Dokki, Giza, 112411, Egypt; sh.elmashed@hotmail.com*

Four difference agricultural by products used usually as roughages in ruminant diets were incubated with five different ZAD concentrations (ZAD is a probiotic containing anaerobic rumen bacteria and cellulytic enzymes) for four different times. Roughages used were rice straw (RS), Wheat straw (WS), corn stalks (CS) and sugarcane bagasse (SCB). By products were chopped at 3.5 cm and analysed for chemical composition as raw material each kind of roughages was then mixed with one concentration of ZAD solution. ZAD concentrations were (0, 0.5, 1, 1.5 and 2 ml/kg) in a solution of water, Urea and molasses. Ensiling incubation periods were (1, 2, 3 and 4 weeks). Effect of ZAD on chemical composition, *in vitro* parameters and bacterial count was recorded for the tasted roughages. Results indicated that: Treatments increased CP, EE, and decreased CF of all roughages used in the present study, Effect of treatments was significant and clear after 3,4 weeks of ensiling. ZAD concentrations had similar effects above 0 ml/kg DM., SCB was the most affected roughage by treatments, In-vitro study showed that SCB was the highest in TGP. The highest DMD was observed for CS with ZAD concentration of 2 ml/ kg and after 48 hrs. of incubation. The highest bacterial count was observed with WS treated with 2 ml/kg of ZAD and incubated for 4 weeks. The results showed a significant interaction between roughage sources and incubation time on *in vitro* Dry matter disappearance (DMD), the highest value were (48 h DMD) for CS 24.03% and lowest value for SCB were 15.16%. A significant interaction was detected (P<0.05) between roughage sources and time on (OMD). The highest value was 28.99 for SCB at 48 h and the lowest were 2.26 for RS at 72 h. The highest bacterial count was recorded with WS with probiotic concentrations 1.5 and 2 ml/kg. (5.93 and 6.03 cfu respectively. It was recommended that ZAD probiotic can be used to treat by products in order to improve its nutritive value as ruminant feed stuff.

Effect of probiotic on productive reproductive and antitoxin activity in lactating Holstein cows

H.M. Gado[1], H.M. Metwally[1], R.M. Abdel Gawad[2], M.A. Kholif[3] and S. Elmashed[4]
[1]*Agriculture college, Ain shams university, Animal production, Hadaeq Shubra, Shubra El-Khima, Cairo, 11351, Egypt,* [2]*National Research Center, Animal production, El tahrir street, Giza, 112411, Egypt,* [3]*Private Farming, Animal Breeding, El Obour, 12681, Egypt,* [4]*Animal Production Institute, Animal breeding, Nadi Elsaid, Dokki, Giza, Cairo, 12681, Egypt; sh.elmashed@hotmail.com*

Forty Eight lactating Holstein Friesian cows (120±3 DIM) in the first season of lactation and averaged 500 kg body weight were divided into three experimental groups (16 cows each). The first group: was the control group offered. NRC requirement 2001, without any supplementation The second group: was supplementation with probiotic (mixture of live bacteria and exogenous enzymes as cellulase, xylanase, alpha amylase in liquid formula prepared to supplement to ruminants diet) (ZAD 1.1 l/ton) The third group: was supplementation with antitoxins product (mixture from different types of clays as well as vitamins) (T5X 1.1 kg/ton) samples: milk, blood and feed samples measured parameters were milk yield, interval time between calving and fertilizing and pregnancy rate. The results indicated that: serum total protein and globulin increased by treatments than control but the difference were not significant (P>0.05). Liver function were not significantly (P>0.05) affected by treatments (P>0.05). Milk yield was significantly (P<0.05) increased about 11.2% by ZAD than control and 11.2% than T5X treatment. Milk total solids, SNF, fat, protein and lactose were significantly (P<0.05) increased by ZAD however it decreased significantly (P<0.05) by T5X when compared with control group the average time from calving to fertilized insemination was not affects significantly by treatments. Whereas, the overall pregnancy rate increased significantly (P<0.05) with ZAD treatment as (69.2%) compared with T5X treatment (53.8%) ZAD treatment has significant (P>0.05) effect for decreased aflatoxins as 32.6 and 24.1% for ZAD and T5X respectively. It was concluded that ZAD has binding properties for mycotoxins in animals feed also, the same effect confirmed for the T5X which contains deactivated yeast and some kind of clays.

Effects of olive leaf or marigold extract on lipid peroxidation in broilers fed n-3 PUFA diets

V. Rezar, A. Levart, R. Kompan, T. Pirman and J. Salobir
University Ljubljana, Biotechnical Faculty, Department of Animal Science, Jamnikarjeva 101, 1000 Ljubljana, Slovenia;
vida.rezar@bf.uni-lj.si

Olive leaves and marigold petals are one of the potent source of plant polyphenols having antioxidant, antimicrobial, and antiviral properties due to its rich phenolic content. In the experiment, 45 one-day old broiler chicks (Ross 308) were randomly divided into 3 groups with two replications and fed *ad libitum* diets rich with linseed oil and not supplemented (Cont) or supplemented with olive leaves (OliveE) or marigold petals (MarigE) extracts. At the end of the experiment at the age of 39 days, birds were sacrificed and blood and liver samples were collected. Concentration of malondialdehyde (MDA) and vitamin E (vitamin E) in plasma, antioxidative capacity of water- (ACW) and lipid- (ACL) soluble antioxidants, SOD and GPx, and liver enzymes in serum were measured. Liver tissue was analysed for MDA, vitamin E, and ACW. Supplementation with extracts had no effects on body weight gain and feed conversion of the birds. Olive leaf extract significantly decreased MDA concentration in plasma for 25%, the 17% reduction in MarigE group was only numerical. Both extracts had no effect on other markers of lipid peroxidation in blood and in the liver. The results indicated that olive leaf extract supplementation had some beneficial effect on lipid peroxidation.

Effect of purified lignin on lambs performance and rumen health

H.V.A. Bezerra, S.B. Gallo, A.F. Rosa, A.C. Fernandes, S.L. Silva and P.R. Leme
University of São Paulo, Department of Animal Science, Duque de Caxias ave. 351, 13630-390, Brazil; prleme@usp.br

Native lignin is a high-molecular weight polymer of phenolic compounds present in the cell wall of plants. Purified lignin (Kraft lignin) differs from native lignin by its low molecular weight with mono-phenolic fragments that have biological characteristics and does not represent a barrier to digestion by animals. Antioxidant, antimicrobial, anti-inflammatory and anti-carcinogenic properties of the phenolic fragments were reported. Therefore, this study was carried out to evaluate Kraft lignin compared with selenium (Se) plus E vitamin, known antioxidants, and monensin, known ionophore, in the performance of lambs fed a high-concentrate diet. Thirty two crossbreed Dorper X Santa Inês with an initial live weight and age of 18 kg and 60 d, respectively, were kept in individual pens and fed a diet composed of coast cross hay (10%) and concentrate (90%). The experimental design was a randomized block with four treatments: control (CTL), without additives, monensin (MON), with 15 mg/kg of the ionophore, Kraft lignin (LKR) containing 12.5 g/kg of the product and selenium and vitamin E (SeE) treatment, with 0.33 mg/kg and 100 IU/kg, respectively. The Se requirement in the literature was 0.26 mg/kg DM, and the diets had respectively 0.13, 0.12, 0.17 and 0.29 mg/kg. Antioxidants enzymes were analysed in blood samples. After 60 days of feeding, the animals were slaughtered and the incidence of ruminitis, carcass traits and meat shelf life were evaluated. The data was analysed with the MIXED model of SAS 9.3 with a 5% level significance. No effect of treatments was observed for the performance or carcass characteristics. The LKR showed four animals (50%) with incidence of ruminitis, presenting lesions in grades 6 or higher. At the 28 days of feed, a higher peroxidases activity was observed in the LKR compared to the MON (P=0.005) and SeE (P=0.015), and tended to be higher than the CTL (P=0.062). The SeE was effective to maintain low lipid oxidation since it presented the lowest values at all times of shelf life (P=0.005). There was no protective action of the Kraft lignin on the mucosa of the ruminal wall, and no positive effect on performance but it increased the activity of antioxidant enzymes.

Efficacy of a probiotic complex containing *Bacillus subtilis* and *Pichia farinose* in broilers

H.Y. Sun, H. Wang, Y.M. Kim, J.Y. Zhang, Y.J. Li, Y. Yang and I.H. Kim
Dankook University, Animal Resources Science, 119, Dandae-ro, Dongnam-gu, Cheonan-si, Chungcheongnam-do, Republic of Korea, 31116, Korea, South; sunhaoyang14@163.com

The objective of the present study was to evaluate the efficacy of probiotic complex supplementation on growth performance and faecal microbial counts in broiler chickens. A total of 768 1-d-old mixed sex Ross 308 broilers were kept in battery cages (124 cm width × 64 cm length × 40 cm height) and fed experimental diets for 35 d. The birds were individually weighed and were assigned to 3 groups (256 birds/group) with 16 replicates (16 birds/cage) each based on similar body weight (42±0.08 g). Treatments consisted of basal diet (CON) and basal diet supplemented with graded levels (0.1 and 0.2%) of probiotic complex consisting of *Bacillus subtilis* (1.0×10^7 cfu/g) and *Pichia farinose* (1.0×10^7 cfu/g). There feed conversion ratio (FCR) was lower (P<0.05) in broilers fed 0.1 or 0.2% and probiotic complex supplemented diets than those fed with CON diet during d 1 to 7; in addition, broilers fed with 0.2% probiotic complex diet had lower (P<0.05) FCR than those fed CON diet during d 21 to 35 and overall d 1 to 35. Supplementation of 0.2% probiotic complex diet had greater (P<0.05) body weight gain (BWG) than CON diet during overall period. In conclusion, dietary probiotic complex supplementation has shown beneficial effect in improving BWG, decreasing FCR, and increasing faecal *Lactobacillus* counts. Additionally, dose responses of probiotics complex were also seen in seen in measured parameters.

A Comparative evaluation of GMO and non-GMO diet in broiler chickens

S. Zhang, X. Ao, S.I. Lee, J. Hu, T. Palanisamy, Y.M. Kim, H.Y. Sun, Y. Yang and I.H. Kim
Dankook University, Animal Resources Science, 119, Dandae-ro, Dongnam-gu, Cheonan-si, Chungcheongnam-do, Republic of Korea, 31116, Korea, South; dydals3561@naver.com

This study was conducted to compare the growth performance, nutrient digestibility, blood profile and meat quality when broiler fed GMO diet and non-GMO diet. A total of 840 broilers (initial BW of 43.03 g) were randomly allocated into 1 of the following 2 dietary treatments (15 broilers per pen with 28 replicates per treatment): (I) Trt 1: GMO maize-soybean meal based Diet; (II) Trt 2: Non-GMO maize soybean meal based Diet. Qualitative analysis, proximate analysis and amino acid analysis of the feed ingredient sample were carried out. The analysis results showed that the total Lys, Met, Thr of non-GMO grains were lower than GMO grains, and the protein content of GMO soybean meal was higher than that of non-GMO soybean meal. Diets were formulated based a nutrient matrix derived from analysis results. Growth performance was measured on day 0, 7, 17 and 32. And all other response criteria were measured on day 32. Feed intake and FCR were greater (P<0.05) in broilers provided with non-GMO diet than that in the GMO group during d 17 to 32. A decrease in FCR was observed in birds fed the GMO diet (P<0.05). No significant impact on blood profile, meat quality and nutrient digestibility was found in response to dietary treatments throughout the experimental period (P>0.05). These results indicated that non-GMO diet showed a negative effect on growth performance without other negative influence.

Antioxidant effects of phytogenic plant additives used in chicken feed on meat quality

Z. Zdanowska-Sasiadek[1], P. Lipinska-Palka[1], K. Damaziak[2], M. Michalczuk[2], A.A. Jozwik[1], J.O. Horbanczuk[1], N. Strzalkowska[1], W. Grzybek[1], K. Justniku[1], K. Kurdru[1], M. Legoda[1] and I. Marchewka[1]
[1]Polish Academy of Sciences Institute of Genetics and Animal Breeding, ul. Postepu 36A, Jastrzębiec, Magdalenka, Poland, [2]Warsaw University of Life Sciences, Department of Animal Science, Nowoursynowska 161, 01-001 Warszawa, Poland; aa.jozwik@ighz.pl

Plants including herbs, vegetables and fruits are important source of antioxidant substances. When fed to production animals they may help to protect the characteristics of the final product. The aim of the current study was to demonstrate the influence of herbal and vegetable blends used in the broiler chickens diet on the oxidative status and physiochemical properties of their breast muscles. Four hundred male broiler chickens (ROSS 308) were randomly divided into four feeding groups (100 birds each) and reared on litter at stocking density of 11 birds/m2 until 42 days of age. Chickens were fed *ad libitum*. The control group (C) received the standard commercial diet, while experimental groups E1, E2 and E3 diet with 2% herbal-vegetables mixtures as an additive. Herbal-vegetables mixtures were composed of 70% onion, 25% thyme, 5% mint or 90% ginger, 7% rosemary, 3% chili or 30% onion, 20% garlic, 25% oregano, 10% fennel, 5% mint, 5% turmeric, 5% ginger in group E1, E2 and E3 respectively. A generalised linear mixed model was applied on all measured parameters, including feeding group as fixed factor using PROC GLIMMIX of SAS v 9.3. Application of the vegetable and herbs mixtures affected meat physiochemical characteristics (pH, WHC, shear force, colour) ($P<0.05$). Chickens receiving herbal and vegetable additives showed higher content of polyphenols, anthocyanins, retinol and α-tocopherol in the breast muscle ($P<0.05$). In all experimental groups, the reduction of TBA was found. Selected herbal-vegetable mixtures applied into broiler diet improved the physiochemical properties of the meat, enriched it with health-promoting substances and protected them against the unfavourable oxidation process.

Standardized natural citrus extract effect on broilers chickens

S. Cisse[1], A. Burel[2], C. Vandenbossche[2] and P. Chicoteau[2]
[1]Laboratoire Commun Feed In Tech, 42 rue Georges Morel, 49070 Beaucouzé, France, [2]Nor-Feed SAS, 3 rue Amédéo Avogadro, 49070 Beaucouzé, France; sekhou.cisse@norfeed.net

The increase of productivity on livestock production has been accompanied by numerous side effect including the increase of stress and the expansion of animal diseases. For preventing these situations, antibiotics have been used as growth promotors for a long time. Nevertheless, due to the development of antimicrobials resistance, this application has been definitely banned in Europe in 2006, giving way to the alternative solutions. Among these solutions, plant extracts are more and more used. Some of them such as citrus extract allow to modulate intestinal microbiota in order to reduce or avoid disease situations and increase animals' zootechnical performances. However, all citrus extract are not the same. Indeed, citrus extract available for feed can vary a lot in terms of composition and concentration of active compounds. These variation can have in important impact on product efficacy and efficiency. In order to assess the effect of a standardized natural citrus extract (SNCE) supplementation on broilers chickens, 17 trials have been realised between 1995 and 2018. Material and SNCE was provided from Nor-Feed SAS Company. Each trial had 2 poultry groups, a control group (CTL group) feed with standard diet without supplementation and a SNCE group (SNCE group) supplemented with SNCE at dose between 250 and 400 ppm. Feed conversion ratio (FCR) and average daily gain (ADG) have been monitored for every trials. Trials have been conducted worldwide. Statistical analyses were performed by Student test (t-test) using GraphPad Prism 7. A meta-analysis was conducted to assess the overall effect of SNCE on broilers. Of the 17 trials, 16 have shown a significant effect on FCR or ADG of broilers chickens ($P<0.05$, t-test). On average, SNCE supplementation on broilers significantly allow to reduce the FCR by 2.3% and increase the ADG by 4.4% ($P<0.05$ Fisher Test). According to these data, SNCE seems to be an effective solution in order to promote zootechnical performances of broilers chickens and offer a good alternative to improve livestock production.

Alkaloids-containing feed additive as rumen fermentation modulator *in vitro*

R. Khiaosa-Ard[1], M. Mahmood[1], M. Münnich[2] and Q. Zebeli[1]
[1]*University of Veterinary Medicine Vienna, Veterinärplatz 1, 1210 Vienna, Austria, [2]Phytobiotics Futterzusatzstoffe GmbH, Wallufer Str. 10 a, 65343 Eltville, Germany; ratchaneewan.khiaosa-ard@vetmeduni.ac.at*

High grain feeding can contribute to acidic and hyperthermal stress in the gut thereby affecting gut health, overall health, and productivity of cattle. It is important to find ways to facilitate rumen fermentation, especially under stress conditions. Anti-inflammatory alkaloids, in particular sanguinarine have been used as a feed additive in swine and poultry diets. Less is known for ruminants and even fewer with regards to its effect during stress conditions. We used a rumen simulation technique to compare the effect of three different doses of Sangrovit® Extra (containing sanguinarine as the major active component): 0% (control), 0.088% of diet dry matter (low), and 0.175% of diet dry matter (high), each was assigned to 2 incubation temperatures: 39.5 °C (normal) and 42 °C (hyperthermal condition) and 2 acidic conditions: normal (target pH 6.6) and mild acidic (target pH 6.0) regulated by the increased proportion of concentrate and a modified buffer. There were 6 experimental runs, each included 5-d adaptation and 5-d measurement period. We found that the low dose was more effective than the high dose in promoting the production of propionate – the key glucose precursor in ruminants – especially under the acidic condition ($P<0.05$). This led to the lowest acetate to propionate ratio with the low dose. A methane mitigating effect of Sangrovit Extra and the effect on reduced ammonia concentration were pronounced with the high dose and under the acidic condition ($P=0.01$). In relation to incubation temperature, an effect of low dose Sangrovit Extra on the increased proportion of propionate was more expressed under the normal rumen temperature compared to that during the hyperthermal condition ($P=0.01$). Addition of Sangrovit Extra contributed minimally to nutrient degradation. In summary, the low dose of Sangrovit Extra was effective to shift the rumen fermentation but the high dose is needed to affect methane formation. The modulatory effect was more pronounced during a mild acidic challenge.

Effects of herbal extract on growth performance and gut health of post weaned pigs

S.K. Park and N.R. Recharla
Sejong University, Food Science & Biotechnology, 209 Neungdong-ro, 05006, Korea, South; sungkwonpark@sejong.ac.kr

In this study, we examined the effects of dietary supplement of herbal extract and multi-enzyme complex on growth performance and gut health of newly weaned piglets. A total of 60 weaned pigs were used during a 4-wk period for this study. There were two dietary treatments: Control (corn soybean basal diet), and HE (basal diet with 1% herbal extract inclusion). The enzyme supplement contained xylanases, β-glucanase, α-amylase and protease. HE treatments did not ($P>0.05$) influence the growth performance of pigs. Pigs fed HE diet showed positive effects on total plasma cholesterol and triglyceride levels and had greater ($P<0.05$) amount of serum IgG than control group. Serum creatinine content was lower ($P<0.05$) in HE fed pigs when compared with control. Based on taxonomic analysis, *Firmicutes* and *Bacteroidetes* phylums were most predominant taxa in all faecal samples, and at the genus level, *Prevotella* genera was more predominant. The genera including *Prevotella* and *Roseburia* were predominant in HE group. These results indicate that the inclusion of herbal extract enhances the health status of piglets during initial post-weaning period, and therefore herbal extract can potentially serve as alternative to antibiotic growth promotors for pigs.

Behaviour of tannin-protein complexes in ruminant digestive tract: influence of pH on complexation

S. Herremans[1], V. Decruyenaere[1], Y. Beckers[2] and E. Froidmont[1]
[1]Centre wallon de Recherches agronomique, Production et filières, Rue de Liroux 8, 5030 Gembloux, Belgium, [2]Université de Liège, Gembloux Agro-Bio Tech, Passage des Déportés 2, 5030 Gembloux, Belgium; s.herremans@cra.wallonie.be

By making complexes with proteins, tannins reduce protein degradability in the rumen. To improve technical performances, nitrogen use efficiency and environmental impacts of ruminants, proteins should be released in the intestine. Tannin-protein complexes are thought to dissociate below pH 3.5 or above pH 8. These conditions could be met in the acidic abomasum or in the alkaline duodenum. However, it is need to better define dissociation pH thresholds as they will determine the fate of the complexes in the digestive tract. The objectives of this study were to determine the these threshold for oak (OT) or chestnut tannin (CT) and four different proteins. A first experiment was performed with bovine serum albumin (BSA) (4.6 g/l) and casein (0.23 g/l). A second experiment used soybean protein (35 g/l; 850 g protein/kg of extract) and pea protein (50 g/l; 850 g protein/kg of extract). Protein precipitation capacity was assessed according to the method adapted from Amory & Schubert (1987). Varying pH conditions (from 1.5 to 9) were tested by adding NaOH or HCl solutions. After incubating tannins with protein solutions, supernatant nitrogen was dosed by the Kjeldahl method and the remaining (precipitated) N was considered as complexed with tannins. The results of the first experiment showed similar BSA precipitation capacity with OT or CT at neutral pH (P>0.05). However, BSA-OT complexes showed a wider pH range of stability than casein-OT complexes (P<0.05). Complexes with casein were stable between pH 2.8 and 6. Dissociation was the highest under pH 2.2 and above pH 7.5. In the case of BSA-OT complexes, stability was observed from pH 2 to pH 6.7. Dissociation mostly occurred under pH 2 and above pH 7.6. These findings suggest that complexes could dissociate, at least partially, in the rumen which usually presents pH between 6 and 7. The results of the second experiment will be available in April 2019 and should provide information on tannin complexes with vegetal proteins. In this experiment, dissociation pH ranges will be more precisely defined.

Effects of supplemental glycerol polyethylene glycol ricinoleate in in broilers

H.Y. Sun, X. Liu, J. Yin, S.I. Lee, J. Hu and I.H. Kim
Dankook University, Animal Resources Science, 119, Dandae-ro, Dongnam-gu, Cheonan-si, Chungcheongnam-do, Republic of Korea, 31116, Korea, South; sunhaoyang14@163.com

This study was conducted to evaluate the effects of glycerol polyethylene glycol ricinoleate (GPGR) supplementation in different energy density diets on growth performance, blood profiles, excreta gas emission, and total tract apparent retention (TTAR) of nutrients in broilers. A total of 544 1-d-old male Ross broilers were used in a 35-day trial. Broilers were allotted into one of four treatment groups in a 2×2 factorial arrangement with two levels of energy density (normal energy or low energy density) and GPGR (0 and 0.035%). From day 18 to 35, GPGR supplemented or normal energy density diet had significant improved (P<0.05) body weight gain (BWG). Meanwhile, GPGR supplemented had significant reduced (P<0.05) feed conversion ratio (FCR) than non-supplemented. During 0 to 35 d, GPGR supplemented diet or the normal energy density diet had significant increased (P<0.01) BWG and reduced (P<0.01) FCR. Moreover, GPGR supplemented tended to increase (P<0.10) the TTAR of dry matter (DM) compared with non-supplemented treatments. Likewise, normal energy density diets had significant improved TTAR of gross energy (GE) (P<0.05) than low energy density diets. No interactive effects were observed between energy density and GPGR supplementation. In conclusion, both dietary GPGR supplementation and normal energy density diets had beneficial effects on growth performance in broiler chickens without any adverse effect on blood profiles and excreta gas emission.

Effect of two sources of Zn at different doses on growth performance in nursery piglets

M. Rodríguez[1], A. Monteiro[2], G. Montalvo[1], J. Morales[1] and S. Durosoy[2]
[1]PigCHAMP Pro Europa, Dámaso Alonso, 14, 40006, Spain, [2]ANIMINE, 335 Chemin du Nover, 74330, France; maria.rodriguez@pigchamp-pro.com

Zinc oxide (ZnO) supplementation in diets of weaned piglets is a current practice which improves growth performance. However, the possibility of reducing those levels without having negatives effects on performance and health should be studied. Therefore, new Zn formulations applied at lower dosages are being developed. The aim of this study (funding from the Eurostars-2 Joint Programme with co-financing from CDTI and the EU H2020 Programme) was to evaluate the efficacy of a potentiated ZnO source (HiZox®, Animine, France; HZ) compared to a conventional ZnO source in piglets' diets (ZO). At 28 days of age, 386 piglets were weaned and allotted to 48 pens (8 piglets/pen) according to body weight (BW) and sex. The nursery phase was divided into 14 d of prestarter and 21 d of starter. In a 2×3 factorial block design, the effects of 2 sources of Zn (HZ vs ZO) and 3 doses (60, 120 and 180 ppm), on performance and clinical signs were studied in prestarter, starter and in the whole nursery period. Performance data were analysed using the MIXED procedure of SAS. Mortality and presence of diarrhoea were analysed as binary variable, using the chi-square test. A source × dose effect ($P<0.05$) was obtained both in average daily gain for the whole nursery period and in final BW, which at the lowest dose supplementation (60 ppm) were higher in HZ than in ZO piglets. However, these differences were not observed at higher doses. Average daily feed intake was not affected ($P>0.05$) by the treatments (neither by the dose nor by the source). At 60 ppm, feed conversion ratio (FCR) did not differ in terms of sources both starter and whole phase. However, at 120 ppm HZ presented lower FCR than ZO, whilst at 180 ppm ZO presented higher FCR than HZ (P source × dose<0.05). Neither source nor doses affected mortality rate ($P>0.05$). Source did not affect presence of diarrhoea; but Zn supplemented at the highest dose (180 ppm) decreased diarrhoea scores compared with lower doses ($P<0.05$). In conclusion, under field condition, Zn dietary supplementation in piglets at low doses (60 ppm) is more effective with HiZox® than ZnO. While at higher doses (120 and 180 ppm), similar growth performance was obtained regardless the source.

The effects of a phytogenic additive on broiler performance, meat quality and liver

A. Möddel, S. Fuchs and M. Wilhelm
Dr. Eckel Animal Nutrition GmbH & Co KG, Im Stiefelfeld 10, 56651, Germany; a.moeddel@dr-eckel.de

The application of phytogenic feed additives is a promising route to increasing performance and promoting health in broilers. A feeding trial with broilers and a negative control was conducted to evaluate the effects of dietary supplementation with a phytogenic feed additive. The carcass quality, vitamin content of the liver and performance were subsequently analysed. A total of 727 five-day-old broiler chicks (Hubbard ISA) were randomly assigned to two diet groups. The dietary treatments consisted of a negative control with a basal diet (NC, n=362) and a positive control with 400 g/t phytogenic feed additive added to the basal diet (PFA, n=365). The birds were slaughtered at 42 days of age, and the composition of the breast muscle tissue as well as the vitamin content of the liver were determined. ANOVA was used to analyse the results. Inclusion of the PFA in the diet resulted in significantly improved performance parameters ($P<0.05$). The average daily weight gain and live weight at the end of the trial were 4% higher (56.5 vs 58.8 g and 2,412.3 vs 2,509.6 g respectively). In addition, the feed conversion ratio improved by 3.9% (1:1.79 vs 1.72) and mortality decreased from 3.6 to 2.6%. These benefits are reflected in an increase in the European Production Efficiency Factor by 6.0% (322.7 vs 304.3) in the PFA group. Supplementation of the diet with the PFA numerically improved the meat quality and vitamin content of the liver. The protein and fat content of the breast muscle increased slightly in comparison to the NC (+0.9% vs +2.7%). Furthermore, the softness of the meat was 4.2% higher, which could be explained by the better intramuscular fat distribution and marbling. The PFA also increased the vitamin A (109.5 vs 103.3 ppm) and vitamin C (4.15 vs 4.03 ppm) content of the broilers' liver. In conclusion, supplementation with Anta®Phyt MO significantly improved broiler performance and promoted the quality of the meat as well as the vitamin content of the liver. Consequently, the application of the PFA increases sustainability and production efficiency naturally.

Plant extracts to strengthen the natural defences of poultry: development of a selection tool

A. Travel[1], F. Alleman[2], A. Roinsard[3], O. Tavares[3], D. Bellenot[4] and B. Lemaire[4]
[1]ITAVI, Centre INRA Val de Loire, 37380 Nouzilly, France, [2]UniLaSalle, 19 rue Pierre Waguet – BP 30313, 60026 Beauvais, France, [3]ITAB, 149, rue de Bercy, 75595 Paris Cedex 12, France, [4]Iteipmai, Melay – BP 80009, 49120 Chemillé-en-Anjou, France; antoine.roinsard@itab.asso.fr

In order to reduce antibiotics use, plant extracts use to strengthen the animal natural defences is becoming increasingly important in poultry production. However, their development as ingredients is hampered, in part, by the lack of robust and related repeatable data. Our aim was to design and test a method to assist in selection, from literature, plant extracts potentially interesting to strengthen natural defences of poultry. We were inspired by Anses methodology that aims to evaluate scientific publications and results. First step consisted in establishing, with help of scientists and practitioners, two reading grids. The first allows level of reliability of bibliographic resources to be noted by verifying 1. that studied extract is correctly characterized and 2. that experimental design and results analysis are relevant to conclude about effect of tested extract. The second grid evaluates effects of studied extract on indicators of immune, inflammatory and antioxidant status of poultry. For each article, reliability score of resource is cross-referenced with effect score of extract, thus highlighting the most effective extracts. Our method has been proven. A fairly extensive bibliographic research phase on plant extracts affecting poultry natural defences, led to the selection of 917 articles, representing 48 plants. 8 of them were selected because: (1) mentioned in book Bruneton 'Pharmacognosie-Phytochimie-Plantes médicinales ' as having effects on immunity, inflammation and oxidative stress; (2) mentioned in at least 5 articles; (3) cultivable in France. Evaluation grid and methodological considerations were allowed to select astragalus, echinacea, ginseng, nigella as interesting plant extracts to strengthen poultry natural defences.

Feedlot performance of Nellore cattle fed different combinations of monensin and virginiamycin

A.L.N. Rigueiro[1], M.C.S. Pereira[1], A.M. Silvestre[1], E.F.F. Dias[2], B.L. Demartini[2], L.D. Felizari[2], V.C.M. Da Costa[2], A.C.J. Pinto[2] and D.D. Millen[2]
[1]São Paulo State University, Rua José Barbosa de Barros, 18610-034 Botucatu, Brazil, [2]College of Technology and Agricultural Sciences, São Paulo State University, Rodovia Comandante Joao Ribeiro de Barros, 17900-000 Dracena, Brazil; danilo.millen@unesp.br

This study, conducted at São Paulo University feedlot, Dracena, Brazil, was designed to evaluate the effects of feeding different combinations of virginiamycin (VM) and monensin (MON), during the adaptation or finishing period on feedlot performance of Nellore cattle. The experiment was designed as a completely randomized block, replicated 6 times (4 animals/pen), in which 120 18-mo-old Nellore bulls (378±24 kg) were fed in 30 pens for 112-d. The adaptation program consisted of feeding three adaptation diets over a 14-d period with concentrate level increasing from 66 to 78% of diet. The finishing program had two finishing diets containing 84% (up to day 72) and 88% concentrate (up to day 112). Cattle were fed *ad libitum* three times daily. Treatments were applied as follows: MON throughout the study; VM throughout the study; MON plus VM during adaptation and only VM when the 84 and 88% concentrate diets were fed; MON plus VM throughout study; MON plus VM during adaptation and while feeding 84% concentrate, and only VM when the 88% concentrate diet was fed. The MON and VM were added into diets at 30 and 25 ppm, respectively (DM basis). Cattle fed MON plus VM during adaptation and finishing with 84% concentrate and only VM when 88% concentrate diet was fed presented greater (P=0.05) final body weight (548 kg) when compared to those fed MON (534 kg) or VM throughout the study (529 kg), and greater (P=0.05) average daily gain than cattle fed VM throughout the study (1.51 vs 1.34). Feeding MON plus VM during adaptation and finishing with 84% concentrate, and only VM in the 88% concentrate diet increased (P=0.05) hot carcass weight (296 kg) compared to those fed MON (289 kg) and VM (289 kg) throughout the study. No effect of treatment was found on dry matter intake (P=0.14) and feed efficiency (P=0.10) Thus, increasing diet energy level and withdrawing MON 40-d prior to slaughter improve feedlot performance of Nellore cattle when virginiamycin is fed throughout.

Effect of a beta-carotene supplementation to dry cows on IgG concentration in the blood of calves

M. Raemy, A. Burren and S. Probst
Bern University of Applied Sciences, School of Agricultural, Forest and Food Sciences, Laenggasse 85, 3052 Zollikofen, Switzerland; stefan.probst@bfh.ch

The reduction in the use of antibiotics through alternatives is a current topic for farmers. A beta-carotene supply to dry cows might contribute to a better calves' health due to a higher antibodies concentration in the calves blood. This hypothesis was verified through a feeding trial with 60 dry cows on 4 different farms in Switzerland. On each farm the dry cows were divided into two groups, one control group without beta-carotene and one test group that was supplied with 600 mg beta-carotene per cow per day during at least the last 14 days of gestation. Calves were fed with colostrum of their mother within 6 hours after birth. Between 24 and 72 hours after birth IgG concentration of the calves' blood was measured using a semi-quantitative quick test (Fassisi Bovine IgG, Fassisi, Goettingen, Germany). Blood IgG concentration was analysed using a generalized linear mixed model with binominal distribution. The probability to attain an IgG concentration in the blood higher than 10 mg/l was 65% in the control group and 93% in the test group, respectively (P=0.012). A concentration higher than 10 mg/l is described in the literature as necessary to ensure an effective immunity transfer and good health of the calves during their first weeks. We conclude that a supply of beta-carotene to dry cows might be a possibility to support calves health during their first weeks of life and therefore reduce the use of antibiotics. Further research is required to identify the optimal amount of beta-carotene supplementation to dry cows.

Effects of 5 different sources of copper and zinc on rumen fermentation *in vitro*

W. Chen[1,2], W.F. Pellikaan[2], L.A. Vande Maele[3] and A. Van Der Aa[3]
[1]Aarhus University, Faculty of Science and Technology, Nordre Ringgade 1, 8000 Aarhus, Denmark, [2]Wageningen University and Research, Department of Animal Sciences, Droevendaalsesteeg 4, 6708 PB Wageningen, the Netherlands, [3]Orffa Additives B.V., Vierlinghstraat 51, 4251 LC Werkendam, the Netherlands; maele@orffa.com

The objective of this study was to investigate effects of different forms of copper (Cu) and zinc (Zn) additives on extent and rate of ruminal gas production (GP, ml/g organic matter) and fermentation end products *in vitro*. Five forms of Cu and Zn supplementations (inorganic: hydroxy and sulphate; organic: glycinate, amino acid complex and proteinate) were evaluated. Supplementations were added to the substrate (grass silage) at inclusion levels of 40 mg/g and 100 mg/g for Cu and Zn, respectively, and incubated (in triplicate) for 72 h in a bicarbonate-phosphate buffered rumen fluid mixture. Total cumulative GP was measured in real time using an automated gas production system. Volatile fatty acid (VFA) concentrations and pH were measured at the end of the experiment. After exclusion of outliers, GP was fitted with a triphasic kinetics model and data were statistically analysed using a mixed model ANOVA procedure and Tukey's multiple comparison test in SPSS. Results show that sulphate and all organic sources of Cu and Zn strongly depressed GP and total VFA compared to the control without added Cu and Zn (P<0.001). Hydroxy source did not inhibit GP and showed a multiphasic response similar to the control, whereas other treatments stopped fermentation within 1.6 h after inoculation and only revealed GP data with a monophasic response. These results suggest that Cu and Zn from sulphate and organic sources depressed rumen fermentation by negatively affecting ruminal microbes. Total VFA concentrations for hydroxy did not differ from control, although a small numerical increase in the molar proportion of propionate could be observed (20.35 and 20.72% propionate for control and hydroxy, respectively). In conclusion, this study shows that hydroxy Cu and Zn had no inhibiting effect on *in vitro* rumen fermentation, whilst sulphate, glycinate, amino acid complex and proteinate Cu and Zn sources all altered and depressed fermentation.

Effects on serum biochemical parameters of tryptophan supplementation in calf milk replacers

N. Yeste[1], L. Arroyo[1], M. Terré[2] and A. Bassols[1]

[1]Universitat Autònoma de Barcelona, Departament de Bioquímica i Biologia Molecular, Facultat de Veterinària, 08193, Bellaterra, Spain, [2]Institut de Recerca i Tecnologia Agroalimentàries, Departament de Producció de Remugants, 08140, Caldes de Montbui, Spain; anna.bassols@uab.cat

Tryptophan amino acid (Trp) is a precursor of serotonin, a neurotransmitter that participates in the control of the affective state of the animal. We studied the effects of supplementation of milk protein-based milk replacer (MR) with 4.5 g/d Trp in 27 Holstein male calves starting at 45 days old. Animals were divided into 2 groups: 'Control' fed MR without supplementation (n=13), and 'Trp' fed with Trp-supplemented MR (n=14). All calves received the same feeding program and were weaned 21 days after the beginning of the study. After weaning, calves were fed starter feed and chopped straw *ad libitum*. From one week before Trp supplementation (t=-1) to one week after weaning (t=4), two blood samples were weekly obtained, one before the morning MR feeding and the other 3 hours after feeding. Several serum biochemistry analytes were determined: those related to energy metabolism (glucose, total protein, triglycerides, cholesterol, creatinine, urea and NEFAs), hepatic enzymes, the stress hormone cortisol, the antioxidant enzymes glutathione peroxidase (GPx) and superoxide dismutase (SOD), and kynurenine (Trp catabolite). Among nutritional parameters, glucose and triglycerides showed the expected increase 3 h after feeding during MR diet but there were no differences once calves were weaned and they were only eating solid feeds. Trp supplementation did not have any effect in glucose concentration. However, animals treated with Trp had lower levels of triglycerides at 3 h after feeding than control calves. The antioxidant enzyme SOD decreased 3 h after feeding, but the difference was less evident at t=4. A tendency for a lower SOD was observed in Trp- supplemented animals. Kynurenine increased in plasma in the Trp-group at 3 h after feeding, but this effect disappears when calves were only fed solid feeds after weaning. Finally, cortisol levels were higher in the group supplemented with Trp than in the control group. In conclusion, supplementation of MR with Trp moderately altered several nutritional, redox and stress markers in serum.

Organic acid based product supplementation in hens challenged with *Salmonella*

V. Ocelova[1], A. Markazi[2], A. Luoma[2], R. Shanmugasundaram[2], G.R. Murugesan[3], M. Mohnl[1] and R. Selvaraj[2]

[1]BIOMIN Holding GmbH, Erber Campus 1, 3131, Austria, [2]The Ohio State University, 209 Gerlaugh Hall, Wooster, OH 44691, USA, [3]BIOMIN America Inc, Overland Park, KS 66210, 10801 Mastin Blvd, Suite 100, USA; vladimira.ocelova@biomin.net

Salmonella as a zoonotic pathogen causing foodborne diseases in humans attracts the attention of the worldwide public. Contaminated meat, eggs, and egg products stand out as source of human infection. *Salmonella* can be present in digestive tract of chicken and shed into the environment, without obvious clinical signs. A trial was carried out to study the effect of organic acid based product (OABP) on caecal *Salmonella* load and relative caecal percentage of Bifidobacterium in hens challenged with *Salmonella enterica* serotype Enteritidis. The product Biotronic® Top3 consisted of propionic, formic and acetic acids, cinnamaldehyde, and a permeabilizing substance, supplemented at 1 kg/ton of feed. Hens were vaccinated with a *Salmonella* vaccine at 14 and 16 weeks of age. Half of the vaccinated birds and all unvaccinated birds were challenged with 250 µl of 1×10^9 cfu *Salmonella* Enterica serotype Enteritidis at 24 weeks of age, resulting in a 3 (vaccinated, challenged, vaccinated + challenged) × 2 (control and product) factorial design. All treatment groups had 8 replicates with 8 birds per replicate. Generated data were analysed using 2-way ANOVA and Tukey's honest significant test when main effects were statistically significant (P<0.05). At 8th dpi (days post challenge), challenged birds supplemented with OABP had significantly lower (P<0.05) caecal *S.* Enteritidis percentage compared to the challenged group without product supplementation. OABP supplementation significantly (P<0.05) increased caecal *Bifidobacterium* percentage at 3, 8, and 17 dpi compared to the un-supplemented groups. The significant main effect (P<0.05) of the OABP on body weights of hens was observed at 3 and 8 dpi. Hens in OABP supplemented groups were significantly heavier in comparison to un-supplemented hens. In conclusion, supplementation with organic acid based product in laying hens significantly decreased *Salmonella* load while increasing *Bifidobacterium* percentage in cecum. Improved intestinal health seemed to have positively influenced body weight of birds.

Effect of digestible phosphorus level on performance and bone mineralization of piglets

C. De Cuyper[1], L. Nollet[2], M. Aluwé[1], J. De Boever[1], L. Douidah[1] and S. Millet[1]
[1]Flanders Research Institute for Agriculture, Fisheries and Food (ILVO), Scheldeweg 68, 9090 Melle, Belgium,
[2]HUVEPHARMA, Uitbreidingstraat 80, 2600 Berchem, Belgium; carolien.decuyper@ilvo.vlaanderen.be

Phosphorus (P) is an essential mineral in pig feed, playing a main role in many metabolic pathways and in the structural compounds of bone and cell membranes. An adequate supply of P is therefore important for optimal growth, development, and welfare of the animal. On the other hand, P fed in excess may lead to avoidable phosphate excretion into the environment. Determining the optimal digestible P level under different circumstances is therefore mandatory. In this study, growth performance and bone mineralization of piglets were evaluated during six weeks at five increasing concentrations of digestible P, obtained by adding increasing levels of phytase (Huvepharma®; 0, 125, 250, 500 or 1000 FTU/kg) to a low P (4.4 g/kg) feed. Each of the five dietary treatments were allocated to six pens housing five piglets (Piétrain boar × RA-SE hybrid sow) per pen. Piglets were weighed individually at four (start), six and ten (end) weeks of age. Feed intake was recorded per pen. Daily feed intake (DFI), daily gain (DG), and feed conversion ratio (FCR) were calculated per pen per period. At nine weeks of age, faecal samples were collected rectally from two pigs per pen during four consecutive days to calculate the digestible P level in each diet. At the end of the trial, one pig per pen was euthanized to determine bone mineralization. Increased phytase levels (i.e. increased digestible P levels) resulted in a better growth performance and a higher bone mineralization: DFI, DG, final bodyweight, and crude ash concentration in bone increased (567 to 657 g/day, 351 to 436 g/day, 23.3 to 26.9 kg and 44.2 to 52.5%, respectively), and FCR decreased (1.61 to 1.51). DFI, DG, and final bodyweight reached a plateau as more digestible P was made available, whereas FCR and crude ash showed a linear response without reaching a plateau in the measured range. In summary, the results of this trial confirm the importance of P and can be used in a meta-analysis to update the recommendations for optimal digestible P levels in piglet diets.

Effect of using levels of Ginger root as feed additives in broiler diets on meat characteristics

A.Y. Abdullah
Jordan University of Science and Technology, Animal Production, Irbid, 22110, Irbid, Jordan; abdullah@just.edu.jo

This study was conducted to evaluate the effect of using ginger root powder (*Zingiber officinale*) in the diets of broiler chickens on growth performance, meat quality and antioxidant status. A total of 400 day-old mixed sex Lohman broiler chicks were used and randomly distributed into 5 dietary treatments; Control: Basal diet only; T1: Basal diet supplemented with 5.0 g/kg ginger from 1-42 d; T2: Basal diet supplemented with 7.5 g/kg ginger from 1-42 d; T3: Basal diet supplemented with 5.0 g/kg ginger from 21-42 d; T4: Basal diet supplemented with 7.5 g/kg ginger from 21-42 d. Each dietary treatment was divided into 4 replicates with 20 chicks each. All diets were isonitrogenous and isocaloric. Body weight and growth performance parameters were recorded weekly for each pen to determine body weight gain (BWG) and feed conversion ratio (FCR). At the end of the experiment, all broilers were slaughtered. Data were analysed by analysis of variance using SAS general linear models. Weekly body weight, gain and ADG were all significantly higher for broiler birds fed basal diets supplemented with ginger compared with the control group with no differences between all treatment groups (T1, T2, T3 and T4). Feed intake and FCR for the broiler birds were not significantly affected by ginger supplementation in the diets compared with control treatment. There were differences ($P<0.05$) between treatments in slaughter live weight, carcass weight, dressing%, and leg cut, liver and heart percentages while other carcass and non-carcass parts were unaffected. Cooking loss% and shear force value were significantly lower and expressed juice% were higher ($P<0.05$) in T4 than the control and other treatment groups. Thiobarbituric acid reactive substances (TBARS) values were increased with increasing days of storage (1-4 d) for each treatment, but the TBARS values decreased significantly in treatment groups compared with control group with lowest in T2 group. This study indicated clearly that rearing chicks on a diet containing ginger give better and higher performance than the control group, but with slightly and more clear effect in groups supplemented with ginger powder for the whole rearing periods (1-42 d) compared with those supplemented from 21-42 d.

Effects of dietary millet grain on performance, plasma metabolites and intestinal health in piglets

N.A. Lefter, M. Hăbeanu, A. Gheorghe, L. Idriceanu and C. Tabuc
National Research&Development Institute for Animal Biology and Nutrition, Calea Bucuresti nr. 1, 077015, Balotesti Ilfov, Romania; nicoleta.ciuca@ibna.ro

The purpose of study was to evaluate the effects of dietary millet grain on performance, plasma metabolites and intestinal health of weaned piglets. Forty weaned piglets (8.14 kg±0.20), 21±3 d of age were assigned into 2 groups (C and E1). The group C (n=20) received a conventional diet based on corn-triticale-soybean meal and in the E1 group (n=20), 25% millet replaced triticale. Both diets were isocaloric and isonitrogenous. The performances (body weight, BW; feed intake, FI; average daily gain, ADG, feed efficiency, FE) were evaluated after 7 and 21 d post-weaning. Blood samples were collected from jugular vein at 21 d. A chemistry analyser was used to determine the plasma metabolites (triglycerides, TG; total cholesterol, TC; high-density lipoprotein cholesterol, HDL-C; total protein, T-Pro; total bilirubin, T-Bil; albumin, Alb; uric acid, UA; creatinine, Cre; urea nitrogen, BUN; aspartate aminotransferase, ASAT; alanine aminotransferase, ALAT; gamma-glutamyl transferase, GGT; calcium, Ca; magnesium, Mg, inorganic phosphorus, IP). The intestinal health was establish using faecal scoring system and by microorganism analysis (total fungal count, TFC; bacteria (*Escherichia coli*, *Staphylococcus aureus* and *Lactobacillus* spp.), expressed as log10. At 7 and 21 d post-weaning, performances of piglets fed either E1 or C diet were comparable (P>0.05). The plasma metabolites were not statistically different among treatments. However, the IP slightly increase (+12%; P<0.07) in the E1 vs C diet. The incidence of diarrhoea did not hint at serious health problems. Except the faecal Escherichia coli where we noticed a slightly decreased (-4.11%, P<0.05) vs C diet, the faecal bacteria and fungus in E1 diet reached similar values to C diet. We can concluded that the dietary millet grain had no adverse effects on performances and health state of weaned piglets and could influence the intestinal health by decreasing the certain pathogenic bacteria such as *Escherichia coli*.

Evaluation of copper sulphate replacement by dicopper oxide in piglet diets

S. De Vos[1], L. Aelbers[1], R. Forier[2] and A. Roméo[2]
[1]INVE Belgie NV, Oeverstraat, 7, 9200 Baasrode, Belgium, [2]Animine, 335, Chemin du Noyer, 74330 Sillingy, France; s.devos@invebelgie.be

Due to recently changed EU legislation with regard to the reduction of maximum copper levels in diets for piglets until 4 weeks (max. 150 ppm) and between 4 and 8 weeks (max. 100 ppm) after weaning, performance parameters in weaner and starter period are expected to become negatively affected. In order to prevent this expected drop in performance, new copper sources such as dicopper oxide (Cu_2O) should be evaluated. Cu_2O is a less water soluble Cu source showing high bioavailability and stronger antimicrobial effects in the gut compared to copper sulphate. The aim of the present trial was to investigate the effect of two copper sources (copper sulphate vs dicopper oxide) on piglet performances. Therefore, a field experiment with 2 treatments was set up with 160 piglets (Topigs 20 × Piétrain) weaned at 21 days of age. Each treatment consisted of two pens of 40 animals each differing in average initial weight of 6.2 kg and 7.8 kg. Piglets were fed a weaner diet and subsequently a starter diet (day 0-18 and day 18-46 after weaning, respectively). Diets were formulated in order to meet the nutritional requirements of the piglets. In both phases, treatment 1 (T1) diets contained 140 ppm added Cu originating from copper sulphate, whereas T2 diets were supplemented with 140 ppm copper from Cu_2O. In the weaner phase piglets from T2 grew 18% more compared to T1, resulting in 540 g higher body weight at day 18. Daily feed intake in T2 was increased by 11% so that the animals showed a 4% better feed conversion ratio (FCR). In the starter phase daily gain in T2 was still 2% increased, whereas feed intake was 2% lower. Also in this phase, FCR in T2 was 4% better and on average the animals weighed 810 g more at day 46 compared to T1. In conclusion, substitution of 140 ppm copper from copper sulphate by Cu_2O in weaner and starter piglet diets resulted in 4% increased growth performance and 4% improved FCR. Using alternative copper sources could be one of the strategies to encounter the expected loss of performance due to reduction of maximum copper levels in diets for weaning piglets.

Ω3-rich natural sources as raw material for livestock feeding: zootechnical response of dairy sheep

N. Mandaluniz, X. Díaz De Otalora, A. García-Rodríguez, J. Arranz, E. Ugarte and R. Ruiz
Neiker, Animal Production, P.O. Box 46, 01080, Spain; nmandaluniz@neiker.eus

Society is increasingly aware of the food it consumes d the way it is produced. The objective of KALIKOLZA project (https://www.kalikolza.com/) is to implement animal feeding systems, mainly by Omega-3 (Ω3) natural sources and to improve the profile of fatty acids (FA). Rapeseed cake and flax are two Ω3-rich sources. A trial was carried out with the experimental flock of NEIKER. Three homogeneous groups of ewes with a single-male-suckling lamb were randomly assigned to 3 concentrates. Each group received a concentrate composed by: control (C, commercial concentrate with soya as protein source and palm as fat source), rapeseed cake (R, 40% of rapeseed cake as protein and fat source) and Rapeseed cake+Flax (R+F, 20% of rapeseed cake and 20% of flax as protein and fat source). Sheep diet was composed by 1 kg of concentrate and 2 kg DM sainfoin. Lambs were individually weighted and slaughtered after one month; sheep were milked individually daily to remove excess milk and individual milk samples and weight data were collected weekly. According to the results, average excess milk, crude fat (CF) and crude protein (CP) contents were: 419.4±281.27 ml/day, 6.3±1.80% CF and 4.6±0.34% CP. In the case of lambs, average birth-weight was 5.21±0.63 kg, average weight to slaughter was 10.9±1.63 kg and average daily growth was 340±92 g/day. According to statistics, R group ewes' produced significantly (P<0.01) more excess milk compared with C and R+F (484a, 432b and 342b respectively). Referring to milk composition, R+F group had significantly (P<0.05) higher protein content compared to R and C (4.82a vs 4.50b and 4.48b). In the case of lambs, even the ewes produced different milk quantity, there were no differences (P>0.05) in their daily growth (297a, 393a and 362a g/day for R+F, R and C groups respectively). Moreover, a parallel study demonstrated that the use of these Ω3-rich sources increase the content of Ω3 in both, milk and lamb meat. As conclusion, the inclusion of Ω3-rich natural sources in dairy sheep feeding has no effects on lambs' growth but has different effects on ewes' milk; the inclusion of R increases milk production while the inclusion of R+F increases its protein content.

Effects of flavour supplements in sheep rations on: feed efficiency for growth and Genes expression

A. Dallasheh
The Hebrew University of Jerusalem, Department of Animal Science, The Robert H. Smith Faculty of Agriculture, Food and Environment, Rehovot, 76100, Israel; areen.dallasheh@mail.huji.ac.il

Flavours modulate the sensory characteristics of feeds to increase voluntary intake, aid in selection for nutritious feed and rejection of feed with low nutrient content or those with toxins in ruminants. Flavours affect sensory characteristics of feeds that might increase voluntary intake including by-product feeds and improve body weight gain. Another factor affecting nutrient intake, is the stimulation of the taste receptors such as the sweet taste receptors. The presence of sugars such as glucose in the intestinal tract increases the expression of the sweet taste receptors, and glucose transporter SGLT1.These stimuli increase functional gene expression and hormones that influence intestinal nutrient uptake glucose metabolism and homeostasis. The aim was to characterize the effect of artificial flavour compounds in diet of ruminants on feed efficiency, and the effect of flavours on the expression of sweet taste receptor, SGLT1 and other functional genes in the small intestines. Eight sheep were randomly assigned to 4 treatments: Sucram, Capsicum, Mix (Sucram and Capsicum at 1:1 ratio) and no flavour as a control. Sheep were offered total mixed ration (TMR), whole corn grain and pellets in 50:25:25 ratio respectively (by weight as is). Intestinal biopsies were taken from the proximal jejunum and later, relative gene expression of: T1R2-T1R3, SGLT1, GLP-1, GLP-2, CCK, AP-N, and ATPase determined by RT-PCR. All data were analysed using the MIXED procedure of SAS in JMP Pro software. Flavours affected average daily weight gain. Sheep fed on Capsicum-flavoured diet gained more weight (16.2 g/MBW) than other treatments.T1R3 expression was greatest with Mix (P<0.001). T1R2 (P=0.026) and SGLT1 (P=0.068) relative genes expression was greater with Mix than with Capsicum treatment. GLP-1(P=0.013), CCK (P=0.05), GLP-2 (P=0.09) relative expression was greatest with Mix. Capsicum increased AP-N and ATPase relative genes expression (P=0.04 and 0.08, respectively) In conclusion, flavours can be used to motivate feed acceptance in ruminants and thereby improve feed intake, and efficiency in sheep.

Vagina-cervix length as a tool to predict a litter size of primiparous sows

M.M. Małopolska[1], R. Tuz[2], T. Schwarz[2] and J. Nowicki[2]
[1]*National Research Institute of Animal Production, Department of Pig Breeding, Krakowska 1, 32-083 Balice, Poland,*
[2]*University of Agriculture in Krakow, Department of Swine and Small Animal Breeding, Al. Mickiewicza 24/28, 30-059 Kraków, Poland; martyna.malopolska@izoo.krakow.pl*

In pig production, achievement of large litters with healthy, vital piglets and optimize sows longevity are the primary goals. The litter size is influenced by ovulation rate, embryos survival, fertility, uterine capacity and intrauterine mortality. The capacity of reproductive tracts seems to be the most limiting factor. Moreover, improvement of reproductive traits is still difficult, so new methods to determine future production potential of gilts are needed. The aim of this study was to determine the relationship between vagina-cervix length (VCL) and litter size. The study was conducted in a large commercial farm located in northern Poland. The research material was 121 hybrid sows, inseminated in the second oestrus. During insemination, the VCL was evaluated for each gilt using foam tip catheter. Animals were divided into 3 groups (percentage distribution of VCL): small (n=23; VCL18,0-23,0 cm), medium (n=65; VCL 23,1-28,0 cm) and large (n=33, VCL 28,1-33,0 cm). The differences between groups in reproductive parameters were analysed using one-way ANOVA followed by Duncan test. The mean VCL differed significantly (P<0.01) among specified groups of gilts. The highest number of live born piglets were noted in a large group (14.12) while the lowest (11.00) piglets in small group (P<0.01). Also, total number of piglets born were the highest in a large group (14.55) in comparison to sows from small (11.30)(P<0.01) and small and medium groups (13.15)(P<0.05). A similar relationship was described by other authors, where the largest litters were noted for gilts with long VCL. Additionally, an increase in litter size caused a slight increase in the number of stillborn piglets. To conclude, the use of VCL evaluation as a potential predictor to litter size or as a method to selection of gilts is possible. Although, more research on different genotypes and herds are necessary.

Different models for estimation of litter size variability phenotypes

J. Dobrzański[1], H.A. Mulder[2], E.F. Knol[3], T. Szwaczkowski[1] and E. Sell-Kubiak[1]
[1]*Poznań University of Life Sciences, Department of Genetics and Animal Breeding, Wołyńska 33, 60-637 Poznań, Poland,*
[2]*Wageningen University & Research, Animal Breeding and Genomics, P.O Box 338, 6700 AH Wageningen, the Netherlands,*
[3]*Topigs Norsvin Research Center, Schoenaker 6, 6641 SZ Beuningen, the Netherlands; jan.dobrzanski@mail.up.poznan.pl*

Selection to increase litter size in pigs has simultaneously increased its variability. In Large White differences in litter size between sows can be larger than 20 piglets, which causes additional labour and affects welfare of animals. Thus there is a need to define phenotypes describing variability of litter size in order to reduce it. The main objective of this study was to obtain and compare new phenotypes of phenotypic variability of total number born (TNB) by using the residual variance of TNB. The dataset contained 246,799 litter observations from 53,803 sows from Topigs Norsvin multiplication farms. Three models were used to obtain estimates of residual variance for TNB in ASReml 4.1: basic (BM), basic with parity as fixed effect (BMP), and random regression (RRM). The within individual's variance of the residuals was calculated and log-transformed, to obtain three new variability traits: LnVarBM, LnVarBMP and LnVarRRM. Then, variance components, heritability (h^2), genetic coefficient of variation (GCV) and genetic correlations (r_g) between LnVar's and between LnVar's and mean TNB were estimated with multi-variate models. Results for additive genetic component (0.026-0.028), residuals (1.23-1.29), h^2 (0.02) and GCV (0.081-0.083) indicated that LnVar's as variability phenotype were robust and not affected by model applied for their estimation. Furthermore, rg between additive genetic components of LnVar's (~0.98) proved that genetically LnVar's are the same trait. The positive r_g between LnVar's and mean TNB (~0.60) confirmed that selection for increased TNB increases its variability. Whereas GCV for LnVar's indicated that TNB variability can be reduced by 1.1 piglet within four generations. These results indicate that phenotypic variability of litter size is under genetic control and could be reduced by selection. The project was financed by Polish National Center for Science (grant no. 2016/23/D/NZ9/00029).

Birth body weight does not always determine subsequent growth performance in grow-finisher pigs

J. Camp Montoro[1,2], E.G. Manzanilla[2], L.A. Boyle[2], O. Clear[2], D. Solà-Oriol[1] and J.A. Calderón Díaz[2]
[1]*Universitat Autònoma de Barcelona, Ciència Animal i dels Aliements, Facultat de Veterinaria, Bellaterra, Barcelona, 08193, Spain,* [2]*Teagasc, Pig Development, Moorepark, Fermoy, P61 C996, Ireland; jordi.montoro@teagasc.ie*

This study investigated the effect of birth and weaning body weight (BW) on performance of grow-finisher pigs. Body weight was recorded within 24 after birth of 144 pigs born within 1 week and pigs were classified as small (S; 0.9±0.13 kg) or big (B; 1.4±0.20 kg). Pigs were weaned at approximately 28 days, weighed and re-classified as S (5.4±1.6 kg) or B (6.3±1.91 kg) yielding a 2×2 factorial arrangement: SS (n=36), SB (n=36), BS (n=35) and BB (n=37). Pigs were fitted with a transponder and housed in mixed sex pens (n=12 pigs/pen) with individual feeding stations. Feed intake was recorded from 2 weeks post-weaning until they reach at least 110 kg of BW when they were sent to slaughter. At slaughter, fat thickness (FT) and lean meat % (LM%) were recorded. Average daily gain (ADG), average daily feed intake (ADFI) and feed conversion ratio (FCR) were calculated for the wean-finisher period. Additionally, days to target slaughter weight were calculated. Data were analysed using mixed model equation methods in PROC GLIMMIX of SAS v9.4. Model included group and sex as fixed effects and pig as random effect. Average daily gain was 76 g lower for the SS pigs versus the other treatments (P<0.001). SS pigs consumed 0.3 kg of feed/day less than SB pigs (P<0.001); however SS pigs had similar ADFI to BB pigs (P>0.05). There was no difference in ADFI between SB and BS pigs (P>0.05). Feed conversion ratio was higher (2.3±0.03) for SB pigs versus the other treatments (2.1±0.03; P<0.05). Days to target slaughter weight was 12.5, 6.5 and 6.4 days longer for SS, SB and BS pigs, respectively compared with BB pigs (P<0.001). Fat thickness was lower for BB pigs than for SS and BS pigs (P<0.001). There was no difference in LM% between groups (P>0.05). Sex was a source of variation (P<0.001) for ADG, FCR, FT and LM%. Pigs with higher birth BW perform better than pigs with lower birth BW even when some of them grow slower during suckling. Some low birth BW pigs have the potential to grow fast during suckling and to have similar growth performance to those born with higher birth BW thereafter.

Redefining the formula for feed conversion ratio in pigs through participatory research

I. Chantziaras[1,2], J. Van Meensel[2], I. Hoschet[1,2], F. Leen[1,2], D. Maes[1] and S. Millet[2]
[1]*Ghent University, Faculty of Veterinary Medicine, Salisburylaan 133, 9820 Merelbeke, Belgium,* [2]*Flanders research institute for Agriculture, Fisheries and Food, B. Van Gansberghelaan 92, 9820 Merelbeke, Belgium; ilias.chantziaras@ilvo.vlaanderen.be*

Feed conversion ratio (FCR) is one of the most important determinants of pig farm profitability and production efficiency. In its simplest form, FCR represents the amount of feed used per unit weight gain of the pig. Yet, this approach entails various limitations hampering its practical applicability such as availability of accurate data and large variation in ways to adapt FCR values for different starting and end weight and mortality rates. Various stakeholders are using their own formulas to determine FCR creating a 'definition nonconformity' when comparing ratios among farms. This study aimed to optimize the calculation of FCR through the use of participatory qualitative research. A multidisciplinary research group of 9 persons (animal scientists, veterinarians, agricultural economists) and a consulting group of 25 stakeholders (representing the Flemish primary sector, feed industry, pharma, genetic companies, large retailers, academia and policy institutions) were involved. The decision problem analysis started with a literature review, followed by in-depth interviews and their analyses (Nvivo 11™). This led to an additional literature review and various focus groups discussing preliminary formulas with the final result being two distinct complimentary formulas that are fit for purpose. Both refer to carcass gain per unit of feed intake. In its most basic form (FCR_1 = (feed) / (nb of delivered pigs × average cold carcass weight – nb of delivered piglets × average piglet weight × 0.72)), it is an objective representation of the animals' performance. For farm benchmarking, a second formula was developed, incorporating a 7-step standardization process that corrects for mortality and bodyweight at start and slaughter. A webtool is designed to ease this standardization process. In conclusion, two different formulas for FCR were developed and proved necessary for practical use: one to follow-up individual farm and one to benchmark farms. The participatory research procedure allowed to develop formulas that are supported by the pig sector and are easy to use.

Mineral precision feeding for gestating sows

C. Gaillard, R. Gauthier and J.Y. Dourmad
PEGASE, INRA, Agrocampus Ouest, 35590 Saint-Gilles, France; charlotte.gaillard@inra.fr

Precision feeding strategy (PF), with the individual and daily mixing of two diets with different lysine content (high or low), was previously reported to reduce protein intake and N excretion by 25% compared to a conventional single diet feeding (CF). The effect of PF on minerals utilization has not been reported yet, even if it seems relevant to maintain sow's productivity and health, as well as for economic and environmental reasons. The objective of this study, realized within the Horizon 2020 EU Feed-a-Gene program (grant agreement n°633531), was to report the variations of digestible P (dP) and total Ca requirements among sows and during gestation and to test if PF based on lysine requirement could also be relevant for P and Ca. A dataset of 2,511 gestations reporting sows' characteristics at insemination and farrowing performance was used as an input for a Python model, based on InraPorc®, predicting nutrients requirements over gestation. Simulations were ran for PF (low: 3.0 g SID lysine/kg, 2.0 g dP/kg; high: 6.5 g SID lysine/kg, 3.3 g dP/kg) and CF (4.8 g SID lysine/kg, 2.5 g dP/kg). The influence of parity (P1, P2, P3+), litter size, and week of gestation on dP and Ca requirements was analysed applying a mixed model. Requirements ranged from 0.88 to 4.14 g/kg for dP and from 2.53 to 13.6 g/kg for Ca. They markedly increased after week 9, increased with litter size and decreased with parity. Compared to CF, average dietary SID lysine and dP content for PF were lower by 27 and 12%, respectively. Over the first 80 days of gestation, all PF and CF sows were fed more than their dP requirement in both strategies. Most of the dP deficits occurred during the last 3 weeks of gestation with 2.9% P deficient sows with PF against 10% with CF. Similar results were obtained for Ca. Mineral PF based on lysine requirements appear less efficient to reduce excessive supplies of P and Ca than for lysine. This is related to differences between P and lysine dynamic of evolution of requirements with time. Nevertheless, the results clearly indicate that precision feeding with the individual and daily mixing of two diets with different P and Ca contents reduced dP intake and consequently P excretion as well as the risk of P deficient sows at the end of the gestation.

The effect of enzymes on nutrient deficient diets on performance, and bone mineralization of pigs

B. Villca[1], G. Cordero[2], P. Wilcock[2] and R. Lizardo[1]
[1]IRTA, Animal Nutrition, Ctrs Reus-el Morell, km 3.8, 43120 Constantí, Spain, [2]AB Vista, Woodstock court, Blenheim Road, Marlborough Business Park, SN84AN Marlborough, Spain; gustavo.cordero@abvista.com

An experiment was carried out to determine the impact of supplementation with phytase and carbohydrases to nutrient deficient diets on growth performance and bone mineralisation in growing pigs. A total of 192 male pigs (Pietrain × Large White×Landrace; initial weight 23.50±3.50 kg) were allocated to 48 pens and distributed in a randomized complete block design comprising 6 treatments with 8 replicates (4 pigs/pen). Diets consisted of a positive control (PC) with no enzyme supplementation, an industry control (IC) with 500 FTU/kg phytase (Quantum Blue, AB Vista), a nutrient (Ca, P, Lys and AA) reduced diet (NC1), a nutrient and energy deficient diet (NC2), and NC1 and NC2 diets supplemented with both phytase (2,000 FTU/kg; Quantum Blue, AB Vista) and carbohydrase (9,600 BXU/kg; Econase XT, AB Vista) (ENZ). The mineral matrix on the NC diets was limited to 0.20% avP and 0.22% Ca due to the phytate level in the diet. Diets were pelleted, and pigs were fed a 2-phase feeding program according to NRC requirements. At the end, one pig from each pen was euthanized, and metatarsus bones were collected for analysis. Data were analysed by GLM procedure of SAS, and means were then separated using Tukey's test. Animals fed IC diet maintained (P>0.05) ADG, FI, FCR, and bone ash whereas those fed NC diets demonstrated a reduction (P<0.05) of ADG, FI and bone ash, and increased FCR relative to those fed the PC diet. Simultaneous inclusion of ENZ in the NC diets improved ADG, FI, FCR and bone ash (P<0.05), restoring performance until to levels similar to the PC. Energy and lysine efficiency were also improved (P<0.05), while phosphorus efficiency in ENZ treatments was greater than the IC diets. These results confirm that supplementation of Quantum Blue and Econase XT in diets limited in Ca, avP, amino acids and energy can result in growth performance and bone ash equivalent to that of pigs fed more expensive nutrient dense diets.

Effect of protein in maternal and weaning diets on performance and faecal consistency of weaned pigs

K.C. Kroeske[1,2], N. Everaert[1], M. Heyndrickx[2], M. Schroyen[1] and S. Millet[2]
[1]Gembloux Agro-Bio Tech, TERRA Teaching and Research Centre, Precision Livestock and Nutrition Unit, Passage des Déportés 2, 5030 Gembloux, Belgium, [2]ILVO, scheldeweg 68, 9090, Belgium; kikianne.kroeske@ilvo.vlaanderen.be

Low protein levels in weaning diets have been associated with lower risk for diarrhoea, which could improve piglets' gut health, and decrease antibiotic use in animal production. The mismatch theory presumes that animals can be prepared during foetal development for certain postnatal situations. Therefore, we investigated the effect of two levels of crude protein in late-gestation diet and piglet post weaning diet on the offspring's susceptibility for diarrhoea post weaning and performance until slaughter. The hypothesis was tested in a 2×2 factorial trial, with high and low protein levels in sow diets (12 vs 17% crude protein (CP)) during the last five weeks of gestation and high or low, one-phase nursery diet (16.5 vs 21% CP). Resulting in four treatments of the four combination of high (H) and low (L) protein diets. All diets were supplemented with amino acids (AA), to reach or exceed the ideal AA-composition. The faecal consistency was measured during the nursery phase of the piglets. Performance of the piglets was registered from birth until slaughter. Statistical analysis was performed using RStudio with linear mixed-effect models (lme4 package), pairwise comparison and estimated measures of means (emmeans package). For performance, no difference was found between treatments. Daily gain (g) did not differ from weaning until 9 weeks (LL 372.00±20.3, LH364.00±7.23, HL370.00±18.8, HH 377.00±4.92), or from 9 weeks until slaughter (LL 766.85±16.62, LH 807.07±32.31, HL 792.11±24.4, HH 804.87±21.15). Feed conversion ratio did not differ from weaning until 9 weeks (LL 1.39±0.04, LH 1.42±0.02, HL 1.42±0.03, HH 1.35±0.02) or from 9 weeks until slaughter (LL 2.45±0.02, LH 2.53±0.05, HL 2.48±0.03, HH 2.39±0.07). For faecal consistency, no interaction between maternal and piglet diet was found. Lower protein in the gestation diet lead to higher diarrhoea scores post weaning, while lower protein in the piglet diets tended to lead to lower diarrhoea scores. In summary, the difference in dietary protein levels for sows and piglets did influence weaning diarrhoea, but did not influence the performance of piglets.

Prebiotic effect of by-products assessed by *in vitro* fermentation combined with porcine immune cells

J. Uerlings[1], M. Schroyen[1], E. Arévalo Sureda[1], J. Bindelle[1], E. Willems[2], G. Bruggeman[2] and N. Everaert[1]
[1]Gembloux Agro-Bio Tech, ULiège, Precision Livestock and Nutrition, Passage des Déportés, 2, 5030 Gembloux, Belgium, [2]Royal Agrifirm Group, Landgoedlaan, 20, 7325 Apeldoorn, the Netherlands; julie.uerlings@uliege.be

The inclusion of prebiotics in the diet of young piglets is generally assumed to be a good strategy to promote intestinal health, and reduce weaning-associated disorders. Therefore, this study investigated the fermentation capacity of some agricultural and industrial by-products, containing significant amounts of dietary fibre, and the impact on the immune response *in vitro*. The ingredients (chicory root and pulp, citrus pulp, rye bran and soy hulls) have been screened in a three-step *in vitro* model of the piglet's gastro-intestinal tract combining an enzymatic hydrolysis and a dialysis step to a batch fermentation. Gas and short-chain fatty acids were analysed and compared to the ones obtained for inulin, considered as positive control. The immunomodulatory effects of the feed hydrolysates and the fermentation supernatants (FS), either sterile-filtered containing only metabolites or complete supernatant containing metabolites and bacteria, were investigated by determination of cytokine production of cultured porcine blood mononuclear cells (PBMCs). The experimental data were subjected to GLM procedures followed by Tukey's HSD. Chicory root displayed the best fermentation characteristics with rapid and extensive gas kinetics and had the highest butyrate ratio after inulin. However, low levels of *Clostridium* clusters IV and XIVa were found for this ingredient. Alternatively, chicory pulp and citrus pulp demonstrated high acetate ratios. Hydrolysates from chicory root, rye bran and citrus pulp displayed TNFα and IL8 cytokine levels in the medium of PBMCs similar to the sham-treated control cells. Remarkably, citrus pulp complete FS reached similar IL8 and TNFα amounts to the control levels. In contrast, sterile-filtered FS induced increased cytokine levels in PBMCs. Therefore, chicory root appears to be a promising ingredient to modulate intestinal fermentation and citrus pulp may exert immunomodulatory effects, due to the feed ingredient itself or by the presence of beneficial bacteria, interacting with the immune cells.

LPS challenge on the blood of intrauterine growth restricted and normal pigs at weaning

C. Amdi, A.R. Williams, J.C. Lynegaard and T. Thymann
University of Copenhagen, Department of Veterinary and Animal Sciences, Grønnegårdsvej 2, 1870 Frederiksberg, Denmark; ca@sund.ku.dk

In modern pig production, large litter sizes result in more piglets being exposed to intrauterine growth restriction (IUGR) and consequently lower growth efficiency than normal piglets. Immune challenges are major obstacles to growth efficiency, as nutrients are diverted away from growth in support of immune-related processes. IUGR piglets may be more susceptible to disease, as their growth is already compromised, and thus more nutrients are portioned to immune defence. The aim of this study was therefore to investigate the cytokine responses in peripheral blood mononuclear cells (PBMC) stimulated with lipopolysaccharide (LPS) of IUGR and normal piglets. Piglets were classified at birth based on their head morphology as either IUGR or normal. At 25 days of age, blood samples were obtained from 20 IUGR and 20 normal piglets. Data was analysed with the statistical program SAS by the GLM procedure. PBMC from IUGR piglets produced significantly less IL-1β following LPS stimulation than PBMC from normal piglets (9.9±1.75 vs 17.3±2.55 ng/ml; P=0.021), as well as numerically lower amounts of IL-6 (929.7±162.13 vs 1,508.2±355.80 pg/ml; P=0.148). Thus, the present study showed a modulation of immune function in IUGR piglets at weaning, compared to normal piglets. These results may suggest a degree of immunological hypo-responsiveness in IUGR piglets which may have implications for resistance to pathogenic challenges in the post-weaning period.

Rapid evaporative ionisation mass spectrometry (REIMS) for at-line boar taint classification

L.Y. Hemeryck[1], A.I. Decloedt[1], J. Balog[2], S. Huysman[1], M. De Spiegeleer[1], J. Wauters[1], S. Pringle[3], A. Boland[4], S. Stead[3] and L. Vanhaecke[1,5]
[1]Ghent University, Lab. of Chemical Analysis, Salisburylaan 133, 9820 Merelbeke, Belgium, [2]Waters Research Centre, Záhony utca 7, 1031 Budapest, Hungary, [3]Waters Corporation, Stamford Avenue, SK9 4AX Wilmslow, United Kingdom, [4]Waters Corporation, Brusselsesteenweg 500, 1731 Asse, Belgium, [5]Queen's University, Institute for Global Food Security, University Road, BT7 1NN Belfast, United Kingdom; lieseloty.hemeryck@ugent.be

Increasing awareness of animal welfare has led to a European incentive to ban the surgical castration of piglets. A valid alternative for castration is the rearing of entire male pigs, but this allows the (re)occurrence of boar taint, an off-odour in meat from entire boars. Hence, due to adverse consumer reactions to pork with boar taint, the rearing of entire boars requires valid boar taint mitigation strategies, which has always been a tough nut to crack. However, the introduction of rapid evaporative ionisation MS (REIMS) offers compelling perspectives for the rapid as well as accurate at-line detection of boar taint by significantly reducing analysis time and workload, yet enhancing research output and efficiency. In this study, REIMS was used as a direct analysis technique to train predictive models for identification of boar taint above the odour threshold (based on sensory (soldering iron method) as well as chemical analysis (UHPLC-HRMS analysis of indole, skatole and androstenone levels). Adipose tissue was sampled using a prototype bipolar handheld sampling device connected directly to a Xevo G2-XS Q-TOF system equipped with REIMS source. The results demonstrate that untargeted mass spectrometric profiling in negative ionisation mode enables the construction of predictive models (Simca, LiveID and AMX) for the classification of carcasses according to boar taint status based on alterations in lipid profiles. As REIMS eliminates sample pre-treatment with analysis taking <10 seconds, it offers significant potential as the first technique enabling accurate *in situ* detection of boar taint. REIMS is a promising and highly innovative tool for several types of food quality and safety applications, furthermore allowing us to move state-of-the-art equipment and applications from bench to production site.

Automatic recognition of feeding and non-nutritive feeding behaviour in pigs using deep learning

A. Alameer[1,2], H.A. Dalton[2], J. Barcardit[1] and I. Kyriazakis[2]
[1]Newcastle University, School of Computing, Newcastle Upon Tyne, NE4 5TG, United Kingdom, [2]Newcastle University, School of Natural and Environmental Sciences, Newcastle Upon Tyne, NE1 7RU, United Kingdom; ali.alameer@newcastle.ac.uk

Automated vision-based early warning systems have been developed to detect behavioural changes in groups of pigs to monitor their health and welfare status. However, automatic feed detection remains a problem in precision pig farming due to problems of light alteration, occlusions and the similar appearances of pigs. Additionally, these systems often overestimate the actual time spent feeding due to the inability to identify and exclude non-nutritive visits (NNV) to the feeding area. To tackle these problems, we developed a robust feed-detection method that is capable of distinguishing between feeding and NNV to the feeding area for group-housed pigs. Our first objective was to demonstrate the ability of this automated method to identify feeding and NNV behaviour with high accuracy. We then tested the system's ability to detect a disturbance in group-level feeding and NNV behaviour due to feed restriction. A GoogLeNet deep learning model was utilised as a base network to accurately identify the feeding (i.e. pig has head inside the feeding trough) and NNV (i.e. pig enters the black mat area with two or more feet with one being a front foot) behaviour of group-housed pigs. The method was designed to monitor a predefined pen area covering two feed troughs and a black mat covering the area in front of feeders using a video surveillance camera. The experimental tests showed that our system could recognise the feeding and NNV behaviour of pigs with an accuracy of 99.4% using 5-folds cross-validation. Following this validation, we tested the method's ability to detect group level feeding and NNV changes from a normal *ad libitum* feeding date and a date of quantitative feed restriction. These experiments demonstrate this method is capable of robustly and accurately monitoring the feeding behaviour of groups of pigs under commercial conditions without the need for additional sensors or individual marking.

Influence of fattening pigs' positive affective state on behavioural and physiological parameters

K.L. Krugmann, F.J. Warnken, J. Krieter and I. Czycholl
Christian-Albrechts-University Kiel, Institute of Animal Breeding and Husbandry, Olshausenstr. 40, 24118 Kiel, Germany; iczycholl@tierzucht.uni-kiel.de

In this study, we analysed in depth multiple behavioural and physiological parameters assessed on-farm on fattening pigs which could be possible indicators for the affective state of animals. Measuring the affective state of animals is challenging as it is not directly measurable but rather constitutes a latent variable. The data collection was conducted on 60 fattening pigs from three different farms with different housing systems. These pigs were assessed with regard to playing behaviour, reactions in behavioural tests as well as body language signals. Furthermore, physiological parameters, such as the diameter and the astroglia cell numbers of hippocampi, salivary Immunoglobulin A (IgA) content or salivary protein composition were examined. Statistical analysis was carried out by partial least squares structural equation modelling (PLS-SEM) with the Software SmartPLS. A hierarchical component model (HCM) was used, which included the pigs' positive affective state as higher-order component (HOC) and the behavioural and physiological parameters as lower-order components (LOC). The validity of the different components of the model proved to be sufficient (average variance extracted (AVE)$\geq$0.5) and split-half consistency was given. The playing behaviour, body language signals and behavioural tests were most influenced by the pigs' positive affective state, as demonstrated by the calculated path coefficients of the model (0.83, 0.79 and 0.62, respectively). Additionally, moderate R^2-values of the endogenous latent variables playing behaviour (69.8%), body language signals (62.7%) and behavioural tests (39.5%) could be detected. Moreover, the indicator 'locomotor play' demonstrated the highest indicator reliability (0.85) for the estimation of the latent variable of pigs' positive affective state. Hence, besides identifying potential indicators for the measurement of positive affective states in pigs, the results of the present study enhanced the comprehension and assessment of pigs' positive affective state in general.

Sow and piglet hair cortisol, a marker for chronic stress in crates and piglets husbandry procedures

L. Morgan[1], J. Meyer[2] and T. Raz[1]
[1]Hebrew University of Jerusalem, Rehovot, 76100 Israel, [2]UMass Amherst, MA, 01003, USA; liat.morgan@mail.huji.ac.il

Food animal welfare is an issue of great importance to society. Still, in many countries sows are restrained during farrowing and lactation, and piglets routinely undergo a set of invasive husbandry procedures (surgical castration; tail docking; teeth clipping). These may have long-term effects, such as high stress, which can potentially lead to impaired health and production. Cortisol, a glucocorticoid produced by the adrenal cortex in response to stress, accumulates in hair and therefore may be a biomarker for chronic stress. Our objectives were: (1) to examine the influence of different restraint periods on hair cortisol, in lactating sows and their piglets; and (2) to examine hair cortisol of piglets and their dams from birth to slaughter when husbandry procedures were gradually avoided. Study1: 80 sows were housed in farrowing crates, while confinement bars were opened randomly after different periods from 3 to 21 days after farrowing. Hair samples were collected on weaning day. Study2: 32 litters were allocated randomly into four groups; G1: all invasive procedures were performed, without environmental enrichment; G2: same as G1, but environmental enrichment was provided; G3: Immunocastration, tail docking, and teeth clipping were performed, with environmental enrichment; G4: Immunocastration, invasive procedures were not performed, environmental enrichment was provided. Hair samples were collected on weaning and slaughter days. In Study 1, sows and piglets hair cortisol concentrations decreased when the restraint period was shortened. Sows hair cortisol was found as a significant mediator between the restraint of the sow and its piglets hair cortisol (Sobel test). For every day of sows' restraint, piglet hair cortisol increased by 1.1 pg/mg (P=0.012). In Study 2, Mixed-effects linear regression analysis revealed that hair cortisol at weaning was lower when husbandry procedures of piglets were avoided (G3: -32.2%, P=0.014; G4: -24.5%, P=0.102). Moreover, higher weaning weight was associated with decreased hair cortisol by 6.64% per additional kilogram of weaning weight (P<0.001). In conclusion, higher hair cortisol was associated with stressful conditions and lower performances in both sows and their piglets.

Enriching neonatal environment: a potential strategy to improve long-term performance in pigs

H.L. Ko[1], Q. Chong[2], X. Manteca[1] and P. Llonch[1]
[1]School of Veterinary Science, Universitat Autònoma de Barcelona, Cerdanyola del Vallès 08193, Spain, [2]Royal (Dick) School of Veterinary Studies, University of Edinburgh, Easter Bush Campus, Midlothian EH25 9RG, United Kingdom; henglun.ko@uab.cat

The aim of the study was to investigate the combined effects of early-socialization and neonatal enriched environment on pig's growth from birth to slaughter. Two treatments were applied on a commercial farm in Lleida (Spain) in June 2017. During suckling period, piglets of control treatment (CON, n=324 from 20 litters) were raised in barren farrowing pens, whereas those of enriched treatment (ENR, n=337 from 23 litters) were co-mingled on D14 (by removing the barrier between adjacent pens) and provided with six objects per pen (2 hemp ropes, 2 commercial rubber chew toys and 2 handmade plastic tube toys). At weaning (D24), animals were regrouped according to treatments. After weaning, both treatments were housed in the same units (nursery unit and fattening unit), fed with identical diets and reared in intensive conditions as usual. Growth performance of 342 pigs (nCON=153; nENR=189) was followed from birth to slaughter. Weighing occurred on D1, 14, 23, 27, 31, 38, 69, 79, and after slaughter (carcass weight). Salivary cortisol was obtained from 6 pre-selected piglets per litter (heaviest, lightest and middle birth weight of male and female) from 17 randomly selected litters on 1 day pre- (-1), 1 day (+1) and 2 days post-weaning (+2). Statistical analyses were performed by R software (3.5.2). Growth performance was analysed with parametric test; slaughter age and salivary cortisol were analysed with non-parametric test. Birth and carcass weight were not different between treatments. Salivary cortisol increased in both treatments after weaning but the rise was significantly higher in CON ((+1, -1), P<0.01; (+2, -1), P<0.01)). ENR weaners showed a tendency (P=0.09) of better average daily gain in a week after weaning (D27). Despite no difference of average daily gain on D79 between treatments, ENR pigs reached the market weight sooner (P=0.07) than CON (ENR: 194.9±1.0 days; CON: 197.7±1.3 days). Results suggest that enriching suckling environment improves weaning adaptability, which may benefit long-term performance by reducing the time to reach slaughter weight.

Complex realities of work in livestock
B. Dedieu
Inra, Sciences for Action and Development, INRA SAD Theix, 63122 Saint Genes Champanelle, France;
benoit.dedieu@inra.fr

Changes to working conditions in agriculture and particularly in livestock farming are investigated by several disciplines that explore different themes. These themes are presented through two areas: one that focuses on the worker (employment, health and skills) and the other on work as a component of farming systems. The analytical frameworks and core research issues are described. The need for a transversal approach of work is argued in order to analyse where the innovating models of livestock farming could lead to.

Approach to the social sustainability of livestock farms
S. Cournut[1], C. Balay[1] and G. Serviere[2]
[1]VetAgro Sup, UMR Territoires, 89 boulevard de l'Europe, 63370 Lempdes, France, [2]Institut de l'Elevage, 9 allée Pierre de Fermat, 63170 Aubiere, France; sylvie.cournut@vetagro-sup.fr

The social dimension of farm sustainability is significantly less documented than the economic and environmental ones whereas it represents an essential piece to understand how livestock farms operate, their territorial and societal role and their evolution facing great socio-economical mutations at local and global scale. Furthermore, with the increase of social expectations concerning livestock activity (animal welfare, quality products and environmental impact of practices), the drop in the agricultural labour force and the enlargement of farm structures, the mutations of farmers expectations concerning their work, the social dimension of sustainable development can no longer be ignored. Our objective was to design a framework to analyse the social sustainability of livestock farms. We chose a comprehensive and non-normative approach to take into account the subjective and context-dependent nature of this dimension. The social sustainability was thereby defined from the expression of actors and farmers interviewed in four French contrasted territories in terms of socio-economic and geographical contexts, but also regarding the type and dynamics of livestock productions. A thematic analysis of interviews allowed us to identify different facets of social sustainability which were organized in seven main axes composing our analytical framework. The four first axes are related to farm-focused sustainability: job meaning, work organization, quality of life and health. The last three axes take into account the embeddedness of farms in a territory and a society: territorial and societal conditions, local and social networks, and contribution to social sustainability of the territory. Our study enlightened the complexity of the social sustainability which refers to different nested organization scales (farmers, associates, employees, family, exploitation, territory), articulates facts and actors' feelings and expectations, deals with both professional and private life, and concerns the farms situation but also its dynamics. The discussion of our framework with farmers, advisors, teachers and local actors confirmed the importance of this social dimension to draw the future of livestock.

Impacts of Dutch livestock production on human health

P.M. Post[1], L. Hogerwerf[1], E.A.M. Bokkers[2], I.J.M. De Boer[2] and E. Lebret[1]
[1]National Institute for Public Health and the Environment, Antonie van Leeuwenhoeklaan 9, 3721 MA Bilthoven, the Netherlands, [2]Wageningen University & Research, Animal Production Systems group, De Elst 1, 6708 WD Wageningen, the Netherlands; pim.post@rivm.nl

Livestock production may affect the health of local residents. Adaptations to reduce such impact, such as installation of air scrubbers and manure fermentation systems, may affect various aspects of human health, thus requiring insight in a variety of health impacts for informed implementation. Building towards such insight, this study aims to present an overview of the impacts of Dutch livestock production on current and emerging zoonotic diseases, on health effects due to particulate matter emissions, on pneumonia, asthma and chronic obstructive pulmonary disease among local residents, on occupational accidents, antimicrobial resistance, odour and noise annoyance and on incidents related to manure fermentation systems. These impacts were assessed quantitatively where possible, using common impact indicators. Burden of disease related to current zoonotic diseases, occupational accidents and health effects due to particulate matter emissions was expressed as disability adjusted life years (DALYs), which summarizes years of life lost and years lived with disease. We found that these health impacts caused about 9×10^3 DALYs, which accounts for about 0.2% of the total Dutch burden of disease and 5% of the total burden of disease related to environmental sources. The estimated burden only partly indicates the total health impact, which we complement by using other indicators like the percentage of persons that perceives severe odour annoyance related to animal husbandry (about 2.5%) or qualitative indications of impact and probability of emerging zoonotic diseases (indicated as somewhat probable, severe impact). The necessity to use such a variety of indicators illustrates the many forms that human health impacts related to livestock production can take, while these are only a selection of positive and negative impacts. What is more, on a local scale some impacts are higher than their contribution to the national disease burden suggest, indicating the importance of taking the local context into account in future developments of the livestock sector.

Assessing and comparing social and biophysical trade-offs in a extensive beef cattle system region

F. Accatino[1], D. Neumeister[2] and M. Tichit[1]
[1]UMR SADAPT, INRA, AgroParisTech, Université Paris-Saclay, 16, rue Claude Bernard, 75005, Paris, France, [2]Institut de l'Elevage, 149 rue de Bercy, 75595, Paris, France; francesco.accatino@inra.fr

Livestock farming systems provide multiple functions that affect a range of stakeholders. Different groups of stakeholders might prefer different functions (social trade-off). Indeed, because of linkages between functions at the biophysical level, a management action aimed at increasing one function might affect another function negatively (biophysical trade-off) or positively (biophysical synergy). Our objective was to assess and to compare the social trade-offs and the biophysical trade-offs (and synergies) in the Massif Central, a region in the center of France characterized by extensive beef cattle systems. For assessing the social trade-off, a workshop was organized where participants were asked to express their preferences for different farming system functions. For assessing the biophysical trade-offs and synergies, we calibrated a statistical model with data of the case study area, linking land cover variables to the provision of three functions (crop production, animal production, and carbon sequestration) and then run a multi-criteria optimization. Results of the participatory assessment showed a clear social trade-off between food-production-related functions (preferred e.g. by member of producer organizations) and natural-resources-related functions (preferred e.g. by members of NGOs). Results of the optimization model revealed a biophysical trade-off between crop production and animal production, as well as between crop production and carbon sequestration; in contrast there was a synergy between animal production and carbon sequestration. Deeper investigation of the results showed that such a synergy could be achieved via increase of grassland surface. The synergy on the biophysical level between animal production and carbon sequestration makes it possible to address the trade-off on the social level, as the biophysical system is able to satisfy two contrasting stakeholder preferences at the same time.

National organizations and management tools for genetic improvement of dairy cattle in Europe

G. Tesniere[1,2,3], J. Labatut[2], V. Ducrocq[4] and E. Boxenbaum[1,5]
[1]*MINES ParisTech, Université PSL, CNRS UMR i3 – CGS, 60 Bd St Michel, 75006 Paris, France,* [2]*INRA, INPT, Université de Toulouse, UMR AGIR, Chemin de Borde-Rouge – Auzeville, 31326 Castanet-Tolosan, France,* [3]*CORAM, National Collective of local Mountain Breeds, BP 42118, 31321 Castanet Tolosan Cedex, France,* [4]*INRA, AgroParisTech, Université Paris Saclay, UMR GABI, Domaine de Vilvert, 78350 Jouy-en-Josas, France,* [5]*CBS Copenhagen Business School, Department of Organization, Kilevej 14A, 2000 Frederiksberg C., Denmark; tesniere.coram@gmail.com*

Since the early 2000s, the development of genomics provides extensive knowledge of the DNA of living organisms. This innovation has transformed the way in which living organisms are evaluated, selected (genomic selection of plants and animals) and marketed. Coupled with political and regulatory changes, this technology contributes to modify the national institutional arrangements in the field of animal genetic improvement as well as actors' practices. The current liberalization process questions both the collective dimension of genetic progress and the property rights of genetic resources. In a comparative perspective involving France, Ireland and The Netherlands, the objective of this overview is to present the plurality of institutional arrangements regarding genomic selection of the Holstein cattle breed. First, it highlights three institutional regimes that reveal different arrangements particularly between public and private organizations. Then, this diversity of arrangements is completed by an analysis of contractual tools. Contracts refer here to formal agreements between breeding companies and farmers through models of strategies aiming at the production and exchanges of genetic resources such as animals, embryos and semen. These models emphasize various forms of property of genetic resources between companies and breeders and also show that the actors' roles in genetic selection activities are redefined. Indeed, the phenomenon of contracting represents one of the major evolutions in the relationship between our societies and nature. These results provide a better understanding of the development of a liberal logic (in The Netherlands) in duality with the reinforcement (in Ireland) or weakening (in France) of a cooperative logic for the production of improved animal genetics.

Farm's social and economic factors and the adoption of agricultural best management practices

T.T.S. Siqueira[1,2] and D. Galliano[1]
[1]*INRA, UMR 1248 AGIR, 24, chemin de Borde-Rouge, CS 52627, 31326 Castanet Tolosan Cedex, France,* [2]*INP-Ecole d'Ingénieurs de PURPAN, 75 Voie du TOEC, 31076 Toulouse, France; tiago.siqueira@purpan.fr*

Exploring the relationship between farm's social and economic factors and the adoption of agricultural best management practices is useful to tackling the environmental challenges faced by the animal production. This study used the data of 47,211 dairy farms from the 2010 French Agricultural Census to study the statistical correlations between these factors and the adoption of nine agricultural best management practices. First, we tested the internal factors related to the characteristics of the farmer, farm's structure, and governance. Second, we tested the external factors related with commercial and regulatory followed by spatial features. The results show that the internal factors like farm size and the contracting any agri-environmental insurance are negatively correlated with the adoption of most of the practices. Communication and information technologies are both positively or negatively correlated with the practices. In terms of governance, farms with individual and corporative legal status have statistically significant adoption behaviour. The share of familial annual working unit is negatively correlated with most of the practices. On the contrary, the diversification has an important positive correlation. Education, age, knowing the succession and the share of land under property have not significant correlation with most of the practices. In terms of external factors, the statistical analysis highlights the significant positive correlation of positioning on alternative markets, short circuits, organic products, or quality markets and the adoption of almost all practices. As the literature commonly suggests, the results show that environmental regulations also drive the adoption. The spatial environment of the farm and, more specifically, the environmental behaviour of neighbouring farms is highly correlated with the adoption. Finally, polices to promote locally farmer's experience exchange and to supporting diversification, high quality products and short circuits can further the adoption of agricultural best management practices reducing the environmental impacts of dairy farms.

Charting challenges of livestock farming using patterns in consumers' reasoning and behaviour
H.J. Nijland
Wageningen University, Communication, Philosophy and Technology, Hollandseweg 1, 6706 KN Wageningen, the Netherlands; hanneke.nijland@wur.nl

The farming and slaughter of animals and the related consumption of meat are increasingly being contested in our current society. However, the ways in which they are called into question are not univalent. The world consists of a variety of (sub-)cultures, in which different arguments are applied by people with different values, norms, and interests, to different animals. In this invited talk, Hanneke Nijland will present patterns stemming from over 15 years of research into the various ways people in everyday-life frame the (non-)acceptability of keeping and killing animals for food. In-depth interpretative analysis of consumers' reasoning and behaviour showed that the long list of topics includes arguments regarding animal welfare and environmental impact, but that many others are involved, such as health, taste, money, and religion – and that there are two implicit values that represent an attitude towards life and death of animals. Moreover, she will explain how there is a clear logic to this, by showing how consumer consideration of the interests of certain parties (including themselves and their loved ones, farmers, animals and the environment) coincide with specific core values and norms for consumption behaviours. She will then zoom in on a new piece of research she is developing, showing how different ways of thinking about animals in philosophy and practice, are linked to specific definitions of their welfare and norms for their treatment, and how these vary in different farming systems contexts. Finally, she will show how these patterns can be used to understand and chart challenges that livestock farming is facing in relation to society, and create space for thought and talk about the intricate issue of farming animals for food.

High school education: a challenge to reduce gap between livestock production and citizen concerns
A. Chouteau[1], G. Brunschwig[2] and C. Disenhaus[3]
[1]IDELE, 149 rue de Bercy, 75595 Paris cedex 12, France, [2] VetAgro Sup, INRA, UMR Herbivores, 89, avenue de l'Europe, 63370 Lempdes, France, [3]UMR PEGASE, INRA, Agrocampus-Ouest, Livestock Science, 65 Rue de Saint-Brieuc, 35000 Rennes, France; alizee.chouteau@idele.fr

In France, as in many other countries, the gap between farmer's reality and citizen's expectations is increasing. As citizens and consumers are more and more concerned with livestock farm evolutions, decreasing this gap is of importance, and this should start at school level. However, information on feelings and knowledge of young people and their teachers about livestock productions are scarce. Two surveys were conducted between March and July 2018 to collect such information's. A closed-questionnaire was full-filled by French secondary school pupils (SSP). SSP were selected because they will soon become adult consumers and citizens. A total of 1,087 SSP from 15 different high school located almost everywhere in France were inquired. Knowledge on livestock originated from media (89%), family or neighbourhood (57%), school (40%), web (40%) or social network, but is anyway very poor. Only 48% know that an ewe need to have a lamb to produce milk. They underline the very important role of livestock farming in providing food (96%), but question the conditions of production, and in particular their negative impact in terms of animal welfare (82%) and on the environment (61%). Only 4% declared to be vegetarian and 0.4% to be vegan, mostly girls. Simultaneously, 28 teachers of life sciences and geography from the same high school answered a semi directive investigation. Many of them (11/28) expressed difficulties in addressing these topics especially for life sciences teachers (10/15). They are indeed poorly informed themselves about livestock and agriculture (8/28), and may have difficulty answering questions from their students. Moreover, this subject is perceived as anecdotic in the training programs (19/28). Finally, they assume to propose a caricatured vision a livestock production proposed by school. The study highlights the necessity to give more information about livestock production to the future citizens and for this purpose to strength the educational resources available for the teachers. This conclusion based on French population need to be enlarged at the European level.

The commitment of Dutch retailers to sustainable egg production

M. Vandervennet and E.M. De Olde

Animal Production Systems group, Wageningen University & Research, P.O. Box 338, 6700 AH Wageningen, the Netherlands; evelien.deolde@wur.nl

Producers and retailers are increasingly expected by consumers, investors, governments and NGOs to monitor and report the sustainability impact of their products. Through the selection of products and commitment to sustainability standards, retailers can play an important role in changing production methods and consumption patterns. Within the Dutch egg production sector, retailer's commitments to animal welfare, environmental issues or the development of new housing systems, are extensively discussed in the media. Decisions of retailers to commit to a certain level of animal welfare certification or sustainability standard can, therefore, have a large impact on the development of the egg production sector. The aim of this study was to analyse the commitment of retailers to sustainability issues in the egg production sector over time. In the Netherlands, consumption eggs are mainly sold through supermarkets. The study focused on the five largest retailers which together have a 76% market share. A systematic analysis of Corporate Social Responsibility (CSR) reports of the retailers and newspaper articles was carried out to analyse which sustainability issues were raised over the past ten years. In addition, interviews with retailers were carried out. Results showed that animal welfare is a key issue in the communication of Dutch retailers related to eggs. In particular, the Dutch animal welfare certification system and organic production were often discussed. Other issues raised included environmental impact, livestock feed (soy), consumer awareness, care farming, food safety, and transparency. The extent to which eggs were discussed in CSR reports varied strongly between retailers and years. While the commitment of Dutch retailers has predominantly focused on animal welfare issues in egg production, attention to other sustainability issues is needed to facilitate sustainable development of the sector. This requires broader sustainability reporting, including the acknowledgement of potential trade-offs between sustainability issues. Further research is needed to evaluate the impact of CSR initiatives on the sector and farmers, and on consumption patterns.

Quantifying unintended macroeconomic consequences of meat taxation

T. Takahashi, G.A. McAuliffe and M.R.F. Lee

Rothamsted Research, North Wyke, Okehampton, Devon, EX20 2SB, United Kingdom; taro.takahashi@rothamsted.ac.uk

As a means to curb GHG emissions associated with ruminant agriculture, a carbon tax against beef and dairy sectors has been proposed in recent years. These proposals are based on modelling studies employing a partial equilibrium framework, a class of economic models that focus on a subset of markets in an economy (food markets in the present case) under the 'separability' assumption that interactions between the studied markets and other markets are negligible. While this condition is largely satisfied for consumers who often budget a proportion of their income for food purchases, its validity on the production side is less clear-cut. A shrinkage of the livestock industry, for example, can result in different, and possibly suboptimal, land use and employment structures beyond the boundary of agriculture. To investigate potential economy-wide impacts of carbon taxation, we set up a GTAP computable general equilibrium modelling framework for the global economy, linked it to global databases of GHG emissions, and imposed an *ad valorem* consumption tax on beef and dairy commodities in the UK and France at the levels identical to those adopted by Springmann *et al.* These rates were 18.6% (UK) / 19.8% (France) for beef and 11.3% / 12.0% for dairy. Our estimation of post-tax food production was largely in agreement with the original study. Primary outputs decreased by 5.9% (UK) / 1.8% (France) for beef and 0.2% / 3.5% for dairy, with on-farm GHG savings of 1.4 Mt (UK) / 1.0 Mt (France) and economy-wide GHG savings of 2.5 Mt / 1.1 Mt CO_2e. The model also revealed, however, that both countries would suffer from large economic losses, with the equivalent valuation (a common indicator to quantify the change in economic welfare post-intervention) amounting to -$310M (UK) / -$232M (France) due to forced reallocation of resources such as transfer of land from livestock farms to arable farms. Importantly, these losses considerably outweighed the social benefit of climate change mitigation ($129M / $58M), evaluated at the carbon price used by Springmann *et al.* to justify the proposed levies ($52/t CO_2e). This finding suggests that the 'optimal' tax being proposed may be underestimating the true value of ruminant agriculture.

Quantifying public attitudes towards consumption of meat produced from genome edited animals

T.W.D. Kirk[1], J. Crowley[1], D. Martin-Collado[2] and C.B.A. Whitelaw[3]

[1]Abacusbio International Ltd, Roslin Innovation Centre, Easter Bush, Midlothian, EH25 9RG, United Kingdom, [2]CITA-Aragon, Avenida Montañana, 930, 50059, Zaragoza, Spain, [3]The Roslin Institute, University of Edinburgh, Easter Bush, Midlothian, EH25 9RG, United Kingdom; jcrowley@abacusbio.co.uk

Novel gene editing technologies offer exciting new opportunities not only to increase the productivity of agriculture, but also to tackle infectious diseases while minimising environmental impact. Thus, gene editing offers a potential solution to several sustainability and food security challenges. Gene editing is being regulated in different ways in different countries, its safety often questioned, and the moral aspects of use in the human food chain regularly politicised. These factors all create controversy and polarised public views, which are not always well informed or quantified. Previous research has focused on public acceptance of gene editing in plants, not animals. Given the potential positive impact of the technology on livestock production, it is important that researchers and potential users understand the level of public acceptance of gene editing animals for the human food chain. Arguably, acceptance of the technology outside the research world is the single largest barrier to commercialisation. We conducted a survey of over 1000 participants to quantify attitudes towards the use of gene edited animals in the human food chain. To gauge consumer's willingness to pay for gene edited meat (relative to 'regular' meat), respondents were asked to choose their preferred option between regular meat products and gene edited meat products at various price levels. Respondents were then asked the same question, but this time specific benefits relating to gene editing were highlighted. The potential characteristics of gene edited meat which were highlighted included improved environmental, health, and animal welfare benefits. Different consumer group's willingness to pay for gene edited meat, both with and without the characteristics described, were compared to the group's attitude towards both gene editing and traditional GMOs.

Farm and abattoir staff opinion regarding pig production systems

D.L. Teixeira[1], L.C. Salazar[2], P. Carrasco[1] and M.J. Hötzel[3]

[1]Universidad de O'Higgins, Instituto de Ciencias Agronómicas y Veterinarias, San Fernando, 00000, Chile, [2]University of Edinburgh, College of Veterinary Medicine, Edinburgh, EH259RG, United Kingdom, [3]Universidade Federal de Santa Catarina, Laboratório de Etologia Aplicada, Florianópolis, 88034, Brazil; dadaylt@hotmail.com

There is growing interest in the use of ante- (AM) and post-mortem (PM) inspection for the detection of animal welfare outcomes. Thus, it is important to understand the views of people directly associated with the pig industry towards the use of such tool. Staff members were recruited at one farm (n=124) and one abattoir (n=106). All participants filled the same survey, which asked their perception about prevalence of tail and ear lesions and severe lameness of pigs on their workplace, what they considered to be the main causes of these conditions, and if they considered that they can be detected at AM and PM inspections. In general, 48% of participants considered the prevalence of tail lesion, ear lesions and severe lameness of pigs to be low at their workplace. Participants considered that these conditions can be detected at AM (77%) and PM (53%) inspections. Density (11.7%), management (7.4%), lack of environmental enrichment (5.2%) and number of pens (4.3%) were considered the main causes of tail lesions. Density (12.7%), management (7.5%), diet (4.8%) and use of antibiotics (3.1%) were considered the main causes of ear lesion. Floor type (16.2%), management (15.7%), density (7.4%) and number of pens (3.9%) were considered the main causes of severe lameness in pigs. Considering the scientific evidence, the findings from this study show that staff members from the pig farm and the abattoir are well informed about these conditions. Therefore, being informed about meat inspection outcome could benefit animal health and welfare management plans on pig farms.

Systems for mother-bonded calf rearing during the milk feeding period in organic dairy farming

L. Mogensen[1], M. Vaarst[2], M.B. Jensen[2] and J.O. Lehmann[1]
[1]Aarhus University, Dept. of Agroecology, Foulum, 8830 Tjele, Denmark, [2]Aarhus University, Dept. of Animal Science, Foulum, 8830 Tjele, Denmark; lisbeth.mogensen@agro.au.dk

Systematic separation of cow and calf early after calving is common practice in both organic and conventional dairy farming. It is broadly accepted by professionals within the dairy sector as a necessity for producing milk for human consumption and justified by the severe reaction of both cow and calf if separation occurs later than 24 hours after calving. The practice of early separation conflicts with the organic principles and is increasingly questioned by consumers, who argue for the development of a dairy production system that supports rearing of cows together with their own calves for an extended time. The main objective of this project is to develop strategies for mother-bonded rearing systems under different Danish organic farming conditions and quantify their effects on farm productivity and economics through analyses of existing dairy cow-calf systems and published experiments. The first step includes a systematic collection of farm data and system descriptions from existing cow-calf systems with focus on housing, management and outcomes of production. Our data collection so far shows that mother-bonded rearing of dairy calves is practiced in a variety of ways with herd size from 30 to 85 cows and milk production delivered to the dairy between 4,000 and 9,500 kg/cow/year. Calves stayed with the dam either full time, part time or full time in the beginning with diminishing time together towards weaning, which led calves to drink between 6-7 l/day and up to 14-16 l/day. Farmers either housed calves and all lactating cows in one group or housed lactating cows with or without calves in two separate groups. Farms used both deep litter and cubicle systems as well as both parlour and robotic milking systems. Finally, calves were either weaned and separated gradually with multiple steps or abrupt at once. The next step will be a synthesis of collected farm data, system descriptions and literature, which will go into developing different strategies for mother-bonded calf rearing in collaboration with Danish organic farmers.

An insight of dairy farming in the Parmigiano Reggiano area

A. Foskolos[1], F. Righi[2], R.G. Pitino[2] and A. Quarantelli[2]
[1]University of Thessaly, Department of Animal Science, Campus Larisa, 413 34 Larisa, Greece, [2]University of Parma, Department of Veterinary Science, Via del Taglio 10, 43126, Parma, Italy, Italy; andreasfosk@hotmail.com

Parmigiano-Reggiano area produces 16% of Italian cattle milk while its cheese production is the second highest in Italian Protected Designation Origin cheeses. With the objective to describe dairy farming in this region, we developed a dataset using recorded information from the Italian association of breeders of 2015, 2016 and 2017, with the following information: location of the farm, cattle breed, number of lactating cows in the herd, annual milk, fat and protein yield (expressed at 305 days), lactation length, average age at 1st calving, average calving age of the herd, and average days open. Statistical analysis was conducted with a mixed effects model in JMP. Fixed effects included farm's location, breed and year. The farm was considered a random effect. A total of 2,351 dairy farms was located in 5 provinces: Bologna (BO; 2.3%), Mantua (MN; 11.2%), Modena (MO; 17.6%), Parma (PR; 29.1%), and Reggio-Emilia (RE; 39.8%). Italian Friesian was the most abundant breed (66.5%) followed by Cross-breeds (21.5%) and Bruna (6.2%). Most intensive dairy farming occurred in MN that was the region with the highest herd size (54 lactating cows), annual milk yield (7,571 kg/year) and lactation length (142 days), and the lowest age at 1st calving (26.4 months) and herd calving age (53.1 months). In contrary, BO was the region with the lowest herd size (25 lactating cows) and milk yield (6,450 kg/year) but the highest fat yield (404 kg/year). On a breed base, Italian Friesian and Jersey seem to be the most industrialized breeds with the highest herd size (115 and 146 lactating cows for Friesian and Jersey, respectively) and annual milk yield (8,396 and 8,126 kg/year for Friesian and Jersey, respectively). Moreover, Angler had the highest fat yield (410 kg/year) with a moderate milk yield (6,891 kg/year). Interestingly, Angler and Rossa Danese had the lowest age at 1st calving among analysed breeds (20.6 and 20.3 months for Angler and Rossa Danese, respectively). Support was provided through the H2020 project CowficieNcy.

Net contribution of dairy production to food supply: impact of intensification and grassland share

C. Battheu-Noirfalise[1], D. Stilmant[1], E. Reding[2], B. Wyzen[2], J. Wester[1], M. Mathot[1], S. Hennart[1] and V. Decruyenaere[1]
[1]Walloon Agricultural Research Centre (CRA-W), 100 rue du Serpont, 6800 Libramont, Belgium, [2]Walloon Breeder Association (AWE), 41 Rue de la Clé, 4650 Herve, Belgium; c.battheu@cra.wallonie.be

Due to their ability to convert cellulose rich resources into high quality animal products, ruminants can play an important role as net food producers. Nevertheless, in order to (1) increase energy and protein density of the diet of high yielding cows and/or (2) to face limited grassland access under low field accessibility and herd size increase, these diets could contain potentially human-edible feed, such as cereals, leading to feed-food competition. In this work we explore the human-edible feed conversion efficiency (heFCE) and the human-edible food production per hectare of fodder area (heFHA), both under gross energy and gross protein forms, for 204 specialised dairy farms, in Wallonia (Belgium). Both milk and meat productions are taken into account. In order to quantify the fraction of human-edible resources in feeds, available literature was used for pure feeds and inquiries were performed near feedstuff producers to take into account the composition of a diversity of concentrates together with their relative importance. First of all, heFCEprotein and heFCEenergy are positively and highly significantly correlated to each other (r=0.71). The results confirm that dairy production intensification both at animal (r=-0.26 between heFCEprotein and milk production per cow; P<0.001) and per surface unit (r=-0.25 between heFCEprotein and the number of livestock units/ha; P<0.001) is negatively correlated to heFCEprotein but positively correlated to heFHAprotein (r=0.37 and r=0.36, respectively (P<0.001)). The proportion of grasslands in the fodder area of the farm is positively and highly significantly correlated to heFCEprotein (r=0.33) with a weak link to heFHAprotein (r=0.15; P<0.05). Nevertheless, all these correlations, even if significant, are weak underlining the need to develop multifactorial approaches to explain the variability observed both for heFCE and heFHA.

Gut microbiota affects behavioural responses of feather pecking selection lines

J.A.J. Van Der Eijk[1,2], M. Naguib[2], B. Kemp[1], A. Lammers[1] and T.B. Rodenburg[1,3]
[1]Wageningen University & Research, Adaptation Physiology, De Elst 1, 6708 WD Wageningen, the Netherlands, [2]Wageningen University & Research, Behavioural Ecology, De Elst 1, 6708 WD Wageningen, the Netherlands, [3]Utrecht University, Department of Animals in Science and Society, Yalelaan 2, 3584 CM Utrecht, the Netherlands; jerine.vandereijk@wur.nl

Early life environmental factors have a profound impact on an animal's behavioural development. The gut microbiota could be such a factor as it influences behavioural characteristics, such as stress and anxiety. Stress sensitivity and fearfulness are related to feather pecking (FP) in chickens, which involves pecking and pulling out feathers of conspecifics. Furthermore, high (HFP) and low FP (LFP) lines differ in gut microbiota composition. Yet, it is unknown whether gut microbiota affects FP or behavioural characteristics related to FP. Therefore, HFP and LFP birds orally received a control, HFP or LFP microbiota treatment within 6 hrs post hatch and daily until 2 weeks of age. FP behaviour was observed via direct observations at pen-level between 0-5, 9-10 and 14-15 weeks of age. Birds were tested in novel object (3 days & 5 weeks of age), novel environment (1 week of age), open field (13 weeks of age) and manual restraint (15 weeks of age) tests. Microbiota transplantation influenced behavioural responses, but did not affect FP. HFP receiving HFP microbiota tended to approach a novel object sooner and more birds tended to approach than HFP receiving LFP microbiota at 3 days of age. HFP receiving HFP microbiota tended to vocalise sooner compared to HFP receiving control in a novel environment. LFP receiving LFP microbiota stepped and vocalised sooner compared to LFP receiving control in an open field. Similarly, LFP receiving LFP microbiota tended to vocalise sooner during manual restraint than LFP receiving control or HFP microbiota. Thus, early life microbiota transplantation had short-term effects in HFP birds and long-term effects in LFP birds. Previously, HFP birds had more active responses compared to LFP birds. Thus, in this study HFP birds seemed to adopt behavioural characteristics of donor birds, but LFP birds did not. Interestingly, homologous microbiota transplantation resulted in more active responses, suggesting reduced fearfulness.

Avoiding feather pecking and cannibalism in laying hens: the dual-purpose hen as a chance

M.F. Giersberg[1], B. Spindler[2] and N. Kemper[2]
[1]Wageningen University & Research, Adaptation Physiology Group, De Elst 1, 6708 WD Wageningen, the Netherlands;
[2]University of Veterinary Medicine Hannover, Foundation, Institute for Animal Hygiene, Animal Welfare and Farm Animal Behaviour, Bischofholer Damm 15, 30173 Hannover, Germany; mona.giersberg@wur.nl

Keeping dual-purpose hybrids is one alternative to the killing of male day-old chickens. However, dual-purpose hens seem to have several additional advantages compared to conventional layers, for instance a lower tendency to develop behavioural disorders, such as feather pecking and cannibalism. In the present study, three batches of conventional layer hybrids (Lohmann Brown plus, LB+) and dual-purpose hens (Lohmann Dual, LD) with untrimmed beaks were observed from 20 to 69(56) weeks of age. Both hybrids were reared under the same conditions and moved to the hen house at about 18 weeks. The animals were kept in a total of four compartments (about 900 hens/compartment) of an aviary system with two replications of each hybrid line per batch. All hens were managed according to standard procedures by the same farm staff. Depending on batch, minor plumage loss on the back was detected in the LB+ flocks at the age of 22 to 25 weeks. Plumage condition deteriorated continuously, so that up to 97% of the LB+ hens showed feather loss to varying extents at the end of the laying period. However, only 4-5% of the LD hens showed slight feather loss on the head and the breast/belly region, which started at the age of 34-41 weeks and remained almost constant until the end of production. Compared to feather loss, integument damage, and thus cannibalism, occurred in the LB+ hens with a delay of 12-21 weeks. With a maximum of 6.5-11.5% injured LB+ hens, a peak was reached at 49-66 weeks of age, depending on batch. In contrast, skin injuries were observed only sporadically in single LD hens (<0.5%). In three consecutive batches, feather pecking and cannibalism occurred in the conventional layer hybrids but not in the dual-purpose hens, though both genetic strains were kept under the same housing and management conditions. Therefore, keeping dual-purpose hens should be considered as an alternative approach to avoid injurious pecking in modern laying hen husbandry.

Network analysis of tail biting in pigs – impact of missed biting events on centrality parameters

T. Wilder[1], J. Krieter[1], N. Kemper[2] and K. Büttner[1]
[1]Christian-Albrechts-University, Institute of Animal Breeding and Husbandry, Olshausenstr. 40, 24098 Kiel, Germany,
[2]University of Veterinary Medicine Hannover, Foundation, Institute for Animal Hygiene, Animal Welfare and Farm Animal Behaviour, Bischhofsholer Damm 15, 30173 Hannover, Germany; twilder@tierzucht.uni-kiel.de

With social network analysis the development of tail biting can be understood by examining the social structure of all pigs instead of focussing on attributes of individuals. Video material of 4 pens (24 undocked pigs/pen) with a tail biting outbreak was continuously analysed for 12 h/day at 2 days before the first large tail lesions, i.e. lesions > diameter of the tail, occurred. Networks consisting of nodes (pigs) and edges (pointing from the biting to the bitten pig) were generated for different aggregation windows (1, 3, 6, 12 h) and centralities were calculated: in- & out-degree, i.e. the node's number of direct in- or outgoing contacts, betweenness, in- & outgoing closeness, i.e. the node's position in the network. Extensive video analysis can provoke distraction leading to missed observations. Thus, the aim of the study was to analyse the impact of missed events on centralities. From all tail biting events, random samples were drawn containing 10-90% of the original observations. Centralities were calculated for the original observations and all samples. To evaluate the changes in the rank order of the nodes Spearman Rank Correlation Coefficients (rs) were measured between the original and each sampled network. The rs values of the centralities (0.23-0.99) increased steadily with increasing sample size. For the 6 & 12 h networks the rs values were >0.9 for the 50% and larger samples for all centralities except for betweenness. For smaller aggregation windows, only sample sizes >60% (3 h networks) or 70% (1 h networks) had rs values >0.9. For the betweenness, only the 80 & 90% samples (for the 1 h network only 90%) had rs values >0.9. Thus, depending on the time aggregation, most of the centralities were quite robust. If these results could be confirmed also for phases with a longer time interval until the first large tail lesions occurred, scan sampling can be used instead of the continuous sampling which implies a reduction of workload without losing significant information.

Genetics of tail-biting receipt in gilts from the Tai Zumu line

L. Canario[1] and L. Flatres-Grall[2]
[1]GenPhySE INRA, INPT ENSAT ENVT, 31326 Castanet Tolosan, France, [2]AXIOM, La Garenne 37210 Azay sur Indre, France; laurianne.canario@inra.fr

This study aims at quantifying the genetic basis of tail-biting receipt in the Tai Zumu population. Tail-biting receipt was recorded at the end of the fattening period on 33,266 gilts that were raised in groups of 6 to 20 females among 3 nucleus herds. The phenotype under measure was the presence of lesion(s) at the end of the fattening period, reflecting that the gilt did not defend itself from tail biters attacks. The objectives were to: (1) compare genetic merit for growth and leanness between non-bitten females and bitten females; (2) quantify the contribution of social genetic effects to the phenotypic variation of the trait 'bitten'; and (3) estimate the effect of the environment on the genetic expression of this trait as interactions between genetics and environment (G×E). The prevalence of tail bites varied from 2.8 to 10.8% among herds that differed according to the groups' characteristics (e.g. size, number of mixings) and feeding system during fattening. Models for tail-biting receipt included the fixed effects of number of group mates, (herd)-year-month, and the random effects of group, litter and single direct genetic effects or also social genetic effects. The model was applied to the complete gilt population and separately to 2 sub-populations, corresponding to the 2 nucleus herds where tail-biting receipt was highest (>5%). Bitten females had a higher genetic merit for growth and leanness than non-bitten females. Tail-biting receipt was analysed as a linear trait, following a normal distribution with the restricted likelihood methodology. Social genetic effects contributed 81 to 93% of total heritable variance, which equalled 40 to 80% of phenotypic variation, whereas only 6-8% was explained with a direct model. Differences in the ranking of connecting sires between herds according to their direct, social and total breeding values yielded rank correlations not different from zero, indicating strong G×E. Selection against tail biting is possible even though the direct genetic contribution for this trait is low. Considering social genetic effects improves quantification of heritable variation, and accounting for them in breeding schemes would increase response to selection against this deviant behaviour.

Genetic analyses of birth weights, tail measurements and scores for tail anomalies in pigs

T. Kunze[1], F. Rosner[2], M. Wähner[1], H. Henne[3] and H.H. Swalve[2]
[1]Anhalt University of Applied Sciences, Strenzfelder Allee 28, 06406 Bernburg (Saale), Germany, [2]Martin-Luther University Halle-Wittenberg, Institute of Agricultural and Nutritional Sciences, Theodor-Lieser-Str. 11, 06120 Halle, Germany, [3]BHZP GmbH, An der Wassermühle 8, 21368 Dahlenburg-Ellringen, Germany; hermann.swalve@landw.uni-halle.de

Tail biting is of ongoing concern in pigs. Environmental factors contribute to this phenomenon but genetic causes also exist. While recording of the trait is difficult as 'biters' and 'those being bitten' have to be identified, the question arises if also size and length of the tail play a role. In this respect, little is known about the genetics of pig tails. Aim of the study was to estimate basic genetic parameters for tail measurements and scores for anomalies of the tail. As individual piglets had to be handled, measurements of birth weight that are very valuable in pig breeding were collected at the same time. Recording was conducted in a Pietrain nucleus line of the German pig breeding company BHZP. The recordings comprised 6,419 records of individual piglets from 582 litters, 410 sows and 68 boars. Univariate and multivariate models were used with and without modelling of random maternal genetic effects in addition to the litter effects and the direct additive genetic effect. Fixed effects considered were parity of sow, time period and sex. Under a multivariate model considering direct, maternal genetic and litter effects, estimates of direct heritabilities were 0.043, 0.306, 0.122, and 0.177 for birth weight, tail length, tail diameter, and body length while maternal heritabilities were 0.137, 0.073, 0.057, and 0.095. Anomaly scores for tails defined as 0/1 were estimated under a univariate threshold model and gave an estimate of heritability of 0.343. Tail diameter, tail length and body length exhibited genetic correlations of direct effects around 0.40 to 0.60. In conclusion, genetic parameters indicate that it is possible to select for shorter tails if correlated effects of this selection can also be taken care of.

Greek farmers' attitudes regarding tail biting and tail docking

M. Kakanis[1], K. Marinou[2], S. Doudounakis[2] and E. Sossidou[3]
[1]*Regional Unit of Pieria, Department of Veterinary Services, 25 Martiou 49, 60132 Katerini, Greece,* [2]*Greek Ministry of Rural Development and Food, Directorate of Animal Welfare, Veterinary Drugs and Veterinary Applications, 2 Acharnon Str, 10176 Athens, Greece,* [3]*Hellenic Agricultural Organization-Demeter, Veterinary Research Institute, Thermi, 57001 -Thessaloniki, Greece; sossidou.arig@nagref.gr*

Although it's been years since the European Commission banned tail docking on a routine basis, a significant percentage of pig farmers apply it as supposedly the only way to minimize the risk of tail-biting. The aim of this study was to obtain a better understanding of Greek pig farmers' attitudes with regard to tail biting. The survey, which is expected to be completed by early July 2019 is based on a questionnaire sent to 120 farmers both through the local veterinary services in hard copy and their national association via an online form. This paper presents the preliminary results collected from 28 respondents to date. More specifically, tail biting is considered to be a significant welfare problem although most of respondents report very low figures in everyday practice (70.3% of respondents. An interesting finding is that enrichment material is still not recognized as an important factor in preventing tail biting (56% of respondents). High incidence of tail biting was anticipated when trying to keep undocked pigs based mostly on previous attempts. This may explain why pig farmers are so reluctant to raise pigs with intact tails. Respondents feel that a curly tail is not an important welfare indicator (55.6% of the respondents) while they firmly believe that there is no way to deal with tail-biting without tail-docking (77.8%). The vast majority of Greek pig farmers (80.8%) identified the length of the tail as a main risk factor for tail biting while 69.2% of them believe pigs don't feel pain when they are docked, a result that should be taken into account when information campaign strategies are developed. As European Commission is urging Member States to fully enforce the implementation of the relevant legislation, it is of utmost importance that farmers get the appropriate education and training in order to share this vision and move towards a new way of breeding pigs with intact tails.

Norwegian pork production – a 'FairyTail' without tail docking

K.H. Martinsen, S.L. Thingnes and T. Aasmundstad
Topigs Norsvin, Storhamargata 44, 2317, Norway; kristine.martinsen@norsvin.no

Tail biting is a serious animal welfare issue and an economic problem for farmers in most countries where pork production is practiced. Removing a significant part of the tail, known as tail docking, have been shown to reduce the risk of tail biting outbreaks. Still, tail docking is not a desirable solution to this problem, because it is a painful procedure for the pig and do not remove the cause of the problem. Tail docking has never been allowed in Norway and still the frequency of tail biting incidences has been low. However, in recent years, statistics from abattoirs and feedback from farmers suggests that tail biting is becoming an increasing problem in Norwegian pork production. Because tail docking is not allowed in Norway, it makes us an ideal country for research on tail biting behaviour. Norsvin has initiated an industry project (FairyTail) addressing tail biting behaviour in Norwegian pork production, with an aim of reducing the frequency of tail biting incidences in Norway and other European countries where Topigs Norsvin genetics are distributed. At the start of the project a survey regarding tail biting incidences in Norway was sent out to all our owners (1,300 herds) and 830 herds answered. The tail biting information is based on the farmers subjective perception (1(never)-4 (often)) of tail biting incidences in their herd. The answers from the survey have been analysed with analysis of variance (ANOVA), in a general linear model with fixed effects. LS-means where compared and preliminary results show there were geographically differences in tail biting incidences within Norway (P<0.0001). Further, no significant difference in tail biting incidences in herds using different terminal lines on TN70-sows was found. Farmers in grower-finisher herds reported that they had significantly more problems with tail biting incidences compared to farmers in piglet producing herds (P<0.0001). Ongoing research in the project focus on use of existing knowledge in advising farmers together with use of new technology, such as genotypes, complex modelling and precision phenotyping.

Identify possible indicators for the early detection of tail-biting in pigs

J. Ahlhorn[1], I. Traulsen[1] and S. Zeidler[2]
[1]*Georg-August-University Göttingen, Livestock Systems, Department of Animal Sciences, Albrecht Thaer Weg 3, 37075 Göttingen, Germany,* [2]*Georg-August-University Göttingen, Breeding Informatics; Department of Animal Sciences, Margarethe von Wrangell-Weg 7, 37075 Göttingen, Germany; juliane.ahlhorn@uni-goettingen.de*

The aim of this study was to investigate reliable indicators for an early detection of tail-biting in undocked pigs. In total, 630 pigs, weaned 27 days p.p., in 10 conventional rearing pens during 9 consecutive batches were examined. The group in the pens were mixed gender groups. During the weaner period of 6 weeks, tail-lesions (0 to 3), losses (0 to 4) and posture (0 to 5) were evaluated animal individual twice a week. In addition, the water and feed consumption per pen, the climatic conditions and the activity behaviour of the individual piglets by video recording were collected as candidates for indicators. In the first 2 weeks of the weaner period 9.4% score 1 (superficial), 8.4% score 2 (small) and 3.2% score 3 (large) of tail-lesions could be observed. Subsequently, the incidence of large lesions increased up to 53.4%. Depending on the batch the amount of tail-losses ranged from 20 to 50% at the end of rearing. The indicator tail-posture is closely connected to the tail-lesions. The average odds ratio for the occurrence of tail-lesions in relation to the tail-posture amounted for hanging tails 1.67 (StdErr 0.18) and for tails jammed between the hind legs the odds ratio is almost twice as high with 2.96 (StdErr 0.40). The odds ratio values for tails jammed between the hind legs show a great variation between the batches and range from 1.06 to 4.83. The average temperature ranged batch-wise between 20.7 °C (StdErr 0.27) and 24.7 °C (StdErr 0.11). In the batch with the highest average temperature the occurrence of tail-lesions was high as well (score 3=31.2%). Water intake changed between batches and pens and increased from on average 4.3-10.9 l at day 1 post weaning up to 24.6-28.4 l at day 40 post weaning. In conclusion results confirm individual tail-posture as an indicator to predict tail-biting. Additionally, climatic variations and changes in activity behaviour seem to be helpful indicators.

Influence of farrowing and rearing systems on tail lesions and losses of docked and undocked pigs

M. Gentz[1], C. Lambertz[2], M. Gauly[3] and I. Traulsen[1]
[1]*Georg-August-University, Livestock Systems, Albrecht-Thaer-Weg 3, 37075 Göttingen, Germany,* [2]*Research Institute of Organic Agriculture (FiBL), Kasseler Strasse 1a, 60486 Frankfurt am Main, Germany,* [3]*Free University of Bozen-Bolzano, Faculty of Science and Technology, Piazza Università 5, 39100 Bolzano, Italy; maria.gentz@uni-goettingen.de*

This study aims to investigate the effects of farrowing and rearing systems on tail lesions and losses of docked and undocked pigs. Therefore, 27-day old weaned pigs (50% docked, 50% undocked) from 3 farrowing systems: (1) conventional farrowing crate (FC); (2) free farrowing (FF); and (3) group housing of lactating sows (GH), were randomly located in two different rearing systems: (1) a conventional system, where they were regrouped and brought to a new barn (CONV); or (2) a wean-to-finish (WTF) system, where the piglets remained in their pens until the end of fattening (120 kg slaughter weight) with higher space allowance. Weekly, the status of tail lesions (Score 0-3) as well as tail losses (Score 0-4) was assessed individually (rearing n=2,951, fattening n=1,252). The incidence of tail lesions was generally higher in CONV compared to WTF (P<0.001). The highest frequency of tail lesions occurred in the CONV system in week 7 (undocked, 61.2%) while the WTF-piglets reached their peak at week 11 (undocked, 42.0%). Compared to undocked animals, docked animals were significantly less affected (P<0.001). They had the highest incidences in week 6 (CONV, 21.7%) and 15 (WTF, 23.6%). Regarding the first tail lesions, the highest hazard rates of new tail lesions were reached before week 8 in both systems for docked and undocked pigs. The highest risk level was reached by the undocked CONV (57.6%). The farrowing system had only a significant effect on tail losses. Pigs raised in FC system had less intact tails and higher incidences of tip losses (73.0 and 20.8%) than those from FF (76.0 and 18.6%) and GH (76.8 and 18.1%). Rearing system had no effect on the frequency of tail losses. In conclusion, the frequency of tail lesions is much higher in undocked compared to docked pigs. WTF rearing reduces the frequency of tail lesions and peak is reached later than in CONV. Early socializing or non-fixation of the sow influenced tail losses significantly but differences were rather low.

A pilot study on tail and ear lesions in suckling piglets

A. Gruempel[1], J. Krieter[2] and S. Dippel[1]
[1]Friedrich-Loeffler-Institut, Institute of Animal Welfare and Animal Husbandry, Dörnbergstraße 25/27, 29223 Celle, Germany, [2]Christian-Albrechts-University, Institute of Animal breeding and Husbandry, Hermann-Rodewald-Straße 6, 24098 Kiel, Germany; angelika.gruempel@fli.de

Tail and ear lesions in suckling piglets are a new topic of interest on piglet producing farms. Our aim was to find risk factors for tail and ear lesions in suckling piglets and investigate a possible relationship with tail and ear lesions in weaner pigs on the same farm. We assessed a median of 10 litters (range 9 to 16; 85 litters total) on each of eight German farms for presence or absence of lesions (injuries, swelling, haemorrhage, necrosis) on tails and ears. Data on management, performance and housing of sows and piglets in the farrowing unit were collected. On the same farm and day, we assessed in median 4 groups with in median 19 weaner pigs for tail and ear lesions. We analysed risk factors for lesions in suckling piglets using classification and regression tree analysis with litter level prevalence of tail or ear lesions as outcome. The relation between lesions in suckling piglets and weaner pigs on the same farm was assessed using Spearman rank correlation. Tail lesions were less prevalent if there was straw in the piglet creep area (6.6 vs 22.0% without bedding). If the creep area was bedded, the prevalence was lower if piglets were docked (3.2 vs.15.0%). Ear lesion prevalence was lower if piglets were <13 days old (31.0 vs 47%). If piglets were older than 13 days, ear lesions were less prevalent if piglet creep area was ≥0.7 m^2 (39.0 vs 50.0%). Tail lesions in weaner pigs negatively correlated with ear lesions (r_s=-0.71; P=0.05) but not with tail lesions in suckling piglets (r_s=-0.20; P=0.64). Ear lesions in weaner pigs did not correlate with tail or ear lesions in suckling piglets (tail: r_s=-0.41; P=0.32; ear: r_s=-0.57; P=0.14). Our results highlighted the importance of piglet creep area quality for tail and ear lesions in suckling pigs. More, especially longitudinal, data are needed for assessing the relationship between tail and ear lesions in suckling piglets and later in life.

Managing tail biting in undocked pigs on fully-slatted floors with different enrichment strategies

J.-Y. Chou[1,2,3], D.A. Sandercock[2], R.B. D'Eath[2] and K. O'Driscoll[3]
[1]University of Edinburgh, Royal (Dick) School of Veterinary Studies, Easter Bush, EH25 9RG, United Kingdom, [2]SRUC, Animal & Veterinary Sciences Research Group, Roslin Institute Building, Easter Bush, EH25 9RG, United Kingdom, [3]Teagasc, Pig Development Department, Animal & Grassland Research Centre, Moorepark, Fermoy, P61 P302, Ireland; jenyun.chou@ed.ac.uk

EU Council Directive (2008/120/EC) prohibits the routine practice of tail docking to control tail biting in pigs, yet most pigs in Europe are still tail-docked. This is due to a lack of effective solutions, especially in fully-slatted systems. This study attempted to use enrichment strategies compatible with fully-slatted floors to rear undocked pigs. Forty-eight pens (12 pigs/pen, n=576) of undocked pigs were followed from birth to slaughter in a 2×3 factorial design. Half the pigs were exposed to enriched environment pre-weaning (paper cup, rubber toy, hessian and bamboo). Three strategies were assigned post-weaning, based on the frequency of enrichment replenishment ('Reduced': on Monday/Wednesday/Friday; 'Medium': daily; 'Optimal': ad-lib). All pens received the same enrichment (8 items/ pen, including an elevated rack supplied with fresh-cut grass). Individual pig weights were obtained at D0, D49, D91 and D113 post-weaning. Direct behaviour observations were conducted twice weekly. Tail and ear lesion scores of individual pigs were recorded every other week. The average daily gain in the finishing stage was higher in Optimal than Reduced groups (1.10±0.01 kg v's 1.07±0.01 kg, P<0.05). Reduced pigs also performed more damaging behaviours combined (tail/ear biting, belly-nosing, mounting, other biting and aggressive behaviours) than Optimal and Medium pigs (P<0.01). Frequency of interaction with the enrichment did not differ between post-weaning treatments. Pre-weaning exposure to enrichment only contributed to a tendency towards higher interaction with grass (P=0.06). No difference in lesion scores was found between treatments. Although sporadic tail biting outbreaks occurred (n=14), they usually resolved within 2 weeks (13.3±4.5 days), and all but one tail-injured pig were successfully reintroduced back to their home pens after removals. Overall, tail biting was manageable, and the optimal enrichment treatment had a positive effect on pigs' growth and reduction in injurious behaviours.

Associations between tail directed behaviour, tail lesions and enrichment type in finisher pigs

L.A. Boyle[1], J.A. Calderon Diaz[1], A.J. Hanlon[2] and N. Van Staaveren[3]
[1]Teagasc, Pig Development Department, Moorepark, P61 C996 Fermoy, Co. Cork, Ireland, [2]UCD, School of Veterinary Medicine, Belfield, D04 V1W8 Dublin 4, Ireland, [3]University of Guelph, Dept. Animal Biosciences, 50 Stone Road East, Guelph, Ontario N1G 2W1, Canada; laura.boyle@teagasc.ie

EC Directive 2001/93 requires that all pigs have access to proper investigation and manipulation materials (i.e. environmental enrichment, EE). However, indestructible materials such as chains (C), hard plastic/rubber (P) or hard wood (W) are the most commonly employed EE. The aim of this study was to investigate associations between tail directed behaviour (TDB), tail lesions and EE type in finisher pigs on commercial farms. In 2015, 31 Irish farrow-to-finish farms were visited as part of larger cross-sectional pig welfare assessment study. Six randomly selected finisher pens were observed for 10 min and the number of pigs affected by mild (superficial bite marks) or severe (blood present) tail lesions was recorded and percentage (median [IQR]) of pigs affected calculated. All-occurrences of TDB were recorded by a second observer during 5 min observation. Associations between EE type and prevalence of tail lesions and TDB were determined using generalized linear mixed models. The most commonly used EE was C (41.4%), followed by P (37.9%) and W (20.7%). The percentage of pigs affected by tail lesions was 10.2% [8.4-12.8%], with most showing mild (8.8% [6.4-11.6%]) rather than severe (0.9% [0-2.2%]) tail lesions. EE type had no effect on overall tail lesion prevalence (P>0.05), however the prevalence of mild tail lesions was lower on farms with W compared to C, with P being intermediate (P<0.05). The amount of TDB observed was lower on farms that provided P or W compared to C (P<0.05). Slightly more destructible EE (i.e. P and W) was associated with less TDB and fewer mild tail lesions. However, in accordance with the literature none were associated with a reduced prevalence of severe tail lesions possibly due to the difficulty in controlling tail biting once it starts. The cross-sectional and associative nature of the study could also mean that the quality of the EE may have been a proxy for different qualities of pig management which may have influenced TDB rather than EE type.

Associations between straw provision, behaviour, ear and tail lesions in undocked pigs

T. Wallgren[1], N. Lundeheim[2] and S. Gunnarsson[1]
[1]Swedish University of Agricultural Sciences, Dept. of Animal Environment and Health, Box 234, 532 23 Skara, Sweden, [2]Swedish University of Agricultural Sciences, Dept. of Animal Breeding and Genetics, Box 7023, 75007 Uppsala, Sweden; torun.wallgren@slu.se

Rearing pigs with intact tails is a large animal welfare concerns in pig production, and the lack of straw as enrichment is a risk factor for tail biting. However, but the behaviour is not yet fully understood. The aim of this study was to investigate associations between pig behaviour and access to straw. A total of 48 pens of growers on 3 farms (n=557), and 64 pens of finishers in 4 farms (n=669), were studied during the grower (10-30 kg) and finishing phase (30-115 kg), respectively. Straw rations range were 4-84 g straw/pig/day. Before daily straw provision, residual straw (RS) in pens was assessed. Every second week pig (PIDB), straw (SDB) and pen directed behaviour (PEDB), as well as, ear (EL) and tail lesions (TL) were scored. Spearman rank correlation was used for statistical analysis. Preliminary results showed that in the beginning of the grower phase TL was negatively correlated to activity (A) (r=-0.38, P<0.01) and RS (r=-0.46, P=0.001), but positively correlated to EL (r=0.38, P<0.01). RS was negatively correlated to A (r=-0.42, P<0.01) and SDB (r=-0.32, P<0.05), PIDB (r=-0.48, P<0.001) and PEDB (r=-0.56, P=0.0001). At end of the grower phase, EL and TL were negatively correlated (r=-0.29, P<0.05). TL was also negatively correlated to A (r=-0.36, P<0.05) and PEDB (r=-0.33, P<0.05), but positively correlated to SDB (r=0.30, P<0.05). In the beginning of the finishing period TL was positively correlated to PEDB (r=0.4, P<0.001) and PIDB (r=0.27, P<0.05), but negatively correlated to SDB (r=-0.32, P<0.05). EL were positively correlated to intact tails (IT) (r=0.27, P<0.05) and PDB (r=0.4, P<0.001), and they were negatively correlated to RS (r=-0.32, P<0.05). In the end of the finisher period, EL was positively correlated to IT (r=0.31, P=0.0124) and PDB (r=0.37, P<0.01) but EL was negatively correlated to RS (r=-0.28, P<0.05). Furthermore, we found that the associations between behaviour, lesions and straw fluctuated over age. The results indicate that residual straw assessment may be a useful in risk analysis of damaging behaviours in growing and finishing pigs.

A longitudinal study of damaging behaviours and their associated lesions in grow-finisher pigs

L.A. Boyle[1], J.C. Pessoa[1,2,3], J. Yun-Chou[1] and J.A. Calderon Diaz[1]

[1]Teagasc, Pig Development Dept., Animal and Grassland Research and Innovation Centre, Moorepark, Fermoy, Co. Cork, P61 C996, Ireland, [2]UCD, Herd Health and Animal Husbandry, School of Vet. Med., D04 V1W8 Belfield, Dublin 4, Ireland, [3]3M3-BIORES, Measure, Model & Manage Bioresponses, KU Leuven, 30, 3001 Leuven, Belgium; laura.boyle@teagasc.ie

Damaging behaviours (DB) in pigs affects welfare and profitability. The aim was to determine the prevalence of DB and the associated lesions in pigs during the grow-finisher period. Weaner (W) pigs (n=1,573, 25±5.3 kg,) were housed in slatted pens (c. 36 pigs/pen) and followed from arrival (T0) for 13 wks until slaughter (114±15.4 kg) in a wean-to-finish farm. Body weight (BW) was recorded at transfer from W to grower (G) and from G to finisher (F) stages when pigs were also inspected for ear (EL), tail (TL) and flank (FL) lesions (scored 0-4). All occurrences of behaviour directed towards the tail (TDB), ears (EDB), belly (BDB) and flanks (FDB) were counted (1×5 min observation/pen/wk). Frequencies of DB were averaged per production stage. TL were re-classified as score 0, 1 and ≥2 and EL as score 0, 1, 2 and ≥3. The prevalence of lesions/score/pen was calculated. All data were analysed by mixed models in SAS including stage as a fixed effect and BW, no. pigs/pen, sex ratio and lesions on arrival (lesions only) as covariates. TDB changed over time (W=2.4±0.23/5 min; G=1.4±0.21; F=2.0±0.21; P<0.01). EDB was highest in W and decreased slightly through G to F (W=6.8±0.36/5 min; G=5.4±0.33; F=5.7±0.42; P<0.01). BDB (W=0.53±0.128/5 min; G=0.91±0.117; F=1.37±0.118) and FDB (W=0.06±0.267/5 min; G=1.04±0.243; F=4.07±0.245) increased significantly over time (P<0.001). BDB (P<0.05), FDB (P=0.06) and EDB (P<0.001) were positively associated with the no. pigs/pen. No FL were observed. There were no stage effects on TL (P>0.05]. Proportion of pigs with Score 0 EL reduced between G (83.6%) and F (53.9%) (s.e. 4.99; P<0.01). High levels of DB were performed throughout the production stages with EDB being the most frequently performed DB followed by TDB. Correspondingly almost 50% of pigs were affected by EL on transfer to the F stage which raises concerns for pig welfare. Reducing the numbers of pigs in the pen could help address the problem.

Raising awareness on the importance of animal genetic resources

G. Leroy

FAO, Animal Production and Health Division, Viale delle Terme di Caracalla, 00153 Rome, Italy; gregoire.leroy@fao.org

The genetic diversity of livestock breed is continually decreasing, threatening long-term food security. Building awareness of animal genetic resources (AnGR) is considered as a key factor for the sustainable management and conservation of those resources, and constitutes a strategic priority of the Global Plan of Action for AnGR. Yet in most countries, including many in Europe, public and private actors rarely realize the important roles and values of livestock breeds, even though support from these actors would be necessary to mobilize resources for enhancing sustainable use, development and conservation. Awareness-raising should not only target multiple actors (e.g. consumers, distributors, tourist offices, media and policymakers), it should also cover various initiatives, such as promoting the development of conservation policies and programmes, value addition to and labelling of breed-related products, recognition and communication on the ecosystem services associated with local breeds and their production systems. The scope of these initiatives should range from local to international. The aim of this session is to provide insight and promote discussion on potential solutions and pitfalls that allow or prevent enhanced communication around AnGR. A diversity of points of view and situations from within and outside Europe will be presented and discussed.

Genomic inbreeding over time for Swedish Red and Swedish Holstein-Friesian cattle

A.M. Johansson[1], E. Strandberg[1], H. Stålhammar[2] and S. Eriksson[1]
[1]Swedish University of Agricultural Sciences, Dept. of Animal Breeding and Genetics, Box 7023, 75007 Uppsala, Sweden, [2]VikingGenetics Sweden AB, Örnsro Box 64, 54321 Skara, Sweden; anna.johansson@slu.se

The aim of this project was to study changes over time in homozygosity and inbreeding in Swedish Red (SR) and Swedish Holstein-Friesian (SH). New techniques and breeding regimes have been implemented for Swedish dairy cattle over the years, such as BLUP evaluation in 1984, joint Nordic evaluation in 2005, and genomic selection from around 2010. We studied if such developments could be detected as changes in homozygosity and inbreeding in these breeds. Genotype data from 50k SNP chip for SR and SH AI-bulls used in Sweden during the last decades were analysed. Early in this time period the transfer from old Swedish Friesian to Holstein was still ongoing. In the SH breed the level of homozygosity and homozygosity excess increased steadily over time since the 1980s. In the SR breed the homozygosity decreased in the 1980-90s, but then increased during the last 15 years, back to the same level as in the 1980s. A similar pattern was seen for inbreeding based on runs of homozygosity: FROH increased steadily for SH from about 2% in the 1980s to about 4% in the latest years, whereas for SR the level was constant at 3% but increased to 3.5-4% in the latest period. The fixation of alleles increased in both breeds in recent years, but the increase was larger in SH. This change might be related to the introduction of genomic selection. The average length of ROH per individual was almost twice as high in the SH as in the SR breed.

Evolution of ROH's distribution along the genome over a decade of genomic selection in dairy cattle

K. Paul[1], A.-C. Doublet[1,2], D. Laloë[1], P. Croiseau[1] and G. Restoux[1]
[1]INRA, AgroParisTech, Université Paris-Saclay, GABI, Allée de Vilvert, 78350 Jouy-en-Josas, France, [2]ALLICE, 149 rue de Bercy, 75012 Paris, France; gwendal.restoux@agroparistech.fr

Previous works showed that genomic selection led to an increase in annual genetic gain in major French dairy cattle breeds but, in some cases, the increase in genetic gain was linked with an accelerated loss of genetic diversity. This could have deleterious consequences on the health and performance of the breed with detrimental economic effects. This accelerated loss of diversity was assessed using Runs of Homozygosity (ROH), which represent contiguous homozygous regions of the genome assumed to result from inbreeding. On the other hand, genomic selection selects targeted regions of the genome much more precisely than genetic selection based on progeny testing only. Therefore the structure of genetic diversity should have evolved differently along the genome. Our objectives were to understand where ROH are located and how they are structured along the genome, to understand what it can tell us about genome structure evolution and gene locations in a selection perspective. To answer these questions, we studied the evolution over time of the positions, lengths and distributions of ROH with respect to the traits under selection. Our dataset consisted of 8,019 sires of three French dairy cattle breeds, Montbéliarde, Normande and Holstein, born between 2005 and 2015 for which we characterized genetic diversity all along the genome. We observed that ROH were increasingly shared among individuals over the years within a breed. In addition, the most shared ROH tended to be older than the least shared ROH. The position of a ROH within the genome could influence their evolution over time, because of differences in recombination/selection rates along the genome. This study shows the opportunity to better understand the mechanisms of the loss of genetic diversity in selection programs and thus to develop efficient and sustainable breeding schemes limiting the loss of genetic diversity and inbreeding depression. Eventually, this new information about genes potentially linked to traits of economic interest may also be used in future genomic selection programs.

Copy number variants identified on the new bovine reference genome partly match known variants

A.M. Butty[1], T.C.S. Chud[1], F. Miglior[1], F.S. Schenkel[1], A. Kommadath[2,3], K. Krivushin[3], J.R. Grant[3], I.M. Häfliger[4], C. Drögemüller[4], P. Stothard[3] and C.F. Baes[1,4]
[1]University of Guelph, Guelph, ON, Canada, [2]Lacombe Research and Development Centre, Lacombe, AB, Canada, [3]University of Alberta, Edmonton, AB, Canada, [4]University of Bern, Bern, Switzerland; buttya@uoguelph.ca

Copy number variants (CNV) form the most common class of structural variants in the animal genome and can be identified as two types of events: copy number loss or copy number gain. Several studies have shown the importance of CNV for economically relevant traits, however, CNV are difficult to identify with high confidence. The trade-off between a high rate of discovery and a low false discovery rate makes CNV identification challenging. Previous studies have shown that low false positive identification rates can be achieved when CNV are identified from two or more sources of information, such as genotype array and whole-genome sequence (WGS) data. The objectives of this study were to identify CNV from 67 Holstein animals with both, array and WGS data, to compare them with previously described bovine CNV, and to describe the possible function of high confidence variants identified. CNV were identified with the software program PennCNV and CNVnator using the signal intensities of updated genotype array variant positions, and on WGS aligned to the latest bovine reference genome (ARS-UCD1.2). In total, 161 CNV were found between methods that had a reciprocal overlap of at least 50%. These CNV were collated and filtered to 36 CNV regions present in more than 5% of our samples. Fifteen regions were copy number losses, seven were copy number gains and 14 regions were found both as losses and gains. Of those CNV regions, four overlapped with previously described variants from literature after remapping breakpoints to ARS-UCD1.2. Of these common variants, one variant on chromosome 16 affects a protein-coding gene of the PRAME family. This family of genes is known to influence immunity and reproduction traits. By developing robust methods to identify structural variants from array data with confidence, we hope to incorporate CNV in addition to single nucleotide polymorphisms into routine genomic selection methods for further genomic improvement of dairy cattle.

High resolution copy number variation analysis using two cattle genome assemblies

Y.L. Lee[1], M. Bosse[1], R.F. Veerkamp[1], E. Mullaart[2], M.A.M. Groenen[1] and A. Bouwman[1]
[1]Wageningen University & Reserch, Animal Breeding and Genomics, P.O. Box 338, 6700 AH Wageningen, the Netherlands, [2]CRV B.V., P.O. Box 454, 6800 AL Arnhem, the Netherlands; younglim.lee@wur.nl

Genetic variations exist in various forms in genomes. Although SNPs have been the choice of variants in numerous studies, due to their abundance, there is a growing body of evidence that copy number variations (CNVs – gain or loss of DNA segments) can have functional impacts. One of the pitfalls in CNV research is that the reference genome quality can affect CNV calling. Here, we performed high-resolution CNV analysis using BovineHD SNP array data (n=476), using two different reference genome assemblies, UMD3.1 and ARS-UCD1.2. One of the main differences between these two assemblies was that the former was built on short-read, whereas the latter used long-read sequencing technologies. We discovered less CNVs using ARS-UCD1.2 (14,272; 31.6 CNVs/animal), compared to UMD3.1 (24,264; 54.3 CNVs/animal). These CNVs spanned ~82 Mb and ~69 Mb, accounting for 3.3 and 2.8% of bovine autosomes in UMD3.1 and ARS-UCD1.2, respectively. The CNV distribution was nonrandom: CNV coverage per chromosome was 0.9-9.6% in UMD3.1 and 0.6-4.9% in ARS-UCD1.2. The most striking difference was seen in BTA12, where 8.7 MB (9.6%) was covered by CNVs based on UMD3.1 while the coverage was only 4.2 MB (4.9%) based on ARS-UCD1.2. This large difference was mainly due to changes of BovineHD SNPs on BTA12:70-77 MB region, where SNPs were sparsely distributed in UMD3.1. In ARS-UCD1.2, many SNPs in this region were replaced to unmapped contigs. This resulted in removing an extremely long CNV (CNV region 1057; 4 MB length) that was detected in this region based on UMD3.1. We corroborate that reduction of CNVs, both in terms of number and length can be a reflection of a sharp decrease in falsely detected CNVs, due to improved quality of ARS-UCD1.2 genome build. Hence, CNVs detected based on UMD3.1 genome build require re-evaluation for better characterisation and understanding of CNVs in cattle genomes.

Improving mating plans at herd level using genomic information

M. Berodier[1,2], P. Berg[3], T. Meuwissen[3], M. Brochard[1,4] and V. Ducrocq[2]
[1]MO3, 01250, Ceyzériat, France, [2]UMR GADI, AgroParisTech, INRA, Université Paris Saclay, 78350, Jouy-en-Josas France, [3]Norwegian University of life Sciences, P.O. Box 5003, 1433 Ås, Norway, [4]Umotest, 01250, Ceyzériat, France; marie.berodier@inra.fr

Since 2011, 189,417 Montbéliarde females have been genotyped at the farmers' request. Farmers are interested in getting GEBV of their young females for selection and for mating purposes. However, more genomic information than only GEBV could be use at mating, such as genomic inbreeding and carrier status for recessive haplotypes or major genes. This study aims at comparing five different on-herd mating strategies for their capacity to maximize expected genetic gain while limiting the expected average progeny inbreeding and the probability to conceive a foetus homozygous for any of three genetic defects segregating in Montbéliarde breed (MH1, MH2 and MTCP). In September 2018, 160 Montbéliarde herds were selected for which more than 80% of the females to be mated were genotyped and where the actual reproducing strategy was known, i.e. the farmer's choice of females to be inseminated with sexed, conventional or beef breed semen. The females of each herd were then mated to 54 purebred top bulls using random mating method (RAND) or two optimized mating methods (LP = linear programming; SEQ = sequential) with or without use of genomic information (PEDI vs GENO). All mating methods were constrained for required type of semen use (sexed or conventional) and for ease at calving for heifers inseminated with conventional semen. A bull could be mated to a maximum of 10% of the females of a herd. In total, 9,655 matings were planned per method. LP_GENO was the best method since it provided the highest average expected genetic gain, the lowest expected progeny (genomic) inbreeding and the lowest risk of conception of a foetus homozygous for any genetic defect. LP_GENO provided on average +12% of economic gain compared to SEQ_PEDI and + 40% of economic gain compared to RAND for the farmer, showing the high value of planned matings compared to random mating. In addition, expected genomic inbreeding was >1% higher when using PEDI rather than GENO coancestry. These results are very promising for an on-herd application since the mate allocation step with an R software was fast (<10 seconds per herd).

Whole genome sequence GWAS reveals muscularity in beef cattle differs across five cattle breeds

J.L. Doyle[1,2], D.P. Berry[2], R.F. Veerkamp[3], T.R. Carthy[2], S.W. Walsh[1] and D.C. Purfield[2]
[1]Waterford Institute of Technology, Cork Road, Co. Waterford, Ireland, [2]Teagasc, Teagasc, Moorepark, Fermoy, Co. Cork, Ireland, [3]Wageningen University & Research Animal Breeding and Genomics, P.O. Box 338, 6700 AH Wageningen, rgw Netherlands; jennifer.doyle@teagasc.ie

While two beef cattle carcasses may have the same weight, they may be morphologically very different; e.g. one carcass may have a better developed loin while the other may have a deeper chest. These morphological differences may contribute to a difference in the overall value of the carcass as primal cuts vary in price. Such differences may hinder successful genomic association studies of carcass weight, especially if the different traits are under different genetic control. Linear type traits, which reflect the muscular characteristics of an animal, have previously been genetically correlated with both carcass weight and conformation and could therefore provide an insight into this hypothesis. Differences in conformation between breeds are also an important differentiator for carcass value. The genetic basis of the differences among breeds is, however, poorly understood. Therefore, the objective of this study was to identify genomic regions associated with 5 muscular linear type traits and to determine if the detected regions are common across 5 breeds. Data on development of hind quarter, development of loin, development of inner thigh, width of thigh and width of withers were available on 1,444 Angus, 6,433 Charolais, 1,129 Hereford, 8,745 Limousin and 1,698 Simmental genotyped beef cattle. Analyses were performed on imputed whole genome sequence data in each of the 5 breeds separately using linear mixed models. Many quantitative trait loci (QTL) were identified for each trait and in each breed and these QTLs were located across the entire genome. Moreover, the only overlap in detected regions among the breeds and traits was a large-effect pleiotropic QTL on BTA2 that contained the *MSTN* gene that was associated with all 5 traits in both Charolais and Limousin. Thus, the results suggest that many QTL are associated with muscularity and these tend to be not only trait-specific but also breed-specific. Knowledge of these differences in genetic architecture among breeds is important to the development of genomic predictions across breeds.

Common genomic regions underlie height in humans and stature in cattle

B. Raymond[1], L. Yengo[2], R. Costillo[2,3], C. Schrooten[4], A.C. Bouwman[1], R.F. Veerkamp[1], B. Hayes[3] and P.M. Visscher[2,5]
[1]Wageningen University & Research Animal Breeding and Genomics, P.O. Box 338, 6700 AH Wageningen, the Netherlands, [2]Institute of Molecular Biosciences, University of Queensland, St. Lucia, 4072, Australia, [3]Queensland Alliance for Agriculture and Food Innovation, University of Queensland, St. Lucia, 4072, Australia, [4]CRV, BV, P.O. Box 454, 6800 AL Arnhem, the Netherlands, [5]Queensland Brain Institute, University of Queensland, St. Lucia, 4072, Australia; biaty.raymond@wur.nl

In this study, we tested the hypothesis that genes containing genetic variants affecting human height also control stature in cattle. For this, we used summary level results of a meta-GWAS for stature in cattle (n=58,265). We also used information on 610 human genes containing eQTLs and 3,920 lead SNPs from a meta-GWAS for height in ~700,000 individuals of European ancestry. First (approach 1), for the 610 human genes with eQTLs, we identified 370 ortholog cattle genes and the number of GWAS hits for stature (SNPs with $P<5\times10^{-8}$) within and around (100 kb either side) the 370 genes. Secondly (approach 2), we mapped the 3,920 lead SNPs from the human study to 2,174 human genes, identified 1,184 cattle ortholog genes and the number of GWAS hits for stature in and around the 1,184 genes. For use in control test, we identified 13,865 cattle genes (after QC) with known orthologs in humans. As control test, we sampled cattle genes (370 for approach1 and 1,184 for approach 2) from the pool of 13,865 cattle genes, with the genes stratified on gene-length, number of SNPs and LD score. In each control test (1000 in total), we compared the number of GWAS hits in and around candidate genes to those observed in and around control genes. Our results show that the number of GWAS hits for stature observed in the 370 candidate genes in approach 1 (1,354) is not more than can be observed by chance, as proportion of samples with significant SNPs equal to or greater than observed in the candidate genes (P) was 0.8. On the contrary, the number of GWAS hits observed in the 1,184 candidate genes in approach 2 (6,419) was more than can be observed by chance (P=0.04). This indicates that, potentially, prior information from human GWAS studies on height can be beneficial in the genetic analysis of stature in cattle.

Session 12   Theatre 8

A fast method to fit the mean of unselected base animals in single-step SNP-BLUP

T. Tribout[1], D. Boichard[1], V. Ducrocq[1] and J. Vandenplas[2]
[1]GABI, INRA, AgroParisTech, Université Paris-Saclay, 78350 Jouy-en-Josas, France, [2]Wageningen University & Research Animal Breeding and Genomics, P.O. Box 338, 6700 AH Wageningen, the Netherlands; thierry.tribout@inra.fr

Single-step GBLUP (SSGBLUP) is the reference method for genomic evaluation. To bypass the inversion of the genomic relationship matrix when many animals are genotyped, equivalent formulations of SSGBLUP have been proposed, predicting the effects of markers rather than breeding values. In such models, missing genotypes are imputed linearly, which requires the centring of the observed genotypes using generally unknown base allele frequencies. Hsu *et al.* proposed to solve this by fitting a covariable to model the mean of unselected base animals. This requires the computation of a covariate vector J with entries equal to -1 for genotyped animals (*ga*) and to $J_n=(A^{nn})^{-1}A^{ng}1$ for ungenotyped animals (*na*), where A^{nn} and A^{ng} are the $na \times na$ and the $na \times ga$ submatrices of the inverse of the pedigree-based relationship matrix. A^{nn} is sparse, so the computations involving $(A^{nn})^{-1}$ can be based on its sparse Cholesky factor L. In dairy cattle populations, the factorization of A^{nn} is fast and L is sparse. However, the factorization is more expensive and L is much denser in beef cattle populations where the number of ungenotyped bulls is large. We propose a simple method to compute the J_n vector at low cost. It requires the following steps: (1) Divide the ungenotyped population into ancestors of *ga* (ANC) and other animals (OTH); (2) Compute L of the A_{ANC}^{nn} submatrix built considering only the *ga* and their ancestors; (3) Compute $J_{ANC}=(A_{ANC}^{nn})^{-1}A_{ANC}^{ng}1$ for ANC animals; (4) for any animal i of OTH, from oldest to youngest, compute $J_{OTH}(i) = 0.5\times(\gamma_{sire}+\gamma_{dam})$, where γ_{parent} is either 0, -1, J_{ANC}(parent) or J_{OTH}(parent) if animal i's parent (sire or dam) is respectively either unknown, genotyped, belongs to the ANC or to the OTH sub-population. This method was tested on a French Charolais breed dataset containing 156,447 ANC, 10,333,930 OTH and 22,449 *ga*. The Cholesky factorization of A^{nn} and A_{ANC}^{nn} using MKL PARDISO and 8 CPUs required 5 h 07 and 33 sec, respectively. The subsequent computation of *Jn* as $(A^{nn})^{-1}A^{ng}1$ or with the proposed method required 27 sec and 1 sec, respectively.

Single-step evaluation for calving traits with 1.5 million genotypes: APY and ssGTBLUP approaches

I. Strandén[1], R. Evans[2] and E.A. Mäntysaari[1]
[1]Natural Resources Institute Finland (Luke), Jokioinen, 31600 Jokioinen, Finland, [2]Irish Cattle Breeding Federation, Highfield House, Newcestown Road, Bandon, Cork, Ireland; ismo.stranden@luke.fi

Single-step genomic evaluation (ssGBLUP) needs the genomic relationship matrix G or its inverse in the mixed model equations (MME). Direct inversion of the G matrix is unfeasible when the number of genotyped animals (n) is large. We investigated ssGTBLUP and APY based ssGBLUP approaches in iterative solving of the MME for large n. In ssGTBLUP, G had form $G=ZZ'+\varepsilon I$ where Z is the (centred and scaled) marker matrix with size n × m (numbers of genotyped and markers), and $\varepsilon=0.01$. Thus, $G^{-1}= I/\varepsilon - TT'$ where $T'=L^{-1}Z'/\varepsilon$ is an m × n matrix and L is the Cholesky decomposition of $(Z'Z/\varepsilon +I)$. Breeding values were estimated for a 6-trait calving performance evaluation for Irish dairy and beef cattle. The model had breed fractions as fixed covariates, and random direct and maternal genetic effects. Pedigree had 9.54 million (M) animals of which n=1.50M were genotyped with 50,240 SNPs. For ssGTBLUP we took the eigenvalues and vectors explaining 98% of variation in G and used a corresponding random core of size 33,636 in APY. Both the preprocessing and the solver kept large matrices in RAM memory to allow efficient parallel computing by 10 cores. Preprocessing was performed using memory efficiency in mind because the computer had 768 GB RAM. Peak RAM need was 592 GB in APY and 371 GB in ssGTBLUP where for ssGTBLUP the computation of $L^{-1}Z'$ was performed in 3 batches of Z matrix to save RAM but increased computing time. Preprocessing time was mainly different in time to invert the G or T matrix: 9.0 h for APY and 3.2 h for ssGTBLUP. In addition, 2.7 h was needed for the eigenvalue analysis. The solver computing times per iteration were 1.34 m for APY and 1.46 m for ssGTBLUP. Number of iterations until convergence were 2,353 for APY and 2,393 for ssGTBLUP. Total solver time was 52.4 h in APY and 58.2 h in ssGTBLUP, and both had peak RAM use of 387 GB when the $(G_{APY})^{-1}/T$ matrices were in double precision. Correlations of estimated breeding values were at least 97.8% for genotyped animals for all traits. In conclusion, large G matrix led to increase in the preprocessing times to save RAM.

Single-step evaluation for calving traits with 1.5 million genotypes: SNP-based approaches

J. Vandenplas[1], R.F. Veerkamp[1], R. Evans[2], M.P.L. Calus[1] and J. Ten Napel[1]
[1]Wageningen University and Research, P.O. Box 338, 6700 AH Wageningen, the Netherlands, [2]Irish Cattle Breeding Federation, Highfield House, Newcestown Road, Bandon, Cork, Ireland; jeremie.vandenplas@wur.nl

The single-step single nucleotide polymorphism best linear unbiased prediction (ssSNPBLUP) method and the single-step genomic BLUP (ssGBLUP), simultaneously analyse phenotypic, pedigree, and genomic information of genotyped and non-genotyped animals. In contrast to ssGBLUP, no genomic relationship matrix is needed for ssSNPBLUP. Instead, SNP effects are fitted explicitly as random effects. In this study we investigated the performance of two ssSNPBLUP methods using data of a 6-trait calving-difficulty evaluation for Irish cattle. For genotyped animals, the first ssSNPBLUP model explicitly estimates residual polygenic effects, while the second model directly estimates aggregate genomic breeding values. The datasets included over 3.5 million records of calving traits, 10.3 million animals in pedigree of which over 1.5 million were genotyped for 50k SNPs. The model included breed fractions as fixed covariates, as well as random direct and maternal genetic effects. Analyses were performed with a multi-threaded Fortran program using 10 CPUs. Mixed-model equations of the two investigated ssSNPBLUP models were solved with the commonly-used preconditioned conjugate gradient (PCG) method, and with a PCG method using a second-level diagonal preconditioner. For time and memory efficiency, all genotypes were kept in memory with a compressed format. Across the different combinations of ssSNPBLUP and PCG methods, maximum amount of RAM memory needed was less than 120 GB, and average time per iteration varied between 66 and 90 seconds. The fastest combination of ssSNPBLUP and PCG methods was the ssSNPBLUP method that directly estimates aggregate genomic breeding values and that was solved with a PCG method using a second-level preconditioner. This combination reached convergence after 2,570 iterations and total time was 64.9 h. In conclusion, it seems possible to perform routine SNP-based single-step evaluations that involve multiple genetic effects and a very large number of genotypes, within manageable amounts of memory and time.

Towards a climate smart European livestock farming – background

J.L. Peyraud

Animal Task Force, 149 rue de Bercy, 75595 Paris, France, Metropolitan; jean-louis.peyraud@inra.fr

International climate agreements, like COP21, have initiated a new era for climate policies. The livestock sector has potential to contributing to mitigating climate impact. In the EU, the sector accounts for 40% of global agricultural emissions or 7% of total emissions, producing about 2,400 Mt of CO_2 equivalent annually, but also methane and NO_2. Enteric emissions, emissions from manure and land use change (LUC) due to deforestation for feed production are among the principal contributors. Thanks to significant efforts, the livestock sector in Europe is starting to contribute to mitigation of climate impacts (SDG 13). R&I, new technologies and relevant incentives to implementation of best practices may enable the livestock sector to come close to CO_2 neutrality for monogastrics and to achieve a 40% reduction for ruminants. Ways to proceed include e.g. implementing mitigation options and enhancing carbon storage under grasslands soils. Climate targets should be integrated into a holistic approach to avoid trade-offs and foster a sustainable use of resources, preservation of biodiversity and improvement of soil quality. Future solutions need to optimise multiple factors through a systems approach, which takes into account the interplay between the system components. The session would like to engage discussion with farmers, industries, scientists, policy makers and with the society. Most important findings will be discussed with a panel. The outcomes of the session will be discussed with a large panel of European stakeholders during the ATF seminar, in Brussels, on 6th Nov. 2019.

Antioxidants in milk and cheese: an insight along the dairy chain stakeholders

G. Niero and M. Cassandro

University of Padova, Department of Agronomy, Food, Natural Resources, Animals and Environment (DAFNAE), Viale dell'Università 16, 35020 Legnaro (PD), Italy; giovanni.niero@phd.unipd.it

Milk and dairy products bioactive compounds take on a key role on animal welfare, technological properties and human health at farm, industry and consumer levels, respectively. These molecules are of great interest for the stakeholders of the entire dairy chain, according a farm to fork insight. On this background, the aim of the presentation is to describe antioxidant molecules in milk and products of different dairy species. Vitamins A, C, E, peptides derived from casein hydrolization and digestion, and whey protein with particular regard to lactoferrin, will be described and discussed in relation to their antioxidant properties. Milk total antioxidant activity will be also presented as a descriptor considering the sum of antioxidant capacity of the aforementioned compounds. Reference analytical methods, including high pressure liquid chromatography (HPLC), gas chromatography (GC), mass spectrometry (MS/MS) and nuclear magnetic resonance (NMR) will be reviewed as for their capability to detect and quantify antioxidants in milk and dairy products. The effectiveness of alternative and fast techniques including spectrophotometric assays, enzyme linked immune sorbent assays (ELISA) and mid-infrared spectroscopy (MIRS) to determine antioxidant compounds will be also assessed. Animal species and breed, days in milk and parity, farming systems and feeding strategies, as well as season and environmental conditions will be examined as sources of variation of milk antioxidants content at farm level. The influence of technological treatments, including skimming and pasteurisation, cheese manufacturing and ripening will be considered for the industry. Finally the main effects of milk and cheese antioxidants will be reviewed to highlight the potential positive role milk and dairy products may have on human health, thus enhancing their added value.

Consumer knowledge and perceptions on milk fat in Denmark, United States and United Kingdom

E. Vargas-Bello-Pérez[1], I. Faber[2], J.S. Osorio[3], S. Stergiadis[4] and F.J.A. Perez-Cueto[2]
[1]University of Copenhagen, Department of Veterinary and Animal Sciences, Grønnegårdsvej 3, 1870 Frederiksberg C, Denmark, [2]University of Copenhagen, Department of Food Science, Rolighedsvej 26, 1958 Frederiksberg C, Denmark, [3]South Dakota State University, Dairy and Food Science Department, 1111 College Ave, Brookings 57007, USA, [4]University of Reading, School of Agriculture, Policy and Development, Earley Gate, Reading RG6 6AR, United Kingdom; evargasb@sund.ku.dk*

This study determined the influence of consumer sociodemographics and country of origin [Denmark (DK), United States (US) and United Kingdom (UK)] on their knowledge and perceptions on milk fat. Participants were recruited through social media. Data was collected online (SurveyXact) from Dec 2018 to Feb 2019, in DK, US and UK. Logistic regressions were used to identify the effect of country of origin and sociodemographic characteristics on consumption, knowledge and perceptions towards milk fat. 501 questionnaires from DK (n=194), UK (n=195) and US (n=112) were answered. Whole milk was more consumed in UK (OR 4.51; 95% CI 2.75-7.56) and US (OR, 4.20; 95% CI 2.39-7.51) than in DK, and this was associated with age (P=0.042) and sex (P<0.001). Respondents in UK and US consumed less skimmed milk than DK (OR, 0.15; 95% CI 0.09-0.23 and OR, 0.10; 95% CI 0.05-0.19 respectively). Saturated fat was a well-known type of milk fat among 48% of the respondents. 64% of DK respondents were concerned about milk fat, while UK (OR, 0.34; 95% CI 0.22-0.52) and US (OR, 0.24; 95% CI 0.14-0.39) were less concerned. Knowledge about saturated fat was associated with age (P=0.012), employment status (P=0.012) and agricultural background (P=0.001). Respondents from UK and US were more likely to perceive milk's fat as a healthy nutrient than those in DK (OR, 1.80; 95% CI 1.18-2.77 and OR, 2.28; 95% CI 1.40-3.74 respectively). Additionally, this perception was associated with employment status (P=0.037). 46% of the participants would be willing to buy milk with a healthier fat content. Results suggest that consumers in DK are different in their purchase behaviour and perception towards milk fat, while consumers in UK and US share common characteristics. These data could be useful for future consumer-sensitive dairy beverage innovation and communication strategies.

Mineral composition of retail goat milk in the UK

S. Stergiadis[1] and M.R.F. Lee[2,3]
[1]University of Reading, School of Agriculture, Policy and Development, P.O. Box 237, Earley Gate, Reading, RG6 6AR, United Kingdom, [2]University of Bristol, Bristol Veterinary School, Langford, Somerset, BS40 5DU, United Kingdom, [3]Rothamsted Research, North Wyke, Devon, EX20 2SB, United Kingdom; s.stergiadis@reading.ac.uk*

The UK dairy goat industry has grown in recent years, reaching a market value of approximately £70 million. In total, 100,000 goats produce 33 million litres of milk per year, of which 52% is consumed as liquid. Although some consumers perceive goat milk as more nutritious and an alternative to cow milk in cases of intolerances or allergies, the nutrient composition of UK goat milk has not been assessed. Since milk is among the main sources of many minerals in human diets, this study investigated the mineral composition of retail goat and cow milk throughout the year. Milk samples (n=84) from 3 goat and 4 cow milk brands were sampled monthly, over 12 months. Mineral profile data were analysed by ANOVA linear mixed effects models, using species, month and their interaction as fixed factors and milk ID as random factor. Goat milk contained more B (+33.0%), Cu (+50.7%), I (+46.1%), K (25.0%), Mg (+21.4%), Mn (+59.2%), P (8.0%), and less Ca (-5.5%), Na (-6.0%), S (-9.1%) and Zn (-15.4%) than cow milk. Mineral composition of milk fluctuates according to stage of lactation, milking methods, environment, season, diet, feeding system, breed and species. At least 60% of UK dairy goats are indoors throughout the year, on diets rich in conserved forage and concentrates; while UK dairy cows are typically fed partly at pasture in the grazing season. This may explain key mineral differences which are driven by soil properties at pasture e.g. high Ca in chalky soils cause poor availability of Mg and application of K fertilisers inhibits uptake of Mg in grass. Differences between species were more pronounced in the grazing period (when cow and goat diets may broadly differ) than in winter (when cows and goats are indoors), thus further highlighting the potential effect of animal diet. Consuming goat milk may increase intakes of B, Cu, I, K, Mg, Mn and P, and reduce intakes of Ca, Na, S, and Zn; but potential effect of these differences on consumer health is unknown.

Milk fatty acid profile of dairy cows is affected by forage species, parity, and milking time

S. Lashkari, M. Johansen, M.R. Weisbjerg and S.K. Jensen
Aarhus University, AU Foulum, Animal science, Blichers Alle 20, 8830 Tjele, Denmark; s.lashkari@hotmail.com

The aim was to study the effect of different forage species on milk fatty acid (FA) profile in dairy cows. Swards with perennial ryegrass (early maturity stage (EPR) and late maturity stage (LPR)), festulolium (FEST), tall fescue (TF), red clover (RC), and white clover (WC) were harvested in the primary growth, wilted, and ensiled without additives. Twelve primiparous (DIM: 59±12, mean ± SD) and 24 multiparous (DIM: 95±64) Danish Holstein cows in an incomplete Latin square design were fed *ad libitum* with total mixed rations with a high forage proportion (70% on DM basis). The TMR differed only in forage source, which was either one of the 6 pure silages or a mixture of LPR silage with either RC or WC silage (50:50 on DM basis). The highest milk proportion of 18:2ω6 was observed in RC (21.7) and WC (21.8 g/kg FA; P<0.001). The highest and lowest milk 18:3ω3 proportion was observed in WC (17.5) and LPR (4.1 g/kg FA), respectively (P<0.001). The highest milk proportion of c9,t11 conjugated linoleic acid (CLA; 3.9 g/kg FA) and c10,t12 CLA (0.2 g/kg FA) was observed in cows fed TF (P<0.001). The ω6 to ω3 ratio were lowest in cows fed WC (1.4) and RC (1.3; P<0.001). Milk yield was highest in cows fed WC (17.2), RC (16.6), and WC-LPR (16.5 kg/milking; P<0.001). Milk protein concentration was highest in EPR (3.6), FEST (3.6) and LPR (3.5%; P<0.001). However, the lowest milk fat concentration was observed in RC (4.2) and WC (4.3%; P<0.001). The primiparous cows showed higher proportion of CLA isomers and C18:1 trans (mainly t11 vaccenic acid) than multiparous cows (P<0.001). Milk yield, milk composition, 18:3ω3, 18:2ω6, ω3 and ω6 proportion were significantly affected by milking time (P<0.001). Concentration of α-tocopherol were highest in cows receiving EPR, LPR, and FEST diets (P<0.001). In conclusion, formation of biohydrogenation intermediates was different in primiparous and multiparous cows, and milk from primiparous cows had higher proportion of biohydrogenation intermediates. The milk FA profile was affected by the forage species fed to the cow and dietary inclusion of RC and WC could be a practical approach to increase the 18:3ω3 and ω3 FA from a human health perspective.

Possibilities for a specific breeding program for organic dairy production

M. Slagboom[1], L. Hjortø[1], A.C. Sørensen[1], H.A. Mulder[2], J.R. Thomasen[1,3] and M. Kargo[1,4]
[1]Aarhus University, Department of Molecular Biology and Genetics, Blichers Allé 20, 8830 Tjele, Denmark, [2]Wageningen University & Research Animal Breeding and Genomics, Droevendaalsesteeg 1, 6708 PB Wageningen, the Netherlands, [3]VikingGenetics, Ebeltoftvej 16, 8960 Randers SØ, Denmark, [4]SEGES Cattle, Agro Food Park 15, 8200 Aarhus N, Denmark; margotslagboom@mbg.au.dk

Breeding programs for organic and conventional dairy production are the same in most countries, despite differences between the two production systems. Breeding goals (BG) might be different for the two production systems and genotype by environment interaction may exist between organic and conventional dairy production, both of which have an effect on genetic gain in different breeding strategies. Other aspects also need to be considered, such as the application of multiple ovulation and embryo transfer (MOET), which is not allowed in organic dairy production. The aim of this research was to assess different environment-specific breeding strategies for organic dairy production. An organic and conventional breeding program were simulated in five different scenarios. Scenario 'Current' had a conventional BG and both MOET and the selection of conventional bulls as breeding bulls in the organic breeding program were allowed. The other four scenarios differed in these two aspects, but all had a specific organic BG in the organic breeding program. Implementation of a specific BG in the organic breeding program with BG weights that were adjusted to the organic production system increased genetic gain (ΔG) in the aggregate genotype slightly. The highest ΔG in the aggregate genotype was achieved in the scenarios where MOET was applied and conventional bulls could be selected for in the organic breeding program. This is therefore a recommendable breeding strategy. Not allowing for MOET or the selection of conventional bulls as breeding bulls in the organic breeding program decreased ΔG by as much as 24% and is not recommendable. ΔG on trait levels showed that a significant increase in functional traits was possible without a decrease in ΔG in the aggregate genotype. Thus, a specific breeding program for organic dairy production gives possibilities to breed in the desirable direction without losing on ΔG in the aggregate genotype.

Consumer views regarding ways to improve animal welfare in beef and dairy production

J.K. Niemi[1], K. Heinola[1], T. Latvala[1], T. Yrjölä[2], T. Kauppinen[3] and S. Raussi[3]
[1]*Natural Resources Institute Finland (Luke), Bioeconomy and environment, Kampusranta 9, 60320 Seinäjoki, Finland,* [2]*Pellervo Economic Research, Eerikinkatu 28 A, 00180 Helsinki, Finland,* [3]*Natural Resources Institute Finland (Luke), The Finnish Centre for Animal Welfare, Latokartanonkaari 9, 00790 Helsinki, Finland; jarkko.niemi@luke.fi*

Previous research shows that the public has concerns on whether animal welfare is satisfactory in modern animal production systems. A labelling scheme can be used to improve animal welfare and to certify that specific requirements are actuarially met in livestock production systems. The aim of this study was to test how the public views strategies to improve animal welfare in beef and dairy cattle production. A survey instrument was developed and administered in Finland in September 2018. The data (n=400 respondents) were a representative sample of population of Finland. We focused on 11 attributes of production: access to pasture or outdoor yard, freedom of movement in dairy cows and beef cattle, extended milk provision to calves and need to suckle, comfort around dairy cows' lying, access to water, leg health, friendly handling of cattle, space allowance, and preventive animal health care. The respondents were clustered into four groups and a multinomial logit regression was used to characterize respondent profiles. Respondents typically considered all 11 attributes as important or very important characteristics of an animal welfare labelling scheme. Dairy cows' access to pasture, continuous access to water, and preventive animal health care were viewed as particularly important characteristics. One of four respondent groups (11% respondents) considered all attributes systematically as quite important characteristics, two other groups (30 and 19%) considered them as very important or important and fourth group (40%) considered all 11 attributes as a very important characteristic of a labelling scheme. Factors such as the respondent living in a city or a suburb, young age and close familiarity with farming through relatives contributed to an increased likelihood of respondent belonging to the fourth group. The results provide guidance on which are the most essential criteria the consumers would like a labelling scheme to address.

Breed differences on sensory characteristics of sheep meat by a taste panel

S. Rodrigues[1], L. Vasconcelos[1], E. Pereira[1], A. Carloto[2] and F. Sousa[1]
[1]*CIMO, Instituto Politécnico Bragança, Campus Sta Apolónia, 5300-253, Portugal,* [2]*ACOB, Largo Coronel Salvador Teixeira, Lote 69/70 R/C Dt, Bragança, 5300-044, Portugal; srodrigues@ipb.pt*

The objective of this work was to determine if there are differences between Churra Galega Bragançana Branca (CGBB) and Churra Galega Bragançana Preta (CGBP) breeds meat. These are Portuguese autochthonous breeds, and CGBB meat has a PDO protection (Regulation EU no. 1151/2012). Samples of lamb loin, with 3 different finishing levels, of the CGBB and CGBP breeds were used. The differences were assessed by a qualified for meat products taste panel with 9 elements plus an unqualified taster. A triangular test was used, with two identical samples and a different one, the experts being obliged to indicate which sample was different and why. The loin samples were separated from the vertebrae, wrapped in aluminium foil and duly encoded. These samples were then placed in a conventional oven and cooked until reaching the temperature of 80 °C in their thermal centre. After confection, the loins were cut into portions with an average thickness of 1 cm, wrapped in foil and duly coded. Each expert was provided with a triad corresponding to meat samples from the two breeds of the same treatment. The procedure was performed three times to test the three treatments of the study. Three replicates of the procedure were performed to test all available samples. The 10 experts participating in the trial performed 90 triangular tests for the difference. In 53 of the answers, the experts correctly identified the different sample. The proportion of assertive answers allows concluding that in the tasting of *longissimus dorsi* the two breeds are differentiable with a level of significance of 0.1%. This differentiation occurred for higher finishing levels, and no significant differences were detected when the degree of finishing was weaker. The descriptors that allowed the experts to identify the different sample were juiciness, hardness, flavour and in smaller number the colour.

Organic livestock farming contentious inputs in France: preliminary results

M. De Marchi[1], H. Bugaut[2], C.L. Manuelian[1], J. Renard[2], R. Pitino[3], M. Penasa[1] and S. Valleix[2]
[1]University of Padova, Department of Agronomy, Food, Natural Resources, Animals and Environment, Viale dell'Università 16, 35020 Legnaro, Italy, [2]VetAgro Sup, ABioDoc Department, 89 avenue de l'Europe – BP 35, 63370 Lempdes, France, [3]University of Parma, Department of Veterinary Science, Via del Taglio, 10, 43126 Parma, Italy; massimo.demarchi@unipd.it

An on-line questionnaire (6 sections, 36 questions) about contentious inputs in organic livestock farming has been developed in English and translated to French based on the Brislin's model. A sample of French organic farmers was contacted via e-mail up to 3 times and through 3 farmers' associations from November 2018 to February 2019. A total of 135 out of 155 questionnaires were usable for the analysis. The men to women ratio were 60/40, primarily from 31 to 50 years old (83/135). In general, the questionnaire was fulfilled by the farm manager (111/135) and most participating farms (90.23%) were small (≤3 employees). In most cases, organic farms reared one (63.70%) or two (22.22%) animal species, mainly beef cattle (38.52%), dairy cattle (27.41%) and sheep (18.52%). In the last year, 48/130 farmers did not treat their animals. When they had to apply a treatment, the use of alternative therapies depended on the health issue they were facing. Between 15.38% (skin problems) and 34.55% (lameness) of the farmers still relied on conventional treatments instead of phytotherapy, homeopathy or probiotics; and between 6.45% (reproductive issues) and 35.29% (mastitis) reported the use of those alternatives as well as conventional treatments. Information about the use of those alternatives was mainly obtained through other farmers (66.42%) and veterinarians (46.27%). Voluntary vaccination of the animals was applied by 30/135 of the farmers. Half of the farmers added vitamins to the animals' diet but 22.39% of them were not aware of the origin of the vitamins. This preliminary analysis suggests the need of further research on alternatives to the use of antibiotics and antiparasitics in organic livestock farming, and underlines that farmers are a key factor for the dissemination/implementation of the results. This project has received funding from the EU Horizon 2020 research and innovation programme under grant agreement No [774340 – Organic-PLUS].

Assessing the main dietary roughage source of dairy systems by milk quality-indicators

G. Riuzzi, V. Bisutti, B. Contiero, S. Segato and F. Gottardo
University of Padova, Animal Medicine, Production and Health, Viale dell'Università 16, 35020, Italy; giorgia.riuzzi@unipd.it

Dairy system authentication is an ongoing topic of interest for producers, consumers and policy makers, with a specific focus to the kind of roughage source used in the ration of lactating dairy cows. A large number of specific milk quality indicators have been proposed to gain insight into dairy systems discrimination. However, these indicators are quite laboratory expensive and/or they do not provide information on the nutritional value. Therefore, this study considered the milk proximate composition coupled with the micro-elements profile with the aim of evaluating an easier traceability, which provides also a nutritional assessment. The study considered 14 dairy farms located in the Veneto region. According to the main roughage dietary sources, the farms were grouped in three experimental thesis: high maize silage (HMS, 6 farms), high permanent meadow hay (HMH, 4 farms) and mixed crop-silage/hay (MSH, 4 farms). Furthermore, the HMH and HMS were split according to the cow's breed: Friesian vs Brown (B-HMII, F-HMH and B-HMS, F-HMS). In 2018, by means of five bulk milk sample collections, chemical composition (crude protein, crude fat, lactose, casein, urea, somatic cells score, cfu) and minerals (TXRF, Total Reflection X-Ray Fluorescence: Ca, Cl, Br, S, P, Cr, Fe, Mn, Zn, Cu, Sr, Rb) were determined. A factorial discriminant analysis was carried out in order to extract the main predictive features, able to discriminate the experimental groups. The stepwise procedure used by the discrimination model selected 5 variables: CP, lactose, SCC, K and Br. Two significant functions (CAN1 and CAN2) accounted for 71.1 and 23.0% of the total variability, respectively. The low value of Wilks' Lambda (0.081, approx. F value=11.51, df1=20, df2=203, P<0.001) and the significant D2 –Mahalanobis distances among the 5 experimental groups indicated that the discriminant model is solid and reliable. Based on the confusion matrix, the correctly classified cases were equal to 73%. An ERA-NET SUSAN project (SusCatt) funded by MIPAAFT. A SAFIL project funded by FONDAZIONE CARIVERONA (call 2016)

Development of a method to estimate the taste of Japanese Black beef based on chemical composition

K. Suzuki[1], T. Komatsu[2], F. Iida[3], Y. Suda[4], K. Katoh[1] and S. Roh[1]
[1]Graduate School of Agricultural Science, Tohoku University, 468-1 Aramaki-Aza-Aoba, Aoba-ku, 980-8572, Japan, [2]Livestock Experiment Station of Yamagata Integrated Agricultural Research Center, 1076 Torigoe-Aza-Itsuponmatsu, Shinjyo, 996-0041, Japan, [3]Japan Women's University, 2-8-1 Mejirodai Bnkyo-ku, Tokyo, 112-8681, Japan, [4]Miyagi University, 2-2-1 Hatatate, Taihaku-ku, 982-0215, Japan; keiichi.suzuki.c6@tohoku.ac.jp

In this study, we developed a method to estimate the taste of Japanese Black beef based on its chemical composition by combining the data of the sensory panel test and the analysis of the chemical components such as amino acids and nucleic acid-related substances. Beef from thirty-five sirloin Japanese black heifers beef graded as more than A4 on the Japanese Meat Grade scale were used in this study. A part of the beef was used for the sensory panel test and another part was used for the chemical composition analysis. The average age of these beef was 34 months old. Ten items were used for the panel test: 'tenderness on the first bite,' 'tenderness while chewing,' 'fibre feeling,' 'juiciness,' 'total texture,' 'sweet scent,' 'off flavour,' 'strength of the aroma,' 'umami intensity' and 'overall evaluation.' First, factor analysis was performed on sensory panel test items to identify latent factors for the sensory panel test. Based on the results of the factor analysis, 10 sensory panel items were classified into three latent factors of texture, taste, and flavour. After that, score diagrams of latent factors were calculated by preparing path diagrams of sensory panels and latent factors by covariance structure analysis. Score diagrams between 10 panel items and 3 latent factors were calculated. Goodness of fit index (GFI), adjusted goodness of fit index (AGFI), and comparative fit index (CFI) of model fitness were estimated to be 0.847, 0.782, and 0.869, respectively. Finally, partial least Squares (PLS) analysis was performed with the latent factor score value as the dependent variable and the chemical composition as the independent variable. As a result of PLS analysis, the taste could be estimated using six chemical components (fat content, inosinic acid, inosine, nucleic acid content, K value, and carnosine + taurine). The contribution rate was 0.80.

Characterization of milk goat composition according to feeding systems in Western France

C. Laurent[1], B. Graulet[1], H. Caillat[2], C.L. Girard[3], J. Jost[4], N. Bossis[4], L. Lecaro[5] and A. Ferlay[1]
[1]INRA, PHASE, UMR Herbivores, 63122 St-Genes-Champanelle, France, [2]INRA, FERLus Les Verrines, 86600 Lusignan, France, [3]Agriculture Agri-Food Canada, Sherbrooke Research and Development Centre, Sherbrooke, J1M 0C8, Canada, [4]Institut de l'Elevage, CS 45002 Mignaloux-Beauvoir, 86550, France, [5]Chambres d'agriculture de Bretagne, CS 14226 Rennes, 35042, France; anne.ferlay@inra.fr

Dairy goat systems are facing a complex socio-environmental context where feed supply is endangered due to climate changes, whereas consumers are increasingly aware of the nutritional quality of the animal products. Thus, increasing the proportion of fresh or conserved grass in the diets could help to increase the self-sufficiency of dairy systems but the impact on milk quality must be assessed. In order to characterize the effects of the level of inclusion and type of grass in the diet on milk composition, bulk milk samplings and surveys on flock characteristics and diets were collected in spring and autumn 2017 29 dairy goat farms from 3 French western regions. Milk samples were analysed for vitamin compounds and fatty acid (FA) composition. A typology of diets was established using PCA and CAH. The effect of diets on milk composition was analysed by variance analysis taking into account dairy goat characteristics and dietary concentrate (XLSTAT 2018.20.12). The diets were characterized according to the main forage (% of dry matter intake): maize silage (MS, 37.4%), hay (H, 46.2%), grass wrapping or silage (GS, 28.5%), and fresh grass (FG, 44.4%). There was no effect of diets on dry matter intake or milk yield. Milk from FG was yellower and richest in retinol, α-tocopherol, vitamin B6 and xanthophylls when compared to milk from the other diets. Milk concentration of vitamin B12 was lower with FG than MS diets. Milk saturated FA concentration was decreased with FG vs other diets, without changes in C6:0-C10:0 concentration. FG increased milk monounsaturated FA when compared to MS and H, value for GS being intermediate. Surprisingly, milk C18:3n-3 concentration was higher for H than MS diets. The FG feeding system seems to enhance nutritional quality of goat milk.

Certification of products as a chance for the development of farms which keep native livestock breed
P. Radomski, J. Krupiński and P. Moskała
National Research Institute of Animal Production, Sarego 2, 31-047 Kraków, Poland; pawel.radomski@izoo.krakow.pl

Research has shown that both consumers and breeders consider promotion and popularization of native breeds necessary as a source of raw materials for production of food with high health-promoting and palatability values. All these circles notice shortcomings in the current activities which promote products from native breeds, indicating large interest in these products in the market at the same time. There is social acceptance and understanding for higher prices of these products and pointing at local markets or directly from producers as the main place of sale and purchase, all perceive it as a great chance in shortening supply chains. New systems solutions are necessary since the market of products from native breeds is a very slight and local niche, but through using promotion activities such as: media campaign presenting the potential of native breeds, appropriate marking of the points of sale and farms participating in the certification scheme, organization of meetings for restaurateurs, food vloggers, hoteliers, there is a chance that its scope and size will increase. The certification system with Native Breed brand could be such a solution. The system of protection and promotion of products from native breeds is a chance for part of homesteads since it will increase their income, it protects cultural heritage, helps in the development of rural tourism and creates additional sources of non-farm income for the farmers who want to deal with processing, as well. The challenge that the certification scheme of Native Breeds faces is surely to convince, on the one hand, the consumers about the quality of products labelled with the certificate, and on the other hand, the breeders about the possibilities the certification scheme provides and through that creating local selling market for the products of native breeds. A vast majority of breeders of native breeds have not benefitted from any forms of promotion so far, which will surely constitute a barrier for them. It is essential to create a chain, which will join all the customer groups i.e. breeders and consumers among which it is also essential to create public awareness about the role and significance of breading and rearing native animal breeds. BIOSTRATEG2/297267/14/NCBR/2016.

Native breeds as a guarantee of the quality of traditional products
P. Radomski, J. Krupiński and P. Moskała
National Research Institute of Animal Production, Sarego 2, 31-047 Kraków, Poland; pawel.radomski@izoo.krakow.pl

The research carried out in many European and worldwide scientific centres indicates that raw materials and processed products obtained from native animal breeds, fed in a traditional way are characterized by more beneficial parameters for processing and at the same time higher contents of biologically active substances, which have a positive impact on human health. The results of the research conducted in the framework of the project BIOSTRATEG indicate good quality of meat of native cattle breeds, including high contents of minerals and favourable fatty acid profile. Moreover, numerous studies have demonstrated that meat obtained from animals kept in a traditional system, based on the maximum use of grazing pasturage is characterized by more beneficial ratio of n-6/n-3 acids and higher contents of desirable polyunsaturated fatty acids (PUFAs). The use of pork and beef meat ingredients for processing to produce high quality charcuterie is characterized by great culinary values. Produced test consignments of processed meat products were widely recognized during presentation and tasting during the fair both in Poland and abroad. Both beef and pork products cured uncooked were held in particular regard. Raw material for manufacturing of these products is obtained from cows grazed on mountain pastures. Excellent traditional preserved meat products are made from meat of Złotnicka pig (White and Spotted) and Puławska pig. High quality of the regional products is also reflected in using traditional methods of production and preservation. One of the oldest methods of preserving meat is smoking, traditionally used in our geographic region. Moreover, thanks to it the distinctive aroma, richer flavours, attractive hue, increased succulence are obtained and nutritional values of food are preserved. The conducted analysis of the traditional products of animal origin from various regions of Poland produced from raw materials derived from native breeds indicates that the products are characterized by high quality and safety particularly in terms of contents of polyaromatic hydrocarbons. The products deserve the recommendation as traditional, high quality products of unique taste. BIOSTRATEG2/297267/14/NCBR/2016.

An improvement in animal welfare in a large Swiss abattoir through the consistent evaluation of data

M.K. Kirchhofer[1] and L.V.T. Von Tavel[2]
[1]Privat, Wohlen, 5033, Switzerland, [2]Swissgenetics, Zollikofen, 3052, Switzerland, lvt@swissgenetics.ch

On the basis of the Animal Welfare Act the 'Directive on the protection of animals at the time of slaughter' governs the handling of animals in Swiss abattoirs. One of its stipulations is that cattle must remain in sheds for a maximum of 4 hours under the reduced shed conditions in abattoirs. However, enforcement and control of this requirement by the veterinary service in large operations with approval to slaughter several hundred cattle per day has proven difficult. In SAP, the software used by the meat processor for data processing, individual animals are registered at three separate time points: by the livestock trade on arrival, by the abattoir at the time of stunning and by Proviande (Swiss Organisation of Meat Marketers) at the time of valuation. The various datasheets were transferred and evaluated by the veterinary service in an Excel spreadsheet and the results were displayed graphically. At 30 minute intervals the graphic shows the frequencies of deliveries, slaughter, shed occupancy and theoretical stall occupancy in the event of a problematic slaughter, valuation and those animals who had to wait more than 4 hours in the abattoir shed. The following parameters were also calculated as key indicators: animals with excess time as a total and in percentages, the cumulative excess time in hours of all animals held too long in sheds and the number of animals listed with excess time after the 30-minute interval of their arrival. The graphic has enabled the dynamism of the abattoir as regards its deliveries, slaughter and animal shed time to be displayed in a clear and understandable way. The key indicators can then be used to objectively evaluate the success of animal welfare improvement measures. The consistent use of these evaluations also enables a fact-based and constructive dialogue between abattoir personnel and the veterinary service. It is also worthwhile to use existing data consistently in an abattoir. This improves animal welfare without impairing the commercial performance of the abattoir. These fact-based evaluations have led to a daily dialogue between veterinary service and abattoir staff and an awareness of animal welfare was promoted by staff in the shed.

Effect of ripening time and salt reduction or substitution in pork sausages sensory characteristics

S. Rodrigues, C. Grando, E. Pereira, L. Vasconcelos and A. Teixeira
CIMO, Instituto Politécnico de Bragança, Campus Sta Apolónia, 5300-253, Portugal; srodrigues@ipb.pt

This work aimed to evaluate the sensory characteristics of meat sausages with 4 different salt formulations (Form1: 2% NaCl, Form3: 1.5% NaCl+0.5% KCl, Form4: 1.5% Sub4 salt+0.5% NaCl, Form5: 0.5% Sub4 salt+1.5% NaCl) and 2 ripening times (6 and 12 days). The evaluated attributes were related to appearance (exterior and interior colour), odour (intensity before and after cutting), taste (salty, bitter and metallic), texture (firmness perceived by thumbs and hardness, juiciness and chewiness in mouth) and flavours (set sensation, intensity and persistence). Five samples of each formulation were evaluated considering the salt formulations in two sessions for a shorter ripening time. The same number of samples were evaluated for a longer ripening time. Samples were evaluated by a 10 elements meat products qualified taste panel, following the Portuguese standards, and using a 7 points scale. Data were submitted to the Product Characterization procedure and to a Generalized Procrustes Analysis (GPA). GPA shows that Form1 had more intense inner odour and flavours persistence, Form3 had a darker interior colour (darker red), the higher odour intensity, hardness and bitter taste, Form4 was where higher values of basic, yet not very high, flavours were noted, and Form5 had the highest values of texture, outer colour and flavours intensity. Results of the characterization of the products indicate that only firmness felt by the thumbs presented a discriminatory power between sausages with shorter ripening time, Form3 had significantly lower firmness than the other formulations. In the sausages with a longer ripening time, GPA indicated Form1 with higher values for hardness, chewing, exterior colour, internal odour, bitter taste and persistence of the flavours, Form3 was considered the most succulent and salty, Form4 was the one that had the most firmness, and Form5 had the highest intensity of interior colour and metallic taste. The characterization of the products identified the external colour as the only attribute with significant discriminatory power. Form1 had significantly darker colour and Form5 had a lighter colour.

Towards value-based marketing of UK lamb

N.R. Lambe[1], N. Clelland[1], E.M. Smith[2], J. Yates[2], J. Draper[3] and J. Conington[1]
[1]SRUC, Roslin Institute Building, Easter Bush, Midlothian, EH25 9RG, Scotland, United Kingdom, [2]The Texel Sheep Society, Stoneleigh Park, Kenilworth, Warwickshire, CV8 2LG, England, United Kingdom, [3]ABP, Birmingham Business Park, Birmingham, B37 7YB, England, United Kingdom; nicola.lambe@sruc.ac.uk

To continue to be a major lamb producer and exporter in Europe, the economic viability of the UK sheep industry must improve. This is achievable by: (a) improving efficiency of production and carcass quality, so that carcasses better meet market requirements; and (b) increasing consumer acceptance by supplying a consistently high quality product. Whole supply-chain solutions are required to enable new advances in breeding, precision livestock management and product quality measurement to be harnessed within a value-based marketing system that will incentivise product quality. Industry-led research projects are underway to investigate the possibilities of breeding for improved lamb carcass and meat quality within the stratified UK sheep industry, and developing industry strategies to allow UK-wide exploitation of the outcomes. Using technologies including Video Image Analysis, Computed Tomography and Near-Infrared spectroscopy, genetic and genomic selection for new product quality traits, relating to lean meat and fat distribution, meat quality and health traits, is being tested. Preliminary results will be presented examining genetic control and relationships amongst product quality traits, across purebred and crossbred sheep populations. Implications and recommendations for industry implementation will be considered.

The putative role of the acetyl-coenzyme A acyltransferase 2 (ACAA2) gene in ovine milk traits

S. Symeou and D. Miltiadou
Cyprus University of Technology, Agricultural Sciences, Biotechnology and Food Science, Anexartisias and Athinon corner, 57, 3603, Cyprus; despoina.miltiadou@cut.ac.cy

The dairy sheep industry in Mediterranean countries is directed by the production of high quality cheese from local breeds. Milk yield (MY), protein, fat and fatty acid (FA) contents of milk determine farmers' income and cheese yield, quality and organoleptic characteristics. Thus, during the last decade, attention has been placed on the identification of genetic loci affecting milk traits in dairy sheep. The *ACAA2* gene encoding an enzyme involved in lipid metabolism is placed on a chromosomal region where QTL for milk, fat and protein yields have been proposed. Molecular characterization of the ovine *ACAA2* gene showed that the gene is monomorphic in Chios sheep, except from a single nucleotide polymorphism (SNP: T>C), in the *3'UTR* of the gene. Mixed model association analysis showed that the SNP was significantly associated with MY in an experimental flock, with the T allele exhibiting an additive effect of 13.4 ± 4.7 kg. Subsequently, association analysis of the identified SNP with MY, fat and protein contents and yields and the FA profile was performed in an extended population of Chios sheep. The SNP was significantly associated with total lactation and protein percentage with respective additive effects of 6.81 ± 2.95 kg and $-0.05\pm0.02\%$, with TT and TC animals exhibiting higher milk yield compared to CC, whereas the latter exhibited lower protein percentage. In addition, dominance effects were detected for milk fat yield, C9:0, C11:0, C12:1 cis-9, C13:0, C15:1 cis-10 FAs and an additive effect was found for the $\omega6/\omega3$ index. Expression analysis from age-, lactation-, and parity-matched animals showed that mRNA expression of the *ACAA2* gene from TT animals was 2.8 and 11.8 fold higher than in CC in liver and mammary gland, respectively. In addition, by developing an allelic expression imbalance assay, it was estimated that the T allele was expressed at an average of 18% more compared to the C allele in the udder, supporting the existence of tissue specific cis-regulatory DNA variation in the ovine *ACAA2* gene.

Selection trace from runs of homozygosity in French dairy sheep

S.T. Rodriguez-Ramilo[1], A. Reverter[2] and A. Legarra[1]
[1]INRA, UMR 1388 GenPhySE, 31326 Castanet Tolosan, France [2]CSIRO Agriculture & Food, Brisbane, QLD 4067, Australia; silvia.rodriguez-ramilo@inra.fr

Runs of homozygosity (ROH) are contiguous homozygous segments of the genome where the haplotypes inherited from each parent are identical. Currently, inbreeding estimated from ROH is considered a powerful approach to distinguish between recent, potentially harmful, from ancient inbreeding, potentially beneficial. Accordingly, inbreeding based on ROH can help to improve the understanding of inbreeding depression. The occurrence of ROH is not randomly distributed across the genome, and islands of ROH across a large number of animals may be the result of selective pressure. The objective of this study is to evaluate whether ROH can be used to explore signatures of selection in French dairy sheep. The data set available included animals from various breeds and subpopulations: Basco-Béarnaise breed (BB); Manech Tête Noire breed (MTN); Manech Tête Rousse breed (MTR); Lacaune Confederation subpopulation (LACCon); and Lacaune Ovitest subpopulation (LACOvi). Animals were genotyped with the Illumina OvineSNP50 BeadChip. After applying filtering criteria, the genomic data included 38,287 autosomal SNPs distributed across 26 chromosomes and 8,700 individuals. One island of ROH was detected on autosome 6 in the same genomic position across animals (between 30-40 Mb). Wright's differentiation coefficients for two SNPs within this island of ROH were high (0.67-0.68). The linkage disequilibrium between both SNPs was also elevated (0.98). The divergence in allele frequencies in those SNPs grouped BB, MTN and MTR breeds in one cluster, and LACCon and LACOvi subpopulations in another cluster. The closest candidate gene is *NCAPG-LCORL*, which has been reported to be under positive selection and suggested to control height and stature in sheep, as well as growth and feed efficiency in cattle. These findings contribute to the understanding of the effects of selection in shaping the sheep genome.

Assessment of methane traits in ewes: genetic parameters and impact on lamb weaning performance

J. Reintke[1], K. Brügemann[1], T. Yin[1], P. Engel[1], H. Wagner[2], A. Wehrend[2] and S. König[1]
[1]Institute Animal Breeding and Genetics, Justus-Liebig-University of Giessen, Ludwigst. 21B, 35390 Giessen, Germany, [2]Clinic for Obstetrics, Gynecology and Andrology of Large and Small Animals with Veterinary Ambulance, Frankfurter St. 106, 35392 Giessen, Germany; jessica.reintke@agrar.uni-giessen.de

There is an increasing interest to reduce methane (CH_4) emissions in sheep via breeding. Furthermore, CH_4 output is associated with improved feed efficiency and decreased dry matter intake. Against this background, the present study focussed on intergenerational aspects and studied the phenotypic and genetic influence of different ewe CH_4 parameters on lamb body weight (LBW). Further traits in association analyses were ewe body weight (EBW), ewe body condition score (BCS) and ewe back fat thickness (BFT). From January 2017 to April 2018, data of 330 ewes (253 Merinoland- and 77 Rhönsheep) and their 629 lambs were recorded. The interval between repeated measurements (for ewe traits and for LBW) was three weeks during lactation. CH_4 concentration (ul/l) in the exhaled air of the ewe was recorded using a mobile laser methane detector (LMD). Afterwards, CH_4 emissions were portioned into a respiration and eructation fraction, allowing the following CH_4 trait definitions: Mean CH_4 during respiration and eructation (amean); Mean CH_4 during respiration (rmean)/eructation (emean); CH_4 sum/minute during respiration (rsum)/eructation (esum); Maximal CH_4 during respiration (rmax)/eructation (emax) and eructation events/minute (event). Increasing CH_4 emissions represent an energy loss for the ewe and were significantly associated with lower LBW, EBW, BCS and BFT. Hence, ewes mobilized body reserves to compensate energy losses. Genetic correlations between CH_4 traits with LBW ranged between -0.33 (esum) and 0.03 (rsum), and from -0.51 (rmax) and -0.08 (rsum) with EBW. Heritabilities for CH_4 parameters were close to zero. Nevertheless, we proved that LMD applications are very valuable for CH_4 recordings under field conditions. Selection on low mean CH_4 emissions (amean) and low CH_4 during respiration (traits: rmean, rmax) will improve feed efficiency of ewes, reduce fodder costs and improve LBW during lactation.

A selection index for flock profitability in an outdoor flock of dairy sheep in New Zealand

N. Lopez-Villalobos[1], M. King[2], J. King[2] and R.S. Morris[1]
[1]*Massey University, Private Bag 11222, 4442 Palmerston North, New Zealand,* [2]*Kingsmeade Cheese Ltd., Olivers Road, 5886 Masterton, New Zealand; n.lopez-villalobos@massey.ac.nz*

Dairy production based on sheep rather than cattle is expanding in New Zealand. The objective of this study was to develop a selection index to rank dairy sheep for contribution to profit in a commercial flock under outdoor conditions of New Zealand with seasonal lambing. A breeding program for this flock was established in 1996, using East Friesian rams with ewes of two New Zealand sheep breeds. Records of weaning weight were used from 1,381 lambs born between 2013 and 2018 to estimate direct (WWd) and maternal (WWm) breeding values (BV) for weaning weight. The pedigree file contained 5,091 animals over 11 generations. Total lactation yields of milk (MILK), fat (FAT) and protein (PROT) were calculated for each ewe-lactation using orthogonal random regression polynomials of order 2. Live weights (LW) of ewes were recorded once in each lactation. Ewes were scored each year for udder support (US). In total there were 445 lactations from 280 ewes recorded from 2016 to 2018. These ewes were the progeny of 26 rams. Breeding values for MILK, FAT, PROT, LW, litter size (LS) and US were estimated with a multi-trait repeatability animal model that included the fixed effects of year, lactation number and deviation from flock median lambing date, and the random effects of animal, ewe permanent and residual error. Variance components required to solve the mixed model equations were derived using phenotypic variances of the data and genetic correlations from published papers. The economic value (EV) for MILK was calculated as the change in ewe profit per litre of milk. The economic values for the other traits were estimated using desired relative economic weights. The index was calculated as $I=2.43 \times BV_{MILK} + 4.42 \times BV_{FAT} + 10.88 \times BV_{PROT} - 4.97 \times BV_{LW} + 75.70 \times BV_{US} + 524.12 \times BV_{LS} + 0.31 \times BV_{WWm}$. The index has been used to select the best ewes and rams as parents of future replacements. The index has also been used for the selection of new-born female lambs. Further work is required to estimate the EVs of the traits using a farm model integrated with a milk processing model.

Redefining key performance indicators for sheep production systems

A.G. Jones, T. Takahashi, H. Fleming, B.A. Griffith, P. Harris and M.R.F. Lee
Rothamsted Research, North Wyke, EX20 2SB, United Kingdom; andy.jones@rothamsted.ac.uk

As with many European countries, sheep farming in the UK has long been associated with low profit margins. To improve the profitability of these systems, various metrics to assist on-farm decisions, generally referred to as Key Performance Indicators (KPIs), have been proposed to date. As the list of KPIs has grown so has confusion amongst commercial producers as to which measurements should be prioritised given their time and budgetary constraints. Using high resolution data from the North Wyke Farm Platform grazing trials in Devon, UK, this study constructed a framework for objectively ranking existing KPIs based on their value for UK sheep producers. A complete list of KPIs in common usage was first compiled; these KPIs were next separated into *performance predictors* and *performance outcomes*, and further segmented into the level at which they are applied; animal level, field level and farm level. Within each level, causal relationships between predictors and outcomes were investigated to assess: (1) how many outcomes could be indicated for each predictor, and (2) how strongly they are correlated, resulting in an *accuracy* score between each predictor and each relevant outcome. Accuracy scores were then adjusted for *impact* (the proportion of farm profitability determined by the relevant outcome) and *ease of application* (how easily the performance predictor can be influenced by changes in management) to produce the overall *benefit* score for each predictor. Finally, these scores were compared against the cost of measuring each predictor, composed of *labour cost* (time taken to collect and evaluate the data) and *physical cost* (financial cost of equipment necessary to collect the data), to rank all predictors according to their cost effectiveness. Our results indicate that performance predictors can provide an efficient way to improve the profitability of sheep systems if the right metrics were selected by commercial farmers. To the best of our knowledge, this is the first study to define the relative values of KPIs based on quantitative evidence.

Effects of nonsynonymous SNPs at GH2-N and GHR genes on coagulation properties of Assaf ewes' milk

M.R. Marques[1,2], J.R. Ribeiro[2], A.T. Belo[2], S. Gomes[3], A.P. Martins[3,4] and C.C. Belo[2]
[1]CIISA, FMV, Avenida da Universidade Técnica, 1300-477 Lisboa, Portugal, [2]INIAV, UEISPSA, Quinta da Fonte Boa, 2005-048 Vale de Santarém, Portugal, [3]INIAV, UTI, Quinta do Marquês, 2780-157 Oeiras, Portugal, [4]LEAF, ISA, Tapada da Ajuda, 1349-017 Lisboa, Portugal; rosario.marques@iniav.pt

Sheep milk coagulation properties are of interest as potential selection criteria once milk is mainly used for cheese yield. The effects of nonsynonymous substitution at the growth hormone (GH2-N copy) and growth hormone receptor (GHR) genes and their haplotypes on milk production, milk composition (fat, protein, lactose, total solids and fat free total solids content), pH, and coagulation properties assessed by Optigraph [clotting time (R), gel firmness after 20 minutes (A20), and after a 2R (AR) period and rate of firming (OK20)] were studied in Assaf ewes. Milk production and composition were evaluated monthly until the sixth month of lactation (184 ewes), and pH and coagulation properties were evaluated at the first and third month of lactation (92 ewes). Data were analysed using a mixed-model procedure with fixed effects of SNP and month of lactation, considering the linear and quadratic effect of ewe' lambing age covariate. In the GH2-N copy gene, two substitutions (X12546 g.597T>C and g.1024T>C, three haplotypes) were studied. GH2-N genotypes and haplotypes had no effect on milk production and composition, however, they affected significantly AR throughout lactation (P<0.01). In the GHR copy gene, SNPs rs1086611503, rs595567866 and rs597181420 and seven haplotypes were studied. GHR genotypes and haplotypes had no effect on milk production and composition, except for SNP rs1086611503, who tended to affect protein and total solids content (P<0.10). Regarding milk coagulation properties, rs597181420 genotypes influenced pH and R, with CC ewes having lower values than TT ewes (P<0.01). The GHR haplotypes influenced significantly R and A20 (P<0.05), and AR (P<0.01) throughout lactation. Effects observed upon gel firmness parameters highlight the usefulness of those SNPs in gene-assisted selection programs for milk coagulation properties. Funding: Project financed by European Fund for Regional Development (ERDF) [ALT20-03-0145- FEDER-000019]

Genetic variability following selection for scrapie resistance in six native Italian sheep breeds

F. Bordin[1], C. Dalvit[1], M. Caldon[1], R. Colamonico[1], L. Zulian[1], S. Trincanato[1], B. Mock[2], F. Mutinelli[1] and A. Granato[1]
[1]Istituto Zooprofilattico Sperimentale delle Venezie, Viale dell'Università 10, 35020 Legnaro, Italy, [2]Federazione Zootecnica Alto Adige, Via Galvani 38, 39100 Bolzano, Italy; fbordin@izsvenezie.it

Scrapie is a neurodegenerative disease of sheep belonging to the group of transmissible spongiform encephalopathies. Susceptibility to scrapie is associated with polymorphisms in the prion protein (PrP) gene. Genetic selection is currently the most effective mean for eradication of the susceptible VRQ allele in favour of resistant ARR allele. The aim of our study was to determine changes of genetic variability in 6 native sheep breeds from autonomous province of Bolzano (northern Italy), following simulation of scrapie selection scenarios. DNA samples from 684 rams were analysed for PrP polymorphisms and for 10 ISAG microsatellite loci to estimate genetic variability (GenAlEx software). The PrP predominant allele was ARQ (51.8%), while the ARR and VRQ allele frequencies were 23.5 and 9.7%, respectively. The ARR/ARR, ARR/ARQ and ARQ/ARQ genotypes represented 6.8%, 23 and 28.6% of all population respectively. Across all microsatellite loci a total of 163 alleles were detected with a mean of 10.4 alleles per locus, and 35 private alleles were found with a frequency below 9%. Average observed (Ho) and expected (He) heterozygosity values overall loci were 0.74 and 0.78 respectively, showing a statistically significant deviation from HWE in all breeds. This heterozygosity deficit is confirmed by positive Fis value, determining a moderate inbreeding rate in each breed. Simulating a mild selection, where only rams having at least a VRQ allele should be excluded from reproduction, Ho, He and Fis remained almost unchanged in each breed, indicating that genetic variability should not be affected by the removal of these individuals. With a moderate selection scenario, considering only rams with at least one ARR allele, we observed a decrease in the mean alleles per breed (8.9) and the maintenance of heterozygosity deficiency except for 2 breeds, where HWE deviation was no longer significant. These results showed that selection strategies, considering only PrP resistant rams, should not dramatically affect genetic variability of these autochthonous breeds.

Across-breed genomic evaluation for meat sheep in Ireland

T. Pabiou[1,2], E. Wall[1], K. McDermott[1], C. Long[1] and A.C. O'Brien[3]
[1]SheepIreland, Highfield house, Bandon, Co. Cork, Ireland, [2]Irish Cattle Breeding Federation, Highfield house, Bandon, Co. Cork, Ireland, [3]Teagasc Moorepark Research Center, Fermoy, Co. Cork, Ireland; tpabiou@icbf.com

The current genetic evaluation for meat sheep run by Sheep Ireland is done within-breed for 6 main breeds (Texel, Suffolk, Charollais, Belclare, Vendeen, and Lleyn) and across 4 modules: growth, birth-related, litter size, and health. Since 2015 Sheep Ireland has procured genotypes on almost 25,000 sheep, through both the OviGen Research project and direct Sheep Ireland initiatives. The objective of the present study was to introduce updates to the current sheep genetic evaluation in Ireland by introducing: (1) across-breed comparisons; and (2) genomic predictions. For all modules, the across-breed model was an animal model accounting for breed by the mean of fixed covariates for 11 breed proportions (coefficient from 0 to 1). The genetic evaluation post-process involved the addition of the breed solutions to the breeding values with respect to the breed composition of each animal to render the breeding values comparable across-breeds. Genomic predictions were undertaken using the single step genomic best linear unbiased prediction method; for accuracy, the two-step method was favoured as to limit potential over-estimation. The across-breed breeding values demonstrate clear differences between terminal and maternal orientated breeds in the Sheep Ireland Terminal and Replacement indexes. The overall correlation for Terminal and Replacement indexes between within- and across-breed evaluation was 0.56 and 0.52, respectively. The gain in accuracy for Terminal and Replacement indexes between within- and across-breed evaluation was mostly due to the growth module genetic parameter re-estimation and was on average 4 and 3%, respectively. For genomic evaluation, all genotypes from the five different genotype panels used in Ireland were imputed to 49,391 single nucleotide polymorphisms using FImpute version 2.2. The breeding value correlations for Terminal and Replacement indexes between the across-breed and the genomic across-breed evaluations were 0.95 and 0.98, respectively. The gain in accuracy for the animals with genotypes is on average 16% for litter size and the genomic information accounts on average for 21% of the published accuracy for litter size.

A new era; the application of technologies to select for lowered methane emissions in New Zealand Sheep.

S.J. Rowe[1], P.L. Johnson[1], S.M. Hickey[2], K. Knowler[1], J. Wing[1], B. Bryson[3], M. Hall[3], P. Johnstone[1], A. Jonker[4], P.H. Janssen[4], K.G. Dodds[1] and J.C. McEwan[1]
[1]Invermay Agricultural Centre, AgResearch Ltd., Private Bag 50034, Mosgiel, New Zealand; [2]Ruakura Research Centre, AgResearch Ltd., Private Bag 3115, Hamilton, New Zealand; [3]Woodlands Research Station, AgResearch Ltd., Woodlands RD1, New Zealand; [4]Grasslands Research Centre, AgResearch Ltd., Private Bag 11008, Palmerston North, New Zealand; suzanne.rowe@agresearch.co.nz

New Zealand is heavily reliant on pastoral based agriculture and has a national sheep flock of ~18.5 million breeding ewes. Maternal sheep production is reliant on the ewe overwintering and successfully rearing at least one lamb. The sustainability and profitability of this system, however, is facing a new threat as awareness grows of the magnitude and impact of ruminant methane emissions on the environment. Ruminant enteric methane emissions are responsible for a third of New Zealand's total greenhouse gas emissions. Strategies limiting methane from ruminants have been put forward to protect the environment and to maintain global food security. Independent breeding strategies exist for increased production and for reduced methane emissions but, to date there has been no data to show whether these breeding objectives might be synergistic, neutral or antagonistic. Ten years ago, a divergent flock of sheep was created to evaluate selection for methane and to monitor the effects of this selection. Initially, sheep were ranked for breeding using methane measures from respiration chambers. Subsequently, a number of proxies have been investigated and effects of selection on methane emissions, production traits, feed intake, carcass and milk quality have been evaluated. Here we describe the main results and describe the flock divergence over the ten-year period. Low methane animals have ~6% less methane than the average sheep, appear economically favourable, grow more wool, have smaller rumens, are leaner, have different microbiomes and differ in fatty acid profiles in muscle.

Assessing the correctness of UK sheep population pedigree using genomic relationship matrix

K. Kaseja[1], G. Banos[1,2], J. Yates[3] and J. Conington[1]
[1]*Scotland's Rural College (SRUC), Roslin Institute Building, Easter Bush, FH25 9RG, Edinburgh, United Kingdom,* [2]*The Roslin Institute, University of Edinburgh, Easter Bush, EH25 9RG, Edinburgh, United Kingdom,* [3]*The British Texel Sheep Society, Stoneleigh Park, Kenilworth, Warwickshire, CV8 2LG, United Kingdom; joanne.conington@sruc.ac.uk*

The UK national genetic evaluation for sheep is based purely on the pedigree provided by breed societies, using conventional Best Linear Unbiased Prediction method. Correctness of pedigree is essential in order to avoid bias in the evaluation. The objective of this study was to determine if the genomic relationship matrix (GRM) can be used to discover and verify pedigree inconsistences. More than 8,500 Texel sheep genotypes were collected over three years using genome-wide DNA arrays, of which 7,995 genotypes passed quality control and were used in the study. For previously confirmed parent-offspring pairs (n=4,994), opposite homozygous verification was performed and minimum, maximum and the average relationship based on GRM was calculated. These values were used as thresholds in order to validate the relationships in the wider population, and were on average 0.50, 0.25 and 0.13 for parent-offspring, grandparent-offspring and great grandparent-offspring relationships respectively; additional thresholds of 0.50 and 0.27 were set for full- and half-siblings, respectively. Unexpectedly high relationship coefficients (r_{ij} equal to 0.78 and 0.70) were discovered between two pairs of half-siblings with a verified common dam, questioning the correctness of the sires. Absence of the respective sire genotypes prevented the use of the opposite homozygote method to check the correctness of these relationships. Further investigation with the use of GRM, revealed that one of the sires was wrongly reported. This finding was based on the relationship coefficients between the animal in question and genotyped potential half-siblings from the reported sire, with r_{ij} indicating a zero relationship between the reported sire and offspring. These results indicate that using GMR with pre-obtained thresholds can indicate, and potentially help to resolve, pedigree errors between pairs of animals using genotypes from the wider population.

Towards the genetic and genomic improvement of British Texel sheep

E.M. Smith[1], J. Conington[2] and J. Yates[1]
[1]*British Texel Sheep Society, Stoneleigh Park, Kenilworth, Warwickshire, CV8 2LG, United Kingdom,* [2]*Scotland's Rural College, Roslin Institute Building, Easter Bush, Edinburgh EH25 9RG, United Kingdom; edsmith@texel.co.uk*

To produce crossbred lambs in Britain, Texel sires are mated to 30% of the national flock (approx. 11M ewes); Texel and Texel-cross ewes also form 12% of this maternal flock. The high level of market share, and accelerating rates of genetic progress, are estimated to contribute over £23M/year to the UK sheep industry. To facilitate the continued progression of the breed, the development and maintenance of a genetic reference population has been necessary to sustainably support the efficient use and commercial application of genomic technology within the wider population. The British Texel Sheep Society began genomic profiling of animals in 2009 with an initial study of 390 rams. Society policy has evolved over the past decade, with embryo donor ewes genotyped (2016 only) and all registered male animals genotyped before their offspring can be registered (n=1,500/year). Building on this foundation of wider population sampling, there has been targeted genotyping of specific animals that have been phenotyped either through CT (computer tomography) scanning (n=300/year) or as part of a footrot and mastitis research programme. The latter has resulted in the collection of hard-to-measure health phenotypes and genotypes from over 4,000 fully performance recorded pedigree ewes over the last four years. In 2017 the Society also genotyped its first Texel-cross slaughter generation animals as part of a meat and carcase quality research project. Support from InnovateUK and collaborations with SRUC, Neogen, AgResearch NZ, Ovita NZ and AbacusBio have helped define commercial policy and research strategy and contributed to this valuable resource. To date, over 10,000 pedigree animals have been genotyped at densities ranging from 15,000 to 606,006 SNPs per animal, representing the largest genomic dataset on a specific terminal sire sheep breed, worldwide. Many of the genotyped individuals and/or their offspring have associated data recorded on traits such as growth rate and carcase quality. The result is a powerful genomic resource that will underpin TexelPlus Genetic Improvement Services and can be exploited for further genomic research purposes.

Genetic parameters for key biomarkers involved in body reserves dynamics: preliminary results

T. Macé[1], E. González-García[2], J. Pradel[3], S. Douls[3] and D. Hazard[1]
[1]INRA, GenPhySE, 24 Chemin de borde rouge, 31326 Castanet Tolosan, France, [2]INRA, UMR SELMET, Place Pierre Viala, 34060 Montpellier Cedex 1, France, [3]INRA, UE321, La Fage, 12250 Roquefort-sur-Soulzon, France; tiphaine.mace@inra.fr

In ruminants, body reserves (BR) are mobilized in situations of negative energy balance, including physiological stages with high metabolic challenges. In contrast, BR accretion occurs when nutrient requirements decrease. This dynamic is partly controlled by metabolic patterns, whereas its genetic determinism is still unknown, at least in sheep. The objective of this preliminary study was to estimate genetic parameters for a few key metabolites and hormones involved in sheep BR dynamic. The BR changes over time were monitored over three years on 318 productive Romane ewes reared exclusively in an extensive rangeland. Plasma non-esterified fatty acids (NEFA), beta-hydroxybutyrate (β-OHB) and triiodothyronine (T3) levels were quantified 2 weeks before and 3 weeks after lambing and 2 weeks before mating. Effects of physiological stage, litter size and a parity-year factor were tested to identify main sources of variation. Genetic parameters were estimated using animal mixed model considering independently measurements at each physiological stage. The least square means for NEFA, β-OHB and T3 were significantly (P<0.001) different according to the physiological stage. The levels of NEFA and β-OHB were higher before lambing (0.44±0.01 mmol/l and 39.83±0.80 mg/l, respectively) and decreased until before mating (0.15±0.01 mmol/l and 23.09±0.95 mg/l, respectively). The level of T3 was equivalent before lambing and before mating (around 1.00 ng/ml) and lower than before mating (1.35±0.02 ng/ml). Heritabilities for NEFA, β-OHB and T3 ranged between 0.11±0.12 to 0.29±0.12, 0.21±0.11 to 0.35±0.17 and 0.16±0.11 to 0.38±0.16, respectively, depending on the physiological stage. Genetic correlations between NEFA and β-OHB before lambing and before mating reached 0.89±0.13, and -0.90±0.25, respectively. The T3 was also highly correlated with β-OHB before lambing (-0.70±0.26). These preliminary results suggest that metabolites and hormones involved in BR dynamics are heritable traits. Further measurements on additional ewes are in progress to confirm these results.

Artificial insemination of indigenous goats in Greece

E. Pavlou, I. Pavlou and M. Avdi
Aristotle University of Thessaloniki, Department of Animal Production, Thessaloniki, 54124, Greece; elefpavlou@gmail.com

Artificial insemination (AI) is applied in developed countries in order to optimize improvement schemes and manage the breeding of goats, targeting to control the kidding period and milk production. Although it's application is widespread in Europe, in Greece is practiced on a small scale. The purpose of the present work is to study the fertility of indigenous Greek goats following the application of A.I. with frozen semen. The experiment was carried out during the non-breeding season. 49 indigenous Greek goats fed and managed under the same conditions in a conventional farm, were used for this experiment. Oestrous was synchronised using intravaginal sponges containing 20 mg flugestone acetate (Chronogest, MSD) for 11 days, followed by intramuscular injection of 380 IU PMSG (Intergonan, MSD) and 50 μg Cloprostenol (Estrumate, MSD) 48 h before sponge removal. Goats were inseminated exo- or intra-cervically, during the induced oestrous at 43 h after sponge removal. The does were inseminated once. Each A.I. dose contained 2.5×10^8 sperm cells, frozen in a 0.2 ml straw. The semen came from Scopelos breed bucks. The influence of age, place of deposition of the semen in the cervix, number of previous kiddings and milk production of the goats, on fertility was analysed using X^2-test. The A.I. has given an average fertility 75.51%. Age of the goat had a significant effect on fertility (P<0.05), with older goats reaching the highest percentage (77.77%). The highest fertility percentage was measured when the AI was intra-cervical (81.81%, P<0,01). Number of previous kiddings had a significant effect on fertility rate (P<0.05). Regarding the milk production (animals with greater than or less than 1.5 l/d), a significant (P<0.05) effect on fertility was observed, with goats producing less milk having higher fertility rates. Results of this experiment demonstrated that the fertility of indigenous Greek goats after A.I. depends on factors, such as age of the goat, place of deposition of the semen in the cervix, number of previous kiddings and milk production of the goats.

Residual intake and body weight gain on the performance and carcass composition of lambs
E.M. Nascimento, C.I.S. Bach, H. Maggioni, W.G. Nascimento, S.R. Fernandes and A.F. Garcez Neto
Federal University of Paraná, Palotina, Paraná, 05.950 000, Brazil; amerieo.garcez@ufpr.br

The aim of this study was to evaluate the relationship between residual intake and body weight gain (RIG) with performance and carcass composition in ½ Dorper × ½ Santa Inês lambs. Nineteen non-castrated male lambs, with four months of age and 24±3 kg of body weight (BW), were fed *ad libitum* in feedlot for 90 days. At the end of trial, lambs were classified as efficient, intermediary and inefficient for RIG based on ±50% of standard deviation for this trait. The final BW, mean metabolic weight, average daily gain and dry matter intake (DMI) not differed (P>0.05) among RIG groups, with mean values of 34.58 kg, 12.41 $kg^{0.75}$, 128.2 g/day and 949.5 g/day, respectively. However, the DMI relative to the BW (DMI_{BW}) presented the lowest and the highest values (P<0.05) in the efficient and inefficient groups, respectively (3.11 vs 3.44% BW/day). The feed conversion ratio reduced (P<0.05) by 0.42% between intermediary and efficient groups (7.15 vs 7.12 kg DM/kg gain), and by 15.94% between inefficient and efficient groups (8.47 vs 7.12 kg DM/kg gain). Conversely, gross feed efficiency increased (P<0.05) by 1.79% from intermediary to efficient group (140 vs 142.5 g gain/kg DM), and by 19.65% from inefficient to efficient group (119.1 vs 142.5 g gain/kg DM). Thus, the efficient lambs consumed 1.35 kg DM/kg gain less and presented 23.4 g gain/kg DM more than inefficient ones. The characteristics of longissimus muscle not differed (P>0.05) among RIG groups, with mean values of 1.76 mm, 2.72 mm, 2.24 mm and 12.22 cm^2 for minimum, maximum and mean backfat thickness measured on the ribeye, and ribeye area, respectively. Differences in body composition are related to variation of residual feed intake (RFI), where low-RFI animals have highest muscle development and lower fat deposition. Since RIG is derived from RFI, it was expected that changes in RIG could be related to differences in body composition. The improvement of efficiency for RIG in lambs leads to better feed utilization, reducing the feed intake relative to body weight and feed conversion ratio, and increasing the gross feed efficiency without affecting the carcass composition.

Seasonal variation of semen characteristics in indigenous Greek goat bucks
E. Pavlou and M. Avdi
Aristotle University of Thessaloniki, Department of Animal Production, Thessaloniki, 54124, Greece; elefpavlou@gmail.com

The aim of this study was to evaluate seasonal variation in ejaculate volume, sperm concentration, total number of spermatozoa per ejaculation, semen's mass and progressive motility, viability of spermatozoa and the percentage of abnormal spermatozoa in indigenous Greek goat bucks. Semen characteristics were assessed in 10 mature bucks, raised on the same farm and under the same conditions, aged 8 months old and weight 42±2,9 kg at the beginning of the experiment. Semen collection and evaluation conducted once a week for a period of 12 months. The influence of seasonality was analysed by Analysis of Variance, using General Linear Models (PASW Statistics 18.0). All bucks examined showed a similar pattern in all the reproductive parameters studied. Seasonal differences (P<0,05) were noted for all study parameters. Semen volume presented the highest value (1,32±0.09 ml) during the early breeding season (summer) and the lowest (1,06±0,08 ml) during the non-breeding season (spring). Mass and progressive motility presented higher values (4,65±0,06 and 4,47±0,07 respectively) during the breeding season and lower values (4,48±0,09 and 4,35±0,09 respectively) during the non breeding season. Viability of spermatozoa showed the highest values (82,16±1,51%) in summer. In addition, semen concentration presented higher values during spring [(6,57±0,17)x109 spermatozoa/ml] and a decrease in values [(5.98±0.14)x109 spermatozoa/ml] during autumn. The total number of spermatozoa per ejaculation was also significantly higher during summer [(7,99±0,63)x109] than in spring [(6.96±0.61)x109]. The percentage of abnormal spermatozoa presented higher values outside of the natural breeding season (16,27±0,56%) and the lowest values during the breeding season (12,03±0,63%). The data presented in this study suggests that indigenous Greek goat bucks' semen quantity and quality exhibit a significant seasonal variation, with the best parameters obtained during the natural breeding season.

Response of dairy ewes supplemented with barley β-glucans to an *E. coli* LPS challenge

A. Elhadi, S. Guamán, E. Albanell and G. Caja

Universitat Autònoma de Barcelona, Group of Research in Ruminants (G2R), 08193 Bellaterra, Spain; gerardo.caja@uab.es

Thirty-six dairy ewes in late-lactation, were used. Ewes were adapted to the experimental conditions and fed alfalfa hay (*ad libitum*) and 350 g/d of conventional barley cv Meseta low in β-glucans (3.8% BG) for 7 d. Thereafter, they were allocated into 3 groups to which the experimental treatments were applied for 14 d. They were: (1) Control (CO), same as adaptation (13.3 g BG/d); (2) Oral (OR), with a new barley variety (cv Kamalamai, with 10% BG (35 g BG/d); and, Injected (IN), same as CO with a single dose of barley BG intraperitoneally injected (140 ml of 1.4% BG; 2 g BG/ewe). After 9 d, ewes were submitted to an intramammary LPS challenge with 1 ml of an *E. coli* endotoxin solution (5 µg/ml, in an udder-half and 1 ml of saline in the other. Rectal temperature, milk yield, milk composition and plasma interleukin 1 (IL-1) were monitored individually for 5 d. Data were treated by PROC MIXED for repeated measurements of SAS and LS means separated by PDIFF at P<0.05. No effects were observed in rectal temperature (P=0.13), milk yield (P=0.29) and α and β IL-1 (P=0.31 to 0.84). However, increasing BG tended (P=0.06) to reduce the decrease of lactose in milk by LPS. This decrease showed the severe affectation of mammary gland by LPS, which was numerically lower in the IN treatment vs the others (3.80 vs 3.57%, respectively). SCC increased due to the effect of LPS, but there were no differences by BG treatment. Plasma α and β IL-1 did not differ by BG nor by time. However, their numerical values increased by time, agreeing with the increase in SCC and inflammation. Moreover, basal IL-1 values were lower for IN, when compared to the others (299 vs 352 pg/ml and 79 vs 152 pg/ml, respectively for α and β IL-1; P<0.01) indicating that BG supplementation may increase the immunity status and the protection of dairy ewes against mastitis. In conclusion, the LPS challenge triggered a short-term immune response in BG supplemented ewes, which was more effective when injected intraperitoneally.

Growth performance and carcass traits of two fat-tailed sheep breeds

M.D. Obeidat and B.S. Obeidat

Jordan University of Science and Technology, Jordan University of Science and Technology, 22110, Jordan; mdobeidat@just.edu.jo

The aim of the current study was to compare growth performance and carcass traits of two fat-tailed sheep breeds. A total of thirty Awassi and Najdi male lambs (15 lambs per breed) were randomly assigned to individual shaded pens (0.75×1.5 m). Lambs were weighed at the beginning of the study and biweekly then after; whereas, feed intake was recorded daily. The study lasted for 77 days (7 days for adaptation and 70 days for data collection). At the end of the study, lambs were slaughtered for the carcass traits evaluation. Data was analysed using the mixed procedure of lamb breed was treated as fixed effect for the growth traits and carcass weight, while lamb was treated as a random effect. For carcass components, lamb breed was the fixed effect, lamb as random effect, and carcass weight as a covariate. No significant differences were detected for initial and final weight, ADG, and feed to gain ratio. Nitrogen intake (g/d) and nitrogen retention (%) were higher (P<0.05) for Najdi breed (34.9 and 27.2 g/d for N intake; 51.3 and 41.7% for N retention, for Najdi and Awassi, respectively). Leg cut percentage was greater (P<0.05) for Awassi lambs, while Shoulder, Loin, rack, fat tail percentages did not differ. In summary, growth performance and carcass traits were comparable between the two breeds.

Forage particle size and starch content in lambs diet – effects on productivity and product quality

J. Santos-Silva[1], A. Francisco[1], M. Janíček[2], J.M. Almeida[1], A.P.V. Portugal[1], E. Jerónimo[3], S. Alves[4] and R.J.B. Bessa[1]
[1]Instituto Nacional de Investigação Agrária e Veterinária, Fonte Boa, 2005-048 Vale de Santarém, Portugal, [2]Slovak University of Agriculture, Trieda Andreja Hlinku 609/2, 949 76 Nitra, Slovak Republic, [3]CEBAL, R. de Pedro Soares, 7800-309 Beja, Portugal, [4]Faculdade de Medicina Veterinária, U. Lisboa, Av. Universidade Técnica, 1300-477 Lisboa, Portugal; jose.santossilva@iniav.pt

Thirty-two Merino lambs were used in experiment designed to evaluate the effects of forage particle size and starch level in complete mixed diets, on growth performance, carcass and meat quality traits. All the diets were supplemented with 6% soybean oil and include 40% of alfalfa. The lambs were randomly allocated to eight diets that combined two forms of presentation of alfalfa hay – chopped or ground – with four levels of barley in the diet – 33.0, 21.3, 11.2 and 0%. The lambs were individually housed and stayed on trial during six weeks until slaughter. Feed intake was controlled daily and live weight weekly. Dry matter intake was 17% higher (1,239 vs 1,061 g/day) and average daily weight gain (ADG) was 10.4% higher (328 vs 297 g/day) when forage was grounded. Carcass and meat quality traits were not affected by treatments except carcass weight that was about 6% higher for ground alfalfa (16.3 vs 15.4 kg). The reduction of barley caused a linear reduction of intake and of ADG. The treatments had no effect on feed conversion ratio. The effects on carcass traits were not relevant but meat shear force decreased and tenderness increased. Fatty acid (FA) composition of m. longissimus thoracis was influenced by treatments. Grinding alfalfa caused a decrease in the sums of saturated, branched chain and n-3 polyunsaturated FA (PUFA) and an increase of the sum of n-6 PUFA and the sum of PUFA. The reduction of barley increased the sums of branched chain, n-6 PUFA and n-3 PUFA. Concerning biohydrogenation intermediates, all the lambs presented meat with high proportion of healthy FA t11-18:1 (6.49% of total FA) and c9,t11-18:2 (1.49% of total FA) and low proportion of unhealthy FA t10-18:1 (2.58% of total FA). The reduction of barley in diet had a moderate positive impact in meat nutritional value decreasing t10-18:1 and this effect has been enhanced by the increase in forage particle size.

Tibetan sheep are better adapted than Han sheep to low energy diets due to more efficient energy use

X.P. Jing[1,2], W.J. Wang[2], Y.M. Guo[2], J.P. Kang[2], J.W. Zhou[2], V. Fievez[1] and R.J. Long[2]
[1]Laboratory for Animal Nutrition and Animal Product Quality, Department of Animal Sciences and Aquatic Ecology, Faculty of Bioscience Engineering, Ghent Universi, Ghent, 9000, Belgium, [2]State Key Laboratory of Grassland and Agro-Ecosysems, School of life sciences, Lanzhou University, Lanzhou, 730000, China, P.R.; xiaoping.jing@ugent.be

Tibetan sheep represent the largest number of indigenous livestock in the Qinghai-Tibetan Plateau (QTP); they play an important role in its ecosystem and are well-adapted to and even thrive under the harsh conditions. The objective of this study was to investigate whether these sheep differed in energy conversion efficiency and regulatory capacity in energy metabolism by feeding them diets with increasing energy density. The trial included 24 Tibetan sheep and 24 Han sheep (a popular sheep breed raised in the northern plains and hilly regions of China) which were used as control. Animals of each species were randomly distributed into one of four groups and were receiving one of the four diets with varying digestible energy (DE) density: 8.21, 9.33, 10.45 and 11.57 MJ DE/kg DM respectively. After 14 d of adaptation, a 42 d feeding experiment and 1-week digestion and metabolism experiment took place. The results revealed that Tibetan sheep have higher apparent digestibility and higher efficiency in dietary energy conversion across treatment which increased linearly as dietary DE level increased, and they require less energy (0.41 vs 0.50 MJ/BW$^{0.75}$/d) for maintenance compared with Han sheep. When supplied with the lowest energy diet (8.21 MJ DE/kg), compared with Han sheep, the higher glucose and glucagon, and lower insulin concentration of Tibetan sheep indicated that they have a higher capacity for gluconeogenesis and regulation ability of glucose metabolism; whereas the higher GH and creatinine, and lower NEFA concentration of Han sheep indicated that they mobilize stored fat more quickly; in addition, the higher NEFA and lower VLDL and TG concentration of Tibetan sheep indicated they have a higher capacity for NEFA oxidation, whereas Han sheep showed a higher TG synthesis ability. Consequently, Tibetan sheep seem to cope better with low energy diets which could indicate an enhanced adaptation to the harsh conditions and challenges of the QTP.

Concentration of IgG in sheep colostrum is improved by melatonin implantation during pregnancy
J.A. Abecia[1], C. Garrido[1], M. Gave[1], A.I. García[1], D. López[1], S. Luis[1], J.A. Valares[2] and L. Mata[2]
[1]IUCA, Universidad de Zaragoza, Miguel Servet, 177, 50013 Zaragoza, Spain, [2]ZEULAB S.L., Polígono PLAZA, Bari, 25 dpdo, 50197 Zaragoza, Spain; alf@unizar.es

This study was designed to study the effect of exogenous melatonin during pregnancy and lamb gender on colostrum quality. Sixty pregnant ewes were used. A group of animals received a subcutaneous melatonin implant (Melovine, CEVA Salud Animal, Barcelona, Spain), at the third month of pregnancy (Group 3M, n=13); another group was implanted with melatonin at the fourth month of pregnancy (Group 4M, n=18); the remaining ewes were non-implanted (Group Control, C, n=29). Immediately after lambing, a sample of colostrum was collected and frozen at -20 °C until analysis. IgG concentrations in ewe's colostrum were analysed using the ELISA test Calokit-Sheep (ZEULAB, Zaragoza, Spain). IgG concentration was evaluated statistically by a multifactorial model with a fixed effect of treatment or not with melatonin, offspring gender (male, female or male+female), and litter size (single vs multiple parturition). Melatonin implantation and gender of the lambs gestated significantly affected colostrum quality (P<0.001), with no effect of litter size. Thus, colostrum of ewes implanted with melatonin had a higher mean (±S.E.M.) IgG concentration than control ewes (55.54±3.09 vs 39.76±3.76 mg/ml, respectively; P<0.05), mainly due to the highest concentration of the 4M group (60.35±3.87 mg/ml) in comparison with 3M (48.32±4.52) and C groups (45.59±3.67) (P<0.001). The effect of lamb gender on IgG levels was evident in singletons (ewes with a male lamb: 57.6±6.14; ewes with a female lamb: 35.13±3.88 mg/ml; P<0.01). In general, the presence of a male in the litter increased colostrum quality (litters with at least on male: 55.18±3.13; litters with no males: 43.65±4.04 mg/ml; P<0.05). In the opposite, the presence of females lambs gave rise to lower levels of IgG in the colostrum (litters with at least one female: 46.25±3.02; litters with no females: 58.96±4.14 mg/ml; P<0.01). In conclusion, implanting ewes with melatonin at the fourth month of pregnancy produces a significant improvement of colostrum quality: moreover when male foetuses are carried by females, IgG concentrations are higher.

Including BMP15 genotype information in genetic evaluation of a prolific line of Barbarine Sheep
C. Ziadi[1,2], J.M. Serradilla[3], A. Molina[1] and S. Bedhiaf-Romdhani[4]
[1]Universidad de Córdoba, Departamento de Genética, Edificio Gregor Mendel, Campus de Rabanales. Ctra Madrid-Cádiz, km 396, 14071 Córdoba, Spain, [2]Institut National Agronomique de Tunisie, 43 Avenue Charles Nicolle, 1082 Tunis, Tunisia, [3]Universidad de Córdoba, Departamento de Producción Animal, Campus de Rabanales. Ctra Madrid-Cádiz, km 396, 14071 Córdoba, Spain, [4]Institut National de la Recherche Agronomique de Tunisie, Laboratoire des Productions Animales et Fourragères, Rue Hédi Karray, 2049 Ariana, Tunisia; ziadichiraz4@gmail.com

The aim of this study was to include the *FecX^bar/BMP15* genotype information in the genetic evaluation of a prolific line of Barbarine Sheep, of which a few number of animals was genotyped. This line is raised at the experimental station of the Tunisian National Institute for Agronomic Research (INRAT) and a selection program for prolificacy is carried out since 1979. The *FecX^bar/BMP15* mutation responsible for prolificacy and ovulation rate was identified, but this information is not considered in actual selection program of this line. Gene content multiple-trait approach was applied to accommodate the large number of missing genotypes in the genetic model. Traits considered were litter size (LS) and gene content (GC), which can be defined as the number of copies of a reference allele in the genotype of an animal. Genetic parameters and breeding values were estimated for both traits with a bivariate animal model using Gibbs sampling methodology, in the case of LS a categorical distribution was considered. Estimates of heritabilities were 0.07 for LS on the threshold scale and 0.99 for GC. Estimates of breeding values obtained with gene content multiple-trait approach were compared with those obtained with a classical animal model without genotype information. Correlation between estimated breeding values in both models was 0.23. This results indicate that including the *FecX^bar/BMP15* effect in the genetic evaluation for this prolific line is possible through the gene content approach. Ignoring the presence of the major gene in this population could generate a wrong ranking of the animals.

Genetic analysis of litter size in a prolific line of Barbarine sheep using random regression models

C. Ziadi[1,2], A. Molina[1] and S. Bedhiaf-Romdhani[3]
[1]Universidad de Córdoba, Departamento de Genética, Edificio Gregor Mendel, Campus de Rabanales, Ctra. Madrid Cádiz, km 396, 14071 Córdoba, Spain, [2]Institut National Agronomique de Tunisie, 43 Avenue Charles Nicolle, 1082 Tunis, Tunisia, [3]Institut National de la Recherche Agronomique de Tunisie, Laboratoire des Productions Animales et Fourragères, Rue Hédi Karray, 2049 Ariana, Tunisia; ziadichiraz4@gmail.com

Litter size records from a prolific line of Barbarine sheep were analysed for the first time using random regression models (RRM) based on orthogonal Legendre polynomials. A total of 2,751 litter records collected between 1989 and 2017 from the first to the sixth parities were provided by the Tunisian National Institute for Agronomic Research (INRAT) and included in the analysis. Estimation of all the covariance components was based on Gibbs sampling technique of Bayesian inference under categorical distribution. Non genetic effects included in the model were: year and month of lambing, and fixed quadratic regression coefficients on litter size with Legendre polynomials. Random additive and permanent environmental effects were modelled by second order Legendre polynomials. Heterogeneity of residual variances, assuming three classes of measurement errors according to parity were considered. Estimates of heritabilities ranged from 0.04 in the first parity to 0.18 in the sixth parity. Estimates of genetic correlations were in a wider range from 0.25 between the second and sixth parities to 0.96 between the third and fourth ones. Due to the changes of variances along the parities trajectory, the use of random regression model to analyse the genetic trajectory of litter size is justified for this prolific Barbarine line.

Is it worth to change from lactational to test-day genetic evaluation model in Latxa dairy sheep?

C. Pineda-Quiroga and E. Ugarte
NEIKER-Tecnalia, Basque Institute of Agricultural Research and Development., Animal Production, Agrifood Campus of Arkaute s/n, E-01080 Arkaute, Spain; cpineda@neiker.eus

Genetic evaluation for milk production and composition in the Latxa breed has been based on 120-days standardized lactation records so far. This methodology requires at least three morning test day records (TD) per ewe taken at different lactation stages, the first of which must have been taken 70 d or less after lambing. A time interval between TD shorter than 35 d is also needed. Otherwise, lactation is not calculated and genetic evaluation of ewes that do not meet these requirements is lost. In the case of the Latxa sheep, these losses represent around 25% on average per year. In contrast, every TD record on each ewe is directly used in TD model analysis after filtering abnormal measures, regardless the recording schemes, making it possible to predict the genetic merit of a greater number of animals. The current study aims at comparing the Estimated Breeding Values (EBV) obtained under lactational (LC) and TD models for milk yield, protein and fat percentage and yield of Latxa Cara Rubia (LCR), Latxa Cara Negra from Euskadi (LCNEUS) and from Navarre (LCNNAF), as well as the reliability (r) of EBV. To do so, 120-day standardized lactational data and TD records from LCR and LCNEUS were collected between 2000 and 2017, while LCNNAF comprised years 2002 to 2017. For TD, milk recordings were considered as a repeated measure of the same trait. The contemporary group herd-year-month of lambing (HYM) and herd-test day (HDT) were respectively fitted for LC and TD models. As expected, TD allowed the evaluation of fairly more animals than LC model: 4.469 additional ewes in LCR, 9.368 in LCNEUS and 1.259 in LCNNAF. As regards EBV, Pearson correlation showed a significant correlation between models for all evaluated traits ($R^2>0.95$, P<0.001 in all ecotypes), indicating that estimated values of both models are highly comparable. It was also noted an increase of about 9% in r of fat percentage, and 4% of protein percentage in both LCR and LCNEUS when TD was tested. In light of the present findings, we might infer that it is suitable to implement the TD model in Latxa genetic evaluations.

The estimation of dispersion parameters for body weight of rams at the end of performance test

M. Simčič[1], B. Luštrek[1], N. Vujasinović[1], M. Štepec[1] and G. Gorjanc[1,2]
[1]University of Ljubljana, Biotechnical Faculty, Department of Animal Science, Jamnikarjeva 101, 1000 Ljubljana, Slovenia, [2]University of Edinburgh, The Roslin Institute and (Dick) School of Veterinary Studies, Easter Bush, Edinburgh, United Kingdom; mojca.simcic@bf.uni-lj.si

The objective of this study was to estimate genetic and environmental dispersion parameters for body weight of rams at the end of performance test in two breeds of sheep in Slovenia, the autochthonous Jezersko-Solčava sheep and the Improved Jezersko-Solčava sheep. Both breeds are widespread in the country and reared mainly for lamb production. The Improved Jezersko-Solčava sheep originated from upgrading the Jezersko-Solčava with Romanov sheep to increase litter size. In these two breed young male lambs are selected from different flocks in the country for a performance test in two test stations with different conditions. Performance test lasts 100 days during which rams are weighted four times. The analysed data collected from 1996 to 2018 had 18,752 body weights on 6,034 rams of both breeds. A pedigree file with 12,092 animals was constructed. Most rams had final test weight around 9 months, so we expressed trait as body weight at 270 days. The fixed part of the model included a breed-test station interaction, litter size, dam parity, season as day of weighing, and age at weighing as a quadratic polynomial nested within breed-test station. The random part of the model included the additive genetic effect of an animal, permanent environment effect and the effect of flock origin-year interaction. Variance components were estimated using REML method implemented in the VCE-6 program. Analysis of variance showed significant differences in body weight at 270 days between breeds and test stations, higher body weight for rams born as single lambs and lower body weight for rams born in the first parity. Heritability estimate for body weight at 270 days was 0.28. The flock of origin-year effect (0.29) and permanent environment effect (0.34) explained similar ratios of variance. The smallest part of the phenotypic variance remained in the residual (0.09). These dispersion parameters will be used in the breeding value prediction from the year 2019 onwards.

Genetic parameter estimates for adult weight and maternal efficiency traits in Santa Inês breed

M. Piatto Berton[1], R.P. Da Silva[2], F.E. De Carvalho[2], P.S. Oliveira[2], J.P. Eler[2], G. Banchero[3], J.B.S. Ferraz[2] and F. Baldi[1]
[1]Faculdade de Ciências Agrárias e Veterinárias – FCAV / UNESP, Via de Acesso Prof.Paulo Donato Castellane s/n, 14884-900 Jaboticabal, SP, Brazil, [2]Faculdade de Zootecnia e Engenharia de Alimentos, FZEA / USP, Av. Duque de Caxias Norte, 225, CEP 13635-900 Pirassununga/SP, Brazil, [3]Instituto Nacional de Investigación Agropecuária -INIA, INIA La Estanzuela Ruta 50, Km. 11, Colonia, Uruguay, Uruguay; mapberton@gmail.com

The aim of this study was to estimate variance components to prolificity, maternal efficiency and ewe adult weight in Santa Inês breed in tropical conditions. The phenotypic records were collected from 700 animals of Santa Inês breed, belonging to four flocks located in Southeast and Northeast states of Brazil. The evaluated traits were: maternal efficiency (ME), metabolic maternal efficiency (MME), ewe adult weight (AW) and metabolic ewe adult weight (MAW). A total of 576 animals were genotyped with the Ovine SNP12k BeadChip (Illumina, Inc.). Quality control consisted of excluding markers with unknown position, on sex chromosomes, monomorphic, with MAF<0.05, call rate≤90%, and with excess heterozygosity. The variance components were estimated using a single animal trait model by the ssGBLUP procedure. The contemporary groups (year and farm) and the covariable body composition and animal age at measurement were included as fixed effects in the model. The analyses were performed using the REMLF90. Moderate heritability estimates were obtained for AW (0.32) and MAW (0.34), and low estimates were obtained for maternal efficiency indicator traits, ME (0.07) and MME (0.08). The genetic correlation estimates between AW, MAW, ME and MME, ranged from -0.58 to -0.77, suggesting that the selection for higher ewe AW would reduce the maternal ewe efficiency. The synchronization of genetic potentials with local environmental conditions, management systems and optimization of reproductive capacity is central for genetic improvement of Santa Inês breed in tropical conditions. Acknowledgments: São Paulo Research Foundation (FAPESP) grant #2014/07566-2 and #2017/10493-5.

Alfa s1 casein (CSN1S1) genetic diversity in two Portuguese goat breeds

M.F. Santos-Silva[1], C. Oliveira E Sousa[1], A.P. Rosa[2], A. Cachatra[3], I. Carolino[1] and N. Carolino[1]
[1]Instituto Nacional de Investigação Agrária e Veterinária (INIAV), EZN-Fonte Boa, 2005-048, Portugal, [2]Associação Nacional dos Criadores de Caprinos da raça Algarvia (ANCCRAL), Azinhal, 8950- Azinhal, Portugal, [3]Associação Portuguesa de Ciprinicultores da Raça Serpentina (APCRS), Rua Diana de Liz, Horta do Bispo, Ap. 194, 7002-503 Évora, Portugal; fatima.santossilva@iniav.pt

Alpha S1 casein as a recognized effect on cheese production processes. Caprine Alfa S1 locus shows high polymorphism, that influence casein level in milk. Certain alleles have being related with high (A, B, C), medium (E) and low or zero (F, D, N, O) levels of casein. Cheese production is very important for caprine production sustainability and its improvement is of main interest. This work developed in collaboration of breeders Associations and ALT-Biotech[RepGeN] project, aimed to analyse genetic variability at alfa S1 casein in two autochthonous goat breeds Algarvia (AL) and Serpentina (SP), from the south of Portugal, for subsequent application in the breeding programs. Blood samples, 178 from AL and 78 from SP, were analysed for alfa S1 polymorphism's. Genotyping was performed by PCR-specific allele supplemented by RFLP's when required. This preliminary results revealed for AL breed the five alleles possible to identify by this methodology (A, B, E, F and N), while SP show those alleles excepting N allele. E allele is the most frequent at the two breeds (65.%- AL) and (73%- SP) as it is seen in other Iberian populations. The A allele is more frequent at AL (18%) than in SP (2.4%), while the B allele is more frequent at SP (21% versus 8%). The F allele has similar representation of approximately 3-4% in the two breeds. The most frequent genotype is EE at the two breeds (43-53%) followed by AE (23%) at AL and BE (33%) at SP. To date that we know, the N allele has not appeared in none of the Portuguese populations excepting AL. The αS1 casein polymorphisms found at the breeds studied make it possible to establish genetic schemes towards the desired alleles, but further studies including more data will be necessary to clarify associations among genotypes and quantitative and qualitative milk traits.

---

n-3 fatty acid level of peripartum ewes diets affect the organ fatty acid profile of suckling lambs

G. Battacone[1], G.A. Carboni[1], G. Pulina[1] and G. Bee[2]
[1]Univerity of Sassari, viale Italia 39, 07100 Sassari, Italy, [2]Agrosocpe Posieux, la tioleyre 4, 1725 Posieux, Switzerland; battacon@uniss.it

The experiment aimed to investigate to what extent maternal dietary n-3 supplementation during gestation and lactation or both affect the fatty acid profile of liver, muscle and brain of suckling lambs. For the study, 2 isocaloric and isoproteic concentrate diets were prepared, control (C) or enriched with extruded linseed diet (L). Twenty-four Sarda breed ewes were fed either the C or L treatment (n=12 each) during the last 8 weeks of gestation. After parturition, 6 ewes of each experimental group continued to receive the same concentrate (C-C; L-L), whereas the other 6 ewes were switched to either the C or L concentrate (L-C; C-L) for 4 weeks of lactation. At birth the 48 lambs were kept with their mothers and fed exclusively maternal milk. All lambs were slaughtered at 4 weeks of age. Fatty acid (FA) profile of the liver and brain tissues was determined by gas chromatography. Compared to the unsupplemented group (C-C), linseed enriched diets in the pre-partum period had no effect on the FA profile of the liver compared to L-C lambs. Suckling lambs from ewes fed linseed after parturition (C-L, L-L) had greater MUFA and lower PUFA level in the liver compared to C-C and L-C lambs. The DHA content of the brain tissues, but not of the liver tissues, was positively affected by the inclusion of linseed in the ewe diets. This effect was greater when offered during both gestation and lactation. In conclusion, dietary 18:3n-3 supply of ewes stimulated DHA production in the brain but not in the liver of suckling lambs.

Total merit index for sheep breeds with focus on meat production

B. Fuerst-Waltl[1] and C. Fuerst[2]
[1]*University of Natural Resources and Life Sciences Vienna (BOKU), Gregor Mendel-Str. 33, 1180 Vienna, Austria,*
[2]*ZuchtData EDV-Dienstleistungen GmbH, Dresdner Str. 89, 1200 Vienna, Austria; birgit.fuerst-waltl@boku.ac.at*

In 2017, more than 11,200 herdbook ewes older than one year belonged to breeds with a focus on meat production, i.e. either a specialized meat breed or the land sheep breed Merinoland, in Austria. Meat performance testing is carried out by trained representatives of the breeding organizations and became mandatory in 2003. Apart from weighing at approximately 40 kg and calculating daily gain, eye muscle depth and fat depth are measured by ultrasound. With regard to fitness-related data, information on lambing intervals, number of lambs born and number of lambs born alive is available due to the routine registration of lambings in the nationwide data base. Along with the introduction of a routine genetic evaluation for the traits mentioned above, a total merit index (TMI) was defined for both breed groups. Economic values were available for daily gain, lambing interval and number of lambs born only. Relative economic weights were hence defined based on the expected selection response. In land sheep, the relative weighting of meat: fitness is 40% (12% daily gain direct, 14% muscle depth, 1% fat depth, 13% daily gain maternal): 60% (21% lambing interval, 20 and 5% number of lambs born maternal and paternal, 11 and 3% number of lambs born alive maternal and paternal)while in specialized meat sheep it is 60% (15% daily gain direct, 26% muscle depth, 1% fat depth, 18% daily gain maternal): 40% (8% lambing interval, 19.6 and 4% number of lambs born maternal and paternal, 6.8 and 1.6% number of lambs born alive maternal and paternal). In the next step, a routine genetic evaluation for longevity will be developed. The trait longevity will also be included in the future TMI.

Relationships between body reserves dynamics and ewes rearing performances in meat sheep

T. Macé[1], E. González-García[2], F. Carrière[3], S. Douls[3], D. Foulquié[3] and D. Hazard[1]
[1]*INRA, GenPhySE, 24 Chemin de borde rouge, 31326 Castanet Tolosan, France, [2]INRA, UMR SELMET, Place Pierre Viala, 34060 Montpellier Cedex 1, France, [3]INRA, UE321, La Fage, 12250 Roquefort-sur-Soulzon, France; tiphaine.mace@inra.fr*

Since a few years, breeders are looking to select more robust animals in order to improve adaptation to more challenging environments. One expression of such robustness is a suitable cyclic alternation of accretion and mobilization of body reserves (BR) i.e. BR dynamics. In a previous study, we characterized several BR trajectories based on body condition score (BCS) and body weight (BW) measurements regularly recorded according to a physiological stages schedule. In addition, in a previous multi-traits approach of BR changes over time we found that BR accretion and mobilization were heritable traits and genetically linked. The objectives of the present study were therefore: (1) to evaluate the link between BR trajectories and ewes rearing performances and; (2) to estimate the genetic correlations between BR changes and ewes rearing performances. By using records from 1,146 productive Romane ewes, their rearing performances were estimated by analysing key parameters like the total weight of the litter at lambing and at weaning, and the average daily gain (ADG) of the lamb during their first, second and third month of life. The effects of the BR trajectories of the dam were also tested. Genetic correlations were estimated between the ewes' BR changes through time and their rearing performances, by conducting animal mixed model analyses. Overall, ewes presenting a stronger decrease in their trajectories during the BR mobilization period, showed superior rearing performances. Negative favourable correlations were found between BR mobilization and the total litter weight at lambing (-0.42±0.15) and between BR mobilization and ADG of lambs during the first and second month of suckling (-0.29±0.14). Positive favourable correlations were found between BR accretion and litter weight at birth (0.42±0.15) and between BR accretion and the lamb ADG during the first month of life (0.37±0.15). Present results suggest that BR dynamics could be considered in genetic selection programs to improve animals' resilience while enhancing ewes' performances when facing external challenges.

Divergent genetic selections for an early behavioural reactivity in meat sheep

D. Hazard, A. Morisset, D. Foulquié, E. Delval, S. Douls, F. Carrière, J. Pradel and A. Boissy
INRA, GENPHYSE, 24 chemin de Borde Rouge, 31326 Castanet Tolosan, France; dominique.hazard@inra.fr

Including behavioural traits in genetic selection could be an advantageous strategy for improving the autonomy of animal to extensive farming systems in an agroecological perspective. More particularly, farm animals are social and gregarious, and relational behaviours are essential for ensuring social cohesion, social facilitation, offspring survival and docility. Breed differences and genetic variation within breed have been reported in sheep for social behaviour and found to be heritable, suggesting such behaviour could be selected. The aim of the present study was to investigate in sheep the efficiency of an early selection for social attractiveness or tolerance to the human. For 7 years, four divergent lines were produced according to the reactivity of lambs to social separation or towards a human. Each year, in average 300 female and male lambs from 250 dams were individually exposed just after weaning to two behavioural tests. The arena test aims at assessing the motivation of the lambs towards conspecifics and the corridor test aims at assessing the reactivity of the lambs to an approaching shepherd. Several behaviours were measured and individual estimated breeding values (EBV) were calculated for selected behaviours for each lamb using an animal mixed model. Extreme animals were chosen according to their high or low EBV either for the frequency of their vocalisations for social reactivity (high vs low frequency), or for the flight distance towards human (short vs long distance). After two generations of selection, the divergent selection led to highly significant line differences for social reactivity or reactivity towards the human. Differences between high and low lines reached 1.8 and 1.0 genetic standard deviation for social reactivity and reactivity towards a human, respectively. Such divergent selections affected significantly locomotor and vigilance behaviours and proximity to flock-mates or the human depending on the lines. Present results confirm that early selection for behavioural traits is feasible. The expected effects on performances and facility to handle flocks will be investigated to assess relevance of such selection for contributing to improve sustainability of livestock.

In breeding, effective population size and coancestry on Latxa dairy sheep breed

I. Granado-Tajada[1], S.T. Rodríguez-Ramilo[2], A. Legarra[2] and E. Ugarte[1]
[1]NEIKER-Tecnalia, Basque Institute of Agricultural Research and Development., Department of Animal production, Agrifood Campus of Arkaute s/n, E-01080 Arkaute, Spain, [2]INRA, UMR 1388 GenPhySE, Chemin de Borde Rouge, 31326 Castanet Tolosan, France; igranado@neiker.eus

Traditionally, inbreeding estimates have been estimated based on pedigree information. However, in sheep there is a considerable proportion of unknown pedigree due to natural mating and limited use of paternity analysis. Therefore, there is an under estimation of inbreeding coefficients based on pedigree. In the genomics era, genomic information can be used to estimate inbreeding. In this study, three different inbreeding estimation methods were assessed (a pedigree-based methodology, a single SNP-based approach and a method based on runs of homozygosity, (ROH) to analyse the genetic diversity of three populations of Latxa dairy sheep: Latxa Cara Rubia (LCR) and Latxa Cara Negra from Euskadi (LCNEUS) and from Navarre (LCNNAF). A total of 981 animals were genotyped with the Illumina OvineSNP50 BeadChip, bringing around 41,200 SNPs and 4,468 animals in pedigree. The results found for LCNEUS and LCNNAF showed an effective population size (Ne) below 100 when inbreeding coefficients were estimated based on pedigree or ROH: Ne PED=64, Ne ROH=86 for LCNEUS, and Ne PED=53, Ne ROH=66 for LCNNAF. Nevertheless, SNP based estimations yielded higher values: Ne SNP=282 and 153 for LCNEUS and LCNNAF, respectively. LCNEUS showed a higher genetic diversity than LCNNAF in any of the evaluated methods. There were differences between pedigree and ROH based results and the SNP based ones, possibly due to the reduced number of genotyped animals. In the case of LCR, which historical importation of semen from the French Manech Tête Rousse (MTR) has avoided the increase of inbreeding per generation, the estimation of effective population size was meaningless. For this breed a study of coancestry between the two breeds has been done, based on pedigree and genomic data, to analyse the evolution of genetic variability. Both methods have reflected, as expected, the important effect on inbreeding and genetic variability of introducing animals from another close breed.

A comparison of traditional and genomic selection schemes in French dairy goats

E. Tadeo-Peralta[1,2], J. Raoul[1,3], I. Palhière[1], S.T. Rodríguez-Ramilo[1], F.J. Ruiz-López[4], H.H. Montaldo[2] and J.M. Elsen[1]
[1] INRA, UMR 1388 GenPhySE, 24 Chemin de Borde Rouge, 31326 Castanet Tolosan, France, Metropolitan, [2]Universidad Nacional Autónoma de México, Facultad de Medicina Veterinaria y Zootecnia, 3000 av. Universidad, 04510 Ciudad de México, Mexico, [3]Institut de l'Elevage, Chemin de Borde Rouge, 31321 Castanet-Tolosan, France, Metropolitan, [4]Instituto Nacional de Investigaciones Forestales, Agrícolas y Pecuarias, Ajuchitlán, 76280 Querétaro, Mexico; elizabeth.tadeo-peralta@inra.fr

Traditionally, selection schemes in French dairy goats were based on genealogical and phenotypic information. However, in 2018, genomic information was introduced to preselect males for progeny testing (PT+GS). This scheme could evolve to a full genomic selection scheme (GS), avoiding the progeny test step in the near future. The objective of this study was to compare the current genomic selection scheme (GS+PT) with a traditional progeny testing (PT) and a full genomic selection scheme (GS) in French Saanen dairy goats. A population, under selection for a production trait (h^2=0.3 and repeatability=0.5) was simulated using a base of real genotypes from 761 proven males. These animals were genotyped with the Ilumina GoatSNP50 chip. After filtering, 46,810 SNP remained for the analyses. One thousand SNP were selected to be QTL. The three selection schemes (PT, GS+PT and GS) were modelled using stochastic simulations to create phenotype and genotype data files to be processed, as in the real life, by single step GBLUP. Results based on 20 replicates were obtained. Compared to the PT scheme, the annual genetic gain was 30 and 35% higher for the PT+GS and GS schemes, respectively. The rate of decrease in genetic variance was the same for the three schemes. Selection accuracy for the young male candidates in the PT scheme was significantly lower (0.26±0.03) than in the GS+PT (0.39±0.02) and in the GS (0.40±0.02). Similar results were observed for young females. However, accuracy values for male candidates were lower than those for young females. Overall, these results confirm the advantage of genomic selection in goat breeding context.

Optimizing nutrition of black soldier fly larvae

T. Spranghers, S. Schillewaert and F. Wouters
VIVES Hogeschool, Agro- and Biotechnology, Wilgenstraat 32, 8800 Roeselare, Belgium; thomas.spranghers@vives.be

Larvae of the black soldier fly (BSF), *Hermetia illucens*, have potential to be reared on a wide spectrum of organic waste streams. On demand of agricultural, food and waste processing companies multiple waste streams were tested as substrates for the larvae in the Insectlab at VIVES (potato peels, tomato plants, brewery waste, fruit and vegetable waste, ….). Most of these waste streams are lacking certain nutrients which are essential to guarantee an optimal growth. Therefore, they were mixed with different rates of high value chicken feed or fed to the larvae in a later stage. In practice, mixtures of different waste streams could be applied to guarantee a sufficient larval biomass production on low value, and consequently low cost, substrates. In order to identify possible interesting mixtures, the nutritional requirements of BSF larvae should be better understood. Therefore, the minimal content of crude protein, crude fat and minerals for a BSF diet were investigated in the lab using artificial diets. The resulting data was combined with data obtained from the above mentioned waste stream assessments and from literature. It was shown that substrates containing 10% protein, 2% fat and 2% minerals on a dry matter basis could result in a sufficient larval growth and favourable feed conversion ratio, as long as sufficient amounts of digestible carbohydrates are present. Protein levels were not directly correlated with larval growth (R^2=0.06), whereas the sum of proteins and non-fibre carbohydrates showed a strong correlation (R^2=0.94). For a BSF substrate, it is advisable that this sum is around 50% of the total dry matter in order to result in a sufficient biomass production. However, more knowledge is necessary about the essential amino acids and digestibility of different proteins and carbohydrates for BSF larvae. Therefore, further research will focus on these topics.

Nutrition of *Tenebrio molitor*: history and novelties for applied research in insect industry
T. Lefebvre
Ynsect, R&D, Genopole Campus1 Bat1 1, rue Pierre Fontaine, 91058 Evry, France; tle@ynsect.com

The yellow mealworm *Tenebrio molitor* is a flagship species in the industrial farming of insects for feed and food. Historically, it has also played an important role as a model species in in research fields like pest control, genetics, physiology, immunology, enzymology and nutrition. The knowledge accumulated over decades of scientific work is an invaluable asset for *T. molitor* in the context of the development of the young field of insect industrial production. Nutrition in particular is a key area of expertise that must be mastered to ensure the success of competitive animal production. Studies on *T. molitor* nutrition started in the 1930s with the work of G. Tessier and were followed by a flourishing period in the 1950s-1970s with studies from entomologists J. Leclercq, G.S. Fraenkel, G.R.F. Davis, which set the first references of the yellow mealworm's nutritional requirements. However, it remains of capital importance for the industry to invest in research and knowledge development, in order to reach the maturity level of traditional animal husbandry. The main fields to be addressed are the estimation of *T. molitor* nutritional requirements, the assessment of digestibility and metabolizability values of feedstuffs and the development of reliable indicators of physiological and economic performances. All these topics are primordial for fine formulation of diets and optimization of economic performance. Ÿnsect has worked in nutrition for several years and has made significant progress in collaboration with French National Institute for Agricultural Research, INRA. After a short review of *T. molitor*'s nutrition science history, the presentation will focus on the methodologies developed to study *T. molitor* populations, illustrated by experimental results.

Biorefinery approach for conversion of organic side-streams into marketable products using insects
L. Bastiaens and Indirect Consortium
VITO NV (Flemish Institute for Technological Research), Sustainable Chemistry, Boeretang 200, 2400 Mol, Belgium; leen.bastiaens@vito.be

'Management of waste as a resource' was described in a European roadmap (COM (2011/571)) as a milestone to be reached by 2020. On the one hand, there are diverse under-spent agrofood side-streams. On the other hand, there is a need for new resources, for instance proteins as alternative for soy proteins. The EU research project InDIRECT (2016-2019) envisioned to develop biorefinery processes as part of new value chains to convert under-spent side-streams from the agro-sector and processing sector into useful marketable products. Cascading processes (multiple compounds from the same feedstock) were envisioned to increase the conversion efficiency and maximise the values of the feedstock. One of the biorefinery approaches considered is a two-step process that can cope with the heterogeneity of side-streams. In a first step the heterogenic feedstock is converted to homogenous biomass using insects. Insects are able to convert a variety of feedstocks into a more homogenous biomass, being their own biomass. In a second step, the insect biomass is further fractionated into a lipid, protein and chitin fraction, that all three have potential towards marketable end products. Fractions are being produced for application tests in feed and chemical applications. An overview of the InDIRECT results will be presented, including growth data of lesser mealworm and black soldier fly larvae on side-streams, correlations between the larvae feed composition and the composition of larvae fed with the feed, and fractionation of insect biomass and processing of chitin. The InDIRECT consortium consists of 9 partners from 4 countries (B, NL, F & I): 2 research organisations (VITO & University of Parma), 5 industrial partners (Nutrition sciences NV – feed additives, two insect farms Millibeter & Protifarm R&D, IMPROVE – plant proteins, and CHEMSTREAM -chemical application), a non-profit organisation linked to farmers and processing industry (Innovatiesteunpunt) and a project supporting SME (Temperio). The project (www. bbi-indirect.eu) is coordinated by VITO and received funding from the Bio Based Industries Joint Undertaking under the European Union's Horizon 2020 research and innovation programme under grant agreement No 720715.

Genotype × environment interactions in black soldier fly larvae grown on different feed substrates
C. Sandrock[1], S. Leupi[2], J. Wohlfahrt[1], F. Leiber[1] and M. Kreuzer[2]
[1]Research Institute of Organic Agriculture, FiBL, Livestock Sciences, Ackerstrasse 113, 5070 Frick, Switzerland, [2]ETH Zurich, Environmental System Science, Universitätstrasse 2, 8092 Zürich, Switzerland; christoph.sandrock@fibl.org

Black soldier fly larvae (BSFL) are promising for producing dietary protein for livestock and aquaculture feeding, as well as for organic waste treatment. Literature indicates life history performance and body composition of BSFL being extremely variable, thus hampering predictions of nutritional quality of insect-based feeds and economic performance in general. BSFL feed substrate is one major driver for the observed variation, as documented in a number of reports. Yet, pronounced differences were still found for comparable BSFL feed material across studies, which have mostly been attributed to different experimental settings. Potential influences of BSF genetics have largely been neglected, due to a hitherto existing lack of appropriate monitoring tools. In a fully crossed factorial design we compared performances of four genetically distinct BSF strains (inferred from newly developed polymorphic nuclear genetic markers) on three feed substrates with six replicates each. Various growth and compositional traits were determined. Both factors, feed substrate and BSF genetics, as well as the interaction between the two, were tested highly significant for virtually all investigated life history traits (mortality, growth dynamics, average larval weight) and body compositional characteristics (dry matter, organic matter, crude protein, ether extract) of harvested larvae. The same applied for feed conversion ratio, nitrogen efficiency, degradation of fibre as well as nitrogen and carbon emissions. Frequently detected stronger differences between genotypes within feed substrates than within genotypes across feed substrates show that BSF genetics is key for interpreting and triggering outcomes. Our results imply that efficiency and sustainability of BSF productions may be substantially increased by considering BSF genetics and interactive effects, depending on available BSFL feed substrates and targeted objectives for implementing BSF products. Ample indication for BSF genotype-specific effects also provides valuable insights for pinpointed BSF breeding.

Survival rate, development time and nutritional features of insect according to rearing substrates
M. Ottoboni, A. Luciano, A. Agazzi, G. Savoini and L. Pinotti
University of Milan, Department of Health, Animal Science and Food Safety Carlo Cantoni, Via Celoria 10, 20133 Milano, Italy; matteo.ottoboni@unimi.it

Larvae of the black soldier fly (BSF), *Hermetia illucens* L. (Diptera: Stratiomydae), are able to convert organic material (i.e. waste and by products) into high-quality biomass, which can be processed into animal feed. Although, several studies have been conducted on evaluation of influence of growing substrate on the nutritional value of BSF larvae and prepupae, this field need more accurate investigation. The intent of this work was to collect, synthesize, discuss and review the available information on the substrates used for rearing BSF larvae in literature. In light of this, a systematic review based on 15 publications published from 2008 to 2019 was carried out in order to evaluate the effects rearing substrate on: Survival rate (%), development time (days) and nutritional features (crude protein – CP; chitin corrected protein; ether extract – EE; ash; and chitin) of deriving insect's biomass. Results obtained indicated that survival rate (%) was negatively correlated (r>-0.5; P<0.01) to EE content in the substrate. A negative correlation (r>-0.5; P<0.01) was also observed between development time (days) and insect's biomass yield (g biomass yield from 100 larvae). When substrates nutritional features were correlate to insect's biomass nutritional feature, EE in the substrate was negatively correlated to CP in harvest larvae (r>-0.5; P<0.01), while EE in the substrate was positively correlated to EE in the larvae (r>0.5; P<0.01). A positive correlation (r>0.5; P<0.01) was also detected for ash in both substrate and larvae. Based on these results it can be suggested that BSF is able to convert organic material (i.e. waste and by products) into high-quality biomass, even though the role and effects of selected nutrients like ether extract/fats in the substrate, merit further investigation.

Producing *Hermetia illucens* larvae with common Western European horticultural residues
C.L. Coudron, D. Deruytter, J. Claeys and S. Teerlinck
Inagro vzw, Insect & Aquaculture, Ieperseweg 87, 8800 Rumbeke-Beitem Belgium; carl.coudron@inagro.be

The production of black soldier fly larvae is often suggested as a sustainable source of protein. It could play a key role in the conversion of low value residual biomass into high value insect protein and thus contribute to an improved usage of all the available biomass. To achieve this, the insect sector must be able to co-exist with other forms of animal production with minimal competition for feed sources (e.g. spent grains), especially if the insects themselves are then used as a feed in animal production. The horticultural sector is an example of a sector with large quantities of residual biomass. Some examples of crops with potential are: (1) Brussels sprouts: 420,000 metric tons of stems and leaves a year (F, N and UK); (2) Chicory: 150,000 tonnes of residual roots (B and N); (3) Tomato: 100,000 tons of fiber rich residual material (B, N and UK); (4) Cauliflower: 154,000 tons of residual foliage each year (B) With B: Belgium, F: Flanders, N: the Netherlands and UK: the united kingdom. Because most of these side streams are only available seasonally, various preprocessing steps were evaluated to stabilize and/or enhance the biomass. Furthermore, due to the varying quality of the residues, a calculated approach is required in order to compose a balanced diet with maximum inclusion of the residual biomass. The minimal nutritional needs of black soldier fly larvae are roughly known. The macro nutrients of different horticultural residues were evaluated based on a proximate analysis and mixtures were made that met the proposed requirements (30% dry matter, 10-20% protein, 2% fat and 30-40% total carbohydrates). The mixtures were evaluated in pilot scale conditions based on the following parameters: (1) harvestability, how easily the larvae and the remaining substrate can be separated from each other; (2) larval growth rate and survival; (3) feed conversion, the amount of feed required for the production of 1 kg of larvae (wet and dry basis); and (4) processing capacity of the larvae, based on weight reduction of the original mixture. Initial results indicate that the pre-fermentation of chicory roots has no negative effect on growth (feed conversion ratio: 1.5-2) but does increase the harvestability of the larvae.

The nutritive value of black soldier fly larvae reared on common organic waste streams in Kenya
M. Shumo[1], I.M. Osuga[2], F.M. Khamis[2], C.M. Tanga[2], K.M. Fiaboe[2], S. Subramanian[2], S. Ekesi[2], A. Van Huis[3] and C. Borgemeister[1]
[1]Center for Development Research (ZEF), University of Bonn, Department of Ecology and Natural Resources Management, Genscherallee 3, 53113, Germany, [2]icipe, Plant Health Unit, P.O. Box 30772, 00100 Nairobi, Kenya, [3]Wageningen University, Department of Plant Sciences, Laboratory of Entomology, Droevendaalsesteeg 1, 6700 AA Wageningen, the Netherlands; mshummo@hotmail.com

In Africa, production is struggling to keep up with the demands of expanding human populations, the rise in urbanization and the associated shifts in diet habits. Insects have been identified as potential alternatives to the conventionally used protein sources in livestock feed due to their rich nutrients content and the fact that they can be reared on organic side streams. Substrates derived from organic by-products are suitable for industrial large-scale production of insect meal. Thus, a holistic comparison of the nutritive value of black soldier fly larvae (BSFL) reared on three different organic substrates, i.e. chicken manure (CM), brewers' spent grain (SG) and kitchen waste (KW), was conducted. Five-hundred (500) neonatal BSFL were placed in 23×15 cm metallic trays on the respective substrates for a period of 3-4 weeks at 28±2 °C and 65±5% relative humidity. The larvae were harvested when the prepupal stage was reached using a 5 mm mesh size sieve. A sample of 200 grams prepupae was taken from each replicate and pooled for every substrate and then frozen at -20 °C for chemical analysis. Samples of BSFL and substrates were analysed for proximate composition, amino acids (AA), fatty acids (FA), vitamins, flavonoids, minerals and aflatoxins. The data were then subjected to analysis of variance (ANOVA) using general linear model procedure. BSFL differed in terms of nutrient composition depending on the organic substrates they were reared on. Crude protein, minerals, amino acids but not vitamins were affected by the different rearing substrates. Low concentrations of heavy metals (cadmium and lead) were detected in the BSFL, but no traces of aflatoxins were found. In conclusion, it is possible to take advantage of the readily available organic waste streams in Kenya to produce nutrient-rich BSFL-derived feed.

Influence of moisture content of feeding substrate on growth and composition of *Hermetia illucens*

L. Frooninckx[1], L. Broeckx[2], A. Wuyts[1], S. Van Miert[1] and M. Van Der Borght[2]
[1]*Thomas More University College of Applied Sciences, RADIUS, Kleinhoefstraat 4, 2400 Geel, Belgium,* [2]*KU Leuven, Lab4Food, Kleinhoefstraat 4, 2400 Geel, Belgium; lotte.frooninckx@thomasmore.be*

Black soldier fly larvae (*Hermetia illucens*) are able to convert organic side streams into high-value biomass, mainly consisting of proteins, fat and chitin. Food waste consisting of pre- and post-consumer products is already being collected separately by several waste management companies. This 'energy mixture' is usually valorised as feedstock for anaerobic digestion. However, it could be better valorised if used as feeding substrate for Black soldier fly larvae. The present study aims to investigate the influence of the chemical composition (moisture, fat, protein, ash, fibre and carbohydrate content) of the 'energy mixture' on growth, feed conversion, waste reduction and composition (moisture, fat, protein, ash, chitin and carbohydrate content) of Black soldier fly larvae and rearing residues. Larvae were daily fed 100 mg of the energy mixture or chicken feed with different moisture contents ranging from 80 to 55%. Larvae fed with substrates containing the highest moisture content grew faster and reached a higher maximal weight per larvae, than larvae fed with less moist substrates. Fat content was lowest in larvae fed with substrates containing the highest moisture content. Protein content did not differ between larvae fed with substrates of high or low moisture content.

Preliminary results on black soldier fly larvae reared on urban organic waste and sewage sludge

M. Meneguz[1], A. Schiavone[1], V. Malfatto[1], F. Gai[2] and L. Gasco[1]
[1]*University of Torino, Department of Agricultural, Forest and Food Sciences, L.go Paolo Braccini, 2, Grugliasco, TO, 10095, Italy,* [2]*Istituto di scienze delle produzioni alimentari, L.go Paolo Braccini, 2 Grugliasco, 10095, Italy; marco.meneguz@unito.it*

The waste management is an issue from an environmental point of view and a challenge for the institutions. The black soldier fly (BSF) larvae rearing could be a valuable solution for this issue. The bioconversion process performed by BSF larvae could enable the reduction of waste and the recovery and conversion of their nutrients into high added value products (fertilizer, protein, fats and chitin). The aims of this research was to evaluate the effects of post treated urban organic waste (PTW) and sewage sludge (SD) on BSF larvae development. Eight PTW-based diets were prepared with different inclusion of SD, with or without probiotic addition. The preliminary results showed as the different inclusion of SD negatively influenced the BSF larvae development while the probiotic did not highlight any effect among diets. The highest level of PTW+SD (30%) with probiotic showed the lower final weights (113.8 mg) in contrast with PTW with probiotic (134.7 mg). The inclusion of SD impacted the waste reduction index, with the lowest reduction operated with the 20% of SD inclusion in contrast with PTW with probiotic (40.4 and 53.8%). Low survival rate was recorded in all treatments (under 53%), arguing due to the bad impact of high level of moisture in diets (>86%).

Investigating the feasibility of household and supermarket organic waste as feed for *T. molitor*

A.M. Meyer[1,2], A. Lecocq[2] and A.B. Jensen[2]
[1]*University of Hohenheim, Societal Transition and Agriculture, Schloss Hohenheim 1, 70599 Stuttgart, Germany;* [2]*University of Copenhagen, Plant and Environmental Sciences, Thorvaldsensvej 40, 1871 Frederiksberg C, Denmark; amezia@vt.edu*

The present global context of the food market and protein consumption is controlled by an immense degradation of the environment. Using insects as food has the potential to become a sustainable, affordable and healthy alternative food source. So far, sustainability remains dependent on finding a suitable organic side stream to rear the insects on. With this in mind, we investigated the potential for the mealworm, *Tenebrio molitor*, to be reared on organic side streams from different sources. We reared 1,500 *T. molitor* larvae on three different types of feed including supermarket organic waste from Nature Energy, household organic waste from BioFos and a control feed of flours with brewery spent grain as a wet feed. The survival and weight gain of the larvae will be measured. In addition, analysis of microplastics, nutritional content of CHN, TOM, and bioaccumulation of heavy metals of the various feed, larvae and faeces will be performed along with expert interviews. The results will allow for an enhanced understanding of the utilization of available food waste within the insect production industry. *T. molitor* should be successfully reared on organic waste and provide a basis for the discussion, implementation and development of insects as a sustainable and nutritious alternative food source that incorporates a circular economy and urban agriculture.

Alternative (commercial) feeds for *Tenebrio molitor*

C.L. Coudron, D. Deruytter, J. Claeys and S. Teerlinck
Inagro vzw, Insect & Aquaculture, Ieperseweg 87, 8800 Rumbeke-Beitem, Belgium; carl.coudron@inagro.be

The choice of feed is amongst the most important decisions a farmer has to make. For mealworms, many breeders work with wheat bran (or wheat flour) in combination with carrots. Although this feed works well, the nutritional quality is not optimal and it is expected that large gains can be made in terms of growth rate, maximum weight and feed conversion rate if the feed is closer to the needs of the mealworm. Furthermore, potentially more ecological or cheaper alternatives may be available when using low value side streams. In this study several feeds, feed sizes, feed additives and moisture sources were investigated to assess if improvements can be made to the standard feed of wheat bran and carrot. In order to ensure that the experiments are immediately relevant for the breeders, all assessed feeds were provided by feed manufacturers or bought and are therefore all commercially available. In the feed trials the feed was added *ad libitum*. In contrast, the moisture sources were given every workday but not *ad libitum* due to mould growth and separation issues. Therefore it was opted to provide less than needed to ensure that it was possible for the mealworms to consume it in 24 h. Previous research (not published) indicated that the feeding radius of small mealworms is only a few centimetres. Therefore the moisture source was provided as evenly as possible not only in the centre. The initial results indicate a better growth performance when 5% of standard chicken meal is provided with a maximum growth enhancement at 25%. Potentially by increasing the amount of carnitine. In contrast to chicken meal and what was expected, ground up leftovers from cookies with a high protein and sugar content did result in a stunted growth and high mortality when provided as feed. This may be due to the presence of cinnamon. Residue from oyster mushroom or shiitake did work well for the mealworms initially, but growth was stunted fast. Finally, mealworms were able to grow on spelt hulls, but at a low rate compared to wheat bran. The first results however indicate a better protein turnover. More feeds and moisture sources will be assessed.

Growing performance of *Hermetia illucens* on lignocellulosic substrates

G. Marchesini, M. Cullere, G. Trevisan and I. Andrighetto
University of Padova, Animal Medicine, Productions and Heath, Viale dell'Università, 16, 35020 Legnaro (PD), Italy;
giorgio.marchesini@unipd.it

In response to the increase in the demand for protein of high biological value, the breeding of insects seems to be a possible alternative source of protein. Since one of the factors that hinder the spread of insects as feed sources is the availability of cheap and healthy food matrices necessary to breed them, the aim of this study was to assess the growing capacity of the larvae of black soldier fly (*Hermetia illucens*), on lignocellulosic by-products. Three g of larvae of about 8 days of age (235 larvae) were bred on dry brewer's grains (Br, NDF=61.8% DM), coffee grounds (Cf, NDF=54% DM) or straw (St, NDF=81% DM), in plastic containers located in the dark, at room temperature (25.3±0.8 °C). For each substrate, 6 replicates were done. The breeding phase lasted 16 days, as the growing cycle was interrupted for all the containers, as soon as the first larvae had come into the pre-pupation phase. At the start and at the end of the trial the larvae were weighed and measured; their number was estimated to evaluate their survival rate and the substrate was weighed and chemically analysed to evaluate the use of its nutrients by larvae. After verifying their normal distribution, all the data were submitted to an ANOVA model, using the substrate as fixed effect. Survival was 81.6, 80.2 and 83.6% for Br, Cf and St, respectively. The larvae achieved a length of 12.5, 11.1 and 12.0 mm, a percentage increase in weight and length of 180, 59.1 and 120% ($P<0.01$) and 80.9, 60.0 and 72.8% ($P<0.01$), respectively. Larvae used 26.8, 12.1 and 8.0 DM g ($P<0.01$) of substrate and led to a reduction of the substrate's NDF equal to 42.0, 5.57 and 9.28% ($P<0.001$) for Br, Cf and St. The selection of the fibrous components by larvae varied with the substrate. To conclude, the larvae of *H. illucens* seemed to live and grow even on lignocellulosic substrates, consuming cellulose and hemicellulose but, as expected, they showed a lower and slower growth when compared to larvae grown on standard substrates such as broiler chicken feed. The research was funded by Fondazione CariVerona, Documento Programmatico 2015, Ricerca Scientifica e tecnologica, Progetto 'Tre poli 5'.

Optimal rearing of *Alphitobius diaperinus* on organic side-streams in the InDIRECT project

N. Gianotten[1], J. Roels[2], M. Lopez[3], S. Sforza[4] and L. Bastiaens[5]
[1]Protifarm, R&D, Harderwijkerweg 141B, 3852 AB Ermelo, the Netherlands, [2]Innovatiesteunpunt, Diestsevest 40, 3000 Leuven, Belgium, [3]IMPROVE, Rue du Fond Lagache, 80480 Dury, France, [4]University of Parma, Department of Food and Pharmacy, Via Università 12, 43121 Parma, Italy, [5]VITO, Flemish Institute for Technological Research, Boeretang 200, 2400 Mol, Belgium; n.gianotten@protifarm.com

Insects are regarded as promising alternative animal based proteins for feed and food. Protifarm focusses on rearing the *Alphitobius diaperinus* or lesser mealworm (LMW). In Ermelo, the Netherlands, a large scale rearing facility houses many billions of larvae, which are reared and processed for food applications. Feed for the LMW larvae is: (1) a large part of the production cost; and (2) has an enormous impact on the sustainability of the rearing system. It is therefore crucial to use as much side-streams as possible in the ration of the LMW. In the InDIRECT project, an EU research project, insects convert under-spent side-streams via bio-refinery into useful products. Insects are able to convert heterogenic side-streams into homogenous biomass. The insect biomass is further fractionated into marketable products like protein, lipid and chitin fractions. Efficiency of rearing the insects on side-streams is one aspect of an economical viable system and is taken into account. Four side-streams (DDGS, corn gluten feed, rapeseed meal and rice bran) were provided by project partners. Each side-stream was combined with standard products available at Protifarm R&D (also side-streams). Combinations were investigated in dose-response studies which were performed according to standard Protifarm protocol. Per treatment, six or eight replicates were used. Yield per tray (experimental unit) and feed conversion ratios are investigated parameters. Chosen side-streams result in marked differences on performance of the LMW larvae, ranging from positive (rapeseed meal) to neutral (rice bran) to negative effects (DDGS) on performance of the larvae. This study received funding from BBI Joint Undertaking in the framework of the European Union's Horizon 2020 research and innovation program, under grant agreement No 720715.

Rearing of *Hermetia illucens* on municipal organic waste: a win-win solution for feed production

S. Bortolini[1], M.R. Alam[1], G. Ferrentino[1], M.M. Scampicchio[1], M. Gauly[1], M. Palmitano[2] and S. Angeli[1]
[1]*Free University of Bolzano, Piazza Università 1, 39100 Bolzano, Italy, [2]EcoCenter S.p.A. via Lungo Isarco Destro 21/A, 39100 Bolzano, Italy; sara.bortolini@unibz.it*

Black soldier fly larvae (BSFL) are well known for their nutritional properties and their capability to convert organic waste into high quality proteins. Bio-waste treatment with BSFL is considered an emerging technology; indeed, BSFL can ingest manure, fruit and vegetable wastes, and carrions, producing insect biomass, rich in proteins and fats. Moreover, recent studies demonstrated that the remaining undigested material has compost-like properties. Currently, the process performance is variable and studies are needed to better understand the decomposition process on complex matrices. The aim of this work was to evaluate the growth parameters of BSFL on the organic fraction of municipal solid waste (OFMSW) of the city of Bolzano (South Tyrol, Italy), at two different temperatures (27 and 31 °C). In a typical experiment, six hundred 4-days-old larvae were reared on 1 kg of OFMSW. To compare the efficacy of OFMSW as growth substrate, 'Gainesville housefly' diet was used as control. Six replicates for each treatment were set up. Larval and prepupal weight was measured three times per week. ANOVA and Tukey test were used for statistics. On OFMSW, maximum average weight of larvae reared was registered after 11 days (0.173 ± 0.024 g at 27 °C, and 0.169 ± 0.025 g at 31 °C). The maximum average weight of prepupae was registered after 14 days (0.128 ± 0.020 g at 27 °C, and 0.141 ± 0.017 g at 31 °C). The OFMSW influenced positively these parameters (P=0.006 and P=0.002). The total yield of prepupae was 75.92 and 57.43%, respectively. The average waste reduction was $57.05\pm0.78\%$ at 27 °C, and $61.25\pm2.69\%$ at 31 °C (wet weight). Based on these achieved results, this research will be further implemented, with the objective to treat in large scale the OFMSW of the entire province, creating at the same time a high-value feed, using BSF meal as a protein constituent. This system might be a win-win solution in waste management because BSFL reduce the amount of organic waste and constitute raw material for high quality feed. The project PROINSECT is funded by 'European Regional Development Fund (ERDF) 2014-2020'.

Side-stream substrate formulations for mealworms in Finland

E. Hirvisalo[1], M. Karhapää[1], M. Mäki[1], P. Marnila[1], B. Lindqvist[1], H. Siljander-Rasi[1], L. Jauhiainen[1], T. Jalava[1], J. Sorjonen[2], H. Roininen[2] and M. Tapio[1]
[1]*Natural Resources Institute Finland (Luke), Production Systems, Myllytie 1, 31600 Jokioinen, Finland, [2]University of Eastern Finland, Department of Environmental and Biological Sciences, P.O. Box 111, 80101 Joensuu, Finland; miika.tapio@luke.fi*

Cultivation of crops used in production substrates is one of the main sources for greenhouse gas emissions in insect production. Replacing them by local side-stream materials is critical for the economic and environmental sustainability of insect production. At the same time, good productivity requires understanding the nutritional requirements of the produced insects. We assessed value of various side streams as substrate ingredients for production of mealworm, *Tenebrio molitor*. It has potential to feed on a wide range of biomasses, but nutritional requirements are inadequately understood. In particular, we aimed to find domestic side streams and other protein-containing raw materials that could substitute imported feed ingredients (soy bean, maize grits) and ingredients directly suitable for food production (wheat bran). We conducted four trials, each with 5-6 treatments. Each treatment was done as 4-5 replicates per trial. Each replicate had 150 larvae and had 300 g feed + fresh vegetables. Larvae were grown for four weeks in controlled conditions (at 24.2 ± 0.2 °C, about $60.3\pm1.4\%$ relative humidity) in row-column experimental design. Growth and viability was measured and recipes were further developed after each trial. Differences in growth and viability were detected and the results confirmed the feasibility of the investigated side streams as feed ingredients for mealworm. Mortality was low through the trials. Potato protein, a side-stream of starch production, was one of the interesting ingredients but is relatively expensive. Local side streams showed their potential, but both nutritional and economical need to be balanced in substrate formulation.

Nutrient composition of three orthopterans as alternatives to 'hard-to-farm' insects in East Africa

T.F. Fombong[1,2], J. Kinyuru[1] and J. Vanden Broeck[2]
[1]*Jomo Kenyatta University of Agriculture and Technology, Food Science, Juja, 00200 Nairobi, Kenya, [2]KU Leuven, Biology, Naamsestraat 59, box 2465, 3000, Belgium; tengweh.fombong@kuleuven.be*

The consumption of insects plays a vital role as a food source across Sub-Saharan Africa, especially East Africa (EA). The most commonly eaten orthopteran insects in EA is the grasshopper, *Ruspolia differens*, despite not yet being farmed and not easy to rear. *Gryllus bimaculatus* (GB), *Locusta migratoria* (LM) and *Schistocerca gregaria* (SG) are three possible orthopteran alternatives that currently are more easy to rear compared to the favourite edible long-horned grasshopper. The study aimed at comparing the nutritional quality of the two locust pests (LM, SG) and a cricket (GB) to the grasshopper, *R. differens,* as a reference in improving food security and livelihood. Therefore, different analytical procedures were employed to investigate their nutritional potentials. Results of the proximate were subjected to one-way ANOVA using the Dunnett's test (with RD as control) in R. This study revealed new protein conversion (Kp) values of 5.31, 5.35, 5.57 and 6.13 for GB, LM, SG and RD, respectively, which are lower than the conventional 6.25 used in our analyses. Protein contents were all above 50% (65, 54 and 61%) for GB, LM, and SG, respectively, with no significant difference to RD. All the insects possessed significant lipid levels ranging between 19-20% in SG and GB and 31% in LM. Oleic acid (22-41%) was the predominant fatty acid while the polyunsaturated arachidonic acid (0.4-1.4%) and docosahexaenoic acid (0.3 and 1.9%) for SG and LM, respectively, were present. All essential amino acids were present with glutamine (120-125 mg/100 g protein) the most abundant. Potassium (796-1,309 mg/100 g) was highest among the minerals and the occurrence of the trace elements iron (4.6-7.3 mg/100 g) and zinc (12.7-24.9 mg/100 g), was also observed. Apart from SG, vitamin B_{12} contents were high (0.22-1.35 µg/100 g). Their nutritional profile compared very well with that of *Ruspolia* suggesting that for instance when a locust outbreak would occur, they could be introduced as alternative nutrient sources in local diets to improve food security and livelihood.

Effects of feeds on the nutritional composition, growth and development of the house crickets

M. Bawa and S. Songsermpong
Kasetsart University, Food Science and Technology, 50 Ngamwonwan Road Ladya, Chatuchak, Bangkok, 10900, Thailand; flierses@yahoo.com

Crickets are sustainable and nutritious future sources of food due to their comparative advantage role in solving global malnutrition and food insecurity problems. The aim of this study was to determine the effects of feeds on the nutritional composition, growth and development of the house crickets being major challenge to crickets' farmers and are important factors in scaling up crickets production in Thailand. The crickets were randomly assigned to 5 experimental feeds from the first day eggs were hatched. Samples were sent for nutritional content analysis at a certified lab, Thai central lab in Bangkok. The nutritional composition of all crickets samples were analysed using official methods, AOAC (2016). The mean adult weight were 0.5826±0.2 g, 0.5326±0.11 g, 0.0.4557±0.009 g, 0.4259±0.017 g and 0.4224 + 0.014 g for crickets on pure pride plus 100 g dry pumpkin per day, pure pride plus 100 g fresh pumpkin per day, pure pride without pumpkin, 50% pure pride plus 50% Betagro chicken feed 215 and Betagro chicken feed 215 respectively. Crickets on a high protein feed, pure pride (20%), supplemented with either fresh or dry pumpkin were more efficient (P<0.05) in converting feed to body size than the other groups. Protein content was high (76.2 g/100 g dry basis) for the group on pure pride feed without pumpkin. The additional of pumpkin to a high protein diet (20%) improved the vitamin contents, but resulted in a low protein content (48.1 g/100 g dry basis), which was lower than expected. Crickets on a high protein diets (20%) recorded the highest (P<0.05) number of females crickets than those on low protein diet (14%), Betagro chicken feed 215. Survival rate and period of eggs laying were also influenced by feed exposures. Crickets can effectively be produced with a high protein feed supplemented with either dry or fresh pumpkin to ensure the inclusion of vitamin benefits of entomophagy.

Food from prey mites affects biological combat with predatory mites

A. Leman[1], G.J. Messelink[1] and O.L.M. Guerin[2]
[1]Wageningen University & Research, Business Unit Greenhouse Horticulture, P.O. Box 20, 2665 ZG Bleiswijk, the Netherlands, [2]Zetadec, Agro Business Park 44, 6708 PW Wageningen, the Netherlands; oriane@zetadec.com

Maintaining the population of predatory mites, used as biological pest control, on ornamentals plants is difficult in case of lack of feed. The purpose of this project was to investigate if predatory mites can be fed on artificial diets. Predatory mites were fed with artificial diets based on the haemolymph of black soldier fly. Effects of diets on oviposition rate were compared using ANOVA tests. Maximum oviposition rate of predatory mites *Amblyseius swirskii* is known to be 2 eggs/day. When fed with artificial diets based on haemolymph, it is 1.5 egg per day. An alternative solution to maintain the population of predatory mites is to feed them with prey mites, with high nutritional values. To optimize the nutritional value of the prey mites, their diets are optimized. When feeding prey mites (as *Tyreophagus entomophagus* and *Acarus siro*) with high-quality protein diets (with ingredients as Typha pollen, bee pollen, dog food, and soy protein concentrate), the population of predatory mites can be up to 5 times larger than when feeding prey mites with diets including wheat bran. To conclude, the establishment and population development of predatory mites in plants can be improved by feeding them with prey mites grown on an optimized protein-rich diet.

The effect of the use of population or hybrid rye in feed compounds on swine fattening performance

T. Schwarz[1], A. Mucha[2], J. Lasek[3], R. Tuz[1], M. Małopolska[2] and J. Nowicki[1]
[1]University of Agriculture in Krakow, Mickiewicza 24/28, 30-059 Krakow, Poland, [2]National Intitute of Animal Production, Sarego 2, 31-047 Krakow, Poland, [3]Experimental Station of the National Institute of Animal Production, Chorzelów Ltd, Chorzelów, 39-331, Poland; rzschwar@cyf-kr.edu.pl

Rye grain is perceived as a feed material of a lower quality, that can negatively influence weight gains of growing pigs as well as feed conversion rate. Last 10 years research showed good productive results of pigs fed compounds containing rye, however, results seem to have variation among experiments. The reason could be different cultivars of rye to be used. The aim of the study was to determine if the use of large amount of population or hybrid rye grain as feedstuff can influence fattening performance of pigs. The study was conducted in Pig Performance Testing Station located in Chorzelow. The research material was 125 polish landrace pigs divided into 5 nutritional groups including control (C), containing equal volume of barley and wheat, and 4 experimental, where equal proportions of barley and wheat were replaced by 40 or 60% of population (E1, E2) or hybrid (E3, E4) rye respectively. Main fattening production parameters were analysed, with estimation of differences among groups using two-way ANOVA followed by Duncan test. There were no significant differences among groups in the most important fattening parameters, however, some of variation are relevant for commercial production. Daily gains were slightly higher in animals fed diet containing hybrid rye, independently of content (934±160; 943±150; 931±115; 989±159 and 984±234 in C, E1, E2, E3 and E4 respectively). However, feed conversion rate was much more similar (2.74±0.42; 2.84±0.50; 2.79±0.52; 2.88±0.52 and 2.76±0.61 in C, E1, E2, E3 and E4 respectively), it was a little better in C and E4 group. To conclude, the use of hybrid rye in feed compounds allowed to increase daily gains in swine fattening, while the use of population rye maintained fattening performance on equal level to controls. Both, hybrid and population rye seem to be good alternative for standard cereal feed materials, creating better economical results thanks to lower market price of grain. NCRD grant ENERGYFEED, contract no. BIOSTRATEG2/297910/12/NCBR/2016.

Effect of feeding bakery by-products on feed intake, milk production and ruminal pH in dairy cows

E. Humer, A. Kaltenegger, A. Stauder and Q. Zebeli
University of Veterinary Medicine Vienna, Institute of Animal Nutrition and Functional Plant Compoun, Department for Farm Animals and Veterinary Public Health, Veterinärplatz 1, 1210, Austria; elke.humer@vetmeduni.ac.at

Bakery by-products (BBP) consist mainly of bakery products not used for human consumption, which are listed as raw feed materials in the positive list. Their availability is increasing during the last years. Yet, information is lacking regarding the effects of inclusion of BBP in diets for dairy cows. Therefore, the aim of this study was to evaluate the effect of graded substitution of grains by BBP on production performance and ruminal pH in high-yielding dairy cows. Twenty-four lactating Simmental cows (149 ± 22.3 days in milk, mean lactation number: 2.63 ± 1.38) were fed a total-mixed ration containing 50% concentrates, 25% maize silage and 25% grass silage (dry matter-basis) throughout the trial. During the first week, all cows received a control (CON) diet without BBP for one week as a baseline. Thereafter, the cows were fed three different diets for four weeks, differing in the BBP-concentration: CON (30% grains, 0% BBP), 15% BBP (15% grains, 15% BBP) and 30% BBP (0% grains, 30% BBP). Feed intake and milk yield were measured daily in all cows and ruminal pH was measured every 10 min using wireless rumen pH-sensors (smaXtec) in 20 cows. Statistical analysis was performed using the MIXED procedure of SAS (9.4). The inclusion of BBP in the diet enhanced the content of ether extract and sugar while reducing the starch and neutral detergent fibre concentration, finally leading to an enhanced energy density. The dry matter intake was enhanced by the inclusion of BBP in the diet by on average 7% ($P=0.01$). A linear increase in the milk yield ($P<0.01$) with BBP-inclusion level was observed (+5% in 15% BBP and +12% in 30% BBP compared to CON). The time the pH fell below 5.8 (used as an index of rumen acidosis) was lowered in cows fed BBP compared to CON (-39% in 15% BBP, -15% in 30% BBP; $P\leq0.05$). Data suggest that changes in the nutrient profile in the diet due to BBP feeding improved production performance and ruminal health in dairy cows. The authors acknowledge Königshofer Futtermittel for providing the concentrates and the H. Wilhelm Schaumann Stiftung for financial support.

Feed design applying circular economy principles: the case of former food products

L. Pinotti, M. Tretola, A. Lucinao and M. Ottoboni
University of Milan, Department of Health, Animal Science and Food Safety, Via Celoria 10, Milano, 20133, Italy; luciano.pinotti@unimi.it

Former Foodstuffs (FFPs) are products that have lost their commercial value on the human consumption market, due to practical, logistical or production issues. However, their nutritional value for animal feed purposes is not at all affected. Consequently, biscuits, bread, chocolate bars, pasta, savoury snacks and sweets, high in energy content (>16.95 MJ/kg of metabolizable energy) in the form of sugar, starch, oil or fat can be considered appealing alternative feed ingredients. The nutritional properties of FFPs are completed by a quite high digestibility and glycaemic index potential, according to the ex-food mixture used in their preparation. Their composition, however, might be variable and some nutrients (i.e. content in free sugars) should be carefully evaluated in order to ensure a proper inclusion in farm animal diets. This implies a functional evaluation with special emphasis on the FFP impact on animal welfare in general and the gastro-intestinal tract (i.e. gut health), in particular. Accordingly, in few studies, conventional ingredients (e.g. cereal grains) have been partially replaced by FFPs in post-weaning piglet's diets in order to investigate the effects of these alternative feed ingredients on growth performance and gut microbiota. Results obtained has indicated that replacing partially conventional cereals with FFPs in post-weaning diets has not detrimental effects on pig growth performance, apparent total tract digestibility of dry matter and on gut microbiota, even though further and larger studies are needed for confirming these observations.

Macauba pulp (*Acrocomia aculeata*) as alternative raw material for growing-pigs

E.F. Dias[1], V.E. Moreira[1], R.P. Caetano[2], A.M. Veira[2], M.S. Lopes[3], R. Bergsma[3], S.E.F. Guimarães[4], J.W.M. Bastiaansen[5], L. Hauschild[2] and P.I.L.D.E. Campos[4]
[1]Universidade Federal dos Vales do Jequitinhonha e Mucuri, Diamantina, 39100-000, Brazil, [2]São Paulo State University, Jaboticabal, 14884-900, Brazil, [3]Topigs Norsvin Research Center, 6641 SZ Beuningen, the Netherlands, [4]Universidade Federal de Viçosa, Viçosa, 36570-900, Brazil, [5]Wageningen University, P.O. Box 338, 6700 AH Wageningen, the Netherlands; paulo.campos@ufv.br

Macauba (*Acrocomia aculeata*) is an oleaginous palm native to tropical America that has received increased attention from the biodiesel industry due to its high productivity and oil quality. Concomitantly, studies have suggested the potential use of macauba by-products in ruminant feeding but with still scarce information on their use on pig-feeding and nutrition. The aim of the study was to evaluate the performance of pigs fed corn and soybean-meal based diets formulated with 0, 50, 100 or 150 g/kg of macauba pulp inclusion. Diets were formulated with similar ME (3.200 kcal/kg), CP (190 g/kg) and digestible Lys (9.7 g/kg). A total of 64 barrows (30.2±1.5 kg initial BW) were assigned to one of the four experimental. The experimental period lasted 35 days and pigs had free access to feed and water. Feed intake (FI) was measured and pigs were weighed weekly to calculate average daily gain (ADG) and feed conversion (FC).Total body minerals, fat and lean content were assessed by dual-energy X-ray absorptiometry at beginning and at the end of the experimental period. Data were analysed using the GLM procedure of SAS including the fixed effects of diet and initial BW as covariate. Pigs fed with dietary inclusion of 150 g/kg of macauba pulp had lower FI than those fed with 0 and 5 g/kg (1,966 vs 2,097 g/d; P<0.01). However, macauba pulp inclusion did not affect neither ADG nor FC. Additionally, pigs fed with 150 g/kg of macauba pulp in the diet had greater backfat thickness than those with no macauba inclusion in the diet (14.9 vs 13.4 mm; P=0.04). According to our results, macauba pulp could be considered as an alternative raw material to be used in diets of growing pigs. However, its inclusion might results in animals with increased backfat thickness.

Nutritive value of an olive cake extracted from a continuous string method for the rabbit

Z. Dorbane[1], S.A. Kadi[2], D. Boudouma[3] and T. Gidenne[4]
[1]Ecole Nationale Supérieure Vétérinaire, ENSV, El-Harrach, Rue Issad Abbes, Algeria, [2]Université M. MAMMERI, Faculté des Sciences Biologiques et Sciences Agronomiques, UN 1501, Tizi-Ouzou, Algeria, [3]École Nationale Supérieure Agronomique, département des productions animales, BP 2700, ex hall de technologie Kharrouba, Mostaganem, Algeria, [4]INRA Occitanie Toulouse, Genphyse, CS 52627, 31326 Castanet, France; zahiadorbane@hotmail.fr

The nutritive value of a sun-dried olive cake extracted from a continuous string 'three stage' method (TSOC) was studied for growing rabbit by comparing 3 diets containing increasing incorporation level of COC (0, 10 and 20%) in substitution to a basal diet (16.9% crude protein, 15.3%ADF). Three groups of 12 rabbits (individually caged) were fed *ad libitum* the three pelleted diets. The faecal digestibility of the diets was measured between 42 and 46 days of age. The chemical composition of TSOC corresponds to a fibre and lignin-rich source: 5.2% of crude protein, 6.4% of ether extract, 67.5% of NDF, 48.2% of ADF and 21.1% of ADL. The substitution of 10 and 20% of the basal diet by TSOC did not significantly affect the weight gain (meanly 35 g/j from 35-46 d old) or the feed intake (meanly 75 g/j from 35-46 d old), but it linearly increased (P=0.043) the feed conversion (2.03-2.30-2.47 respectively for the 0-10-20% TSOC incorporation rate). TSOC incorporation reduced the digestibility of organic matter, crude protein and NDF from 67.8 to 55.3%, 80.4 to 75.3%, 31.5 to 18.4% (P<0.001) respectively. The digestible energy (DE) content of the olive cake estimated by regression was 2.94±0.52 MJ DE/kg DM. Protein digestibility of TSOC was estimated at 37.0%, corresponding to a digestible crude protein concentration of 22.4±6 g DP/kg DM. Crude olive cake could be considered as a moderate source of nutrients for growing rabbit but a good fibre and lignin source.

Energy and protein values of lucerne leaf meal in growing pigs
D. Renaudeau[1], M. Duputel[2], E. Juncker[3] and C. Calvar[4]
[1]PEGASE, INRA, Agrocampus-Ouest, 16, le clos, 35590 Saint-Gilles, France, [2]Chambre d'Agriculture de Bretagne, Avenue du chalutier ''Sans pitié', 22195 Plérin, France, [3]TRUST'ING, 16 Rue du Chêne Cartier, 44300 Nantes, France, [4]Chambre d'Agriculture de Bretagne, Avenue Borgnies Desbordes, 56009 Vannes, France; david.renaudeau@inra.fr

Forage legumes efficiently utilise the growing season and represent a relevant way to ensure soil cover and to produce high protein yields. Among the range of forage, leaf fraction of lucerne would be considered as a valuable protein source for swine. Two experiments were thus conducted to determine the energy and the protein values of lucerne leaf meal (LLM) in growing pigs. In the first experiment, total tract digestibility coefficient (TTD) of energy was measured on a total of 25 pigs (60 kg BW) randomly allotted to five different dietary treatments: 0, 5, 10, 15 and 20% of LLM was incorporated in a wheat-soybean meal diet. On a dry matter basis, the average crude protein (CP), lysine, crude fibre and gross energy contents of LLM were 25.3%, 1.09%, 15.5% and 19.3 MJ/kg, respectively. After 10-days adaptation period, faeces and urines were collected during seven consecutive days. In the second experiment, the apparent (AID) and standardized (SID) ileal digestibility coefficient of AA were measured on five pigs (32.4 kg) fitted with ileo-rectal anastomosis and arranged in a 5×5 Latin square design with 6 periods and five diets containing 0, 5, 10, 15 and 20% of LLM. Pigs were fed each of the five diets during one 7-days period, and ileal digesta were collected on d 5, 6 and 7. On the 6th period, all the pigs were fed with a free protein diet for the measurement of the basal ileal endogenous. As expected, the TTD of energy linearly decreased with increasing inclusion of LLM in the diet. On average, the TTD value and the digestible energy content of leaf meal were 45.4% and 8.5 MJ/kg. Increasing the rate of incorporation of LLM reduced the AID for crude protein and amino acids. This reduction was linear (P<0.05). On average, the AID and SID of lysine were 57.7 and 62.5%, respectively. These low digestibility of energy and lysine are probably explained by the presence of anti-nutritional factors (saponins, polyphenols, etc.) at a high concentration in the LLM.

***Nicotiana tabacum* L. cv. Solaris: an innovative forage for dairy heifers**
A. Fatica[1], F. Di Lucia[2], F. Fantuz[3], H.F. De Feijter[2], B.P. Brandt[2] and E. Salimei[1]
[1]Università degli Studi del Molise, Dip. Agricoltura, Ambiente, Alimenti, via de Sanctis, 86100, Campobasso, Italy, [2]Sunchem BV, Johann Siegerstraat 20, 1096 BH Amsterdam, the Netherlands, [3]Università degli Studi di Camerino, Scuola di Bioscienze e Medicina Veterinaria, Via Gentile III Da Varano, 62032, Camerino, Italy; a.fatica@studenti.unimol.it

The chemical composition of *Nicotiana tabacum* L. cv Solaris has been studied and its biomass has been successfully conserved by ensiling (SiloSolaris). The aim of this trial was to investigate the effects of the innovative ensiled forage in heifers' diet. Sixteen growing heifers of Friesian breed, weighing 220-360 kg and aged 8-16 months, were selected and divided in two homogeneous groups, Control (CTR) and SiloSolaris (SS) Group. Balanced and nutritionally equivalent rations were developed based on the nutritional needs, age and weight: CTR Group received 2.3 kg of concentrates mixture/head and 5-6 kg of hay/head daily, while SS Group was daily fed with 2.3 kg of concentrates mixture/head, 3.25-5.5 kg of hay/head and 0.5 kg (in the adaptation phase, 20 d) to 5 kg of SiloSolaris/head. The trial lasted 49 days. Body condition (BCS), faecal consistency (FS) and locomotion (LS) were scored on each animal, according to the literature. In addition, the investigated consumption of feedstuffs (offered-refusals), body weight (WG) and average daily gain (ADG) were monitored. The effect of the dietary treatment on dry matter (DM) consumption, WG, ADG, BCS, FS and LS were processed by analysis of the variance, also considering the covariate effect of values at d 0, when significant. Significance was declared at P<0.05. In average, DM intake was 5.8 and 6.5 kg, respectively for SS and CTR heifers. ADG was 0.96 vs 0.88 kg/d for SS and CTR group (P>0.05). Taken together, the results suggest that *N. tabacum* L. cv. Solaris ensiled biomass is appreciated by Frisian heifers and did not negatively affect both growth performances and animal welfare. In a view of circular and green economy, this study represents a further successful proof of this innovative multitasking (energy and animal feeding) crop, able of recovering the 'know-how' of the traditional tobacco cultivation in Centre-South Italy.

Evaluation of the nutritional value of poultry by-products in the diet of growing pigs

P. Bikker[1], F. Molist[2] and R. Davin[2]
[1]*Wageningen University & Research, Wageningen Livestock Research, P.O. Box 338, 6700 AH Wageningen, the Netherlands, [2]Schothorst Feed Research, Meerkoetenweg 26, 8218 NA Lelystad, the Netherlands, rdavin@schothorst.nl*

Processed animal proteins (PAPs) obtained as by-product from the slaughter of pigs and poultry are valuable protein sources in animal diets. PAPs can make an important contribution to a sustainable animal production to meet the increasing global demand for animal sourced food while reducing Europe's dependency on imported soybean meal. Processing of PAPs has been improved in the last decades to assure optimal biosecurity and nutritional value of the products, but adequate data on the nutrient digestibility are lacking because of the EU-wide ban to use PAPs in farm animal diets. To facilitate rational future use of PAPs, this study was conducted to determine the nutritional value of five poultry by-products (low-, medium- and high-ash poultry meal, feather meal and blood meal) in 48 growing pigs (Tempo × (LR×LW), BW 45.4 kg) in two periods. Six diets (basal and PAPs) were formulated to evaluate Ca and P digestibility in period 1 (day 0-21), and ileal amino acid digestibility and total tract digestibility of proximate components and energy in period 2 (day 21-35). Feed was supplied at 3.0 × maintenance in 2 meals per day, faeces were collected the last 4 days in each period and chyme was collected after sacrifice of the pigs on the final day. Data were analysed by ANOVA and post-hoc testing of treatment differences. Preliminary data for daily gain and FCR (P<0.001) indicate important differences in nutritional value of the products and an effect of the ash content within the group of poultry meal products. PAPs inclusion also affected the pH in the stomach (P<0.001), but not in other segments of the digestive tract, which may influence physiological processes. Ileal and excreta samples are under analyses; it is expected that PAPs AA profile and ash content will impact mineral, protein and energy digestibility values. Inclusions of up-to-date values for nutrient content and digestibility of these poultry by-products in feedstuff tables will allow their optimal use in pig diets when EU legislation allows the use of PAPs in farm animal diets.

Rumen-bypass nanoparticles: between ingestion and milk production

J. Albuquerque[1,2], S. Casal[1], I. Van Dorpe[3], A.J.M. Fonseca[4], A.R.J. Cabrita[4], A.R. Nunes[5] and S. Reis[1]
[1]*LAQV, REQUIMTE, FFUP, Department of Chemical Sciences, Rua Jorge Viterbo Ferreira no. 228, 4050-313, Porto, Portugal, [2]Instituto Ciencias Biomédicas Abel Salazar, Universidade do Porto, Rua Jorge Viterbo Ferreira no. 228, 4050-313, Porto, Portugal, [3]PREMIX-Especialidades Agrícolas e Pecuárias. Lda, Parque Industrial II – Neiva, 4935-232, Viana do Castelo, Portugal, [4]LAQV, REQUIMTE, ICBAS, UP, Rua Jorge Viterbo Ferreira no. 228, 4050-313, Porto, Portugal, [5]CQM – Centro de Química da Madeira, Universidade da Madeira, Campus da Penteada, 9020-105 Funchal, Portugal; joao.albuquerque.costa@gmail.com*

The main objective of this work was to assess if solid lipid nanoparticles (SLNs), already proven to be able to resist ruminal digestion, were able to resist the digestive processes of remaining bovine digestive track and deliver lysine (Lys) into the bloodstream. SLNs were produced by an organic solvent-free method based in sonication and using only natural occurring lipids and surfactants already used in food applications. The nanoparticles (NPs) were then submitted to different biological media, designed to mimic the *in vivo* conditions of the abomasum, the small intestine and the blood stream as well as fresh bovine blood. Dynamic light scattering (DLS) analyses were performed to compare the NPs prior to and after incubation in the respective media to determine their stability in the latter. Two-tailed student's t-test and one-way analysis of variance (ANOVA) were performed to compare dependent samples and multiple groups of independent samples, respectively. When the groups presented significant statistical difference (P≤0.05), the differences between the respective groups were compared with a post-hoc test (Tukey, P≤0.05). The results showed that the proposed SLNs were able to resist digestion in both the abomasum and the small intestine, while appearing to be slowly degraded in the bovine blood. This ensures that NPs are able to transport Lys across the entire bovine digestive track and release it in the blood where it can be used by the animal. To conclude the tested NP formulations are capable of resisting digestion in the bovine digestive track and of delivering Lys in the blood. Showing high promise for future applications in the Lys supplementation of dairy cows.

The influence of large amounts of rye grain in feed compounds on gilts and barrows fattening perform

R. Tuz[1], T. Schwarz[1], M. Małopolska[2] and J. Nowicki[1]
[1]University of Agriculture in Krakow, Mickiewicza Ave. 24/28, 30-059 Krakow, Poland, [2]National Intitute of Animal Production, Sarego 2, 31-047 Krakow, Poland; rztuz@cyf-kr.edu.pl

Rye grain is not popular in swine fattening as a feedstuff, although the research of last 10 years showed very good productive and economical results of pigs fed mixtures containing large amount of rye. Metabolism of growing barrows is different than gilts, causing disparity in gains and carcass quality, with clearly larger adiposity. Low content of fat in rye grain could potentially influence composition of carcass. The aim of this study was to determine if the use of large amount of rye grain as feedstuff can influence fattening and slaughter performance of pigs, according to sex. The study was conducted in commercial farm located in central Poland. The research material was 500 DanBred pigs divided into 2 nutritional groups (C: control, containing equal volume of barley and wheat; and E: experimental where equal proportions of barley and wheat were replaced by 30 and 60% of rye in grower and finisher respectively). Each group was additionally divided according to sex (G – gilts and B – barrows). Main production parameters were analysed, with estimation of differences among groups using two-way ANOVA followed by Duncan test. Mean daily gains were higher (P<0.01) in barrows independently from nutritional group (1,202±134; 1,181±133; 1,155±107; 1,138±113 in CB, EB, CG and EG respectively), however, feed conversion rate was worse (2.53; 2.63; 2.41; 2.41 in CB, EB, CG and EG respectively). Backfat thickness in CB was larger (P<0.01) in comparison to gilts and barrows fed rye compounds (16.4±3.1; 14.5±3.4; 14.2±3.1; 13.9±2.9 in CB, EB, CG and EG respectively). Consequently lean meat content in CB was lower (P<0.01) in comparison to other groups (57.4±2.6; 58.7±2.7; 59.0±2.4; 59.1±2.3 in CB, EB, CG and EG respectively). To conclude, the use of rye in feed compounds significantly decreased the level of adiposity and increased meatiness of barrows, not influencing slaughter parameters of gilts. Addition of rye to swine diet seem to be a good method to improve barrows performance. Financial source NCRD grant ENERGYFEED, contract No BIOSTRATEG2/297910/12/NCBR/2016.

Nutritional valorisation of *Arundo donax* by treatment with urea

B. Rocha, C.S.A.M. Maduro Dias, C.F.M. Vouzela, H.J.D. Rosa, J.S. Madruga and A.E.S. Borba
University of the Azores, Institute of Agricultural and Environmental Research and Technology (IITAA), ECOFIBRAS (MAC/4.6D/040), Rua Capitão João d'Ávila, 9700-042 Angra do Heroísmo, Açores, Portugal; alfredo.es.borba@uac.pt

The *Arundo donax*, commonly known as cane, is a robust perennial herb, used in agriculture since ancient times, in hedges and slope safety. It can be found from coastal dunes, cliffs and coastal wetlands, to bushes, permanent and semi-natural pasture, as well as agricultural areas and anthropic vegetation. It's one of most widespread Macaronesia island invasive plants. In this work, we propose to increase *A. donax*'s low nutritional value, through a treatment with urea, with the goal of making it a sustainable and environmentally-friendly animal feed alternative and fighting its spread. *A. donax* plants were collected and dried at 60 °C in an oven with controlled air circulation. They were then dried and sprinkled with a 3.5 DM% urea solution and placed in a leak-proof container for 4 weeks. The chemical and *in vitro* digestibility properties of both treated and untreated (control) *A. donax* samples were analysed in triplicate. The data was analysed according to the t-test for independent samples, with significant differences whenever P<0.05. The obtained results indicate that the urea treatment leads to a significant increase (P<0.05) of crude protein contents (from 8.56 to 11.59 DM%) and produce a non-significant decrease (P>0.05) in NDF (from 77.73 to 76.89 DM%), ADL (6.83 to 6.29 DM%) and DM digestibility (43.31 to 40.83%). We can conclude that the urea treatment in a 3.5 DM% concentration has no significant effects in *A. donax*'s nutritional value, so we intend to explore other treatments, which can include increasing the urea concentration or applying treatments with NaOH.

The potential of atella to replace industrial brewers' grain in dairy rations in Northern Ethiopia

A. Tadesse[1,2], B. Rata[3], Y. Tesfay[4] and V. Fievez[1]
[1]Ghent University, Animal Sciences and Aquatic Ecology, Coupure Links 653, 9000 Gent, Belgium, [2]Mekelle University, Animal, Rangeland and Wildlife Sciences, Mekelle, 231, Ethiopia, [3]Ghent University, Food Safety and Food Quality Coupure Links 653, 9000 Gent, Belgium, [4]International Livestock Research Institute, Addis Ababa, 5689, Ethiopia; alextmu@yahoo.com

The study was conducted to compare the nutritive value of atella (fermentation residue of local beer or tella) and industrial brewers' grain (BG) in terms of rumen crude protein degradability and intestinal digestibility as assessed by an *in vitro* simulation of the digestive process. In this study the most common type of atella in Tigray region, originating from a tella brewing process, was prepared from ingredients composed of 0.25:0.03:0.02:0.70 maize flour:barley malt:local hops:water. During tella preparation, maize flour dough was baked on a steel plate, kept on a three-stone wood fire at 113±14.6 °C for 10±2.2 minutes. Both atella (n=3) and BG (n=2) samples were incubated in an *in vitro* rumen simulation system for 2, 4, 6, 10, 24, 48 and 72 hours to determine the rumen degradability of DM and CP. In this laboratory simulation experiment, the nylon bag technique was used to determine rumen degradability and intestinal digestibility. The non-parametric Mann-Whitney test revealed that there was no difference (P>0.05) in rumen degradability and intestinal digestibility between atella and BG. The CP content was 199 and 214 g/kg DM for atella and BG respectively. Atella and BG were characterised by a positive degradable protein balance (OEB). Effective rumen protein degradability (ERPD), calculated based on the degradation kinetics assessed through the nylon bag technique in combination with a fixed rumen outflow rate (kp=0.03/h) was 0.425 g/g CP for atella and 0.451 g/g CP for BG. CP intestinal digestibility determined on the 10 h rumen incubated residues, was 0.837 and 0.858 g/g BPCP (bypass crude protein) for atella and BG respectively. The estimated true protein digested in the small intestine (DVE) was 0.518 and 0.500 g/g CP for atella and BG respectively. From this study, it can be concluded that the nutritive value (CP composition, in sacco *in vitro* simulation characteristics, ERPD and DVE) of atella is comparable to that of BG.

Study on *Nicotiana tabacum* L. cv Solaris as a source of biomass for animal feeding

A. Fatica[1], A. Alvino[1], S. Marino[1], F. Di Lucia[2], H.F. De Feijter[2], B.P. Brandt[2], F. Fantuz[3] and E. Salimei[1]
[1]Università degli Studi del Molise, Dip. Agricoltura, Ambiente, Alimenti, via de Sanctis, 86100, Campobasso, Italy, [2]Sunchem BV, Johann Siegerstraat 20, 1096 BH Amsterdam, the Netherlands, [3]Università degli Studi di Camerino, Scuola di Bioscienze e Medicina Veterinaria, Via Gentile III Da Varano, 62032, Camerino, Italy; a.fatica@studenti.unimol.it

While a deep crisis is affecting the smoke tobacco cultivation in Italy, the non GMO *Nicotiana tabacum* L. cv. Solaris has been developed (PCT/IB/2007/053412) as 'energy crop', since its maximized production of flowers/seeds are rich in oil used as biofuel. In certain climatic conditions, after the harvest of the inflorescences, a second harvest of biomass is possible. The aim of this study is to assess the use of this biomass as an innovative forage. In the triennium 2016-2018, Solaris biomass samples (n=15) were collected in three experimental sites (Vicenza, Chieti and Perugia), dried and chopped at 4 cm and analysed for chemical constituents (humidity, crude protein, crude oil and fat, crude ash, neutral detergent fibre, acid detergent fibre and acid detergent lignin), sugars (glucose, fructose, sucrose), starch, total alkaloids (expressed as nicotine), minerals (Ca, P, Mg, K, Na and chlorides), according to official analytical methods. Descriptive statistical analysis of data was performed and results are reported on a dry matter basis (mean ± standard deviation, per 100 g DM). Results allow to define the Solaris biomass as a good quality forage, in terms of crude protein (19.0±2.4 g), crude oil and fats (7.9±4.06 g) and fibre contents (46.6±4.1 g NDF; 37.5±5.0.g ADF), despite high levels of lignin (12.9±3.9 g ADL), ash content (17.2±2.4 g) and calcium content (2.5±0.6 g Ca). The content of non-structural carbohydrates (sugars and starch) allows the use of Solaris biomass as silage. The average content of total alkaloids was 350±185 mg, which is much lower than values reported for smoking tobacco varieties. The results confirm the use of cv. Solaris from the second harvest biomass as a good quality forage, adding value to the Solaris multitasking attitude as a source of renewable energy (oil from seed and bio-methane from biomass) and animal feed protein (Solaris seed cake).

Nutritive value and *in vitro* organic matter digestibility of dried white grape pomace
P. Vašeková, M. Juráček, D. Bíro, M. Šimko, B. Gálik, M. Rolinec, O. Hanušovský and R. Kolláthová
Slovak University of Agriculture in Nitra, Department of Animal Nutrition, Trieda Andreja Hlinku 609/2, 949 76 Nitra,
Slovak Republic; vasekovap@gmail.com

The aim of experiments were determined organic matter (OM), dry matter (DM), crude fibre (CF), acid detergent fibre (ADF), neutral detergent fibre (NDF), crude protein (CP), *in vitro* digestibility of organic matter (IVOMD), macro and microelements in dried white grape pomace (*Vitis vinifera* L., variety *Pinot Gris*). Chemical analysis was determined with standard laboratory procedures and principles in the Laboratory of Quality and Nutritive Value of Feeds (Faculty of Agrobiology and Food Resources, Slovak University of Agriculture in Nitra). *In vitro* digestibility was determined by pepsin-cellulase method. In grape pomace samples average DM 942.05 g/kg and OM 907.25 g/kg content were found. All samples had an adequate fat (98.90 g/kg) and CP (99.80 g/kg) content. However, CF (184.50 g/kg) with their fractions NDF (433.13 g/kg) and ADF (358.88 g/kg) were very high. These values correspond to the quantity of nutrients in forages. Coefficient of digestibility OM was 42.37%. The highest content from macroelements in the content of K (14.45 g/kg) and from microelements in the Fe concentration (90.55 mg/kg) was determined. On the other side the lowest amount from macroelements in Na (0.58 g/kg) and microelements in Mn content (12.95 mg/kg) was observed. Results show that dried grape pomace as by-product of winery has potential as nutritional source for animal nutrition. This work was supported by the Slovak Research and Development Agency under the contract no. APVV-16-0170.

Effects of feeding alternative feeds on nursing performance of Awassi ewes and their suckling lambs
B.S. Obeidat, M.K. Aloueedat and M.S. Awawdeh
Jordan University of Science and Technology, Department of Animal Production, P.O. Box 3030, 22110 Irbid, Jordan;
bobeidat@just.edu.jo

Two experiments were conducted to assess the effects of replacing conventional feeds with alternative feeds (AF) on nutrient digestibility and milk components and body weight (BW) changes in lactating ewes and their lambs. Experiment 1 (Exp. 1) evaluated nursing ewe performance while experiment 2 (Exp. 2) evaluated nutrient digestibility and N balance. In both experiments, the same diets and AF (dried distillers grains with solubles, carob pods, olive cake and bread by-product) were used. The isonitrogenous dietary treatments were: (1) no AF (CON); (2) 200 g/kg AF (AF200); or (3) 400 g/kg AF (AF400) of dietary dry matter (DM). Animals in both experiments were housed in individual shaded pens. In Exp. 1, twenty-seven Awassi ewes (initial BW=51.6 kg; age=3 to 4 years) were randomly distributed into the corresponding diets (9 ewes/diet) and fed *ad libitum*. Nutrient intake was evaluated daily during the experimental period, which lasted for 8 weeks. In Exp. 2, 18 Awassi ewe lambs (BW=38.1 kg; age=7 to 8 months) were used and assigned to the same dietary treatments (6 ewe lambs/diet) during a 21-day trial (14 housed individually in shaded pens and 7 days in metabolic cages). Feed intake and faecal and urinary outputs were recorded. In Exp. 1, no differences in animal BW were detected. With the exception of neutral detergent fibre (NDF), which was lower (P=0.05) in the AF-containing diets compared with the CON diet, DM, crude protein (CP) and acid detergent fibre (ADF) intake was not affected by treatment. Milk yield (950, 929 and 893 g/d for the CON, AF200 and AF400, respectively) and composition and feed to milk ratio were comparable among treatment diets. Cost of milk production was lower (P<0.05) in the AF diets than in the CON diet (1.11, 0.94, and 0.86 US$/kg milk, for the CON, AF200 and AF400, respectively). In Exp. 2, DM, CP, NDF and ADF digestibility as well as N balance were not affected by the treatment diets. Results of the current study indicate the possibility of including different alternative feeds in Awassi ewe diets (either at 200 and 400 g/kg of dietary DM) to reduce the cost of production but without negatively affecting production parameters.

Alternative feedstuffs had no effects on haematological and biochemical parameters of Awassi lambs

M.S. Awawdeh[1], B.S. Obeidat[2] and H.K. Dager[2]
[1]Jordan University of Science and Technology, Faculty of Veterinary Medicine, Jordan University of Science & Technology, 22110, Jordan, [2]Jordan University of Science and Technology, Faculty of Agriculture, Jordan University of Science & Technology, 22110, Jordan; dr.awawdeh@gmail.com

Dietary inclusion of alternative feedstuffs (AF) could impact animal performance, ruminal ecosystem, and/or blood parameters. The objective of the current trial was to evaluate the effects of dietary inclusion of mixed AF (dry bread, carob pods, olive cake, and sesame meal) on haematological and biochemical parameters of growing lambs. Male Awassi lambs (n=27, average BW=20.0±0.5 kg) were randomly assigned to one of three dietary groups (9 lambs/group). Diets were formulated to contain (on DM basis) 0 (CTL), 250 (25A), or 500 (50A) g/kg of mixed AF. Blood samples were collected from each lamb before morning feeding at the beginning (day 7), middle (day 35), and end (68 day) of the study. All haematological parameters (white blood cell count; WBC, red blood cell count; RBC, platelets, packed cell volume; PCV, haemoglobin; Hb, mean corpuscular volume; MCV, mean cell haemoglobin; MCH, and mean cell haemoglobin concentration; MCHC) were not ($P \geq 0.11$) substantially affected with treatment diets. Except for urea and cortisol, dietary treatments had no significant ($P \geq 0.07$) effects on levels of all of the measured metabolites (glucose, total proteins, albumin, globulins, calcium, triglycerides, total cholesterol, low density lipoproteins; LDL, high density lipoproteins; HDL, and creatinine) and enzymes (creatine kinase; CK, alkaline phosphatase; ALP, alanine aminotransferase; ALT, and aspartate aminotransferase; AST). Lambs fed the 50AF diet had the lowest blood levels of urea and cortisol. Under condition of the current study, dietary inclusion of AF up to 500 g/kg did not negatively impact lambs health assessed with changes in haematological and biochemical parameters.

***In vitro* dry matter and crude protein digestibility in bones and raw food (BARF) diets**

M. Ottoboni, E. Fusi, A. Luciano, C. Coppola and L. Pinotti
University of Milan, Department of Health, Animal Science and Food Safety Carlo Cantoni, via Celoria 10, 20133 Milano, Italy; matteo.ottoboni@unimi.it

Feeding a bones and raw food (BARF) diet has become an increasing trend in canine nutrition. These diets try to imitate the feeding behaviour of the wolf. Thus, BARF diets contain a high amount of animal components like meat, offal, and raw meaty bones, (representing the wolf's prey) combined with comparatively small amounts of vegetable ingredients. This study evaluated the *in vitro* dry matter and crude protein digestibility in BARF diets. For these purpose eight samples of dogs BARF diets and two commercial dog food, used as reference materials, were analysed and tested in the assay. The BARF diets were based on raw beef and poultry by-products, while the commercial pet food were one dry and one wet. All samples were analysed for dry matter (DM), crude protein (CP), ether extract (EE) and ash content. Furthermore, using an *in vitro* assay, simulating gastric and small intestinal digestion, both dry matter digestibility (IVD-DM) and crude protein digestibility (IVD-CP) have been measured. All BARF diets and wet pet food were characterized by high moisture content (DM 380 g/kg), while in the case of dry pet food DM content was 920 g/kg. On average BARF diets and commercial diets were characterised by the following values, on dry matter basis: CP 368 g/kg; EE 442 g/kg; ash 52 g/kg. All BARF samples and reference materials were characterized by high digestibility values. Both IVD-DM and IVD-CP reached values higher than 80%. Of note in the case of IVD-DM a substantial variability within samples has been observed (SD ±5.5). While in the case of IVD-CP value observed presented less variability (SD ±1.4). In light of these results, it can be concluded that proposed IDV method is adequate in determining protein digestibility in BARF diets, while the assay probably needs some optimisation for DM digestibility in order to reduce its variability.

Detection of reciprocal translocations in pigs using short read sequencing

A.C. Bouwman[1], M.F.W. Te Pas[1], M.F.L. Derks[1], M.A.M. Groenen[1], R.F. Veerkamp[1], M. Broekhuijse[2] and B. Harlizius[2]
[1]*Wageningen University and Research, P.O. Box 338, 6700 AH Wageningen, the Netherlands,* [2]*Topigs Norsvin Research Centre, P.O. Box 43, 6640 AA Beuningen, the Netherlands; aniek.bouwman@wur.nl*

In pig breeding, male selection candidates are routinely screened for karyotype abnormalities in several countries worldwide. Abnormal karyotypes have a frequency of 0.4%. The most common abnormality seen in live animals is reciprocal translocations (RT). Although the individuals carrying RT are healthy and viable, they show subfertility. Karyotyping is costly, labour intensive, and difficult. Envisioning a further reduction of sequencing cost in the future and more extensive use of sequence data, sequencing selection candidates might become a routine. Being able to use these sequences for screening of karyotype abnormalities would be cost effective and might even be a more sensitive screening than the visual inspection after chromosome painting currently applied. To evaluate such an approach, nine pigs with a positive RT test based on chromosome painting were sequenced at 30× coverage using Illumina paired end short reads. Discordant read pairs and split reads were used by Delly v2 to identify inter-chromosomal translocations in the sequence data, returning a long list including many false positives. Results were filtered using sequence data from control animals including TJ tabasco (the animal used for the pig reference genome assembly) to filter on genome build issues and repetitive DNA elements or other commonly observed issues. Additional filtering on mapping quality, number of supportive reads, as well as, presence of a reciprocal match was performed. The final set of possible reciprocal breakpoints was visually inspected to discriminate further between false and true breakpoints. The visual inspection was successful in eliminating false positives, but not all true positives where detected. Currently we are developing an analysis pipeline to replace the visual inspection. In addition, we will investigate the specificity and sensitivity of our RT detection pipeline with 30× coverage, as well as, with 15× coverage to reduce sequencing cost.

Genome-wide association analyses and fine mapping of immune traits in dam lines using sequence data

C.M. Dauben[1], E.M. Heuß[1], M.J. Pröll-Cornelissen[1], K. Roth[1], H. Henne[2], A.K. Appel[2], K. Schellander[1], E. Tholen[1] and C. Große-Brinkhaus[1]
[1]*University of Bonn, Institute of Animal Science, Endenicher Allee 15, 53115 Bonn, Germany,* [2]*BHZP GmbH, An der Wassermühle 8, 21368 Dahlenburg-Ellringen, Germany; cdau@itw.uni-bonn.de*

Health and immune traits are linked to robustness of piglets. To prevent performance reduction through health impairment, it could be useful to enhance robustness due to immune competence. Indicators triggering a health-promoting immune competence need to be specified regarding negative impacts and interactions. However, the genetic background of the porcine immune system still remains unclear. So far, a limited number of immune relevant QTL has been analysed. Therefore, this study aims to identify biological relevant SNPs associated with health and immune traits in pigs. A total of 535 Landrace (LR) and 461 Large White (LW) piglets were included in this study. Phenotypes were recorded for the complete and differential blood count as well as for eight cytokines and haptoglobin. All animals were genotyped by using the Illumina PorcineSNP60 BeadChip. Based on genotype and pedigree information, a subset of 57 animals, which represent the genetic structure of the populations, was selected for whole genome sequencing. A breed-specific GWAS was performed with a generalized linear model. Using sequence data for fine mapping, significant QTL regions associated with immune regulative mechanisms are analysed in more detail. In total, the univariate GWAS revealed 355 significant SNPs in LR and 96 significant SNPs in LW. Numbers and positions of the SNPs vary between the traits and breeds. However, across both breeds, we identified a region on SSC9 associated with IL1β, IL4, IL6, IL10 and TNFα, suggesting pleiotropic effects. First results of the fine mapping show 15 variants located close to a specific SNP within this region. All of the variants occur in LR and LW, indicating an across-breed effect of this region on cytokines mediating in the immune response. In conclusion, along with the knowledge about the genetic potential of the traits analysed, we are able to clarify mechanisms interacting in the porcine immune system. This information may be used to improve breeding strategies and thereby enhancing robustness of piglets.

Selection of pigs for social genetic effects improves growth in crossbreeds

L.V. Pedersen[1], B. Nielsen[1], O.F. Christensen[2], S.P. Turner[3], H.M. Nielsen[1] and B. Ask[1]
[1]Danish Pig Research Centre, Axelbo.. J, 1600 Cph, Denmark, [2]AU Foulum, Blichers Alle 20, 8830 Tjele, Denmark, [3]SRUC, West Mains Road, Edinburgh EH9 3JG, United Kingdom; livp@seges.dk

The average daily gain (ADG) of animals housed in groups are affected by heritable social genetic effects (SGE). However, reported selection responses are inconsistent, probably because of small sample sizes and selection environments differing from experimental or production environments. Therefore, we aimed to: (1) show that selection for SGE in purebreds in a nucleus environment can improve ADG in crossbreds in a production environment; and (2) investigate whether social genetic parameters are different for ADG between purebreds in a nucleus environment and crossbreds in a production environment. A selection experiment was conducted on sows from one Danish nucleus herd. High SGE litters were made by mating high SGE DanBred Yorkshire boars with high SGE DanBred Landrace sows, and vice versa for low SGE litters. In total, 4,728 crossbred castrates from 1,171 litters were housed on a production farm in 135 high and 138 low social groups with group sizes varying from 12 to 19 pigs. The average estimated SGE was 1.12 for high and -0.84 for low social groups. Pigs were weighted at 24 hours and ~7 weeks after entering the finisher unit. A linear regression model was applied to ADG, which included the complementary social genetic effect (cSGE, sum of the pen mates' SGE) as a covariate. Preliminary results showed that a 1 unit increase in the cSGE resulted in an increased ADG of 0.59 g/day per pig. With a pen size of 18 pigs/group, this corresponds to an additional ADG of 10.6 g/day for the whole group. Direct and social genetic (co)variances for ADG were estimated in a multivariate analysis using a social genetic animal model for purebred relatives and a social genetic sire-dam model for crossbreds. The results showed direct genetic variances of 3,873 and 2,770 and social genetic variances of 54 and 119 for the sire and sow components on the crossbreds, respectively. The estimated genetic (co)variances indicated that SGE differ between purebreds and crossbreds for the sire component, but not the sow component. In conclusion, selection for pigs with higher SGE for ADG increases ADG in crossbreds in a production environment.

Verifying the existence of second litter syndrome in pigs

M. Pszczola[1], E.F. Knol[2], H.A. Mulder[3] and E. Sell-Kubiak[1]
[1]Poznan University of Life Sciences, Department of Genetics and Animal Breeding, Wojska Polskiego 28, 60-637 Poznan, Poland, [2]Topigs Norsvin Research Centre, Schoenaker 6, 6641 SZ Beuningen, the Netherlands, [3]Wageningen University & Research, Animal Breeding and Genomics, P.O. Box 338, 6700 AH Wageningen, the Netherlands; mbee@up.poznan.pl

It has been indicated that fertility performance of some sows in 2nd parity is lower than in the 1st one – a so-called 2nd litter syndrome (SLS). Interestingly, SLS sows and non-SLS sows have similar performance in latter parities. The SLS has been associated with negative weight loss after the 1st gestation and too early 1st insemination of young gilts. However, it was not tested whether the presence of SLS is purely due to the properties of the trait distribution and possible just by chance (i.e. SLS is non-existent). We attempted to verify the existence of SLS. Data set with records of total number born piglets (TNB) of 50,000 sows and 10 parities was simulated. Simulation had 100 replicates and used means, and SD's of TNB in subsequent parities from the real data. To check the frequency of occurrence whether TNB in the subsequent parity is lower than in the preceding one, we divided the data set based on the following condition:, where i is the parity number. Further, data from 67 herds were used to indicate the existence of SLS. The simulations showed that average expected frequency across all the parities was 0.48 (±0.06), and average observed frequency was 0.44 (±0.04). Data from individual herds showed that the expected and observed frequencies agreed well. The average difference between the observed and expected frequencies was 0.05 (±0.03) and correlation between them was 0.93. The results of this study showed that numerous instances of $TNB_i > TNB_{(i+1)}$ observed in herds are likely due to statistical properties of the analysed trait and as long as its frequency is not higher than the expectation based on the distribution it should not be considered an abnormality or syndrome. Financed by the Polish National Science Centre NCN SONATA grant no. 2016/23/D/NZ9/00029.

Looking at nutrition related traits in pigs through genetic-tinted glasses

L.M.G. Verschuren[1,2], M.P.L. Calus[2], D. Schokker[2], R. Bergsma[1], F. Molist[3] and A.J.M. Jansman[2]
[1]Topigs Norsvin Research Center B.V., P.O. Box 43, 6640 AA Beuningen, the Netherlands, [2]Wageningen Livestock Research, Droevendaalsesteeg 4, 6708 PD Wageningen, the Netherlands, [3]Schothorst Feed Research, Postbus 533, 8200 AM Lelystad, the Netherlands; lisanne.verschuren@wur.nl

To improve pig performance, especially feed efficiency, faecal nutrient digestibility is an important trait of feed ingredients and complete diets. However, feeding the same quantity of diet to pigs with a similar body weight can result in a different performance. This study investigated the variation in grower-finisher performance explained by the variation in faecal nutrient digestibility, independent of feed intake. Data were collected of 114 three-way crossbreed grower-finisher pigs (65 female and 49 male) which were either fed a diet based on corn/soybean meal or on wheat/barley/by-products. Faecal samples were collected on the day before slaughter (mean body weight 121 kg) and used to determine faecal digestibility of dry matter, ash, organic matter, energy, crude protein, crude fat, crude fibre and non-starch polysaccharides. The effects of diet, sex and feed intake on faecal nutrient digestibility were estimated with a mixed model and were highly significant for most nutrients. The residuals of the model created new variables, corrected faecal nutrient digestibility, and were used to estimate their association with performance traits. In the first phase of the experiment (mean body weight <70 kg), up to 7.7% of the variation in feed conversion rate was explained by the corrected nutrient digestibilities. Corrected digestibility of dry matter, organic matter and crude protein explained up to 6.1% of the variation in backfat thickness and crude protein digestibility explained 3.6% of the variation in average daily gain. There was no relationship between corrected faecal nutrient digestibility and performance of the pigs in the second phase of the experiment (mean body weight >70 kg). In conclusion, during early stages of the growing-finishing period, part of the variation in performance was explained by variation in faecal digestibility of nutrients, with highest contribution of crude protein.

Using life cycle assessment to estimate environmental impacts from correlated genetic traits in pigs

M. Ottosen, S. Mackenzie, M. Wallace and I. Kyriazakis
Newcastle university, School of Natural and Environmental Science, T&W, NE1 7RU Newcastle Upon Tyne, United Kingdom; m.ottosen2@newcastle.ac.uk

Pig performance has seen major improvements during the last decades, but it is unknown which traits are most influential on the environmental impacts of pig systems. A central element in livestock breeding is that many traits are correlated, and this complicates the estimation of the environmental impacts through life cycle assessment (LCA). An animal model was built for an indoor pig production system using typical diets, cumulative feed intake and performance from the Danish pig production industry. Twelve genetic traits were used as independent variables spanning growth, energy use, reproduction and mortalities. Feed intake for all stages of the production system was adjusted to the needs of the animal, which changed according to changes in the genetic traits. Genetic variance and correlations were collected from modern pig literature, and a correlation matrix was compiled from the different studies, weighted by sample size and strength of the correlation. Model sensitivity was tested in such a way that higher degree of correlations could be gradually taken into account: (1) one at a time sensitivity analysis (OATSA) was performed by changing all traits by +/-2 SD; (2) traits were clustered by upwards hierarchical clustering based on the Euclidean correlation distance between the traits, and each cluster was varied according to the internal correlations in a Cluster of Traits sensitivity analysis (CoTSA); (3) Finally, all traits were randomly sampled with Latin Hyper Cube sampling according to the correlation matrix to calculate the total sensitivity index (TSI). The OATSA suggested that energy maintenance, large pig growth rate and a few sow growth traits were the most sensitive traits for all investigated impacts. The CoTSA suggested that the model is most sensitive to a cluster of growth and energy traits, about half as sensitive to a cluster of reproductive traits and not sensitive to a cluster of sow performance traits. The TSI showed high sensitivity to energy maintenance and lean meat content, but nearly no sensitivity to large pig growth rate. This study demonstrated a novel method for compiling and applying correlations in genetic trait to LCA of livestock systems.

Back to the future: virtual concepts for sustainable pig production systems

M. Von Meyer-Höfer, H. Heise, C. Winkel and A. Schütz
Georg-August University of Göttingen, Agricultural Economics and Rural Development, Platz der Göttinger Sieben 5,
37073 Göttingen, Germany; mvonmeyer@uni-goettingen.de

As providing an adequate amount of meat is not enough anymore to legitimize any production methods, future pig production systems must not only take economic but also ethical and public demands into consideration. Previous research mainly focused on technological, scientific and economic optimisation, which is not fully adopted in practice yet. However, including public demands has played only a minor role so far, for sustainable pig farming broad public acceptance is indispensable. In this context, it is particularly important to develop production systems that are realisable, acceptable and above all, also consider intra and inter sustainability goal conflicts. Against this background we will present our current (October 2017-May 2019) transdisciplinary project where within a series of professionally moderated future workshops innovative systems of pig production are identified for all production phases. The suggested systems are developed in a multi-stakeholder discourse including farmers, scientists and divers other stakeholders and evaluated from different perspectives (animal welfare, agribusiness, emissions). Moreover, their public acceptance is tested via surveys including f-NIRS measurements. The jointly developed standards underlying the production systems from sow to pork production are above all, but not limited to: more space, natural light / air, straw litter, roughage, roofed exterior climate area, wallows, structured pens. Additionally, wood is suggested as an innovative material to achieve more aesthetic and sustainable buildings. The final models are presented in computerized 3D visualisations to ensure a broad accessibility and dissemination. The here presented work of the project group is supposed to provide a solid basis for the conceptualisation of investments to be made by individual farmers, the development of future research strategies and the deduction of political implications in EUs core pig producing countries.

Environmental impacts of housing and manure management in European pig production systems

G. Pexas[1], S. Mackenzie[1], M. Wallace[1,2] and I. Kyriazakis[1]
[1]Newcastle University, Agriculture, School of Natural and Environmental Sciences, King's Rd, NE1 7RU, Newcastle upon Tyne, United Kingdom, [2]University College Dublin, School of Agriculture and Food Science, Belfield, 4, Dublin, Ireland; g.pexas2@newcastle.ac.uk

The potential of modifications to pig housing and manure management to reduce the environmental impact of European pig production systems was evaluated using life cycle assessment. Initially, we performed a local sensitivity analysis on pig housing and manure management related parameters, to identify environmental hotspots in these components of the system. The environmental impacts of in-house slurry acidification, screw-press slurry separation and anaerobic-digestion of slurry were compared and benchmarked using parallel Monte Carlo simulations. Further to this, we evaluated each manure management system's impacts against scenarios that described modifications in pig housing, through a 4 (the above plus the base line) × 5 pig housing scenario analysis. The functional unit was 1 kg of live weight pig at the farm gate. The impact categories assessed were: Non-Renewable Resource Use and Non-Renewable Energy Use (NRRU, NREU), Global Warming Potential (GWP), Acidification Potential (AP) and Eutrophication Potential (EP). Danish pig systems were used as a case in point, with data provided by the Danish Pig Research Centre. The environmental impact hotspots identified, and their relative contribution to each impact category they significantly affected were: level of slurry dilution (AP, 15% and EP, 10%), slurry removal regime (AP, 3.8% and EP, 2.4%) and ventilation system efficiency (NREU, 4% and GWP, 4.6%). The alternative scenario analysis suggested that in-house slurry acidification was the only manure management system that systematically, significantly reduced environmental impacts compared to the average manure treatment, for all impact categories. Additionally, it was the least sensitive to changes in pig housing related parameters (e.g. fan efficiency, barn insulation). Modifications in manure management have the potential to significantly reduce the environmental impact of pig production. The interactions between pig housing and manure management should be considered if aiming for the most environmentally optimal system configuration.

Cost and benefit evaluation of genomic selection in Canadian swine breeding schemes

Z. Karimi[1,2], M. Jafarikia[1,2] and B. Sullivan[2]
[1]*University of Guelph, Animal Biosciences, 50 Stone Rd E, N1G 2W1 Guelph, Canada, [2]Canadian Centre for Swine Improvement, 960 Carling Avenue, Ottawa, ON K1A 0C6, Canada; zahra@ccsi.ca*

Replacing traditional genetic evaluation by genomic evaluation may result in higher selection accuracy and, accordingly, higher genetic gain. However, this would generate additional costs due to genotyping of animals. The goal of this study was to investigate the net economic gain of genomic selection compared to traditional selection. Scenarios considered were based on the simulation study of Lillehammer *et al.* with adjustments to match a typical pre-genomics selection program used in Canadian dam lines. Economic analysis considered that genotyping costs would be incurred at the nucleus while benefits would accrue at nucleus, multiplication and commercial levels. Four different cases were evaluated with varying heritability and genetic correlation assumptions between reproduction and production traits, and with varying numbers of males and females genotyped, as simulated in Lillehammer *et al*. The cost of 50k SNP genotyping was considered 25 Canadian dollars per animal. Economic benefit was calculated as a function of the increased accuracy of genetic evaluation and the current economic gains estimated in Canada and published in the Annual Report of the Canadian Centre for Swine Improvement (2018). In summary, there was a relative increase in genetic gain from 0.09 to 0.57 depending on the scenario, with higher costs for more accuracy gain. If we only consider benefit at the nucleus level, the payback would be from 26 to 42 years, depending on the scenario. Considering the extra improvement at the multiplication level, the pay back period was only 33 to 47 months. At the commercial level, the pay back period was just 13 to 20 weeks. Thus, for a nucleus operation, it would be difficult to justify the investment in any of these scenarios without considering the benefit from the multiplication or commercial levels. However, it is worth noting that the benefits at the multiplication level alone would justify the investment for all scenarios evaluated, even if cost was not shared by the commercial producers.

The intestinal, metabolic, inflammatory and production responses to heat stress

L.H. Baumgard[1], R.P. Rhoads[2], J.W. Ross[1], A.F. Keating[1], N.K. Gabler[1] and J.T. Selsby[1]
[1]*Iowa State University, Department of Animal Science, 313 Kildee Hall, Ames, IA 50011-3150, USA, [2]Virgina Tech University, Department of Animal and Poultry Sciences, 3080 Litton Reaves Hall, Blacksburg VA, 24061, USA; baumgard@iastate.edu*

Heat stress (HS) compromises efficient animal production and reduces livestock output. Suboptimal production during HS was traditionally thought to result from hypophagia. But recent discoveries indicate that heat-stressed animals utilize homeorhetic strategies to modify metabolic and fuel selection priorities independently of reduced feed intake. With regards to lactation, reduced nutrient intake only explains about 50% of the decrease in milk synthesis. We believe that the epicentre of the remaining 50% originates at the gastrointestinal tract. For a variety of reasons, HS compromises intestinal barrier integrity in ruminant and non-ruminant animals. Intestinal hyperpermeability to luminal contents can result in local and systemic inflammatory responses. Consequently, heat-stressed animals are simultaneously confronted with environmental hyperthermia and endotoxemia-induced inflammation. Ironically, both HS and endotoxemia (catabolic conditions) either cause insulin (a potent anabolic hormone) secretion or markedly enhances glucose-stimulated insulin secretion. The reason for the hyperinsulinemia is unclear, but leukocytes are insulin sensitive and most activated immune cells initiate a metabolic switch from oxidative phosphorylation to aerobic glycolysis (the Warburg Effect), thus requiring glucose as their primary fuel and as a biosynthetic precursor. In fact, we have recently demonstrated *in vivo* that a lipopolysaccharide (LPS)-intensely activated immune system consumes ~1 g glucose/kg BW0.75/h in growing pigs and cattle and lactating dairy cows. This translates into ~2 kg of glucose/d for a lactating dairy cow. The HS-induced hyperinsulinemia explains why HS animals do not mobilize adipose tissue despite being in both a negative energy balance and catabolic state. Heat stress and endotoxemia also negatively affect multiple reproductive parameters and, coupled with hyperinsulinemia, likely explains decreased farm animal fertility during the warm summer months.

Future heat stress risk in European dairy cattle husbandry

S. Hempel[1], C. Menz[2] and T. Amon[1,3]
[1]Leibniz Institute for Agricultural Engineering Potsdam-Bornim (ATB), Engineering for Livestock Management, Max-Eyth-Allee 100, 14469, Germany, [2]Potsdam Institute for Climate Impact Research (PIK), Climate Resilience, Telegraphenberg A 31, 14473 Potsdam, Germany, [3]Free University Berlin, Veterinary Medicine, Institute of Animal Hygiene and Environmental Health, Robert-von-Ostertag-Str. 7-13, 14163 Berlin, Germany; shempel@atb-potsdam.de

In the last decades, an exceptional global warming trend was observed. Along with the temperature increase, modifications in the humidity and wind regime may amplify the regional and local impacts. Modifications in housing management are the main measures taken to improve the ability of livestock to cope with the resulting climatic stress conditions. Measures and systems that balance welfare, environmental and economic issues are, however, barely investigated in the context of climate change and are thus almost not available for commercial farms. In Europe, cows are economically highly relevant and are mainly kept in naturally ventilated buildings that are most susceptible to climate change. We used a modelling chain to estimate future heat stress risk in dairy cattle husbandry. Meteorological data was collected inside three reference barns in Central Europe and the Mediterranean region. An artificial neuronal network (ANN) was trained to relate the outdoor weather conditions to the indoor microclimate. Subsequently, this ANN model was driven by regional climate model projections. For the evaluation of the heat stress risk, we considered the amount and duration of heat stress events, which we defined as hours of at least moderate heat stress. We found that by the end of the century the number of annual stress events can be expected to increase by up to 2,000 hours while the average duration of the events increases by up to 22 hours relative to a reference period 1971 to 2000. Although the degree of severity of the projected increase of heat stress risk varies depending on the region, the climate model and the anticipated greenhouse gas concentration, there was an overall increasing trend. This implies strong impacts on animal welfare, milk yield and emissions and an urgent need for mid-term adaptation strategies.

Environmental effects on feed consumption and feeding behaviour of cattle with known feed efficiency

Y.R. Montanholi[1], M. Sargolzaei[2], J.E. Martell[3], A. Simeone[1], V. Beretta[1] and S.P. Miller[4,5]
[1]Universidad de la Republica Uruguay, Facultad de Agronomía, Av. Gral. Eugenio Garzón 780, Montevideo, 12900, Uruguay, [2]University of Guelph, Ontario Veterinary College, Pathobiology, 50 Stone Road East, Guelph, ON, N1G2W1, Canada, [3]Independent Researcher, 207 Farnham Road, Bible Hill, NS, B2N2X6, Canada, [4]University of Guelph, Centre for the Genetic Improvement of Livestock, 50 Stone Road East, Guelph, ON, N1G2W1, Canada, [5]Angus Genetics Inc., 3201 Frederick Avenue, Saint Joseph, MO 64506, USA; yuri.r.montanholi@gmail.com

The efficiency of feed utilization for growth is influenced by the capacity of cattle to cope with environmental conditions while maximizing retained energy. Our objective was to assess the relationships of weather parameters with feed efficiency, measured via residual feed intake (RFI; DM kg/d). Crossbred feedlot steers (n=104; 239±21 days of age) were performance tested in a covered barn during winter/spring. Assessments of individual feed intake on a corn-based ration, and every 28-day measures of body weight and composition were computed to determine RFI. Weather parameters (temperature, humidity, pressure sunshine, precipitation and radiation) were recorded in a station 800 meters away. Feeding behaviour measures (time at the feeder, TF; intake per visit, IN and; number of visits, NV) were calculated using data from the automated feeding system. Regression and categorical analyses were conducted, with steers divided into differing RFI groups (Efficient=26, -0.82; Average=52, 0.00 and; Inefficient=26, 0.85 DM kg/d) in consideration to periods of the day (day, night, sunrise, sunset) and feed freshness (new, old). Preliminary analysis indicate that a rise in temperature and humidity were associated with a decrease in TF and NV of inefficient cattle, but not for Efficient. Efficient cattle seem to increase IN at a higher magnitude in response to temperature when compared to Inefficient. The rise in pressure was associated with a similar increase in TF for all cattle. Efficient cattle had lower NV and increased TF over the day, and increased TF with new feed. In conclusion, feed efficiency relates to phenotypes that are less sensitive to temperature and humidity variation. Feed efficiency may also relate to a potential dominance of efficient cattle.

What is an outlier telling us – hot summers might affect reproductive performance of sheep in autumn

M.J. Rivero[1], G.B. Martin[2], A. Cooke[1] and M.R.F. Lee[1,3]
[1]Rothamsted Research North Wyke, Okehampton, EX20 2SB, United Kingdom, [2]The University of Western Australia, Crawley, WA, Australia, [3]University of Bristol, Bristol Veterinary School, Langford, BS40 5DU, United Kingdom; jordana.rivero-viera@rothamsted.ac.uk

Effects of stress on reproductive performance in ewes may have severe consequences for fecundity and fertility, and therefore reduce lambing percentage. We hypothesised that heat stress (HS) on summer may affect fertility of sheep and we explored this using the Comprehensive Climate index – CCI. We studied the CCI from post-weaning (Jul-1) to the end of summer (Sep-30) and tested whether it was linked to reproductive performance at mating (late Oct – mid Dec) determined at scanning (Jan) for a UK sheep flock. A regression analysis was performed using the sum of hours CCI was above 25 °C (I25) and the ram percentage (%R) as predictors of the pregnancy scanning (%S) of Suffolk × mule ewes (average 412 ewes). The data comprised records for six seasons (2013-2018) at Rothamsted Research-North Wyke. For 2013 to 2018, %R were 3.85, 2.98, 3.47, 2.91, 3.28 and 3.01%, I25 were 233, 126, 15, 68, 82 and 218 h, and %S were 207, 192, 199, 193, 196 and 182%. Over the six seasons, %R explained 70% of the variation in %S (P=0.04) whilst I25 had no significant effect (P=0.88). However, by removing the outlier year with a high %R (2013), the analysis changed markedly, with I25 explaining 92% of the variation in %S (P<0.01), and %R became non-significant (P=0.25). It might be hypothesised that the potential negative effect of the HS occurred in summer 2013 was overcome by the use of a high %R (extra sperm in the system). In other words, the effects of HS might be only expressed when the %R is limiting. HS might affect male performance by disrupting spermatogenesis, with the 60 d spermatogenic cycle explaining the delay between the period of stress and the mating outcome. In the females, HS might affect the number of follicles entering the committed population and thus becoming susceptible to atresia, but the delay (≤130 days before ovulation) might be too long. Other factors also need to be considered (forage quality, ewe age, lambing status), but the evidence suggests that the previous summer weather may affect reproductive performance of sheep in temperate climates.

Heat stress during early-mid gestation causes placental insufficiency and growth restriction in pigs

W. Zhao[1], F. Liu[2], J.J. Cottrell[1], B.J. Leury[1], A.W. Bell[3] and F.R. Dunshea[1]
[1]The University of Melbourne, Faculty of Veterinary and Agricultural Sciences, Parkville, 3010, Australia, [2]Rivalea (Australia) Pty Ltd, Corowa, 2646, Australia, [3]Cornell University, Department of Animal Sciences, Ithaca, 14853, USA; weichengz@student.unimelb.edu.au

Pregnant gilts suffering from seasonal infertility often lead to compromised foetal growth and placental insufficiency, which are causative factors for altered postnatal growth. The objectives of this study were to quantify the effect of early-mid gestational heat stress (GHS) on placental efficiency and foetal morphology in pigs. Fifteen Large White × Landrace gilts were randomly allocated to either cyclical heat stress conditions (HS; 8 h/d, 33 °C/ 28 °C; n=8) or thermo-neutral control conditions (TN; 20 °C; n=7) during d40 to 60 of gestation. All pigs were fed 2.0 kg daily of a gestational diet containing 13 MJ/kg DE. Gilts were killed on the last day of thermal exposure. Foetal weights, crown-rump length (CRL), and head circumference were measured in all foetuses (n=203). The body mass index (BMI; weight:CRL^2) and the ponderal index (PI; weight:CRL^3) were calculated. The placentae were separated for placental weight, surface area and efficiency (foetal: placental weight) calculations. Three foetuses per litter, one each from tip, mid and cervical sections of one uterine horn were dissected for organ weights. Data were analysed using Linear Mixed Models in GenStat 18th. Gestational HS increased placental weights by 20% (86.6 vs 103.7 g, P=0.006), while the foetal weights were not significantly different (95.7 vs 92.0 g, P=0.12), meaning that GHS reduced placental efficiency (1.17 vs 0.91, P<0.001). Furthermore, the GHS tended to increase placental surface area (719 vs 792 cm^2, P=0.067). The GHS foetuses had lower BMI (58.3 vs 48.9, P<0.001), PI (4.68 vs 3.60, P<0.001) and head circumference (10.15 vs 9.50 cm, P=0.005), but higher CRL (13.42 vs 13.74 cm, P=0.032). In summary, GHS caused placental insufficiency in pigs, as evidenced by enlargement of the placentae without a commensurate increase in foetal weights. Although the GHS foetuses were not lighter, morphological differences were evident in that they grew disproportionately, increasing the probability of growth restriction later in life.

Hepatic expression of thermotolerance related genes and early thermal conditioning in broilers

M. Madkour[1], M. Aboelenin[2], H. Hassan[1], O. Aboelazab[1], N. Abd El-Azeem[1] and M. Shourrap[3]
[1]National Research Center, Animal Production, El Buhouth St., Dokki, Cairo, Egypt, 12622, Egypt, [2]National Research Center, Cell Biology Department, El Buhouth St., Dokki, Cairo, Egypt, 12622, Egypt, [3]Ain-Shams University, Department of Poultry Production, P.O. Box, 68 Hadayek Shoubra, 11241, Egypt; mahmoud.madkour9@gmail.com

Early-life thermal conditioning in broiler chickens has long lasting impacts on later life. Therefore, the aim of this study was to clarify the molecular mechanism by which early thermal conditioning (TC) impacts the hepatic expression of heat shock proteins (HSPs), glucocorticoid receptor (GR), and oxidant NADPH oxidase 4 (NOX4) as well as antioxidant enzyme response of broiler chicks at six weeks of age. Two hundred one-day-old broiler (Cobb 500) chicks were allocated into four equal treatments. The first group was let under optimal brooding conditions (control, CE), while the 2nd, 3rd and 4th groups were exposed to a thermal conditioning (TC) at 39±1 °C for 6 h/day through the third day post hatch (TC3), or exposed to a thermal conditioning at the fifth day post hatch (TC5), or exposed to a thermal conditioning at the seventh day post hatch (TC7), respectively. At 42 days of age, chicks of all groups were exposed to 40±1 °C for 6 hours and 8 birds from each group were rapidly sacrificed for liver sampling. Results showed that, hepatic expression of NOX4 and GR decreased (P<0.05) by 50 and 58% respectively in chicks of the TC5 compared with the CE group. The TC5 group had the lowest level of SOD expression with a 30% decrease compared with the control group (CE). TC5 had the lowest (P≤0.05) hepatic expression of CAT and HSPs (60, 70 and 90). While the highest values of these traits were recorded for TC7 with the exception of HSPs that were increased (P<0.05) with TC3 comparable to the other groups including the CE group. It can be concluded that thermal conditioning of broiler chicks at day five post hatch is the most effective to induce thermotolerance acquisition at later ages as evidenced by the lowest expression of HSP genes and stress indicators.

Saliva as a tool to detect chronic stress in piglets

S. Prims, M. Dom, C. Vanden Hole, G. Van Raemdonck, S. Van Cruchten, C. Van Ginneken, X. Van Ostade and C. Casteleyn
University of Antwerp, Faculty of Pharmaceutical, Biomedical and Veterinary Sciences, Universiteitsplein 1, 2610 Wilrijk, Belgium; sara.prims@uantwerpen.be

Monitoring chronic stress is advantageous for the pig farmer because stress influences the animals' welfare and as such their zootechnical performances and susceptibility to infectious diseases. Salivary analysis could serve as a non-invasive, fast and objective evaluation tool. Cortisol is predominantly used as a salivary biomarker to assess chronic stress. However, interpreting salivary cortisol levels has certain constraints since they are affected by various factors, including a circadian rhythm. Therefore, there is still a need for a reliable set of salivary biomarkers to detect chronic stress in pigs. To reach this goal, twenty-four 7-day-old female piglets were allocated to either a control group or a stressed group (overcrowding and frequent mixing with unfamiliar animals). At day 28, weight gain was determined and saliva was collected from each animal using a Micro·SAL™ (Oasis Diagnostics®). The salivary protein composition and protein-specific ratios were determined using an isobaric labelling method (iTRAQ) for tandem mass spectrometry. Additionally, cortisol levels of hair and saliva were assessed using a commercial ELISA (IBL-International). After 3 weeks, stressed animals gained less weight than the control animals (mixed model). Additionally, hair from the stressed group contained the highest concentration of cortisol, whereas salivary cortisol levels did not differ between groups. Weight gain and hair cortisol levels correlated, while none of these parameters correlated with salivary cortisol levels (Spearman's assay). Of a total of 596 salivary protein identifications, 8 proteins showed a significant fold change (>1.4). The relative abundance of 2 of these proteins, lipocalin 1 (LCN1_PIG) and an uncharacterised protein (I3LL32_PIG), significantly correlated with weight gain and hair cortisol levels. A multiple-reaction monitoring (MRM) analysis is currently being performed to further validate these possible biomarkers.

Effect of initial udder's health status on inflammatory response to once-daily milking in dairy cows
J. Guinard-Flament[1], S. Even[2], P.A. Lévêque[1], C. Charton[1], M. Boutinaud[1], S. Barbey[3], H. Larroque[4] and P. Germon[5]
[1]PEGASE, INRA, Agrocampus Ouest, 35590, Saint-Gilles, France, [2]UMR STLO, INRA, 35000 Rennes, France, [3]UE du Pin, INRA, 61310 Exmes, France, [4]UMR GenPhySE, INRA, 31326 Castanet Tolosan, France, [5]UMR ISP, INRA, Université François Rabelais, Tours, France, INRA, 37380 Nouzilly, France; jocelyne.flament@agrocampus-ouest.fr

Once-daily milking (ODM) is known to induce udder inflammation, and could lead to a higher mastitis incidence due to increased milk leakage and extended milk stasis that could promote penetration and multiplication of pathogens. As a fact, ODM increases milk somatic cell counts (SCC) with or without an increased number of clinical mastitis. However, bacteriological analyses are often lacking to describe udder inflammation. Our study aimed to describe the effect of initial health status of the udder on the inflammatory response observed when implementing ODM. The trial used 63 cross-bred Holstein × Normande dairy cows at 78±5 days in milk and comprised a 1-wk control period (cows milked twice daily) and a 3-wk period of ODM (d1, 1st d of ODM). The udder and animal inflammatory status were assessed by measuring SCC 4 d/wk in whole-udder milk, interleukins (IL) 6 and 8 in quarter milk, and haptoglobin in plasma on d0, 1, 2, 3 and 14. The infectious status of quarters was determined at d0, 3 and 14 by milk bacteriological analysis and identification of minor and major pathogens by 16S DNA sequencing. Four initial infectious and inflammatory udder's classes (57 of the 63 cows) were determined: no infected and inflamed quarters (n=13), no infected quarters but at least one inflamed quarter (n=15), at least one quarter inflamed and infected with a minor pathogen (n=14), and at least one quarter inflamed and infected with a major pathogen (n=15). Milk yield losses due to ODM were higher for cows with inflamed udders compared to non-inflamed udders (25-27% vs 17%, respectively). Implementing ODM increased SCC and IL8. The increase in IL8 was higher for cows infected with a major pathogen. Cows with no infected but inflamed quarters contrasted by showing only a slight increase in SCC with no change in IL8 and haptoglobin. Inflammatory and milk yield responses to ODM thus varied according to the initial udder health status.

Effects of gestational heat stress on sow and progeny performance at farrowing
M.C. Père, A.M. Serviento, E. Merlot, A. Prunier, H. Quesnel, L. Lefaucheur, I. Louveau and D. Renaudeau
PEGASE, INRA, Agrocampus Ouest, 35590 Saint-Gilles, France; isabelle.louveau@inra.fr

Heat stress during gestation could negatively affect pregnancy and alter foetal development, resulting in long-term developmental damage in the offspring. The effects of two thermal environments for mixed-parity gestating sows on morphological and physiological traits in newborn piglets were investigated. A total of 24 Landrace × Large White sows (12 Primiparous and 12 Multiparous) were distributed to a 2×2 experimental design considering the effect of parity (Primiparous vs Multiparous) and gestational thermal environment (Heat Stress (HS): diurnal 28 to 34 °C vs Thermo-Neutral (TN): diurnal 18 to 24 °C). Sows were fed a standard gestation diets. Farrowing was supervised to allow sampling and morphometric analyses from piglets at birth. Piglets born alive and stillborn were counted and weighed for each sow. Blood samples were collected from umbilical vein immediately after delivery (10 piglets per litter on average; n=245 in total). Data were analysed using SAS Proc MIXED with environment, parity, and their interactions as fixed effects. The gestational thermal environment and the parity of the sow did not influence the duration of gestation (P<0.10). A significant parity × environment interaction was found for litter size (P=0.002) and litter weight (P=0.02). At birth, no significant difference was observed between HS and TN piglets for body weight (1,349±291 g), cranio-caudal length (25±3 cm) and biparietal diameter (5.3±0.4 cm). Plasma concentrations of glucose, the main energetic substrate of the foetus, were not influenced by the gestational HS and the parity of the sow (3.2±1.0 mM on average). Concentrations of fructose, an indicator of foetal maturity, were not affected by the thermal environment, but were lower (P=0.02) in piglets born from Multiparous (1.8±0.7 mM) than Primiparous sows (2.4±0.7 mM), suggesting a lower maturity in piglets from Primiparous sows. Albumin concentration, considered as a foetal liver maturity marker, did not differ between groups (6.0±1.0 g/l on average). Altogether, the current findings show no significant effect of heat stress during gestation on piglet morphological and physiological traits at birth.

Differential expression of heat shock protein #70 gene in PBMC of beef calves exposed to heat stress

W.S. Kim[1,2], J.G. Nejad[1,2], D.Q. Peng[1,2], Y.H. Jo[1,2], J.H. Jo[1,2], J.S. Lee[1] and H.G. Lee[1,2]
[1]Konkuk University, Department of Animal science & Technology, Seoul, 05029, Korea, South; [2]A team of Educational Program for Specialists in Global Animal Science, Brain Korea 21 Plus Project, Seoul, 05029, Korea, South; hglee66@konkuk.ac.kr

The objective of this study was to investigate the heat shock protein (HSP) gene expression in peripheral blood mononuclear cells (PBMC) of beef calves during and after sudden exposure to heat stress conditions, compared with calves in thermoneutral conditions. Data were collected from sixteen Korean native calves (BW 166.9±6.23 kg and age 169.6±4.60 d), which were kept at four designated temperature humidity index (THI) levels based on ambient temperature and humidity: threshold (22 to 24 °C, 60%; THI=68 to 70), mild (26 to 28 °C, 60%; THI=74 to 76), moderate (29 to 31 °C, 80%; THI=81 to 83) and severe (32 to 34 °C, 80%; THI=88 to 90) stress level. Calves were subjected to ambient temperature (22 °C) for 7 days (thermoneutral; TN), after which chamber temperature and humidity were raised to each THI level for 21 days (heat stress; HS). For PBMC isolation, blood samples were collected every three days from the jugular vein of four beef calves per experimental group during the TN and HS periods. Physiological parameters of heat stress were measured every three days including rectal temperature (RT) and heart rate (HR). The results showed that RT (P=0.001) and HR (P=0.001) were increased in the HS period compared with the TN period. In PBMC, HSP70 gene expression was increased after sudden exposure to heat stress conditions (moderate and severe levels). However, as the high-level temperature condition was maintained, HSP70 gene expression returned to a normal range after three days of continuous heat stress. There were no differences in HSP90 and HSPB1 gene expression during the heat stress conditions. In conclusion, the HSP70 gene plays a pivotal role in protecting cells from being damaged and reacts sensitively to heat stress in regard with immune cells compared to other HSP genes. In addition, this study suggests that calves have actively integrated physiological mechanisms to adapt themselves to heat stress after three days of continuous exposure by regulating the HSP70 gene expression.

Individual thermal response of Avileña Negra-Ibérical calves at feedlot

M. Usala[1], M. Carabaño[2], M. Ramon[2], C. Meneses[2], N. Macciotta[1] and C. Diaz[2]
[1]University of Sassari, Agricultural Sciences, Viale Italia 39 Sassari, 07100, Italy, [2]INIA, Mejora Genética Animal, La Coruña Km 7.5, Madrid, 28040, Spain; cdiaz@inia.es

The increasing demand for meat in developing countries, together with the rapidly increasing change of the world's climate conditions has made the thermal tolerance one of the most important adaptive aspects for cattle. In this study, the aim was to investigate the genetic component of thermal response in male calves of Avileña-Negra Ibérica (ANI) as a first step for an eventual genetic evaluation to improve thermal tolerance in this breed. A total of 29,591 weight records taken during the fattening period belonging to 5,876 animals collected between 2005 and 2017 were used. Pedigree file included 15,418 animals. The range of average daily temperatures at the location of the feedlot on the day of recording (Ta) ranged from a minimum of -4 to a maximum of 30 °C. Three random regression models (RRM), one with no temperature effect and two with second or third order of Legendre polynomials on Ta for additive genetic and permanent environmental effects were fitted to estimate the variability in individual reaction to increasing thermal loads of weight in feedlot. For each model, the random effects included direct genetic, permanent environmental effect of the animal, and residual component. Fixed effects included a quadratic age covariate nested to age-of-entry class (4 levels), and contemporary group (47 levels). Estimated breeding values for changes in weight associated to increase of one degree of Ta at different values of Ta were estimated for all animals. The model with cubic random regressions on Ta showed the best goodness of fit. The ranges of estimated breeding values for weight in feedlot under cold [-4, 2.5] °C, comfort [2.5, 23.5] °C and hot [23.5, 29] °C regions were [-129.96, 85.67] kg, [-112.35, 82.57] kg and [-103.79, 85.05] kg, respectively. Along the temperature scale, heritability estimates decreased from 0.59 for the coldest temperatures to 0.52 to raise beyond 10 °C up to 0.64 and slightly decrease when Ta was above 26 °C. Heterogeneity of variances along the Ta scale suggest the existence of a G×E related to thermal stresses.

Physiological processes as linked to heat stress effects on production traits of Holstein cows

H. Amamou[1], M. Mahouachi[2], Y. Beckers[1] and H. Hammami[1]
[1]Gembloux Agro-Bio Tech, University of Liège, AgroBioChem, Passage des Déportés 2, 5030, Belgium, [2]High School of Agriculture of Kef, University of Jendouba, Le Kef, 7119, Tunisia; hamamou@doct.ulg.ac.be

Production losses due to heat stress (HS) are increasing concerns in the world and the impacts of climate change are expected to worsen the situation. This study aimed to investigate the effect of HS on production and physiological responses of Holstein cows. Two experiments each lasted 6 weeks were conducted in Tunisia, the first during summer under HS (n=80) and the second during autumn under thermoneutral (n=80) conditions. Respiration rate (RR), skin temperature (ST), rectal temperature (RT) and milk yield were measured, and milk samples were collected on 2 days every week during each experimental period. Ambient temperature (T, °C) and relative humidity (RH, %) were measured inside the barn at 10-minute intervals during the experimental periods to calculate the temperature-humidity index (THI). Reaction norm models were applied to quantify the individual responses of cows across the trajectory of THI. A clustering methodology was developed to identify tolerant and sensitive cows to HS based on their slope for response of physiological and production traits during HS period. Analysis of the pattern of individual responses across the THI scale shows that cows exhibit similar reaction norms for all traits marked with positive slope for milk yield suited with similar positive slopes for physiological traits. It could be stipulated that the maintenance of the milk yield level was associated with higher RR. On the other hand, other cows exhibit different reaction norms across the traits (e.g. positive slopes for physiological parameters and in opposite negative slopes for production traits or vice versa). The largest proportions of cows (82%) showed responses with steeper negative slopes for all studied traits, and were qualified as sensitive. In contrast, 18% of cows were qualified as heat tolerant as they show positive slopes for response for HS in production and physiological traits at high THI values. Therefore, RR could be used as an indicator for thermotolerance. The results of this study deepen our understanding of different aspects of HS resilience.

Relationships between body temperature, respiration rate and panting score in lactating dairy cows

P.J. Moate, S.R. Williams, J.L. Jacobs, M.C. Hannah, W.J. Wales, L.C. Marett and T.D.W. Luke
Agriculture Victoria Research, Dairy Production, 1301 Hazeldean Road, 3821, Australia; tim.luke@ecodev.vic.gov.au

A simple method to assess the thermal status of lactating dairy cows during periods of high heat load is required for on-farm situations where measurement of body temperature (BT) may be impractical. Strong relationships have been documented between BT and respiration rate (RR), and between BT and panting score (PS) for beef cattle, but not for lactating dairy cows. This research established relationships between RR and BT and between PS and BT of lactating dairy cows. Three feeding experiments were conducted with lactating Holstein cows offered various dietary treatments (feed supplements of either wheat, barley, corn, canola-meal, canola oil or betaine). In each experiment, 2-day heat challenges that mimicked an actual heat wave, were conducted in controlled-climate respiration chambers. The first experiment involved 2 treatments and 18 cows while the second and third experiments each involved 24 cows and 4 treatments. During the night, the temperature was set at 25 °C, relative humidity (RH) at 60%, and Temperature Humidity Index (THI) was 74, while in the day, temperature was 33 °C, RH was 50% and THI was 84. A total of 620 observations on BT, RR and PS (0, no panting to 4.5 open mouth, excessive drooling) were recorded by a trained observer. Measurements on individual cows were made at 07:00 h and 15:30 h and BT was measured using a rectal probe thermometer. Relationships were analysed using mixed models containing a linear fixed-effect for the independent variable (RR or PS) and random effects for experiment, treatment within experiment, and cow by day split for time of day. Mean BT was 39.1 °C, and ranged from 36.9 to 41.3 °C. Mean RR was 73 and ranged from 18 to 129 breaths per minute, while mean PS was 1.2 and ranged from 0 to 3.5. The relationships between RR and BT and between PS and BT were both highly significant (P<0.001): BT = 39.14 +- 0.064 + 0.0184 +- 0.00101 x (RR – 72.53) R^2=0.47, RMSEP=0.57 BT = 39.31 +- 0.0400 + 0.212 +- 0.0400 x (PS -1.137) R^2=0.18, RMSEP=0.76. We conclude that while both RR and PS can be used to predict BT, on the basis of the smaller root mean square prediction error (RMSEP), the equation involving RR is to be preferred.

Fibre quality influences faecal glucocorticoid metabolites of South African Mutton Merino sheep

H.A. O'Neill, F.W.C. Neser and B.T. Elago
University of the Free State, Department of Animal, Wildlife and Grassland Sciences, 205 Nelson Mandela Drive, Park West, Bloemfontein 9301, South Africa; neserfw@ufs.ac.za

In South Africa, heat, drought and lack of adequate pasture may predispose sheep to nutritional stress. The objective of this study was to determine whether fibre source quality will influence stress responsiveness in sheep, measured through faecal glucocorticoid metabolite concentration. A nutritional stress response test was performed on 20 sheep randomly divided into four treatment groups [T1(n=5); T2(n=5); T3(n=5) and T4(n=5)]. Sheep were individually kept in metabolic crates for thirty days. Rations were nutritionally balanced but different fibre sources were used. T1 were fed Alfalfa Hay; T2 were fed Maize Stover; T3 were fed Soya Hulls and T4 were fed Eragrostis Tef. Faeces was digitally removed at 5:00 am and 19:00 pm from the caudal rectum of each animal. Samples were frozen at -20 °C until analyses. Heart rate of each animal was measured immediately after faeces removal. A total of 520 faecal samples were collected. Faecal glucocorticoid metabolite (fGCM) concentrations were measured by an enzyme immunoassay. Faecal metabolite concentrations were expressed as mass/g dry weight. Baseline values of fGCM concentrations were determined; values greater than the mean plus 2 standard deviations were removed. Baseline values were compared among the 4 treatment groups using one-way ANOVA analysis from R statistical package. There were no significant differences in heart rate between treatment groups. Higher variations were observed in overall morning fGCM concentrations among all treatment groups compared to those in the evening. An overall morning median fGCM concentration of 172.9±235.7 ng/g DW and 191.7±98,5 ng/g DW for the evenings were found. T4 had significantly higher fGCM concentrations for both mornings (302.0±192.4 ng/g DW) and evenings (237.0±107.6 ng/g DW) compared with other treatment groups. Fibre quality may induce a stress response in sheep. Different levels of fGCM concentration in sheep could be directly or indirectly related to fibre quality. It will be suitable to use alfalfa hay, maize stover and soya hulls as fibre sources, as they are more available and cheaper in South Africa.

Detection of fin cortisol concentration in sturgeon

M. Ataallahi[1], J. Ghassemi Nejad[1,2], M.H. Salmanzadeh[3] and K.H. Park[1]
[1]Kangwon National University, Animal Life Sciences, Chuncheon, Gangwon, 24341, Korea, South, [2]Konkuk University, Animal Science, Seoul, 05029, Korea, South, [3]Persian Gesture, No.1/167, 3rd majd, Sanabad-Mashad, Khorasan Razavi, 9184668477, Iran; ataallahim@kangwon.ac.kr

Cortisol is known as an indicator of stress in animals, and thereby provides a valid framework for researchers to monitor the extent of stress and the general pattern of stressors in fishes. Whereas blood cortisol provides information about acute stress, chronic stress from poor water quality, reduced oxygen level, and rapid changes in salinity might be detected by the accumulative concentration of cortisol in fin tissue. Therefore, we examined cortisol in fin of three sturgeon species to evaluate whether fin can be used as a basic stress indicator. Nineteen sturgeons, including seven Beluga (*Huso huso*; 18±2.1 months of age, weight: 2,700±300 g, length: 55±5 cm), seven Siberian (*Acipenser baerii*; 9.6±2.4 months of age, weight: 1,750±250 g, length: 45±5 cm), and five Sevruga (*Acipenser stellatus*; 14±1.3 months of age, weight: 1000±100 g, length: 65±5 cm), were used for fin cortisol extraction. Fins of fish were harvested from the dead bodies. Fin cortisol concentrations were determined by ELISA assay. These species were kept in the Persian Gesture farm in Iran. The temperature of water was 19.5±0.5 °C and oxygen of water was 6.5±1.5 mg/l. Fish were fed twelve times daily with commercial diets. Data were analysed using the GLM procedure of SAS. Results showed the existence of cortisol in the fin of the sturgeon species. There were no significant differences in fin cortisol concentrations between the sturgeon species, and the concentrations of cortisol were 3.45±0.33, 2.92±0.43, and 3.34±0.47 pg/mg for *Huso huso*, *Acipedeadlinser baerii*, and *A. stellatus*, respectively. The current study indicates that fin cortisol concentrations are detectable but further studies are needed to show that fin cortisol concentration may be a relevant indicator of chronic stress.

Capsaicin enhances thermophysiological performance of ergotamine-aggravated heat stress in rats

H. Al-Tamimi[1] and R. Halalsheh[2]
[1]Jordan University of Science and Technology; Faculty of Agriculture, Animal Production, Irbid, Amman Street, 22110, Jordan, [2]The Hashemite University; Faculty of Applied Health Sciences, Medical Laboratory Sciences, Zarqa strategic site, Amman-Mafraq international highway, 13116, Jordan; hjaltamimi@just.edu.jo

Compromised cardiovascular hemodynamics (e.g. persistent vasoconstriction) can hinder thermoregulatory pathways especially during heat stress. Ergotamine (ERG), an alkaloid found in some medications, is used to simulate fescue toxicosis in farm ruminants, induces peripheral ischemia, and exacerbates hyperthermia. Capsaicin (CAP) is a monoterpene that enhances peripheral blood flow. A trial using 32 male Sprague Dawley rats was conducted to examine changes in thermophysiological parameters by CAP along with ERG under chronic heat stress (HS; air temperature; Ta=31.6±0.5 °C). All animals (chronically-biotelemetered to continuously monitor core body temperature [Tcore] and heart rate) were initially kept at thermoneutrality (TN; Ta=22.7±0.3 °C) for 4 days, followed by a 5-day HS, and received either ERG (0 or 0.5 mg/kg BW/day; oral gavage) and/or CAP (0 or 50 mg/l drinking water) within the last 3 days of the trial, in a 2×2 factorial arrangement of treatments. At the end of the trial, blood samples were drawn to measure selected hepatic enzymes (aspartate aminotransferase [AST] and alkaline phosphatase [AlkP]), triglycerides (TriG) and total antioxidant activity (TAO). No dissimilarities in radiotelemetric parameters were noticed throughout the TN period among treatment groups. However, switching to HS induced a hyperthermic response in all rats. With the onset of ERG/CAP treatments, a clear ERG by CAP by time interaction (P<0.01) was prominent, such that the ERG group had higher Tcore than the control group (more than 0.5 °C surplus), while CAP resulted in hyperthermia alleviation when used with ERG. CAP also caused compensatory tachycardia in ERG-treated rats by the end of the trial. This suggests that the potent vasoconstrictive effect of ERG was mitigated by the vasodilatory potential of CAP (which also succeeded in normalization of ERG-prompted hindrances of AST, AlkP, TriG and TAO). Treatment with CAP mitigates thermophysiological perturbations acutely caused by ERG, including hyperthermia and some compromised hepatic enzymes.

Multilevel evaluation of prebiotic supplementation and heat stress effects in laying hens

S. Borzouie[1], B. Rathgeber[1], C. Stupart[1], I. Burton[2] and L.A. Maclaren[1]
[1]Dalhousie University, Department of Animal Science and Aquaculture, 58 Sipu Awti, Truro, Nova Scotia, B2N 4H5, Canada, [2]National Research Council of Canada, 1411 Oxford St, Halifax, Nova Scotia, B3H 3Y8, Canada; borzouie@dal.ca

This study was planned to investigate the effects of prebiotic seaweed supplementation and heat stress on haematology, plasma chemistry and metabolites as indicators of hen health. In a short-term trial, Lohmann Brown and LSL White laying hens in month 10 of the lay cycle were supplemented or not with 3% red seaweed *Chondrus crispus* (CC) for 21 days prior to sampling. In a long-term trial, Brown and LSL laying hens were assigned to 0 or 3% CC from week 34 to 74. Long-term trial included a 21-day heat stress period or not beginning at week 71. Blood samples were collected at slaughter for assessment of plasma chemistry, leukocyte count and metabolomics analysis. Plasma metabolites were characterized using 1H nuclear magnetic resonance spectroscopy (NMR). Multivariate statistical methods (principle component analysis and heat map) revealed considerable differences between the metabolic fingerprint of hens fed 0 and 3% CC as well as brown and LSL hens. Metabolites were identified and quantified with an average of 50 high-concentration compounds for each sample (>1 µM). Heterophil/lymphocyte (H/L) ratios were assessed in blood samples from long-term trial birds as an indicator of health status. Prebiotic intake (CC vs control) was associated with an increase in H/L ratios (P<0.05) and higher levels of H/L ratio were observed in Brown strain compared with LSL. Heat stress increased H/L ratio in Brown line (P<0.05). In the short-term trial, LSL hens responded to heat stress with increase in plasma alanine aminotransferase (P<0.05). Significant interactions were observed between heat-stress and seaweed treatments for plasma aspartate aminotransferase, glutamate dehydrogenase and cholesterol levels (P<0.05). Overall, the blood multi-level evaluation indicates that prebiotic seaweed intake moderates metabolism and its response to heat stress.

Advanced bone measurements show clear differences between laying hens kept in cages or floor pens

A. Kindmark[1], A.B. Rodriguez-Navarro[2], F. Lopes Pinto[3], E. Sanchez-Rodriguez[2], N. Dominguez-Gasca[2], C. Benavides-Reyes[2], H.A. McCormack[4], R.H. Fleming[4], I.C. Dunn[4], H. Wall[3] and D.J. De Koning[3]
[1]Uppsala University Hospital, Department of Medical Sciences, Akademiska sjukhuset, Ing.40, 3 tr, 75185 Uppsala, Sweden, [2]Universidad de Granada, Dpto. Mineralogia y Petrología, 18002 Granada, Spain, [3]Swedish University of Agricultural Sciences, P.O. Box 7070, 750 07 Uppsala, Sweden, [4]University of Edinburgh, The Roslin Institute and R(D) SVS, Easter Bush, Roslin, Midlothian, EH25 9RG, United Kingdom; dj.de-koning@slu.se

Poor bone quality is a major welfare problem in laying hens, which can be improved by genetic selection. We previously showed that birds housed in floor pens showed significantly higher bone strength (P<0.001) than birds housed in furnished cages. Here we investigate changes in bone properties between birds kept in floor pens or furnished cages. From around 850 birds, the right tibia was analysed using peripheral quantitative computed tomography (pQCT) and breaking strength, while the left tibia was analysed using thermogravimetric analysis (TGA). TGA measures different bone components by weighing the bone during a constant heating ramp and determining the weight loss at characteristic temperatures. Analyses focused on Pearson correlations between TGA measures, pQCT measures, and breaking strength. Differences between bone properties of birds from furnished cages or floor pens were analysed with two-sided t-tests. Fifteen out of 16 pQCT-derived traits showed significant differences (P<0.001) between birds from furnished cages and floor pens. Twelve pQCT traits correlated strongly with breaking strength (r>0.75). TGA data show highly significant (P<0.001) differences in the composition and degree of mineralization of cortical bone. The TGA analyses show that the differences in breaking strength can be better studied focusing on the cortical bones. The amount and composition of medullary bone shows clear differences between furnished cages and floor pens, but these properties do not correlate with breaking strength and only sparsely with pQCT measures. The cortical TGA data is more informative because: a) clear differences between furnished cages and floor pens b) some significant correlations with breaking strength and c) unique correlation structures with pQCT data.

Effect of day length and inhibition of prolactin on the mammary epithelial cell exfoliation

L. Herve, V. Lollivier, M. Veron, P. Debournoux, J. Portanguen, P. Lamberton, H. Quesnel and M. Boutinaud
PEGASE, INRA, Agrocampus Ouest, 35590, Saint-Gilles, France; marion.boutinaud@inra.fr

In dairy cows, a long-day photoperiod (LDP) is known to increase milk yield. This increased milk yield is associated with greater concentrations of prolactin (PRL). In contrast, a short-day photoperiod (SDP) was shown to increase the rate of mammary epithelial cell (MEC) exfoliation as also observed after an experimentally-induced inhibition of PRL. The aim of this study was to investigate if PRL mediates, at least in part, the effect of photoperiod on the MEC exfoliation process. Eight Holstein dairy cows producing 42±3.7 kg of milk per day were submitted to 4 treatments according to a 4×4 Latin square design: 2 day lengths (8 h of light/d for SDP or 16 h of light/d for LDP) for 20 d and to a pharmacological treatment (a single i.m. administration of 5.6 mg of cabergoline, an inhibitor of PRL secretion, or vehicle) on d10. Blood samples were collected before, during, and after milking to measure plasma concentration of PRL. Milk samples were collected for milk MEC purification using an immunomagnetic method. Milk protein mRNA levels were analysed in milk purified MEC by q-PCR to evaluate MEC metabolic activity. Data were analysed by analysis of variance using SAS and the effect of photoperiod, cabergoline treatment, and the interaction between photoperiod and cabergoline treatment were tested. Photoperiod had no effect (P>0.05) on PRL secretion, milk yield, MEC exfoliation rate and metabolic activity. It was therefore impossible to draw conclusion regarding the potential role of PRL as a mediator of the effect of photoperiod. Cabergoline induced a decrease in PRL secretion, by decreasing both basal concentration and the surge at milking, and induced a decrease in milk yield (P<0.01). This decrease in milk yield in response to cabergoline was associated with a decrease in the metabolic activity of MEC associated with a down-regulation of the expression of κ-casein and α-lactalbumin gene (P<0.01). Cabergoline had no effect on MEC exfoliation rate showing that in the present study, the MEC exfoliation process did not participate in the decrease in milk yield in response to the inhibition of PRL by cabergoline.

Underfeeding increases *ex vivo* blood cell responsiveness to bacterial challenges in dairy cows

M. Boutinaud[1], E. Arnaud[1], R. Resmond[1], P. Faverdin[1], G. Foucras[2], P. Germon[3] and J. Guinard-Flament[1]
[1]*PEGASE, INRA, Agrocampus Ouest, 35590, Saint-Gilles, France,* [2]*UMR IHAP ENVT INRA, 23 Chemin des Capelles – BP 87614, 31076 Toulouse, France,* [3]*UMR ISP, INRA, Université François Rabelais, Centre de Tours, 37380 Nouzilly, France; marion.boutinaud@inra.fr*

In dairy cattle, underfeeding is known to influence cow's homeostasis by inducing body reserve mobilization and by reducing milk production. However, its impact on cow's immune responses is less known. This study aimed to examine blood cell responsiveness to *ex vivo* bacterial challenges in cows fed a standard or a low energy diet. Fifty-four Prim'Holstein at mid lactation received a standard diet (based on corn silage) from calving to 201 days in milk, and then were fed a lower energy diet (14.5% concentrates replaced by 11.5% straw and 3% dehydrated alfalfa) for 5 wk. Milk yield was recorded daily. Blood samples collected at wk -2, -1 and 1, 2 and 5 after diet change were incubated *ex vivo* with heat-killed *Escherichia coli* or *Staphylococcus aureus* bacteria, or saline as a control for 23 h at 38.5 °C. Plasma Interleukin 8 (CXCL8), an indicator of the inflammatory response, was then measured by ELISA. Plasma haptoglobin, glucose, urea and NEFA concentrations were determined. Data were analysed using a mixed model of variance testing the effect of week with a cow week random effect. The decrease in energy diet content was associated with a loss of 3 kg/d milk yield (P<0.001) and reduced glucose and urea plasma concentrations (P<0.01) while plasma NEFA and haptoglobin concentrations did not vary. A higher production of CXCL8 was observed after *E. coli* and *S. aureus* stimulations during the underfeeding period (P<0.01), being maximal 2 and 5 wk after diet change, for *E. coli* and *S. aureus* respectively. Thus, a moderate underfeeding of dairy cows could modify the cow's inflammatory response.

The acute phase proteins ITIH4 and haptoglobin as markers of metritis in Holstein dairy cows

J.A. Robles[1], R. Peña[1], A. Bach[2], M. Piñeiro[3] and A. Bassols[1]
[1]*Departament de Bioquímica i Biologia Molecular i Servei de Bioquímica Clínica Veterinària, Facultat de Veterinària, Universitat Autònoma de Barcelona, 08193 Bellaterra, Spain,* [2]*Institut de Recerca i Tecnologia Agroalimentàries, ICREA and Departament de Producció de Remugants, Caldes de Montbui, 08140, Barcelona, Spain,* [3]*Acuvet Biotech, C/ Bari 25, 50197 Zaragoza, Spain; mpineiro@acuvetbiotech.com*

Metritis is a frequent disease in post-partum Holstein dairy cows caused by delayed uterine involution, persistent inflammation, and consequent bacterial colonization of endometrial mucosa. Alternative diagnostic tools are needed for the efficacious treatment of the disease and as a mean to decrease antibiotic treatments. Acute-phase proteins (APPs) are released in plasma mainly by liver at early stages of illness to promote the immune response. Herein, we evaluated two APPs: haptoglobin (Hp) and ITIH4, as potential biomarkers of inflammation in cows suffering metritis. A total of 139 Holstein dairy cows were studied in the periparturient period and examined for the occurrence of metritis symptoms and their severity after parturition. Blood samples were collected at 10 and 5 d pre-partum and at 1, 3, 7, and 10-14 d post-partum. Species specific immunoturbidimetric assays were used for APP analysis. Data were analysed using an analysis of variance accounting for the fixed effects of presence or absence of disease, time, and their 2-way interaction, and cow entered the model as a random effect. ITIH4 and Hp concentrations in serum were similar in pre-partum samples of healthy and metric cows. However, a noticeable increase in post-partum cows was detected, with this increase being more pronounced in cows with metritis. Furthermore, the increase in ITIH4 and Hp serum concentrations was proportional to the severity of the disease. Serum ITIH4 and Hp concentrations decreased 10 d after calving in healthy cows and remained elevated in metritic cows. Hp serum concentrations changed more markedly than ITIH4, but both biomarkers followed a similar pattern. In conclusion, Hp and ITIH4 serum levels are useful biomarkers for diagnosis and monitoring bovine metritis during the onset of the disease.

A new approach to predict the success of insemination of cows using a systematic heat-assessment

L. Von Tavel, F. Schmitz-Hsu, U. Witschi and M. Kirchhofer
Swissgenetics, Meielenfeldweg, 12, 3052, Switzerland; lvt@swissgenetics.ch

The success or failure of an insemination is mainly dependent on the cow. For this purpose, a simple but effective system is needed to judge the cow in order to predict the success of insemination. The collected data will contribute to a more reliable estimate of the bull's non-return-rate. We designed a system with which the technician records the relevant data about the cow during each insemination process. The goal was to collect data to get a more reliable estimate of the bull's non-return-rate. Based on literature, six factors were chosen to assess the status of the cow and were evaluated in a field test: position of the vulva, quantity of mucus, uterus tonus, size of the uterus, cervix passage (easy/tight), insemination timepoint (early/late in heat). These six factors were scored with three levels each. To simplify data entry and to increase the tangibility for the technician, pictograms were shown on his tablet. Additionally, the technician enters a number between 1 and 6 indicating his personal prediction of the probability of success of the insemination. The collected data were used in the statistical analysis of the insemination result using the Non-Return Rate at 56 days (NRR56). For two months, 8,184 inseminations done by 27 technicians were evaluated. The acceptance was satisfactory. Each cow-heat was scored. The variability of the system was used. The evaluation of the combination of the factors, which were used more than 100 times, showed that the proposed heat-assessment scores were positively associated with the probability of success of the insemination. There was a difference of 34.7 points in the NRR56 between the subset with the best level of all factors and the subset with the worst level of all factors. The personal prediction of the technician (1-6) at the time of insemination was consistent with the NRR56. Compared to the previous system, the new approach allowed a much more differentiated prediction. The proposed approach will be very helpful to evaluate the cow during the routine insemination process. The obtained data will also improve the evaluation of the bull's NRR56, which is essential for artificial insemination organisations in terms of quality control of the semen straws released on the market.

Histomonosis – new approaches on disease prevention

D. Liebhart[1] and M. Hess[1,2]
[1]University of Veterinary Medicine, Clinic for Poultry and Fish Medicine, Veterinärplatz 1, 1210, Vienna, Austria, [2]Christian Doppler Laboratory for Innovative Poultry Vaccines (IPOV), Veterinärplatz 1, 1210, Vienna, Australia; dieter.liebhart@vetmeduni.ac.at

The flagellate *Histomonas meleagridis* is the cause of histomonosis (syn.: blackhead disease or histomoniasis) and can cause typhlohepatitis in turkeys as well as in chickens. Recently used drugs cannot be applied anymore against the disease in many countries worldwide due to concerns on product safety. As a consequence, reports on outbreaks of the disease in poultry flocks increased indicating a re-emergence of histomonosis. The infection of host birds with *H. meleagridis* can occur directly or via the vector *Heterakis gallinarum* from contaminated faeces. We demonstrated in recent experiments that the infection with histomonads via the oral or cloacal route can led to fatal histomonosis in turkeys whereas in chickens it causes milder clinical signs. However, both host species can show severe inflammation of caecum and liver. For specific detection and localization in tissue samples, we established an *in situ* hybridization to distinguish *H. meleagridis* from other protozoa like *Tetratrichomonas gallinarum*. So far, different prophylactic approaches against histomonosis were experimentally investigated including chemicals and plant derived compounds with limited effectivity or no approval to use in poultry flocks. In recent years, experiments investigating vaccination of turkeys and chickens were performed due to the demand to control histomonosis. We found that both host species can be protected from fatal histomonosis, respectively a drop in egg production by vaccination using clonal *in vitro* attenuated histomonads. Investigations on the cellular immune reaction of vaccinated chickens and turkeys showed the importance of Interferon gamma positive cells as well as a significant reduction of B cells, T cells and monocytes/macrophages following challenge compared to non-vaccinated and challenged birds. In conclusion, histomonosis is a re-emerging disease in many countries. Recent investigations demonstrated that vaccination using *in vitro* attenuated *H. meleagridis* effectively protects turkeys and chickens from histomonosis and is safe in use, arguing for the most promising approach to prevent histomonosis in poultry.

Anthelmintic resistance in ruminants in Europe: challenges and solutions

J. Charlier[1], E. Claerebout[2], D. Bartley[3], L. Rinaldi[4], G. Von Samson-Himmelstjerna[5], E. Morgan[6], H. Hoste[7] and S. Sotiraki[8]
[1]Kreavet, Hendrik Mertensstraat 17, 9150, Belgium, [2]Ghent University, Salisburylaan 133, 9820, Belgium, [3]Moredun Research Institute, Pentlands Science Park, Edinburgh EH26 0PZ, United Kingdom, [4]University of Naples Federico II, Via Delpino 1, 80137 Napoli, Italy, [5]Freie Universität Berlin, Robert-von-Ostertag-Str. 7-13, 14168 Berlin, Germany, [6]Queen's University Belfast, Lisburn Road, Belfast, BT9 26 7BL, United Kingdom, [7]INRA / ENVT, 23 Chemin des Capelles, 31076 Toulouse, France, [8]Veterinary Research Institute HAO-DEMETER, Campus Thermi 57001, Thessaloniki, Greece; jcharlier@kreavet.com

Cattle, sheep and goats are parasitized by various helminth species, the most important being the gastrointestinal nematodes, lungworms and liver fluke. These pathogens can cause severe disease and affect productivity in all classes of stock and are amongst the most important production-limiting diseases of grazing ruminants in Europe and globally. Essentially, all herds/flocks in a grass-based production system are affected. While in specific cases, mortality can be high, the major economic impact is due to sub-clinical infections causing reduced growth, milk production and fertility. A recent comprehensive study, funded by the EU-GLOWORM project, estimated the economic burden in the EU at an annual cost of €1 B in dairy cattle and € 0.3 B in the dairy cattle and meat sheep sector, respectively. A major constraint on the control of helminth infections in livestock is treatment failure due to anthelmintic resistance (AR). Frequent, indiscriminate or inappropriate use of anthelmintic drugs to control these parasites has resulted in selection of drug-resistant helminth populations. AR is now widespread in all the major GIN of sheep and in liver fluke and is an emerging problem in cattle parasites. The economic impact of treatment failures today is unknown, but if no alternatives to current control options become available, pasture-based livestock industries are likely to suffer major economic losses. Here, we review the current knowledge on the development, occurrence, impact and envisaged solutions of AR in ruminant livestock in Europe.

Sheep, strongyles and sequencing for sustainability: investigating ivermectin resistance on UK farms

J. McIntyre[1], K. Maitland[1], J. McGoldrick[1], S. Doyle[2], J. Cotton[2], N. Holroyd[2], E. Morgan[3], H. Vineer[4], K. Bull[5], D. Bartley[6], A. Morrison[6], K. Hamer[1], N. Sargison[7], R. Laing[1] and E. Devaney[1]
[1]University of Glasgow, Glasgow, G61 1QH, United Kingdom, [2]Wellcome Sanger Institute, Hinxton, CB10 1SA, United Kingdom, [3]Queen's University Belfast, Belfast, BT7 1NN, United Kingdom, [4]University of Liverpool, Liverpool, L69 7ZX, United Kingdom, [5]University of Bristol, Bristol, BS8 1TH, United Kingdom, [6]Moredun Research Institute, Edinburgh, EH26 0PZ, United Kingdom, [7]Royal (Dick) School of Veterinary Studies, Edinburgh, EH25 9RG, United Kingdom; j.mcintyre.1@research.gla.ac.uk

Parasitic gastroenteritis is a major production limiting disease of small ruminants worldwide. The abomasal nematode *Teladorsagia circumcincta* is the primary pathogen on most sheep farms in the UK during the summer months, when lambs are in their primary growth phase. Expertly adapted to temperate climates, its small ruminant hosts and capable of surviving year round, *T. circumcincta* is challenging to control, particularly with the rapid development of resistance to all widely used anthelmintic classes. Ivermectin is the most commonly used anthelmintic in the UK but resistance is highly prevalent; recent studies have demonstrated rising numbers of farms with detectable ivermectin resistance by faecal egg count reduction test (FECRT), a concern for sustainable control of parasites in the future. There are many limitations to the FECRT, and such limitations, combined with a highly complicated disease, make understanding the effect of different management practices on the development of anthelmintic resistance difficult. Molecular tests are needed for sensitive diagnosis of resistance, but such tests require knowledge of the regions of the genome conferring resistance. In this study we undertook genome-wide sequencing of two UK farm populations of *T. circumcincta* pre- and post-ivermectin treatment using next generation sequencing techniques (ddRAD-Seq and Pool-Seq) to identify regions of the genome under selection. Despite the fragmented nature of the *T. circumcincta* draft genome assembly, we found a single large locus to be under ivermectin selection, in addition to many smaller loci. Though many selected loci appear farm specific, several are under selection on both farms.

First non-chimeric nicotinic acetylcholine receptor from invertebrates

K. Kaur[1], L. Rufener[2] and T.E. Horsberg[1]

[1]Norwegian University of Life Sciences, Ullevåls veien 72, 0454, Norway, [2]INVENesis, Rue de la Musinière 8, 2072 St-Blaise, Switzerland; kiran.kaur@nmbu.no

Nicotinic acetylcholine receptors (nAChRs) are ligand-based cation channels found in the central and peripheral nervous system, muscle, and other tissues of many organisms. These receptors respond to the neurotransmitter acetylcholine or drugs like nicotine. nAChRs are also the target proteins for a large group of chemicals, the neonicotinoids, being used worldwide against various arthropod parasites (on humans, livestock and in aquaculture), pests (agriculture), and vectors for diseases (mosquitoes). In-depth studies on the function of these channels in arthropods have to date been almost impossible, as no research group managed to express a functional nAChRs in an *ex vivo* system using genes only from the target arthropod species. So far, only chimeric nAChRs have been obtained from insects and other arthropods, by co-expressing nAChR subunits from insects and vertebrates (mainly chicken) in *Xenopus laevis* oocytes. This hugely limits our understanding of the assembly and function of nAChR in arthropods, and their sensitivity to various parasiticides and pesticides. Besides, this poses a major hurdle in designing novel tools and compounds targeting these receptors in various parasites and pests. We have overcome this hurdle by obtaining a first non-chimeric, fully functional nAChR in an arthropod (salmon louse) in *Xenopus laevis* oocytes. The present study will provide a better understanding about the assembly and basic function of arthropod nAChRs using this non-chimeric channel. Besides, it will serve as screening tool to determine the effect of various existing and novel compounds on salmon louse. The present study will lead to a paradigm shift in the current ways of obtaining nAChRs from arthropods with a switch to develop species-specific nAChRs screening models. This will provide a novel platform to develop better screening tools, e.g. novel compounds targeting a specific arthropod, which in turn will lead to better management of various parasites and pests.

***In vivo* assessment of the anthelmintic effects of by-products (peels) from the chestnut industry**

S. Marchand[1], S. Ketavong[1], E. Barbier[2], M. Gay[3], H. Jean[3], V. Niderkorn[4], S. Stergiadis[5], J.P. Saliminen[6] and H. Hoste[1]

[1] INRA, UMR 1225 IHAP. INRA ENVT, 23 chemin des capelles, 31076 Toulouse Cedex, France, [2]Société MG2Mix, Chateaubourg, 35220 Chateaubourg, France, [3]Société Inovchâtaigne, St Médard Mussidan, 24400, France, [4]INRA, UMR Herbivores, Saint-Genès-Champanelle, 63122, France, [5] University of Reading, School of Agriculture, Policy and Development, 1 Earley Gate, P.O. Box 236, United Kingdom, [6]University of Turku, Dptt of Chemistry, Lab. of Organic Chemistry & Chemical Biology, 20014, Turku, Finland; h.hoste@envt.fr

Gastro-intestinal nematodes (GINs) remain a major threat for small ruminants when bred outdoors. For >50 years, synthetic anthelmintics (AH) have been the cornerstone to control GINs. However, the spread of resistance to commercial AHs is nowadays worldwide. There is an urgent need to explore alternative solutions to AHs to control GINs. The use of Legume forages containing condensed tannin (CT) forages as nutraceuticals has been widely explored. The current *in vivo* study aimed at examing the AH effects of other resources, e.g. by-products of the chestnut industry. This study aimed at comparing the effect of 4 diets differing by their CT contents and/or nature on the parasitological and pathophysiological parameters in 4 groups of 7 lambs, experimentally infected with *Haemonchus contortus* and *Trichostrongylus colubrifomis*. The total tannin (TT) and polyphenol contents (PT) were measured for each diet. The lambs received 4 diets made isoproteic based on dehydrated pellets containing CTs differing by their contents and nature. These groups were (1) a control group (C) (CT = TT content % = 0); (2) sainfoin (S) (TT= 2, 03); (3) chestnut (Ch) (CT=3,12); (4) chestnut + sainfoin (Ch + S) (CT=1,4). The results of faecal egg counts per group showed significant differences in the group Ch+S compared to the 3 others. The necropsy results also indicate a 50% reduction in the number of *Haemonchus* in the Ch + S group when compared to the 3 other groups. No differences were observed for the *T. colubriformis* populations. These *in vivo* results tend to confirm previous *in vitro* studies on the interest of peels of chestnuts to help in the control of GINs in small ruminants. They also underlined that besides the quantity of CT their nature can influence the AH effects.

COST Action COMBAR: combatting anthelmintic resistance in ruminants in Europe

S. Sotiraki[1], E. Claerebout[2], D. Bartley[3], L. Rinaldi[4], G. Von Samson-Himmelstjerna[5], E. Morgan[6], H. Hoste[7] and J. Charlier[8]
[1]Veterinary Research Institute HAO-DEMETER, Campus Thermi 57001, Thessaloniki, Greece, [2]Ghent University, Salisburylaan 133, 9820 Merelbeke, Belgium, [3]Moredun Research Institute, Pentlands Science Park, Edinburgh EH26 0PZ, United Kingdom, [4]University of Naples Federico II, Via Delpino 1, 80137 Napoli, Italy, [5]Freie Universität Berlin, Robert-von-Ostertag-Str. 7-13, 14168 Berlin, Germany, [6]Queen's University Belfast, Lisburn Road, Belfast, BT9 26 7BL, United Kingdom, [7]INRA/ENVT, 23 Chemin des Capelles, 31076 Toulouse, France, [8]Kreavet, Hendrik Mertensstraat 17, 9150 Kruibeke, Belgium; smaro_sotiraki@yahoo.gr

Anthelmintic resistance, part of the larger antimicrobial resistance phenomenon, is a growing concern in the control of ubiquitous parasites in grazing livestock. The COST Action COMBAR (2017-2021) aims at coordinating research at the European level to advance knowledge in this area, develop and promote new solutions. COMBAR is a growing network, uniting 177 researchers from 31 countries with expertise in diagnostics, vaccine development, targeted selective treatment strategies and decision support with the aim to integrate the various disciplines and propose new control options. Collaboration with economists helps to understand the financial trade-offs of implementing new strategies to fight AR, while the adoption of social sciences helps understand and overcome the socio-psychological barriers to the uptake and maintenance of sustainable control approaches. The solution to AR will not be a single new drug, but is based on a diagnosis before treatment approach as well as a broader panel of control options including vaccines and nutraceuticals. This will require more veterinary guidance in helminth control at farm level and increased collaboration and support from regulatory authorities to enhance the market access and uptake of sustainable control approaches and new solutions.

Acaricidal activity of plant-derived essential oil components against *Psoroptes ovis*

Z.Z. Chen[1], W.V. Mol[1], M. Vanhecke[1], L. Duchateau[2] and E. Claerebout[1]
[1]Laboratory of Parasitology, Faculty of Veterinary Medicine, Ghent University, Salisburylaan 133, 9820 Merelbeke, Belgium, [2]Department of Nutrition, Genetics and Ethology, Faculty of Veterinary Medicine, Ghent University, Salisburylaan 133, 9820 Merelbeke, Belgium; zhenzhen.chen@ugent.be

Psoroptes ovis is a major health problem in beef cattle. Treatment is limited to local administration of amitraz or pyrethroids or systemic administration of macrocyclic lactones. Treatment failures with macrocyclic lactones have been reported in recent years. To investigate potential alternative treatments, the acaricidal activity of four plant-derived essential oil components, i.e. geraniol, eugenol, 1,8-cineol and carvacrol against *P. ovis* was assessed *in vitro* and *in vivo*. Three components showed a concentration-dependent acaricidal activity in a contact assay, with LC50 of 0.56, 0.38 and 0.26% at 24 h for geraniol, eugenol, and carvacrol, respectively. In a fumigation bioassay, carvacrol demonstrated the best efficacy as it killed all mites within 50 min after treatment, whereas geraniol, eugenol, and 1,8-cineol needed 90, 150 and 90 min, respectively. Following a 72 h incubation period in a residual bioassay, eugenol and carvacrol killed all mites after 4 h of exposure to LC50 and LC90, while geraniol killed all mites only after 8 h exposure at LC50. Topical treatment with 2% carvacrol in Tween-80 of six calves with experimental *P. ovis* infestations, reduced mean mite counts by 98.48±2.36% at 6 weeks post treatment. In the control group which was treated with Tween-80 only, the mite population increased with similar kinetics as a typical experimental mite infestation. Topical application of carvacrol on shaved skin caused mild and transient erythema 20 min after treatment. No other side effects were observed. Considering the strong acaricidal activity of carvacrol *in vitro* and *in vivo* and the mild and transient local side effects after topical treatment, carvacrol shows potential as an acaricidal agent in the treatment of *P. ovis* in cattle.

Chamomile extract mitigates bowel inflammation and ROS-overload related to nematode infections

S.H. Hajajji[1,2], M.A.J. Jabri[1], M.R. Rekik[2] and H.A. Akkari[1,2]
[1]Institut Supérieur de Biotechnologie de Beja, Avenue Habib Bourguiba, 9000, Tunisia, [2]Ecole Nationale de Médecine Veterinaire, Sidi thabet, 2020, Tunisia; hafidh_akkari@yahoo.fr

Chamomile (*Matricaria recutita* L.) is a plant which has been reported to be effective in treating several parasitic and digestive diseases. The present study was conducted to evaluate the anthelmintic activity of chamomile methanolic extract (CME). *In vitro*, the anthelmintic activities of CME were investigated on the L3 larvae of *Heligmosomoides polygyrus* in comparison to albendazole. *In vivo*, Swiss albino mice were infected with infective third (L3) larval stage of *H. polygyrus* by intragastric administration. Moreover, the effects of CME and albendazole on worm eggs, adult worms, serum cytokine production and oxidative stress were studied. All used doses of CME showed a potent anthelmintic activity both *in vitro* and *in vivo* and the effect being similar to treatment with albendazole. Moreover, *H. polygyrus* infestation was accompanied by an intestinal oxidative stress status characterized by an increased lipoperoxidation, a depletion of antioxidant enzyme activity, as well as an overload of hydrogen peroxide. We have also recorded an increase of pro-inflammatory mediator (TNF-α, IL-6, and IL-1β) levels after treatment with CME (14 ± 0.8; 41 ± 2; 58 ± 4 pg/mg protein, respectively, with the concentration 800 mg/kg BW) when compared with infected control mice (20 ± 1; 59 ± 2, and 83 ± 4 pg/mg protein, respectively). However, extract treatment alleviated all the deleterious effects associated with *H. polygyrus* infection. These findings suggest that CME can be used in the control of gastrointestinal helminthiasis and associated oxidative stress.

Do nematode infections impair vaccine-induced humoral immune responses in chickens?

G. Das[1], M. Stehr[1], C.C. Metges[1], M. Auerbach[2], C. Sürie[2] and S. Rautenschlein[2]
[1]Leibniz Institute for Farm Animal Biology (FBN), Institute of Nutritional Physiology, Wilhelm-Stahl-Allee 2, 18196, Dummerstorf, Germany, [2]University of Veterinary Medicine Hannover, Clinic for Poultry, Bünteweg 17, 30559, Hannover, Germany; gdas@fbn-dummerstorf.de

Nematode infections with *Ascaridia galli* and *Heterakis gallinarum* are highly prevalent in chickens kept in free-range and organic systems. Chicken's immune system deals with intracellular (e.g. viruses) and extracellular (e.g. nematodes) pathogens mainly through Th-1 and Th2-type immune responses, respectively. Because Th1- and Th2-type immune responses may be traded-off, we investigated whether nematode infections impair vaccine-induced humoral immune responses in chickens. During the growing period (17 weeks), 180 pullets of two genotypes, namely Lohmann Brown Plus (LB) and Lohmann Dual (LD) were subjected to a vaccination program, including immunization against major bacterial and viral diseases and coccidiosis. At 24 weeks of age, the hens received either a placebo or an oral inoculation of 1000 *A. galli* and *H. gallinarum* eggs. Starting from 2 weeks post infection (wpi) infected and uninfected hens were necropsied at timed intervals (i.e. 2, 4, 6, 10, 14 and 18 wpi) to quantify worm burdens. Plasma concentrations of vaccine-induced antibodies against Newcastle disease virus (ND), infectious bronchitis (IB) and avian metapneumovirus (AMVP), and non-specific immunoglobulins (i.e. IgY, IgM, IgA) were analysed by ELISAs. On average each hen harboured 14 ± 13.5 *A. galli* and 139 ± 103 *H. gallinarum* which did not differ between the two host genotypes ($P>0.05$). There were no effects of infections on ND and IB titers ($P>0.05$). Overall, LB hens had higher AMPV titers than LD hens ($P<0.05$). Uninfected LD hens had lower AMPV titers than their infected counterparts at 6 and 14 wpi ($P<0.05$). Infections had no effect on the concentrations of non-specific immunoglobulins ($P>0.05$). LB hens had higher levels of IgY and IgA ($P<0.05$) and tended to have higher IgM levels than LD hens ($P=0.056$). Our results collectively suggest that nematode infections have no adverse effects on vaccine-induced or non-specific humoral immune responses in chickens.

Gastro-intestinal nematode infections and preventive measures in Flemish organic dairy herds

L. Sobry[1] and J. Vicca[2]
[1]*Inagro, Ieperseweg 87, 8800 Rumbeke-Beitem, Belgium, [2]Odisee, Hospitaalstraat 23, 9100 Sint-Niklaas, Belgium; luk.sobry@inagro.be*

In organic dairy herds a close monitoring of infections with gastro-intestinal nematodes (GIN) is important in developing a preventive strategy. Maximal access to pasture is an important principle in organic dairy systems but implies a higher infection risk with GIN. The abomasum nematode *Ostertagia ostertagi* is considered as the most important GIN infection in cattle in Flanders with considerable financial implications. Monitoring of the infection level in herds is fairly simple by determining antibody level on bulk milk. During the years 2016-2018 organic dairy herds in Flanders were monitored in this way. In 2018, 27 farms making up more than 90% of Flemish organic dairy herds were monitored. The average optical density ratio (ODR) was 0.69 and ranged between 0.48 and 0.96. An ODR>0.5 is considered to have a negative effect on milk production, with raising ODR the daily milk loss per cow increases. According to our data the average organic dairy cow loses an estimated amount of 0.6 litre per day as a result of infection with *O. ostertagi*. Preventive treatments with anthelmintics are not allowed on organic farms. Prevention of infection with GIN focuses on developing a good host resistance and thus lowering infection pressure on the pasture. The development of host resistance should be focussed on grazing management of calves entering pastures for the first time. In addition, there is currently a growing focus on the use of nutraceuticals. In the fight against GIN, especially condensed tannins appear to have an influence because they greatly reduce the vitality of the GIN nematodes. Sainfoin is a leguminous crop rich in these tannins. On an organic dairy farm an adapted pasture management was combined with a supplement of either lucerne or sainfoin pellets in the calves' feed during their first pasture season. In September and October, serum samples were analysed with ELISA for pepsinogen, an abomasum enzyme, as a measure of abomasum damage. The feeding of sainfoin, as compared to lucerne, had no influence on daily growth in the calves (P=0.80), did not suppress FEC (P=0.57) and also did not protect against abomasum damage (P=0.13). Abomasum damage clearly increased from September to October.

Does garlic change the taste of cow milk when fed as a feed supplement?

S. Probst[1], M. Aarts[2] and D. Hartig Hugelshofer[1]
[1]*Bern University of Applied Sciences, School of Agricultural, Forest and Food Sciences, Laenggasse 85, 3052 Zollikofen, Switzerland, [2]HAS University of Applied Science, Onderwijsboulevard 221, 5223 DE 's-Hertogenbosch, the Netherlands; stefan.probst@bfh.ch*

Flies and other parasites are commonly known pests for cows. Garlic (*Allium sativum*) might be a solution for the reduction of flies as it is included for its function as insect repellent in equine supplements. Until now it is unknown if consumers could taste a difference of cow milk when cows get garlic as a feed supplement. This study was conducted to answer this question. Nine Simmental cows where divided into a control group (n=4) and a test group (n=5). The test group received 10 g of a garlic supplement per cow per day in experiment 1 (T1) and 50 g garlic supplement in experiment 2 (T2). After an adaption period of 7 days for each experiment, the milk of every cow was collected individually. The milk was blended to reach the equal fat content in control group and test group. Then the milk was homogenized and pasteurized. A sensory difference test (triangle test based on ISO 4120:2004) was conducted to identify whether consumers could taste a difference between the milk samples (test set up: one sample is different and two are equal; H0: There is no perceptible difference between the samples). The test was carried out with 56 consumers (T1) and 55 consumers (T2), respectively. They were asked to identify the different milk sample (with or without garlic taste). The minimum correct responses for a perceptible difference are calculated with $x=(n/3)+z\times\sqrt{(2\times n/9)}$ with a z-value of 1.64 for P=0.05. If the correct responses are greater than or equal to the calculated value there is a rejection of the assumption of 'no difference'. The calculated value was 25 for both tests. Therefore, 25 correct answers are needed to conclude there is a statistically significant difference between the two milk samples. 27 (T1) and 26 (T2) consumers did correctly detect the different sample during the sensory test (P<0.05). Based on this study, consumers recognized a difference between the milk of cows fed with garlic, independent of the garlic concentration in the feed. Further research is required to identify different garlic substances for feed which might not result in a taste difference of milk.

In vitro efficacy of kefir produced from camel, goat, ewe and cow milk on *Haemonchus contortus*

D.A. Alimi[1], M.R. Rekik[2] and H.A. Akkari[1]
[1]Ecole Nationale de Medecine Veterinaire, Sidi thabet, 2020, Tunisia, [2]International Center for Agricultural Research in the Dry Areas (ICARDA), Amman 11195, 950764, Jordan; hafidh_akkari@yahoo.fr

One of the great challenges of veterinary parasitology is the search for alternative methods for controlling gastrointestinal parasites in small ruminants. Milk kefir is a traditional source of probiotic with great therapeutic potential. The objective of this study was to investigate the anthelmintic effects of kefir on the abomasal nematode *Haemonchus contortus* from sheep. The study used camel, goat, sheep and cow milk as a starting material to produce species-specifics kefir. All kefirs showed a significant concentration-dependent effect on *H. contortus* egg hatching at all tested concentrations. Compared to others, kefir from camel milk showed higher inhibitory effects on egg hatching ($P<0.001$). In relation to the effect of kefirs on the survival of adult parasites, all kefirs induced concentration-dependent mortality in adults, with variable results. The complete mortality (100%) of adults of *H. contortus* occurred at concentrations in the range 0.25-2 mg/ml. The highest inhibition of motility (100%) of worms was observed after 8 h post exposure with camel milk kefir at 0.25 mg/ml. These findings indicate that kefir can be considered a potential tool to control haemonchosis in sheep. Further investigations are needed to assess the active molecules in kefir responsible for anthelmintic properties, and to investigate potential *in vivo* effects.

Potential risks of nematode parasitism in farmed snails in Greece

K. Apostolou[1], M. Hatziioannou[1] and S. Sotiraki[2]
[1]University of Thessaly, Ichthyology & Aquatic Environment, Fytoko Street, 38445, Nea Ionia Magnesia, Greece, [2]National Agricultural Research Foundation, Veterinary Research Institute, NAGREF Campus, 57001 Thermi, Greece; apostolou@uth.gr

The aim of this study was to classify heliciculture farms, as well as register parasite species in farmed snails (*Cornu aspersum maximum or Cornu aspersum*) in Greece. Farms located in Central, Western and North Greece were classified into four different types i.e. open field (35%), net-covered greenhouse (38%), mixed system with net-covered greenhouse or open field (19%) and elevated sections (8%). Farmers were surveyed on a designated questionnaire concerning welfare problems of livestock. The questions included presence of other animals in the farm, preventive spraying and neighboring farms. A total of 1,300 snails (14 ± 2.4 g) from 26 farms (2017) and 900 snails (13.2 ± 2.7 g) from 18 farms was collected (2018) during a two-year experiment. Parasitological examinations through standard dissection and faecal egg counting techniques were performed. The most common problem was the entry of other animals in the farm with 77% of farmers being unable to deal with the issue. Additionally, 85.7% of farms had mice issues. Only 23% of snail farmers practiced spraying against pathogens. Approximately 43% of farms abut another animal farming. Results of 2017 showed presence of adult nematodes in all farm systems, with higher values in elevated sections (50%). A total of 19.2% of snail farms were positive for adult parasitism. In 2018, all farms were positive and the average number of eggs per gram of faeces was $2,535\pm3,272$. Adult nematodes appeared in only two systems [(i.e. open field (42.8%), net – covered greenhouse (40%)]. In both years, three nematode species were encountered. The species were *Phasmarhabditis hermaphrodita* (20-40%; intestine), *Muellerius capillaris* (40%; foot) and *Alloionema appendiculatum* (20-60%; intestine). Two farms were positive for adult parasites in both years. The increase in percentage of positive snail farms with adult parasites in second year (38.8%) suggests that preventive measures should be taken in order to protect farmed snails and restrict the spread of parasites.

Schooling philosophy and horse welfare

F.O. Ödberg[1] and L. Gombeer[2]
[1]Ghent University, Nutrition, Genetics and Ethology, Heidestraat 19, 9820 Merelbeke, Belgium, [2]Belgian Equitation Academy, Chaussée de Mons, 41/405, 1400 Nivelles, Belgium; frank.odberg@ugent.be

In the antiquity Xenophon dealt explicitly with schooling, promoting kindness and understanding behaviour. Literature in Roman times and early and late middle ages is scarce. Linear progress in schooling cannot be found before the transition between the renaissance and baroque periods. Writings from renaissance masters show a generally violent way of schooling, strangely contrasting with appeals for kindness. The 17th century and 18th centuries showed a gradual improvement: moving (1) from anthropomorphism to understanding horse behaviour, (2) from coercion to suppleness obtained through gymnastics. Many baroque and early 19th century masters had understood intuitively conditioning laws discovered scientifically in the 20th century by the Pavlovian school and behaviourists. The essence of schooling was settled in the baroque period. It subsided from the 19th on because of the French revolution that closed academies and more interest in racing and hunting (more weight on the forehand and less collection). Pressure of competition in the 20th nearly eradicated academic tradition. Some fundamental riding principles will be translated into scientific terminology. 'Independence of aids' is based on discrimination of stimuli and overshadowing. 'Discretion of aids' on generalization and second-order conditioning. 'Descente de mains et de jambes' on avoidance of habituation, of confusion, and of the impossibility to apply negative reward. 'Legs without hands, hands without legs' on avoidance of experimental neurosis due to contradictory stimuli. The above mentioned principles are seldom applied nowadays. The FEI holds a large responsibility in the tolerance of a brutal way of riding. Protest is increasing. Scientists however seek facts. Frequency of conflict behaviours is an indicator at ethological level. However, some horses could be in learned helplessness and there is no litmus test for that. There is a need for physiological stress parameters standards according to breed, age, sex. Horses schooled according to different philosophies are often from different breeds making matched-pairs studies difficult and standardization of ontogeny is impossible outside laboratory conditions.

Educating young horses: the role of voice in a worrying situation

S. Barreau[1] and J. Porcher[2]
[1]Ecole Blondeau, La Grande Pièce, Saint Hilaire St Florent, 49400 Saumur, France, [2]INRA, SAD, 2 Place Viala, 34060 Montpellier, France; sophie.barreau@ecoleblondeau.com

The main objective of the Chevaleduc project – which is the framework for the results presented here – is to show that breaking-in constitutes an initial vocational training for young horses. Our hypothesis is that this training allows the acquisition of the basic skills needed by a race horse throughout its career. Our field research was conducted at Ecole Blondeau (Saumur, France). In a stable dedicated to breaking-in, where 240 foals are accommodated every year, all foals go through a standardized twenty-step process implemented by all riders. The majority of the 20 steps of the Blondeau method are real-life work situations. One hundred thoroughbred male and female sport horses and foals were filmed going through those 20 steps with 7 trainers. The videos were then analysed using Observer xt software (Noldus). The sixth step is 'the van': at that stage, roughly 30 min after the beginning of the training, the horse learns how to load into a van, often a dark and narrow space. Our observations concerning that step were conducted from August 28th to December 12th 2017. During the study of the 81 videoclips dealing with that stage, we observed 53 instances of a particular gesture from the foal (its lips delicately resting on the trainer's neck), 52 of which were in direct response to the trainer's voice. There were two particular forms of expression: satisfaction: 'C'est bien! (Well done!)' uttered in an exclamatory and rather strong way- and encouragement: 'Tu vas y arriver ma fille! (You can do it, girl!)', uttered in a soft and caring voice. It is in response to these words of encouragement that the aforementioned behaviour appears. We show that, initiated in a context of great physical proximity with the trainer and of strong emotional support, in the face of a worrying situation, the horse's gesture appears to be one of reassurance and appeasement, aiming to convey the animal's trust in the trainer, thus enabling it to confront the situation. This gesture of contact can also be understood in connection with two social behaviours in horses, mutual grooming and mutual greetings.

Does a self-loading positive reinforcement-based training improve loading procedures in meat horses?

F. Dai[1], G. Di Martino[2], L. Bonfanti[2], C. Caucci[2], A. Dalla Costa[2], B. Padalino[3] and M. Minero[1]
[1]University of Milan, DiMeVet, via Celoria 10, 20133 Milan, Italy, [2]IZSVe, Viale dell'Università 10, 35020 Legnaro PD, Italy, [3]University of Bari, DiMeV, Piazza Umberto I, 70121 Bari, Italy; francesca.dai@unimi.it

Self-loading techniques are proven to reduce the likelihood of sport horses to show behavioural problems and the need of human intervention (e.g. using aids to load the horse) at loading, consequently reducing time needed and decreasing injuries occurrence both to horses and handlers. Horses kept for meat production are generally transported to the slaughterhouse without any training. It was hypothesized that self-loading training would reduce stress-related behaviours and human intervention needed during loading. Thirty-two meat horses (M=18; F=14; 6 months-old) were included in the study. Animals had limited interactions with the farmer and were not used to being restrained, lead with a lead rope, nor transported. Horses were randomly divided in two groups: Control Group (C; n=14) and Training Group (T; n=18). T horses were trained to self-load, applying a target training protocol using a shaping process. Training sessions were performed three times per week (9:30-13:30 and 14:30-16:30), for six weeks; training was then repeated once a week to maintain the memory until the transport toward a slaughterhouse. The loading phase was video-recorded and loading time was directly recorded using a stopwatch. All horses were transported to the slaughterhouse on fourteen different days using the same truck. Behaviour was then analysed, using Solomon Coder software, with a focal animal continuous recording method. T horses required a shorter loading time (mean 44.44±47.58 sec) than C horses (mean 463.09±918.19 sec) (T test; P<0.05) and needed less human intervention to load (T: mean 2.28±1.70; C: mean 11.86±23.45. T test; P<0.1); T horses also showed more forward locomotion towards the truck than C horses (T test; P<0.05). Although not statistically significant, stress-related behaviours (e.g. rear, kick, mount, paw, defecate) were shown mainly by C horses. Our preliminary results suggest that self-loading training may be useful to improve loading procedures in meat horses, mitigating stress behaviours, reducing time needed and decreasing the need of human intervention for loading.

Equine Blink rate, reactivity and learning: can one predict the other?

R.C. Clarke[1] and A. Hemmings[2]
[1]University Centre Myerscough College, Equine, St Michaels Rd, Bilsborrow, Preston, PR3 0RY, United Kingdom, [2]Royal Agricultural University, Equine, Stroud Rd, Cirencester, GL7 6JS, United Kingdom; rcclarke@myerscough.ac.uk

Horses are an animal in which not only superior physical performance but also good temperament is needed in order to achieve optimal performance and as such this has been an important selective characteristic in breeding. In horses it has been found that blink rate in the left eye is associated with central dopamine activity within the brain and as such can be used as a behavioural parameter to measure this. As a link between novel object testing and reactivity and equine learning ability, it has been shown that high levels of fearfulness can negatively impair learning ability in the horse. This therefore suggests that learning ability could be detected via measurement of eye blink rate. The objective of this study was to investigate possible links between dopamine release in the brain via eye blink rate, equine reactivity and learning. A total of 20 horses of mixed age, experience, breed and sex were used where all horses were habituated to the testing areas. Equine learning was assessed through the use of a reversal discrimination task and fearfulness/reactivity was assessed through the use of a novel object test. The current study revealed there was a link between dopamine release via eye blink rate and learning ability where, as eye blink rate and consequent level of dopamine release increased, learning ability in the reversal discrimination task decreased. There was also a link demonstrated between learning ability and reactivity responses where, as reactivity in the novel object test increased so too did learning ability, with less mistakes to the reversal criterion being made. These results suggest that if one of these traits is known, the ability to predict the other is now available. This in turn will allow horse owners and trainers to tailor their programmes to match the complexity, levels of comprehension and learning intensity that is innate to the horse, which will improve the welfare of all horses involved in some aspect of the equine industry.

Influence of intrinsic effects on effort and recovery stress assessed with infrared thermography

D.I. Perdomo-González, E. Bartolomé, M.J. Sánchez-Guerrero and M. Valera
Universidad de Sevilla, Departamento de Ciencias Agroforestales, ETSIA, Ctra. Utrera Km 1, 41013 Sevilla, Spain;
daviniapergon@gmail.com

The Spanish Sport horse (CDE) is a composite breed used worldwide for sport performance. During the sport events, horses are exposed to several environmental stimuli that could affect their sport performance as a result of the stress developed. Furthermore, as important as the stress level reached during the exercise, is the effort developed to achieve the adequate physiological condition to perform and how the animal recover from it, as this will condition its adaptation to the sport events. Thus, the aim of this study was to assess the influence of different intrinsic effects on the effort and recovery stress suffered by CDE during a standard warming exercise. In this study, eye temperature (ET) assessed with infrared thermography was collected from 495 CDE while performing a regular warming exercise. During the exercise, 3 ET samples per animal were obtained one hour before (BE), just after (JAE) and one hour after (AE) the warming exercise. In order to evaluate effort and recovery levels, a previous ET increase (PETI=JAE-BE) and a later ET increase (LETI=JAE-AE) were calculated, respectively. According to the composite nature of the CDE breed, 5 genetic lines were assessed according to their origin: 'German' (L1); 'Thoroughbred' (L2); 'Trotter' (L3); 'Pura Raza Español' (L4) and 'Others' (L5). In order to consider the influence of effects related to the animal itself (intrinsic effects) on the effort (PETI) and recovery (LETI) stress suffered by the CDE, sex, age and breed composition (genetic line) were assessed with a Multiple-trait ANOVA test, a Least Square Means Test and a Duncan's Test. Results showed significant differences in PETI and LETI for sex and genetic line effects, while only in PETI for the age effect. As regards to sex and genetic line, males and L3 CDE showed a higher effort stress and recovered almost completely after 1 h from the exercise. As regards to age, the youngest animals showed a significantly higher effort stress than older animals. Our results highlight the importance of animal's intrinsic effects when developing the CDE Breeding Program, as it would condition the stress suffered by the animal.

Network analysis of the group structure of horses on pasture by using GPS data

F. Hildebrandt, K. Büttner, J. Salau, J. Krieter and I. Czycholl
Institute of Animal Breeding and Husbandry, Christian-Albrechts-University, Olshausenstr. 40, 24098 Kiel, Germany;
fhildebrandt@tierzucht.uni-kiel.de

The herd structure and hierarchy of large groups (>20 horses) is not yet adequately described. Therefore, the aim of this study is to characterize group structure, behaviour and social interactions between horses using network analysis. The study assessed 53 horses held in an open stable system in Northern Germany. In addition to a permanently available paddock area, the horses had pasture access for one hour each day. To record the position, each horse was equipped with a GPS-sensor (sampling frequency: 0.1 Hz) attached to a nylon collar. In total, the horses were observed for 9 months (Jun 2018 – Feb 2019). The pasture time (9.00-10.00 am) on 30 days in October was analysed and divided into 10-minute intervals. For each time interval, undirected networks were generated containing nodes (horses) which are combined by edges (defined as an approach closer than 6 m in any coordinate). Density and fragmentation for each interval were calculated. The density is the ratio of actual edges in the network to all possible edges, whereas the fragmentation measures the network components in relation to all nodes in the network. The density ranged between 0.01-0.33 and the fragmentation varied from 0.33-0.98. Higher density was obtained in the first and last 10 minutes of the daily-analysed hour, but not in the intervals between. This might be caused by a bottleneck-effect caused by the gate during pasture entry and exit. Furthermore, the higher density in the last 10 minutes might be explained by horses waiting at the drinking troughs, as water was only available in the paddock area. Additionally, the first and last 10 minutes showed a tendency toward lower fragmentation. This can be explained by larger group movements yielding to larger network components. In conclusion, density and fragmentation can be used as an indicator for group dynamics in horses. Next evaluation steps include the complete observation period to investigate the usage of and group formation in the complete housing environment.

Horse housing: active stable and tracks systems – management and practices in Belgium and France

M. Humbel, J.F. Cabaraux and M. Vandenheede
University of Liège Veterinary Faculty, Veterinary Management of Animal Resources, Dpt. D43 Ecologie de la santé et des productions animales (DPA) Quartier Vallée 2 avenue de Cureghem, 4000 Liège, Belgium; mhumbel@uliege.be

In Western countries, horses are frequently kept in individual stable despite related health and welfare concerns. Alternative systems for group housing are of growing interest among horse's owners (e.g. 'active stable', 'Paddock trail', 'equi-tracks', 'paddock tracks'). These organised outdoor areas intend to provide day long feeding, herd life and to enhance movement. There is a lack of information on the characteristics of those housing systems and their eventual effect on horse welfare. Owners of such systems were recruited via social media, website and peers recommendation in Belgium and France. Data were collected from 56 herds. The questionnaire gathered information on management choices, available areas, sheltering, feeding practices, enrichment, workload, type of horses and group composition. Trends in management choices were identified by using descriptive statistics. Artificial shelters are more frequent than natural ones only. Forage is most often distributed via hay rack or individual nets. Strategies to enhance locomotion rely on dispersal of resources along the tracks or automated processes. Toys are almost never used to provide enrichment while the use of hedgerow or trees is often considered. Work time per horse varies with the level of mechanisation. Horses are mostly used for leisure purpose but competition use is also reported. Behaviour observations and collection of welfare indicators in those systems will then allow us to evaluate their suitability for horses.

Use of alkanes and alcohols for hay intake estimation in stabled horses

M. Maxfield, E. Andrade, A.S. Santos and L.M.M. Ferreira
CITAB-UTAD, Qta Prados, 5001-801, Portugal; assantos@utad.pt

Horses are herbivore animals that spend most of their time in foraging activities. When stabled, this behaviour should be kept, and hay intake is one way to promote this. Knowledge on the amount of hay that a stabled horse consumes is important to adequately manage diet to the animal's nutritional requirements. Given the wide variety and importance of hay in horse feeding, the correct evaluation of its nutritional characteristics, including voluntary intake, is important. However, knowing the accurate amount of voluntary intake of hay can be complicated when horses are stabled in groups or outdoors. Internal markers such as alkanes and alcohols have been successfully used for estimation of pasture intake in previous studies. In this study, we intended to assess the use of alkanes and alcohols to estimate hay intake in stabled horses. For this purpose, 3 diets varying concentrate and hay proportion where used: D1: 5% concentrate; D2: 10% concentrate; and D3: 20% concentrate. Concentrate was marked with beewax in order to ensure a known profile of alkanes and alcohols. Hay intake was estimated on a 10 day trial period, with total faecal collection on the last 4 days. Results showed significant differences for estimate methodologies ($P<0.001$), with the best estimates being obtained when alkanes or the combination of alkanes+alcohols where used. Accuracy of estimate was higher in D3, this was expected since the amount of alkanes is very low in hay, thus the higher amount of concentrate, the best estimates would be expected. This study indicates that alkanes can be used for hay intake estimation in stabled horses.

Serum malondialdehyde in horses supplemented with selenium and vitamin E moderately exercised

E. Velázquez-Cantón, A.H. Ramírez-Pérez, J.C. Ángeles-Hernández and J.C. Ramírez-Orejel
Universidad Nacional Autónoma de México, Facultad de Medicina Veterinaria y Zootecnia, Av. Universidad 3000, 04510, Mexico; rpereza@unam.mx

Twenty-four unexercised healthy horses were used in an 77-days trial to evaluate the relationship between serum malondialdehyde (MDA) concentrations and supplementation with selenium (Se, Se-methionine) and vitamin E (E, α-tocopheryl) in horses moderately exercised in a particulate matter (PM_{10}) polluted environment (Data from the Ministry of the Environment (SEDEMA, Mexico city). Horses were individually housed, randomly allotted to 4 treatments in a factorial trial (2×2, Se, E levels) with repeated measures (n=6). Antioxidant supplementation levels were low (L) and high (H). Experimental treatments were LSeLE, HSeLE, LSeHE and HSeHE (LSe, 0.1; HSe, 0.3 mg Se/kg of DM and LE, 1.6; HE, 2 IU vitamin E/kg of BW). Se and E were fully supplemented because the diet was poor (Se, undetectable, <2 μg/kg DM; E, 14.4 IU/kg DM). D0 corresponded to serum MDA baseline value before starting the experimental supplementation. The experimental days were as follows: d0 to d21, adaptation period; d28 to d49, exercise period (3 consecutive days per horse; 5-20-5 minutes, including: warm up-cantering-cool down); d56, rest; d63, end of supplementation. Jugular blood samples were taken every week (d28-d49, at the end of last exercise day). Data were analysed by a mixed model for the design described above. MDA was quantified as Gutteridge (1975). The highest (P<0.05) MDA was at d7 (1.8±0.09 nmol/ml) Supplementation (Se, E, Se × E) did not affect MDA. The relationship between MDA, PM_{10} and day was analysed and visualized using multiple sequence regression to test linear and nonlinear effects by RSM package in R. Relationship was: MDA μg/ml = 5.1 + 0.05 × day + 0.12 × PM_{10} – 0.001 × PM_{10} × day + 0.0002 × day^2 – 0.0005 × PM_{10}^2 (P<0.001, adjusted r^2=0.3). MDA would be higher on the first days of exposure to high PM_{10} concentrations and later decrease over time, even though PM_{10} increased. This could be an adaptation response; however, these results require further research.

The effects of athletic kinesio taping technique on locomotion in sport horses

S. Biau, L. Tourneix and I. Burgaud
IFCE, Pôle Développement, Innovation et Recherche, Ecole Nationale d'Equitation BP 207, 49411 Saumur Cedex, France; sophie.biau@ifce.fr

The effects of kinesio taping (KT) are being investigated in human sport medicine. However only a few studies have been performed on horses. The objective of this study was to assess the immediate effects of athletic Kinesio taping on locomotion of horses at walk and trot, with facilitation application method on abdominal muscles, essential for the function of the spine during locomotion Twelve healthy sport horses (ages: 10.83±3.23 years old and weight: 589±47.5 kg) were tested without KT and with KT. Their locomotion was tested with the Equimetrix® system on a straight line at walk and trot. They were first measured without KT. Then they returned to the box for 30 min. During these 30 minutes, the KT was fixed 15 minutes before the second test in hand. The KT was glued according to the muscle facilitation application method: Vetkintape® 50 mm with 25% of tension on rectus abdominis muscles, and ktape® 60 mm with 100% of tension, cross obliquus externus abdominis muscles. A statistical test of Wilcoxon compared the locomotor parameters calculated (propulsion, dorsoventral activity, dorsoventral displacement, symmetry, regularity) without and with tape (P<0.05) The results showed a significant difference at the trot with KT vs without KT. The propulsion, the dorsoventral activity and displacement were higher with KT (6.7±2.4 W/kg vs 8.2±2.7 W/kg, P=0.005; 16.8±3.8 W/kg vs 18.2±3.5 W/kg, P=0.02; 11.4±2.2 cm vs 12.4±1.7 cm, P=0.03). This increase of trot activity was probably related to the ventral line sheathing stimulated by KT (eccentric contraction of abdominal muscles), resulting in a decrease in thoraco-lumbar extension and an increase in lumbosacral flexion, especially at the stand phase, producing a good propulsion. No difference was shown at walk probably related to the under activity of abdominal muscles at walk vs trot. The athletic Kinesio Taping on abdominal muscles stimulated immediately the activity of trot in hand. These results could find applications in the conduct of a training, (propulsion and vertical displacement are expected qualities at trot) or rehabilitation phase, following a convalescence, or in addition to a treatment of back pain.

Friends, family or food: principles of human-animal relationships and their implications for animals

C. Croney
Purdue University, Center for Animal Welfare Science, 625 Harrison St, West Lafayette, IN 47907, USA; ccroney@purdue.edu

The idea that humans are innately drawn to nature, and particularly to animals, as outlined in the biophilia hypothesis, is now well established. Undoubtedly, this attraction provides a basis for the human-animal bond and other human-animal interactions. The quality of these relationships and their implications for animals, however, is of increasing concern in developed western nations. The term, 'human-animal bond' connotes a close connection and a mutually beneficial relationship that implies a special duty of human care or obligation to the animals with whom such bonds exist. While such verbiage is commonplace in discourse related to companion animals, the term 'human–animal bond' is not typically used in production animal agriculture. This exclusion, in conjunction with industry terminology that frequently describes agricultural animals as commodities, potentially reinforces the notion that particularly in large-scale, intensive production systems, close connections to farmed animals are precluded, unrealistic, or non-existent. Consequently, existing concerns about the quality of care and life (welfare) experienced by livestock and poultry animals may be escalated. Yet, accumulating evidence illustrates the existence and effects of human-animal interactions on agricultural animal behavioural and physiologic state as well as their related impacts on animal production and profitability of farms. While such interactions with farmed animals may not perfectly mirror common conceptions of bonded-relationships between humans and other species, their nature and implications for animal treatment should be explored, especially because many of the factors that elicit attraction to and care-giving behaviour toward animals apply to those reared for food production. The interactions between language, culture, science, evolving societal experiences with and beliefs about animals, and our perceived obligations to them are therefore relevant to discussions about the ethics of contemporary agricultural animal production.

Different religions, different ethics

L. Caruana
Gregorian University, Piazza della Pilotta 4, 00187 Rome, Italy; caruana@unigre.it

Many people assume that serious reflection on animal ethics arose because of recent technological progress, the sharp rise in human population, and consequent pressure on global ecology. They consequently believe that this sub-discipline is relatively new and that traditional religions have little or nothing to offer. In spite of this however, we are currently seeing a heightened awareness of religion's important role in all areas of individual and communal life, for better or for worse. As regards our relations with nature in general and with animals in particular, and as regards the foundational idea of creaturehood, religious traditions have played, and are still playing, a central role in moulding the subliminal conscience of billions of people, guiding their moral dispositions that often remain unarticulated. This paper therefore explores our relation to animals by referring not only to the binary conceptual structure animality-humanity, as philosophers often do, but by referring also to the triple conceptual structure animality-humanity-divinity. After critically evaluating some of the relevant attitudes that derive from the major world religions, the paper tries to determine the extent to which the doctrine of these religions converge on some useful central principles regarding animal ethics and animal production. The result of this research supplies added support to the claim that the study of religious outlooks in this area serves to rediscover neglected perspectives and thereby to enlarge the horizon of current philosophical work.

The philosophy of animal rights

G.L. Francione[1,2]
[1]University of East Anglia, School of Politics, Philosophy, Language and Communication Studies, Norwich Research Park, Norwich NR4 7TJ, United Kingdom, [2]Rutgers University, Rutgers Law School, 123 Washington Street, Newark, NJ 07102, USA; gary.francione@rutgers.edu

A right is a way of protecting an interest. An interest is something we prefer, desire, or want. There are two basic ways in which we can protect interests. Those who subscribe to a consequentialist moral theory (of which there are different sorts) support protecting interests depending on whether the consequences of protecting those interests are better than the consequences of not protecting them. Those who subscribe to a rights theory claim that some interests are so important that we should protect them irrespective of consequences. To say that an interest is protected by a right does not mean that we protect those interests absolutely. There is a great deal of disagreement about what rights humans should have. But we all agree that all humans should have the right not to be chattel property. That is why we reject human slavery. If humans are property, all of their interests, from the most minor to the most fundamental, can be valued by an owner who has property rights in the human. Nonhuman animals, like humans, have interests. Depending on the species we are talking about, and depending on individuals within that species, those interests will vary. All sentient beings have at least two interests: the interest in not suffering and the interest in not dying. That is, although not all sentient beings may think about their lives in the same way, all of them desire or want to remain alive. The use of animals as property for food, clothing, and other purposes implicates at least two related, but different, interests that animals have. Using animals in the ways that we use them involves doing things to animals that they want, desire, or prefer that we not do: we cause them suffering and we kill them. Just as in the case of human chattel slavery, if animals are property, all of their interests, from the most minor to the most fundamental, can be valued by a human owner who has property rights in the animal. If animals matter morally, then we must follow the basic moral rule – the principle of equal consideration – and treat like cases alike. We must accord to animals the same right that we accord to all humans unless there is a good reason not to do so. We do not have a good reason to not accord all sentient beings the right not to be property. I will discuss the notion of animal rights as part of what I call the abolitionist approach to animal rights.

Modifying animals: beyond welfare

B. Bovenkerk
Wageningen University and Research, Hollandseweg 1, 6706 KN Wageningen, the Netherlands; bernice.bovenkerk@wur.nl

When discussing the humane treatment of animals, people tend to focus on the implications of our actions for animal welfare. However, in the areas of selective breeding and genetic modification of animals, arguments and concerns often go beyond welfare considerations. While less consensus exists about such considerations than about the importance of welfare, and they often go back to fundamental value differences, it is still important to publicly discuss them. Public discussion will help opponents understand the position of the other side and will bring to the fore considerations that are easily overlooked. Common objections to selective breeding and genetic modification of animals are that they violate animal integrity, that 'tinkering' with animals' genetic make-up is unnatural, and that it amounts to playing God. These objections are often dismissed out of hand – by scientists and ethicists alike – but there is more behind these concerns than meets the eye. The examination of three case-studies will make this clear: biotechnological procedures with animals for biomedical research, pedigree dog breeding, and genetic technologies in breeding of livestock.

Session 22

Theatre 5

Sustainable intensification of animal production: what does it mean?

P. Garnsworthy

University of Nottingham, School of Biosciences, Sutton Bonington, Loughborough LE12 5RD, United Kingdom; phil.garnsworthy@nottingham.ac.uk

The Foresight (2011) report on The Future of Food and Farming concluded that to support the growing population, 'global food supply will need to increase without the use of substantially more land and with diminishing impact on the environment: sustainable intensification is a necessity'. Sustainable intensification of animal production has to embrace the three main pillars of sustainability: economic, environmental, and social. Sustainable intensification has been interpreted differently by different stakeholders, who usually focus on social or environmental aspects. People who oppose animal production say that a sustainable food strategy should rely on intensive crop production, and that the grains, protein crops and fish products fed to animals should be diverted for direct human consumption. Such a strategy ignores the essential role of animals in global food security, whereby animals utilise non-arable land and agrifood byproducts to generate nutritious products. There are also ethical and environmental consequences of relying solely on plant-based diets. Many people believe that sustainable animal production means 'natural', 'free-range' or 'organic', which limits intensification. Often such beliefs derive from perceived ethical and welfare detriments of intensive livestock systems, whereas in reality these issues are independent of intensity. Livestock farming receives negative publicity regarding environmental impact. Impacts per unit area can increase with intensification, but recent studies have shown that boosting yields on existing farmland greatly reduces impacts on wild populations, and decreases environmental impact per unit of product. Intensification does not mean that future animals will only be produced in intensive systems. Intensification means increasing output per unit of input by making more efficient use of resources. Greater efficiency of any system usually leads to improved economics, lower environmental impact, and more food to meet societal demands, thus addressing all three pillars of sustainability. In conclusion, sustainable intensification of animal production means increasing efficiency of converting feed into high quality animal products, increasing profit, with high standards of animal welfare and minimal environmental impacts.

Session 23

Theatre 1

Exotic breeds and crossbreeding in developing countries

G. Leroy[1], C. Pena[1], R. Baumung[1], P. Boettcher[1], F. Jaffrezic[2] and B. Besbes[1]
[1]Food and Agriculture Organization, Viale delle Terme di Caracalla, 00153 Rome, Italy, [2]INRA, UMR 1313 GABI, Domaine de Vilvert, 78352 Jouy en Josas cedex, France; roswitha.baumung@fao.org

Many developing countries view introgression and crossbreeding with highly productive exotic breeds an interesting solution to improve the performance of local livestock. This approach can be faster and may demand less investment in capital, infrastructure and technical know-how than within-breed selection. In parallel, increased use of exotic breeds and indiscriminate crossbreeding are considered as major threats for local genetic resources. Nevertheless, little has been quantified on the extent of those two phenomena. Based on national censuses and breed population data from the Domestic Animal Diversity Information System, considering 10 species and 21 developing countries, we observed that the share of local breeds within species decreased by an average of 1% per year during the last 20 years. The corresponding increase in non-local animals was distributed between pure exotic breeds and crossbred animals. This tendency was significantly greater in Asia and Latin America than in African countries. On one hand, in several countries, increased use of exotic breeds and crossbreeding has been accompanied by increased production, in milk yield for instance. On the other hand, the cause and effect is unknown; It is speculated, that the tendencies observed may contribute to an increased proportion of locally adapted livestock breeds with declining population trends. Numerous differences were observed across countries and species basis of how crossbred animals are defined and monitored. This heterogeneity must be considered when assessing impacts of germplasm imports.

SimHerd Crossbred for estimating the economic effects of crossbreeding

M. Kargo[1,2], J.B. Clasen[3], J. Ettema[4], R.B. Davis[1] and S. Østergaard[2,4]
[1]SEGES, Agro Food Park 15, 8200, Denmark, [2]Aarhus University, Blickers alle 20, 8830, Denmark, [3]Swedish University of Agricultural Sciences, Ulls väg 26, 756 51, Sweden, [4]Simherd A/S, Niels Pedersens alle 2, 8830, Denmark; morten.kargo@mbg.au.dk

Crossbred dairy cows are often more robust and more profitable compared to purebred cows. However, the interest in crossbreeding dairy cattle has until recently been limited in developed countries in the Northern hemisphere. One of the reasons is assumed to be the lack of economic figures for the overall herd profit when applying systematic crossbreeding programs. SimHerd Crossbred is a supplement to the SimHerd simulation software. SimHerd is a mechanistic, dynamic and stochastic model that simulates the production and state changes of dairy cows and young stock in a dairy herd. In SimHerd Crossbred each animal is described by the breed composition and the degree of heterozygosity. The effect of crossbreeding is estimated based on a dominance model. Different systematic crossbreeding systems such as two- or three-breed rotational crossbreeding and terminal crossbreeding can be simulated. Also, the effect of using different breeds for crossbreeding can be evaluated. This requires knowledge of the breed level for a large range of traits where production, fertility, health, calving ease, and mortality are among the most important. Furthermore, the F_1-heterosis estimates for all traits and breed combinations are needed. The SimHerd Crossbred software has been used to estimate the effect of systematic rotational crossbreeding between Holstein, Nordic Red and Montbeliarde – the so-called ProCross. Compared to a Holstein herd, the ProCross herd produced 140 kg ECM less per cow year, had 7.5 percentage points lower replacement rate, 10% fewer mastitis treatments per cow and 7 percentage points higher conception rate. This increased the yearly herd profit by 130 Euro per cow in the ProCross herd. In addition, examples of other crossbreeding programs including other breeds will be presented.

Effect of crossbreeding and using new techniques on genetic merit and economy in Swedish dairy herds

J. Clasen[1], M. Kargo[2,3], S. Østergaard[4], W.F. Fikse[5], L. Rydhmer[1] and E. Strandberg[1]
[1]Swedish University of Agricultural Sciences, Department of Animal Breeding and Genetics, Box 7023, 75007 Uppsala, Sweden, [2]SEGES, Danish Agriculture & Food Council, Agro Food Park 15, 8200 Aarhus N, Denmark, [3]Aarhus University, Department of Molecular Biology and Genetics, 20 Blichers Allé, 8830 Tjele, Denmark, [4]Aarhus University, Department of Animal Science, 20 Blichers Allé, 8830 Tjele, Denmark, [5]Växa Sverige, 26 Ulls väg, 75651 Uppsala, Sweden; julie.clasen@slu.se

Crossbreeding, genotyping, sexed semen and beef semen can improve herd economy in different ways. The economic and genetic consequences of combining them are complex. We investigated 22 scenarios of average Swedish herds having only purebred Holstein or managing systematic terminal crossbreeding with Holstein purebreds and terminal F1 Swedish Red × Holstein crossbreds. The scenarios had different combinations of strategic use of sexed semen with or without genotyping of purebred heifers to select future replacement heifers based on genomic breeding values. Use of beef semen was included in all scenarios to produce calves for slaughter and limit the surplus of replacement heifers. All dairy crossbred females were serviced with beef semen, i.e. they had no influence in breeding. The scenarios were simulated using a combination of two simulation models: SimHerd Crossbred that simulates herd dynamics and ADAM that simulates the additive genetic herd level. Crossbreeding had positive effects on the herd profit compared to purebreeding, due to heterosis and increased income from beef × dairy slaughter calves. Interestingly, when large amounts of sexed semen were used, the purebred cows in the crossbreeding herds had higher genetic merit than in the purebreeding herds. Genotyping and sexed semen had clear positive effects on the genetic merit. Use of sexed semen allowed for more use of beef semen, which had positive effects on the herd profit. Crossbreeding reduced the number of purebred heifer calves to be genotyped, which reduced the genotyping costs. In general, any combination of the breeding tools were economically beneficial, compared to not using any of them. The most profitable scenario included crossbreeding, sexed semen in 50% of heifers and in 25% of cows, and genotyping of all purebred heifers.

A selection index to select cows for farm profitability in a commercial grazing herd of Uruguay

D. Laborde[1] and N. Lopez-Villalobos[2]
[1]El Pedregal, Cusuglionl 453, 0500 Ti nidad, Uruguay; [2]Massey University, Private Bag 11222, 4442 Palmerston North, New Zealand; dlaborde@adinet.com.uy

The objective of this study was to develop a selection index to rank Holstein-Friesian (HF), Jersey (J) and HF×J crossbred cows for farm profit in a commercial herd under grazing conditions of Uruguay with a seasonal calving. In the year 2017 the herd was comprised of 127 HF, 9 J and 1,207 HF×J crossbred cows, which were herd-tested, in average, three times during the year. Total lactation yields of milk (MILK), fat (FAT) and protein (PROT) were calculated for each cow-lactation using orthogonal random regression polynomials of order 3. Somatic cell score (SCS) was calculated as log2(somatic cell count) and an average value is obtained for each cow-lactation. Mating records were used to calculate days from start of mating period to conception (SMCO). Live weight (LW) of cows were recorded once in each lactation. Cows were scored each year for udder depth (UD), udder support (US), front teat placement (FT) and rear teat placement (RT) by a qualified inspector. In total there were 4,081 lactations from 2,371 cows recorded from 2015 to 2017. The cows were the progeny of 97 sires. Breeding values (BV) for MILK, FAT, PROT, LW, SMCO, SCS, UD, US, FT and RT were estimated with a multi-trait repeatability animal model that included the fixed effects of year, lactation number, breed group and deviation from herd median calving date, and the random effects of animal, cow permanent and residual error. The economic values for MILK, FAT, PROT and LW were derived from a farm model that evaluates the change in cow profit per unit change of the trait, considering revenues from milk and beef, and feed and farm costs. The economic values for the other six traits were estimated using desired relative economic weights. The index was calculated as $I=-0.074×BV_{MILK}+2.13×BV_{FAT}+7.25×BV_{PROT}-1.48×BV_{LW}-3.60×BV_{SMCO}-36.84×BV_{SCS}+30.86×BV_{UD}+19.67×BV_{US}+21.10×BV_{FT}+20.07×BV_{RT}$. The index has been used to select the best cows as mothers of future replacements and cull the cows with the lowest selection index value. The index has also been used for the selection and culling of heifer replacements.

Milk characteristics and cheese yield of Holsteins and 3-breed rotational (ProCross) crossbred cows

S. Saha, G. Bittante, N. Amalfitanò and L. Gallo
University of Padova, Department of Agronomy, Food, Natural resources, Animals and Environment (DAFNAE), Via dell'Università 16, 35020, Legnaro, Padova, Italy; sudeb.saha@studenti.unipd.it

Interest in crossbreeding of dairy cattle has grown in commercial dairy farm to improve the fertility, health and longevity respect to purebred Holstein (HO). Conversely, information on milk properties important for rennet behaviour, such as the size of fat globule and casein micelle, and on cheese yield properties of milk from crossbred (CROSS) cows are sparse. The aim of the present study (University of Padova, BIRD188213) was to investigate the effects of 3 breed rotational cross breeding (ProCross) of Holstein (HO) with sires of Viking Red in first (F1: VR×HO), Montbéliarde in second (F2: MO×F1), HO in third (F3: HO×F2) and again VR in fourth generation (F4: VR×F3) on milk characteristics and cheese yield (CY). Evening milk samples (n=102, 2,000 ml/cow) were collected from multiparous HO (n=33) and CROSS (n=69: 17 F1, 18 F2, 17 F3, 17 F4) cows reared in a commercial farm located in Northern Italy. Individual milk samples were assessed for milk composition, fat globule and casein micelle size and processed through a laboratory model-cheese manufacturing procedure to determine CY. Data were analysed using a linear mixed model including the random effect of date of sampling (5 dates) and the fixed effect of breed combination. Compared to HO, CROSS cows yielded 10% less milk, but showed increased protein and casein content (+3% and +6%, respectively) and lower SCS (2.53 vs 3.35). Milk from CROSS cows showed smaller average casein micelle size (131.0 vs 133.4 nm, P<0.01), and smaller volume-surface average diameter (d3,2: 3.23 vs 3.41 μm, P=0.02) and volume-weighted average diameter (d4,3: 3.76 vs 3.93 μm, P=0.10) of fat globules than HO cows. CY averaged 17.7% and was comparable for milk from HO and CROSS cows, likewise the proportion of solids and water in the curd and the recovery of nutrients from milk to curd. In conclusion, when compared to that from HO, milk from CROSS cows showed some advantages for milk quality and fat globule and casein micelle size, without alteration of CY. The trial is in progress to increase cow sample size.

Use of mate allocation in pig crossbreeding schemes: a simulation study

D. González-Diéguez[1], L. Tusell[1], A. Bouquet[2,3] and Z.G. Vitezica[1]
[1]GenPhySE, INRA, Université de Toulouse, 31326 Castanet-Tolosan, France, [2]IFIP Institut du Porc, BP 35104, 35651 Le Rheu, France, [3]France Génétique Porc, BP 35104, 35651 Le Rheu, France; david-omar.gonzalez-dieguez@inra.fr

One of the main goals in a crossbreeding scheme is to improve the performance of crossbred population by exploiting heterosis and breed complementarity. Dominance is one of the likely genetic bases of heterosis and, nowadays, estimating dominance effects in genetic evaluations has become feasible in a genomic selection context. Mate allocation strategies that account for inbreeding and/or dominance can be of interest for maximizing the crossbred performance. The objective of this study was to simulate scenarios including or not mate allocation strategies in two-breed pig crossbreeding schemes. The different crossbreeding scenarios have been compared in terms of genetic gain (within-breed) and total genetic value in crossbred populations. The benchmark scenario is a crossbreeding scheme where within-line selection is performed on purebred genomic estimated breeding values and crossbreds come from random matings of the best purebreds. The other subsequent scenarios are conceived to evaluate the potential benefits of accounting for inbreeding, dominance and crossbred performances in the genetic evaluation model. Genomic mate allocation is a promising strategy to improve the crossbred performance.

Meta-analysis for selection of feed efficiency in crossbred pigs

M.N. Aldridge[1], R. Bergsma[2] and M.P.L. Calus[1]
[1]Wageningen University & Research, Animal Breeding and Genomics, P.O. Box 338, 6700 AH Wageningen, the Netherlands, [2]Topigs Norsvin, P.O. Box 43, 6640 AA Beuningen, the Netherlands; michael.aldridge@wur.nl

We reviewed the literature and completed a meta-analysis of parameter estimates to determine which current and novel traits are promising indicator traits to select for feed efficiency of crossbred pigs. Feed conversion ratio (FCR) was the key trait of interest, average daily gain and daily feed intake were also included. For each following category a single trait was selected; digestibility (dry matter), microbiota (Alloprevotella), feeding behaviour (eating time per day), group records (group daily feed intake), welfare (joint lesions or total lesion count), indirect genetic effects (growth rate with social effect), and biomarkers (nitrogen excreted). A genetic correlation matrix was built with published results, and unknown correlations assumed as low (0.10). The nearest positive definite matrix and published genetic standard deviations were used to calculate a covariance matrix. To determine which traits have the most potential to improve FCR, 100% of selection was placed on each trait separately, until the selected trait increased by one genetic standard deviation. When selection was placed on FCR, it decreased by 7.1%, from 2.52 (kg/kg) to 2.34 (kg/kg). The other traits reduced FCR in descending order: dry matter digestibility (4.5%), daily feed intake (2.2%), average daily gain (1.9%), eating time per day (1.3%), nitrogen excreted (1.2%), group daily feed intake (0.8%), total lesion count (0.8%), Alloprevotella (0.7%), joint lesions (0.7%), and growth rate with social effect (0.7%). Current traits are likely to have the largest impact on feed conversion ratio. Including novel traits in selection, is likely to have a positive impact on FCR, particularly from digestibility, feeding behaviour, biomarkers, and group records. To determine an optimal index to select for feed efficiency, further analysis will use selection index theory with multi-trait selection. It will include additional traits from each category (including perturbations as a new category), estimate missing correlations with industry data, account for differences in genetic correlations between purebreds and crossbreds, and take into consideration ease and costs of measuring to prioritise traits.

Crossbred evaluations using ssGBLUP and algorithm for proven and young with distinct sources of data

I. Pocrnic[1], D.A.L. Lourenco[1], C.Y. Chen[2], W.O. Herring[2] and I. Misztal[1]
[1]University of Georgia, Animal and Dairy Science Department, 425 River Road, Athens, GA 30602, USA, [2]Genus PIC, 100 Bluegrass Commons Blvd, Hendersonville, TN 37075, USA; ipocrnic@uga.edu

Genomic selection is routinely applied to many purebred farm species but can be extended to predictions across purebreds as well as for crossbreds. This is useful for swine and poultry, for which selection in nucleus herds is typically performed on purebreds, whereas the commercial products are crossbreds. Single-step genomic BLUP (ssGBLUP) is a widely applied method that can use algorithm for proven and young (APY), that allows for greater computing efficiency by exploiting the theory of limited dimensionality of genomic information and chromosome segments (Me). This study investigates the predictivity as a proxy for accuracy across and within two purebred pig lines and their crosses, under the application of APY in ssGBLUP setup, and different levels of Me overlapping across populations. The data consisted of approximately 210k phenotypic records for two traits and more than 720k animals in pedigree. Genotypes for 43k SNP were available for 46k animals, from which 26k and 16k belong to purebreds, and 4k to crossbreds. The models included bivariate animal model with three lines evaluated as one joint line, and for each trait individually a three-trait animal model with each line treated as a separate trait. Both models provided the same predictivity across and within the lines. Using either of the pure lines data as a training set resulted in a similar predictivity for the crossbreeds. Across-line predictive ability was limited to less than half of the maximum predictivity for each line. For crossbreds, APY performed equivalently to direct inverse when the number of core animals was equal to the number of eigenvalues explaining 98-99% of the variance of G including all lines. Predictivity across the lines is achievable because of the shared Me between them. The number of those shared segments can be obtained via eigenvalue decomposition of genomic information available for each line.

Relevance of genotyping crossbred pigs for selection of nucleus purebred pigs for finisher traits

C.A. Sevillano[1], M.P.L. Calus[1], A. Neerhof[2], J. Vandenplas[1], E.F. Knol[2] and R. Bergsma[2]
[1]Wageningen University & Research Animal Breeding and Genomics, P.O. Box 338, 6700 AH Wageningen, the Netherlands, [2]Topigs Norsvin Research Center, P.O. Box 43, 6640 AA Beuningen, the Netherlands; claudia.sevillanodelaguila@wur.nl

Adding crossbred (CB) genomic information might improve estimation of breeding values (BV) of purebred (PB) individuals for CB performance, as it allows taking into account genetic differences in PB and commercial CB performance. Given that multiple breeding lines are involved in genomic prediction for CB performance, properly aligning pedigree and genomic relationships in single step approaches is especially important. A single-step approach that uses the concept of metafounder (MF) might be a potential approach to achieve this, and might help to improve the prediction of CB performance. We performed an analysis using a large dataset from a sire line, and their three-way CB offspring to show the added value of using CB information in the training population with or without the MF approach. There were 47,257 (6,670) PB pigs from the sire line, and 53,625 (4,186) CB offspring with phenotypic (genomic) information. The accuracy of the EBV of PB individuals for CB performance for average daily gain (ADG), average daily feed intake (ADFI), back fat (BF), and loin depth (LD) were calculated using different information sources for training: (1) PB phenotypes, genotypes, and CB phenotypes; (2) PB and CB phenotypes and genotypes; and (3) PB and CB phenotypes and genotypes using MF. To test the predictive ability of the models, we ignored the phenotypes from 142 youngest sires and their 15,137 CB offspring. For the 142 validation sires, we calculated the average of the individual CB offspring deviation to be used as true BV. When using only genomic information from PB, the accuracies were around 0.46-0.61. After including CB genomic information, prediction accuracy for ADG increased 1% while accuracies for ADFI, BF, and LD decreased 2 to 10%. When CB genomic information was included using MF, prediction accuracies for ADG, BF, and LD increased 3-4%, and decreased 2% for ADFI. Therefore, including CB genomic information seems to be beneficial for prediction accuracy, when genomic and pedigree information are properly aligned as achieved with the MF approach.

Optimizing genomic breeding program designs to improve crossbred performance

Y.C.J. Wientjes, P. Bijma and M.P.L. Calus
Wageningen University & Research, Animal Breeding and Genomics, P.O. Box 338, 6700 AH Wageningen, the Netherlands; mario.calus@wur.nl

The commercial production animals in pig and poultry systems are generally crosses between 3 or 4 purebred lines. Genetic improvement in those systems is generated by selecting purebred (PB) animals, although the breeding goal is focussing on crossbred (CB) performance. The aim of this study was to investigate whether a PB or CB reference population is optimal for genomic selection of the PB selection candidates for CB performance, and how this depends on: (1) the genetic correlation between PB and CB performance (r_{pc}); and (2) the genetic distance between the CB reference population and the PB selection candidates. We simulated a PB population consisting of 40 sires and 400 dams with a litter size of 6 per generation. This population was under genomic selection for PB performance for 8 generations. We compared the accuracy of predicting breeding values for CB performance of PB individuals in generation 9, using either a PB or 4-way CB reference population, each of 2,400 individuals. The CB reference population was generated after applying two, one or zero multiplication steps. This resulted in a minimum genetic distance of 9, 7, or 5 steps through the pedigree connecting the CB reference population to the PB selection candidates. We either used the same sires (top) or different (subtop) sires as used in the PB population to generate the CB populations. The accuracy of the PB reference populations was 0.31 at high r_{pc} (0.75), and 0.19 at low r_{pc} (0.5). The accuracy of the CB reference populations was similar for both r_{pc} values. Among the CB reference populations, the accuracy was always highest for the population generated using zero multiplication steps and using top sires (0.27). The other CB reference populations showed a similar accuracy (0.25), although the genetic distance with the PB selection candidates did differ across those populations. Altogether, those results indicate that the optimal design of a reference population to improve CB performance depends on the r_{pc} and on the genetic distance of the CB reference population to the PB selection candidates.

Potential of DNA pooling for the inclusion of commercial slaughterhouse data in genetic improvement

M.N. Aldridge[1], B. De Klerk[2], Y. De Haas[1] and K.H. De Greef[1]
[1]Wageningen University & Research, Animal Breeding and Genomics, P.O. Box 338, 6700 AH Wageningen, the Netherlands, [2]Cobb Vantress B.V., Koorstraat 2, 5831 GH Boxmeer, the Netherlands; michael.aldridge@wur.nl

Health traits are recorded in slaughterhouses on commercial animals for welfare and food safety. Here, methods are examined to link such health data to the breeding population, but it is not economical or logistically viable to identify or genotype individual crossbred broilers. One way to achieve this link between commercial health data and breeding animals, is with DNA pooling (genotyped sample which multiple individuals provided equal contributions of DNA). Footpad dermatitis of broilers is being used as a case study to test this. Before experimental work begins we want to ensure DNA pools are representative of individuals used to build the pool and that pools are sufficiently different from each other. A typical broiler population was simulated, with pooled genotypes (estimated from allele frequencies) to model DNA pools. From a four-way cross, 25 animals were randomly selected from each the top and bottom 10% of phenotypes. High and low genotype pools of 25, ten, and five animals were built. Non-segregating SNPs and SNPs with the same allele frequency in both pools were filtered. To estimate relatedness within and between pools a G matrix was built. Preliminary analysis indicates a high relatedness of individuals within a pool and with the corresponding pooled genotype, with a lower relatedness between the high and low pools. Pools of five or ten individuals, had a 100% accuracy when matching individuals to the correct genotype pool. In the pools of 25 individuals, 92% were matched to the correct genotype pool. Genotype pools are representative of individuals used to build them, and there is a higher relationship within pools than between pools. Smaller pools are suggested but the optimum size is to be determined. Relatedness between pools and ancestors will be used to identify pure line sires. We are confident that these results will be repeatable with real DNA pools, built from samples we collected based on footpad dermatitis. This assumes equal contributions of DNA, but we will also explore the effect of unequal contributions. A slaughter trial will be conducted to test genomic prediction using DNA pools.

Genomic predictions in Norwegian and New Zealand sheep breeds with similar development history

H.R. Oliveira[1], J. McEwan[2], J. Jakobsen[3], T. Blichfeldt[3], T. Meuwissen[4], N. Pickering[5], S. Clarke[2] and L.F. Brito[1]
[1]Purdue University, 270 Russell St, IN 17907, USA, [2]AgResearch, 176 Puddle Alley 9092, New Zealand, [3]NSG, Box 104, 1431 Ås, Norway, [4]Norwegian University of Life Sciences, Universitetstunet 3, 1430 Ås, Norway, [5]Focus Genetics, 17 Mahia St, 4144, New Zealand; hrojasde@purdue.edu

The prediction of accurate molecular breeding values (mBVs) is dependent on various parameters, including the size of training population and genetic relationship between training and prediction populations. Various Norwegian (NOR) and New Zealand (NZ) sheep populations were developed by crossing multiple breeds, with overlapping founder breeds. The main goal of this study was to describe the genetic diversity and connectedness between NOR and NZ sheep populations, and validate mBVs predicted for NOR animals using the NZ training population. A total of 47,056 NZ animals were either genotyped with a high-density (HD; 606k SNPs) panel or with a low-density (LD) panel and imputed to HD. 828 Norwegian White sheep rams were genotyped on the HD panel. A PCA-plot based on the HD genotypes showed no clear evidence of diverging clustering among animals, suggesting a genetic similarity between populations from both countries. Estimated breeding values (EBVs) for weaning weight (WW_{NOR}), carcass weight (CW_{NOR}), EUROP carcass classification and EUROP fat grading were predicted for NOR animals. In addition, mBVs for WW_{NZ}, pre-slaughter weight, CW_{NZ}, x-ray CW_{NZ}, carcass fatness at the GR site, ultrasound fat depth and ultrasound eye muscle depth were predicted for NOR animals using the NZ training population. EBVs and mBVs for NOR animals were compared based on the Pearson correlation coefficient adjusted for the average EBV reliability. Pearson correlation coefficients ranged from 0.19 (between WW_{NOR} and WW_{NZ}) to 0.40 (between CW_{NOR} and x-ray CW_{NZ}). Considering the differences in trait definition between countries, these findings indicate a promising opportunity to perform across country genomic predictions in composite sheep populations. Further analyses will be performed to better understand the genetic relationship among the aforementioned populations, as well as alternative prediction methods.

Analysis of effective population size and inbreeding in composite beef cattle

B. Abreu Silva, F. Bussiman, F. Eguti, B. Santana, A. Acero, L. Grigoletto, J.B. Ferraz, E. Mattos and J.P. Eler
University of São Paulo, Av. Duque de Caxias Norte,225, 13635900, Brazil; jbferraz@usp.br

The Montana beef cattle is a composite breed developed with the objective to make use of the heterosis and complementarity advantages of adapted animals breeds to the tropics, making production systems more efficient. The aim of this study was to determine its effective population size (Ne) and the rate of inbreeding. The data contains pedigree information of 1,397,180 animals born between 1970 and 2018. The analyses were performed in a software system (PopRep). Five methods were used to assess the Ne estimates (based on (1) census information; (2) the rate of inbreeding; (3) parents generation; (4) logarithmic regression of one minus inbreeding; (5) equivalent complete generations). The rate of inbreeding per generation was -0,10% per year, which evidences the heterosis retention across the generations since the transmission of inbreeding was negative. Ne estimates were 3,768, 1,858, 440, 953, 400 for the 5 methods, respectively; these estimates may indicate the true Ne ranges near to 1000 effective animals. Surprisingly, although the pedigree is quite large, in some generations, the observed completeness can be influenced by the entrance of animals with unknown genealogy, since in the formation of the composite the inclusion of several breeds is allowed, which can overestimate the effective size. Other studies are being developed to determine the population structure of this composite beef cattle.

Predicted performance for pre-weaning traits and cow-calf efficiency in a beef crossbreeding program

G.M. Pyoos[1,2], M.M. Scholtz[1,2], M.D. Macneil[2], A. Theunissen[2,3], P.D. Vermeulen[2] and F.W.C. Neser[2]
[1]Agricultural Research Council, Animal Production, Private Bag X02 Irene, 0062, South Africa, [2]University of the Free State, Animal Wildlife and Grassland Sciences, P.O. Box 339, 9300 Bloemfontein, South Africa, [3]Northern Cape Department of Agricultural, Land Reform and Rural Development, Private Bag X9, Jan Kempdorp, 8550 JanKempdorp, South Africa; neserfw@ufs.ac.za

Crossbreeding systems allow for use of favourable effects of heterosis and the opportunity to use breeds whose additive effects offset each other's weaknesses. The objective of this study was to predict levels of cow-calf production that may be achieved through top-cross, two-breed rotation, and terminal sire crossbreeding systems. The use of systems analysis to examine these mating systems, taken to equilibrium breed composition and stabilized heterosis, replaced a much lengthier and more costly experiment that would be otherwise required. Calves, sired by Afrikaner (AF), Bonsmara (BN), Nguni (NG), Angus and Simmental bulls and out of AF, BN, and NG dams, were weaned from 2015 to 2017. Predicted performance of the various crossbreeding systems followed standard regression theory. At birth, the Nguni calves weighed less than either Bonsmara or Afrikaner calves. The Nguni calves also grew slower from birth to weaning than Bonsmara calves. As a consequence of these effects, Bonsmara calves were heavier at weaning than either Afrikaner or Nguni calves. The calves from Nguni dams were older at weaning than those from Bonsmara dams (210 days versus 198 days). The Nguni-Afrikaner criss-crosses produced approximately 12 kg less weaning weight relative to the other pairs of breed combinations. Overall, the straightbred breeding system was least efficient, followed by the criss-cross system (+2%) and the terminal sire system that utilized Simmental (+4%), with the terminal sire system utilizing Angus being on average most efficient (+8%). The severe drought and extreme heat of the 2015/2016 summer season had a marked effect on the pre-weaning performance of the calves (preceding year ADG=879±27 g/d., 2015/2016 ADG=687±28 g/d, following year ADG=770±28 g/d). Calves sire by Afrikaner bulls were less affected by the year effects than their contemporaries sired by Bonsmara, Nguni, Angus and Simmental bulls.

Additive and dominance genomic parameters for backfat thickness in purebred and crossbred pigs

M. Mohammadpanah[1], M. Momen[2], H. Gilbert[3], C. Larzul[3], M.J. Mercat[4], A. Mehrgardi[1] and L. Tusell[3]
[1]Shahid Bahonar University of Kerman, Dept of Animal Science, Faculty of Agriculture, 76169133 Kerman, Iran, [2]Virginia Polytechnic Institute and State University, Department of Animal and Poultry Sciences, VA, 24061 Blacksburg, USA, [3]INRA, GenPhySE, 24 chemin Borde Rouge, 31326 Castanet-Tolosan, France, [4]IFIP/ALLIANCE R&D, La Motte au Vicomte, 35651 Le Rheu, France; llibertat.tusell-palomero@inra.fr

In pig crossbreeding programs, genetic evaluation has been based predominantly on purebred data accounting only for additive genetic effects, whereas improving crossbred performance is the ultimate goal. Theoretically, a combined crossbred and purebred selection method is advised if genetic correlation between purebred and crossbred populations differ from unity. If dominance effects are large enough, assortative mating strategies can enhance the total genetic values of the offspring. Hence, estimates of genetic parameters for purebreds and crossbreds are needed to assess the best selection crossbreeding scheme strategies. In this study, additive and dominance genetic variance components and additive and dominance genotypic correlations between a Piétrain and a Piétrain × Large White populations were estimated for backfat thickness (BFT). A total of 607 purebreds and 620 crossbred BFT records were analysed with a genotypic bivariate model that included hot carcass weight and inbreeding coefficient as covariates, an additive and a dominance genotypic effects, and a pen nested within batch random effect. Genetic parameters were estimated with EM-REML plus an additional iteration of AIREML to obtain the asymptotic standard deviations of the estimates. The additive genotypic correlation between purebreds and crossbreds was high, 0.82, indicating that the genetic progress attained in the purebreds can mostly be transferred to the crossbreds. Dominance genetic variance represented about 10% of the BFT phenotypic variance in both populations, suggesting that assortative matings could slightly enhance both purebred and crossbred performances. However, the underlying genetic mechanisms responsible for the dominance effects could differ between populations since dominance genotypic correlation was 0.49.

Effects of aging on epigenetics of immune cells in dairy cattle

H. Jammes[1], A. Chaulot-Talmon[1], C. Pontelevoy[1], L. Jouneau[1], C. Richard[1], G. Foucras[1] and H. Kiefer[2]
[1]UMR1198 BDR, INRA, ENVA, Université Paris Saclay, 78352 Jouy en Josas, France, [2]UMR1225 IHAP, Université de Toulouse, ENVT, INRA, 31076 Toulouse Cedex 3, France; helene.jammes@inra.fr

Today Holstein dairy cows have an average number of lactations of only 2.5. infertility and higher susceptibility to mastitis remains the main reasons for culling, highlighting the existing trade-offs between production, health and fertility in dairy farms. The immune ability depends on a long process of cell differentiation, involving epigenetic mechanisms, such as DNA methylation, and activities of specific transcription factors to timely and precisely regulate the gene expression. In this study, genome-wide DNA methylation of purified monocytes was studied in relation to the age of cows. To control for the variability due to genetic background of animals, and thus precisely characterize the contribution of epigenome alteration, we benefited from the exceptional resources of cloned cows of different ages obtained from the same cell line (unique genotype; 2 groups, 6-7 year-old cows and 15-17 years-old cows; n=7 per group). All cows were born in an experimental INRA farm, raised under the same conditions limiting environmental changes, and all animal husbandry information was collected. Blood samples were collected 15 days after ovulation. The monocyte subpopulation was purified using positive selection with CD14 magnetic beads. Reduced Representation Bisulfite Sequencing was performed and sequences analysed using our dedicated pipeline. Using the set of 14 libraries, we identified original differentially methylated cytosines (DMCs) highlighting methylation differences induced by aging and associated with relevant genes for immune function. In the long-term, identification of epigenetic marks could serve to reveal new indicators of the immune ability of animals in relation to intrinsic (health, fertility, age) and extrinsic factors (infection, nutrition, breeding practices, climate). These markers may help also to understand a part of missing causality in the relationship between genetic information, environment and immunity and to better characterize the robustness of animals.

Long-term impact of early life nutrition on DNA methylation patterns in bull sperm

J.P. Perrier[1], D.A. Kenny[2], C.J. Byrne[1], A. Chaulot-Talmon[3], E. Sellem[4], L. Schibler[4], H. Jammes[3], P. Lonergan[5], S. Fair[1] and H. Kiefer[3]
[1]University of Limerick, Laboratory of Animal Reproduction, Department of Biological Sciences, Schrodinger Building, V94 T9PX Limerick, Ireland, [2]Teagasc, Animal and Bioscience Department, Grange, C15 PW93 Dunsany, Ireland, [3]INRA, UMR1198, Biology of Development and Reproduction, Domaine de Vilvert, batiment 230, 78352 Jouy-en-Josas, France, [4]Allice, 149 rue de Bercy, 75595 Paris, France, [5]University College Dublin, School of Agriculture and Food Science, Belfield, D04 F6X4 Dublin, Ireland; helene.kiefer@inra.fr

Recent data from our group demonstrated that offering calves a high plane of nutrition in early life hastens the age at puberty onset by approximately 1 month, without resulting in any latent effects on measures of post-pubertal semen quality. As diet can influence DNA methylation profiles in sperm, the goal of this study was to ascertain if early life calf nutrition could influence DNA methylation patterns in sperm produced post-puberty. Holstein-Friesian calves were assigned to a high (H) or low (L) plane of nutrition for the first 24 weeks of age, then reassigned to the L diet until puberty, resulting in LL and HL groups. Sperm DNA methylation patterns from a sub-group of bulls in the HL (15 months of age; n=9) and in the LL (15 and 16 months of age; n=7 and 9, respectively) were obtained using Reduced Representation Bisulfite Sequencing. Both 15 and 16 months were selected in the LL treatment as these bulls reached puberty approximately 1 month after the HL treatments. Principal Component Analysis and hierarchical clustering demonstrated that inter-individual variability unrelated to diet or age dominated DNA methylation profiles. While the comparison between 15 and 16 months of age revealed almost no change, 2,932 Differentially Methylated CpGs (DMCs) were identified between the HL and LL groups. DMCs were mostly hypermethylated in the HL group, and enriched in introns, intergenic regions and CpG shores and shelves. Moreover, DAVID analysis using the genes associated with DMCs revealed a significant enrichment for the ATP-binding process. Together, these results demonstrated that enhanced plane of nutrition in pre-pubertal calves induced moderate but long-lasting changes in sperm DNA methylation profiles after puberty.

Genetic and non-genetic factors determine DNA methylation patterns in bull spermatozoa

J.-P. Perrier[1], E. Sellem[1,2], L. Jouneau[1], M. Boussaha[3], C. Hozé[2,3], S. Fritz[2,3], A. Aubert[1], C. Le Danvic[2,4], D. Boichard[3], L. Schibler[2], H. Jammes[1] and H. Kiefer[1]

[1]INRA, ENVA, Université Paris Saclay, UMR BDR, Domaine de Vilvert, 78350 Jouy-en-Josas, France, [2]ALLICE, 149 rue de Bercy, 75012 Paris, France, [3]INRA, AgroParisTech, Université Paris Saclay, UMR GABI, Domaine de Vilvert, 78350 Jouy-en-Josas, France, [4]CNRS/USTL, UGSF – UMR8576, 50 avenue de Halley BP 70478, 59658 Villeneuve d'Ascq Cedex, France; helene.kiefer@inra.fr

Spermatozoa have a remarkable epigenome in line with their degree of specialization, their unique nature and different requirements for successful fertilization. In humans and model species, alterations of sperm DNA methylation have been reported in cases of spermatogenesis defects, male infertility and exposure to toxics or nutritional challenges, suggesting that a memory of environmental or physiological changes is recorded in the sperm methylome and could potentially be transmitted to the embryo. On the other hand, DNA methylation is directly affected by the CpG content of the genome and its alteration by DNA sequence polymorphism. While bull semen is widely used in artificial insemination (AI) in a context of genomic selection of sires, there are no data about the respective contributions of genetic and non-genetic factors to the determination of DNA methylation patterns in bull spermatozoa. We characterized the sperm DNA methylome of 55 AI bulls using reduced representation bisulfite sequencing (RRBS). Five breeds were represented (Holstein, Montbéliarde, Normande, Abondance, Charolais). For the Montbéliarde breed, we investigated whether fertility biomarkers could be identified in the sperm methylome. Based on field fertility indicators, we compared fertile ejaculates, subfertile ejaculates and ejaculates from bulls with a disappointing career, with genomic predictions (n=10/group). The phenotypic characterization revealed no significant difference except for the proportion of motile sperm, which was reduced in disappointing bulls. We identified 4,823 fertility-related differentially methylated CpGs (DMCs) and 37,962 breed-related DMCs. An integrative approach was conducted to identify CpGs and SNPs in correlation, which revealed a complex interplay between the genome and the epigenome.

Gene expression of differentially-methylated genes associated with fertility trait in sheep

O.E. Othman and H. Shafey

National Research Centre, Cell Biology Department, El Buhouth St., Dokki, 12622, Egypt; othmanmah@yahoo.com

Fecundity is one of the most important economic traits for sheep meat production. Therefore, the increasing in the sheep fecundity will dramatically increase the economic benefits. The reproductive traits have low heritability and do not show a remarkable response to phenotypic selection. So the investigation of genetic information associated with reproductive ability can enhance the selection. DNA methylation is an epigenetic factor plays an essential role in mediating biological process including the gene expression of fecundity-related genes. This work aimed to assess the gene expression of some differentially-methylated genes associated with fertility trait in sheep. Total RNA was extracted from ten ovarian samples – five from high fecundity and five from low fecundity sheep groups – using the TRIzol reagent and dissolved in RNase-Free water. The quality and quantity of RNA were determined using a NanoDrop instrument. cDNA were synthesized from total RNA using a reverse transcription reagent kit with gDNA Eraser. The synthesized cDNA was subjected to qRT-PCR to assess the gene expression level of fifteen genes using specific primers and SYBR Green Master mix. The expression level of each gene was standardized to GAPDH as a housekeeping gene and calculated using $2^{-\Delta\Delta Ct}$ method. The results showed that ten out of fifteen tested differentially-methylated genes; SCY1 like pseudokinase 1 (SCY1), mammalian target of rapamycin (mTOR), ATP binding cassette subfamily G member 2 (ABCG2), Proteasome 26S subunit, Non-ATPase 7 (PSMD7), Serine/Threonine kinase 3 (STK3), Neurotensin (NTS), Urocortin-3 (UCN3), Gonadotropin releasing hormone 1 (GnRH1), Corticotropin releasing hormone (CRH) and Gonadotropin releasing hormone receptor (GnRHR) were significantly highly expressed in ovaries of high fecundity than those in low fecundity sheep groups. This results confirmed that role of DNA methylation as an epigenetic factor affecting on fertility trait in sheep.

Genetic determinism of DNA global methylation rate in sheep

L. Drouilhet[1], F. Plisson-Petit[1], D. Marcon[2], F. Bouvier[2], C. Moreno[1], S. Fabre[1] and D. Hazard[1]
[1]INRA, UMR1388 GenPhySE, 24 chemin de Borde Rouge, 31320 Castanet Tolosan, France, [2]INRA, UE0332 Domaine de la Sapinière, 18390 Osmoy, France; stephane.fabre@inra.fr

Recent studies highlighted that DNA or histone biochemical modifications, called epigenetic marks, influence adaptation and production traits. However, whether the epigenetic variations are under a genetic determinism or not remains unknown. Thus, we have considered DNA global methylation rate (DGMR) as a new phenotype in sheep. DGMR was obtained by LUminometric Methylation Analysis (LUMA) and its variability was explored in blood cells and other various tissues, within breed and between breeds. Thirty lambs from two sheep breeds (Romane and Martinik Black Belly; 16 males and 14 females) were bred simultaneously to share a common environment. At the age of 5 months, they were slaughtered and seventeen tissues per animal were collected. Concerning the non-reproductive tissues, the sex, breed and tissue effects and their interactions were tested. The tissue effect was highly significant ($P<0.0001$), the sex effect was only significant for the pancreas and the kidney cortex (sex×tissue interaction; $P=0.03$), whereas the breed effect was never significant. Concerning the reproductive tissues, the breed and tissue effects were significant ($P=0.02$ and $P<0.0001$, respectively), the DGMR in Martinik Black Belly being higher than in Romane, whatever the tissue. The correlations between the DGMR measured in blood cells and other tissues were low or not significant, indicating that blood DGMR would not be a predictor of the DGMR of tissue less accessible. The proof of concept of the existence of a genetic determinism of DGMR considered as a quantitative trait was obtained through LUMA measurement of 940 DNA blood samples from Romane lambs genotyped on the 50k SNP chip. Blood DGMR was variable between animals with an average of $70.7\pm6.0\%$. Among the tested effects, the sex and the ram effects were significant. Blood DGMR was heritable (0.20 ± 0.05) and joint analyses combining linkage disequilibrium and linkage revealed two main genomic regions influencing this new phenotype. Such results indicate that blood DGMR could be used in genetic selection, but before, further analyses are in progress to investigate relationship between blood DGMR and productive or adaptive traits.

Genome-wide DNA methylation profiles of pig testis reveal epigenetic markers/genes for boar taint

X. Wang and H.N. Kadarmideen
Technical University of Denmark, Department of Applied Mathematics and Computer Science, Richard Peterson Plads, Building 324, 2800 Kongens Lyngby, Denmark; hajak@dtu.dk

Boar taint (BT) is an offensive flavour observed in non-castrated male pigs (boars) that reduces the market price of the carcass. Surgical castration effectively avoids the taint but is associated with animal welfare concerns. The functional annotation of farm animal genomes (FAANG) for understanding the biology of complex traits and diseases can be used in the selection of breeding animals to achieve favourable phenotypic outcomes. The characterization of pig epigenomes/methylation changes between animals with high and low boar taint and genome-wide epigenetic markers that can predict boar taint are lacking. Reduced representation bisulfite sequencing (RRBS) of DNA methylation patterns based on next generation sequencing (NGS) is an efficient technology used to identify candidate epigenetic biomarkers associated with BT. Three different BT levels were analysed using RRBS data to calculate the methylation levels of cytosine. The co-analysis of differentially methylated cytosines (DMCs) and differentially expressed (DE) genes identified 32 significant co-located genes. The joint analysis of GO terms and pathways revealed that the methylation and gene expression of seven candidate genes were associated with BT; in particular, FASN plays a key role in fatty acid biosynthesis, and PEMT might be involved in oestrogen regulation and the development of BT. This study is the first to report the genome-wide DNA methylation profiles of BT in pigs using NGS and summarize candidate genes associated with epigenetic markers of BT, which could contribute to the understanding of the functional biology of BT traits and selective breeding of pigs against BT based on epigenetic biomarkers.

Influence of genetic background on the bovine milk microRNA composition

S. Le Guillou[1], J. Laubier[1], A. Leduc[1], F. Launay[2], S. Barbey[2], R. Lefebvre[1], M.-N. Rossignol[1], S. Marthey[1], D. Laloë[1] and F. Le Provost[1]
[1]INRA, UMR1313 GABI AgroParisTech Université Paris-Saclay, 78350 Jouy-en-Josas, France, [2]INRA, UE 326 Le Pin au Haras, 61310 Exmes, France; sandrine.le-guillou@inra.fr

The concept of milk as a healthy food has opened the way of studies on milk components, including macro- and micronutrients, as well as a novel class of molecules with broad regulatory properties, the microRNAs. Their presence in large quantities in milk has led to focus attention on their potential role on health. Some studies suggested that microRNAs present in milk could: (1) affect functions such as immunity, growth, development, cell proliferation and apoptosis; (2) be transmitted from mother to infant; (3) have a potential action across species; and (4) influence milk effects on consumer health. Intrinsic and stable components of milk, their comprehensive identification and characterization have been performed in several species, such as human and bovine, and depends on the stages of lactation. In that context, we have studied milk microRNA composition variations according to genetic variables. We have compared milk miRNomes of Holstein and Normande dairy cows, two breeds with contrasted lactation performances. We have performed high throughput small RNA sequencing on milk samples from 10 mid-lactation primiparous cows for each breed. microRNAs contributing more than 1% of the milk miRNomes are the same in the two breeds. Our study reveals the presence of 182 microRNAs with a significantly differential level between breeds. We previously showed a direct relation between microRNAs expression in the mammary gland and their level in milk. The milk miRNomes have been compared to mammary miRNomes already performed on dairy (Holstein, Montbéliarde) and beef (Limousine) cattle breeds. Variations according to breed in milk were also observed in mammary gland for five microRNAs. For the first time, this study allowed to evaluate genetic variations of the bovine milk miRNomes according to the breeds. These variations may have several impacts on offspring and consumers and thus requires particular attention. The presence of microRNAs in milk opens a line of investigation to use them as biomarkers for downstream consequences of health, physiological or metabolic status, in animals and humans.

RumimiR: a detailed microRNA database focused on ruminant species

C. Bourdon, P. Bardou, E. Aujean, S. Le Guillou, G. Tosser-Klopp and F. Le Provost
INRA, UMR GABI, INRA, Domaine de Vilvert, 75338, France; fabienne.le-provost@inra.fr

In recent years, the increasing use of Next Generation Sequencing technologies to explore the genome has generated large quantities of data. For microRNAs, more and more publications have described several thousand sequences, all species included. In order to obtain a detailed description of microRNAs from the literature for three ruminant species (bovine, caprine and ovine), a new database has been created: RumimiR. To date, 2,887, 2,733 and 5,095 unique microRNAs of bovine, caprine and ovine species, respectively, have been included. In addition to the most recent reference genomic position and sequence of each microRNA, this database contains details on the animals, tissue origins and experimental conditions available from the publications. Identity with human or mouse microRNA is mentioned. The RumimiR database enables data filtering, the selection of microRNAs being based on defined criteria such as animal status or tissue origin. For ruminant studies, RumimiR supplements the widely used miRBase database by browsing and filtering using complementary criteria, and the integration of all published sequences described as novel. The principal goal of this database is to provide easy access to all ruminant microRNAs described in the literature.

Modulation of gene expression by CNVs

M. Mielczarek[1,2], M. Frąszczak[2] and J. Szyda[1,2]
[1]National Research Institute of Animal Production, Krakowska 1, 32-083, Poland; [2]Wrocław University of Environmental and Life Sciences, Department of Genetics, Kozuchowska 7, 51-631, Poland; magda.mielczarek@upwr.edu.pl

Genetic polymorphisms may contribute to the variation of gene expression in multiple ways including gene dosage effects, gene coding region disruption and deletion or duplication of regulatory sequences. This project aimed to characterize the impact of copy number variations (CNVs) located within coding sequences on the level of gene expression. For a single Duroc × Pietrain female pig, genomic DNA extracted from white blood cells and RNA obtained from skeletal muscle were taken from the SRA database (PRJNA354435). Additionally, RNA-seq for 168 Duroc × Pietrain individuals were investigated (PRJNA403969) to measure the baseline gene expression level (expression of transcripts not affected by CNVs). Two bioinformatics pipelines were performed. (1) CNV detection pipeline applied for genomic data included: (i) the raw data editing based on reads quality (ii) alignment of reads against the reference genome; and (iii) CNV detection and validation. (2) RNA-seq data were (i) edited based on reads quality; and (ii) the level of gene expression was quantified. As a final step, regression models were used to investigate the impact of CNV on gene expression. Depending on the genomic location of CNV (coding or non-coding regions) and the length of polymorphism occupied those regions we found significant differences in gene expression levels. In conclusion, CNVs are important modulators of gene expression.

A generalized relationship matrix suitable for parent-of-origin analyses

N. Reinsch, I. Blunk and M. Mayer
Leibniz-Instiutute for Farm Animal Biology, Institute of Genetics and Biometry, Wilhelm-Stahl-Allee 2, 18196 Dummerstorf, Germany; reinsch@fbn-dummerstorf.de

The gametic relationship matrix has been shown to be a useful tool for modelling the presence of genomic imprinting effects, or, more generally, parent-of-origin effects. Gametic models that account for all possible kinds of genomic imprinting – maternal or paternal, full or partial – require four gametic effects per pedigree member. In cases where phenotypes are only recorded in non-parents, as frequently the case in meat animals, a substantial smaller number of equations for genetic effect are sufficient for the same purpose, if a special kind of reduced model is applied. In all other circumstances, however, the gametic model is the only available choice that is associated with a high number of equations for random genetic effects. As a general alternative that is designed to alleviate the curse of high equation numbers and which can favourably adapted to a variety of analysis needs we propose a share of the individuals from a given pedigree to be represented by a single genetic effect – their transmitting ability – and all others by two gametic effects. The kind of representation therein can freely be chosen by the data analyst according to current needs. The covariance of the resulting mixture of genetic effects can then be represented by a generalized kind of relationship matrix. Rules for setting up the inverse of this matrix directly from a pedigree file are presented in detail. These rules in particular have to take into account through what kind and by how many genetic effects the parents of an individual are represented in the model. Along with details on direct inversion illustrations of cases where the application of this matrix is thought to be beneficial in the context of imprinting analyses – including maternal effects – are given.

Transgenerational benefits of extruded linseed supply to dairy cows on the reproductive performance

J.M. Ariza[1], T. Meignan[1,2], A. Madouasse[1], G. Chesneau[2] and N. Bareille[1]
[1]BIOEPAR, INRA, Oniris, La Chantrerie, 44307, Nantes, France, [2]Valorex, La Messayais, 35210, Combourtillé, France; nathalie.bareille@oniris-nantes.fr

Future dairy cows are conceived during the maternal lactation, a productive period characterized by a large demand for energy and nutrients. Indeed, during this period embryo and foetal development closely compete with the large metabolic requirements of mammary gland for milk production. Supplementing extruded linseed (EL), a feed rich in n-3 fatty acids, to pregnant dairy cows may help to enhance their performance and to prevent negative metabolic imbalances. This retrospective study investigated whether the reproductive performance of primiparous cows is affected by the environment of their dam during pregnancy. To investigate the potential dose-effect relationship during pregnancy, we defined a proxy of EL intake per day by using deliveries of EL based feeds from 22 companies in the study period 2008-2015 in France. Based on recommendations for EL use on the field, 2 levels of exposures of the dam during the first pregnancy semester were created: low [<300] (n=7,186 AIs) and high [>300] (n=10,845 AIs) g/cow/d. The reference population was composed of primiparous cows issued from pregnant dams that did not receive EL in the same herds than exposed dams (n=25,557 AIs). Mean daily EL intake in exposed pregnant dams population was 324 g/cow/d (±226.1). Reproductive performance was studied on 43,588 AIs from 1,042 herds and 29,335 primiparous cows using Cox models for days to first AI and days to conception, and logistic regression models for risk of return-to-service, adjusted for factors likely to influence the reproductive performance and for a herd random effect. Risk of return-to-service between 18 and 78 days after first and second AI did not differ between cows issued from exposed dams or from reference dams. Days to first AI was significantly reduced in cows from exposed dams, whatever the exposure level (HR: 1.09 for the low EL level and HR: 1.08 for the high EL level). The effect was the same for time from calving to conception (HR: 1.10 for the low EL level and HR: 1.07 for high EL level). This original large-scale epidemiological study provides new insights into potential epigenetic effects of feeding EL during the foetal development of dairy cows.

Hypoxia during embryo development induced phenotypic plasticity and affected post hatch performance

A. Haron[1,2], D. Shinder[2] and S. Druyan[2]
[1]The Hebrew University of Jerusalem, The Robert H. Smith Faculty of Agriculture, Food and Environment, Rehovot, 76100, Israel, [2]ARO The Volcani Center, Institute of Animal Sciences, Dept. of Poultry and Aquaculture Sciences, P.O. Box 6, Bet-Dagan 50250, Israel; shelly.druyan@mail.huji.ac.il

The continuous genetic selection for performance traits resulted in a considerable enhancement of daily feed consumption, leading to alterations in growth mechanisms and development. These developments were not accompanied by the necessary increases in the size of the cardiovascular and respiratory systems, nor sufficient enhancements of their functional efficiency. Thus, modern broilers have an elevated metabolic rate and consequently elevated internal heat production that leads to insufficient maintenance of dynamic steady-state of thermoregulation processes, resulting in enhancement of body temperature fluctuations. The scientific assumption of this study was that a preferred and efficient adaptive response to oxygen shortage during the critical phases of embryonic development may induce alterations in the thermoregulatory control system. The aim of this study was to investigate the effect of hypoxic exposure, from E16 to E18 (the plateau period) for 12 h/d or constant 48 h on metabolism and growth in the embryo and in the post-hatch chick up to marketing age. Daily hypoxic exposures of 12 h or continuously for 48 h to hypoxia of 17% oxygen during the Plateau period (E16-E18), caused a decrease in the metabolic rate of the embryo. Lower metabolic rate leads to lower heat production and enables the growing broiler to invest less energy in maintenance and to allocate the remaining metabolic energy to growth. In addition, this adaptation had a long-lasting effect by improved thermotolerance of broiler chickens exposed to either long term high environmental temperature (32 °C) or cycling chronic hot conditions (32/24 °C).

Prenatal environment affects preweaning milk intake in dairy cattle: preliminary results
K. Verdru, M.L. De Vuyst, M. Hostens and G. Opsomer
Faculty of Veterinary Medicine, Department of Reproduction, Obstetrics and Herd health, Salisburylaan 133, 9820 Merelbeke, Belgium; karel.verdru@ugent.be

Optimizing the productivity of dairy cattle has become more important than ever. Nulliparous heifers are being bred at a young age and multiparous cows generally produce large amounts of milk during gestation. Evidence that the prenatal environment is an important determinant of postnatal health and productivity in dairy cattle has recently been published. Aim of this study was to identify prenatal environmental factors that affect preweaning milk intake in dairy calves. Female dairy calves (n=131) from 1 herd were measured and weighed within 7 days after birth. Their milk intake and drinking behaviour at the automated feeder were closely monitored during the 70-day preweaning phase. Calves were offered *ad libitum* starter and hay and could consume up to 8 litres of milk replacer. We hypothesized that prenatally challenged calves would consume more milk replacer, because they are programmed to live in an environment where nutrients are scarce. We could not find a difference in milk consumption between calves originating from primi- versus multiparous dams, nor could we find an effect of parity within the multiparous group. Calves consuming more milk were heavier at birth (P<0.05), indicating that nutrient requirements increase with increasing weight. Calves that had a higher milk replacer intake also visited the automated feeder more frequently (P<0.05), indicating a higher urge to nurse in these animals. Calves from primiparous dams, born in winter consumed less milk replacer (P<0.05) than those born in any other season. We observed no effect of age at first calving in primiparous dams. Offspring born out of multiparous dams with higher milk production during gestation, higher 305-days production and higher maximum daily milk yield consumed more milk (P<0.05) during the preweaning phase. Whereas calves originating from dams with delayed post-parturient rise in milk production and higher milk yield at the end of lactation, consumed less milk at the automated feeder (P<0.05). These results indicate that the prenatal environment affects early-life nutrient intake in dairy calves. Currently, more data is being analysed to further substantiate these preliminary results.

Expression of genes involved in the immune mammary gland response potentially regulated by miRNAs
E. Kawecka[1,2], M. Rzewuska[2], E. Kościuczuk[1], M. Zalewska[3], K. Pawlina-Tyszko[4], D. Słoniewska[1], S. Marczak[1], T. Ząbek[4], L. Zwierzchowski[1] and E. Bagnicka[1]
[1]*Institute of Genetics and Animal Breeding PAS, Animal Improvement, Postępu st. 36a, 05-552 Magdalenka, Poland,* [2]*Warsaw University of Life Sciences, Veterinary Medicine, Nowoursynowska st. 159, 02-776 Warsaw, Poland,* [3]*Warsaw University, Department of Applied Microbiology, Miecznikowa st.1, 02-096 Warsaw, Poland,* [4]*National Research Institute of Animal Production, Department of Animal Molecular Biology, Krakowska st. 1, 32-083 Balice, Poland; e.kawecka@ighz.pl*

The miRNA, small non-coding RNA, regulates expression of genes mainly by transcript degradation or inhibition of translation. The aim of the study was to analyse expression of the selected genes possibly regulated by miRNAs (*MAPK*1, *MEF2C*, *MEF2A*, *LR4* at mRNA level) in the udder secretory tissue of cows infected with coagulase-positive (CoPS) or negative staphylococci (CoNS) vs the non-infected cows (H). The miRNAs differentially expressed in the udder during mastitis in our previous microarray analysis were selected and their target genes were found in silico using the TarBase v.8 software. The genes that were targeted for the largest number of miRNAs were selected. The 55 udder quarter samples were collected from 40 slaughtered HF cows. Six sample groups were distinguished: infected with CoPS, parities 1-2 (n=13); CoPS, parities 3-4 (n=14); CoNS, parities 1-2 (n=5); CoNS, parities 3-4 (n=9); H, parities 1-2 (n=9); H, parities 3-4 (n=4). The mRNA levels were estimated using the qPCR technique. The *GAPDH* and *HPRT* were used as reference genes. There was no difference in *MEF2C* gene expression between groups. A lower *MEF2A* expression in CoPS1-2 group vs the CoPS3-4 and higher expression in CoPS3-4 vs CoNS3-4 and H3-4 groups (P≤0.01) was stated. Moreover, the higher *TLR*4 expression in CoPS3-4 than in H3-4 and CoNS1-2 compared to H1-2 groups (P≤0.05) were found. All these genes appeared potentially regulated by miRNAs, but their activity depends also on the type of pathogen and parity. Financed by the NSC, No. 2015/17/B/NZ9/01561 and KNOW Scientific Consortium 'Healthy Animal – Safe Food', MSHE No. 05-1/KNOW2/2015.

DNA methylation changes with differentiation of intramuscular preadipocyte cells from J. Black cattle

Y. Suda[1], H. Aso[2], Y. Hirayama[1], H. Kitazawa[2], K. Kato[2] and K. Suzuki[2]
[1]Miyagi University, 2-2-1 Hatatate, Taihaku, Sendai, Miyagi, 982-0215, Japan, [2]Tohoku University, 980-8572, 468-1 Aramaki Aza Aoba, Aoba, Sendai, Japan; suda@myu.ac.jp

Meat quality of Japanese Black cattle (JB) are known as a excel character in Wagyu, and has become very popular worldwide. However the main mechanism of developing their characteristic features is not well understood although many researchers have them reported SNPs and QTLs which may be concerning with them. DNA methylation to C in CG rich of the upstream region of start codon controls the expression of many genes on a genome wide level with reflecting to environmental effects. This study aims to clarify regions modified by methylation or demethylation in genomic DNA followed to induction of differentiation in bovine intramuscular preadipocyte cells from Japanese Black cattle. Bovine intramuscular preadipocyte cells, BIP established from Japanese Black cattle was cultured and induced the differentiation by following to a methods by Aso et.al. Genomic DNA from BIP was extracted by using an extraction kit and one for lipid, Takara Bio. And Methylation analysis was performed by using Infinium Methylation EPIC Bead Chip in Takara Bio. 860,000 of regions in genomic DNA of BIP cells were detected. 183,815 regions of 860,000 were estimated as CG rich region. DNA methylation in 120 of regions changed significantly according to the induction of differentiation. Genes controlled by the promoters which had modification changed in CpG might relate to lipid accumulation or differentiation.

Potential role of epigenetics to improve utilisation of feeds low in n-3 LC-PUFA in aquaculture

S. Turkmen, H. Xu, F. Shajahan, M. Zamorano and M. Izquierdo
Universidad de Las Palmas de Gran Canaria, Research Institute in Sustainable Aquaculture and Marine Conservation (IU-ECOAQUA), Crta. Taliarte s/n, 35214, Telde, Spain; serhat.turkmen@fpct.ulpgc.es

There is an increasing number of evidences regarding how dietary interventions during the early stages of development can have long-term on the metabolism of different organisms, including fish. In the recent years, in commercially important fish species, it has been demonstrated the possibility of routing the metabolism for better usage of more sustainable formulated diets. In this review, the effects of early dietary interventions, focusing on improving the utilisation of the novel feeds on growth of gilthead sea bream will be presented. Our studies showed offspring's utilisation of low n-3 LC-PUFA improved if nutritional clues were supplied during the spawning period, not during early or late larval stages in gilthead sea bream. Obtained progenies from different fed brood fish showed altered gene expression of lipid metabolism gene such as *fads2*, at the larval stage even fed with the same commercial diets. This altered metabolism led better utilisation of low n-3 LC-PUFA diets at 6 and 18 months of juveniles; we observed lower feed conversion ratios thus higher final weight if fishes were nutritionally challenged with low n-3 LC-PUFA diets in later life stages. Following studies showed the possibility of improving the growth of the offspring by selection of the broodstock by *fads2*expression levels at eight-months-old juveniles if fed with low n-3 LC-HUFA diets. On the other hand, our recent studies showed a genetic selection of the broodstock could contribute better utilisation of the vegetable meal and oil-based diets and affects progenies' performance. This novel feeding strategy may lead to better utilisation of low n-3 LC-HUFA, vegetable-based diets this, in turn, can contribute the sustainable growth of the aquaculture sector.

Session 25

How to make sure that PLF truly improves animal welfare?

F. Tuyttens

Flemish Research Institute for Agriculture, Fisheries and Food (ILVO), Scheldeweg 68, 9090 Melle, Belgium; frank.tuyttens@ilvo.vlaanderen.be

PLF is often advocated for its opportunities to improve animal welfare, but potential threats are rarely mentioned. I discuss several such threats. First, animal welfare problems that are expressed unordinary may not trigger a warning and may thus go unnoticed. This can partly be dealt with by focusing on multiple predictor variables that are generic welfare indicators, and by proper algorithm validation to produce accurate alerts in a wide variety of conditions. Another part of the solution is to ensure that PLF does not replace direct problem detection by caretakers. However, the rise of PLF may reduce the likelihood of caretakers directly noticing welfare problems because of the other two threats: reduced contact with the animals and reduced skills to detect problems during such contacts. With labour heavily affecting production costs it is likely that a rise in PLF will facilitate on-going trends of increased herd size and animals per stockperson. This poses risks as deep knowledge of the normal behaviour of all animals in the herd is needed for detecting early signs of abnormalities. Furthermore, social psychology studies indicate that intimate contact with animals strongly affects people's attitude and hence behaviour towards them. Moreover, with the rise of PLF the personality profile of a successful farmer will likely shift from animal to technology-centred. The consequences of less animal contact and of being more technology-minded on farmers' care, attitude and behaviour towards their animals have hardly been studied. Further industrialization facilitates the instrumentalization of animals which increasingly conflicts with citizens' moral concerns for sentient beings. The final threat concerns the vulnerability of animals kept in highly automated, minimally staffed, large-scale farms to technological break-downs. Back-up plans would need to safeguard the interests of the animals but may be expensive to install, maintain and implement. To conclude: PLF is booming in animal production science and marketing, partly by emphasizing the opportunities for animal welfare. However, the potential threats of PLF for animal welfare and for the social license to produce food from animals need to be acknowledged and studied as well.

Session 25

Automated assessing social networks and daily barn activities of dairy cows from video surveillance

J. Salau and J. Krieter

Institute of Animal Breeding and Husbandry, Kiel University, Hermann-Rodewaldt-Straße 6, 24118 Kiel, Germany; jsalau@tierzucht.uni-kiel.de

Network analysis has become a valuable tool in animal science, as it provides a lot of parameters to analyse herd structure, individual behaviour, and animal interactions. Networks consist of nodes and edges. In this social network the animals serve as nodes, and contacts between the animals are regarded as edges. The contacts within a herd of 36 Holstein Friesian cows are obtained from the video recordings of 8 cameras AXIS M3046-V. Daily recordings take place between morning and afternoon milking. The barn is located in Northern Germany, has dimensions 12×26 m and provides 37 lying cubicles and 18 feeding places. All cameras are installed above the centre line in 3.5 m height and are facing downwards. Degree and betweenness are centrality parameters associated with individual nodes of the network. Degree counts the number of edges connected to a node. Betweenness is the number of shortest paths that pass through a node. Diameter and density are parameters regarding the whole network. Diameter is defined by the longest of all shortest paths between two nodes, and density is the ratio of the number of edges to the number of possible edges. The recorded image material is stitched to achieve images covering the complete barn. For these images convolutional neural networks will be trained to automatically find and track the cows. In preparation for the automation step, animal contacts have been specified manually regarding recordings from 9 am to 9:15 and 3 pm to 3:15 for one day. Degree and betweenness ranged from 0-5 (2.1±1.4) and 0-0.2 (0.03±0.05) before noon compared to a range from 0-8 (2.5±2.1) and 0-0.3 (0.5±0.8) in the afternoon. Diameter and density were 8 and 0.11 compared to 9 and 0.13. A clique is a fully connected group of cows within the herd. The maximal number of cliques larger than 3 cows and the size of the largest clique increased from 4 to 7, respectively, from 4 to 5. Higher centrality parameters as well as more and larger cliques might be explained by unrest prior to afternoon milking (4 pm) leading to more contacts within the herd. With the aspired automation longer time periods can be analysed, and comparisons between social interaction and management data get possible.

Association between feeding behaviour and wellness scores in Jersey cows using triaxial accelerometer

D. Du Toit, G. Esposito, J.H.C. Van Zyl and E. Raffrenato
Stellenbosch University, Animal Sciences, Merriman and Bossman Ave, 7602 Stellenboch, South Africa; emiliano@sun.ac.za

Non-invasive methods to asses behaviour (e.g. triaxial accelerometer) have been validated mainly in Holstein-Friesian. The study aimed at investigating feeding behaviour of transition Jersey cows and its interaction with wellness scores, using a commercial triaxial accelerometer. The cows (n=145) were fitted with the Silent Herdsman neck collar (Afimilk, Israel). From -21 to 21 DIM, eating and rumination time were monitored hourly; rumen fill, faecal, locomotion, leg hygiene, body condition score (BSC) were assigned every two days (1 to 5 for all scores). Feeding behaviour data were provided by Afimilk (Israel) before algorithmic transformation. Pearson correlations evaluated associations between collar data and scores. Feeding behaviour was also analysed with GLIMMIX procedure of SAS, with cow and DIM as random and repeated variables, respectively. Groups of cows within the specific score value, parity, DIM and the respective interactions were included as fixed factors. BCS was included as a continuous variable. Feeding and rumination times were positively correlated with fill score and BCS (0.34 to 0.55; P<0.01), and negatively correlated with leg hygiene and locomotion scores (-0.32 to -0.12; P<0.01). Parity didn't affect daily rumination time, but older cows ruminated more (P<0.001) around parturition. DIM was significant (P<0.0001) resulting in cows reducing rumination time from 423±52 min at -21 DIM to 197±8 min at calving. Moreover, a drastic reduction of eating time was observed after calving, from 635±33 min, at -21 DIM, to 542±10 min for the first 10 DIM. Only from 20 DIM, cows ruminated at least 400 min/day. Locomotion had a significant effect on rumination (P<0.0001), with a decrease of 1 hr from score 1 to 4. When including eating time as response variable, younger cows were spending more time at the manger (599 vs 538 min; P<0.0001). Locomotion represented a limitation to normal eating resulting in a drastic reduction of time spent eating from 583 min, for score 1, to 512 min, for score 4. The data suggest that accelerometers might need to have Jersey cows-specific algorithms and that, relatively to the scores analysed, at least locomotion is associated with feeding behaviour and may be early-detected.

Selecting variables from sensor data using principal components and partial least squares

I. Dittrich[1], M. Gertz[1], B. Maassen-Francke[2], K.H. Krudewig[3], W. Junge[1] and J. Krieter[1]
[1]Kiel University, Institute of Animal Breeding and Husbandry, Olshausenstr. 40, 24098 Kiel, Germany, [2]GEA Farm Technologies GmbH, Siemensstr. 25, 59199 Bönen, Germany, [3]365FarmNet Group GmbH & Co. KG, Hausvogteiplatz 10, 10117 Berlin, Germany; idittrich@tierzucht.uni-kiel.de

Early sickness detection is a main challenge for farmers. The implementation of sensor systems and development of detection algorithms are a possible aid in finding sickened animals as early as possible. Nevertheless, sensor systems collect many variables describing behavioural patterns like activity or indexes of milk production (e.g. milk yield), which is not necessarily helpful in algorithm development due to e.g. autocorrelations or too many parameters. Therefore, the aim of the study was to identify useful variables from sensor data by using principal component analysis (PCA) and partial least squares (PLS) for later employment in an automated sickness detection approach. The specific sensor data were collected from September to December 2018 (n_{cows}=480). After selection of 140 cows (>100 observations) the dataset consisted of 57 healthy and 83 lame cows. Daily values were calculated and z-transformed for 23 variables to ensure the same preconditions. The PCA identified 7 principal components (PCs), which accounted for 79% of the observed variation. The PLS model extracted three factors (Fs), which explained 92% of the dependent variable's (health status) variation. The extracted PCs and Fs were implemented in multivariate cusum control charts (MCUSUM-PCA; MCUSUM-PLS) for the detection of lameness. To evaluate the classification accuracy (lameness detection) of the MCUSUM, sensitivity (SE-PCA; SE-PLS), specificity (SP-PCA; SP-PLS) and false positive rates (FPR-PCA; FPR-PLS) were calculated and compared regarding the method of variable selection. Both approaches showed at least acceptable sensitivities (SE-PCA=85%; SE-PLS=65%) and specificities (SP-PCA=70%; SP-PLS=95%), while the false positive rates were acceptable for the MCUSUMPLS (FPR-PLS=5%) and less acceptable for the MCUSUM-PCA (FPR-PCA=30%). The used methods PCA and PLS for variable selection showed acceptable results for the classification accuracy parameters when using the extracted variables in MCUSUM chart to detect lameness.

Validation of a locomotive detection system in cows with the help of direct behavioural observation

B. Kulig[1], G. Grodkowski[2], R.J. Redlin[3] and T. Sakowski[4]
[1]*Kassel University, Nordbahnhofan 1a, 37213 Witzenhausen, Germany, [2]Warsaw University of Life Sciences, Ciszewskiego 8, 02-786 Warszawa, Poland, [3]University of Applied Sciences, Philipp-Müller-Straße 14, 23966 Wismar, Germany, [4]Institute of Genetics and Animal Breeding of the PAS, Postępu 36A, Jastrzębiec, 05-552 Magdalenka, Poland; bkulig@uni-kassel.de*

The locomotives detection system SensOor Ear Tag combined with the herd management system of CowManager B.V. was used for directly recorded accelerations in all spatial directions triggered by movements of the animal. The aim of this investigation was to test whether such system, mainly adapted for the detection of heat and health problems in dairy cows primarily in stable housing, can also be used for grazing systems. The behavioural observations on 57 cows for a total of 129 hours in 2016 and 2017 were carried out at the two farm in Poland and in Germany. With regard to the peculiarities of grazing the following behavioural patterns were added to the observational catalogue: grazing standing and walking, ruminating lying and standing, ruminating walking, standing, lying/resting and active/walking. The measurement system analysis, which takes into account the recording of the different behaviour patterns in stable and pasture was developed. A partial least squares regression (PLS-R) solved that matching problem. The pasture behaviour was used in the PLS-R as the x-side of the equation and the patterns identified by the sensor were predicted. The behaviour estimated by the PLS-R was compared with the behaviour indicated by the sensor using a classical statistical method comparison. Automatic sensor systems produce masses of data without any additional effort and the statistical law of large numbers becomes therefore effective. The measurement results reveal the individual animal differences by increasing the sample size. For this reason, it can be recommended to investigate the use of locomotion recordings for animal breeding more closely. The available results are promising enough. The author(s)/editor(s) acknowledge the financial support for this project provided by transnational funding bodies, being partners of the FP7 ERA-net project, CORE Organic Plus, and the cofound from the European Commission.

Step recognition and feature extraction for turkey gait measured with IMU sensors

A.C. Bouwman[1], W. Ouweltjes[1], J. Mohr[2] and B. Visser[3]
[1]*Wageningen University & Research, P.O. Box 338, 6700 AH Wageningen, the Netherlands, [2]Hendrix Genetics North America, Riverbend Dr Suite C 650, N2K 3S2 Kitchener, Canada, [3]Hendrix Genetics Research & Technology Centre, P.O. Box 114, 5830 AC Boxmeer, the Netherlands; aniek.bouwman@wur.nl*

In livestock production locomotion is an important welfare and health trait. Breed4Food aims to improve locomotion through breeding and herd management using sensor technology for precision phenotyping. In most breeding programs locomotion is recorded once in a lifetime based on subjective observations by human experts. In addition, locomotion issues like lameness are often recorded when observed in daily farm management. While larger animals like cows are often equipped with accelerometers to keep track of their behaviour, this type of sensor recording is underdeveloped in poultry. Therefore, the aim of this study was to use inertial measurement units (IMUs) to record the locomotion of turkeys. In turkey breeding programs the walking ability of breeding candidates is subjectively scored on a scale from 1 (poor) to 6 (good) by an expert while the birds walk through a corridor one by one. During this routine scoring procedure three IMUs were attached to each bird with velcro straps while walking, one on each leg and one on the neck. The IMU provide acceleration, gyroscope, and magnetic field data in x, y, and z-axis, as well as, pitch, roll and yaw output. Steps were annotated by hand in the acceleration patterns from the leg IMUs. The machine learning technique gradient boosting was used for step recognition based on the leg IMUs. Results showed that with a very limited training set of 25 steps from one leg IMU from 3 animals we could get very good step recognition. The step recognition was even better than our own annotation. Based on the steps, we will extract features such as stride cycle, stance phase, swing phase. The IMUs provide information on pitch, and we aim to extract additional features related to balance, and leg strength. By annotating the extracted features with walking ability scores we aim to predict the walking ability score of turkeys using machine learning techniques. Positive results will lead to an objective scoring system, which could also lead to more frequent scoring or even continuous monitoring of the locomotion of turkeys.

Testing the potential of the SowSIS for on-farm automatic lameness detection in breeding sows

P. Briene[1], O. Szczodry[1], P. Degeest[1], A. Van Nuffel[1], J. Maselyne[1], S. Van Weyenberg[1], J. Vangeyte[1], B. Ampe[2], F.A.M. Tuyttens[2,3] and S. Millet[2]
[1]*ILVO (Flanders Research Institute for Agriculture, Fisheries and Food), Technology and Food Science Unit, Burgemeester Van Gansberge 115 bus 1, 9820 Merelbeke, Belgium, [2]ILVO (Flanders Research Institute for Agriculture, Fisheries and Food), Animal Science Unit, Scheldeweg 68, 9090 Melle, Belgium, [3]Ghent University, Faculty of Veterinary Sciences, Salisburylaan 133, 9820 Merelbeke, Belgium; petra.briene@ilvo.vlaanderen.be*

Lameness is a very common problem in breeding sows, which often goes undetected for long periods of time. This has severe consequences on the welfare and performance of sows. Automatic lameness detection could help pig farmers to recognize and treat the problem sooner. The Sow Stance Information System (SowSIS) consists of 4 force plates, built into an electronic sow feeder, providing non-invasive 'stance'-output for each leg. Stance information variables can be extracted from the data. Stance data was automatically collected for 53 gestating group-housed sows for 74 days per sow. The gait of these sows was visually scored twice a week using a 150 mm tagged visual analogue scale to determine their lameness status. This gait score was used as a reference and compared to the SowSIS measurements. First, significant variables were used in a multilevel linear model to predict lameness. With the model estimates, it was determined whether the model would classify a sow as lame or non-lame using gait score ≥55 mm as the cut-off value. The model could detect lameness with 71.7% sensitivity, 89.7% specificity, 76.7% lame predictive value and 87.0% non-lame predictive value. The 5 severely lame sows (gait score >90 mm) were all correctly classified as lame. Secondly, different linear models were tested to determine the lame leg on a sub-dataset. The support vector machine model and random forest model could predict the lame leg correctly by 100% when fitted to a validation dataset. The SowSIS shows great promise as an on-farm lameness detection system.

Water usage as health indicator in pig production

A.H.M. Van Der Sanden and M.A. Van Der Gaag
Connecting Agri & Food, Oostwijk 5, Postbus 511, 5400 AM Uden, the Netherlands; a.vandersanden@connectingagriandfood.nl

Water usage of sows, piglets and fattener pigs was measured at four farms. Water consumption was estimated by measuring the water usage. The water measuring device was installed at the drinking water line and measured the water flow rate of a compartment with up to 10 farrowing sows, 135-230 piglets or 290 fatteners respectively. Water intake was measured every five minutes during at least 12 consecutive months. The water intake of 20 farrowing sows was measured at the individual level on two farms with each one compartment. Indoor temperature was measured real time and data were online accessible. Additionally online available data on outdoor temperature and relative humidity from nearby weather stations were included. Specific circumstances were registered, like veterinary treatments, entry and delivery date, occurrence of health problems, abnormalities in faeces. Sow reproduction data were registered. Normal drinking water patterns for sows, weaners and fattener pigs were estimated. Actual water usage was compared with the normal drinking water pattern. Actual daily water usage deviating more than 10% from the normal value was considered as a change in the expected drinking pattern. Unified analysis was conducted of abnormal drinking water behaviour and moment of vaccination, occurrence of health problems including diarrhoea, and increase or decrease in outdoor temperature by more than 11 °C. Relations were identified between water usage and specific climate or health related circumstances. Sick animals showed decreased water intake before showing symptoms. In conclusion, daily water usage can be used as a predictive parameter for health status. Further research is under way.

First validation of a prototype of multi-spectral camera to quantify skin lesions in pigs

A. Prunier[1], L. Delattre[2], J.M. Delouard[2], B. Früh[3] and C. Tallet[1]
[1]PEGASE, INRA, Agrocampus Ouest, 16 Le Clos, 35590, Saint Gilles, France, [2]3D Ouest, 5 rue Louis de Broglie, 22300 Lannion, France, [3]FIBL, Ackerstrasse 113, 5070 Frick, Switzerland; armelle.prunier@inra.fr

The aim of the present work is to detect and quantify skin lesions with fresh or coagulated blood in pigs. If successful, this objective technics will replace on-farm detection by an observer as described in scientific papers or in the Welfare Quality® protocol. Therefore, it could be used for welfare assessment in scientific experiments, for welfare certification, for genetic selection against aggressiveness or for farmers to have an objective evaluation of behavioural problems occurring in their farms. The method is based on a method described by Goel *et al*. Six images are acquired by an active camera with projection of light characterized by 6 different wavelengths. These wavelengths were chosen so that haemoglobin could be specifically absorbed, and hence detected, at least by one length. The images are taken successively in a short interval of time so that that they are well superposed. These images are analysed by a software which allowed, in a first step, to produce 2 images: a well-defined image of the pig (left image below), a white and black image (right image below) where pixels in black corresponds to skin lesions with blood and pixels in grey to the skin without blood. An area of the body with skin lesions is depicted inside the white circle in both images. Some work is still undertaken to reduce the time interval between images, improve the software and validate the method.

Monitoring behaviour of fattening pigs using low-cost RGB-depth cameras under a practical condition

S. Zhuang[1], J. Maselyne[1], J. Vangeyte[1] and B. Sonck[1,2]
[1]Flanders Research Institute for Agriculture, Fisheries and Food (ILVO), Burg. van Gansberghelaan 92, 9820 Merelbeke, Belgium, [2]Ghent University, Department of Animal Sciences and Aquatic Ecology, Faculty of Bioscience Engineering, Coupure links 653, 9000 Ghent, Belgium; shaojie.zhuang@ilvo.vlaanderen.be

Indoor housed fattening pigs sometimes exhibit undesirable lying habits in partially slatted pens, and continuous observations on the animals are needed to identify the causal factors and evaluate the effectiveness of the corresponding solutions. Low cost RGB-depth cameras have been shown to be a promising tool in monitoring farm animals, but application in practice remains a challenge due to insufficient sensor performance and variable farm conditions. The aim of this study was to test the applicability of a relatively recent low-cost RGB-depth camera model for monitoring the behaviour of fattening pigs under a practical condition. Pigs housed in two separate climate controlled compartments with eight 2×2 m pens each (six pigs per pen) were monitored with the camera system. Eight RGB-depth cameras were mounted above the pens to obtain a top-down view (two pens per camera). In each compartment a mini-PC was used to log the depth and RGB videos. A depth-based algorithm was developed to identify the pigs and their standing/lying positions. In the test environment, the mean absolute error of the camera on the depth measurement was <0.03 m, with median absolute deviation <0.013 m. Preliminary results based on 1,761 frames randomly sampled from a full fattening period showed that the location of pigs could be detected most cases. The sensitivity and specificity for identifying pigs in standing position were respectively 92 and 99.4%. The camera satisfied the general requirements for full-time observations on the posture and spatial preference of the pigs. However, the depth value could not be properly obtained at the locations exposed to excessive sunlight, which occasionally near the barn windows during certain hours of the day. Other practical issues and prospects that were related to the specific camera model and the indoor environment of the pig barn are discussed.

Development and validation of an embedded tool to measure postural activity of lactating sows

L. Canario[1], Y. Labrune[1], J.F. Bompa[1], Y. Billon[2], L. Ravon[2], S. Reignier[2], J. Bailly[2], M. Bonneau[3] and E. Ricard[1]
[1]INRA, INPT-ENSAT, INPT-ENVT, UMR1388 GenPhySE, 24 chemin de Borde Rouge, 31326 Castanet-Tolosan, France, [2]INRA, GenESI, Le Magneraud, 17700 Surgères, France, [3]INRA, URZ UR143, Petit-Bourg, 97170 Guadeloupe, France; laurianne.canario@inra.fr

The objective was to develop a tool to measure sow postural activity during lactation while kept between fences in the farrowing crate. Several trials were carried out to select sensors and test their ability to detect 5 different positions of the sow. First, several sensors, placed in a single metal box, were attached to the upper part of the back of sows with a custom-built belt that passes underneath the belly, just behind the front legs. For validation of the positions predicted from sensor data, sow behaviour was recorded with use of a digital camera and video records were analysed by a single observer. Prediction ability was tested with machine learning applied to random forests for the three-axis data provided by the sensors in the X, Y and Z dimensions. When comparing information provided by three sensors, i.e. two accelerometers (sensors A and B) and an application developed for a smartphone (sensor C), one sensor detected five positions more accurately than the other two. The error rate of prediction from 30% of the data was 2.1% with sensor A, 3.1% with sensor B, and 4.8% with sensor C. Additional trials focused on the ability of sensor A to distinguish among the five positions, using five sows. Statistical sensitivity was 0.94 and 0.95 for the right and left lateral positions, respectively; 0.79 for the ventral position; 0.75 for the standing position; but only 0.33 for the sitting position (because it represented only 4% of sows' time budget). One single axis was sufficient to detect the two lateral positions. Next, we concentrated our interest on the advantages and drawbacks of 2 accelerometers: we ran new trials to compare outputs from sensor A with those obtained with a sensor D. We used acquisition speeds varying from 1 to 16 obs/sec to optimize the sampling rate to detect positions accurately and to test if postural changes at risk for the piglets can be detected. Analyses are in progress. The use of accelerometers to measure finely the postural activity of lactating sows is promising.

Stray and abandoned dog overpopulation management – Israel's unique system

L. Morgan[1], B. Yakobson[2] and T. Raz[1]
[1]The Hebrew University of Jerusalem, Rehovot, 76100, Israel, [2]Ministry of Agriculture and Rural Development, Rishon Lezion, 50200, Israel; liat.morgan@mail.huji.ac.il

Dog overpopulation is a major problem worldwide, which impacts animal welfare and health, as well as public health. Furthermore, it may be related to disease transmission and attacks of livestock, wild animals and human. In Israel, a unique governmental database is managed in order to increase responsible ownership and to enforce mandatory dog registration and rabies yearly vaccination policy. In addition, since 2012, a unique online searchable database has been gathering most homeless pets offered for adoption from non-profit organizations and municipal shelters (http://Yad4.co.il). Our objectives were: (1) to investigate the registered dog population in Israel and to assess its association to the abandoned dog population; and (2) to reveal the risk factors for a dog to be adopted or to stay at the shelter. Data analyses included 758,288 registered dogs and 22,545 adoptable dogs. Analyses revealed that only 214,101 out of the 343,872 dogs that are registered as 'active' are known to have owners. Approximately 40,000 dogs 'disappear' from the database every year, which means that some of them may be abandoned. Multi Variate Linear regression strengthened it by revealing that the number of abandoned dogs increased by the numbers of active registered dogs and dogs which had 'disappeared' from the database in the previous year (P<0.05). Among the registered dogs, 8% were puppies (<1 year old); however, only 1.7% of the abandoned dogs were puppies (Mean ± SD age of dogs when abandoned: 1.7±2.12 years). Among these dogs, the risk factors to stay at the shelter were when the dogs were described as: 'mix-breed', 'male', 'suitable for senior' or 'for athletes' (P<0.05). In conclusion, a governmental national database is an important dog population management tool, with a potential to predict the number of abandoned dogs and to lead for better decisions and regulations. Since most abandoned dogs are older than one year, promoting responsible ownership would be vital. Moreover, a national online database, such as Yad4 website, may successfully improve dog adoption rate; however, dog characteristics described in the website have a significant impact on the chance for adoption.

Calf feeding behaviour in a cow-calf contact system with access to automatic milk feeders

K. Piccart[1], S. Van Weyenberg[1], D. Weary[2] and J. Føske Johnsen[3]
[1]*Flanders Research Institute for Agriculture, Fisheries and Food (ILVO), Burg. Van Gansberghelaan 115 bus 1, 9820 Merelbeke, Belgium,* [2]*University of British Columbia, Vancouver Campus 248-2357 Main Mall, Vancouver, BC V6T 1Z4, Canada,* [3]*Veterinærinstituttet, Ullevålsveien 68, Pb 750 Sentrum, 0106 Oslo, Norway; kristine.piccart@ilvo.vlaanderen.be*

Automatic milk feeders (AMF) generate a large amount of data on feed intake and feeding behaviour of individual calves, which in turn might be useful for monitoring their health and welfare. AMF have also been used to provide supplementary nutritional support for suckling dairy calves in a cow-calf contact system. In this research, AMF data were analysed to characterize the feeding behaviour of dam-reared dairy calves (n=20) with and without the opportunity to suckle the dam. In the first phase, calves were housed in a calf creep next to the cow pen during the day, allowing visual and auditory contact with the dam. During the night, calves had access to the adjacent cow pen. Ten calves could suckle the dam (AMF + dam), while the other 10 calves could not (AMF only) because access to the dam's udder was blocked using an udder net. All experimental calves had access to 12 l of milk from the AMF. After 6 weeks, the calves were moved to an adjacent separation pen. The separation was divided into 2 phases: partial separation from day 43 to 46 (i.e. nose-to-nose contact with dam) and total separation from day 47 to 50 (i.e. no more contact with dam). All calves continued to have access to the AMF 24 h/d during this period. During the final phase, the calves were gradually weaned by reducing milk allowance by 1.5 l/day. A mixed model was used to describe the relationship between treatment (AMF + dam vs AMF only), phase and health status (healthy vs sick) with the outcome variables body weight, milk intake, rewarded and unrewarded AMF visits. Results showed that the AMF + dam calves preferred suckling when given the opportunity, but switched to the AMF during the separation phases. Throughout the study, calf body weight did not differ between treatments (interaction age × treatment: P=0.78). Regardless of treatment group, the number of unrewarded AMF visits decreased when calves were sick (P<0.05), suggesting that a decline in these visits may be a useful indicator of calf health.

Microbiability of faecal nutrient digestibility in pigs

L.M.G. Verschuren[1,2], D. Schokker[2], A.J.M. Jansman[2], R. Bergsma[1], F. Molist[3] and M.P.L. Calus[2]
[1]*Topigs Norsvin Research Center B.V., P.O. Box 43, 6640 AA Beuningen, the Netherlands,* [2]*Wageningen Livestock Research, Droevendaalsesteeg 4, 6708 PD Wageningen, the Netherlands,* [3]*Schothorst Feed Research, Postbus 533, 8200 AM Lelystad, the Netherlands; lisanne.verschuren@wur.nl*

Microbes play an important role in nutrient digestion, especially when fibrous diets are fed to the pigs. The aim of this study was to use microbial relationships to predict faecal nutrient digestibility. Data were collected of 114 three-way crossbreed grower-finisher pigs (65 female and 49 male) which were either fed a diet based on corn/soybean meal or on wheat/barley/by-products. On the day before slaughter (mean body weight 121 kg) faecal samples were collected and used to determine faecal digestibility of dry matter, ash, organic matter, crude protein, crude fat, crude fibre, and non-starch polysaccharides. The faecal samples were also sequenced for the 16S hypervariable region of bacteria (V3/V4) to profile the faecal microbiome, resulting in 3,041 operational taxonomic units after filtering. With these data, we calculated the proportion of variance in faecal nutrient digestibility explained by variation in the faecal microbiome, i.e. the 'microbiability'. The microbiabilities ranged from 1.3% for ash digestibility to 93.2% for crude protein digestibility. Using leave-one-out cross-validation, we estimated the accuracy of predicting individual digestibility based on the microbiome profile. The accuracies of prediction for crude fat and ash digestibility were virtually 0, and for the other nutrients, the accuracies ranged from 0.42 to 0.63. In conclusion, the faecal microbiome profile gave high microbiabilities for faecal digestibility of dry matter, organic matter, crude protein, crude fibre, and non-starch polysaccharides. The accuracies of prediction are relatively low if the interest is in precisely predicting individual digestibility, but looks promising from the perspective of ranking animals in a genetic selection context; Especially considering that the prediction accuracy is expected to increase if the training data increases and comprises more variation in individual digestibility.

Effect of gut microbiota composition on carcass and meat quality traits in swine
P. Khanal[1], C. Maltecca[1], C. Schwab[2], J. Fix[2] and F. Tiezzi[1]
[1]North Carolina State University, Department of Animal Science, Raleigh, NC, 27695, USA, [2]The Maschhoffs LLC, Carlyle, IL, 62231, USA; cmaltec@ncsu.edu

To date studies on the effect of host gut microbiome on carcass and meat quality traits of swine remain limited. The objectives of this study were: (1) to estimate the microbiability and investigate its impact on heritability estimates of meat quality and carcass traits; and (2) to estimate the microbial correlation between the meat quality and carcass traits in the commercial swine population. The study population consisted of 1,123 three-way cross individuals genotyped with 60k SNP chips and phenotyped for carcass and meat quality traits. Fecal 16S microbial sequences for all individuals were obtained at three different stages: weaning (WEAN: 18.64±1.09 days); 15 weeks post weaning (W_15: 118.2±1.18 days); and off-test (OT: 196.4±7.80 days)). Data were analysed using a single and multi-trait model, which included fixed effects of dam line, contemporary group, and sex, as well as random effects of pen, animal additive genetic and microbiome. The last two were modelled with the use of relationship matrices among individuals obtained by genomic and microbial information. Analyses were conducted in ASREML v.4. The contribution of microbiome to all traits was significant, although it varied over time, with an increase from WEAN to OT for most traits. Adding microbiome information did not affect the estimates of genomic heritability of meat quality traits but changed the estimates of carcass composition traits. A decrease in heritability was observed for OT. The decrease in heritability ranged from 2% for carcass average daily gain (CADG) to 10% for fat depth. Microbial correlations ranged from -0.93±0.11 between firmness and shear force to 0.97±0.02 between CADG and loin weight. Results suggest gut microbial composition can contribute to the improvement of complex traits in swine. These results may lead to establishing a newer approach of genetic evaluation process through the addition of gut microbial information.

Piglets infected with ETEC F4 and F18: effect of MUC4 and FUT1 genotypes
F.R. Massacci[1,2,3], S. Tofani[2], M. Tentellini[2], S. Orsini[2], C. Lovito[2], C. Forte[2], D. Luise[3], C. Bevilacqua[1], L. Marchi[2], M. Bertocchi[3], C. Rogel-Gaillard[1], G. Pezzotti[2], J. Estellé[1], P. Trevisi[3] and C.F. Magistrali[2]
[1]INRA, GABI, Domaine de Vilvert, 78352, France, [2]Istituto Zooprofilattico Umbria e Marche, SC3RS, G.Salvemini, 06121, Italy, [3]University of Bologna, DISTAL, G. Fanin, 40127 Bologna, Italy; francesca.massacci2@unibo.it

Enterotoxigenic *Escherichia coli* (ETEC) is the etiological agent of post-weaning diarrhoea (PWD) in piglets. The *MUC4* and the *FUT1* genes have been associated with the susceptibility to ETEC F4 and F18, respectively. The aim of this study was to investigate the effects of the genotype of *MUC4* and *FUT1* in piglets naturally infected with ETEC F4 and F18. 71 piglets were divided into 3 groups based on two antimicrobial administration routes: A) parenteral antibiotic, B) oral antibiotic and C) control group without antibiotic. Animals arrived in the facility on weaning day (T0). For groups A and B, at T0 amoxicillin was administered during 5 days either parentally or orally. Animals were evaluated at the end of the amoxicillin administration (T1) and 7 days after (T2). At each time point, faecal scores and body weight were recorded, and the presence of ETEC F4 and F18 in faecal samples was assessed by PCR. Results revealed that 50/71 piglets were naturally infected by ETEC F4 and 21/71 by ETEC F18 at T0. Only F18 was detected at T1. At T0, Fisher tests showed that *MUC4* genotype was significantly associated with the presence of ETEC F4 and the faecal scores (P<0.05). Intriguingly, the *MUC4* resistant genotype was associated with ETEC F4 absence but also with a higher diarrhoea score. At T1, *FUT1* was associated with the presence of ETEC F18 (P<0.05) but not with the diarrhoea scores. Antibiotic administration was significantly associated with the presence/absence of F18 and the diarrhoea score at T1 and T2 (P<0.05). Our results confirm that *MUC4* and *FUT1* genotypes are associated with the susceptibility to ETEC F4 and F18 infection. Production of the gut microbiota data is ongoing and the results will be correlated with the piglets' genotypes. Next step will thus be to study how the gut microbiota evolves in relation to ETEC infection, genotypes and antibiotic administration.

The association of the host genome with microbiome composition and growth traits in pigs

C. Maltecca[1], D. Lu[1], C. Schillebeeckx[2], N.P. McNulty[2], C. Schwab[3], C. Shull[3] and F. Tiezzi[1]
[1]NC State University, Animal Science, Raleigh, 27695 0001, USA, [2]Matatu Inc, Saint Louis, 63108, USA, [3]The Maschhoffs LLC, Carlyle, 62231, USA; cmaltec@ncsu.edu

This work aimed at evaluating the potential interaction between the host genome and its gut microbiome in swine. Three thousand and one pigs, originating from 28 founding Duroc Sires, were used in the study. Half of the pigs were generated through purebred matings while half by matings with F1 Landrace × Large White dams. All pigs were genotyped with the Illumina PorcineSNP60 Beadchip while sires were sequenced at an average depth of 10×. Faecal microbiome samples were collected at weaning, week 15 and week 27 of the feeding trial, along with backfat thickness (BF), live weight (WT), and loin depth (LD) measures. We estimated heritability for taxa of the gut bacteria, identifying taxa that are important for the phenotypes, as well as scanning the host genome for single nucleotide polymorphisms (SNPs) potentially influencing the gut microbiome composition in both crossbred and purebreds. Forty-eight taxa from week 15 and 20 at week 27, were significantly associated with backfat, weight and loin depth at an FDR of 5%. Taxa with a substantial contribution to traits variation included *Succinivibrio dextrinosolvens, Lactobacillus reuteri, Prevotella copri, Peptococcus niger, and Oxalobacter formigenes*. Heritability estimates for the significant taxa at weaning, week 15, and week 27 ranged from low to moderate (0.01 to 0.34) and were heterogeneous between crossbred and purebreds. Several SNPs were found significantly associated with OTUs at week 15 and 27. These SNPs located in chromosome regions that contained a total of 378 genes, belonging among others, to pathways that protect epithelial cells in the host gut, transfer substances transmembranes, promote cell proliferation, differentiation, motility, and survival.

To what extent rumen microbiota composition is host-driven – an experimental approach in goats

V. Berthelot[1], P. Gomes[1,2], P. Mosoni[3] and L.P. Broudiscou[1]
[1]UMR Modélisation Systémique Appliquée aux Ruminants, INRA, AgroParisTech, Université Paris-Saclay, 75005 Paris, France, [2]Neovia, Rte Talhouët, 56250 Saint-Nolff, France, [3]Université Clermont Auvergne, INRA, UMR 0454 MEDIS, 63000 Clermont-Ferrand, France; valerie.berthelot@agroparistech.fr

Ruminants are products of the symbiosis between the host organism and its rumen microbiota. Both entities co-evolve from birth to adulthood and are subjected to environmental factors like feed and breeding conditions. It is now well established that each adult animal harbours its own microbiota even when animals are fed and bred identically, suggesting that the host has a significant impact on its microbiota composition. Our objective was to determine to what extent the host controls microbiota composition. We conducted an original experimental approach in which the link between microbiota and host was disrupted. Twelve 3-year-old rumen cannulated dry goats raised in similar conditions from birth were fed a similar diet rich in fibre during 8 months. After 3 months, the link between host and microbiota was experimentally disrupted by emptying the rumen of each animal, pooling the 12 rumen contents, and finally refilling each empty rumen with the 12 homogenized microbiota on a day referenced as d0. Rumen samples were taken before (d-53) and after disruption (d3, d35, d114, and d141) to analyse the rumen microbiota composition using 16S metabarcoding. Richness and Shannon diversity indices underlined a significant increase in microbial α-diversity on d3, as expected. It stabilized from d35 to d141 at higher levels than on d53. Regarding β-diversity, community structures were compared between goats and days. Clear clustering patterns based on days were observed with communities on d114 and d141 being closer to each other than to the previous days post-disruption. However, communities did not return to the original state before the disruption. Goats also impacted the microbial composition but to a lower extent. From the disruption step up to five months later, the 12 ecosystems mostly evolved following a common pattern in which the hosts played a minor, though definite, role. Our data suggest that the rumen microbiota composition was more environmentally- and diet-driven than host-driven in goats fed high fibre diets.

Identification of host-characteristics influencing rumen microbiota composition in dairy goats

P. Gomes[1,2], L.P. Broudiscou[2], P. Mosoni[3] and V. Berthelot[2]
[1]*NEOVIA, Route de Talhouët, 56250 Saint-Nolff, France, [2]UMR Modélisation Systémique Appliquée aux Ruminants, INRA, AgroParisTech, Université Paris-Saclay, 75005 Paris, France, [3]Université Clermont Auvergne, INRA, UMR 0454 MEDIS, 63000 Clermont-Ferrand, France; pierre.gomes@agroparistech.fr*

Rumen microbiota is involved in host feed valorisation. Animals bred in the same environment still have the variability of microbiota composition, suggesting a significant impact of the host. In this study, we challenged the link between rumen microbiota composition and host to evaluate how the host can influence its rumen microbiota composition through its characteristics and performances. For this, we used ten 4-year-old rumen cannulated lactating goats raised in similar conditions from birth. The original link between microbiota and host was experimentally disrupted by collecting all the rumen contents, pooling and mixing contents, and refilling goat rumens with the same homogenized microbiota on a day referenced as d0. We studied the evolution of their rumen microbiota composition with 16S metabarcoding before (d-8, d-5) and after (d2, d139, and d148) disruption. At the same time, host animal characteristics (body weight, dry matter intake) and performance (peak and persistence of milk yield) were measured. Rumen flow kinetics were characterized by the intake rate, its variability, between-bout intervals, rumen liquid, and solid turnover rates. Prior to disruption, the Bray-Curtis dissimilarity between goats was 0.336 at the OTU level. The breed, body weight, dry matter intake, and rumen flows partly explained the variability of microbiota composition. On d2, the dissimilarity between goats increased to 0.443 without any significant correlation with the goat variables. On d139 and d148, the dissimilarities between goats were 0.333 and 0.363 respectively. As prior to disruption, breed, body weight, intake rate and between-bout intervals partly explained the variability of microbiota composition. The dissimilarity between the initial (d8 and d5), and the final samples (d139 and d148) was 0.343. It seems that the goat rumen microbiota gets back closer to the initial state than just after the disruption. Animal characteristics linked to feeding behavior seem to be a determinant of microbiota composition.

Rumen fatty acid metabolism and bacterial community

L. Dewanckele[1], L. Jing[1], B. Stefanska[2], B. Vlaeminck[1], J. Jeyanathan[1], W. Van Straalen[3], A. Koopmans[3] and V. Fievez[1]
[1]*Ghent University, Coupure Links 653, 9000 Gent, Belgium, [2]Poznan University of Life Sciences, Wojska Polskiego 28, 60-637 Poznan, Poland, [3]Schothorst Feed Research B.V., Meerkoetenweg 26, 8218 Lelystad, the Netherlands; lore.dewanckele@ugent.be*

Distinct blood and milk 18-carbon fatty acid proportions and buccal bacterial populations in dairy cows differing in reticulorumen pH response to dietary supplementation of rapidly fermentable carbohydrates Nine Holstein dairy cows were fed diets with increasing proportions of rapidly fermentable carbohydrates (RFCH) to investigate the effect on reticular pH, milk fat content (MFC), plasma and milk 18-carbon fatty acid (FA_{C18}) proportions and buccal bacterial community. Cows were fed a TMR throughout the experiment, which consisted of four periods: i/ adaptation (d0-4), ii/ low (d5-18), iii/ increasing (d19-24) and iv/ high RFCH (d25-28). During the increasing period, the standard concentrate was gradually and partly replaced by a concentrate high in RFCH. The reticular pH was measured using a bolus. Plasma and milk samples were collected and analysed for FA_{C18} proportions, and buccal swabs were collected for bacterial community analysis based on 16S rRNA gene amplicon sequencing. Inter-animal variation was observed in terms of reticular pH, which allowed to divide the cows into two groups: tolerant ($t_{pH<6.0} \leq 0.1$ h/d) and susceptible cows ($t_{pH<6.0} \geq 1.26$ h/d). The lower reticular pH of susceptible cows was accompanied by lower MFC. Both groups already differed in reticular pH and MFC during the low RFCH period. Even more, higher RFCH amounts did not decrease the reticular pH in any of the two groups. Nevertheless, MFD was observed in both groups during the high compared to the low RFCH period. Lower MFC in animals with lower reticular pH or during the high RFCH period was associated with a shift in FA_{C18} toward *trans*-(*t*)10 at the expense of *t*11 intermediates, and shifts in the relative abundance of specific bacteria. Genera *Dialister*, *Sharpea*, *Carnobacterium* and *Acidaminococcus* were more abundant in situations with greater *t*10 proportions.

Gut microbiome and incidence of foodborne pathogens are affected by diet in pasture-raised chickens

J.M. Lourenco[1], M.J. Rothrock Jr.[2], Y.M. Sanad[3], F.L. Fluharty[1] and T.R. Callaway[1]
[1]University of Georgia, Athens, GA, 30602, USA, [2]USDA ARS, Athens, GA, 30605, USA, [3]University of Arkansas, Pine Bluff, AR, 71601, USA; jefao@uga.edu

Using a farm-to-fork approach, this study aimed to examine the effects of feeding two distinct diets – one containing soy (SB), and one soy-free (SF) – on the gut microbiome of pasture-raised chickens. Microbial DNA was extracted, and 16r RNA gene sequencing was performed on caecal contents of 1-day-old birds, faeces collected from their pastures (at 4, 7 and 12-weeks-old), caecal contents obtained at the slaughterhouse, whole carcass rinses (WCR) obtained at the slaughterhouse, and WCR obtained from final products (after carcasses were frozen for 1 month). Results showed that the number of observed OTUs significantly increased (P<0.001) as birds fed the SF diet aged (from 1-day-old to 12-weeks old); however, no significant differences (P=0.11) in the number of OTUs were observed as SB-fed birds aged. Relative abundance of the genera *Oscillospira*, *Faecalibacterium* and *Ruminococcus* increased as birds aged (P≤0.001), whereas the abundance of *Lactobacillus* decreased (P≤0.02) in both SB and SF diets. For bacterial groups regarded as foodborne pathogens, there was no difference (P≥0.15) in *Salmonella* abundance between SB and SF in any sample type; however, SF broilers had a lower abundance of *Campylobacter* during their entire lifecycle. This effect was particularly pronounced in the faecal material collected when birds were 12-weeks-old (P=0.003) as well as in the WCR obtained from the final product (P=0.04). *Acinetobacter* abundance was also lower (P=0.05) in the WCR from birds consuming SF. These findings demonstrate that the feed offered to pasture-raised broilers can significantly affect their gut microbial populations. Moreover, they suggest that the use of soy-free diets may represent a viable strategy to reduce contamination of carcasses in pasture-raised chicken production systems since the incidence of foodborne pathogens such as *Campylobacter* and *Acinetobacter* were reduced in broilers fed SF.

Selection for intramuscular fat modifies microbial genome for energy metabolic routes in rabbit gut

M. Martínez-Álvaro, A. Zubiri-Gaitán, C. Casto-Rebollo, A. Blasco and P. Hernández
Institute for Animal Science and Technology, Department of genetics, Camino de Vera s/n, 46022, Spain; marinamartinezalvaro@gmail.com

High intramuscular fat content (IMF) of meat has an economic interest, as it improves meat quality. An exceptional experimental material to study the genetic basis of IMF deposition was developed in rabbits, consisting of two lines from a common genetic origin and divergently selected for IMF. In this study we investigate the changes on the cecum microbes genome, identifying the metabolic routes modified by the selection. A correlated response on the microbiome would imply a link between host genome and its metagenome. Cecum content samples were collected from 16 and 17 rabbits from each of the two selected lines for high (H) and low (L) IMF, in the 10th generation of selection. Samples were sequenced with an Illumina NextSeq instrument. Metagenomic data were pre-processed using the centred log ratio transformation due to its compositional nature. Distinct microbial genes between the two lines were identified using Projection to Latent Structures Discriminant Analysis (PLS-DA), selecting variables with a variable importance for projection (VIP) >1. The final model included 105 microbial genes, showing a classification ability after cross-validation (Q^2) of 91.6%. These 105 microbial genes coded for proteins involved in several metabolic pathways, being one of the most represented the energy metabolism pathway (18 genes). These genes showed different relative abundance in H and L lines. For instance, two genes involved in methane metabolism, and two involved in the metabolism of mannose and fructose specific carbohydrates were more abundant in the H line, whereas genes related lipopolysaccharides biosynthesis were more abundant in L line. Besides, when analysing only these 18 genes in PLS-DA model, its explanatory power was high (Q^2=55.7%). This preliminary analysis highlights the importance of the gut microbiome in the muscular lipid deposition in rabbits and shows that selection for IMF led to a correlated response in their metagenomics profile, particularly in the energy metabolic routes. These results imply a link between the genes of the individual and the genes of its gut microbes.

Rabbit microbiome analysis by using compositional data analysis techniques

A. Zubiri-Gaitan, M. Martínez-Alvaro, A. Blasco and P. Hernández
Institute for Animal Science and technology, Universitat Politècnica de València, 46022, Valencia, Spain;
azubirigaitan@gmail.com

The objective of this study was to analyse the effect of selection for intramuscular fat content (IMF) in the cecum microbiota composition of rabbits, taking into account the compositional nature of microbiome data. The animals used for this study came from two rabbit lines divergently selected for IMF. Cecum samples were taken from 16 and 17 rabbits from the high (H) and low (L) IMF lines, respectively. The metagenome was obtained by Illumina NextSeq sequencing, and taxonomic classification of the reads was performed using Kaiju. 2,653 operational taxonomic units at genus level were identified. All genera with zero counts in more than 25% of the individuals were discarded and 1,435 genera were kept. The remaining zero counts were replaced by imputed values using a Bayesian-multiplicative treatment. The vector of counts of each rabbit was normalized to unit constant sum. For this study, the compositional nature of microbiome data was considered in order to compare the cecum microbiota of the high and low IMF lines. Compositional data is defined in a sample space called simplex with Aitchison geometry, which is different from the Euclidean geometry of the real space. Compositional data can be transformed into real space coordinates by an isometric log-ratio transformation (ilr), where standard statistical techniques can be applied. Ilr transformation is defined as the logarithm of the ratio between the geometric means of different variables (microbial genera) combinations. We call these ratios 'balances'. In the present study, an algorithm called *selbal* was applied to the microbiome dataset in order to find the optimal balance able to discriminate the lines. The balance selected included four bacterial genera, grouping *Firmicutes Caldicoprobacter* and *Firmicutes Dendrosporobacter* on one side, and *Candidatus Vecturithrix* and *Acidobacter_na* on the other side. The balance was able to successfully discriminate the lines, thus indicated by a correlation of 0.98 between predicted and observed values. When testing the ratio H/L for these microbial genera, L showed greater relative abundance of *Firmicutes Dendrosporobacter* but lower of *Candidatus Vecturithrix* and *Acidobacter_na* in the high line.

Colonic microbiota of weanling piglets fed diets containing rice distiller's by-product

O. Nguyen Cong, B. Taminiau, B. Jérôme, T. Vu Dinh and J.-L. Hornick
University of Liège, Veterinary Management of Animal Resources, Quartier vallée 2, avenue de Cureghem 6, B43a, 4000
Liège, Belgium; ncoanh@gmail.com

This study was conducted to evaluate the effects on gut microbiome of different levels of rice distiller's by-product (RDP) in diets of weaned piglets. A total of 48 weaned castrated male crossbred pigs, initial body weight (IBW) 7.5±0.97 kg, and age about 4 wks, were used. They were randomly allocated into 3 diet groups of 4 blocks (4 pigs per block) according to similar IBW and sow origin by treatment. The piglets were fed either a control diet, a diet with 15% RDP, or a diet with 30% RDP for a total of 35 days experiment. On days 14 and 35, eight piglets per treatment (2 per block) were exsanguinated and eviscerated for the collection of digesta samples at 6 hours after feeding. Individual samples of each colonic digesta were collected in sterile tubes (PSP@ Spin Stool DNA Plus Kit, Germany), and DNA extracted in Biotechnology laboratory and stored at −20 °C. PCR-amplification of the V1-V3 hypervariable region of bacterial 16S rRNA and sequencing on a MiSeq sequencer using v3 reagents (ILLUMINA, USA) was performed. Gene amplicon profiling of 16S rRNA revealed that colonic microbiota composition of weaned pigs differed between treatment groups and over periods, especially at family and genus levels. On day 14, a decreased proportion of *Lachnospiraceae_ge*, *Ruminococcaceae_ge*, *Ruminococcaceae_UCG-005*, *Bacteroidales_ ge*, and *Prevotella_1*, and an increased proportion of *Prevotellaceae_ge*, *Prevotella_2*, and *Prevotella_9* were observed with inclusion of RDP, while opposite effect was found on day 35. Additionally, the proportion of genera *Lachnospiraceae_ge*, *Ruminococcaceae_ge*, *Ruminococcaceae_UCG-005*, and *Bacteroidales_ge* decreased over periods in the control diet but increased largely in RDP30, which was considered as beneficial to the host. In conclusion, RDP in a favourable way modulated gastrointestinal microbiota composition in piglets.

Impact of early life feeding management on fattening calves ruminal metagenome

S. Costa[1], G. De La Fuente[1], M. Blanco[2], J. Balcells[1], I. Casasús[2] and D. Villalba[1]
[1]University of Lleida, Av. Alcalde Rovira Roure, 191, 25198, Lleida, Spain, [2]CITA, Av. Montañana 930, 50059, Zaragoza, Spain; scosta@ca.udl.cat

The aim of this work (GENTORE H2020) was to study the impact of feeding management in early life on fattening calves ruminal metagenome. Eight *Parda de Montaña* male calves were raised with their dams from birth to weaning (156 days [d] of age); they were fed on their dams' milk and had free access to hay and straw (DAM). Ten Holstein male calves were separated from their dam at birth and they received milk replacer and a starter concentrate until weaning (56 d of age) (REP). From weaning, they were fed cereal-based concentrate and barley straw *ad libitum*. Ruminal fluid was obtained via oesophagus tube twice: in growing (GRO; 172 d of age, 241 kg body weight [BW]) and finishing periods (FIN; 295 d of age and 438 kg BW). Rumen bacterial and archaeal community composition was analysed by taxonomic profiling of 16S ribosomal RNA V3-V4 variable regions. Venn diagram, Shannon index values, multivariate analysis and Adonis test were obtained and analysis of variance of microbial abundance data (including treatment, period and their interaction as effects) was performed. Core bacterial/archaeal community (shared taxa by all individuals) was comprised of 87 OTUs (out of 459), which gathered the 92.0% of analysed sequences. Contrarily, specific OTUs of each treatment and period clustered few sequences (between 0.06-0.25%, within groups). Ruminal biodiversity, in terms of Shannon index values, was affected by treatment by period interaction (P=0.003) and it was higher in DAM animals in both GRO (2.45 in DAM *vs* 1.03 in REP) and FIN (2.32 in DAM *vs* 1.86 in REP), showing different microbial development in both groups. Multivariate analysis plot showed clear clustering of animals according to treatment and period, which was confirmed by the Adonis test (P<0.001). *Bacteroidetes*, *Firmicutes*, and *Actinobacteria* were the main phyla in both treatments. DAM animals had higher titers of *Selenomonas, Olsenella* and *Methanobrevibacter* genus, whereas REP ones had more abundant *Prevotella* and *Agathobacter* genus. The feeding management in early life clearly affected the calves' ruminal metagenome and this effect lasted over all their fattening period.

Effects of dietary capsicum supplementation on growth performance and gut microbiota of weaned pigs

J.H. Cho[1], R.B. Guevarra[1], J.H. Lee[1], S.H. Lee[1], H.R. Kim[1], J.H. Cho[2], M. Song[3] and H.B. Kim[1]
[1]Dankook University, Department of Animal Resources Science, Cheonan, 31116, Korea, South, [2]Chungbuk National University, Department of Food and Animal Sciences, Cheongju, 28644, Korea, South, [3]Chungnam National University, Department of Animal and Dairy Science, Daejeon, 34134, Korea, South; hbkim@dankook.ac.kr

Dietary capsicum has been scientifically recognized to improve the gut health of weanling pigs due to its anti-inflammatory functions, which may reduce the negative effects of pathogen infection. Nowadays, it is urgent to develop alternatives to antimicrobials in livestock to address the problem of antimicrobial resistance. However, there have been no reports to observe the composition of the gut bacterial communities in response to feeding dietary capsicum extracts in weanling pigs. Therefore, the current study was conducted to determine the effects of capsicum on growth performance and intestinal microbiota of weaned piglets using high-throughput next generation sequencing. A total of 12 Landrace-Yorkshire-Duroc (LYD) pigs at 4 weeks of age were randomly assigned to 2 dietary treatments (2 pigs/pen; 3 pens/treatment). The dietary treatments were: (1) control diet based on corn and soybean meal (CON);and (2) capsicum group (CAP) (CON + 300 ppm capsicum extract). Faecal samples were collected at 0 and 6 weeks after feed supplementation. Total DNA extracted from fresh faecal samples were used to amplify the 16S rRNA gene for Illumina MiSeq sequencing. Sequence quality filtering was performed using Mothur and analysis of the faecal microbiota was performed using the QIIME pipeline. Data for growth performance was analysed using the PROC GLM procedure of SAS. Capsicum extracts showed an increase in average daily gain (ADG) (14 to 29 g/day) and feed efficiency (0.591 vs 0.631 g/g) as compared with CON group throughout the experimental period. Taxonomic analysis revealed a higher relative abundance of *Prevotella* and *Methanobrevibacter* in capsicum fed piglets than in the CON group. Principal coordinate analysis (PCA) showed clear separation of the microbiota from the CAP group at week 6 as compared to CON. In conclusion, dietary capsicum extracts can improve the growth performance by shifting the gut microbial community composition of weaned piglets.

Intestinal microbiota and growth performance of weaned pigs fed with dietary garlic and turmeric

R.B. Guevarra[1], J.H. Cho[1], J.H. Lee[1], S.H. Lee[1], H.R. Kim[1], J.H. Cho[2], M.H. Song[3] and H.B. Kim[1]
[1]Dankook University, Department of Animal Resources Science, Cheonan, 31116, Korea, South, [2]Chungbuk National University, Department of Food and Animal Sciences, Cheongju, 28644, Korea, South, [3]Chungnam National University, Department of Animal and Dairy Science, Daejeon, 34134, Korea, South; robin.becina.guevarra@gmail.com

Garlic and turmeric are spices that can be considered as best disease-preventing foods which have received increased clinical attention. Insufficient data exist on the influence of garlic and turmeric extracts on the gut microbiota and such studies are lacking in pigs. A total of 18 Landrace-Yorkshire-Duroc (LYD) pigs at 4 weeks of age were randomly assigned to 3 dietary treatments (2 pigs/pen; 3 pens/treatment). The dietary treatments were: (1) control diet based on corn and soybean meal (CON); (2) turmeric group (CON + 300 ppm turmeric powder); and (3) garlic group (CON + 300 ppm garlic powder). Faecal samples were collected at 0 and 6 weeks after feed supplementation. Total DNA extracted from fresh faecal samples were used to amplify the 16S rRNA gene for Illumina MiSeq sequencing. Sequence quality filtering was performed using Mothur and analysis of the faecal microbiota was performed using the QIIME pipeline. Data for growth performance was analysed using the PROC GLM procedure of SAS. Taxonomic analysis revealed that garlic fed piglets showed a higher relative abundance of genera *Parabacteroides*, *Methanobrevibacter*, *Oscillospira*, and *Ruminococcus* compared to control pigs. In addition, analysis of the microbiota of pigs fed with turmeric extract showed an increase in the relative abundance of genus *Methanobrevibacter*. Principal component analysis showed a clear separation of intestinal microbiota between control and turmeric fed piglets at week 6. Pigs fed with garlic and turmeric diets showed higher average daily gain and gain to feed ratio than in the control group (P<0.05). Collectively, these results suggest that dietary garlic and turmeric supplementation can improve the growth performance by shaping the intestinal microbial community composition of weaned piglets. Here we present the effects of dietary garlic and turmeric extracts on intestinal microbiota for the nutritional management of gut health in weaned pigs.

Development of the microbiota along the gastrointestinal tract of calves in early postnatal life

L. Arapovic[1], B.P. Thu Hang[2], C.E. Hernandez[1], M.A. McGuire[3], B.O. Rustas[1] and J. Dicksved[1]
[1]Swedish University of Agricultural Sciences, Department of Animal Nutrition and Management, Box 7024, SE750 07 Uppsala, Sweden, [2]An Giang University, Department of Animal Sciences and Veterinary Medicine, No 18, Ung Van Khiem street, Dong Xuyen ward, Long Xuyen, Viet Nam, [3]University of Idaho, Department of Animal and Veterinary Science, 875 Perimeter Drive MS 4264, Moscow, USA; johan.dicksved@slu.se

Microbial colonization of the gastrointestinal tract (GIT) is crucial for proper development of the mucosal immune system in calves. The species composition of the initial colonizers is likely important for health and growth. In ruminants, most studies of the microbiota have primarily focused on the developed rumen due to the high importance in feed digestion and climate footprint. Few studies have focused on the early colonization of the microbiota in the pre-ruminant period. This study aimed to characterize the microbiota along the GIT of calves in early postnatal life. Intestinal contents and mucosal scrapings were collected separately from six GIT segments (rumen, abomasum, duodenum, ileum, cecum, and colon) from 13 2-d-old and 12 7-d-old calves after euthanasia. Calves received colostrum as their first meal (an amount equal to 8.5% birth weight) and were then fed whole milk. The composition of the microbiota in the collected samples was assessed by sequencing 16S rRNA gene amplicons using Illumina MiSeq. Clustering of bacterial communities among samples was assessed with principal coordinate analysis (PCoA) and analysis of similarity (ANOSIM). Sequencing revealed that the microbiota composition was linked to the intestinal segment with large differences between the proximal and distal parts of the GIT, but with higher similarity in segments with close proximity to each other. Clustering by segment became more evident in the 7-d old calves. Even though there was a large overlap in composition between mucosal scrapings and intestinal content, specific microbial groups were enriched in either mucosal scrapings or intestinal content. Results demonstrate that already at 2 d of age a niche adaptation of the microbiota is evident.

Effect of 9c,11t and 10t,12c conjugated linoleic acid on bovine PBMC apoptosis and viability

G. Ávila, C. Catozzi, D. Pravettoni, G. Sala, C. Lecchi and F. Ceciliani
Università di Milano, Dipartimento di Patologia Animale, Via Celoria,10, 20133, Milano, Italy, gabriela.avila@unimi.it

The conjugated linoleic acid (CLA), a group of naturally occurring isomers of the essential fatty acid linoleic acid, are receiving special attention in animal and in human nutrition. Anticarcinogenic, anti-atherosclerotic, antidiabetic and immunomodulatory effects of CLA were reported. However, the effects of CLA on immune cells remain undisclosed. Results in ruminants are contradictory, reporting either increased proliferation and viability of peripheral mononuclear cells (PBMC), or a decrease, or no effect at all in these and other immune functions. Apoptosis is an effective strategy to modulate mononuclear cells' lifespan. No information about how CLA modulates apoptosis in ruminant PBMC is available. In this preliminary study we sought to determine the *in vitro* effects of two CLA isomers, namely 9c,11t and 10t,12c (individually and in combination) on apoptosis and viability of bovine PBMC. PBMC from 6 healthy animals were isolated from whole blood and incubated with increasing concentrations of both isomers (10, 50, 100, and 500 µM). Apoptosis was assessed by measuring the fluorescence produced after specific cleavage of DEVD by Caspase-3 and -7, while the viability was quantified using a colorimetric MTT based assay. Statistical analyses were performed in GraphPad Prism 8.0.2 using repeated measures one-way ANOVA, followed by Tukey's multiple comparison for the different CLA concentrations and paired t-tests for the isomers' effects. Apoptosis rate was not affected by any of the CLA isomers, while the viability of cells was reduced (P<0.05) when cells were incubated with 100 µM and 500 µM CLA. Also, no differences between isomers were observed. These preliminary results suggest that 9c,11t and 10t,12c CLA isomers do not affect apoptosis of bovine PBMC and that viability at concentrations compared to the physiological ones (50 µM) is not impaired. These results are a starting point for further analyses of the *in vitro* impact on other immune related functions like chemotaxis, oxidative burst and phagocytosis, and to compare the differential effects of saturated and unsaturated fatty acids on immune functions.

Risk factors for intestinal health problems in broilers in Belgium: a cross-sectional study

M. Ringenier, N. Caekebeke, F. De Meyer, V. Eeckhaut, R. Ducatelle, F. Van Immerseel and J. Dewulf
Faculty of Veterinary Medicine, Ghent University, Salisburylaan 133, 9820 Merelbeke, Belgium; moniek.ringenier@ugent.be

Objective Dysbiosis (DB) is a main indication for the use of antimicrobials in broilers. The only method to diagnose DB, as described by Terilynck, is relatively subjective and a more objective method is needed to determine the presence of DB in an early stage to reduce antimicrobial usage. Therefore, we wanted to determine the prevalence of DB and coccidiosis in broilers in Belgium. Also, the aim was to identify risk and protective factors for the development of DB. Methods 50 Belgian broiler flocks were randomly selected and visited when the animals were at 28 days of age. Farm-specific information was collected and determination of the health status of the flock was done by performing necropsies on 10 broilers. Macroscopic lesion scoring for DB and coccidiosis was performed and several intestinal segments were collected for determination of morphological parameters. Water and feed samples were collected for analysis. Results The average DB score on farm level was 3,45±1.07 [0-5.2] and on animal level was 3.53±1.58 [0-9]. The average coccidiosis score on animal level for all three *Eimeria* species was <1 (*E. acervulina* 0.74±0.97 [0-4]; *E. maxima* 0.75±0.66 [0-3]; *E. tenella* 0.21±0.41 [0-2]). No significant associations were found between the DB score and the coccidiosis score. Broilers with footpad lesions had significant shorter villi in comparison with animals without footpad lesions. Against expectations, a higher total DB score was significantly negatively correlated with a lower food conversion rate of the farms. A significant negative correlation was found between the length of the villi and the amount of neutral detergent fibre present in the feed. The average number of days that the farms treated the animals with antibiotics was 5.96±2.97 [0-14] (not significantly associated with DB score). Conclusions The correlations found in this study were not strong, which raised the question if the DB score on day 28 is the right parameter to determine risk and protective factors for the development of DB. Further research is needed to see if stronger correlations are found at a younger age.

A holistic One Health perspective on future agricultural systems with full nutrient recycling
B. Amon and Optinutrient Consortium
Leibniz Institute for Agricultural Engineering and Bioeconomy (ATB), Max-Eyth-Allee 100, 14469 Potsdam, Germany;
bamon@atb-potsdam.de

Agriculture as currently practiced causes a significant amount of environmental damage, with consequent negative effects on health and health-related costs. In order to prevent these problems in future, our 'OptiNutrient' concept uses a true *One Health* approach to place agriculture in a new context and expand its range from the narrow task of producing biomass to include a comprehensive role in protecting and promoting health, thus taking responsibility for reducing environmental and health risks. Agriculture will play an integral role in protecting and promoting health within a comprehensive *One Health* policy. We suggest to set up five clusters for research and development to achieve efficient recycling of nutrients, and improve human and animal health. The Crop Cluster will research concepts for self-sufficiency in nutrients from the biological cycle. Regionally adapted crop rotations integrate legumes, and use innovative sensor technologies and improved plant growth models to utilize nutrients from a wide spectrum of bio-based fertilizers. The Animal Cluster will investigate innovations for animal health and more efficient feed utilization. Regional breeding strategies and measures that respond to individual genetic dispositions will be developed. Integrating the Nutrition Cluster ensures a holistic system perspective. This will require concepts for comprehensive nutrient recycling of urban material streams including solutions for potential hygienic risks. The Nutrient Cluster will break new ground by bringing together all the relevant actors and nutrient flows in the *Bio-based Nutrient Platform*, identifying nutrient streams and ensuring their distribution according to needs with high temporal and spatial precision. Innovations for processing of organic waste will be investigated. Methods for intelligent data analysis and information flows will be used for processing and analysing data gathered in all clusters. The Impact Cluster will ensure close cooperation of all stakeholder groups to make sure that innovations are implemented in practice. The acceptance of selected technologies by consumers will be assessed. OptiNutrient experiments and results will be expanded by means of European and global modelling.

The One Health European Joint Programme (OHEJP): a new European initiative in One-Health
J. Hoorfar
DTU, FOOD, Kemitorvet, 2800 Kgs Lyngby, Denmark; jhoo@food.dtu.dk

The One Health EJP is a €90 million project under Horizon 2020 that comprises a network of 39 national public organisations, including reference centres for animal and public health infectious diseases in 19 member states of the EU. The overall objective is share knowledge to enhance prevention, detection and control of zoonoses and antimicrobial resistance. The project concept recognises that human health is tightly connected to the health of animals and the environment, for example that animal feed, human food, animal and human health, and environmental contamination are closely linked. Therefore the study of infectious agents that may cross species and environmental barriers to move between these compartments is imperative. In line with the 'Prevent-Detect-Respond' concept, the main focus of the new OHEJP is to reinforce collaboration between institutes by enhancing transdisciplinary cooperation and integration of activities by means of dedicated Joint Research Projects, Joint Integrative Project and through education and training in the fields of Foodborne Zoonoses (FBZ), Antimicrobial Resistance (AMR) and Emerging Threats (ET). The OHEJP is an exemplar of the 'One Health' concept, and boasts a landmark partnership among 39 acclaimed food, veterinary and medical laboratories and institutes across 19 member states in Europe, and the Med-Vet-Net-Association. An interdisciplinary, integrative and international approach to One Health is essential to address the existing and emerging threats of zoonotic disease and antimicrobial resistance. Through the OHEJP there are opportunities for harmonization of approaches, methodologies, databases and procedures for the assessment and management of foodborne hazards, emerging threats and AMR across Europe, which will improve the quality and compatibility of information for decision making. The joint research projects (JRPs) and joint integrative projects (JIPs) are key instruments to facilitate partner organizations working together and aligning their approaches, increasing their knowledge base of host-microbe interactions, and improving epidemiological studies and risk assessments which ultimately equip risk managers with the best tools for intervention measures.

Upcycling food leftovers and grass resources through farm animals

O. Van Hal[1], I.J.M. De Boer[1], A. Muller[2], S. De Vries[3], K.-H. Erb[4], C. Schader[2], W.J.J. Gerrits[3] and H.H.E.
Van Zanten[1]
[1]Wageningen University & Research, Animal Production Systems group, De Elst 2, P.O. Box 337, 6700 AH Wageningen, the
Netherlands, [2]Research Institute of Organic Agriculture (FiBL), Ackerstrasse 113, 5070 Frick, Switzerland, [3]Wageningen
University & Research, Animal Nutrition group, De Elst 2, P.O. Box 337, 6700 AH Wageningen, the Netherlands,
[4]University of Natural Resources and Life Sciences, Institute of Social Ecology, Schottenfeldgasse 29, 1070 Vienna,
Austria; hannah.vanzanten@wur.nl

Although animal production currently competes for resources with food production, it is increasingly acknowledged that farm animals can contribute to a resource efficient food system by upcycling 'low-opportunity-cost feed' (LCF, defined as food leftovers and grass resources), into nutritious animal-source food. To fully exploit the benefits of such animal production, available LCF should be fed to that combination of farm animals that contributes most to human nutrition. We, therefore, optimise the use of LCF available in the EU using a model that distributes LCF to those animal production systems that maximise animal protein production. We included EUs most common livestock systems (pigs, laying hens, broilers, dairy cattle and beef cattle), considering their nutrient requirements under low mid and high productivity. Available LCF included food processing by-products and food waste (derived by combining current food supply with data on wastage and processing) and currently available grass resources. Optimal conversion of these LCF provides 31 g animal protein/(cap*d), requiring a variety of animal production systems and productivity levels. Most appropriate production systems had a high conversion efficiency (laying hens, dairy cattle), were able to valorise specific LCF (dairy cattle for grass; pigs for food waste), or were able to valorise low quality LCF due to a low productivity. Using only highly productive livestock reduced animal protein supply by 16 to 26 g/(cap*d). The estimated animal protein supply from animals fed solely on LCF was, furthermore, sensitive to assumptions regarding the availability and quality of LCF, especially grass resources. Our model provides valuable insights into how animals can efficiently use LCF, essential for a transition towards a resource efficient food system.

Environmental impacts and their association with performance and excretion traits in growing pigs

A.N.T.R. Monteiro[1], L. Brossard[1], H. Gilbert[2] and J.Y. Dourmad[1]
[1]PEGASE, Agrocampus Ouest, INRA, 16 le Clos, 35590 Saint-Gilles, France, [2]GenPhySE, Université de Toulouse, INRA,
INPT, ENSAT, Castanet-Tolosan, France; jean-yves.dourmad@inra.fr

The selection of animals for improved production traits has been, for a long time, the major driver of pig breeding. More recently, because of the increasing concern with environment, new selection criteria have been explored, such as nitrogen (N) excretion which is related to both feed efficiency and environmental impact. However, many studies indicate that life cycle assessment (LCA) provides much better indicators of environmental impacts. In this context, the objective of this study was to investigate, using a modelling approach, the relationships between production traits and LCA impacts of individual growing pigs. The LCA considered the process of pig fattening, including production and transport of feed ingredients and complete feeds, the raising of pigs, and manure management. Impacts were calculated at farm gate and the functional unit was one kg of body weight gain over fattening. Performances of pigs were simulated for two feeding programs, 2-phases and precision feeding, using InraPorc population model (2,000 pigs per scenario) and considering between-animal variability. The LCA calculations were performed for each pig according to its own performance and N excretion. The CORR procedure (SAS version 9.2) was used to investigate the correlations between LCA impacts and performance and excretion traits. The results indicated that N excretion was positively correlated with feed conversion ratio (FCR; r=+0.94), climate change (CC; r=+0.94), acidification potential (AC; r=+0.96), and eutrophication potential (EU; r=+0.96), whatever the feeding program. Surprisingly, FCR appeared the best indicator of LCA impacts with a very high and positive correlation (r>+0.99) with CC, AC, and EU for both feeding programs. Despite the lower CC, AC, and EU of pig production in a precision feeding program (3, 16 and 11% lower than the 2-phases, respectively), the correlation within each outcome were very similar. It was concluded that the use of FCR as a selection criterion in pig breeding is the most effective way to associate improved performance and low environmental impact of pig fattening.

Nutrition and antimicrobial resistance issues with zinc oxide in pigs

J. Zentek

Institute of Animal Nutrition, Veterinary Medicine, Königin-Luise-Str. 49, 14195 Berlin, Germany;
juergen.zentek@fu-berlin.de

Zinc is one of the essential trace elements that is needed in pigs at a relatively low concentration (50-80 mg/ kg of feed). Zinc has numerous metabolic functions, which, due to its structural and functional significance, can very quickly lead to clinical deficiency symptoms. In addition to obvious changes in the skin integrity, numerous other effects have been described, such as immunological deficits. Due to the positive effects of a high, so-called pharmacological dosage of zinc oxide, especially in weaned piglets, higher concentrations, which clearly exceed the requirements, are used worldwide. The European Union has set the maximum value in feed for piglets at 150 mg/kg. This is higher than the physiological requirement and should be sufficient even under the most unfavourable conditions to guarantee a physiological zinc supply of pigs. Pharmacological dosages of zinc oxide are increasingly viewed critically, especially in Europe, as there are negative repercussions on the environment due to increased zinc excretion via manure as well as an induction of antimicrobial resistance of gastrointestinal bacteria. This has been clearly demonstrated in a number of studies. In this respect, recent studies focus on the question of how a lower concentration of zinc in feed and the targeted use of more readily available zinc sources can achieve a safe supply while at the same time ensuring the health of the animals.

Contribution of animal breeding to reduce environmental impact of animal products

H. Mollenhorst and Y. De Haas

Wageningen University and Research, Animal Breeding and Genomics, P.O. Box 338, 6700 AH Wageningen, the Netherlands; yvette.dehaas@wur.nl

Animal production is responsible for 14.5% of total anthropogenic greenhouse gas (GHG) emissions. Approximately half of these emissions originate directly from animal production, whereas the other half comes from feed production. Animal breeding aims at improving animal production and efficient use of resources, which results in a reduction of environmental impact. The objective of this study was to quantify the contribution of animal breeding to reducing the environmental impact of the four major livestock species in the Netherlands (with their main product), namely broilers (meat), laying hens (eggs), pigs (meat) and dairy cattle (milk). A literature review was performed to assess historical trends in environmental impact, mainly focussed on GHG emissions, and general performance criteria, like feed efficiency and lifetime production. Furthermore, we quantified current environmental impact as well as the result of recent genetic improvements of the four animal products based on data provided by animal breeding companies. For eggs and broiler and pig meat, we focussed on GHG emissions and nitrogen and phosphorus efficiency, whereas for dairy we focussed on enteric methane emissions, an important contributor to GHG emissions. Results showed that selection on increased (feed) efficiency indirectly reduces environmental impact per unit of animal product by about 1% per year. If the aim is to directly select on environmental traits, recording of new traits is required; e.g. nitrogen and phosphorus contents of meat and eggs and methane emission of individual dairy cows.

One Health aspects of antimicrobial resistance

S. Schwarz
Institute of microbiology and Epizootics, Freie Universität Berlin, Department of Veterinary Medicine, Robert-von-
Ostertag-Str. 7-13, 14163 Berlin, Germany; stefan.schwarz@fu-berlin.de

'One Health' is the evolution of the earlier used term 'One Medicine' which recognizes that humans do not exist in isolation, but are a part of a larger whole, a living ecosystem, and that activities of each member of this ecosystem affect the others. The 'One Health' concept recognizes that the health of people is connected to the health of animals and the health of the environment. Interdisciplinary collaborations and communications in all aspects of healthcare of humans, animals and the environment are indispensable. Although the term 'One Health' is fairly new, the concept has long been recognized, both nationally and globally. Since the 1800s, scientists have noted the similarities in disease processes among animals and humans, although human and animal medicine were practiced separately until the 20th century. Under the 'One Health' umbrella, many different topics can be summarized, one of which is antimicrobial resistance. Antimicrobial agents are used extensively in human medicine, veterinary medicine, but also in aquaculture and horticulture. In all sectors, bacterial resistance to antimicrobial agents occurs and the resistant bacteria as well as their resistance properties are interchanged. Examples will be provided that refer to emerging resistance properties seen recently among bacteria from humans and animals, including transferable resistance to: (1) carbapenems; (2) polymyxins; and (3) oxazolidinones, all of which are antimicrobial agents listed as 'critically important' by the WHO. The examples presented will illustrate that antimicrobial resistant bacteria and their resistance genes have evolved in different sectors and are spread within and between these sectors via variable transmission routes.

The past and future of nitrogen use efficiency in agriculture

J.W. Erisman
Louis Bolk Institute, Kosterijland 3, 3981 AJ Bunnik, the Netherlands; j.erisman@louisbolk.nl

For a long time there was enough naturally occurring nitrogen (N) to provide food for the world's peoples. Then we decided to live in settlements and in cities and while the population started to grow nitrogen management became important. Still there was no shortage. Major breakthroughs in nitrogen science was more driven by warfare and market mechanisms than for food. When population increased further nitrogen became limited for our food. With the invention of the chemical fixation of N the scarcity problem was solved. But this transition from enough, to scarcity, to plenty has come with a tremendous environmental cost. It has changed the physical, societal, industrial and economic landscapes in the world. The history of nitrogen (N) use by people is a field well-fertilized with materials detailing the evolution of knowledge about nitrogen, the development of the Haber-Bosch process and the extent and impact of N cycle alteration by people. This presentation provides an historical overview of the growth of knowledge about N and about its impacts, both positive and negative. It marks the major needs for nitrogen and the accumulation of N knowledge with respect to its benefits (feeding the wars) and its problems (specialization in agriculture and cascading and accumulating N impacts). By focusing on the major drivers for managing nitrogen in the future we explore different potential solutions of how to feed the world's population while preserving the landscape. Furthermore, we will identify the need for continuous knowledge and technology development.

European research and innovation for a sustainable and competitive livestock production sector

J.-L. Peyraud[1] and B. Amon[2]
[1]Animal Task Force, Maison Nationale des Eleveurs, 149, rue de Bercy, 75595 Paris Cedex 12, France, [2]Leibniz Institute for Agricultural Engineering and Bioeconomy (ATB), Max-Eyth-Allee 100, 14469 Potsdam, Germany; jean-louis.peyraud@rennes.inra.fr

The Animal Task Force (ATF) is a leading body of expertise linking European industry and research providers for developing innovation in the livestock sector. ATF members are European research providers, and industry representative bodies that support the interests of Europe's livestock industries. The ATF encourages the future development of livestock production systems using holistic agriculture approaches to optimise synergies in production systems. Integration of livestock and plant sectors could help to maximise circularity in production ecosystems, evaluate and protect the properties of agro-ecosystems, optimise the use of biomass, minimise damage to the environment, improve the efficiency of livestock production and provide ecosystem services. In March 2018, the ATF published a position paper for research and innovation (R&I) within Horizon Europe supporting a sustainable and competitive terrestrial livestock production sector in Europe, including herbivores and monogastrics. ATF fully acknowledges that interrelations between green and blue economies should be carefully considered to achieve a sustainable European food and non-food production. Part 1 of the Vision Paper stresses the importance of the European livestock sector in a global context. Part 2 describes the new paradigm shift and expected R&I support required to address the main challenges facing livestock farming. Part 3 proposes a framework for R&I around 4 main levers that may be combined: agro-ecological practices, circularity, innovative (bio)technologies and governance. Finally, part 4 outlines the main impacts expected from European R&I. Bringing together research groups with complementary expertise (feeding, animal breeding, reproduction, nutrition, physiology and health, new technologies (incl. ICT, -omics approaches and New Breeding Techniques, food evaluation, modelling, economics, sociology, multi-criteria evaluation, etc.) is necessary to address the complexity of issues related to societal challenges and to consider the huge diversity of production systems and various scales.

Control of biotrickling filter efficiency on NH_3 emitted by piggeries

E. Dumont[1], S. Lagadec[2], N. Guingand[3], L. Loyon[4], A. Amrane[5] and A. Couvert[5]
[1]IMT Atlantique, UMR CNRS 6144 GEPEA DSEE, 4 rue Kastler CS 20722, 44307 Nantes, France, [2]CRAB, rue Le Lannou CS 74223, 35042 Rennes, France, [3]IFIP Institut du Porc, La Motte au Vicomte, 35651 Le Rheu, France, [4]IRSTEA, UR OPAALE, 17 avenue de Cucillé, 35044 Rennes, France, [5]Université de Rennes ENSCR, CNRS. ISCR-UMR 6226, 11 allée de Beaulieu, 35000 Rennes, France; nadine.guingand@ifip.asso.fr

Ammonia emitted by piggeries can be removed using biotrickling filtration. As ammonia is a very soluble compound in water, removal efficiency (RE) around 70-80% can be expected. However, the accumulation of nitrogen salts in water leads to a decrease in RE. Consequently, there is need to improve the management of equipment by controlling the amount of water which has to be discharged and replaced by fresh water in order to limit the accumulation of nitrogen salts. Such an improvement is based on the knowledge of the nitrogen mass balance between the gas phase and the liquid phase. The objective of this study was to establish the nitrogen mass balance of a pilot-scale biotrickling filter treating ammonia emitted by a pig house. The experiment was carried out for 14 weeks on a French pig farm. The biotrickling filter installed to treat the polluted air generated by 54 fattening pigs was filled with a structured plastic packing WAT NET 150 NC 20/48 (0.9×0.9×0.45 m). The airflow rate was of 1,350 m^3/h corresponding to an Empty Bed Residence Time (EBRT) of 1 s. In the gas phase, temperature, relative humidity, ammonia and nitrous oxide concentration were hourly measured. In the liquid phase, temperature, pH, electrical conductivity, and concentration of nitrogen salts (NH_{4+}, NO_{2-}, NO_{3-}) were weekly measured. Results showed that nitrogen mass balances carried out on both phases are in agreement (5% difference). A steady transfer rate of ammonia from the gas phase to the liquid phase was obtained (3.22 gN/h corresponding to 10.0 gN/pig/week). From the measured concentrations of nitrogen salts in the liquid phase, it was calculated that the nitrogen mass transfer was 9.5 gN/pig/week. Moreover, it was also evidenced that the amount of nitrogen salts dissolved in water could be correlated to the water conductivity. As a result, the measurement of this parameter could be a useful tool to determine the amount of ammonia removed from the gas phase.

Ruminant milk fatty acids and human health: effects of species in interaction with diets

H. Fougère and L. Bernard
INRA, Theix, St Genès-Champanelle, 63122, France, h.fougere@yahoo.fr

Dairy products are the main vector of lipids contained in human diets: they provide fatty acids (FA) with positive or negative effects on human health. Among breeding factors, nutrition, as the use of lipid supplements, is a rapid and efficient tool to modulate milk fat yield and FA composition, to improve the health value of milk for the consumer. Indeed, certain FA found in milk may exert negative effects (12:0, 14:0, 16:0) when consumed in excess, whilst others have potentially (4:0, 18:1, ω3 FA) or putative (11 CLA) beneficial effects on human health. A comparative study on the dietary regulation of lipid metabolism in dairy cows and goats was performed using various dietary lipids to clarify the specificities of response to diets of these 2 ruminant species to control milk fat yield and composition. The effects of diets containing no additional lipid (CTL) or supplemented with corn oil (5% dry matter intake (DMI)) and wheat starch (COS), marine algae powder (MAP) (1.5% DMI), or hydrogenated palm oil (HPO) (3% DMI), on milk fat content and composition were studied in cows and goats (n=12 per species) conducted simultaneously according to a 4×4 Latin square design. Data were subjected to ANOVA using the mixed procedure of SAS. Dietary treatments had no significant effects on milk yield in both species. In cows, milk fat content decreased with COS and MAP (-45% and -22%, respectively; P<0.001) and increased with HPO (+13%, P<0.001) compared with CTL. In goats only MAP decreased (-15%, P<0.001) milk fat content. The major effects of COS compared to CTL were, in cows, a decrease in saturated FA concentration (4:0, 6:0, 8:0, 10:0, 14:0, 16:0) as observed in goats but to a lesser extent, and an increase in trans-FA in both species (+340% in cows, +97% in goats) including c>9, t>11 CLA and t>10, c>12 CLA. MAP increased long-chain PUFA ω3 in both species (+ 500% in cows, +267% in goats) whereas HPO had no effect on milk FA with potential effects on human health. These data highlighted differences in ruminant milk FA composition depending on the diet interacting with the species.

Emissions from various systems of dairy cows and fattening pigs housing systems

J. Walczak[1], W. Krawczyk[1] and P. Sendor[2]
[1]National Research Institute of Animal Production, Kraków, Poland, Krakowska 1, 32-083 Balice, Poland, [2]Agency for Restructuring and Modernisation of Agriculture, Promienistych 1, 31-481 Kraków, Poland; jacek.walczak@izoo.krakow.pl

According to the NEC Directive and EU CAP, it is necessary to determine efficient methods for reducing NH_3 and GHG emissions from animal production. The aim of the research was to determine NH_3, N_2O and CH_4 emission level from different housing systems of dairy cows and fattening pigs. The experimental material consisted in 48 heads of HF dairy cows (DC) and 240 head crossbred fatteners (CF). Animals were housed over 2 months in 6 climatic respiration chambers, each of which was fitted out with a different housing systems. Standard cubicle housing (SCH), SCH with frequent removal of slurry (FRS), SCH with 8 hours of pasturing (PC) and grooved floor (GF) were used for DC. FC was housed on the fully slatted floor (FSF), part (30%) slatted floor (PSF), FSF with frequent removal of slurry (FSR) and PSE with slurry channel sloped walls SCS. The study showed the significant reduction effect for DC and NH_3 when the slurry is removed 4 times a day (FRS) in relation to the SCH (16.9 vs 21.1 kg/place/year). But the highest difference was in GF (14,79 kg/place/year). In PC, CH_4 emission was the lowest of all DC compared group (73.8 kg/place/year). In the case of N_2O, the reduction result (0,004 kg/pc/year) was comparable to the FRS group. For FP the use of PSF resulted in a 60% reduction of NH_3 compared to FSF (0.84 vs 2.1 kg/place/year). It was a highly statistically significant difference. However, no differences were observed in CH_4 emissions. For NH_3, FSR method allowed to obtain about 20% emission reduction, similarly to the SCS (1,69 and 1,70 kg/place/year). FSR also reduced CH_4 emission compared to the FSF group (1.98 vs 2.25 kg/place/year). Only in terms of N_2O, no significant differences were found here. The study shows that it is possible to obtain even an over 20% reduction of GHG and NH_3 emissions from animal housing systems. Unfortunately, not all methods work in the same way on CH_4 and NH_3, N_2O emission due to different oxygen requirements in the formation.

Efficiency of biogas production partly using share solid manure from various livestock

J. Walczak[1], M. Sabady[2], K. Prochowska[1] and P. Sendor[3]
[1]National Research Institute of Animal Production, Krakowska 1, 32-083 Balice, Poland, [2]Veterinary Inspectorate, Wilsona 21, 97-500 Radomsko, Poland, [3]Agency for Restructuring and Modernisation of Agriculture, Promienistych 1, 31-481 Kraków, Poland; jacek.walczak@izoo.krakow.pl

Maize silage and slurry are the most common feedstock for biogas production. However, maize is a valuable feed and food material that should be replaced with waste, i.e. '2nd generation substrates'. The aim of the study was to formulate fermentation beds in which part of maize was replaced with solid manure derived from different livestock. The experiment used 8 mixtures (experimental groups, EG) of substrates containing different proportions of solid manure collected from: laying hens (17% in EG1, 23% in EG2), dairy cows (20% in EG3, 22% in EG4), mink (17% in EG5, 20% in EG6), and foxes (21% in EG7), 60-72% maize silage and 8-21% cattle slurry. The control group (CG) was a substrate composed of maize silage (40%) and cattle slurry (60%) alone. Chemical analysis of the individual components was followed by formulation of substrate mixtures to ensure a C/N ratio of 26:1 and a dry matter content of 13%. The mixtures were fermented using micro fermenters equipped with substrate dispensers and mixers, corresponding to wet fermentation, under mesophilic conditions (37 °C) with complete control over the direction and parameters of the ongoing processes (pH, amount and composition of biogas, etc.). The experiment was conducted in 4 replicates. The highest CH_4 concentration in biogas was determined for EG5, EG6 and EG7 groups (65.1; 63.6; and 63.5%) compared to CG (57.4%). The best results in terms of methane yield per unit of substrate organic dry matter (ODM) were obtained for EG7 (304.1 m^3 CH_4/t) and EG1 (298.4 m^3 CH_4/t) which was similar to CG (330.3 m^3 CH_4/t). The same groups differed in respect to chemical oxygen demand (COD) reduction (57.58% vs 72.08% and 66.03%). Except for lower COD in EG2, all other groups had similar CH_4 production rates than the CG. The results obtained showed that solid manure could be a partial substitute (up to 21%) for maize silage in methane fermentation.

Economic evaluation of broiler litter by the nutrient balance

R.A. Nacimento, M.F. Silva, J.M.B. Andreta, L.F. Araújo, C.S.S. Araújo and A.H. Gameiro
University of Sao Paulo, Department of Animal Science, Av. Duque de Caxias Norte, 225, 13.635-900 Pirassununga SP, Brazil; gameiro@usp.br

The intensification of animal production, among other causes, has increased drastically the use of feed. The manure produced has become an environmental issue. However, manure can be used as crop fertilizer in local farms or be sold as organic fertilizers. But for this, it is fundamental to know the manure chemical composition, contributing to decision-making on the best agricultural and livestock manage practices, seeking for a more sustainable production system. However, there are few scientific studies of manure evaluation. Thus, this study aims to use the NB as a tool to broiler litter economic valuation. We used previous University of Sao Paulo experimental data from broiler production, submitted to conventional feed program ('CON') or reduced nutritional matrix feed program supplemented with 1,500 FTU/kg of phytase ('1500'). Data on dietary nutrient composition, performance and carcass yield were considered. The data were extrapolated to a representative broiler production facility at Sao Paulo State (29,000 animals/flock, six flocks per year and rice hull as litter). The litter represented input of 8.3, 1.5 and 5.8% of N, P and K in the studied broiler systems, respectively. The nutrient use efficiency of CON and 1500 was highest for N (49.84 and 51.18%), followed by P (17.84 and 24.90%) and K (15.94 and 16.02%), respectively. The annual broiler litter potential revenue equivalent to chemical fertilizer would be US$ 35,158.81 and US$ 28,576.65 for CON and 1500, respectively. The phytase utilization increase P use efficiency by 7.1% of in representative broiler system, with no effect in animal performance. However, the litter provided by 1500 situation was of lower commercial value than CON. In conclusion, NB is a tool feasible to estimate the broiler litter economic value.

AniFair – a GUI based tool for assessing animal welfare using multi criteria analysis
J. Salau, L. Friedrich, I. Czycholl and J. Krieter
Institute of Animal Breeding and Husbandry, Kiel University, Hermann Rodewald Straße 6, 24118 Kiel, Germany,
jsalau@tierzucht.uni-kiel.de

In assessing animal welfare, observers gather multiple indicators which need to be converted into an objective overall welfare score to provide comparable classification of farms. The indicators are both collected as qualitative judgements and measured on various numerical scales, so that no comparability among the decision criteria is given. Studies have proven, that with the current aggregation system, some indicators affect more than one category and are, thus, weighted stronger in the calculation of the welfare score. As adaptions of the aggregation system have been discussed, the Multi-criteria decision analysis software 'AniFair' was developed. An approach to achieve comparability between criteria is Measuring Attractiveness by a Categorical Based Evaluation TecHnique (MACBETH). Here, the decision maker can judge the differences of attractiveness qualitatively in terms of attributes ranging from 'none' to 'extreme' instead of having to name quantitative differences. With the criterion 'colour' the decision maker could exemplarily evaluate the difference of attractiveness between 'blue' and 'red' as 'weak'. From these judgments, comparable numerical scales are calculated. Using all criteria in a final decision, aggregation needs to take place. 'AniFair' uses the Choquet integral for this step, which enables the modelling of criteria interaction, in contrast to the weighted mean, which considers the criteria mutually independent. Hereby, the decision maker can influence interaction between and relative importance of criteria according to own preferences and knowledge of the decision problem. All information can be entered into 'AniFair' via Graphical User Interfaces (GUI). 'AniFair' was implemented in R and can be downloaded at www.tierzucht.uni-kiel.de. It is introduced via an example from the Animal Welfare Assessment Protocol for pigs and was applied to real data collected from 13 farms in Northern Germany by an animal welfare expert. As result the farms could be ranked according to their welfare status, and as criteria have been made comparable, aimed advice could be given on how to improve the welfare situation.

Subclinical doses of the mycotoxin deoxynivalenol induce rumen epithelial cell remodelling in cattle
J.R. Bailey[1], H.R. Fleming[2], A. Christou[1,2], V. Morris[1], M. Bailey[1], T.A. Cogan[1] and M.R.F. Lee[1,2]
[1]University of Bristol, Bristol Veterinary School, Langford, BS40 5DU, United Kingdom, [2]Rothamsted Research, North Wyke, Okehampton, Devon, EX20 2SB, United Kingdom; michael.lee@rothamsted.ac.uk

Mycotoxins are fungal metabolites frequently present in silage, which have been implicated in reduced livestock performance due to altered nutrient metabolism, endocrine malfunction and immune-suppression. These effects have been well documented at high doses, but the effect of lower doses is not well understood. The aim of this work was to evaluate the effect of low doses of the common mycotoxin, deoxynivalenol (DON), on the architecture of the rumen epithelium as part of a larger study to better understand the impact of subclinical doses on animal health and rumen function. Holstein steers (n=24) at five months of age were separated into four groups of six and acclimatised to a 60:40 grass silage:concentrate diet over five weeks. Following this, DON was included in the daily feed at: 0, 0.5, 1.5 or 4.5 mg/kg dry matter (DM). DM and water intake were monitored daily, and animals were weighed twice weekly. Following seven days of DON exposure, animals were euthanised and samples of rumen epithelium collected to determine histopathological changes by haematoxylin and eosin staining and expression of Ki67, indicating proliferating cells, and cleaved caspase-3, indicating apoptotic cells by immunofluorescence. Data were analysed by Kruskal-Wallis test with a Dunn's multiple comparisons test. DM and water intake, and weight gain were not different across treatments with no manifestation of clinical disease. Animals exposed to 4.5 mg DON/kg DM had significantly thickened rumen epithelia when compared to control animals (0 mg DON/kg DM) (P=0.023). Similarly, the proportion of proliferating cells within the rumen epithelium was significantly reduced in animals which ingested 4.5 mg DON/kg DM (P=0.0239) when compared to the control. Rumen epithelial cell apoptosis was not significantly affected by presence of lower levels (<4.5 mg/kg DM) of DON. We conclude that short-term exposure to subclinical doses of DON induces significant remodelling of the rumen epithelium. If exposure is prolonged, this has potential to detrimentally affect rumen epithelial function.

Differences of essential minerals and trace elements in cow and buffalo milk

F. Fantuz[1], S. Ferraro[2], L. Todini[1], A. Fatica[3] and E. Salimei[3]
[1]Università di Camerino, Scuola di Bioscienze e Medicina Veterinaria, via Gentile da Varano, 62032 Camerino, Italy,
[2]Università di Camerino, Scuola di Scienze e Tecnologie (sezione Chimica), via Gentile da Varano, 62032 Camerino, Italy,
[3]Università del Molise, Dip. Agricoltura, Ambiente, Alimenti, via De Sanctis, 86100 Campobasso, Italy; salimei@unimol.it

Although most of the global milk production is accounted for by cow milk (83%), buffalo milk is the second largest type of milk produced worldwide accounting for 13% of global milk production, but only few information is available on the mineral fraction in buffalo milk. Aim of this trial was to study the Ca, P, K, Na, Mg, Zn, Fe, Cu, Mn, Co, Se, and Mo concentrations in cow and buffalo milk raised in the same farm. Thirty lactating Italian Friesian cows and thirty lactating Mediterranean buffaloes were used to provide milk samples. Cows and buffaloes were machine milked twice a day and individual milk samples were collected during the afternoon milking. Experimental animals were group fed a total mixed ration and samples of feedstuffs were collected for analysis. Milk samples and feedstuffs were analysed by Inductively Coupled Plasma-Mass Spectrometry. Differences in milk elements concentration between the two species were determined by t-test. The concentration of minerals (% dry matter) and trace elements (mg/kg dry matter) in total mixed ration, respectively for cows and buffaloes, was Ca 0.78 vs 0.88; P 0.29 vs 0.24; K 1.47 vs 1.81; Na 0.29 vs 0.24; Mg 0.19 vs 0.18; Zn 143.1 vs 127.2; Fe 161.3 vs 149.8; Cu 19.3 vs 17.6; Mn 75.6 vs 67.1; Co 0.43 vs 0.41; Se 0.38 vs 0.35; Mo 3.60 vs 5.53. Significantly higher concentration (mg/l for minerals; μg/l for trace elements) was observed in buffalo milk for Ca (1,598 vs 1,153), P (1,340 vs 975), Mg (154.6 vs 104.0), Zn (4,978 vs 3,435), Fe (321.9 vs 212.7) and Cu (109.8 vs 56.7). On the contrary, significantly higher concentration was observed in cow milk for K (1,461 vs 1,021), Na (391.7 vs 333.0) and Mo (73.8 vs 49.9). The milk Ca/P ratio (1.18 cow vs 1.21 buffalo milk), Na/K ratio (0.29 cow vs 0.35 buffalo) and the milk Mn (25.8 cow vs 30.1 buffalo), Se (25.4 cow vs 26.9 buffalo) and Co (0.78 cow vs 0.94 buffalo) concentration were not significantly different between the two species.

Tackling matrix effects of feed extracts for the detection of ruminant by-products by targeted MS

M.C. Lecrenier[1], L. Plasman[1], R. Sanna[2], J. Henrottin[3] and V. Baeten[1]
[1]Walloon Agricultural Research Center (CRA-W), Valorisation of agricultural products, Chaussée de Namur, 24, 5030 Gembloux, Belgium, [2]Haute Ecole Louvain en Hainaut (HELHa), rue Trieu Kaisin, 136, 6061 Montignies sur Sambre, Belgium, [3]CER Groupe, Rue du Point du Jour, 8, 6900 Marloie, Belgium; m.lecrenier@cra.wallonie.be

Animal by-products are an interesting source of feed materials. However, since the bovine spongiform encephalopathy ('mad cow disease') crisis, their use has been strictly regulated. Official controls are based on a combination of light microscopy and PCR but sometimes these methods are unable to distinguish unauthorised and authorised materials. Ultra-high performance liquid chromatography coupled to tandem mass spectrometry (UHPLC-MS/MS) method was therefore developed to address the analytical gaps. This method has the advantage to be species and tissue specific. The sample preparation and the analytical method were designed to provide a fast, simple and powerful method suitable for routine. But an evaluation of the method on a large number of samples in order to assess its robustness has highlighted that some feeds have major effects on peptide signal (ion suppression). The subject of this study was to evaluate various types of SPE cartridge in order to optimise the purification step regarding the matrix effects. Recovery rates were also compared. For matrix effect assessment, commercial pig feeds and aquafeeds were artificially adulterated with 1% (w/w) of bovine blood meal and 0.1% (w/w) of milk powder. For the recovery evaluation, a mixture of 10:1 of blood meal and milk powder was prepared. All samples were extracted in triplicate. Proteins were extracted in a buffer containing 200 mM TRIS-HCl pH 9.2, 2 M urea followed by trypsin digestion and purification with SPE cartridge. Sep-Pak tC18, Oasis PRiME HLB and Oasis PRIME MCX cartridges were selected for this study. Analyses were performed by liquid chromatography coupled with a triple quadrupole mass spectrometer using MRM (multiple reaction monitoring) mode. Labelled peptides were used as internal standard in order to compare the results independently of the retention time variation due to matrix effect.

The impact of different types of bacteria on the technological parameters and quality of cow's milk

P. Brodowska[1], K. Kawecka[1,2], M. Rzewuska[2], M. Zalewska[3], D. Reczyńska[1], D. Słoniewska[1], S. Marczak[1], S. Petrykowski[1], L. Zwierzchowski[1] and E. Bagnicka[1]

[1]Institute of Genetics and Animal Breeding PAS, Department of Animal Improvement, Postępu 36, 05-552 Magdalenka, Poland, [2]Warsaw University of Life Sciences, Faculty of Veterinary Medicine, Nowoursynowska 159, 02-776 Warsaw, Poland, [3]Warsaw University, Department of Applied Microbiology, Miecznikowa 1, 02-096 Warsaw, Poland; p.brodowska4@ighz.pl

Over 150 species of bacteria causing mastitis including coagulase-positive (CPS), coagulase-negative Staphylococci (CNS), and environmental and opportunistic pathogens have been identified. The aim of the study was to analyse the impact of the presence of bacteria in cow's milk on its technological parameters. The study was carried out on 304 Polish HF cows. The dataset comprised the 547 records on daily milk yield, protein, fat, casein, lactose, dry matter, free fatty acid, citric acid contents, FPD, and somatic cell count (SCC), clotting time, whey quality, and cheese yield and quality. Whey and cheese quality was assessed in arbitrarily six-point scale, with 1 as the highest. The microbiological analysis was done using VITEK2 equipment. The milk samples collected three times during two lactations were divided into 5 groups: 1st – bacteria-free group (n=60), 2nd – bacteria not causing or occasionally causing mastitis (n=52), 3rd – opportunistic bacteria (n=185), 4th – CNS (n=205), 5th – CPS (n=45). The MIXED procedure of SAS package was used for variance analyses. The SCC was transformed into natural logarithm scale (lnSCC) before analysis. As expected, the lowest lnSCC, the highest daily milk yield, and casein content were stated for 1st and 2nd groups. The worse technological milk parameters were observed in 3rd, 4th and 5th groups which covered the milk samples containing the pathogenic bacteria than in 1st and 2nd groups. However, the worst technological milk parameters were stated in CNS group, even worse than in CPS. It means that coagulase-negative staphylococci, until recently considered less pathogenic bacteria, deteriorate the milk technological parameters to a greater extent than highly pathogenic bacteria from CPS group. Funded by Scientific Consortium KNOW 'Healthy Animal-Safe Food', MSHE decision No. 05-1/KNOW.2/2015.

Resource conversion by black soldier fly larvae: towards standardisation of methods and reporting

G. Bosch[1], D.G.A.B. Oonincx[1], H.R. Jordan[2], J. Zhang[3], J.J.A. Van Loon[4], A. Van Huis[4] and J.K. Tomberlin[5]

[1]Wageningen University & Research, Animal Nutrition Group, De Elst 1, 6708 WD Wageningen, the Netherlands, [2]Mississippi State University, Department of Biological Sciences, 295 Lee Blvd, MS 39762, USA, [3]Huazhong Agricultural University, State Key Laboratory of Agricultural Microbiology, F9G3+36 Hongshan, Wuhan 430070, China, P.R., [4]Wageningen University & Research, Laboratory of Entomology, Droevendaalsesteeg 1, 6708 PB Wageningen, the Netherlands, [5]Texas A&M University, Department of Entomology, TAMU 2475, TX 77843-2475, USA; guido.bosch@wur.nl

The use of larvae of the black soldier fly (*Hermetia illucens*; BSF) to convert low-value residual organic resources into high-value products is developing into a global industry. The number of publications dealing with diet conversion efficiency by BSF larvae also greatly increased over the last years. It appears that research methods and the information reported vary considerably among studies, which hamper comparison of the nutritional quality of organic resources across studies. For livestock species there are internationally accepted standardised procedures developed to evaluate nutritional quality of ingredients. This has resulted in publicly available feeding tables describing species-specific nutritional qualities of ingredients, which are instrumental for formulating diets supporting optimal animal performance and use of resources. In this presentation procedures are discussed and proposed with the aim to arrive at standardisation of the research methods employed in future resource conversion studies with BSF larvae and how results are reported. Aspects covered are fly colony origin, rearing procedures, reference and experimental feeding substrates, sample preparation including microbiota and chemical analyses.

What synchronization efforts are worth the investment in mass insect rearing facilities

N. De Craene
Circular Organics, Slachthuisstraat 120/6, 2300 Turnhout, Belgium; nouchka.decraene@circularorganics.com

In natural populations, high variabilities in the timing of life events is a sound evolutionary strategy. Having a temporal dispersion in the stages of the life cycles of individuals within a population, prevents the entire population being decimated by a singular disastrous event. In mass production systems, however, this natural variability is detrimental to the reliability and predictability of production outputs. As insect mass production deals with very large numbers of individuals, the resulting inconsistency of output is even greater. Even more than traditional mass production system concerning live organisms, insect production requires precautionary measures to minimize natural variability. We developed several procedures to enable the synchronization of life events in the mass production of black soldier fly (*Hermetia illucens*), including predictable survivability, growth curves and bioconversion rates, to achieve greater consistency in the production system. In this presentation, we discuss our findings regarding these synchronization efforts and whether they are worth the investment.

Development and evaluation of real-time PCR targets for the authentification of insect flours

F. Debode[1], A. Marien[1], D. Akhatova[2], C. Kohl[1,3], H. Sedefoglu[1,4], L. Mariscal-Diaz[1,4], A. Gérard[1,5], O. Fumière[1], J. Maljean[1], J. Hulin[1], F. Francis[5] and G. Berben[1]
[1]Walloon Agricultural Research Center (CRA-W), Unit Authentication and Traceability, chaussee de Namur, 24, 5030 Gembloux, Belgium, [2]University of Chemistry and Technology (UCT), Department of Biochemistry and Microbiology, Technická 5, 166 28 Prague 6, Czech Republic, [3]Haute école de la Province de Liège, quai Gloesener 6, 4020 Liège, Belgium, [4]Haute école Louvain-en-Hainaut, rue Trieu Kaisin 136, 6061 Montignies-sur-Sambre, Belgium, [5]University of Liège (ULg), Gembloux Agro-Bio Tech, Functional and Evolutionary Entomology, Passage des Déportés 2, 5030 Gembloux, Belgium; f.debode@cra.wallonie.be

Insects are rich in proteins and could be an alternative source of proteins to feed animals. In recent years, numerous companies have started producing insects for feed purposes. In Europe, the processed animal proteins obtained from seven insect species have been authorised for aquaculture by Commission Regulation (EU) 2017/893. The stakeholders are now hoping future modifications of the legislation to allow the feeding of poultry and pig. Methods to detect if a product really contains insects and to authenticate insect products will be mandatory. To fulfil the need in authentication methods, we started to develop real-time PCR methods specific to the insects. Methods for *Tenebrio molitor* and *Hermetia illucens* were already developed and published. Methods are now ready for *Alphitobius diaperinus*, *Gryllus assimilis*, *Acheta domesticus* and *Musca domestica*. These methods were checked with success for their specificity, sensitivity and robustness. The applicability of the methods was also tested on real industrial insect flours at the exception of *M. domestica* as industrial samples were not available. An interrogation also remains for *M. domestica* as many close species are erroneously considered to be *M. domestica*. In addition to real-time PCR, approaches based on metagenomics and high throughput sequencing are ongoing in order to provide a more complete overview of the insects in presence.

An optimised test for recording hygienic behaviour in the honeybee

E. Facchini[1], P. Bijma[2], G. Pagnacco[3], R. Rizzi[3] and E.W. Brascamp[2]
[1]Hendrix Genetics Research Technology & Service B.V., P.O. Box 114, 5830 AC Boxmeer, the Netherlands, [2]Wageningen University & Research, Animal Breeding and Genomics, P.O. Box 338, 6700 AH Wageningen, the Netherlands, [3]University of Milano, Department of Veterinary Medicine, via G. Celoria 10, 20133 Milan, Italy; elena.facchini@hendrix-genetics.com

Hygienic behaviour (HB) in honeybees reflects social immunity against diseases and parasites. Young bees showing HB detect, uncap, and remove infested brood from a colony. We developed a new variant of freeze-killed brood (FKB*) test to optimise the duration of the HB test, the costs, and safety for the operator. In 2016, we performed a comparison between traditional FKB and FKB* on 25 unselected and unrelated colonies in the apiary of the University of Milano. To estimate repeatability and heritability, in 2017 and 2018, FKB* was used to phenotype, respectively, 56 and 95 colonies twice, in the context of a breeding programme. FKB* took less time and required a smaller amount of liquid nitrogen. The two methods showed a correlation between colony effects of 0.93, indicating that they measure the same trait. For single records, the phenotypic correlation between both methods was 0.64. Estimated heritability and repeatability for single records HB were 0.23 and 0.24, respectively, whilst heritability for the average HB value of two records was 0.37.

Investigating genetic and phenotypic variability of honeybee queens' individual traits

E. Facchini[1], F. Turri[2], F. Pizzi[2], D. Laurino[3], M. Porporato[3], R. Rizzi[4] and G. Pagnacco[4]
[1]Hendrix Genetics Research Technology & Service B.V., P.O. Box 114, 5830 AC Boxmeer, the Netherlands, [2]National Research Council, Institute of Agricultural Biology and Biotechnology, via Einstein, 26900 Lodi, Italy, [3]University of Turin, Department of Agricoltural, Forest and Food Sciences, Largo Paolo Braccini 2, 10095 Grugliasco (Torino), Italy, [4]University of Milano, Department of Veterinary Medicine, Via Celoria 10, 20133 Milan, Italy; elena.facchini@hendrix-genetics.com

The quality of the queen may have an important effect on a colony's development, productivity and survival and queen failure or loss is considered a cause for colonies' decline worldwide. The queen quality, resulting from her genetic background, developmental condition, mating success and environment, can be assessed by some morphological measures. The aim of this study was to obtain preliminary genetic parameters of some traits that describe the quality of bee queens. An Italian breeding and beekeeping company provided 147 related queens with known pedigree bred during spring/summer of 2017 and 2018 using a standardized production system. The queen quality traits considered were body weight, weight and width of the tagmata (head, thorax and abdomen), wing length, diameter and volume of spermatheca, number of ovarioles and number of sperms in the spermatheca. An animal model was applied in univariate and bivariate analysis to obtain (co)variance estimates. Heritability of body and tagmata weight ranged from 0.46 (±0.34) to 0.54 (±0.34), whereas lower estimates were found for the tagmata width and wing length (from 0.13±0.26 to 0.42±0.32). Heritabilities estimated for the spermatheca diameter and volume, number of ovarioles and number of sperms were 0.17±0.34, 0.88±0.39, 0.70±0.35 and 0.57±0.35, respectively. Many phenotypic correlations related to size (weight and width) were high and positive, while we found weak correlations between morphology and reproductive traits. Genetic correlations showed large standard errors and are to be considered with caution.

A cross-laboratory study on analytical variability of amino acid content in three insect species

D.G.A.B. Ooninex[1], G. Bosch[1], M. Van Der Borght[2], R. Smets[2], L. Gasco[3], A.J. Fascetti[4], Z. Yu[4], V. Johnson[5], J.K. Tomberlin[5] and M.D. Finke[6]
[1]Wageningen University & Research, Animal Nutrition Group, De Elst 1, 6708 WD Wageningen, the Netherlands, [2]Faculty of Engineering Technology, Department of Microbial and Molecular Systems, Kleinhoefstraat 4, 2440 Geel, Belgium, [3]University of Turin, Department of Agricultural, Forest, and Food Sciences, largo P. Braccini 2, 10095 Grugliasco, Italy, [4]University of California, School of Veterinary Medicine, Department of Molecular Biosciences, 1089 Veterinary Medicine Drive, CA 95616, Davis, USA, [5]Texas A&M University, Department of Entomology, 370 Olsen Boulevard, College Station, TX 77843, USA, [6]Mark Finke LLC, 17028 E Wildcat Dr, AZ 85263, Rio Verde, USA; guido.bosch@wur.nl

Insects are primarily considered as a protein source. Hence, protein content and amino acid (AA) composition, which indicates protein quality, are important characteristics. Protein content is commonly estimated by multiplying N content by a conversion factor. Conventionally 6.25 was used, but more recently factors between 4.67 and 4.86 were reported for specific insect species. Both the conversion factor and the protein quality depend on the accuracy of AA analyses. Such analyses normally consist of two steps; first the matrix is destroyed and peptide bonds are broken via hydrolysis to liberate the AAs. Then, these AAs are quantified via ion exchange or high performance liquid chromatography. Methodological differences among laboratories can lead to variable results, which we evaluated in this study. House crickets (HC), yellow mealworms (YMW) and black soldier fly larvae (BSFL) were freeze-dried and ground, and subsamples were distributed over seven laboratories that used different methods to determine N and AA contents. Coefficient of variation (%) was low for N content (HC 2.8, YMW 3.0, BSFL 5.0) and considerably higher for AAs (HC 11.2-30.6, YMW 2.9-50.2, BSFL 1.4-34.7). Among the AAs low CV values where found for Aspartic acid (3.8-11.2) and Glutamic acid (1.4-15.8), whereas high CV values where found for Cysteine (22.6-34.7 and Methionine (30.6-50.2). The analytical variability in AA values warrants more standardisation of methodologies among laboratories as this will improve the validity of reported protein quality of insects.

Current uses and challenges of microalgae in animal feeding

J.A.M. Prates
CIISA, Faculty of Veterinary Medicine, University of Lisbon, Avenida da Universidade Técnica, 1300-477 Lisboa, Portugal; japrates@fmv.utl.pt

Microalgae are mainly composed by proteins, carbohydrates, lipids, vitamins, minerals and bioactive compounds, such as carotenoids. The effect of dietary microalgae on production and meat quality of livestock species are discussed here. Research evidence so far has shown that the inclusion of microalgae in animal diets could improve growth and meat quality in ruminants, pigs, poultry and rabbits. These findings are highly dependent on microalgae composition and their amount in the diet. In a general overview, the inclusion of *Arthrospira platensis* in pig and poultry diets increases average daily gain but negatively affects feed conversion ratio. Regarding *Schizochytrium* sp., this microalga improves fatty acid composition in pork and poultry meat, essentially due to its high content in docosahexaenoic acid (DHA). Chlorella, at very low percentages in feed, benefits growth performance parameters of poultry. The use of microalgae as feed ingredients is very promising as an alternative to corn and soybean, thus mitigating the current competition among food-feed-biofuel industries. In addition, microalgae contribute to the protection of environment and natural resources, namely land degradation and water deprivation. Microalgae also provide a sustainable alternative for n-3 LCPUFA availability, thus protecting worldwide fatty fish stocks. However, the cost-effective production and use of microalgae is a major challenge in the near future. In fact, the current microalgae cultivation technology should be improved to reduce their production costs. In addition, we foresee that the efficiency of microalgae incorporation in monogastric diets could be largely improved by the use of carbohydrate-active enzymes (CAZymes). CAZymes will allow the increase of nutrients bioavailability, as a consequence of recalcitrant microalgae cell wall degradation. Overall, the inclusion of microalgae in feed represents a very promising strategy for the maintenance and development of livestock sector, as an environmental friendly alternative to balance food-feed-biofuel industries. Supported by PTDC/CVT-NUT/5931/2014 and UID/CVT/00276/2019 (FCT, Portugal) and 08/SI/2015/3399 (P2020, Portugal).

Microalgae are suitable protein feeds for lactating dairy cows

M. Lamminen, A. Halmemies-Beauchet-Filleau, T. Kokkonen, S. Jaakkola and A. Vanhatalo
University of Helsinki, Department of Agricultural Sciences, P.O. Box 28, 00014 University of Helsinki, Finland;
marjukka.lamminen@anis.au.dk

The potential of non-defatted microalgae as protein feeds for dairy cows was evaluated in a series of four Latin square experiments conducted with 4-8 lactating cows on grass-silage based diets. In the experiments: (1) mixture of *Chlorella vulgaris* and *Spirulina platensis;* or (2) *S. platensis* were compared to rapeseed meal (RSM); (3) *S. platensis*, *C. vulgaris* and mixture of *C. vulgaris* and *Nannochloropsis gaditana* to soya bean meal (SBM); and (4) *S. platensis* to RSM and faba beans (FB). Microalgae replaced the conventional protein feeds partially [50% of crude protein (CP)] in Exp. 1, 2 and 4, or completely (100% of CP) in Exp. 1-3. Data was analysed by analysis of variance. In this abstract, the main results with significance (P<0.05) or tendency towards significance (0.05<P≤0.10) are summarised. The CP concentration was on average 690, 597 and 385 g/kg dry matter (DM) for *S. platensis*, *C. vulgaris* and *N. gaditana*, respectively. Cows showed reduced palatability of microalgae in three experiments. When forage and concentrates were fed separately, cows compensated the decreased intake of microalgae containing concentrates by increasing silage intake, leading to unaffected DM intake (DMI). When fed as total mixed ration, DMI was decreased on average 0.65 kg/d with microalgae. The slight inferiority of microalgae to RSM was demonstrated in one out of three experiments by lower milk yield, and in two experiments by lower N use efficiency (NUE) and human-edible protein efficiency (HEPE), as well as tendency towards lower protein yield on microalgae than RSM diets. Microalgae seemed to be suitable substitute for grain legumes. Microalgae resulted in similar milk and protein yields and NUE as SBM, and tended to increase HEPE in comparison to SBM. Compared to FB, microalgae increased milk yield and NUE, tended to increase protein yield, and did not affect HEPE. In conclusion, the results showed no biological or physiological constraints for protein feed use of microalgae. The greatest challenge limiting the feed use of microalgae for lactating cows is the poorer palatability relative to conventional feeds.

The effects of *Schizochytrium* sp. lipid extract or fish oil on lamb meat fatty acids

A. Francisco[1], S.P. Alves[2], J. Santos-Silva[1] and R.J.B. Bessa[2]
[1]Insituto Nacional de Investigação Agrária e Veterinária (INIAV), Estação Zootécnica Nacional, Fonte Boa, 2005-048 Vale de Santarem, Portugal, [2]Faculty of Veterinary Medicine, Universidade de Lisboa, CIISA, Polo Universitário do Alto da Ajuda., 1300-477 Lisboa, Portugal; rjbbessa@fmv.ulisboa.pt

Thirty-six ram lambs born with approximately 60 days of age were randomly allocated to 3 complete ground diets based on dehydrated lucerne (70% dry matter) and supplemented with either; 6% of sunflower oil (S diet); 4% of sunflower oil and 2% of fish oil (FS diet); or 4% of sunflower oil and 3.53% of *Schizochytrium* sp lipid extract – Trevera® (TS diet). Lambs stay on trial for 6 weeks until being slaughtered and longissimus dorsi muscle and subcutaneous fat (SCF) sampled for fatty acid (FA) analysis. The TS diet strongly decreased (≈ 50%) the feed intake and growth performance compared to the other treatments. The FA profile of muscle lipids was profoundly affected by treatments. The deposition of 20:5n-3, 22:5n-3 and 22:6n-3 (i.e. n-3 LC-PUFA) increased in muscle polar lipids (PL) from 4.3% FA in S to 9.0% with FS and 6.6% with TS. Notably, n-3 LC-PUFA were deposited in muscle neutral lipids (NL) and SCF following a different pattern, being highest with TS rather than with FS diet: 0.18, 0.99 and 1.53% FA respectively for S, FS and TS in muscle NL and 0.16, 1.2 and 2.1% FA respectively for S, FS and TS in SCF. Compared to other diets, the TS diet depressed the stearoyl-CoA desaturase derived FA like *c*9-16:1, *c*9-17:1 and *c*9-18:1 in muscle NL but not on muscle PL. The proportion of *t*11-18:1 in muscle NL increased sharply from 3.8% FA with S diet to circa 13% with both FS and TS diets, whereas the *c*9,*t*11-18:2 increased from 1.4% with S, to 1.7% with TS and 2.4 with FS diet. The same response pattern was observed in muscle PL and SCF. The *t*10-18:1 remained quite low in all treatments and the *t*10-/*t*11-18:1 ratio was always below 0.4 with S diet and below 0.05 with FS and TS diets. Concluding, both fish oil and Trevera promoted the deposition of a large proportion of biohydrogenation intermediates and some deposition of n-3 LC PUFA. Trevera differs from fish oil as it favoured the deposition of n-3 LC PUFA in muscle NL and SCF but also it strongly depressed the feed intake.

Dietary algal β-(1,3)-glucan and its role in modulating the immune system in pigs and poultry

N. Smeets, R. Spaepen and V. Van Hamme
Kemin Europa NV, Toekomstlaan 42, 2200 Herentals, Belgium; natasja.smeets@kemin.com

B-glucans are known for their immune modulating effects and are added to feed to enhance resistance to disease and consequently improve animal health. This study describes two trials that studied the effect of an algal derived β-(1,3)-glucan on 1. the inflammatory response and cell-mediated immune system in pigs and 2. the vaccination efficiency against Infectious bursal disease (IBD) in poultry. In a first trial, piglets of 6 supplemented (1 g/sow/day β-(1,3)-glucan -Aleta™- starting 3 w before lactation) and 6 non-supplemented mother sows, were divided into a supplemented (200 g/T β-(1,3)-glucan) and a non-supplemented piglet group, resulting in 4 groups with 30 piglets/group. Blood was taken from the piglets 42 d after weaning and analysed for T-lymphocyte counts and haptoglobin. In a second trial, 96 male Ross 308 broilers were divided into 3 treatments (8 replicates): a positive control group (vaccination with a live IBD vaccine, no treatment), a negative control group (no vaccination, no treatment), and a treatment group (vaccination and treatment with 50 g/T β-(1,3)-glucan). Blood samples were taken at 18 d and 35 d to measure antibody titers (IgG) against IBD. The first trial showed that β-(1,3)-glucan supplementation significantly decreased haptoglobin level in piglets originating from supplemented mother sows, indicating a decreased level of inflammation. Additionally, a significantly decreased population of CD4-CD8- T-lymphocytes and significantly increased populations of CD4+CD8lo and CD4-CD8+ T-lymphocytes in piglets originating from supplemented mother sows were observed, indicating an increased lymphocyte proliferation and activity. The second trial showed that in the positive control group 25% of all susceptible birds had a positive response to vaccination with an average IgG titer of 3,906 ELISA units. In the β-(1,3)-glucan treated group, 52% of all susceptible birds (without maternal antibodies) reacted positive to the vaccination with an average IgG titer of 4,564 ELISA units, indicating an increased IBD vaccination efficiency. In conclusion, the trials showed the immune modulating effects of β-(1,3)-glucan: an alleviation of inflammation, enhancement of cell-mediated immune responses and an increased vaccination efficiency.

Algae as animal feed – poster pitches

R.M.A. Goselink
Wageningen Livestock Research, Animal Nutrition, De Elst 1, 6700 AH Wageningen, the Netherlands;
roselinde.goselink@wur.nl

In this block of 15 minutes, the presenting authors of the posters submitted to the session 'Algae as animal feed' will shortly present their poster. Posters are shown in the session room and after each poster pitch, there is room for a question from the audience in the session room. Abstracts of the posters can be found in the proceedings.

Seaweed as a potential feed resource for farmed animals

M. Øverland, L.T. Mydland and A. Skrede
Norwegian University of Life Sciences, Animal and Aquacultural Sciences, Aboretveien 6, 1432 Ås, Norway;
margareth.overland@nmbu.no

The ocean offers large opportunities for cultivating seaweed to provide renewable biomass as a sustainable alternative to animal feed ingredients, and there is great interest in exploring seaweed as a novel feed resource. Brown seaweeds are a promising biomass that contain variable levels of nutrients depending on species, season of harvest, geographic origin, and environmental conditions. The nutritional value is, however, highly variable. The content of water and ash is high, the content of essential amino acids and protein can be low, and a high content of indigestible carbohydrates can adversely affect the energy value for monogastric animals and fish. Optimal use of seaweed in feeds requires suitable processing, and biorefinery approaches may increase protein content and improve nutrient availability. Brown seaweeds can be used as a source of fermentable sugars and essential nutrients in the fermentation media for the production of microbial feed ingredients such as yeast. Enzymatic saccharification of the brown seaweed *Saccharina latissima* to produce a high sugar yield has been investigated using commercial and in house enzymes such as alginate lyases. Seaweeds are also rich in unique bioactive components and there is growing interest in the potential health beneficial effects of compounds such as laminarin and fucoidan in different seaweeds and seaweed products. A general overview of seaweed as a feed resource, including nutrient composition, factors affecting nutrient availability, and optimal processing of seaweed for use in feeds, will be discussed.

Polysaccharides in seaweeds: necessary steps in order to use seaweeds in animal feed

W. Muizelaar[1], J. Krooneman[2,3] and G. Van Duinkerken[1]
[1]Wageningen Livestock Research, De Elst 1, 6708 WD Wageningen, the Netherlands, [2]Carbohydrate Competence Center, Nijenborgh 7, 9747 AG Groningen, the Netherlands, [3]University of Groningen, Engineering and Technology Institute Groningen, Nijenborgh 4, 9747 AG Groningen, the Netherlands; wouter.muizelaar@wur.nl

In recent years the scientific interest to use seaweeds (SW), or extracts thereof, in animal feed is rising. Using land plants in animal diets is the current standard and many analytical methods and knowledge on animal digestibility and physiology are based upon this practice. Marine plants, like SW, do share some characteristics with land plants, but also fundamentally differ with respect to the composition of their cell walls. Differences in cell wall polysaccharides (PS) possibly influence analytical methods, experimental design, data analysis, and the interpretation of results. Seaweeds are commonly sub-divided in three groups of red, green and brown, usually referred to as *Rhodophyta*, *Chlorophyta*, and *Phaeophyceae*. Within these groups there is also a great variety in the cell wall composition. Evolutionary red and green SW are more closely related to land plants than brown SW. However, all SW share a common characteristic which land plants seem to have lost, the ability to incorporate sulphated PS into the cell wall. Besides having potential bioactive properties in the animal (for animal health), these and other SW PS might not be effectively broken down by the animal digestive system. Recent observations during the chemical characterisation of the SW also showed different trends in the NDF, ADF and ADL content when determined by the *van Soest* method. Generally, in brown SW, NDF is higher than ADF, which should not be possible by definition. The NDF (g/kg DM) and ADF (g/kg DM) content in *Ascophyllum nodosum*, *Fucus serratus*, *Laminaria digitata*, *Saccharina latissima* and *Undaria pinnatifida* are respectively 168.6 NDF and 254.0 ADF, 162.1 NDF and 277.6 ADF, 132.2 NDF and 173.3 ADF, 161.2 NDF and 168.7 ADF and 155.1 NDF and 165.2 ADF. We hypothesize that the PS alginate ends up in the ADF fraction instead of the NDF fraction. In conclusion, re-assessment of current analytical methods or the development of new methods with respect to the PS fraction of SW is needed in order to incorporate them in animal diets.

In vitro gas production of 11 seaweeds as indicator for rate and extent of rumen fermentation

V. Perricone[1,2], W. Muizelaar[1], G. Van Duinkerken[1], W. Pellikaan[3] and J. Cone[3]
[1]*Wageningen Livestock Research, De Elst 1, 6708 WD Wageningen, the Netherlands,* [2]*Univeristy of Milan, Department of Health, Animal Science and Food Safety (VESPA), Via Celoria 10, 20133 Milano, Italy,* [3]*Wageningen University & Research, Animal Nutrition Group, De Elst 1, 6708 WD Wageningen, the Netherlands; vera.perricone@unimi.it*

The European dairy cattle sector is exploring novel feed materials to increase protein self-sufficiency and mitigate environmental impact of animal production (i.e. CH_4 emissions). In this direction, marine biomass has potential as alternative feed for ruminants, not competing for land. A study was executed to evaluate 11 North West European seaweed species as alternative feed material for dairy cattle. Proximate composition, phenols, and bromoform content were analysed. *In vitro* total gas production (TGP) was assessed as indicator for rumen fermentation, and CH_4 production was determined. Grass silage was used as reference. After 72 h of incubation, *Palmaria palmata*, *Laminaria digitata* and *Saccharina latissima* had similar TGP as grass silage (334.9 ml/g OM), suggesting that rumen fermentation was not affected (P>0.05). CH_4 production of *P. palmata* did not differ from grass silage as well. *Undaria pinnatifida*, *Porphyra umbilicalis* and *Ulva lactuca* were responsible for a mild reduction in TGP (234.8, 221.4 and 212.4 ml g/OM, respectively; P<0.05). Their CH_4 production, along with *L. digitata* and *S. latissima,* was lower than grass silage (P<0.01). *Chondrus crispus, Gracilaria gracilis, Ascophyllum nodosum. Asparagopsis armata* and *Fucus serratus* had the strongest effect, severely inhibiting both TGP and CH_4 production (P<0.01). Assuming that a mild reduction in TGP might be due to a lack of adaptation of rumen microflora to seaweeds, *P. palmata*, *L. digitata, S. latissima, U. pinnatifida, P. umbilicalis* and *U. lactuca* could be considered as possible substitute to grass silage. However, according to their protein content, only *P. palmata*, *U. pinnatifida, P. umbilicalis* might be considered as potential alternative feedstuff for dairy cows. Further investigations are required to better understand their nutritive value, their potential effect on animal health and the sustainability of their cultivation.

Macro algae *Ulva lactuca* in broiler feed: *in vitro* and *in vivo* digestibility of protein extracts

L. Stokvis[1,2], M.M. Van Krimpen[1] and R.P. Kwakkel[2]
[1]*Wageningen Livestock Research, Animal Nutrition, P.O. Box 338, 6700 AH Wageningen, the Netherlands,* [2]*Wageningen University and Research, Animal Nutrition Group, P.O. Box 338, 6700 AH Wageningen, the Netherlands; lotte.stokvis@wur.nl*

To feed the growing world population living a life of increasing welfare, novel sources of feed for the production of protein in the form of meat, milk and eggs need to be found, that are not competing for resources currently in use. Seaweed could be such a novel protein source. Using broilers, an experiment was conducted to determine the digestibility of seaweed protein extracted from the seaweed *Ulva lactuca*, where potential bioactive effects on bird health were studied simultaneously. Three products were tested, differing in the process of extraction, leading to differences in inclusion of only soluble or also insoluble protein like cell wall bound protein. Chemical composition and *in vitro* organic matter and nitrogen digestibility were analysed next to the *in vivo* experiment. The seaweed products were added to a standard diet, at a concentration of 10%. The study was conducted with 288 14-day old Ross 308 broilers in 24 floor pens (0.96 m^2) with 12 birds per pen. Three experimental diets were fed with the different protein extracts diluting the standard diet, versus 1 control diet, with 6 replicates (pens) per treatment. Feed and water were available *ad libitum* and intake was monitored. Birds were weighed at day (d) 0, 14, 21 and 28, faeces was collected qualitatively at pen level at d 26-27-28 and litter quality was determined by visual scoring and analysing dry matter content at d 14, 21 and 28. At d 28, birds were sacrificed and ileal content was collected to determine digestibility of the seaweed protein products. Ileal tissue samples were collected and analysed for morphology and gene expression. Blood samples were collected and analysed for cytokines and chemokines, and caecal content was analysed for microbial composition. With the results of this study, which will be presented at the EAAP in Ghent, we will be able to determine the digestibility of the seaweed protein products, and get more insight in the bioactive properties of the protein extracts, which contributes to the future use of seaweed protein in animal diets.

The effect of feeding long-chain polyunsaturated fatty acids to lambs in the rumen lipid metabolism

S.P. Alves[1], A. Francisco[2], J. Santos-Silva[2] and R.J.B. Bessa[1]
[1]CIISA, Faculdade de Medicina Veterinária, Universidade de Lisboa, Av. da Universidade Tecnica, 1300 477, Portugal, [2]Instituto Nacional de Investigação Agraria e Veterinaria, Fonte Boa, Vale de Santarem, 2005, Portugal; susanaalves@fmv.ulisboa.pt

This work aims to study the effect of *Schizochytrium* sp. lipid extract or fish oil on fatty acid (FA) and dimethyl acetals (DMA) composition in the rumen of lambs. Thus, 36 ram lambs with approximately 60 days of age at start of the experiment were randomly allocated to 3 complete ground diets based on dehydrated lucerne (70% dry matter) and supplemented with: 6% of sunflower oil (S diet); 4% of sunflower oil plus 2% of fish oil (FS diet); or 4% of sunflower oil plus 3.53% of *Schizochytrium* sp. lipid extract – Trevera® (TS diet). After 6 weeks on feed, lambs were slaughtered, and rumen contents were sampled for FA and DMA analysis. FA and DMA were prepared from freeze-dried rumen samples by direct transesterification and analysed by gas chromatography and mass spectrometry. Ruminal biohydrogenation (BH) and completeness estimation were also calculated. The TS diet increased (P<0.001) the biohydrogenation of 18:2n-6 (72% vs 92%) and 18:3n-3 (66% vs, 90%) compared to S and FS diets. The estimated completeness for the C18 unsaturated FA decreased from 69% with S diet to 18% with FS diet; and was lowest (P<0.05) with TS (12%). The very incomplete BH observed with long chain polyunsaturated FA (LC-PUFA) supplementation led to a strong accumulation of t11-18:1, which increased from 3.4% of total FA in S to 24.5% of total FA with both FS and TS diets. The completeness of C20 FA was greater with the TS diet compared with the FS, while the opposite was observed for the completeness of the C22 FA. The proportion of the sum of branched chain FA and DMA were similar among treatments, but its profile differ markedly. Although the DMA-16:0 was the main DMA in all treatments, it comprised only 39% of total DMA in S diet, increasing to 53% with FS and increasing further to 61% with TS. In summary, the source of the LC-PUFA (*Schizochytrium* sp. lipid extract vs fish oil) affected differently the FA and DMA composition in the rumen of lambs.

Effect on milk quality of replacing soybean meal with spirulina in a hay-based diet for dairy cows

E. Manzocchi, M. Kreuzer and K. Giller
ETH Zurich, Institute of Agricultural Sciences, Universitätstrasse 2, 8092 Zurich, Switzerland; elisa.manzocchi@usys.ethz.ch

Alternative protein sources produced on non-arable land are needed to be able to reduce the exploitation of the latter for feed production. Cyanobacteria like spirulina (*Arthrospira platensis*) are an efficiently growing source of proteins, which might be used in dairy cow nutrition. The aim of this study was to assess the effect of substitution of soybean meal by spirulina on feed intake, milk yield, milk quality, as well as on the milk coagulation properties. An experiment was conducted with 12 late-lactating Brown Swiss and Holstein cows (2±1 lactations, 329±69 days in milk). There were divided into two groups (n=6) and fed with two isoenergetic (5.3 MJ/kg dry matter (DM)) and isonitrogenous (20 g N/kg DM) hay-based diets containing, on a dry matter basis, either 5% spirulina dry powder (SPI) or 6% soybean meal (SOY). The diets were homogeneously mixed in order to avoid feed selection. After 2 weeks of adaptation to the diets, 3 weeks of sampling followed. Feed intake and milk yield were recorded daily. Milk composition and colour were assessed four times during the sampling period, whereas rennet coagulation properties were measured twice. The data were analysed with the Mixed procedure of SAS with feed, breed and their interaction as fixed factors. The substitution of soybean meal with spirulina did not affect the DM intake (SPI 17.6, SOY 17.2 kg DM/day). This indicates that spirulina has a high palatability in dairy cows. Moreover, milk yield (SPI 18.1, SOY 16.7 kg/day), milk composition and rennet coagulation properties (rennet coagulation time, rate of firming and curd firmness) were not affected either. However, the milk from spirulina-fed cows had a significantly higher yellowness (b*, SPI 14.9 vs SOY 13.8, P=0.013) at unchanged brightness (L*, SPI 75.1, SOY 75.1) and redness (a*, SPI -5.65, SOY -5.65). The yellow colour could have resulted from a transfer of carotenoids from spirulina to the milk. This might improve its oxidative stability. If confirmed for β-carotene, a precursor of vitamin A, spirulina might be nutritionally valuable for cows and consumers. Furthermore, the increased supply with carotenoids might improve also the antioxidant status of the cows. Sensory assessments of the milk are ongoing.

Animal performance and milk fatty acid profile of ewes fed algae oil

T. Manso[1], B. Gallardo[1], P. Gómez-Cortés[2], M.A. De La Fuente[2], P. Lavín[3] and A.R. Mantecón[3]
[1]ETS Ingenierías Agrarias, Universidad de Valladolid, Avd. Madrid s/n, 34004 Palencia, Spain, [2]Instituto de Investigación en Ciencias de la alimentación (CSIC-UAM), Nicolas Cabrera, 9, 28049 Madrid, Spain, [3]Instituto de Ganadería de Montaña (CSIC-ULE), Finca Marzanas, 24346 Grulleros (León), Spain; tmanso@agro.uva.es

The aim of this study was to evaluate the effect of dietary inclusion of algae oil on milk FA profile of lactating ewes. Thirty-six Churra ewes (58.4±3.26 kg BW) in mid lactation (42±2.3 days in milk at the beginning of the experiment) were randomly distributed in 6 lots (6 ewes/lot), which were assigned (3 lots/treatment) to two experimental diets consisting of a TMR based on alfalfa hay and a concentrate (40:60) supplemented with 20 g of hydrogenated palm oil/kg of DM (CON) or 20 g/kg of algae oil (ALG). Fresh diets were offered daily *ad libitum* at 09:00 and 18:00 h, and clean water was always available. Intake of DM was recorded daily for each experimental lot. Ewes were milked twice daily throughout the experimental period (6 weeks) and individual milk production and composition was registered once a week. On day 42 of the experiment, a pooled sample of each lot was collected for FA analyses. The intake and milk production and composition were analysed using the MIXED procedure and the milk FA composition was subjected to ANOVA using the GLM procedure of SAS. Supplementation with ALG decreased DM intake (-41%), milk (-30%) and fat yield (-31%) ($P<0.001$). The ALG diet caused a reduction in milk saturated FA as proportion of total fatty acids (SFA) ($P<0.001$). The *trans* monounsaturated FA were increased with ALG diet ($P<0.001$), but there were not differences ($P>0.05$) in some putative fat synthesis inhibitors, especially *trans*-10 18:1. Additionally, ALG diet enhanced the proportion of *trans*-11 18:1, *cis*-9, *trans*-11 18:2, and C20-22 n-3 polyunsaturated FA, mainly 22:6 n-3 which was increased from 0.03 to 2.26% ($P<0.001$). A lower milk n6/n3 ratio (3.89 *vs* 0.66, $P<0.001$) was found when algae oil was used. It can be concluded that supplementing ewes' diet with algae oil could be a promising way to improve milk FA composition from a human health point of view. New techniques of rumen protected fat should be developed to avoid the negative effects of algae oil on animal performance.

Changes in milk-fed lamb performance in response to marine algae meal in the diet of dairy sheep

B. Gallardo[1], P. Lavín[2], A.R. Mantecón[2] and T. Manso[1]
[1]ETS Ingenierías Agrarias, Universidad de Valladolid, Avd. Madrid s/n, 34004 Palencia, Spain, [2]Instituto de Ganadería de Montaña (CSIC-ULE), Finca Marzanas, 24346 Grulleros (León), Spain; tmanso@agro.uva.es

The aim of this study was to evaluate the effects of supplementing lactating ewe diets with marine algae meal on ewe milk and meat performance of their suckling lambs. After lambing, twenty-four pregnant Churra ewes were fed daily a TMR containing lucerne and concentrate at a 40:60 ratio (forage:concentrate). Each ewe was randomly assigned to one of two treatments: Control (with 2% of fat from calcium salts of palm oil, Magnapac®) and MA (with 2% of fat from marine algae meal, Biomega Tech® Feed). Feed intake was recorded daily and milk yield and composition were registered weekly during the first month of the lactation. Milk production was recorded by the oxytocine technique. All lambs were reared exclusively on milk and were slaughtered when they reached 11 kg live weight. Data were evaluated by the GLM procedure of SAS. According to results, dry matter intake was reduced when ewes were supplemented with marine algae meal ($P<0.05$). MA treatment caused a decreased in milk (-33.8%), fat (-42.1%) and protein (-27.0%) yield ($P<0.05$). However, there were no significant differences in milk fat percentage ($P>0.05$) due to dietary treatment and the milk protein percentage was higher (+9.5%) in Control than in MA ewe milk ($P<0.05$). According to the experimental design, there were no difference in birth and lamb slaughter weights ($P>0.05$). However, lamb growth (305 vs 227 g/d), carcass dressing percentage (53.0 vs 49.8%) and kidney knob fat (212 vs 100 g) were also affected ($P<0.05$) by diets (Control vs MA, respectively). In conclusion, supplementing ewes' diet with marine algae meal reduce milk and meat performances and further research would be necessary to elucidate whether marine algae meal enhances the nutritional quality of ewe milk fat and of suckling lamb´s meat.

Estimation of direct and maternal genetic responses based on a piglet post-natal survival selection

T.Q. Nguyen[1], R.J. Dewhurst[1], P.W. Knap[2], S.A. Edwards[3], G. Simm[4] and R. Roehe[1]
[1]Scotland's Rural College, Edinburgh, EH25 9RG, United Kingdom, [2]Genus PIC, Genus PIC, Schleswig, Germany, [3]Newcastle University, School of Natural and Environmental Sciences, Newcastle, NE1 7RU, United Kingdom, [4]University of Edinburgh, The Royal (Dick) School of Veterinary Studies, Edinburgh, EH25 9RG, United Kingdom; tuan.nguyen@sruc.ac.uk

To genetically improve piglet survival is a challenge due to its low heritability, binary observation and being controlled by both genes of the piglet (direct effects) and genes affecting the dam's mothering performance (maternal effects). The objective of this research was to evaluate the effect of two generations of selection on piglet survival in both piglet and sow performance, using data from 22,481 piglets raised under outdoor conditions. The first generation animals were produced from 26 Landrace boars selected in groups of high and average estimated breeding values (EBV) for maternal genetic effects on piglet postnatal survival. From each group, 285 and 283 daughters were raised and mated with groups of Large White boars (n=42) divergently selected for direct genetic effects of postnatal survival to obtain the second generation animals. The piglet data included survival at birth (SVB), survival at weaning (SVNP) and individual birth weight (IBW). The sow data comprised of number born piglets (NB), average piglet birth weight, standard deviation of piglet birth weight within litter (SDBW), survival ratio at birth and at weaning (SVNPL). Genetic parameters and breeding values were estimated by Bayesian multivariate threshold model at piglet level and Bayesian multivariate linear model at sow level. At piglet level, maternal and direct selection responses respectively (calculated as groups' EBV differences) were for SVB $0.11\pm0.02\%$ and $0.89\pm0.03\%$, SVNP $0.21\pm0.05\%$ and $0.45\pm0.08\%$, and IBW -0.009 ± 0.001 and 0.02 ± 0.003 kg. At sow level, there was an increase in SVNPL of $1.46\%\pm0.004$, a reduction in NB of 0.14 ± 0.04 piglets and a decrease in SDBW of 0.008 ± 0.003 kg. The obtained selection responses indicate that improvement of piglet survival through genetic selection based on direct and maternal EBV was successful and revealed a negatively correlated response in litter size.

Validation of a piglet vitality score for maternal pig breeds in Austria

K. Schodl[1], B. Fuerst-Waltl[1], R. Revermann[1], C. Winckler[1], A. Willam[1], C. Leeb[1], P. Knapp[2] and C. Pfeiffer[1]
[1]University of Natural Resources and Life Sciences Vienna (BOKU), Gregor-Mendel Str. 33, 1180 Vienna, Austria, [2]Schweinezucht und Besamung Oberoesterreich, Waldstr. 4, 4641 Steinhaus, Austria; birgit.fuerst-waltl@boku.ac.at

The strong focus of breeding goals for sows on prolificacy negatively affects welfare of sows and piglets. Individual birth weights of piglets from large litters are lower and vitality is impaired which causes higher mortality rates. Thus, Austrian pig breeders aim to include a piglet vitality index (PVI) into routine genetic evaluation. PVI should comprise mean individual birth weights, standard deviation of birth weights and piglet vitality (PV). PV should be qualitatively assessed by breeders using a score from 1 (not vital) to 4 (vital) at litter level. Although other studies have used similar approaches to assess PV, this has not been validated so far. The aim of this study was to validate the PV score using liveborn preweaning mortality rate (PWMR) and to estimate genetic parameters for PV, total number of piglets born (TNPB) and PWMR. For this purpose, stockpersons on 23 farms assessed 2,323 litters after training for PV assessment. A linear mixed model was fitted for PWMR with farm, obstetrics, year-season, sow breed, PV score and parity as fixed effects and sire and sow nested within farm as random effects. To describe the relationship between PV and PWMR, Spearman rank correlations were calculated. The genetic analysis was performed for a subsample of 2,900 records from 22 farms fitting a trivariate linear animal model. The PV score 1, 2, 3 and 4 were used for applied in 0.68, 4.02, 25.7 and 69.6% of the litters, respectively. LS-Means for PWMR [%] were 23.8, 22.3, 15.1 and 7.3 for PV score 1, 2, 3 and 4, respectively. Except for score 1 and 2 PWMR differed significantly between scores. Heritability for PV was 0.11 ($\pm$0.04) and genetic correlations were -0.68 ($\pm$0.16) and -0.65 ($\pm$0.18) between PV and TNPB and PV and PWMR, respectively. The medium strong relationship between PV and PWMR and the differences between PWMR in the score categories suggest that PV may be routinely recorded given regular training. Genetic correlations indicate that breeding for large litters may reduce PV whereas breeding for PV may reduce PWMR.

Genome-wide association studies on stillbirth, birth weight and postnatal growth in pigs

C. Große-Brinkhaus[1], E. Heuß[1], M.J. Pröll-Cornelissen[1], H. Henne[2], A.K. Appel[2], K. Schellander[1] and E. Tholen[1]
[1]University of Bonn, Institute of Animal Science, Endenicher Allee 15, 53115 Bonn, Germany, [2]BHZP GmbH, An der Wassermühle 8, 21368 Dahlenburg-Ellringen, Germany; cgro@itw.uni-bonn.de

Piglet mortality has a negative impact on animal welfare, public acceptance and decreases the subsequent viability of pig performance. Moreover, the profitability of piglet production is mainly determined by the number of weaned piglets per sow. This situation is intensified by an increasing litter size which influences birth weight and piglet survival negatively. The objectives of this study were to evaluate stillbirth (SB), birth weight (BW) and postnatal growth on individual piglet level using genome-wide association studies (GWAS) in Landrace (LR) and Large White (LW). SB (n_{LR}: 911, n_{LW}: 796) was analysed using a logistic regression investigating full sibs. BW, intermediate weight (IW, three weeks after birth) and daily gain (DG) (N_{LR}: 2.290, N_{LW}: 2.291) were analysed as normally distributed traits. In order to identify the best model, results using raw phenotypes and various mixed model approaches as well as breeding values were compared according to their applicability for the respective traits. First results showed eight and two significant SNPs for SB in LR and LW, respectively. However, overlapping regions were not identified comparing the results of both breeds. Identified chromosomal regions (SSC 1, 3, 6, 7, 8, 13, 16) comprising significant SNPs have been already described in literature for total number of stillborn piglets on litter level in maternal lines. Potential candidate genes (e.g. UNC13A, KIF17, THSD4, MAPK10, CAPSL) were found for SB in LR. Other regions identified here were often associated with traits like teat number, birth weight, amount of corpus luteum and litter size. Generally, BW, IW and DG seem to be polygenically determined traits with only chromosome-wide significant SNPs found in both breeds. Especially in LW interesting QTL regions were found for IW and DG including candidate genes associated with growth, feed intake and cellular development. Interestingly, many significant associations identified here, are in relation with the immune system.

Preliminary genomic analysis for birth weight homogeneity in mice

N. Formoso-Rafferty[1], J.P. Gutiérrez[1], F. Goyache[2] and I. Cervantes[1]
[1]Facultad de Veterinaria, Universidad Complutense de Madrid, Avda. Puerta de Hierro s/n, 28040, Madrid, Spain, [2]Área de Genética y Reproducción Animal, SERIDA-Deva, Camino de Rioseco 1225, 33394, Gijón, Asturias, Spain; n.formosorafferty@ucm.es

A 20-generations divergent selection experiment, including two lines for high (H-line) and low (L-line) variability for birth weight in mice informed that homogeneous animals were more productive and healthy. Moreover, L-line individuals had more robust reproductive performance (litter size and survival of the pups) even under environmental challenge conditions such as food restriction during growth and reproduction periods. A total of 384 DNA samples belonging to 212 L-line and 172 H-line females were genotyped using the Affymetrix Mouse Diversity Genotyping Array to identify genomic regions associated to environmental variability for birth weight in mice. A preliminary analysis has been performed using a single marker regression approach (224,415 SNPs kept after quality control) for individual (BW) and standard deviation (SD) of birth weight within litter. Phenotypes were corrected for the effect of parturition number (and pup sex for BW). L-line and H-line females were analysed independently and -log10(P-value) was used as a parameter to select SNPs with important effect, selecting only those with values higher, in both lines, than half the global maximum of the parameter across lines, which resulted in a threshold of 39 for BW and 7 for SD. Gene-annotation enrichment and functional annotation analyses were performed with BioMart Software by Ensembl Genes 95 and DAVID 6.8 bioinformatics tools. Functional clusters with enrichment score higher than 1.3 (equivalent to Fisher P-value of 0.05) were identified: 13 for SD, 8 for BW and 2 for both traits. Most genes included in functional clusters for SD and BW were involved in response to stimulus, regulation of biological process, development and reproduction functions. For BW genes related with the immune system were identified as well. Selection can have selected genes involved in a better immune response and, consequently, in robustness. This is a preliminary research investigating the genetic basis of robustness determination in *Mus musculus* using genomic data.

Artificially reared piglets exhibit gut dysfunction at maternal separation and at weaning

J. Degroote[1], C. Van Ginneken[2], S. De Smet[1] and J. Michiels[1]
[1]Ghent University, Department of Animal Sciences and Aquatic Ecology, Coupure Links 653, Ghent, 9000 Belgium, [2]University of Antwerp, Department of Veterinary Sciences, Universiteitsplein 1, Wilrijk, 2610, Belgium; jerdgroo.degroote@ugent.be

Artificial rearing (AR), i.e. maternal separation of piglets at an early age (d3) and feeding a milk replacer until 3 weeks of age, is being practiced to handle supernumerary piglets. Fragmentary data are available how gut function matures pre-and post-weaning in AR piglets, hence this study was designed to deliver a more comprehensive insight, including a comparison between LBW and normal birth weight (NBW) piglets. Here, the histo-morphology and mucosal fluxes for FITC-dextran 4 kDa (FD4) and horseradish peroxidase (HRP) were determined in the jejunum of LBW and NWB piglets. Per birth weight category, 6 piglets were sample at d5, d8, d20 (weaning), d22, d25, d32 and d48 of age in AR piglets. These were compared with age-matched controls that remained with the sow until weaning (CON), which were sampled at d1, d3, d8, d20 (weaning), d22, d25, d32 and d48 of age. Data were separately analysed per day by a general linear model with rearing system and birth weight category as fixed factors, together with their interaction. Additional t-tests were done to evaluate specific hypotheses. Following maternal separation at d3 in AR piglets, villus length was significantly reduced at d8 (versus CON). Irrespective of rearing system, villus length was reduced at d22 and d25 compared to d20 (day of weaning). Yet, villi of AR piglets were also significantly smaller at d22 (versus CON). Villus width was significantly increased at d20 and d22, and crypts were deeper at d8, d20, and d22 in AR piglets (versus CON). Weaning also resulted in an increased FD4 flux at d22 compared to d20 (day of weaning). Nevertheless, the FD4 and HRP fluxes were significantly higher at d20, d22, d25, and d32 in AR animals (versus CON). Crypt depth, FD4 fluxes and HRP fluxes were higher, and villus length was reduced on d5 in AR piglets compared to CON piglets at d3. Conclusively, AR induced villus atrophy and a decreased barrier function both at maternal separation and at weaning. Most remarkably, AR piglets were characterized by long-lasting barrier malfunction in the post-weaning phase.

Influence of a supplementation in probiotics and vitamins on postnatal development of piglets

M. Girard, L. Duval, G. Michel, S. Dubois and G. Bee
Agroscope, Route de la Tioleyre 4, 1725 Posieux, Switzerland; marion.girard@agroscope.admin.ch

The selection for sow hyperprolificacy increased the number of less viable slow-growing low birth weight (BtW) piglets. Improving their survival and growth is therefore primordial. This study aimed to establish the impact of a supplement containing probiotics, vitamins A and E and selenium on piglet growth from birth to 2 weeks post-weaning and on the occurrence of diarrhoea. A total of 361 piglets were allocated in 2 groups: 175 piglets from 14 litters were administrated 2 ml of supplement (S+) the day of birth while 186 piglets from 14 litters received 2 ml of water (S-). Piglets were further grouped into 3 BtW-categories (Cat): low (L; ≤1.2 kg), medium (M; 1.2 to 1.7 kg), high (H; >1.7 kg). They were weighed at day (d) 0 (birth), 2, 5, 9, 16, 21, 25 (weaning) and 7 and 14 d post-weaning (PW). From d 21 onwards, diarrhoea occurrence was monitored. In addition, on d 16 and 14 d PW, faeces samples were collected from 2 females per litter to determine the volatile fatty acids (VFA) profile. They were selected according to their BtW and average daily gain (ADG) from birth to d 16, resulting in 2 groups: SLOW (BtW: 1.27±0.25 kg, ADG: 177±53 g/d) and FAST (BtW: 1.76±0.18 kg, ADG: 313±68 g/d). Piglets of the L/S- group but not of the L/S+ had a lower ($P<0.05$) body weight at 14 d PW compared to those of the M/S- ad M/S+ groups. The supplement did not affect the ADG from birth to weaning nor the VFA level in the faeces. Diarrhoea occurrence was 4%-units lower ($P=0.02$) in S+ than S-piglets 7 d PW. These findings suggest that the supplement might improve health status of L piglets. The production of VFA in faeces was similar ($P>0.10$) between SLOW and FAST piglets, but on d 16, the relative abundance of propionate was lower whereas that of butyrate, isobutyrate and isovalerate was greater in FAST compared to SLOW piglets. These results suggest that differences in microbiota colonization early in life within litter and their metabolic products directly impact growth performance of the host.

The effects of maternal hydroxytyrosol supplementation on the early postnatal growth of offspring

M. Vázquez-Gómez[1], C. García-Contreras[2], L. Torres-Rovira[1], J.L. Pesantez-Pacheco[3,4], P. Gonzalez-Añover[1], S. Astiz[4], C. Óvilo[2], B. Isabel[1] and A. Gónzalez-Bulnes[4]
[1]Faculty of Veterinary Medicine, UCM, Animal production Department, Av.Puerta de Hierro, 28040 Madrid, Spain, [2]SGIT-INIA, Department of Animal Breeding, Ctra.La Coruña, km7.5, 28040 Madrid, Spain, [3]School of Veterinary Medicine and Zootechnics, UCuenca, Av. Loja, Cuenca, Ecuador, [4]SGIT-INIA, Department of Animal Reproduction, Av. Puerta de Hierro, 28040 Madrid, Spain; martavazgomez@gmail.com

Polyphenols have great antioxidant activity and anti-inflammatory and immuno-modulatory properties, such as hydroxytyrosol. The goal of this study was to assess the effects of hydroxytyrosol supplementation of maternal diet during the last two-thirds of pregnancy on prenatal and early postnatal growth patterns of offspring. From Day 35 of gestation to delivery, 20 control (C; 10) and treated (HT; 10) Iberian sows were fed with a restricted diet (50% amount) to increase the IUGR process risk, while HT sows daily received 1.5 mg of hydroxytyrosol. Their offspring were weighed at birth, 15 days-old and weaning, and the average daily weight gain (ADWG) was calculated to evaluate its development. Moreover, mortality rates were analysed. Some traits were evaluated by litter size, categorized as 2-6, 7-8 & 9-10 piglets/litter. The results support the usefulness of the maternal supplementation with hydroxytyrosol to improve prenatal development in a swine model of IUGR pregnancies. The group HT showed by a higher mean birth-weight than group C (P<0.001) and lower incidence of low birth-weight piglets (P<0.05), although there were no differences in litter size or neonatal mortality. On the other hand, the effects of hydroxytyrosol treatment on birth-weight were determined by offspring sex and litter size. Females were lighter than females (P<0.05) at birth, but there was no difference at weaning due to a catch-up process. Moreover, HT piglets from the largest litters showed higher body-weight and ADWG than C piglets. In fact, there was a difference of around 1 kg between these piglets in the groups HT and C at weaning. In conclusion, hydroxytyrosol supplementation of maternal diet during pregnancy enhances prenatal and early postnatal developmental patterns, particularly in the largest litters.

Effect of glutamine supplementation in the neonatal phase on growth of low birth weight piglets

Z. Li, Q.L. Sciascia, J. Schregel, S. Goers, A. Tuchscherer and C.C. Metges
Leibniz Institute for Farm Animal Biology (FBN), Wilhelm-Stahl-Allee 2, 18196 Dummerstorf, Germany; li@fbn-dummerstorf.de

Low birth weight (L) piglets suffer from higher rates of mortality, and delayed development and growth compared to normal birth weight piglets. Maternal milk is thought to provide insufficient glutamine (Gln) for maximal growth of suckling piglets. Thus, additional Gln supply may improve piglet growth. We investigated the effect of oral Gln supplementation, during the early neonatal period, on growth and plasma metabolite profiles of L piglets. At birth, 48 male German Landrace piglets (0.8-1.2 kg,<25% percentile of the birth weights in FBN pig breeding facility), born to first parity sows, were selected and suckled by their dams in standardised litters (12 piglets/dam). Litter mate piglets randomly received an oral dose of Gln (1 g/kg body weight (BW)/day, Gln-L, n=24) or alanine (Ala, 1.22 g/kg BW/day; isonitrogenous to Gln, Ala-L, n=24), from 1 to 12 days of age (d). Piglets were weighed daily and crown-rump length (CRL)/ abdominal circumference (AC) were measured at birth, 5 (n=24/group), 7 and 12 d (n=12/group). Piglets received an i.p. injection of D2O (0.2 ml/kg BW, 99% atom-% D) at 11 d to measure milk intake (n=12/group). Plasma was collected to analyse milk intake (12 d) and metabolites (5, 12 d). Data was analysed using the MIXED procedure of SAS. Least square means were separated using the Tukey test (P<0.05). Gln-L piglets were heavier than Ala-L piglets at 10 (P=0.07), 11 (P=0.03) and 12 d (P=0.02). Gln-L piglets had higher milk intake (375.9 vs 327.6 g/kg BW/d, P=0.02), were longer at 12 d (CRL: 32.9 vs 31.4 cm, P=0.05) and were wider at 5 d (AC: 28.8 vs 26.8 cm, P=0.01) than Ala-L piglets. Plasma triglycerides (TG) were lower (5 d: 0.9 vs 1.2 mM, P=0.06) and plasma urea and non-esterified fatty acids (NEFA) were higher (12 d: 3.4 vs 2.7 mM, P=0.07 and 401 vs 310 μM, P=0.06) in Gln-L. We conclude that oral Gln supply in the early neonatal period improved growth of L piglets. The different plasma TG and NEFA levels suggest an altered lipid metabolism, whilst the increased plasma urea might be due to the higher milk intake of Gln-L piglets.

Sow lying behaviour, crushing losses and nest acceptance in free farrowing systems in early lactation

S. Ammer[1], C. Sprock[1], M. Fels[2], N. Kemper[2] and I. Traulsen[1]
[1]Department of Animal Sciences, Georg-August University, Albrecht-Thaer-Weg 3, 37075 Goettingen, Germany, [2]University of Veterinary Medicine Hannover, Foundation, Institute for Animal Hygiene, Animal Welfare, Bischofsholer Damm 15, 30173 Hannover, Germany; stefanie.ammer@uni-goettingen.de

The objectives of this study were to analyse the sow lying behaviour, crushing losses and piglets' nest acceptance in free farrowing systems during early lactation. Therefore, 36 sows were kept in free farrowing systems with 2 different piglet nests: (1) a box with small entry (BOX, n=8 sows) and (2) a covered one with large entry (OPEN, n=28 sows) and were observed within the first 72 hours postpartum using video records. The assessment included each lying down event per sow: (1) fully controlled (CON); (2) controlled forehand in prone position (PRONE); (3) controlled forehand in side position (SIDE); (4) uncontrolled (UNCON) as well as the location of the events inside the pen and each crushing loss of a piglet caused by: (1) a lying down event (LOSSLY) or (2) a change in lying position (LOSSPOS). In addition, the time until the first piglet and 80% of the piglets used the piglet nest was recorded. In total, 783 lying down events were observed with 28.4% for CON, 40.6% for PRONE, 25% for SIDE and 6% for UNCON. The causes for crushing losses (n=92) amounted 33.7% for LOSSLY and 66.3% for LOSSPOS. Most of LOSSLY occurred due to SIDE (48.4%) and UNCON (35.5%) followed by PRONE (16.1%), while none of the losses was caused by CON. Increasing litter sizes led to rising crushing losses. The mean time until the first piglet used the nest after the onset of farrowing was higher for BOX (335 min; Min: 131 min; Max: 788 min) compared to OPEN (161 min; Min: 8 min; Max: 619 min). Additionally, after on average 2,720 min 80% of the litter were in the nest for OPEN, while only one litter for BOX could reach this parameter within the first 72 h at 4,281 min after farrowing. In conclusion, free farrowing systems combined with increasing litter sizes bear the risk of rising piglet mortality due to crushing. Most of the crushing losses were caused by lying position changes of the sow, fewer by lying down events. Open structured piglet nests increase the acceptance for early utilization from the piglets.

Hut climate impacts piglet survival

S.-L.A. Schild[1,2], L. Foldager[1,3], M.K. Bonde[4], H.M.-L. Andersen[5] and L.J. Pedersen[1]
[1]Aarhus University, Department of Animal Science, Blichers Allé 20, 8830, Denmark, [2]Swedish University of Agricultural Sciences, Department of Biosystems and Technology, Sundsvägen 16, 23053 Alnarp, Sweden, [3]Aarhus University, Bioinformatics Research Centre, C.F. Møllers Allé 8, 8000 Aarhus C, Denmark, [4]Center of Development for Outdoor Livestock Production, Marsvej 43, 8960 Randers, Denmark, [5]Aarhus University, Department of Agroecology – Agricultural Systems and Sustainability, Blichers Allé 20, 8830, Denmark; sarah.lina.schild@slu.se

In outdoor production, piglet mortality varies across the year. This variation could be related to seasonality in hormones affecting parturition and lactation, but may also be related to changes in farrowing hut climate. The aim of the current study was to quantify the effect of yearly variation in farrowing hut climate on piglet mortality in organic sow herds. Data were collected at five organic sow herds in which sows gave birth in insulated A-frame huts. Hut temperature (T) and humidity were recorded using data loggers and farmers recorded production results. Stillbirth risk (n=646 parturitions) was analysed using a binomial generalized mixed model (GLMM). The rate of liveborn mortality until castration (median day 5; n=568 parturitions) was analysed using a negative binomial GLMM. Sows gave birth to 18 piglets (median). Stillbirth risk depended on time of year ($F_{3,635}$=4.4, P=0.004), the lowest risk being in winter. In winter, a higher T day 1 *pre partum* than at parturition increased stillbirth risk ($F_{1,99}$=6.39, P=0.013). The remaining part of the year, T category (three levels: T≤22 °C, 22 °C<T<27 °C; T≥27 °C) affected stillbirth risk ($F_{2,511}$=6.46, P=0.002) with an increased risk at T≥27 °C compared to 22 °C<T<27 °C. Time of year also affected liveborn mortality ($F_{3,561}$=3.9, P=0.009), the lowest rate being in spring. No effect of hut temperature was found for liveborn mortality. Hut humidity (<100% or =100%) affected neither stillbirth nor liveborn mortality significantly. In conclusion, high T increased stillbirth risk, suggesting, high T may challenge sows in outdoor production. Contrary, low T did not appear to affect liveborn mortality, indicating, that in insulated farrowing huts with proper management, sows are able to maintain a sufficient nest microclimate.

Locomotion as a possible indicator of vitality in the newborn pig

C. Vanden Hole, P. Aerts, S. Prims, M. Ayuso, S. Van Cruchten and C. Van Ginneken
University of Antwerp, Universiteitsplein 1, 2610 Wilrijk, Belgium; charlotte.vandenhole@uantwerpen.be

In the domestic pig, large litter sizes, combined with a limited uterine capacity (i.e. uterine, placental and foetal function), lead to intrauterine competition. As such, intrauterine crowding (IUC) causes an increased heterogeneity in within-litter birth weights and a larger number of small piglets. These piglets often exhibit a low vitality (physical strength), causing them to lose competition for a functional teat with their heavier littermates. This difficulty in retrieving milk from the sow makes small piglets more prone to starvation. In addition, by being less vital, they get crushed by the sow more frequently. Consequently, mortality rates in low birth weight/low vitality piglets (L piglets) around the time of birth get quite high. We attempted to unravel vitality-related differences by comparing locomotor development between L piglets and piglets with a normal birth weight/normal vitality (N piglets). To this end video recordings and associated digitization of the footfalls were made of walking piglets (n=25) at ten time points during early development (0, 1, 2, 4, 6, 8, 24, 26, 28 and 96 h after birth). By means of mixed models, motor performance, neuromotor maturation and gait variability were studied from spatio-temporal gait variables (e.g. stride and step length, stride frequency and duty factor) and asymmetry indices. Both absolute and normalized data (according to the dynamic similarity concept) were included, to distinguish neuromotor maturation from effects caused by growth. A lower motor performance (measured by absolute self-selected speed and its components) in L piglets and the increase of these variables in function of body mass, indicates that IUC affects locomotion mainly by impairing growth in utero. A difference in neuromotor skill was observed at birth, though only for swing and stance duration. In addition, neuromotor maturation afterwards was similar for L and N piglets. From this study, we conclude that L piglets might be considered a smaller and fictitious younger version of N piglets.

Genetic parameters for litter traits at farrowing and weaning in Landrace and Large White pigs

S. Ogawa[1], A. Konta[1], M. Kimata[2], K. Ishii[3], Y. Uemoto[1] and M. Satoh[1]
[1]Graduate School of Agricultural Science, Tohoku University, 468-1 Aramaki Aza Aoba, Aoba-ku, Sendai 980-0845, Japan, [2]Cimco Corporation, 2-35-13 Kameido, Koto-ku, Tokyo 136-0071, Japan, [3]Division of Animal Breeding and Reproduction, Institute of Livestock and Grassland Science, NARO, 2 Ikenodai, Tsukuba, Ibaraki 305-0901, Japan; shinichiro.ogawa.d5@tohoku.ac.jp

Increasing litter size at weaning (LSW) is required for maternal pig breeds and the nursing ability of dam is also important to keep the litter size at farrowing until weaning. In this study, genetic parameters were estimated for litter traits at farrowing (total number born (TNB), number burn alive (NBA), number still born, total litter weight at birth (LWB), mean litter weight at birth (MWB defined as LWB/NBA), and survival rate at birth (SVB as NBA/TNB)) and those at weaning (LSW, total litter weight at weaning (LWW), mean litter weight at weaning (MWW), and survival rate from birth to weaning (SVW as LSW/NBA)) in Landrace and Large White pigs. The numbers of litters for traits at farrowing and weaning were 62,534 and 7,274 for Landrace, respectively, and 49,817 and 6,144 for Large White, respectively. Pedigree data for Landrace and Large White pigs included 79,224 and 68,615 animals, respectively. Piglets were weaned at 21 days of age and the records without cross-fostering were analysed. Single-trait and two-trait repeatability models were exploited to estimate the heritability and the genetic correlation, respectively. Heritability estimates of SVB and SVW were 0.1 or lower in both breeds. The genetic correlation between NBA and LSW was estimated to be higher than 0.8. Estimated genetic correlations of NBA with MWW and SVW were about –0.5 and –0.3, respectively. The genetic correlation between NSB and SVB was almost –1. Genetic correlations of LWW with NBA, MWW, and SVW were all positive. The results indicate that increasing NBA solely is expected to increase LSW and LWW but decrease MWW and SVW, that decreasing NSB contribute to the improvement of SVB, and that adding LWW in breeding objective as an indicator for nursing ability of dam could be effective in avoiding the decrease of MWW and SVW while not disturbing the increase of LSW.

Plasma metabolites and amino acids in low and normal birth weight piglets at birth and 4 hours later

Q.L. Sciascia, Z. Li, J. Schregel, S. Görs and C.C. Metges
Leibniz Institute for Farm Animal Biology (FDN), Institute of Nutritional Physiology, Wilhelm-Stahl-Allee 2, 18196
Dummerstorf, Germany; sciascia@fbn-dummerstorf.de

As litter size has increased so has piglet mortality, due mostly to the higher prevalence of low (L) birth weight (BiW) piglets. At birth, it is evident that L piglets are at a disadvantage compared to their normal (N) littermates – disadvantages that point to developmental retardation potentiated by changes in nutrient metabolism. Therefore, the aim of this study was to measure plasma amino acid (AA) and metabolite concentrations in the first 4 hours of life, and determine if differences exist between L and N piglets. Male German Landrace littermate piglets (L; BiW 0.8-1.2 kg, n=15) and (N; BiW 1.4-1.8 kg, n=15) were sourced from gilts with litter sizes between 10-20 piglets. At birth (before first suckling) and 4 hours post-birth (4 h), plasma samples were collected via venepuncture and plasma free amino acids, amino-metabolites (AA-m) and metabolites were measured. Data was evaluated using the MIXED procedure of SAS and where appropriate, with repeated measures. Least square means were separated using the Tukey test (P<0.05). At birth, the concentration of AA β-alanine was lower (P=0.02) and carnosine was higher (P=0.08), and the metabolites inositol (P=0.001) and lactate (P=0.06) were higher in L pigs. At 4 h, all essential AA, except threonine, were lower, the non-essential AA asparagine tended to be lower (P=0.07) and glycine was higher (P=0.01) in L pigs. The AA-m α-aminoadipic acid (P<0.01) and citrulline (P<0.01) were lower and carnosine (P=0.02) and taurine (P=0.03) were higher in L pigs at 4 h. At 4 h plasma glucose (P=0.09) tended to be lower and inositol (P<0.01) was higher in L pigs. The concentration of all AA (P<0.01), except alanine; and all AA-m, except hydroxy-proline increased from 0 to 4 h both in L and N piglets. The metabolites fructose and lactate were lower and glucose higher (P<0.01) at 4 h compared to birth in L pigs. Similar changes in metabolites were observed in N piglets from birth to 4 h, except inositol, which was higher (P<0.01). Results show, that from birth, there are metabolic differences between L and N piglets that may underlie the differences in post-natal growth potential.

Birth weight and glutamine supplementation effects on early postnatal muscle development of piglets

Y. Zhao[1], E. Albrecht[1], Z. Li[2], J. Schregel[2], Q. Sciascia[2], C.C. Metges[2] and S. Maak[1]
[1]Leibniz Institute for Farm Animal Biology, Institute of Muscle Biology and Growth, 18196 Dummerstorf, Germany,
[2]Leibniz Institute for Farm Animal Biology, Institute of Nutritional Physiology, 18196 Dummerstorf, Germany;
elke.albrecht@fbn-dummerstorf.de

Piglets with low birth weight (LBW) usually have a lower survival rate and diminished overall growth and performance. Piglets can partially compensate the growth retardation with adapted nutrition. Since milk is thought to provide insufficient glutamine (GLN) for LBW piglets, the current study investigated the influence of GLN supplementation during the early suckling period on development and lipid deposition in the M. longissimus. Four groups were generated consisting of LBW male piglets and their normal birth weight (NBW) littermates. Piglets were supplemented between 2 and 11 dpn with either 1 g GLN/kg body weight or an isonitrogenous amount of alanine (ALA). Twelve piglets per group were slaughtered at 5, 12 and 26 dpn and tissue of M. longissimus was snap frozen in liquid N2. Serial cryostat sections were stained using either Oil Red O or the alkaline phosphatase reaction to quantify structural changes including intramyocellular lipid droplets, adipocytes, capillaries, and muscle fibres. The abundance of myosin heavy chain (MYH) isoforms 1, 2 and 7 was quantified with western blots. Significant effects of age and BW on intramyocellular lipid droplets were observed (P<0.01), but no supplementation effect. LBW piglets had more lipid droplets than NBW piglets at 5 dpn (P<0.01). The differences decreased with age. Adipocyte development increased with age, but was not influenced by BW or supplementation. There was a trend for larger muscle fibres in NBW and GLN supplemented piglets (P<0.1). Furthermore, there was a trend for fewer capillaries in LBW piglets (P<0.1). Abundance of MYH7 decreased with age, whereas MYH1 increased. Supplementation with GLN promoted fibre conversion from slow to fast myosin as indicated by less MYH7 protein in GLN supplemented piglets (P<0.01). The results indicate that BW differences were accompanied by differences in lipid deposition, capillarization, and muscle fibre structure, suggesting a delayed development in LBW piglets. Those disadvantages can partly be ameliorated by GLN supplementation.

Drenching low birth weight piglets: friend or foe?

K. Van Tichelen[1], M. Ayuso[1], S. Van Cruchten[1], J. Michiels[2] and C.J.D. Van Ginneken[1]
[1]Antwerp University, Veterinary Sciences, Universiteitsplein, 2610 Wilrijk, Belgium, [2]Ghent University, Animal Sciences and Aquatic Ecology, Coupure, 9000 Ghent, Belgium; chris.vanginneken@uantwerpen.be

Farmers face difficulties in redeeming their investment in larger litters since this comes with a higher litter heterogenicity and mortality due to the higher prevalence of pigs with a low birth weight (BW) in these larger litters. The major problem these small pigs face is ingesting adequate amounts of colostrum and milk. Thus, farmers practice different rearing strategies for granting these small pigs access to functional teats and increasing their chances to win the competition for food with littermates. One of such strategies is drenching with an enriched formulated milk replacer. However, handling of the pigs during the drenching act could provoke stress and possibly counteract the beneficial effect of the supplement. In this study the effect of drenching (sham (n=31); 2 ml milk replacer (n=31)) for 7 days after birth vs non-drenching (n=27) on skin lesion score (0-3), weight gain, and survival of small pigs (mean BWlitter – 2.5×SD < BW < mean BWlitter – 1×SD) was examined on days 1, 3, 9, 24 (weaning) and 38 (post-weaning). Mixed models were fitted to analyse the effects of drenching vs non-drenching on weight gain, Pearson correlation examined the effect on the maximum skin lesion per pig, and Kaplan Meier tested the effect on survival. The lesion score was not affected by treatment (9 pigs scored 0, 26 pigs scored 1; 32 pigs scored 2; 22 pigs scored 3) (P=0.2). Nor did treatment affect weight gain during the drenching period (P=0.7; 0.89±0.46 kg), during the suckling period (P=0.09; 2.87±1.23 kg) and after weaning (P=0.7; 1.63±1.11 kg). Surprisingly, drenching lowered the mortality when compared to the pigs that were not handled (82%) (P=0.02). There was no difference with regard to mortality between the pigs that were sham drenched (45%) or those that were given 2 ml of milk replacer (55%) (P=0.9). Thus, drenching did not impose a significant stress to the pigs and induced some changes that led to an improved survival. This should be considered when evaluating the effects of the formulae that are supplemented via drenching.

Effect of age on the functional glycosidase activity in the small intestinal mucosa of piglets

N. Van Noten[1], E. Van Liefferinge[1], J. Degroote[1], S. De Smet[1], T. Desmet[2] and J. Michiels[1]
[1]Ghent University, Department of Animal Sciences and Aquatic Ecology, Coupure Links 653, 9000 Gent, Belgium, [2]Ghent University, Department of Biotechnology, Coupure Links 653, 9000 Gent, Belgium; noemie.vannoten@ugent.be

Recently, phytochemicals have gained interest in animal production because of their potential beneficial effects on animal health. However, these compounds might occur as their glycoconjugated form in plants, or glycosides might be constructed for specific purposes, which affects their bioavailability. After ingestion, the aglycone (i.e. the free phytochemical) is split from its sugar moiety by glycosidase enzymes present along the gastro-intestinal tract (GIT). Yet, to better predict the fate of glycosides in the GIT of the young pig, we aimed to gain insights in the effect of animal age and intestinal site on a selected number of glycosidases. For this purpose 36 median weight piglets originating from 6 litters were sampled at 6 time points: 10 days before weaning (corresponding to 13 days of age), on the weaning day, and 2, 5, 14 and 28 days after weaning. Each sampling day 6 piglets (one per litter) were humanely sacrificed and mucosal scrapings were obtained from the small intestine (SI) at 2.5 (SI1), 25 (SI2) and 75% (SI3) of total length. The α-glucosidase (α-gluc), β-glucosidase (β-gluc) and β-galactosidase (β-gal) activities of mucosal extracts were measured by the release of p-nitrophenol (pNP) from its respective p-nitrophenyl-glycoside. Results are expressed in units (1 U = 1 μmol pNP released per min per g protein). The data were analysed by a general linear model with age and small intestinal site as fixed factors. The α-gluc activity (mean value 1.40 U) was not affected by age or SI site (P>0.05). The β-gluc activity in SI1 (17.1 U) and SI2 (16.3 U) was significantly higher than in SI3 (11.9 U). The same pattern was found for β-gal activity with values of 25.8, 24.9 and 19.4 U for SI1, SI2 and SI3, resp. The age effect was reflected by a significant drop in β-gluc (-44.7%) and β-gal (-71.5%) activity between weaning and 2 days after weaning. In conclusion, the piglets' mucosa harbours glycosidase activities to hydrolyse phytochemical glycosides. Weaning results in a drop in β-gluc and β-gal activity, whereas α-gluc activity remains low across time.

Consequences of low birth weight on porcine intestinal maturation and growth

M. Ayuso, S. Van Cruchten and C. Van Ginneken
University of Antwerp, Veterinary Sciences, Universiteitsplein, 1, 2610 Wilrijk, Belgium;
miriam.ayusohernando@uantwerpen.be

Proper gut development is needed for nutrients' digestion and assimilation. The normal maturation of the gut may be altered when challenges occur during prenatal development, for example in intrauterine growth restricted pigs. However, the extent of this delayed maturation and its implications during the first weeks of life is not fully understood. This study was conducted to characterize the postnatal growth and intestinal development patterns of 24 normal (NBW) and 24 low birth weight pigs (LBW). Four NBW and 4 LBW pigs were sacrificed at 0, 8 and 96 h and 10, 28 and 56 days of age and BW, gross and microscopic morphometry of the small intestine (SI), and ALPi and PCNA protein abundance were determined at different time points as indicators of cell proliferation (PCNA) and enterocyte differentiation (ALPi). Body weight was used to normalize the SI weight and length; the SI weight was similar, but the SI length was significantly higher in LBW. However, NBW showed a higher SI weight/SI length ratio, corresponding to a thicker and thus more developed intestine than LBW. With regard to the microscopic morphology, LBW piglets had shorter and narrower villi and shallower crypts than NBW, leading to a smaller intestinal surface in this weight category (in agreement with their lower SI weight/SI length ratio). In both groups, villus height increased during the first hours after birth and dropped to about 45 of its maximum height after weaning. Villus width remained constant and crypt depth increased after weaning in both groups. Age did not affect ALPi protein abundance in LBW but it did in NBW, with the lowest abundance observed at 96 h. At this time point, the ALPi levels were higher in LBW pigs. The percentage of PCNA-positive cells varied with age, with the highest value at the 96 h time point. Moreover, the percentage of positive cells was significantly higher in NBW than in LBW at 96 h of life. Since PCNA and ALPi-staining represent opposite cell populations (i.e. proliferating vs differentiating cells), our results highlight the 96 h time point as a highly proliferative and sensitive to BW stage. Also, we report a marked impairment of intestinal development in LBW pigs that leads to decreased intestinal surface and that extends beyond the suckling period.

Head morphology and body measurements of IUGR and normal piglets – a pilot study

M.N. Nielsen, J. Klaaborg and C. Amdi
University of Copenhagen, Department of Veterinary and Animal Sciences, Grønnegårdsvej 2, 1870 Frederiksberg,
Denmark; ca@sund.ku.dk

In Danish pig production, approximately 30% of all pigs have been exposed to some level of intrauterine growth restriction (IUGR) during gestation. Piglets suffering from IUGR have a higher risk of mortality as they are smaller and weaker at birth compared to normal piglets. However, with the right management, such as grouping them at a nurse sow suitable for small piglets, they can easily reach weaning. Piglets suffering from IUGR can easily be identified by their headshape, as the brain sparing effect during intrauterine growth retardation has resulted in a larger head to body ratio compared to a normal piglet. It is therefore an easier on-farm tool for the farmer to find and select them by their headshape than body weight alone. The aim of this pilot study was make further measurements to quantify the headshape identification method. The study included eight piglets from eight different sows. They were divided into two groups; IUGR and normal based on their body weight and head-shape. Registrations were carried out on newborn piglets and included several measurements of the head and body that were then analysed by proc glm in SAS. Results showed that measurements of the head were significantly smaller in IUGR piglets compared to normal piglets including height (6.5 vs 7.1 cm SEM 0.21; $P<0.018$), length (6.9 vs 8.8 cm SEM 0.29; $P<0.005$) diameter of the head (18.6 vs 22.3 cm SEM 0.60; $P<0.01$) and the angle of the nasal-bridge (45.0 vs 31.5 ° SEM 3.52; $P=0.003$) and the angle of the head-nasal bridge (20.7 vs 22.3° SEM 0.60; $P=0.002$), respectively. Additionally, length of the snout to eye (3.8 vs 5.1 SEM 0.20; $P=0.003$) and ear to ear (4.2 vs 5.4 SEM 0.15; $P=0.003$) was smaller in IUGR piglets, respectively. Furthermore, measurements of the body were also significantly smaller in IUGR piglets compared to normal piglets including length between the shoulders and hips and the length from snout to tail ($P<0.05$). In conclusion, there is a significant difference in the morphological features between IUGR and normal piglets that are most likely attributed to the brain sparing effect.

Effects of IFNG and IFND in porcine conceptus and uterine cells
S.O. Jung, S.H. Paek and J.Y. Kim
Dnakook University, 119, Dandae-ro, Dongnam-gu, Cheonan-si, Chungcheongnam-do, Republic of Korea, 31116, Korea,
South; jso7000@naver.com

Large numbers of embryos and foetuses are lost during the peri-implantation period (Day 12-30 of pregnancy) because there is disturbance in coordinated cross talk between the conceptus and maternal endometrium. One of the messages that the placenta of pigs sends to the uterus comes in the form of secretion of interferon gamma and interferon delta. We hypothesized that interferons change the endometrial immune environment, increasing maternal immunity for protection and achieving maternal immune tolerance to the semi-allograft conceptus. In this study we examined the effects of IFNG and IFND within porcine uterine cells and conceptus to determine if: (1) the endometrial cells exhibits characteristics of an inflamed state; (2) induce apoptosis of endometrial cells; (3) IFNG increases the specific expression of chemokines and their receptors that potentially recruit immune cells to the endometrium. We treated endometrial explants tissues from on Day 12 of the oestrous cycle and conceptus from gilts on Days 12 pregnancy with increasing dose of IFNG and IFND. After specific treatment, those tissues were analysed by real-time PCR, immunofluorescence and TUNEL assay. Our results revealed that combination treatment of IFNG and IFND increased expression of hypoxia-inducible factors (HIF1-alpha and HIF1-beta) and T-cell co-signalling receptors as well as chemokine ligand 12 (CXCL12) and CXC chemokine receptor type 4 (CXCR4) in endometrial cells and conceptus. Moreover, IFNG induce apoptosis of uterine cell to favour endometrium remodelling in the presence of PBMCs. These system-level analyses will provide a more comprehensive understanding of the specific biological pathway to improve embryonic survival and successful pregnancy in the commercial swine industry.

Validation of refractometry as a fast method to quantify IgG in serum of newborn piglets in farm
G. Álvaro-Minguito[1], E. Jiménez-Moreno[1], M. Fornós[1], B. Isabel[2], C. López-Bote[2] and D. Carrión[1]
[1]Cargill Animal Nutrition, Mequinenza, 50170 Zaragoza, Spain, [2]Faculty of Veterinary Medicine, UCM, 28040 Madrid,
Spain; marta_fornos@cargill.com

Sow's colostrum play an important role in the immunity for the newborn piglets. Immunoglobulin (Ig) G protects against possible infections and improves the viability of piglets. There is a direct correlation between this concentration of IgG in colostrum and the serum IgG concentration in newborn. The quality and quantity of colostrum is directly related to the IgG concentration and the survival of piglets. The aim of this study was to validate the refractometry technique as a quick and economical analytical procedure to determine the IgG concentration in serum of piglets. Sampling was carried out in 3 farrowing batches and 120 serum samples were obtained from 40 litters from of different parity. Blood samples were taken from three 4 days old piglets per litter and stored at 4 °C during 18 h until analysis. In each case, three piglet birth sizes were selected (the lightest, 1.04 ± 0.19 kg, medium; 1.50 ± 0.10 kg, and the heaviest, 1.93 ± 0.20 kg per piglet). Serum samples were analysed for the determination of IgG concentration using Optical or Digital Brix® refractometers and both readings were compared with the IgG concentration measured by radial immunodiffusion (IDR) method. A minimum concentration of 1000 mg of IgG/dl is needed to ensure a proper transfer of passive immunity. Seventy percent of the samples overcame 1000 mg of IgG/dl and had a Brix® readings between 6 and 12%. Contrary, 30% of the samples contained IgG concentrations lower than 1000 mg of IgG/dl and had Brix® readings between 2 and 5.9%. In this study, serum IgG concentration averaged $1,451\pm672$ mg/dl. Optical and Digital Brix® refractometer readings of serum were highly correlated ($r=0.954$; $P<0.001$). The correlation between Optical Brix® readings and the corresponding IgG concentration was of 0.593 ($P<0.001$) while with Digital Brix® readings, was of 0.541 ($P<0.001$). Therefore, both Brix® refractometers may be used indistinctly as a fast, economical and reliable method of estimating IgG concentration, providing a practical and differential indicator between good and poor IgG concentration.

Housing of farrowing sows in pens with opening crates – piglet losses in relation to pen dimensions

K. Menzer, M. Wensch-Dorendorf and E. Von Borell
Martin Luther University Halle-Wittenberg, Institute of Agricultural and Nutritional Sciences, Theodor-Lieser-Straße 11, 06120 Halle, Germany; katja.menzer@gmail.com

The permanent restriction of movement of sows during the entire suckling period is increasingly criticized. While labour productivity and avoidance of losses through crushing (overlying) by the sow have been the main focus of development of farrowing pens, efforts are being made today to reduce the sows' confinement to a tolerable minimum. Opening crates (OC) allow sows to move after a short period of confinement post-farrowing, preventing piglets to be crushed and overlayed. The OC Systems available on the market differ drastically in their design features. Supported by the European Innovation Partnership (EIP) initiative, 4 variants of OC pens (all crates opened 7 days after farrowing) were tested in 11 trials on one farm (using Topigs Norsvin genotype), with a special focus on piglet losses during the suckling period. Two Modifications (MOD) of the OC had a narrow pen design: MOD 1 (2.9 m^2 floor space allowance for each sow, n=100 sows/1,397 piglets) and MOD 2 (3.1 m^2 floor space allowance for each sow, n=32 sows/420 piglets) and two types had a square base area: MOD 3 (3.0 m^2 floor space allowance for each sow, n=30 sows/390 piglets) and MOD 4 (3.7 m^2 floor space allowance for each sow, n=32 sows/413 piglets). The statistical analysis was carried out with the procedure GLIMMIX (SAS 9.4, 2012, SAS Institute Inc., Cary, NC). The piglet losses were modelled as binomially distributed, using a logit link function. F-test and adjusted t-test were used to analyse the fixed effects pen design/ pen modification, number of litters and litter size. As a result, less piglet losses were seen in OC pens with a narrow design (9.5% SE=0.8%) compared to those with a square base area (13.4%, SE=1.3%, P=0.003). It can be concluded, that pens with a square base and greater space allowance for sows provoke higher piglet losses MOD 1: 10.2% (SE=0.92%) and MOD 2: 7.2% (SE=1.2%), MOD 3: 12.7% (SE=1.7%) and MOD 4: 13.7% (SE=1.7%). Piglet losses in pens with the smallest width (MOD 2) were significantly lower compared to the two types of pens with a square base (MOD 3, P=0.026 and MOD 4, P=0.007).

Piglet's vocalizations during castration and vaccination

M.C. Galli[1], A. Scollo[2], S. Minervini[3], G. Riuzzi[1] and F. Gottardo[1]
[1]University of Padova, Department of Animal Medicine, Production and Health, Viale dell'Università 16, 35020, Italy, [2]Swivet Research snc, Via Che Guevara 55, 42123 Reggio Emilia, Italy, [3]MSD Animal Health, Via Fratelli Cervi, 20090, Segrate, Italy; giorgia.riuzzi@unipd.it

Public opinion considers fundamental to safeguard the welfare of food-producing animals. Consequently, the reduction of pain during routine procedures gains increasing importance. Aim of the study was to evaluate piglets' distress and pain using their vocalizations during castration and vaccination, the most common stressful farm practices. In the study, the vocalizations (numbers and peak value in dB) of 173 piglets were recorded during three different procedures: (1) castration (excising each testis by cutting the spermatic cord); (2) intramuscular vaccination using a needle syringe (IM); and (3) intradermal needle-free vaccination using the IDAL® device (MSD Animal Health). Eighty-eight 7-day-old male piglets were assigned to castration, while 43 and 42 14-day-old piglets were assigned to IM and ID respectively. Vocalizations were recorded by a digital sound level meter (T Tocas® Professional Sound Pressure Level Meter) placed approximately 50 cm away from the pig's head and analysed by a SoundLabsound capture software. A chi square test was applied to identify differences in the number of piglets that produced at least one vocalization during the procedures, whereas the intensity of the vocalizations was analysed using a one-way ANOVA Model. All piglets vocalized during castration, while 61.9% of those vaccinated IM and 37.2% of those vaccinated ID did it. A statistically significant difference was observed between castration and vaccination (P<0.001). Considering the peak value, the highest was measured in the IM group (101.9 dB), whereas the lowest one in the ID group (90.8 dB) and that recorded for castration was 97.9 dB (P<0.001). The research highlights how the different procedures can affect the vocalizations repertoire reflecting distress and likely pain in piglets. Moreover, vocalizations could be a promising objective tool for evaluating and classifying the state of pain and distress. This could contribute to improve and develop less painful practices for pigs.

How to handle trade-offs and synergies in our search towards a sustainable food system?

I.J.M. De Boer, A. Van Der Linden and E.M. De Olde
Animal Production Systems group, Wageningen University & Research, P.O. Box 338, 6700 AH Wageningen, the Netherlands; imke.deboer@wur.nl

Food systems are central to several pressing and existing challenges of our times, such as climate change, biodiversity loss, and malnutrition. The key challenge of the coming decades, therefore, will be to produce enough safe and nutritious food in a sustainable way. In our search towards sustainable food systems, however, we inherently face trade-offs that emerge in the development and implementation of interventions and policies. A sustainability trade-off occurs if a gain in one issue (e.g. animal welfare) comes at the expense of another issue (e.g. climate change or farm profitability). Trade-offs can occur within and across scales (e.g. farm, region, nation, chain) and time frames. To enable transparent decision-making in food systems, therefore, we first need to understand the complexity of the food system, and the multi-dimensional benefits and costs of interventions (including trade-offs and synergies), across scale and time. Using examples, we demonstrate that identification of trade-offs depends, among others, on the scale of analysis (e.g. farm or region), the boundary of the system (e.g. chain or food system), and the time frame chosen (e.g. one or multiple years). This knowledge then can feed into multi-stakeholder dialogues about future options and limitations of sustainable food systems. In such a dialogue, multiple stakeholders aim to identify shared ambitions and dreams, and explore underlying perspectives and values. Insight in these underlying values can help to find consensus and jointly identify so-called small wins, i.e. meaningful and feasible small steps that can contribute to a more sustainable food system.

A multidimensional decision support tool to assess the sustainability of livestock farming systems

F. Accatino[1], M. Zehetmeier[2], J.P. Domingues[1], E.M. De Olde[3], A. Van Der Linden[3] and M. Tichit[1]
[1]UMR SADAPT, INRA, AgroParisTech, Université Paris-Saclay, 16, rue Claude Bernard, 75005, Paris, France, [2]Bavarian State Research Center for Agriculture, Menzinger Straße 54, 80638, München, Germany, [3]Animal Production Systems group, Wageningen University & Research, De Elst 1, 6708 WD Wageningen, the Netherlands; francesco.accatino@inra.fr

The sustainability of livestock sector in Europe has a complex nature. The portfolio of benefits and costs of animal production systems spans environmental, economic and social dimensions that can be measured at different levels, i.e. from farm to region and country. Indeed, there are trade-offs and synergies between dimensions and levels, meaning that an improvement in one indicator might have unintended consequences on other indicators. Our objective was to build a Decision Support System (DSS) to evaluate the sustainability of animal production systems accounting for its complexity. Such a tool can be used to better understand trade-offs and synergies between sustainability indicators in current animal production systems and to better understand the impacts of innovation practices. The DSS was grounded on a set of indicators of the three dimensions of sustainability on multiple levels. Indicators can be of three different types: directly collected, computed from collected data, or simulated by models. In particular, the simulated indicators constitute the modelled consequence of an innovation practice. Our application of the DSS regards dairy and beef cattle in France, dual-purpose cattle in Germany, and laying hens in the Netherlands, with data extracted from existing databases or via direct on-farm surveys (between 10 and 20 farms were considered per case study). Results, obtained comparing farms within each case study, showed trade-offs both within sustainability dimensions (e.g. between nitrogen conversion efficiency and nitrogen feed self-sufficiency in French dairy farms) as well as between dimensions (e.g. between social and environmental indicators). The DSS is envisaged to take the form of an interactive web-based dashboard and will support a more complete understanding of trade-offs and synergies between sustainability indicators as well as the selection by end-users of practices that better soften trade-offs.

Sustainability performance of grass-based beef production systems in the Bourbonnais area of France
A. Van Der Linden[1], E.M. De Olde[1], C. Mosnier[2], C. Pineau[3], M. Tichit[4] and I.J.M. De Boer[1]
[1]Wageningen University & Research, Animal Production Systems group, De Elst 1, 6700 WD Wageningen, the Netherlands, [2]INRA, UMR 1213-Herbivores, 63122, Saint-Genès Champanelle, France, [3]Institut de l'Elevage, Economie des exploitations, 9 Allée Pierre de Fermat, 63170 Aubière, France, [4]INRA/AgroParisTech, UMR 1048 SAD APT, 16 Rue Claude Bernard, 75005 Paris, France; aart.vanderlinden@wur.nl

A lack of insight in trade-offs among environmental, economic, and social consequences of interventions hampers decision-making about future options for sustainable development of livestock farms. Benefit-cost portfolios can be used to show the environmental, economic, and social consequences of an intervention at multiple scales. The objective of this study is to assess the consequences of various interventions in beef production systems in the Bourbonnais region of France by comparing benefit-cost portfolios. We used the bio-economic optimization model Orfee to quantify environmental (n=11), economic (n=4), and social indicators (n=4) of three farm types representative for the Bourbonnais region (155-165 ha, two farm types sell weaned calves, one sells fattened calves). Besides the current farm configuration, we simulated the following interventions separately: mechanisation of feeding and spreading straw in stables, cultivating a mix of cereals and protein crops for silage (5 ha), and increasing the area for wrapped bales instead of hay making (10 ha). Under the current configuration, the farm type fattening male calves had a lower net economic result (-44%) than the two farm types selling weaned calves. A higher level of mechanisation resulted in a trade-off between the economic net result (on average -17% for all farm types) and labour requirements (-14%). Cultivating a mix of cereals and protein crops for silage resulted in a synergy between concentrate intake (-32%) and the net result (+7%). Increasing the area for wrapped bales slightly decreased concentrate intake (-4%). Innovations hardly affected greenhouse gas emissions per kg live weight (-2.1 up to +0.6%). We conclude that combining farm models and benefit-cost portfolios is useful to get insights in trade-offs and synergies among sustainability indicators, and can support decision-making on implementation of interventions by stakeholders.

Assessing current livestock-induced biomass flows in Europe's NUTS2 regions for evaluating feasible
M.C. Theurl, A. Mayer, L. Kaufmann, S. Matej and K.-H. Erb
Institute of Social Ecology, Department of Economics and Social Sciences (WiSo), Schottenfeldgasse 29, 1070 Vienna, Austria; michaela.theurl@boku.ac.at

The consistent high demand for livestock products in Europe is putting increasing pressures upon the global environment. Livestock production has huge implications on resource use and contributes substantially to global greenhouse gas (GHG) emissions, e.g. livestock consumes 60% of global biomass harvest and contributes 15% to annual global anthropogenic GHG emissions. While livestock can in theory be fed from grasslands, residue and waste flows, competition between food and feed production on croplands is increasing, while environmental costs and benefits of intensified grassland harvest are highly debated. A systematic understanding of livestock-induced biomass flows allows to assess future scenarios of food and feed production with consumption, but also to detect sub-national import dependencies of regional livestock systems and leakage effects in terms of land use or GHG emissions. In this study, we use the widely used Material and Energy Flow Assessment (MEFA) framework. We built a comprehensive set of biomass flow and land use data, encompassing information on ecological productivity of agricultural areas, primary production on cropland and grasslands, spatially-explicit information on land-use patterns, livestock activities and information on the final consumption of biomass products. Data origins from previous research, Eurostat, FAOSTAT and CAPRI. First, we calculated the extent of biophysical biomass flows of each NUTS2 region. We found, that high-producing regions (above 12 Mt DM/yr of biomass) were North-Western parts of France, central and Southern Spain, Denmark, and the UK and Ireland. In the majority of NUTS2 regions, the share of biomass harvested from grazing land (i.e. grass and hay) is smaller than cropland production (DM/yr). Since a lot of sustainability issues are related to these flows, we secondly, assessed their environmental effects in terms of C-stock changes, GHG emissions, and HANPP, and analysed trade-offs. Our assessment serves as a starting point which allows to model the territorial impacts of e.g. the implementation of livestock-related innovations and to assess limitations in land availability and emissions.

New EU27 livestock typology reveals areas for targeted innovation

J.P. Domingues[1], C. Perrot[2], M. Theurl[3], S. Matej[3], K.H. Erb[3] and M. Tichit[1]
[1]INRA/SAD-APT, 16 rue Claude Bernard, 75005, Paris, France, [2]Institut de l'élevage, 149 Rue de Bercy, 75595 Paris cedex 12, France, [3]Institute of Social Ecology (SEC), University of Natural Resources and Life Sciences, Schottenfeldgasse 29, 1070 Vienna, Austria; domi.joaopedro@gmail.com

The European Union is characterized by a large diversity of livestock farming systems and steering them towards more sustainable patterns of production requires targeted strategies and an improved knowledge on their diversity and spatial distribution. Such information is, for instance, key for identifying relevant areas to which innovations towards more sustainable production forms can be applied and to assess their implications. During the 2000s, many livestock typologies were developed, but none of these explicitly integrates consistent information on land use patterns, livestock characteristics such as animal density and co-localization, price competitiveness as well as non-price competitiveness. We developed a livestock typology for EU27 at the resolution of NUTS3 regions (n=986). This approach allows to identifying areas where livestock play an important role for ecosystems and the regional economy, and where similar agro-ecological conditions occur. We combined expert based knowledge with statistical multivariate methods based on various datasets. Results revealed that articulating land use and livestock to price and non-price competitiveness variables enable to highlight areas where livestock play important role: These can be: (1) very densely populated areas where both ruminants and monogastrics are concentrated, e.g. driven by economies of scale; or (2) areas where grassland based ruminant systems are key in aspects of high quality products and play a major role in landscape conservation. We discuss the suitability and contributions of such a typology to address the manifold sustainability challenges related to livestock systems in Europe.

Innovations to enhance benefits and limit costs in two livestock territories in France

D. Neumeister, E. Dolleans, C. Pineau, S. Fourdin, A.C. Dockes and C. Perrot
Institut de l'Elevage, 149 Rue de Bercy, 75595 Paris Cedex 12, France; delphine.neumeister@idele.fr

In the framework of the ERANET Animal Future project, two in-depth analysis were carried out in North of France (Boulonnais – dairy production) and Central France (Bourbonnais – suckling cows' production). The analysis aimed at identifying innovations that improve livestock systems sustainability, by fostering the benefits that livestock offers to the territories and by reducing its costs. In each territory, two sets of qualitative interviews were conducted: firstly a listening tour of agricultural stakeholders to identify innovations and to understand what's at stake in the territory (24 interviews: 19 in Bourbonnais, 5 in Boulonnais); Secondly an in-depth analysis of these innovations thanks to individual qualitative interviews with the drivers of each innovation in order to qualify them: successes and defeats, benefits and trade-offs, breaks and motivations, needs and partners, etc. (30 interviews: 17 in Bourbonnais, 13 in Boulonnais). Based on the Duru *et al.* concept, the analysis was translated into an 'innovation flower concept'. Results were discussed with local stakeholders during a participative workshop in the two regions. The territories are both characterised by a large part of grasslands, hence a majority of innovations includes grass and hedges valorisation. Despite different territorial contexts (touristic sea side territory versus low density area), innovations can be divided into the three pillars of sustainability: environmental issues (increase in grazing, grazing fattening, solar panels...), economical issues (outlet diversification, contracts...) or social issues (producers' organisations, on farm transformation, farm tourism...). This study shows the diversity of solutions that farmers can develop, whatever their convictions, skills and livestock dimensions. Motivations to develop an innovation are mostly personal (increasing profit, securing sales, improving working conditions, etc.) but its impact can be beneficial for the whole territory. The practical implementation of these innovations is often dependent on the local policies support. This same analysis is currently carried out in 8 other case studies in Europe. A cross synthesis will be available next year.

Landscape measures to improve livestock systems nitrogen management and planning

T. Dalgaard

Aarhus University, Department of Agroecology, Blichers Allé 20, 8830, Denmark, tommy.dalgaard@agro.au.dk

The objective of this paper is to demonstrate how new landscape level management practices can help to increase the use efficiency of nutrients, and in particular help to mitigate environmental and climate effects from intensive livestock production systems. The experimental setup is a real landscape, where both arable and livestock production systems have been digitized in a Geographical Information System, together with a linked database of output and input factors, and their nitrogen (N) content. Based on these data, the www.farm-N.dk model has been used to quantify annual N flows, and distribute farm N surpluses into different types of N losses and the soil N balance. Statistical confidence intervals show significant differences between farm types, with higher N losses at livestock compared to arable farms, and a relatively higher nitrate leaching at pig farms, as compared to a relatively higher ammonia emission part at the cattle farms. In general, the livestock farms showed a positive soil N balance while the arable farms showed a negative balance. A joint analyses of the farm level N balances and the related types of N losses, showed a non-linear relationship between livestock density and N losses, with relatively higher losses at the most intensive livestock farms. Consequently, optimising the distribution of nitrogen between farms could be used to mitigate N losses, including the potential water pollution from nitrates, air pollution from ammonia and the emissions of nitrous oxides and other greenhouse gasses. Furthermore, landscape level planning of the agricultural production, and combination with measures to reduce N pollution can be used to reach local objectives, as for instance protection of especially sensitive biotopes or drinking water reservoirs. Such options are collected in a catalogue with recommendations for such landscape level joint up management of water, air and greenhouse reduction measures, and the potentials of the approach is discussed based on experiences from a set of study landscapes across Europe.

Assessing the optimality and robustness of ecosystem services trade-offs: a probabilistic approach

F. Joly, M. Benoit and B. Dumont

Université Clermont Auvergne, INRA, VetAgro Sup, UMR Herbivores, Theix, 63122 Saint-Genès-Champanelle, France; frederic.joly@inra.fr

Sustainability science proposes different methods to quantify trade-offs and synergies between ecosystem services (ES). These methods are usually based on spatial correlations, temporal correlations, or production possibility frontiers (PPF). The latter is inspired by the Pareto optimality front, defined as the limit of the area where a criterion cannot be increased without reducing another one. On the PPF a given ES can be favoured compared to another one, e.g. a regulating ES over a provisioning ES or vice-versa, but the loss incurred by the other ES is minimized. A PPF hence represents the optimal forms of ES bundles a system can provide. The concepts of ES bundles and trade-offs can be mobilised to manage livestock agroecosystems. In this context, we considered relevant to add a probabilistic component to PPF based methods. Obtaining an optimal solution may indeed be difficult for a farmer, owing to the complexity of biological processes involved in animal production. Fine-tuned management decisions may have to be made to provide ES bundles belonging to a PPF, and slight deviations from these decisions may strongly reduce the probability of reaching it. Here, we introduce a robustness-based framework to take into account this probabilistic component. We provide an example of application of the framework to a pasture grazed by a mixed sheep/cattle herd, and analyse the trade-off between a provisioning service (meat production) and two regulating services (climate regulation and erosion control). We simulated management decisions, i.e. combinations of various stocking rates and cattle/sheep ratios, and modelled the resulting ES bundles, i.e. the sets of ES values. We then quantified the probability of obtaining each of the bundles through the number of management decisions leading to them. This number is indeed variable as a given set of values can be delivered by one or several management decisions. We finally identified the PPF and observed that the probability of obtaining bundles belonging to the PPF, or close to the PPF, are amongst the highest. This result indicates that mixed grazing has the potential to optimize the delivery of several ES in a robust way.

Food and biodiversity impacts of conservation scenarios on Dutch agricultural land

V.J. Oostvogels, A. Kok, E.M. De Olde and R. Ripoll-Bosch
Wageningen University & Research, Animal Production Systems group, P.O. Box 338, 6700 AH Wageningen, the
Netherlands; vincent.oostvogels@wur.nl

Biodiversity declines at an alarming rate and agriculture is an important driver of this loss. To reduce agriculture's negative impact on biodiversity, the European Commission (EC) proposes to increase the agricultural area under conservation measures. Such measures often go however at the expense of food production. The magnitude of this trade-off depends on the type of measures and the extent to which they need to be implemented, but the EC remains rather unclear on these details. We therefore aimed to explore how conservation measures on agricultural land could take shape, and, based on that, to assess potential land-use implications for food production and biodiversity. We derived claims on the required conservation measures from recent publications and complementary author interviews. Claims aimed to conserve a subset of declining species or to make future nature conservation more robust. Based on the measures they described, we defined four scenarios: MB, MB+, WA, and WA+. The first two consisted of targeted changes in agricultural practice, the second two of generic changes in both agricultural practice and land-use. Scenarios MB+ and WA+ were more elaborate variants of MB and WA. We applied these scenarios to a fixed area of agricultural land. We assessed impacts on food production based on energy and protein yield as feed for dairy production, and impacts on biodiversity based on potentially disappeared fraction (PDF) of plant species richness. Relative to the baseline, MB and WA reduced total energy yield by 3 and 41%, respectively, and total protein yield by 4 and 35%, respectively. In the baseline, mean PDF of the total case study area was 17% (i.e. 83% of reference plant species richness). In MB and WA, mean PDF improved to 16 and 10%, respectively. In MB+ and WA+, effects were more pronounced. This confirms that conservation measures on agricultural land, while increasing biodiversity on the area to which they are applied, can substantially decrease the food output of that area. Understanding of this trade-off is critical when conservation measures on agricultural land are part of a strategy to reduce agriculture's negative impact on biodiversity.

Tools for evaluating trade-offs between robustness to price and yield variations in dairy goat farms

L. Puillet[1], V. Verges[1,2], P. Blavy[1], P. Chabrier[2] and V. Picheny[2]
[1]UMR MoSAR, INRA, AgroParisTech, Université Paris Saclay, 75005 Paris, France, [2]UR MIAT, INRA Toulouse, 31326
Castanet Tolosan, France; laurence.puillet@inra.fr

In the context of agroecological transition, the challenge for dairy goat farms is to decrease the use of purchased concentrates and increase the use of grass in diets, with forages or grazing. Such changes potentially expose the system to different constraints: purchasing feedstuff exposes the system to price fluctuations, relying on more grass-based diets exposes to forage yield fluctuations. The system can express different types of robustness in face of these perturbations and trade-offs may appear between them. Objective of this study was to quantify the effects of price and forage yield fluctuations on robustness indicators with a global sensitivity analysis (GSA) of a herd model. Contrary to a one factor at a time approach, GSA allows the exploration of the whole space of parameters and their interactions. With this tool, we explored which management strategies better balance robustness indicators and if they are the same depending on feeding system. Seven parameters were studied: herd milk production potential, breeding season length, reproductive success, replacement rate, culling criteria, extended lactations management and concentrate supplementation level. A simulation plan was defined to optimize parameters space exploration and computation time. Simulations were run for contrasting feeding systems (e.g. maize silage, alfalfa hay, pasture grazing). To reduce computational cost and allow GSA, Kriging models were fitted to simulation data to obtain cheap-to-evaluate surrogates of model outputs. Sensitivity indices of outputs, related to robustness to price and forage yield fluctuations, were then computed using the kriging models. They quantified the contribution associated to each input (management parameters) to the variance of robustness indicator. As a complementary tool, a user-friendly interface was developed to allow the visual exploration of robustness indicators depending on feeding systems and management parameters. This study emphasizes the benefits of combining mathematical and computer methods with systemic modelling to deepen our understanding of trade-offs and synergies and quantify them.

Integrating animal welfare into sustainable development of livestock systems

E.A.M. Bokkers
Wageningen University & Research, Animal Production Systems, P.O. Box 338, 6700 AH Wageningen, the Netherlands;
eddie.bokkers@wur.nl

Assessing the impact of animal welfare innovations on sustainable development of livestock systems requires an integrated research approach including at least the key themes: environmental impact (e.g. use of resources and emissions to air, water or soil) and profitability. With an integrated analysis trade-offs and synergies can be identified, which will prevent unforeseen consequences and can stimulate desired directions. In several projects we worked on improving animal welfare while reducing the environmental impact of livestock systems and ensuring economic viability. We studied trade-offs between animal welfare and the environment in organic and conventional pig production and found that land use and contribution to climate change related to feed showed to be higher (83% resp. 36%) in organic pig production, while welfare indicators gave a variable picture. Different production systems for laying hens were compared for various sustainability issues resulting in that systems with a potential better welfare performed economically better but had a higher contribution to greenhouse gas emissions and acidification and needed more land. An integrated analysis of veal diets to improve calf welfare, while taken into account environmental and economic consequences showed that a diet containing more roughage improved welfare, was economically attractive and neutral for the environment. A study in dairy cows showed that subclinical ketosis and mastitis proportional have a smaller impact on climate change (2.3 and 6.2%) than on economic revenues (5.6-12.4%). One big challenge is to collect data for animal welfare, environmental and economic indicators from the same farm at the same time point, which is the best way for identifying trade-offs and synergies, and therefore essential to significantly contribute to sustainable development of livestock systems. Describing and analysing livestock systems based on more than one theme of sustainability creates complex, but more complete pictures and avenues for the future. It is critical that the impact of animal welfare innovations are considered integrated with environmental impact and economic performance if these innovations are to be long-lasting.

Assessing the effects of mature size on efficiency of beef and sheep breeding enterprises

T.J. Byrne[1], T.W.D. Kirk[1], P.R. Amer[2], K. Matthews[3], J. Richards[4] and D. Bell[5]
[1]AbacusBio International Limited, Roslin Innovation Centre, Easter Bush, Midlothian EH25 9RG, United Kingdom,
[2]AbacusBio Limited, P.O. Box 5585, Dunedin 9016, New Zealand, [3]AHDB, Stoneleigh Park, Kenilworth, Warwickshire
CV8 2TL, United Kingdom, [4]Hybu Cig Cymru – Meat Promotion Wales, Tŷ Rheidol, Parc Merlin, Aberystwyth SY23
3FF, United Kingdom, [5]Quality Meat Scotland, Rural Centre, West Mains, Ingliston, Newbridge EH28 8NZ, United
Kingdom; tbyrne@abacusbio.co.uk

In recent years strong selection for growth traits in beef cattle and sheep genetic improvement programs and improvements in animal management and feeding have resulted in increased breeding female mature weight globally. This has positive and negative effects on farm efficiency and financial performance. These effects include changes to fertility, increased replacement and mature female maintenance costs, faster growing progeny which can be slaughtered heavier (increasing carcase revenue and feed costs) or slaughtered earlier (decreasing feed costs), and differences carcase quality (meat and fat yield). An understanding of the implications of increasing breeding female mature weight is required in order to inform best practice approaches to genetic improvement and management. How costs and revenues change with respect to variation in breeding female mature weight determines the optimum mature breeding female mature weight, where the farm system is most efficient. This study focused on UK farming systems, accounted for differences in feeding between intensive and extensive production, and identified optimum breeding female mature weights under several scenarios. For a typical UK farm, we identified optimum breeding female mature weights of 740 kg for beef cows and 70 kg for ewes. The implications of changes to breeding female mature weight, in terms of carcase weight produced, feed consumed, and gross margin were modelled at the level of the UK industry. We also consider genetic and non-genetic tools which can employed by the industry, for commercial farmers and breeders, in order to bring about a shift in mature weights towards the optimum level. The outcomes represent key knowledge for genetic improvement programs, breeder and commercial farmers, and other industry stakeholders in the UK.

Estimation of labour requirement for feeding of dairy cows with loose barn dried hay

J. Mačuhová, V. Stegmann, B. Haidn and S. Thurner
Institute for Agricultural Engineering and Animal Husbandry, Vöttinger Str. 36, 85354 Freising, Germany;
juliana.macuhova@lfl.bayern.de

The interest of dairy farmers in feeding of cows with hay and especially with barn dried hay is increasing in the last years. However, less information is available about labour requirement for feeding of barn dried hay by recently used techniques. The aim of this study was to create calculation models for the estimation of labour requirement for selected tasks during feeding of loose barn dried hay. Therefore, labour studies were performed on Bavarian dairy farms to determine standard times for single work elements. For most investigated standard times, the required accuracy (ε) of 10% was already reached, for the others, further investigations will be performed. First calculation models were created for the estimation of labour requirement for two feeding systems: telehandler with gripper and forage wagon. In both cases, picking up of hay from the storage place is performed with a hay crane. The influence of many variables on labour time requirement can be tested in calculation models (e.g. herd size, amount of hay per cow and feeding, capacity of forage wagon, gripper, and grabber, distances, etc.). So e.g. by feeding once daily, the labour required per cow and year for distribution of hay at the feed table is between 2.33 and 1.36 h and between 2.56 and 0.77 h for herds between 20 and 120 cows (with decreasing tendency with increasing herd size) by using a telehandler (capacity of gripper 200 kg) and forage wagon (capacity 30 m^3), respectively. The labour required per cow and year for the filling of an intermediate storage for telehandler and the filling of a loader wagon with a hay crane (capacity of grabber 200 kg) decreases for small herd sizes with increasing herd size (from 1.19 to 1.12 h, herd sizes between 20 and 40 cows). However, for larger herd sizes, it increases from 1.21 to 1.62 h by herd sizes between 60 to 120 cows, due to increasing distances needed to be driven to pick up the hay from the drying/storage boxes when constant width of boxes is considered. The calculation models should support farmer in choosing of a feeding system and the right capacity of applied techniques for his farm conditions.

Capturing impacts of innovations in livestock farming using a vision design and assessment framework

V. Karger, M. Zehetmeier, G. Dorfner and A. Reindl
Bavarian State Research Center for Agriculture, Menzinger Straße 54, 80638 Munich, Germany;
vanessa.karger@lfl.bayern.de

Innovations implemented in livestock farming systems take effect on a complex set of sustainability indicators. Alongside the desired impacts of an innovation, side- and displacement-effects emerge that may result in synergies or trade-offs in an economic, ecologic or social dimension. Farmers and decision makers need to be able to assess sustainability effects of an innovation on all sustainability dimensions in order to make educated decisions regarding the implementation of an innovation. While there are simulation and optimization models that allow scientists to estimate the effects of innovations on various sustainability indicators – especially within the economic and environmental dimensions – the complexity of impacts triggered by an innovation can hardly be captured. In order to enrich bio-physical models and also take better account of social and animal welfare aspects that are difficult to compute, we adapted a vision design and assessment framework developed by Halbe and Adamowski that integrates stakeholder and expert knowledge. Vision design and assessment are accomplished via a participatory approach using functional analysis, causal loop diagrams and fuzzy cognitive mapping. We apply the methodology on the case studies of dairy cow and pig farming in Germany, sheep farming in Spain and Scotland and cattle farming in Portugal and France. In interviews with stakeholders and experts innovations for the examined livestock farming systems are used as start variables. In the further course of the interviews functions, needs and other influencing factors (such as impediments) within the system are added and causal linkages between these variables are drawn and weighed by the interviewees. The resulting mind-maps – one for each interviewee – are then merged in order to create a comprehensive system structure. The final diagram is a visualization of trade-offs and synergies induced by an innovation in an agricultural system that allows for a comprehensive assessment of sustainability effects of an innovation.

Modelling the impact of climate change on the yield of European pastures

M. Dellar[1,2], C.F.E. Topp[2], A. Del Prado[3], G. Pardo[3], G. Banos[1,2] and E. Wall[2]
[1]University of Edinburgh, The Roslin Institute, University of Edinburgh, Easter Bush Campus, Midlothian, EH25 9RG, United Kingdom, [2]SRUC, Peter Wilson Building, King's Buildings, West Mains Road, Edinburgh, EH9 3JG, United Kingdom, [3]Basque Centre for Climate Change (BC3), Alameda Urquijo, 4, 48-1a /48008 Bilbao, Spain; martha.dellar@sruc.ac.uk

With rising CO_2 concentrations, increasing temperatures and changes in rainfall patterns, the European climate is undergoing significant changes. These changes have the potential to dramatically affect pasture growth and therefore grazing livestock. Our research uses two different modelling approaches to predict the impact of climate change on the productivity of European pastures: (1) an empirical approach based on linear regression of climatic and managerial variables; and (2) a dynamic approach using the Century ecosystem model. Each approach has its benefits and limitations. By using both we can mitigate the drawbacks of each and get an improved picture of how pastures are likely to change in the future. We looked at different geographic regions across Europe and found that pasture yields will increase in Alpine and northern regions, and either stay constant or slightly increase in the Atlantic region (depending on fertiliser use). For continental Europe, yields were predicted to increase for permanent grasslands but decrease for temporary ones. For southern Europe the trend was less clear, with the different methodologies producing conflicting predictions. It was clear from the results that management practices had considerably more effect on yields than climate change, suggesting that it should be possible to mitigate any negative impacts of climate change through changes in management practices. In most regions these results are good for grazing livestock, with yields either staying constant or increasing. In areas which may see decreases in yields it could become necessary for farmers to increase their use of bought-in-feed and/or implement changes in their management practices.

SusSheP – sustainable sheep production in Europe

C. Morgan-Davies[1], J. Kyle[1], P. Creighton[2], I.A. Boman[3], N. Lambe[1], A. McLaren[1], H. Wishart[1], A. Krogenaes[4], E. Wall[5], T. Pabiou[5], M.G. Diskin[2], K.G. Meade[6], N. McHugh[6], L. Parreno[7], E. Caldas-Silveira[8], X. Druart[8] and S. Fair[7]
[1]SRUC, Hill & Mountain Research Centre, Kirkton, Crianlarich, FK20 8RU, United Kingdom, [2]Teagasc, Athenry, Co. Galway, Ireland, [3]NSG, P.O. Box 104, 1431 Ås, Norway, [4]Norwegian University of Life Science, Ullevaelsveien, 72, Oslo, Norway, [5]SheepIreland, Shinagh, Bandon, Ireland, [6]Teagasc, Moorepark Fermoy, Co. Cork, Ireland, [7]University of Limerick, Castletroy, Limerick, Ireland, [8]INRA, Val de Loire, 37380 Nouzilly, France; claire.morgan-davies@sruc.ac.uk

Sheep systems in Europe vary in production, breed types and management systems. To be sustainable, they must be underpinned by environmental and welfare friendly practices, as well as being profitable and labour efficient. SusSheP is a research project funded under the European Research Area Network, Sustainable Animal Production Framework that aims to address some of these sustainability and profitability issues. The project is set across four countries (Ireland, France, Norway and UK) and focusses on key industry problems, namely the length of a ewe's productive life, farm labour and carbon hoofprint, and welfare-friendly sheep artificial insemination (AI) methods. The key objectives of the project are to provide new genetic tools for farmers to increase ewe longevity, quantify labour input and carbon hoofprint in contrasting sheep systems, develop other methods of AI to replace laparoscopic insemination by characterising ewe breeds differences in cervical function, and maximise knowledge transfer and uptake of these alternative approaches by the farming community. This paper will present some preliminary results of the project, and, in particular, will focus on the carbon hoofprint (using AgriCalc) and labour efficiency of three contrasting production systems in the UK and Ireland. It will also present preliminary results of breed effects on cervix characteristics to develop cervical AI, as well as common factors affecting ewe longevity in an attempt to determine the best longevity traits for inclusion in breeding indexes. SusSheP will ultimately provide farmers with better information and tools to enable them to make their own system more sustainable, thus ensuring uptake across the farming industry in Europe.

Heat stress response in sheep populations under different climatic and productive schemes

M. Ramón[1], C. Pineda-Quiroga[2], M. Serrano[3], E. Ugarte[2], C. Díaz[3], M.A. Jiménez[4], M.D. Pérez-Guzmán[1], R. Gallego[2,5], F. Freire[4], L. Mintegui[2] and M.J. Carabaño[3]
[1]IRIAF-CERSYRA, Avda. del Vino 10, 13300 Valdepeñas, Spain, [2]NEIKER-ARDIEKIN, Campus Agroalimentario de Arkaute, 01080 Vitoria-Gasteiz, Spain, [3]INIA, Mejora Genética Animal, Ctra. de A Coruña km 7.5, 28040 Madrid, Spain, [4]ASSAF.E, Granja Florencia, 49800 Toro, Spain, [5]AGRAMA, Avda. Gregorio Arcos, 19, 02005 Albacete, Spain; m.ramon.fernandez@gmail.com

One of the challenges of the project 'Innovation for sustainable sheep and goat production in Europe (iSAGE)' is to deal with extreme weather events. In particular, heat stress is expected to have a growing negative impact on sustainability because of climate change. In this study, we aimed at the quantification of response to extreme weather events in three dairy sheep breeds, raised under different climatic conditions and productive systems in Spain. Historic milk recording and AI results data (from 10+ years) provided by Assaf (As), Latxa (Lx) and Manchega (Mn) breed associations were joined with weather information from close to farms weather stations. Statistical models used to quantify productive and reproductive response to extreme thermal loads (TLs) included cubic regressions of daily milk yields or AI results on the date of recording and environmental noise effects (Flock and year of lambing, parity number and lactation stage at recording and prolificacy). Comfort regions differed for breeds from temperate vs warmer climate conditions and across traits. Heat stress thresholds were highest for milk yield and for Mn and As breeds (25 °C in Mn and Ass vs 20 °C in Lx) and lowest for fat yield (22 °C for Mn vs 18 °C in Lx). For the heat stress region, maximum losses for milk/fat/protein yields ranged from -22/-3/-2 g/day and °C over max daily TL (30 °C) for the highest producing, intensively raised As breed to -6/-1.1/-0.3 g/day and °C above max daily TL (25 °C) for Lx local sheep breed raised under semi-extensive conditions. Fertility loss was around -1.6% unit of AI success/°C above max TL for both As and Mn breeds. Both productive level and climate conditions of origin of the breed seem to determine the response to high heat loads in sheep breeds.

Persistency of lactation yields of milk, fat and protein in New Zealand dairy goats

M. Scholtens, N. Lopez-Villalobos, D. Garrick and H. Blair
Massey University, School of Agriculture and Environment, Private Bag 11-222, 4442 Palmerston North, New Zealand; m.scholtens@massey.ac.nz

Lactation persistency can be defined as the ability of the doe to maintain production following peak yield. Knowledge of the variation of persistency between animals is important to farmers so that they can make informed management decisions to improve farm productivity. The objective of this study was to describe the lactation curves of daily yields of milk, fat and protein in New Zealand dairy goats. A total of 113,895 herd-test records from 14,187 does were used to predict 36,195 individual lactation curves at each parity using random regression. Lactation curves were modelled using third-order orthogonal polynomials as a random effect for each animal-lactation with an unstructured covariance matrix. Persistency was measured as the area under the lactation curve from day 181 to 270 divided by the area under the curve from day 1 to 90, multiplied by 100. Peak and day of peak were determined for each lactation. Total yields, persistencies, peak yields and days at the peak yield were analysed with a linear model that included fixed effects of breed group (Saanen and crossbred), parity number, breed-by-parity number interaction, deviation from median kidding date for a given herd and year as a covariate and the random effects of herd and year. Total 270-day milk yield (MY) ± SD were 894.7±224.4 kg with a peak of 3.7±0.9 kg at day 92 and milk persistency of 84±8%. Peak yield and MY increased as does become older, until the fourth parity, then they decreased. The day of peak MY was later in lactation and milk persistency was greater for younger does than for older does. Effect of breed group was significant (P<0.001), with crossbred does producing greater yields with higher persistency than Saanen does. Similar results were found for FY and PY. Phenotypic correlations between 270-day yields and persistency were 0.26 (MY), 0.01 (FY) and 0.03 (PY), between peak yield and persistency were 0.19 (MY), -0.39 (FY) and -0.16 (PY), and between the day of peak and persistency were 0.95 (MY), 0.60 (FY) and 0.78 (PY). Knowing the shape of the lactation curve will assist farmers to devise better feeding and management strategies to optimise their production system.

Prevalence of hyperketonemia in Australian commercial periparturient dairy goats
F. Zamuner[1], K. Digiacomo[1], A.W.N. Cameron[2] and B.J. Leury[1]
[1]The University of Melbourne, Faculty of Veterinary and Agricultural Sciences, Building 184 Royal Parade Level 1, Rm 108, Parkville, Victoria 3010, Australia, [2]Meredith Dairy Pty Ltd, 106 Cameron Rd, Meredith, Victoria 3333, Australia; fzamuner@student.unimelb.edu.au

β-hydroxybutyrate (BHB) is regularly used to estimate herd's energy status and estimate the prevalence of hyperketonemia (HYK) in dairy cows. Sustained HYK is a strong indicator of ketosis, the primary metabolic disease affecting periparturient goats. However, analytic studies describing the patterns of HYK occurrence in dairy goat herds are scarce in the literature. This cross-sectional survey aimed to characterize HYK occurrence according to goat's parity and time of sampling by analysing weekly prevalence of HYK (BHB>0.8 mmol/l) in primiparous (PRIM; n=319) and multiparous (MULT; n=621) clinically healthy periparturient Saanen does (1-7 years, LW; 66±17.0 kg, and BCS; 2.5±0.3) from a commercial dairy farm in Australia. Blood was harvested via jugular venepuncture, once-weekly, from -21±3 d antepartum (AP) to 21±3 d postpartum (PP). Blood concentrations of BHB were measured using a ketone meter (FreeStyle Optium Precision Neo, Abbott Ltd., UK). Weekly prevalence (%) of HYK were; 0.7, 1.7, 2.6, 11.0, 14.1, 12.8, 12.2 vs 0.8, 0.0, 0.0, 0.7, 3.0, 4.3, 4.5, from -3 to 3 weeks from delivery, in MULT vs PRIM, respectively. Further, 24 PRIM does (8%) and 177 MULT does (29%) had at least one HYK event PP. From those, 13 PRIM does (54%) and 90 MULT does (51%) had the first HYK event between 3 to 14 DIM. Our results suggest that MULT goats are 3 times more likely to develop HYK than PRIM does. Additionally, PP prevalence of HYK increased 7-fold in MULT does and 11-fold in PRIM does compared to AP prevalence. It is concluded that; (1) 3 to 14 DIM is the best sampling window for early detection of lactational ketosis in dairy goat herds, (2) PRIM does should be evaluated separately from MULT does for greater accuracy estimating NEB and true prevalence of HYK at the herd level.

Barymetric predictions of body weight in goats
J. Viel[1], A. Ringuet[1], J. Mauclert[1], E. Laclef[1] and S. Jurjanz[1,2]
[1]ENSAIA, Sustainable Development of Agriculture, BP 20163, 54500 Vandoeuvre-lès-Nancy, France, [2]Université de Lorraine-INRA, UR AFPA, BP 20163, 54500 Vandoeuvre-lès-Nancy, France; juliettevielh4@gmail.com

Body weight (BW) of livestock animals is a key parameter to determine actions of herd management as reproduction or sales. In some cases, measuring weight with scales is difficult to implement, especially when animals are grazing on plots with a difficult access. Barymetric methods represent an alternative, as BW can be estimated from body measurements of animals. Such methods exist for cattle and horses but there is a lack of recognized barymetric formula for goats. Therefore, establishing a barymetric model for BW in goats is a real stake for breeders, especially in extensive rearing systems. As a result, 28 adult dairy goats (50-100 kg BW), 16 primiparous and 12 multiparous, were used for the study. These goats from the schoolfarm 'La Bouzule' near Nancy (France) had a strong body development. After milking and feeding time, goats were isolated. Then, the following measurements were carried out on each animal by 4 different operators: height at withers (HW; height of the animal from the ground to withers taken at the front legs), chest circumference CC (circumference of the animal just behind the front legs) and length of back LB (distance between the shoulder blades and the base of the tail). The LB and CC were determined with a tape measure but the HW using a measuring rod. Finally, all animals were weighed on a scale. The link between these measurements and the BW was analysed by a multiple regression with the software R. The best prediction was obtained using the CC: BW= 2.02 CC – 121.55, with R^2=0.88. This model predicts the BW of adult goats at ±8.5 kg or 12% of the weight of the animal. This preciseness could be improved by a larger sample size. Besides, this prediction could be slightly improved by adding LB: BW = 1.53 CC + 0.83 LB – 137.40 with R^2=0.91. However, this small improvement of BW predictions seem not to justify the addition of a 2nd parameter. BW predictions in adults only via body volume as CC was expected but in growing animals, adding a measurement reflecting body height seems useful. Finally, this model shows a good potential of application in field conditions to precise visual evaluation by the breeders.

Milk yield and composition of Local Balady and Boer Goats in Egypt

M.T. Sallam[1,2] and A.A. Abd El-Ghani[1,2]
[1]Minia University, Animal Production, Faculty of Agriculture, 111144, Egypt, [2]Minia University, Animal Production, Faculty of Agriculture, 111144, Egypt; akrum312@hotmail.com

Forty eight does, 25 from Egyptian local breed Balady and 23 from the exotic Boer breed, were used to examine differences in total milk yield (TMY), milk composition, lactation length (LL), lactation curves and the correlation coefficients among the milk yield and some body and udder measurements. Breed of does and type of rearing had strong (P<0.01) effects on TMY, LL, and average daily milk yield (ADMY). Boer does presented the highest (P<0.01) TMY and longest LL (116.0 kg milk in 112 days) when compared with the Egyptian Balady does (74.0 kg milk in 98 days). Also, breed of does and type of rearing had strong (P<0.01) effects on milk fat and lactose content. Milk production peaked at the 3rd week for the Balady does and at 5th week for the Boer does. The volume of filled udder and the volume difference (filled-empty) were greater (P<0.05) for Boer does. Also, Boer does presented higher (P<0.05) body length, average body weight, chest girth, abdominal girth and height at withers than Balady does. The volume of the filled udder, body length and height at withers in both Balady and Boer does were positively (P<0.01) correlated with TMY. Based on positive and high (P<0.01) correlations among volume of filled udder, body length, height at withers, and milk yield, it was suggested that use of these measurements to select indirectly for milk production in non-dairy goats would be effective.

Lactational effects of melatonin during autumn in two breeds of dairy ewes

A. Elhadi, A.A.K. Salama, X. Such and G. Caja
Universitat Autònoma de Barcelona, Group of Research in Ruminants (G2R), 08193 Bellaterra, Spain; abdelaali.elhadi@uab.cat

A total of 72 adult dairy ewes, 36 Manchega (MN; 72.4±1.9 kg BW) and 36 Lacaune (LC; 77.7±2.3 kg BW) breed animals were used to evaluate the effect of melatonin implants in the lactation performance. Ewes were penned in 12 balanced groups of 6 ewes, fed a TMR *ad libitum* (forage:concentrate = 55:45) and randomly assigned to a 2×2×3 factorial design (treatment×breed×replicate). The performance trial occurred in early lactation (d 28-114) during autumn. Treatments were: Control (CO; n=36), that did not receive any melatonin treatment; and, Melatonin (MEL, n=36), that were implanted in the ear base (18 mg/ewe) at wk 1 after the weaning of the lambs (35±1 d). Milk yield was recorded daily using automatic milk meters and ruminal bolus transponders, and milk composition were measured biweekly. Data were analysed by the PROC MIXED for repeated measurements of SAS and least squares means separated by PDIFF at P<0.05. No MEL effects were detected on DM intake (P=0.12) and milk yield (P=0.60), although they were 51 and 87% greater (P<0.001) in LC ewes when compared to MN, respectively. Milk composition varied by breed, the LC producing lower (P<0.001) fat and protein contents than the MN ewes, but no CO vs MEL effects were detected on milk composition (P=0.25 to 0.99) nor milk-fat standardized milk (P=0.73). Nevertheless, numerical greater persistence coefficients of milk yield were estimated for the group CO vs MEL of both breeds (LC, –0.015 vs –0.013 l/d; MN, –0.012 vs –0.010 l/d, respectively), the MEL treatment producing a numeric decrease of 7 and 3% in the milk of MN and LC ewes, respectively. On the other hand, BW and BCS did not vary by MEL treatment or by breed throughout the experiment. In conclusion, the use of exogenous MEL implants, together with the endogenous MEL produced under decreasing photoperiod conditions, had no effects on the early lactation performances of dairy sheep, despite of their milk production potential.

Productive and reproductive aspects of two different sheep breeds under Egyptian conditions

F. Younis[1], N. Ibrahim[2], T. Mostafa[3], A. Abd Elsalaam[3], O. El-Malky[3] and H. Tag Eldin[4]
[1]Desert Research Center, Animal physiology, 1 Mathaf El mataria, Cairo, 11753, Egypt, [2]Faculty of Agriculture, Beni Suef University, Animal Production Department, Beni Suef, 62511, Egypt, [3]Animal Production Research Institute, Dokki, Giza, 12618, Egypt, [4]Animal Health Research Institute, Chemistry Department, Dokki, Giza, 12618, Egypt; alphayounis@gmail.com

Livestock undergo various kinds of stresses such as: physical, nutritional, chemical, psychological and thermal. The aim of this study was to determine the influence of type of breed and physiological status on productive, blood metabolic profile, some of hormones concentration in the blood and reproductive performance of ewes. Investigations were carried-out on 40 healthy ewes belonging to the Bedouin flock in pastoral areas of Burj Al Arab region, aged 3-5 years old and with average weight of 35.0-40.0 kg. Ewes were grazing on the natural pasture (*Acacia saligna* and *Atriplex halimus* plants) and supplemented with a concentrate feed mixture. The effect of breed was strong (P<0.001) for RBCs, MCV and MCH, while, effect of physiological status was strong (P<0.001) for RBCs, Ht % and MCH. Ossimi ewes had higher level (P<0.05) of insulin concentrations than Barki ewes. The breed type and physiological status had a significant (P<0.05) effect for thyroid hormones, especially for thyroxin concentrations, which Ossimi ewes had higher values for T3 and T4 (1.76 and 40.14 µg/dl) than Barki ones (1.67 and 38.20 µg/dl), respectively. Ossimi ewes presented higher (P<0.05) leptin concentrations than Barki ewes during different physiological status. Ossimi ewes showed better reproductive performance than that Barki ewes, namely: oestrus duration, oestrus rate, day of oestrous, non-return rate, dominant follicle diameter, CL diameter, conception rate, fecundity, kidding rate, reproductive ability, kids born per ewes joined, ewes aborted/ewes conceived, twining frequency and kids weaned/kids kidded. Altogether these physiological, and biochemical responses makes the native breed to survive in harsh environment.

β-lacto globulin, kappa casein and prolactin genes effects on milk production of Awassi sheep

K.I. Jawasreh[1] and A.H. Al Amareen[2]
[1]Jordan University of Science and Technology, Department of Animal Production, Irbid- Jordan, 22110, Jordan, [2]National Agriculture Research Center, Livestock Department, Baqa- Jordan, 19381, Jordan; kijawasreh@just.edu.jo

The aim of this study was to assess the influence of genotypes of PRL, β-LG, and CSN3 genes on milk production and composition of Awassi ewes. A total of 167 genotypes were determined by PCR amplification followed by sequencing. The Mixed Model of SAS was used for performing the analysis. β-LG gene had a greater prevalence of the B allele (0.58), with BB and AB genotypes predominating in the population, with frequencies of 0.32 and 0.51, respectively. The frequencies of alleles A and B at locus PRL were 0.82 and 0.18, resulting in the higher frequency of the AA genotype (0.73). CSN3 gene polymorphisms were more restrict due to the higher frequency of the T allele (0.92), leading to very high frequencies for the TT (0.85) genotypes compared to TC (0.15) genotypes. The β-LG and CSN3 genotypes had no affect (P>0.05) on milk production. For the PRL gene, the AA genotype presented higher (P≤0.05) milk production than BB and AB genotypes. The AB genotype from the β-LG gene was associated with lower (P≤0.05) milk fat content. The BB genotype produced higher (P<0.05) SNF, protein and lactose content when compared with the AA and AB genotypes. The AA and BB genotypes, of PRL gene were associated with higher (P<0.05) levels of SNF and protein content, while the higher (P<0.05) lactose content was recorded for the BB genotype. For the CSN3 gene, the TT genotype was associated with the higher (P<0.05) levels of SNF, while the genotype TC was associated with the lower (P<0.05) levels of SNF content. However, the β-LG*PRL genotype AABB was associated with higher (P<0.05) milk production. The AABB genotype (β-LG*PRL) interaction was also associated with the higher (P<0.05) fat percentage compared to other genotypes. Higher values (P<0.05) of SNF and protein and lactose contents were recorded for the BBBB genotype of β-LG*PRL gene, compared to the other genotypes. Higher values (P<0.05) of fat and SNF percentage were recorded for the BBTT genotype of PRL*CSN3 gene, compared to the other genotypes. It can be concluded that the studied genes can be incorporated in breeding programs for improving milk production and composition traits in Awassi sheep.

Relationship of bodyweight and BCS with backfat and longissimus dorsi thickness in four dairy sheep

S.-A. Termatzidou, N. Siachos, G. Valergakis, K. Lymperis, M. Patsikas and G. Arsenos
Aristotle University, Faculty of Veterinary Medicine, School of Health Sciences, Box 393, 54124, Thessaloniki, Greece;
stermat@vet.auth.gr

The objective of this study was to use the ultrasound measurements of backfat thickness (BFT) and longissimus dorsi depth (LDD) to evaluate body condition score (BCS) and bodyweight (BW) changes in four different dairy breeds of sheep. Thirty randomly selected and clinically healthy non-pregnant, non-lactating ewes (Chios, n=8;Frizarta, n=8;Lacaune, n=7 and Assaf, n=7) were used. The experimental treatment involved a 6-weeks period of over-feeding and a 4-weeks period of under-feeding, considering the nutrient requirements for ewe's maintenance. BCS (scale 1-5; increments of ¼ units) was assessed by palpation. BW of individual ewes was measured weekly; the same time BFT and LDD were assessed with ultrasound (5.0 MHz linear transducer) at the transversal processes of the 3rd-4th lumbar vertebrae. BFT and LDD values and BCS values were analysed with repeated measures mixed models (IBM SPSS v.25). Breed, treatment and their interaction (B×T) were set as fixed effects and ewe was used as random effect. The most appropriate covariance structure was selected. Additionally, pairwise correlations between BW, BCS, BFT and LDD were calculated within each breed. BFT and LDD were strong predictors of BCS (BFT: β=0.255, se=0.023, P<0.001; LDD: β=0.057, se=0.006, P<0.001). The latter was affected by Breed (F=12.72, P<0.001) and Treatment (F=164.99, P<0.001). The highest correlation between BW and BCS (r=0.876) was observed in Lacaune ewes; lower correlations were observed for Chios, Frizarta and Assaf ewes (r=0.312, 0.363 and 0.137, respectively). Correlation coefficients between BCS and BFT (0.588, 0.594, 0.684 and 0.357 for Chios, Frizarta, Lacaune and Assaf ewes, respectively) and between BCS and LDD (0.373, 0.515, 0.584, and 0.647 for Chios, Frizarta, Lacaune and Assaf ewes, respectively). Correlations between BCS and BW, BFT and LDD differed (P<0.05) among the four breeds. At the same BCS, the depth of subcutaneous adipose and muscle tissues were different (P<0.05) among the four breeds. In conclusion, the existing BCS scale does not describe accurately the depletion or restoration of body reserves in different breeds of dairy sheep.

Evaluation of udder morphology and milk production in prolific and meat ewes

L. Del Pino[1], E. Salazar-Diaz[2], L. Rodríguez-Arias[2], C.B. Marques[1] and G. Ciappesoni[1]
[1]INIA, Meat & Wool, Ruta 48 km 10, 90200, Uruguay, [2]Universidad de Ciencias Aplicadas y Ambientales, Calle 222 55 – 37; Bogotá, 111166, Colombia; gciappesoni@inia.org.uy

Currently in Uruguay, several experiments of crossbreeding and synthetic breeds are being developed with the inclusion of East Friesian (EF), Finnsheep (F) and Texel (T) meat breeds. The milk production and udder morphology of 57 ewes (8 F, 24 EF and 25 T breeds) from three flocks were evaluated. A total of 161 records (24 F, 68 EF and 69 T breeds) were collected in three test-day controls with average days in milk (DIM) 21, 40 and 60. The ewes were randomly assigned to four experimental groups and separated from the lambs during 4 hours in a good pasture. Before that, it was allowed the nursing of the lambs and the ewes were manual milked to empty their udders. Glandular cistern area (CA) were recorded by ultrasonography (from the side area of the udder) after intramuscular injection of 1 I.U. of synthetic oxytocin, followed by mechanic milking to record the milk weight (MW) and quality (protein-P% and fat-F% percentages). All the traits were adjusted for the effects age of the ewe (1-3), type of lambing (1-2), breed (F, EF, T) and DIM and the square of DIM (both as covariables). Correlation between the residuals of CA and MW from these models were computed. EF showed a wider CA (35.7±2.0 cm2) than F (25.0±3.4) and T (28.4±2.6 cm2) breeds. This was expected because EF is a dairy breed selected to be tolerant to long milking intervals. However, no differences (P>0.05) were observed for milk production (F 382±47; EF 430±29; T 392±37 ml). The F and T breeds showed higher (P<0.0001) P% (5.6±0.1 and 5.6±0.1%, respectively) than EF (4.9±0.1%) and the similar (P>0.05) F% (7.0±0.4; 6.9±0.3 and 6.5±0.2, respectively). The correlation between CA and MW residuals were 0.48 F, 0.42 T and zero (P>0.05) for EF breeds. These findings would indicate a greater proportion of alveolar production in the EF breed and/or a better response to the release of milk by the oxytocin injection than the non-dairy breeds. Though, further research would help to determine the breed's differences between milk production and quality, as their conversion into kilograms of lambs weaned.

Qualitative behaviour assessment in intensively and extensively reared lambs

R. Bodas[1], M. Montañés[1], V. Cadavez[2], T. Peric[3], M. Baratta[4], N. Ko[5] and J.J. García-García[1]
[1]Instituto Tecnológico Agrario (ITACyL), Junta de Castilla y León, Subdirección de Investigación y Tecnología, Ctra, Burgos, km 119, 47071 Valladolid, Spain, [2]Mountain Research Centre, Polytechnic Institute of Braganza, Braganza, Portugal, [3]Univerza v Novi Goric, Nova Gorica, Nova Gorica, Slovenia, [4]University of Turin, Dept. Veterinary Science, University of Turin, Turin, Italy, [5]University of Stuttgart, Department Life Cycle Engineering, Stuttgart, Stuttgart, Germany; bodrodra@itacyl.es

Qualitative behaviour assessment (QBA, from the AWIN welfare assessment protocol for sheep) relies on the ability of humans to integrate perceived details of behaviour into descriptors with emotional connotation that can be scaled and added to other quantitative indicators. The complete AWIN protocol was performed in 14 groups (6 extensive, 2 semi-intensive and 6 intensive rearing system) of 15 young lambs (2 months of age) participating in the EcoLamb project (ERA-Net SusAn funded), which aims to holistically evaluate lamb production sustainability (meat quality, ecological footprint and animal welfare). Data on QBA (items were being scaled from 0 –absence- to 10 -all the animals fully expressed the evaluated item-), familiar approach and fleece quality tests were subjected to descriptive statistics analyses and showed herein. Extensively reared lambs scored higher in descriptors such as aggressive, defensive, physically uncomfortable or apathetic, whereas intensively reared lambs showed higher values in descriptors such as agitated and fearful but also in other as active, sociable, vigorous, subdued, calm, inquisitive and assertive. Semi-intensively reared lambs scored in between. All the animals ruminated to some extent (when observed), the quality of the fleece was always acceptable and no stereotypes were recorded. Minor lameness problems were observed in one extensive farm. Regarding familiar approach, all animals (except in one farm) voluntarily approached to human, despite a flight distance of 2 to 4 m was observed at the beginning. Although animal welfare is a real complex matter to be assessed, do animals from different rearing systems express different degree of positive emotional state?

Milk energy output and Carbon footprint from sheep and goats in different forage systems

A.S. Atzori, M.F. Lunesu, F. Correddu, F. Lai, A. Ledda and A. Cannas
University of Sassari, Dipartimento di Agraria, Viale Italia 39, 09030 Sassari, Italy; asatzori@uniss.it

The aim of this work was to study the relationship among carbon footprint (CF) and average production level of a sample of 21 Sardinia dairy sheep and goat farms belonging to different forage systems (FS). Sheep farms (n=15) belonged to lowland (L; n=5), to Hill (H; n=5) and to Mountain (M; n=5) systems. Goat farms belonged intensive indoor (I; n=3) management and semi-extensive (SE; n=3) systems. Farm data were gathered with face to face surveys to farmers, for a complete life cycle inventory (LCI) of farm production processes from October 1st 2016 to September 30th 2017. The LCI included information on flock, animal diets, feed purchases, crops, farm stocks, and an energy use audit. Data were analysed with a modified Tier 2 of the IPCC (2006), by using locally developed equations for animal energy requirements, coefficients of IPCC for estimation of animal and manure CH_4 and N_2O emissions and literature coefficients for purchased feeds and for energy. Emissions were allocated 100% to milk yield considering fat and protein corrected milk (FPCM) adjusted to 6.5% fat, 5.8% protein (1,047 cal/kg) for sheep and 3.5% fat and 3.1% protein (671 cal/kg) for goat. Average FPCM production level of the flock was equal to 198±49, 173±43, 217±90 kg head/year/ for sheep raised in L, H and M forage systems, respectively. Average production level for goats was 711±205, 581±415 kg/head/year for I and SE systems, respectively. Carbon footprint resulted equal to 3.15±1.41 and 3.72±0.80, 1.94±0.15 and 3.03±1.93 kg of CO_2 equivalents per kg of FPCM in L, H, M, I and SE systems. Production level of all the surveyed sheep and goat farms, expressed in terms of annual milk energy output (PLE; Mcal of milk energy/head/year), explained the most part of the variability of the emission intensities measured as CF = $84.4 \times PLE^{0.614}$; R^2=0.87) and as enteric methane (kg of CH_4/kg of FPCM) = $69.74 \times PLE^{0.711}$; R^2=0.92. In conclusion, similar environmental polices of emission mitigations can be formulated for sheep and goat systems using the milk energy output per head as proxy of environmental performances. The authors gratefully acknowledge the Europe Union for the financial support (Project Forage4CLimate LIFE 15 CCM/IT/000039).

Microbiological quality of lamb meat raised in intensive and extensive systems in Italy

F. Chiesa, P. Morra, I. Viola, G. Perona, T. Civera and M. Baratta
University of Turin, Veterinary Science, Largo Braccini 2, 10095, Italy; francesco.chiesa@unito.it

In the framework of Ecolamb, a project funded under the ERA-NET Cofund SusAn programme aimed to enhance the profitability and competitiveness of the European sheep production sector, the microbiological quality and the shelf-life of lamb meat was investigated. A total of 48 lambs (aged 4 months), 28 Biellese and 20 Sambucana breed, raised in intensive and extensive systems, respectively, were sampled at slaughter. Total Viable Count (TVC) was evaluated according to Regulation (EC) 1441/2007, using the abrasive sponge method on four different sites of 100 cm^2. Samples were also tested for the presence of common meat-borne microbial pathogens (*Salmonella*, *Listeria monocytogenes* and *E. coli* O:157). Longissimus dorsi muscles were used to determine the shelf life of fresh, chilled lamb meat stored at +4 °C, under vacuum packaging. The loins were collected at boning 1 day post slaughter and divided into four pieces that were vacuum packaged (VP) and stored at 4 °C. VP Samples were then analysed for TVC, Lactic Acid Bacteria (LAB) and Psychrotrophic Plate Count (PPC) at 3, 9 and 15 days after slaughter. None of the pathogens was detected. The TVC on carcasses resulted constantly under the satisfactory limit set by Regulation 1441/2007 (3.5 log ufc/cm^2). After two weeks of vacuum packaged chilled storage at +4 °C the counts of all the parameters were within the limit for acceptable quality (7 log_{10} cfu/g). A statistically significant difference (Mann-Whitney test, P<0.05) between the groups was shown for TVC after 15 days post-slaughter (Sambucana > Biellese) and LAB and PPC after 3 days post-slaughter (Biellese > Sambucana). The results from the present study suggest that storage time for vacuum packaged fresh, chilled lamb meat can reach 2 weeks at a temperature of +4 °C. The differences observed between the two groups may reflect different microbial population derived from different breeds and breeding systems. All VP meat samples were of satisfactory microbiological quality at the end of shelf life despite the differences described above.

An investigation into the development of screen-printed biosensors for fatty acid analysis in meat

A. Smart, O. Doran, A. Crew and J.P. Hart
University of the West of England, Health and Applied Sciences, Coldharbour Lane, Bristol, BS16 1QY, United Kingdom; amy.smart@uwe.ac.uk

The composition of fatty acids (FAs) in meat is an important indicator of quality; particularly relative concentrations of saturated, mono- and poly- unsaturated FAs. Strategies have been designed to improve the FA composition of meat, however current methods of measuring FAs to verify anticipated improvements are impractical; gold standard chromatographic methods are time consuming multi-step processes, requiring highly skilled personnel, and are expensive. In contrast, the novel biosensor approach based on screen-printing technology is rapid, reagentless, user-friendly and inexpensive with potential for mass production. Developing biosensor technology for rapid measurement of FAs in meat could streamline abattoir processing. The aim of this project is to develop novel screen-printed biosensors for the direct measurement of FAs in meat. Screen-printed carbon electrodes were drop-coated with a selected enzyme to measure FAs. Applied voltage was optimised for measurement of α-linolenic acid using hydrodynamic voltammctry, and calibration and precision studies were performed using amperometry in stirred solution. The applied voltage for the operation of the biosensor was optimised. From the position of the plateau, +0.5V vs Ag/AgCl was selected, which provides maximum sensitivity without excessive voltage (using lower voltages minimises the possibility of interferences). A calibration study was carried out at the optimised voltage. The steady state current was proportional to concentration up to 200μM α-linolenic acid. This indicates that the biosensor holds promise for the measurement of the selected FA. The reproducibility of the biosensor was assessed by examining the response of three individual biosensors to additions of α-linolenic acid. The co-efficient of variation was found to be 2.77%, which demonstrates good precision. This investigation demonstrates that the novel amperometric screen-printed biosensor can successfully measure α-linolenic acid. This technology holds promise for the measurement of other fatty acids present in meat; future work will involve developing biosensors for other FAs based on this platform technology.

Effects of whole, cracked or steam-flaked corn on growth performance in feedlot Damascus goats

M.C. Neofytou[1], D. Sparaggis[2] and O. Tzamaloukas[1]
[1]Cyprus University of Technology, Department of Agricultural Sciences, Biotechnology and Food Science, P.O. Box 50329, Limassol, Cyprus, [2]Agricultural Research Institute, P.O. Box 22016, Nicosia, Cyprus; ouranios.tzamaloukas@cut.ac.cy

The effects of corn processing and the type of concentrate mixture were studied on weaned Damascus breed kids of 70 days of age. Sixty animals were divided on the basis of live-weight and age into four groups and randomly allocated to the following treatments type of concentrate mixture: (1) pelleted concentrates (P), where all ingredients were ground and pelleted in 5-mm cubes; (2) dry-rolled corn grains (DRC); (3) whole grain corn (WGC); or (4) steam-flaked corn (SFC), mixed with pellets made from the other ingredients of the concentrate mixture. Diets were formulated to be iso-energetic with corn comprised 707/kg of the concentrate mixture, and offered *ad libitum* along with 0.1 kg per head of Lucerne hay. Treatments effects on feedlot kid performance were evaluated in a seven-week trial, with measurements of live weight gain, feed consumption and rumen pH. Animals fed SFC presented lower (P<0.05) feed consumption and lower (P<0.05) average daily gain when compared to the other three groups. Animals fed WGC, DRC and P diet showed similar (P>0.05) growth rates and feed consumption. Feed efficiency and pH measurements were similar (P>0.05) across treatments and therefore the lower live-weight gain in SFC animals was attributed to lower feed consumption.

Slaughter and carcass characteristics of suckling lambs of Bardoka local sheep breed

B. Markovic, M. Djokic and D. Radonjic
University of Montenegro, Biotechnical Faculty, Mihaila Lalića 1, 81000, Montenegro; bmarkovic@t-com.me

The objective of this study was to investigate the growth performance and carcass quality of suckling lambs of Bardoka breed. This is a local sheep breed traditionally reared in Montenegro, Albania, Serbia and Kosovo. Since this breed is characterized by the relatively good milk production, the lambs are fattened only during the suckling period and slaughtered at about 3 months of age, and after that ewes are milked. During first two months lambs were feed with milk, hay and concentrate feed add libitum. During the third month lambs were fed with milk, hay in combination with pastures and 0.150 kg/day of concentrate feed. At the end of the feeding trial, 12 lambs (6 males and 6 females) with age between 85 and 95 days were randomly selected. The following traits were investigated: live weight before slaughtering, weight and dressing percentage of warm and chilled carcasses with and without offal, weight and share of edible and non-edible by-products of slaughtering and skin, feet, head. Lambs average live weight before slaughtering were 25.4 for males and 24.4 kg for females. During the suckling period, male lambs presented higher (P<0.05) average daily gain than female lambs (0.246 versus 0.223 kg/day). Dressing percentage of warm carcasses were 57.8 for male and 57.6% for female lambs, while for chilled carcasses were 56.5 and 55.7%, respectively. However, it is important to stress that carcasses of suckling lambs usually include head and edible offal (heart, liver, lung and kidney). The average weight of chilled carcasses, without head and edible offals, were 11.3 kg for male and 10.9 kg for female lambs. No differences (P>0.05)were found between sexes regarding weight and dressing percentage, while the male lambs presented higher (P<0.05) weight of head, feet and skin (0.98, 0.88 and 3.23 kg, respectively) than female lambs (0.84, 0.88 and 2.97 kg, respectively). Considering carcass joints proportions, no differences (P>0.05) were found between male and female lambs: the proportion of first category joints were 36.0% in both genders; the proportion of second category joints were 31.5% versus 32.0%; and the proportions of third category joints were 32.5% versus 31.0%, for male and female lambs, respectively.

Breeding practices applied in Skopelos goat farms under the collective organization scheme

D. Tsiokos[1], F. Kanellidou[2] and C. Ligda[2]
[1]Hellenic Agricultural Organization – Demeter, Research Institute of Animal Science, Paralimni Giannitsa, 58100
Pella, Greece, [2]Hellenic Agricultural Organization – Demeter, Veterinary Research Institute, P.O. Box 60 272, 57001
Thessaloniki, Greece; chligda@otenet.gr

Within the PERFORM (ARIMNET funded) project, breeding and management practices of local sheep and goat breeds are examined in connection with their farming systems. The specific case of the Skopelos goat breed was studied, as a local breed valued by the breeders for its excellent characteristics and high milk production. The work aimed to provide an insight on the current breeding practices of goat farmers under the milk recording scheme. For this purpose, data collected in the frame of the scheme from 2008 until 2016 were analysed. The number of farms participating in the scheme varied through the years, mainly due to organizational gaps. Currently, 43 Skopelos goat farms are included in the milk recording scheme. For the period analysed, on average females are kept until the age of 6 to 7 years, while males are used for 2 years maximum. The 92% of the bucks used originate from the same farm and very little exchange between farms is practiced. However, differences on breeding practices between the farms are observed, due to the size of the farm and the location. Besides the data available through the milk recording scheme, additional information was collected with interviews to the farmers. A questionnaire aiming to a first appraisal of the existing collective organization was also used. The questionnaire addressed to the Regional Centre of Genetic Resources, the Skopelos Breeders' Association and goat farmers and included parameters related with the level of development of the collective organization, i.e. recorded traits, bucks' exchange practices, setting of breeding goals, other activities of the Association, etc. This assessment provided the basis for the debate with the Breeders Association aiming to identify the means to enhance the efficiency of the breeding program (i.e. revising the goals, implementing new tools and alternative selection schemes).

Effect of the dietary level of protein on feed intake and performance of Assaf heavy fattening lambs

C. Saro, S. Andrés, P. Oviedo-Luengo, I. Mateos, J. García-Rodríguez, M.J. Ranilla and F.J. Giráldez
Instituto de Ganadería de Montaña. CSIC-Universidad de León, Grulleros, 24346 León, Spain; cristina.saro@unileon.es

Thirty Assaf lambs, divided into three groups, equilibrated for live body weight (29.8±1.88 kg), were used to study the effect of different dietary levels of protein on feed intake, animal performance and carcass characteristics. Three total mixed rations including barley straw, barley, corn, soybean meal, urea, molasses, soy oil, bicarbonate, vitamin-mineral premix, with a 15:85 forage:concentrate ratio, were formulated in order to supply 135 (LP group), 155 (MP group) and 175 (HP group) g of crude protein/kg dry matter. Animals were housed individually and had free access to fresh water. Feed was offered *ad libitum* and daily intake was recorded. Lambs were weighed once a week and they were slaughtered at the end of the experimental period (63 days). Carcasses were weighed before and after chilling for 24 hours at 4 °C and the following parameters were recorded or calculated: hot and cold carcass weight (HCW, CCW), chilling losses, carcass commercial yield and proportions of carcass joints. Data were subjected to analysis of variance, considering the protein content in the diet as fixed effect. Linear and quadratic trends were evaluated using orthogonal polynomial contrasts. Dry mater intake (DMI) was not affected (P>0.05) by dietary protein level, although DMI was numerically greater in HP than in LP group (1,256, 1,401 and 1,464 g DM/day for LP, MP and HP groups, respectively). Average daily gain (221, 268 and 297 g/day for LP, MP and HP, respectively) and feed to gain conversion rate (5.82, 5.32 and 4.97 g DMI/g ADG for LP, MP and HP, respectively) increased (P<0.01) and decreased (P<0.01), respectively, with increasing level of dietary protein. There was a linear effect (P<0.05) of dietary protein level on CCW, with the highest (26.1 kg) and lowest (22.8 kg) values for HP and LP groups, respectively. Neither chilling losses (2.0±0.29%) nor commercial yield (50.8±0.83%) or proportion of different carcass joints were affected (P>0.05) by dietary treatment. Results suggest that increasing dietary protein content from 13.5 to 17.5% enhances performance throughout the fattening period.

Session 33

Theatre 1

Stakeholder involvement and ethics in conservation of animal genetic resources in gene banks

W. Kugler[1], E. Charvolin[2], N. Khayatzadeh[3], G. Mészáros[3], B. Berger[4], E. Martyniuk[5], J. Sölkner[3], M. Wurzinger[3] and A. Dore[1]

[1]SAVE Foundation, Neugasse 30, 9000 St. Gallen, Switzerland, [2]INRA, Rue de l'Universite 147, 75338 Paris, France, [3]Universitaet fuer Bodenkultur, Gregor Mendel Strasse 33, 1180 Wien, Austria, [4]HBLFA Raumberg-Gumpenstein, Department for Biodiversity of the Institute for Organic Farming and Biodiversity, Austrasse 10, 4601 Thalheim, Austria, [5]Warsaw University of Life Sciences, Szkoła Główna Gospodarstwa Wiejskiego, Nowoursynowska 166, 027787 Warszawa, Poland; office@save-foundation.net

Cryoconservation of reproductive triggers ethical concerns of stakeholders amongst others for the choice of the breed to be conserved. To mitigate any negative responses, stakeholders need to be consulted. 'Dialogue fora' were implemented by the Innovative Management of Genetic Resources (IMAGE) H2020 project to strengthen communication, knowledge exchange and to get direct feedbacks from stakeholders. The topics covered during these fora were: (1) the general role of gene banks; (2) sanitary regulations; and (3) economics of conservation. During the first meeting opinions were shared about gene bank collaborations, material exchange, ethical issues in biotech and standardization of gene bank management. The sanitary regulation meeting resulted in a clear call to identify and list exceptions for derogations from the EU Animal Health, specifically considering the needs of gene banks. The awareness regarding these needs has to be raised also within the European Commission, with the help of European gene bank network (EUGENA). Discussions about economics in conservation made clear that there is a need to support the exchange of material through public funding, but management should run through (public) research and breeder societies. Anonymous surveys were conducted among different actors, such as breeders and NGOs, commercial breeding and governmental bodies. The questions focused on the motivations for breed conservation, criteria of breed choice and funding sources for cryobanks. Innovations in bio-banking methods and trade-offs like the acceptance of further research on improvement of semen freezing, support for cloning methods and the sacrifice of animals for cryoconservation were asked in another part of the survey.

Session 33

Theatre 2

Rationalization and further development of European livestock genebank collections

S.J. Hiemstra[1], P. Boettcher[2], D. Moran[3] and C. Danchin-Burge[4]

[1]Wageningen University & Research, Centre for Genetic Resources, the Netherlands (CGN), Droevendaalsesteeg 1, 6700 AH Wageningen, the Netherlands, [2]FAO, 00153, Rome, Italy, [3]University of Edinburgh, EH25 9RG, Midlothian, United Kingdom, [4]Institut de l'Elevage, 75012, Paris, France; sipkejoost.hiemstra@wur.nl

Many local breeds in Europe are at risk of extinction (FAO, 2019). Hence a genetic reserve in the form of gene bank collections is crucial to be able to adapt to future changes in markets, climate or production systems. European countries have developed animal germplasm collections as a security back up and genetic reserve for breeding and research. The challenge ahead for such an *ex situ in vitro* conservation strategy is demonstrated by SDG indicator 2.5.1 'the proportion of local breeds for which sufficient genetic material is stored in gene bank collections for long term conservation'. IMAGE survey data shows that for many local breeds the amount of genetic material stored is still limited. The IMAGE project has undertaken rationalization case studies, in order to evaluate national gene bank strategies and objectives, to assess the genetic diversity captured in gene bank collections compared to diversity in live breeding populations, and to monitor changes in genetic diversity over time. In addition to national gene banks, universities and research institutes harbour a large variety of DNA and tissue collections. Future research and breeding would benefit from better integration and access to data of both germplasm and DNA/tissue collections. From an economic point of view, Europe should aim for an optimized European genetic reserve, minimizing costs, maximizing genetic diversity stored, and maximizing the options for future benefits. Within IMAGE, we developed a mathematical model to illustrate alternative scenarios. IMAGE also demonstrates differences between gene banks regarding their state of development. Quality management principles should be implemented by all gene banks, considering sampling strategies, conservation methods, legal aspects, data management and use policies. IMAGE results will be integrated in future work of the European Gene Bank Network for Animal Genetic Resources (EUGENA). The IMAGE project received funding from the EU H2020 Research and Innovation Programme under the grant agreement n° 677353.

Rationalization and characterization of gene bank collections: a case study

C. Danchin
Institut de l'Elevage, 149 rue de Bercy, 75012 PARIS, France; coralie.danchin@idele.fr

The French national cryobank is holding cryoconserved semen from six rare goat breeds. Most males were sampled when the breeds' conservation programs started and very limited information was available at the time. Pedigree data were scarce or even unknown, and bucks were often chosen from information based on their phenotypes and region of origins. The IMAGE project presented an opportunity to better characterize these animals thanks to genotype analysis. DNA was extracted from the frozen semen of the oldest bucks and genotyped with a 54k SNP chip. Their data were compared with animals from the live populations. An analysis was performed by using the PLINK, Admixture and TreeMix software. Our study revealed what was the real kinship between animals, including the ones without pedigrees, pinpointed the animals that were crossbred, and helped us target the strains present in the live population without representatives in the French cryobank. On a different perspective, since some of the goat breeds were genotyped for the first time with SNP data, it allowed us to have more insights on how the French goat breeds were related between each other. Finally, our study showed that the various diversity indicators calculated, based on SNP data, still need a better consensus from the research community. Indeed, most indicators and parameters were defined with population genetics hypothesis fit for Human and/or wild species and do not correspond well to rare livestock breeds. The small number of animals, recent bottlenecks and short generation intervals do not comply with some of the hypothesis used in various SNP analysis programs. Depending on the parameters used, results could vary in a quite wide range, which could lead to erroneous answers. Some tips learnt from our experience are described during the process analysis but further researches are still needed in this area.

Optimising *ex situ* genetic resource collections for Spanish livestock conservation

R. De Oliveira Silva[1], B. Vosough Ahmadi[2], O. Cortes Gardyn[3] and D. Moran[1]
[1]The University of Edinburgh, GAAFS, Easter Bush Campus, Midlothian, EH25 9RG, United Kingdom, [2]FAO, EuFMD, Viale delle Terme di Caracalla, 00153 Roma RM, Italy, [3]University Complutense of Madrid, Faculty of Veterinary, Av. Séneca, 2, 28040 Madrid, Spain; rafael.silva@ed.ac.uk

The global existential challenges for *in situ* livestock breed conservation are well documented, and while *ex situ* collections may have some limitations they nevertheless offer the potential to reduce extinction risks, affording option value to society in terms of maintaining future breeding opportunities for productivity and heritage traits. However, how much should we be seeking to collect and conserve in *ex situ* collections and where? A collaborative aim might be to harmonise collections to avoid costly overlaps in collecting and storage of genetic materials, keeping the most diverse collection at least cost. At a national scale, this managerial challenge entails selecting genetic or reproductive materials from several hundred candidate breeds from various species differentiated by location, collection costs, budgetary constraints and further limitations in terms of storage capacity, and specialised labour availability. To address this specific objective we developed a mathematical model to optimise breed conservation choices, and to evaluate alternative scenarios for efficiently re-allocating genetic materials currently stored in different Spanish cryogenic banks, allowing for cross-country gene collection, cost and cryogenic capacity differentials. We show how alternative allocations could reduce overall conservation costs, and illustrate the diversity-cost relationship using notional supply curves. These provide guidance for both policy makers and conservation scientists seeking to define objectives for the design of efficient *ex situ* conservation. We acknowledge funding from the IMAGE (Innovative Management of Animal Genetic Resources) project funded under the European Union's Horizon 2020 research and innovation program under grant agreement No 677353.

Breakthrough to improve the reproductive capacity of gene bank material

E. Blesbois[1], K. Liptoi[2], H. Woelders[3], B. Bernal-Juarez[4], S. Nandi[5], A. Thelie[1], F. Pizzi[6], E. Varkonyi[2], F. Guignot[1], M. McGiou[1] and J. Santiago Moreno[4]

[1]INRA, UMR-PRC, 37380 Nouzilly, France, [2]HaGK, Isaszegi u. 200, 2100 Godollo, Hungary, [3]University of Wageningen, P.O. BOX 338, 6700 AH Wageningen, the Netherlands, [4]INIA, Puerta de Hierro k 5.9, 2804 Madrid, Spain, [5]University of Edinburgh, Easter Bush Campus, EH64SX Midlothian, United Kingdom, [6]PTP, via Einstein, 26900 Lodi, Italy; elisabeth.blesbois@inra.fr

The aim of the WP 3 of IMAGE is to improve the efficiency of the reproductive cells to be stored in the cryobanks for the conservation of domestic animal genetic resources. Different cells and tissues are under study in major production species. Two tasks (3.1, 3.2) are focused on improving sperm fertilizing ability, the task 3.3 is turned on the study of gonad transfer, the task 3.4 on the study of Primordial Germ cells (PGCs) and the task 3.5 on pig embryo vitrification. Many advances were obtained in each Task. In the Tasks focused on semen (3.1, 3.2), different improvements of sperm freezing process were suggested in chickens and rams. New sperm quality criteria are in progress through the studies of DNA damages, proteic and mi RNA actors of fertility in chickens or bulls. Gonad transfer studies in the chicken (3.3) allowed to methodologic progress, to the identification of efficient breed couples of donor-recipient animals and to the production of healthy progeny after transfer. Improvements in PGCs biotechnologies (3.4) allowed to the production of a serum free standard medium for cryopreservation, to improvements in the knowledge and control of quality of cultured PGCs, and to the identification of interspecies chicken-guinea fowl sterile hybrids as recipient animals for donor PGCs. In the task 3.5 focused on pig embryo vitrification, a mathematic model simulating the osmotic events occurring during vitrification indicated adequate conditions reached in only seconds of embryos exposure to cryoprotectants. This is now used as a basis for experimental vitrification. All of these progresses allow enhancing the methodologies of conservation, characterization and evaluation of reproductive collections, and the quality of European *ex situ* collections. Funding: IMAGE project from the EU H 2020 Research and Innovation Program, agreement no. 677353.

Freezing poultry semen: effects of CPA concentration × cooling rate, and other factors

H. Woelders[1], A. De Wit[1], I. Grasseau[2], E. Blesbois[2], B. Bernal[3] and J. Santiago Moreno[3]

[1]Wageningen UR, Animal Breeding and Genomics, P.O. Box 338, 6700 AH Wageningen, the Netherlands, [2]INRA, UMR PRC, INRA Centre Val de Loire, Nouzilly, 37380, France, [3]INIA, Reproducción Animal, Av. Puerta de Hierro, 18, 28040 Madrid, Spain; henri.woelders@wur.nl

Cryopreservation of chicken semen is used in gene banks for *ex situ* conservation of genetic diversity, or in the breeding industry to conserve selection lines. However, optimization of freezing medium and freezing protocol is necessary as cryo-resistance may be low in specific breeds. In the present study a number of CPAs were compared, and effects of DMA concentration, cooling rate, and other variables were studied. The effect of osmolality of the base extender (no CPA) on sperm cells longevity during 5 °C storage was first tested using extenders with equal composition in terms of solute ratios, but having osmolalities ranging from 290-410 mOsm/kg. Higher osmolalities had a strong negative effect on sperm motility, which was only partly reversible, indicating permanent injury. Six related CPAs (methylformamide, methylacetamide, dimethylformamide (DMF), dimethylacetamide (DMA), propane-1,2-diol, and diethylformamide) were first pre-screened at 0.6 M for freezing semen from individual cocks (n=10) in 0.25-ml straws at a cooling rate of 250 °C/min. Post-thaw % motile and % live sperm were highest with DMA and DMF. Finally, semen from individual cocks or pooled semen was frozen in 0.25-ml straws, using cooling rates (CRs) of 4, 50, 250, and 440 °C/min and [DMA] of 0.4, 0.6, 1.0, and 1.5 M. Results of microscopical sperm assessment were presented elsewhere. Data from flow cytometry, Tunel analysis, and CASA motility analysis will now be presented. Results show clear effects of both CR and [DMA]. Percentage motile and % live sperm were highest for CRs 50-250 °C/min. Higher DMA concentrations gave better post-thaw sperm survival. However, longevity of the sperm at 1.0 and 1.5 M DMA was compromised. Therefore, [DMA] may best be 0.6-1.0 M at a CR of 50-250 °C/min. This work was part of the IMAGE project which received funding from the European Union's Horizon 2020 Research and Innovation Programme under the grant agreement no. 677353.

The use of genomic variation in European livestock genebank collections

R.P.M.A. Crooijmans[1], L. Colli[2], A. Stella[2], P. Ajmone-Marsan[2], S.J. Hiemstra[3] and M. Tixier-Boichard[4]
[1]Wageningen University and research, Animal Breeding and Genomics, Droevendaalsesteeg 1, 6708 PB Wageningen, the Netherlands, [2]Universita Cattolica del S. Cuore, via Emilia Parmense 84, 29122 Piacenza, Italy, [3]Wageningen University and Research, CGN, Droevendaalsesteeg 1, 6708 PB Wageningen, the Netherlands, [4]INRA, UMR GABI, 78352, Jouy-en-Josas, France; richard.crooijmans@wur.nl

The majority of the SNP arrays developed so far from whole genome sequencing data are biased towards genomic variation found in a few commercial populations. Moreover, the set of markers contained in the SNP arrays is changing often in time and overlap between the different versions and platforms is not always guaranteed. Furthermore, in several cases arrays are not freely available without permission of the consortia involved in their development. In the framework of the Horizon2020 IMAGE project, we have identified a large pool of SNP variants by whole genome sequencing of over 500 individuals from a large variety of local breeds and species and combined this with publicly available SNPs to develop panels to be used for different applications and analyses. One of these applications is the development of a multi- species SNP array (that might include cattle, sheep, goat, pig chicken, turkey, buffalo, horse, duck, rabbit, camel, camelids, honey bee) containing 10k SNPs per species selected from the total SNP pool. This low cost SNP array will make it possible to genotype a large number of samples of the different species collected in gene banks worldwide. Opening the molecular reservoir of variation in these gene bank collections will add value to the material stored, detect the uniqueness of the collections and monitor variation changes over time. The generated whole genome sequences will also be used in genome wide association studies (GWAS) for specific traits present in traditional breeds, like e.g. ptilopody and bantam traits in chicken. The variants related to these and other traits of interest, detected in the course of IMAGE project, or variants presented in literature for specific traits, will also be selected and added on the multi-species SNP array. The IMAGE project received funding from the EU H2020 Research and Innovation Programme under the grant agreement n° 677353.

Genetic diversity and inbreeding in Dutch cattle breeds based on gene bank collections

A.E. Van Breukelen, H.P. Doekes and J.K. Oldenbroek
Wageningen University & Research, Animal Breeding and Genetics, Droevendaalsesteeg 1, 6708 PB Wageningen, the Netherlands; anouk.vanbreukelen@wur.nl

To preserve genetic diversity within and between domestic breeds in the Netherlands, the Centre for Genetic Resources (CGN) collects *ex situ* genetic material and maintains gene bank collections. In this study, we investigated genetic diversity within and between bulls from Dutch cattle breeds and the Holstein Friesian breed based on material stored in the gene bank. Analyses were performed using 37k SNP data of in total 900 bulls. Based on genetic distances the most distinct breeds were the Dutch Friesian (DF; including Dutch Friesian Red and White (DFR)), the Dutch Belted (DB), the Groningen White Headed (GWH), the Dutch Red and White (MRY), the Improved Red and White (IRW) and the Holstein Friesian (HF). Overall, Dutch HF gene bank bulls are the least genetically similar to bulls of the other native Dutch breeds. Between breeds large differences in genomic inbreeding levels (F_{ROH}) were observed, GWH bulls were the most inbred and IRW bulls the least. When performing optimal contributions selection, with all bulls as selection candidates and maximizing diversity, the largest contribution is allocated to HF bulls. Additionally, when applying the method of Eding *et al.*, HF bulls were allocated the highest unique diversity. Nonetheless, Dutch breeds contain diversity additional to diversity in HF bulls and are therefore important for conservation. These results provide insight into the genetic diversity currently stored in the Dutch gene bank and help to determine conservation strategies (e.g. to compose a core set and safe set).

Optimization of introgression breeding programs with MoBPS

T. Pook[1,2], A. Ganesan[1,2], N.T. Ha[1,2], M. Schlather[1,3] and H. Simianer[1,2]
[1]University of Goettingen, Center for Integrated Breeding Research, Albrecht-Thaer-Weg 3, 37075 Goettingen, Germany,
[2]University of Goettingen, Animal Sciences, Albrecht-Thaer-Weg 3, 37075 Goettingen, Germany, [3]University of Mannheim,
Stochastics and Its Applications Group, A5,6 B 117, 68161 Mannheim, Germany; torsten.pook@uni-goettingen.de

Performing introgression of one or more beneficial QTL that are only present in a wild type or gene bank material into an elite breeding population is a complex optimization problem. On the one hand, a high genetic share of the elite line is to be maintained to regain its productivity level and economic competitiveness. On the other hand, fast introgression can cause inbreeding depression with resulting lower long-term gain and potentially causing health and animal welfare issues. Depending on the species and its reproduction biology, the amount of resources and resulting number of animals in the program, optimal mating strategies can be different. We focused our analysis on introgression schemes in chicken but also considered other species. Introgression schemes were evaluated based on the speed of introgression, short and long-term performance on simulated traits and inbreeding levels. In terms of the design of the introgression scheme, we considered different mating strategies (e.g. optimum genetic contribution selection) and assessed potential benefits of genotyping. This is not only relevant for genomic selection but also to track the share of the genome originating from the elite line. Additionally, one might consider putting an emphasis on regions physically close to the beneficial QTLs, as recovering the elite line genotype in those regions is most challenging. Finding a closed solution to such a complex allocation problem with uncertain mating strategy is virtually impossible. Instead, we propose the usage of the Modular Breeding Program Simulator (MoBPS). The software is an R-package openly available at https://github.com/tpook92/MoBPS which supports the simulation of complex and large scale breeding programs with focus on livestock and crop populations. User-input can be inserted in a web-based application including a tool to visualize the mating scheme between different breeding groups.

Introgression of blue eggshell colour from a gene bank collection into a White Leghorn breeding line

C. Dierks[1], N.T. Ha[2], H. Simianer[2], D. Cavero[3], R. Preisinger[4] and S. Weigend[1]
[1]Friedrich-Loeffler-Institut, Institute of Farm Animal Genetics, Höltystrasse 10, 31535 Neustadt, Germany, [2]University
of Goettingen, Center for Integrated Breeding Research, Albrecht-Thaer-Weg 3, 37075 Goettingen, Germany, [3]H&N
International, Am Seedeich 9, 27472 Cuxhaven, Germany, [4]EW GROUP GmbH, Hogenbögen 1, 49429 Visbek, Germany;
claudia.dierks@fli.de

The aim of the present study is to demonstrate the efficient transfer of a specific monogenic trait maintained in gene bank into a contemporary high performing white egg layer chicken line. The project is part of the EU project IMAGE (Innovative Management of Genetic Resources) under the umbrella of Horizon 2020. The trait of interest, blue eggshell colour is inherited in a dominant way and the causal mutation is a large insertion on chromosome 1 upstream of SLCO1B3. In 2016, six Araucana cocks were mated with ten White Leghorn (WL) hens. Two marker-assisted backcross generations (BC1 and BC2) followed by an intercross-generation (IC) will be generated, aiming at a high performing homozygous blue layer WL-like line. Based on genotype data of the founder animals, we identified 24 highly informative SNPs on chromosome 1 bracketing the insertion site. Additionally 13 SNPs from a customized 52k SNP chip were suitable to distinguish between the Araucana and WL genome in this region. These markers were used to detect recombinant animals of BC1 that carried the blue eggshell mutation on the one hand and had highest content of WL genome on the other hand with regard to the inserted region on chromosome 1. Out of these recombinants, animals with the highest proportion of the recipient genome and the highest degree of diversity were selected for mating of the BC2. The IC is planned for end of 2019. Performance data of the BC1 and commercial WL were compared. Analysis of performance data of the BC2 is still under progress. The results are promising as the mean laying rate of BC1 animals (81.2% in carriers and 83.6% in non-carriers) was 5.4% lower than in the commercial line (87.8%). In BC1 animals, the eggshell strength was on average 40 N and significantly lower ($P<0.0001$) compared to the commercial line (45 N). The mean egg weight was measured four times and was 61.6 g in BC1 hens and 62.2 g in the commercial line.

Annotation of selection signatures in the bovine breed Asturiana de Valles

C. Paris[1], S. Boitard[1], B. Servin[1], N. Sevane[2] and S. Dunner[2]
[1]INRA, GenPhySE, 24, chemin de Borde-Rouge, Auzeville Tolosane, 31326 Castanet-Tolosan, France, [2]Universidad Complutense de Madrid, Facultad de Veterinaria, Avda. Puerta de Hierro, 28040 Madrid, Spain; simon.boitard@inra.fr

Past events of positive selection leave characteristic signatures in the genetic diversity of a population, which can be detected by genome-wide scans based on present time molecular data. However, determining the adaptive trait or the onset and intensity of selection at a given locus is often difficult from such data. By providing direct access to the temporal evolution of allele frequencies, the analysis of genomic data extracted from gene banks might significantly improve our understanding of selection history in livestock species. The aim of this study is to evaluate whether the analysis of genomic samples collected at different times in the recent past allows: (1) detecting recent selection events; and (2) annotating selection signatures found by classical approaches based on present time data only. To answer these questions, we considered the case study of the Spanish bovine breed Asturiana de los Valles (RAV), for which genotyping data was available for 137 animals with birth dates between 1980 and 2010. Fifteen additional RAV animals born in 2008-2013 were sequenced at ~8X coverage. These data were used to detect historical selection signatures in RAV using a classical statistic (nSL) based on a single sampling time. A new statistical approach allowing detection of selection from genomic time series was developed and applied to the combined dataset including nine distinct generations. This approach pointed out several candidate regions with a clear shift in allele frequencies over the few last generations. Besides, the analysis of recent allele frequency trajectories at nSL candidate SNPs outlined that selection at these SNPs was likely completed before 1980. Combining the time series and nSL approaches, 13 candidate regions under selection in RAV were detected, including genes related to carcass and meat traits (such as MSTN, RBPMS2 or OAZ2), immunity (GIMAP7, GIMAP4, GIMAP8), olfactory receptors (OR2D2, OR2D3, OR10A4, 0R6A2) and milk traits (ARFIP1). In conclusion, these approaches were found complementary and the time series approach provided relevant new information concerning selection history in Asturiana de los Valles.

Dutch gene bank for farm animals: beyond conserving genetic material

A. Schurink, B. Hulsegge, J.J. Windig, A.H. Hoving-Bolink and S.J. Hiemstra
Centre for Genetic Resources, the Netherlands (CGN), Wageningen University & Research, P.O. Box 338, 6700 AH Wageningen, the Netherlands; sipkejoost.hiemstra@wur.nl

The Centre for Genetic Resources, the Netherlands (CGN) focuses on the conservation and use of vegetable crops, farm animal breeds and autochthonous forest species. One of the main activities is cryopreservation of farm animal genetic diversity for long term conservation purposes. From the 1970s onwards, genetic material (mostly semen) has been cryopreserved and stored in the Dutch gene bank for farm animals. Nowadays, the gene bank contains just over 347k insemination doses of 8,457 donors from 139 breeds, belonging to 11 farm animal species. Main focus is on the economically most important farm animal species in the Netherlands (cattle, pig, sheep, goat, horse and chicken). The gene bank is a valuable source of genetic diversity for long term conservation, an insurance for future catastrophes and (near) extinction of breeds. At the same time, gene bank material is already used today to support breeding programmes of Dutch endangered breeds. For example, gene bank material has been and still is an important source for breeding Red and White Friesian cattle, as each year dozens of insemination doses are provided. Moreover, genetic material from old breeding lines conserved in the gene bank is used to support the Dutch Landrace pig breeding population for several years now. Without the possibility of using gene bank material, these breeds may become extinct. The gene bank also forms a unique archive of genentic diversity, representing four decades of farm animal breeding. Genomic characterization of gene bank collections shows the impact of changing breeding goals, consolidation of breeding lines, and introduction of genetic management and genomic selection methodologies. Further use of modern genomic techniques to characterize gene bank collections may result in the identification of valuable genetic variation that is no longer present in current breeding populations. The Dutch gene bank will continue to enhance its collections, creating a unique genetic diversity reservoir for future breeding and for research.

Value of gene bank material for the commercial breeding population of Dutch Holstein Friesian cattle

J.J. Windig[1,2], H.P. Doekes[1,2], R.F. Veerkamp[2], P. Bijma[2] and S.J. Hiemstra[1,2]
[1]Wageningen UR, Centre for Genetic Resources, the Netherlands (CGN), P.O. Box 338, 6700 AH Wageningen, the
Netherlands, [2]Wageningen UR, Animal Breeding and Genomics, P.O. Box 338, 6700 AH Wageningen, the Netherlands;
jack.windig@wur.nl

Each year samples from Holstein Friesian bulls used for breeding have been stored in the Dutch national gene bank
since the 1990s. Using the optimal contribution method we investigated whether old samples may contribute to the
current breeding program. Optimal contributions can maximise genetic merit while constraining relatedness levels
and loss of genetic diversity. Inclusion of cryobank bulls in the current breeding program, did not provide additional
merit when total merit was maximised and relatedness restricted at high levels. At low levels of restricted relatedness
inclusion of cryobank bulls did provide additional merit. For sub-indices yield, fertility and udder health additional
merit was provided at any level. Optimal contributions were extended to constrain relatedness differently at different
parts of the genome, or fix allele frequencies of single alleles, such as the allele for polledness in cattle, to a certain
level. Using regional constraints we were better able to limit the loss of genetic diversity at parts of the genome further
away from the allele under selection. Gene bank collections can thus be a valuable resource for commercial breeding
populations, especially when loss in genetic diversity is to be constrained or in case of changes in the breeding goal.

Characterization genetic and evaluation of diversity in Spanish chicken breeds based on MHC class II

S.G. Dávila, M.G. Gil, P. Resino-Talaván, O. Torres and J.L. Campo
I.N.I.A., Animal Breeding, Crta. Coruña, km 7, 28040 Madrid, Spain; sgdavila@inia.es

Spain have a rich genetic diversity of native chickens, nowadays under threat of extinction because the
utilization of industrial hybrids to produce eggs. Chickens major histocompatibility complex (*MHC*) is located
on microchromosome 16 and contains class *II* (*B-LB*) gene, which is similar to mammalian *MHC* class II. The
B-LB II gene has been intensively studied for its extensive polymorphism and its important role in presentation of
extracellular antigen peptides to initiation of immune response. The objective of this study were: (1) to characterize
the genetic polymorphisms of chicken *MHC B* class *II* genes in 12 Spanish breeds and a White Leghorn population;
(2) to evaluate the genetic diversity within and between populations. A fragment of 267 pb was amplified in 221
birds revealing the sequence alignment of the alleles across populations a total of 116 variable sites, of which
97 parsimony informative and 19 singleton variables sites, respect the sequence of *Gallus gallus*. Among the
polymorphic sites, the nucleotide substitutions in the first, second and third positions of the fragment was 29.3, 30.2
and 40.5%, respectively. Alignment of the deduced amino acid sequence revealed 59 nosynonymous substitutions,
being significantly higher than that of synonymous. These results indicated a high genetic variation within chicken
MHC B II that could be related to an association with traits of disease resistance. The average diversity within of
Spanish breeds ranging from 0.105 (Blue Andaluza) to 0.061 (White-Faced Spanish). Estimates of evolutionary
divergence between groups showed a range from 0.186 (between Black-barred Andaluza and White Leghorn) to
0.086 (between Buff Prat and Black-Breasted Red Andaluza) being 0.128 the average distance. A phylogenetic tree
constructed by the NJ method, showed a clear separation between White Leghorn and the remaining breeds. Within
the Spanish breeds, one of the most evident cluster was found for the breeds carrying the dominant alleles at the *E*
locus (extended black and birchen). The results indicate that the *MHC B* class *II* gene was useful in studying the
genetic diversity of chicken breeds.

Inventory of relic Romanian native cattle breeds

D. Gavojdian
Research and Development Institute for Bovine Balotesti, sos. Bucuresti-Ploiesti, km 21, Balotesti, Ilfov, 077015, Romania;
gavojdian_dinu@animalsci-tm.ro

The aim of the study was to evaluate the current status of relic cattle breeds in Romania, in order to aid the development of future conservation strategies and revival plans. Until mid-18th century the Grey cattle represented over 95% of the cows reared. The Romanian Grey had 4 recognised varieties: Transylvanian (now extinct); Moldovan (83 cows remaining); Ialomiteana (now extinct) and Dobruja (declared extinct). To improve the sub-optimal production of Grey cattle, Simmental, Schwyz, Pinzgauer and Red cattle were imported. This led to the formation of the improved/modern Romanian breeds such as Romanian Spotted, Maramures Brown, Transylvanian Pinzgauer and Dobruja Red. After 1960, the Dobruja Red breed was absorbed by the Friesian breed, leading to the development of Romanian Black and White HF population. Due to the absence of herd books and national conservation programs, the Dobruja Grey and Dobruja Red cattle breeds are nowadays regarded as officially extinct. However, on remote islands of the Danube Delta populations of Dobruja Grey and Dobruja Red cattle were identified, given the isolation of these islands (no AI practiced) and the harsh rearing conditions, some populations remained secluded in the delta, with census estimates of 70-80 Dobruja Red cows and 120-150 Dobruja Grey cows. Urgent *in situ* and biobanking conservation efforts are needed for the Dobruja Grey and Dobruja Red breeds, in order to preserve FAnGR and genetic diversity.

Population structure and genetic diversity analysis in five rare endangered goat breeds in Austria

N. Khayatzadeh[1], M. Klaffenböck[1], G. Mészáros[1], M. Wurzinger[1], B. Berger[2] and J. Sölkner[1]
[1]BOKU, Department of Sustainable Agricultural Systems, Gregor-Mendel-Straße 33, 1180 Vienna, Austria, [2]ÖNGENE, Austraße 10, 4600 Wels, Austria; negar.khayatzadeh@boku.ac.at

Maintenance of genetic diversity within and between breeds plays an important role for designing sustainable conservation programs. Seven autochthonous goat breeds in the Alpine region were identified by the Austrian national association for gene conservation of farm animals (ÖNGENE) as Austrian endangered goat breeds, which are part of a national conservation program. A set of 47,652 single nucleotide polymorphisms (SNP) for 145 goats from five of these breeds (34 Blobe, 22 Chamois Colored Alpine, 28 Pinzgau, 33 Styrian Pied and 28 Tauern Pied) were used to infer population structure and estimate the genetic diversity, using heterozygosity, average pairwise genetic distance, linkage disequilibrium (LD), effective population size (N_e) and runs of homozygosity (F_{ROH}) in order to evaluate the genetic variability within and between breeds. According to the principal component analysis, a quite clear separation for Tauern Pied and more relation between the other breeds were detected. Six ancestries were identified for these five breeds by ADMIXTURE, where Blobe goats showed two different ancestries in their genetic background. We added the genotype information on 22 Passeier goats from South Tyrol, Italy, whose main habitat along the alpine range is very close to that of the Blobe breed. The Passeier goats showed similar background with some Blobe goats, indicating migration between these two populations. According to the genetic diversity measurements, the highest heterozygosity (0.418) and average pairwise genetic distance (0.320) were observed for Styrian Pied goats and the lowest values were observed for Tauern Pied goats; 0.376 and 0.275. Likewise, highest LD (0.070) with smallest N_e (104) were estimated for Tauern Pied goats. In total, a reasonable amount of genetic diversity and good separation between breeds were observed for the breeds. The low levels of genetic diversity in the Tauern Pied goats were confirmed by the highest levels of genomic inbreeding (0.107 ± 0.033) in this breed, due to a more homogenous founder population compared to the other breeds.

IOF2020: accelerate innovation and Internet of Things for agriculture and food

J. Maselyne, E. Lippens and J. Vangeyte
ILVO (Flanders research institute for agriculture, fisheries and food), Technology and Food Science Unit, Burg. Van Gansberghelaan 115 bus 1, 9820 Merelbeke, Belgium; jarissa.maselyne@ilvo.vlaanderen.be

The Internet of Things (IoT) is a powerful driver to transform the farming and food sector into a fully connected web of objects that are context-sensitive and generate decisions, perform actions and provide alerts. As such, IoT will be a real game changer in agriculture that drastically improves productivity and sustainability. The IoF2020 project (Internet of Food and Farm, 2017-2020, H2020 Innovation Action, grant nr. 731884) fosters a large-scale take-up of IoT in the European farming and food domain. The project revolves around several use cases in specific domains of agriculture (arable, dairy, fruit, vegetables, meat): 19 original use cases (2017-2020) and 14 additional use cases (2019-2020) that originated from the open call. Support for technical, business and ecosystem building is provided by the Work Packages. Overall, the project will generate over 50 products tested at 100 demonstration sites and testbeds with expected 10-15% productivity growth and 20-25% sustainability improvement. Results from the dairy and meat use cases are amongst others dairy cow health and fertility applications, beef cattle tracking products, pig farm intelligence dashboards, milk quality testing, chain transparency and traceability solutions and logistics optimization products. From the supporting Work Packages re-usable components, KPI catalogues, business models and ethics in data management are some of the generic results that can be used within and outside of the project. User acceptance testing, value proposition exercises and impact assessment are put in place to guarantee the sustainability of the products beyond the project. The project clearly shows the potential of IoT and precision livestock farming solutions and accelerates novel technologies in the agrifood domain.

Stickntrack solution for tracking grazing dairy cows and cattle in nature areas

O. Guiot[1], L. Claeys[2], E. Eygi[2], E. Cannière[3], K. Piccart[1], T. Van De Gucht[1], E. Lippens[1], J. Maselyne[1] and S. Van Weyenberg[1]
[1]Institute for agriculture, fisheries and food, Burg. Van Gansberghelaan 92, 9820 Merelbeke, Belgium, [2]Sensolus, Rijsenbergstraat 148D, 9000 Gent, Belgium, [3]Inagro, Ieperseweg 87, 8800 Rumbeke-Beitem, Belgium; kristine.piccart@ilvo.vlaanderen.be

Technological solutions for cow tracking are useful for several applications, such as proving and certifying milk produced from grazing cows, digital logging of Flemish ammonia emission reduction measures, tracking cattle in natural reserves, etc. The Stickntrack solution, developed by the company Sensolus, is a robust GPS tracker for tracking of physical assets (containers, airport material, etc.). During the IOF2020 project it was adjusted for livestock usage and implemented at different test farms for dairy and beef cattle in order to test it's potential for the different applications. It proved especially useful for cattle tracking in natural reserves. Stickntrack is an easy to implement location tracker with long battery life, using GPS signals and SigFox communication. Tests were performed in 2018 at Saeftingherhof (border of Belgium and the Netherlands), where beef cows are grazing from May till October in a natural reserve, a delta of many mud flats and salt marshes with trenches that fill up twice a day with the salty water of the Schelde river. Under these conditions, cows can get stuck in one of the trenches or get lost in the reserve, leading to large search actions for the farmers and possible welfare and health issues, loss of production or even death as a consequence for the cow if they are not found in time. The GPS tracker optimized time management and reduced stress for the farmers, as they can monitor their cow's activities and spend less time searching for cows (several people searching multiple hours vs a 'find my cow' navigation application). The return on investment was very clear. Within the test period, two cows that were trapped, have been saved. This means a recovered potential revenue loss of around € 5,000 compared to a small investment for the Stickntrack solution. The Stickntrack is being further optimized for the cattle application, for all these farmers in need of a 'life-saving' solution.

Detection of rumination in cattle using bolus accelerometer sensors and machine learning

A.W. Hamilton[1], C. Davison[1], C. Michie[1], I. Andonovic[1], C. Tachtatzis[1] and N. Jonsson[2]
[1]University of Strathclyde, Electronic and Electrical Engineering, 99 George Street, Glasgow, G1 1RD, United Kingdom,
[2]University of Glasgow, School of Veterinary Medicine, 464 Bearsden Rd, Glasgow, G61 1QH, United Kingdom;
andrew.w.hamilton@strath.ac.uk

Reticuloruminal function is central to the digestive efficiency in dairy and beef cattle, with many technologies such ear tags, activity collars, bolus sensors being used to monitor the time spent ruminating. Cattle that are ill or injured feed less and so ruminate less, therefore evaluating the time spent ruminating provides insights into animal health. pH boluses are used to provide information on digestive function, associating pH values with clinical conditions such as sub-acute rumen acidosis. These boluses can incorporate accelerometers to provide an approximate measure of animal activity. By deriving hyper-parameters from the acceleration activity data and using machine learning methods, it is possible to derive rumination periods in cattle from the bolus sensor, potentially leading to the reduction in the number sensor systems required for optimal cattle monitoring. The study created hyper-parameters based on inter contraction interval and the underlying signal strength and show that it is possible to classify periods of rumination. Performance was benchmarked using accelerometer based collars to provide truthing and validation data.

Internet of Things and blockchain to create shared value in the beef supply chain

F. Maroto-Molina[1], I. Gómez-Maqueda[2], J.E. Guerrero-Ginel[1], A. Garrido-Varo[1], C. Callejero-Andrés[2], R. Blanco-Carrera[2] and D.C. Pérez-Marín[1]
[1]ETSIAM University of Cordoba, Animal Production, Madrid-Cadiz Rd. km 396, 14071 Cordoba, Spain, [2]Sensowave SL, Avenida de Castilla 1, 28830 San Fernando de Henares, Spain; g02mamof@uco.es

The beef supply chain is a complex system, involving grasslands, crop farms, livestock farms, feedlots, transporters, slaughterhouses, retailers and consumers. Events occurring at farm level have huge effects on industry, consumers and other stakeholders, and vice versa. Nowadays, traceability systems collect data, mostly by hand, from some segments of the supply chain, mainly to assure food safety to consumers. However, the potential for data sharing along the beef supply chain (upstream and downstream) is much bigger than that. Shared value systems based on data integration can allow every segment of the beef supply chain to improve efficiency and product quality, to certify production conditions, etc. Regarding data availability, the Internet of Things (IoT) is a paradigm shift, as it allows to automatically record objective data from every segment of the supply chain. On the other hand, blockchain technologies represent an opportunity to share key data among beef supply chain stakeholders in a secure and verifiable way, allowing the design of utilities based on data sharing, such as smart contracts. ShareBeef, a project funded by the last IoF2020 Open Call, is working on the development of a technological framework, based on IoT devices, Open Data, cloud platforms and blockchain services, to facilitate data sharing along the beef supply chain. Demonstration testbeds are currently being deployed in six countries: Bulgaria, Croatia, Ireland, Italy, Portugal and Spain. More than 1000 animals will be monitored during the project, including grazing cows, lactating calves and fattening calves. The improvement of decision-making on the basis of data sharing is expected to result in a optimization of crop production; a reduction of water and fertilizer use; a reduction of calving interval and calf losses; an improved weight daily gain; a reduced farmer work effort; and an improved information for consumers regarding production conditions.

Unlocking the value of data via Internet of Things in the fattening pig sector

C. Vandenbussche[1], F. Renzi[2], T. Montanaro[3], K. Mertens[2], D. Conzon[3], P. Briene[1], K. De Wit-Joosten[4], D. Aarts-Van De Loo[4], R. Klont[5] and J. Maselyne[1]
[1]*ILVO, Burg. Van Gansberghelaan 115, 9820 Merelbeke, Belgium,* [2]*Porphyrio, Warotstraat 50, 3020 Leuven-Herent, Belgium,* [3]*Links Foundation, Via Pier Carlo Boggio 61, 10138 Torino, Italy,* [4]*ZLTO, Onderwijsboulevard 225, 5223 DE 's-Hertogenbosch, the Netherlands,* [5]*Vion Food Group, Boseind 15, 5281 RM Boxtel, the Netherlands;* chari.vandenbussche@ilvo.vlaanderen.be

As part of the IOF2020 project, the Pig Farm Management use case focuses on Internet of Things (IoT) in the fattening pig sector. The pig sector is facing challenges of high costs, difficult economic situation and increasing demands on topics like animal welfare and emission reduction. To keep up with these requirements, an ever-increasing monitoring-efficiency is needed. In fact, a pig farmer has to take into account many different aspects going from health inspection, managing stocks and personnel to optimizing feeding whilst also following the latest developments in the sector. The present-day pig farmer is already assisted in his job as caretaker and farm manager by technological installations (e.g. feeding and climate systems), but the data that these systems generate is most often under-used. Consequently, a clear overview of the on-farm data linked with technical results from the slaughterhouse and automatic problem detection, combined with tracking of actions and treatments is needed more than ever. To this aim, the Pig Business Intelligence Dashboard was developed as one easy tool that provides the overview that is needed to effectively steer the farm management. In this work, we will present the Pig Farm Management Use case, the development of the Pig Business intelligent Dashboard, the encountered challenges and the solution applied to overcome them. A data sharing agreement, for instance, was introduced to assure data privacy in the integration of heterogeneous IoT devices/platforms. While, the combination of an IoT platform, the FIWARE Orion Context Broker and a real-time warning system allowed the integration of IoT devices and platform. To conclude, we will report the experience of the 5 practical farms, where the product is tested and further developments are fostered by continuous feedback and recommendations.

Development and testing of individual level warning systems for fattening pigs

C. Vandenbussche, P. Briene, B. Sonck, J. Vangeyte and J. Maselyne
ILVO, Burg. Van Gansberghelaan 92, bus 1, 9820 Merelbeke, Belgium; chari.vandenbussche@ilvo.vlaanderen.be

As part of the IOF2020 project, the Pig Farm Management use case focuses on the use of Internet of Things in the fattening pig sector. While the Pig Business Intelligence Dashboard combines the technology that is available on the farm today, the use case also focuses on the development of future solutions. Traditionally, pigs are monitored on a daily basis by a caretaker. This is often combined with basic checks of the functionality of feeding, drinking and climate regulation and cleaning activities. However, detecting individual problems in a group of pigs can be challenging. Individual pig behaviour might also be difficult to interpret by the farmer during a short observation time. RFID transponders (electronic identification) added to the pig ear tags can be used to identify individual pigs in feeding stations or to monitor presence near feeding and drinking facilities. Combining these data with smart algorithms allows early detection of abnormal behaviour and informs the farmer in case of problems. On the ILVO farm, a fattening pig compartment with 8 pens (15 pigs per pen of 19 m^2) was equipped with two different feeding systems. Four pens had a feeding station with weighing module and four other pens had a single round feeder with high frequency RFID antenna. Pigs were fed *ad libitum* with two-phase dry feed. Algorithms were developed based on two fattening rounds of 120 pigs each and tested during a third round. To estimate the effect of alerts on the treatment of pigs, the pig caretakers received daily alerts for 50% of the pigs. For the other pigs, alerts were generated, but not communicated towards the caretakers. Three times a week, all pigs and all alerts were checked by experts. A total of 91 alerts (excluding 4 pigs with continuous false alerts due to too strict starting conditions) were generated by the algorithm during a period of 75 days. In the same period, the caretakers reported 48 problem pig-days and supplied 20 treatments during their daily checks. In this session we answer the following questions. What is the performance of the alert system? How did they experience the daily list of alerts? Did the alerts correspond with their own observations? What are future steps to be taken to improve the algorithm for a pig farm?

Responsible governance of data in smart farming: results of reflective stakeholder workshops

S. Van Der Burg, M.J. Bogaardt and E.B. Oosterkamp
Wageningen Economic Research, P.O. Box 35, 6700 AA Wageningen, the Netherlands; elsje.oosterkamp@wur.nl

Sensors, drones, weather satellites and robots are examples of technologies that make farming 'smart'. A reading of the literature suggests that current ethical discussion about smart farming circles around three themes: (1) data ownership and access; (2) distribution of power; and (3) impacts on human life and society. All of these themes relate to the governance of farm data. Discussions that fall under these themes have however not yet reached a satisfying conclusion, thus leading to different opinions regarding who the owner of farm data is, who should have access to data, who should be entitled to use data and for what purposes. At the moment of writing this abstract, we're in the middle of shaping workshops, which will be used to enhance reflection of stakeholders, that started in January 2019 and last till April 2019. Participants will be farmers, tech service providers, policy makers, insurance companies, banks and NGO's. The content of the workshops is shaped on the basis of a literature study as well as qualitative interviews with stakeholders. Based on this background, we shaped future scenarios which present different ways to govern farm data in the future. During the workshops participants are invited to reflect on the pros and cons of these scenarios. This reflection is structured in a layered fashion: first, we ask for an intuitive reaction to the scenarios, then we ask participants to reflect on the basis of personal values (which we derived from the interviews), and lastly we ask them to reflect on their preferences as members of society who consider societal issues related to the scenarios. During the session we intend to present the results of these workshops: what scenario is the preferred scenario and why? These results are generated by the ethical work package of the IOFH2020 project. In the session we also give an introduction to the IOFH2020 project and share some use case stories and achievements. The session also provides insight into how ethical reflection with stakeholders about a technological future can be shaped, and reflects on the possibilities it offers as well as its limitations.

Networking European Farm to Enhance cRoss ferTilisation & Innovation uptake Through demonstratIon

A.G. Adrien Guichaoua
ACTA, European and Regional Affairs, 149 Rue de Bercy, 75595 PARIS Cedex 12, France; adrien.guichaoua@acta.asso.fr

NEFERTITI is a unique project (H2020, Societal Challenge 2, RUR 12-2017 call) that establishes 10 thematic networks that bring together regional clusters (hubs) of Demo-activities and the involved actors. NEFERTITI focuses on creating added value from the exchange of knowledge, actors, farmers and technical content over the networks in order to boost innovation, to improve peer to peer learning and improve network connectivity between farms actors over Europe. NEFERTITI will also pay attention to transfer the knowledge and know-how gained by the project to a wide range of stakeholders, from farmers and members of the farming community; state, private and public advisory service; commercial companies within the agriculture and Agri-forestry sectors; policy makers at national and European levels; education and agricultural schools; research and higher education institutes, European Network for Rural Development (ENRD) and National Rural Networks (NRNs), Operational Groups and Thematic networks and enterprises. NEFERTITI project is a continuity of two Horizont 2020 projects: (1) peer-to-peer learning: accessing innovation through demonstration (PLAID project) and (2) building an interactive agridemo-hub community enhancing farmer to farmer learning (AgriDemoF2F project). The objectives of the project are: (1) Establishing 10 interactive thematic networks in the 3 main agricultural sectors: animal production, arable farming and horticultural production (2) Bringing together 45 regional hubs of demo-farmers and R&I actors, advisors, cooperatives, NGOs, industry, education, researchers and policy makers (3) Supporting 3 annual campaigns of 'on farm demonstrations' (4) Boosting innovation uptake, improving peer to peer learning and network connectivity between farming actors (5) Communicating and disseminating the practical-oriented outcomes adapted at local level through different channels, and production of relevant material in each language of the partners (6) Improving the policy dialogue and networking for sustainability. During the first year of the project the activities are oriented to the establishment of hubs and networks, setting up a platform, gather best practices and prepare methodologies for the demo activities in the next three years.

SmartAgriHubs project: outline of 2 flagship innovation experiments

D. Van Damme[1], G. Bruggeman[2], H. Gerhardy[3], P. Rakers[1], J. Vangeyte[1] and A. De Visscher[1]
[1]Flanders research institute for agriculture, fisheries and food, Technology and Food Science Unit, Burg. Van Gansberghelaan 115 bus 1, 9820 Merelbeke, Belgium, [2]Nuscience, Booiebos 5, 9031 Drongen, Belgium, [3]Marketing Service Gerhardy, Am Stahlbach 17, 30826 Garbsen, Germany; geert.bruggeman@nusciencegroup.com

The SmartAgriHubs project aims to accelerate the digital transformation of the European agrifood sector. A broad network of Digital Innovation Hubs (DIHs) will be built, expanded and connected with other local DIHs and competence centres throughout Europe. This multi-layer approach is supported by 28 Flagship Innovation Experiments (FIE), part of 9 regional clusters, in which innovative ideas, concepts and prototypes are further developed and introduced into the market. The North-West European cluster exists of 5 FIEs, of which two intend to lower the abundant use of antimicrobials in monogastric animals by improving animal health and welfare, i.e. DIG-ITfarm and SmartPigHealth, finally leading to a reduction in antimicrobial (cross) resistance. In the FIE DIG-ITfarm novel synergistic feeding concepts and innovative precision livestock farming technologies (PLF) will be combined. Sensor prototypes will be tested in barns of both broilers and fattening pigs to develop proven PLF based predicting and monitoring services. SmartPigHealth on the other hand only concerns pigs and aims both for an increase in the transparency on how pigs are kept and pork is produced and for a decrease in the antimicrobial use and resistance. Predictive analytic methods will be developed based on applied sensors, enabling a continuous health assessment of the pigs and giving a prognosis of upcoming diseases and mal-conditions. The development of a sustainable ecosystem 'pork production' is one of the extra aims. Data of sensors, information on biological and economic parameters and assessments of organs in the slaughterhouses will be merged to achieve this goal. Both FIEs will help to tackle the main issue of antimicrobial (cross) resistance, a threat for both human and veterinary medicine.

Transparent data sharing in the dairy production

S. Van Weyenberg, P. Ilias, S. Bosch, D. Van Damme and J. Vangeyte
Institute for agriculture, fisheries and food, Burg. Van Gansberghelaan 92 bus 1, 9820 Merelbeke, Belgium; stephanie.vanweyenberg@ilvo.vlaanderen.be

The data revolution in the agrifood sector is a common opportunity for all stakeholders along the entire food chain. The ever-increasing amount of data collected at dairy farms, enriched with data further in the chain, offers a potential for new data driven business models. But data will only become valuable if integrated throughout the entire production chain. Yet, all stakeholders depend on the willingness of data sharing and trust and recognition of data ownership is critical. Sustainable digital innovation ecosystems with respect for privacy, ownership and added values for all stakeholders need to be developed. DataHub4Agrofood, a joint initiative of ILVO, Smart Digital Farming, leading companies from the Belgian dairy industry (AVEVE, Boerenbond, CRV, DGZ, Innovatiesteunpunt, Milcobel) and the European Regional Development Fund (EFRO), aims to develop such an ecosystem. The first aim is to build a digital cloud platform able to support data sharing between participants having multiple business roles. The hub wants to achieve horizontal stakeholder integration along the supply chain to secure data exchange for the future and stimulate innovation by creating trust among partners. Clear agreements about data access, use, management and ownership are made. All parties must be able to rely on the agreements being kept so that they will continue to share their data. The farmer is recognized as the owner of the raw data. He determines which other parties may have access to his data. Only authorized data can be further exchanged and combined. The potential benefits of data sharing are demonstrated through use cases in dairy farming. These cases will illustrate how integrated data will enable the diverse stakeholders to optimize their activities (farm management, administrative workload, disease detection and prevention, planning and delivery forecasts, feed and medicine traceability, product traceability). Ongoing data feedback between partners will allow improved 'cause-effect' management at dairy farms and in companies further down the supply chain. The effort of the whole dairy sector in terms of sustainability and animal welfare can be shared transparently with retail, consumers and government.

Challenges in the implementation of cloud-based precision livestock management services provision

C. Davison, A. Hamilton, R. Atkinson, C. Tachtatzis, C. Michie and I. Andonovic
University of Strathclyde, Department of Electronic and Electrical Engineering, 204 George Street, Glasgow G1 1XW,
United Kingdom; i.andonovic@strath.ac.uk

The presentation initially details the features of a platform with the capability to provision of a range of services that implement Precision Livestock Farming management. The platform elements comprise robust, high node count sensor networks gathering a range of data from individual animals integrated into a cloud based software environment. The analysis of the data yields actionable information that optimise the practices of a mix of stakeholders including, for example, farmers/herdsmen, animal fertility specialists, feed specialists over various mediums such as smartphones. The presentation will then focus on the challenges in the evolution to new business models based on provisioning a range of services – 'software-as-a-service' – at both the technical and just as importantly commercial levels. For example, all data streams must be synchronised to enable impactful derivation of information; all data streams must be structured to enable cleansing, mining and analysis of the combined data through advanced (artificial intelligence) software; all data streams must be secured. At the commercial level the partnering of standalone system providers is mandatory for the creation of the spectrum of services that can be envisaged. The establishment of standards and data ownership is a major goal.

FITPig – pigs on internet – test-bed for online monitoring

A. Herlin[1], C. Verjus[2], S. Dasen[2], C. Callejero[3], A. Jara[4], A.P. Verdu[4] and I. Cuevas Martínez[4]
[1]Swedish University of Agricultural Sciences, Biosystems and Technology, P.O. Box 103, 23053 Alnarp, Sweden, [2]Centre Suisse d'Electronique et de Microtechnique, Rue Jacquet-Droz 1 Switzerland, 2002 Neuchâtel, Sweden, [3]Digitanimal SL, C/Federico Cantero Villamil 2B, Móstoles (Madrid), Spain, [4]HOP Ubiquitous S.L., Luis Buñuel 6, 30562, Ceutí, Murcia, Spain; anders.herlin@slu.se

The structural, genetic and technological development in pig production has resulted in a substantially improved efficiency of pig production. However, there is a concern on animal welfare due to a perceived lack of attention of the individual animal and a move to reduce antibiotics. The use of the potential of every animal and avoid losses is the key to profitability. Monitoring the individual animal by the use of IoT sensors, sending online information, farmers will have automated alerts, early warnings of disease; they can respond quickly to problems. The pig sector has very few commercially available sensor-based systems and it will be a new technology step upwards to provide such technologies for the sector. In the use-case project, FITPig, we are using a farm test-bed to monitor physiological signs (activity and heart rate) using a sensor embedded in an ear tag. For the end-user, it will provide a context-based alarm and decision support. This sensor includes features that are able to monitor health, farrowing and behaviour of the animals which will be further developed and evaluated during the project. The core of the FITPig use case consists of: (1) a physical layer, an IoT device on the animal: a photoplethysmographic sensor as well as an accelerometer that is integrated into an ear tag attached to the ear of the pigs. These sensors will allow continuous monitoring of the heart rate and activity of the animal; (2) a connectivity and service layer with gateways in the farm; (3) information management and application layers where data will be collected and analysed and information provided to the farmer in near real-time. By making use of sensors to monitor animal health and behaviour in pigs, we expect early and improved illness detection, reduced mortality, reduced use of antibiotics and overall enhanced animal well-being which will benefit pig farmers and society.

An agent-based model to evaluate the performance of reproductive technologies in beef cattle

O.A. Ojeda-Rojas[1], D.B. Coral[1], G.L. Sartorello[1], M.E.Z. Mercadante[2] and A.H. Gameiro[1]
[1]University of Sao Paulo, Department of Animal Science, Av. Duque de Caxias Norte, 225, 13.635-900 Pirassununga SP, Brazil, [2]Instituto de Zootecnia, Centro Avançado de Pesquisa de Bovinos de Corte, Rodovia Carlos Tonanni, km 94, 14160-900 Sertãozinho SP, Brazil; gameiro@usp.br

This study aimed to develop a model of a synthetic population to estimate the technical and economic performance of two scenarios composed of natural matting (NM) and NM + Timed Artificial Insemination (TAI). An agent-based stochastic simulation model was constructed and parameterized using a combination of field data (collected at the Centro APTA Bovinos de Corte, Sao Paulo State, Brazil) and data from scientific literature. The farm area was 500 ha and the initial population of agents in the model was 400 heifers and 14 bulls. After weaning, all calves were sold. The Effective Operational Cost (EOC), Total Operating Cost (TOC) and Total Cost (TC) per kg produced annually for each scenario were compared by ANOVA using the SAS GLM procedure. The results of the model were evaluated using a 5-year time horizon at one-day time intervals, across 5 stochastic model replicates. The NM+TAI scenario results in a higher weight at weaning in both female (4.70% equivalent to 8.68 kg) and male calves (5.17% equivalent to 10.53 kg). Additionally, calf productivity per unit area increased 18.75%, from 0.48 calves/ha to 0.57 calves/ha. Similarly, when heifers, lactating cows, and dry cows were included, the weaning weight per exposed cow was higher in the NM+TAI scenario (155.41±7.57 kg) than in the NM scenario (128.07±4.97 kg). The EOC was higher (P<0.001) in the scenario that used only NM (R\$ 4.36±0.32) than in the NM+TAI scenario (R\$ 3.83±0.18). Similarly, the TOC was higher (P<0.001) for NM (R\$ 6.84±0.49) than for NM+TAI (R\$ 5.86±R\$ 0.28). Likewise, TC was higher (P<0.001) for the herd under the NM scenario (R\$ 14.97±1.09) than for the herd that used NM+TAI (R\$ 12.48±0.59). Note: € 1.00 = R\$ 4.24. Under the simulated conditions, the model indicated that the NM+TAI scenario presented better technical and economic performance. The present study demonstrated the potentiality of an agent-based simulation model within the scope of a cow-calf production system allowing different technological combinations without altering the real world.

Low protein feeding as a potential measure to reduce ammonia emissions from beef cattle barns

K. Goossens, E. Brusselman, S. Curial, S. De Campeneere, P. Van Overbeke and L. Vandaele
Flemish institute for Agricultural, Fisheries and Food Research (ILVO), Scheldeweg 68, 9290 Melle, Belgium; karen.goossens@ilvo.vlaanderen.be

Nitrogen depositions in Flanders exceed the critical threshold of what native vegetation can bear, resulting in a loss of vegetation biodiversity. Ammonia (NH_3) is the major source of nitrogen coming from agriculture. Since 2014, the Flemish agricultural sector was targeted by the Programmatic Approach on Nitrogen (PAN), aiming at decreasing the nitrogen deposition on protected nature areas by implementing NH_3 reducing measures. For beef cattle farmers only grazing has been approved as an NH_3 reducing measure. The aim of this experiment was to evaluate low protein feeding as an NH_3 reducing measure for beef cattle. The experiment was performed in the ILVO emission barn, with 4 identical and mechanically ventilated deep litter compartments. The NH_3 concentration in the outgoing air was analysed using a cavity ring-down spectroscope (Picarro G2103) and the airflow rate was constantly logged. The trial was performed with 28 Belgian Blue heifers (age 12-20 months, weight 350-550 kg), divided over 4 groups. Two different TMRs were fed, equal in energy and intestinal digestible protein content (PDI) but varying in crude protein levels (CP 11.6 vs 14.2%) and in rumen degradable protein content (RDP -15 vs +15 g/kg DM). The 2 TMRs were switched over the 4 groups in 4 consecutive periods of 3 weeks each. A significant effect of the CP intake on the NH_3 concentrations was found. The NH_3 concentrations were on average 49% higher in the groups fed the high protein diet (1,385±261 vs 2,759±487 ppm; P<0.001). DM intake was not different between treatments (P>0.05), resulting in an equal energy intake (P>0.05), an equal PDI intake (P>0.05) but a significant difference in CP intake (933±42 vs 1,147±61 g CP/day; P<0.001). No effects of the diet on growth performances were found in this trial. The results of this study prove that reduction in dietary protein intake can be an effective source-oriented measure to reduce NH_3 emissions. In 2 ongoing trials long term effects of low protein feeding on growing Belgian Blue heifers and steers are under investigation.

The role of beef production in reducing the environmental burdens from livestock production

A. Foskolos[1,2], A.D. Soteriades[3], D. Styles[3] and J. Gibbons[3]
[1]University of Thessaly, Department of Animal Science, Campus Larisa, 41110 Larissa, Greece, [2]Aberystwyth University, Institute of Biological Environmental and Rural Sciences, Gogerddan, SY23 3EE Aberystwyth, United Kingdom, [3]Bangor University, School of Natural Sciences, Deiniol Road, LL57 2UW, Bangor, United Kingdom; andreasfosk@hotmail.com

Milk and beef production generate environmental burdens both globally and locally. Across many regions a typical dairy intensification pathway is for dairy farms to specialize on milk production and reduce the co-production of beef. Dairy-beef thus reduces and beef needs to be produced elsewhere if beef production is to be maintained. Life Cycle Assessment (LCA) studies quantifying the environmental implications of dairy and beef production have largely focused on the farm level and not captured system connections. Further LCA work has generally represented the 'average' farm of a region, consequently ignoring the range in farm management observed in practice and few studies consider a range of LCA environmental footprints other than carbon footprints. Here we present a 15-year LCA assessment of a total of 738 dairy (3,624 data points) and 1,887 suckler-beef (10,340 data points) commercial UK farms by considering five major LCA footprints. We also explore the footprint implications of compensating for reduced dairy-beef through producing this 'displaced' beef on suckler-beef farms. We found a substantial variation in farm footprints not captured in 'average farm' studies. Dairy-beef was much more efficient than beef produced on suckler-beef farms in terms of footprints per unit of beef output. Reducing dairy-beef and replacing it on a suckler-beef farm generally significantly increased environmental burdens. A reduction in carbon footprint was also associated with a reduction in other burdens suggesting no trade-off between local and global emissions. Increasing dairy farm diversification via higher dairy-beef output per unit of milk reduced burdens by up to 40%. We conclude that overspecialization of dairy farms in milk production increases the combined burdens from beef and milk, and that more intensive beef systems that make more efficient use of forage land play a crucial role in mitigating these burdens. Support was provided through the Sêr Cymru NRN-LCEE project Cleaner Cows.

Performances and greenhouse gas emissions of alternative beef production schemes in organic farming

M. Mathot, A. Mertens, D. Stilmant and V. Decruyenaere
Walloon Agricultural Research Centre, rue du Serpont, 100, 6800 Libramont, Belgium; m.mathot@cra.wallonie.be

Organic farming systems are currently strongly encouraged according to environmental and societal considerations. In these systems, herbivores must graze and valorise local resources. Under such constraint, what is the best beef production scheme? With which animal breed? To explore these questions, we compared the animal and environmental performances of steers and bulls of dual-purpose Belgian Blue (DPBB) or Limousine (LIM) breeds, with a maximization of the proportion of grazed grass and silage in their diet. Three repetitions in time of four groups ((DPBB or LIM) × (steer [S] or bull [B])) of 2 heads were raised in the purpose to be slaughtered while reaching around 700 kg of liveweight. After common grazing periods, bulls and steers received, in barns, contrasted diets with more grass based feed (less concentrates) in the steers diets, this under a two phases (growing [G] and finishing [F]) fattening scheme. Liveweight gains (lwg) were recorded through monthly weighing. Individual enteric methane emissions were quantified at pasture with GreenFeed (gf). In barn, individual ingestion and enteric CH_4 emissions (with gf) were also recorded. During some periods, CH_4, N_2O and NH_3 emissions from animal-manure system, using chambers for two animals, and from the stored manures produced in chambers were also followed. Amongst the information collected and that we will present, we pointed out that, in average, cattle weighed 491±44 kg at start of growing phase, 610±45 kg at the beginning of the finishing phase and were slaughtered at 732±63 kg. However main differences were observed in phases duration between steers and bulls due to large variation in lwg/day (during G phase: S=0.9±0.1, B=1.32±0.2; during F phase: S=1±0.3, B=1.5±0.2 kg lw/day). Contrasts in emissions were also observed. Indeed, enteric methane emissions were higher in G and F phases for steers (262±31 g CH_4/lwg and 341±121 g CH_4/lwg, respectively) compared to bulls (171±21 g CH_4/lwg and 135±15 g CH_4/lwg, respectively). In order to perform a systemic comparison of the different meat production alternatives, the impact of feed production schemes will be included in the environmental footprint definition.

LIFE BEEF CARBON Project – effectiveness of the mitigation strategies

S. Carè, G. Pirlo and L. Migliorati
CREA, Research Centre for Animal Production and Aquaculture, Via A. Lombardo 11, 26900 Lodi, Italy;
sara.care@crea.gov.it

At COP21, nations have committed to keep the increase of the temperature from the preindustrial age lower than 1.5 °C and to cut at least 30% of the greenhouse gas (GHG) emissions in 1995 by 2030. Beef production is also under pressure to reduce GHG emissions, but there is a long list of strategies for reducing GHG emissions and increasing soil carbon sequestration. One hundred and seventy two European beef farmers engaged in the LIFE BEEF CARBON project have committed to reduce GHG emissions by 15% within 10 years, by adopting one or more of available strategies. In Italy, carbon footprint (CF) of 20 intensive fattening farms in Veneto and Piemonte was estimated before and after the application of a series of mitigation strategies, most of which aimed to improve animal welfare, increase renewable energy, reduce synthetic fertilizers, increase animal performances or increase soil carbon stock. The environmental performances were evaluated through a life cycle assessment approach, using the CAP2ER® (Niveau 2). The functional unit was 1 kg of live weight gain (LWG) and the environmental categories were: global warming (GW), acidification (AC), eutrophication (EU) and energy consumption (EC). The difference of GW, AC, EP and EC before and after the introduction of the mitigation strategies were statistically evaluated using Wilcoxon Signed Ranks Test (R, software version 3.4). At the beginning of the project, the farms' mean size was 66±66 ha with a capacity of 468±334 animal places; the average LWG was 1.22±0.21 and the average live weight production was 213±159 T/year. After the introduction of the mitigation strategies GW, AC and EC decreased significantly (P<0.05) from 8.60±1.54 to 7.97±1.45 kgCO$_2$ eq/kg LWG, from 0.04±0.01 to 0.03±0.01 (kgSO$_2$eq/kg LWG) and from 28.3±7.74 to 25.8±7.75 (MJ/kg LWG) respectively. No difference was observed for EU. The farms that combined more than one mitigation strategies have already achieved the goal of the project to reduce the CF. The activity was carried out in collaboration with ASPROCARNE and UNICARVE and was funded by European Commission area Environment (LIFE BEEF CARBON project).

Assessing and predicting carcass and meat quality traits

M.-P. Ellies-Oury, V. Monteils, A. Conanec and J.-F. Hocquette
INRA, VetAgro Sup, UMRH 1213, Herbivore Unit, Theix, 63122 Saint Genes Champanelle, France;
marie-pierre.ellies@inra.fr

At the meat production chain level, numerous beef carcass classification systems have been implemented throughout the world to evaluate their economic value and facilitate their commercialization. Each of them calls on a set of indicators to evaluate carcass quality. At the same time, at the consumer level, different factors appear to be important for quality assessment, especially intrinsic qualities (beef taste and especially flavour, juiciness, tenderness, freshness, leanness, healthiness and nutritional value) but also extrinsic qualities (such as the impact of farming on the environment or animal welfare concerns). Nevertheless, it is difficult to reach the expectations of all the stakeholders, and a trade-off needs to be found to satisfy retailers and consumers at the same time. In this context, the present paper aims to review current knowledge about factors of importance (with special emphasis on breeding practices) that could be considered to predict and manage carcass and meat quality all over the meat production chain. The first part of this paper deals with retailers' and consumers' expectations and describes how to assess carcass and meat quality. Then, the second part is about the changes in livestock practices that affect the characteristics of carcass and indictors of meat quality. After presenting the methodology of integration of the various variation factors, different models of quality prediction developed recently are then proposed. The last part presents the current state of knowledge concerning meat quality. At the present times, although the objectives of quality management are clearly established, the predictive models are, however, limited and sometimes incomplete and further developments need to be proposed for instance for nutritional quality. Indeed, the laws of response of quality indicators to breeding factors seem to vary significantly (especially according to breed and muscle). Generic models have been developed to integrate all factors regulating meat sensory quality, such as the Meat Standards Australia grading scheme.

Genetic parameters for whether or not cattle reach a desired carcass quality and yield specification

D.A. Kenny[1,2], C. Murphy[1], R.D. Sleator[1], R.D. Evans[3] and D.P. Berry[2]
[1]Cork Institute of Technology, Bishopstown, Co. Cork, Ireland, [2]Teagasc, Moorepark, Co. Cork, Ireland, [3]Irish Cattle Breeding Federation, Bandon, Co. Cork, Ireland; davida.kenny@teagasc.ie

The ability of the beef industry to provide consumers with consistent, high-quality products is paramount to ensuring viability within an ever-more challenging and diverse market-place. The objective of the present study was to estimate the contribution of genetics in determining whether or not cattle achieved desirable carcass quality and yield specifications. A range of combinations of carcass specifications were evaluated, namely a carcass weight between 270 kg and 380 kg, an age at slaughter $\leq$30 months, a EUROP fat score between 2+ and 4= (i.e. 6 to 11 on a 1 to 15 scale) and a EUROP conformation score $\geq$O= (i.e. $\geq$5 on a 1 to 15 scale). Carcass records were available on 58,868 cattle finished from 2,923 herds. Whether or not a carcass fulfilled the overall specification (i.e. all 4 metrics combined) or different combinations of the trait-specific specifications was defined as a series of binary traits. Variance components were estimated using linear mixed models. Fixed effects included dam parity, birth-herd type (i.e. beef or dairy), coefficients of heterosis and recombination, birth type (i.e. singleton or twin) and contemporary group of finishing herd, year, season and sex (i.e. bull, heifer or steer). An additive genetic effect was fitted as a random term. Heritability estimates for direct measures of carcass conformation, fat and weight were 0.73 (0.02), 0.63 (0.02) and 0.68 (0.02), respectively. The heritability of achieving single desired specifications for carcass conformation, fat or weight were 0.12 (0.01), 0.11 (0.01) and 0.18 (0.02), respectively. The heritability of achieving the overall desired specification was 0.18 (0.02). The genetic standard deviation of the single and overall specification traits ranged from 0.08 to 0.17. The genetic correlations between the continuous measures of carcass conformation, fat or weight with the achievement of the overall specification were -0.29 (0.04), 0.29 (0.06) and -0.60 (0.03), respectively. The genetic parameters presented indicate that potential exists for genetic selection to improve the quality and yield specifications of cattle carcasses.

Cattle welfare self-assessment and benchmarking tool for farmers via mobile application

M. Thys[1], A. Watteyn[1], R. Vaes[2], G. Vandepoel[2], B. Sonck[1] and F. Tuyttens[1]
[1]Flanders Research Institute for Agriculture, Fisheries and Food (ILVO), Scheldeweg 68, 9090 Melle, Belgium, [2]Boerenbond, Diestsevest 40, 3000 Leuven, Belgium; mirjan.thys@ilvo.vlaanderen.be

Monitoring animal welfare is valuable for farmers to raise awareness about why and how to assess welfare, to identify points requiring attention, and to evaluate effects of farm management modifications over time. Therefore, a mobile application has been developed for farmers to self-assess the welfare of their livestock with instant feedback on their status (including potential risk factors, historical scanning results and benchmarking). Similar to the Welfare Quality© and KTBL Protocol, the tool primarily includes animal-based 'outcome' indicators (e.g. directly related to body condition, injuries, behaviour). Indicators were carefully selected to ensure that the main welfare issues are covered, and to allow straightforward and robust scoring of a limited but illustrative number of animals in a time-span of 1.5 to 2 hours. Additionally, key questions on farm management, housing and production parameters are included to allow for benchmarking with comparable farms. Self-scans, allowing offline data recording, have been developed for beef cattle, lactating cows, dry cows and dairy calves. Within each animal category, separate scans are provided based on type of housing, to allow for appropriate data collection. After online submission of a completed scan, a report is automatically generated calculating scores for each key welfare indicator on a 0-100 scale. In addition, scores are graphically depicted over time and benchmarked anonymously with those of comparable farms. In order to assess whether the tool was comprehensible, inviting and feasible for farmers to include in their routine management, the self-scan was tested by regional study groups of farmers on 8 farms. As a result the number of animals to be individually scored was reduced and several questions were rephrased for clarity. By scanning and benchmarking animal welfare periodically, farmers are encouraged to address points of attention, and able to monitor effects of measures taken over time. Effects of implementing the tool on the actual animal welfare status on farm, as well as the comparability of data both within and between farms still need to be assessed.

Technologies to predict eating quality in Australian beef carcasses

S.M. Stewart[1], P. McGilchrist[2], G.E. Gardner[1], L. Pannier[1] and D.W. Pethick[1]
[1]Murdoch University, School of Veterinary and Life Sciences, Murdoch, WA 6150, Australia, [2]University of New England, School of Environmental and Rural Science, Armidale, NSW 2351, Australia; l.pannier@murdoch.edu.au

Within the Australian beef industry visual marbling score is used as one of the key inputs into the Meat Standards Australia (MSA) carcass grading system. It acts as a proxy for intramuscular fat (IMF %), which is a strong driver of eating quality in beef, positively influencing tenderness, flavour, juiciness and overall liking. However, as visual marbling assessment is subjective, it may be susceptible to a degree of imprecision and inaccuracy. Alternatively, current methods to determine IMF% are typically destructive and may be arduous, most requiring laboratory measurement. Recent work indicates that marbling and IMF independently contribute to palatability scores in Australian beef, suggesting that marbling score describes aspects of eating quality beyond just the percentage of fat in meat. Thus new technologies that aim to predict both of these traits may further improve the prediction of beef eating quality. The Advanced Livestock Measurement Technologies project was initiated to accelerate the development of devices that can measure traits that predict eating quality at chain-speed within Australian abattoirs. Several photonics devices are currently being investigated and include hyperspectral and RGB type cameras and probes, as well as computed tomography. In addition to testing the robustness of these technologies, and comparing their relative precision and accuracy, industry systems are being established regulate their calibration and auditing. Accurate and precise objective measurement of IMF % and marbling will improve the prediction of consumer eating quality, delivering a more consistent product to consumers. This will enhance the transparency of price signalling within the supply chain, capturing value in both the processing and the retail sector. More accurate feedback will enhance management decisions on-farm, and accelerate genetic gain.

SmartCow: fostering synergies through mapping cattle research infrastructures and research

M. O'Donovan[1], R. Baumont[2], E. Kennedy[1], C. Guy[1], C. Fossaert[3], P. Madrange[3] and B. Esmein[4]
[1]Animal and Grassland Innovation Centre, Teagasc, Moorepark, Fermoy, Co Cork, Ireland, [2]INRA, UMR Herbivores, Theix, 63122 Saint-Genès-Champanelle, France, [3]Institut de l'Elevage, 149 rue de Bercy, 75012 Paris, France, [4]EAAP, Via Giuseppe Tomassetti 3 A/1, 00161 Roma, Italy; rene.baumont@inra.fr

The European Strategy Forum on Research Infrastructures (ESFRI) roadmap clearly identified the need for improved coordination, harmonisation and access to European research infrastructures (RIs) of farm animals. One of the main objectives of the SmartCow project (www.smartcow.eu) funded by the European Commission for 4 years from 1st February 2018, is to link the RIs, resource capability, and research needs across Europe. There is huge research capabilities within the research communities, but there is very little known on what resources, research activities and databases are in use or available within RIs. We inventoried European animal RIs, databases and existing sample banks, and catalogued available equipment and related techniques, which are in use across the SmartCow consortium: France (INRA – Herbipôle and Le Pin), Ireland (Teagasc – Moorepark and Grange), England (University of Reading), Scotland (SRUC – Dairy and Beef Centres), Spain (IRTA), Netherlands (Wageningen University and WUR), Belgium (CRAW) and Germany (FBN Leibniz). Twelve of the centres are dairy based with a further 5 stations beef focussed. Dairy cow number totalled 2,540 across the RI's, the breed type dominated by Holstein Friesian cows. Total beef cow numbers were 740, dominated by Limousin, Angus, Simmental and Charolais breeds. The calving spread was spring calving in Ireland, all other countries had calving split between spring and autumn. Animal measurement capability was very standard across RI's. The consortium has excellent expertise in animal methane methodology, with 6 RI's having enclosed chambers, 7 RI's having capability using the SF6 technique and 8 RI's having the Green Feed – methane measurement technology. An interactive map of the RI's and their capabilities has been generated and is available on the project website. The mapping of RIs will be completed in SmartCow participating countries and in the rest of Europe through a questionnaire sent through EAAP network.

What prospective scenarios will be compatible with sustainable crop livestock systems?

C. Mosnier[1], I. Abdouttalib[1] and N. Dubosc[2]
[1]*Université Clermont Auvergne, Inra, VetAgro Sup, UMR Herbivores, Theix, 63122 Saint-Genès-Champanelle, France,*
[2]*Chambre Régionale d'Agriculture, chemin de Borde Rouge, 31320 Auzeville-Tolosane, France; claire.mosnier@inra.fr*

Crop-livestock systems (CLS) can use complementarities between crops and livestock to reduce their input consumption. Focus groups have been organized in four French regions to discuss how crop-livestock systems might evolve according to three contrasting prospective scenarios. To support these discussions, the impacts of key drivers of each scenario have been simulated with the bioeconomic farm model Orfee to one typical crop-livestock farm of each region. The liberal scenario is characterized by the largest acreage increase, the highest prices variability, the lowest taxes and no subsidies (model inputs). The territorial scenario has the lowest variability of prices, livestock and legume crop coupled payments are maintained, a high share of grazing and low share of concentrate feed in animal diets are imposed. The environmental scenario introduces a tax on GHG emissions and favours organic farming. To adapt to the new context, the simulated farms can modify herd size, animal diets, production per animal, crop acreage and can introduce legume crops. The results show that CLS will be maintained in all scenarios but with different levels of sustainability. In the liberal one, dairy production increases and becomes more intensive per hectare due to higher average milk price, while suckler cow production decreases (cut of coupled subsidies). Incomes are high but variable. Environmental indicators are weaker. In the territorial scenario, the share of livestock in total sales is close to the current situation, incomes are low but more stable, social and environmental indicators are intermediate. In the ecological scenario, livestock is maintained owing to its capacity to produce organic fertilizers and to value crops that store carbon and that are good precedent for cash crops. Livestock production is more extensive per hectare. Environmental indicators are the best but social indicators are intermediate. The future will probably be at the crossroad of these three scenarios. To promote CLS and agroecology, we have to develop incentive tools targeting agricultural industries, consumers and policymakers.

Evaluation of nitrogen excretion equations for pasture-fed dairy cows

C. Christodoulou[1], J.M. Moorby[2], E. Tsiplakou[1] and A. Foskolos[2,3]
[1]*Agricultural University of Athens, Department of Nutritional Physiology and Feeding, Iera Odos 75, 11855, Athens, Greece, [2]Aberystwyth University, Institute of Biological, Environmental and Rural Sciences, Gogerddan, SY23 3EE, Aberystwyth, United Kingdom, [3]University of Thessaly, Department of Animal Science, Campus Larisa, 41110 Larissa, Greece; andreasfosk@hotmail.com*

Prediction of nitrogen (N) excretion of dairy cattle is a direct tool to improve N management and thus, to mitigate environmental N pollution. Empirical equations of N excretion have been evaluated for indoor dairy cattle but there is no evaluation for cows fed high amounts of fresh forage. Therefore, the objective of the study was to evaluate N excretion equations with a unique dataset of zero grazing trials. A literature search identified 138 equations predicting N excretion by 18 studies. These were evaluated with an independent dataset derived from 55 lactating Holstein-Friesian cows that were part of 7 zero-grazed trials. Squared sample correlation coefficients reported were based on either the best linear unbiased prediction (R^2_{BLUP}) or model-predicted estimates (R^2_{MP}). Root mean square prediction error (RMSPE) was used to determine model accuracy, and the concordance correlation coefficient (CCC) to simultaneously assess accuracy and precision. Results showed that N intake was related to the prediction of urinary N (UN; R^2_{MP}=0.74, R^2_{BLUP}=0.98, RMSPE=16.5, CCC=0.82), faecal N (R^2_{MP}=0.53, R^2_{BLUP}=0.97, RMSPE=7.5, CCC=0.72), and feed N apparently digested (R^2_{MP}=0.98, R^2_{BLUP}=0.99, RMSPE=10.2, CCC=0.95). Milk N was better predicted using milk yield as a single independent variable (R^2_{MP}=0.77, R^2_{BLUP}=0.96, RMSPE=10.6, CCC=0.74). Additionally, dry matter intake was a good predictor of UN, Milk N, faecal dry matter excretion and total manure excretion. However, predictive equations for milk urea N did not perform satisfactorily. In conclusion, we identified several equations from the literature that can be used to predict N excretion from dairy cows fed grass pasture. Support was provided through the H2020 project CowficieNcy.

Indirect estimation of milk production of beef cows of submitted to foetal programming

M.H.A. Santana, F.J. Schalch Junior, A.C. Fernandes, G.P. Meciano, I.M. Ruy, J.N.P. Leite, N.P. Garcia and G.H.G. Polled
FZEA/USP, Animal Science, Avenida Duque de Caxias Norte, 225, 13635-900, Brazil; miguel-has@hotmail.com

Foetal programming has been the subject of many studies in beef cattle. This strategy is especially interesting for tropical regions where there is a dry season during gestation of the matrix that can affect progeny performance. The aim of this study was to evaluate milk production indirectly from cows that were submitted to foetal programming during their pregnancy. A total of 69 Nellore cows were divided into three groups of gestational nutritional plans: NP – only mineral supplementation (0.03% BW), PP – protein and energy supplementation in the final third of gestation (0.3% BW), and PC – protein and energy supplementation all gestation (0.3% BW). After parturition, all cows were submitted to the same food protocol (mineral supplement only) and the same environmental conditions. At 120 days postpartum the indirect estimation of the milk production was made, in brief the calves were separated from the cows for 16 hours and were weighed, after weighing 45 minutes with the matrices and were weighed again. The experimental design was a randomized block design where the treatments were nutritional plans and the blocks were considered the birth years of the matrices (2010-2015), the data were analysed by PROC GLM of SAS. The average milk yield among the three treatments was 2.84±1.43, 3.02±1.65 and 3.32±1.91 for NP, PP and PC, respectively. There was no significant difference between the treatments (P>0.05) and there was also no difference in the age of the matrix in its milk production. The foetal programming of Nellore cows did not affect the milk production of the matrices at 120 days of lactation.

Gestational weight gain in Nellore cows in foetal programming

M.H.A. Santana, F.J. Schalch Junior, A.C. Fernandes, G.P. Meciano, E. Furlan, I. Mortari, L.F.M. Azevedo and R.C. Rochetti
FZEA/USP, Animal Science, Av. Duque de Caxias Norte 225, 13635-900, Brazil; miguel-has@hotmail.com

Cow nutrition during gestation is very important not only to the animal, but also to the health and productivity of her offspring. Additional care is required in the tropical regions during the middle and late thirds of gestation where, generally, the quality and amount of pasture is lower due to the dry season. The objective of the present study was to examine the weight gain of cows during their gestation that were subjected to three different nutritional plans of foetal programming. A total of 126 Nellore cows during their entire gestation were evaluated in a tropical environment, under pasture of *Brachiaria brizantha* cv. Marandu. The matrices were also divided into three nutritional plans of foetal programming that are considered the treatments: NP – only mineral supplementation (0.03% BW), PP – protein and energy supplementation in the final third of gestation (0.3% BW), and PC – protein and energy supplementation all gestation (0.3% BW). Randomized block design was used as experimental design where the treatments were nutritional plans (NP, PP and PC) and the blocks were considered the birth years of the cows (2010-2015), the data were analysed by PROC GLM of SAS. The mean weight gain was 12.4, 34.6 and 41.3 for treatments NP, PP and PC, respectively. As expected, cows that were supplemented with energy and protein throughout pregnancy and those supplemented only in the final third had greater weight gain than those supplemented only with minerals. The effects of this nutritional will be evaluated in the future in their progenies by omics and economic approaches to verify the effects of this foetal programming in the system of production of beef cattle in tropical environments.

Prediction of response to selection on principal components for growth traits in Nellore cattle

G. Vargas[1,2], F.S. Schenkel[1], L.F. Brito[3], L.G. Albuquerque[2], H.H.R. Neves[4], D. Munari[2] and R. Carvalheiro[2]
[1]University of Guelph, Department of Animal Biosciences, 50 Stone Road East, N1G 2W1, Guelph, ON, Canada,
[2]Sao Paulo State University, Department of Animal Science, Via de Acesso Prof. Paulo Donato Castelane, 14884-900, Jaboticabal, SP, Brazil, [3]Purdue University, Department of Animal Sciences, 610 Purdue Mall, 47907, West Lafayette, IN, USA, [4]GenSys Associated Consultants, R. Guilherme Alves 170, 90680-000, Porto Alegre, RS, Brazil; vargasg@uoguelph.ca

The selection of breeding animals based on combined traits has been primarily done according to the Selection Index (SI) theory. However, SI derived from Principal Components (PCs) was suggested as an alternative for combining different selection criteria. This study was carried out to compare the genetic progress expected by selecting animals based on PCs or SI used by Nellore cattle breeding programs in Brazil (Final Index, FI). The PCs were obtained from growth traits, i.e. weaning and yearling weight gain and visual scores (conformation, precocity and muscling), and scrotal circumference (reproductive-performance indicator trait). The dataset included phenotypic and pedigree information from 355,524 animals. Variance components and EBVs were obtained from a multivariate analysis using a mixed linear animal model. PCs were obtained from the eigen-decomposition of the additive genetic (co)variance matrix among traits. Standardized genetic gains per generation were predicted for each trait. The PC1 showed higher overall genetic progress (expected genetic gain in each trait ranged from: 0.68 to 1.21) when compared to PC2 and PC3. However, lower than that obtained for FI (range: 0.98 to 1.45). The PC2 showed negative gains for growth traits and conformation (range: -0.85 to -0.03) and positive gains for the other traits (range: 0.14 to 0.52). For PC3, genetic gains ranged from -0.28 to -0.10 for weaning traits and from 0.03 to 0.81 for yearling traits. The first PC and the FI yielded to favourable genetic gains for all traits, while the PC2 and PC3 would lead to a change in animal biological type (early or late growth biotypes and weaning or yearling performance, respectively) when used as criteria for selection in Nellore cattle.

Effects of gluconeogenic supplements on energy metabolism and fertility in suckler cows

S. Lobón[1], J. Ferrer[1], M.A. Gómez[2], A. Noya[1], I. Casasús[1], C. Mantecón[2] and A. Sanz[1]
[1]CITA de Aragón – IA2, Montañana 930, 50059 Zaragoza, Spain, [2]NOVATION 2002 SL, C/ Marconi 9, Coslada, 28820 Madrid, Spain; asanz@aragon.es

It was hypothesized that the intake of gluconeogenic supplements around fecundation has a flushing effect on energy metabolism and fertility in suckler cows. The effect of three gluconeogenic supplements on insulin, metabolites, productive and reproductive parameters was studied in 39 lactating Parda de Montaña cows (6.8±2.6 years; 568±65 kg live-weight (LW); 2.8±0.2 body condition score (BCS)) during a 24-day trial. Cows were fed daily 9 kg alfalfa hay and 1 kg triticale meal. At 50±15 days postpartum, cow-calf pairs were distributed into 4 treatments: (1) PRO, 228 g supplement/cow/day (containing 65% propilenglycol); (2) GLY-PRO, 366 g/cow/day (33% glycerine + 9.5% propilenglycol); (3) GLY, 375 g/cow/day (40% glycerine); (4) CONTROL. Supplements were offered from day 1 to 8 (period 1) and from day 16 to 24 (period 2) of the study. Before and after each period, two blood samples (in 48 h) were collected to determine the concentration of insulin, glucose, β-hydroxibutyrate, urea and non-esterified fatty acids (NEFA), and LW and BCS were registered. Concurrently with period 2, cows were involved in a Cosynch protocol and were inseminated on day 24 (pregnancy was diagnosed by ultrasound 35 days later). During the trial, GLY-PRO cows lost less LW than PRO and GLY groups (-8 vs -19 and -16±2.8 kg; P<0.05), and similar to CONTROL (-13 kg). Calves from GLY group had lower LW gain than GLY-PRO and CONTROL calves (0.536 vs 0.866 and 0.804±0.083 kg/day; P<0.05), and similar to PRO calves (0.704 kg/day). Insulin and metabolite concentrations were affected by the interaction between treatment and time (P<0.05), except for glucose (P>0.05). GLY-PRO cows showed the highest values of insulin and NEFA after the first period of gluconeogenic supplementation (P<0.05). Fertility rate was 50% in PRO, GLY-PRO and CONTROL groups, and 60% in GLY (P>0.05). In conclusion, the intake of gluconeogenic supplements during two short periods prior to insemination had slight effects on cow-calf weight gains and energy metabolism, but did not improve fertility in suckler cows. Further studies with longer supplementation times should be evaluated.

Surface methane and nitrous oxide emissions from a commercial beef feedlot in South Africa
K. Lynch, C.J.L. Du Toit, W.A. Van Niekerk and R. Coertse
University of Pretoria, Animal Sciences, Hatfield, 0028, South Africa; linde.dutoit@up.ac.za

The objective of this study was to quantify methane (CH_4) and nitrous oxide (N_2O) emissions from feedlot pen surfaces and manure management systems under commercial conditions over three prominent seasons experienced in the Mpumalanga province of South Africa. Methane and N_2O emission fluxes were quantified in a randomized block design during 2015 in the dry and cold season (DC), dry and hot season (DH) and the wet and hot season (WH). Emission fluxes were measured using closed static chambers over a 21 day period on 3 separate sampling days at midday during each season. Three chambers were randomly placed per sample site with 3 replicates per emissions surface (starter pens, grower pens, finisher pens, effluent dams and manure piles). Headspace samples were collected from the chambers at 0, 10, 20 and 30 minutes stored in 5 ml evacuated vials, and analysed using a SRI 8610c chromatograph. Data was analysed with the general linear model of SAS. The seasonal rainfall and temperature ranged from 6 mm and 16.9 °C in the DC season to 129 mm and 31.6 °C in the WH seasons. Pen surface CH_4 flux rates ranged from 0.5 to 4.9 mg CH_4/m^2/hr, but were not influenced by season except in the grower pen treatment which had a lower ($P<0.05$) CH_4 flux in the DC season of 0.5 mg CH_4/m^2/hr compared to the DH and WH seasons of 4.9 and 4.6 mg CH_4/m^2/hr respectively. The WH season resulted in the highest CH_4 flux ($P<0.05$) in both the effluent dams and the manure piles with flux rates of 345.9 and 249 mg CH_4/m^2/hr respectively. No seasonal or within season differences were observed ($P>0.05$) in N_2O flux rates at surface emission sources within the feedlot. The grower pens had the highest ($P<0.05$) surface CH_4 flux rates within a season during the DH and WH seasons with 4.9 and 4.6 mg CH_4/m^2/hr, respectively followed by the finisher pens with 2.7 mg CH_4/m^2/hr during the DC season. The DH season recorded the highest CH_4 flux rate from pen surfaces followed by the WH and DC seasons with the highest CH_4 emissions flux recorded in the WH season from the effluent dams and manure piles. On average the grower pen surfaces resulted in the highest CH_4 flux rates between the feedlot pens.

HealthyLivestock: a Chinese-European project to reduce the need for antimicrobials
H.A.M. Spoolder[1], S.M. Yang[2], A. Ayongxi[3], J. Vaarten[4] and B. Kemp[5]
[1]Wageningen Livestock Research, Animal Welfare, P.O. Box 338, 6700 AH Wageningen, the Netherlands, [2]Institute of Quality Standards and Testing Technology for Agro-Products (IQSTAP) of CAAS, No.12 Zhongguancun South St., Haidian District, Beijing, 100081, China, P.R., [3]The International Cooperation Committee of Animal Welfare, No.30 Building of CAAS, No.12 Zhongguancun South Street, Haidian District, Beijing, 100081, China, P.R., [4]Federation of Veterinarians of Europe, Avenue de Tervueren 12, 1040 Brussels, Belgium, [5]Wageningen University, Adaptation Physiology group, P.O. Box 338, 6700 AH Wageningen, the Netherlands; hans.spoolder@wur.nl

In the HealthyLivestock project Chinese and European scientists and industry partners work together to reduce antimicrobial (AM) use by the pig and broiler industries, by improving animal health & welfare without compromising productivity. The project investigates and develops 4 interlinked strategies to reduce AM need: (1) Biosecurity: reducing risk of pathogen presence within a farm through zoning-based Health & Welfare plans, including animal based indicators of success. (2) Resilience: increasing ability of animals to cope with endemic diseases, through novel stress-reducing housing systems and probiotic improvement of gut health. (3) Rapid detection: applying precision farming techniques to facilitate early detection, diagnosis and intervention of health & welfare problems. (4) Precision medication: using pharmacokinetics to target AM to only individuals or groups in need. In a second phase the project will validate the technical innovations by establishing their societal acceptability and economic viability. It will also assess the relationships between the Health & Welfare plans, the level of pathogens on the farm and AM residues in product and manure. In phase 3 the project's industrial partners dedicate their network and expertise to knowledge exchange, including the further development of quality assurance schemes in China and the EU. The project is funded by the European Commission through Horizon2020, and by the Chinese Ministry of Science and Technology.

Strategies to reduce the need and use of antimicrobials in animal production: a European perspective

S.J. More
University College Dublin, UCD School of Veterinary Medicine, Belfield, Dublin, D04 W6F6, Ireland; simon.more@ucd.ie

Antimicrobials are a global common good, and prudent use, both in humans and animals, is critical to their long-term effectiveness. This paper will provide an overview of efforts within Europe to reduce the need for and use of antimicrobials in animal production. The review will consider three broad themes. Firstly, an overview of the science as presented in the 2017 RONAFA joint opinion of EMA (the European Medicines Agency) and EFSA (the European Food Safety Authority) and subsequent relevant publications from Europe. Secondly, relevant legislative changes, in particular the new European Regulation on veterinary medicines and medicated feed. And finally, progress that has been made and practical initiatives that have emerged to reduce the need for and use of antimicrobials in animal production in Europe.

Reduce antimicrobial use through better animal welfare practices in China

A. Ayongxi and A.T. Liu
The International Cooperation Committee of Animal Welfare, No.30 Building of CAAS, No.12 Zhongguancun South Street, Haidian District, Beijing, 100081, China, P.R.; ayxiccaw@163.com

Livestock raised at high stocking densities in close confinement have high levels of stress, and such conditions favour occurrence and spread of infectious diseases. Consequently, antimicrobials are routinely administered to animals for treatment and prevention of diseases. Along with the expansion of intensive farming in China, the consumption of veterinary antimicrobials has been growing fast. Improved animal welfare and health management can significantly improve the immunity and resilience of livestock, reduce infectious diseases and the need for veterinary antimicrobials. Smooth transition toward rational use of antimicrobials in China requires joint efforts of all the relevant stakeholders on improving animal welfare and health management practices of intensive animal farming. In China, the International Cooperation Committee of Animal Welfare (ICCAW) is leading the efforts in improving animal welfare with a holistic multi-stakeholder science-based approach, thus reducing antimicrobial use in livestock production. In the industry, we develop animal welfare standards and conduct training in animal welfare practices. We improve animal welfare practices through modelling of demonstration farms and acknowledgement of animal welfare awards. Furthermore, we promote the consumption of animal welfare-friendly products through assessment and labelling. In the academia, we promote relevant research and innovation programs. Among the public, we survey public attitudes on animal welfare and promote animal welfare concepts through promotion videos. Most importantly, we engage all the relevant stakeholders, share experiences and build consensus through the organization of annual animal welfare conferences. The comprehensive and holistic work led by ICCAW continues to widen the reach of our impact and to build an animal welfare development model with Chinese characteristics, which has significance in animal welfare improvement and antimicrobial reduction in China, as well as being an important reference for animal welfare development in other areas with relatively inadequate resources.

Biosecurity challenges to reduce the needs of antimicrobials in pig and broiler farming systems

C. Fourichon[1], W. Song[2], M. Bokma-Bakker[3], G. Kefalas[4], X. Yi[5], C. Belloc[1], P. Ferrari[6] and L. Xuielian[2]
[1]Oniris & INRA, BIOEPAR, CS 40706, 101 route de Gachet, 44307 Nantes, France, [2]DBN, 14F, no. 27, Zhongguancun Street, Beijing 100080, China, P.R., [3]Stichting Wageningen Research, Wageningen Livestock Research, De Elst 1, 6708 WD Wageningen, the Netherlands, [4]VTN, Propylaion – Viomichaniki Periochi 18, Lefkosia 2033, Cyprus, [5]New Hope Liuhe Limited Liability Company, T3 11, block A, Wangjing center, Wangjing stree, Beijing 100102, China, P.R., [6]CRPA, Viale Timavo 43/2, Reggio Emilia 42121, Italy; christine.fourichon@oniris-nantes.fr

Biosecurity is pivotal to keep under control the exposure of farm animals to pathogens. Therefore, it is considered as a strong lever to reduce the needs of antibiotic use in farm animals. In the HealthyLivestock project, we develop and evaluate innovative approaches to support improvements in biosecurity in pig and broiler farms. Based on existing knowledge, we develop a risk assessment tool to evaluate biosecurity performance in a farm. We further identify biomarkers that can be used to monitor the results of biosecurity and that enable to detect early a failure in biosecurity procedures. This approach is original because it is result-oriented, whereas existing tools to assess biosecurity are resource-based. The innovative tools and biomarkers will be used to support the definition and implementation of health control plans which are tailor-made to the specific risks identified in given farms. The consequences of implementing the tools and plans on animal health, antibiotic use, and performance will be assessed in a variety of farming systems and potential exposure in 7 countries.

Critical fail and success factors for reduced use of antibiotics in veal calves

M.H. Bokma-Bakker, J.W. Van Riel, C.C. De Lauwere, A.F.G. Antonis and M. Kluivers-Poodt
Wageningen Research, P.O. Box 338, 6700 AH Wageningen, the Netherlands; martien.bokma@wur.nl

The Dutch veal calf sector has initiated various activities since 2007 to reduce antibiotic use within the sector, resulting in a 47% reduction to date. To obtain guidance for further improvements, in 2016/2017 research was carried out to identify critical success factors for low antibiotic use in white-veal calves at the request of the Dutch ministry of Agriculture and the veal industry. By statistical analysis of industry data-bases, associations were determined between farm and herd characteristics and antibiotics use. Significantly more antibiotics were used for larger than for smaller veal calves herds (up to 26% more antibiotics; herd sizes ranging from <400 calves (reference) to >1,200 calves). In case of more nationalities in the herd, the difference between large and small herds was even more extreme. A higher proportion of heifer calves in the herd was associated with lower antibiotic use (over 13% lower with >80% heifers in the herd). This effect was even stronger after weight correction. Within the studied average calf weight range upon arrival (40-53 kg), antibiotic use was inversely related to calf weight: approximately 1% less use of antibiotics (in daily dosages per animal) per kg. Herds with predominantly Irish calves used over 30% less antibiotics (after weight correction) than comparable herds from other origins. Herds where all calves were brought in on the same day had a lower antibiotic use (5%) than herds with 7 or more days between entry of the first and last calf. In 2018-2019, a more in-depth study on the possible reasons behind these relationships is carried out, using comparable large white-veal farms with a consistent low or high use of antibiotics.

Effective waterlines cleaning protocols: a new way to reduce antibiotic usage?

M. Leblanc-Maridor[1], S. Brilland[2], C. Belloc[1] and P. Gambade[2]
[1]Oniris & INRA, BIOEPAR, CS 40706, 101 route de Gachet, 44307 Nantes, France, [2]Sanders, tbc, Pontivy, France;
mily.leblanc-maridor@oniris-nantes.fr

To guarantee the best quality of water from the source to the animal troughs, it's important to be aware that water quality can be adversely affected by the formation of biofilms in distribution systems, which represent persistent reservoir for potentially pathogenic bacteria. In addition, the presence of biofilm in water distribution systems makes disinfection difficult or it can decrease the efficacy of oral treatments administered to the animals like vaccines, antibiotics or nutritional factors. In pig husbandry, weaning is a critical management period since piglets become exposed to social, environmental as well as nutritional changes which might be regarded as stressful events. In this study, we have chosen this sensitive period to evaluate in pig farms the effects of different mechanical and chemical waterlines cleaning protocols, similar to those used in poultry farms. Two different protocols were tested. They combined: (1) the mechanical action of draining; (2) one detergent (either an alkaline or an enzymatic one); (3) another draining state; and finally (4) one acid used at an antibacterial concentration. The experiment was set up during the down period in post-weaning rooms. To follow the bacteriological quality of water during protocols, we counted the total flora at 22 and 37 °C in water. Before and after the experiment, cotton swabs were applied into the pipes to evaluate biofilm. Bacterial concentration in water increased along the pipelines: total flora was higher at watering place than at the entry of the building. Both protocols combining mechanical and chemical procedures reduced total flora, improved water quality and cleanliness of pipes. Our results show that waterlines cleaning protocols (transposed from those used in poultry farms) reduce water's total flora and could be part of the health prevention measures for troubles which are linked to a poor water quality. The improvement of water management could be also used to reduce antibiotic consumption during this period. It would be interesting to measure the re-contamination of water flowing in pipes in order to adapt protocols mixing optimization of water quality for animals and convenience for farmers.

Levers to reduce antibiotic use in French beef farming: a multilevel perspective

F. Bonnet-Beaugrand[1], S. Assié[1], A. Poizat[1], A. Rault[1], L. Hervé[2], P. Loiseau[2] and N. Bareille[1]
[1]UMR BIOEPAR INRA ONIRIS, CS40706, 44307 Nantes Cedex 3, France, [2]Terrena Innovation, La Noëlle BP 20199, 44150 Ancenis, France; nathalie.bareille@oniris-nantes.fr

The beef farming industry encounters large health risk, mainly due to the separation of fattening operations from cow-calf operations. Antibiotic use has proved to be an efficient solution on an economical and epidemiological perspective, but can no longer be considered optimal considering public health perspectives. Although prudent practices have already been identified, they remain niches of innovation and have not been implemented at a large scale. In this communication, we aim at identifying the main barriers and levers for the adoption of these new practices. We draw on a range of studies embedded in a multilevel perspective research program. Presentation of risk factors or niches of innovation are then exemplified through assessment studies concerning the consequences of batching on growth and health, coordination in the young bulls industry or the conception of a risk assessment tool to avoid systematic antibiotic use. The identification of sociological and economical barriers and levers relies on a qualitative approach (54 interviews and 6 field observations) among various French stakeholders of beef farming. The guidelines focused on the value chain organisation and the way the interviewees take weanling's health and especially BRD risk management into account. The results show that the barriers are more or less similar for middle men and fatteners and highlight logistics and information transmission among the value chain, whereas cow-calf producers are more concerned by the balance between the valuation of weanlings and the workload induced by better practices. We then discussed incentives regarding barriers and levers identified at different levels of the value chain: valuation of weanlings or work organisation at the farm level, logistics reconfiguration, information transmission or insurances at the industry level, marketing at the retailers' level, regulation on price or antibiotic limitations at the national or EU level. The efficacy of scaling up may depend on a combination of those incentives.

Early life experiences affect the adaptive capacity of animals to cope with challenges in later life

B. Kemp[1] and H.A.M. Spoolder[2]
[1]Wageningen University, Adaptation Physiology group, P.O. Box 338, 6700 AH Wageningen, the Netherlands, [2]Wageningen Livestock Research, Animal Welfare, P.O. Box 338, 6700 AH Wageningen, the Netherlands; bas.kemp@wur.nl

Health, welfare and productivity are influenced by the animal's capacity to adapt to stressors that it is exposed to. Stress is not necessarily detrimental, but a 'load' of simultaneous, persistent or severe challenges exceeding the adaptive capacity of farm animals may lead to reduced resilience and culminate in behavioural and physiological disturbances, as well as increased disease susceptibility. Therefore, there needs to be a proper balance between the challenges an animal is exposed to and its adaptive capacity. Evidence is accumulating that early-life experiences influence this balance. In broilers, high incubation temperatures were found to increase the mortality due to ascites during the production phase. Further work showed that high incubation temperatures hamper heart development during incubation, which may reduce their cardiovascular capacity later in life. Secondly, young laying hens benefit from feeding immediately after hatch, as this has long term effects on intestinal microbiota composition and increases systemic humoral T-cell dependent immune responses. In pigs, enrichment during lactation reduces feed neophobia and results in better pre-weaning growth performance and higher feed intake. Furthermore, loose housing during lactation allows the sow to teach piglets to eat, which has been shown to reduce food neophobia and result in higher pre-weaning growth and less damaging behaviours after weaning. Finally, enriched housing versus a barren environment for weaner pigs affect their response to a PRRSv (Porcine Reproductive and Respiratory syndrome) and APP (Actinobacillus Pleuro-Pneumoniae) challenge. The percentage of piglets with APP induced lung lesions and the extent and severity of lung lesions was substantially lower in enriched housed piglets as compared to barren housed piglets. These and other examples will be presented to support the importance of early life experiences when shaping the animal's adaptive capacity.

Ewe prepartum supplementation with polyunsaturated fatty acids affects lamb immunity and behaviour

X. Averós[1], I. Granado-Tajada[1], I. Beltrán De Heredia[1], J. Arranz[1], A. García-Rodríguez[1], L. González[2], N. Elguezabal[3], R. Ruiz[1] and R. Atxaerandio[1]
[1]Neiker-Tecnalia, N-104, km 355, 01992 Arkaute, Spain, [2]Centr. Inv. Agr. Mabegondo, AC-542, Km 7 Mabegondo, 15318 A Coruña, Spain, [3]Neiker-Tecnalia, Parq. Tec. Bizkaia, Berreaga 1, 48160 Derio, Spain; xaveros@neiker.eus

Prenatal interventions may help reducing antimicrobial use in animal production. Effects of prepartum supplementation with PUFA-ω-3 on ewe colostrum composition and quality, lamb passive transfer immunity, and lamb behaviour and immune response after post-weaning aversive handling (AH) were studied. Colostrum was collected from 20 Latxa ewes after birth, among 48 ewes split into 2 groups fed 2 diets during 5 weeks pre-lambing (CO: 500 g control feed/1 kg corn silage/*Festuca pratensis ad libitum* per ewe and day; L: 500 g feed rich in PUFA-ω-3/1 kg corn silage/*Onobrychis viciifolia ad libitum* per ewe and day). Colostrum physicochemistry, fatty acid composition, and IgG concentration were analysed. Twenty lambs were separated from ewes and fed colostrum (10% BW until 18 h postpartum). Lamb serum IgG concentration and apparent efficiency of absorption (AEA) were studied until 21 d of age. Lambs were weaned (17.0±0.4 kg) and subjected to AH (30 min immobilization twice a day) during 12 d. Then their response to social isolation and social motivation was tested, and they were subjected to a phytohaemagglutinin (PHA) inflammatory challenge (IC). Colostrum fat % tended to be lower in L (P=0.085), although biologically relevant essential fatty acid concentrations were higher (P<0.05). Colostrum IgG concentration and AEA until 48 h were not affected by diet, but serum IgG concentration tended to be higher at 21 d of age in L regarding CO lambs (P=0.055). BW after AH was affected by treatment×sex, L males being heavier than L females (P=0.026). During social motivation tests, treatment×sex affected time spent close to other lambs with L females spending longer periods than CO females (P<0.001). Lambs fed L diet presented lower cortisol (P=0.034) and higher IL-2 levels (P=0.030) during IC. Fatty acid colostrum composition can be modified with prepartum diet without affecting IgG concentration. Prepartum diet supplementation influences lamb behaviour and inflammatory response after aversive handling postweaning, with sex modulating the effect.

Impact of interventions during the neonatal development of piglets

D. Schokker[1], A. De Greeff[2] and J.M.J. Rebel[1]
[1]Wageningen Livestock Research, Animal health and welfare, De Elst 1, 6708 WD Wageningen, the Netherlands,
[2]Wageningen Bioveterinary Research, infection biology, Houtribweg 39, Lelystad, the Netherlands; annemarie.rebel@wur.nl

Health and resilience are important aspects for future livestock farming. Genetics, nutrition, and management are key factors that influence these traits. One of the key tissues that affects health is the gastro-intestinal tract (GIT), including the immune system and the resident microbiome. In the GIT the interaction between microbiota and intestinal tissue contributes to nutrient uptake, as well as health of the host. Neonatal development is an important period where the microbiota colonizes the GIT and simultaneously development of the host immune system occurs. Furthermore, the immune system is being programmed to react against certain antigens or not (immune reaction vs tolerance). In previous work we and others have shown that nutrition in early life has impact on this immune system programming. Nevertheless, the precise biological mechanisms of the development of a 'healthy' gut remain to be elucidated. Recently, we have performed multiple studies to identify to what extent this immune development can be modulated by interventions. We focused on early life interventions and different routes of intervention, i.e. maternal via the sow, as well as neonatal in pre-weaned piglets. We observed effects of these different (routes of) interventions on the GIT system, i.e. on microbiota composition, host gene expression and histology. Nutritional interventions (i.e. oligosaccharides, beta-glucans, or medium chain fatty acids) led to effects on intestinal immunological development and microbiome composition, depending on the route of intervention. These (explorative) studies pave the way for further research to identify the optimal intervention route and age. With these results we could determine the window-of-opportunity to modulate the GIT development in a beneficial way for health and resilience traits in order to reduce the antibiotic usage.

Automated monitoring of broilers from different hatching conditions: the HealthyLivestock approach

M.F. Giersberg[1], R. Molenaar[1], I.C. De Jong[2], C. Souza Da Silva[2], H. Van Den Brand[1], B. Kemp[1] and T.B. Rodenburg[1,3]
[1]Wageningen University & Research, Adaptation Physiology Group, P.O. Box 338, 6700 AH Wageningen, the Netherlands,
[2]Wageningen Livestock Research, Animal Welfare and Health, P.O. Box 338, 6700 AH Wageningen, the Netherlands,
[3]Utrecht University, Faculty of Veterinary Medicine, Department of Animals in Science and Society, Yalelaan 2, 3584 CM Utrecht, the Netherlands; mona.giersberg@wur.nl

The early peri-hatching environment has effects on the health, resilience and welfare of broiler chickens in later life. Optimal early and later life conditions in combination with reliable automated flock monitoring systems will likely contribute to a reduction in the use of antimicrobials in broiler production. The aim of this study is to investigate if recently developed hatching systems promote health and welfare in broiler chickens, and to validate a PLF tool to monitor these parameters automatically. Therefore, a grow-out experiment with chickens from three different hatching conditions is performed. The chickens are hatched either conventionally (no light, feed or water in the hatcher), in a system which provides feed and water in the hatcher (HatchCare, HatchTech) or on-farm (X-treck, Vencomatic), where feed and water is available after hatch and, in addition, transport of day-old chickens from hatcher to farm is not necessary. The animals are housed in a total of 12 floor pens (1,150 animals/pen) with four replications of each treatment. Each pen is equipped with a camera of the eYeNamic system (Fancom) taking top-view images of the flocks. A customized software program translates these images into indices for animal distribution and activity, which are valuable indicators for leg and food pad health. Furthermore, changes in flock activity after inspections rounds by farm staff may be used to measure the animals' fear responses towards humans automatically. In order to validate this PLF tool and to investigate whether it is capable to detect also small differences between the treatment groups, the broilers' health and welfare are assessed manually by means of established scoring protocols. The experiment will be finished at the end of Spring 2019. All data analyses will be completed, and the results will be presented at the EAAP conference.

Posters on strategies to reduce the need for antimicrobials

S. Messori
OIE, 12 Rue de Prony, 75017, Paris, France; stefano.messori@yahoo.it

At the end of this session we will invite the presenters of poster on antimicrobials to show one slide and try to get the audience interested to visit it during the poster session. A good 'teaser' will make everyone curious to see the poster in real life!!

Effects of oral administered garlic on postweaning pig's health and performance

H. Ayrle[1], H. Nathues[2], A. Bieber[1], M. Mevissen[3], M. Walkenhorst[1] and A. Maeschli[1]
[1]Research Institute of Organic Agriculture FiBL, Department of Livestock Sciences, Ackerstrasse 11, 5070, Switzerland, [2]Vetsuisse Faculty, University of Bern, Department of Clinical Veterinary Medicine, Clinic for Swine, Bremgartenstrasse 109a, 3012 Bern, Switzerland, [3]Vetsuisse Faculty, University of Bern, Veterinary Pharmacology & Toxicology, Department Clinical Research and Veterinary Public Health, Laenggassstrasse 124, 3012 Bern, Switzerland; ariane.maeschli@fibl.org

Effects of dried garlic as feed additive (Allium sativum L. of European Pharmakopoeia quality, AS) on health and performance of postweaning (pw) pigs were investigated. Piglets (n=600, all half-siblings) of a Swiss farm were within each litter randomly assigned to three treatment groups and observed from birth until three weeks pw. For the first two weeks pw piglets received either 0.3 g dried AS-powder/kg body weight (BW)/day, 6 mg colistin-sulphate/kg BW/day or a placebo (PL). Clinical score (CS), BW and daily weight gain (DWG) were measured weekly. Data were analysed using generalized mixed effect models. Due to antibiotic pen treatment because of severe pw diarrhoea 3 pens (53 and 54 piglets) of each the PL- and AS-treatments had to be excluded from the analysis. BW of the AS group was significantly higher compared to PL (0.99 kg; P=0.003) and by trend higher compared to colistin (0.63 kg; P=0.064) in the end of the third week pw. DWG of PL piglets was significantly lower compared to AS (61 g; P=0.008) and colistin piglets (61 g; P=0.001). CS of AS-treated animals was significantly lower compared to the colistin-group (P=0.04). In conclusion supplementation of dried garlic powder shows a high potential as a feed additive to improve performance and overall clinical health of postweaning pigs but does not save them from severe diarrhoea.

Effects of creep feed copper level on the health and ionomic profiles of sucking piglets
F. Zhang, W.J. Zheng, Y.Q. Xue and W. Yao
Nanjing Agricultural University, Weigang#1, Xuanwu District, Nanjing, China, 210095, China, P.R.;
zhengweijiang@njau.edu.cn

High-level copper in the diet may further enhance the co-selection toward antibiotic resistant bacteria and promote the spread of antibiotic resistance. We are required to assess the safety and necessity of this excess supplementation of Cu in feed again. Eighteen Suhuai sows at their second parity were allocated to 3 experimental treatments in which they 172 suckling piglets could access to antibiotic free creep feed including different copper level: 6, 20, and 300 mg/kg Cu ($CuSO_4$), and offered *ad libitum* from d 14 until weaning at 40 days of age. In our study, the growth performance, hair, serum, and faecal ionomic profiles and the correlations with serum biochemical parameters of suckling piglets were all analysed, our results suggested that dietary high-level (300 mg/kg) copper could only promoted growth in short term, but lack of a significantly growth response over the entire experimental period. Also, the hepatorenal function and redox balance were impaired. The feed conversion rate (FCR) was increased in CON (20 mg/kg Cu) group, which could effectively improve the antioxidant ability and protect tissues from oxidative damage compared with LC (6 mg/kg Cu) and HC (300 mg/kg Cu) groups, the diarrhoea rate was in the acceptable range. The unbalance of Na, K and Fe, Zn in the HC group, which leading to redox unbalance and hepatorenal dysfunction. Taken these results together, dietary 20 mg/kg Cu seems suitable to meet the needs and maintain the health of suckling piglets compared with 6 and 300 mg/kg Cu in diet.

Effect and mechanism of FAPS on intestinal mucosal immunity in mice
K. Zhang, Z.J. Liang, L. Wang, J.Y. Zhang, X.Z. Wang, K. Zhang and J.X. Li
Lanzhou Institute of Husbandry and Pharmaceutical Sciences, Chinese Academy of Agricultural Sciences, TCVM DEP., #335, Jiangouyan, Qilihe, Lanzhou, Gansu, 730050, China, P.R.; fyjohn@sina.com

In this study, we aimed to investigate the effect of fermented astragalus polysaccharide (FAPS) on intestinal mucosal immune suppression induced by cyclophosphate, and to provide solutions for reducing the dosage of chemical drugs and antibiotics. Animal models of intestinal mucosal immunosuppression were established with cyclophosphamide. Normal control (NC), Model contrl (MC), Levamisole hydrochloride (LH) and low, medium and high dose of FAPS groups were set up. After FAPS gradient dose gavage, the successful establishment of the animal model was evaluated by changes in body weight, immune organ index, blood routine index, intestinal tissue morphology and SIgA secretion. The model was established to further determine the gavage dose of FAPS, and the optimal dose was used to study whether it could improve the intestinal mucosal immune suppression induced by cyclophosphamide in mice. Results showed that compared with model group, medium and high dose group, its weight rebound faster, especially in high dose group, thymus index, spleen index, WBC, intestinal SIgA level and C/V value of medium and high dose group significantly increased (P<0.01). Intestinal morphology has improved significantly, intestinal villus in neat rows, slender compact, length is significantly enhanced (P<0.01). It suggested that FAPS can prevent intestinal tissue injury caused by cyclophosphamide, protect the integrity of the intestinal tissue morphology. Compared with the model group, IgA, IgG, IgM and cytokines IL-2 and IFN-γ were significantly increased (P<0.01) in serum. SIgA -chain and pIgR as the components of SIgA, the gene expression of SIgA -chain and pIgR also increased significantly (P<0.01) in the high and medium dose groups of FAPS. FAPS can significantly improve intestinal mucosa at a gavage dose of 600 mg/kw. FAPS can improve intestinal mucosal immune suppression by regulating the secretion of SigA, so as to enhance animal immunity and provide scientific basis for solutions to reduce the dosage of chemical drugs and antibiotics.

The effect of Shegan Dilong granule on the prevention of infectious bronchitis of chicken

R.H. Xin, J.S. Xie, Y.J. Luo, G.B. Wang and J.F. Zheng
Lanzhou Institute of Husbandry and Pharmaceutical Sciences of CAAS, 335jiangouyuan, qiliho, 730050 Lanzhou, China,
P.R.; xinruihuamys@126.com

Infectious bronchitis (IB) of chicken is an acute infectious disease caused by infectious bronchitis virus (IBV) in chickens. Shegan Dilong granule is a new prescription of traditional Chinese medicine, which is mainly composed of belamcanda chinensis, lumbricus, rhizoma menispermi, smoked plum, etc. It has the functions of clearing pharynx and improving larynx, resolving phlegm and relieving cough. This study will determine whether the drug has a protective effect on IB. Methods: three experimental groups were inoculated with M41 strain of chicken embryo allantoic fluid virus by nasal drip. After challenge, the incidence of chicken was observed every day and clinical symptoms were recorded. Death of chickens autopsy in time, observe the pathological changes and record the morbidity and mortality. At the same time, the experimental chickens in each group were weighed, the average weight was calculated, and the weight growth and food intake of the experimental chickens were compared. Blood samples were collected from the wing vein 14 days after infection. All the test chickens were killed after the experiment, The filtrate of lung and tracheal was inoculated in the allantoic cavity of 10 day old SPF chicken embryo, 37 °C for 48 h, the allantoic fluid were collected for IBV RT-PCR detection. The positive rate was compared between each group. Results: the incidence of IB was reduced by 20 and 23.3% when adding shegan dilong granules to the feed at the dose of 0.5 and 1.0 g/kg body weight, respectively, the positive rates of IBV RT-PCR in the test group chickens were 60 and 56.7%, significantly lower than the positive rates in the control group (86.7%). During the whole trial period, the clinical symptoms and pathological changes of test chickens in the medium and high dose groups were relatively low, and the onset peak was delayed 1-2 days. During the experiment, no adverse drug reactions were observed. Shegan dilong granule has a significant effect on the prevention and treatment of IB. It is recommended to add 0.5 g/kg of body weight to the feed for 7 consecutive days, which can be used for the prevention and treatment of IB.

Functional biological feed by microbial fermentation: a way to improve nutrition value of feed

Y. Fan, X.M. Bai, W. Hao, X.L. Liu and W.P. Song
State Key Laboratory of Direct-Fed Microbial Engineering/Beijing DBN Technology Group, No.66 Xixiaokou Road,
Haidian District, 100193, Beijing, China, P.R.; swping@163.com

Antibiotics are used in swine production for long periods, resulting in drug-resistant strains emergence and environmental pollution, also posing health hazards for both animals and humans. Microbial fermented feed maybe a good choice to reduce antibiotics. The aim of this study was to investigate the nutrition value of functional biological feed and its effect on production performance of lactating sows. The corn-soybean meal-bran mixed feed was fermented with lactic acid bacterium and Saccharomyces cerevisiae combined with multiple enzymes, at 30 ℃ for 96 h (water: feed ratio of 2:5). A total of 20 lactating sows were randomly divided into 2 treatments, with 10 replicates per treatment and 1 sow per replicate. Sows in control group were fed the basal diet (with antibiotic), and sows in experimental group were fed the experiment diet which contained 20% fermented feed (without antibiotic) and 80% basal diet. The experiment lasted for 30 days. The results showed that: (1) Lactic acid bacterium DBN01 could efficiently synthesize acid. Lactic acid bacterium DBN01 and *Saccharomyces cerevisiae* SC01 could obviously inhibit the growth of pathogen. (2) After fermentation for 96 h, the pH of the mixed feed decreased from 6.40 to 4.29, the total acid content rose from 0.42 to 2.82%, and the acid-soluble protein content (dry matter) increased from 2.07 to 4.88%. The SDS-PAGE analysis indicated that some macromolecular peptides were degraded into small molecular peptides. More abundant metabolites were observed in the functional biological feed after fermentation determined via HPLC. (3) Compared to control group, the average daily gain of weaned piglets and feed intake of lactating sows in experimental group were improved by 1.63 and 3.68%, respectively. In addition, the mortality of weaned piglets in experimental group was reduced by 75.43%. In conclusion, functional biological feed has a positive effect on the nutrition value of feed and production performance of lactating sows, with a promising potential in reducing antibiotics in swine production.

Water quality: differences of perception and management between poultry and pig producers

M. Leblanc-Maridor[1], S. Brilland[2], P. Gambade[2] and C. Belloc[1]
[1]Oniris & INRA, BIOEPAR, CS 40706, 101 route de Gachet, 44307 Nantes, France, [2]Sanders, rue Monge, 22600 Loudeac, France; mily.leblanc-maridor@oniris-nantes.fr

Drinking water is an essential nutrient for animals. Indeed, when the physiological animal's requirements are not satisfied, performances can decrease and/or diseases may appear, both having an economic impact for pig or poultry productions. In order to update data concerning the different practices and perception for water management in pig and poultry farms, a study has been conducted to compare water supplies, their optimization and the different management practices for piglets in post weaning rooms and broiler chickens. A survey was carried out in pig and poultry farms located in the West region in France. Twenty-five pig producers and 25 poultry farmers have been selected. A questionnaire was completed during an interview. The association between practices and production characteristics was analysed with khi-deux tests. Water mainly comes from drilling. For continuous water disinfection, 60% of pig farmers use chlorination whereas 80% of poultry farmers perform it with different disinfectants and the remaining 20% used tap water. Water is also an administration route for antibiotics, anthelmintics, vaccines and nutritional factors. Regarding sanitary status of animals, some recurrent diseases can be linked to unadapt water quality, especially digestive disorders during the post-weaning period (82%) and lameness of birds (90%). For many criteria poultry farmers are more aware of water quality than pig farmers. The main differences in their practices concern the monitoring of water consumption and the water pipe maintenance (including cleaning measures to eliminate the biofilm). None pig farmers perform pipes' draining while 72% of poultry farmers do. During the down period, all the poultry farmers set up protocols with mechanical and chemical procedures: 44% use flushing, draining, base, acid and disinfectant whereas 24 out of 25 pig farmers only clean the troughs in post-weaning rooms. This study underlined that the control of water management is more settled in poultry farming compared to the pig industry. These different treatments and maintenance practices could help to prevent digestive disorders in weaners and thus to reduce antibiotic consumption.

Effect of an advisory project in veterinary homeopathy for livestock on health and antibiotic use

A. Maeschli[1], M. Jakob[1,2] and M. Walkenhorst[1]
[1]Research Institute of Organic Agriculture FiBL, Ackerstrasse 113, 5070 Frick, Switzerland, [2]Technische Universität München, Arcisstrasse 21, 80333 München, Germany; ariane.maeschli@fibl.org

To reduce the use of antimicrobials is an indisputable need in livestock production. Complementary and alternative veterinary medicine (CAM) may contribute to this. Even if farmers demand in CAM for livestock is high only few veterinarians or animal health practitioners (HP) specialised in this field are available in Switzerland. The 2012 founded project 'Kometian' offers telephone CAM-advice by veterinarians and HP. We evaluated: (1) the development of the health of individual animals after farmers getting advice; (2) the development of udder health and the yearly rate of antimicrobial treatments on dairy farms during their first three years of Kometian participation. In the end of 2018, 510 farms joined the project. After getting advice in CAM for individual animals, all farmers were contacted again to gain information about the health development of the animals. In the years 2017 and 2018 we could evaluate 947 cases, 836 (88%) related to cattle. Main indications were mastitis, gastrointestinal and respiratory disorders and fertility problems. In 656 cases (69%) farmers reported a substantial improvement only by CAM treatment. In 765 cases (81%) no further conventional treatment was used. On 24 dairy farms the treatment journal and the monthly milk recording of the year before 'Kometian' starts (Y0) and the third project year (Y3) was compared. The rate of antimicrobial treatments in dairy cows declined significantly ($P<0.001$) from 56.2 to 32.8 treatments per 100 animals and year, while the udder health remains stable (35.4 and 34.1% milk recordings with Somatic Cell Counts >100,000/ml). In calves the rate of antimicrobial treatments per 100 animals and year declined from 11.4 (Y0) to 1.4 (Y3). The advisory program seems to support the recovery of individual animals. Furthermore it seems to enhance dairy farmers to reduce the use of antimicrobials in cows and calves on farm level, without a deterioration of udder health.

AMPs as active components of insect protein
A.J. Józefiak
Poznan University of Life Sciences, Department of Preclinical Sciences and Infectious, Wolynska 35, 60-637, Poland;
agata.jozefiak@up.poznan.pl

Insects at all life stages are rich sources of protein, fat and many other important nutrients. Together with the extraction of first insect antimicrobial protein (AMP) in 1980, more than 1,500 proteins with antimicrobial activity have been identified in different organisms. The interest in insect AMPs is based on the knowledge from ancient times, where different insects were used in the treatment of a number of different ailments. Harbouring approximately 1.5 to 3 million different species, insects are a significant potential source of AMPs and knowledge on their antimicrobial effects is continuously increasing. It is known that in addition to their antimicrobial effect, they boost host specific innate immune responses and exert selective immunomodulatory effects involved in angiogenesis and wound healing. In general, they are characterized as heat-stable with no adverse effects on eukaryotic cells. Depending on their mode of action, insect AMPs may be applied as single peptides, as a complex of different AMPs and as an active fraction of insect proteins in the nutrition of different livestock. The great potential for the use of AMPs in animal production is primarily associated with the growing problem of antibiotics resistance including the support of animal growth and health, treatment of infections and in the preservation of food. Studies on antimicrobial peptides (AMPs) and their applications have become one of the hot topics in the areas of animal science and the food and feed industries. The mechanism of insect AMPs actions has been shaped over hundreds of years of evolution, and it is very conservative, suggesting the risk of bacterial resistance to be low. Moreover, bactericidal insect AMPs may also protect organism against viruses and fungi to be considered as immune modulators have modulating effect on animal microbiome.

The ENTOBIOTA project: added value for *Hermetia illucens* by exploring its microbiota
L. Van Campenhout, J. De Smet and M. Van Der Borght
KU Leuven, Kleinhoefstraat 4, 2440 Geel, Belgium; leen.vancampenhout@kuleuven.be

Black soldier fly larvae are mainly sold as feed ingredient. To be economically valuable, the feed industry requires BSFL to be available at a high volume, constant quality and low cost. The substrates allowed for rearing are legally limited and exclude low-cost alternatives. These aspects hinder the use of BSFL as a competitive protein source in feed. A new research project ENTOBIOTA (coordinator KU Leuven; partners UAntwerp, Thomas More, Inagro; 01/2019-12/2022; 3 PhDs) aims to upgrade the exploitation potential of BSFL by studying their microbiota, to increase their growth yield and direct their use also to the pharma and waste sector. The first work package of the project will optimize the microbial gut equilibrium by adding (a) microbial symbiont(s) during rearing, for better growth and competitive exclusion of human food pathogens towards authorization of more substrates. Preliminary results from literature and own research will support the isolation of those symbionts. Larvae will be challenged with food pathogens during rearing to assess the capacity of the isolates to reduce pathogen count. The second work package will study the fermentation of substrates to increase their digestibility for the BSFL and to enhance e.g. the ω-3 fatty acid content of the larvae. In the third work package, BSFL will be explored as a source of novel antimicrobial peptides (AMPs) for a number of problematic human pathogens (multi-resistance). The larvae will be challenged with these pathogens during rearing, AMPs will be identified in larval extracts with antimicrobial activity, and finally the AMP encoding genes will be expressed in an expression system to produce and characterise the peptides. The last work package will investigate BSFL as supermarket waste converter. In a circular economy, supermarket waste is a highly valuable substrate for BSFL, but it is frequently contaminated with (micro)plastics during automatic unpacking of foods. In continuation of preliminary experiments, larvae will be reared on food waste spiked with plastics and additives, and the bioaccumulation, biotransformation and excretion of contaminants will be monitored. Also, microbial symbionts will be isolated and their role in the biotransformation potential of BSFL will be assessed.

A new wave of insect products

C.S. Richards
AgriProtein, Research and Development, 1 Rochester Rd, Philippi, 7750, Cape Town, South Africa;
cameron.richards@agriprotein.com

The notion of using protein and fats refined from insects as ingredients in animal feeds has been of commercial interest for many years and there are now more than 50 companies worldwide working to industrialize this opportunity. Along with these two products, there are several emerging value-added products that have the potential to offer insects as more than just a source of protein and fat. We will present the mechanisms and potential applications of these new value-added products and highlight associated product safety measures.

Antimicrobial activity of insects' fats

D. Nucera[1], M. Meneguz[1], S. Piano[1], S. Dabbou[2], I. Biasato[1], A. Schiavone[2] and L. Gasco[1]
[1]Turin University, Department of Agricultural, Forest and Food Sciences, largo braccini 2, 10095 Grugliasco, Italy,
[2]Turin University, Department of Veterinary Sciences, largo braccini 2, 10095 Grugliasco, Italy; daniele.nucera@unito.it

Insects are considered a promising protein sources for feed. To produce insect-protein meals, lipid fraction of larvae is extracted. Research showed these fats as antibacterial and valuable ingredients. Hence, this study is aimed to assess the antibacterial activity of fats extracted from *Tenebrio molitor* (T) and *Hermetia illucens* (H). Methods: T and H fats were mixed with bacterial pure cultures in a final concentration of 3 log(cfu)/ml. These mixes were incubated 24 hours at 37 °C. Every 2 hours of incubation, cultures were streaked. *Salmonella* Tiphymurium, *Salmonella* Enteritidis, *Yersinia enterocolitica* (Y.e.), *Pasteurella multocida* (P.m.) and *Listeria monocytogenes* (L.m.) were tested. Each strain/fat combination was analysed in triplicates, an aliquot of 3log(cfu)/ml of broth was used as negative control (A), whereas a mix of soy oil and bacterial broth (3log(cfu)/ml) was used as standard oil control (B). Growth curves were graphed using means and standard errors (M±SE). Results: Controls (A,B) reached final concentrations of 7-10log(cfu)/ml. Fats on Salmonellae' growth did not show any effects. Y.e. showed lower growth when challenged with insects oils than in controls: 8±2log(cfu)/ml (T) and 4±2log(cfu)/ml (H) compared to 10±0.3log(cfu)/ml (A,B). For L.m. growth observed after 6 h showed: 4.6±0.8log(cfu)/ml (T) and 3.7±0.8log(cfu)/ml (H) compared to 5.6±0.4log(cfu)/ml (A,B); after 24 h, lower counts were detected only in cultures challenged with H fat (4.8±1log(cfu)/ml) compared to 9.5±0.03log(cfu)/ml (T) and 10±0.3log(cfu)/ml (A,B). P.m. was inhibited by both insects oils after 10 h of incubation, reaching at 24 h 4±1.2log(cfu)/ml (T), 2.1±0.4log(cfu)/ml (H), compared to 7±0.4log(cfu)/ml (A,B). Conclusions: These results disclose the potential of insect fats as antibacterial ingredients: the activities against gut pathogens may indicate a potential for impairing their growth/colonization (i.e. *Pasteurella*, *Yersinia*) in farmed animal, fed with diets containing oils. The anti-*Listeria* activity could indicate a potential application in the food industry for preserving the safety of RTE foods.

Evaluation of insect derived functional feed ingredients in poultry diets

S. Verstringe[1], G. Bruggeman[1] and L. Bastiaens[2]
[1]Nutrition Sciences N.K., Booiebos 5, 9031 Drongen, Belgium; [2]VITO NV Flemish Insititute for Technonolgical Research, Boeretang 200, 2400 Mol, Belgium; stefanie.verstringe@nusciencegroup.com

An indirect biorefinery approach was developed converting side-streams into crude extracts via a two-step process consisting of: (1) conversion of biomass by insects; and subsequent (2) biorefinery of the insect biomass. The aim of the first step is to convert heterogenic feedstock into a homogenous biomass via insects. Insects are able to convert a variety of feedstock into a more homogenous biomass, being their own biomass enriched in functional feed ingredients. Black soldier fly biomass as well as biorefinery products thereof were evaluated for their nutritional value in poultry trials. Diets were formulated in iso-nutritional and iso-energetic feed formulation, based on linear programming software (WinMix™). The birds were one-day-old-chickens, and were grown for 42 days. At the start of the trial, the birds were allocated to the different pens by weight. This allocation was made in order to have an equal average weight and an equal standard deviation around the average weight for each treatment and pen. Feed intake was registered. During the trial, a veterinarian and a Felasa D certified person supervised the performed poultry experiment according to the international guidelines described in law EC/86/609 and further amendments. During the whole trial period the birds were fed *ad libitum*. A dose-response of the retained insect fractions was tested, and following parameters were analysed: Daily weight gain, daily feed intake, feed conversion ratio, microbial and histochemical analysis in the gastrointestinal tract. These parameters enable to determine the efficacy of the insect fractions on zootechnical performance, and in particular the feed conversion ratio as well as on survival rate and health status of the animal, as a measure of antibiotic use.

Insects in poultry feed: a meta-analysis of the effect on average daily gain

N. Moula, E. Dawans and J. Detilleux
University of Liege, Farah, Bld de Colonster, 4000 Liege, Belgium; jdetilleux@uliege.be

Today, insects are receiving a great attention as a potential source of poultry feed due to the high costs and limited future availability of conventional feed resources. A great number of studies have evaluated the effect of their inclusion on poultry average daily gain (ADG) but results are not always consistent and the statistical power of these individual studies are often not enough to put in evidence significant association even though it exists. Thus, our goal was to summarize and critically analyse published individual studies within the framework of a meta-analysis. We searched Pubmed, Medline and Google Scholar for reports published up to January 2019. Two independent reviewers assessed the eligibility of each study based on predefined inclusion criteria (study design, inclusion rate of insects, poultry species). We recorded ADG for all reported rates of insects' inclusion in poultry feed, computed the difference in ADG with respect to a reference isoproteic and isoenergetic diet containing no insects. Next, we performed a meta-analysis to calculate a pooled difference in ADG (DIFF) using a random effects model and a meta-regression to explore any potential source of heterogeneity between studies (R 'rma' package). Of the 75 studies reviewed, 55 met the inclusion criteria and had appropriate data for the meta-analysis. They account for a total of 116 trials with inclusion rates going from 0 to100%. Studied insects included meal worms, black soldier fly larvae, maggots, grasshoppers, crickets, locusts, and caterpillars. Animals were layers, broilers, quails and guinea fowls. With respect to the reference diet, the pooled DIFF (95% CI) was -0.10 (-0.81 to 0.62) with marked heterogeneity ($I2=99.2\%$). Heterogeneity between studies was mostly explained by the insect species (lowest DIFF for grasshopper), inclusion rate (decreasing DIFF with increasing inclusion rate) and country of publication (highest reported DIFF in Asia). Overall, our results show not too high inclusion rate of insects, other than grasshopper, can be recommended in poultry feed.

Effect of different insect meals on performance and gut health of monogastric animals

M.E. Van Der Heide, R.M. Engberg and J.V. Nørgaard
Aarhus University, Department of Animal Science, Blichers Allé 20, 8830 Tjele, Denmark; marleen.vanderheide@anis.au.dk

Two experiments are designed with the objective to compare animal performance and gut health of piglets and broilers fed insect meal or soybean meal: (1) 600 one-day old broiler chickens are housed in 24 pens for 35 days; and (2) 96 piglets (weaned at 28 days) are pair-housed during three weeks. Pens are assigned to one of four isonitrogenous and isoenergetic dietary treatments: de-hulled soybean meal (control), or 10% meal (94-96% dry matter) of *Tenebrio molitor* or *Hermetia illucens* or *Alphitobius diaperinus*. Feed intake and animal weight are recorded weekly (pigs) or biweekly (chicks) on pen level. Total excreta collection from chicks placed in metabolic cages for five days, starting from day 13, is conducted to determine apparent total tract digestibility of fat, dry matter, ash, methionine, cysteine and nitrogen retention. Clinical parameters including diarrhoea (piglets), and footpad dermatitis and litter quality (broilers) are recorded. After killing, samples from crop (chicks), stomach (piglets), small intestine and ceca are obtained at day 22 (broilers) and at the end of experimental periods (both species). The pH, selected intestinal bacteria (enterobacteria and lactobacilli) and short chain fatty acids are measured in digesta samples. Segments of the digestive tract as well as some major lymphoid organs are weighed. Monogastrics fed diets containing insects are hypothesized to have similar performance to those fed the control diet, due to the high protein and fat content in insect meal, which is comparable to commonly used soybean meal. Apparent nutrient digestibility in broilers is hypothesized to be negatively affected by the chitin content in the insect meals. A potential prebiotic effect of chitin may be reflected by increased SCFA production in particular in the ceca. The easily available medium short chain fatty acids in insect meals may provide a highly available source of energy. Antimicrobial peptides and chitin may have antimicrobial and immunomodulatory effects resulting in significant improvement of gut health reflected by reduced occurrence of clinical findings (diarrhoea and footpad lesions).

Effects of replacing soybean with black soldier fly in broiler feeding programs

D. Murta[1,2,3], M.A. Machado[4], R. Nunes[3], J.M. Almeida[5] and O. Moreira[5]
[1]Faculty of Veterinary Medicine – ULHT, Campo Grande 376, 1749-024 Lisboa, Portugal, [2]CIISA – FMV, ULisboa, Av. Universidade Técnica, 1300-477 Lisboa, Portugal, [3]EntoGreen, Ingredient Odyssey, Quinta das Cegonhas, 2001-907 Santarém, Portugal, [4]ISA, ULisboa, Tapada da Ajuda, 1349-017 Lisboa, Portugal, [5]INIAV, Santarém, Fonte Boa, 2005-048 Vale de Santarém, Portugal; olga.moreira@iniav.pt

Although the acceptance of insects, such as black soldier fly (*Hermetia illucens*; BSF) as a novel feed ingredient rich in protein and lipids is increasing, it is ultimately important to assess if it might affect the animal's production potential. This study aims to evaluate the effect of the partial replacement of soybean meal and soybean oil by insect meal (BSF) in broiler diets on the metabolic and productive performances. 48 1-day-old Ross 308 broiler were individually allocated in metabolic cages and randomly distributed by four dietary treatments formulated to be isoproteic and isoenergetic: control diet (BSF0) and partial replacement of soybean meal and soybean oil by 25% (BSF25), 50% (BSF50) and 75% (BSF75) of dry BSF larvae meal. Live body weights (LW), average daily gain (ADG) and feed conversion ratio (FCR) were estimated weekly. Feed and water intake and excreta were measured daily. The animals were slaughtered at 28 days of age. Diets and excreta were chemically analysed and apparent digestibility of Dry Matter (DMD), Nitrogen (ND) and Energy (ED) were calculated. The statistical analysis was based on the post-hoc DMS test and was performed by SPSS. LW, ADG and FCR were significantly different (P<0.05) at weeks 2 and 3, with the highest FCR (P<0.05) observed for diets BSF75 and BSF50. However, when considering the trial to its full extent there were no differences between diets. Broilers in diet BSF75 presented the lowest values of DMD, ND and ED, with a mean decrease of 9% in ND, compared with BSF0. That diet showed the highest levels of crude fibre and lignin compared with the control diet: 6.04 vs 3.88 and 1.61 vs 0.55%DM, respectively. Animals in diet BSF25 showed superior DMD, ND and ED: 79.4, 75.7 and 82.1% respectively. It was concluded that the inclusion of BSF larvae in diets to partially replace soy (at 25 and 50%), as alternative protein source, might be a sustainable nutritional solution in broiler feeding.

Black soldier fly larvae feed the future of laying hens: benefits for welfare and circularity

M.A.W. Ruis[1], J.L.T. Heerkens[2], J. Katoele[3] and L. Star[2]
[1]VHL University of Applied Sciences, Department of Animal Management, Agora 1, 8934 CJ Leeuwarden, the Netherlands,
[2]Aeres University of Applied Sciences, De Drieslag 4, 8251 JZ Dronten, the Netherlands, [3]Wadudu Insect Centre,
Noordveen 1, 9412 AG Beilen, the Netherlands; marko.ruis@wur.nl

Black soldier fly (BSF) larvae are very suitable insects for poultry feed: they provide a sustainable source of protein and contain substances that work as antimicrobial agents. It is allowed to feed them alive, or as lipids, to poultry. In previous research, an indication has been obtained that production of laying hens can be guaranteed if 20% of the diet is replaced by larvae from the BSF. The BSF can also play a major role in making the food chain more circular: BSF larvae can grow on residual flows from the food chain, but also on manure. However, both residual flows and manure are currently not permitted as rearing substrates for insects; the safety is yet not clear enough. That inclusion of live BSF larvae in the diet may also benefit the welfare of poultry, is substantiated and discussed on the basis of results of two experiments, each with 400 Bovans Brown laying hens, in 20 units having 20 hens (n=5-7/treatment). BSF larvae were fed alive to these hens from 20-36 wks of age, additionally to unlimited mash feeding. In both experiments, a 'self-harvesting system', in which BSF larvae are reared on poultry manure in the hen house, was one of the feeding treatments. Young BSF larvae grew well on the manure, and reduced the amount of manure by 20%. Overall activity and foraging behaviour increased, and feather damage from injurious pecking behaviour decreased, when compared to treatments were no larvae were fed, or where hens consumed larvae in a short time, e.g. by spreading them in the litter. Even better results were obtained when larvae were fed through a perforated PVC tube. The uptake of larvae was also prolonged, but more controlled with this treatment. Production performance was ensured, despite an increased behavioural activity. It is concluded that live BSF larvae may enrich the life of laying hens with prolonged time of foraging. Rearing of BSF larvae on manure is promising, but its implementation depends on adaptation of legislation.

Effects of dietary black soldier fly larvae oil on ileal microbiome and transcriptome in growing pig

S. Ji, S. Lee and K. Kim
National Institute of Animal Science, Animal Nutrition & Physiology Team, 1500, Kongjwipatjwi-ro, Iseo-myeon, Wanju-gun, Jeollabuk-do, (55365), Korea, South; syjee@korea.kr

Black soldier fly larvae (BSFL), *Hermetia illucens*, has been considered as a sustainable source of protein and fat for poultry, fishes, and pigs. Its lipid fraction contains high amounts of lauric acid, which has an antimicrobial effect on a wide range of microbes. This study is for clarifying the effect of BSFL oil supplementation on the growth performances, gut microbes and ileal mucosal gene expression of growing pigs. Twenty-four growing pigs(17.4±2.8 kg) were allocated into four groups(Con(-), Con(+), 0.5T, and 1.0T) and fed with BSFL oil(C12:0, 25.4%DM) at three levels(0, 0.5 and 1.0% per kg feed) for 28 days. All groups except Con(-) were intraperitoneally injected with LPS(O111:B4) 100 µg/kg BW except the Con(-) group on the final day. After 24 hours, their ileal digesta and mucosa were collected for microbiome and RNAseq analysis. And the length of their villi and crypts were also measured microscopically. The average daily gain was higher in the 0.5T(685.7±140.5 g) than the control(608.9±105.3 g) and 1.0T(542.9±101.6 g) groups. In ileal villi: crypt ratio, the 0.5T(0.84±0.22) had higher value than the remaining groups(Con(-), 0.94±0.09; Con(+), 0.80±0.26; 1.0T, 0.75±0.12). In the ileal digesta, microbiome of the 0.5T group contained the lowest ratio of Proteobacteria among all the groups. However, there were no significant difference among groups. We could not find any significant differences among the groups in differentially expressed genes of ileal mucosa. Interestingly, NOS2 expression results were similar to those of growth performances in spite of no statistical correlation. In conclusion, BSFL oil could be regarded as an alternative fat source, but further researches are needed to make use of BSFL oil as the antimicrobial agent for pigs.

Insect oil as an alternative to palm oil in broiler chicken nutrition
A. Benzertiha[1,2], B. Kierończyk[2], M. Rawski[3], A. Józefiak[4] and D. Józefiak[1,2]
[1]*HiProMine S.A., Poznańska 8, 62-023 Robakowo, Poland,* [2]*Poznań University of Life Sciences, Department of Animal Nutrition, Wołyńska 33, 60-637, Poznań, Poland,* [3]*Poznań University of Life Sciences, Division of Inland Fisheries and Aquaculture, ul. Wojska Polskiego 71c, 60-625, Poznań, Poland,* [4]*Poznań University of Life Sciences, Department of Preclinical Sciences and Infectious Diseases, Wołyńska 35, 60-637 Poznań, Poland; damjo@up.poznan.pl*

The study was conducted to evaluate the effect of total replacement of palm oil and poultry fat with *Tenebrio molitor* oil (TM oil) in broiler chicken diet on the growth performance and lipid fatty acid composition of liver and breast muscle tissues. A total of 72 seven-day-old female Ross 308 were used. The birds were randomly distributed to 3 different groups, 12 replicates per group and 2 birds per replicate. The experiment was 30 days in metabolic cages. The basal diet was formulated on maize and soybean meal basis. 5% of palm oil, poultry fat or TM oil was added to the diet. The growth performance parameters (BWG, FI, and FCR) were measured during day 7, 14, 21, and 30. Immediately after slaughter, samples from breast muscle and liver tissue were into dry ice for analyses. Data were tested using the General Linear Models procedure of SAS software. Means were separated using Duncans' tests following one-way ANOVA. In all periods of the experiments, BWG, FI or FCR were not affected by dietary treatments. The liver tissue of the chickens fed a diet supplemented with TM oil showed the lowest value of SFA (P=0.004) and the highest of UFA (P=0.004). In addition, TM oil significantly decreased MUFA (P<0.001) and increased PUFA (P<0.001) in comparison to PF and PO. Furthermore, SFA and UFA profile of the breast tissue was not affected by any of the dietary fat sources. However, TM oil reduced MUFA content (P<0.001) and increased PUFA (P<0.001). Moreover, n-3 and n-6 fatty acids were significantly increased in case of TM oil supplementation (P=0.006; P<0.001 respectively). In conclusion, the use of TM oil in broiler chicken diet did not show any effect on the growth performance. Moreover, TM oil supplementation improved the fatty acids profile of the liver tissue and the breast muscle.

Insects as functional feed additives affect broiler chickens' growth performance and immune traits
A. Benzertiha[1,2], B. Kierończyk[2], P. Kołodziejski[3], E. Pruszyńska–oszmałek[3], M. Rawski[4], D. Józefiak[2] and A. Józefiak[1,5]
[1]*HiProMine S.A., Poznańska 8, 62-023 Robakowo, Poland,* [2]*Poznań University of Life Sciences, Animal Nutrition, Wołyńska 33, 60-637 Poznań, Poland,* [3]*Poznań University of Life Sciences, Animal Physiology and Biochemistry, Wołyńska 33, 60-637 Poznań, Poland,* [4]*Poznań University of Life Sciences, Division of Inland Fisheries and Aquaculture, ul. Wojska Polskiego 71c, 60-625 Poznań, Poland,* [5]*Poznań University of Life Sciences, Department of Preclinical Sciences and Infectious Diseases, Wołyńska 35, 60-637 Poznań, Poland; damjo@up.poznan.pl*

Two independent experiments were conducted to investigate the effect of insect full-fat meals (*Tenebrio molitor* and *Zophobas morio* larvae), added 'on top' of a complete diet (Exp. 1) or calculated into diets (Exp. 2), on the growth performance, and immune system traits of broiler chickens. 1000 one-day old female Ross 308 broiler chicks were used in both experiments. In the first trial, the birds were randomly assigned to 6 treatments i.e. NC (negative control) – no additives; PC (positive control) – NC + salinomycin (60 mg/kg diet); TM02 – NC + 0.2% TM full-fat meals; ZM02 – NC + 0.2% ZM full–fat meals; TM03 – NC + 0.3% TM full–fat meals; and ZM03 – NC + 0.3% ZM full-fat meals. In the second trial, 4 treatments were set: NC – no additives; PC – NC + salinomycin (60 mg/kg diet); TM03 – NC with 0.3% TM full fat meals; and a ZM03 – NC with 0.3% ZM full-fat meals. Data were tested using the GLM procedure of SAS software. Means were separated using Duncans' tests following one-way ANOVA. Insect full-fat meals increased the BWG and FI (Exp. 1; P=0.024, P=0.022, respectively), and no effect on the FCR was recorded. In addition, in groups fed insect full-fat meals and PC comparing to NC decreased the IgY (P=0.045) and IgM, (P<0.001) levels significantly. In second experiment, IgM levels were also decreased (P<0.001) in groups fed insect meals. Moreover, the IgM levels were negatively correlated to the BWG (r=–0.4845) and the FI (r=–0.4986), with statistically significant values (P<0.001). In conclusion, the current results confirmed that both *T. molitor* and *Z. morio* may be applied as functional feed additives to achieve improved growth performance and changes in the immune system.

Effects of replacing soybean with black soldier fly (*Hermetia illucens*) in broiler meat quality

J.M. Almeida[1], M.A. Machado[2], R. Nunes[3], D. Murta[3,4,5] and O. Moreira[1]
[1]*INIAV, Santarém, Fonte Boa, 2005-048 Vale de Santarém, Portugal, [2]ISA, ULisboa, Tapada da Ajuda, 1349-017 Lisboa, Portugal, [3]EntoGreen, Ingredient Odyssey, Quinta das Cegonhas, 2001-907 Santarém, Portugal [4]Faculty of Veterinary Medicine, ULHT, Campo Grande 376, 1749-024 Lisboa, Portugal, [5]CIISA, FMV, ULisboa, Av. Universidade Técnica, 1300-477 Lisboa, Portugal; joaoalmeida@iniav.pt*

The worldwide increase in broiler production has a considerable environmental and economic impact, increasing the urgency to find sustainable alternative feed ingredients. Recent studies indicate that some species of insects have great production potential and may be a source of both protein and lipids, which is the case of black soldier fly (BSF) (*Hermetia illucens*). However, it is necessary to evaluate possible factors that might alter meat quality parameters, since those are critical factors for the market success of the final product. The objective of the current study was to evaluate the influence of replacing soy (soybean meal and soybean oil) with BSF meal in three different iso-energetic and isoproteic broiler feeding programs, and how this will influence meat quality. A total of 48 1-day-old Ross 308 broiler were fed four different treatments that consisted of a basal diet where soybean meal and soybean oil were replaced by 0% (BSF0), 25% (BSF25), 50% (BSF50) and 75% (BSF75) of dry BSF larvae. The animals were slaughtered at 28 days of age. Meat quality and appearance were evaluated six days storage at 2 ± 0.2 °C in a portion of pectoral major muscle. It was observed that BSF50 and BSF0 had higher values for Chroma and Hue than the other regimes ($P<0.05$). pH average values are slightly higher than normal (6.18-6.26, for BSF0 and BSF25, respectively) in all treatments. On the contrary, drip loss mean values are lower or under normal standards (six-day storage: higher and lower values belong to BSF50 and BSF75, 3.28 g and 2.79 g, respectively). Also, nutritional composition of meat is within normal values. It was concluded that the used replacements of soybean by BSF meal don't affect poultry meat quality.

Chitin does not impair protein and lipid digestion in monogastric animals

A. Caligiani[1], L. Soetemans[1,2], G. Leni[1], L. Bastiaens[2] and S. Sforza[1]
[1]*University of Parma, Food and Drug Department, Parco Area delle Scienze, 27/A, 43124 Parma, Italy, [2]VITO, Flemish Insititute for Technonolgical Research, Boeretang 200, 2400 Mol, Belgium; stefano.sforza@unipr.it*

Insect larvae, flour or by-products are more and more used as ingredients in the diets of some animal species. Their use is currently regulated in the European Union by the Regulation EC No 893/2017. Chitin, one of the main insect biomolecules, is a long-chain polymer of N-acetylglucosamine constituting the exoskeleton. In literature very little it is known about the influence of chitin on feed digestibility and some studies have reported that chitin might reduce the feed intake and nutrient availability. In the present study, we investigated the effect of purified chitin on the digestibility of proteins and lipids in model systems. Chitin from *Hermetia illucens* was mixed with a model lipid (lard, from 0 to 40% chitin in the mixture) and a model protein (milk serum protein concentrate, MSPC, from 0 to 30% chitin in the mixture). Samples were than subjected to a standard protocol of *in vitro* static digestion that mimics the gastro-intestinal digestion of monogastrics. The digested mixtures from chitin/lard were analysed by ^{1}H NMR for studying the composition of lipid fraction. The recorded spectra did not show significant differences between samples with and without chitin, all showing a comparable ratio of triglyceride and free fatty acids. These results demonstrated that the co-presence of purified chitin, until the 40% of the total mixture, does not have implication on the digestibility of lipids. The digested mixtures from chitin/MSPC were analysed with UPLC/ESI-MS for the peptides characterization. Results showed that samples containing different amounts of purified chitin released during digestion not only the same amount of peptides, but also the same peptide sequences as those released in the control (100% protein). In conclusion, the presence of purified chitin seems not to affect the digestibility of lipid and protein, even if further *in vivo* studies are necessary to confirm these preliminary results. This study received funding from BBI Joint Undertaking in the framework of the European Union's Horizon 2020 research and innovation program, under grant agreement no. 720715.

Do *Hermetia illucens* larva substrates affect sensory traits of quail breast meat?

A. Dalle Zotte[1], Y. Singh[1], E. Pieterse[2], L.C. Hoffman[2,3] and M. Cullere[1]
[1]University of Padova, Animal Medicine, Production and Health, Agripolis, Viale dell'Università, 16, 35020 Legnaro (PD), Italy, [2]University of Stellenbosch, Department of Animal Sciences, Matieland, 7602, South Africa, [3]Queensland University, Centre for Nutrition and Food Sciences, QAAFI, 39 Kessels Rd, 4108 Brisbane, Australia; antonella.dallezotte@unipd.it

The present research studied a 10% dietary inclusion of full-fat *Hermetia illucens* (HI) dried larvae, reared on different substrates, on the sensory profile of quail (*Coturnix coturnix japonica)* breast meat. HI larvae were reared either on a substrate made up of 100% layer mash (HI1) or 50% layer mash: 50% fish offal (HI2), rich in n-3 fatty acids. Ten-day old chicks were randomly allocated to 3 different dietary treatments: a Control diet (C) with soybean meal as main protein source, and two C diets including either 10% HI1 or 10% HI2, respectively. At 28 days of age, quails were slaughtered, and 52 breasts/treatment were dissected. Thirty-six breasts were cooked in a water bath up to 74 °C core temperature, cooled and Warner-Bratzler Shear Force (WBSF) was measured. Sixteen breasts were subjected to a descriptive sensory analysis to detect possible differences: C vs HI1 vs HI2. Samples were evaluated by an eight-member trained panel using a list of descriptors. Off-odours and off-flavours characterization was also performed. WBSF data were analysed by a one-way model with diet as fixed effect, whereas sensory data by a mixed model with diet and panellist as fixed and random effect, respectively. A Chi^2 test was performed on off-odours and off-flavours characterization. The dietary treatments did not affect neither the WBSF of quail breasts, nor their overall sensory profile. The only exceptions regarded two textural attributes: juiciness which was significantly higher (P<0.001) in C (5.90) and HI1 (4.90) groups compared to HI2 (4.53), and fibrousness which was the highest (P<0.001) in HI1 (5.08) and HI2 (5.28) quail breasts compared to C breasts (4.79). HI larvae rearing substrate containing only layers mash or additional fish waste did not influence the sensory characteristics of quail meat. The research was supported by Padova University (Italy): funds ex 60%, 2016 DOR1603318, Senior Researcher Scholarship (Prot. n. 1098, 2015).

Effect of the inclusion of *Calliphora* sp.-derived flour in the diet of organic slow-growing chicken

J.A. Abecia[1], C. Palacios[2], I. Revilla[2] and A. Sarmiento[2]
[1]University of Zaragoza, Animal Production, Miguel Servet, 177, 50013, Sri Lanka, [2]University of Salamanca, Costruction and Agronomy, Avda. Filiberto Villalobos, 119, 37007, Spain; alf@unizar.es

Organic farms cannot always ensure that the ideal temperature recommendations for broiler breeding are met scrupulously during the first week of life. On the other hand, new alternative protein sources are explored for feeding livestock and poultry. In fact, we have previously tested the inclusion of an insect-derived flour in the diet of organic chickens during the starting period, obtaining a significant increment of daily growth rate (DGW) and live weight (LW) when 10-15% of flour was added. The aim of this work was to determine the effect of the inclusion of an insect-derived flour in the diet of organic chicken subjected to lower temperatures than the usually recommended during the first week of life. Slow-growing broilers (RedBro) were bred under organic farming conditions and subjected to 25 °C during the first week of life. Four groups (n=12 per group) were compared, according their diets during the first 30 days of life: fed 100% commercial organic chicken mixture (control, C); fed 95% chicken mixture+5% *Calliphora* sp.-derived flour (F5); fed 90%+10% flour (F10); or fed 85%+15% flour (F15). The animals were weighed weekly from the first day of life. A trend to significant differences (P=0.06) between mean LW (±S.E.M.) of the F15 (222±0.02 g) and C (145±0.012 g) groups was observed (P=0.06). DGR between day 7 and 14 was higher (P<0.05) in the F15 group (17.78±0.002 g/d) than in the C group (7.02±0.002 g/d). Mortality rates at day 30 were high (C:59.3; F5:33.3%; F10:25.0%; F15: 30.8) with no differences among groups. In conclusion, housing temperature of chickens during the first week of life is crucial for their survival. The addition of an insect-derived flour in the diet only achieved a higher daily growth rate during the second week of life, so that it was unable to compensate the low temperatures during the first week.

Insects for animal feed – a project for Poland

T. Bakuła, Ł. Zielonka, K. Obremski and J. Kisielewska
University of Warmia and Mazury in Olsztyn, Faculty of Veterinary Medicine, Department of Veterinary Prevention and Feed Hygiene, Oczapowskiego 13, 10-718, Poland; lukaszz@uwm.edu.pl

At present in Poland we do not have any traditions in the consumption of insects, we do not have national available technologies for their breeding and processing. However, we see opportunities and the need to develop and implement technology for the production of processed insect protein for animal feeding. We started a project funded by the National Centre for Research and Development aimed at developing a strategy for the development of insect protein production. The project will last until the end of 2021. We chose two species of insects considered to be the most suitable for small and large-scale breeding in the local climatic and environmental conditions of Poland, it is black soldier fly (BSF; *Hermetia illucens*) and mealworms. (*Tenebrio molitor*). Laboratory analyses of nutrients from selected insect species, including proteins and fats and other desirable and undesirable ingredients will be performed. Feeding trials involving poultry will be conducted on the laboratory and industrial scale. Social surveys on attitudes towards replacing genetically modified protein with insect protein in animal feed will be performed with the use of innovative modelling methods. Various groups of food consumers will be surveyed. In the last stage of the project, a strategy for industrial-scale insect farming and insect protein utilization in animal nutrition will be developed. The principles of insect breeding and farming will be established taking into account quality control and assurance (HACCP). Commercial insect breeding can be combined with environmental protection because food-processing wastes and by-products can be effectively managed and utilized in all stages of the production process. The strategy will also promote new industrial and business activities among both individual farmers and animal feed mills, thus creating new job opportunities. In the next few years, the implementation of the novel technology for processing insect protein may exert a beneficial influence on the feed protein balance in Poland.

Effect of heating and high-pressure processing on the potential of black soldier fly larvae as feed

J. Ortuño[1], M. Campbell[1], A. Stratakos[2], M. Linton[2], N. Corcionivoschi[2] and K. Theodoridou[1]
[1]Queen's University, Institute for the Global Food Security, University Road, BT7 1NN Belfast, United Kingdom, [2]Agri-Food and Biosciences Institute, Newforge Lane, BT9 5PX Belfast, United Kingdom; j.ortunocasanova@qub.ac.uk

The EU is moving towards considering insects, including black soldier fly larvae (BSFL), as a source of alternative protein for animal feed. However, more studies are needed to explore effective processing methods to certify insect feed as a safe and suitable ingredient in animal diets. The study assessed the effect of heating and high-pressure processing (HPP) on the hygienisation of the whole BSFL, the *in vitro* true dry matter digestibility (IVTDMD) in ruminants and monogastrics, and the relationship with the chitin content. BSFL reared on by-products from the brewer's industry were supplied by Hexafly. BSFL were sacrificed by freezing at -18 °C after 24 h-fasting. BSFL samples (n=9) were thawed, vacuum packed, and 7 different treatments were applied combining heating (90C) or HPP (400 and 600 MPa) at different times. Microbiological analyses included Total Viable Counts, Lactic acid bacteria, Enterobacteria, yeasts and moulds. The chitin content was estimated by the difference between ADF and ADL. The IVTDMD for ruminants was determined by incubating the larvae in rumen fluid followed by NDF digestion in a Fibre Analyser; for monogastrics, was established by simulating the 3 phases of their *in vivo* digestion. The MIXED procedure of SAS was used for comparing treatments means. Pearson's correlations were applied to establish the relationship between IVTDMD and chitin. Both heating and HPP reduced the BSFL microbial load of all the species evaluated in comparison with control samples. However, heating showed a higher antimicrobial effect than HPP, and, within the latters, the mildest one (400 MPa × 1.5 min) was less effective. Then, heating for 10 min was the most time-efficient treatment for a suitable hygienisation of the whole BSFL. The chitin was unaffected by heating but was increased after HPP. The IVTDMD was positively and negatively correlated with the chitin, for ruminants and monogastrics, respectively. These findings suggest a potential increase of the BSFL chitin content during HPP. Further studies are required to clarify this issue.

Using insect meal in the diet of Atlantic salmon decreases accumulation of arsenic in the fillet

I. Biancarosa[1], V. Sele[1], I. Belghit[1], R. Ørnsrud[1], E.-J. Lock[1] and H. Amlund[2]
[1]Institute of Marine Research, Postboks 1870 Nordnes, 5817 Bergen, Norway, [2]Technical University of Denmark, National Food Institute, Kemitorvet, Bygning 202, 2800 Kgs. Lyngby, Denmark; ibi@hi.no

In the European Union (EU), the use of insect meal (IM) and insect oil is permitted in feed for fish. However, research on feed and food safety of using insect-based ingredients in aquaculture is scant. Arsenic is a contaminant of major concern, as it can be very toxic to living organisms. In the EU, maximum levels (MLs) are set for arsenic in feed materials and complete animal feed (EC Directive 2002/32 and amendments). Arsenic is present in different chemical forms, or arsenic species, with differences in bioavailability and toxicity. In general, organoarsenic species are less toxic than inorganic arsenic forms. The main sources of arsenic in fish feed are marine ingredients, mainly fish meal (FM) and fish oil. The dominant form of arsenic in FM is arsenobetaine (AB), whereas arsenolipids are the major forms in fish oils. To our knowledge, there is no data on arsenic species in IM. In this study, seawater Atlantic salmon (*Salmo salar*) were fed diets where FM was replaced with 33, 66 and 100% IM. The IM was from black soldier fly larvae (*Hermetia illucens*) fed on media containing 60% seaweed. Seaweed can contain high levels of arsenic. Our main aim was to assess the potential transfer of arsenic from feed to fish. Speciation analyses were also performed to determine the major arsenic species present in IM, the diets and in the fish fillets. The concentrations of total arsenic in the four diets were similar and all below the EU ML for complete feed. Speciation analysis of the IM showed that AB accounted for only 1.2% of total arsenic, while a major fraction of arsenic in the IM were unidentified species. Transfer of arsenic from feed to fillet occurred, although there was a clear reduction of arsenic concentrations in the fillet of Atlantic salmon fed diets with increasing inclusion of IM. We assume that the IM in the diets contributes with arsenic species which are not as readily available and/or not accumulating in the muscle tissue of the fish.

Comparison of the process parameters during pelleting in different feed formulation

V. Böschen
Forschungsinstitut Futtermitteltechnik, Frickenmuehle 1A, 38110 Braunschweig, Germany; iff@iff-braunschweig.de

Insects are in the focus of feed and food production as sustainable protein sources. With respect to the industrial scale production, the yellow mealworm is a suitable option. Pelleting is an important process in feed industry and more and more in aquaculture industry, too. A lot of parameters influence the pelleting process, e.g. raw material composition, meaning raw fibre, raw protein, fat or starch content, as well as the kind of starch or technological parameters like pellet diameter, the length of the press channel or the gap between roller and die. At present, little is known about the processing features or the pellet quality of feed formulations including insect meal. Such information is crucial for adaption of techniques with respect to similar pellet quality with changing formulation. Therefore, the aim of the investigations is to compare the impact of increasing content of insect meal on pellet characteristics without changing nutrition properties of the whole formulation. The different characteristics of these pellets for fish and pork feed are presented. The IGF Project 28 LN of the International Research Association Feed Technology e.V. (IFF) was funded by the Federal Ministry of Economics and Technology via the AiF as part of the program for the promotion of industrial joint research and development (IGF) on the basis of a resolution of the German parliament.

Welcome and opening

J. Fink-Gremmels[1] and J. Seifert[2]
[1]Utrecht University, P.O. Box 00152, 2584 CM Utrecht, the Netherlands, [2]FEFANA, Rue de Treves 45, 1040 Brussels, Belgium; j.fink@uu.nl

Facing the global challenges associated with an increasing world population and the growing demand for animal proteins, climate changes and declining resources, and last but not least the concerns of consumers about antimicrobial resistance and animal health and wellbeing, the feed industry strives to offer concepts and novel approaches to improve feed efficacy, reduce the environmental impact of the livestock industry, and to increase the resilience of animals to infectious diseases and physiological stressors of daily life. Therefore, FEFANA, the EU Association of Specialty Feed Ingredients and their Mixtures, took the initiative to organize a one-day symposium with the aim to provide a platform to address and discuss these challenges. Invited key note lectures will summarize recent insight in crop production and the central role of the soil microbiome in healthy plant production, strategies to improve animal resilience reducing the need to use antibiotics by means of speciality feed ingredients, and a meta-analysis assessing the impact of potential dietary feed additives. Along with the key note lectures individual short communication, some of which are already presenting success stories, will complete the insight in new developments and solutions. The day-programme includes two round table discussions, allowing a fruitful interaction between the speakers from academia and industry with the audience.

A healthy plant microbiome for healthy food and feed

L.S. Van Overbeek
WUR, Droevendaalsesteeg 1, 6708 PD Wageningen, the Netherlands; l.s.vanoverbeek@wur.nl

Plants live in association with micro-organisms. The combination of micro-organisms and plants living together as one single living unit is called the 'plant microbiome'. From the plant microbiome perspective, plants must be considered as 'super organisms', where micro-organisms contribute to plant health, growth (yield) stimulation and nutrient acquisition, but also can cause food devastation, either by destroying plants (plant pathogens), or by contaminating food and feed materials derived from arable plants (food spoiling and human pathogenic micro-organisms). A positive balance between beneficial micro-organisms on the one hand side and plant pathogenic and food spoiling micro-organisms on the other side in the plant microbiome is essential for safe and sustained food and feed production. The concept of 'microbiomes' is new and would provide basic knowledge to existing measures in plant production, such as breeding and agronomic practices. In fact breeding and agronomic practices are influencing plant microbiome composition, but our current knowledge on that aspect is still at an immature stage. Technical innovations in the area of high throughput DNA, protein and metabolic compound analyses revolutionize our basic knowledge in this area, leading to new insights and concepts in plant microbiome functioning. In this lecture I will give an overview on the concepts and technical innovations in plant microbiome research. The impact of plant treatments, e.g. with plant-beneficial microbes and/ or with chemical compounds or extracts derived from compost, plants and algae will be discussed. The interaction of these plant additives with the physiological and metabolic status of plants and microbiome composition and functioning will be discussed. For example, it is a well-known fact that micro-organisms, chemical compounds and extracts can stimulate plant hormone pathways, and on their turn, plant hormones stimulate plant resilience against biotic and abiotic stresses. The interaction between plant hormone-stimulated pathways and plant-associated micro-organisms in arable plants is still in an exploratory stage. The role of the microbiome in stimulation of plant resilience is a topic that is currently under investigation within my research group.

Circadian variation in methane emission by sheep fed ryegrass-based pasture

A.A. Biswas and A. Jonker
AgResearch Limited, Grassland Research Centre, Animal Nutrition and Physiology Group, Tennent Drive. 11 Dairy Farm Road, Palmerston North-4442., New Zealand; ashraf.biswas@agresearch.co.nz

A large body of information is available on daily enteric methane (CH_4) emissions from sheep, but little information is available on circadian emission profiles, which is important information for developing successful measurement protocols for alternative rapid CH_4 spot-sampling methods. The objective of the current study was to determine the circadian variation in CH_4 emission from sheep fed ryegrass-based pasture. Data of nine respiration chamber experiments with growing Romney wether or ewe lambs or yearlings offered cut pasture twice daily were used for the current analysis. In all experiments, CH_4 emissions were measured within animal at approximately every 6 minutes over two consecutive days and each data point was expressed as g/d before summary statistics and correlations were analysed using R version 3.4.2. Among the 9 trials, minimum CH_4 production rate ranged from 8.7±0.10 g/d to 17.3±0.18 g/d and maximum CH_4 production rate ranged from 20.0±0.17 g/d to 26.2±0.19 g/d. Maximum CH_4 emission rate was reached at approximately 2 hours after each morning and afternoon feeding and minimum CH_4 emission rate in the day occurred before morning feeding. The absolute circadian variation indicated by the ratio of maximum to minimum CH_4 production rate ranged from 1.6±0.02-fold to 2.6±0.04-fold among the 9 trials. Dry matter intake (kg/d) correlated negatively with the maximum/minimum CH_4 ratio (R^2= -0.31). When the CH_4 emission rates over 24 h were expressed as a daily mean, the actual CH_4 emission rates were within 15% of the mean for 16 to 21 hours. In summary, the magnitude of circadian variation in CH_4 emissions in sheep fed ryegrass-based pasture decreased with increasing dry matter intake.

Round table: Strategies to reduce the environmental impact of animal farming

L.S. Van Overbeek
Wageningen University Research, Wageningen Plant Research, Droevendaalsesteeg 1, 6708 PB Wageningen, the Netherlands; l.s.vanoverbeek@wur.nl

Round table with L.S. van Overbeek, A. Burel, F. Garcia-Launey, A.A. Biswas, X.W. Zhou and J. Fink-Gremmels. Societal concerns related to the livestock sector address environmental sustainability, the efficiency of crop production and the use of biocidal agents in agricultural practice. Moreover, the contribution of the livestock sector to the production of green-house gasses and its contribution to climate changes is controversially discussed. This Round Table brings together scientists from different disciplines to identify tools and solutions from the field, which can reduce the environmental impact of animal farming while assuring feed safety and security.

Role of specialty feed ingredients in tackling societal challenges
S. Deschandelliers and P. Doncecchi
Adisseo France, R&D, 10, place Général de Gaulle, 92160 Antony, France; paolo.doncecchi@adisseo.com

Providing Food and nutrition security is becoming more complex with respect to the long-term environmental sustainability, especially taking into account the interconnected challenges of natural resources scarcity, climate change and population growth, as well as slow motion emerging threats as antimicrobial resistance. Even if the use of antimicrobials for growth promotion has significantly declined, still key antimicrobials continue to be used routinely in several regions for this purpose. Today it is acknowledged that animal nutrition plays an important role to increase livestock production efficiency while minimizing its environmental impact, and to keep animal healthy and feeling well. Specialty Feed Ingredients, including feed additives, have a key role to play in the fight against Antimicrobial Resistance as part of an integrated multi-stakeholder approach for developing innovative solutions. Indeed, our ingredients are really playing at the edge of nutrition and health, helping animal health by strengthening the animals' intestinal tract and therefore increasing their resilience or immune competence to stressors. E.g. through the maintenance of a proper balance in the gut microbiome limiting the development and action of potential pathogens in the gut, or the increase of the capacity of the body to repair from mucosa lesions, and as well limiting the risk of gut inflammation by improvement of feed digestibility. Innovation is key for development of solutions, and we think that we need a revised enabling global policy environment for innovation, allowing new ingredients to translate from R&D to delivery into use. For that, communication and collaboration along the whole value chain is critical, to raise awareness of the benefits of new technologies and ingredient, to fill information gaps, and highlight the common challenges of the sector.

The effect of two probiotics on the prevalence of antimicrobial resistance of *E. coli* in broilers
A. Lee[1], M. Aldeieg[2], M. Woodward[2] and C. Rymer[1]
[1]University of Reading, Agriculture, Policy and Development, Reading, RG6 6AR, United Kingdom, [2]University of Reading, Food and Nutritional Sciences, Reading, RG6 6AP, United Kingdom; ahran.lee@pgr.reading.ac.uk

This study was undertaken to determine the effect of a yeast (*Candida famata*) and a bacterium (*Lactobacillus plantarum*), administered alone or in combination in the drinking water, on the population of *Lactobacillus* sp. and coliforms, and of the prevalence of antimicrobial resistance (AMR) in *Escherichia coli* isolated from caecum taken throughout the life of broiler chickens. Male (Ross 308) day-old chicks (220) were used. *C. famata* (isolated from a chicken) and *L. plantarum* (isolated from a pig) was administered via the drinking water. Water was provided either untreated (CON) or with *C. famata* (CF; 10^8/ml), *L. plantarum* (LP; 10^5-10^8/ml), or a combination of CF and LP (CFLP; 10^6-10^8/ml) in water hoppers on two days each week for 35 d. Birds were sacrificed every week and samples of duodenal, ileal and caecal digesta were taken to determine the population of *Lactobacillus* sp. Coliform populations were determined in caecal digesta and a colony of *Escherichia coli* taken from each sample to determine its resistance to ampicillin (AMP), tetracycline (TET), nalidixic acid (NAL) and chloramphenicol (CAM, all 50 µg/ml). The effect of bird age, administration of CF or LP and the interaction between all main effects on the *Lactobacillus* sp. and coliform population was determined by ANOVA. Chi square analysis was used to determine the association between either bird age or treatment and AMR status. No significant interactions were observed between main effects, and neither CF nor LP had any effect on the population size of *Lactobacillus* sp. or coliforms. The population of *Lactobacillus* sp. and coliform decreased with age (P<0.001). Resistance to NAL and CAM was low (7.8 and 3.9% respectively). TET resistance was high (78%) at 8 d but then declined to 21% at 35 d (P<0.001) with no effect of treatment. AMP resistance also declined from 100% at 8 d to 38% at 35 d (P<0.001) with a tendency (P=0.097) for CF to reduce it from 58% (CON) to 32% (CF). AMP and CAM resistance is overcome to some extent as birds age, and the administration of CF may further reduce the prevalence of AMR.

Effect of riboflavin dosage and source on growth, slaughter traits and welfare of broilers

C. Lambertz[1], J. Leopold[1], K. Damme[2], W. Vogt-Kaute[3], S. Ammer[4] and F. Leiber[4]
[1]Research Institute of Organic Agriculture (FiBL), Kasseler Strasse 1a, 60486 Frankfurt am Main, Germany, [2]Poultry Competence Centre of the Bavarian Institute for Agriculture, Mainbernheimer Str. 101, 97318 Kitzingen, Germany, [3]Naturland e.V., Kleinhaderner Weg 1, 82166 Gräfeling, Germany, [4]Research Institute of Organic Agriculture (FiBL), Ackerstrasse 113, 5070 Frick, Switzerland; christian.lambertz@fibl.org

The availability of GMO-free riboflavin (vitamin B_2) is a major issue in organic production. Effects of riboflavin-enriched feed produced with the unmodified yeast *Ashbya gossypii* by fermentation as an alternative to conventional (GMO-based) riboflavin on performance traits and welfare indicators were investigated at graded dosages. In a 4×2 factorial design, 1,600 slow-growing broilers were fed four dietary treatments in two runs. Starter diets included a basal diet: (1) without any riboflavin supplementation (N-C); (2) with conventional riboflavin supplementation at 9.6 mg/kg (P C); (3) with riboflavin supplementation from the non-GMO source at 3.5 mg/kg (A-low); and (4) with riboflavin supplementation from the non-GMO source at 9.6 mg/kg (A-high). For the finisher diet, P-C and A-high were supplemented with 8.0 mg/kg and A-low with 3.5 mg/kg. Diets were formulated according to organic regulations. Broilers were kept in pens of 20 animals in floor husbandry. Body weight, feed and water consumption were recorded weekly, welfare indicators bi-weekly. Slaughter traits were assessed for five males and females per pen at 62/63 days of age. Final body weight of A-high differed from N-C and A-low, but not from P-C. From week two until six of age, A-high had a higher daily weight gain and dressing percentage (74.4%) was better in A-high compared with all other groups (73.3%). Breast percentage of A-low was lower than that of both control groups but did not differ from A-high. The highest frequency of scores indicating fatty liver syndrome was found in P-C, followed by N-C and A-low. A score indicating light footpad dermatitis occurred most frequently in A-high, however at a low prevalence. In conclusion, the tested riboflavin-enriched feed derived from fermentation of *A. gossypii* can be used at levels of commercial recommendations as alternative to riboflavin produced from GMO.

Mechanisms involved in heat stress response and characterization of intervention strategies

S. Varasteh, S. Braber and J. Fink-Gremmels
Utrecht University, Faculty of Veterinary Medicine, Institute for Risk Assessment Sciences (IRAS) and Institute for Pharmaceutifal Sciences (UIPS), Yalelaan 104, 3584 CM Utrecht, the Netherlands; j.fink@uu.nl

Under current predictions of a climate change, it is assumed that the intensity and frequency of heat stress (HS)-related disorders are increasing in humans and animals. One of the primary organs which is affected by HS is the gastrointestinal tract, where splanchnic ischemia is leading ultimately to a 'leaky gut syndrome. In a series of experiments, various *in vitro* models were validated, using human intestinal epithelial cells (Caco-2 cells) as the best characterized intestinal epithelial cells. Non-permeable, confluent monolayer of Caco-2 cells were exposed to different conditions of hyperthermia to evoke HS. Validation parameters included the expression of heat shock proteins and markers of oxidative stress and inflammation. Transepithelial resistance, as an indicator of barrier integrity, the paracellular transport of marker substances, as indicators of a disrupted barrier, and the expression of tight junction proteins were measured. Subsequently these models were used to test various compounds, which might be given as feed additives, and which were selected according to their different mechanisms of action. The list of products included: (1) galacto-oligosaccharides, known to stabilize intestinal integrity by supporting the expression and re-assembly of claudins and occludins; (2) the amino acid L-arginine assumed to protect epithelial integrity by stabilizing the HS-induced reduction of the nitric oxide synthesis; and (3) α-lipoic acid, recognized for its anti-inflammatory properties and positive effect on cell proliferation. Results showed that all three test compounds conveyed a certain degree of protection to epithelial cells and stabilized the intestinal integrity. These measurable effects were dose-dependent and in some cases, such as for L-arginine, a clear dose optimum was observed, with adverse effects at higher concentrations. It could be concluded that the established experimental models are a very useful tool for the rapid identification of potential protective compounds. However, the experiments also demonstrated that it remains necessary to establish an optimal *in vivo* dosing regimen as demonstrated in experiments with chickens.

Relationships between heavy metals in milk, soil and water from agricultural and industrial areas

X.W. Zhou[1,2], H. Soyeurt[2], N. Zheng[1], C.Y. Su[1] and J.Q. Wang[1]
[1]Chinese Academy of Agricultural Sciences, Institute of Animal Sciences, No. 2 Yuanmingyuan West Road, Beijing, China, P.R., [2]Gembloux Agro-Bio Tech/University of Liège, Passage des déportés 2, 5030 Gembloux, Belgium; xuewei.zhou@student.uliege.be

Environmental pollution from various industrial activities may result in the presence of heavy metals in the diet of dairy cows and their potential transfer into milk. This study investigated the relationship between the heavy metals content in the milk of individual cows and the concentrations in soil and water from industrial and agricultural areas. Sixty milk samples (10 per farm) were collected from the udder during milking procedure. Six underground water and six soil samples were collected simultaneously. Concentrations of Pb, As, Cr and Cd in raw milk and water samples were measured by ICP-MS. Lead, Cr and Cd was measured by AAS in soil samples, and As was detected by AFS. Kruskal-Wallis tests were performed to study the area effect. Spearman correlation was calculated between Pb, As, Cr and Cd contents in milk with those in water and soil. Ranges of Pb, As, Cr and Cd in milk were 0.025-10.45, 0.002-1.53, 0.017-5.01 and 0.006-0.27 µg/l. Levels of Pb and Cd in milk from agricultural area were significant lower (P<0.01) than that from industrial area. Significantly (P<0.01) higher As residues were recorded in milk from the agricultural area. There was no difference between areas for Cd.. Lead, As and Cd from the farm in agricultural area showed difference with other farms, while no difference existed for Cr. Weak positive correlations for the levels of Pb (r=0.03) and As (r=0.37) were found between milk and water. Strong correlations were found for Cr (r=0.60) and Cd (r=0.65) between milk and soil. The results suggested that Pb and As were not mainly introduced into milk from drinking water. Contents of Cr and Cd in milk may be derived from soil. Negative correlations between milk and water were found for Cr and Cd. Lead and As levels in milk showed a negative correlation soil concentrations. The results indicated that heavy metal residues in milk have a complex origin, and water and soil are not the only contributors to this contamination.

Effect of a natural repellent on dust mite

A. Burel[1], R. Dao[1], X.N. Nguyen[2], P. Coquelin[2] and P. Chicoteau[1]
[1]Nor-Feed, 3 rue amedeo Avogadro, 49070 Beaucouze, France, [2]Nor Feed Vietnam, 34 Hoang Viet Str., Tan Binh District, HCMC, Vietnam, Viet Nam; anne.burel@norfeed.net

Mites are common pests of stored cereals and oilseeds (Collins, 2005). The most abundant and frequently reported mites are *Acarus siro* and *Tyrophagus putrescentiae*. Heavy mite infestations can taint grain, making it unsuitable for milling and unpalatable to livestock. Infestation can also reduce zootechnical performances such as growth rates. Moreover, mites are a source of clinically important allergens and cause diseases like asthma, dermatitis and allergic rhinitis. Initially, a natural in-feed ingredient composed of essential oils from lemongrass (*Cymbopogon nardus*), clove (*Eugenia caryophyllus*), oregano (*Oreganum vulgare*) and citrus extract, was developed to be a repellent against red-mites, a poultry ectoparasite. In order to develop a new application, this natural repellent was tested, on stored feed, infested with dust mites, to evaluate its potential repellent effect on this kind of arthropods. To this end, two trials were conducted in our lab, with respectively 9 and 10 Petri dishes. Each Petri dish was divided in two zones. Infested feed (coming from the field, mites could be *A. siro* or *T. putrescentia* according to Kaur & Dhingra) was dropped on the control zone and infested feed supplemented with the natural repellent at 500 ppm was dropped on the natural repellent zone. At the end of a 24 h period, 5 counts per zone and per dish were done, these counts were summed to have the total number of mites per dish. Data were analysed by Shapiro-Wilk test to check the normality. A Khi^2 test was done on every dish box. A significant (P<0.001) lower number of mites were observed in the natural repellent zone compared to the control zone. On both trials we observed around 70% less of mites on the natural repellent zone. In conclusion, a high repellent effect of Nor-Mite on mites usually found in stored feed was observed when Nor-Mite was added to the infested feed. The observations will be completed with a trial conducted under common storage conditions to confirm these promising first results.

Multiobjective feed formulation for pig: methodological approach and application

F. Garcia-Launay[1], C. Crolard[1], E. Teisseire[2], M. Davoudkhani[1] and J. Aubin[2]
[1]PEGASE, INRA, Agrocampus Ouest, 35590 Saint-Gilles, France, [2]SAS, Agrocampus Ouest, INRA, 35042 Rennes, France; mohsen.davoudkhani@inra.fr

Animal production is responsible for several environmental impacts, to which feed production has usually the major contribution. However, the traditional least-cost feed formulation (LCF) method minimizes the cost without consideration of its environmental impacts. Multi-objective feed formulation has already been proposed to reduce the environmental impacts of pig feeds. It includes an objective function which is a weighted sum (WS) of the normalised values of feed cost, and four environmental impacts of the feed (climate change, land occupation, phosphorus demand and cumulated energy demand) calculated by Life Cycle Assessment. Normalised values are calculated dividing them by their reference value (REF) obtained with LCF (method Norm1). An additional factor α, ranging from 0 to 1 to explore the space of optimal solutions, is then used to weight the relative influence of feed cost and environmental impacts. The aim of this study was to explore potential improvements of the multi-objective method, with applications to pig feeds. We compared WS method with the ε-constraint method, which consists in minimizing a single objective while setting a maximum constraint for another one. In our case, the objective function was the weighted sum of the environmental impacts and the constraint was applied on feed cost. Then, we compared the behaviour of the model with Norm1 and with a new normalization method (Norm2), which subtracts the minimum criterion value to the criterion calculated and divides it by the difference between the REF and the minimum criterion values. WS and ε-constraint methods produced some common optimal solutions but only ε-constraint method allowed to obtain all the solutions of the Pareto front of the problem. Norm2 allowed defining solutions, which reduce consistently all environmental impacts whereas Norm1 allowed compensations between the impacts, which can potentially lead to increase of some impacts. Multi-objective feed formulation method can be substantially improved by implementing Norm2 to avoid possible compensations between impacts, and by applying the ε-constraint method to have access to the whole set of feasible optimal solutions.

Will the real Pietrain please stand up? Genomics reveal Pietrain pig diversity and substructure

W. Gorssen[1], N. Buys[1], R. Meyermans[1], J. Depuydt[2] and S. Janssens[1]
[1]Livestock Genetics, Department of Biosystems, KU Leuven, Kasteelpark Arenberg 30 – bus 2472, 3001 Leuven, Belgium, [2]Vlaamse Piétrain Fokkerij vzw, Deinse Horsweg 1, 9031 Drongen, Belgium; wim.gorssen@kuleuven.be

The Pietrain is a black-spotted pig breed named after Pietrain, a Belgian village where it originates between 1920-1950. The superior conformation and the spread of artificial insemination made the Pietrain develop into Belgiums' most popular terminal sire line. Driven by this success, the breed quickly conquered Europe and also spread worldwide. Due to the narrow genetic origin and subsequent selective breeding, genetic diversity was questioned. Although Pietrain pigs are bred worldwide in closed populations, the Pietrain is regarded as one breed. We questioned to what extent these populations have genetically diverged in time, giving rise to subpopulations. Medium density genotypes were available from 1,771 Pietrain pigs originating from 8 populations worldwide. Analyses were performed via PLINK 1.9, Admixture and the *RzooRoH* R-package. Runs of homozygosity (ROH) analysis showed that on average 26% of the Pietrain genome was identified as a ROH. Most inbreeding (18%) originated from events more than 8 generations ago. For the most recent 8 generations, inbreeding was clearly lowest in the Belgian (5.9%) and highest in the USA (10.2%) population. ROH islands were identified on SSC8. At this region more than 90% of all Pietrain pigs had ROH. This selection signature is most likely caused by close inbreeding during breed formation, because populations became rather independent from the 1970s on. Principal component analysis revealed three clusters, wherein the Belgian and a French population clearly deviated from other Pietrain populations. Admixture analysis confirmed the existence of three genetically distinct subgroups. Our results show a clear historical pattern of selection on SSC8 in Pietrain pigs. Nevertheless, independent breeding has resulted in at least three genetically distinct subgroups. These findings raise the question if we can still consider the Pietrain as one unique breed.

Genome-wide association study (GWAS) for fat depth and muscle depth in commercial crossbred Piétrain

M. Heidaritabar[1], M. Bink[2], A. Huisman[3], P. Charagu[4], M.S. Lopes[5] and G. Plastow[1]
[1]Livestock Gentec, University of Alberta, Edmonton, Canada, [2]Hendrix Genetics, Research, Technology & Services B.V, Boxmeer, the Netherlands, [3]Hendrix Genetics, Swine Business Unit, Boxmeer, the Netherlands, [4]Hendrix Genetics, Swine Business Unit, Regina, Canada, [5]Topigs Norsvin, Research Center, Beuningen, the Netherlands; marzieh.heidaritabar@mbg.au.dk

The popularity of Piétrain pigs, originating from Belgium, is because of their exceptional muscularity and leanness. Fat depth and muscle depth are economically important traits and good indicators of carcass lean content, which is one of the main breeding objectives in pig breeding programs. Our aim was to detect genomic regions associated with these two traits in a sample of 1,849 commercial crossbred Piétrain pigs genotyped with a custom 50k Illumina SNP panel. After quality control, 44k SNPs remained for further analyses. We performed a standard genome-wide association study (GWAS) using single-marker association analysis with a false discovery rate of 0.05. Phenotypic records were pre-corrected for sex as fixed effect and the random effect of litter. In total, 22 significant SNPs in three potential regions (two on SSC1 at 55-57.6 Mbp and 154.7-159.5 Mbp; and one on SSC18 at 10.2-10.7 Mbp) associated with fat depth and 18 SNPs in one potential region (on SSC17 at 13.7-18.3) associated with muscle depth were detected. The associated SNPs are located in or near several potential candidate genes. Among these genes, *ANK1*, *CDH20* and *MC4R* have been previously reported as being associated with fat deposition traits. Our current findings provide insights into the genomic regions influencing fat and muscle depth in crossbred pigs. We are also investigating the importance of dominance in the Piétrain crossbred population, as non-additive genetic effects are of particular importance in crossbred populations. GWAS to investigate the role of dominance effects in the genetic architecture of fat depth and muscle depth is underway. Evaluation of dominance effects for the associated regions may provide a better understanding of the complex genetic architecture of leanness in crossbred pigs.

Genetic analyses of boar taint and reproduction traits in Landrace and Large White

I. Brinke[1], K. Roth[1], M.J. Pröll-Cornelissen[1], C. Große-Brinkhaus[1,2], I. Schiefler[2] and E. Tholen[1]
[1]University of Bonn, Institute of Animal Science, Endenicher Allee 15, 53115 Bonn, Germany, [2]Association for Bioeconomy Research (FBF e.V.), Adenauerallee 174, 53113 Bonn, Germany; ibri@itw.uni-bonn.de

Although the decision about the ban of surgical castration without anaesthesia is extended for two more years in Germany, discussion about suitable alternatives has to be more intensified. Fattening of entire young boars is the most suitable alternative when it comes to the integrity of the animals, which is a crucial aspect for consumer acceptance. Possible antagonistic relationships between boar taint compounds androstenone (AND) and skatole (SKA) and reproduction, caused by common synthesis pathway, can lead to a relevant risk if breeding against boar taint. To reveal unfavourable relations between AND, SKA and several maternal/paternal reproduction traits, data from Landrace (LR) and Large White (LW) population was used for a multivariate variance component estimation with a polygenetic model. Heritabilities (h^2) and genetic correlations (r_g) showed breed specific differences. Whereas the r_g of -0.18 between SKA and age at first insemination (AFI) in LW showed an unfavourable relationship between boar taint and reproduction traits, SKA and AFI in LR tend to be favourable correlated (r_g=0.35). Additionally, all animals were genotyped to perform a genome wide association study (GWAS) for AND, SKA and reproduction traits. Number of significant markers with possible pleiotropic effects differed between trait and line. One exception was observed for SKA on SSC 14 in the gene CYP2E1. In addition, different scenarios to evaluate the accuracy of genomic selection (GS) strategies were performed. First results indicate moderate accuracies in a range between 0.55 to 0.83 applying a 5-fold cross validation for AND and SKA. As shown in variance component estimation, the antagonistic relationships between boar taint and reproduction traits have to be considered. Results from GWAS and GS confirmed the need for breed specific analysis. Further steps include the addition of hormone profiles to the multi-trait analyses to reveal genes with possible pleiotropic effects on boar taint and reproduction traits to exclude them from further GS.

Genetic parameters for reproductive and longevity traits in Bísaro pigs

G. Paixão[1], A. Martins[1], A. Esteves[1], R. Payan-Carreira[1,2] and N. Carolino[2,3]
[1]Universidade de Trás-os-Montes e Alto Douro, CECAV, Vila Real, 5000-801, Portugal, [2]Universidade de Évora, Departamento de Medicina Veterinária, Évora, 7002-554, Portugal, [3]INIAV, Strategic Unit for Biotechnology and Genetic Resources, Vale de Santarém, 2005-048, Portugal; gus.paixao@utad.pt

The Bísaro pig has gained popularity in recent years reflecting the success of the conservation program. Nevertheless, no data is available for animal genetic evaluation in this breed. Therefore, this study aimed to estimate genetic parameters and trends for reproduction related traits in Bísaro pigs. Through a REML procedure applied to mixed linear models, 27,844 farrowing records, from 1995 to 2017, were used to analyse total number of pigs born per litter (NBT), number of pigs born alive (NBA), number of stillborn (NSB), number of pigs weaned per litter (NBW), age at first farrowing (AFF), farrowing interval (FIT), length of productive life (LPL), lifetime number of litters (LNL), lifetime pig production (LTP) and lifetime efficiency (LTP365). The heritability estimates for litter size traits were low and ranged from 0.007±0.004 to 0.015±0.006. Differently, the heritabilities for traits related to longevity and lifetime production traits were moderate (0.078±0.026 to 0.121±0.030). AFF registered the highest heritability value (0.345±0.028). NSB and FIT presented high values of additive genetic coefficient of variation (0.177 and 0.271) in contrast with low heritability estimates (0.007±0.004 and 0.002±0.005). Very tight genetic correlations were found between NBT and NBA (0.968), NBW and NBT (0.974), and NBW and NBA (0.945). Weak genetic correlations were found between both NBT and NSB (0.352) and between NBA and NSB (0.107). Longevity and lifetime production traits presented high positive genetic correlations (0.811-0.969) and moderate to high phenotypic correlations (0.266-0.946). No major genetic changes were registered over time for most of the analysed traits, except for AFF and LPL, having registered an overall decreased of mean estimated breeding values (21.3 and 17.5) and negative genetic trends of -0.6 and -0.4 (P<0.001), respectively.

Prediction of individual breeding values from group recordings

B. Nielsen[1], B. Ask[1], H. Gao[2] and G. Su[2]
[1]SEGES, Pig Research Centre, Breeding and Genetic, Axeltorv 3, 1609 Copenhagen, Denmark, [2]Aarhus University, Center for Quantitative Genetics & Genomics, Blichers Allé 20, 8830 Tjele, Denmark; bni@seges.dk

Selection for low feed consumption per body weight gain is one of the most important traits in pig breeding programs. Feed intake in pigs is often recorded during a well-defined period of performance test. Pigs are kept in pens and electronic feeders allow for feed intake recorded on individually. However, the use of electronic feeders makes recorded feed intake recording very costly. To reduce the phenotyping costs of feed intake, feed intake recorded at pen level can be an alternative approach, however group size in practice are typically large. To investigate if group feed records of large groups (20 pigs per group) are feasible in a practical selection program, an experiment was conducted with 6,478 purebred DanBred Landrace male pigs growing from approximately 30 kg to 100 kg in 644 pens. Each of two pens was connected to one feeding unit (the two pens was referred to a 'double pen' in the context) recording the amount of feed intake of all the pigs in the two pens. In total, feed intake of 322 feed groups were recorded. All pigs were given dry pelleted feed *ad libitum*. During the test period the aggregated feed intake of all pigs in each double pen was recorded. At the start and end of the experiment, the body weight of all pigs was recorded and the aggregated weight gain of all pigs in each double pen was calculated including weight gain of the pigs that dropped out before end of test. Thereby, group records of feed intake in the 322 feed units and individual record of body weight gain were obtained. An animal model for group records was used to obtain individual breeding values for each of the 6,478 pigs. The model includes genotype information of 6,172 male pigs which cover 95% of the experimental pigs. The results showed that group records can be implemented into practical breeding programs. For the animals without individual feed record, group pooled records increases the accuracy of breeding values for feed conversion rate, however individual recorded feed intake lead to highest accuracy.

Analysis of group recorded feed intake and individual records of body weight and litter size in mink

M.D. Madsen[1], T.M. Villumsen[1], M.S. Lund[1], B.K. Hansen[2], S.H. Møller[3] and M. Shirali[1]
[1]*Aarhus University, Molecular Biology and Genetics, Blichers Alle 20, 8830 Tjele, Denmark, [2]KopenhagenFur, Langagervej 60, 2600 Glostrup, Denmark, [3]Aarhus University, Animal Science, Blichers Alle 20, 8830, Denmark; mette.madsen@mbg.au.dk*

The aims of this study were: (1) to model the genetic heterogeneity of male and female in group recorded feed intake (FI) in mink; (2) to perform combined analysis of group recorded FI with single recorded body weight (BW) and litter size (LS); and (3) to derive residual feed intake (RFI) as measure of feed efficiency separately in males and females using group records of FI. Cumulated FI, recorded during the growth period for 9,936 cages with one male and one female per cage, and BW, recorded at the end of the test period for 7,813 male and 8,140 female mink from 2013 to 2016, were available for analysis from research centre Foulum. LS records from 6,445 yearling females from 2006 to 2015 was also included. The pedigree contained 29,597 animals covering 2001 to 2016. The data was analysed in a Bayesian setting with Gibbs sampling. A multivariate model was developed with group recorded FI, modelled to obtain genetic parameters for male and female separately, BWs for male and females along with LS. The genetic parameters of RFI of male and female were obtained simultaneously by conditioning FI for BW using genetic partial regression coefficients. Results are presented as means and SD of the posterior distribution of each parameter. The heritability of RFI was 0.16 (0.02) in males and 0.06 (0.02) in females. The heritability of BW was 0.54 (0.03) in males and 0.62 (0.03) in females, and the heritability of LS was 0.12 (0.02). Genetic correlation between sexes in RFI (0.84 (0.08)) and BW (0.81 (0.03)) were both high but significantly different from unity. The genetic correlation between LS and BW in males and females were -0.48 (0.09) and -0.69 (0.08), respectively; while the genetic correlations between LS and RFI in either sex were not significantly different from zero. Here we presented a model for combined group and single recorded trait analysis that obtained genetic heterogeneity across sexes. When feed efficiency was modelled with RFI no unfavourable genetic correlation with litter size and body weight was observed.

Comparison of statistical parameters from genetic and genomic analyses of indirect genetic effects

B. Poulsen[1], B. Ask[2], T. Ostersen[2] and O.F. Christensen[1]
[1]*Center for Quantitative Genetics and Genomics, MBG, Aarhus University, Blichers Allé 20, 8830 Tjele, Denmark, [2]Danish Pig Research Centre, Axelborg, Axeltorv 3, 1609 København V, Denmark; bgp@mbg.au.dk*

It is generally acknowledged that pigs influence the phenotypes of their pen mates. The effects of interactions between pen mates are partially genetically determined. Genetic merits for own growth and growth of pen-mates are referred to as Direct Genetic Effects (DGEs) and Indirect Genetic Effects (IGEs), respectively. In theory, inclusion of IGEs in genetic evaluation and selection may improve genetic progress. However, the superiority of models with both DGEs and IGEs over models with DGEs only is inconsistent with regards to predictive performance. This may be due to sensitivity to data structure of such models, and if so then using the Genomic Relationship Matrix (GRM) instead of the Numerator Relationship Matrix (NRM) may offer a solution. However, experiences from models with DGEs indicate that the choice of relationship matrix influences parameter estimation. Therefore, the objective of this study was to compare statistical parameter estimates and predictive ability of social interaction models using either NRM or GRM by analysing average daily gains of Danbred Landrace pigs. A total of 11,437 pigs entered test. At end of test, 10,136 pigs were phenotyped and 10,998 pigs had genotype information. A Single-step Relationship Matrix (SSRM) combining pedigree and genomic information was used instead of a GRM to account for some animals not being genotyped. Model parameters were estimated using average-information restricted maximum likelihood. IGE variance components estimated with the SSRM were smaller than those estimated with NRM (21 ± 7.4 vs 87 ± 22.8). The correlation between DGEs and IGEs was -0.2 ± 0.15 when estimated with the SSRM and 0.12 ± 0.13 when estimated with NRM. The predictive performance of DGEs and IGEs increased when using SSRM over NRM as covariance structure. For SSRM, the predictive performance decreased when including IGEs in prediction. The opposite was true for NRM. Collectively, the predictive performance was best with SSRM as covariance structure and without IGEs. In conclusion, SSRM decreased both the estimated variance and predictive performance of IGEs.

Using meat inspection data to improve pig health traits by breeding

A. Horst[1], J. Krieter[1], B. Voß[2] and J. Krieter[1]
[1]Institute of Animal Breeding and Husbandry, Christian-Albrechts-University, Olshausenstr. 40, 24098 Kiel, Germany,
[2]BHZP GmbH, An der Wassermühle 8, 21368 Ellringen, Germany; ahorst@tierzucht.uni-kiel.de

The extensive data-pool of organ lesions documented during meat inspection may provide an opportunity to integrate health-related phenotypes into the breeding-program. However, several factors influence the documentation of organ lesions causing inter-abattoir and intra-abattoir variations. To consider for such factors statistically, specific requirements have to be met: first, a satisfactory data quality (amount of observations per day and farm) and second, a well-linked data structure (connection between abattoirs). The aim of this study is to analyse the data structure and to investigate what conditions are necessary to establish a valuable data-pool for breeding purposes. The study assessed animal specific data from the German breeding organization BHZP (130,569 fattening pigs, 25 farms, 7 abattoirs). To increase the amount of animals and the connection between abattoirs, the BHZP data was expanded by organ lesions documented by the German Quality and Safety database (975,374 pigs, 57 farms, 7 abattoirs). For this purpose, following criteria where determined referred to the farm: at least 40 pigs per delivery, 10 deliveries during data period and delivery to at least two abattoirs. Network analysis was performed on both data sets to investigate the data structure. A generalized linear mixed model including abattoir (AB1-AB7) and slaughter day (SY), nested in AB, was applied to prevalence of lung and liver lesions. The LSMeans (LSM) for SY vary within AB (coefficient of variation (CV) for e.g. AB2 CV-lung=79.1%, CV-liver=92.2; AB7 CV-lung=47.9%, CV-liver=59.8%) and between ABs (e.g. lung lesions: AB2 14.5%±11.5% vs AB7 17.2%±8.27%). The LSM-SY for liver lesions showed less deviation than for lung lesions (e.g. AB2 6.68%±6.16% vs 14.5%±11.4%; AB7 9.21%±5.50% vs 17.2%±8.28%). Similar characteristics were found in LSM-AB in both data sets, concluding that improving connectedness provides an opportunity to estimate abattoir-specific effects. Further research on data structure, including farm as random effect to the model and linking pedigree information to estimate breeding values is in progress to draw a definitive conclusion.

Significant dyads in agonistic interactions and their impact on centrality parameters in pigs

K. Büttner, I. Czycholl, K. Mees and J. Krieter
Institute of Animal Breeding and Husbandry, Christian-Albrechts-University, Olshausenstr. 40, 24098, Germany;
kbuettner@tierzucht.uni-kiel.de

Dominance between two animals is characterised by the consistent outcome of the agonistic interactions to the advantage of one animal. It is claimed that the significant asymmetric outcome has to be tested before further sociometric measures are calculated. Thus, two calculation methods considering pen or dyad individual limits for significant dyads were suggested (resulting in the data sets PEN and DYAD) and their impact on centralities in newly mixed weaned pigs (93 pens; 8.91±0.58 animals/pen) were compared to all dyadic interactions (data set ALL). Directly after mixing, all animals were video recorded for 3 days documenting the agonistic interactions. For all data sets, networks were built based on the information of initiator and receiver with the pigs as nodes and the edges between them illustrating attacks. Centralities were calculated describing the position of each animal in the network. To evaluate the relation between the centralities for different observation times or data sets Spearman rank correlation coefficients (rs) were calculated. The comparison between day 2 and 3 showed the highest rs values independent of the centrality and data set considered, i.e. focussing on only two days is sufficient to get reliable results. Comparing the centralities between the data sets, the rs values showed a wide variation with regard to time and centrality. The rs values between ALL and PEN showed always higher values compared to the rs values between ALL and DYAD. Here, the centralities focussing on direct and indirect outgoing agonistic interactions (initiation of fights), i.e. out-degree and outgoing closeness, showed the highest rs values (0.60-0.75). The centralities focussing on ingoing agonistic interactions (being attacked), i.e. in-degree and ingoing closeness, showed lower rs values (0.45-0.65). The lowest rs values had the betweenness which measures the extent to which an animal lies on paths between other animals (0.1-0.5). It can be concluded that focussing on significant dyads has an immense effect on the calculated centralities, whereby centralities focussing on active agonistic interactions (i.e. initiation of fights) showed the most robust results.

Meat and fat quality of pigs intended for Spanish cured ham: effect of male castration and feeding

L. Pérez-Ciria[1], F.J. Miana-Mena[1], G. Ripoll[2] and M.A. Latorre[1]
[1]IA2- Universidad de Zaragoza, C/ Miguel Servet 177, 50013, Spain, [2]IA2-CITA de Aragón, Av. Montañana 930, 50059, Spain; leticiapcgm@gmail.com

Currently, the castration is necessary in heavy male pigs such as those intended for dry-cured ham elaboration. Immunocastration could be an alternative to surgical castration, considering the animal welfare, but the maintenance of the product quality should be also guaranteed. A total of 90 Duroc × (Landrace × Large White) male pigs was used to assess the impact of the type of castration (surgical castration vs immunocastration) and of different diets on meat and fat quality of pigs intended for Teruel ham, which is a Spanish label of high quality dry-cured hams. Surgical castration was carried out at the first week of age and immunocastration consisted of three injections of Improvac® at 56, 101 and 122 days of age. The diets tested were: A=control, B=high net energy level (NE) and C=low standardized ileal digestible Lysine level (Lys SID). During the growing period (80 to 109 kg body weight-BW), the diet A contained 2,330 kcal NE/kg and 0.77% Lys SID, the diet B included 2,480 kcal NE/kg and 0.77% Lys SID and the diet C provided 2,330 kcal NE/kg and 0.67% Lys SID. During the finishing period (109 to 137 kg BW), the diet A contained 2,330 kcal NE/kg and 0.63% Lys SID, the diet B included 2,480 kcal NE/kg and 0.63% Lys SID and the diet C provided 2,330 kcal NE/kg and 0.54% Lys SID. A sample of meat from each carcass (n=15) and 48 samples of subcutaneous fat chosen at random (n=8) were taken to be analysed. Meat from immunocastrated males (ICM) showed lower intramuscular fat content and lightness, but higher moisture than that from surgical castrated males (SCM) (P<0.05). The diet C carried out the highest cooking losses (P=0.003). Fat from ICM presented a lower proportion of total monounsaturated fatty acids than that from SCM (P=0.028). Besides, in ICM, diets B and C decreased the total polyunsaturated fatty acids (P=0.012). We can conclude that immunocastration of male pigs provides lower intramuscular fat content and fat less monounsaturated than surgical castration. Also, the diet has to be considered in ICM, because those with high energy content or low Lys level can affect the fat composition and therefore the product quality.

Modelling growth performance of pigs and within-room thermal balance in different local conditions

N. Quiniou[1], A. Cadero[1,2], M. Marcon[1], A.C. Olsson[3], K.H. Jeppsson[3] and L. Brossard[2]
[1]IFIP-Institut du Porc, BP 35, 35650 Le Rheu, France, [2]Pegase, INRA, Agrocampus Ouest, 35590 Saint-Gilles, France, [3]Swedish University of Agricultural Sciences, Biosystems and Technology, P.O. Box 103, 23053 Alnarp, Sweden; nathalie.quiniou@ifip.asso.fr

A model has been used to assess both direct and indirect consumption of energy by growing pigs fed *ad libitum*, housed in fattening rooms with various insulation characteristics, and under different outdoor temperatures. This model combines a growth model and a bioclimatic model. It simulates thermal exchanges at the room level, based on interactions between the insulation of the room, available equipment (fans, heaters…), the parameters of the climate control box, the characteristics of pigs, and the feeding strategy. Heat sources are the animals (sensible heat) and heaters when available. Heat losses are due to insulation characteristics of the room and air renewal. The model has been evaluated from data collected simultaneously on pigs, diets, indoor and outdoor temperature (T) during a trial, and the error of prediction of indoor T was below 0.5 °C on an hourly basis. Thereafter, simulation will be performed, based on the same population of pigs and feeding strategy under different outdoor T, different heater powers in the room combined with different insulation level of wall material. For this purpose, four time series of outdoor T have been collected over 12 months (one in France and three from South to North in Sweden), as well as building characteristics in both countries (heater power: from 0 to 26 Watt/fattening place; 1 or 3 insulated layers). For each combination of climate and building, indoor T and pig performance will be simulated as well as total energy consumption and its partition between direct and indirect components. This research was part of the Pigsys ERA-Net project, co-funded under European Union's Horizon 2020 RI program (from SuSan, www.era-susan.eu, Grant Agreement n°696231) by the French ANR (grant n°ANR-16-SUSN-0003-02).

Incidence of heating the liquid feed on performance of fattening pigs

E. Royer[1], N. Lebas[1,2], G. Roques[2] and L. Alibert[1,2]

[1]Ifip-institut du porc, 34 bd de la gare, 31500 Toulouse, France, [2]GIE Villefranche Grand Sud, 258 rte de la Mathébie, 12200 Villefranche-de-Rouergue, France; eric.royer@idele.fr

Pig farms producing their own renewable energy could improve animal health or performance by the controlled heating of liquid feed. This study evaluated the effects of different liquid feed temperature on performance of growing-finishing pigs. For the fattening period (27.5 to 112 kg), a total of 144 pigs received, according to plan, a growing then a finishing diet mixed with water in a ratio of 2.8:1 l/kg and distributed at 10, 20 or 30 °C. Pigs of each treatment were allocated to 8 pens of 6 animals each, spread throughout the liquid feeding facility and received 13 times per week a liquid feed distributed by an automated supply system. Environmental temperature was maintained at 24 °C during the whole period. After distribution, pigs fed the 30 °C meal ate faster than pigs delivered 20 or 10 °C meals. It was visually observed that the latter ones agitated the mixture before eating. From d 0 to 52, pigs receiving 30 °C meal had better ADG and FCR than pigs given the 20 °C meal, pigs of the 10 °C meal being intermediate (ADG: 773, 740 and 764 g/d, respectively; P=0.02; FCR: 2.58, 2.69 and 2.61 kg/kg, respectively; P=0.03). Performance was similar among treatments from d 52 to 100 (P>0.10). Overall from d 0 to 100, FCR of pigs distributed the 30 °C feed was 2.2% lower than pigs given the 20 °C or the 10 °C meals (2.69 vs 2.77 and 2.74 kg/kg, respectively; P=0.04). Live weight at d 100 was higher for pigs given 30 °C meal than for pigs fed at 20 °C, pigs with a 10 °C meal being intermediate (110.4, 108.7 and 109.6 kg, respectively; P=0.04). 30 °C pigs tended to have a higher back fat depth than 20 and 10 °C pigs (13.7 vs 12.8 and 13.0 mm, respectively; P=0.07). These results are in agreement with the amount of heat exchange energy required to raise a liquid feed to body temperature. An additional profit of 1.1 € per pig for heated meal vs temp/cold meal was calculated on the basis of the study technical results. Under the economic conditions of 2016, a 2.2% decrease of the FCR could amortize an investment in a renewable energy up to € 5,000 per year in the case of a fattening unit with capacity for 1,500 pigs.

Digestive efficiency is a heritable trait to further improve feed efficiency in pigs

V. Déru[1,2], A. Bouquet[3], E. Labussière[4], P. Ganier[4], B. Blanchet[4], C. Carillier-Jacquin[2] and H. Gilbert[2]

[1]France Génétique Porc, La Motte au Vicomte, 35651 Le Rheu Cedex, France, [2]INRA, GenPhySE, Chemin de Borde Rouge, 31320 Castanet-Tolosan, France, [3]IFIP-Institut du Porc, La Motte au Vicomte, 35651 Le Rheu Cedex, France, [4]INRA, UMR PEGASE, 16, Le Clos, 35590 Saint-Gilles, France; vanille.deru@inra.fr

The use of diets with dietary fibres from alternative feedstuffs less digestible for pigs is a solution considered to limit the impact of increased feed costs on pig production. This study aimed at determining the impact of an alternative diet with fibres on individual digestive efficiency coefficients, and to estimate their heritabilities and genetic correlations with other production traits. A total of 480 Large White pigs were fed a high fibre diet (FD) and 547 of their sibs were fed a conventional diet (CO). For each animal, digestibility coefficients (DC) of energy, organic matter, and nitrogen were predicted from faeces samples analysed with near infrared spectrophotometry. Individual daily feed intake (DFI), average daily gain (ADG), feed conversion ratio (FCR) were recorded as well as lean mean percentage (LMP), carcass yield (CY) and meat quality traits. The FD pigs had significantly lower DC than CO pigs (-5 to 6 points). The DC were moderately to highly heritable, with heritabilities ranging from 0.41±0.14 to 0.50±0.15 in CO, and from 0.62±0.17 to 0.70±0.17 in FD. Genetic correlations between DC and ADG (from -0.65 to -0.52), FCR (from -0.75 to -0.33), and DFI (from -0.83 to -0.57) were high and negative in both diets. The DC were slightly unfavourably correlated with CY (from -0.24 to -0.11) and favourably correlated with LMP (from 0.03 to 0.29). Genetic correlations were generally unfavourable with meat quality traits (from -0.75 to 0.09). Genetic correlations of DC between diets were close to 1, so no interaction between feed and genetics could be evidenced for these traits. To conclude, DC measured in farm conditions are interesting criteria for selection to account for animal digestive capacity, due to moderate to high heritabilities and high genetic correlations with FCR. However, according to these first results, it would have to be selected together with carcass yield and meat quality to avoid adverse genetic trends on the latter traits.

Effect of genotype and dietary protein on digestive efficiency and protein and fat metabolism

L. Sarri, A.R. Seradj, M. Tor, G. De La Fuente and J. Balcells
Universitat de Lleida, Agrotecnio Center, Departament de Ciència Animal, Av. Rovira Roure 191, 25198 Spain;
laura.sarri.espinosa@gmail.com

This assay aims to analyse the influence of the genotype/productive variety and dietary crude protein (CP), on digestive capability and protein deposition in growing pigs, in order to improve further precision feeding and productive efficiency. For this purpose, 16 growing pigs of two different varieties were used: entire hybrids (F2) progeny of [F1: Duroc × Landrace] ♀ × Pietrain ♂, and castrated purebred Durocs (D) (30.5±1.27 and 25.19±0.91 kg of BW, respectively). Half of the pigs of each genotype were fed with diets which differed on CP content (LP 15% vs NP 17%). Performance and digestive efficiency parameters were determined; protein fractional synthesis rate (FSR) was analysed using the flooding dose technique, which consisted of an intravenous infusion of phenylalanine (containing 15% of (2H5-Phe)) with subsequent blood sampling from 12 to 40 min after the start of the infusion and sacrifice of the animals. Fat incorporation rates (IR) were analysed comparing the concentration of deuterated stearic acid (added to the diet during the experimental period) in tissues and plasma. F2 pigs showed a higher feed intake (P<0.05) compared to Duroc, although their growth didn't differ (P>0.05). The protein content didn't influence the feed intake, except in the intake of crude protein (CP). Average daily gain was higher in animals fed LP diet compared with NP (P<0.05); this effect was especially noticeable in the F2 variety, whose growth (kg/day) was (0.33±0.08 NP vs 0.53±0.08 LP), with respect to Duroc (0.38±0.04 NP vs 0.36±0.05 LP). Apparent dry matter (DM) digestibility was higher in Duroc pigs, and in pigs fed NP diet (P<0.05); pigs fed NP diet also showed a higher real CP digestibility. The FSR was not affected either by the productive variety or by the level of CP in the diet. Fat IR was affected by the animal genetics, presenting Duroc animals higher rates of fat incorporation in liver and muscle (P<0.05). No differences were found between both varieties in subcutaneous tissue, probably due to the high individual variability in fat deposition observed in young pigs. This study is part of the Feed-a-Gene project and received funding from the European Union's H2020 program under grant agreement no. 633531, as well as Spanish National funding by MINECO (AGL2017-89289-R).

Effect of sampling time on apparent digestibility of organic matter and net energy of pig feed

L. Paternostre, J. De Boever and S. Millet
Flanders Research Institute for Agriculture, Fisheries and Food (ILVO), Animal Sciences Unit, Scheldeweg 68, 9090
Melle, Belgium; louis.paternostre@ilvo.vlaanderen.be

A low feed conversion ratio (FCR) in pig production is important for both economical and environmental reasons. The FCR is influenced by many factors including the feed and its net energy (NE) content. The NE can be estimated from the composition and the digestibility of its nutrients. Digestibility can be measured with an indirect method, using an indigestible external marker. This method allows spot sampling of the faeces which is less laborious than total collection. In a study with sows at our institute, it was shown that the faecal sampling time affected the results. However, the sows were fed restrictedly, while most fattening pigs in Belgium are fed *ad libitum*. The aim of this study was to compare the digestibility of organic matter (DC_{OM}) and NE of pig feed from samples taken at either 9 am or 2 pm. For this experiment, 4 compound feeds were fed *ad libitum* to 24 pens (6 pens/feed) with 3 fattening pigs per pen. After 17 days of adaptation, faeces (from minimum 2 pigs per pen) were sampled twice a day for five consecutive days. The five samples were pooled per pen and per collection time, freeze dried and ground. Feeds (n=4) and faeces samples (n=48) were analysed by accredited ISO-methods. The DC_{OM} was derived from acid insoluble ash as marker and NE was calculated according to the Dutch system (CVB, 2016). The data were compared with a student t-test. The DC_{OM} of the 4 feeds amounted (mean ± sd) to 82.7±2.0 and 82.9±1.8% for the morning and afternoon collection, respectively and the NE to 9.59±0.38 and 9.61±0.36 MJ/kg, respectively and were not significantly (P>0.05) different. Thus, it appears that taking one spot faeces sample per day during 5 consecutive days from fattening pigs fed *ad libitum* is sufficient for reliable digestibility measurements.

Selection of genetically diverse animals from pedigree in Croatian autochthonous pig breeds

N. Karapandža, Ž. Mahnet, V. Klišanić, Z. Barać and M. Špehar
Ministry of Agriculture, Vukovarska 78, 10000, Croatia; nina.karapandza@mps.hr

The selection of the animals for genotyping is rarely described in the genetic diversity studies. The aim of this study was to combine various indicators of genetic diversity and population structure based on pedigree data to find the genetically most diverse animals for SNP genotyping in Croatian autochthonous pig breeds. Pedigree analysis obtained animals of Black Slavonian (BLS; 6,445), Turopolje (TUP; 632), and Banija spotted (BSP; 179) pig breeds. Based on the financial resources, genetically diverse groups in the living population per breed have been chosen according to the following criteria: largest number of offspring by sire or dam (25% of selected animals), the cumulated proportion of genetic variance explained by the selected ancestors (additional 25% of animals), and the average relationship among animals (remaining 50% of animals). Data were prepared using SAS statistical package while genetic diversity parameters were estimated using ENDOG and CFC software. Average number of offspring per sire or dam was 12.5 in BLS, followed by 10.3 in BSP, and 7.6 in TUP, respectively. The cumulated proportion of genetic variance explained by the selected ancestors ranged from 0.952 to 0.999 in BLS, from 0.994 to 0.999 in TUP, and from 0.951 to 0.987 in BSP. Average relationship among selected animals ranged from 0.002 to 0.029 in BLS, from 0.029 to 0.104 in TUP, and from 0.160 to 0.238 in BSP. In all breeds, average inbreeding coefficient was lower in the sample groups compared to the total alive populations as follows: 0.006 vs 0.063 for BLS, 0.021 vs 0.024 for TUP, and 0.052 vs 0.118 for BSP. Maximum value for the relationship among individuals of 0.125 as a threshold value could not be applied in BSP due to a high relatedness in this small population. When genotypes for selected animals will be available, genomic parameters will be compared with those obtained from pedigree data and the usefulness of this approach can be tested and upgraded.

Association between PLIN2 polymorphism and carcass and meat quality traits in pigs

D. Polasik[1], E.M. Kamionka[2], M. Tyra[3], G. Żak[3] and A. Terman[1]
[1]West Pomeranian University of Technology in Szczecin, Al. Piastów 17, 70-310 Szczecin, Poland, [2]University Hospital Heidelberg, Im Neuenheimer Feld 365, 69120 Heidelberg, Germany, [3]National Research Institute of Animal Production, ul. Sarego 2, 31-047 Kraków, Poland; grzegorz.zak@izoo.krakow.pl

Perilipin 2 plays role in lipid droplets formation, which are the storage of lipids in cells. Expression of PLIN2 gene is positively correlated with backfat thickness and is higher in fat-type pigs. Polymorphism in 3'-UTR (GU461317:g.98G>A) of PLIN2 through abolishing a binding site for GATA-1 transcription factor (G allele) may be associated with its differential expression and thus with phenotypic traits. Therefore the aim of this study was to determine possible relationships of PLIN2 variants with carcass and meat quality traits in pigs reared in Poland. Study included 3 breeds of pigs: Polish Landrace (n=261), Polish Large White (n=189) and Puławska (n=67). Nine carcass traits and nine meat quality traits were determined. g.98G>A polymorphism was detected by means of ACRS-PCR method. Association analysis was performed for each breed separately using GLM procedure. It was shown that AA genotype was present only in Puławska breed, but due to very low frequency (0.01) it was excluded from analysis. During testing of carcass traits we found associations between PLIN2 genotypes with weight of loin without backfat and skin and height of the loin eye in Polish Large White and Puławska breeds (P≤0.05). In case of the first trait, pigs with GG genotypes were characterized by its highest values, however in case of the second by lowest. In Polish Large White we also noticed strong correlations (P≤0.01) for loin eye area with GG genotype being favourable. Analysing meat quality traits we obtained consistent results for all 3 breeds in relation to intramuscular fat content and water holding capacity. We noticed that pigs with GG genotypes represented highest values of these traits (P≤0.05). We also found correlation (P≤0.05) between PLIN2 genotypes and luminosity in Puławska breed.

Frequency of genotypes associated with development of the uterus in sows
A. Mucha[1], K. Piórkowska[1], K. Ropka-Molik[1], M. Szyndler-Nędza[1], M. Małopolska[1] and R. Tuz[2]
[1]National Research Institute of Animal Production, Sarego 2, 31-047 Kraków, Poland; [2]University of Agriculture in Krakow, Mickiewicza Ave. 24/28, 30-059 Krakow, Poland; aurelia.mucha@izoo.krakow.pl

In studies in mice, it was shown that the HOXA10 and HOXA11 genes are expressed along the embryonic axis of the Muller's duct. The HOXA10 gene plays an important role in the development of the uterus, while HOXA11 of the uterus and cervix. The aim of the study was to identify polymorphisms of *HOXA10* and *HOXA11* genes in 372 Polish Large White (PLW) and Polish Landrace (PL) pigs. New polymorphisms were identified in the HOXA10 gene by High Resolution Melting (HRM) screening method using KAPA HRM FAST PCR Kits (KAPA Biosystem, USA) and the Eco Real-Time PCR System (Illumina, USA). The mutation types were identified by Sanger sequencing using the GenomeLab DTCS-Quick Start Kit (Beckman Coulter, USA) on a Beckman Coulter sequencer. The polymorphisms in *HOXA11* gene were genotyped by PCR-RFLP. The restriction enzyme used in this analysis was *HpaII*. As a result, 4 mutations were identified in the *HOXA10* gene: *g.45398502C>A*, *g.45398508_45398509delinsCC*, *g.45398714C>G* and *g.45398767G>A*. Within the mutation g.45398508_45398509delinsCC, only two genotypes were found (del / ins – 0.27 and ins / ins – 0.73). Three polymorphic forms of this gene were identified in the locus *g.45398714C>G* (GG -0.67, GC – 0.32, CC – 0.01). For the other two locus, three polymorphic forms were also detected. For the mutation at the locus g.45398502C>A the frequency of AA and AC genotypes was similar to those at the locus *g.45398767G>A* of genotypes AA and AG (0.44 and 0.48, 0.50 and 0.44, respectively). Mutations within the *HOXA11 g.3786A> G* gene have been identified as a result of previous studies. The frequency of genotypes was GG – 0.22; GA – 0.50 and AA – 0.28. Hardy-Weinberg disequilibrium was found for the *HOXA10 g.45398767G>A* and *HOXA11 g.3786A>G genes*. This study was supported by statutory activity of the National Research Institute of Animal Production, project no 01-4.06.01.

Differences in removed testicles and spermatic cords after castration with different techniques
S.M. Schmid and J. Steinhoff-Wagner
Institute of Animal Science, University of Bonn, Katzenburgweg 7-9, 53115 Bonn, Germany; simone.schmid@uni-bonn.de

According to EU Directive 2008/120/EC, only the 'castration of male pigs by other means than tearing of tissues' is allowed, but tearing of spermatic cords is still common practice in many countries. An ongoing survey shows that 21.5% of questioned German pig farmers (n=65) use tearing during castration. Previously we found that testicle weight was higher after tearing off the spermatic cords when piglets were anesthetized. The aim of this study was to (1) test the hypothesis that lack of body tension in anesthetized piglets enables to tear out more tissues during castration and (2) describe anatomical details of removed testicles. In study 1, 49 piglets were weighed and grouped: control piglets castrated after administration of analgesia (C, n=13), piglets castrated under general anaesthesia (IN, n=12), and piglets castrated with local anaesthesia (LIN, n=25). In 26 piglets, the severing of the spermatic cords was done by tearing (routine procedure at this farm), while in 23 piglets, cords were cut off with a scalpel. Testicles of all piglets were weighed. In study 2, piglets (n=8) were castrated by tearing and cutting one testicle of each piglet. Testicles were weighed separately and tissue dimensions measured. Effects were estimated by linear models with SAS 9.4. Piglet weight was not different. Torn testicles' weight (study 1 4.32 g; study 2 1.84 g) was numerically higher in all groups than weight of cut ones (study 1 3.86 g; study 2 1.55 g), and significantly higher as relative to body weight (P<0.05). Minimum and maximum length of spermatic cords were higher in torn testicles (P<0.02 and P<0.001, respectively). Fat tissue surrounding the testicles was present in all cut but only in some torn testicles, since the fat skin slipped off and remained in the body due to twisting while severing. This may explain the larger range of weights in piglets with torn cords (2.51 g). The hypothesis that testicle weight is higher in anesthetized piglets after tearing the spermatic cords could not be proven in this study and requires a larger test population. Considering the castration technique more research and knowledge transfer will help to avoid unnecessary removal of tissues and improve animal welfare.

Mathematical model of the behaviour of adult stem cells in skeletal muscle and adipose tissue of pigs

M.H. Perruchot[1], O. Duport[1], E. Darrigrand[2], I. Louveau[1] and F. Mahé[2]
[1]PEGASE, INRA, Agrocampus Ouest, 35590, Saint-Gilles, France, [2]Rennes 1 University, IRMAR, Campus de Beaulieu, 35000 Rennes, France; isabelle.louveau@inra.fr

Adult stem cells are recognized for their key role in tissue development, growth and body homeostasis throughout life. They may represent a relevant level of tissue adaptation in response to different factors inducing changes in growth and body composition. Recently, their manipulation has been considered as a relevant strategy to manage the number and types of differentiated cells within tissue. In this context, a better understanding of the biology of these cells in skeletal muscle and adipose tissue is needed to control the variability of body composition in growing pigs. The involvement of these cells in growing animals remains largely unexplored. The objective of the current study was to model the behaviour of adult stem cells during proliferation and differentiation by continuous compartmental models. The parameters of the mathematical models were estimated by inverse problem using data previously obtained and data from additional experiments. Data were obtained from primary culture of stromal-vascular cells derived from skeletal muscle and subcutaneous adipose tissue of growing pigs (1 and 42-day-old pigs). Cells were cultured in different nutritional conditions (glucose and methionine as sources of variation in the media). An analysis of the complete model provided insight into the behaviour of these cell populations in terms of proliferation and differentiation. Such an analysis was impossible *in vitro* because both phenomena cannot be studied at the same time. With the generated model, additional information on cell division was obtained. For instance, the global cell division time of adult stem cells in muscle was estimated at 8h30. The current mathematical model is a relevant tool to investigate the dynamics of cell proliferation and differentiation. In further studies, it will facilitate the identification of key factors that govern precursor cell fates in skeletal muscle and adipose tissue.

Indicators animal behaviour in growing hybrid sows subjected to castration by biological immunization

R. Braun, S. Pattacini and V. Franco
Universidad Nacional de La Pampa, Facultades de Agronomía y Ciencias Exactas y Naturales, Ruta 35, km 334, 6300 Santa Rosa, La Pampa, Argentina; braun1816@gmail.com

Immunocastration in pigs inhibits the action of pituitary gonadotrophins and, consequently, reduces sexual activity in the fattening stage. In males the sexual odour decreases due to the absence of the secretion of the hormone testosterone and in females the reduction of oestrogen, attenuating sexual libido, fights and mounts during oestrus. Males and females were treated with commercial immunizing vaccines and productive variables and animal behaviour patterns were compared with non-immunocastrated females. Twenty-four groups of pigs of 60 individuals each were evaluated on average ±5, from weaning to slaughter (6.9±0.2-129±6 kg), with the following distribution: 6 groups with immunocastrated male pigs, 8 groups with immunocastrated females and 10 groups of females without treatment. The application of the gonadotropin inhibitor was given to those treated at the same age and in two doses. The treatments were ordered in a completely randomized, unbalanced experimental design. In each treatment, the average quantitative values of daily weight gain from slaughter (kg/day) were measured – DWG, food consumption (kg/day), feed conversion efficiency (consumption / DWG), age at slaughter (days) and quality of beef to slaughter (% of Lean). Ethological studies were carried out after the second vaccination for the observation of stereotypes and to assess the normality or not of the behaviours, the social patterns of organization and hierarchies, the motivation and the expression of preferences of the animals housed in groups. There were no significant differences between the treatments of females in the productive variables, but they presented different behaviours during fattening, resulting in very aggressive non-immunocastrated sows. The immunocastrated males gained more weight, but they were less lean than the females, regardless of whether they were castrated or not.

Performance traits of Puławska pigs depending on polymorphism in the RYR1 gene (c.1843C>T)

M. Szyndler-Nędza[1], K. Ropka-Molik[1], A. Mucha[1], T. Blicharski[2] and M. Babicz[3]
[1]National Research Institute of Animal Production, Sarego 2, 31-017 Kraków, Poland, [2]Institute of Genetics and Animal Breeding, Jastrzębiec, ul. Postępu 36A, 05-552 Magdalenka, Poland, [3]University of Life Sciences in Lublin, Akademicka 13, 20-950 Lublin, Poland; aurelia.mucha@izoo.krakow.pl

The objective of the study was to determine the relationship between polymorphism in the RYR1 gene (rs118192172) and on-farm test results (fattening, slaughter and reproductive traits) of pigs maintained in conservation herds. The experiment covered 76 boars and 291 gilts kept in 33 herds. Animals aged 150-210 days were weighed and P2 and P4 backfat thickness as well as P4 loin muscle depth (P4M) were measured with an ultrasonic device. Standardised daily gain (g/day) of the animals and carcass meat percentage (%) were determined. The following reproductive traits of sows were also collected: number of teats, age at first farrowing, dates of next farrowing, number of piglets born alive and weaned at 21 days of age. Hair roots were sampled from the gilts, sows and boars to determine the RYR1 gene polymorphism. It was found that only 3% of the animals in the analysed population had TT genotypes, whereas 28.34% were heterozygous (CT). This polymorphism (CT) was carried by 28.34% of the animals. When analysing the effect of this polymorphism on fattening and slaughter traits in live animals, it was found that sows with TT genotype, compared to the others, were characterised by higher backfat thickness only (P≤0.05). In the boars with CC and CT genotypes, no significant differences were noted between the values of the analysed traits. For reproductive traits of the sows, it was observed that females with TT genotype weaned more piglets until 21 days of age (P≤0.05). Although the analysed polymorphism had no negative effect on reproductive traits of the sows in the studied population, it is suggested that boars carrying the T allele should be eliminated first from breeding due to the possible increase in the number of stress-susceptible pigs in the population and the incidence of malignant hyperthermia (TT), and thus the increased risk of PSE meat defects in fattening pigs. Financed by the National Centre for Research and Development BIOSTRATEG2/297267/14/NCBR/2016.

Cost-effectiveness of environmental impact mitigation strategies in European pig production

G. Pexas[1], S. Mackenzie[1], M. Wallace[2] and I. Kyriazakis[1]
[1]Newcastle University, Agriculture, School of Natural and Environmental Sciences, King's Rd, NE1 7RU, Newcastle upon Tyne, United Kingdom, [2]University College Dublin, School of Agriculture and Food Science, Belfield, 4, Dublin, Ireland; g.pexas2@newcastle.ac.uk

The growing demand for pork meat on a global scale implies an increase in the potential environmental impacts associated with pig supply chains. Emerging technologies, novel construction materials and alternative farm management practices are called to improve the sustainability and efficiency of pig production systems. However, implementing 'green' solutions is not always an economically viable option. In this study, we assessed the cost – effectiveness of modifications in pig housing and manure management systems targeting environmental impact reductions in European pig production. To achieve this, we performed a Marginal Abatement Cost (MAC) analysis for the implementation of alternative pig housing technologies (e.g. air purification systems), farm management strategies (e.g. alternative slurry removal regimes) and manure management practices (e.g. anaerobic digestion, in-house acidification). We evaluated the performance of the impact mitigation strategies on the following environmental impact categories: Non-renewable resource use (NRRU), Global Warming Potential (GWP), Acidification Potential (AP) and Eutrophication Potential (EP). Additionally, we performed a sensitivity analysis to assess the robustness of each technology under varying financial scenarios (e.g. variations in interest rates). The functional unit was the production of one kilogram of live weight pig (1 kg LW) at the farm gate. Danish pig production systems were used as a case in point, with data provided by the Danish Pig Research Centre. By employing the MAC methodology, we assessed the impact on farm profitability of innovations that aim to improve sustainability in pig production systems. Moreover, we identified potential economic impact hotspots and low-risk areas of the production system regarding investments for environmental impact reductions. Finally, we provide a framework that can potentially aid decision making in the choice of an environmentally sustainable pig system configuration that does not sacrifice the profitability of the farm.

Upgraded EU rapeseed meal improves growth performance and nutrient digestibility in growing pigs

A.D.B. Melo[1], T. Oberholzer[2], E. Royer[3], E. Esteve-García[1] and R. Lizardo[1]
[1]IRTA, Animal Nutrition, Ctrs Reus-el Morell, km 3.8, 43120 Constantí, Sri Lanka, [2]Bühler AG, Gupfenstrasse 5, 9240 Uzwil, Switzerland, [3]IFIP-Institut du Porc, 34 bd de la gare, 31500 Toulouse, France; rosil.lizardo@irta.es

Rapeseed meal (RSM), a by-product of oil industry, contains large amounts of protein but also contains a high fibre content that limit its use to non-ruminant animals. Tail-end dehulled RSM presents high protein content and reduced fibre fraction, resulting more attractive for animal feeding. Also, pelletizing process could contribute to improve the nutritive value of dehulled RSM. The aim of the present study is evaluating the growth performance and nutrient digestibility of growing pig´s diet based on dehulled RSM, combined with die size and steam from pelletizing process. A $2\times2\times2$ factorial design was used with conventional (35% CP) or upgraded RSM (40% CP), 4×40 or 4×60 mm of die size, processed with or without steam. Conventional or upgraded RSM were included in similar proportion (22.5%) as the only protein source and no adjustment of synthetic amino acids (AA) was done. Diets were provided *ad libitum* overall the trial, which lasts 49 days. One-hundred forty-four pigs weighing 27.55 kg were allocated at 72 pens, 2 pigs for pen (male and female), totalling 9 blocks of live weight per treatment. At the last week of experiment, faecal samples were collected for 3 days and lyophilized before lab analyses. Growth performance and nutrient digestibility were used as responsive criteria. Pig´s weight gain and feed conversion ratio were improved by dehulled RSM. Feed efficiency is improved in pigs fed dehulled RSM, 4×60 mm die size with steam on pelleting process. The digestibility of energy, crude protein, lysine, methionine, threonine and valine is improved in diet based on dehulled RSM and no affected by die size. A greater nutrient content and availability such as found in the diet based on dehulled RSM is crucial to improve animal performance and it was enhanced after pelleting process.

Impact of prebiotics or enhanced housing system on entire male pigs' growth towards improved welfare

M.A. Ramos[1,2], M.F. Fonseca[2], R.P.R. Da Costa[1,2], R. Cordeiro[3], M.A.P. Conceição[1,2], A. Frias[2] and L. Martin[2]
[1]CERNAS, ESAC Bencanta, 3045-601 Coimbra, Portugal, [2]ESAC/IPC, Coimbra Agriculture School, Polytechnic of Coimbra, ESAC Bencanta, 3045-601 Coimbra, Portugal, [3]Uzaldo Lda, R Balastreira 8, 3090-649 Coimbra, Portugal; monica.fonseca@esac.pt

In swine industry, the finishing of heavy entire males presents difficulties given the aggressive behaviour among animals reaching puberty and the possibility of boar taint in meat. The inclusion of selected raw-materials rich in non-digestible oligosaccharides (NDOs) such as chicory inulin or beet pulp in finishing diets has been pointed as helpful in reducing boar taint (highlighting inulin). Also, housing systems or management practices promoting the cleanliness of animals and the reduction of livestock density, could promote less skatole absorption from faeces. It is also known that these NDOs, can act as prebiotics, improving intestinal mucosa integrity, impacting positively the general health status and even the growth performance of animals. Improving the housing systems with more space/animal and providing manipulative objects (toys) could, theoretically, reduce aggression, promoting exploratory behaviour instead. In that sense, a field trial was conducted for 7 weeks, to assess the impact of diets containing NDOs and housing improvement on growth performance, animal behaviour, health status and boar taint of 60 entire male pigs (23 weeks of age, weighing 94.5±5.7 kg). Six groups of 10 animals were randomly assigned and received beet pulp (10%), inulin + beet pulp (4%+5%, respectively) or control diet, when housed in standard (1 m^2, 1 feeder, 1 drinker, 1 toy) or improved (1.25 m^2, 2 feeders, 2 drinkers, 3 toys) pens. Only zootechnical data is available at the moment, the remaining analyses are being carried out. Regarding final live and carcass weights, no significant differences were observed among groups. The same was also noticed for the average daily gain, though the animals fed on beet pulp diet and raised in improved housing presented a numerically superior performance (1.02±0.17 kg/day) when compared to remaining groups (0.90±0.14). We did not observe significant influence on growth by the tested diets or housing systems. Further analyses will elucidate their impact on welfare and boar taint.

Aggressive behaviour and body lesions in juvenile boars of 6 to 8 months age

M. Fonseca[1], M.A. Ramos[1,2], M.A.P. Conceição[1,2], L. Martin[1], A. Frias[1], R. Cordeiro[3], N. Lavado[4] and R.P.R. Du Couto[1,2]

[1]Coimbra Polytechnic, ESAC, Animal Science, Bencanta, 3045-601, Coimbra, Portugal, [2]Coimbra Polytechnic, CERNAS, Bencanta, 3045-601, Coimbra, Portugal, [3]Uzaldo Ltd, R Balastreira 8, 3090-649 Coimbra, Portugal, [4]Coimbra Polytechnic, ISEC, R. Pedro Nunes, 3030-199 Coimbra, Portugal; mocasfon@gmail.com

With the onset of puberty boars tend to demonstrate aggressive behaviours to pen mates. Breeding intact male pigs beyond 23 weeks of age could lead to loss of carcass value due to external scar lesions present on the carcasses. This study aimed to understand the effect of aggressive behaviour on the occurrence of skin lesions. Sixty pigs (Pietrain × F1 (Landrace × Large White)) were allocated into groups of 10, fed ad libitum with a control diet 0% inulin (Pen A & D), and diets of 3% (Pen B & E) and 6% (Pen C & F) prebiotic inulin. The animals were maintained under conventional housing (1 m²/pig; Pen ABC) with one watering point and one entertainment item and under improved housing (1,9 m²/pig; Pen DEF), 2 water points and 3 environmental enrichment items, of which 2 were permanent and one item exchanged at different phases. Six CCTV video cameras were used to record behaviour for 24 hours/day and for 48 days. Of these only 28 days of 7h30/day/video diurnal activity were subjected to behavioural analyses using BORIS software. Lesions were scored using a modified Quality Welfare assessment protocol and conducted every 10 days by the same person. Statistical analysis of data was done using R software. In conventional housing (A, B, C) the inclusion of inulin reduced the level of lesions, significantly at 6% towards to Control (0%). Globally, improved housing conditions also reduced the lesion score. Pen F (6% improved) presented the lowest lesion score. A significant difference was found between D and A. A positive correlation was found when comparing the frequency of aggressive behaviour with number of lesions from 23 weeks to 29 weeks of age. These results demonstrate the possibility of prebiotic inulin in reducing aggression and lesions, thereby improving animal welfare and carcass value.

Genetic parameters for traits derived from behavioural tests for lactating sows

J. Kecman[1], J. Neu[2], N. Göres[2], F. Rosner[1], B. Voss[3], N. Kemper[2] and H.H. Swalve[1]

[1]Martin-Luther University Halle-Wittenberg, Institute of Agricultural and Nutritional Sciences, Theodor-Lieser-Str. 11, 06120 Halle, Germany, [2]University of Veterinary Medicine Hannover, Foundation, Institute for Animal Hygiene, Animal Welfare and Farm Animal Behaviour, Bischofsholer Damm 15, 30173 Hannover, Germany, [3]BHZP GmbH, An der Wassermühle 8, 21368 Dahlenburg-Ellringen, Germany; jelena.kecman@landw.uni-halle.de

Aim of the present study was to estimate genetic parameters for different behavioural traits in lactating sows that characterize their reaction towards humans during routine management activities. The study was carried out on a nucleus farm (BHZP GmbH) with purebred landrace sows (db.01), kept in single housing free-movement pens. Data was collected from October 2016 until December 2018. Three behavioural tests were performed. The 'Dummy Arm Test' (DAT, 770 sows, 1,444 observations) evaluated the aggression of sows towards humans when their piglets were handled. The 'Towel Test' (TT, 772 sows, 2,847 observations) was used to analyse the sow's response to novel objects. The 'Trough Cleaning Test' (TCT, 772 sows, 2,805 observations) assessed the sow's reaction on routine management procedures. Estimation of variance components was performed univariately using an animal model. The following fixed effects were considered in addition to the random animal effect and random permanent environmental effect: batch, parity of sows, observer and status of pen (open/closed). Estimates of heritability (SE) were $h^2=0.17$ (0.05) for aggressive behaviour of sows towards humans (DAT), $h^2=0.19$ (0.04) for response of sows in relation to novel objects (TT) and $h^2=0.13$ (0.04) for reactions during routine activity (TCT). Additionally, genetic correlations between behavioural traits were estimated using a multivariate animal model. Additive genetic correlations (SE) ranged from $r_g=0.59$ (0.37) between TT and TCT to $r_g=0.77$ (0.30) between TT and DAT. These behavioural traits of lactating sows could be used as new phenotypes for the genetic selection on gentle and easy to handle sows. The genetic correlations point in the same direction, giving a hint for related reaction patterns. Further investigations will include an analysis of the relationship between behavioural traits and rearing performance of sows.

Assessing pubertal age through testicular and epididymal histology in Bísaro pig
G. Paixão[1], A. Esteves[1], N. Carolino[2,3], M. Pires[1] and R. Payan-Carreira[1,3]
[1]Universidade de Trás-os-Montes e Alto Douro, CECAV, Quinta dos Prados, Vila Real, 5000-801, Portugal, [2]INIAV, Strategic Unit for Biotechnology and Genetic Resources, Santarém, 2005-048, Portugal, [3]Universidade de Évora, Departamento de Medicina Veterinária, Évora, 7002-554, Portugal; gus.paixao@utad.pt

Bísaro pig (BP) had grown in numbers in the last decade, representing one of the most important native Portuguese breed. This study aims to estimate the age of puberty in male BP through testicular and epididymal morphometry. Fifty-six pairs of testis and epididymis were collected from male BP ranging in age from 1 to 8 month-old. Samples were collected post-mortem (n=26) or from surgical castration (n=30), from May 2017 to April 2018, sourced from six different farms. After collection, testis and epididymis were trimmed, weighed and measured. Tissue samples were processed for paraffin embedding and routine haematoxylin–eosin staining. Studied parameters included spermatogenesis scoring (SS), the diameter of seminiferous tubule (DST), the density of Sertoli (DS) and Leydig (DL) cells, the diameter of Leydig cells (DLC) and nucleus (DLN), and the ratio between tubular/interstitial areas (RTI). Correlations between testicular and epididymal length, width, depth, weight and volume, were highly positive (r: 0.866-0.997; P<0.001; n=56). Positive correlation was also found between DLC and DLN (r: 0.732; P<0.001; n=52). Differently, DST increased proportionally to the animal's age (R2: 0.69; P<0.001; n=52) varying from 52.91 μm to 241.95 μm. RTI acted similarly increasing in older animals (R2: 0.76; P<0.001; n=52). It varied from 25.36 to 77.56%. On average, tubules have 17.52(4.27) sertoli cells with a density varying from 2.32 to 86.45 cells/μm. While DS decreases (R2: 0.72; P<0.001; n=52), DL increases (R2: 0.32; P<0.001; n=52) as the animals gets older. A GLM model was used to predict average testis dimensions and animal's age at pre-defined stages; when SPZ is found in the epididymis, testis are 6.88 cm length and 4.49 cm width, at 118.21 days. At this age, the most likely median SS is 6 and the mean predicted SS is 5.29. When SPZ is found in the vas deferens testis are 7.46 and 4.82 cm, at 145.74 days. The most likely median SS is also 6 and the mean predicted SS is 6.07.

Genetic parameters of feeding behaviour traits in Finnish pig breeds including social effects
A.T. Kavlak and P. Uimari
University of Helsinki, Agricultural Science, Koetilantie 5, 00014, Helsinki, Finland; alper.kavlak@helsinki.fi

Pigs are housed in groups during the test period. Social interactions between pen mates may affect average daily gain (ADG) and feeding behaviour traits of pigs sharing the pen. The aim of this study is to estimate the genetic parameters of ADG and feeding behaviour traits with social interaction model in combined Finnish Yorkshire (n=2,586), Finnish Landrace (n=2,961) and crossbred (n=829) data. Feeding behaviour traits included the number of visits per day (NVD), time spent in feeding per day (TPD), daily feed intake (DFI), time spent feeding per visit (TPV), feed intake per visit (FPV), and feed intake rate (FR). The test period was divided into 5 periods of 20 days. The number of pigs in each pen varied from 8 to 12, but only 8 random pigs per pen were included in this analysis. The mixed linear model included sex, breed, and herd×year×season as fixed effects, and batch×pen, litter, the animal itself (direct genetic effect), and pen mates (indirect genetic effects) as random effects. The estimated total heritable variation (the ratio of total heritable variance to the phenotypic variance, T2) was 0.26 for ADG. Excluding the social interaction from the model yielded h2 of 0.26, thus the effect of social interaction in this data was minimal for ADG. For feeding behaviour traits (period 1) estimates of total heritable variation were 0.44, 0.40, 0.48, 0.54, 0.63, and 0.65 for NVD, TPD, DFI, TPV, FPV, and FR, respectively. These are considerably higher than h2 (varied from 0.14 to 0.36). Especially, for TPV, FPV and FR social interaction was important, while other feeding behaviour traits (NVD, TPD, DFI) demonstrate a small increase in the classical heritability. Similar differences between T2 and h2 were also obtained for period 4. The results indicate that social interactions have a considerable effect on feeding behaviour of pigs but not on ADG.

Economic optimization of feeding strategy in pig-fattening units with an individual-based model

M. Davoudkhani[1], F. Mahé[2], J.Y. Dourmad[1], E. Darrigrand[2], A. Gohin[3] and F. Garcia-Launay[1]
[1]PEGASE, INRA, Agrocampus Ouest, 35590, Saint Gilles, France, [2]IRMAR, Université de Rennes I, CNRS, UFR Campus de Beaulieu, 35042, Rennes, France, [3]SMART LERECO, INRA, 35011, Rennes, France; mohsen.davoudkhani@inra.fr

Economic results of pig farming systems are highly variable and depend on the prices of feed ingredients used to formulate the diets as well as on pig performance and pork price. Thus, feeding strategies in fattening units are critical factors in the economic outputs of pig production and major levers for improvement. This work aims at improving the profitability of pig farms by proposing the best compromise between the cost of feeding and the performance of animals. We rely on an individual-based bioeconomic model, which calculates the average gross margin per fattened pig and the environmental impacts of the production according to biological traits of each pig, such as feed intake and protein deposition potentials, and the feeding strategy. The objective function maximizes the average gross margin per pig using an optimization procedure regarding the feeding strategy. The optimization problem is solved with an evolutionary algorithm. The behaviour of this process was investigated, in one-phase and two-phase feeding programs, using two among three feeds A, B, and C, formulated to achieve 110, 90 or 90% of digestible lysine requirements of an average pig at 30, 65 and 120 kg of body weight, respectively. We studied as decision variables the percentage of the two feeds in the blend at each phase and the average live weight of the pen at the end of the first phase. With this optimization strategy, it is possible to optimize simultaneously the feed mixture to be distributed at each phase and the pig weight at feed change. In two-phase feeding program, for the pig population tested, optimal pen average pig weight at diet change was lighter than common practice (50 vs 65 kg), and optimal feed digestible lysine contents amounted 8.5 and 6.3 g/kg in phase 1 and 2, respectively. Further investigations will extend the range of available decision variables and consider multi-objectives optimization to account for environmental impacts.

Resilience of livestock farming systems: concepts, methods and insights from case studies on organic

G. Martin, M. Bouttes and A. Perrin
INRA Occitanie-Toulouse, UMR AGIR, CS 52627, 31326 Castanet Tolosan, France; guillaume.martin@inra.fr

The context of agricultural production is unstable as a result of complex and interrelated factors that are beyond farmers' control such as climate, markets and public policy. These factors give rise to a range of hazards and changes: global warming, extreme climatic events, volatility of prices... Farmers have no alternative than to develop the resilience of their livestock farming systems, i.e. their capacity to remain against these hazards and changes. Resilience is being promoted by three capabilities: (1) buffer capability: the system is able to tolerate hazards and changes without departing from its routine regime; (2) adaptive capability: the system is able to implement technical, organizational or commercial adaptations to cope with hazards and changes and quickly return to its routine regime; (3) transformative capability: the system is able to transform to remain. Several factors have been shown to promote the resilience of livestock farming systems: diversification of pastures, crops and livestock within farms, rusticity of pasture and crop cultivars and livestock breeds, dual-purpose crops, pluriactivity of farmers... We focus on organic dairy farming and present the insights from several studies conducted in France. In these case studies, resilience was either assessed through classical technical and economical indicators or through farmers' perceptions. The most resilient livestock farming systems were those that displayed a consistent dimensioning (animal number to land potential balance, animal number to workforce balance, etc.), that strongly relied on grazing and made thrifty use of inputs to promote self-sufficiency. We present a four-step iterative cycle to support farmers in developing their farms' resilience: realizing farm exposure to hazards and changes, assessing farm sensitivity against these hazards and changes, designing and implementing alternative scenarios and monitoring their impacts. We conclude by emphasizing that resilience thinking constitutes a break from traditional analytical and managerial approaches by promoting the adaptability and transformability of systems rather than seeking optimal solutions. It involves thinking about livestock farming systems with more dynamic and holistic perspectives.

How much is enough – the effect of nutrient profiling on carbon footprints of 14 common food products

G.A. McAuliffe, T. Takahashi and M.R.F. Lee
Rothamsted Research, Sustainable Agriculture Sciences, North Wyke, Okehampton, EX20 2SB, United Kingdom;
graham.mcauliffe@rothamsted.ac.uk

Life cycle assessment (LCA) of agri-food systems has received criticism on functional units in recent years; namely, those based on mass (e.g. environmental impacts per kg product) fail to reflect the nutritional value of individual commodities. Consequently, a wave of novel research has materialised over the last decade, with a shifting focus from product quantity to quality. Although no single methodological solution has been agreed upon nor uniformly adhered to, one of the more popular options is using nutrient profiling to estimate environmental impacts per proportion of daily nutritional requirements satisfied by a commodity. Derivation of nutrient indices, however, necessitates a selection of nutrients to be included, and the impacts of this decision on LCA results are not generally well-understood. The aim of this study, therefore, was to examine the effect of adopting four different nutrient density scores (NDS) on the relative carbon footprints (CF) amongst 14 food products commonly consumed as protein sources. Mass-based CF from 737 production systems around the world were sourced from a recently-published meta-analysis and recalculated using nutritional data obtained from USDA. NDS were calculated using either 6, 9, 11 or 15 nutrients to encourage and, in all cases, 3 nutrients to discourage (saturated fat, sodium and total sugar). Under the mass-based functional unit (100 g product), animal-derived products almost always showed higher CF than plant-based products. When nutritional quality was accounted for, however, product rankings became less clear-cut. For example, pork and tofu generated global averages of 1.141 and 0.324 kg CO_2-eq/100 g product, yet 0.145 and 0.149 kg CO_2-eq/1% NDS under the 15-3 scoring. This reversal of rank results from superior nutritional composition of pork over tofu and suggests that, if consumed according to optimal dietary intakes, less pork would be required than tofu to achieve the same uptake of nutrients. As more nutrients were added to the NDS, animal-based products tended to perform more favourably, indicating that mass-based evaluation of CF may be biased in favour of plant-based products.

Farm resilience: a farmers' perception case study

E. Muñoz-Ulecia, A. Bernués, I. Casasús and D. Martin-Collado
CITA-Aragón, Animal Production and Health Unit, Avda. Montañana 930, 50059 Zaragoza, Spain;
emunnozul@cita-aragon.es

In Europe, the number of mountain farms is decreasing due to various socioeconomic drivers. Although mountain livestock farming systems are generally considered as extensive, they are actually very diverse, influenced by both internal (use of natural resources, purchased feedstuffs, farmer's age, etc.) and external factors (agricultural policy, socioeconomic context, environmental conditions, etc.). In addition, farmers need to adapt to crucial challenges that affect agriculture globally, e.g. increasing risk of droughts due to climate change and higher prices of inputs due to market dynamics. Understanding farmers' views on the relevance of actions and strategies to face these challenges is key to study mountain farming resilience. The aim of this work was to analyse: (1) farm resilience strategies according to farmer response to climate and market changes; and (2) the influence of farms and farmer characteristics on those strategies. We carried out a survey on 54 beef farmers in the central Pyrenees (Spain), gathering information about farm structure, management and economic performance. We also measured farmers' perception on the importance of different actions to deal with: (1) 2-year-long drought; and (2) rise of input prices, using a Likert scale from 1 (not important) to 5 (extremely important). Specifically, we considered actions related to pastures and feed management, reproductive management, herd size, external advice, development of quality brands, diversifying farm activity or seeking for other sources of income outside farming. According to farmers, the most relevant actions to face droughts were using new areas of pasture (average relevance of 3.4) or reducing herd size (3.3), in contrast with the lower relevance of seeking for external advice (2.4). Regarding the increase of inputs' price, the highest importance was given to using new areas of pasture (4.2) and extending the grazing season (4.2), as opposed to developing a quality brand (2.6) and seeking for external advice (2.4) that had the lowest importance. Several farm and farmer profile characteristics influenced their views on the relative importance of actions to face these challenges; e.g. farmer age, size of utilized agricultural area, or farm type (fattening on-farm or not).

Resilience of yak farming in Bhutan

N. Dorji[1,2,3], M. Derks[3], P. Dorji[1], P.W.G.G. Koerkamp[3] and E.A.M. Bokkers[2,3]
[1]College of Natural Resources, Animal Science, Royal University of Bhutan, Lobesa, Punakha, P.O. Box 1264, Bhutan, [2]Wageningen University and Research, Animal Production Systems group, P.O. Box 338, 6700 AH Wageningen, Netherlands, [3]Wageningen University and Research, Farm Technology group, P.O. Box 16, 6700 AA Wageningen, Netherlands; nedup.dorji@wur.nl

Yak-based transhumant systems are influenced by socioeconomic developments, policies, and environmental changes. Little is known about the impact of these factors on the resilience of yak farming practices among different regions in Bhutan. The yak farming practice changes were assessed through interviews with yak herders in three regions (west, 22; central, 20; east, 25) and with livestock professionals (28). Our results show that at present forage shortage in the rangeland (herders, 93%; livestock professionals, 96%), yak mortality (herders, 96%; livestock professionals, 96%) and to a lesser extent labour availability (herders, 30%; livestock professionals, 96%) are the main concerns in all yak farming regions. These concerns have increased due to socioeconomic developments (e.g. education) and strong conservation policy. Some factors causing forage shortage, however, are specific to certain regions, e.g. competition with the horse population (west, 91%), cattle and cattle-yak hybrid (east, 72%; central; 70%), cordyceps collection (west, 55%; central, 60%), and prohibited burning of rangelands (central, 80%; east, 76%). According to the respondents, the market to sell yak products and livestock extension services has improved, while forage shortage and yak mortality has increased over the years. In addition, family labour available to herd yaks, as well as the number of young family members (successors) to take over yak herding over the years was perceived as slightly decreased. Based on the experiences and perceptions of yak herders and extensionists, we conclude that increasing forage shortage in the rangelands, decreasing numbers of successors, and increasing yak predation (wild carnivores) are the major threats to the resilience of yak farming in Bhutan.

Measurement enhanced the operationalization of resilience concept applied to livestock farms

M.O. Nozières-Petit[1], E. Sodre[2], A. Vidal[1], S. De Tourdonnet[2] and C.H. Moulin[1]
[1]UMR SELMET, Montpellier Université, INRA, CIRAD, Montpellier SupAgro, 2 place Viala, 34060 Montpellier, France, [2]UMR Innovation, Montpellier Université, INRA, CIRAD, Monptpellier SupAgro, 2 place Viala, 3460 Montpellier, France; charles-henri.moulin@supagro.fr

Resilience is a multi-faced concept, with many definitions, but also difficulties to operationalize it. Carpenter et al. proposed to measure the resilience of what to what. Darnhofer operationalized the concept, considering three capabilities for farm management. Our hypothesis is that on-farm approaches on practices and performances enables highlighting those capabilities. Through surveys in French Mediterranean, we studied how 8 meat sheep farmers dealt with economic hazards, from 2009 to 2013, and how 11 dairy sheep farmers coped drought from 2015 to 2018. Most of dairy farmers sought for the stability of the animal outputs, delivering the same annual quantity of milk. Meat farmers were far more flexible, with inter-annual variability of the patterns of lamb types, sold to various operators. Seeking stability, farmers used buffer capabilities, relying on excess capacities, redundancy in resources, and adjustment and innovation on crop and livestock operations. For flexibility, farmers also used adaptive capabilities, as changing mating session schedules and feeding system, to catch opportunities with downstream operators. Finally, some farmers were engaged in radical transformation, mobilizing transformative capabilities. In combination with comprehensive interviews, the quantification of several variables through a mid-term of 5 five year is very relevant to understand what is at stake for the farmers (resilience of what) and the levers they used in their situation. In that way, we show that the notion of production goals, classically used in farm management, is not relevant in all situations.

Environmental footprint and efficiency of mountain dairy farms

M. Berton[1], S. Bovolenta[2], M. Corazzin[2], L. Gallo[1], S. Pinterits[3], M. Ramanzin[1], W. Ressi[3], C. Spigarelli[2], A. Zuliani[2] and E. Sturaro[1]
[1]University of Padova, DAFNAE, Viale dell'Università 16, 35020 Legnaro, Italy, [2]University of Udine, DISAA, Via Sondrio 2, 33100 Udine, Italy, [3]Umweltbüro, Bahnhofstraße 39, 9020 Klagenfurt, Austria; marco.berton.1@unipd.it

The study presents the results of a project (IR VA Italia-Österreich 'TOPValue') aiming at identifying the added value of livestock 'mountain products' in terms of multifunctionality. The specific aim of this paper is to analyse the environmental footprint (Life Cycle Assessment method) and production efficiency (gross energy and potentially human-edible conversion ratios, ECR and HeECR respectively). Data originated from 75 farms (38±25 LU, 20.9±5.4 kg fat protein corrected milk – FPCM/cow/day), associated to 9 cooperative dairies in the eastern Alps. Herd and manure management, on-farm feedstuffs production, purchased feedstuffs and materials were included into the system boundaries. Impact categories assessed were Climate Change, Cumulative Energy Demand, Land Occupation (LO). Two functional units were used: 1 kg of FPCM and 1 m^2 of farming land. Milk vs meat allocation (IDF method) was used. Mean impact values were 1.2±0.2 kg CO_2-eq, 3.3±1.6 MJ, 2.3±1.0 m^2/y per 1 kg FPCM, and 0.5±0.2 kg CO_2-eq, 1.4±0.7 MJ per 1 m^2. Mean ECR was 6.5±0.9 MJ feed/MJ milk, with 93% of gross energy deriving from non-human-edible feedstuffs, nearly totally produced on-farm. We tested the effect of herd size (3 classes) and management strategies (use of pasture and/or summer farms). Herd size did not affect impact categories, ECR and HeECR. Farms using pasture and/or summer farms for lactating cows showed significant greater values for LO per 1 kg FPCM and lower values for impact categories per 1 m^2, probably because of a lower stocking rate (-35%). Besides, farms with pastures showed a greater diet self-sufficiency ratio (+32%) and a lower HeECR (-41%) due to lower use of potentially human-edible concentrates. The results evidenced that the traditional managing options in the mountain dairy farming system (small-scale farms using pasture and summer transhumance) generally do not worsen the environmental footprint indicators but enhance the decoupling of milk production from crop production intended for direct human consumption.

Enhancing resilience of EU livestock systems; what is the role of actors beyond the farm?

M. Meuwissen[1], W. Paas[1], G. Taveska[2], E. Wauters[3], F. Accatino[4], B. Soriano[5], M. Tudor[6], F. Appel[7] and P. Reidsma[1]
[1]Wageningen University, Hollandseweg 1, 6706 KN Wageningen, the Netherlands, [2]Sveriges Lantbruksuniversitet, Almas Alle 8, 750 07 Uppsala, Sweden, [3]ILVO, Burgemeester van Gansberghelaan 92 bus 1, 9820 Merelbeke, Belgium, [4]INRA, Rue de l'Universite 147, 75338 Paris Cedex 07, France, [5]CEIGRAM, Calle Ramiro de Maeztu 7 Edificio Rectorado, 28040 Madrid, Spain, [6]Institute of Agricultural Economics, Calea 13 Septembrie 13, 050711 Bucharest, Romania, [7]IAMO, Theodor Lieser Strasse 2, 06120 Halle Saale, Germany; miranda.meuwissen@wur.nl

In dealing with challenges, such as reduced societal acceptance and power issues along the chain, multiple actors from livestock systems play a role. Yet, most research on enhancing the resilience of systems focuses on the farm level. This paper also considers the role of other actors. The aim of the paper is to: (1) identify system actors who have strong patterns of mutual dependence with livestock farmers; (2) elicit the perceived resilience of livestock systems in terms of robustness, but also with regard to capacities to adapt and transform the system; and (3) identify individual and collective competences to enhance resilience. The paper builds on participatory research in six case study regions including intensive livestock systems in Sweden and Belgium, extensive livestock production in two regions of France and Spain, and mixed systems in Romania and Germany. Preliminary findings show that capacities of robustness and adaptability are perceived to be higher than the transformative capacity of livestock systems. In the paper, differences between systems and regions will be elaborated including resilience enhancing (and constraining) strategies and competences of farmers and other actors the system. The work presented in this paper is part of the Horizon 2020 SURE-Farm project (www.surefarmproject.eu).

Organization of an alfalfa hay sector between cereal farms, livestock farms and a local cooperative

E. Thiery[1,2], G. Brunschwig[1], P. Veysset[1] and C. Mosnier[1]
[1]Université Clermont Auvergne, VetAgro Sup, Inra, UMR Herbivores, INRA Saint-Genès-Champanelle 63122 Saint-Genès-Champanelle, France, [2]AgroSup Dijon, 26, Boulevard Docteur Petitjean-CS 87999, 21079 Dijon Cedex, France; eglantine.thiery@agrosupdijon.fr

The dependence of livestock farms on protein and crops farm on nitrogen is increasingly questioning farmers and stakeholders in the supply chains about the value of legume crops. However, for growers, legume crops require technical skills and know-how, particularly for forage harvesting, which are becoming increasingly rare in the agricultural landscape. In addition, markets for protein-rich forages such as alfalfa hay remain disorganized and prices fluctuate widely from year to year and even from one cut to another. This fluctuation in prices, as well as in the quality of fodder, does not encourage farmers to turn away from oilseed and protein crop cakes. The first studies on crop and livestock group initiatives for alfalfa hay transactions show that these different approaches could not be sustained. The participation of an actor organizing the sector and the market, (a cooperative) can be a promising way of implementation. Our study will focus on the approach of a cooperative group setting up an alfalfa industry between lowland cereal growers and mountain breeders in the same region. We will analyze the sensitivity of the price of alfalfa hay by taking into account agronomic gains (nitrogen fixation in particular) using a bio-economic optimization model. Our objective is to determine under which agronomic and economic conditions, the alfalfa hay sector can be created sustainably to link mountain dairy farms and lowland cereal farmers in the Rhône-alpin context.

Performance, longevity and financial impacts of removing productive ewes early from mountain flocks

H. Wishart, C. Morgan-Davies, A. Waterhouse and D. McCracken
SRUC, Hill & Mountain Research Centre, Kirkton, Crianlarich, FK20 8RU, United Kingdom; claire.morgan-davies@sruc.ac.uk

Traditionally in the UK many breeding ewes are sold from mountain farms, after their fourth or fifth annual crop of lambs, to lower altitude upland farms where they are mated with longwool breeds to produce more prolific crossbred ewes. This is a key element of the UK stratified sheep industry. However culling mountain ewes at a fixed age from the flock limits longevity and may not be in the financial interests of the farm. To understand whether culling on age from a mountain farm still occurs, a questionnaire was carried out with 115 farmers. Results showed that there was a difference ($P<0.05$) between system type (mountain, upland and lowland) for those flocks where ewes were culled on age with the largest majority being mountain (55%), then upland (41%) and finally lowland (29%). However not all mountain farms culled on age which could indicate a change in culling practices from our previous understanding. Actual results from a research mountain flock was used to model the financial implications of culling based on age. Results showed that a flock where productive ewes were allowed to remain after their fourth lamb crop had the potential for a greater margin compared to a flock that culled ewes after their fourth crop. This was largely a result of more replacement ewe lambs being sold and reduced costs for the first unproductive year before lambing at two years old. However, financial viability and sensitivity were highly dependent upon relative sale value of retained ewes, and input costs for replacement females up until their first lamb crop. Furthermore, performance of older ewes (those aged 5.5 years old or older at the start of the production year) was compared to all other ages. Results showed that performance was similar between older ewes and all other age groups of ewes, in terms of: survival, number of lambs born and weaned, and weight of lambs weaned. Therefore, culling practices in the UK mountain flocks appear to be changing. Retaining older ewes within the flock could provide potential benefits as they are able to perform as well as younger ewes and could increase overall longevity for these flocks. This also brings potentially greater profits and resilience for mountain farmers.

How can we better support the future in dairy farmers from the point of view of the stakeholders?

A.-L. Jacquot[1], F. Kling-Eveillard[2] and C. Disenhaus[1]
[1]PEGASE, INRA, Agrocampus Ouest, 35590, Saint-Gilles, France, [2]Institut de l'élevage, 149 rue de Bercy, 75595 Paris Cedex 12, France; anne-lise.jacquot@agrocampus-ouest.fr

The context of milk production is changing in France. The operations are expanding and specializing, support to the farmers must evolve. During 2017, dairy farmers and advisors were brought together in focus groups to reflect and discuss the support needs in these dairy farms, as they felt they were for themselves, in the case of farmers, or among the farmers they work with in their activity for the advisors. Farmers in dairy farms were selected on specific criteria: more than 100 cows, more than 3 persons in the workforce or with a high productivity per labour unit (more than 500,000 litres per labour unit). Two focus groups of farmers took place, 12 farmers in Western France and 14 in Eastern France. The focus group of advisors took place in Brittany with 18 advisors from various trades and structures (banks, veterinarians, milk recording, food, chambers of agriculture, strategy consulting, National Social Insurance for agricultural workers, …). The farmers and advisors gathered expressed their strong interest of this purpose for these dairy farms. Farmers identify specificities in the management of large dairy herds 'an essential organization of work, more time spent in the office than in the stable, business management and the management at herd-scale and not per cow'. Western farmers have highlighted their need for support around the management of human relations, and in particular communication within the work collective, while Eastern farmers testify to needs on employee management (training, sustainability of the workforce, labour legislation,...) and how to adapt their farm to a changing context (market uncertainties). The advisors insisted on the need to develop long-term support measures (preventive rather than curative actions), whose added value is sometimes difficult to have recognized and sold to farmers. They insist on the need to acquire specific references on large herds, as well as on the sharing of various data between advisors in order to design an approach adapted to each farm.

How agro–ecological transition could sustain goat keeping in nomadic systems of Iran

F. Mirzaei
Animal Science Research Institute of Iran, Animal Production Management, 31585 Karaj, Iran; fmirzaei@gmail.com

Pastoralism is a type of animal production characterized by the use of spontaneous food resources. Since 50 years, the dominant paradigm of agriculture has been based on specialization and intensification of production systems by maximizing the issue of production factors including higher yields, higher inputs, higher dependence to food industry. Its positive effect is more food safety and its negative effect is environmental impacts and decreasing of the farmers. Five principles of Agroecology for livestock are as follows: the integrated management of animal health, the reduction of inputs by using ecological processes, the reduction of pollutions by controlling the biological cycles, the use of diversity in production systems to increase their resilience and the preservation of biodiversity(of pasture, of landscapes, of local populations) by adapting practices. To define agroecological characteristics for goat keeping system of Iran following items should be considered and developed; animal nutrition, sustainable pasture management, crops and forage practices, disease prevention, breeds and reproduction, animal welfare, food safety and hygiene, marketing and management, conditions of social and economic sustainability, environmental sustainability and societal contribution. The productivity of Iranian goats is low, with a national average estimated at 150 kg of milk and 20 kg of meat per goat per year. The Iranian goat sector is dominated by extensive herds dedicated to the production of goats for meat and milk, which have very seasonal sales concentrated mainly during the feast of sacrifice and Ramadan. The production of goat milk is undergoing an important development, especially in the centre of the country, and is allowing for a significant improvement in the profitability of goat operations.

Italian ryegrass yield prediction for forage supply to ruminant livestock farming in South Korea

J.L. Peng and L.R. Guan
Kangwon National University, Room 308, Building 1, College of Animal Life Sciences, 1 Gangwondaehak-gil, 24341, Chuncheon-si, Gangwon-do, Korea, South; pengjinglungood@163.com

The ruminant livestock farming system in South Korea consumes lots of imported forages which financially costs a lot and remains the risks of forage quality and supply capacity in the context of climate and market uncertainties. Therefore, high quality domestic forage supply draws great concerns of the interested parties. As a fundamental study of developing the precision production system for high quality forages using agricultural big data in South Korea, the objective of this research was to construct the yield prediction model of Italian ryegrass using climatic, soil and cultivar information. The forage cultivation and cultivar data were obtained from the national forage cultivation experiment reports, the climatic and soil data were gathered from the Korean Meteorological Administration and Korean Soil Information Center, respectively. Stepwise regression was used to select the optimal climatic variables which are autumnal growing days (AGD), mean temperature in January (MTJ), spring growing days (SGD), period to accumulated temperature 150 °C (PAT150), and spring amount of precipitation (SAP). The calculation criteria were developed to generate the soil suitability score. The cultivars were classified into early-, middle-, and late-maturity groups. With dry matter yield (DMY) as the dependent variable, the yield prediction model of Italian ryegrass was developed via general linear model considering the selected climatic variables and soil suitability score as the quantitative variables and the cultivar maturity as the qualitative variable as follows: DMY (kg/ha) = 7,555.16 + 70.03AGD − 49.76MTJ + 78.90SGD − 67.06PAT150 + 1.85SAP − 89.71Soil + Maturity Category (early=864.23, medium=0, late=428.25). The R-squared of the model was 45%. The residual diagnostics and 3-fold cross validation showed the fitness and the validity of the model was in the good level. The model will be further improved and used for Italian ryegrass yield prediction after merged into the precision production system for high quality forages in South Korea.

Characteristics of carcass traits and meat quality of broilers chickens reared underconventional and

A. Ehsani
Tarbiat modares university, Jalal Al-Ahmad highway, Nasr Bridge, Tehran, 1497713111, Iran; alireza.ehsani@modares.ac.ir

Alternative chicken production systems becoming popular in recent years due to animal welfare criteria's and consumer's perceptions. General beliefs indicate that the quality of chicken meat reared under free range is higher than that from conventional production systems. The aim of this study was to compare the quality and quantity of carcass traits produced under conventional and free-range systems. Data included a total of 74 day-old chickens of slow-growing Ja57 strain until the age of 78 days. Birds were randomly divided into two groups, one was reared in a conventional closed house and the other was moved to the free-range until the end of the fattening period. A sample of 6 male chickens selected in each system with a target living weight of 3.2 to 3.5 kg were randomly chosen to provide material for analysis. The meat quality characters, pH at 45 minutes, ultimate pH, colour coordinates, drip loss, cooking loss, and water-holding capacity were measured and also proximate parameters such as, crude protein, total fat, and crude ash were measured. There were no significant differences in main carcass yield and breast muscles, however, colour values dramatically were influenced by rearing systems. The absence of significant differences in the carcass yield and other under study traits between two different systems are may be due to short rearing period with outdoor access. Samples were taken from breast muscle and differences were seen in colour for birds reared under the conventional system having a smaller hue angle and saturation value than those grown in the free-range. Moreover, the drip loss parameter was significantly higher in free-range chickens. The Ash and protein contents of breast muscles were similar, although, raw breast meat from free-range birds had significantly lower fat content. The results proves that free-range can modify the appearance, colour values and fat content of chicken meat and it can be a part of the interests of meat production consumers.

Determinants of the sustainability of dryland sheep farming: preliminary results

P.M. Toro-Mujica[1,2], R.R. Vera[2], C. Arraño[2] and L. Robles[3]
[1]*Instituto de Ciencias Agronómicas y Veterinarias. Universidad de O'Higgins, Ruta 90, Km.2. San Fernando., 3070000, Chile,* [2]*Pontificia Universidad Católica de Chile, Ciencias Animales, Vicuña Mackenna 4860. Macul. Santiago, 8430613, Chile,* [3]*Universidad Autónoma del Estado de México, Departamento de Nutrición Animal, Toluca, México, 50200, Mexico; pmtoro@uc.cl*

The study corresponds to a descriptive exploratory analysis that addresses the existing relationship between characterization variables and their influence on the sustainability of sheep farming of Chile. During the second semester of 2018, 55 surveys were carried out in farms of the Central zone of Chile. As a criterion for the selection of farms, it was considered that sheep farming should be one of the three main economic activities of the farm. The survey included closed and open questions, focused on learning the composition of the family, the diversity of activities that take place on the farm, and the number of animals among others. The proximity to urban centres was a variable considered for the selection of farms, selecting two subzones: farms within 80 km or less of an important urban centre (considering as such the capital of the country or cities with more than 1 million inhabitants) (Zone 1), and farms located at a distance greater than 120 km. of large urban centres (Zone 2). In the analysis of the information, relative and absolute frequencies of qualitative characterization variables were determined; also, the relationship between variables was analysed through contingency tables, categorizing quantitative variables; descriptive statistics were calculated for quantitative variables. Among the main differences observed between both types of farms are the average area, herd size, and the number of activities carried out. The overall average farm area was 38 ha (Zone 1: 13 ha, Zone 2: 59 ha;), while herd size was 87 sheep (Zone 2: 48 heads, Zone 1: 120 heads); 44% of the interviewees developed another activity (non-agricultural or agricultural) (Zone 1: 61%, Zone 2: 31%). Among the causes responsible for the abandonment of sheep production are its low profitability, the lack of complementary employment opportunities for young people, and the advanced age of a large part of the farmers (70% with more than 60 years).

Productive evaluation of tropical livestock agribusinesses based on most probable producing ability

H. Estrella-Quintero[1], V. Mariscal-Aguayo[1] and E. Salas-Barboza[2]
[1]*Universidad Autonoma Chapingo, CRUOC, Rosario Castellanos 2332, 44950 Guadalajara, Mexico,* [2]*Freelancer, Floresta 1366, 44530 Guadalajara, Mexico; estrellaqh@hotmail.com*

Data records are fundamental for the information structure of any livestock farm. Records allow to make the ensuing technical and managerial decisions, and are the base for determining the genetic progress of the herd. The aim of this work was to assess the productive performance of the dual-purpose farms from a southern region of Mexico with regards to its most probable producing ability (MPPA). The study was undertaken during 2018, with double-purpose farms located in Huimanguillo, Tabasco, with an average of 51 adult cows, which were artificially inseminated and natural mated, and with no health issues. MPPA for milk production was obtained from the reports generated by the software DOBLE AGROPEC Star®. Five farms with MPPAs estimated for 256 lactations, 19 genotypes, and data from calved cows were considered. Data were studied with an analysis of covariance using a completely randomised design with the genotype and the lactation number as the covariate, using the statistical software SAS. In the first analysis without including the covariates effects, statistical differences were obtained on the MPPA values among the different farms (P<0.01), obtaining the maximum value (2,017.53) in farm number 3. A significant effect of the covariable genotype was found (P<0.01) but not for the number of lactation. Genotypes Swiss American × Sardinian Black and Brahman × Sardinian were significantly different from the remaining genotypes, and with the former having the highest estimated value (2,366.81). When the covariate genotype was added to the model, statistical differences (P<0.01) on the MPPA values for the five farms were found, with the farms 2 and 3 showing the highest values (1,773.89 and 1,995.82, respectively) and being different to the remaining farms. The productive assessment through the most probable producing ability allowed to identify the farm with the best cows on the basis of production and genotype.

Potential intake of snails and slugs by free-ranged livestock

K. Germain[1], C. Collas[2], M. Brachet[1] and S. Jurjanz[2]
[1]INRA, Le Magneraud, 17700 Surgères, France, [2]Université de Lorraine, INRA, 54505 Vandoeuvre, France;
karine.germain@inra.fr

Free-ranged livestock may ingest invertebrates (IV) during plot exploration. IV can be a protein supply especially interesting in organic systems but can also transfer soil-bound contaminants. It is thus necessary to know how many IV are present on the plot at a given period and how many would be ingested by free-ranged livestock. If the biological cycles of IV are well described, the interactions with livestock can modify their population dynamics as shown by earlier studies on earthworms. As no study on the dynamics of snails and slugs (SaS) on outside runs exist, their number and mass have been determined during spring and autumn on outside runs and controlled conditions of organic chicken at the experimental station 'INRA Le Magneraud' (West France). The IV are caught with Barber-traps on five placements on each plot (50×50 m): two located 2 m in front of the poultry house, two others at the end of the plot (2 m before the fence) and one in the middle of the run. During the spring campaign (April-May), SaS were revealed on two types of plots (grass covered or under trees) every 2 weeks starting the day before giving access to the chicken (d35) and up to the week before slaughter (d86). The autumn campaign (October-November) was carried out on the same runs on three times: before access of chicken, just before withdrawal of birds (slaughter) and an intermediate point. Results showed huge variations between the different runs but the highest count, made on a grass covered run, revealed only 50 slugs (8 g) and 170 snails (11 g). Generally, more slugs than snails were present on the plots (in number and mass). Initial number of SaS decreased along the exploration period to reach nearly 0 at the end of May. As similar results were got in autumn (grass plots > plots under trees, snails > slugs, beginning > end), it seems likely that predation is a central factor of disappearance of SaS even if dry and hot weather can amplify this phenomenon. Moreover, livestock ingest gastropods available on outside plots decreasing their occurrence but further work is necessary to understand dynamics of IV population on plots of livestock.

Productive behaviour of dual-purpose cows in the region of Tierra Blanca, Veracruz, Mexico

V. Mariscal-Aguayo[1], H. Estrella-Quintero[1] and E. Salas-Barboza[2]
[1]Universidad Autonoma Chapingo, CRUOC, Rosario Castellanos 2332, 44950 Guadalajara, Mexico, [2]Freelancer, Floresta 1366, 44530 Guadalajara, Mexico; valmara@hotmail.com

Dual-purpose production systems are very heterogeneous in management, which makes them vulnerable to factors such as changes in production, quality and availability of forage. The objective of the present study was to evaluate the productive behaviour of 20 agribusinesses with commercial dual-purpose cattle, located in the tropical region of Tierra Blanca, Veracruz, Mexico during the years 2014 and 2015. The feeding was based on grazing forages supplemented with concentrates. The information was obtained from the global report generated by DOBLE of AGROPEC Star® software. Data of average daily milk production per cow were analysed using SAS. Significant differences were obtained in the average milk production (kg/cow/day) of the agribusinesses evaluated (P<0.01), observing the highest values in agribusinesses 2 and 15, and the lowest ones in agribusinesses 9 and 10. The analysis shows that there are differences in productivity per animal among the ranches evaluated. Only five groups were specified to show these differences, which explains about 80% of the variation observed; nevertheless, the location (municipality) of the agribusinesses, was not determinant in the production level, therefore, neither in the groupings obtained. Once again, agribusinesses 2 and 15 stand out, with the maximum value (10.25 in ranch 15) and, 9 and 10 with the lowest value (2.57 in ranch 10); the average milk production (kg/cow/day) obtained for the 20 agribusinesses evaluated was 5.83. In the analysis of the average behaviour and trend of the evaluated variable, an irregular curve was obtained from the third order regression which explains the 86.27% variation observed in the agribusinesses under study, with a maximum value (7.7) in January and a minimum (4.25) in May. The productive behaviour of the cows that are in the agribusinesses located in the region of Tierra Blanca, Veracruz is typical of the tropical livestock of Mexico, managed under an extensive system where the main source of food is the forage, which is a function of the greater or lesser availability according to year season, which finally determines the level of milk production.

Linking horse production and leisure activities: drivers, responses and consequences for the sector

R. Evans
Norwegian Univ Coll of Agriculture and Rural Development (HLB), Arne Garborgsveg 22, 4340 Bryne, Norway;
rhys@hlb.no

Much of the horse industry does not resemble other sectors of animal production. In most animal production sectors, the production is dominated by producing food. Other sectors also include other uses of animals, but the balance between food production and production for other purposes leans overwhelmingly towards food. The horse industry flips this balance over. Horses are indeed bred for food in Europe, in significant numbers. But the vast majority of horses are bred for leisure uses. This paper begins to assemble a picture of the horse industry in these terms and asks about the implications for horse production within a European system of animal production.

Towards a reconstitution of riders' paths and a typology of riding schools and clients

C. Eslan[1,2], C. Vial[1,2], S. Costa[1] and P. Rollet[1]
[1]INRA, MOISA, CIHEAM-IAMM, CIRAD, Montpellier Supagro, Univ Montpellier, -, 34000 Montpellier, France, [2]IFCE,
Pôle développement innovation et recherche, 61310 Exmes, France; eslan.camille@gmail.com

Since 2012, the evolution of the French Equestrian Federation (FFE) licensees has reversed after 70 years of continuous progression. In addition, customer versatility is intensifying. Thus the FFE policy targets riders' loyalty and the conquest of new audiences. However, the horse market is still not studied enough by marketers and there is today a real need to understand riders' motivations and profiles. This would allow equestrian managers to better promote and target their offer, retain their current customers, and attract new ones. Therefore, a quantitative survey was conducted online in France among 771 actual riders and 243 former riders, focusing on their paths, practices, and wishes. All respondents are aged 15 and over and non-owners of horse(s). Almost all French riders came to horse riding being motivated by the link with the horse. They started in a riding school (being more or less loyal to their club) but their paths lead them either to stop riding or to continue in riding schools, livery yards or at horse owner friends' home. More than half of current riders would like to become horse owners in the future. Former riders stopped riding due to lack of time, financial reasons or dissatisfaction but almost all of them (92%) would be open to practice again. For both current and former riders, Olympic disciplines is the most wished activity, followed by 'taking care of a horse' and 'groundwork', whereas competition is the most unwanted one. Using a hierarchical classification with R, we proposed a typology of current riders. We identified 5 types, some of whom never wish to practice competition or Olympic disciplines which are the main activities recorded in horse riding schools (90%). 6 types of horse riding school were also identified according to their offer. The results show that riding schools 'offer is very standardized in spite of clients' different expectations leading sometimes to dropping out of horse riding. These results offer an opportunity for equestrian managers to target their activities according to their clients' needs and expectations, in particular making better use of riders' desire to create links with horses.

A model of loyalty to services: an example from recreational riders in French riding schools

C. Vial[1,2], C. Eslan[1,2], S. Costa[1] and P. Rollet[1]
[1]MOISA, INRA, CIHEAM IAMM, CIRAD, Montpellier Supagro, Univ Montpellier, Montpellier, France, 2 place Pierre Viala, 34060 Montpellier, France, [2]Ifce, pôle développement innovation et recherche, Ifce, 61310 Exmes, France, celine.vial@inra.fr

Adopting a consumer's behaviour marketing approach, the study questions the mechanisms of service loyalty to a sporting club in the field of horse riding. In a national context of high versatility of horse riders and of a recent drop in the number of French Equestrian Federation licensees, we are interested in riders' relationship with their riding school. The loyalty concept has been widely studied in the literature. This study adopts a dynamic, relational and situational approach, assuming an influence of satisfaction, community and emotional commitments, trust and switching barriers on attitudinal loyalty. A quantitative survey conducted online in France among 630 riders (aged 15 and over, practicing in riding schools, and non-owners of horse(s)) enabled to build a structural equation model. This one shows that rider's attitudinal loyalty to the club is directly influenced by their satisfaction, their relationship with their instructor, their community commitment toward their riding school, and switching barriers (that are perceived switching costs and the attractiveness of alternatives). Emotional commitment to the group of friends only influence community commitment. The results highlight the key role of the instructor in client satisfaction, community commitment, perceived switching costs and consequently in loyalty. In contrast to what we expected, the relationship with a specific horse does not directly influence attitudinal loyalty but increases perceived switching costs. The study also seems to underline a lack of customer knowledge about alternative offers. Multi group structural equation modelling points out differences according to the social profile of the rider. For instance, students and young workers are the only group whose attitudinal loyalty is influenced by switching costs. Another example is that community engagement does not seem to be important in older (and upper social class) clients' attitudinal loyalty. Finally, these results provide evidence for equestrian structures to adapt their offer and communicate better to build customer loyalty.

Horse assisted activities as a treatment for participants with Parkinson's disease

A. Kaić[1], B. Luštrek[2], C.P. Likar[3], M. Martini Gnezda[3], D. Flisar[4] and K. Potočnik[2]
[1]University of Zagreb, Faculty of Agriculture, Department of Animal Science and Technology, Svetošimunska 25, 10000 Zagreb, Croatia, [2]University of Ljubljana, Biotechnical Faculty, Department of Animal Science, Jamnikarjeva 101, 1000, Slovenia, [3]Parkinson's Disease Society of Slovenia, Šišenska cesta 23/I, 1000 Ljubljana, Slovenia, [4]University Medical Center Ljubljana, Department of Neurology, Zaloška cesta 2, 1000 Ljubljana, Slovenia; klemen.potocnik@bf.uni-lj.si

Horse assisted activities (HAA) as an effective treatment could improve function in daily activities and quality of life in individuals with a variety of neurological conditions. The study was conducted to assess the effects of HAA in people with Parkinson's disease (PD). Seven participants (4 men and 3 women) took part in all nine meetings that were held during the study. The average age of the participants was 65 years. Before and after the study, participants responded to the questionnaire on health and mood. In addition, at the end of the last day of the study implementation, participants were asked about the observed changes according to the situation before the start of the study. With general questionnaires and a small group, it was not able to demonstrate statistically significant differences according to the state of the participants before they were included in the study and upon its completion. However, a very important finding is that with all the participants, regardless of age, condition of disease, and the fact that they have practically no experience with animals, it was achieved the same level of knowledge and skills. After nine meetings, all the participants were able to independently lead the horse and, with the help of the guide, sit on a horse with outspread arms and with no stirrups us and follow the horse's movement while walking. Despite the fact that due to the above reasons it was not possible to estimate the statistically significant effects of horse assisted activity on patients with PD, the positive effects of the activity with horses on the participants were noticed. During the study we adopted project proposal for each of 9 meetings, with detailed program and expected achievements of each meeting. Such program will be tested in near future with new group of participants with PD.

Soil intake in grazing sport horses

S. Jurjanz[1], C. Collas[1], C. Quish[2], B. Younge[2] and C. Feidt[1]
[1]*Université de Lorraine-INRA, UR AFPA, BP 20163, 54505 Vandoeuvre, France, [2]University of Limerick, Department of Biological Science, Faculty of Science and Engineering, Limerick, Ireland; stefan.jurjanz@univ-lorraine.fr*

Soil intake is the main exposure to environmental contaminants in free-ranged animals. In horses, the main concern of soil intake is reduced welfare by pathologies especially parasites, as their meat is hardly consumed. However, no data are available. Three levels of Daily Herbage Allowance (DHA: 2, 3, 4% of the body weight BW) were studied in 6 grazing sport horses via a Latin square design 3×3 performed in East Limerick, Ireland. Each period consisted in 10 d of adaption of the horses to the DHA on the new plot followed by 6 d to carry out all the measurements. The surface offered to the horses was adjusted to reflect the experimental DHA, and was estimated every 2 days measuring the herbage mass on the next surface to be grazed. The plots allocated were made using temporary electric fencing, and the two horses assigned during one period to a given treatment moved every 2nd day with pre and post-grazing sward heights (SH) measurements. Individual faecal samples were collected from the centre of the dung to avoid soil or dust contamination. Soil and grass were sampled from each plot at each period. Soil intake was estimated by a method using an internal marker (acid insoluble ash AIA) and an estimate of dry matter (DM) digestibility of the diet. Therefore, AIA contents were analysed in soil, feed and faeces and digestibility of ingested grass was estimated according to Mésochina *et al*. When DHA decreased from 4 to 2% BW, daily soil intake increased from 3.8 to 4.6% of total DM intake ($P<0.05$) corresponding to 547 and 653 g soil ($P<0.05$) respectively. The DHA of 3% BW resulted in an intermediate soil intake (in proportion and amount). Pre-grazing SH being similar between treatments (11.9 cm), post-grazing SH was significantly shorter for the lowest DHA (3.1 cm) compared to the highest DHA (4.4 cm). The DHA of 3% BW enhanced a post-grazing SH of 4.1 cm. The strong relationship to low SH has been shown in cattle and sheep but the enhanced soil intake in horses is lower. Indeed, in ruminants, a sparse grass availability reflected by post-grazing SH<4 cm enhanced soil intakes >10% what could be explained by different grass prehension.

The effect of different herbage allowances on dry matter intake and digestibility in grazing horses

C. Quish
University of Limerick, Biological Sciences, Schrodinger Building, Limerick, Ireland; carol.quish@ul.ie

Grass is a valuable forage for horses and well-managed can provide an excellent source of nutrients and assist the horse's overall health and well-being. This study investigated the impact of three different daily herbage allowances on grass dry matter intake (GDMI) and grass dry matter digestibility (GDMD) in grazing horses. Six Irish sport horses were assigned to three daily herbage allowances using a double Latin-square experimental design, T1: 2% BW/d, T2: 3% BW/d and T3: 4% BW/d. Each experimental period consisted of ten days of adaptation to the herbage allowance and six days of measurement. Daily herbage allowances were achieved by adjusting the grazing area offered to the horses every two days. Herbage mass disappearance was calculated using pre-grazing herbage mass relative to post-grazing herbage mass. Apparent digestibility of DM and CP was estimated using total faecal collection method. Faecal output was individualised for each horse by feeding indigestible synthetic markers daily. Two naturally occurring indigestible markers AIA and ADIA were used to estimate DM digestibility. GDMI on T1 (13.0±0.75 kg DM/d) was significantly higher compared to T2 (17.5±0.82 kg DM/d) and T3 (20.6±0.93 kg DM/d) ($P<0.05$). Both DMD (T1: 40.6±0.01, T2: 53.9±0.09 and T3: 62.10±0.07%) and apparent digestibility of protein (T1 40.2±0.03, T2 59.8±0.2 and T3 72.3±0.01%) were significantly different between the treatments ($P<0.05$). Daily herbage allowances presented significantly different digestible DM and CP intakes. Estimated digestibility coefficients using AIA (T1 65.6±3.57, T2 65.8±3.38 and T3 61.2±2.56%) and ADIA (T1 69.2±6.08, T2 68.2±4.67 and T3 64.9±6.53%) were not significantly affected by daily herbage allowances ($P>0.05$). This study describes nutrient availability on a grass-based diet for horses using three different digestibility methods at different grass allowances. All methods highlight the potential source of digestible nutrients available from grass for horses.

Influence of mare body conformation on growth and development of Lusitano foals until weaning

M.J. Fradinho[1], D. Assunção[2], A.L. Costa[2,3], C. Maerten[2], V. Gonçalves[2], P. Abreu[2], M. Bliebernicht[2,3] and A. Vicente[1,4]

[1]CIISA-Centro de Investigação Interdisciplinar em Sanidade Animal, Faculdade de Medicina Veterinária, UL Islam Lisboa, 1300-477 Lisboa, Portugal, [2]Pôle Reproduction, Haras de la Gesse, 31350 Boulogne-sur-Gesse, France, [3]Embriovet, Lda, Muge, 2125-348 Muge, Portugal, [4]ESAS, IPS, Santarém, 2001-904 Santarém, Portugal; mjoaofradinho@fmv.ulisboa.pt

Embryo transfer has been widely used as a reproductive technique to support equine breeding and selection. However, in the Lusitano breed, the use of this technique is fairly recent. Because maternal uterine capacity was recognized as one of the factors that may influence foetal and post-natal growth of the foals, the main objective of this work was to study the influence of body conformation of the dams on growth and development of their foals, from birth to weaning. Fifty Lusitano foals, born between 2017 and 2018 on a Lusitano studfarm in France, were weighed and measured (height at withers – HW, girth – G and cannon circumference – CC) within the first 24 hours after birth, and then every two weeks up to weaning. For data analysis, foals were grouped according to mare status and body conformation into: foals from Lusitano biological mothers (LM; n=24); foals born from Lusitano recipient mares (RLus; n=17) and foals born from warmblood recipient mares (R; n=9). Throughout gestation and lactation, husbandry conditions and management were identical for all animals, and diets were adjusted according to individual needs. Linear and quadratic effects of time and the effects of group, year of birth, gender and their interactions were evaluated through a mixed model, considering repeated measures. The effect of group was significant for body weight (BW), HW and G (P<0.05), with foals from R mothers presenting higher values when compared to LM. Considering BW and G, there was an interaction with the year of birth (P<0.001), with higher values for foals born in 2017. The effect of gender was not significant. Besides being innovative for the Lusitano breed, the obtained results, shows the importance of the selection of recipient mares with regard to body shape and the inherent uterine capacity. Project CIISA UID/CVT/00276/2019.

Meeting nutritional needs to ensure dairy donkeys' welfare

F. Raspa[1], D. Vergnano[1], L. Cavallarin[2], A. McLean[3], M. Tarantola[1] and E. Valle[1]
[1]University of Torino, Veterinary Sciences, L. Braccini 2, 10095 Grugliasco, Italy, [2]ISPA-CNR, L. Braccini 2, 10095 Grugliasco, Italy, [3]UC Davis, One Shields Av., Davis, CA 95616, USA; dianavergnano@libero.it

The increasing scientific interest on donkey milk in paediatric nutrition has led to an enhanced number of dairy donkey farms in Europe. Amongst other aspects, there is the need of a better comprehension of dairy donkeys' nutritional welfare with the aim to satisfy their nutritional needs. Few attempts have established the nutritional needs of donkeys during lactation. In order to provide an overall nutritional assessment, management-based indicators should be evaluated as hazards that may affect dairy donkeys' nutritional welfare, since they are essential to correctly manage dairy animals. Moreover, animal-based indicators should be assessed to explore conditions that may influence feed intake, feeding behaviour and nutritional requirements. Currently, just BCS and skin tent test are proposed to evaluate the 'Good Feeding' Principle. However, BCS is not a measurement of past nutrition and doesn't evaluate current nutritional status or needs. Therefore, more studies are necessary to comprehend the nutritional requirements of the jenny. Some information can be extrapolated from the available literature about the average daily milk yield produced and the foal's daily weight gain in the first six months of life. Based on this information it is possible to perform some proposals and calculate the energy requirements (ER) for milk production adding it to the maintenance ER. For example, the ER for 5 litres of milk produced by a jenny of 300 kg BW could be 3.04 Mcal NE. No data are available for dairy donkeys' protein requirements; but some indications can be found from the 2015 INRA system, estimating the protein requirement 33 g MADC/kg milk produced, in addition to the maintenance requirement. Vitamins and mineral requirements for lactating donkeys are unknown, but it seems that feeding management influences the fat-soluble vitamin concentration of milk. Understanding nutritional requirements of dairy donkeys is necessary to meet their welfare. The suggestion to evaluate BCS is not sufficient and nutritional requirements should be taken into account.

Analysis of the composition of frozen and lyophilisate mare milk for differences in composition

G.M. Polak

National Research Institute of Animal Production, Department of Horse Breeding, Sarego, 2, 31-047 Cracow, Poland; grazyna.polak@izoo.krakow.pl

The aim of the study was to analyse the composition of mare's milk in order to detect the difference in the composition of frozen and lyophilized milk. Forty-four samples, from the cold-blooded Sokólska mares (26) and the Polish Halfbred Horse (18) was analysed. The milk was obtained from May to early September 2018 once a day. Every time two identical samples were taken from each mares. Immediately after milking they were cooled to a temperature of 3-6 °C, frozen to -20 °C, and then sent to a laboratory, where one of pair identical samples was thawed and analysed under fat, protein and physicochemical properties. The second was lyophilised and analysed under the same aspects. The results of analysis were compared in pairs (for twin samples) for differences in the content of ingredients in frozen and lyophilized milk. For statistical calculations they were used the Shapiro-Wilk and Student's t-test. In two cases, there were statistically significant differences ($P=0.05$): (1) the amount of lactoferrin was higher in milk than in lyophilizate (from 0.17 to 0.41%); and (2) the amount of β-IG was lower in milk than in the lyophilizate (from 0.22 to 0.56%), which may indicate that process of lyophilization has a significant effect on its composition.

Technological approach to donkey milk cheesemaking and sensory characterization of the product

M. Faccia[1], G. Gambacorta[1], G. Martemucci[2] and A.G. D'Alessandro[2]

[1]University of Bari, Department of Soil, Plant and Food Sciences, Via Amendola 165/A, 70126 Bari, Italy, [2]University of Bari, Department of Agricultural and Environmental Science, Via Amendola 165/A, 70126 Bari, Italy; michele.faccia@uniba.it

Making cheese from equide milk is a difficult task, due to lower casein concentration and different structure of the micelle with respect to ruminants. Recently, some protocols have been reported for donkey milk that could open new perspectives for the valorisation of this livestock sector. The present communication aimed to: summarize the current knowledge about this topic and present the results of an experimentation, in which two innovative cheesemaking protocols were tested. So far, the information available in the scientific literature indicate that, even though encouraging results have been obtained, several issues are still present, mostly regarding cheese structure and yields. Our experimentation had the objective of contributing to resolve such issues by developing a suitable processing protocol. A series of cheesemaking trials were performed at laboratory level on Martina Franca donkey milk taken from a local farm. The technological scheme adopted was based on the protocol reported in a previous research, used as control, with two modifications. During the cheesemaking trials, samples of milk, whey and fresh cheese were taken and subjected to analysis of the gross composition, electrophoresis of the protein fraction, and sensory evaluation by a panel of experts. The results showed that some improvements were obtained: in one case cheese had better texture, in the second case, the yield was improved. The electrophoresis study gave punctual indications about the changes of the protein matrix in cheese, and allowed to connect them to the technological results. Finally, the sensory evaluation supplied an interesting description of the main organoleptic features of the cheeses, and demonstrated their acceptability.

Aging of different horse muscles: a proteomic approach

P. De Palo[1], R. Marino[2], J.M. Lorenzo[3] and A. Maggiolino[1]
[1]University of Bari A. Moro, Department of Veterinary Medicine, SP per Casamassima, km 3, 70010 Valenzano, Italy, [2]University of Foggia, Department of the Science of Agriculture, Food and Environment (SAFE), Via Napoli 25, 71121 Foggia, Italy, [3]Meat Technology Center of Galicia, Avenida de Galicia no. 4, Parque Tecnolóxico de Galicia, San Cibrao das Viñas, 32900 Ourense, Spain; pasquale.depalo@uniba.it

The aim of this study was to evaluate the effect of post-mortem aging on tenderness development and on proteolysis of myofibrillar proteins in 3 different horse muscles. A total of thirty-six horse samples from longissimus lumborum (LL), semitendinosus (ST) and semimembranosus (SM), respectively, were purchased from a local slaughterhouse. Each muscle was removed from the left side of twelve Heavy Draft Horse carcasses 24 hours post-mortem and transferred under refrigerate condition to the laboratory. All animals were slaughtered at about 12 months of age and were reared in the same farm. Each muscle was divided into three sections resulting in six samples for each animal and was stored at 2 °C under vacuum packaging and analysed at 1, 3, 6, 9 and 14 days of aging; cranial and caudal sections were randomised across aging time. Instrumental tenderness, MFI, total collagen, SDS-PAGE and Western blot were analysed using the GLM procedure of the SAS statistical software (SAS Institute, 2013). The mathematical model included fixed effect due to muscle, aging and muscle × aging, and random residual error. Warner-Bratzler shear force decreased during aging in all muscles, showing the lowest values in longissimus lumborum (LL) during all post-mortem period. Myofibril fragmentation index significantly increased in LL and semimembranosus (SM) muscles throughout aging time whereas in semitendinosus (ST) it increased only at 14 days of aging. Proteomics analysis revealed the major content of intact myofibrillar proteins with high molecular weight in ST muscle in the first phase of aging, while, at 14 days a greater accumulation of TnT-derived polypeptides and spots isoforms ascribed at MLC2 and MLC1 proteins was found. Data suggest that an extending aging time nullified the tenderness differences between LL and ST muscles in horse meat and highlight the more extensive post mortem proteolysis occurring in ST muscle.

Volatile compounds profile of donkey meat during aging

A. Maggiolino[1], R. Marino[2], J.M. Lorenzo[3] and P. De Palo[1]
[1]University A. Moro of Bari, Department of Veterinary Medicine, S.P. per Casamassima km 3, 70010, Valenzano, Italy, [2]University of Foggia, Dipartimento di Scienze Agrarie, degli Alimenti e dell'Ambiente, Via Napoli 25, 71122 Foggia, Italy, [3]Centro Tecnológico de la Carne de Galicia, Avenida de Galicia 4, Parque Tecnolóxico de Galicia, 32900 San Cibrao das Viñas, Province of Ourense, Spagna, Spain; aristide.maggiolino@uniba.it

The aim of the work is to assess the ageing effect on the volatile compounds profile of vacuum packaged donkey meat stored at 4 °C for 14 days. Ten Martina Franca breed male donkeys, 12 months old, were included in the trial. Samples of longissimus lumborum (LL) muscle were taken, sliced in 20 mm thick steaks, vacuum packed, stored at 4 °C and analysed at 1, 3, 6, 9 and 14 days after slaughtering. At each time, samples were grill cocked and Volatile Organic Compounds profile was performed by solid-phase microextraction and gas chromatography-mass spectrometry. VOC profile was analysed using a one-way ANOVA, where ageing time was set as independent variable. Means were compared using the Tukey´s test for repeated measures. Pentane and octane showed decreasing trend, with the lowest values at 9 and 14 ageing days (P<0.01), with no differences showed in the total hydrocarbons (P>0.05). Aromatic hydrocarbons have an important contribution to meat flavour, and they showed a decreasing trend, with the lowest values observed from the 6th to the 14th ageing day (P<0.01); furan 2-penthyl showed lowest values from the 6th ageing days until the trial end (P<0.01). Total aldehydes did not show to be affected by ageing time, although benzaldehyde was the only one that showed at 14 ageing days the highest values than other days (P>0.01). Aldehydes were the most produced by donkey meat and they are often associated to tallowy and meaty odours. Hexanal, that is considered a good oxidation indicator, characterized by high aromatic potential, providing freshly cut-grass and green aroma notes, is the most present compound. Total alcohols, and hexanol, showed lowest values from the 6th to 14th ageing days (P<0.01). The aging of donkey meat under vacuum conditions for 14 days leads to small changes in the volatile compounds profile after grilling cooking. However, the volatile compounds that showed some variation included primarily those indicative of lipid oxidation.

Young rider's satisfaction and loyalty to a club: factors and influences

C. Eslan[1,2], C. Vial[1,2], S. Costa[1] and O. Thomas[3]
[1]INRA, MOISA, CIHEAM-IAMM, CIRAD, Montpellier Supagro, Univ Montpellier, 2 place Viala, 34000 Montpellier, France, [2]IFCE, Pôle développement innovation et recherche, La jumenterie, 61310 Exmes, France, [3]UniLaSalle, Campus de Rouen, 3 Rue du Tronquet, 76130 Mont-Saint-Aignan, France; eslan.camille@gmail.com

Since 2012, the evolution of the French Equestrian Federation licensees has reversed after 70 years of continuous progression. In addition, customer versatility is intensifying. These trends are especially pronounced among children under 13, who are the main target clientele of riding schools. The services offered by riding schools have evolved but still remain too standardized and seem to have difficulties to adapt to a changing demand. In this context, this research focuses on the expectations of these young customers, but also on their satisfaction and loyalty to the club. It assumes that parent and child have complementary influences in the assessment and decision processes. The analysis is based on face-to-face surveys among 87 young riders between 7 to 12 years of age and their parents, in two French regions. The results highlight the influence of family members, especially the mother, in the initial choice for horse riding and for a riding school. However, once the practice is engaged, child's satisfaction and desire to stay or not in the club seem to take precedence over parent's opinions. From the point of view of the child and the parent, loyalty process towards riding school is largely influenced by the tripartite relationship that exists with the teacher. For the child, there is also an important role of the child's emotional commitment to one specific pony or horse and to the group of friends in the club. These two elements influence child's satisfaction but can also represent switching barriers. Equestrian disciplines that are proposed can also represent a source of dissatisfaction for the child. For the parent, there is also an emphasis on their own reception conditions. These results invite equestrian structures to question building and promotion of their offer in order to adapt it to the specificities of this market, while taking into account the distinct and complementary roles of young children and their parents in loyalty process.

The horsemeat market in France: first overview and research perspectives

C. Vial[1,2] and S. Costa[1]
[1]MOISA, INRA, CIHEAM-IAMM, CIRAD, Montpellier Supagro, Univ Montpellier, Montpellier, France, 2 place Pierre Viala, 34060 Montpellier, France, [2]Ifce, pôle développement innovation et recherche, Ifce, 61310 Exmes, France; celine.vial@inra.fr

In France, before the 20th century, horses were used for the army, transport and agriculture. Therefore, horsemeat came from aged or injured adult horses (dark red meat) and was mainly consumed by poor people or in crisis periods. Since the 20th century, France has become both producer and consumer of horsemeat. However, draft foals bred for this market (producing light red meat) are 80% exported, mainly to Italy. Whereas France imports reformed horses to satisfy 80% of the national consumption that remains oriented towards dark red meat (adult horses). This imported meat raises the question of its quality and traceability. Moreover, horsemeat consumption is decreasing in France (-46% in the 10 last years) despite the sustainable characteristics of this product: its high quality (more iron, zinc, better fats, good taste) could reduce the total amount of consumed meat, its production in extensive conditions respects animal well-being, horses emit less methane than cattle, it enables the maintenance of mountain grassland areas and of nine endangered local heavy horse breeds… Nevertheless, one can notice horsemeat consumption peaks consequently to sanitary crisis as the one of the 'mad cow', or when horse meat was found into beef 'lasagna', which suggest a consumption potential to exploit. In this context, this poster will first present the specific characteristics of the French horsemeat market. First data will be collected in the beginning of 2019 through a literature review, the analysis of databases on food consumption and first exploratory surveys among consumers and non-consumers of horsemeat. The long-term aim of this project is to obtain a typology of horsemeat eaters, potential eaters, and refractory eaters. Our goal is to understand people representations, consumption motivations, but also barriers that limit this consumption, in order to find potential new markets for this production and to better adapt its promotion.

Influence of temperature and storage time on oxidation and fatty acids in foal meat

M.V. Sarriés[1], M. Benavent[1], E. Ugalde[1], M. Ruiz[1], K. Insausti[1], M.J. Beriain[1], J.M. Lorenzo[2] and A. Purroy[1]
[1]IS-FOOD. Research Institute for Innovation & Sustainable Development in Food Chain. ETSIA. Universidad Pública de Navarra, Campus de Arrosadía, Pamplona, Spain., 31006, Spain, [2]CTC. Centro Tecnolóxico da Carne, Rua Galicia, 4, Parque Tecnológico de Galicia, San Cibrao das Viñas, Ourense, 32900, Spain; vsarries@unavarra.es

The purpose of this work was to study the effect of temperature (2 vs 8 °C) and storage time (0, 4, 8 and 12 days) of foal meat under vacuum packaging conditions on protein oxidation (carbonyl compounds) (nmol/mg protein), on lipid oxidation (malonaldehyde content (MDA) (mg MDA/kg of meat)) and on the quality of fatty acids (mg of FA/100 g of meat) from 22 Galician Mountain and Burguete crossbreeding foals slaughtered at 13 months of age. Neither temperature nor the interactions between temperature × storage time were significantly different in any of these variables ($P>0.05$). However, there were differences between the storage time for carbonyl compounds ($P<0.01$) and for MDA value ($P<0.001$). Carbonyl compounds and MDA content achieved the lowest values after 4 days of storage and thereafter they increased reaching the highest levels on the 12th day (carbonyl 'day 0': 2.21; 'day 4': 1.94; 'day 8': 1.97; 'day 12': 2.3 and MDA 'day 0': 0.49; 'day 4': 0.39; 'day 8': 0.44; 'day 12': 0.47). In terms of fatty acid content, storage time affected significantly the values of saturated fatty acids (SFA) ($P<0.01$) and polyunsaturated fatty acids (PUFA) ($P<0.001$), so that as storage time increased the main SFA and PUFA content decreased. In this regard, palmitic fatty acid (C16:0) ($P<0.01$) and stearic fatty acid (C18:0) ($P<0.01$) were the SFA most affected (C16:0 'day 0': 598.42; 'day 4': 461.04; 'day 8': 397.68; 'day 12': 346.75 and C18:0 'day 0': 191.11; 'day 4': 149.7; 'day 8': 138.87; 'day 12': 123). As for PUFA, linoleic fatty acid (C18:2n6c9c12) and linolenic fatty acid (C18:3n3c9,c12,c15) were the most affected (C18:2n6c9c12 'day 0': 377,71; 'day 4': 304.51; 'day 8': 292.53; 'day 12': 290.41 and C18:3n3c9,c12,c15 'day 0': 275.04; 'day 4': 196.07; 'day 8': 176.11; 'day 12': 147.31). To conclude, the preservation temperature did not affect the foal meat quality, and on the contrary, storage time had a great impact on protein and lipid oxidation, so prolonged storage is not recommended.

High protein intake does not increase activation of mTOR compared to low protein intake in horses

C.M.M. Loos, S.C. Dorsch, A.M. Gerritsen, K.R. McLeod and K.L. Urschel
University of Kentucky, W.P. Garrigus Bldg, Lexington, 40546, USA; carolineloos00@gmail.com

Research in non-equid species has shown that consumption of a protein-rich meal activates the skeletal muscle mechanistic target of rapamycin (mTOR) pathway in a dose-dependent manner and that stimulation above maximal threshold levels does not result in further increase in activation. We have previously shown that high levels of dietary protein also activate mTOR pathways in horses, however there is a paucity of data regarding the level of dietary protein necessary to achieve maximal activation. Thus, the objective of the current study was to evaluate activation of muscle protein synthetic pathways following consumption of either a low (LO) or high (HI) protein pellet in mature, sedentary horses. In a randomized, crossover design, six mature, thoroughbred mares were fed 0.2% of BW/day of each treatment pellet for 6 d. On d7, venous blood and gluteus medius muscle samples were collected before and 90 min after feeding the low (0.2 g protein/kg BW) or high (0.6 g protein/kg BW) protein pellet. Plasma glucose, insulin, urea nitrogen (PUN) and amino acid concentrations and the relative abundance and phosphorylation of mTOR pathway signalling molecules were determined. Data were analysed using mixed procedures in SAS with treatment, time and their interaction as the fixed effects and horse as the random subject. Regardless of treatment, feeding increased ($P≤0.05$) plasma insulin, PUN and amino acid concentrations and activation of mTOR, Akt and rpS6. There was no effect of level of protein intake on postprandial plasma glucose and insulin concentrations or activation of mTOR, Akt and rpS6 ($P>0.05$). Horses on HI had higher ($P<0.05$) levels of PUN, valine and leucine and greater ($P<0.1$) postprandial concentrations of plasma asparagine, arginine, isoleucine, ornithine and lysine compared to LO. These results illustrate that the consumption of moderate amounts of protein is sufficient to stimulate mTOR pathways and that elevated protein intake does not result in further increase in activation of muscle protein synthetic pathways in mature, sedentary horses.

Study on selected minor trace elements in donkey milk: first results

F. Fantuz[1], S. Ferraro[2], L. Todini[1], A. Fatica[3] and E. Salimei[3]
[1]Università di Camerino, Scuola di Bioscienze e Medicina Veterinaria, via Gentile da Varano, 62032 Camerino, Italy,
[2]Università di Camerino, Scuola di Scienze e Tecnologie (sezione Chimica), via Gentile da Varano, 62032 Camerino, Italy,
[3]Università del Molise, Dip. Agricoltura, Ambiente, Alimenti, via De Sanctis, 86100 Campobasso, Italy; salimei@unimol.it

Aiming to provide more in-depth data on the mineral content of donkey milk, the diurnal and nocturnal concentration of Lithium (Li), Boron (B), Titanium (Ti), Rubidium (Rb) and Strontium (Sr) and their distribution in different milk fractions have been studied. In experiment 1, four pluriparous jennies (Martina Franca derived population), 235 kg average body weight and in mid lactation (100-120 d), were machine milked during the day (at 10:00 and 18:00 h) and during the night (at 23:00 and 02:00 h) to provide individual milk samples. In experiment 2, individual milk samples from four donkeys were collected during the afternoon milking and aliquoted. One aliquot was ultracentrifuged to provide milk serum. Inductively Coupled Plasma-Mass Spectrometry was used for the analysis of the mentioned elements in whole milk and milk serum. Milk yield averaged 713.4 (±97.0, s.d.) ml/ milking. The individual daily intake (feedstuffs and water) accounted in average for 2.6 mg Li, 246.2 mg B, 48.3 mg Ti, 178.2 mg Rb, 250.2 mg Sr. Results from milk samples obtained at 10:00 and 18:00 h were grouped as day milk whereas samples obtained at 23:00 and 02:00 h were grouped as night milk. In experiment 1 data were processed by one way ANOVA. The average concentrations of the investigated trace elements did not differ significantly between milk obtained during the day and during the night (Li 6.54 vs 6.65 µg/l; B 308.88 vs 315.65 µg/l, Ti 130.66 vs 128.27 µg/l, Rb 909.69 vs 884.67 µg/l, Sr 485.53 vs 510.05 µg/l). First results on trace elements distribution indicated that all the milk Li (102%), B (99%) and Rb (98%) are associated with the serum fraction, whereas donkey milk serum contains only 44 and 28% of total Ti and Sr respectively, suggesting that the most of these latter elements are likely associated with the casein fraction.

Milk production quantity model evaluation in saddle horse Anglo-Arabian type lactating mares

L. Wimel, J. Auclair-Ronzaud and C. Dubois
IFCE, DIR Station Expérimentale, La Valade, 19370 Chamberet, France; laurence.wimel@ifce.fr

Quantification of mare milk production is a great challenge. Indeed, be able to evaluate milk yield can lead to selection based on maternal qualities and to a better understanding of the foal nutritional environment in the first months of its life. Milk yield was evaluated over three years (2016-2018) on 97 Anglo-Arabian mares. Were studied 40 primiparous with an average weight of 530 kg (426 to 635 kg) and mean age of 5.5 years old (5 to 6 yrs), and 57 multiparous with an average weight of 577 kg (480 to 724 kg) and a mean age of 9.7 years old (6 to 14 yrs). From day 3 after parturition to day 180, they had a free access to pasture and water *ad libitum*. Forage availability in pasture was managed through a rotational grazing system. We follow 3 indicators (M1, M2, WSW1) at 5 lactation points (d3, m1, m2, m3, m6). At each measuring point, foals are not separated from their dams. M1 and M2 measures are milking, both realized without Oxytocin injection and using an Udder pump (Udderly EZ ™ Pump). First milk yield (M1) occurs after a 3 hrs period during which the foal does not suckle. It reflects maximal storage capacity of the udder. The second milk yield (M2) occurs 30 minutes after M1 and is an estimate of udder filling capacity in 30 minutes. The third indicator, WSW1 (weigh-sucking-weigh1) is monitored 30 minutes after M2 and measures foal ingestion after 4 hrs separation from the mare and 30 minutes udder filling. In order to evaluate milk yield through M2 and WSW1 we used Wood's model as well as a mixed linear model based on mares' live weight, parity, days of lactation, M1 and year. Linear mixed model seems to fit better our data than Wood's curve does, based on AIC values. However, both models give similar results, which are consistent to those find in literature. Indeed, we observed a greater production for multiparous than for primiparous (P<0.001). We noticed as well that multiparous mares weighing more than 600 kg at foaling produced more milk than those weighing less than 600 kg at foaling (P<0.05).

Session 43

Theatre 1

The IMAGE unified data portal to integrate and represent European gene bank data

P. Cozzi[1], J. Fan[2], A. Sokolov[2], O. Selmoni[3], E. Vajana[3], S. Joost[3], E. Groeneveld[1], P. Flicek[2], P. Harrison[2] and A. Stella[1,4]

[1]UCSC, Via Emilia Parmense, 84, 29100 Piacenza, Italy, [2]EMBL-EBI, Wellcome Genome Campus, Hinxton, Cambridge, CB10, United Kingdom, [3]EPFL, Route Cantonale, 1015 Lausanne, Switzerland, [4]CNR, IBBA, Via A. Corti 12, 20133 Milano, Italy; paolo.cozzi@ibba.cnr.it

The Innovative Management of Animal Genetic Resources (IMAGE) project was established to improve animal gene bank management within Europe. One key challenge that this project addresses is the integration and transparent use of the vast information stored within more than 60 gene banks/genetic collections spanning 20 European countries, together with the collection of newly generated data. This imposes a major strategic challenge for the management and conservation of animal genetic resources due to the huge amount of heterogeneous data distributed across gene banks in different locations, storage formats and languages. The solution we present here comprises: (1) a well-defined metadata rule set ensuring high quality and comparable data across the diverse collections originating in different storage formats and languages; (2) development of a single Inject tool helping gene bank managers to enhance, standardise, tag and submit their gene bank data to a common data pool that integrates all gene bank records from across Europe; (3) the sustainability offered by archiving of this data within the EBI BioSamples public archive; and (4) a bespoke data portal that integrates gene bank metadata with generated 'omic datasets from within IMAGE and cross referencing to other gene bank and breeding database resources from across Europe such as those hosted by the FAO. Within the data portal, a Geographic Information System tool is included to assist the user in identifying/storing the geographical origin of the samples as well as displaying individual/population genetic parameters and biological attributes through interactive maps. Querying across all types of data is also expected to facilitate targeted search to identify genetic material of interest residing somewhere in the partner gene banks and collections.

Session 43

Theatre 2

SmartCow: integrating European cattle research infrastructures to improve their phenotyping offer

R. Baumont[1], R. Dewhurst[2], B. Kuhla[3], C. Martin[1], L. Munksgaard[4], C. Reynolds[5], M. O'Donovan[6] and B. Esmein[7]

[1]INRA, UMR Herbivores, Theix, 63122 Saint-Genès-Champanelle, France, [2]SRUC, West Mains Road, EH9 3JG Edinburgh, United Kingdom, [3]FBN Leibniz, Wilhelm-Stahl-Allee 2, 18196 Dummerstorf, Germany, [4]Aarhus Universitet, Nordre Ringgade 1, 8000 Aarhus, Denmark, [5]University of Reading, Whiteknights Campus, RG6 6AH Reading, United Kingdom, [6]Teagasc, Moorepark, Fermoy, Co Cork, Ireland, [7]EAAP, Via Giuseppe Tomassetti 3 A/1, 00161 Roma, Italy; rene.baumont@inra.fr

The sustainability of cattle production relies on improved resource use efficiency, reduced GHG emissions, and improved animal health and welfare. Facing this challenge require far more complex animal traits than previously that need to be assessed under a range of conditions. It is now time to look for means to phenotype complex animal traits using smart technologies and rapid analytical methods in a standardised way applied in many contexts. The SmartCow infrastructure project (www.smartcow.eu) was selected by the European Commission for 4 years from 1st February 2018. The project will ensure that existing guidelines are adopted (e.g. ICAR guidelines) and work towards the use of unified measurement methods when no international standard is existing. The development of the cattle ontology of traits (ATOL) and the ontology of environmental conditions (EOL) will be an important step to unify research methodologies. Refining *in vivo* methods to evaluate feed efficiency and gas emissions will generate innovations in experimental design for more accuracy. The development of new biomarkers (proxies) that can be easily measured in milk, faeces, urine, or blood through rapid analytical methods (e.g. infrared spectroscopy) will bring new phenotyping capacities. The development of tools to generate new and improved information from animal sensors and other routinely collected data (e.g. prediction of individual cow status in terms of health and welfare) will also enable a more efficient phenotyping and genetic selection of cattle. Finally, transnational access to 11 major research infrastructures will provide access to around 2,500 dairy and 1000 beef cattle and generate new phenotyping opportunities in research projects to be financed by the SmartCow project. Eleven projects have already been selected after the first call.

GplusE: Mid-infrared milk analysis based technologies adding value to gene banks

N. Gengler[1], C. Grelet[2], H. Soyeurt[1], F. Dehareng[2] and GplusE Consortium (www.gpluse.eu)[3]
[1]ULiège-GxABT, Passage des Déportés, 2, 5030 Gembloux, Belgium, [2]Walloon Agricultural Research Centre, Chée de Namur, 24, 5030 Gembloux, Belgium, [3]Lead Partner UCD, Belfield, 4 Dublin, Ireland; nicolas.gengler@uliege.be

Fitter farms through the use of animal gene banks need an excellent level of characterization of the genetic material represented in these banks. If genomic characterization is straight forward the phenotypic characterization of specific traits of interest is less. Even worse, some traits may even not be known as of interest when the material was conserved. Here mid-infrared (MIR) spectra based milk analysis strategies as studied during the recently finished EU FP7 project GplusE would allow to add value to gene bank collection. Required for this would be the recording of spectra for individuals of preserved breeds, if possible from spectrometers which were at the same time participating to a standardization procedure, and the preservation for at least a part of these MIR records the corresponding preserved (frozen) milk. There are at least four ways MIR based technologies can add value. First, MIR prediction equations developed well after the conservation of the MIR data can be used to determine a posteriori novel phenotypes for these old animals. Prediction would be more reliable when made on standardized spectra and can be done with the sole knowledge of the spectra so even for larger groups of animals then those preserved (e.g. daughters of preserved sires). Also, second, preserved frozen samples associated to some MIR records, can be used to validate, or even improve these equations adding variability that has maybe disappeared since preservation. These frozen samples can also be useful because novel reference methods (e.g. proteomics) may appear and these samples can contribute to better novel equations because they increase variability of reference calibration dataset. Finally, MIR spectra data, but also MIR based predictions can be used to establish breed differences. Using directly MIR has the advantage of a phenotypic characterization (i.e. milk phenome) close the genome. As for the use of the genome for selection of candidates for gene banks, the use of this milk phenome could be a strategy to assess and to cover the existing variability in a given breed and among breeds to preserve.

Responses of pigs divergently selected for cortisol level or feed efficiency to a challenge diet

H. Gilbert[1], E. Terenina[1], J. Ruesche[1], L. Gress[1], Y. Billon[2], P. Mormede[1] and C. Larzul[1]
[1]GenPhySE, INRA, INPT, ENSAT, Université de Toulouse, 31326 Castanet Tolosan, France, [2]GenESI, INRA, 17700 Surgères, France; helene.gilbert@inra.fr

Selection for feed efficiency is questioned as reducing the ability of farm animals to face stress and overcome challenges. The hypothalamic-pituitary-adrenocortical (HPA) axis activity, measured by the cortisol level in plasma after ACTH injection, has been proposed as an indicator of robustness. The objectives of this study were to evaluate (1) if a modified cortisol level in pigs alters their feed efficiency and their production performance; and (2) if alternative dietary resources would affect these responses. Parallel trials including divergent lines for plasma cortisol 1 hour after ACTH injection and divergent lines for residual feed intake (RFI) were run during growth with a conventional diet and a diet with high fibres, low energy and low protein content. Selection for higher or lower plasma cortisol levels after stimulation of the adrenal gland did not impact growth and feed intake traits, but it had a significant impact on body composition and carcass yield, with improved composition in the high cortisol animals. The two lines had similar responses to an alternative diet, with decreased growth rate and feed intake, and increased feed conversion ratio. On the other hand, lines selected for divergent RFI had different responses to the alternative diet, the more efficient line having a more reduced growth rate with the diet with lower energy and AA contents. However, in terms of FCR it remained more efficient. The initial hypothesis of decreased efficiency associated to increased cortisol was not validated. This project has received funding from the European Union's Horizon 2020 research and innovation programme under grant no. 633531.

H2020 IMAGE project strategy to investigate local adaptation in European sheep

L. Colli[1], O. Selmoni[2], M. Barbato[1], M. Del Corvo[1], E. Vajana[2], E. Eufemi[1], B. Benjelloun[3], S. Joost[2], P. Ajmone Marsan[1] and IMAGE Consortium[4]

[1]Università Cattolica del S. Cuore, Department of Animal Science, Food and Nutrition (DIANA), via Emilia Parmense, 84, 29122 Piacenza, Italy, [2]École Polytechnique Fédérale de Lausanne (EPFL), Laboratory of Geographic Information Systems (LASIG), School of Architecture, Civil and Environmenta, Route Cantonale, 1015 Lausanne, Switzerland, [3]Institut National de la Recherche Agronomique Maroc (INRA-Maroc), Centre Régional de Beni Mellal, Beni Mellal, Beni Mellal, Morocco, [4]INRA, Animal Genetics and Integrative Biology Unit, Domaine de Vilvert, Bât 211, 78352 Jouy-en-Josas, France; paolo.ajmone@unicatt.it

The IMAGE H2020 project objective is the valuation of livestock and poultry gene banks and biodiversity. Among the case studies run within the project, one uses landscape genomics (LG) to investigate selective sweeps associated to adaptation to climatic conditions in local sheep populations and genetic collections. To select sheep individuals to analyse, 965 geo-referenced biological samples have been split into groups based on the environmental conditions at their sites of origin. Animals living in the most diverse environmental conditions have been identified by a Principal Component Analysis followed by hierarchical clustering performed on 24 climatic and bioclimatic variables, altitude as well as land cover. The target area covered Europe and the Mediterranean region, including Northern Africa and South-western Asia. Subsequently, each of the eight main environmental clusters we identified was further subdivided into 4 sub-clusters to allow a finer mapping of the diverse environmental conditions. The final set of 672 animals was selected so to ensure both the representativeness of the environmental sub-clusters and the within-breed variation. For all animals, high-density SNP genotypes have been obtained with the Illumina Ovine HD BeadChip and merged with data from 742 geo-referenced individuals available from IMAGE partners, previous research projects and public databases. The final dataset including 1,514 individual genotypes will be analysed by landscape genomics, selection signature and population structure approaches to describe the distribution of diversity across the study area and to detect signals of environment-driven selection.

Economic resilience and efficiency indicators of conventional and organic dairy farms across Europe

C. Grovermann, S. Quiedeville, M. Stolze, F. Leiber and S. Moakes

FiBL, Research Institute of Organic Agriculture, Ackerstrasse, 5070 Frick, Switzerland; florian.leiber@fibl.org

The Horizon 2020 project GenTORE is investigating proxies for and improvement of resilience and efficiency traits in dairy and beef cattle. In this context also the farm level is assessed, first, as the environment for the animal phenotypes, and second, with regard to the farms' economic resilience and efficiency itself. The results presented here show the effect of compliance with organic production standards on dairy farms, which was assessed in terms of their economic efficiency and profitability across four geo-climatic classes in Europe, using FADN data from 39,000 dairy farms. Substantial evidence shows that organic certification contributes to environmentally safe production conditions. Less is known about its effect on the economic performance and resilience of farms. While yields might be lower in several instances, costs and benefits behave differently. The effect of certification is estimated through entropy balancing, an innovative method to achieve covariate balance in observational studies. To balance treatment and comparison, a range of control variables are used: Farm size, economic size, degree of specialisation, share of family labour, the share of forage and fodder areas and altitude. To also account for the heterogeneity of dairy farming in Europe, a class splitting model is used to allocate farms to four distinct groups, each exhibiting more similar characteristics in terms of production technology and climate than the entire sample. While differences in efficiency between conventional and organic certified farms are relatively small across the 4 classes, ranging from no significant effect to a positive effect of 2%, impacts of organic certification on profitability are more pronounced, ranging from no significant effect to a positive effect of 47%. Besides environmental gains, organic certification can thus have potential as an institutional innovation for a profitable dairy economy However, the balanced data shows slightly higher variance for the profitability of organic farms in three classes, which might suggest that economic resilience of organic farms is more limited than in conventional systems.

Capacity building to enhance the use of animal genetic resources in a multinational context

L.T. Gama[1], P. Boettcher[2], O. Cortes[3], M. Naves[4], M.R. Lanari[5], L.M. Melucci[5], C. Lucero[6], S.J. Hiemstra[7], A. Elbeltagi[8], B. Benjelloun[9], A. Bouwman[7], J.J. Windig[7], A.J. Amaral[1] and M. Tixier-Boichard[4]

[1]FMV-Univ. Lisboa, Av. Univ. Técnica, 1300-477 Lisboa, Portugal, [2]FAO, AnGR, V. Terme Caracalla, 00153 Rome, Italy, [3]Univ. Complutense Madrid, Av. Puerta Hierro, 28040 Madrid, Spain, [4]INRA, UMR GABI, 78352 Jouy-en-Josas, France, [5]INTA, Modesta Victoria 4450, 8400 Bariloche, Argentina, [6]AGROSAVIA, vía Bogotá-Mosquera, 250047, Colombia, [7]Wageningen Univ. Research, Box 338, 6700 Wageningen, the Netherlands, [8]APRI, Nadi El Said, Dokki, Giza, Egypt, [9]INRA-Tadla, 567 Beni-Mellal, Tadla, Morocco; ltgama@fmv.ulisboa.pt

Management of genetic diversity is crucial to develop programs aimed at the conservation and sustainable utilization of Animal Genetic Resources (AnGR). The IMAGE project was developed with the goal of enhancing the use of genetic collections and upgrading the management of animal gene banks, by developing genomic methodologies, biotechnologies, and bioinformatics for better knowledge and exploitation of AnGR. One key component of IMAGE has been the development of capacity building programs designed to strengthen the activity of gene bank managers and other stakeholders involved in managing AnGR. Three postgraduate training courses were organized in 2018, in Argentina, Colombia and The Netherlands. These one-week courses involved lecturers from several IMAGE partners, and aimed to build capacity in conservation and management of AnGR using novel technologies. The course programs were designed with high quality standards, engaging international lecturers and post-graduate students and professionals, with the goal of developing the skills of researchers and managers engaged in conservation genetics and population management, providing the environment and framework for participants to apply the acquired knowledge in their projects. Nearly 100 participants from 17 different nationalities (in Europe, Africa and America) attended the courses, which covered topics such as the role of AnGR in sustainable development, assessment of genetic diversity using different tools, management of small populations, development and management of conservation programs, and reproductive technologies. Additional training sessions are planned before the project ends in early 2020.

Burning issues in biodiversity 2: fitter livestock farms from better gene banks?

M. Tixier-Boichard and IMAGE Consortium

INRA, AgroParisTech, University Paris-Saclay, GABI, 78352 Jouy-en-Josas, France; michele.tixier-boichard@inra.fr

Animal gene banks are an investment for countries and research institutions as well as an asset for the livestock sector. A better knowledge of these collections is key to stimulate their use. Since information about gene bank collections is often hard to find for a breeder, the IMAGE H2020 project is developing a web portal to connect all data pertaining to gene bank collections. Unfortunately, data on gene bank collections are scarce, except for cattle. IMAGE is producing molecular data with the aim to correlate them with adaptation or specific phenotypic traits of breeds being present in gene banks, but it does not focus on phenotyping, except for semen. Cooperation between EU-funded projects on livestock is supported by the Common Dissemination Booster dedicated to fitter animal farms. This 'CDB' gathers GenTORE, FEED-A-Gene, SAPHIR, GplusE, SmartCow and IMAGE projects. This session will feature results and approaches of these projects that could be used to better document gene bank collections. The SmartCow infrastructure project is devoted to phenotyping: animals with original phenotypes would be a target for gene bank collection, and resources from gene banks could be used to produce animals from past genotypes to be compared with current ones. IMAGE is also using time series of gene bank collections to identify selection signatures in a cattle breed. GplusE developed mid-infrared milk analysis based technologies that could be used in different ways to add value to gene banks. FEED-A-GENE is identifying criteria and populations related to a better feed efficiency in pigs and chickens fed alternative feedstuffs. GenTORE will illustrate impacts of geoclimatic classes and farm type (organic or conventional) on economic resilience of dairy farms. Association between environmental variation and genetic diversity will be illustrated by IMAGE with a landscape genetics approach in sheep. Cooperation between these projects should increase our knowledge of phenotype-genotype relationships. Refined characterization of gene bank samples would improve the prediction of animal performance and enhance the complementarity between gene banking strategies and management of on-farm populations. This approach could then be disseminated in training programs of IMAGE.

Generalizing the construction of genomic relationship matrices

L. Gomez-Raya[1], W.M. Rauw[1] and J.C.M. Dekkers[2]

[1]*Instituto Nacional de Investigaciones Agroalimentarias, Mejora Genética Animal, Carretera de la Coruña km 7, 28040, Spain, [2]Iowa State University, Ames, IA, Department of Animal Science, 239D Kildee Hall, 50011 Ames Iowa, USA; gomez.luis@inia.es*

The construction of genomic relationship matrices (GRM) for genomic breeding value estimation is based on the assumptions of random mating and no selection. However, those conditions do not hold in farm animals. Populations of pigs are often not in Hardy-Weinberg equilibrium (HWE) because of crossing to take advantage of heterosis. Methods to generalize the construction of a GRM under no HWE conditions are proposed using a biological parameterization. For each locus not in HWE equilibrium with genotypes AA, Aa and aa, the additive GRM is centred by $1-f_{AA}+f_{aa}$, $-f_{AA}+f_{aa}$, and $-1-f_{AA}+f_{aa}$, respectively. Similarly, the centring of the dominance relationship matrix for the same locus is $-f_{Aa}$, $1-f_{Aa}$, and $-f_{Aa}$ for individuals with genotypes AA, Aa and aa, respectively. Formulas for the scaling of the GRM under no HWE were also derived. Illumina SNP arrays were used to construct additive and dominance relationship matrices for two data sets of approx. 200 Landrace × Large White crossbreds from a field test for natural infections for Porcine Reproductive and Respiratory Syndrome Virus (PRRSV). Departures from HWE were observed for a majority of SNPs. For the two data sets, the diagonals of the additive GMR had an average of 0.84 and 0.87 when assuming HWE but were nearly 1.00 after using the scaling proposed in this study. Dominance GMR were 1.03 for both trials when assuming HWE but 1.00 without this assumption. Components of variance with and without the assumption of HWE were estimated for the area under the viremia curve, maximum viremia and average daily gain. Estimates of variance of additive components for the three phenotypes analysed were biased upwards by about 10% when assuming HWE. In conclusion, the proposed methods should be used instead of the methods assuming HWE because they are more general to any type of population.

meBLUP: a new single step gBLUP combining pedigree and SNP information without genomic inverse

E. Groeneveld[1] and A. Neumaier[2]

[1]*Institute of Farm Animal Genetics (FLI), Höltystrasse 10, 31535, Germany, [2]Faculty of Mathematics, University of Viennna, Oskar-Morgenstern-Platz 1, 1090 Vienna, Austria; eildert.groeneveld@gmx.de*

The solution of Henderson's mixed model equations (MME) becomes computationally tractable only when the inverse of the numerator relation matrix (NRM) is set up directly. On the other hand, for a genomic (SNP) based relationship matrix (GRM), an actual inverse needs to be computed. Based on an augmented model equivalent to the MME, with model equations (ME) of the simple form Ax=b (where A is the incidence matrix for fixed and random factors, x a vector of unknowns, and b a right hand side), we developed a new single step gBLUP combining pedigree and SNP data. One set of ME explains the additive genomic effects by an infinitesimal model with one regression coefficient per animal for each SNP in the panel. A second set of ME has one trivial equation for each SNP random effect. A joint pedigree and SNP based genetic evaluation called meBLUP is done by combining these two set of ME to Ax=b and solving for x by preconditioned conjugate gradients (PCG). In this formulation no inverse of either a NRM and GRM appears. meBLUP was implemented in VCE and PEST2 and tested with simulated data based on the correlation between true breeding value and the model equation based gBLUP. Compared to the ssBLUP as implemented in blupf90, meBLUP typically showed equal or higher correlations for smaller, while ssBLUP gave higher correlations for larger panel sizes. CPU time increases linearly with the number of individuals and the panel size: going from 2,500 to 40,000 SNPs increases the CPU time per iteration on 200,000 phenotypes and 102,902 SNP records from 0.08 sec to 12.7 sec, by a factor of 16 each. In contrast, inversion time in ssBLUP grows cubically with the size of the matrix inverted. Performance was assessed by a 3 trait dairy data set of 10 mio milk records and 3 mio pedigree, resulting in 7 mio unknowns. One PCG iteration took 1 sec for the pedigree only model, while adding a 30k panel for 10,000 animals increased this to 1.7 sec. Convergence was reached after 34 iterations for the meBLUP model while it took 749 iterations for the AM resulting in a total processing time of 58 vs 743 sec. Using meBLUP often leads to a much lower total time.

Factors affecting accuracy of estimated effective number of chromosome segments for small breeds

J. Marjanovic and M.P.L. Calus

Wageningen University & Research, Animal Breeding and Genomics, P.O. Box 338, 6700 AH Wageningen, the Netherlands; jovana.marjanovic@wur.nl

Small dairy cattle breeds often produce less than mainstream breeds, which endangers their existence. Genomic selection (GS) may significantly increase the genetic improvement per year and can be used to improve long-term perspectives of small breeds by making them more lucrative for farmers. GS requires a sufficiently large breed-specific reference population for genomic prediction (GP), which is the main obstacle for numerically small breeds. This may be overcome by adding individuals from other breeds to the reference population, however, the benefit strongly relies on relatedness between the breeds, which can be inferred through the effective number of chromosome segments (M_e). In fact, M_e is the key parameter in determining the accuracy of GP. Our objective was to investigate the number of individuals needed to accurately estimate M_e within and between populations. In addition, we studied the effect of marker density on the estimates of within and between population M_e. Two populations of 5,000 individuals separated for 100 generations were simulated to reflect selection history and LD structure of current domestic cattle breeds, using QMSim software. We tested five sample sizes of 10, 50, 100, 500, 1000 individuals on the accuracy of within population M_e and combinations of these sample sizes from each breed on the accuracy of between population M_e. M_e is presented as the average of estimates from 50 replicates. The initial number of SNPs was 720k. We selected every 2^x-th SNP, where x ranged from 1 to 6, to test the effect of marker density. The M_e was computed as the variance of genomic relationships within or between populations. Our results show that at least 50 individuals are needed to accurately estimate within population M_e, regardless of SNP density. Lower sample size resulted in significant oscillations of the estimates. However, SNP density was important for between population M_e. For example, 45k SNPs underestimated between population M_e value by 20% compared to 720k. Importantly, for estimates of between population M_e, both populations need to have at least 100 genotyped individuals.

Impact of data sub-setting strategies on variance component estimation for Interbeef evaluations

R. Bonifazi[1], J. Vandenplas[1], J. Napel[1], A. Cromie[2], R.F. Veerkamp[1] and M.P.L. Calus[1]

[1]Wageningen University & Research, Animal Breeding and Genomics, 6700 AH Wageningen, P.O. Box 338, the Netherlands, [2]Irish Cattle Breeding Federation, Highfield house, Bandon, P72 X050, Co Cork, Ireland; renzo.bonifazi@wur.nl

Interbeef (International beef) cattle evaluations are required to predict breeding values on a national scale while using data from multiple countries. Accurate international evaluations depend on accurate estimation of genetic correlations (rG) among countries. Despite the availability of many individual phenotypes for international beef cattle evaluations, the low usage of artificial insemination in beef compared to dairy leads to poor genetic connectedness among countries. This makes the estimation of genetic parameters challenging, both in terms of obtaining accurate and consistent estimates, as well as requiring long computational time. The objective of this study was to investigate and compare data sub-setting strategies in beef low-connected populations to obtain more efficiently accurate estimates of genetic correlations across countries. Age-adjusted weaning weight phenotypes and pedigree information were available for 3,128,338 Limousine beef cattle males and females, born between 1972 and 2017. The phenotypes were distributed across 10 European countries and 19,357 herds. A total of 1,441 bulls had offspring in more than one country (common bulls). Variance component estimation for a multi-trait animal model, where weaning weight is modelled as a different correlated trait for each country, was performed including the full dataset using Monte Carlo EM REML as implemented in the MiX99 software. Estimated rG, for both direct and maternal genetic effect, and the required computational time of the full data analysis will be compared with those from three data sub-setting strategies: (1) random selection of herds from the biggest country; (2) reduction of data based on genetic connectedness measurements; 3) herd selection targeting common bulls and using country-specific contemporary group size. It is expected that a balanced distribution of records across countries using common bulls will provide more accurate and more computationally efficient rG estimation for international beef cattle evaluations.

Genome-wide association study for natural antibodies in colostrum of Swedish dairy cattle

J. Cordero-Solorzano[1,2], J.J. Wensman[2], M. Tråvén[2], J.A.J. Arts[1], H.K. Parmentier[1], H. Bovenhuis[1] and D.J. De Koning[2]
[1]Wageningen University & Research, P.O. Box 338, 6700 AH Wageningen, the Netherlands, [2]Swedish University of Agricultural Sciences, Ultuna, 750 07, Uppsala, Sweden; dj.de-koning@slu.se

Colostrum with sufficient antibodies is essential for the newborn calf, as it requires this passive immunity to survive until weaning. High variation in the amount of colostrum antibodies in Swedish dairy cows has been reported, with a large proportion having low antibody levels. Natural antibodies (NAb) are produced without any antigenic stimulation and target self-antigens and pathogen-associated molecular patterns (PAMPs). Our objective was to estimate genetic parameters and detect quantitative trait loci (QTL) for two NAb isotypes (IgG, IgM) in colostrum binding keyhole limpet hemocyanin (KLH) and muramyl dipeptide (MDP). Three experimental farms were included in the study, 1,719 colostrum samples from 1,313 cows between 1 to 6 parities, calving from January 2015 to April 2017 were collected. 70% of the animals were Swedish Red and 30% Swedish Holstein. Antibodies were measured from colostrum using indirect ELISAs. To estimate genetic parameters, a linear mixed model with repeated measures (different calvings from the same cow) was run using ASReml 4, correcting for cow parity number, time from calving to colostrum sampling and breed, including Herd-Year-Season of calving and sample storage plate as random effects. An imputed 50k SNP array from a LD 7K array was used for the Genome-wide association study (GWAS), running the same model but including the SNP genotype as a fixed effect. Heritabilities for colostrum NAbs ranged from 0.15 to 0.27, with a permanent environment effect for IgG isotypes accounting for 30% of the variance and for IgM ranging from 15 to 19%. Genetic correlations between IgG and IgM ranged from 0.1 to 0.4. The GWAS revealed one QTL on BTA3 for IgM (KLH and MDP), the latter comprised of 7 SNPs (-log10(p)=4.4), two significant and five suggestive, ranging from 80-105 Mbp and another QTL on BTA7 for IgG (KLH and MDP) consisting of 3 SNPs (-log10(p)=4.1), from 85-113 Mbp. Our results suggest that natural antibodies can potentially provide an effective tool to improve colostrum quality using genetic selection.

WGBLUP model improves accuracy of breeding values prediction in a commercial line of broilers

H. Romé[1], T.T. Chu[1], R. Hawken[2], J. Henshall[2] and J. Jensen[1]
[1]Aarhus university, Blichers Allé 20, 8830, Denmark, [2]Cobb-Vantress, P.O. Box 1030, 72761-1030 Siloam Springs, Arkansas, USA; helene.rome@mbg.au.dk

To maximize genetic gain in broilers, the main lever is to improve accuracy of breeding values prediction. Genomic best linear unbiased prediction (GBLUP) model increased accuracy of predicted breeding values in poultry. Nevertheless, this accuracy could be improved by considering information on genetic architecture of traits of interest. In this study, we compared accuracy and inflation of breeding value prediction using GBLUP and WGBLUP model in commercial line of males and females broilers. This population was genotyped with an Illumina 50k array and their body weight (BW) was recorded for five successive weeks. Weeks of age and sexes were studied separately so that ten traits were analysed. After quality control 18,096 individuals genotyped and phenotyped for 46,502 SNP were kept for analyses. For cross-validation, dataset was split in two groups. Animals were assigned randomly to one group. Within each group, SNP effects were estimated using a single regression marker. Variance component were estimated using REML module in DMU on full dataset. Breeding values were computed with a GBLUP and WGBLUP models on both reduced datasets. For WGBLUP model, to estimate one group, SNP weight estimated within the other group were used to compute the weighted genomic matrix. Accuracies were computed as the correlation between the 'true' phenotype and the estimated breeding value based on reduced dataset divided by the square root of heritability. The inflation was estimated via the slope of the regression (b) of the true phenotype on the estimated breeding value. Gain of accuracy from GBLUP to WGBLUP ranged from 0.0 to 4.2%. An important inflation of predicted breeding values were observed with WGBLUP model (b=0.7 in average) whereas no inflation was observed with GBLUP (b=1.0 in average). In conclusion, taking into account genetic architecture of trait could lead to a higher predictive ability but leads to an overestimation of predicted breeding values.

Impact of preselection on genetic evaluation of selection candidates using single step GBLUP

I. Jibrila, J. Ten Napel, J. Vandenplas, R.F. Veerkamp and M.P.L. Calus
Wageningen University and Research, Animal Breeding and Genomics, Droevendaalsesteeg 1, 6708 PB Wageningen, the Netherlands; ibrahim.jibrila@wur.nl

This study was conducted to investigate the impact of preselection on genetic evaluation of selection candidates using single step GBLUP (ssGBLUP). A breeding programme with a single-trait breeding goal was simulated, with features of pig and poultry breeding programmes. A recent population of 15 generations with selection using pedigree BLUP was simulated from a historical population of 3,000 generations of random mating without selection. In selecting the parents of the 16th generation from the individuals of the 15th generation, 8 preselection scenarios were tested. In any scenario, the simulated heritability was 0.1, 0.3 or 0.5; the form of preselection was genomic, phenotypic, parental average, random or no preselection; and the preselection intensity was either high (10% of males and 15% of males preselected) or very high (5% of males and 12.5% of females selected). After each preselection scenario, breeding values of all animals were predicted using ssGBLUP, and the scenarios were evaluated for accuracy, bias and ability to correctly estimate genetic trends. Initial results are available for three scenarios when using a heritability of 0.3, which are no preselection, high genomic preselection and very high genomic preselection. For the 3 scenarios, mean Pearson's correlation coefficient (SEM of 10 replicates in brackets) between true breeding values (TBV) and genomic estimated breeding values (GEBV) of selected animals in the 15th generation (a measure of accuracy) was 0.63 (0.01), 0.59 (0.01) and 0.61 (0.01), respectively. Also for the selected animals in the 15th generation, mean regression coefficient of TBV on GEBV (a measure of bias) was 1.00 (0.01), 0.97 (0.01), and 0.97 (0.02), respectively. Mean realised genetic gain was 2.62 (0.04), 2.34 (0.03) and 2.23 (0.03) additive genetic standard deviations, respectively. Although mean realised genetic gain is significantly higher without preselection, results for accuracy and bias confirm the ability of ssGBLUP to reasonably handle the impact of genomic preselection, and that genomic preselection can greatly reduce cost of phenotyping with minor impact on accuracy and bias of genomic prediction of selection candidates.

Genetic analysis of different categories of fertility disorders in Holstein cows

V. Müller-Rätz[1], R. Schafberg[1], F. Rosner[1], K.F. Stock[2] and H.H. Swalve[1]
[1]Martin-Luther University Halle-Wittenberg, Institute of Agricultural and Nutritional Sciences, Theodor-Lieser-Str. 11, 06120 Halle, Germany, [2]IT Solutions for Animal Production (VIT), Heinrich-Schröder-Weg 1, 27283 Verden, Germany; renate.schafberg@landw.uni-halle.de

Diagnoses of health disorders and diseases were made available within the project GKUHplus for Holstein dairy cattle for the period from January 2013 to May 2015. In a first analysis, a total of 35,325 completed lactations in parities 1 to 3 from 27,389 cows in 44 herds could be included. In general, traits were defined as binary traits considering the entire lactation or as continuous traits counting the number of independent periods of disease. The focus of the study was on fertility disorders, either summarizing all kinds of disorders or specifically addressing different forms of disorders. Analyses were performed in multiple trait linear and threshold animal models using ASReml software. Estimates of heritabilities ranged between 0.03 and 0.12 with higher estimates found for binary traits analysed under threshold models. Fertility disorders exhibited positive, i.e. antagonistic genetic correlations with milk yield in the range of 0.29 to 0.43. Relationships with udder health traits and foot and leg disorders were more complex. Sterility, defined as total, and sterility of uterine or ovarian origin showed positive genetic correlations with foot and leg disorders of between 0.28 and 0.34. Genetic correlations of sterility (total) with clinical mastitis had positive values only for later parity cows. Uterine sterility showed negative genetic correlations with clinical mastitis, while ovarian sterility was uncorrelated with mastitis cases. Post-partum fertility disorders had positive genetic correlations with clinical mastitis of 0.43 and also with foot and leg disorders. Continued documentation of health events in the dairy farms (approx. 66,000 lactations until 2017) provides the basis of further analyses which contributes to improved understanding of correlation patterns among disorders to be addressed in future dairy cattle breeding programs.

Modelling fertility trait as censored observations in Finnish Ayrshire data

K. Matilainen[1], G.P. Aamand[2] and E.A. Mäntysaari[1]
[1]Natural Resources Institute Finland (Luke), Myllytie 1, 31600 Jokioinen, Finland, [2]Nordic Cattle Genetic Evaluation, Agro Food Park 15, Skejby, 8200 Århus N, Denmark; kaarina.matilainen@luke.fi

Currently observations for the interval from first to last insemination (IFL) in Nordic fertility evaluations are penalized for animals with uncertain pregnancy status after last recorded insemination. Another option would be to use right censored Gaussian model, where unknown observations are replaced by observations generated based on current solutions for model effects and time of censoring. The trait, IFL in first parity cows, was used to investigate three approaches: (1) Linear model with original observations (LM); (2) Linear model with penalized observations (LMp) and 3) Right censored Gaussian model (CM). Data contained 1.7 million IFL records for Finnish Ayrshire cows from 1992 to 2018, with 14% of the observations censored. For CM analysis, 0.6% of the observations were removed to prevent problem with fixed effect classes that have all observations censored. Pedigree contained 3 million animals. All analyses had the same heritability 0.05 and the same model effects as in the Nordic evaluations. The CM method was implemented in MiX99. Both linear model analyses converged in 5 minutes, whereas time needed for non-linear CM analysis was half an hour. Correlation between estimated breeding values (EBV) by LM and LMp for all RDC sires was high (0.99) and EBV by CM had higher correlation between EBV by LMp compared to EBV by LM (0.96 vs 0.94). Among the top 5% of these sires, 79% (70%) were the same based on CM and LMp (LM). Validation of the models was based on the regression of deregressed proofs from full data on EBV from reduced data. Validation reliabilities (regression coefficients) for parent average were 0.16 (0.71), 0.19 (0.81) and 0.21 (0.76) for LM, LMp and CM. In conclusion, CM was feasible although more time consuming compared to linear models.

Identification of genomic regions associated with bovine fertility

H. Sweett[1], E. Ribeiro[2], A. Livernois[1,3], S. Nayeri[1], M. Romulo Carvalho[2], J. Felipe Warmling Sprícigo[2], F. Miglior[1] and A. Cánovas[1]
[1]University of Guelph, Centre for the Genetic Improvement of Livestock, Department of Animal Biosciences, 50 Stone Road W, N1G 2WE, Guelph, Canada, [2]University of Guelph, Department of Animal Biosciences, 50 Stone Road W, N1G 2W1, Guelph, Canada, [3]University of Guelph, Department of Pathobiology, Ontario Veterinary College, 50 Stone Road W, N1G 2W1, Guelph, Canada; hsweett@uoguelph.ca

Fertility plays a key role in the success of calf production, but there is evidence that reproductive efficiency in beef and dairy cattle has decreased during the past half century worldwide. Pregnancy failure has been attributed to early losses after fertilization or failure to conceive, due to poor bull and/or cow fertility. Identifying animals with superior fertility could significantly impact cow-calf production efficiency. The objective of this research was to combine structural and functional genomic data to identify candidate regions affecting fertility in beef and dairy cattle. A genome-wide association study (GWAS) was performed using GenABEL on a population of 329 crossbred (mainly, Angus × Simmental) beef bulls to identify genomic regions associated with scrotal circumference (SC). Quality control retained SNPs on autosomes, with a call rate $\geq 90\%$, a P-value for Hardy-Weinberg Equilibrium test $>10^{-6}$, and a minor allele frequency $>5\%$ for further analyses. Preliminary results identified two significant SNPs on BTA16 (FDR<0.05) associated with SC that were previously found to play a role in spermatogenesis. Regarding cow fertility, 82 Holstein cows were artificially inseminated (AI) and vaginal swabs, uterine biopsies, and transcervical flushing samples were collected on Day 15 post AI. A conceptus was recovered from 16 cows and uterine flushes were tested for interferon-tau signalling for pregnancy confirmation, revealing a total of 37 pregnant cows (14 primiparous, 23 multiparous) and 45 non-pregnant cows. Further analysis will be performed using RNA-Sequencing to compare the uterine transcriptome of pregnant (n=35) and non-pregnant (n=35) cows to identify key regulator genes and structural variations associated with pregnancy. The study of both bull and cow fertility will lead to a more complete understanding of the reproductive performance of cattle.

Genotype imputation on several SNP-chip panels for genomic selection on Norwegian White Sheep

X. Yu[1], J.H. Jakobsen[2], T. Blichfeldt[2], J.C. McEwan[3] and T.H.E. Meuwissen[1]
[1]*Norwegian University of Life Sciences, Department of Animal and Aquacultural Sciences, Arboretveien 6, 1432, Ås, Norway,* [2]*Norwegian Association of Sheep and Goat Breeders, Box 104, 1431, Ås, Norway,* [3]*AgResearch Limited, Invermay Agricultural Centre, 19053 Mosgiel, New Zealand; jj@nsg.no*

Genotyping costs are considerable for implementation of genomic selection (GS). Cost of genotyping has therefore been one of the reasons for slower implementation of GS in small ruminants compared to dairy cattle. Genotyping with low density chips and imputation to higher density can greatly reduce the costs. Although sparser chips are cheaper, imputation to higher density is less accurate than actual high-density genotypes. In a project of genomic selection in Norwegian White Sheep (NWS), we genotyped 828 rams with 606k SNP chip (HD). These rams were selected to better impute newer born in the complete NWS population. All other rams in this project have so far been genotyped with 8k chip (LD). The imputation was done with Beagle 5. Of the genotyped animals, 345 were genotyped with both LD and HD panels. Three methods were used to determine imputation quality at various levels. Firstly, imputation rates were tested in the 345 ID who were genotyped with both HD and LD chips, with HD genotypes as control. The error rates were also examined against allele frequencies and genomic position. Secondly, G matrices constructed with various (imputed) panels were QQ-plotted one by one in order to compare their relationship estimates. In general, diagonals of imputed genotypes deviated from diagonals of non-imputed genotypes. Thirdly, single-step GBLUP runs were conducted including various G-matrices in addition to pedigree-based relationships. Genetic trends of progeny tested rams were compared to genetic trends of comparative BLUP run excluding the genomic information. In general, trends were upwardly biased when including imputed genotypes in the ssGBLUP, which was not observed using non-imputed genotypes. Generally, the results show that our imputation directly from LD to HD does not work well. We hypothesize that the gap to impute directly from 8k LD to 606k HD is too large, and the use of some more intermediate density SNP chips is recommended.

SNP prioritisation in GWAS with dense marker sets

B. Czech[1], T. Suchocki[1,2] and J. Szyda[1,2]
[1]*Wroclaw University of Environmental and Life Sciences, Biostatistics group, Department of Genetics, Kozuchowska 7, 51-631 Wroclaw, Poland,* [2]*National Research Institute of Animal Production, Krakowska 1, 32-083 Balice, Poland; bartosz.czech@icloud.com*

The development of modern methods of SNP genotyping – whole genome sequencing (WGS), targeted amplicon sequencing (TAS), and SNP array, allow to use constantly increasing number of markers for testing their association with phenotypes. However, such increasing number of (often highly correlated) markers that are statistically tested poses a big challenge in controlling the I type error. The aim of this study is to compare different multiple testing correction methods, proposed by Bonferroni, Holm, Hochberg, Benjamini & Hochberg, and Benjamini & Yekutieli, and compare distribution of P-values received from GWAS applied to different real data structures. In addition we compare different SNP prioritization approaches to GWAS data. The material comprises a set of nominal P-values resulting from GWASs: (1) 2,977 cows genotyped by the GeneSeek® Genomic ProfilerTM HD panel; (2) 288 chicken genotyped using targeted amplicon sequencing in candidate gene regions; and (3) 32 cows with whole genome sequences. The estimated proportions of true null hypotheses amounted to 0.999 (for both traits) for the 2,977 cows genotyped by SNP chip. The Bonferroni correction selected 3 (0.005%) and 2 (0.003%) SNPs for both traits genotyped using microarray significant at the 5% level, while the estimated false discovery rates resulted in 3 (0.005%) and 2 (0.003%) SNPs with the maximum discovery rates of 5%.

Using genomics to understand the genetic architecture of tolerance to Aleutian disease in mink
Y. Miar
Dalhousie University, Department of Animal Science and Aquaculture, 58 Sipu Awti, Truro, B2N5E3 Nova Scotia,
Canada; miar@dal.ca

Aleutian disease (AD) caused by the Aleutian mink disease virus, is the most economically significant disease impacting mink (*Neovison vison*) production in Canada and worldwide. The costs of AD to the Canadian mink producers is over a million dollars annually in testing and loss of herds. Clinical outcomes following infection include reproductive failure, reduced growth performance and increased mortality. Although this disease has no vaccine or treatment, genetic variation in AD tolerance exists in mink due to the variability in their immune response to infection. Genomics technology provides tremendous opportunities to elucidate the biology of disease tolerance but the potential of this technology has not yet been investigated in mink health. The goals of this project are to use genomics and bioinformatics approaches paired with analysis of immunity to advance our understanding of the genetic mechanisms underlying AD tolerance in mink. This project will seek to: (1) create the first genome assembly of mink using next generation sequencing and design a robust and reliable SNP assay (50-60k) for genomics discovery in mink; (2) discover genome structure and signature of selection as well as identify new genetic variants explaining variation in AD tolerance; and (3) identify the genetic relationships of AD tolerance with other economically important traits such as feed efficiency, reproductive performance, fur quality, growth rate and pelt size. The AD tests were performed on 1,212 mink from the Canadian Centre for Fur Animal Research (CCFAR) at Dalhousie Agriculture Campus (Truro, Nova Scotia) and Parkinson Fur Farm (Rockwood, Ontario) using iodine agglutination test, pack cell volume, counter-immunoelectrophoresis, and quantitative enzyme-linked immunosorbent assay. These animals will be genotyped using the developed assay for subsequent genomic analyses. The information gained in this project will contribute novel insights into genetic architecture of AD tolerance in mink and how this may impact mink health and productivity. This work will help reduce the costs of AD to the mink industry.

SNP discovery and association study with milk fat traits in Mediterranean river buffalo Leptin gene
G. Cosenza[1], D. Gallo[1], G. Chemello[2], G. Gaspa[2], L. Di Stasio[2] and A. Pauciullo[2]
[1]University of Naples Federico II, Agriculture, Via Università 100, 80055 Portici (NA), Italy, [2]University of Turin,
Agricultural, Forest and Food Sciences, Largo Paolo Braccini 2, 10095 Grugliasco (TO), Italy; alfredo.pauciullo@unito.it

Leptin is a 16 kDa protein highly expressed in the adipose tissue, pituitary gland, brain and mammary gland. Leptin is responsible for the regulation of body weight and energy homeostasis and is one of the potential biomarkers for high animal performances, leading to better adaptability and productivity. In cattle, many studies have been reported on associations between Leptin gene (*Lep*) and traits like meat quality, carcass characteristics, fertility, feed intake, energy balance, body weight, etc. Other studies have also reported that this hormone can influence milk quality in dairy cattle, sheep and goats. On contrary, there has been limited research regarding association studies between SNPs at this *locus* and milk traits in buffalo. *Lep* gene in buffalo consists of three exons and two introns (GenBank: AH013754.2) and it has been mapped on chromosome 8 (BBU8q32). The aim of the present study was to identify and analyse the variability of the *Lep locus* in Italian Mediterranean river buffalo and to test possible associations with milk fat traits. Individual fresh milk and blood samples were collected from 766 unrelated lactating buffaloes reared in 14 farms of Southern Italy. 306 milk samples were used for fat analysis (fat content, fatty acid composition and fatty acid classes), whereas blood samples were treated for genomic DNA isolation. We sequenced the exon 3 (over 1.8 kb) of 8 buffaloes randomly chosen. Nucleotide alignment showed a synonymous transition G>A at the 231st nucleotide of exon 3 that identifies a new allele named A1. The entire panel of DNA samples was genotyped in outsourcing (KBiosciences, UK). The allele frequency of the common allele (*Lep A*) was 0.84 (genotype distribution: 547 A/A, 196 A/A1, 23 Lep A1/A1). The statistical analysis, using a mixed linear model, showed no direct effect of *Lep* g.231G>A genotype on milk fat traits. However, this result does not exclude the existence of an association of this SNP with other traits of economic interest.

Leptin receptor LEPR gene is associated with reproductive seasonality in Rasa Aragonesa sheep breed

K. Lakhssassi[1], M. Serrano[2], B. Lahoz[1], P. Sarto[1], L. Iguacel[1], J. Folch[1], J.L. Alabart[1] and J.H. Calvo[1,3]
[1]Centro de Investigación y Tecnología Agroalimentaria de Aragón – IA2, Zaragoza, 50059, Spain, [2]INIA, Madrid, 28040, Spain, [3]ARAID, Zaragoza, 50018, Spain; klakhssassi@cita-aragon.es

Leptin gene and its receptor have been strongly associated with production and reproduction traits. The aim of this research was to perform an association study among some polymorphisms detected in the *LEPR* gene and three reproductive seasonality traits: the total days of anoestrus (TDA), based on weekly individual plasma progesterone levels and defined as the sum of days in anoestrus; the progesterone cycling months (P4CM), defined for each ewe as the rate of cycling months between January and August, and also based on progesterone determinations; and the oestrus cycling months (OCM) defined for each ewe as the rate of months cycling between January and August, and based on oestrus records. A total of 239 Rasa Aragonesa ewes from one flock were controlled from January to August 2012. We sequenced the exons 4 and 20 of the *LEPR* of 20 ewes with extreme phenotype values for TDA and OMC to search polymorphism. Four non-synonymous SNPs were selected for genotyping the whole population (n=203): one in exon 4 (rs411478947), and 3 in exon 20 (rs412929474, rs428867159, and rs405459906). Only, the interaction between the SNP and age affected the TDA, P4CM and OMC traits for SNP rs412929474 in exon 20, showing different effects between mature and young ewes. After Bonferroni correction, the TDA phenotype differed among genotypes in young ewes, finding significant differences between the GG and AG genotypes. The haplotype and the interaction haplotype × age, affected also the TDA and OMC traits. Two haplotypes were associated with OMC variability considering the whole population. Thus, ewes with 0 copies of these haplotypes showed more (h2) or less (h4) oestrous cycling months (depending on the haplotype) that those with 1 copy. Also, other haplotype was associated with TDA phenotype in both young and adult ewes, finding significant differences between animals carrying 0 or 1 copies. This genetic information address for the first time the role of the *LEPR* gene in ruminant's reproductive seasonality.

Sequence-based GWAS on carcass traits and organ proportions in Charolais cows

P. Martin[1], S. Taussat[1,2], D. Krauss[3], D. Maupetit[3], A. Vinet[1] and G. Renand[1]
[1]Institut National de la Recherche Agronomique, UMR1313 Génétique Animale et Biologie Intégrative, Domaine de Vilvert, 78352 Jouy en Josas, France, [2]Allice, 149 rue de Bercy, 75012 Paris, France, [3]Institut National de la Recherche Agronomique, UE332 Unité expérimentale de Bourges, Domaine de la Sapinière, 18390 Osmoy, France; pauline.martin@inra.fr

Slaughter traits have a main economic value in beef cattle. They are also an important information source on the physiological variability among animals. To investigate the genomic control of some of these traits, a sequence-based GWAS was performed on 578 adult Charolais cows. Animals were slaughtered at seven years of age, after a fattening period. They were phenotyped for 18 traits: empty body weight (EBW), carcass yield, percentage of muscle, percentage of fat, 5th quarter, 5th quarter's fat, leather, rumen, omasum, abomasum, intestines, total digestive tract, total reproductive tract, liver, lungs, heart, kidneys and spleen. The carcass yield and all 5th quarter and organ traits were expressed relative to the EBW. Cows were genotyped with the BovineSNP50 SNPchip (Illumina Inc., San Diego), and their genotypes were imputed up to the sequence level based on the run6 of the 1000 bull genomes project (2,333 animals including 128 Charolais bulls), using successively the FImpute and Minimac software. GWAS analyses were performed using the GCTA software. All traits together, we identified 19 significant QTL at the stringent sequence based Bonferroni threshold (-logP>8.19). The carcass yield, the percentages of muscle and fat, the 5th quarter and the 5th quarter's fat were all found highly associated with the same region of the BTA2. These QTL reflect the well-known effect of the myostatin major gene on the carcass composition. Even though the relative weight of the 5th quarter shows a strong association with this gene, no QTL was observed at this position for the organs taken individually. Other QTL were detected in association with the kidneys (several QTL on BTA 18 and 24), with the lungs (two QTL on BTA 19), with the rumen (BTA 2 and 28), with the heart (BTA5) and with the abomasum (BTA 28). Further investigation will be made in parallel with other phenotypes such as feed efficiency.

Genome-wide association study for backfat thickness trait in Nellore cattle

R.P. Silva[1], M.P. Berton[2], F.E. Carvalho[1], L. Grigoletto[1], J.B.S. Ferraz[1], J.P. Eler[1], R.B. Lobo[3] and F.S. Baldi Rey[2,3]
[1]Univ. of Sao Paulo/FZEA, Veterinary Medicine, Rua Duque de Caxias Norte, 225, Campus da USP, Pirassununga, SP, 13635-970, Brazil, [2]UNESP/FCAV, Animal Science, Via de Acesso Prof.Paulo Donato Castellane s/n, Jaboticabal, SP, 14884-900, Bulgaria, [3]ANCP, Rua João Godoy, 463, Jardim América, Ribeirão Preto, SP, 14020-230, Brazil; presidencia@ancp.org.br

This study was carried out to estimate genetic parameters and to identify genomic regions associated with backfat thickness (BF). Animals were scanned at a point three-fourths the length ventrally of the Longissimus muscle area. Data from ~60,000 animals participating in the Nellore Brazil breeding program of National Association of Breeders and Researchers (ANCP) were used. A total of 8,545 animals were genotyped using the Bovine HD Beadchip 777k SNPs. The analyses were performed using the BLUPF90 family of programs. The SNP substitution effects were estimated using a single-step procedure (ssGWAS). The model included the fixed effect of the contemporary group and animal age at evaluation (linear covariable), the additive genetic effect and residual effects were considered as random. The GWAS results were reported based on the proportion of additive genetic variance explained by windows of 10 adjacent SNPs. The identification and positioning of the SNPs in the bovine genome were performed using the NCBI, Ensembl and DAVID database. The heritability was moderate for BF (0.17±0.06). The genomic regions located on chromosomes 1, 2, 6, 10 and 22 explained 7.3% of the additive genetic variance. Surveying candidate genes in those regions associated with BF revealed genes such as TKT, FNDC5, CHRND and PLCL2. The TKT gene was associated with intramuscular fat in cattle. The FNDC5 gene was related with lipid and carbohydrate metabolism. The CHRND gene was associated with muscle mass reduction during ageing, influencing the final percentage of fat and muscle in beef cattle. The PLCL2 gene was related with tenderness and is highly linked to fat accretion and body composition. The identification of these regions would contribute to a better understanding and evaluation of backfat thickness in Nellore cattle.

Effect of lactose percentage in milk on longevity in Polish Holstein-Friesian cows

M. Morek-Kopec[1] and A. Zarnecki[2]
[1]University of Agriculture in Krakow, Department of Genetics and Animal Breeding, Al. Mickiewicza 24/28, 30-059 Krakow, Poland, [2]National Research Institute of Animal Production, ul. Krakowska 1, 32-083 Balice k. Krakowa, Poland; rzmorek@cyf-kr.edu.pl

Lactose percentage in many countries is recorded together with other milk constituents but rarely included in breeding programs. Different degrees of association between lactose content and fertility, health and longevity have been reported. To analyse the effect of lactose percentage in milk on longevity a Weibull proportional hazard model was applied, using two definitions of longevity: true longevity as length of productive life measured by number of days from first calving to culling (uncensored records) or last test day (censored records), and functional longevity as length of productive life adjusted for production. First lactation test-day lactose percentage data (TD_LP1) from 2,577,222 cows in 23,018 herds, calving for the first time between 1995 and 2017 were collected from the SYMLEK recording system. Mean length of productive life was 1,123.3 days for 1,432,259 uncensored records (56.6%) and 954.8 days for 1,144,963 censored records. Overall mean TD_LP1 was 4.85% with SD 0.22%. Based on these values, cows were assigned to five classes (LP1) according to mean lactose percentage deviation from overall mean. For functional longevity the statistical model included time-independent fixed effects of lactose percentage classes, age at first calving, and time-dependent fixed effects of year-season, parity-stage of lactation, annual change in herd size, relative fat and protein yield, time-dependent random herd-year-season effect and sire effect. The same model without fat and protein effects was used for true longevity. Likelihood ratio tests showed a highly significant overall impact of LP1 on culling risk. Increased culling risk was associated with low lactose in milk. Compared to cows with medium and high lactose percentage, the relative risk of culling cows in the lowest-lactose class was 1.4 times higher when the effect of protein and fat was included in the model, and almost twice higher when those effects were not included. Higher risk of culling cows with low lactose percentage might be related to its possible role as indicator of health and fertility, and indirectly survival.

Genetic parameters for workability and conformation traits Polish Holstein-Friesian dairy cattle

B. Szymik, P. Topolski, K. Żukowski and W. Jagusiak
National Research Institute of Animal Production, Department of Genetics and Cattle Breeding, Krakowska 1, 32-083 Balice near Cracow, Poland; bartosz.szymik@izoo.krakow.pl

The aim of this study was to estimate genetic parameters between workability traits (WT) and conformation traits with and without genotypes sires of cows in Polish Holstein-Friesian dairy cattle. The analysis includes all conformation (descriptive and specific) traits from Polish dairy cow evaluation system. Records of 13,280 cows (milking from 2007 to 2014) were collected in the database system SYMLEK belonging to Polish Federation of Cattle Breeders and Dairy Farmers. The cows were scored for WT at the second test-day of the first lactation. Genetic parameters between WT and conformation traits were estimated using Gibbs sampling method which has been implemented in the BLUPF90 package. Multi-trait linear model of observation was applied to estimate (co)variance components. The model included the fixed effects of HYS and lactation stage, fixed regressions on percent of Holstein-Friesian genes and age of calving and random genetic effect. Genetic and phenotypic correlation coefficients between WT and conformation traits were estimated first only with pedigree data information. Next, data has been extended with genomic information about 2,228 sires of cows. For milking temperament (MT) and milking speed (MS) we estimated heritability coefficients 0,12 and 0,25, respectively. In the group of descriptive traits heritability ranged from (0.25) for calibre to (0.08) for legs and feet. Among the specific traits the highest heritability was found for length of teats (0.32), and the lowest for the legs and feet (0.08). Genetic correlation coefficients between MS and conformation traits ranged from -0.37 for the udder structure to 0.13 for the calibre. Genetic correlation coefficients between MT and milk conformation traits varied from -0.49 for the calibre to -0.37 for legs and feet. Phenotypic correlation coefficients between MS and conformation traits ranged from –0.01 for legs and feet to 0.04 for teat length, and between MT and conformation traits were slightly lower (from -0.02 for udder width to 0.01 for teat position).

Construction and evaluation of economic selection indices for Japanese Black cattle

H. Hirooka[1], K. Oishi[1], E. Muraki[2] and K. Inoue[3]
[1]Kyoto University, Graduate School of Agriculture, Oiwake-cho, Kitashirakawa, Sakyo-ku, Kyoto, 606-8502, Japan, [2]Gifu Prefectural Livestock Research Institute, Takayama, Gifu, 506-0101, Japan, [3]National Livestock Breeding Center, Nishigo, Fukushima, 961-8511, Japan; hirooka@kais.kyoto-u.ac.jp

In Japan, breeding values for carcass and reproductive traits of Japanese Black (Wagyu) cattle have been estimated using animal model BLUP methods since 1990s and the estimated breeding values have been used in selecting sires and cows for breeding purpose on national, prefecture and farm levels. Although carcass traits, in particular marbling score, have been historically emphasized in selecting breeding animals in Japanese breeding schemes, reproductive traits are also receiving increased attention in recent years because reproductive traits are an important determinant of farmers' profitability. The objectives of this study were to construct economic selection indices based on both carcass and reproductive traits for Japanese Black cattle and demonstrate usefulness of the economic selection indices. The estimated breeding values of sires and cows for two carcass traits (marbling scores and carcass weight) and one reproductive trait (calving interval) in Gifu prefecture were used as example data in this study. The economic values of the three traits were calculated using a bio-economic model for life cycle cow-calf-feedlot production systems for Japanese Black cattle. The calculated economic values (yen per year for life-cycle cow basis) were 70,449, 941 and -2,655 for marbling score, carcass weight and calving interval, respectively (US\$1=110 yen). The pseudo selection of sires and cows were conducted based on selection by each single trait and selection by economic indices (based on all three traits and two traits except marbling score) which are expressed by a linear combination of the estimated breeding value weighted by the economic value of the traits. Results of pseudo selection demonstrated that the economic selection index based on all the three traits can provide highest increase of net merit and realize genetic superiority of selected sires and cows for the three traits. It was suggested from the result that constructing economic selection index would contribute to greater profitability on national, prefecture and farm level.

Genetic relationships between growth curve parameters and reproductive traits in Wagyu cattle

K. Inoue[1], M. Hosono[1], H. Oyama[1] and H. Hirooka[2]
[1]National Livestock Breeding Center, Nishigo, Fukushima, 961-8511, Japan, [2]Kyoto University, Sakyo, Kyoto, 606-8502, Japan; k1inoue@nlbc.go.jp

In recent, the price of calves at the market has been rising and thus pressing feedlot farmers' businesses due to decreasing the number of calves in Japan. Genetic improvement for reproductive traits is therefore required to increase the cow population and their producing calves. For production side, body size of cows has been increasing as a result of improving carcass weight, weight gain and so on. It is known that body size and growth pattern of cows may affect reproductive traits. The objective of this study was to estimate genetic parameters for growth curve characteristics and reproductive traits including conception rate (CR) at the first service, age at first calving (AFC) and gestation length (GL) in Japanese Black female cattle. Gompertz growth function was fitted to body weight and age data to obtain mature weight (MW) and rate of maturing (ROM) of female cattle. The records of CR, GL and AFC for the first calving were also collected. Records from total of 3,204 animals were used in this analysis. Genetic parameters were estimated using AIREML method with a linear bivariate animal model. The average heritability estimates for the growth curve parameters were moderate (0.29 for ROM) and high (0.58 for MW), whereas the estimates for reproductive traits were low (0.02 to 0.10). The negative genetic correlation (-0.26) was found between MW and ROM, suggesting that a heifer with faster ROM has smaller MW. The genetic correlation of CR with MW was negative (-0.42) and with ROM was strongly positive (0.91). The results suggested that a heifer with faster ROM and smaller MW has better CR. On the other hand, a heifer with smaller MW would increase AFC, because the significantly negative genetic correlation (-0.49) was found between AFC and MW. The genetic correlations of GL with MW and ROM were 0.09 and -0.20, respectively.

Evaluation criterion for response to selection with constraint

M. Satoh, S. Ogawa and Y. Uemoto
Tohoku University, Aramaki-Aoba 468-1, Sendai, 980-8572, Japan; masahiro.satoh.d5@tohoku.ac.jp

The aims of the present study are to represent the concept of restricted breeding values algebraically and to propose a criterion for evaluating the genetic responses achieved by using a restricted selection procedure. An additive genetic mixed model characterized by multiple traits with constraints was assumed. If the random errors approach zero and the fixed effects can be completely estimated correctly in the model, the restricted best linear unbiased predictor of breeding values (u_R) is equal to $[[_q\text{-}G_0C_0(C_0{}'G_0C_0)^{-1}C_0{}']\otimes I]u$, where G_0, C_0, and u are the additive genetic variance-covariance matrix for the q traits, the matrix for restriction, and the vector of breeding values, respectively. Therefore, if we want to evaluate the response to restricted selection, such as by a stochastic computer simulation study with known breeding values, we can use u_R as only one criterion.

Bayesian response to selection on linear and ratio traits of feed efficiency in dairy cattle

M.S. Islam[1], J. Jensen[2], P. Karlskov-Mortensen[1], P. Løvendahl[2] and M. Shirali[2]
[1]Division of Animal Genetics, Bioinformatics and Breeding, University of Copenhagen, 1870 Frederiksberg, Denmark, [2]Center for Quantitative Genetics and Genomics, University of Aarhus, 8830 Tjele, Foulum, Denmark; nvf800@alumni.ku.dk

The aims of the current study were to: (1) estimate genetic parameters of linear trait of residual feed intake (RFI) and ratio trait of feed conversion ratio (FCR), along with dry matter intake (DMI), and energy sinks (energy corrected milk (ECM), body weight (BW) and body condition score (BCS)) at different periods across the first lactation; and (2) to derive the Bayesian responses to direct selection on RFI or FCR. The study included repeated recordings of feed intake and energy sink traits on 847 primiparous Danish Holstein cows during 44 lactation weeks. Entire lactation was grouped into 6 periods, each of the first 5 periods was 8 weeks in length and 6th period was 4 weeks in length. Bayesian multi-variate repeatability animal model was used for DMI, ECM, BW, BCS in each period separately. RFI was obtained by conditioning DMI for ECM, BW, and BCS using partial genetic regression coefficients. The posterior distribution of the FCR's breeding values was derived from the posterior distribution of 'fixed' environmental effects and additive genetic effects on DMI and ECM. Response to selection was defined as the difference in additive genetic mean of the selected top individuals, expected to be parents of the next generation, and the total population after integrating genetic trends out of the posterior distribution of selection responses. Posterior means of heritability were 0.15 to 0.5 for DMI; 0.32 to 0.42 for ECM; 0.54 to 0.64 for BW; 0.28 to 0.43 for BCS and 0.06 to 0.10 for RFI in period 1 to 6. Selection against RFI showed a direct response of -0.69 to -1.34 kg/d and correlated responses of -0.004 to 0.41 kg/kg for FCR and -0.69 to -1.35 kg/d for DMI in period 1 to 6 with no significant effect on production traits. Selection against FCR showed a direct response of -3.54 to 0.01 kg/kg and correlated responses of -0.62 to -0.27 kg/d for RFI; -2.20 to -0.20 kg/d for DMI; -4.28 to 5.50 kg/d for ECM; -44.17 to -10.85 kg/d for BW and -0.14 to 0.08 for BCS in period 1 to 6. Selection for FCR might result in unexpected response to selection.

Estimation of genomic breeding values for three milk traits in the Frizarta dairy sheep

A. Kominakis[1], G. Antonakos[2] and A. Saridaki[3]
[1]Laboratory of Animal Breeding & Husbandry, Department of Animal Science and Aquaculture, Iera Odos 75, 11855 Athens, Greece, [2]Agricultural and Livestock Union of Western Greece, 13th Km N.R. Agrinio-Ioannina, 30100 Lepenou, Agrinio, Greece, [3]School of Environmental Engineering, Technical University of Crete, University Campus, 73100, Chania, Greece; acom@aua.gr

The aim of the present study was threefold. First, to estimate genomic breeding values (GEBVs), second to compare GEBVS with 'classical' estimated breeding values (EBVs) and third to predict phenotypic records based on GEBVs (and fixed effects) of three milk traits i.e. milk yield (MY), fat content (FC) and protein content (PC) in the Frizarta dairy sheep. A total number of 485 Frizarta dairy ewes genotyped with the ovine 50k SNP chip array with available pedigree data and phenotypic records on the three traits were used. 'Classical' EBVs were estimated using the BLUP method while estimation of GEBVS was performed using the GBLUP method and 46,232 SNPs passing typical marker quality criteria (call rate>0.95, minor allele frequency $\geq$0.05 and Fisher's Hardy-Weinberg equilibrium test $P<10^{-4}$). Prediction of the phenotypic values for the three traits was made via a k-fold cross-validation approach. Pearson correlations between EBVs and GEBVs was as high as 0.51, 0.53 and 0.61, for MY, FC and PC, respectively. The respective correlations between EBVs and phenotypic values were 0.46, 0.40 and 0.43, for MY, FC and PC, respectively. Finally, correlations between GEBVS and phenotypic values were as high as 0.69, 0.58 and 0.70 for MY, FC and PC, respectively. Current results suggest that genomic selection might be efficient in this specific population. However, they need to be verified in a larger sample before a clear conclusion could be drawn.

Autorregressive and random regression models in multiple lactations of Holstein cattle

D.A. Silva[1], J. Carvalheira[2], P.S. Lopes[1], A.A. Silva[1], H.T. Silva[1], F.F. Silva[1], R. Veroneze[1], G. Thompson[2] and C.N. Costa[3]

[1]*Universidade Federal de Viçosa, Animal Science, Avenida P.H. Rolfs, s/n, Viçosa, 36570-900, Brazil,* [2]*University of Porto, CIBIO-InBio/ICBAS, Rua Padre Armando Quintas, Crasto, n/a, Vairão, 4485-661, Portugal,* [3]*Empresa Brasileira de Pesquisa Agropecuária, Embrapa Gado de Leite, Rua Eugênio do Nascimento, 610, Dom Bosco, Juiz de Fora, 36016240, Brazil; plopes@ufv.br*

Autoregressive (AR) and Random regression (RR) models were fitted to test-day (TD) records from first three lactations of Brazilian Holstein cattle. Data comprised 4,142,740 milk yield (MY) and somatic cell score (SCS) records registered between 1994 and 2016 from 2,322 herds. MY and SCS additive genetic variances estimates using AR model were, respectively, 7.0 and 22.2% higher than those obtained from the RR model. Residual variances of MY (SCS) in the first, second and third lactation were, respectively, 15.5 (6.6), 39.2 (18.9) and 39.8% (19.1%) lower when estimated using the AR model. MY heritabilities (h^2) were similar using both models and ranged from 0.17 to 0.23. For SCS, the h^2 in the first, second and third lactation were, respectively, 0.17 (0.11), 0.16 (0.14) and 0.14 (0.14) using the AR (RR) models. MY and SCS genetic correlations between TD estimated using the RR model ranged from 0.68 to 0.90 and from 0.71 to 0.92, respectively. The rank correlation between EBV obtained from AR and RR models were 0.96 and 0.94 for MY and 0.97 and 0.95 for SCS, respectively, for bulls with 10 or more daughters and cows. Annual genetic gains for bulls (cows) obtained using AR model were of 46.11 (49.50) kg for MY and -0.019 (-0.025) score for SCS; whereas which for RR model were of 47.70 (55.56) kg for MY and -0.022 (-0.028) score for SCS. Akaike information criterion (AIC) values were lower for the AR model and for SCS, they were, on average, 52.9% (P<0.05) lower. Both models performed well and may be used for genetic evaluations of production traits of the Brazilian Holstein cattle. Given the lower number of parameters to estimate the AR model is more parsimonious and would be a reasonable choice to be used in genetic evaluations.

Genetic parameters for lactation curve traits in Holstein dairy cows

M. Salamone[1], H. Atashi[1,2], J. De Koster[1], G. Opsomer[1] and M. Hostens[1]

[1]*Ghent University, Department of Reproduction, Obstetrics and Herd Health, Salisburrylaan 133, 9820, Belgium,* [2]*Shiraz University, College of Agriculture, Departement of Animal Science, Shiraz, Iran; matthieu.salamone@ugent.be*

The graphical representation of milk yield during lactating period is lactation curve. Typical lactation curve of a dairy cow shows a peak occurring during first weeks after calving, followed by a daily decrease in milk yield until the cow is dried-off. The objectives of this study were to estimate (co)variance components and genetic parameters for lactation curve traits in Holstein dairy cows. The records on 9,917 primiparous Holstein cows with calving year of 2010 to 2018 distributed over 91 herds in four countries (the Netherlands, Belgium, Great Britain and Denmark) were used. To describe the lactation curve, the MilkBot model was fitted on milk recording events from each animal. Pedigree information was collected for all phenotyped animals which included 36,681 animals. The genetic analyses were carried out through restricted maximum likelihood (REML) method. (Co)variance components for milk yield, and lactation curve parameters including ramp (controlling the rate of rise in milk production in early lactation), scale (representing the theoretical maximum daily yield) and decay (representing the rate of senescence of production capacity) were estimated using single and multiple-trait animal models. The estimated heritability for milk yield and lactation curve parameters ranged from 0.35 (milk yield) to 0.12 (ramp parameter of lactation curve). Genetic correlations among milk and lactation curve traits ranged from 0.10 (between the scale and ramp parameter of the lactation curve) to 0.97 (between milk yield and scale parameter of lactation curve).

Preliminary longevity analysis for birth weight homogeneity in mice

N. Formoso-Rafferty, J.P. Gutiérrez and I. Cervantes
Facultad de Veterinaria, Universidad Complutense de Madrid, Avda. Puerta de Hierro s/n, 28040, Madrid, Spain;
n.formosorafferty@ucm.es

Longevity is a desirable trait in breeding programs because it increases the profitability of the farm. Selecting for homogeneity has been recently started to be included as one of the selection objectives in genetic breeding programs because it has been shown to relate to higher robustness. A divergent selection experiment for birth weight variability in mice during 20 generations has shown that homogeneous animals would be less heavy but more effective in terms of litter size at birth and weaning, survival and feed efficiency. The objective of this work was to compare the longevity of the two divergent lines. A current experiment with an initial number of 43 females and 43 males from both variability lines were mated one to one, and stayed together from the last birth in order to have several parturitions until the end of the productive life. Females are discarded when the time elapsed from the last parturition was higher than 63 days. The number of animals reaching consecutive parturitions was recorded and compared based on a single statistical Chi-square test between lines. Interval between parturitions and litter size at birth and weaning were also compared between lines via a generalized linear model with the line and number of parturition as fixed effects and the female as random effect besides residual. There were significant differences (P<0.001) between lines in litter size at birth and at weaning and in the interval between consecutive parturitions. Regarding litter size, low line presented 3 pups more than high line and their mean interval between parturitions were 28.0 and 36.8 days respectively. In addition, the low variability line performed 35% more first births and twice second ones than the high variability line. Although the experiment continues developing, the percentage of females that were discarded was 25.6% in low line and 65.1% in high line. According to these preliminary results, the line selected for low variability presented important reproductive advantages. Higher longevity was found in homogeneous animals suggesting higher robustness and animal welfare.

GWAS for six milk production traits in Spanish Churra Sheep using deregressed EBV

M. Sánchez-Mayor[1], R. Pong-Wong[2] and J.J. Arranz[1]
[1]Universidad de León, Producción Animal, Facultad de Veterinaria. Campus de Vegazana s/n, 24071. León, Spain,
[2]The Roslin Institute and R(D)SVS, Genomics and Genetics, University of Edinburgh, Roslin, Midlothian., EH25 9PS,
Scotland, United Kingdom; msancm@unileon.es

Genome-wide association studies were carried out for six milk production traits in a commercial population of Spanish Churra dairy sheep: Milk yield (MY, grams), Protein yield (PY, grams), Fat yield (FY, grams), Protein percentage (PP, %), Fat percentage (FP, %) and Total solids percentage (TSP, %). The analysis considered the phenotypic test-day records for these traits available in the breeder association' database from 1970 to 2016. A total of 2,368 individuals of the population were genotyped with two medium-density SNP chips (50k-Chip) from Illumina and Affymetrix. After a quality control, 39,470 autosomal SNPs common to both chips remained available for analysis. The GWAS was done using deregressed EBVs of genotyped individuals. First, a univariate REML/BLUP analysis was performed on each trait, including records from all genotyped individuals plus those from their flock contemporaneous and ancestors. A repeatability model was used to account for the individuals' permanent environment effect as well as their polygenic effects (modelled using the NRM calculated with pedigree information). Thereafter the EBVs from genotyped animals were deregressed as EBV/r2 and the GWAS was carried out on them using an in-house built software. The model used in the GWAS analyses included an additive polygenic effect (modelled using a GRM) plus the effect of the SNP, with the heritability being reestimated when testing each SNP. Bonferroni corrected significance thresholds were used to account for multiple testing and determine genome- and chromosome-wise thresholds. At the genome-wise significance level ($P<1\times10^{-6}$), we found 8 significant SNP markers, located on Chromosome 3 between 134.9 and 140.6 Mb, where it corroborated the LALBA gene found in previous studies and two other genes POU6F1 and ADCY6. Other chromosome-wise significant associations were identified (76 SNPs) on chromosomes 1, 3, 4, 6, 8, 13, 20 and 26. The results reported here suggests that the methodology used is appropriate for GWAS analysis.

A locus on BTA5 with sex-dependent effect on horn growth in Norwegian Red cattle

A.B. Gjuvsland, H. Storlien and A.G. Larsgard
Geno SA, Storhamargata 44, 2317 Hamar, Norway; arne.gjuvsland@geno.no

The main determinant of the presence of horns in cattle is the Polled locus, with the polled allele *P* dominant over horned *p*. The prevailing hypothesis of horn genetics includes two other loci Scurs (*Sc*) and African horns (*Ha*), which are both epistatic to Polled and show sex-dependent inheritance in heterozygous polled animals. The Polled locus has been mapped to BTA1 and causal variants identified. For *Sc* and *Ha* the situation is different, with lack of replication of mapping results for *Sc* and no reports on *Ha*. We extracted register data on hornbud and dehorning for genotyped Norwegian Red calves (39,932 females, 19,206 males) with *P* allele frequency 0.165. Conditioning on sex and Polled genotype we found that male and female *PP* animals had similar polled registration rates (F:0.77, M:0.77), while more females than males were registered as polled among *Pp* (F:0.75, M:0.66) and *pp* (F:0.08, M:0.04) animals. Dehorning rates in *pp* animals were slightly higher in females (F:0.66, M:0.62) and low for both sexes in *PP* animals (F:0.01, M:0.02). However, for *Pp* animals dehorning rates were much higher in males (F:0.03, M:0.19). We grouped animals by sex and Polled genotype (F-*pp*, F-*Pp*, M-*pp*, M-*Pp*) and performed GWAS (GCTA-MLMA-LOCO) on imputed 777k genotypes with hornbud and dehorning registrations as binary traits. For dehorning of *Pp* males we found a clear GWAS peak on BTA5 around 70 Mb. The top snp BovineHD0500019653 ($P<10^{-21}$) has a MAF of 0.129 and a considerable effect on dehorning rate in *Pp* males (CC:0.21, CT:0.11, TT: 0.05). For hornbud registrations in *pp* and *Pp* males we also found suggestive GWAS peaks on BTA5, while none of the analyses on females showed a peak in this area. Thus, our analysis reveals a locus on BTA5 with gene action resembling that of the hypothesized *Sc* and *Ha* loci. The routine registrations in Norwegian Red do not include explicit recording of scurs. To determine whether the locus on BTA5 gives scurs or horns we will follow up with registrations of scurs in selected calves.

Routine approach to identification of mastitis using copy number variations in dairy cattle

K. Żukowski, P. Topolski and M. Skarwecka
National Research Institute of Animal Production, Department of Cattle Breeding, Balice, 32-083, Poland;
kacper.zukowski@izoo.krakow.pl

Mastitis, is one of the most common and expensive disease in dairy cows, implies significant losses in the dairy farms. The aim of this work was to confirm links between copy number variation (CNVs) in selected and previously reported DNA sequences and the incidence of clinical mastitis. Consequently, to use this information to routine identification of mastitis in Polish dairy cows based on CNVs. As a material we used cows genotypes deposited in Polish cattle SNP genotypes database (cSNP) including veterinary history of mastitis. The phenotype information was collected on four experimental farms belonging to National Institute of Animal Production. More than 1000 cows previously reported with mastitis were genotyped on low density EuroG LD chip extended with previously reported association of SNPs and CNVs with mastitis. The proposed approach was based on identified CNVs affecting mastitis and comparison of the results with veterinary history of each cow in reference group. Next, the cross-validation was performed. As a result, we observed a significant correlation between the cows which were carrying CNVs affecting mastitis in validation group. However, not all CNVs regions were confirmed to be positive in the wide cow population. As a result, we focused on regions covered APP, SSFA2, TXNRD2, ADORA2A, NDUFS6 and genes localized on BTA28.

KASP assay of 89 SNPs in Romanian cattle breeds and their association with mastitis and lameness

D.E. Ilie[1], D. Gavojdian[2], R.I. Neamt[1] and L.T. Cziszter[3]
[1]*Research and Development Station for Bovine, Arad, 310059, Romania,* [2]*Research and Development Institute for Bovine, Balotesti, 077015, Romania,* [3]*Banat's University of Agricultural Sciences and Veterinary Medicine 'King Michael I of Romania', Timisoara, 300645, Romania; danailie@animalsci-tm.ro*

Mastitis and lameness are costly production diseases influencing farming efficiency and animal welfare. Given the limited number of researches done up-to-date concerning genetic studies of autochthonous cattle breeds reared in Romania, the objective of our study was to analyse the missense SNPs belonging to 6 genes (CXCR2, CXCL8, TLR4, BRCA1, LTF, BOLA-DRB3) which were found to be associated with genetic resistance or susceptibility in mastitis and lameness. A total of 298 cattle reared in Western Romania (250 Romanian Spotted and 48 Romanian Brown) were included in the research herd. Bovine SNP data were collected from the dbSNP (NCBI). Kompetitive Allele Specific PCR (KASP) genotyping assay was used for the bi-allelic discrimination of the selected 89 SNPs. A total of 35 SNPs (39.3%) among the selected 89 SNPs were successfully genotyped of which 31 SNPs were monomorphic. The polymorphic SNPs were found in two genes: TLR4 (rs460053411) and BOLA-DRB3 (rs42309897, rs208816121, rs110124025). Thereby, a total of 35 SNPs (monomorphic and polymorphic) were analysed from a total of 10,430 genotypes assayed. A total of 8 alleles and 11 genotypes were found at 4 polymorphic SNPs located in exon regions. The polymorphic SNPs with MAF<5% and call rate <95% were used further for the association study. The results showed that rs42309897 and rs110124025 were significantly associated with both lameness and mastitis prevalence (P≤0.05) and rs110124025 was significantly associated with mastitis prevalence (P≤0.05) in both breeds. Based on current results the identified SNPs could be used as candidate genetic markers for lameness and mastitis resistance in Romanian Spotted and Romanian Brown cattle.

Benchmark of algorithms for multiple DNA sequence alignment across livestock species

A. Bąk[1], G. Migdałek[2] and K. Żukowski[1]
[1]*National Research Institute of Animal Production, Department of Cattle Breeding, Balice, 32-083, Poland,* [2]*Pedagogical University of Cracow, Department of Plant Physiology, Cracow, 30-084, Poland; kacper.zukowskI@izoo.krakow.pl*

Due to the growing amount of biological data, it is often necessary to select the most optimal estimation method. One of the most important benchmark of genomics is to modelling homology between considered DNA sequences. Multiple sequence alignment is a highly powerful tool for molecular and evolutionary biology and there are several programs and algorithms applicable for this purpose. The purpose of this work was to study the most commonly used DNA alignment algorithms to select the optimal tool dedicated for short sequences. The bioinformatics pipeline considered steps: (1) selection of reference genome sequences of ARS1.2 for cattle, EquCab3.0 for horse and vicPac2 for alpaca with a low E-value using TBLASTn; (2) removing gaps for these sequences; (3) alignment of obtained sequences using examined algorithms; (4) matching the quality of aligned sequences with sequences of reference genomes by mscore software. The time of computation was archived for whole analysis. The seven programs, each based on different alignment algorithms were used, namely: ClustalO, ClustalW, Kalign, MAFFT, MUSCLE, Probcons and T-Coffee. The result obtained in this study showed that the fastest are progressive algorithms such as Kalign or MUSCLE-FAST. Moreover, the iterative algorithms like MAFFT and MUSCLE revealed a higher quality of alignment. The T-Coffee and Probcons programs were computational cost-effective, simultaneously they were generating a medium-quality calculation in a relatively long time. The best quality of alignment was shown by iterative variants of the MAFFT program, however the speed of the calculations was relatively low. The fastest algorithm was Kalign, making alignment much faster than the competitors, but achieving average results in the quality of the alignment. The average speed ratio in relation to the quality of the analysed algorithms was obtained by the progressive version of MAFFT, NS1. The results of this study will be used to re-alignment of variant primers in new livestock genome releases.

Traces of Carpathian wolf in genome of dogs

N. Moravčíková, D. Simonová and R. Kasarda
Slovak University of Agriculture in Nitra, Tr. A. Hlinku 2, 94976 Nitra, Slovak Republic; nina.moravcikova1@gmail.com

The aim of this study was to identify the footprints of selection across genomes of the Czechoslovakian Wolfdog (CWD), the German Shepherd Dog (GSD) and wild wolf (WW) as outgroup species. Due to the fact that modern Wolf Dog breeds such as the Czechoslovakian Wolfdog represent results of crossing between wolf-like or ancient breeds (the German Shepherd Dog, the Alaskan Malamute or Siberian Husky) with wild wolves, the analysis of their genome can be used to decipher the impact of intensive anthropogenic hybridisation on the genomic architecture as well as to identify specific genetic variants, unique for each of the species. In this study, the genomic data for overall 32 animals that were genotyped using Illumina CanineHD Whole-Genome Genotyping BeadChip (LUPA project) were used. After SNP pruning the database stored information for 127,397 SNPs markers distributed across the autosomal genome. The footprints of selection were identified based on the scan of variation in genome-wide linkage disequilibrium (LD) patterns among selected breeds. The comparison of LD patterns between CWD and GDS showed 60 genomic regions, between CWD and WW 94 regions, and between GSD and WW 76 genomic regions under strong selection pressure. In case of all breeds under consideration the longest region were identified on chromosome 19. Most of the signals were found directly or very close to the sequence of genes responsible for canine colour patterns (e. g. MC1R, ASCL1, FAT1, CRSP1) which confirm generally accepted fact that the coat colour is very good trait to reveal the power of artificial selection.

Genetic parameters for semen production traits in Swiss dairy bulls

A. Burren[1], H. Joerg[1], M. Erbe[2], A.R. Gilmour[3], U. Witschi[4] and F. Schmitz-Hsu[4]
[1]Bern University of Applied Sciences, School of Agricultural, Forest and Food Sciences HAFL, Länggasse 85, 3052 Zollikofen, Switzerland, [2]Bavarian State Research Center for Agriculture, Institute of Animal Breeding, Prof.-Duerrwaechter-Platz 1, 85586 Poing-Grub, Germany, [3]VSNi, Orange, NSW 2800, Australia, [4]Swissgenetics, Meielenfeldweg 12, 3052 Zollikofen, Switzerland; alexander.burren@bfh.ch

The objective of this study was to estimate variance components (VC) for the semen production traits ejaculate volume, sperm concentration and number of semen portions in the Swiss cattle breeds Brown Swiss (BS), Original Braunvieh (OB), Holstein (HO), Red-Factor-Carrier (RF), Red Holstein (RH), Swiss Fleckvieh (SF) and Simmental (SI). For this purpose, semen production traits from 2,617 bulls with 124,492 records were used. The data were collected in the years 2000-2012. The model for genetic parameter estimation across all breeds included the fixed effects age of bull at collection, year of collection, month of collection, number of collection per bull and day, interval between consecutive collections, semen collector, bull breed as well as a random additive genetic component and a permanent environmental effect. The same model without a fixed breed effect was used to estimate VC and repeatabilities separately for each of the breeds BS, HO, RH, SF and SI. Estimated heritabilities across all breeds were 0.42, 0.25 and 0.09 for ejaculate volume, sperm concentration and sperm motility, respectively. Different heritabilities were estimated for ejaculate volume (0.42; 0.45; 0.49; 0.40; 0.10), sperm concentration (0.34; 0.30; 0.20; 0.07; 0.23) and number of semen portions (0.18; 0.30; 0.04; 0.14; 0.04) in BS, HO, RH, SF and SI breed, respectively. The phenotypic and genetic correlations across all breeds between ejaculate volume and sperm concentration were negative (-0.28; -0.56). The other correlations across all breeds were positive. The phenotypic and genetic correlations were 0.01 and 0.19 between sperm motility and ejaculate volume, respectively. Between sperm motility and sperm concentration the phenotypic and genetic correlations were 0.20 and 0.36, respectively. The results are consistent with other studies and show, that genetic improvement through selection is possible in bull semen production traits.

Genetic parameters for reproductive traits of Nellore cattle using a threshold-linear model

N.G. Leite[1], P.V.B. Ramos[1], T.E.Z. Santana[1], A.L.S. Garcia[2], H.T. Ventura[3], S. Tsuruta[2], D.A.L. Lourenco[2] and F.F. Silva[1]
[1]Universidade Federal de Viçosa, Animal Science, PH Rolfs Ave no number, 36570900, Brazil, [2]University of Georgia, Animal and Dairy Science, 425 River Road, 30602, USA, [3]Associação Brasileira dos Criadores de Zebú, Fernando Costa Park no number, 38022330, Brazil; nataliabaraviera@gmail.com

Reproductive performance is one of the most important features of beef cattle, especially in tropical environments. Stayability (STAY) is a relevant trait used in the selection index to improve female fertility and profitability in Nellore cattle. Although STAY is easily measured, it is recorded late in the animal's life, leading to an increase in generation interval. Thus, age at first calving (AFC) can be an alternative indicator trait of female fertility, which can be recorded earlier and can be used for the indirect selection of STAY. The aim of this study was to estimate genetic parameters for both AFC and STAY. In our analyses, STAY was considered a binary trait, and was defined as the probability that a cow will remain in the herd until 76 months of age given she calved 3 times. The pedigree and phenotypic datasets were provided by the Brazilian Association of Zebu Breeders. Pedigree was composed by 1,188,542 animals and the phenotypic dataset by 592,057 and 508,238 records for STAY and AFC, respectively. Genetic parameters were estimated using the Gibbs-sampling program THRGIBBS1F90, from the BUPF90 family, which allows the combination of linear and binary/categorical traits in a single analysis. The model included the mean as systematic effect and animal, contemporary group, and residual effect as random. Contemporary groups were defined as a combination of year and season of birth and heard of the first calving. The convergence was checked using the software POSTGIBBSF90. Heritability estimates were low for both traits, with posterior means of 0.075 ± 0.003 and 0.071 ± 0.002 for STAY and AFC, respectively. The STAY showed a negative correlation with AFC of -0.28 ± 0.026. The present results indicate that selection for AFC as an earlier reproductive trait for STAY will promote a slow improvement, even though favourable.

Autoregressive model in genetic evaluation for interval reproductive traits of Holstein dairy cattle

H.T. Silva[1], C.N. Costa[2], D.A. Silva[1], A.A. Silva[1], F.F. Silva[1], P.S. Lopes[1], R. Veroneze[1], G. Thompson[3,4] and J. Carvalheira[3,4]
[1]Univ. F. Viçosa, Campus Univ., 36571-000 Viçosa, MG, Brazil, [2] Embrapa Gado de Leite, R. E. Nascimento, 610, 36038-330 Juiz de Fora, MG, Brazil, [3]CIBIO, Univ. Porto, R. Padre Armando Quintas, 4485-661 Vairão, Portugal, [4]ICBAS-Univ. Porto, R.Jorge Viterbo Ferreira, 228, 4050-313 Porto, Portugal; jgc3@cibio.up.pt

The objectives of this study were to evaluate the autoregressive (AR) repeatability model for genetic evaluation of interval reproductive traits of Portuguese Holstein cattle comparing with the classic repeatability (REP) model. The data used for this study included records from the first four calvings of cows occurred between 1995 and 2018. Previous analyses indicated that the voluntary post-partum waiting period was 42 days and we assumed a standard gestation length of 280 ± 15 days. The reproductive traits considered were Interval from Calving to 1st service, Days Open, Calving Interval and Daughter Pregnancy Rate, with 416, 766, 872 and 766 thousand records, respectively. Both models comprised fixed (month and age classes associated to each calving order) and random (herd-year-season, animal and permanent environmental) effects. The difference between the 2 models was on the structure of the (co)variance matrices for the herd-year-season and permanent environmental effects: the AR model was fitted with first-order autoregressive (co)variance structures while the REP models included only a simple IID (co)variance structure. For all traits the AR models presented the best fit for the data with the smallest Akaike Information Criteria and the Akaike Weights were also close to unity, indicating that the AR models were better in explaining the traits variability. EBV rank correlations for the top-100 and the total population of bulls ranged from 0.71 to 0.88 and from 0.95 to 0.97, respectively. These results indicate that the re-ranking observed at the top sires probably means more opportunities for selection. The difference between EBV accuracies obtained from the 2 models for the 1, 10 and 50% best bulls was significant ($P<0.05$) for all traits favouring the AR approach. Overall, the AR performed better than the REP models and is recommended for genetic evaluation of longitudinal interval reproductive traits.

Genome-wide association for milk production, lactation curve and calving interval in Holstein cows

H. Atashi[1,2], M. Salavati[3], J. De Koster[2], J. Ehrlich[4], M.A. Crowe[5], G. Opsomer[2] and M. Hostens[2]
[1]*Shiraz University, College of Agriculture, Department of Animal Science, Shiraz, Iran,* [2]*Ghent University, Department of Reproduction, Obstetrics and Herd Health, Salisburylaan 133, 9820 Merelbeke, Belgium,* [3]*University of Edinburgh, The Roslin Institute and Royal (Dick) School of Veterinary Studies, Easter Bush, Midlothian, United Kingdom,* [4]*Dairy Veterinarians Group, 832 Coot Hill Rd., Argyle, NY 12809, USA,* [5] *University College Dublin, Dublin, Ireland; hadi.atashi@ugent.be*

Genomic-wide association studies (GWAS) are a powerful tool for detecting genomic regions explaining variation in phenotype. The objective of this study was to detect genomic regions associated with milk yield, lactation curve parameters and calving interval in 11,262 Holstein dairy cows. A genome-wide association study was performed for each trait, using a single SNP (single nucleotide polymorphism) regression mixed linear model and imputed high-density panel (777k) genotypes. Deregressed estimated breeding values of the animals were used as the response variable. The number of significant SNPs identified varied from 534 for milk yield to 18 for calving interval. The most significant SNP for milk yield, lactation curve parameters and calving interval was BOVINEHD1400000216 (rs134432442) mapped inside CPSF1 gene followed by ARS-BFGL-NGS-4939 (rs109421300) mapped inside DGAT1 gene. Investigating overlapping regions among milk yield and lactation curve parameters resulted in identifying 173 significant SNPs on BTA14 at position of 1.3 to 3.4 Mb. The only overlapping peak among the milk yield, lactation curve parameters and calving interval was the region identified on BTA14 at position of 1.7 to 2.1 Mb includes 18 significant SNPs. Previously proposed candidate genes for milk yield supported by this work include DGAT1, GRINA, CPSF1, SLC2A4RG, GHR and THRB, whereas new candidate genes are ARC, TOP1MT (BTA14), FBXO4, and PRKAA1 (BTA20). The chromosomal region identified in this study not only confirm several pervious findings but also detected new regions that may contribute to genetic variation in milk production, lactation curve parameters and calving interval.

Establishing a female reference sample for susceptibility to digital dermatitis in Holstein cows

H.H. Swalve[1], M. Wensch-Dorendorf[1], G. Kopke[1], F. Rosner[1], D. Oelschlägel[1], B. Brenig[2] and D. Döpfer[3]
[1]*Martin-Luther University Halle-Wittenberg, Institute of Agricultural and Nutritional Sciences, Theodor-Lieser-Str. 11, 06120 Halle, Germany,* [2]*University of Göttingen, Institute of Veterinary Medicine, Burckhardtweg 2, 37077 Göttingen, Germany,* [3]*University of Wisconsin, Food Animal Production Medicine Section, School of Veterinary Medicine, Madison 53706, USA; hermann.swalve@landw.uni-halle.de*

Digital dermatitis (DD) is an infectious claw disease in cattle. Genetic effects of the cow contribute to susceptibility. Most often, recording and analysing DD is done using data from hoof trimmers about active, ulcerative stages of DD. For the purpose of this study, inactive DD stages were used as well. Data consisted of 13 large dairy farms visited three times at intervals of three weeks. Herds were enrolled in the study if at least a third of the animals had been genotyped (SNP-array) under a random scheme within herd or when entire birth years had been genotyped. The resulting data set comprised 5,040 cows with 50-k genotypes or 10-k genotypes imputed to 50-k that had been scored for DD at least 2 or 3 times. Phenotypes were defined as 0 for completely healthy cows and as 1 for cows that showed either ulcerative, active lesions or inactive DD. Data was analysed by single-step GBLUP using BLUPF90. SNP effects estimated were used to derive genomic breeding values for A.I. sires and a completely independent sample of hoof trimmer data including records for 31,000 first lactations was used for validation. The results show that sires with a lowest gEBV showed incidences of 0.21 (s.e. 0.009) for DD in their daughters while top sires had a daughter average of 0.13 (s.e. 0.008). Former genome-wide association studies and gene expression analyses have provided initial insights into the identification of chromosomal regions and potential candidate genes. So far, especially regions on the chromosomes 6, 8, 16, 23, and 26 have been published and the derived candidate genes include IL-8, IL-19, IL-24, TLR4, HTRA1, MHCII as well as keratins and keratin-associated proteins. We were able to confirm the relevance of these known chromosomal regions and additionally identified regions of interest on BTA 3, 11, 14, 19, 25, and 28, with largest effects of a SNP on BTA 19.

Improved cow claw health by selection – a Nordic experience

H. Stålhammar[1] and C. Bergsten[2]
[1]VikingGenetics, Box 64, 53221 Skara, Sweden, [2]Swedish University of Agricultural Sciences, Department of Biosystems and Technology, Box 103, 23053 Alnarp, Sweden; hasta@vikinggenetics.com

The objective of this study was to evaluate the phenotypic and genetic trends in claw health traits recorded by claw trimmers in Sweden, Denmark and Finland. The phenotypic data was collected and processed by Växa Sverige and the breeding values were estimated by Nordic Cattle Genetic Evaluation. Annual trends for and correlation between Claw Health Index (CHI introduced in Sweden 2005) and the Nordic Total Merit (NTM, updated in November 2018) were estimated. The CHI is weighted and favours the individually most costly claw disorder, sole ulcer including white line disease, although several other traits are also included. The number of cows recorded and the average number of claw records per cow per year have increased considerably during the last 10 years. Especially when the welfare reimbursement for claw trimming (claw coin) was introduced in Sweden in 2016 the number of records raised. The prevalence of sole ulcer, sole haemorrhage, heel horn erosion has noticeable been reduced in the Swedish cow population during the period, while the prevalence of digital dermatitis has increased slightly. Moreover, the number of cows with multiple remarks has decreased. This favourable development is supported by the genetic trends in CHI for both males and females. The positive gain has been larger in Swedish Holstein than in Swedish Red, which is supported by the size of correlations between CHI and NTM in respective breed. The study clearly shows that by using bulls with high merit in CHI the expected prevalence in future daughters is largely reduced. The favourable development in claw health have thus improved cow welfare as well as improved the profit for the dairy herd, by less lameness causing claw disorders.

Association of genomic and parental breeding values with cow performance in Nordic dairy cattle

C. Bengtsson[1,2], H. Stålhammar[1], E. Strandberg[2], S. Eriksson[2] and W.F. Fikse[3]
[1]VikingGenetics, VikingGenetics Sweden AB, 53294 Skara, Sweden, [2]Swedish University of Agricultural Sciences, Dept. of Animal Breeding and Genetics, Box 7023, 750 07 Uppsala, Sweden, [3]Växa Sverige, Växa Sverige, Box 288, 751 05 Uppsala, Sweden; chben@vikinggenetics.com

Genotyping females in the Nordic countries started on a large scale in 2012. The main reason was to include genotyped females into the reference population and thereby increase the accuracy of genomically enhanced breeding values (GEBV). Since the start, over 250,000 females have been genotyped and information from these animals' lactations has been recorded. The purpose of this study was to compare GEBV and parent average breeding values (PA) for females regarding their ability to predict the cows' future phenotype. Validation of GEBV and illustrating the relationship between genomic prediction and the future phenotype is key to increasing confidence for the genomic technology among farmers. Data were collected from the Swedish, Danish and Finnish milk recording schemes. Production, conformation, fertility and other functional traits, adjusted for systematic environmental effects, were used as measures of cow performance. GEBV and PA were collected from the Nordic Cattle Genetic Evaluation. Preliminary results for the Swedish Red showed that GEBV and PA breeding values worked best to predict future phenotypes for high heritability traits. For protein yield the difference between the bottom and the top quartile was 21 kg fat for PA prediction and 24 kg fat for genomic prediction. The correlation between PA yield index and adjusted protein yield was 0.194 (0.155-0.232) and the correlation between genomic yield index and adjusted protein yield was 0.214 (0.177-0.252). There were tendencies that GEBV functioned better than PA breeding values for most of the analysed traits. Except for milkability there were no significant differences between indexes and their prediction of future phenotypes. The preliminary results also showed that the conventional genetic evaluation, without genomic information, works well for many of the studied traits.

Patterns of DNA variation between autosomes and sex chromosomes in *Bos taurus* genome

B. Czech[1], B. Guldbrandtsen[2] and J. Szyda[1,3]
[1]*Wroclaw University of Environmental and Life Sciences, Biostatistics group, Department of Genetics, Kozuchowska 7, 51-631 Wroclaw, Poland, [2]Aarhus University, Center for Quantitative Genetics and Genomics, Department of Molecular Biology and Genetics, Blichers Allé 20, Postboks 50, Tjele, DK, 8830, Denmark, [3]National Research Institute of Animal Production, Krakowska 1, 32-083 Balice, Poland; bartosz.czech@icloud.com*

The new ARS-UCD1.2_Btau5.0.1Y assembly of the bovine genome includes considerable improvements. We assume that a more accurate identification of patterns of genetic variation can be achieved with it. Sex chromosomes in mammals originate from autosomes, but over time, the Y chromosome has degenerated losing the majority of its genes. The aim of this study was to explore differences in genetic variations between autosomes, the X chromosome, and the Y chromosome. In particular, SNP and InDel densities and annotations were compared between chromosomes. Furthermore, InDel lengths and nucleotide divergence, and Tajima's D were compared between chromosomes. Whole-genome DNA sequences of 217 individuals representing different cattle breeds were examined. The analysis included alignment to the new reference genome and variant calling. 23,655,292 SNPs and 3,758,781 InDels were detected. BTA25 had the highest concentration of SNPs (2.58%), while the Y chromosome had the lowest (0.1%). Both sex chromosomes had very similar average lengths of insertions, but the Y chromosome contained deletions with the shortest average length. The X and the Y chromosomes had the highest density of variants at the end of the chromosome, while the autosomes had quite uniform distributions of variants. In contrast to autosomes, both sex chromosomes had negative values of Tajima's D and lower nucleotide divergence. That implies a correlation between nucleotide diversity and recombination rate, which is obviously reduced for sex chromosomes. Also, the relatively lower N_e for sex chromosomes leads to a lower expected density of variants.

Expression of genes involved in the cows' mammary gland defence during staphylococcal mastitis

E. Kawecka[1,2], E. Kościuczuk[1], M. Zalewska[3], D. Słoniewska[1], M. Rzewuska[2], S. Marczak[1], L. Zwierzchowski[1] and E. Bagnicka[1]
[1]*Institute of Genetics and Animal Breeding PAS, Animal improvement, Postępu st 36a, 05-552 Magdalenka, Poland, [2]Warsaw University of Life Sciences, Veterinary Medicine, Nowoursynowska ST. 159, 02-776 Warsaw, Poland, [3]Warsaw University, Department of Applied Microbiology, Miecznikowa st.1, 02-096 Warsaw, Poland; e.kawecka@ighz.pl*

Staphylococci are the main group of pathogens causing udder inflammation. The aim of the study was to analyse the selected genes expression at mRNA and protein levels (C-C motif chemokine 2 – *CCL2*, C-C motif chemokine ligand 5 – *CXCL5*, interleukin 1β – *IL-1β*, interleukin 8 – *IL8*, tumour necrosis factor α – *TNFα*, serum amyloid A – *SAA*, haptoglobin – *Hp*, alpha-S1-casein – *CNS1S1*) in the udder secretory tissue (parenchyma) of cows infected with coagulase-positive (CoPS) or negative staphylococci (CoNS) vs non-infected cows (H). The 55 udder quarter samples were collected from 40 slaughtered Polish Holstein-Friesian cows. Three sample groups were distinguished: infected with CoPS (n=24), CoNS (n=14), or non-infected (H; n=13). The mRNA level was estimated using the qPCR technique, while protein concentration was measured using ELISA tests. The *GAPDH* and *HPRT* were used as reference genes. At mRNA level the differences were found only for genes: *SAA* between CoPS and H, *Hp* between CoPS and CoNS and *CXCL5* between CoPS, and H groups. However, there were no differences in the level of protein encoded by *SAA* and *Hp* genes. The higher concentrations of four cytokines: Il-8, TNFα, CCL2, CXCL5 in CoPS vs CoNS group were found, with no differences with H group. As expected, CSN1S1 had a higher concentration in H vs CoPS group, but only at the tendency level (0.05).

Association of TGFB1 and TNF-α gene SNPs with dairy cattle production and udder health state traits

P. Brodowska[1], E. Kawecka[1,2], M. Zalewska[3], D. Reczyńska[1], D. Słoniewska[1], S. Marczak[1], S. Petrykowski[1], L. Zwierzchowski[1] and E. Bagnicka[1]
[1]Institute of Genetics and Animal Breeding PAS, Department of Animal Improvement, Postępu 36A, 05-552 Magdalenka, Poland, [2]Warsaw University of Life Sciences, Faculty of Veterinary Medicine, Nowoursynowska 159, 02-776 Warsaw, Poland, [3]Warsaw University, Department of Applied Microbiology, Miecznikowa 1, 02-096 Warszaw, Poland; p.brodowska4@ighz.pl

Somatic cell count (SCC) as a genetic marker of the udder health state has a low heritability coefficient (<0.1). To accelerate their genetic progress SNPs could be used in selection. One of the groups of proteins involved in the udder inflammation are those encoding cytokines, e.g. tumour necrosis factor alpha (TNF-α) and transforming growth factor (TGFB1) as proinflammatory agents. The aim of the study was to analyse the association between SNPs: DdeI-g.3960C>T of TGFβ1 and RsaI-g.1844C>T of TNF-α genes with production traits. The study was conducted on 271 HF dairy cows being from 1st to 7th lactations. Altogether, 8,515 observations on daily milk yield, protein, fat, casein, lactose, dry matter contents, fatty acid profile, SCC as well as the pedigree information from 10 generations obtained from Polish Federation of Cattle Breeders and Dairy Farmers were used to analysis. The REML method with one-trait mixed repeatability animal model was used to assess the associations. The SCC was transformed into natural logarithm scale (lnSCC) before analysis. The following genotype frequencies were stated – for DdeI-g.3960C>T (TGFβ1 gene): CC (0.27), CT (0.69), TT (0.04) and for RsaI-g.1844C>T (TNF-α gene): TT (0.19), TC (0.56), CC (0.25). The TT homozygous of both genes was associated with the lowest lnSCC and highest protein and lactose contents, however, they had the lowest frequency. The expected genotype frequencies of both genes deviate from the observed ones, which means that this population is not in equilibrium for these genes, i.e. traits which undergo selection are probably associated with studied SNPs. Thus, both genotypes could be useful markers in the marker assisted selection of dairy cows with higher resistance to mastitis.

Use of repeated group measurements with drop out animals for genetic analysis

H. Gao[1], B. Nielsen[2], G. Su[1], P. Madsen[1], J. Jensen[1], O.F. Christensen[1], T. Ostersen[2] and M. Shirali[1]
[1]Aarhus University, Center for Quantitative Genetics and Genomics, Department of Molecular Biology and Genetics, Blichers alle 20, Foulum, 8830 Tjele, Denmark, [2]2SEGES, Pig Research Centre, Axelborg, Axeltorv 3, 1609 Copenhagen, Denmark; hongding.gao@mbg.au.dk

Under commercial animal breeding, it is more cost-effective to measure performance on group level compare to individual level for some economically important traits. The objective of this study was to develop a random regression model based on repeated group measurements with a consideration of drop out animals for estimation of variance components (VC) and prediction of breeding values. Data were simulated based on individual feed intake in a pig population with records available at six time points during the test period. The simulated phenotypes consisted of additive genetic, permanent environment, and random residual effects. Additive genetic and permanent environmental effects were both simulated and modelled by first order Legendre polynomials. Three grouping scenarios based on genetic relatedness of group members were investigated: (1) medium within genetic relationship ($\text{Group}_{3\times4}$): group consisted of 4 different families with 3 pigs from each family; (2) high within group relationship ($\text{Group}_{6\times2}$): all animals from 2 different litters were allocated to a group; (3) low within group relationship ($\text{Group}_{1\times12}$): all animals were from different litters. To investigate the effects of the drop out animals during test period, two drop out strategies within each grouping scenario were assessed: (1) Drop_{ran}: animals dropped out randomly; (2) Drop_{phe}: animals were ranked based on their phenotypes at each testing time point and then inferior animals dropped out. Using group measurements yielded similar VC estimates but with larger SDs compared with those of using individual measurements in all scenarios. Similar VCs estimates were observed from Drop_{ran} compare to no drop out scenario. Dropping out animals by Drop_{phe} led to underestimated VCs. Different grouping scenarios produced similar VCs estimation. The results show that, the developed model can properly handle group measurements with drop out animals, and can achieve comparable prediction accuracy for traits measured at the group level.

Genetic evaluations for dam type specific calving performance traits in a multi breed population

R. Evans, A. Cromie and T. Pabiou
Irish Cattle Breeding Federation, Genetics, Highfield House, Bandon, Cork, Ireland, N/A, Ireland; revans@icbf.com

The objective of this study was to derive multiple breeding values for direct calving difficulty based on records on four defined cow types namely dairy heifers (DH), dairy cows (DC), beef heifers (BH) and beef cows (BC). Calving difficulty is assessed by Irish dairy and beef farmers on a 4-point scale covering no assistance (1) to veterinary assistance (4). Birth weight (kg) and birth size (1 to 5 scale) were also available as potential predictor traits. Genetic parameters were estimated with DMU and Estimated Breeding Values (EBVs) were estimated in MiX99. Model fixed effects included breed proportions and specific heterosis effects as fixed covariates and birth year, sex of calf and age of dam as fixed effect classes. Contemporary groups were defined as 3-month windows within herd and treated as random effects. Random effects included dam permanent environment effect for traits with repeated dam records and direct and maternal genetic effects were included for all six traits. Direct heritability, maternal heritability and direct-maternal genetic correlations were 0.16, 0.04 and -0.07 for DH, 0.08, 0.02 and -0.07 for DC, 0.17, 0.09 and -0.26 for BH, 0.15, 0.08 and 0.01 for BC, 0.41, 0.09 and -0.48 for birth weight and 0.24, 0.05 and -0.39 for birth size, respectively. Genetic correlations for direct genetic effects across the four calving difficulty traits averaged 0.75 with the lowest correlation between DC and BC and the highest between BH and BC. Correlations between birth weight and birth size and the four calving traits averaged 0.63 and 0.82, respectively. Validation of these new EBVs was undertaken by omitting the phenotypes for 2018 born animals. Accuracy of prediction was assessed by comparing parental average EBVs with yield deviations (YDs) for these animals. Comparing against the current official calving EBV (a single calving difficulty trait), the new EBVs displayed increased correlations with the YDs (0.18 vs 0.23 for DH; 0.21 vs 0.23 for DC; and 0.20 vs 0.21 for BH and BC. Regression slopes indicated more bias in the old *vs* new EBVs particularly for DH (1.6 v 0.91) and BH (1.7 vs 0.91). This indicates the current single trait EBVs are underestimating the actual phenotypic calving difficulty in these dam types.

Integration of external information from multiple traits into the national multitrait evaluations

T.J. Pitkänen[1], M. Koivula[1], I. Strandén[1], G.P. Aamand[2] and E.A. Mäntysaari[1]
[1]Natural Resources Institute Finland, Myllytie 1, 31600 Jokioinen, Finland, [2]NAV Nordic Cattle Genetic Evaluation, Nordre Ringgade 1, 8200 Aarhus N, Denmark; timo.j.pitkanen@luke.fi

Integration of external information from multiple traits, e.g. international estimated breeding values (EBVs) from Interbull or Interbeef, into domestic multitrait evaluations, is presented. The method requires EBVs and reliabilities for each trait for internationally evaluated animals (hereinafter external animals). The method has three steps. First, the multitrait reversed reliability approximation is used to obtain effective record contributions (ERC) for external animals. ERC is approximated first using reliabilities from the domestic data only, and then from reliabilities from international evaluations. To avoid double counting of information, the correlation(s) between the evaluated traits are accounted during ERC approximation. The difference between the international and the domestic ERCs estimates the amount extra information due to the foreign data in international evaluation. In the second step, multitrait deregressed proofs (DRPs) are calculated for the domestic and international evaluations, separately based on corresponding EBV and ERC data. The correlation between the traits is taken into account also in this step. Finally, based on two DRPs and two ERCs, pseudo-observation for external animal is calculated. The pseudo observations will approximate the external records, which, given the weight $ERC_{int} - ERC_{dom}$, will give national EB_{Vblend} that are the same as the EBV_{int} in joint evaluations. The method was tested on an artificial setup where EBVs for milk, protein, and fat yields from joint evaluation model for Nordic countries were blended to fictitious domestic evaluation for Denmark alone. Both evaluation models were for 305 d yields for milk, protein, and fat in 3 parities. Hence, in total BVs for 9 traits were estimated in each model. To mimic blending of MACE proofs, yield indices for milk, protein, and fat were calculated by weighted average of EBVs over lactations. Three yield indices were then integrated to the domestic evaluations using the proposed method.

TrackLab 2: automatic recording and analysis of the behaviour of animals kept in groups

L.P.J.J. Noldus[1], W. Ouweltjes[2], T.B. Rodenburg[3,4], E.K. Visser[5], B.J. Loke[1] and A. Van Gijssel[1]
[1]*Noldus Information Technology BV, Nieuwe Kanaal 5, 6709 PA Wageningen, the Netherlands,* [2]*Wageningen Livestock Research, De Elst 1, 6708 WD Wageningen, the Netherlands,* [3]*Utrecht University, Yalelaan 2, 3584 CM Utrecht, the Netherlands,* [4]*Wageningen University, P.O. Box 338, 6700 AH Wageningen, the Netherlands,* [5]*Aeres University of Applied Sciences, De Drieslag 4, 8251 AJ Dronten, the Netherlands; lucas.noldus@noldus.nl*

With the trend towards larger group housing systems in farm animals, it becomes increasingly important to be able to monitor the behaviour and welfare of individual animals housed in groups. Here we present TrackLab 2, a new software package and integrated system for the acquisition and analysis of location, activity and social behaviour of group-housed animals. It is the successor of TrackLab 1, which has been used in a wide variety of livestock research projects on cattle, pigs, poultry and sheep. TrackLab 2 offers several functional and technical innovations relative to its predecessor: the movement analysis and behaviour assessment routines are compatible with both indoor (ultra-wideband, UWB) as well as outdoor (GPS) tracking; the tracking hardware is customized for animals of different sizes; the accuracy achieved with UWB-based localization allows for analysis of behaviours that cannot be achieved with other positioning techniques (e.g. RFID, WiFi): social behaviour/network analysis, accurate activity classification and place-preference analysis; the analysis includes results per individual animal as well as statistics for experimental groups; it has a distributed and scalable client-server architecture, supporting multiple concurrent users and measurements at multiple locations (e.g. barns). Practical trials and validation studies with dairy cattle and poultry have demonstrated the performance of the system under demanding conditions, during prolonged tracking of large numbers of animals, both in barns and in the field. We will discuss the findings and their implications for practical application of the system. We hope that TrackLab 2 will contribute to livestock research (behavioural phenotyping, testing different diets, housing and management systems) as well as precision livestock farming (monitoring individual animal health and welfare).

Application of a precision feeding program in growing pigs: effect on performance and nutrient use

L. Brossard[1], M. Marcon[2], J.Y. Dourmad[1], J. Van Milgen[1], J. Pomar[3], V. Lopez[3] and N. Quiniou[2]
[1]*PEGASE, INRA, Agrocampus-Ouest, 35590 Saint-Gilles, France,* [2]*IFIP-Institut du Porc, BP 35, 35650 Le Rheu, France,* [3]*Department of Agricultural Engineering, Universitat de Lleida, Av. Alcalde Rovira Roure 191, 25198 Lleida, Spain; ludovic.brossard@inra.fr*

Improvement of feed efficiency in growing pigs is a key issue for the economic and environmental sustainability of livestock production. This can be achieved with novel techniques such as precision feeding (PF). Within the Horizon 2020 EU Feed-a-Gene program (grant agreement n°633531), a decision support system (DSS) was developed to implement precision feeding in commercial pig farms. This study aimed to test the functioning of the DSS in practical conditions and the consequences on performance and nutrient use of growing pigs fed *ad libitum*. Sixty-four pigs were reared from 77 to 161 days of age (34 to 109 kg body weight, BW) in a single pen equipped with an automatic weighing-sorting system and eight automatic feeders allowing to register feed intake and deliver a tailored blend of two diets (A and B with respectively 1.0 and 0.4 g SID Lysine (Lys)/MJ NE, and 9.7 MJ NE/kg) to individual pigs. Pigs of the control group received a blend providing 0.9 g SID Lysine (Lys)/MJ NE until the average group BW was 65 kg (growing phase) and 0.7 g SID Lys/MJ NE thereafter (finishing phase). For the PF group, the assessment of the SID Lys requirement was performed individually and on a daily basis, based on up to 20 previous records of BW and feed intake. Feed composition was changed accordingly by blending diets A and B in appropriate proportions. Daily feed intake, average daily gain, and feed conversion ratio did not differ between treatments for the overall period or per period. During the growing period, the SID Lys intake and the nitrogen intake and excretion were respectively 10.8, 8.8, and 14.4% lower in the PF group compared to the control group (P<0.05). During the finishing period, these values were only numerically lower (difference <2%; P>0.68). This could result from a slightly higher feed intake (+100 g/d, P=0.24) in PF group combined with a SID Lys supply already low in control group. A second experiment will be performed in the same conditions to confirm the potential of the PF using the developed DSS.

Using a commercial precision livestock farming sensor to record dairy cows' behaviour at pasture

M. Bouchon, L. Rouchez and P. Pradel
INRA Herbipole PHASE Theix 63122 Saint-Genes-Champanelle, France; matthieu.bouchon@inra.fr

Precision livestock farming system play an increasingly important role in farms and represent new way of acquiring data for research. This study aimed at validating activities (eating, ruminating, resting) and posture (standing or lying) data output from a commercial PLF device (Axel, Medria®) mounted on dairy cows' neck, against direct visual observation at pasture. The device is a three-axial accelerometer commonly used to heat and feed monitoring that sends raw data to a server. The data are automatically proceeded to give cow's activity and posture at 5 minutes interval. The experimentation was carried out at INRA Herbipole Marcenat experimental facility in June 2018. We observed six dairy cows in mid lactation (3 Prim'Holstein and 3 Montbeliarde) by continuous focal sampling for 180 hours, i.e. 30 hours per cow, split into five periods of six hours. Theses direct observations were the gold standard. Cows were at pasture, with no other feed, in order to record grazing behaviour only. The performance of the sensor varied between behaviours. Ruminating and grazing showed good sensitivity (77%, 81%) and specificity (92%, 93%); resting presented a lower sensitivity (65%) but a very good specificity (93%), and posture a lower specificity (62%) but a higher sensitivity (89%). The overall accuracy was also good, from 83% for the posture to 90% for resting. There were no significant effect of the cow nor of the breed. Regarding time-budget analysis, the linear regressions for all activities showed good results with slopes close to 1 and R^2 over 0.76 (except for the posture, R^2=0.63). Gathering the individual data in a mean herd activity by six hours period lead to an improved agreement for the three activities (R^2=0.9 for grazing, ruminating and resting). Results found in the literature are slightly better, but most of the experiments aimed at calibrating specific models from raw data to detect a given behaviour in experimental condition. In conclusion, the Axel Medria sensor, initially designed to help farmers taking decision on animal breeding, show good concordance with visual observations of dairy cattle activities and postures at pasture. It also opens the possibility of studying dairy cattle behaviour at the group/herd level with very low constraints.

Opportunities and challenges of data integration with focus on claw health for decision support

C. Egger-Danner[1], M. Suntinger[1], F. Grandl[2], P. Majcen[3], F. Bapst[4], O. Saukh[4], M. Fallast[5], A. Turkaspa[6], J. Duda[2], C. Linke[1], F. Steininger[1], T. Wittek[7], B. Fuerst-Waltl[8], F. Auer[9] and J. Kofler[7]
[1]ZuchtData, Dresd. Str.89/19, Vienna, Austria, [2]LKV Bavaria, Landsb.Str.282, Munich, Germany, [3]LK Austria, Schauflerg.6, Vienna, Austria, [4]TU Graz/CSH, Inffeldg.16/I, Graz, Austria, [5]smaXtec, Belgierg.3, Graz, Austria, [6]SCRbyAllflex, Hamelacha St.18, Netanya, Israel, [7]Univ.Vet.Med., Veterinaerpl.1, Vienna, Austria, [8]BOKU, Greg.-Mend.-Str.33, Vienna, Austria, [9]LKV Austria, Dresdner Str.89/19, Vienna, Austria; suntinger@zuchtdata.at

New technologies are revolutionising the dairy industry. Instead of punctual measurements, sensors record animal behavioural patterns that can imply on the animal well-being and welfare. The large amounts of data generated by monitoring and the integration of various data sources promise completely new insights into animal health. Traditional data pipelines with information from performance recording, genetics, feeding in combination with environmental effects have existed for some time. For claw health, information from claw trimming, veterinary diagnoses and lameness scoring has only been partly made available. Sensor technology provides alarms based on irregularities of normal behaviour for early detection of disorders. Advanced methodology offers the possibility to combine various environmental information and genomic background to get new insights in the occurrence of or susceptibility to disorders. To explore these opportunities the big challenge is the integration of different data sources. In practice, data are often generated by different hardware and software products, which makes data integration more difficult due to different data exchange formats of the communication partners involved. Traits are defined differently by different products. It is therefore necessary to create structures to bring these data sources together in order to provide farmers with maximum support for herd management. Another challenge of data integration from different sources is compliance with legal data protection regulations. Cooperation between different partners and integrating different data is the precondition for successfully applying advanced data technologies based on complex trait definitions. Based on the project D4Dairy steps to overcome these challenges are presented.

Detecting perturbations in dairy cows liveweight trajectories

O. Martin and A. Ben Abdelkrim

INRA, UMR MoSAR Agroparistech 16 rue Claude Bernard, 75005 Paris, France; olivier.martin@agroparistech.fr

Today, managing individual variability in dairy farming systems is a potentially innovative way to face the challenges of efficiency and robustness in a changing and uncertain environment. Precision farming has primarily developed through the automation of data acquisition but interpretative tools to capitalize on this raw material are lacking. The challenge is to design tools as translators of individual time series data on animal performance into phenotypic information providing quantification on variability and further useful benchmarks for decision support. In this study, we present a model of body weight trajectory with explicit representation of perturbations, and a fitting procedure on data to infer a theoretical unperturbed weight trajectory and extract perturbation features. This model-based interpretation of data provides estimates of the timing and intensity of perturbations affecting cows. The model is based on generic models describing growth and liveweight changes over repeated reproductive cycles throughout lifespan, associated with a model detecting individual perturbations. Model parameters provide quantification of unperturbed and perturbed patterns of liveweight trajectories with an estimate of the timing and intensity of perturbations affecting cows. The model was implemented in R and fitted to individual liveweight time series data recorded in commercial French Holstein herds. We performed a statistical analysis on the estimated model parameters to provide an overview of the variability between cows related to liveweight trajectories and detected perturbations. The present communication describes the model and its fitting procedure, presents results of the fitting procedure and puts into perspective the use of this method and more generally of model-based approaches as management tools in the context of precision farming.

Using nonlinear state space models for robust short-term forecasting of milk yield

M. Pastell

Natural Resources Institute Finland (Luke), Latokartanonkaari 9, 00790 Helsinki, Finland; matti.pastell@luke.fi

State space models are used widely in animal science for state estimation of physiological and behavioural measurements. They provide a powerful tool to combine biological knowledge in the state equation combined with statistical methods for parameter estimation. Short term forecasting of milk yield is a useful tool in determining whether changes in production have occurred due to e.g. disease or management changes. Particle filters are a Sequential Monte Carlo (SMC) method to estimate state space models that can't be estimated using traditional linear or linearized approaches. We aimed to develop a robust nonlinear particle filter model for short term (up to 14 days ahead) forecasting dairy cow lactation curves forecasting. Lactation curves were modelled using the Wilmink in difference equation form: $Y_t = Y_{t-1} + \beta_1 \Delta t + \beta_2 e^{kt} \Delta t$ Student distribution was used as error distribution for observations to reduce the effect of outliers in state estimation. A dataset of over 800 lactation curves recorded in milking parlours from 2 research farms was used in modelling. A Julia library for fitting the models was developed and optimization was used to estimate model parameters. I compared 14 different global optimization methods and they all converged to similar results, but with significant differences in required iterations. I estimated the number of particles is required for repeatable results. We determined that 10,000 particles were needed to obtain repeatable results on the estimation of used model on the used dataset. Further the performance of maximum likelihood estimation (MLE) was compared to optimizing the model for mean square error of 7 and 14 days ahead forecasting. Details of the optimization method will be presented together with forecasting results of models optimized for different prediction horizons. The results indicate that optimizing the model for specific forecasting horizon yields low mean square error of prediction and smoother parameter estimates compared to MLE estimates leading to more accurate forecasts. The used modelling method is very flexible and additional predictive parameters such as feed intake will be added to the model in future work.

Promising parameters to foresee intake and feed efficiency at pasture – a meta-analysis approach

M. Boval and D. Sauvant
INRA PHASE INRA AgroParisTech Université Paris-Saclay, 75005, Paris, France, France;
maryline.boval@agroparistech.fr

In a context of major advances in electronic/computer technologies, identifying key and routinely measurable criterial to detect animals that are likely to be more efficient in various feeding environments, is essential. For that purpose, a meta-analysis carried out on 110 publications (n) and gathering 905 experiments (nexp) involving feeding behaviour of grazing cattle and small ruminants, stressed the importance of the bite mass (BM, mg DM/bite/kg BW, 0.11<1.79±1.27<7.41, n=582), as an essential unitary criterion for dry matter intake (DMI, 0.25<2.79±1.06<8.01% BW, n=238). The intra-experiment equation is DMI = 2.16 + 0.56 BM (RMSE=0.42, n=180, nexp=67). This meta-analysis also highlighted the importance of the incisive arcade (IA, 2.22<5.23±2.11<8.60 cm, n=112), which is measurable and related both to BW (IA = 0.91 BW $^{0.346}$, RMSE=0.27, n=20) and to BM (BM = 0.015 IA$^{1.88}$, RMSE=0.10, n=45, nexp=21). In addition, we have tested several predictive regressions of DMI and notably one based on BM combined with ruminating time (RU, 0.52<6.06±2.15<9.57 h/day, n=265). The equation is DMI= 1.395 + 0.652 BM + 0.097 RU, (RMSE=0.19, n=87, nexp=33). It must be emphasized that grazing time (GT, 3.33<8.79±2.43<18.00 h/day, n=142) has no influence in the latter equation; moreover, the association of BM and GT in order to predict DMI is less accurate than the association of BM and RU (RMSE of 0.28 vs 0.19, respectively). Otherwise, biting rate (BR, 11.18<46.89±14.65<106.71 bites/min, n=559), which is also an easily measurable parameter, presents no intra experiment relationship with DMI. As the BM can be estimated from the IA, measurable with a simple Vernier caliper, and that RU is easily measurable with various technologies such as acoustic recordings or accelerometers, these findings are very promising for assessing feed efficiency and individually.

Validation of sensor on rumination and feeding behaviour of dairy heifers in two feedlot systems

R.D. Kliemann[1], E.M. Nascimento[1], S.R. Fernandes[1], M.M. Campos[2], T.R. Tomich[2], L.G.R. Pereira[2] and A.F. Garcez Neto[1]
[1]Federal University of Paraná, Palotina-Paraná, 85.950-000, Brazil, [2]EMBRAPA, Dairy Cattle, Juiz de Fora-Minas Gerais, 36.038-330, Brazil; americo.garcez@ufpr.br

The aim of this study was to validate the sensor HEATIME® HR SYSTEM for rumination activity and evaluate the behaviour traits of dairy heifers in tie-stall and loose-housing systems. Eleven Gir heifers with 10 months of age and 179±26 kg of body weight were housed in both systems following a crossover design with two periods of five consecutive days for data recording. A total mixed ration was offered *ad libitum* in both systems. Before each period, the animals were adapted to the collars with the sensors and systems during seven days. Each day of evaluation regarded 8 hours of observation (08:00-12:00 and 14:00-18:00). The time spent in feeding, water intake, rumination in standing and lying, and idleness in standing and lying were assessed by visual evaluation, whereas the sensor monitored only the rumination activity. In the visual evaluation the activities were recorded every three minutes, and the data collected by the sensor were recorded in 2-h interval. For the sensor validation, the data from visual evaluation were adjusted to 2-h interval. The regression of rumination recorded by the sensor to the visual evaluation was significant only for loose-housing system (P=0.0002), but the Pearson correlation between both measures was negative and low (r=-0.25; P=0.0002). The sensor overestimated the rumination by 27.3% in loose-housing (28 vs 22 min/2 h) and 38.5% in tie-stall (36 vs 26 min/2 h) and, therefore, the sensor was not validated. Regarding the systems comparison by visual evaluation, the time spent drinking, idleness in standing, rumination in standing and lying, and the total rumination time not differed between systems. However, the heifers in tie-stall spent more (P<0.05) time in feeding (41 vs 29 min/2 h), idleness in lying (19 vs 15 min/2 h) and total idleness (22 vs 17 min/2 h), whereas those in loose-housing spent more (P<0.05) time in other activities (50 vs 30 min/2 h). The sensor is not effective to record the rumination activity in Gir heifers. Loose-housing leads animals to greater activity not related to feeding compared to tie-stall.

Sensing the individual animal feed efficiency

I. Halachmi[1], N. Barchilon[1,2], R. Bezen[1], V. Bloch[1], O. Geffen[1], A. Godo[1], J. Lepar[1], H. Levit[1], E. Vilenski[1], E. Metuki[1,3] and S. Druyan[1,2]
[1]*PLF Lab., Institute of Agricultural Engineering, Agricultural Research Organization (A.R.O.) – The Volcani Centre, P.O. Box 15159, HaMaccabim Road 68, Rishon LeZion, Israel,* [2]*Animal Science, Agricultural Research Organization (A.R.O.) – The Volcani Centre, P.O. Box 15159, HaMaccabim Road 68, Rishon LeZion, Israel,* [3]*Spark.co.il, Israel; halachmi@volcani.agri.gov.il*

Precision livestock farming (PLF) might be defined as 'real-time monitoring technologies aimed at managing the smallest manageable production unit, otherwise known as the 'sensor-based' individual animal approach'. The first widely adopted application of PLF, years before the term PLF was coined, was the Israeli individual electronic milk meter for dairy cows in the 1970s and early 1980s, followed by commercialized behaviour-based oestrus detection, rumination tags and online real-time milk analysers. However, the dairy cow is not the only animal species in the PLF arena; it was applied to other species at approximately the same time. Undoubtedly, these technologies will continue to change the way that animals are managed. Moving forward, this technological shift provides reasons for optimism regarding improvements in both animal and farmer well-being. Producers, animal-feed suppliers, milk and meat processors and other animal-products stakeholders can examine real-time data organized in reports to identify abnormal deviations from a baseline. However, the data themselves are meaningless unless they are transformed into information that can be used in a good decision-making program. Precision livestock monitoring technologies will never replace producers' intuition. This review lecture will explore four research case studies (dairy, poultry, sheep and goats) – animal husbandry developments; currently running in the Israeli PLF lab, from research to on-farm application; improved production efficiency, animal welfare and GHG emission reduction.

An automatic method of forage intake estimation for Tibetan sheep based on sound analysis

H. Sheng[1], L.S. Zuo[1], G.H. Duan[1], H.L. Zhang[1], S.F. Zhang[2], M.X. Shen[1], W. Yao[1], L.S. Liu[1] and M.Z. Lu[1]
[1]*Nanjing Agricultural University, 1# Weigang Nanjing, Jiangsu province, China, 210095, China, P.R.,* [2]*Qinghai Nationalities University, No.3 Bayizhonglu, Chengdong District, Xining City, Qinghai Province, China, 810007, China, P.R.; lmz@njau.edu.cn*

Forage intake is one of the most important indicators for health evaluation of Tibetan sheep. How to monitor the herbivorous quantity of free-grazing Tibetan sheep automatically is a big challenge. In this paper, a method to estimate the forage intake of Tibetan sheep based on feeding audio analysis is studied. Oat was provided and 114 pieces of herbivorous audio were collected from 8 female Tibetan sheep (body weight 50±1 kg) using a voice recorder. The fresh weight of oat before and after a Tibetan sheep eating was weighed, and the difference of the two values was used as the gold standard of the intake of a single feeding. Each piece of the herbivorous audios was preprocessed firstly and a 32-dimensional Mel-frequency cepstral coefficients (MFCC) feature vector was obtained. The principal component analysis (PCA) method was introduced to map the MFCC feature vector into 10-dimensional P-MFCC feature parameters. The Gaussian kernel-based support vector machine (SVM) was applied to identify the chewing audios based on the extracted P-MFCC feature parameters, with an accuracy of 97.055%. The behaviour measurements (BMS) and acoustic measurements (AMS) were extracted from the chewing audios, and the relation between the forage intake and BMS, AMS parameters were calculated. A linear relation was found among these variables. The total energy flux density in chews (a parameter of ASM), chewing duration (belongs to BMS) and number of chews (belongs to BMS) were the most important explanatory variables. A linear regression test was performed to define the coefficient of determination between the total energy flux density of chewing and the amount of forage intake of sheep, which resulted in $R^2=0.842$. Furthermore, the relation between the amount of forage intake and the total energy flux density in chews, chewing duration, number of chews was investigated and the coefficient of determination R^2 was 0.893. The model established based on the integrated metric had the best estimation accuracy, which can explain 89.3% of forage intake changes.

High performance computing developments for precision pig farming, open sea fishing and aquaculture

J. Maselyne[1], D. Jensen[2], K. Sys[1], H. Polet[1], M. Aluwé[1], R. Van De Vijver[1], R. Klont[3], M. Bouwknegt[3] and K. Bovolis[4]
[1]ILVO, Burg. Van Gansberghelaan 92, 9820 Merelbeke, Belgium, [2]University of Copenhagen, Department of Veterinary and Animal Sciences, Grønnegårdsvej 2, 1870 Frederiksberg C, Denmark, [3]Vion Food Group, Boseind 15, 5281 RM Boxtel, the Netherlands, [4]I2S (Integrated Information Systems SA), Mitropoleos 43, 15124 Maroussi, Athens, Greece; jarissa.maselyne@ilvo.vlaanderen.be

In agriculture and livestock farming, big data, advanced analytics and robotics are trending. There is an increased need to fuse multiple data sources and deploy advanced algorithms that require parallel execution. The H2020 project CYBELE (IA, 2019-2021, grant nr. 825355) aims to demonstrate how the convergence of high performance computing (HPC), big data, cloud computing and Internet of Things can revolutionize farming. Nine HPC-enabled demonstrators in various fields of agriculture are selected to show the potential impact, four of which are in animal production: (1) 'Pig weighing optimization', where the live weight of pigs is estimated based on images, and subsequently combined with time series of drinking and temperature data to predict undesired events. (2) 'Sustainable pig production', based on carcass data in the slaughterhouse, meat processing is optimized and trends in health issues are detected whilst also on-farm sensors are deployed to alert when a pig is ill. (3) 'Open sea fishing', includes the vision-based automatic detection of fish species, the optimization of operational decision on the fishing boats based on sensor fusion and better monitoring of the North Sea ecosystem via an integrated modelling approach. (4) 'Aquaculture monitoring and feeding optimization', where drone images, weather, production and sensor data are used to optimize feeding, prevent fish diseases, monitor the environment and identify algae blooms. CYBELE will deliver secure access to large-scale HPC infrastructures that can handle the demonstrators' data and algorithms efficiently and timely. At this moment, the requirements of the system are being defined based on internal and external stakeholders' insights. These requirements will flow into the CYBELE technical architecture that will be built during the project and upon which the demonstrators will perform their analysis and develop their solutions.

SEM-EDAX analysis used as a tool to investigate dairy cows' stored feed resources

C.V. Mihali[1,2], D.E. Ilie[1], R.I. Neamt[1] and A.E. Mizeranschi[1]
[1]Research and Development Station for Bovine Arad, Romania, Research, 32, Bodrogului st., Arad, 310059, Romania, [2]Vasile Goldis Western University from Arad, Faculty of Medicine, Romania, Life sciences, 86, Liviu Rebreanu st., Arad, 310414, Romania; mihaliciprian@yahoo.com

The most used fodders during the cold season are corn silage, concentrate, hay, alfalfa hay. These forages must be evaluated qualitatively and analysed from the point of view of origin, aspect, colour, smell, botanical composition, purity. To achieve these goals, financial resources and time-consuming analyses could be performed. A less used method is elemental and morphology analysis of forages. Scanning Electron Microscopy (SEM) with Secondary Electrons (SE) and Dispersive Spectrometry and X-ray Fluorescence (EDAX) represents a powerful tool for characterizing and analysing preserved feed resources that compose the dairy cows' diet. In the current research we used a new and rapid analysis of forage by SEM/EDAX quality and imagistic analysis. The fodder included in the dairy cow diet during the cold season in the Research and Development Station for Bovine Arad was used for our analyses. The analyses were performed with a SEM Quanta 250 microscope with SE and EDAX detections. The forages considered in the current analysis were randomly sampled from different points of storage: peripheral/central area of corn silage, mixed fodder (wheat mix, maize, sunflower, wheat bran), alfalfa hay, ensilage hay. In SEM/EDAX analyses, the next chemical elements were found and expressed (weight percentage Wt% or atomic percentage/ At%): Mg, Si, K, Ca, C, O, Al, Pb, S, P, Cl, Na. Analysis of fodder morphology revealed vegetal structures with different degrees of degradation and fungus hyphae were found in one of the sample. All samples showed a large number of starch grains. No parasitic invertebrate or protozoa with aetiologies in various parasitic pathologies have been identified. To conclude, SEM/EDAX analyses represents a quick and accurate method that can be employed to identify the chemical elements and characterize the quality of plant morphology from different types of fodder and also to identify the presence of any foreign bodies that may appear in forage.

On-farm estimation of claw traits related to health and milk production
R. Kasarda[1], R. Žídek[2], J. Candrák[1], M. Vlček[3] and N. Moravčíková[1]
[1]Slovak University of Agriculture, Department of Animal Genetics and Breeding Biology, Tr. A. Hlinku 2, 949 76 Nitra, Slovak Republic, [2]Slovak University of Agriculture, Department of Food Hygiene and Safety, Tr. A. Hlinku 2, 949 76 Nitra, Slovak Republic, [3]SEPRED Ltd., Parkové nábrežie 877/17, 949 01 Nitra, Slovak Republic; radovan.kasarda@uniag.sk

Functional traits of cattle and feed efficiency have significant impact on profitability, because influencing the cost in production process. Claw and leg disorders are recognized as one of the three most important reasons of culling in today's dairy herds. The Aim of the study was to analyse digital images of dairy cow claws after functional trimming by use of photometry and estimate genetic parameters of claw traits and claw disorders as well as their association with production traits, reproduction traits and metabolic status by use of multi-trait mixed models. Collection of data about claw traits and claw health was made in 4 selected herds of Holstein cattle (TOP10 Slovakia) after functional trimming during fixation in the fixation cage. Cows in high production group were considered. Total number of observations was 860 (716 animals) with 3 generation pedigree information, both red and black and white Holstein cows were used in evaluation. Right hind claw was used. Claw angle, claw length, heel depth, claw height, claw diagonal and claw width were measured, and two measures of area were considered: total and functional claw area, as well. Presence of interdigital dermatitis and heel erosion (IDHE) with evaluation of degree of erosion, digital dermatitis (DD) with degree evaluated as well as overall presence of sole ulcers. Metabolic status (acidosis/ketosis) of cows was evaluated based on fat/protein ratio based on test-day recording data obtained from official milk recording made by Breeding Services of Slovak Republic s. e. Photometry results are comparable with direct measurement of traits on animals and models with use of photometry obtained measures are suitable for management of animals. Directional selection is only effective tool in case of claw health improvement.

Faecal-NIRS for predicting digestibility and intake in cattle: efficacy of two calibration strategy
D. Andueza[1], P. Nozière[1], S. Herremans[2], A. De La Torre[1], E. Froidmont[2], F. Picard[1], J. Pourrat[1], I. Constant[1], C. Martin[1] and G. Cantalapiedra-Hijar[1]
[1]INRA-UMRH, Centre de Theix, 3122 Saint-Genès-Champanelle, France, France, [2]Walloon Agricultural Research Center, Production and Sectors Department, Rue de Liroux, 8, 5030 Gembloux, Belgium; donato.andueza@inra.fr

Diet digestibility and intake are animal traits difficult to measure in practice. Thus, several indirect methods including faecal visible/near infrared (VIS/NIR) spectroscopy, have been developed for predicting these traits. The development of this technology requires great number of samples, which are not always obtained in standardized conditions among experiments. Consequently, the lack of standardization could be associated to a loss of precision of VIS/NIR models. In order to test this hypothesis, we used 88 VIS/NIR spectra of faeces sampled on cattle beef and dairy cows from 3 different experiments. Experiments included measurements of the individual organic matter digestibility (OMD) and daily dry matter intake (DI) of animals. Two calibration strategies were compared for predicting these measurements: (1) specific experiment calibration; and (2) a global calibration procedure. Models were assessed by standard error of cross-validation (SECV) and coefficient of determination of cross-validation (R2CV). The SECV was separated into bias and SECV corrected by bias SECV(C). Bias and SECV(C) for both calibration strategies were compared using the procedure of Fearn in 1996. Data sets gave value ranges of 0.60-0.78 for OMD and 53-181 g/kg body weight (BW)0.75 for DI. For OMD, the SECV and R2CV were 0.02 and 0.86 for both specific experiment calibration and global calibration procedure. For DI, the SECV and R2CV were 6.90 g/kg BW0.75 and 0.96 respectively, for the specific experiment calibration and 7.62 g/kg BW0.75 and 0.95, respectively, for the global calibration procedure. For both, OMD and DI, no significant differences between calibration strategies were found for bias and SECV(C). From the results of this preliminary study, we conclude that mixing data from experiments conducted in very different conditions does not degrade the precision of the obtained VIS/NIRS models to predict OMD and DI in cattle.

Experts' opinion on how to validate an individual welfare monitoring system for fattening pigs
P. Briene, C. Vandenbussche, B. Sonck, J. Vangeyte and J. Maselyne
ILVO (Flanders Research Institute for Agriculture, Fisheries and Food), Burgemeester Van Gansberge 115 bus 1, 9820, Belgium; petra.briene@ilvo.vlaanderen.be

The pig production sector today is a fast-developing and challenging sector. Pig farmers try to improve production while keeping animal health and welfare standards high. For the IOF2020 project, the Pig Farm Management use case focuses on the use of Internet of Things (IoT) technology in pig production. Automated, individual monitoring of the health and production of fattening pigs is an example of IoT technology which can help pig farmers to optimize their time management and resources by helping them to identify pigs with health problems. Additionally, health, welfare and also productivity can be increased via early treatment of individual pigs where problems otherwise might be detected late or not at all. An additional aim of the use case is to develop an individual pig monitoring dashboard. To develop this, feeding and drinking data of individual pigs have been collected using RFID ear tags to identify each individual pig in combination with RFID antennas at the feeders and the drinkers. Three fattening rounds have so far taken place, each with 120 pigs, divided into eight pens. Algorithms were designed to detect abnormal deviations in individual pig's behaviours, which then result in an alert. The process of designing, testing and validating the detection system is a very challenging task which includes many decisions to be made. During an interactive session which will be held at a PLF workshop organised by the University of Copenhagen coming April, expert opinions from the participants will be collected. During this interactive session, using example cases, different steps in this process will be visualised and discussed such as: what problems should this system detect, how to categorize health problems whilst also considering severity, how to handle the observers' bias, time difference between alerts and observations, etc. These expert opinions will be used to identify challenges, investigate the different approaches to validation and identify possible needs of the potential users of a detection system. The results of this workshop will be presented at the conference.

Tool for dairy farmers to predict on site the composition of their forage with NIR spectroscopy
N.C. Chamberland
Walloon Agricultural Research Center, Food and Feed Quality Unit, Chaussée de Namur, 24, 5030 Gembloux, Belgium; n.chamberland@cra.wallonie.be

The Walloon Agricultural Research Center (CRA-W) is currently testing a new approach using NIR spectroscopy applied to the context of dairy farming in Wallonia, Belgium. One of the aims of this project is to develop and validate new analytical methods for predicting quality parameters such as dry matter (DM), chemical composition (Starch, Crude Protein, ADF, NDF and Ash) and digestibility of wet forages directly on farm (specifically maize silage, grass silage, fresh grass and hay). Fifteen dairy farms have been selected in Wallonia in 2018 to collect fresh samples. NIR spectra were measured directly on site with three portable NIR spectrometers: the FieldSpec 4 from ASD (350-2,500 nm), the NIR4FARM from AUNIR (950-1,650 nm) and the flameNIR from OceanOptics (940-1,665 nm). These instruments allow acquiring NIR spectra directly on farm. Moreover, samples were also measured with a benchtop XDS instrument from FOSS (400-2,498 nm) and a Bruker MPA instrument (800-2,500 nm) in the Food and Feed Quality Unit of the CRA-W. Reference values were obtained by prediction with a FOSS DS2500 on dried and ground samples. In 2018, 150 fresh maize silage samples and 106 fresh grass silage samples were collected and analysed on site and on benchtop instruments (FOSS XDS and Bruker MPA). Results obtained with portable instruments are still in treatment and will be presented in further studies. Regarding the first results, we can observe than models for fresh grass silage are mostly better than models for fresh maize silage based on the observation of RPD of cross-validation (Ratio of Prediction) for both FOSS XDS and Bruker MPA benchtop instruments. The low RPD of the models for fresh maize silage are mainly linked to the high heterogeneity of this matter. Models for both fresh grass silage and fresh grass maize will be improved and updated with new samples collected in 2019. These results are the first step to develop an user-friendly tool for dairy farmers to predict on site the composition of their forage. enabling the calculation of their nutritional value and the adaptation of animal's feeding for a better sustainability.

Assessing sustainability trade-offs between pig welfare, economy of production and farmer wellbeing

J. Helmerichs[1], S.J. Hörtenhuber[2], A.K. Ruckli[3] and S. Dippel[1]
[1]Friedrich-Loeffler-Institut, Dörnbergstr. 25/27, 29223 Celle, Germany, [2]FiBL Deutschland e.V., Kasseler Str. 1a, Postfach 90 01 63, 60486 Frankfurt am Main, Germany, [3]University of Natural Resources and Life Sciences (BOKU), Gregor-Mendelstraße 33, 1180 Vienna, Austria; juliane.helmerichs@fli.de

Public demand for more sustainable pig production systems with high animal welfare standards is increasing. Thus, farmers face the challenge to improve pig welfare while at the same time generate an adequate income and have satisfying working conditions. Our contribution gives an overview of tools for assessing pig welfare, economic sustainability and farmer wellbeing. We illustrate differences and agreements between methods and show results obtained with those tools, based on a review of scientific and generic publications in English. While there are some tools for assessing pig welfare only (e.g. Welfare Quality®), others combine pig welfare and environmental sustainability (ProPig), environmental, economic and social sustainability (SAFA, Public Goods Tool), or economic sustainability and farmer wellbeing (RISE, SMART). However, there are no tools yet that combine the three areas pig welfare, economic sustainability and farmer wellbeing. Studies assessing pig welfare mostly use comparable indicators, while indicators for economic sustainability and farmer wellbeing are more variable between tools. Moreover, assessment of economic sustainability sometimes extends beyond the farm level and includes aspects of the local economy. Several studies show strong relations between economy, farmer wellbeing and pig welfare. Higher pig welfare strongly correlates with higher production costs. Farmers are more satisfied with their job, if they have more autonomy in their work, and pigs of farmers with higher job satisfaction are more productive. Farmers are facing these trade-offs when trying to increase the sustainability of their farm. More data from different production systems are needed to better understand the relationships and support farmer decision making as well as sustainability improvement policies.

Developing a methodology for sustainability scoring of pig farms

R. Hoste
Wageningen Economic Research, P.O. Box 35, 6700 AA Wageningen, the Netherlands; robert.hoste@wur.nl

Given the search for added value in pig meat supply chains in the Netherlands, the Dutch government and industry signed a covenant in 2009 to support this development. Meat processors are in search of tools to measure sustainability of pig farms. A tool has been developed in Excel to quantify a sustainability score for pig farms. Sustainability is based on four Components: Pig (animal health and welfare), Planet, Profit and People (labour and society), with underlying Aspects. Aspects were defined, as well as scoring algorithms and weighting of Aspects within Components. For each component the score was in the range of 0 to 9, with the highest value considered to mostly contribute to sustainability. Scores of the Components were added, which resulted in a Farm score in the range of 0 to 36. Scoring was performed for piglet production and growing-finishing separately. The tool has been applied on 7 farms. The Farm score varied among farms between 11.6 and 19.3 for sows and between 10.6 and 19.7 for growing-finishing pigs. On most farms the Profit Component scored lower than the Planet Component. For application in practice, the definition of Aspects, their algorithms and the weighting should be discussed with stakeholders and experts to increase the value for users. Sustainability scoring can be used by farmers to improve insight into sustainability issues in their farm management, and for discussion among farmers. Theoretically it can be used for settling between meat processors and their supplying farmers if sustainability goals are met. However, guarantees for correct application of the tool need then to be further elaborated.

Farmers' and other stakeholders' view on sustainable pig production systems

S. Hörtenhuber[1], R. Chapman[2], P. Ferrari[3], M. Gebska[4], J. Helmerichs[5], C. Hubbard[2], C. Leeb[6], C. Munsterhjelm[7], A.K. Ruokli[9], K. Swan[7], H. Vermeer[8] and S. Dippel[5]
[1]FiBL Deutschland, Kasseler Str. 1a, PF 90 01 63, 60486 Frankfurt am Main, Germany, [2]Newcastle University, NE1 7RU, NE1 7RU Newcastle, United Kingdom, [3]Fondazione CRPA Studi e Ricerche, Viale Timavo 43/2, 42121 Reggio Emilia, Italy, [4]Warsaw University of Life Sciences, Nowoursynowska 166, 02-787 Warszawa, Poland, [5]Friedrich-Loeffler-Institut, Dörnbergstraße 25/27, 29223 Celle, Germany, [6]University of Natural Resources and Life Sciences Vienna, Gregor-Mendel-Str. 33, 1180 Vienna, Austria, [7]University of Helsinki, Viikintie 49, P.O. Box 57, 00014 Helsinki, Finland, [8]Wageningen University & Research, De Elst 1, 338, 6700 AH Wageningen, the Netherlands; stefan.hoertenhuber@fibl.org

Various approaches exist for assessing sustainability of pig production systems, however, few are feasible for on-farm application. Therefore, the SusPigSys project consortium developed an indicator set based on literature, stakeholder and expert opinion in seven countries (AT, DE, FI, IT, NL, PL, UK). This contribution describes farmers' and stakeholders' view on sustainability of pig production systems as a basis for a practical assessment protocol. In each country, a stakeholder workshop was conducted, which included in total 90 scientists, representatives of farmer associations and allied industries, farmers, NGOs, consumers and policy makers. Furthermore, 67 volunteering farmers were interviewed during farm visits. The general view on sustainability of pig production, as well as specific indicators were collected in group and individual interviews. Stakeholders mentioned mostly environmental aspects regarding atmospheric emissions, nutrient cycles or feeding and economic aspects, e.g. profit and fair prices. Concerning animal welfare, clinical and behavioural aspects, housing systems and breeding were often named. Regarding social aspects, acceptance of pig farming by society and farmer livelihood were most important. In two countries (UK, PL) farmers mostly indicated interest in economic issues (e.g. sufficient income), while farmers in the other countries focussed on environmental aspects (e.g. nutrient efficiency). However, also animal health and welfare topics (e.g. absence of disease) and social aspects (e.g. succession) were mentioned.

Pig farmers' willingness to participate in animal welfare programs

A. Kuhlmann, S. Schukat and H. Heise
University of Göttingen, Department of Agricultural Economics and Rural Development, Agribusiness Management, Platz der Göttinger Sieben 5, 37075 Göttingen, Germany; sirkka.schukat@uni-goettingen.de

The demand for products with higher animal welfare (FAW) standards is growing within the European Union. In response to the growing public debate, representatives from agricultural associations, the slaughtering industry and food retailing started the industry solution 'Initiative Animal Welfare' (IAW) in 2015 with the aim of establishing higher FAW standards on a broad basis in both, poultry and pig production in Germany. In contrast to other animal welfare programs (AWPs), the IAW receives considerable support from farmers. Farmers are considered a very important stakeholder group for the successful implementation of higher FAW standards but so far little is known about their attitudes and the determinants of participation in programs that request higher FAW standards. For this study, pig farmers in Germany were questioned via a large-scale online survey in summer of 2018 (n=239). Based on the unified theory of acceptance and use of technology by Venkatesh *et al.*, a partial least squares path modelling was run. Results show that the expected performance as well as the expected costs associated with the IAW influence pig farmers' behavioural intention to participate in the IAW substantially. Furthermore, the decision is influenced by social determinants and facilitating conditions, such as deadweight effects. Farmers' hedonic motivation, fair remuneration and previous experiences with the establishment of higher FAW standards can influence their intention to take part in the IAW. In addition, farmers' trust in the program constitutes a major determinant. There are also moderating variables such as age and work experience that influence farmers' intention to take part in the IAW. Our results have important managerial implications for the IAW and can help to design further tailor-made AWPs that fulfil the requirements of both pig farmers and the broader public. Participating in AWPs could create an opportunity for pig farmers to escape from the pressure to produce at the level of world market prices and instead take advantage of a more differentiated market segment for meat produced with higher FAW standards which is accepted and financially rewarded by society.

Welfare assessment of pigs using a multi-factorial approach (MulTiViS)

H. Gerhardy[1], H. Meyer[2], H. Plate[2], I. Spiekermeier[3], B. Wegner[4], J. Grosse-Kleimann[5], H. Nienhoff[3] and L. Kreienbrock[5]
[1]*Marketing Service Gerhardy, Am Stahlbach 17, 30826 Garbsen, Germany, [2]VzF GmbH Erfolg mit Schwein, Veerßer Str. 65, Uelzen, Germany, [3]Swine Health Service, Chamber of agriculture Lower Saxony, Mars la Tour Str. 1-13, Oldenburg, Germany, [4]University of Veterinary Medicine Hanover, Foundation, Germany, Institute for Animal Hygiene, Animal Welfare and Farm Animal Behaviour, Bünteweg 2, Hannover, Germany, [5]University of Veterinary Medicine Hannover, Department of Biometry, Epidemiology and Information Processing, Bünteweg 2, Hannover, Germany; hubert.gerhardy@t-online.de*

Farm animal welfare attracted political, social and consumer attention. At the moment there is a lack of information to assess pig welfare comprehensively on farm level taking into account various issues at any one time. A tool to monitor the development of animal health on national level is missing. To provide information on indicators to estimate animal welfare, the project MulTiViS has been launched. Using various data sources of 200 commercial farms in the North of Germany (biological and economic data, use of antibiotics, information on assessments in slaughterhouses as well as on pig behaviour and clinical assessments of pigs on farms) indicators should be developed using multi-factorial analysis. Based on the description of the actual situation the challenges of the diversity of production methods and differences in assessments of organs between slaughterhouses will be presented. The project is supported by funds of the Federal Ministry of Food and Agriculture (BMEL) based on a decision of the Parliament of the Federal Republic of Germany via the Federal Office for Agriculture and Food (BLE) under the innovation support programme.

Pig welfare self-assessment and benchmarking tool – a mobile application for farmers

A. Watteyn[1], M. Thys[1], W. Wytynck[2], G. Vandepoel[2], B. Sonck[1] and F. Tuyttens[1]
[1]*ILVO, Scheldeweg 68, 9090 Melle, Belgium, [2]Boerenbond, Diestsevest 40, 3000 Leuven, Belgium; anneleen.watteyn@ilvo.vlaanderen.be*

Today's society requires the farmer to focus much more on animal welfare. Although farmers believe they already pay a lot of attention to animal welfare, they could benefit from extra information about why and how to assess welfare, identifying points requiring attention, and evaluating effects of farm management modifications. Until now there has been no tool for monitoring animal welfare by the farmer. Therefore, a mobile application has been developed to self-assess facultative the welfare of their livestock with instant feedback which includes potential risk factors for important welfare issues, previous results and benchmarking. Similar to the Welfare Quality© and KTBL Protocol, the tool primarily includes animal-based 'outcome' indicators (e.g. body condition, injuries, behaviour). Indicators were carefully selected to ensure that the main welfare issues are covered, and to allow straightforward and robust scoring of a limited but illustrative number of animals in a time-span of 1.5 to 2 hours. Additionally, key questions on farm management, housing and production parameters are included to allow for benchmarking with comparable farms. Self-scans, allowing offline data recording, have been developed for pregnant sows (shortly after insemination and after rehousing), lactating sows and their litters, growing pigs and finishing pigs. After online submission of a completed scan, a report is automatically generated with scores for several welfare indicator on a 0-100 scale, where 0 is the worst and 100 is the best score. In addition, scores of the own farm are graphically depicted over time and benchmarked anonymously with those of comparable farms. In order to assess whether the tool was comprehensible, inviting and feasible for farmers to include in their routine management, the self-scan was tested by three farmers. This pilot trial resulted in a further reduction of the number of animals to be scored individually and in the rephrasing of several questions for clarity. By assessing and benchmarking animal welfare periodically, farmers together with their veterinarians and/or consultants are encouraged to address points of attention, and able to monitor effects of measures taken over time.

Applying life cycle assessment to sparse historical data from pig systems
M. Misiura, M. Ottosen, S. Mackenzie, J. Filipe and I. Kyriazakis
Newcastle University, School of Natural and Environmental Science, Tyne and Wear, NE1 7RU Newcastle Upon Tyne,
United Kingdom; m.ottosen2@newcastle.ac.uk

Feed inputs account for a large part of environmental impact from pig systems. However, within life cycle assessment (LCA), genetics and management of crop production systems are also considered as important determinants of the environmental impact of livestock systems. As such, for rigorous comparisons of livestock production systems from different eras with respect to environmental impact, a robust method is needed to estimate diet composition based on animal requirements. The objective of this study was to compare the environmental impacts of British indoor and outdoor bred pigs based on growth rates, feed intake and of sow reproductive performance from 2003 and 2017 systems. Animal daily requirements were estimated by modelling the need for energy and protein based on equations for conservation of mass, energy and protein. To accommodate lactation nutrient dynamics, a sow growth model modified to impose conservation of protein was implemented to estimate sow requirements. Least cost diets, based on historical prices of feed ingredients, were formulated to meet the estimated energy, amino acids and mineral requirements of all life stages. Storage and spreading of manure were modelled based on excreted nutrients, and artificial fertilizers were replaced with typical ratios reported in literature. Global Warming Potential (GWP), Terrestrial Acidification Potential (TAP), Land use (LU) and Fossil Resource Scarcity (FRS) were calculated with SimaPro, linking the sow reproductive cycle to the grower pig via piglet production. We found that impacts changes per functional unit of 1 kg of live pig at farm gate from 2003 to 2017 respectively for indoor and outdoor systems were 8.8 and 5.0% lower for GWP, 2.1% higher and 6.1% lower for TAP, 13.2 and 16.6% lower for land use, and 86.9 and 83.7% higher for FRS. The major increase in FRS are accounted for by higher inclusion of artificial amino acids in 2017 and higher fertilizer replacement in 2003. This method can be applied to all systems where basic industry data are available, and thus is a promising tool for estimating the implications of industrial development for the sustainability of livestock systems.

The carbon footprint, nitrogen and phosphorus efficiency in boars, barrows and immunocastrates
S. Millet[1], C. De Cuyper[1], V. Stefanski[2], K. Kress[2] and A. Van Den Broeke[1]
[1]ILVO (Flanders Research Institute for Agriculture, Fisheries and Food), Scheldeweg 68, 9090 Melle, Belgium,
[2]Behavioral Physiology of Livestock, University of Hohenheim (UHOH), Garbenstr. 17, 70599 Stuttgart, Germany;
sam.millet@ilvo.vlaanderen.be

With the pressure to ban surgical castration, more male pigs are kept entire or are immunocastrated. In the SuSi (Sustainability in pork production with immunocastration) project, trials are carried out in different European countries to assess all aspects of sustainability with immunocastrates. Two studies, one in Germany (UHOH) and one in Belgium (ILVO) have been finished. In this abstract, we present the effect of castration strategy and trial location on nitrogen (N) and phosphorus (P) efficiency and on the carbon footprint of the ingested feed (CFP) per kg of bodyweight (BW) gain in boars (BO), barrows (BA), and immunocastrates (IC). In each trial, BO, BA and IC received the same diet. Animals were followed up from 10 weeks to slaughter in a three-phase-feeding regime. At ILVO, N efficiency of BO (51±1%) was significantly higher than IC (48±1%), which in turn was higher than BA (44±1%). At UHOH, BO (38±1%) and IC (39±1%) had a significantly higher N efficiency than BA (34±1%). In both trials, P efficiency was significantly higher in BO and IC compared to BA (48±1, 47±1 and 43±1% at ILVO and 36±1, 37±1 and 33±1% at UHOH for BO, IC and BA, respectively). Both at ILVO and at UHOH, the CFP per kg BW gain was significantly lower in BO and IC compared to BA. Both trials confirm that the better feed efficiency of BO and IC results in a lower environmental impact compared to BA. However, the differences between trials were by far exceeding the differences between sexes. This indicates that the castration decision may affect environmental sustainability, but farm management (including diet) may be more important. The research was done within ERA-Net Cofund SusAn (696231) Sustainability in pork production with immunocastration (SuSI).

Tailored phase-feeding program for liquid-fed growing pig towards a reduced use of protein rich diet

F. Maupertuis[1], D. Olivier[1] and N. Quiniou[2]
[1]Chambre d'Agriculture des Pays de la Loire, Ferme expérimentale, Les Trinottières, 49140 Montreuil-sur-Loir, France,
[2]IFIP-Institut du Porc, La Motte au Vicomte, BP 35, 35650 Le Rheu, France; nathalie.quiniou@ifip.asso.fr

The growth profile and amino acid requirements of crossbred pigs were assessed with the InraPorc® software and used to investigate the possibility of implementing different phase-feeding strategies using a liquid feeding system. Two batches (16 pens × 14 pigs per pen) were studied over the 44 to 120 kg body weight (BW) range. Two diets with 1.0 g (A) or 0.5 g (B) of digestible lysine per MJ of net energy (LYS/NE) were blended in proportions that depended on the strategy and the BW range. If using the 2-phase (2P) strategy, LYS/NE was 0.9 up to 65 kg BW and 0.8 afterwards. If using the 5-phase strategy (5P), LYS/NE was 0.90 (BW<50 kg), 0.85 (50-65 kg), 0.80 (65-85 kg), 0.70 (85-105 kg), and 0.60 (BW>105 kg). The daily supply of feed increased weekly by around 0.2 kg/pig, up to a plateau (gilts: 2.8 kg/d, barrows: 2.6 kg/d). The change from one phase to another was undertaken by the average BW per pen (measured every 3 weeks) and the minimum amount of feed the liquid-feeding system could deliver. Due to delivery problems, data of two pens were removed from the trial. No significant difference were observed between 2P and 5P pigs on daily feed intake (respectively: 2.38 and 2.42 kg/d, P=0.47), average daily gain (822 and 814 g/d, P=0.64), and feed conversion ratio (2.93 and 3.00, P=0.26). Diet A represented on average 65.4 and 48.1% of the total feed intake for 2P and 5P pigs (P<0.01), respectively. Based on the difference between N intake calculated from the crude protein (CP) content of diets A (15.6%) and B (10.7%) and N retention assessed from carcass leanness (P=0.37), N output was reduced by 6% with the 5P strategy (P=0.04). The decrease would have reached 19% with a two-phase strategy based on typical growing-finishing diets (16 and 15% CP) instead of the low-protein blends (14.6 and 13.6% CP) used in this study. Based on global environmental impacts of feedstuffs incorporated into diets A and B, the 5P strategy may contribute to a reduced consumption of phosphorus and climatic change. This research was part of the DY+ project, co-funded under the European Union's Feader program.

Effect of dietary energy level on performance and environmental sustainability in male pigs

A. Van Den Broeke, C. De Cuyper, M. Aluwé and S. Millet
Flanders Research Institute for Agriculture, Fisheries and Food (ILVO), Animal Sciences Unit, Scheldeweg 68, 9090 Melle, Belgium; alice.vandenbroeke@ilvo.vlaanderen.be

One of the possible alternatives for surgical castrated male pigs (SC) is the production of immunocastrated pigs (IC). The scientific knowledge on optimal management of IC, however, is limited. While it is generally assumed that pigs eat according to their energy needs, it is unsure whether this is still valid for IC. Therefore, a study was designed to test the effect of dietary energy concentration in late finishing on performance, carcass quality and environmental sustainability in SC and IC in a 2×2 factorial design. Ninety-six IC and 96 SC were housed in single-sex group pens of 6 animals per pen and fed *ad libitum* with a three phase diet. All animals received the same diet in phase 1 and 2. In the last phase (from 2nd IC injection to slaughter), pens were divided into 2 groups, receiving a high energy (HE; 9,6 MJ/kg) or a low energy (LE; 8,5 MJ/kg) diet. Crude protein and digestible amino acid level content of both diets were equal. No interactions between sex and diet were observed on any of the measured parameters. Energy concentration of the diet did not affect average daily feed intake (DFI) significantly. As a result, daily gain in finishing was lower (1,004±36 vs 1,066±36 g/day) and feed conversion ratio higher (2.99±0.04 vs 2.76±0.04 g/g) in LE compared to HE. The average daily lean meat gain did not differ between HE and LE (P=0.664) indicating that excess energy intake, lead to a higher fat deposition and therefore a lower lean meat percentage (61.7±0.2% HE vs 62.5±0.2% LE). The nitrogen efficiency of HE (46±1%) was significantly higher compared to LE (44±1%) but the carbon footprint of ingested feed per kg of bodyweight gain was also higher in HE compared to LE (2,826±27 vs 2,658±27 g CO_2 eq/kg), giving contrasting effects of diet on environmental sustainability. This trial could not confirm our hypothesis that IC and SC would react differently on dietary energy level but shows however that DFI in (immuno-)castrated pigs is not solely based on fulfilling their energy needs but also on other parameters like gastro-intestinal capacity and satiety of the diet. The research was done within SuSi (ERA-Net Cofund SusAn, 696231).

SusPigSys: assessment and feedback of sustainability of pig production systems

A.K. Ruckli[1], C. Leeb[1], K. De Roest[2], M. Gebska[3], J. Guy[4], M. Heinonen[5], J. Helmerichs[6], S. Hörtenhuber[7], H. Spoolder[8], A. Valros[5] and S. Dippel[6]
[1]BOKU, Gregor-Mendel-Straße 33, 1180 Vienna, Austria, [2]FCSR, Viale Timavo 43/2, 42121 Reggio Emilia, Italy, [3]SGGW, Nowoursynowska 166, 02-787 Warszawa, Poland, [4]NU, Agriculture Building, Newcastle Univ., NE1 7RU Newcastle upon Tyne, United Kingdom, [5]HU, Viikintie 49, P.O. Box 57, 00014 Helsinki, Finland, [6]FLI, Dörnbergstr. 25/27, 29223 Celle, Germany, [7]FiBL, Kasseler Str. 1a, PF 90 01 63, 60486 Frankfurt am Main, Germany, [8]WUR, De Elst 1, 338, 6700 AH Wageningen, the Netherlands; antonia.ruckli@boku.ac.at

Despite the need for pig farmers to have recommendations on how to optimise the balance between the apparently conflicting areas of sustainability (economy, environment, pigs, farmers, society), very little on-farm data is available for supporting informed holistic decisions. The SusPigSys project therefore aims to collect, summarise and disseminate evidence-based information on successful strategies for improving sustainability in typical as well as niche pig production systems. Project outcomes will include an integrative on-farm assessment and feedback tool to help pig farmers compare and improve economic and environmental performance as well as animal welfare and their own wellbeing. Various stakeholders are involved throughout the project. They identified key issues for assessing sustainability on-farm in workshops in 7 countries represented in the SusPigSys consortium (AT, DE, FI, IT, PL, NL, UK). This was combined with existing protocols (e.g. SAFA guidelines) to develop a detailed sustainability assessment protocol, which was subsequently applied on a total of 68 farms (2 person days/farm). These data are the basis for generating a valid condensed protocol, which will then be applied to a larger number of farms (1 person day/farm). Data will be used for developing an integrative analysis toolbox for summarising farm data, which will be linked with an existing international pig production database to allow enhanced benchmarking. Furthermore, the toolbox will be integrated in a software to form a farmer decision support tool ('app') with farm-specific feedback on how to improve sustainability of pig production.

Sustainability of pig production through improved feed efficiency

W.M. Rauw[1], E. Gómez Izquierdo[2], L.A. García Cortés[1], E. De Mercado De La Peña[2], J.M. García Casco[1], J.J. Ciruelos[2] and L. Gomez-Raya[1]
[1]INIA, Dept de Mejora Genética Animal, Crta de la Coruña km 7.5, 28040 Madrid, Spain, [2]ITACYL, Centro de Pruebas de Porcino, Ctra. Riaza-Toro S/N, 40353 Hontalbilla, Spain; rauw.wendy@inia.es

The human population is projected to rise to over 11 billion people by the year 2100, resulting in a concomitant increase in meat production and consumption. However, because livestock production has been linked to deforestation and climate change, there is a need for improvements in the efficiency of food and feed production. Genetic improvement of livestock is key to this development, and has resulted in an unprecedented increase in lean growth rate and feed efficiency in commercial pig lines. Improved pig production levels must be supported by improved quality of resources that allow for the expression of production traits, or may otherwise result in a reduction in the animal's resilience to environmental stressors and the genesis of production diseases. Therefore, modern livestock diets consist of feed ingredients with high nutritional and commercial value that is sourced from international markets. For example, soybean meal is a major ingredient in livestock feeds because of its relatively high protein content, however, Europe is heavily reliant on its imports from countries in which soybean production has expanded into their natural ecosystems. The notion of the unsustainability of this heavy dependency of the EU to soybean meal imports emphasizes the need for using local feed resources and feedstuff co-products. However, since the quantity and quality of feed resources limits productive output, this may require a different type of animal with different performance characteristics than those currently selected in intensive, high quality input – high output production systems. This is supported by research that shows a negative relationship between potential for high production and resilience to environmental stressors. Evaluation of the environmental, social, and economic impact of improved feed efficiency on local feed resources and feedstuff co-products may facilitate the development of new sustainable pig production systems. This research is part of the Eranet SusAn 'SusPig' project 'Sustainability of pig production through improved feed efficiency' (www.suspig-era.net).

Evaluating the environmental impacts of selection for residual feed intake in pigs

F. Soleimani Jevinani and H. Gilbert
GenPhySE, INRA, INPT, ENSAT, Université de Toulouse, 31326 Castanet-Tolosan, France; faezeh.soleimani-jevinani@inra.fr

To appoint proper strategy for prospective feed efficient pig farming it is wise to shed light on ongoing selection scenarios. Selection based on residual feed intake (RFI) has been proposed to improve feed efficiency, and potentially reduce the environmental impacts accordingly. Data were collected for pigs from the 5th generation of lines divergently selected for RFI (low line, more efficient pigs, LRFI; high line, less efficient pigs HRFI). Individual records for daily feed intake, body weights and body composition were available for about 60 pigs per line. The averages of feed conversion ratio and daily feed intake in the LRFI pigs were 7% lower than the average of the HRFI pigs according to the records. To compare the effects of the difference in feed efficiency on environmental impacts, a parametric model for life cycle assessment (LCA) was developed based on the net energy fluxes. A nutritional growth pig tool, InraPorc® was integrated as a module into the model to embed the flexibility for change in feed, traits and housing conditions, along with simulating individual pig performance. The comparative LCA showed 6% lower environmental impacts for LRFI on average relative to HRFI, considering climate change (CC), acidification potential (AP), eutrophication potential (EP), land occupation (LO) and water depletion (WD). Impacts of CC, AP, EP, LO and WD for 1 kg live weight of pig as the functional unit were 2.61 kg CO_2-eq, 44.8 g SO_2-eq, 3.37 g PO_4-eq, 4.208 m^3, 0.0448 m^3 for LRFI, and 2.76 kg CO_2-eq, 48 g SO_2-eq, 3.62 g PO_4-eq, 4.44 m^3, 0.047 m^3 for HRFI, respectively. Parallel Monte Carlo simulation on input parameters demonstrated that LRFI pigs would have a lower environmental impact than HRFI pigs in more than 70% of the calculations, except for water depletion (55% of the calculations). A once-at-a-time sensitivity analysis revealed that environmental impacts are highly sensitive to protein content of the body and protein deposition. This correspondence between improvement of feed efficiency and reduction of environmental impacts is very promising for prospective sustainable breeding objectives.

Estimated cost savings from reduction of ammonium-salt based inorganic second aerosol from livestock

S. Lee, M. Kim and J.H. Hwang
Jeonju University, Agricultural Conference, 303 Cheonjam Ro Wansan Gu North Jeolla, 56093, Korea, South; soilsang@gmail.com

Disturbance of nitrogen (N) cycle is receiving more attention, while the interests of domestic contribution of ammonia emission for releasing of fine particulate matter in South Korea. Of course, the impact from China has mainly focused on climate situation depending on regional temperature. The N cycle related with animals and various industrial activities influenced in a complicate manner based on reactive forms of Ammonia (NH_3) and NOx. Atmospheric transfer of reactive N compounds (NH_3, NO_3, NO_2, NO, N_2O), especially NH_3 and the relations with balance of fine particulate matter (PM) are mainly focused on the domestic effect from the agriculture in South Korea while the external impact on particulate matter is mainly affected from China. Additionally, the methodology of dry/wet deposition network was engaged in our research site with help of US National Atmospheric Deposition Program (NADP) to develop a comprehensive multi-pollutant approach for N, with special attention to the fate of emissions from agriculture. The reduction technology of ammonium-salt based inorganic second aerosol from livestock manure composting process was applied using the cattle manure-based biochar as the compost ingredients of moisture control mixing material. This work was carried out with the support of 'Cooperative Research Program for Agriculture Science & Technology Development (Project No. PJ014297)' Rural Development Administration, Republic of Korea.

Measuring particles in pig housing
S. Lagadec[1], N. Guingand[2] and M. Hassouna[3]
[1]CRAB, rue le Lannou, 35042 Rennes, France, [2]IFIP institut du Porc, La motte au vicomte, 35651 Rennes, France, [3]INRA, UMR SAS, 65 rue de St Brieuc, 35000 Rennes, France; nadine.guingand@ifip.asso.fr

Livestock contributes to the emission of particles in the atmosphere. Literature shows drastic differences between emission factors by animal category related to the measurement methodologies applied. Based on this observation, a project involving research and development organizations has been developed in order to develop a measurement protocol strictly adapted to pig building conditions. The project is organized in 3 steps: (1) identification of specific conditions related to pig building; (2) analysis of metrology equipment able of adapting to these conditions; and (3) development of a protocol adapted to the equipment identified in the previous step. Conclusions of the first step (1) are the following ones: ammonia concentration inside piggery vary between 0 and 50 ppm, relative humidity between 70 and 100% and temperature between -10 and +40 °C. TSP, PM_{10} and $PM_{2.5}$ should be measured continuously in the ambience, in the extracted duct and outside. Massic concentrations but also concentration in number of particles per volume unit should be measured. In order to analyse morphology of particles, sampling should be possible. Analysis of measurement equipment – step (2) – led to choose the optical measurement methodology applied in the GRIMM 1.109 (Intertek). Nevertheless, in order to validate collected data, gravimetric method with simple filter will also be applied. For particle measurements, 24 h sampling period should be achieved in the middle of the corridor (1-1.50 m high). This duration has been chosen in order to integrate diurnal and nocturnal changes inside piggeries. To calculate the emission factor per fattening pig, three periods (between 14-18 days, 45-50 days and 78-82 days) were identified for measuring particles. The project is currently in progress with the second phase consisting of the implementation of this protocol in commercial pig farms in Western France.

Effects of soil spreading of swine manure on bioavailability of trace elements in contaminated soils
M. Kim[1,2] and S. Lee[1]
[1]Jeonju University, Agricultural Convergence, 303 Cheonjam Ro Wansan Gu North Jeolla, 55069, Korea, South, [2]Korea University, O-Jeong Eco-Resilience Institute, Seoul, 02841, Korea, South; soilsang@gmail.com

This study evaluated the effects of spreading of swine manure (SM) to soil on bioavailability of trace elements, depending on the initial soil pH in acidic, neutral, and alkali conditions. Swine manure was obtained from a commercial fertilizer vendor and air dried, ground and sieved to less than 0.5 mm. Three types of trace elements contaminated soils were collected from different pH soils. After air-dried and passes through a 2-mm sieve. Swine manure ground was applied to each soil at a 2% w/w ratio and mixed for homogeneity. After 4 weeks, chemical and biological assessments were conducted using Mehlich-3 solution and phytotoxicity test, respectively. pH, electrical conductivity, and loss on ignition of SM were 7.8, 9.8 ds/m and 58.5%, respectively. Application of SM changed soil pH from 5.9, 7.2, 8.3 to 6.36, 7.32, and 8.39, respectively and significant difference was observed in only acidic soil. Water holding capacity, which is important index for soil fertility, was significantly increased in all types of soil by applying SM mainly composited by organic matter. In all kinds of soil, bioavailability of trace elements was not increased by SM. However, SM treatments could not promote root and shoot elongation but decreased the absorption of trace elements in their shoots. These results suggested that the application of organic amendments including SM in agricultural soil requires further study. This research was supported by Technology Development Program for Agriculture and Forestry [318014], Ministry for Food, Agriculture, Forestry and Fisheries, Republic of Korea.

Pig manure economic valuation by nutrient balance

R.A. Nacimento[1], M.F. Silva[1], E.R. Afonso[1], O.A. Ojeda-Rojas[1], J.C.P. Palhares[2] and A.H. Gameiro[1]
[1]University of Sao Paulo, Department of Animal Science, Av. Duque de Caxias Norte, 225, 13.635-900 Pirassununga SP, Brazil, [2]Embrapa, Pecuária Sudeste, Rodovia Washington Luiz, km 234 s/n, Fazenda Canchim, 13.560-970 São Carlos SP, Brazil; gameiro@usp.br

Nutrient balance is a method used to evaluate the environmental performance of farms and other activities, through inputs and outputs of nutrients such as nitrogen (N), phosphorus (P) and potassium (K). This work evaluated the economic value of the manure to an intensive system for finishing pigs. The evaluation was made using the nutrient balance approach. For the estimates, research data were used and extrapolated to a commercial intensive Brazilian system of pigs in the finishing phase with 773 animals, submitted to corn-based diets and soybean co-products for 117 days. The inputs were the amount of N, P and K consumed from the diet. The outputs were N, P and K in animal product (pig meat) and inedible parts. The final weight of animals was ~135 kg and carcass yield 80.23%. The estimated price of the manure was calculated based on the average market price of chemical fertilizers: urea (45% N) US$ 0.51 /kg, superphosphate (18% P_2O_5) US$ 0.38/kg, and potassium chloride (60% K_2O), US$ 0.51. Estimated loss of N was 50% due to volatilization of ammonia. The balance was positive for the three nutrients, the inputs were larger than the outputs. The difference between inputs and outputs remained in the system as manure. The nutrient balance per production cycle was 2,476.17 kg N, 474.24 kg P and 1,037.40 kg K. The nitrogen showed the highest efficiency use 38.57%, followed by P (20.14%) and K (16.19%). The valuation of manure produced in equivalence to commercial fertilizers was US$ 4,944.41 per production cycle, corresponding to US$ 6.40/animal and US$ 14,554.12/year. The results emphasize the importance of the management of pig manure as well as its commercialization as a suggestion for the correct destination of this product.

Applied analytical hierarchical processing in a social sustainability study of pig farming in Sweden

S. Zira[1], E. Ivarsson[2], E. Röös[3] and L. Rydhmer[1]
[1]Swedish University of Agriculture, Animal Breeding and Genetics, Box 7023, 75007, Sweden, [2]Swedish University of Agriculture, Animal Nutrition and Management, Box 7024, 75007, Sweden, [3]Swedish University of Agriculture, Energy and Technology, Box 7032, 75007, Sweden; stanley.zira@slu.se

Poor conditions for human and animals are antecedents of low social sustainability in pig production. Efforts meant to improve social sustainability should prioritize these antecedents in their order of importance. In this pilot study we tested analytical hierarchical processing (AHP) as a method of prioritization of the antecedents of low social sustainability. Five issues for pig farm workers, five for pigs and three for consumers were identified in a literature review including 7 published articles. We used three experts in Swedish pig production as judges in pairwise comparisons of these issues. The R package AHP was used for the weighting and analysis of judges' consistency. All the three judges had an internal consistency <0.2 for pig issues, whilst for worker issues only one judge had an internal consistency <0.2. For consumer issues, internal consistency ranged from 0.28 to 1.3 and the weights are thus not presented. For worker issues, low income had the highest relative weight (0.43) followed by low work pleasure (0.27), poor worker health (0.15), long work hours (0.09), and poor communication between farm owners and workers (0.06). For pig issues, poor health (0.30) had the highest weight followed by poor ambient environment (0.24), poor handling at slaughter (0.22), lack of freedom to express normal behaviour (0.20), and weak animal welfare law (0.04). To conclude, the pilot study shows that AHP can be used to weight social sustainability issues using experts but consistency needs to be considered. There is ongoing work to see the effect of increasing the number of judges and also improving the questioning pattern. Based on this pilot study the highest priority for worker issues should be given to income and the least to communication between farmer and workers. For pig issues similar priority should be given to pig health, pig ambient environment, handling at slaughter and freedom to exhibit natural behaviour.

Genetic parameters for feed intake and growth curves in three-way crossbred pigs

R.M. Godinho[1], R. Bergsma[1], S.E.F. Guimarães[2], F.F. Silva[2], E.F. Knol[1], J. Van Milgen[3], H. Komen[4] and J.W.M. Bastiaansen[4]

[1]*Topigs Norsvin Research Center B.V., Schoenaker 6, 6640 AA Beuningen, the Netherlands,* [2]*Universidade Federal de Viçosa, Animal Science, Avenida Peter Henry Rolfs, s/n, 36570-900 Viçosa, Brazil,* [3]*INRA, Agrocampus Ouest, 35590 Saint-Gilles, France,* [4]*Wageningen University & Research, Animal Breeding and Genomics, Droevendaalsesteeg 1, 6708 PB Wageningen, the Netherlands; rodrigo.godinho@topigsnorsvin.com*

Feed intake (FI) curves can be fitted using non-linear models (NLMs) which are based on a small number of parameters that usually have biological interpretations. NLMs are implemented with functions that mimic the behaviour of the continuous trait along a dependent variable. In this sense, a Gamma function of the maintenance energy expenditure was proposed. Describing FI for growth and maintenance throughout the whole growing-finishing period, this curve expresses fully the pig's intake capacity and energy expenditure. Using an NLM approach, breeding goals can be defined aiming to change the shape of the curves by treating the estimated parameters as phenotypic observations in genetic models. We recommend fitting curves that describe FI as a function of body weight given the strong interrelationship between FI and body weight. We also recommend selecting pigs with flatter curves, as they will eat less at the same stage of growth than pigs with steeper curves, and pigs with higher FI precocity, i.e. higher FI in early stages of growth associated with a higher growth maturation rate and a consequent lower FI later on the finishing period. Phenotypes of 2,230 crossbred pigs were available for this study. Medium to high heritability estimates for all curves' parameters, indicate that these traits are a feasible alternative for pig breeding programs aiming to change the shape of the FI and the growth curves in crossbred pigs. Selection for higher growth maturation rate, flatter FI curves and higher FI precocity is antagonistic to selection for heavier adult weight in crossbred pigs. As the common slaughter weight in pork production systems is around half the mature weight predicted in this study, selection for growth maturation rate, FI precocity and flatter FI curves might be advantageous.

Economic weights of sperm quality traits for sire breed using the gene flow methods

Z. Krupová, M. Wolfová, E. Krupa and E. Žáková

Institute of Animal Science, Přátelství 815, 10400 Prague Uhříněves, Czech Republic; krupova.zuzana@vuzv.cz

When breeding of pigs the increase of lean meat content has long been a goal of pig breeders, especially in sire breeds or lines. However, negative correlation between the meat production and sperm quality traits (SQTs) was found. Some authors reported that SQTs are heritable with genetic variation. Therefore, the objective of this study was to incorporate the SQTs into the comprehensive bio-economic model and to calculate the marginal economic values and economic weights for the new trait complexes in the Pietrain breed involved in the three-way crossing system of The Czech National Program for Pig Breeding. The numbers of discounted gene expressions of selected parents of each breed for all evaluated traits were summarized within all links of the crossbreeding system during the 8-year investment period using discounting with the half-year discount rates of 5%. Sperm volume, sperm concentration, progressive motion of spermatozoa in percent (motility) and percentage of abnormal spermatozoa were defined as the main characteristics for sperm quality. The bases assumptions when calculating the economic values of SQTs was that improving of SQTs lead to a higher number of sperm doses produced per boar ejaculate and decrease the costs per sperm dose and thus decrease the costs for insemination of sows. Number of sperm doses available from boar ejaculate was a function of boar age. From the total importance of the 18 traits in Pietrain breed the most important traits were health traits represented by survival rates of pigs in different rearing periods followed by feed efficiency and carcass traits. The relative economic weight of the four SQTs made 8%, therefore it is recommended to include SQTs into the breeding goal for sire breeds. The study was supported by project MZE-RO0719 and QK1910217 of the Czech Republic.

Effect of sow lines and type of Belgian Piétrain sire line on carcass and meat quality

E. Kowalski[1,2], E. Vossen[1], M. Aluwé[2], S. Millet[2] and S. De Smet[1]
[1]*Ghent University, Department of Animal Sciences and Aquatic Ecology, Coupure links 653, 9000 Ghent, Belgium,*
[2]*1Flanders research institute for agriculture, fisheries and food (ILVO), Scheldeweg 68, 9090 Melle, Belgium;*
eline.kowalski@ilvo.vlaanderen.be

For decades, pig breeding in Flanders has focused strongly on a high carcass lean meat content and low feed conversion ratio. However, the selection towards high carcass quality has unintendedly led to a lower meat quality in terms of taste, juiciness and PSE-like meat. Indeed, with higher lean growth and decreased fat deposition, lower intramuscular fat (IMF) level can be anticipated due to the genetic correlation between lean meat percentage and IMF level. As not only sire line, but also sow line may influence the outcome, the aim of this study was to evaluate the crossbred offspring of three most common sow lines in Flanders with two types of Belgian Piétrain sire lines on carcass quality (lean meat content) and meat quality (pH1, drip loss and IMF) of the loin. Three hybrid sows were compared: Topigs 20, TN70 (Topigs Norsvin), Mira (RA-SE genetics). The first type of sire line was selected for a higher growth rate (Optimal) and the second type of sire line for a higher carcass quality (Premium). Across three rounds, the lean meat percentage of 270 pigs (135 gilts and 135 immunocastrates) were evaluated. A total of 216 pigs (18 pigs/crossing/sex) were selected to evaluate meat quality. In the preliminary results (means of round 1), a higher lean meat content was observed by the offspring of TN70 (66.0%) compared to Mira (64.4%) and Topigs 20 (64.5%). No differences between the two sire lines were observed (65.3% Optimal, 65.2% Premium). Regarding the initial pH no big differences can be observed between the three sow lines and two sire lines, the mean varied between 6.34 and 6.44. Between the three sow lines, the drip loss percentage was the highest by Mira (10.2%), followed by TN70 (9.2%) and the lowest by Topigs 20 (7.9%). The Optimal sire line showed a higher drip loss percentage (9.2%) compared to Premium (8.8%). Regarding the IMF, Topigs 20 had the highest IMF (1.83%) compared to the two other (Mira:1.68% and TN70: 1.68%). The IMF content of the was 1.60% for optimal and 1.84% for the Premium sire line. The complete dataset will be available at the end of March 2019.

Dietary inclusion of ensiled avocado (Persia Americana) oil cake on pig growth performance

M.L. Seshoka[1], I.M.M. Malebana[1], P.J. Fourie[2], A.T. Kanengoni[3] and B.D. Nkosi[1,4]
[1] *Animal Production Institute, Division for Animal Nutrition, P/Bag X2, 0062, Irene, South Africa, [2]Central University of Technology, Department of Agriculture, P/ Bag X20539, 9300, Bloemfontein, South Africa, [3]Joburg Zoo, P/ Bag X 13, 2122, Parkview, South Africa, [4]University of Free State, Centre for Sustainable Agriculture, P.O. Box 339, 9300, Bloemfontein, South Africa; dnkosi@arc.agric.za*

This study was done to evaluate effects of different dietary inclusion levels of avocado (Persia Americana) oil cake silage on the growth performance of pigs. Avocado oil cake (AOC) contains 24% dry matter (DM) and was mixed with wheat bran and sugarcane molasses at a ratio of 70:25:5 respectively. The mixture was ensiled in 210 l drums for 90 days. Diets that contained 0, 3 and 5% AOC silage were formulated to be iso-nitrogenous and iso-caloric and fed to 27 large white × landrace (LW × LD) pig crosses of 22-24 kg live weight. Pigs were randomly blocked by weights to each dietary treatment, making 9 pigs/diet. The pigs were housed individually in single pens in a barn. An adaptation period of 21 days was allowed and the experiment took 60 days. Diets were fed ad-lib, pigs were weighed at the start and continued weekly till the end of the experiment. Daily feed intake was measured, average daily gains (ADG) and feed conversion rate (FCR) were measured. The addition of AOC silage did not (P>0.005) affect the daily feed intake, ADG and FCR. This suggest that silage from AOC can be incorporated at not more that 5% in the diets of growing pigs. Further work to evaluate this silage on its effects on the carcass traits of pigs is warranted.

Italian ryegrass improves growth performance with increased gut *Lactobacillus* in pigs

S.K. Park[1], N. Recharla[1] and D.W. Kim[2]

[1]Sejong University, Food Science & Biotechnology, 209 Neungdong-ro, 05006, Korea, South, [2]National Institute of Animal Science, Swine division, Sunghwan, 31000, Korea, South; sungkwonpark@sejong.ac.kr

This study investigated the effect of dietary Italian ryegrass (IRG) on growth performance of growing-finishing pigs. A total of 20 pigs {Duroc × (Landrace × Yorkshire)} were randomly assigned to control (normal diet) or IRG group. Supplementation of 1% IRG reduced the feed intake from 41.24 kg to 39.85 kg. Average daily gain was increased (P<0.05) in IRG treatment group by 8% when compared with control. Furthermore, when compared with control group, population of *E. coli* was decreased but *Lactobacillus* increased in the faeces from pigs fed IRG (P<0.05). Microbial community analysis showed that *Lactobacillus* and *Treponema* spp. were increased(P<0.05) in faeces from IRG feeding group but pathogenic bacteria including *Clostridium*, *Roseburia*, and *Streptococus* showed a tendency to decrease compared to the control. Results from the present study suggest that pig diet supplement with IRG might improve growth performance of pigs with increased beneficial microbes in porcine gut.

Effect of phytase on performance and bone mineralization in growing pigs fed a deficient P diet

B. Villca[1], G. Cordero[2], P. Wilcock[2] and R. Lizardo[1]

[1]IRTA, Animal Nutrition, Ctrs Reus-el Morell, km 3.8, 43120 Constantí, Spain, [2]AB Vista, c/Peru, 6, 2ª Planta Oficina 4, Edificio Twin Golf A, 28290 Las Rozas, Spain; rosil.lizardo@irta.es

An experiment was conducted to determine the effect of dietary supplementation with phytase on growth performance, bone mineralisation and bone strength in growing pigs. One hundred eighty male and female Pietrain×(Landrace×Large White) pigs of 20 kg liveweight were used. They were distributed into 9 blocks according to initial body weight and allocated at 4/pen in 45 pens for a 42-d trial. Intra-block pigs were randomly distributed to the 5 treatments corresponding to positive (PC: 0.58 and 0.35% total and digestible P, respectively) and negative (NC, 0.28 and 0.11% total and digestible P, respectively) control diets or the NC supplemented with 300, 600, 1,200 FTU/kg of phytase, respectively. At the end of performance trial, one pig per pen was euthanized and metatarsi bones were collected for bone strength and mineralisation analysis. Regardless the dose used phytase inclusion statistically improved weight gain, feed intake and feed efficiency (P<0.01) relatively to the NC diet. No differences were observed between the PC diet and those containing phytase. Bone ash, and P and Ca contents were reduced (P<0.01) on pigs fed the NC relatively to PC diet. Bone ash and P content respond proportionally to increasing levels of phytase supplementation. Metatarsi of pigs fed PC and NC diets showed the highest and lowest bone stiffness (P<0.001), respectively. Inclusion of phytase progressively increased bone stiffness and ultimate force. It can be concluded that supplementation of diets with Quantum Blue phytase improved growth performance and bone mineralization of pigs fed low-P diets.

Steroidome and metabolome analysis in gilt saliva to identify biomarkers of boar effect receptivity

G. Goudet[1], P. Liere[2], L. Nadal-Desbarats[3], D. Grivault[4], C. Douet[1], J. Savoie[1], S. Ferchaud[4], F. Maupertuis[5], A. Roinsard[6], S. Boulot[7] and A. Prunier[8]

[1]INRA, PRC, 37380 Nouzilly, France, [2]INSERM, U1195, 94276 Kremlin Bicêtre, France, [3]INSERM, U1253, 37044 Tours, France, [4]INRA, GENESI, 17700 Surgères, France, [5]Chambre d'agriculture, Pays de la Loire, 44150 Ancenis, France, [6]ITAB, ITAB, 49105 Angers, France, [7]IFIP, Institut du Porc, 35650 Le Rheu, France, [8]INRA, PEGASE, 35650 Le Rheu, France; armelle.prunier@inra.fr

Our objective was to develop alternatives to hormonal treatments to synchronize oestrus of gilts. Before puberty gilts exhibit a pre-puberty period during which boar exposure could induce and synchronize first ovulation. To develop practical non-invasive tools to identify this period and improve detection of the gilts to stimulate, we searched for salivary biomarkers of the pre-puberty period. Saliva samples were collected from 30 Large-White × Landrace crossbred gilts from 140 to 175 days of age. Gilts were exposed to a boar twice a day and subjected to oestrus detection from 150 to 175 days of age. Among the 30 gilts, 10 were detected in oestrus 4 to 7 days after introduction of the boar and were considered receptive to the boar effect, 14 were detected in oestrus more than 8 days after boar introduction, 6 did not show oestrus and were considered non-receptive. Saliva samples from 6 receptive and 6 non-receptive gilts were analysed for steroidome using GC-MS/MS and for metabolome using 1H-NMR spectroscopy. Four saliva samples per gilt were analysed: 26 days and 11 days before boar introduction (BI-26 and BI-11), the day of boar introduction (BI), 3 days later for receptive gilts (BI+3) or 7 days later for non-receptive gilts (BI+7). Data were analysed using repeated measures one-way ANOVA and orthogonal partial least squares discriminant analysis. Thirty steroids and 35 metabolites were detected in gilt saliva. The concentrations of 6 steroids were higher ($P<0.05$) in receptive gilts than in non-receptive gilts at BI-26, BI-11 and BI. The concentration of 2 metabolites were lower ($P<0.05$) in receptive gilts than in non-receptive gilts at BI-11. These candidates could be potential salivary biomarkers to detect receptive gilts. However, their low and variable concentrations in saliva require expensive analysis and limit their use in pig farms.

Welfare of entire males, immunocastrates and surgical castrated pigs in socially unstable groups

L. Wiesner, K. Kress, U. Weiler and V. Stefanski

University of Hohenheim, Behavioral Physiology of Livestock, Garbenstr. 17, 70599 Stuttgart, Germany; linda.wiesner@uni-hohenheim.de

For animal welfare reasons, stakeholders committed to ban surgical castration of male pigs without anaesthesia in future. Therefore, alternatives such as fattening entire males or immunocastrates are evaluated. The goal of the study was to compare those alternatives in terms of animal welfare problems in socially unstable environments. Behaviour of 72 pigs (surgical castrates, SC, n=24; immunocastrates, IC, n=24, entire males, EM, n=24) were analysed. Animals (Piétrain × German Landrace) were fed *ad libitum* and housed in groups of 6 animals/pen under socially unstable conditions, induced by two mixing phases (interchanging pen mates twice a week) during two weeks around 1st and three weeks around 2nd Improvac vaccination of IC animals. The study was conducted in two consecutive trials. Aggressive (fights, head-knocks, biting) and sexual behaviours (mounting with different intensities, anal-genital nosing) were evaluated. The animals were observed on three days in the second mixing phase and on three days shortly after (continuous sampling for one hour per group and day). Penis biting in EM and IC was validated by physical examination of penile injuries at slaughter. Behavioural data from first trial are analysed so far. Statistical analysis was performed by using Wilcoxon signed rank test (RStudio 1.1.463), values are shown as mean ± SEM. Preliminary analysis indicates that the highest number of aggressive behaviours per animal and hour occurred in groups of EM (3.90±1.53), followed by IC (2.56±1.05), both significantly different ($P<0.01$) from lowest numbers shown by SC (1.25±0.60). Sexual behaviours occurred the least often in IC (0.94±0.43) and SC (0.47±0.23) groups and were different from EM (2.48±1.01; $P<0.001$). Final data of both trials will be presented at EAAP. Penile injuries were analysed for both trials (n=48) and revealed significantly more penile injuries per animal in EM than IC (EM: 5.19±0.98; IC: 1.60±0.39; p±0.001). Immunocastration is a promising tool to improve the welfare status of fatting pigs regarding social conflict, especially problems with non-reproductive sexual behaviour. These results were accomplished within SuSi project, financed by SusAn ERA-NET.

Profile and consumer acceptance of salami from immunocastrated, castrated and entire male pigs

E. Kostyra[1], S. Żakowska-Biemans[1], M. Aluwé[2], A. Van Den Broeke[2], M. Candek-Potakar[3] and M. Škrlep[3]
[1]Warsaw University of Life Sciences (WULS-SGGW), Faculty of Human Nutrition and Consumer Sciences, ul. Nowoursynowska 159c, 02-776 Warszawa, Poland, [2]Flanders Research Institute for Agriculture, Fisheries and Food (ILVO), Animal sciences unit, Scheldeweg 68, 9090 Melle, Belgium, [3]Agricultural Institute of Slovenia, Hacquetova ulica 17, 1000 Ljubljana, Slovenia; marijke.aluwe@ilvo.vlaanderen.be

The increasing concern of some consumer segments for animal welfare has led to consider alternatives to surgical castration (SC), like raising entire males (EM) or immunocastration. However, the social acceptance of immunocastration or the possible presence of boar taint in pork from EM rises concerns. Therefore, the main aims of the study were: (1) determine sensory characteristics of traditional salami from IC, SC and EM using trained assessors sensitive to boar taint using Quantitative Descriptive Profiling (QDA); (2) evaluate consumers' acceptance of key attributes and willingness to buy with 9-point scale; and (3) assess consumers' sensory and emotional characteristics of salami using Check-All-That-Apply (CATA) questionnaire. The QDA data showed differences for salami, in particular in the intensity of some odour and flavour (e.g. sweat, manure and persistent impression). Salami from EM revealed the lowest overall sensory quality, whereas salami from IC and SC showed relatively similar sensory characteristics. The consumers' acceptance did not significantly differ regardless of the production system. CATA results confirmed profiling characteristics and showed some differences, namely EM salami was associated with non-intense meat flavour, bitter taste, spicy impression and feelings like persistent taste, disappointment and emotion related to being negatively surprised. Generally IC salami was perceived mainly as having familiar flavour, whereas SC products were identified with sour, meat, fatty and positive attributes (e.g. delicate, pleased and positively surprised). The combination of different sensory methods can provide meaningful insights for product evaluation and understanding more deeply consumers response to IC/EM/SC products. The research was done within ERA-Net Cofund SusAn (696231) Sustainability in pork production with immunocastration (SuSI).

Sustainable precision livestock farming: real-time estimation of body protein mass in pigs

A. Remus[1], S. Méthot[1], L. Hauschild[2] and C. Pomar[1]
[1]Agriculture and Agri-Food Canada, Sherbrooke, QC, J1M0C8, Canada, [2]University of São Paulo State, FCAV, Jabotical, SP, 14883108, Brazil; alnremus@gmail.com

This study aimed to predict real-time individual body protein mass (BP, kg), protein deposition rate (PD, g/d) and the proportion of PD in the body weight (BW, kg) gain (BWG, kg/d) over the growing-finishing period, as required for precision feeding. Data from 81 pigs (21 gilts and 60 barrows; 35-135 kg BW) were used. BP was estimated using Dual X-ray absorptiometry (GE Lunar Prodigy Advance) on days 1, 28, 56 and 84 of trial. Individual BP was regressed over BW assuming a null intercept and quadratic response (BIC=3,371). The linear and quadratic parameters were equal between trials and sexes (P>0.05; GENMOD procedure of SAS). The general regression estimating average population BP was obtained using the MIXED procedure of SAS, considering within pig repeated measures in time and the regression parameters as fixed effects: $BP = 0.221601 \times BW - 0.0004076 \times BW^2$ (1) The coefficients of concordance and determination were both 0.99, and the root mean square error was 0.19. The PD/DG was obtained by deriving equation 1, as follows: $PD/BWG = 0.221601 - 0.0008152 \times BW$ (2) Observed BW was regressed over time (day), and BWG was obtained using the first derivative of the BW equation. Average PD input was 163 g/d for gilts and 160 g/d barrows (observed values). This information which was used to simulate average population PD/BWG with the InraPorc and NRC (2012) models comparing to Equation 2 values. Equation 2 predicts a linear PD/BWG decrease (interval: 35 to 135 kg BW) whereas InraPorc and NRC predict a quadratic decrease. Equation 2 estimations of PD/BWG at 30 and 124 kg were respectively, 19.7 and 11.8%, whereas InraPorc estimations were 15.4 and 12.5%, respectively. Equation 2 accuracy was evaluated through the mean absolute percentage error (MAPE) using observed average BW and PD/BWG of 9 trial (292 pigs). Equation 2 was more accurate (MAPE=-3.22%) compared to the NRC model for barrows (MAPE=-4.39%) and gilts (MAPE=-6.64%). Equation 2 is able to predict real-time PD/BWG with small error, allowing for more precise estimations of nutrient requirements in pigs fed with daily tailored diets as required for precision feeding.

Regulatory framework for production and use of insects as feed in the EU

W. Trunk

European Commission, DG SANTE / Animal Nutrition, Veterinary Medicines, B232 04/23, 1049 Brussels, Belgium; wolfgang.trunk@ec.europa.eu

Insect derived products can play an important role, both to contribute to the protein supply of the EU livestock and in the margins of the EU Action Plan for the Circular Economy. Therefore, Directorate General SANTE of the European Commission elaborated some years ago a Strategic Safety Concept for Insects as Feed in order to allow a sound development of the production and use of insects in the EU while simultaneously ensuring the safety of the food chain. More precisely, Commission Regulation (EU) 2017/893 (in application since July 2017) authorised the use of processed insect protein in feed for aquaculture animals. Based on the description of the current situation, the presentation will elaborate on the next steps, which are envisaged to further expand the insect sector, such as the authorisation of processed insect protein as feed to other non-ruminant, food-producing animals or the assessment of new substrates for the insects.

Cross-sectoral matchmaking and networking opportunities for further scaling of the insect sector

M. Peters and P. Eyre

New Generation Nutrition, Deutersestraat 12, 5223 GV 's Hertogenbosch, the Netherlands; marianpeters@ngn.co.nl

The global demand for animal products is expected to more than double between 2000 and 2050. Animal feed production is increasingly competing with resources for human food and fuel production. Insects can play an important role in fulfilling the demand for animal protein through the use of alternative resources that are considered currently as bio-waste. Insect protein is already approved for fish farming in the EU and could provide a solution to feeding other livestock in the future. The market for insect production is expected to expand by 20% annually over the next five years. In terms of up-scaling, the insect sector faces both challenges and opportunities as it grows. To maximise up-scaling activities, the sector will require the transfer of knowledge, techniques and systems from other livestock sectors, that will facilitate the success of insects as a viable feed and food source. Common practices and standards applied in other sectors such as breeding systems, automation, climatization and downstream processing can be newly adapted to insect production. This knowledge and technical transfer must be combined with the positioning of insect production as a sustainable entrepreneurship opportunity. To enable this, it is vital to bring together entrepreneurs, governmental bodies, research institutes and value chain stakeholders to discuss, exchange and learn in order to move forward. Therefore, a matchmaking session will be held to involve these key actors and to facilitate discussion on central themes suggested by delegates. For this matchmaking session, all participants of the EAAP meeting with an interest in the up-and-coming insect sector are invited to connect with sector experts and vice versa. The EAAP meeting is an excellent opportunity to link diverse stakeholders to disseminate cross-sectoral knowledge and technology, initiate research and development projects, and to connect to relevant actors. Participants have the opportunity to arrange face-to-face meetings with people from a variety of different institutions to discuss preferred topics of interest. By facilitating these linkages, a robust network across the insect sector can be built to bolster the position of insects for feed and food applications.

Dietary feed additives with antibacterial effects as replacers of antibiotics for weaned piglets

J. Dewulf[1], W. Vanrolleghem[2], S. Tanghe[2], S. Verstringe[2], G. Bruggeman[2], D. Papadopoulos[3], P. Trevisi[4], J. Zentek[5] and S. Sarrazin[1]
[1]*Universiteit Gent, Faculty of veterinary Medicine, Salisburylaan 133, 9820, Merelbeke, Belgium,* [2]*Nutrition Sciences, Booiebos 5, 9031 Drongen, Belgium,* [3]*Aristotle University, Thessaloniki, Thessaloniki, Greece,* [4]*University of Bologna, Department of Agricultural and Food Science, Viale G. Fanin, 46, 40127 Bologna, Italy,* [5]*Freie Universität Berlin, institute of animal nutrition, Königin-Luise-Str. 49, 14195 Berlin, Germany; jeroen.dewulf@ugent.be*

This meta-analysis evaluated the use of potential dietary feed additives (pDFA) as alternatives for antibiotics and zinc oxide in piglets at weaning. Twenty-three peer-reviewed *in vivo* studies were identified between January 2010 and January 2017. The pDFA in these studies could be grouped into 5 classes: plant extracts, antimicrobial peptides, lysozyme, medium chain fatty acids/ triglycerides and chitosan. Mixed-effect meta-analyses with the type of pDFA as fixed effect were performed for the growth parameters 'average daily gain' (ADG) and 'feed conversion ratio' (FCR), which are two important economic performance parameters for farmers. For each class of pDFA, the results of the meta-analysis showed a significantly higher average daily gain in the group with pDFA compared to the negative control group, while no significant difference with the positive control group was observed. Furthermore, a positive effect on FCR was found, i.e. significantly less feed was needed to gain 1 kg of body weight in the group with pDFA compared to the negative control group. No significant difference with the positive control group was observed for each class of pDFA, except for plant extracts, where also the FCR was significantly reduced in the treatment group. These results suggest that pDFA could reduce the use of antimicrobials without significant negative effects on the performance indicators.

Inhibitory effect of SCFA and MCFA on contaminants of liquid pig feed and intestinal bacteria

M.M.J. Van Riet[1,2], S. Vartiainen[3], G. Jurgens[3], A. Seppala[1], K. Rikkola[1], S. Vermaut[1] and I. Peeters[1]
[1]*Eastman Chemical Company, Technologiepark 21, Zwijnaarde, Belgium,* [2]*Ghent University, Heidestraat 19, Merelbeke, Belgium,* [3]*Alimetrics Research, Koskelontie 19, Espoo, Finland; mir.vanriet@eastman.com*

Microbial contaminants present in liquid pig feed seem to be susceptible to short chain fatty acids (SCFA) and medium chain fatty acids (MCFA). Adding these acids triggers growth inhibition and this would be valuable to optimise production, storage and hygiene of liquid pig feed. Similarly, intestinal microbes found in pigs seem to be susceptible to SCFA and MCFA but may depend on bacterial strain and other conditions. The objective of this *in vitro* study was to determine the growth inhibitory effect of several SCFA and SCFA+MCFA combinations on 8 bacterial and 2 yeast strains. The SCFA and/or MCFA products (1 l/t) were added to a liquid bacterial growth media after which the media was inoculated with 9% v/v overnight cultured microbial inocula. Growth media pH and temperature were adjusted for tested strains as follows: *Bacillus cereus, Lactobacillus salivarius, Escherichia coli, Salmonella enterica, Staphylococcus aureus*: TSGY, pH 5, 37 °C; *Clostridium perfringens*: TSGY, pH 6.2, 37 °C, anaerobic; *Streptococcus suis*: TSB with 2% serum, pH 7, 37 °C microaerophilic; *Campylobacter jejuni*: BHI, pH 6.5, 37 °C, microaerophilic; *Candida humilis* and *Saccharomyces cerevisiae*: YM, pH 5, 25 °C. Growth inhibition was determined measuring culture optical density at 600 nm at multiple time events over a 24 h period, except for *C. jejuni*. Growth inhibition (%) per product was statistically analysed relative to the negative control using a two-tailed Student t-test. All products inhibited the growth of tested bacterial strains dependent on time after inoculation, except for *C. jejuni*. The SCFA+MCFA combination showed superior growth inhibition, whereas lactic acid had the lowest efficacy. Only the SCFA+MCFA combination was able to inhibit the growth of yeast cells effectively. *E. coli* showed a time-dependent susceptibility to SCFA and MCFA. In conclusion, SCFA and SCFA+MCFA inhibited growth of several bacterial strains. The SCFA+MCFA also inhibited yeast cell growth.

Dietary protein optimization using a rabbit model: towards a more sustainable production in nitrogen

P.J. Marín-García, M.C. López, L. Ródenas, E.M. Martínez-Paredes, E. Blas and J.J. Pascual
Instituto de Ciencia y Tecnología Animal, Universitat Politècnica de València, Animal Science, Camino de Vera, 46022, Valencia, 46022, Spain; pabmarg2@upv.es

Livestock is one of the sources of N pollution worldwide. Formulating feed according to the requirements of animals at the ileal real level would greatly reduce the excretion of this contaminant. Plasma urea N level (PUN) could be a good indicator of the protein optimization in the diet. The objective of this work will be to propose a model (using growing rabbits) to optimize protein nutrition stepwise. In a first trial (Trial 1), using 918 animals, it was evaluated which of the 27 combinations minimized the PUN −3 levels of inclusion [M, medium (current); H, high (+ 15%); L, low (-15%)], for the first 3 limiting amino acids (AA) in rabbits (lysine, sulphur AA and threonine), at faecal apparent level−. In the second trial (Trial 2), using 116 animals, the combination of Trail1 that minimized PUN was compared with the current recommendations for growth performance and apparent ileal digestibility of feeds. From the results of Trial1, it was observed that the combination of AA that minimized the PUN values was MHL (for lysine, sulphur and threonine, respectively). In addition, in the Trial2 it was found that, animals with the MAB feed, both the growth rate and the feed conversion ratio were improved ($P<0.05$). Therefore, in growing rabbits it is recommended 5.2, 4.7 and 3.0 g/kg of lysine, sulphur and threonine digestible at the ileal level, respectively. This model can be used to optimize the diets of other zootechnical species and reduce N contamination.

Galactomannan fenugreek extract as proposed alternative to antibiotics in young rabbits nutrition

J. Zemzmi[1], L. Rodenas[1], E. Blas[1], E. Martinez-Paredes[1], M.C. López-Lujan[1], J. Moya[1], J.J. Pascual[1] and T. Najar[2]
[1]University Politécnica de Valencia, Institute for Animal Science and Technology, C.V, 46022 Valencia, Spain, [2]National Agronomic Institute of Tunisia, Animal Science, 43 Av C.N, 1082 Tunis, Tunisia; zemji@doctor.upv.es

Antimicrobials resistance has become one of the greatest threats to global health. Rabbit production, despite its small productive size, has a high dependence on the use of antimicrobials. Galactomannan (GM) has been proposed as a soluble fibre that may have a prebiotic effect in different species. To be qualified as a prebiotic GM must satisfy three conditions: to not be digestible by the gastric and small intestine enzymes, be highly fermented in the caecum and be selectively utilized by host microorganisms conferring a health benefit. Initially, purity and galactose/mannose ratio of a GM extracted from Tunisian fenugreek seeds was evaluated. Subsequently, five increasing levels of GM (0, 0.5, 1, 1.5 and 2%) were included in two commercials rabbits feeds (one enriched in soluble fibre and other in lignin), and the 10 feeds were subjected to pepsin and pancreatin digestion followed by fermentation using a caecal inoculum. Finally, two diets differing in soluble fibre (high and low) were formulated with and without 1% of GM, and were used to evaluate growth performance, health status, nutrient digestibility and caecal activity in growing rabbits. Fenugreek GM was characterized by purity of 69% and galactose/mannose ratio of 1.06. Most of the ingested GM (83.6%) escapes *in vitro* digestion by pepsin and pancreatin. GM inclusion seems to increase dietary neutral detergent fibre digestion. GM was fully fermented, increasing production of volatile fatty acids, reducing ammonia nitrogen. Respect to caecal activity, a significant increase in caproic acid concentration in GM groups ($P<0.01$) was observed, which suggests a different microbial activity. Despite showing the characteristics to be considered as a prebiotic of interest in rabbits, the dietary inclusion of 1% of GM did not have any significant effect on the health status and growth performance of growing rabbits. Future interpretation of its effect on caecal microbiota will provide more knowledge about the potential of this GM as a prebiotic for rabbits.

Vitafibra, a unique health fibre, as a successful part of zinc oxide replacement

M. De Vos, R. D'Inca, K. Lannoo and J. Vande Ginste
Nutrition Sciences, Booiehos 5, 9031 Drongen, Belgium; maartie.devos@nusciencegroup.com

Enterotoxigenic *Escherichia coli* F4 is associated with post-weaning diarrhoea in piglets. The initial step of an *E. coli* infection is to adhere to host-specific receptors present on the enterocytes and thereby immune responses are initiated[1]. The aim of this study was to investigate the potential of a health fibre (Vitafibra) on the prevention of adhesion by agglutination of *E. coli*. To study the binding capacity of Vitafibra, *E. coli* F4 strains were mixed into a 100 ml solution together with 0.2 g of Vitafibra (except for the negative control). *E coli* strains were allowed to adhere to Vitafibra. Afterwards, samples were filtrated (4-7 µm Whatmann) and filtrates were plated to check the amount of *E. coli* that could be recovered. In addition, the remaining residue was cultured to check how much of the residual *E. coli* was still able to grow. The amount of culturable *E. coli* was determined on the start culture, on the filtrate and on the residue by classical laboratory methods. The addition of Vitafibra to the bacterial strain, results in a strong reduction of *E. coli* in the filtrate. Only 1% of *E. coli* passes through the filter, the remainder (99%) was found in the residue and thus considered to be captured by Vitafibra. To show *E. coli* was disactivated, a cultivation method was performed on the residue. This method showed that only 21% of the original *E. coli* could still be cultivated, meaning that 78% was bound and prohibited to grow. Vitafibra has the ability to capture 99% of *E. coli* in an *in vitro* suspension. Moreover, the major part of this *E. coli* was unable to be cultivated after the agglutination process. To conclude, Vitafibra reduces the risk of pathogen overgrowth by capturing *E. coli* bacteria. This makes Vitafibra a successful feed additive in the battle against *E. coli* diarrhoea.

Round Table: implementing innovative solutions in animal nutrition

J. Dewulf
Ghent University, Faculty of Veterinary Medicine, Salisbury Laan 153, 9820 Merelbeke, Belgium; jeroen.dewulf@ugent.be

Round Table: implementing innovative solutions in animal nutrition: tools and success stories from the field to reduce the need of antibiotic use in animal farming. J. Dewulf, P. Doncecchi, A. Slawinska, P.L. Marin-Garcia M., M.M.J. Van Riert. The global concern about the increase of antimicrobial resistance has stimulated the development of innovative solutions in animal nutrition aiming at the reduction of the need to use antibiotics in livestock. Various specialty feed ingredients and feed additives have been developed, considering the specific needs of animal species and categories. This Round Table brings together scientists and industry representatives to identify and discuss critical success factors for the implementation of such new strategies at the global level.

A blend of active compounds to optimize growth in finishing pigs exposed to heat stress

E. Janvier, F. Payola, C. Launay, E. Schetelat and A. Samson

Neovia, Talhouët, Saint Nolff, B.P. 80234, 56006, Vannes Cedex, France; ejanvier@neovia-group.com

Heat stress in pigs occurs in tropical countries as in European countries during summer. As a result, pigs have a lower feed intake and growth is sometimes severely affected. There is therefore room for nutrition to develop solutions to overcome poor feed intake in pigs exposed to heat stress. For this purpose, a trial was designed to evaluate a blend of plant extracts, micro and macro minerals and flavouring agents (Fresh'up, Neovia) targeting inflammation, gut permeability, acid-base homeostasis and nutrient absorption. In total, 72 pigs (73.7 ± 0.6 kg BW) housed in eight pens of nine individuals (females and barrows) were assigned to the experiment. Pigs were divided into two groups from 111 d of age until slaughter: control group (CON) fed a conventional finisher diet (10 MJ NE/kg, 14.5% CP, 0.80% dLys) and experimental group (FRESH'UP) fed the same finisher diet but supplemented with the blend of active compounds (Fresh'up, Neovia). Pigs were exposed to heat stress during the experimental period (d 111 – d Slaughter). Temperature was set at 32 °C during the day and at 28 °C during the night with a relative humidity ranging between 40 and 50%. Pigs were fed *ad libitum* all over the trial. Results showed a tendency for higher average daily feed intake (ADFI) during the finishing phase for pigs in the FRESH'UP group compared to the CON group ($2,578\pm355$ g/d and $2,461\pm283$ g/d respectively, P=0.09). Consequently, the average daily gain (ADG) tended to be improved for the FRESH'UP group compared to the CON group (875 ± 146 g/d and 814 ± 131 g/d respectively, P=0.10). The feed conversion ratio (FCR) was numerically lower for the FRESH'UP group (2.98 ± 0.35 and 3.07 ± 0.42 respectively, P>0.10). Additionally, there was no interaction Dietary treatment × Sex (P>0.10). To conclude, a holistic approach dealing with the different damages induced by heat stress seems beneficial to optimize growth performance in finishing pigs.

Assessment of immunomodulatory role of probiotics for chickens using *in vitro* test

A. Slawinska[1], A. Dunislawska[1], S. Powalowski[2] and M. Siwek[1]

[1]UTP University of Science and Technology, Department of Animal Biotechnology and Genetics, Mazowiecka 28, 85-084 Bydgoszcz, Poland, [2]JHJ sp. z o.o., Nowa Wies 11, 63-308 Gizalki, Poland; slawinska@utp.edu.pl

The immunomodulatory role of probiotics can be utilized to condition intestinal health and host defence against enteric diseases. For an efficient colonization of the gastrointestinal tract, probiotics should be applied in perinatal period. At the moment, the optimal way to deliver immunomodulatory probiotics in chickens is through *in ovo* technology. Probiotics that exert immunomodulatory properties have to be first recognized by the immune system of the host. Host immune cells that are involved in antigen recognition are mainly macrophages and dendritic cells, which are part of the host innate immune system. Those cells are able to phagocyte bacteria and present their peptides on the surface. For this reason, the *in vitro* test that we propose is based on the chicken macrophage-like cell line (HD11). The test is based on a co-culture of the HD11 cells with live probiotic bacteria and subsequently the assessment of the specific gene expression signatures using RT-qPCR method. Earlier, we have developed a list of immune-related gene expression signatures using whole-genome expression microarrays for chickens (Affymetrix). They allow us to distinguish between potent and poor immunomodulatory probiotics. These immune-related gene expression signatures include cytokines (*IL-6, IL-12b* and *LIF*), chemokines (*CCL1* and *CCL20*), heat shock proteins (*HSP25* and *HSPA2*) and matrix metaloproteinases (*MMP1, MMP9* and *MMP10*). Up-regulation of those genes indicates higher immunomodulatory properties of the probiotic under consideration. In this study, we aim to validate the *in vitro* test for the assessment of immunomodulatory probiotics for poultry. The validated *in vitro* platform will be used to establish novel immunomodulatory probiotics for *in ovo* applications for poultry. The National Centre for Research and Development in Warsaw, Poland (OVO-FOOD, ID:433847).

Determination of the requirements of digestible amino acids

P.J. Marín-García, M.C. López, L. Ródenas, E.M. Martínez-Paredes, E. Blas and J.J. Pascual
Instituto de Ciencia y Tecnología Animal, Universitat Politècnica de València, Animal Science, Camino de Vera, 46022,
Valencia, España., 46022, Spain; pabmarg2@upv.es

Ileal digestibility is the best indicator in order to know the degree of nutritional use. However its knowledge is scarce for growing rabbits. In a previous trial, it was studied the best total amino acid (AA) combination (Lysine = Lys, sulphur AA = sAA and Threonine = Thr) which improved the protein synthesis and maximized the productive parameters. The objective of this work will be to determine the requirements at ileal level of AA of this new combination (MHL) and of the current recommendations (MMM). A bait test (between 28 and 63 days of life) was performed on 30 rabbits of the R line (selected by growth rate). Diets were formulated starting from the same basal mixture, but only the levels of Lys, sAA and Thr were different. Rabbits were randomly assigned to one of the two diets (marked with yterbium) and slaughtered (at 63 day of life). Ileal content was collected and analysed. The coefficients of ileal digestibility were equal between the experimental diets, but a tendency was observed (P>0.05) when the AA levels were higher (the addition was made with highly digestible AA). Finally, the current digestible recommendations at ileal level are established as 5.2, 3.6 and 4.3 g/kg of DM of Lys, sAA and Thr, respectively. While the new combination, which improved production traits, health and protein utilization, is situated with values of 5.2, 4.7 and 3.0 g/kg of DM of Lys, sAA and Thr, respectively.

Synbiotics for *in ovo* aplication – *in vitro* design

K. Stadnicka[1], P. Gulewicz[2], M. Succi[3], G. Maiorano[3] and M. Bednarczyk[1]
[1]UTP University of Science and Technology, Animal Biotechnology and Genetics, Mazowiecka 28, 85-084 Bydgoszcz,
Poland, [2]Poznan Science and Technology Park of the AMU Foundation, Biotechnology Centre, Rubiez 46, 61-612 Poznan,
Poland, [3]University of Molise, Agricultural, Environmental and Food Sciences, Via Francesco De Sanctis 1, 86-100
Campobasso, Italy; katarzyna.stadnicka@utp.edu.pl

Synbiotics injected *in ovo* can beneficially modulate microbiome of the embryo long prior to hatch and allow to mitigate peri-hatch stresses. In automated hatcheries, the *in ovo* modulation of gastrointestinal tract is possible as early as on 12th day of egg incubation by injecting 0.2 ml of solubilized stimulants. Synbiotic must have optimal arrangement of two interacting components: probiotic/s species and prebiotic. A question that needs answer is how the *in vitro* image of a synbiotic composition should look to premise a desirable physiological effect and bioavailability during *in ovo/ in vivo* application? We investigated the interaction between selected prebiotic carbohydrate substrates: ALGAE 1 (AL1), ALGAE 2 (AL2), ALGAE 3 (AL3), BITOS (BI), Inulin (IN) and different probiotic bacteria: *Lb. rhamnosus* GG (L1), *Lb. rhamnosus* FL2 (L2), *Lb. rhamnosus* FL3 (L3), *Lb. rhamnosus* FL4 (L4), *Lb. rhamnosus* FLC5 (L5), *Lb. rhamnosus* AT194 (L6), *Lb. rhamnosus* AT195 (L7), *Lb. rhamnosus* 39 (L8), *Lb. rhamnosus* H25(L9) and *Bifidobacterium* ATCC (B1). The probiotics were grown on 1/ a standard De Man, Rogosa and Sharpe agar (MRS) with glucose as a reference substrate, 2/ on modified MRS media where glucose was replaced with selected prebiotics as carbohydrate sources. The changes in optical density (OD) were plotted in real time using Automated Microbiology Growth Curve Analysis System, Bioscreen C (Oy Growth Curves Ab Ltd) and compared with one-way ANOVA and Sheffe's pairwise tests. As a result, the specific mutual effects (P≤00.1) of prebiotic and probiotic were observed: combination of AL3 × L9 (OD 2.82±0.06) and BI × L3 (OD 1.94±0.44) were found to be optimal *in vitro*. Interestingly, the AL3 prebiotic (OD 2.26±0.15) acted as a 'universal' substrate for all the probiotics tested. This protocol allows for low cost preselection of potentially unlimited number of pro/prebiotic combinations prior to using them in the *in vivo* poultry trials.

Effect of chronic exposure to endotoxins on respiratory health of broilers

M. Kluivers-Poodt, J.M. Rommers, N. Stockhofe, A.J.A. Aarnink and J.M.J. Rebel
Wageningen University & Research, De Elst 1, 6708 WD Wageningen, the Netherlands; marion.kluivers@wur.nl

Human health is affected by the inhalation of endotoxins from poultry farming. The aim of this study was to investigate the effect of chronic aerosol exposure to endotoxins on respiratory health of broilers. One-day-old broilers were housed in two separate rooms where a continuous low (LE, 119 EU/m^3) or high (HE, 6,203 EU/m^3) endotoxin (*Escherichia coli* O55:B5) level was created. At D34 broilers were either culled (Con), intranasally challenged (IBC) or vaccinated (IBV) with Infectious Bronchitis virus. IBV and IBC were culled at D37. Bodyweight and feed intake were measured weekly. Behavioural observations were performed in week 3 and 5. Blood, trachea and lung tissue were collected at D34 (Con) or D37 (IBC and IBV). In blood IgM and IgG natural antibody titres (NAB) were measured, in lung tissue cytokines and TLR-4 mRNA expression, in the upper, middle, and lower part of the trachea ciliary movement. In LE resp. HE, body weight at D34 was 2.09 (±0.35) and 2.14 (±0.26) kg, average feed intake 85 and 86 g/day. HE showed numerally more passive behaviour at 5 weeks of age. No differences were found IgM and IgG NAB. IgM NAAB levels tended to be higher in LE, IgG NAAB tended to be higher in LE-males than HE-males. No differences were found between HE and LE in mRNA expression of IFN-α, IFN-β and IL-10. However, TLR4 mRNA expression significantly differed between Con-HE and Con-LE. IBC had higher TLR4 than Con and IBV in HE and LE. No effect was found of endotoxin exposure on ciliary activity, histopathological lesions in the trachea and lungs and inflammatory response in the lung. However, after IBV and IBC challenge histological differences were found in the beak between HE and LE. Chronic exposure to high levels of airborne endotoxin did not affect production performance, but induced behavioural changes. Respiratory health of broilers was affected as shown by differences in TLR 4 expression and histology of the beak.

The influence of laying performance and estradiol-17ß on keel bone fractures in laying hens

S. Petow, B. Eusemann, A. Patt and L. Schrader
Friedrich Loeffler Institut, Institute of Animal Welfare and Animal Husbandry, Dörnbergstraße 25/27, 29223 Celle, Germany; stefanie.petow@fli.de

Keel bone fractures belong to the most severe animal welfare problems in the egg production industry. It is likely that egg production influences keel bone health due to the high demand of calcium for the eggshell. This is in part, taken from the skeleton. The high plasma concentrations of oestrogens, which are linked to the high laying performance, may also affect the keel bone as sexual steroids have been shown to influence bone health. The aim of the current study was to investigate the influence between egg production, genetically determined high laying performance, and estradiol-17ß on keel bone characteristics. A high performing layer line (WLA) and a low performing layer line (G11) were examined. After onset of lay, 100 hens of each layer line were divided into four treatment groups: Group S received a deslorelin acetate implant to inhibit follicle maturation. Group E received an implant of estradiol-17ß, group SE received both implants, and group C were control hens. At regular intervals, the keel bone of all hens was radiographed and estradiol-17ß plasma concentrations were assessed. For statistical analysis, generalized linear mixed-effects models were used. Prevalence of keel bone fractures was lower in non-productive hens compared to productive hens (2.3% in group S and 1.9% in group SE vs 48.3% in group C and 48.1% in group E). Furthermore, the high performing layer line WLA showed more keel bone fractures compared to the low performing layer line G11 (43.1 vs 29.5%). Radiographic density, reflecting bone mineral density, increased throughout the entire experimental period in non-laying hens but not in laying hens (treatment×layer line×sampling period: P<0.05). Treatment with estradiol-17ß did not seem to have an effect on keel bone measures. Although SE hens showed estradiol-17ß concentrations similar to C hens, keel bone measures were comparable to S hens. We assume that egg production plays a major role in the etiology of keel bone fractures, possibly by affecting bone mineral density. Moreover, our results indicate that the selection for high laying performance may negatively influence keel bone health.

First approach: validation of a scoring system to assess pododermatitis in Pekin ducks

L. Klambeck[1,2], J. Stracke[1], B. Spindler[1], D. Klotz[3], P. Wohlsein[3], H.-G. Schön[4], F. Kaufmann[2], N. Kemper[1] and R. Andersson[2]
[1]University of Veterinary Medicine Hannover, Foundation, Institute for Animal Hygiene, Animal Welfare and Farm Animal Behaviour, Bischofholer Damm 15, 30173 Hannover, Germany, [2]University of Applied Sciences Osnabrueck, Department of Animal Husbandry and Poultry Sciences, Am Kruempel 31, 49090 Osnabrueck, Germany, [3]University of Veterinary Medicine Hannover, Foundation, Department of Pathology, Buenteweg 17, 30559 Hannover, Germany, [4]University of Applied Sciences Osnabrueck, Faculty of Agriculture and Landscape Architecture, Am Kruempel 31, 49090 Osnabrueck, Germany; l.klambeck@hs-osnabrueck.de

As visual scoring is influenced by subjective bias of the observers, the aim of the study was to validate a visual scoring system for assessing the occurrence and the severity of pododermatitis in Pekin ducks. It should be investigated, whether the visual scoring system detects ulcerations in duck feet in actual prevalence, as ulcerations are reported as an indicator for severe and potentially painful lesions in poultry. At an abattoir, n=100 Pekin duck feet were collected representing each level of a five-point scoring system (n=20 feet per score level). An investigation of size of footpads and lesions as well as the proportion of lesions was performed using digital images of the feet and an image analysis software. As a reference method for the visual scoring system, each foot was examined histopathologically regarding occurrence, severity and type of pododermatitis. With the aim of assessing the reliability of the visual scoring system, histopathological findings and results of image analysis were compared with the results of visual scoring. According to our results, the visual scoring system did not fully correspond with the prevalence of ulcerations, as 47 feet histopathologically classified as ulcerative pododermatitis could not be detected macroscopically. Moderate correlations between size of lesion and ulceration (0.43) and respective percentage of lesion and ulceration (0.46) were found, respectively. Our findings suggest that the examined parameters visual scoring system, size of lesion and percentage of lesion are not appropriate to detect ulcerative pododermatitis in Pekin ducks in actual prevalence when performing visual inspection.

G×E interactions of body weight for broilers raised in bio-secure and commercial environments

T.T. Chu[1], H. Romé[1], E. Norberg[1,2], D. Marois[3], J. Henshall[3] and J. Jensen[1]
[1]Aarhus university, Center for Quantitative Genetics and Genomics, Blichers Allé 20, 8830 Tjele, Denmark, [2]Norwegian University of Life Sciences, Department of Animal and Aquacultural Sciences, P.O. Box 5003, 1432 Ås, Norway, [3]Cobb-Vantress, P.O. Box 1030, 72761-1030 Siloam Springs, AR, USA; helene.rome@mbg.au.dk

In genetic improvement programs for broilers, purebred birds are kept under strict bio-secure environmental conditions (B), while under a commercial production environment (C), the hygienic conditions are at a lower level. In this study, we explored the genotype by environment interactions (G×E) for broiler body weight (BW) in environment B and C. A pedigree-based BLUP multivariate model was used to estimate genetic parameters of BW traits at week five and six measured in environment B and C. The model accounted for heterogeneous variances between sexes by applying standardization to male and female BW differently. We found that the average performances of birds in B were higher than those of birds in C. Genetic correlations between BW traits measured in B and C environments were in the range of 0.479-0.535, which indicate that it will be a significant re-rankings of birds in terms of breeding values between the two environments. The genetic correlations was decreasing as week of age increased from five to six. Additive genetic variances of BW traits measured in B were more than two times larger than for the traits measured in C. Heritability of BW traits measured in B (0.274-0.301) was lower than the heritability of the traits measured in C (0.305-0.366). Additive genetic variances of BW traits for males were 1.37-1.66 times higher than the variances for females. In addition, differences between B and C environments also had effects on permanent environmental maternal effects. The permanent environmental correlations between BW traits measured in B and C was in the range of 0.589-0.723. In conclusion, environmental differences between B and C led to several indications of considerable G×E interactions for BW traits including different average performances, significant re-rankings, heterogeneous variances and different heritability for BW measured in B or C environments.

Session 49 Theatre 6

SNP prioritisation for immune responses traits of hens
T. Suchocki[1,2], B. Czech[2], M. Siwek[3] and J. Szyda[1,2]
[1]*National Research Institute of Animal Production, Krakowska 1, 32-083 Balice, Poland,* [2]*Wroclaw University of Environmental and Life Sciences, Department of Genetics, Kozuchowska 7, 51-631 Wroclaw, Poland,* [3]*University of Technology and Life Sciences, Animal Biotechnology Department, Mazowiecka 28, 84-085 Bydgoszcz, Poland; tomasz.suchocki@upwr.edu.pl*

The aim of this study was to create the list of genes influenced the most immune responses traits of hens. Immune responses are the category of complex or quantitative traits and they are controlled by multiple genes with different magnitudes of phenotypic effects. In our study adaptive immunity was represented by the specific antibody response toward keyhole limpet hemocyanin (KLH); innate immunity was represented by natural antibodies toward lipopolysaccharide (LPS) and lipoteichoic acid (LTA). Analysis was carried out in an experimental population, created by crossing two breeds of hens: Green-legged Partridgelike and White Leghorn. The final F2 generation consisted of 506 individuals obtained in six hatches. The regions with significant QTLs for immune responses presented by Siwek *et al*. were resequenced and finally we obtained 195 SNPs with minor allele frequency higher than 0.05 and call rate higher than 95%. Based on mixed model we calculated additive and dominance effect of each SNP and then we used the SNP prioritization procedure which allow us to create the list of genes with the highest impact of selected immune responses.

Session 49 Theatre 7

Optimising enrichment use for commercial broiler chickens
M. Baxter and N.E. O'Connell
Queen's University Belfast, Institute for Global Food Security, Malone Road, Belfast BT7 1NN, United Kingdom; m.baxter@qub.ac.uk

Environmental enrichments usually refer to physical or sensory changes made to an animal's environment with the intention of improving their health, behavioural repertoire or mental well-being. These additions are becoming more common in farms supplying certain retailers, with research showing multiple benefits of increasing the complexity of barren housing. However, using enrichment to improve the welfare of commercially housed broiler chickens presents a particular challenge. Their intensive genetic selection for production traits and a lack of environmental stimulation has often translated into high levels of inactivity. While their Red Jungle Fowl ancestors spend the majority of their time foraging, broiler chickens now spend a comparable amount of time sitting down. This modern behaviour pattern is linked to several welfare concerns, including poor leg health, infections and contact dermatitis. However, our research has shown a willingness among broilers to engage with their environment when provided with appropriate enrichments. For example, broilers will readily make use of a dustbathing area throughout their production cycle and show clear preferences for dustbathing materials. Broilers supplied with dustbathing areas also had better gait scores in their final week compared with those housed in a control treatment, suggesting that dust baths may have a role in improving leg health. Perches are also often implicated in improved broiler leg health, however the typical single bar perches are often unsuitable for broilers. Suspended platform perches, that increase in height as birds age, are more attractive to broilers and may have positive effects on behavioural repertoire and fearfulness. We have also investigated the effectiveness of straw bales, a common addition in both laying hen and broiler enriched housing. Although stimulating pecking and foraging is the intuitive aim of straw bale enrichments, they are typically used as resting areas by broilers and may serve a separate function by providing protective cover. With pressure on global poultry production continuing to increase, making efficient changes to broiler housing may be a practical way of maintaining their health and welfare in intensive systems.

Production performance of two dual-purpose chicken breeds in a mobile stable system

F. Kaufmann, U. Nehrenhaus and R. Andersson
University of Applied Sciences Osnabrück, Animal Husbandry and Poultry Sciences, Am Krümpel 31, 49090 Osnabrück,
Germany; f.kaufmann@hs-osnabrueck.de

Debates on culling day-old male egg-type chicks are constantly growing. Besides the *in-ovo* sex determination and the fattening of male chicks of layer lines, the use of Dual Purpose Breeds is considered as a strategy to avoid killing of day-old male egg-type chicks. The current study investigated the performance of two Dual Purpose Breeds in a mobile production system under organic conditions. 509 Lohmann Dual (LD) and 505 Lohmann Dualexperimental (LDex) day old chicks were reared in two mobile stable barns as hatched. The cockerels were slaughtered on day 85, the hens remained in their respective mobile barn for a prolonged laying period (74 weeks, including molting). Body weight (BW) development, group feed consumption and mortality was recorded in frequent intervals. At time of slaughter of the males, randomly selected birds (n=30/group and genotype) were dissected to determine slaughter weight (SW) and carcass composition. Additionally, laying performance of the hens was recorded in frequent intervals during the laying period. Average BW and SW of cockerels at d 85 was 2,650 (SD: 262.1) and 1,786 g (SD: 185.3) in LD and 2,176 (SD: 263.9) and 1,449 g (SD: 179.3) in LDex, respectively. Over the fattening period, average daily weight gain was 34.6 g in LD and 29.7 g in LDex. Percentage of breast muscle (16.3%) and legs (32%) did not differ between genotypes. In 68 weeks of production, average laying performance (excluding molting) was 62.2% in LD and 66.9% in LDex; an average LD hen produced 289 eggs whereas it was 307 eggs in LDex. Average egg weight during 68 weeks of production was around 60.5 g in both genotypes, whereas LD hens had a higher feed consumption per kg egg mass (X:1) when compared with LDex (2.9 kg vs 2.4 kg). The performance of both Dual Purpose Breeds is not comparable with those of specialized hybrids. The use of such breeds may still be sufficient if the products realize higher prices at the market as their production is more expensive and less efficient when compared to specialized hybrids. However, when taking animal welfare and ethical aspects into account, the use of Dual Purpose Breeds may still be considered as one potential strategy to avoid culling of day olds.

Genetics of the relationships between immunocompetence, resilience, and production in chickens

T.V.L. Berghof[1], K. Peeters[2], J. Visscher[2] and H.A. Mulder[1]
[1]Wageningen University and Research Animal Breeding & Genomics, P.O. Box 338, 6700 AH Wageningen, the Netherlands, [2]Hendrix Genetics B.V., P.O. Box 114, 5830 AC Boxmeer, the Netherlands; tom.berghof@wur.nl

It is hypothesized that selection for production traits in animals has resulted in allocation of resources to production at the expense of other physiological processes, such as immune function: This resource allocation led to trade-offs between production and other life functions. The aim of this study is to dissect the genetic basis of resource allocation between immunocompetence, resilience, and production. Immunocompetence is defined as the level of natural antibodies (NAb). Breeding for higher NAb levels was previously shown to increase avian pathogenic *Escherichia coli* resistance in chickens, and was hypothesized to improve humoral adaptive immunity. Resilience is the capacity of an animal to be minimally affected by disturbances or to rapidly return to the state pertained before exposure to a disturbance. We hypothesize that less resilient animals are more susceptible to environmental perturbations and will consequently show more fluctuations in day-to-day production than more resilient animals. A daily-measured and relevant production trait in chickens is egg production, which was therefore selected for further investigation for both production and resilience. The study population consists of one genetic line with three subpopulations: (1) a population of approximately 3,000 purebred chickens with observations for NAb (i.e. immunocompetence); (2) a population of approximately 18,500 purebred chickens with daily observations for egg production (i.e. resilience and production); and (3) a population of approximately 1,200 crossbred sire-families with daily observations for egg production (i.e. resilience and production). By estimating heritabilities and genetic correlations with double hierarchical generalized linear models, we will investigate and quantify relationships between immunocompetence, resilience, and production. This leads to better understanding of resource allocation, and gives opportunities for livestock breeding organizations to breed for livestock with good immunocompetence, resilience, and production.

Contribution of the chair: selected posters + relevance of other species

K. Stadnicka[1] and E.F. Knol[2]

[1]Animal Biotechnology and Genetics, UTP University of Science and Technology, Mazowiecka 28, 85-084 Bydgoszcz, Poland, [2]Topigs Norsvin Research Center, Research, Schoenaker 6, Beuningen, the Netherlands; katarzynakasperczyk@utp.edu.pl

Currently there is no Poultry nor Rabbit Commission within EAAP. The Pig Commission volunteered to organize a poultry session, which resulted in 29 abstracts, collected from different commissions and free communications. 13 abstracts were chosen for presentation, with 15 posters on display. In this timeslot the chair and the president of the pig commission want to highlight some posters and want to share some data on the presence of poultry, rabbit and probably mice and other species abstracts over the past years to explore an extension of the pig commission to 'pig and poultry commission', 'monogastric without horses commission', or another format. This same discussion will be/was held (again) during the pig commission business meeting on Thursday morning.

Evaluating body weight and egg number per bird per week for four tropically adapted chicken breeds

S.W. Alemu[1], O. Hanotte[1,2], R. Mrode[3,4] and T. Dessie[1]

[1] Live Gene, International Livestock Research Institute, Addis Ababa, Ethiopia, [2] Cell, Organisms and Molecular Genetics, School of Life Science, University of Nottingham, United Kingdom, [3]Animal Veterinary Science, Scotland Rural College, Edinburgh, United Kingdom, [4]Animal Biosciences, International Livestock Research Institute, Nairobi, Kenya; s.worku@cgiar.org

The production system of Ethiopia 60 million chicken is largely free range and based on the smallholder system. Upgrading smallholder poultry to small-scale commercial framework using more productive and tropically adapted breeds is central to rural development. With the aim of providing access to improved chicken by rural farmers in Ethiopia and other African countries, a project called African chicken genetic gain has been testing the performance traits (e.g. live body weight at week 20 (lbw) and number of eggs per bird per week (epbpw) in four tropically adapted chicken in two stations in Ethiopia for 64 weeks. The breeds included indigenous Horro, Koekoek, Kuroiler, Sasso_RIR. This paper evaluates the performance of these breeds for lbw and epbpw in the two stations (Bishoftu/Debrezeit and Haramaya). A linear mixed model was fitted with week (for epbpw only) and pen (for both traits) fitted as random effects, and station and breed as fixed effects. The result showed us that for epbpw, both breed and station were significant and more importantly the interaction between breed and station was significant, which implies that the breed performance depends on the station. For example, at Bishoftu/Debrezeit station, Koekoek (4.2±0.22) yielded the highest epbpw followed by Sasso_RIR (3.9±0.19), Horro and Kuroiler yield the least number of eggs per week (3.5±0.24). However, at Haramaya station, Koekoek and Sasso_RIR yielded the large epwpb (3.75±0.25) followed by Horro (3.7±0.26) and Kuroiler (3.6±0.26). The difference in performance in the two stations might be related to the climate of the stations. With respect to lbw, there was no interaction between station and breed but both factors were significant. Overall, Koekoek and Sasso_RIR performed best in epbpw and Kuroiler and Sasso_RIR in lbw. Thus, there is a very good prospect in producing chicken that yields large lbw and epbpw if it is coupled with a right breeding programme and management system.

Benefits of using genomic information for broiler breeding program in presence of G×E interactions

T.T. Chu[1,2], E. Norberg[2,3], C. Huang[4], J. Henshall[4] and J. Jensen[2]
[1]Wageningen University & Research, Animal Breeding and Genomics, 6709 PG Wageningen, the Netherlands, [2]Aarhus University, Center for Quantitative Genetics and Genomics, Blichers Allé 20, 8830 Tjele, Denmark, [3]Norwegian University of Life Sciences, Department of Animal and Aquacultural Sciences, 1432 Ås, Norway, [4]Cobb-Vantress Inc., Siloam Springs, Arkansas 72761-1030, USA; chu.thinh@mbg.au.dk

An increase in accuracy of prediction by using genomic information in animal breeding programs has been well documented. However, benefits of genomic information and methodology for evaluation are missing for breeding programs in the presence of genotype by environment interactions. This study explored benefits of using genomic information in a broiler breeding program when selection candidates were raised and tested in a bio-secure environment (B) and close relatives were tested in a commercial production environment (C). A pedigree-based BLUP model was used to estimate variance components and predict breeding values (EBVs) of body weight (BW) traits at week five and six, measured in environment B and C. A ssGBLUP model that combined pedigree and genomic information was also used to predict EBVs of those traits. Cross-validations were based on statistics of correlation, mean difference and regression slope for EBVs estimated from full and reduced datasets. Those statistics were estimators of population accuracy, bias and dispersion of EBV prediction. Validation animals were genotyped and non-genotyped birds in B environment only. It was found that the genetic correlations between BW traits measured in environment B and C were in the range 0.479-0.535. The use of combined pedigree and genomic information in ssGBLUP substantially increased population accuracy of EBVs, and reduced bias of EBV prediction for genotyped birds compared to the use of only pedigree information. The increase in accuracy of EBVs also occurred for non-genotyped birds, however bias of prediction also increased for non-genotyped birds. The relative increase in population accuracy was from 31.7-73.1% for genotyped birds and from 6.3-14.9% for non-genotyped birds. In conclusion, the use of combined pedigree and genomic information increases population accuracy of EBVs for genotyped birds in B compared to the use of pedigree only.

Comparison of broiler productivity fed different types of diet at commercial farm in Northern Japan

A. Tozawa and S. Sato
Teikyo University of Science, Department of Animal Sciences, 2-2-1 Senjusakuragi, Adachi, Tokyo, 120-0045, Japan; akitsu-t@ntu.ac.jp

The differences of broiler productivity fed a commercial diet with antibiotics (Type A), a commercial diet with brown rice and rye without antibiotics (Type B), and a commercial diet with vitamins and *Lactobacillus*-based probiotic without antibiotics (Type C) were studied. Data of 909 broiler flocks kept in 41 broiler farms in Northern Japan from Apr. 2017 to Feb. 2018 were used. There were 56 flocks for Type A, 178 flocks for Type B, and 675 flocks for Type C. Mortality, feed conversion ratio (FCR) and daily gain (DG) were calculated as productivity. Steel-Dwass test were performed for the statistical analysis for comparing between types of the diet. Mortality was higher in Type C (mean ± S.D.=6.57±3.44%) than in Type A (5.08±2.28%) and Type B (4.71±2.31%) (P<0.01). Also, FCR was higher in Type C (1.80±0.08) than in Type A (1.78±0.06) and Type B (1.78±0.08) (P<0.05), and DG was lower in Type C (64.4±2.3 g/day) than in Type A (65.5±1.8 g/day) and Type B (65.3±2.3 g/day) (P<0.01). These results indicate that the broiler fed a diet with vitamins and *Lactobacillus*-based probiotic without antibiotics in conventional farm had lower productivity than the broiler fed a commercial diet or a commercial diet with brown rice and rye without antibiotics. We are now continuing analysis of differences in season, feeding company, strain, flock size and hatchery in each type.

L-selenocystine and L-selenomethionine and their deposition in broiler muscle tissue
S. Van Beirendonck[1,2], B. Driessen[1], G. Du Laing[3], B. Bruneel[4] and L. Segers[4]
[1]Dier&Welzijn, Kleinhoefstraat 4, 2440 Geel, Belgium, [2]KU Leuven, Faculty of Engineering Technology, Kleinhoefstraat 4, 2440 Geel, Belgium, [3]Ghent university, Faculty of Bioscience Engineering, Coupure Links 653, 9000 Gent, Belgium, [4]Orffa, Vierlinghstraat 51, 4251 LC Werkendam, the Netherlands; bruneel@orffa.com

The aim of this study was to investigate the bioavailability of specific organic selenium (Se) compounds on their Se deposition capability in broiler muscle tissue. Three different Se products (inorganic and organic) were added at 0.2 mg Se/kg feed, including two organic sources: L-selenomethionine (L-SeMet) and L-selenocystine (L-SeCys). One day-old male broilers were fed for 14 days control or one of three treatment starter diets. A wheat-corn-soybean meal basal diet containing no added Se was used for the production of all starter diets. Each treatment consisted of 4 pens with 5 animals per pen. Three broilers per pen were sacrificed on d14 and representative samples of the breast muscle were taken and analysed for Se content by ICP-MS. Data were analysed using a linear mixed model with treatment as a fixed effect and pen as a random effect, in the statistical software program SAS 9.4. Results show that the addition of L-SeMet resulted in the highest Se content in muscle (293 µg Se/kg) compared to the basal diet without added Se (P<0.05). The addition of L-SeCys and sodium selenite resulted in a significantly lower deposition of 169 µg/kg and 194 µg/kg, respectively, compared to the addition of L-SeMet (P<0.05). In conclusion, this study shows that L-SeMet is incorporated to a high extend into muscle proteins. L-SeCys, on the other hand, has no benefit over inorganic selenium to increase the Se content in broiler muscle tissue. In literature, L-SeMet is known to be deposited to a high extend in (muscle) proteins.

Fluctuation of chicken meat price in European Union in 2007-2018
K. Utnik-Banaś
University of Agriculture in Kraków, Institute of Economics and Enterprises Management, al. Mickiewicza 21, 31-120 Kraków, Poland; rrbanas@cyf-kr.edu.pl

The aim of this study is analysis of price level and influence of particular kind of fluctuation on variability range of chicken meat price in European Union as a whole market and chosen particular countries of EU. Research material consist of monthly time series of chicken meat price in whole European Union and particular countries for the years 2007 -2018. Data come from Integrated System of Agricultural Information supervised by Ministry of Agriculture in Poland. Information about level of chicken livestock production was taken from FAOSTAT database. Time series of price were described by multiplicative model and seasonal decomposition was performed using Census X-11 method. Price of broiler chicken meat in European Union average in year 2007 was on the level of 176 EURO per 100 kg. During twelfth years period meat prices were subject different fluctuations and finally increased to the level of 186 EURO in year 2018. Prices of chicken meat in EU were significantly influenced by seasonal fluctuation: generally higher prices are in summer (102.6% in June) and lower in winter (98.5% in December). In the one year time horizon variability of prices was shaped in 5,6% by irregular changes,42.1% by seasonal and 52.3% by cyclical fluctuation. The highest prices of chicken meat were in Germany (254 Euro/100 kg as average in analysed period), followed by: Finland (246 €), Cyprus (242 €), and Denmark (226 €). The lowest prices were in Poland (131 €), Great Britain (148 €), Bulgaria (149 €) and Portugal (160 €). The paper was created with the grant support project NCBiR no. BIOSTRATEG2/297910/12/NCBR/2016.

Relationship between the number of hatchings and the pre-incubation conditions of poultry eggs

D.I. Carrillo-Moreno[1], J.H. Hernandez[1], C.A. Meza-Herrera[2], J.A. Beltran-Legaspi[3], E. Carrillo-Castellanos[3], J.Z. Ordoñez Morales[1] and E.G. Véliz-Deras[1]
[1]Universidad Autonóma Agraria Antonio Narro, Torreón Coahuila, 27054, Mexico, [2]Universidad Autónoma de Chapingo. Unidad Regional Universitaria de Zonas Áridas, Durango, 35230, Mexico, [3]Instituto Tecnológico de Torreón, Torreón, Coahuila, 27170, Mexico; dalia.ivettecm@gmail.com

The aim of the present study was to evaluate if the conditions of the incubation period (storage time and temperature) of eggs from backyard flocks influences the hatching average. We analysed data from eggs from different domestic bird backyard flocks, data of 60,347 eggs was collected from January 2010 to May 2014. The variables considered were average of hatchings at different storage times (days from laying to day of entry to the incubator) and egg temperature art reception. A vertical multi-stage incubator with a forced air hatcher and a capacity of 750 eggs at 37 °C and a relative humidity of 70% was used. Data was analysed with the statistical software SPSS 25. From the analyses of the 60,347 eggs, the following means (± standard deviation) were obtained: storage days 7.80 (±3.39), egg temperature at the time of reception 24.7 °C (±6.18), hatchings 37.4%. The reception temperature and storage time are negatively correlated with the average of hatchings (R^2=0.691, P<0.01, R^2=0.569, P<0.001 respectively). The results grouped into ranks indicate that the average of hatchings decreases considerably in eggs stored more than 9 days (29.4%) compared to those with a time less than 6 days (42.2%). The range of egg temperature at reception of 20.1-25 °C had the highest average of hatchings (47%) while the range where the lowest average was obtained was 30.1-35 °C (24.4%). The analyses indicate that the pre-incubation management of eggs from domestic poultry in backyard flocks, specifically storage time and egg temperature, correlates negatively with the average of hatchings. Suggesting that a storage greater than nine days at a temperature higher than 30 °C affects the average of hatchings, it is recommended to continue studies to establish these associations.

Effects of de-oiled lecithin in laying hens

Y. Yang[1], D.H. Nguyen[1], S.B. Yoon[2], W.J. Seok[1], K.A. Kim[1] and I.H. Kim[1]
[1]Dankook University, Animal Resources Science, 119, Dandae-ro, Dongnam-gu, Cheonan-si, Chungcheongnam-do, Republic of Korea, 31116, Korea, South, [2]Feedbest., Ltd, 354-58, Mojeon 1-gil, Seonggeo-eup, Seobuk-gu, Cheonan-si, Chungcheongnam-do, Republic of Korea, 31116, Korea, South; yangyi1104657897@163.com

The objective of the present study was to evaluate the effects of de-oiled lecithin on egg production, cracked egg rate, egg quality and blood profiles in laying hens. A total of 420 Hy-line brown laying hens was randomly allotted into 1 of the following 3 treatments: CON (basal diet), TRT1 (CON + 0.1% DOL-60), TRT2 (CON + 0.1% DOL-97). During the week 1 to week 6, there was no significant difference (P>0.05) on egg production and cracked egg rate among treatments. Yolk colour was significantly higher (P<0.0001) in TRT2 than in other treatments. Huaugh units showed improvement (P<0.05) in TRT2 than in other treatments at 4th week. At the 3rd week, shell colour was improved (P<0.05) in the TRT2, compared to those in treatments fed with CON diet. The serum total cholesterol and Triglyceride concentrations showed no difference compared with all treatments at week 6, but the TRT2 showed significant higher HDL-cholesterol (P=0.0456) and lower LDL-cholesterol (P=0.0404) compared with TRT1. Taken together, our observations suggest that de-oiled lecithin benefits egg quality and reduce LDL-cholesterol in laying hens.

Productive response in broilers fed with rape seed (*Brassica napus*)

R. Braun, J. Chapado, M. Magnani and S. Pattacini
Universidad Nacional de La Pampa, Facultades de Agronomía y Ciencias Exactas y Naturales, Ruta 35, km 334, 6300
Santa Rosa, La Pampa, Argentina; silca1816@hotmail.com

Rapeseed grains with normal levels of glucosinolates and erucic acid can be used to feed farm birds by including them in their diets. The objective of this research was to determine if the rapeseed grain prior to inclusion in the balanced diets during the growth-fattening period, allow to improve the composition of the carcasses and the quality of the meat and its fatty acid composition. The treatment received an initiation diet until 30 days and another termination diet until 51 days that included 6% rapeseed in its composition. The age at slaughter was at 51 days. It was measured through a calibrated electrode peachimeter with buffer solutions in the muscles of the breast. The weight losses by cooking (%) were determined by a thermal bath at 72 °C for 50 minutes in samples of 100 g of the breast muscle. In this investigation the tenderness of the meat was measured with Warner-Bratzler. The iodine index was also detected at the time of slaughter by the Hanus method on peritoneal fat samples. Experimental carcasses were taken samples of peritoneal fat, and by gas chromatography, the abundance of fatty acids was determined. The correlation of the weights accumulated on each individual was analysed using R^2. The average data obtained from the weekly and final weights of males and females of both treatments did not present significant differences, but important biological differences in the difference of weight by sex. Regarding the muscle, protein and fat values of the breast (%); the control obtained significantly greater differences in fat %. Significant differences were observed in cooking losses and cutting force favourable to treatment. Without statistical significance in pH. The II_2 of the fat of rape-eating chickens is higher and is reflected in higher % unsaturated fatty with three and more than three double bonds in the carbon chain (ω 3). In any case, the II_2 of the fat refers to the amount of halogen (Iodo) that is added to the double bonds, and in both treatments it is high because the unsaturated exceed 70%. The ratio $\omega6/\omega3$ that for the control was 10.69: 1 and for the treatment 8.30: 1.

Animal welfare and productivity of hosted hens in a cage-free system

R. Braun, S. Pattacini and J. Sosa Bruno
Universidad Nacional de La Pampa, Facultades de Agronomía y Ciencias Exactas y Naturales, Ruta 35, km 334, 6300
Santa Rosa, La Pampa, Argentina; braun1816@gmail.com

The production of eggs in cage-free floor systems is diametrically different from production in confinement, especially in cage batteries in ships populated by thousands of hens in high densities, not only in management issues, but also in sanitary aspects and of animal behaviour, as well as all the systems to floor maintain high productive standards little away from the confined systems. The purpose of this work was to determine if a productive system of cage-free layers. In this work the factors related to the productive system to cage-free floor will be studied, comparing their results with those of the conventional methods of production in cages. For the freedom of hunger and thirst, the indicators of live weight, growth and eating behaviour, productive of eggs along the posture and feed conversion efficiency were evaluated. Regarding freedom of discomfort, the indicators were animal density, temperature, relative humidity, air flow and light intensity; for the freedom of pain, injury and disease, the indicators of body condition (live weight), plumage status, degree of dirtiness, presence of wounds and injuries, nail length and mortality were evaluated; For the freedom to express normal behaviour and freedom from fear and distress, the indicators of: absence of abnormal behaviours such as stereotypies, cannibalism and fear reaction test were determined. The productive indicators with reference to live weight at 18 weeks, feed intake and feed conversion were acceptable for floor productions. Only the percentage of laying was relatively lower, settling in the order of 85-90% at the peak of posture and maintenance during the laying cycle, 5% lower than in confined hens. In this experience it was difficult to adapt the hens to the nesting, especially because they were not suitable at the beginning to conditions of certain intimacy that the hen requires at the time of laying. This meant that 50% of the eggs were collected from the floor. The situation involved protecting the nests with black nylon to darken and place blind floors with chips in the nest boxes, this only resolved 30% of the problem, the habit of laying on the floor prevailed, an aspect that complicated the harvest and increased the breakage of eggs.

European guide of good and better practices for poultry transport

L. Warin[1], J. Litt[1], C. Mindus[1], E. Sossidou[2], H.A.M. Spoolder[3] and L. Bignon[1]
[1]ITAVI, 7, rue du Faubourg Poissonnière, 75009 Paris, France, [2]Hellenic Agricultural Organization – Demeter, Veterinary Research Institute, 57001 Thermi Thessaloniki, Greece, [3]Wageningen Livestock Research, P.O. Box 338, 6700 AH Wageningen, the Netherlands; litt@itavi.asso.fr

The 'Animal Transport Guides' project, funded by the DG SANTE of the European Commission, aimed to develop guides for good practices for all types of animal transport (short or long, to the farm or to the slaughterhouse) and took place from 2015 to 2018. The guides facilitate the understanding and practical implementation of Regulation 1/2005 on the transport of animals. Fourteen European partners worked on all main livestock species: sheep, cattle, pigs, horses and poultry. A first comprehensive bibliographical scientific studies and available guides of good practices was conducted to identify as many good practices as possible (197 for poultry). Several consultations of stakeholders (European professionals, scientists, animal protection associations) were done to establish a consensus on practices that should be included on the final guide. 163 good and better practices were included in the final guide. Good practices are defined as procedures and processes that ensure compliance with requirements of legislation or regulations and better practices are defined as providing additional guidance on how procedures and operations can be improved to exceed any legally defined minimum welfare requirement. Teaching materials (e.g. illustrated fact sheets and videos) on key practices were also developed. The impact of the guides on the behaviour of stakeholders was also evaluated through questionnaires. 411 stakeholders were questioned. 39% indicate that they apply the EU rules, but are interested to do more and could use the new guidelines and 17% indicate they already apply more than the EU rules, but are interested to learn about other practices. These results indicate a positive impact of this work and a good implementation of these guides. It is expected that the guides will support the industry in improving their practices and are widely disseminated.

Influence of humic acids on ammonia concentrations, final live weight and mortality of broilers

Z. Palkovičová, M. Uhrinčať and J. Brouček
NPPC, Hlohovecká 2, 951 41 Lužianky, Slovak Republic; palkovicova@vuzv.sk

In this study, we investigated the influence of humic acids on ammonia concentrations, final live weight and mortality of broilers. We applied a preparation containing humic acids (>45%) into the litter (140 g per m^2) in the experimental hall. The second hall was control. Ammonia concentration (mg/m^3), temperature (°C) and humidity (%) were measured every 10 minutes for 35 days. The measured data were evaluated by the statistical program Statistix 10. We found that the average ammonia concentrations were lower in two locations (fans: 0.59±0.01, the middle of the hall: 1.15±0.01) and higher in one location (flaps: 0.86±0.02) of the experimental hall than in the control hall (fans: 0.79±0.02, the middle of the hall: 1.40±0.01, flaps: 0.41±0.01) with significance $P<0.001$. The average temperatures were lower in two locations (flaps: 22.53±0.06, fans: 25.30±0.06) and higher in one location (the middle of the hall: 27.78±0.04) of the experimental hall than in the control hall (flaps: 22.64±0.08, fans: 27.16±0.07, the middle of the hall: 26.10±0.04) with significance $P<0.001$ for the middle of the hall and fan location. The average humidity values were higher in all locations of the experimental hall (flaps: 44.04±0.15, fans: 40.11±0.13) than in the control hall (flaps: 39.79±0.20, fans: 35.30±0.16) with significance $P<0.001$. Within the individual halls, we detected very high significant ($P<0.001$) differences in the values of ammonia concentration, temperature and humidity. The average NH$_3$ concentration was lower in the experimental hall (0.95±0.01) than in the control hall (1.01±0.01). Conversely, the average temperature and humidity were higher in the experimental hall (25.84±0.03; 42.08±0.10) than in the control hall (25.57±0.04; 37.55±0.13). During the study, 946 and 869 broilers died in the experimental and control hall, respectively, but the differences in broiler mortality were not significant ($P=0.06$). The average final live weights of broilers were 2.05 and 1.94 kg in the experimental and control hall, respectively. The applied preparation significantly decreased the ammonia concentrations in the housing area and increased the final broiler weight by 110 g on average. This article was possible through APVV 15-0060 of the SRDA Bratislava.

Chestnut tannins in laying hen feed and their effect on metabolism

K. Buyse[1,2], B. Wegge[3], E. Delezie[2], G.G.P. Janssens[1] and M. Lourenço[2]
[1]*Ghent University, Department of Nutrition, Genetics and Ethology, Heidestraat 19, 9820 Merelbeke, Belgium,* [2]*ILVO, Scheldeweg 68, 9090 Melle, Belgium,* [3]*Sanluc International NV, Schoolstraat 49, 9860 Oosterzele, Belgium; kobe.buyse@ilvo.vlaanderen.be*

It seems that small doses of chestnut tannins can be beneficial for poultry gut health and overall performance, but the working mechanism is not well known. This study assessed the effect of chestnut tannins on nutrient digestibility, performance and metabolism in laying hens (Lohmann Brown) of 32 weeks of age. Two basal diets were tested namely a widely used corn-soy based (C) diet and a wheat-rapeseed-palm oil based challenge diet (W) to which chestnut tannins were added (500 mg/kg) (T+) or not (T-) and two levels of vitamin E were tested (50 mg/kg: E50 and 25 mg/kg: E25), making a total of six different treatments (C/T-/E50, C/T+/E50, C/T+/E25, W/T-/E50, W/T+/E50, W/T+/E25). The vitamin E dose was included to evaluate the antioxidant capacity of the tannins. In total 54 laying hens were housed in digestibility units for 12 days, after which gut content and faeces were sampled. No significant increases in digestibility of gross energy and crude protein were seen in groups fed chestnut tannins compared to control groups (C/T-/E50 and W/T-/E50). Crude fat digestibility was significantly better in groups fed corn-soy based diets with added chestnut tannins (C/T+/E50 and C/T+/E25) compared to the control group (C/T-/E50) but no effects were seen in hens fed the (W) diet. A better crude fat digestibility was observed in groups supplemented with tannins and normal vitamin E dose (C/T+/E50) compared to tannin supplemented feed with low dose vitamin E (C/T+/E25). In (W) diets the metabolizable energy was significantly higher when supplemented with tannins (W/T+/E50 and W/T+/E25) compared to the control diet (W/T-/E50). Production traits, feed retention and viscosity of digesta showed no significant differences between diets. It can be concluded that low doses of chestnut tannins improved fat digestibility and metabolizable energy, but that effect depends on the composition of the basal diet. Furthermore obtained results are valid for the specific tested chestnut extract and can't be extrapolated to similar products.

Improving disease resistance in chickens: divergent selection on natural antibodies

T.V.L. Berghof[1,2], J.J. Van Der Poel[2], J.A.J. Arts[1], H. Bovenhuis[2], M.H.P.W. Visker[2] and H.K. Parmentier[1]
[1]*Wageningen University & Research Adaptation Physiology, P.O. Box 338, 6700 AH Wageningen, the Netherlands,* [2]*Wageningen University and Research Animal Breeding & Genomics, P.O. Box 338, 6700 AH Wageningen, the Netherlands; tom.berghof@wur.nl*

Natural antibodies (NAb) are antibodies recognizing antigens without previous exposure to this antigen. Keyhole limpet hemocyanin (KLH)-binding NAb titers in chickens are heritable, and higher KLH-binding NAb titers have been associated with higher survival. This suggests that breeding for higher NAb titers might improve general disease resistance. A purebred White Leghorn chicken line was divergently selected and bred on total KLH-binding NAb titers at 16 weeks of age for 6 generations, and resulted in a High and Low line. The average estimated breeding value differences in KLH-binding NAb titers increased with 0.36 for total, 0.40 for IgM, and 0.32 for IgG per generation. Generations 4 and 6 of the selection lines were inoculated with an avian pathogenic *Escherichia coli* (APEC) at 8 days of age. Mortality and morbidity after 1 week were significantly reduced in the High line compared to the Low line, which suggests a higher APEC resistance in the High line compared to the Low line. To investigate possible correlated responses on the immune system, several traits were measured at different ages in several generations: the High line showed higher different NAb titers at different ages, antibody concentrations, percentage of antibody-producing B cells, and bursa weight at young age compared to the Low line. This suggests that KLH-binding NAb selection has a favourable correlated response on the humoral adaptive immune system. No line differences were observed for T cells, γδ T cells, natural killer (NK) cells, and antigen-presenting cells (APC). This might indicate that the selection had no unfavourable correlated responses on other parts of the immune system. This selection experiment shows that selective breeding on total KLH-binding NAb titers at 16 weeks of age is possible, and that selection for higher NAb has a beneficial effect on resistance to APEC infection. In addition, the selection experiment suggests a promising opportunity for improving general disease resistance without unfavourable correlated selection responses.

How accurate is keel bone damage assessment by palpation and how does experience affect this?

S. Buijs[1,2], J.L.T. Heerkens[2,3], B. Ampe[2], E. Delezie[2], T.B. Rodenburg[4] and F.A.M. Tuyttens[2]
[1]Agri Food and Biosciences Institute, Large Park, BT26 6DR Hillsborough, United Kingdom, [2]Flanders Research Institute for Agriculture, Fisheries and Food (ILVO), Scheldeweg 68, 9090 Melle, Belgium, [3]Aeres University of Applied Sciences, De Drieslag 4, 8251 JZ Dronten, the Netherlands, [4]Utrecht University, Yalelaan 2, 3584 CM Utrecht, the Netherlands; stephanie.buijs@afbini.gov.uk

Palpation is the most commonly used method to evaluate keel bone damage in living hens. Although the importance of experience with this technique is commonly emphasized, knowledge on its effect is scarce. Ten assessors with or without prior experience palpated the same 50 laying hens twice for the presence of deviations, medial fractures, and caudal fractures. Each assessor's accuracy was determined by comparing palpation results to post-dissection assessment. To determine intra-assessor consistency (i.e. test-retest reliability) Cohen's Kappa was calculated for each assessor. Linear models were used to compare both statistics between experienced and inexperienced assessors, and accuracy between the 1st and 2nd assessment session. Prior experience led to more accurate results when assessing deviations (% accurate assessments: experienced 83% vs inexperienced 79±1%, P=0.04) and caudal fractures (41 vs 29% ±3, P=0.03). Experienced assessors were also more accurate when assessing medial fractures in the 1st session (82 vs 68 ±3%, P=0.04), but not in the 2nd one, by which time the inexperienced assessors' accuracy had increased to 77±4% (P=0.04). Thus, both prior experience and experience gained during the experiment contributed to improved accuracy. However, effect sizes were small for deviations and all assessors lacked accuracy when assessing caudal fractures. Intra-rater consistency of caudal fracture assessment was poor (0.36-0.44) and unaffected by prior experience (P=0.55). The true prevalence of caudal fractures (85%) was markedly underestimated by both experienced (37%) and inexperienced (16%) assessors. In conclusion, experience improves accuracy but does not guarantee high accuracy for all types of damage. Future research should determine if other training methods (e.g. comparison to post-dissection scores or to radiographs) improve accuracy further.

Using mixed rice bran in laying hen diets

U.O. Emegha
Ebonyi State College of Education, Ikwo, Animal Science, 13 Daniel Street Aguogboriga, Abakaliki, Nigeria; pstemeghaudu52@gmail.com

Rice brain is energy and protein rich ingredient used in poultry diets to balance energy and protein requirements. The purpose of this study is to examine the effects of rice bran on performance and egg quality during peak production of a commercial white laying strain of 22 week of age. Dietary treatments were consisted by inclusion of rice bran 0, 5, 10 and 15% Levels. Each treatment had 6 reps in which 12 birds were randomly assigned in wired floor battery cages equipped with nipple drinkers and through feeders. Layers accessed to feed and water freely. Lighting regime was adjusted to 16 hour light/8 hour dark. The experiment lasted for 10 weeks. Overall results of the present experiment indicated that rice bran could be included up to 10% without any adverse effect on laying performance, egg quality and digestive organs. The increasing price of feed stuffs used for poultry diets demands nutritionists to look for new cheaper ingredients able to replace the expensive feeds. Use of mixed rice bran in layer ration was studied during the period of 13 weeks in an experiment in Abakaliki, Ebonyi State, Nigeria with completely randomized design (CRD). After determination of chemical composition, six diets were formulated by using 0, 5, 10, 15, 20 and 25% of mixed rice bran (diets 1-6). During the experiment, feed intake, egg production, egg mass, egg weight, egg specific gravity, haughty unit score, body weight changes and mortality were measured. The results indicated that use of mixed rice bran up to 25% is profitable and had no significant effect on each of above traits except for egg mass (P<0.05). Diets containing 10 and 15% of mixed rice bran (diets 3, 4) had the best values for egg weight, egg mass and feed conversion ratio. It is concluded that feeding laying hen with maize-soya beans meal diets containing 25% mixed rice bran was practically feasible without compromising performance.

Impact of a litter amendment on welfare indicators and litter quality in a Turkey husbandry

K. Toppel[1], H. Schoen[1], F. Kaufmann[1], M. Gauly[2] and R. Andersson[1]
[1]University of Applied Sciences Osnabrueck, Department of Animal Husbandry and Poultry Sciences, Am Kruempel 31, 49090 Osnabrueck, Germany, [2]Free University of Bozen, Faculty of Science and Technology, Piazza Università 5, 39100 Bozen-Bolzano, Italy; k.toppel@hs-osnabrueck.de

Litter moisture content as well as the pH-value of poultry litter is discussed to significantly affect foot pad health (FP) in turkeys. The objective of this study was to evaluate effects of a bisulphate litter treatment (SBS) on foot pad lesions (FPD) and litter parameters in European turkey production systems. The study was performed on a research farm, two groups of 142 male turkey poults were raised for 147 days. Birds were placed on wood shavings from day 1 to 35 (3.4 kg/m^2), then kept on chopped straw until the end of the trial (day 147). Before placement of day olds, a total of 4 g SBS was spread on 100 g bedding material, whereas no litter treatment occurred in the second group (Con). Dispersing of litter was repeated in the experimental group on day 15 and 22. Starting on day 35 additional SBS application was performed together with every third re-bedding. The experiment was repeated four times, whereas the first two trials were a pre-study (S1) and repetition 3 and 4 were the main study (S2). During S1 mortality rate, body weight (BW; biweekly, n=40 birds/group and sampling date) and FPD (post mortem, n=220 feet/group) were investigated, whereas pH and dry matter content (DM) of the litter as well as BW and FP were recorded biweekly within S2 (n=60 birds/group and sampling date). SBS had no influence on mortality rate (8.0% vs Con 11.6%) and BW (P>0.05) in S1, however, FPD was significantly reduced in the experimental group when compared to Con (P<0.05). In S2 SBS neither affected mortality rate (12.7% vs Con 12.0%) nor BW (P>0.05), however, SBS significantly reduced FPD incidence compared to Con (P<0.05). No differences were found in DM of the litter (P=0.342) but in pH-value (P=0.000). SBS reduced the initial litter pH to 2.8 in the experimental groups (6.3 in Con). The results indicate positive effects of SBS on FP without influencing mortality or BW. Further studies are needed to confirm the results under commercial field conditions in order to improve health and welfare conditions in turkey production systems.

A genomic single-step random regression model for egg production in turkeys (*Meleagris Gallopavo*)

H. Emamgholi Begli[1], B.J. Wood[1,2], E.A. Abdalla[1], O. Willems[2], F. Schenkel[1] and C.F. Baes[1,3]
[1]University of Guelph, Centre for Genetic Improvement of Livestock, Department of Animal Biosciences, Guelph, Ontario, N1G 2W1, Canada, [2]Hybrid Turkeys, A Hendrix Genetics Company, Kitchener, Ontario, N2K 3S2, Canada, [3]Institute of Genetics, Vetsuisse Faculty, University of Bern, Bern, 3001, Switzerland; hemamgho@uoguelph.ca

There is increasing interest in using single-step GBLUP in animal genomic evaluation because of its improved accuracy and simplicity of routine application. In statistical analyses of longitudinal traits, random regression models have advantages over alternative models, because variances and covariances are more accurately modelled over time and, consequently, they provide more accurate breeding value predictions. The objective of this study was to evaluate the potential improvement in accuracy and reduction in bias of estimated breeding values by using a genomic single-step random regression (RR-ssGBLUP) model compared to a pedigree-based (RR-PBLUP) model in a commercial turkey line. Egg production over five consecutive months from age at first egg was recorded for 5,547 birds, of which 2,269 were genotyped. Breeding values for each of the five months were estimated using best linear unbiased prediction, where hatch-week was included as a fixed effect, and both additive genetic and permanent environment were random effects. Accuracy was assessed as the correlation between genomic or pedigree estimated breeding values and the corresponding phenotypes adjusted for environmental effects divided by the square root of the heritability. Prediction accuracies for RR-PBLUP ranged from 0.11 to 0.44, while for RR-ssGBLUP accuracies were between 0.24 and 0.45. Predictions based on RR-ssGBLUP were slightly less biased than those based on RR-PBLUP, except for the first and 3rd months. RR-ssGBLUP yielded higher prediction accuracies than RR-PBLUP in all months, except for the first one. The results of this study showed that using RR-ssGBLUP to estimate the breeding values for egg production improved prediction accuracy compared to RR-PBLUP over the five months, with the only exception being the first month. Therefore, the use of RR-ssGBLUP for improving accuracy of prediction of breeding values for egg production in turkeys may be warranted.

Should I stay or should I go – pros and cons of alternative career options for young scientists

C. Lambertz[1], M. Åkerfeld[2], N. Buys[3], G. Loh[4] and J. Van Arendonk[5]

[1]YoungEAAP, Via G. Tomassetti 3 A/1, 00161 Rome, Italy, [2]Swedish University of Agricultural Sciences (SLU), Ulls väg 26, 75007 Uppsala, Sweden, [3]KU Leuven, Kasteelpark Arenberg 20, 3001 Leuven, Belgium, [4]Evonik Industries, Goldschmidtstraße 100, 45127 Essen, Germany, [5]Hendrix Genetics Research, Technology & Services B.V., Spoorstraat 69, 5831 CK Boxmeer, the Netherlands; christian.lambertz@unibz.it

This debate on the pros and cons of staying in academia or changing to companies working in the livestock sector targets young scientists during their early stages of their scientific career independent whether at master's, PhD or Postdoctoral level. Whenever graduations approach, important decisions with impact on our future career have to be taken. For young scientists a variety of options arise with job opportunities becoming more and more diverse. Similarly, job requirements change constantly and may vary even more dramatically in the future, for example in view of developments such as digital farming. In this session, several speakers will highlight their career paths, including both changing from academia to the livestock industry, and the other way round. The first is for most of the young and early career scientists the usual way. However, also the way back to universities or research institutions might interest people who left academia for a certain time. Accordingly, livestock scientists with a variety of career pathways moving from research to livestock industries or vice versa will present us their motivations and discuss skills and expertise that are helpful and needed for successful job changes in the livestock sector.

Improvements in larviculture of *Crangon crangon*: steps towards its commercial aquaculture

B. Van Eynde[1], D. Vuylsteke[1], O. Christiaens[2], K. Cooreman[1], G. Smagghe[2] and D. Delbare[1]

[1]Flanders research institute for agriculture, fisheries and food, Animal Sciences Unit-Fisheries, Ankerstraat 1, 8400 Oostende, Belgium, [2]Faculty of Bioscience Engineering Ghent University, Department of Plants and Crops, Coupure Links 653, 9000 Ghent, Belgium; Daan.delbare@ilvo Vlaanderen.be

The European brown shrimp, *Crangon crangon*, is a highly valued commercial species in Europe. The fisheries are characterized by a strong seasonal variation of average landings. Attempts to 'catch and hold' wild adults in land-based rearing systems have been proven to be very difficult. Preliminary experiments showed that efficient larval rearing is a first essential step to domesticate this species and to establish a commercially viable and sustainable culture of the European brown shrimp. This study describes the design and operational procedures of a small-scale rearing system for the larviculture of *C. crangon* with emphasis on high survival. The focus of this research is on the selection of an adequate feed and the optimization of the feeding regime. One-day old larvae were reared in Erlenmeyer flasks with natural seawater (300 larvae/l). Different diets were tested namely, *Artemia* nauplii, *Artemia* nauplii in combination with micro-algae and one-day old enriched *Artemia* nauplii. After 21 days, the survival rates were 67.6% (*Artemia*), 75.6% (combination) and 61.3% (enriched *Artemia*). Although the combination of *Artemia* nauplii and micro-algae resulted in a higher survival rate, the difference with the other diets was not significant. Also the feeding rate of the larvae was determined with individual trials. Three feeding regimes were tested namely, 10, 30 and 50 *Artemia* nauplii per day per larva. At the end of the experiment a feeding regime was suggested based on the major moulting periods during the larval development. It was proposed that the amount of nauplii per larvae (per day) must be increased with 5 to 10 nauplii per larvae after each moulting event. Based on these results, we took the first steps to rear *C. crangon* larvae successfully in captivity. This research will enable us to provide an effective larval rearing system to maximize production of healthy post-larvae for further potential grow-out settings.

Understanding the regulatory mechanisms for sex determination and colour pattern genes in zebrafish

S. Hosseini, M. Gültas, L. Hahn, A.O. Schmitt, H. Simianer and A.R. Sharifi

University of Goettingen, Department of Animal Sciences, Albrecht-Thaer-Weg 3, 37075, Goettingen, Germany; shahrbanou.hosseini@uni-goettingen.de

The transcriptional regulation of gene expression in higher organisms is essential for different biological processes. These processes are regulated by transcription factors (TF) and their combinatorial interplay, which are essential for complex genetic programs and regulation of the transcriptional machinery in tissue. The role of specific TF interactions and pathways in gene regulatory mechanisms has not been studied for sex determination and sex-associated colour pattern genes in zebrafish with respect to sexual attractiveness, which is relevant for fish species with sexual diversity and plasticity in nature during global climate change. In order to address this limited knowledge, we first performed the transcriptome analysis (RNA-Seq) of 48 gonad and caudal fin tissue samples of adult zebrafish for understanding the underlying molecular mechanism of association between sex and colour genes. We then performed a systematic approach for the detection of specific and shared regulatory mechanisms between sex and colour patterns, which combines identification of: (1) sex determination and colour pattern candidate gene sets; (2) gene set-specific TF interactions by analysing the promoter region; (3) overrepresented pathways and master gene regulators to investigate the upstream processes; and (4) comparison of the results of both phenotypes (sex and colour) for their shared regulatory mechanisms; (5) monitoring the significant TF interactions using their expression level in RNA-Seq dataset. We found 27 and 15 TF interactions as significant for sex and colour patterns, respectively. From these, seven interactions have been identified as shared regulatory elements for both phenotypes, in which SOX9, GATA4, SP1 and STAT6 are known to play an important role for sex determination and coloration. Furthermore, the investigation of upstream processes provides pivotal knowledge about a shared master regulator (PKACA) for both phenotypes, which plays a fundamental role for proper explanation of our main hypothesis about the association between sex determination and colour patterning genes in zebrafish.

Quantification of non-additive genetic effects in Nile tilapia using both pedigree and genomics

R. Joshi[1], J.A. Woolliams[2,3], T.H.E. Meuwissen[3] and H.M. Gjøen[3]

[1]GenoMar Genetics AS, Bryggegata 3, 0250, Oslo, Norway, [2]University of Edinburgh, Midlothian, EH25 9RG, United Kingdom, [3]NMBU, Arboretveien 6, 1433, Ås, Norway; rajesh.joshi@genomar.com

The increasing availability of cost-efficient genomic resources allows the dissection of the phenotypic variances observed in breeding populations into different genetic components. In this study we compare these estimates to pedigree-based ones in Nile tilapia. The objective of this study was to partition the observed variances into additive and non-additive genetic components in a Nile tilapia sub-population designed to separate these components. Modifications were made in a commercial Nile tilapia selective breeding population through factorial mating with reciprocal cross and artificial fertilisation with communal rearing of the hatched larva to minimise the confounding of effects common to fullsibs. 6 different phenotypes [Body weight at harvest (BWH), body depth (BD), body length (BL), body thickness (BT), fillet weight (FW) and fillet yield (FY)] were available for 2,524 individuals, with pedigree over 22 generations. Among them, genotypes (n=39,927 SNPs after filtration) were available for 1,119 individuals. Univariate pedigree and genomic based BLUP were compared for the 6 traits using alternative univariate models. The basic model used was an animal model, which was gradually expanded to a model with additive, dominance, maternal and first order epistatic interactions effects by adding each effect as random effects in a heuristic approach. The non-additive genetic variance was found to be almost entirely additive by additive epistatic variance, although in pedigree studies this source is commonly assumed to arise from dominance. The estimates of the additive by additive epistatic ratio (P<0.05) was found to be 0.15 and 0.17 in the current breeding population using genomic data. In addition, the study found maternal variance (P<0.05) for BD, BWH, BL and FW explaining approximately 10% of the observed phenotypic variance. The study also estimated negative effects of inbreeding, with 1.1, 0.9, 0.4 and 0.3% decrease in the trait value with 1% increase in the individual homozygosity for FW, BWH, BD and BL, respectively. The inbreeding depression and lack of dominance variance was consistent with an infinitesimal dominance model.

Genomic selection in Nile tilapia increases prediction accuracy and gives unbiased estimates of EBVs

R. Joshi[1], A. Skaarud[1], M. De Vera[1], A.T. Alvarez[1] and J. Ødegård[2]
[1]GenoMar Genetics AS, Bryggegata 3, 0250, Oslo, Norway, [2]AquaGen AS, Torgard, 7462 Trondheim, Norway; rajesh.joshi@genomar.com

Genomic selection has shown to increase the predictive ability and genetic gain, across all the species, including the aquaculture. Hence, the aim of this paper is to compare the predictive abilities of pedigree and genomic based models in univariate and multivariate approaches, with the aim to utilise genomic selection in Nile tilapia breeding program. 48,960 SNP genotypes on 1,444 animals phenotyped for body weight at harvest (BW), fillet weight (FW) and fillet yield (FY) with 14 generations of pedigree was available for analysis after quality control. Estimated breeding values (EBVs and GEBVs) were obtained using traditional(P) BLUP and genomic(G) BLUP models, using both univariate and multivariate approaches. Prediction accuracy and biasness was evaluated using 5 replicates of 10-fold cross-validation using different cross-validation methods with both univariate and multivariate approaches. Further, impact of these models and approaches on the genetic evaluation was assessed based on the ranking of the selected animals. GBLUP univariate models were found to increase the prediction accuracy and reduce biasness of prediction compared to other PBLUP and multivariate approaches. GBLUP models were found to increase the prediction accuracy by ~20% for FY, >75% for FW and >43% for BW, compared to PBLUP models. Using GBLUP models caused major re-ranking of the selection candidate, with no significant difference in the ranking due to GBLUP univariate and GBLUP multivariate approach. Genomic selection is beneficial to Nile tilapia breeding program as it increases prediction accuracy and gives unbiased estimates of the breeding values compared to the pedigree. It is recommended to use GBLUP univariate approach in the routine genetic evaluation for the commercial traits in Nile tilapia.

GWAS and accuracy of predictions for resistance against IPN in rainbow trout

M.L. Aslam[1], S. Valdemarsson[2], A.A. Berge[2] and B. Gjerde[1]
[1]Nofima, AS, Breeding and Genetics, Osloveien 1, 1433 Ås, Norway, [2]Osland Stamfisk AS, Osland brygge, 5962 Bjordal, Norway; luqman.aslam@nofima.no

Infectious pancreatic necrosis (IPN) is one of the most prevalent and contagious disease which causes huge economic lose in rainbow trout production. Current study was designed to estimate genetic variation for resistance against IPN, map quantitative trait loci (QTL), and explore potential of marker assisted/genomic selection against IPNV. To reach these specific objectives, a rainbow trout population of 119 full-sib families (the offspring of 56 sires and 119 dams, ~25 sibs per family) was produced from the breeding nucleus of Osland Stamfisk. The fry were challenged using bath challenge model at VESO Vikan with IPNV-R-L5 strain of IPNV. Mortalities were recorded during the course of challenge test which was terminated when asymptote in cumulative mortalities was reached at day 25 with 65.2% observed mortalities. Tissue samples were collected for all the recorded fish (dead/live) for DNA extraction, genotyping, and genomic analyses. Estimation of genetic parameters, genome-wide association analysis (GWAS), and predictions of EBVs were performed using a linear mixed animal model. The estimated heritability for resistance to IPNV using genomic information was moderate (0.32). GWAS revealed three QTLs on three different chromosomes OMY1, OMY6, and OMY13 including 9, 5, and 10 SNPs respectively, presenting significant association to the trait with P-value crossing genome-wide Bonferroni corrected threshold $P \leq 2.80 \times 10^{-07}$. The proportion of total genetic variance explained by the three SNPs (with single top most SNP taken from each of the three QTLs) was ~37% of the total genetic variance. Accuracies of prediction of EBVs using models with genomic vs pedigree information exhibited 44-54% increase in accuracy. These results show that marker assisted selection using the significant SNPs may prove an efficient cost effective strategy to develop a rainbow trout population with higher survival and robustness against IPNV. Genomic selection gives opportunity to perform both within and between family selection with both higher selection accuracy and intensity and thus higher genetic gain.

Effects of group mate relatedness on body weight and variability of body weight in Nile tilapia

J. Marjanovic[1,2], H.A. Mulder[2], H.L. Khaw[3] and P. Bijma[2]
[1]*Swedish University of Agricultural Sciences, Department of Animal Breeding and Genetics, P.O. Box 7023, 75007 Uppsala, Sweden, [2]Wageningen University & Research, Animal Breeding and Genomics, P.O. Box 338, 6700 AH Wageningen, the Netherlands, [3]WorldFish, Jalan Batu Maung, 11960 Bayan Lepas, Malaysia; jovana.marjanovic@wur.nl*

In a wide range of animal species relatives show better social behaviour to each other than to unrelated conspecifics, for example food sharing and reduced aggressiveness. Such behaviour has the potential to increase fitness and may have evolved through the process known as kin selection. In addition to fitness benefits, reduced competition in agricultural populations may also increase performance. In aquaculture populations, competition for feed is an issue, as it reduces growth and inflates phenotypic variability among individuals. In domestic Nile tilapia the coefficient of variation (CV) of body weight is ~60%, which is high and undesirable. A potential way to reduce competition, and increase yield and uniformity of trait values in Nile tilapia, is to utilize the consequences of past kin selection, i.e. the potential evolution of kin discrimination and cooperative behaviour among relatives. However, it is almost entirely unknown whether relatedness among group mates in Nile tilapia leads to higher growth rates and better uniformity. In this study we compared two experimental treatments: rearing of fish in kin groups vs rearing in non-kin groups. In the kin groups all individuals were full-sibs. In the non-kin groups half of the individuals were full-sibs from the same family as the individuals in the kin group, while the other half were unrelated individuals. Six families were used in the experiment. We analysed average body weight (BW), and standard deviation and CV of BW, between the two treatments. Results of our study show that individuals had significantly higher BW in groups composed of kin (8.6±2.6 g), indicating that domestic Nile tilapia may exhibit kin-biased behaviour. The effect of relatedness on BW was higher for males than females, beyond the scaling effect, which may be related to a higher level of competition among males. There was no difference in variability of BW between both treatments. Aquaculture farming may benefit in yield by rearing individuals in groups composed of relatives.

Genetic variation for thermal sensitivity in growth and lice tolerance of Atlantic salmon

H.L. Khaw[1], S.A. Boison[2], L.-H. Johansen[3], M.W. Breiland[3] and P. Sae-Lim[4]
[1]*Nofima AS, Breeding and Genetics, P.O. Box 210, 1431 Ås, Norway, [2]Mowi ASA, P.O. Box 4102 Sandviken, 5835 Bergen, Norway, [3]Nofima AS, Fish Health, P.O. Box 6122, 9291 Tromsø, Norway, [4]Pathum Thani Aquatic Animal Genetic Research & Development Center, 39 mu 1, Klong Ha, Klong Luang, 12120 Pathum Thani, Thailand; hooi.ling.khaw@nofima.no*

Climate change is one of the major global concerns with no exception to aquaculture. In Atlantic salmon, growth is very much dependent on ambient temperature. Thermal stress can result in growth reduction when the ambient temperature is outside the optimum range (11-14 °C). Apart from growth, sea lice infection is also of great concern in the salmon industry. Studies showed that an increase in sea surface temperature could change the spatial distribution of sea lice outbreak. In literatures on other fish species, researchers show that there is possibility for selective breeding on environmental sensitivity. Thus, the objective of this study is to quantify the heritable variation for thermal sensitivity in growth and lice tolerance of Atlantic salmon. A sea lice challenge test was setup, where 50 full-sib families of post-smolt were tested at temperature of 5, 10 and 17 °C (two replicates/temperature; 20 fish/ family/replicate), with equal sea lice density. Fin-clips were taken prior to the experiment. Thermal growth coefficient (TGC = [(end weight$^{1/3}$) – (initial weight$^{1/3}$)]/day×degree) and log lice count (ln(1 + lice count)) were calculated and used as the phenotypes. Multivariate model treating phenotypes at each temperature as different traits was fitted to estimate the genetic parameters using restricted maximum likelihood in WOMBAT. Genomic relationship matrix was used in the estimation. Heritabilities for TGC and log lice count ranged from 0.21±0.05 to 0.37±0.06 and 0.20±0.05 to 0.30±0.05, respectively. Results showed moderate to strong re-ranking for growth (r_g of 0.57 to 0.79) and lice tolerance (r_g of 0.41 to 0.66) at different temperatures, which indicated that there is genetic variation in thermal sensitivity. Therefore, there is possibility to select for thermal sensitivity in growth and lice tolerance in Atlantic salmon.

Haplotype-based genomic predictions in Atlantic salmon
S.A. Boison, S. Gonen, A. Norris and M. Baranski
Mowi ASA, P.O. Box 4102, 5835 Bergen, Norway; salomon.boison@mowi.com

The use of genomic information for selection in the Atlantic salmon has been reported to result in substantial increases in accuracy of selection and selection intensity. In several Atlantic salmon populations, genomic predictions have been based on single nucleotide polymorphism (SNP) alleles, however the use of haplotype alleles in genomic prediction may increase accuracy of selection. Therefore, the objective of this study is to evaluate the accuracy (correlation between estimated breeding values and phenotype) and bias of genomic prediction in Atlantic salmon using pedigree, SNP or haplotypes with fixed length in kilobase-pairs (100 kb, 250 kb, 500 kb, 1000 kb) or number of SNPs (5, 10, 20) of non-overlapping moving windows. Genomic best linear unbiased predictions (GBLUP) and Bayesian methods (BayesCπ with π=1 and 5%) were tested. A total of 907 fish generated from 32 sires and 73 dams were challenge tested for cardiomyopathy syndrome and genotyped with a non-commercial 55,735 SNP Affymetrix array. The phenotype used in this study was histology score (0 to 3) of the atrium at 45-days post-challenge test with the CMS virus. Parental genotypes were added to the offspring genotypes before phasing with Beagle v4.1. A family-based validation scheme where two full sibs are randomly selected from each full-sib family as validation (n=150) and the rest of the siblings as training (n=757) was repeated 20 times. Accuracy of genomic prediction using SNP alleles were significantly higher (18% – GBLUP and 28% – BayesCπ=5%) and genomic breeding values were less biased than using pedigree information. In general (SNP and haplotype allele-based models), BayesCπ models resulted in higher accuracy (~14%) than GBLUP models and BayesCπ=5% gave the highest prediction accuracy. Haplotypes constructed from 100 kb windows were marginally more accurate than SNP allele-based models (0.459 vs 0.462 for GBLUP; 0.526 vs 0.535 for BayesCπ=5%) and similar results were obtained from haplotypes constructed with 5 SNP markers. In conclusion, genomic information resulted in higher prediction accuracy than pedigree information, BayesCπ models gave significantly more accurate breeding values than GBLUP and finally using haplotypes in genomic prediction for Atlantic salmon resulted in marginal increase in prediction accuracy.

A mega-analysis using microbial prediction of individual performance across aquaculture species
G.F. Difford and E.H. Leder
Nofima, Breeding & Genetics, P.O. Box 210, 1431 Ås, Norway; gareth.difford@nofima.no

In recent years, a substantial number of small-scale studies have attempted to characterise microbial composition in aquaculture and wild species such as Atlantic Salmon and Atlantic Cod. Typically, these studies focus on effects of dietary interventions, life stages or anatomical sampling sites as well as differences between distinct populations. Often the focus is on describing microbial composition and finding a 'core microbiome' common to all individuals. Key limitations to microbial surveys are the use of correlation and lack of causation, so determinist factors shaping host microbial interactions can only be implied. Here we investigate the host microbiota interactions in shaping host response to external factors and differences between fish in phenotypic variation. We compile a database of more than 100+ freely available studies on key aquaculture species and use microbial relationship matrices to predict host phenotypes. The cross-validation prediction accuracy of microbial composition across vast spatial and temporal ranges with independent validation sets, shows promising and robust results ranging from (R^2 0.23-0.65) across host associated traits like (length, weight and age) as well as environmental traits (diet, salinity and thermal tolerance). These findings indicate that diet, environment and host shape host microbe interactions.

Imputation and genomic prediction accuracies for pancreases disease in Atlantic salmon

B.S. Dagnachew[1], M. Baranski[2] and S.A. Boison[2]
[1]Nofima, Breeding and Genetics, Osloveien 1, 1433 Ås, Norway, [2]Mowi ASA, Sandviksboder 77AB, 5835 Bergen, Norway;
binyam.dagnachew@nofima.no

Genomic selection (GS) can increase the rate of genetic gain in aquaculture breeding. Unfortunately, the cost of GS is high because a large number of individuals need to be genotyped in aquaculture breeding schemes. However, strategical imputation from lower to higher marker density could be considered for reducing cost. In this study, we quantified the potential of imputation from low marker density for cost effective genomic selection in Atlantic salmon breeding. A dataset from Mowi breeding population was obtained from a pancreases disease (PD) challenge trial. Genotypes and phenotypes from 7,645 individuals were available. From each family, parents were genotyped with ~ 53k SNP and the genotypes of the offsprings were strategically reduced to 0.5k SNP. The genotypes of the offsprings were imputed from 0.5k to ~53k SNPs and imputation accuracy was calculated as a correlation between true and imputed genotypes. Genomic breeding values (GEBV) were predicted using true and imputed genotypes. From each family, the phenotypes of 15% of the sibs were masked (i.e. validation individuals) and accuracies of prediction were calculated as correlations between GEBV and the masked phenotypes. The correlations were weighted by the inverse of the square root of heritability. The imputation accuracies were 0.66 when no parents were genotyped and 0.81 when only one of the parents (either dam or sire) was genotyped. When both parents were genotyped, the imputation accuracy increased to 0.90. There was no difference in imputation accuracy between genotyping only dams and only sires. The results also showed that there no difference in imputation accuracy between male and female offsprings. Genomic prediction accuracy was 0.85±0.03 when using true genotypes compared to 0.79±0.03 when using imputed genotypes. Similar trends were also observed for three other disease traits and two weight related traits. The results showed that imputation can be leveraged to minimize genotyping cost without losing too much prediction accuracy. It also observed that both parents needed to be genotyped to get a reasonable imputation accuracy.

Regression equation for loin eye area prediction through morphometric traits in *Colossoma macropomum*

C.A. Perazza[1], F.O. Bussiman[2], A.W.S. Hilsdorf[1] and J.B.S. Ferraz[3]
[1]Universidade de Mogi das Cruzes, Av. Dr. Cândido Xavier de Almeida e Souza, 200 Mogi das Cruzes, SP, 08780-911, Brazil, [2]Universidade de Sao Paulo/FMVZ, VNP, Rua Duque de Caxias Norte, 225, Campus da USP, Pirassununga, SP, 13635970, Brazil, [3]University of Sao Paulo/FZEA, Veterinary Medicine, Rua Duque de Caxias Norte, 225, Campus da USP, Pirassununga, SP, 13635970, Brazil; wagner@umc.br

The main commercial cut of Tambaqui (*Colossoma macropomu*m, Cuvier, 1818)) comprises the loin and ribs, so a new phenotype, Loin Eye Area (LEA) can become a target in breeding programs. Ultrasound tools allow to accurately collect that information, however the high cost of the equipment make it difficult to use on a large scale in small-sized properties. Studies with regression equations for farmed tropical fishes traits are still scarce and regression equations in aquaculture are limited to Nile tilapia, rainbow trout and salmon. This work aims to establish an equation that efficiently estimates LEA by morphometric body measurements. The study was performed in Biofish Aquaculture Company located northern Brazil. Animals were tagged with a Passive Integrated Transponder. The LEA data were measured using a portable digital ultrasound machine and the morphometric traits Body Weight (BWe), Standard Length (SL), Head Length (HL), Body Width (BWi), Head Height (HH), and Body Height (BH) were collected. Akaike Information Criterium (AIC) and the coefficient of determination (R^2) were performed to check the model adjustment. Some assumptions were verified, like verification of residual normality, multicollinearity effects, and homoscedasticity of error variance. The estimated regression equation was assessed using a different database (evaluation population). The most adjusted model (AIC value and R^2 0.841) included the morphometric trait BWe, BH and BWi with coefficients 2.075, 0.284 and 0.556, respectively. In the evaluation population, mean values of LEA from ultrasound (2.9716) and those estimated by the equation (2.962) were statistically identical (P=0.832). Spearman's coefficient using these two LEA estimates was 0.87, suggesting significant correlation between them. The BWe, BWi, and BH data that is already collected by Tambaqui fish farmers can be a straightforward way to predict LEA.

Development and application of a duplex PCR assay for detection of viruses in *Crangon crangon*

B. Van Eynde[1,2], O. Christiaens[1], D. Delbare[2], K. Cooreman[2], K.S. Bateman[3], G.D. Stentiford[3], A.M. Dullemans[4], M.M. Van Oers[5] and G. Smagghe[1]
[1]Faculty of Bioscience Engineering, Ghent University, Department of Crop Protection, Coupure Links 653, 9000 Ghent, Belgium, [2]Flanders research institute for agriculture, fisheries and food, Animal Sciences Unit-Fisheries, Ankerstraat 1, 8400 Oostende, Belgium, [3]Cefas, European Union Reference Laboratory for Crustacean Diseases, Barrack Road, The Nothe, Weymouth, Dorset, DT4 8UB, United Kingdom, [4]Wageningen Research, Business Unit Biointeractions and Plant Health, P.O. Box 16, 6700 AA Wageningen, the Netherlands, [5]Wageningen University, Laboratory of Virology, P.O. Box 16, 6700 AA Wageningen, the Netherlands; benigna.vaneynde@ilvo.vlaanderen.be

Disease outbreaks are a major problem associated with the development of shrimp aquaculture industry worldwide. Especially viruses can cause growth retardation and/or high mortality, resulting in major economic losses. During research to evaluate the potential of *Crangon crangon* as suitable candidate for aquaculture, histology of hepatopancreatic tissue of *C. crangon* caught along the Belgian Coast, revealed a typical pathology as described by Stentiford *et al*. This study describes a duplex PCR assay developed for the detection of CcBV, based on amplification of the CcBV lef-8 gene and E75 as an internal amplification control gene. PCR primers were designed based on preliminary genome sequencing information from the virus and transcriptomic data available for *C. crangon*. Sequencing of the resulting amplicon confirmed the specificity of this PCR assay. Finally, the duplex PCR assay was applied to samples of *C. crangon* hepatopancreas collected from several sites across its distribution range to screen for the presence of CcBV. Based upon PCR, the prevalence of CcBV was on average 87%, comparable to previous reports of high prevalence, based upon histology, in shrimp collected from UK sites. The development of this detection tool fo CcBV offers the ability to rapidly and reliably screen wild *C. crangon* to select healthy and specific pathogen-free shrimp and it will also provide a way to monitor th presence of CcBV in cultured *C. crangon* populations.

Seasonal abundance of the *Sparicotyle chrysophrii* and haematocrit values of the farmed sea bream

S. Colak[1], R. Baric[1], M. Kolega[1], D. Mejdandzic[1], A. Doric[2], B. Mustac[2], B. Petani[2], I. Zupan[2] and T. Saric[2]
[1]Cromaris d.d., Gazenicka ul. 4b, 23000 Zadar, Croatia, [2]University of Zadar, Department of ecology, agronomy and aquaculture, Trg kneza Viseslava 9, 23000 Zadar, Croatia; bpetani@unizd.hr

Gilthead sea bream *Sparus aurata* (Linnaeus, 1758) is one of the most important species in Croatian mariculture production. The increase in sea bream production in the last few years has been supported by the massive establishment of sea cages, although it has also contributed to the extension of pathogenic parasitic diseases. Monogeneans are ubiquitous and abundant in the aquatic environment. They are often responsible for economic losses in fish farming. Polyopisthocotylean monogenean *Sparicotyle chrysophrii* parasitise as ectoparasite on the gills of wild and cultured gilthead sea bream and it has already caused lethal epizootics in sea cages in the Mediterranean area. Disease is manifested by anxiety, loss of appetite, weight loss and severe fish anaemia. In this research we have monitored the occurrence of *S. chrysophrii* and the influence of parasite infestation on fish survival, growth, and health. The experiment was carried out at fish farm Cromaris d.d., within 7 commercial net cages 9×5 m and per 10,000 fish individuals of the same origin and initial weight. Sampling was performed once a month, which included biometric analyses of the total fish length and weight. Additionally, the prevalence of the infested sea bream with *S. chrysophrii* and overall *S. chrysophrii* abundance were determinated. Sea surface temperature and oxygen were also recorded during the experiment. The results showed that the number of *S. chrysophrii* was significantly higher ($P<0.05$) in May (2.2 ± 0.5), July (2.7 ± 0.6) and August (3.2 ± 0.6) than in December (0.9 ± 0.1). Also, haematocrit values were significantly higher in August (44.4 ± 2.1) and September (42.9 ± 1.2) than in period from December to April (December 28.1 ± 1.9; February 27.3 ± 1.2; April 28.5 ± 1.2) despite the overall parasite abundance. Possible causes of lower haematocrit values and lower parasite abundance may be due to the lower sea temperature in winter period but further analyses are needed.

Non-aerated ponds reduces variances and heritabilities compared to aerated ponds in Nile tilapia

S.B. Mengistu[1], H.A. Mulder[1], J.A.H. Benzie[2,3], H.L. Khaw[4] and H. Komen[1]
[1]*Wageningen University & Research, Animal Breeding and Genomics, P.O. Box 338, 6700 AH Wageningen, the Netherlands,* [2]*WorldFish, Jalan Batu Maung, 11960 Batu Maung, Malaysia,* [3]*University College Cork, School of Biological Earth and Environmental Sciences, Distillery Fields, North Mall, Cork, T23 N73K, Ireland,* [4]*Nofima, P.O. Box 210, 1431 Ås, Norway, Nofima, P.O. Box 210, 1431 Ås, Norway; samuel.mengistu@wur.nl*

Nile tilapia has been selectively bred for intensive production systems while most of the small-holder production still takes place in non-aerated ponds where large diurnal oxygen fluctuations are apparent and dissolved oxygen levels can drop to almost zero during the night and early morning. The objective of this study was to estimate heritabilities and genotype by environment interaction between aerated and non-aerated ponds for body weight in Nile tilapia. For this experiment, about 3,000 fish were produced by mass spawning and PIT tagged. Then, they were randomly distributed to an aerated (high oxygen) and a non-aerated pond (low oxygen). The grow-out period was 218 days. Body weight was recorded two times, at 163 days and at 218 days (harvest weight). Among the 1,500 stocked fish in each pond, 1,026 fish from aerated pond and 1,038 fish from non-aerated pond were genotyped by sequencing (GBS). Genomic relationship matrix was built using 11,929 SNPs. Heritabilities and genetic correlations were estimated using a bivariate mixed animal model. The heritabilities for harvest weight were 0.24±0.06 and 0.17±0.06 for aerated and non-aerated ponds, respectively. Additive genetic and residual variances for harvest weight in non-aerated pond were 73 and 56% lower, respectively, compared to additive and residual variances for harvest weight in aerated pond. Genetic correlations for harvest weight between the two environments were not estimable. Instead we estimated the genetic correlations between harvest weight and body weight at 163 days of age when growth in both environments was still comparable. The genetic correlations between harvest weight with body weight at 163 days were 0.92±0.06 and 0.87±0.06 in aerated and in non-aerated ponds, respectively. No aeration leads to large differences in variance and lower heritability compared to aerated ponds.

Use of oil, solvent-extracted meal and protein concentrate from *Camelina sativa* for rainbow trout

J. Lu[1], S.M. Tibbetts[2], S.P. Lall[2] and D.M. Anderson[1]
[1]*Dalhousie University, Faculty of Agriculture, Department of Plant and Animal Sciences, 58 Sipu (River) Road, Haley Institute, Truro, Nova Scotia, B2N 5E3, Canada,* [2]*National Research Council of Canada, Aquatic and Crop Resource Development Research Centre, 1411 Oxford Street, Halifax, Nova Scoita, B3H 3Z1, Canada; jinglu@dal.ca*

A 98-day feeding trial was conducted in a 24-tank freshwater flow-through system (14 °C) to evaluate the effects of using dietary camelina oil (CO), solvent-extracted camelina meal (SECM), and camelina protein concentrate (CPC) in rainbow trout fry diets (1.0 g/fish, 50 fish/tank). Triplicate tanks were randomly fed one of eight diets including a Camelina sative-free control diet, two diets containing CO at either 50 or 100% replacement of fish oil, three diets containing 6, 12, or 18% SECM, and two diets containing 6 or 12% CPC. Up to 100% of added fish oil can be replaced with CO in diets without negatively affecting growth performance, nutrient utilization and proximate whole-body carcass composition of rainbow trout fry after 98 days of feeding. Dietary inclusion of up to 18% of SECM or 12% of CPC did not have negative effects on growth performance, feed efficiency or proximate whole-body carcass composition. However, fish fed diets containing 18% SECM, required a longer acclimation period to the feed and, ultimately, gained less weight, presumably due to a poorer protein utilization efficiency in the early days of the trial.

Session 52

Theatre 1

Humans and horses: a community of destiny

V. Deneux
INRA, UMR Innovation, Bat 27, 2 place Viala, 34060 Montpellier Cedex 2, France; vanina.deneux@inra.fr

By bringing animals into the domus, we didn't only share our home, our life with them, we have set up a complex system of interactions that generate societal and even civilizational achievements. Our relations with animals are labour relationships: humans and animals, we subjectively invest in the production of goods or services. Also domestic animals are not part of our culture, they are agents of culture. The aim of this communication is to show, from the case of horses, how our relations of life and work with them have created and shaped our European societies. The horse is with the dog, the domestic animal with which we have the most diversified working relations. One of the first contributions of horses, its velocity, allowed the first reduction of time and space between human communities. He helped us in the fields, in the cities, he participated in the war effort and was an actor of European industrialisation. Today our societies are service societies therefore our working relationships with horses have adapted in this sense: sport, leisure, mediation, care, social links. Our urban heritage testifies to the importance of horses in cities until the Second World War: road sizes, old stable of forge buildings, hodonymia… Moreover, with each major societal change, our collaborations have adapted or even reinvented. Furthermore we can draw a parallel between the living and working conditions of humans and horses. For example, in the 19th century, the industrialisation of Europe led to the proletarianisation of both humans and horses (mines, transport, etc.). Similarly, we can draw a parallel between our current lifestyles and their consequences: urban and sedentary life, which have among other negative effects, obesity, digestive disorders, depression. In a process of socio-historical dynamics, I will present the major evolutionary phases, the breaks and the reinventions of our relations of life and work with horses. After having shown the great types of collaboration between humans and horses in history, I will address the societal functions of the horse in our current society. Finally, while there are threats to the future of our relations with domestic animals, I will address the political interests and issues of horse work in tomorrow's European societies.

Session 53

Theatre 1

Developing the wool sector as a tool for local/regional development

Y. Dupriez and K. Thollier
Filière laines, 21 rue de Transinne, 6890, Belgium; ygaelle@laines.be

The first questions about the use of local wool in Belgium (Wallonia) were raised in 2010; when several discussions pointed out the contradiction between the concern of local customers/investors in wool, and the fact that local sheep farmers had to send their underpaid wool to China to 'get rid' of it while consumers and enterprises of the sector import merino wool from Australia/New Zealand. The question was asked: would it be possible to use Belgian wool as a tool for local development, and how? Based on these facts, our project began developing activities around local wool with 5 main goals: raise the customer/citizen's awareness about the value of wool and its sector; protect the work of sheep farmers; protect the artisans' know-how and techniques; make wool a definitely modern resource on top of being natural and sustainable. In 2017, together with 11 partners, we launched an European-Interreg project to widen the scope of our activities to the neighbouring regions: France (Lorraine) and Luxembourg. The partners are working with farmers, shearers, crafts(wo)men, companies, designers, schools, local initiatives to: organise the sector and create a transnational cluster to raise awareness around wool; innovate in the Great Region and launch new products through: a wide market survey, the individual coaching of nearly 80 project leaders and the creation of more than 60 new products based on local wool, 2 design competitions, the use of sheep wool from natural reserves, a study and experiment for sound and heat insulation in public buildings, a design workshop, etc. – train on sorting, shearing, knowledge about wool for professionals, individuals, students; and implement a 'virtual collection system' communicate: raising awareness on wool and its qualities for the general public and for companies, create a touristic route, organise events, create a 'house of wool' etc. The wool is strong, breathable, thermoregulatory, absorbent of VOCs, insulation of hot, cold and noise, light, healthy, etc. The wool is noble, natural, renewable, durable, allowing multiple uses. The wool meets the challenges of today.

How to assess sustainability of animal fibre production – a case study on Cashmere in Mongolia

M. Wurzinger[1], L. Supper[1] and M. Purevdorj[2]
[1]*BOKU-University of Natural Resources and Life Sciences, Gregor-Mendel-Strasse 33, 1180 Vienna, Austria, [2]University of Agriculture and Life Science, Zaisan-53, Ulaanbaatar, Mongolia; maria.wurzinger@boku.ac.at*

There is an increasing awareness, but also demand from consumers for sustainably produced animal fibre products. Several methods to assess sustainability of agricultural production have been developed in the last years. There also some specific tools available for pastoralist systems, but not addressing all four dimensions of sustainability (environmental, social, economic and cultural) at the same time. A large proportion of wool and animal fibre is produced in low-input, extensive pasture-based systems by smallholder farmers all over the world – highly specialized systems with some unique characteristics. The aim of the study is to develop a first draft for a comprehensive framework to assess the four dimensions of sustainability (environmental, social, economic and cultural). In a first step a possible set of indicators for the four dimensions of sustainability are selected based on an extensive literature review. In a second step, a participatory, action-research approach is applied. The proposed framework is tested in two regions of Mongolia, where Cashmere goats are kept by nomadic livestock keepers. This step is a validation, but also a further development of the assessment tool. Integration of opinion from various stakeholders like livestock keepers, government representatives, but also processing industry is essential for the identification of useful and cheap assessment indicators. Therefore, individual interviews with different stakeholders are performed to get detailed information on current management system of Cashmere goats and about the value chain of Cashmere. In addition, focus group discussions are carried out to cross-validate obtained information and explore further opinions on the future of Cashmere production. In a next step, the assessment framework will be tested for other production systems to validate the applicability of the framework under other conditions.

A new wool-shedding sheep breed

D. Allain[1], B. Pena-Arnaud[1], D. Marcon[2], C. Huau[1], D. François[1] and L. Drouilhet[1]
[1]*INRA, UMR GenPhySE, Chemin de borde Rouge, Auzeville, 31326 Castanet Tolosan, France, [2]INRA, UE0332, Domaine de la Sapinière, 18390 Osmoy, France; daniel.allain@inra.fr*

In Europe, there is a growing interest for the use of breeds that shed their wool naturally. Indeed the weak relative value of meat and wool and the increasing shearing costs lead breeders to favour breeds that shed naturally annually. By using an original selection strategy, a new wool-shedding sheep breed was created in two consecutive steps. Firstly, a new genotype was created by introgressing the gene pool from the Martinik Hair, a hairy sheep known to shed, into the Romane breed through 4 consecutive backcrossing generations. For those animals, ability to shed and shedding extent increased by 75.4 and 130.2%, respectively, at 7 months of age compared to the Romane breed. Then the introgressed animals were selected during 8 generations on shedding extent over the body. To do so, a population of 150 ewes and 10 rams was selected on shedding extent estimated breeding values (EBV) under an annual breeding system with lambing period in early March. Fleece shedding extent was measured at mid-June in both adult ewes, and all lambs i.e. at the age of 3.5 months in lambs and once a year in ewes. All rams were replaced each year while ewe replacement varied from 30 to 50%. After 8 generations of selection of introgressed animals, all ewes showed a complete or nearly complete shedding of their fleece at mid-June: mean shedding extent was 89.3±17.9% in ewes indicating that annual shearing was not required anymore. In young lambs, mean fleece shedding extent measured at 3.5 months of age was 61.6±35.6%. Heritability estimate of shedding extent was high (0.38±0.04) and a genetic gain of 3 σg was observed after 8 generations of selection in this new breed deriving from the French Romane breed which has a high production potential. Breeding performances as well as lamb growth performances of this new breed were similar to those of the French Romane breed. Our selection strategy based on an introgression process of the gene pool of fleece shedding from a hairy moulting sheep is an interesting strategy to lead to a new shedding breed and/or improving adaptive traits in breeds having a high production potential. Current investigations are in progress to identify causal mutations in wool-shedding.

Application of the resistance test in assessing changes in lamb's wool depending on age

P. Cholewińska, A. Wyrostek, K. Czyż and D. Łuczycka
Wrocław University of Environmental and Life Sciences, Norwida 25, 51-625, Poland; paulina.cholewinska@upwr.edu.pl

The aim of the study was detected the differences between lambs wool depending on age using the electrical properties of wool. Animal material were lambs of the Olkuska breed from twin litters (n=10). The sample of wool was collected three times: 0 days old, 28 days old and 56 days old. The samples were analysed to check differences between electrical properties, heat resistance and thickness. The results of electrical properties based on resistance show the significant between wool on each ages. The highest resistance was characterized from 56 days old lambs wool. However, lamb wool from day 0 was characterized by the lowest resistance. The results obtained have a bearing on the remaining wool properties examined, such as heat resistance and thickness. Wool from 56 days old lambs had the highest diameter (25,57 μm) of fiber and heat resistance (WPC) at level 0,432 compare to 0 days old lambs with diameter 15,84 μm and heat resistance at level 0,219. Differences in the results obtained in this case show the influence of factors such as filling the hair with keratin, the amount of secreted sebum translates into the obtained results. In addition, when analysing wool with respect to age, one can notice a gradual increase in its thickness, heat-protection parameters which are influenced by factors mentioned earlier. However, the very low resistance of wool in the day 0 of lambs' life may also be the result of deposition of foetal remains and saliva of mothers, which contain a large amount of mineral salts increasing the electrical conductivity of hair. Examination of the coat in the future after proper refinement of algorithms can facilitate the assessment of the quality of the fibre, but also capture changes taking place under the influence of various factors such as infection and diet, because the study of electrical features detects changes at the molecular level.

Energetic and thermal adaptations of llamas (*Lama glama*) in the High Andes of Peru

A. Riek[1], A. Stölzl[2], R. Marquina Bernedo[3], T. Ruf[4], W. Arnold[4], C. Hambly[5], J.R. Speakman[5,6] and M. Gerken[2]
[1]Friedrich-Loeffler-Institut, Institute of Animal Welfare and Animal Husbandry, Dörnbergstr. 25/27, Celle, Germany, [2]University of Göttingen, Department of Animal Sciences, Albrecht-Thaer-Weg 3, Göttingen, Germany, [3]Centro de Estudios y Promoción del Desarrollo del Sur, Umacollo, Arequipa, Peru, [4]University of Veterinary Medicine Vienna, Research Institute of Wildlife Ecology, Savoyenstraße 1, Vienna, Austria, [5]University of Aberdeen, Institute of Biological and Environmental Sciences, AB24 2TZ, Aberdeen, United Kingdom, [6]Chinese Academy of Sciences, Institute of Genetics and Developmental Biology, 100101, Beijing, China, P.R.; alexander.riek@fli.de

The aim of our long-term study was to determine if the llama, exhibits seasonal and/or daily adjustment mechanisms in energy expenditure (EE) and body temperature (T_b) in its natural habitat of the high Andes in South America. For that purpose, seven non-pregnant llama dams were kept under a traditional Andean herding system at the High Andean Plateau (approx. 4,500 m a.s.l.) in Peru. We used a telemetry system continuously measuring T_b, activity and distances travelled over a period of 308 days. Additionally, we measured EE using the doubly labelled water method at four different time points for two weeks each. The EE varied between the four different measurement periods. The lowest and highest individually recorded EE were 11.6 MJ/d and 28.3 MJ/d, respectively. EE was correlated with daily distances travelled (r=0.48, P<0.001) and was substantially lower compared to true ruminant species. The average daily T_b amplitude was very variable and reached on some days more than 3 °C. There was a significant positive relationship between T_b and ambient temperature (T_a) over the entire study period (R^2=0.40, P<0.001) as well as between T_b and T_a amplitude (R^2=0.63, P<0.001). In conclusion, our study provides evidence that llamas kept at the Andean High Plateau have an exceptionally low EE compared to true ruminant species. Furthermore, llamas seem to adjust their T_b according to T_a which must involve some trade-offs that allow them to save energy instead of keeping their T_b constant.

Calving date as a potential breeding objective to manage the reproductive seasonality in alpacas

A. Cruz[1], I. Cervantes[2], R. Torres[1], N. Formoso-Rafferty[2], R. Morante[1], A. Burgos[1] and J.P. Gutiérrez[2]
[1]*INCA TOPS S.A, Miguel Forga 348, Arequipa, Peru,* [2]*Universidad Complutense de Madrid, Departamento de Producción Animal, 28040 Madrid, Spain; gutgar@vet.ucm.es*

The reproductive seasonality is the norm in the production of alpacas in Peru. It conditions the female reproductive performance, as a female not becoming pregnant within the reproductive period will remain open for a year, with the corresponding losses in the productivity of the farm. Births are preferred to occur in the middle of the reproductive season as there will not be sufficient food resources if they are too soon, and the animals will lose the opportunity of becoming pregnant for a year if they occur too late, having in addition an increase in the probability of enterotoxemia. Management of this scenario has been traditionally treated by analysing the age at first calving and calving interval traits, but these traits do not account with the search of the optimal date. Thus, the objective of this work was to estimate genetic parameters for the calving date trait to be considered as a possible selection objective itself or combined with its variability. Data from complete reproductive campaigns from Pacomarca experimental farm from second half of 2001 to 2018 were used, counting with 6,533 records of calving date from huacaya genetic type that were recorded as the number of days elapsed from the previous reproductive interseason date as reference. There were 9,575 animals in the pedigree file. Homogeneity and heterogeneity models were fitted both including sex, year and colour as systematic effects, age of the female as covariate and genetic and permanent environmental random effects. The calving date heritability become 0.09 (sd=0.02) with a repeatability of 0.20 (sd=0.01), suggesting the possibility of breeding animals to delay or advance the date. The genetic coefficient of the variance for the variability was 0.10 (sd=0.05), a value that suggests also the possibility of selecting animals performing the birth preferably close to the middle of the season. Genetic correlation between the mean and the variability was -0.84 (sd=0.12), suggesting that concentrating births would delay them within the season.

Status of the alpaca breeding in Poland

A. Slawinska[1], A. Zmudzinska[1] and M. Wierzbicki[2]
[1]*UTP University of Science and Technology, Department of Animal Biotechnology and Genetics, Mazowiecka 28, 85-084 Bydgoszcz, Poland,* [2]*Polish Alpaca Breeders Association, Nowogrodzka 31, 00-511 Warszawa, Poland; slawinska@utp.edu.pl*

Alpaca (*Vicugna pacos*) is gaining recognition as a valuable livestock species in Poland. Rapidly growing interest in alpaca breeding has been driven by their unique properties as fibre-producing as well as companion and therapy animals. The standards and guidelines for alpaca breeding in Poland are under ongoing legislative procedure. Polish alpaca breeders are associated in the Polish Alpaca Breeders Association (PABA, http://pzha.pl/). The breeders associated in PABA own approximately 1,500 animals, 700 of which are currently registered and included in the database. The vast majority of the alpacas (91%) registered in PABA belong to huacaya breed and the remaining 9% are Suri. The demographic structure indicates that the majority of the animals are young females. Sex proportion is approximately two females to one male. 80% of alpacas were born in 2012 or later. The fleece colour distribution ranges from white (37%), different shades of beige and fawn (27%), brown (21%) to black (10%) and grey (6%). The appaloosa and multicolour alpacas are extremely rare in Poland (about 1%). In the near future, PABA will continue registering animals in the database. The animals registered in PABA database will be characterized with respect to their genetics, fitness and the fibre quality. In this paper we are going to present the case studies regarding health and reproduction that were described in Polish alpaca herds. At the early stages of alpaca breeding in Poland fostering collaboration between breeders and scientist plays a critical role. Strategic implementation of the research program on Polish alpacas will contribute to the knowledge on camelids and their adaptation to European environment.

A mixed-model analysis including hidden inbreeding depression load

L. Varona[1,2], J. Altarriba[1,2], C. Moreno[1,2] and J. Casellas[3]

[1]*Instituto Agroalimentario de Aragón, c/ Miguel Servet 177, 50013 Zaragoza, Spain,* [2]*Universidad de Zaragoza, Anatomía, Embriología y Genética Animal, c/ Miguel Servet 177, 50013 Zaragoza, Spain,* [3]*Universitat Autònoma de Barcelona, Departament de Ciència Animal i dels Aliments, Plaça cívica, 08193 Bellaterra, Spain; lvarona@unizar.es*

Inbreeding is caused by the mating of related individuals and the most common consequence is inbreeding depression. Studies have detected heterogeneity in inbreeding depression among founder individuals, and a recent study has developed a procedure for predicting individual inbreeding depression load. The objectives of this study were to develop a computational strategy and a software program that can analyse large data sets and to extend the model by including the genetic covariance between additive and inbreeding depression load effects, which were applied to two data sets of weaning weight from Spanish beef cattle populations (Pirenaica and Rubia Gallega). The datasets comprised 35,126 and 75,194 live weights between 170 d to 250 d of age, and pedigrees of 308,836 and 384,434 individual-sire-dam entries for Pirenaica and Rubia Gallega, respectively. A FULL model that included heterogeneous individual inbreeding loads whose mean may differ from zero and covariance between additive and inbreeding load effects, and five reduced models (R1 to R5) were implemented and compared based on the logarithm of the conditional predictive ordinate (logCPO). The reduced models set to zero the mean (R1, R3 and R5) and the variance (R4 and R5) of the inbreeding depression loads and the covariance with the additive breeding values (R2, R3, R4 and R5). The models that had the best fit were FULL in Pirenaica and R3 in Rubia Gallega. The model comparisons and the variance component estimations confirmed heterogeneous inbreeding depression loads in both populations, with a posterior mean (and standard deviation) percentage of variance explained by inbreeding depression load associated with an inbreeding of 0.10 of 0.028 (0.010) and 0.043 (0.008) in Pirenaica and Rubia Gallega, respectively. A strong negative correlation (-0.566±0.077) between additive effects and inbreeding depression loads was detected in the Pirenaica, but not in the Rubia Gallega population.

Genomic and pedigree methods to analyse inbreeding depression in Basco-Béarnaise rams

Z.G. Vitezica[1], I. Aguilar[2], J.M. Astruc[3] and A. Legarra[1]

[1]*INRA, GENPHYSE, 31326 Castanet-Tolosan, France,* [2]*Instituto Nacional de Investigación Agropecuaria, (INIA), 90200, Canelones, Uruguay,* [3]*Institut de l'Elevage, 149 rue de Bercy, 75595 Paris, France; zulma.vitezica@inra.fr*

Inbreeding depression is caused by increased homozygosity of individuals and it is involved in the decrease of performance and fitness of the animals. The goal of this study is to estimate inbreeding depression for semen traits in Basco-Béarnaise rams. Genomic (e.g. using SNP-by-SNP approach) or pedigree-based inbreeding can be used to obtain inbreeding coefficients (f). It is possible to detect inbreeding depression (b) from the regression of the phenotype (y) on f, as $y=X\beta \pm fb+u+e$, where u is the vector of breeding values. In sheep, inbreeding from pedigree tends to be underestimated due to missing pedigrees. Inbreeding coefficients, f, were estimated using four methods. The first method used the traditional pedigree-based inbreeding. The second used pedigree-based inbreeding but accounting for non-zero relationships for unknown parents. The third method used the metafounder relationships in pedigree-based inbreeding. The fourth method calculates inbreeding from a combined relationship matrix using pedigree and genotypes with metafounder relationships. Breeding values, u, for the first three methods were estimated with the classical BLUP using phenotypes and pedigree information and the fourth method was Single Step GBLUP with metafounders.

Identification of homozygous haplotypes associated with reduced fertility in Finnish Ayrshire cattle

K. Martikainen and P. Uimari
University of Helsinki, Department of Agricultural Sciences, Koetilantie 5, 00014 University of Helsinki, Finland;
katja.martikainen@helsinki.fi

Inbreeding depression, the reduction in phenotypic performance due to inbreeding, is assumed to be mostly caused by homozygosity for deleterious recessive alleles. Inbreeding gives rise to continuous segments of homozygous genotypes known as runs of homozygosity (ROH), which are shown to be enriched for these deleterious recessive alleles and can therefore be linked to inbreeding depression. However, within a region, the unfavourable effect is likely due to a single unique ROH genotype while the other ROH genotypes may have a neutral or even a favourable effect on the phenotype. Our goal was to identify homozygous haplotypes within ROH with an unfavourable effect on female fertility traits in the Finnish Ayrshire population. Data were available for 13,712 Finnish Ayrshire females that were genotyped with low density panel (8k) and imputed to 50k density. Haplotypes within ROH were detected using the Unfavourable Haplotype Finder software, which first scans the genome using a sliding window approach to identify haplotypes within ROH associated with an unfavourable effect on the phenotype. Then, the haplotypes identified are tested for their significance using a linear mixed model. Multiple haplotypes with statistically significant association with reduced fertility were detected across the genome. The haplotypes with the highest statistical significance were associated with 9.2, 9.5, and 17.1 days longer interval from calving to first insemination for first, second, and third parity cows, respectively. Corresponding results for the interval from first to last insemination were 12.8, 13.2, 29.6, and 37.7 days longer interval for heifers and for cows of first, second, and third parity, respectively. Similarly, for the number of inseminations the effect was 0.3, 0.5, 0.6, and 1.0 more inseminations, respectively. Haplotypes identified in this study with detrimental effect on fertility when homozygous can be used to more efficiently control the effects of inbreeding in the Finnish Ayrshire breeding program.

Comparative approach of missing homozygosity and GWAS in Brown Swiss cattle

F.R. Seefried[1], I.M. Häfliger[2], M. Spengeler[1] and C. Drögemüller[2]
[1]Qualitas AG, Chamerstrasse 56, 6300 Zug, Switzerland, [2]Institute of Genetics, Vetsuisse Faculty, Bremgartenstrasse 109A, 3001 Bern, Switzerland; franz.seefried@qualitasag.ch

Ten years ago the framework of genomic selection has been implemented in the two Braunvieh populations (Original Braunvieh and Brown Swiss) leading to large-scale genotype data. Recently, genotyping of females has been increased in numbers. In addition, the number of re-sequenced animals has been increased as well. Both significant changes enable now a comparative approach between identification of haplotypes with reduced or missing homozygosity and GWAS for specific traits like female fertility, birth, and growth related traits, which are closely related to the missing homozygosity hypothesis. Analyses of missing homozygosity were performed based on the recent cow reference genome (ARS-UCD1.2). Genome-wide scans and sliding window approach were applied using genotyped trios on ~47k density. A total number of 41 regions were found with significant reduced homozygosity. For each region the longest haplotype with significant lack of missing homozygosity was chosen for association analysis. Additive genetic effects were estimated for these haplotypes using de-regressed EBVs. In a parallel approach, QTL regions were identified during GWAS runs using imputed WGS data using de-regressed EBVs as phenotypes and based on the former assembly (UMD3.1). Dosage data, summing up the genotype probabilities were used as genomic input data. GWAS was performed using EMMAX applying the mixed model approach and a significance threshold at $P<1\times10^{-6}$ for QTL detection. In order to minimise false positives due to the low threshold, we excluded QTL with a single associated SNV. This led to the detection of 62 putative QTL. The QTL were distributed across 20 chromosomes and 28 population-trait combinations. Since some QTL-regions were overlapping between different traits, a final number of 42 different QTL regions was obtained. Finally, bringing both approaches together revealed 6 regions that have been detected as well as during homozygosity screens and during GWAS runs. Genome sequencing is ongoing in order to detect the underlying causative sequence variants. Tools for avoiding risk matings on farm level will follow subsequently after causative variants have been detected.

Missing ROH – recommendations for tuning PLINK in ROH analyses
R. Meyermans, W. Gorssen, N. Buys and S. Janssens
Livestock Genetics Department of Biosystems, KU Leuven, Kasteelpark Arenberg 30 box 2472, 3001 Leuven, Belgium;
roel.meyermans@kuleuven.be

PLINK is probably the most used program for SNP genotype analyses and runs of homozygosity (ROH) studies, both in human and animal populations. ROH analyses have become the state-of-the-art method for inbreeding assessment. In PLINK, ROH analyses require the --homozyg flag, with the specification of several input parameters. Often, the choice of the parameter values seems not been thought through and default settings are used whereas the choice of some parameters can strongly affect the results. We investigated the effects of pruning of SNPs (both for linkage disequilibrium and for low minor allele frequencies) prior to the ROH analysis and examined the effect of the --density parameter (the minimal average SNP density in a ROH) on over thirty populations of pigs, cattle, sheep and horses. An improper use of these parameters can have severe impact on the outcome of the ROH analysis. For example, we found that a badly chosen SNP density parameter can underestimate genomic inbreeding by a fourfold. We recommend the examination of these criteria and to report them in publications. We also encourage the use of the proposed genome coverage parameter to examine the ROH analysis' validity.

Dissection of inbreeding in line 1 Hereford
E.A. Hay[1], P. Sumreddee[2], S. Toghiani[1], A.J. Roberts[1] and R. Rekaya[2]
[1]USDA Agricultural Research Service, Fort Keogh Livestock and Range Research Laboratory, 243 Fort Keogh, Miles City, MT, 59301, USA, [2]University of Georgia, Department of Animal and Dairy Science, 425 River rd, Athens, Ga, 30602, USA; elhamidi.hay@ars.usda.gov

Line 1 Hereford cattle were linebred at the USDA Agricultural Research Service station in Miles City, Montana since the early 1930s. This population was started by 2 paternal half sib bulls and 50 cows. The objective of this study is to evaluate the extent of inbreeding in this close population and assess its effect on growth and fertility traits. Inbreeding was estimated based on pedigree (FPED) and genomic information. Using 785 animals genotyped for 30,810 single nucleotide polymorphisms (SNP), three estimates of genomic inbreeding coefficients were derived based on the diagonal elements of the genomic relationship matrix using estimated (FGRM) or fixed (FGRM0.5) minor allele frequencies or runs of homozygosity (ROH) (FROH). The pedigree consisted of 10,186 animals and was used to calculate FPED. The traits analysed were birth weight (BWT), weaning weight (WWT), yearling weight (YWT), average daily gain (ADG), and age at first calving (AFC). The number of ROH per animal ranged between 6 and 119 segments with an average of 83. The shortest and longest segments were 1.36 and 64.86 Mb long, respectively. The average inbreeding was 29.2, 16.1, 30.2, and 22.9% for FPED, FGRM, FGRM0.5, and FROH, respectively. Inbreeding depression analyses showed that a 1% increase in an animals pedigree based inbreeding resulted in a decrease of 1.20 kg, 2.03 kg and 0.004 kg/d in WWT, YWT, and ADG, respectively. Using genomic inbreeding, similar effects on growth traits was observed although with varying magnitude of the effects across the methods. Across all genomic inbreeding measures; WWT, YWT, and ADG decreased by 0.21 to 0.53 kg, 0.46 to 1.13 kg, and 0.002 to 0.006 kg/d for each 1% increase in genomic inbreeding, respectively.

G×E and selection response for type traits in a local cattle breed using a reaction norm approach

C. Sartori[1], F. Tiezzi[2], N. Guzzo[3] and R. Mantovani[1]
[1]*University of Padova, DAFNAE, Viale dell'Universita 16, 35020 Legnaro (PD), Italy,* [2]*NC State University, Department of Animal Science, 120 W Broughton Dr, 27607 Raleigh, NC, USA,* [3]*University of Padova, BCA, Viale dell'Universita 16, 35020 Legnaro (PD), Italy; cristina.sartori@unipd.it*

This study aimed to investigate G×E and response to selection (R) for type traits under diverse environmental conditions (EC) in which the local Rendena cattle is reared. Target traits were 24 linear type traits (LTT) routinely collected in primiparous cows by trained classifiers, and 3 factor scores (FS: muscularity, udder conformation, udder volume) obtained from factorial analysis of 10 LTT. Traits referred to 15 years and belonged to 8,538 cows sired by 807 bulls. The EC considered were: (1) farm altitude (plain/hill/mountain); (2) housing (tie-stall/loose housing); (3) feeding system (hay based/total mixed ration); (4) summer pasture (yes/no). The G×E term was valued under the reaction norm approach, i.e. a first step based on a single trait animal model (M1), and a subsequent random regression sire model step (M2) accounting the solutions of the herd-year-classifier (HYC) effect of M1. These latter were used as environmental covariate to assess the intercept G and the slope G×E across HYC levels. A multivariate response to selection (R) in different EC was computed for the FS and the other traits under the actual selection program: milk, fat and protein yield, and performance test traits in young candidate bulls. Solutions of M1 showed that sires' EBVs varied in different EC; a correlation from 83 to 93% was found, e.g. between the average EBVs of daughters feed with the hay based system vs the total mixed ration. The sum of G, G×E and their covariances (M2) was on average the 25% of the total phenotypic variance (P), with the sole G×E term accounting from 6 to 41% of P. A greater R for muscularity and udder conformation was found in farms in plain and without summer pasture, under loose housing and total mixed ration as feed; udder volume showed opposite results. Considering all traits used for R, a greater response was found for mountain farms, loose housing, the hay based feeding and no summer pasture. Results of the study suggested to consider the G×E in genetic evaluation of local breeds like Rendena.

Investigations on G×E interactions at single trait and index level in Brown Swiss dairy cattle

M. Schmid[1], A. Imort-Just[1], R. Emmerling[2], C. Fuerst[3], H. Hamann[4] and J. Bennewitz[1]
[1]*University of Hohenheim, Institute of Animal Sciences, Garbenstrasse 17, 70599 Stuttgart, Germany,* [2]*Bavarian State Research Centre for Agriculture, Institute for Animal Breeding, Prof.-Duerrwaechter-Platz 1, 85586 Poing, Germany,* [3]*ZuchtData EDV-Dienstleistungen GmbH, Dresdner Strasse 89/19, 1200 Vienna, Austria,* [4]*State Office for Spatial Information and Land Development Baden-Wuerttemberg (LGL), Stuttgarter Strasse 161, 70806 Kornwestheim, Germany; markus_schmid@uni-hohenheim.de*

This study was conducted to infer whether genotype by environment (G×E) interactions affect German-Austrian Brown Swiss sire breeding values for milk production and functional traits and hence may lead to a re-ranking of bulls in different environments. The evaluations were based on phenotypes derived from the routine genetic evaluation and pedigree information including 185,439 animals. Bivariate sire models were applied to assess the effect of the (binary scaled) environments farming system (organic, conventional) and farm location (altitude threshold: 800 m above sea level). G×E effects were analysed using the results of the bivariate models on the trait level as well as on the selection index level, which considered also different relative economic weights. In a further approach, milk energy yields were calculated from first lactation performances of daughters to be a continuously scaled environmental descriptor of the herd production level and fitted in random regression reaction norm sire models. Depending on the trait, environment and model, the datasets contained phenotypes of 44,648 to 56,759 cows which descend from 56 to 638 sires. Bivariate analyses of a respective trait measured in different environments resulted in genetic correlations between 0.87 on the trait level (protein yield depending on the farm location) and 0.95 (milk yield and fat yield depending on the farm system), which indicate no significant G×E interactions on the trait level. G×E effects on the index level were more pronounced, but still on a low level. In agreement with this, the reaction norm models revealed no substantial G×E effects. These findings imply that breeding values obtained from the routine German-Austrian Brown Swiss sire evaluation might be valid in all sorts of production environments.

Genomic selection breeding program benefits more than traditional one in presence of G×E

L. Cao[1], H. Liu[1], H.A. Mulder[2], M. Henryon[3,4], J.R. Thomasen[5], M. Kargo[1,6] and A.C. Sørensen[1]
[1]Aarhus University, Molecular Biology & Genetics, Blichers Alle 20, 8830 Tjele, Denmark, [2]Wageningen University
& Research Animal Breeding and Genomics, P.O. Box 338, 6700 AH Wageningen, the Netherlands, [3]School of Animal
Biology, University of Western Australia, 35 Stirling Highway, Crawley, WA 6009, Australia, [4]Seges, Danish Pig Research
Center, Axeltorv 3, 1609 Copenhagen V, Denmark, [5]VikingGenetics, Ebeltoftvej 16, 8960 Randers SØ, Denmark, [6]Seges,
Agro Food Park 15, 8200 Aarhus N, Denmark; cao.lu@mbg.au.dk

We tested the hypothesis that cooperation between breeding programs that use genomic selection (GS) in the presence
of genotype by environment interaction (G×E) has larger benefits in terms of selection response and rate of inbreeding
than traditional pedigree-based selection (PS). We tested this hypothesis by simulating two breeding programs,
each with their own cattle population and environment. Cooperation referred to combined genetic evaluation and
across-environment selection of bulls. Four scenarios either with genomic-based or pedigree-based evaluation or
within-environment or across-environment selection of bulls were performed. The genetic correlation (r_g) was varied
between 0.5 and 0.9. Scenarios were compared for selection response, rate of inbreeding, proportion of foreign bulls,
and the split-point genetic correlation (SP-r_g). If r_g was lower than SP-r_g, then each breeding program selects foreign
bulls only in the short-term, while if r_g is higher than SP-r_g then each breeding program selects foreign bulls in the
short and long-term. Cooperation of GS breeding programs resulted in 60% more selection response, 24% reduction
in rate of inbreeding and a lower SP-r_g (0.85-0.875 versus 0.9) than cooperation between PS breeding programs.
The minimum rate of inbreeding appeared at the SP-r_g. This study shows the benefits of cooperation of genomic
breeding programs in the presence of mild G×E.

Genotype by environment interaction using reaction norms for milk yield in Brazilian Holstein cattle

V.B. Pedrosa[1], H.A. Mulim[1], A.A. Valloto[2], L.F.B. Pinto[3], A. Zampar[4] and G.B. Mourão[5]
[1]Ponta Grossa State University, Department of Animal Science, Av General Carlos Cavalcanti 4748, 84030-900 Ponta
Grossa Paraná, Brazil, [2]Paraná Holstein Breeders Association, Rua Professor Francisco Dranka, 608, 81200-404
Curitiba Paraná, Brazil, [3]Federal University of Bahia, Av. Adhemar de Barros s/n, 40170-115 Salvador Bahia, Brazil,
[4]State University of Santa Catarina, Rua Beloni Trombeta Zanin, 680-E, 89815-630 Chapecó Santa Catarina, Brazil,
[5]University of São Paulo, Av. Pádua Dias, 11, 13418-900 Piracicaba São Paulo, Brazil; vbpedrosa@uepg.br

This study aimed to verify the genetic behaviour of a Holstein cow population in the country's highest productivity
milk basin, located in southern Brazil. It was evaluated the animal's genetic response to the variations in environmental
temperature, through the analysis of the effects resulting from the genotype by environment interaction, based on
reaction norms. Therefore, milk production data was collected for 67,360 primiparous cows from the database of the
Paraná Holstein Breeders Association in Brazil, with the purpose of evaluating the temperature effect, considered
as an environmental variable, distinguished under six gradients over the region. The random regression model was
adopted, utilizing the fourth order under the Legendre polynomials, applying the mixed models of analysis by the
REML method, and using the WOMBAT software. The total milk production on average was estimated at 8,374.82
kg, signifying the superior productivity of the animals evaluated, against the Brazilian output level. The coefficients
of heritabilities were found in the low to moderate range, from 0.18 to 0.23, displaying a decline with a rise in the
temperature, highlighting the influence it exerted on the heritabilities. However, all the genetic correlations between
the gradients were above 0.80, in the range of 0.873 to 0.998, revealing no remarkable interaction between the
genotype and environment. Based on the genetic correlation results, it can be concluded that the application of the
temperature effect in the models of genetic analysis of the evaluated population is not required. Funded by Fundação
Araucária, CAPES and CNPq.

Genotype by environment interaction for fat yield in Brazilian Holstein cattle using reaction norms

V.B. Pedrosa[1], H.A. Mulim[1], A.A. Valloto[2], L.F.B. Pinto[3], A. Zampar[4] and G.B. Mourão[5]
[1]Ponta Grossa State University, Animal Science, Av General Carlos Cavalcanti 4748, 84030-900 Ponta Grossa Paraná, Brazil, [2]Paraná Holstein Breeders Association, Rua Professor Francisco Dranka, 608, 81200-404 Curitiba Paraná, Brazil, [3]Federal University of Bahia, Av. Adhemar de Barros s/n, 40170-115 Salvador Bahia, Brazil, [4]State University of Santa Catarina, Rua Beloni Trombeta Zanin, 680-E, 89815-630 Chapecó Santa Catarina, Brazil, [5]University of Sao Paulo, Av. Pádua Dias, 11, 13418-900 Piracicaba Sao Paulo, Brazil; vbpedrosa@uepg.br

In order to verify the genetic behaviour of Holstein cattle in different environmental conditions, data of 67,360 primiparous cows from Paraná Holstein Breeders Association in Brazil was used, considering six gradients of temperature in the Southern part of the country. The assessed region is the main productivity area for milk and solids yields in Brazil, but, on the other hand, is one of the most diverse in terms of climate. For that reason, a genotype by environment interaction study for milk fat was conducted, based on reaction norms, to analyse the impact of the temperature over genetic expression of cows. The random regression model was adopted, utilizing the fourth order under the Legendre polynomials, applying the mixed models of analysis by the REML method, and using the WOMBAT software. The fat yield average was estimated at 277,81 kg, demonstrating the superior milk fat productivity of the animals evaluated, against the Brazilian output level. The coefficients of heritabilities were moderate, ranging from 0.21 to 0.27, where the higher heritability was achieved in a lower temperature region, and the lower heritability was found at the higher temperature region. However, all the genetic correlations between the gradients were above 0.80, in the range of 0.89 to 0.99, revealing no remarkable interaction between the genotype and environment. This result indicates that the application of the temperature variable in the models of genetic analysis for fat yield in the population evaluated is not required. Funded by Fundação Araucária, CAPES and CNPq.

Pedigree-based genetic relationships and genetic variability in Estonian Holstein population

T. Kaart[1,2], K. Smirnov[3] and H. Viinalass[1,2]
[1]BioCC OÜ, Kreutzwaldi 1, 51006 Tartu, Estonia, [2]Institute of Veterinary Medicine and Animal Sciences, Estonian University of Life Sciences, Kreutzwaldi 62, 51006 Tartu, Estonia, [3]Institute of Mathematics and Statistics, University of Tartu, J. Liivi 2, 50409 Tartu, Estonia; tanel.kaart@emu.ee

In 2018 there were 515 dairy farms with 81,821 dairy cows (96% of all dairy cows) in milk recording in Estonia. (Results of Animal Recording in Estonia 2018). To achieve faster genetic progress and due to economic reasons lot of sperm of the same bulls is imported and still existing own bulls in Estonia are actually sons of foreign top bulls. This in turn may reduce the genetic variability and selection possibilities. The task of present study was to assess the bulls' use in Estonian Holstein population, estimate the genetic relationships between more frequently used bulls and study the genetic variability among heifers. Two different databases obtained from the Estonian Livestock Performance Recording Ltd were used: (1) the whole Estonian dairy cows' pedigree (as of March 2018) containing more than 1.7 million animals; and (2) the pedigree-based estimated milk production breeding values of all heifers born in 2017 and 2018 (49,522 animals). Based on pedigree data the inbreeding coefficients of all animals and the additive genetic relationship coefficients between more frequently used nowadays bulls were calculated using R package pedigreemm. Based on heifers´ data the genetic variability between and within farms was estimated. The pedigree analysis indicated, that the average inbreeding of animals born in last decade was 0.023, however, among 325 bulls with more than 100 daughters (with at least one daughter born after 2013) there were 2,122 pairs of bulls with their additive genetic relationship higher than 0.125 and 420 pairs of bulls with their additive genetic relationship higher than 0.25. Heifers born in last two years were daughters of 693 bulls in total, whilst 18.2% of heifers were daughters of only four different sires and 31 bulls were sires of 50% heifers. Analysis of variance of farms with more than 200 heifers showed, that even there existed quite large variability between farms in heifers estimated breeding values, the within farm variability was much higher considering 85% of the total variability.

Intensity of selection in the genome dual-purpose cattle breeding in Slovakia

R. Kasarda, B. Olšanská, A. Trakovická, O. Kadlečík and N. Moravčíková
Slovak University of Agriculture, Department of Animal Genetics and Breeding Biology, Tr. A. Hlinku 2, 949 76 Nitra, Slovak Republic; radovan.kasarda@uniag.sk

The aim of this study was to analyse the distribution of runs of homozygosity (ROH segments) across cattle genomes in Slovakia. We analysed Slovak spotted and Slovak Pinzgau cattle as two Slovak autochthonous breeds, as well as Holstein, Simmental and Ayrshire, which were used for grading -up of these two breeds in the past. The analysis was implemented up to 12 generations (>4 Mb). The genome-wide data for in total of 420 animals were obtained by two platforms: Illumina Bovine SNP50v2 BeadChip and ICBF International Dairy and Beef v3. After quality control, the database consisted of 35,934 common SNPs across all of the analysed breeds. Selection signals were determined by application of genome-wide significance threshold limit set on 10%. The average length ROH on autosomal genome was: for Slovak Spotted 7.12 Mb, Slovak Pinzgau 8.59 Mb, Holstein 10.45 Mb, Simmental 8.8 Mb and for Ayrshire 9.1 Mb. Largest ROH segment 15.62 Mb was observed in Ayrshire and shortest ROH segment was observed in Holstein (6.11 Mb). Presence of selection signals in Slovak Spotted cattle on BTA 5 and 6, Slovak Pinzgau cattle on BTA6, Holstein on BTA13, Simmental on BTA9 and for Ayrshire on BTA19 in regions of QTL. Generally, the result show that the regions with selection signals contained the QTLs affecting dairy production, beef production, reproduction and muscle development.

Genomic characterisation of Czech Holstein cattle

L. Vostry[1], H. Vostra-Vydrova[2], N. Moravcikova[3], B. Hofmanova[1], J. Pribyl[2], Z. Vesela[2] and I. Majzlik[1]
[1]Czech University of Life Science Prague, Kamycka 129, 16500 Prague, Czech Republic, [2]Institut of Animal Science, Pratelstvi 815, 10400 Prague, Czech Republic, [3]Slovak University of Agriculture in Nitra, Tr. A. Hlinku 2, 94976 Nitra, Slovak Republic; vostry@af.czu.cz

Recent fast development of molecular techniques has enabled estimation of a genomic inbreeding coefficient (F_{ROH}), which reflects realized autozygosity and can be further partitioned to chromosomes and chromosomal segments. In this work, the genetic diversity based on the runs of homozygosity in local bulls of Holstein breed was described. Further the genetic diversity of local Holstein bulls was describe by potential autozygosity islands as regions with extreme runs of homozygosity frequency. A total of 30 bulls were analysed. All of animals have been genotyped using Illumina BovineSNP50 BeadChip V2 and after quality control the final dataset was composed of 52,530 autosomal loci. The inbreeding coefficient (F_{ROH}) was expressed as the length of the genome covered by runs of homozygosity divided by length of the autosomal genome covered by all SNPs (2.5 Gb). The runs of homozygosity segments greater than 4 Mb (F_{ROH}>4 Mb) cover in average 9.3% of the genome, the average runs of homozygosity segments greater than 8 MB was 6.4%, whereas inbreeding estimates based on runs of homozygosity >16 Mb (F_{ROH}>16 Mb) achieved 3.6%. This value signalized high recent inbreeding in analysed population. The autozygosity islands based on the 4 MB runs of homozygosity threshold resided in eleven regions on chromosomes 2, 4, 8, 7, 10, 13, 16, 19, 20, 21 and 25. The strongest signals was found on chromosomes 10 and 20. The high values of genomic inbreeding coefficient indicated high loss of genetic diversity in local subpopulation of Holstein cattle breed, which is caused probably by common ancestry used in breeding program at farms.

Characterization of quarter milk yield losses during clinical mastitis in dairy cows

I. Adriaens[1], L. D'Anvers[1], B. De Ketelaere[1], W. Saeys[1], K. Geerinckx[2], I. Van Den Brulle[3], S. Piepers[3,4] and B. Aernouts[5]

[1]KU Leuven, Mechatronics, Biostatistics and Sensors, Kasteelpark Arenberg 30, 3001 Heverlee, Belgium, [2]Hooibeekhoeve, Province of Antwerp, Hooibeeksedijk 1, 2440 Geel, Belgium, [3]M-Team UGent, Salisburylaan 133, 9820 Merelbeke, Belgium, [4]Mexcellence BVBA, Gontrode Heirweg 168, box 01.02, 9820 Merelbeke, Belgium, [5]KU Leuven, Livestock Technology, Kleinhoefstraat 4, 2440 Geel, Belgium; ines.adriaens@kuleuven.be

Clinical mastitis has a huge impact on the profitability of modern dairy farms, not in the least because of its negative effect on milk production. Due to the lack of a conclusive system for automated detection, clinical mastitis is often noticed later on farms equipped with an automatic milking system (AMS) compared to conventional farms. This has an impact on the production recovery and clinical and bacteriological cure. As a first step towards an automated monitoring of recovery and cure through the measurement of various milk parameters, this study aimed to characterize milk yield dynamics during clinical mastitis in AMS farms. Using a quarter level approach thereby allows to discriminate between the systemic (e.g. due to general illness, energy needed for the immune response and decreased feed intake) and local (e.g. caused by the direct pathogen effects in the affected udder quarter) impact of the intramammary infection. First, the effect of severity of the clinical mastitis and the causal pathogen on milk yield losses and yield dynamics during recovery was studied. To this end, all clinical mastitis cases were closely followed on the experimental farm 'de Hooibeekhoeve' in Flanders, Belgium over a period of 3 years. For each detected case, the severity was estimated via clinical scores of the milk, udder and cow, while the presence of a mastitis pathogen in quarter milk samples taken at day 0, 7, 14 and 21 after detection of the infection was assessed through bacteriology. Furthermore, weekly DHI data were collected for monitoring of the somatic cell count. Milk losses were calculated using a linearized mixed model and Bayesian predictions. The dataset consisted of 38 cases, for which the quarter level losses showed a clear distinction between the infected and non-infected quarters, strongly dependent on the causal pathogen and severity of the infection.

Timing and consistency of luteolysis detection using two algorithms for milk progesterone monitoring

I. Adriaens[1], W. Saeys[1], N.C. Friggens[2], O. Martin[2], K. Geerinckx[3], B. De Ketelaere[1] and B. Aernouts[4]

[1]KU Leuven, Mechatronics, Biostatistics and Sensors, Kasteelpark Arenberg 30 box 2456, 3001 Heverlee, Belgium, [2]INRA, Modélisation Systémique Appliquée aux Ruminants, 16, Rue Claude Bernard, 75231 Paris, France, [3]Province of Antwerp, Hooibeekhoeve, Hooibeeksedijk 1, 2440 Geel, Belgium, [4]KU Leuven, Livestock Technology, Kleinhoefstraat 4, 2440 Geel, Belgium; ines.adriaens@kuleuven.be

Milk progesterone can be used to closely monitor the fertility status of dairy cows as it allows identification of prolonged postpartum anoestrus, oestrus, cysts, embryonic mortality and pregnancy. Although milk progesterone can be measured quite reliably on farm, the measured concentration is subject to sampling errors, assay variability, calibration issues and varying concentrations of milk components, all influencing the determined concentration. In order to interpret the values correctly, and subsequently provide valuable fertility information to the farmers with a high sensitivity and specificity, an appropriate data processing and monitoring algorithm is required. In this study, we compared two algorithms to monitor fertility based on milk progesterone, using both on-farm and simulated data. The first algorithm is considered as the current state of the art and consists of a smoothing multiprocess Kalman filter combined with a fixed threshold for decision making (MPKF+T). The second algorithm, progesterone monitoring algorithm using synergistic control (PMASC), models the progesterone time series using two sigmoidal growth functions, after which decisions are made based on individual process control charts. The on-farm data included 1,843 oestrus alerts, and number and timing of these were compared for both algorithms. Because of the time-lag inherent to the smoother, alerts by MPKF were given 20±16 h later than the first out-of-control measurement identified by PMASC. The analysis on well-controlled simulated data, in which variability was included both at cycle and at measurement level, showed that PMASC was less dependent on the actual level of progesterone during oestrus, and that using a model-based indicator calculated from PMASC allowed for a reliable and consistent estimation of true luteolysis time, even when the sampling rate during luteolysis was decreased with 66%.

Sensor fusion for early detection of mastitis in dairy cattle
C. Davison, C. Michie, A. Hamilton, C. Tachtatzis and I. Andonovic
University of Strathclyde, Electronic and Electrical Engineering, 204 George Street, G1 1YW, United Kingdom;
christopher.davison@strath.ac.uk

Mastitis can have a significant effect on both animal health and farm profitability. Electrical conductivity (EC) measurements have been studied in order to determine their suitability to accurately determine the onset of mastitis. Here we examine the combination of EC measurements with milk constituent measurements and activity based sensors in order to assess the potential for providing an early indication of the onset of mastitis. The study was carried out on an Agri-EPI satellite farm, chosen as an exemplar Scottish commercial herd. The herd consisted of 200 Holstein-Freisian cows of varying parity and stage of lactation, each fitted with an Afimilk behavioural monitoring collar, providing aggregate measures of rumination and eating time budgets. The cows had ad-libitum access to 4 Fullwood M^2erlin automatic milking robots, enabling the cow to self-regulate her milking schedule. The robots provided per-quarter conductivity and yield measurements per milking, and bulk fat, protein, and lactose milk content. Trained observers monitored the herd for visual signs of mastitis, and these observations were contrasted with automated alerts from activity based sensors, milk constituents, and from EC measurements. Activity based measurements were found to alert before changes were detected using either milk constituents or EC measurements.

NIR sensor for online quality analysis of raw milk: from concept to validation
B. Aernouts, J. Diaz-Olivares, I. Adriaens and W. Saeys
KU Leuven, Biosystems, Kleinhoefstraat 4, 2440, Belgium; ben.aernouts@kuleuven.be

As milk contains valuable information on the cow's metabolic status, regular analysis of the produced milk is a very efficient way to monitor cow and udder health. Near infrared (NIR) spectroscopy has been shown to be a valuable technique for rapid, non-destructive and on-line analysis of the raw milk composition. NIR reflectance is relatively easy to measure in-line, but provides insufficient information on the lactose in the milk serum. The NIR transmittance results in superior prediction of milk fat, protein and lactose, but is challenging because of fat globules and the water fraction (±86%) respectively scattering and absorbing the NIR light. Accordingly, this measurement can only be performed when the milk is bypassed and presented in a thin cuvette (≤2 mm). Based on this knowledge, an NIR (950-1,650 nm wavelength range) sensor prototype was designed and built to measure the NIR reflectance and transmittance simultaneously on a raw milk sample in a 2 mm flow-through cuvette. The sensor was connected to an automatic milking system (AMS) that collected a representative milk sample (±50 ml) for each individual milking and shared information on its status: teat cleaning, teat cup attaching/removing, milking, system cleaning and status of milk pump and bypass-valve. As such, the sample handling and NIR measurements by the sensor were synchronized with the AMS (VMS, DeLaval) and measurements were performed without any human intervention. The sensor was automatically cleaned together with the scheduled cleaning cycles of the AMS. The sensor collected the raw milk spectra of 55 cows for each milking (150 milkings per day on average) over a period of 10 weeks. Calibration models (partial least squares) were built on the first 150 milk samples for which reference analyses were available. The robustness of the original calibration was evaluated every week against the reference analysis of 2 samples per cow. Online evaluation of the Hotelling T^2 and Q-residual statistics was used to identify sample spectra which could potentially contribute to the (re)calibration of the NIR sensor. This resulted in an overall prediction error (RMSE) smaller than 0.1% (w/w) for milk fat (range 1.5-8%), crude protein (2.7-4.5%) and lactose (4-5.2%).

Online quality analysis of raw milk: potential of miniature spectrometers
B. Aernouts, J. Diaz-Olivares, I. Adriaens and W. Saeys
KU Leuven, Biosystems, Kleinhoefstraat 4, 2440, Belgium; ben.aernouts@kuleuven.be

Over the past 50 years, genetic selection and improved feed and management practices resulted in an increased milk production per cow lactation. As these modern cows are prone to production-related disorders, they need to be monitored closely to guarantee animal health and welfare. On the other hand, the increasing size of dairy farms results in less time available for each individual animal. Therefore, tools are needed to provide reliable and useful information on individual cows and help the dairy farmer to optimize the animal management while reducing the work load. Milk contains valuable information on the metabolic and nutritional status of dairy cows. Therefore, regular analysis of the produced milk is an efficient way to monitor cow health and welfare. Frequent milk analysis is only feasible if it is performed on the farm with a minimal investment of labour and resources. Different studies, both in the lab and on the farm, indicate that near infrared (NIR) spectroscopy holds the potential for rapid, non-destructive and on-line analysis of the raw milk composition. Nevertheless, commercial NIR detectors are typically costly (>10k euro) and relatively large, not allowing for implementation in existing milking systems. In this study, we evaluated the ability of an affordable Micro-Electro-Mechanical-System (MEMS) based NIR sensors to analyze raw milk and predict the fat, protein and lactose concentration. These NIR MEMS sensors use Fabry-Perot Interferometers for wavelength scanning, which enables compact sensor packaging and fast signal collection. The performances of the miniature spectrometers (different wavelength ranges) were compared against a benchtop (diode array) NIR spectrometer in an on-farm validation study. In total, 200 raw milk samples with a varying composition were measured. Calibration models (partial least squares) were built on the first 100 milk sample, while the other samples were used as independent validation set. Although the prediction error for the main milk components was significantly lower in the benchtop instrument (RMSE<0.1) compared to the miniature spectrometers (RMSE=0.11-0.13), the latter already provides a valuable indication of the milk composition.

Milk urea: a promising non-invasive indicator of SARA in dairy goats in the context of PLF?
S. Giger-Reverdin and C. Duvaux-Ponter
UMR Modélisation Systémique Appliquée aux Ruminants, Inra, AgroParisTech, Université Paris-Saclay, 16 rue Claude Bernard, 75005 Paris, France; sylvie.giger-reverdin@agroparistech.fr

Sub-acute ruminal acidosis (SARA) is a major nutritional disease occurring in intensive ruminant production due to the use of diets rich in highly fermentable ingredients in order to meet requirements. Its detection is not easy, because with a same diet, there is a large among-animals variation in rumen fermentation and pH values. In the context of precision livestock farming (PLF), it is worthwhile to find a non-invasive indicator of SARA. Eight mid-lactating cannulated goats received successively two diets, in a cross-over design, differing by their percentage of concentrate (30 (30D) vs 60% (60D) on a dry matter basis). After two weeks of adaptation, rumen juice was sampled five times, on two different days and for each diet: before the morning feed allowance and every two hours thereafter. Rumen pH was measured immediately. For each goat, and with each diet, the minimum pH value (pHmin) was calculated. Uraemia was measured on one plasma sample taken before feed allowance. Milk urea was determined by mid-infrared spectroscopy (MIR) on the days of rumen sampling. pHmin varied from 5.60 to 6.36 for 30D and from 5.14 to 6.08 for 60D with a large among-animals variation within a diet. pHmin was highly correlated to uraemia (pHmin=6.81-1.70 Uraemia (g/l), n=16, ngoats=8, r^2=0.90, RSD=0.156) and to milk urea (pHmin=6.97-1.84 milk urea (g/l), n=16, ngoats=8, r^2=0.83, RSD=0.206). Uraemia and milk urea increased when pHmin decreased. Within a diet a decrease in pH is linked to an increase in fermentations with higher VFA and ammonia concentrations that might be different between animals. The flux of ammonia through the rumen wall increases with VFA concentration and is transformed into urea in the liver which diffuses passively to the mammary gland and is transformed into milk urea. Thus, when rumen pH decreases, milk urea increases. Moreover, milk urea depends on the diet composition (crude protein, fermentable organic matter or excess of degradable nitrogen in the rumen). Milk urea measured by MIR is a promising criterion in the context of PLF because it is an individual non-invasive criterion potentially able to detect SARA at a low cost within a herd.

Evaluation of factors affecting passive transfer of immunity to calves using a Brix refractometer

A. Soufleri[1], G. Banos[1,2], N. Panousis[1], N. Siachos[1], G. Arsenos[1] and G.E. Valergakis[1]

[1]Faculty of Veterinary Medicine, School of Health Sciences, Aristotle University of Thessaloniki, Box 393, 54124, Thessaloniki, Greece, [2]Scotland's Rural College and Roslin Institute University of Edinburgh, EH25 9RG, Midlothian, United Kingdom; nsiachos@vet.auth.gr

The objective was to evaluate the factors affecting passive transfer of immunity to newborn calves, by estimating serum total protein with a Brix refractometer. The study was conducted in 10 commercial dairy herds in Northern Greece from February 2015 to September 2016. One thousand and thirteen Holstein cows and their newborn calves were used. Colostrum quality, estimated as total solids concentration was assessed with a digital Brix refractometer and expressed in % Brix values. Calves were blood sampled 24-48 hours post-calving and serum total protein was estimated in % Brix values (>7.8%: indicative of adequate passive transfer of immunity) using the same refractometer. Season, quantity of 1st colostrum meal fed to calves and interval between calving and 1st meal were recorded. Calf serum Brix values were analyzed using a univariate general linear model with season, colostrum quality, quantity fed and interval between calving and 1st meal as fixed effects, and farm as random effect. Moreover, the effect of colostrum quality and quantity fed on calf serum Brix values was assessed with a logistic binary regression. Mean Brix values for colostrum and calf serum were 25.8±4.6% and 8.7±1.0%, respectively. Farm, season, colostrum quality and quantity fed had statistically significant effects on calf serum Brix values (P<0.01). The latter were higher in summer and autumn (P<0.05); higher colostrum quality and quantity fed also resulted in higher calf serum Brix values (P<0.001). A 1% increase in colostrum Brix value, increased the probability of calf serum Brix value to exceed the target threshold of 7.8% by 12.9% (P<0.001). Moreover, a 1 kg increase in colostrum quantity fed, increased the probability of calf serum Brix value to exceed the above target threshold by 98.0% (P<0.001). Even moderate improvements in colostrum quality and quantity fed greatly enhances passive transfer of immunity.

Phenotypic analysis of blood β-hydroxybutyrate predicted from cow milk mid-infrared spectra

A. Benedet, M. Franzoi and M. De Marchi

University of Padova, Department of Agronomy, Food, Natural Resources, Animals and Environment (DAFNAE), Viale dell'Università 16, 35020 Legnaro (PD), Italy; anna.benedet@phd.unipd.it

Hyperketonaemia is one of the most costly metabolic disorders affecting dairy cows in early lactation and it is commonly diagnosed through the concentration of β-hydroxybutyrate (BHB) in blood. Although the reference test relies on the laboratory quantification of BHB in blood, several cow-side ketone tests are available for an easier and semi-quantitative measurement of this ketone body in field conditions. However, blood sampling and analysis are time-consuming and expensive, and the procedure is stressful for the animal. Recently, mid-infrared spectroscopy (MIRS) prediction models for blood BHB have been developed. Mid-infrared spectroscopy is commonly used to determine milk composition in routine milk recording systems and, thus, at population level. The present study aimed to investigate phenotypic variation of predicted blood BHB in the most important Italian cattle breeds. The first test-day record (between 5 and 35 days in milk) of Holstein-Friesian, Brown Swiss and Simmental cows reared in multi-breed herds was considered from a large spectral dataset collected between 2011 and 2016. Blood BHB prediction model previously developed was applied to spectral data. Sources of variation of predicted blood BHB were investigated using a linear mixed model, including the fixed effects of breed, herd, year and month of sampling, stage of early lactation, parity and interactions between the main effects. Random factors were cow and residual. Blood BHB concentration was greatest for Holstein-Friesian and lowest for Brown Swiss cows. Multiparous cows had greater blood BHB concentration than primiparous cows. On average and especially for multiparous animals, the greatest BHB concentration was detected between 5 and 15 days in milk, except for Simmental cows, which exhibited a quite flat trend of BHB concentration between 5 and 35 days in milk. The greatest BHB values were observed during spring and early summer for all considered breeds. Environmental effects identified in this study will be included as adjusting factors in within-breed genetic evaluations of predicted blood BHB.

Capitalizing on European collaboration for large-scale screening for ketosis in dairy cows

C. Bastin[1], M. Calmels[2], A. Werner[3], U. Schuler[4], X. Massart[5], C. Grelet[6], M. Gelé[7], D. Glauser[8] and L.M. Dale[3]
[1]Walloon Breeding Association, 5590, Ciney, Belgium, [2]Seenovia, 72000, Le Mans, France, [3]LKV – Baden-Wuerttemberg, 70190, Stuttgart, Germany, [4]Qualitas AG, 6300, Zug, Switzerland, [5]European Milk Recording EEIG, 5590, Ciney, Belgium, [6]Walloon Agricultural Research Centre, 5030, Gembloux, Belgium, [7]French Livestock Institute, 75012, Paris, France, [8]Suisselab AG, 3052, Zollikofen, Switzerland; cbastin@awenet.be

This study presents: (1) the development of calibration equations to predict non-esterified fatty acids (NEFA) and β-hydroxybutyrate (BHB) blood concentrations from bovine milk mid-infrared (MIR) spectral data; and (2) the practical use of such equations to implement large-scale screening for ketosis in dairy herds. For the calibration of blood BHB and NEFA, the trial dataset contained standardized milk spectral data associated with blood NEFA content (n=1,516) and blood BHB content (n=735). Data were collected in France, Germany and Switzerland during the OptiMIR and Acetone projects. The calibration model showed intermediate coefficients of determination (R^2=0.73 for BHB and R^2=0.79 for NEFA). The R^2 of cross-validation and the ratio of standard deviation to standard error of cross-validation were respectively 0.68 and 1.76 for BHB whereas 0.77 and 2.06 for NEFA. These calibration models were released into the European Milk Recording network but used differently by its members. In Germany, predictions of these models were used to validate the ketosis risk indicator obtained from the KetoMIR2 model (D4Dairy project). Correlations of this ketosis risk indicator with blood BHB predictions and blood NEFA predictions were 0.59 and 0.78 respectively. In the Walloon Region of Belgium, the currently used ketosis risk indicator combines 4 MIR predicted traits (fat to protein ratio, contents in milk of C18:1 cis-9, acetone and BHB) and classifies the cows in 4 groups of risk. It showed a specificity and a sensitivity slightly higher than 70%. MIR predictions of blood NEFA and BHB will be used to reinforce this indicator and improve its performances. In conclusion, European collaboration permitted the development, the release and a country specific implementation of MIR predictions in order to achieve large-scale screening for ketosis in dairy cows through milk recording.

Indicators of subacute ruminal acidosis in early lactating dairy cows in commercial dairy farms

M. Zschiesche[1], A. Mensching[1], A.R. Sharifi[1], H. Jansen[1,2], D. Albers[2] and J. Hummel[1]
[1]Georg-August-University Goettingen, Animal Sciences, Kellnerweg 6, 37077 Goettingen, Germany, [2]Agricultural Chamber Lower-Saxony, Mars-la-Tour-Str. 6, 26121 Oldenburg, Germany; marleen.zschiesche@uni-goettingen.de

Subacute ruminal acidosis (SARA) is a digestive disease that affects animal welfare. Because of feeding changes and general stress especially early lactating cows are at a high risk of developing SARA. Definitions rely on thresholds derived from pH measurements in the ventral rumen. In recent years, the continuous recording using intra-reticular pH boluses has become increasingly common, but this on-farm measurements are cost-intensive. Thus, indicators associated to the pH are in the focus of research. In this study, reticulum-pH was measured on ten commercial dairy farms, and associated with potential indicators. On each farm ten early lactating cows were chosen [20±4 days in milk (DIM), 37.3±8.9 kg milk/d, 4.4±0.8% milk fat, 3.2±0.3% milk protein (mean ± SD)]. Data recording was on 11 consecutive days per farm. Reticular pH and chewing behaviour were measured continuously. Milk, blood, urine, and faecal samples were collected four times per cow. As potential indicators milk, chewing, and blood data were tested. The first step of the statistical evaluation consisted in testing the indicators individually for their ability to predict the daily mean pH (pHmean) and the standard deviation of daily pH (pHsd) with linear mixed models. In a second step, a backward selection was performed for both pHmean and pHsd using multiple linear mixed models. On average 31% of the examined cows were below the pH-threshold for SARA. Applying linear mixed models, only β-hydroxybutyrate, the number of ruminate boluses and rumination chews showed a significant effect, for pHsd only the milk lactose. Within backward selection for pHmean, the DIM, milk lactose and eating time remained in the model. DIM, milk lactose, number of rumination chews and glucose were selected for pHsd. Still, only 4 and 14% of the variance could be explained with the help of the fixed part of the model, respectively. In particular, the selection of some unexpected indicators and the large animal-specific variance underline that general thresholds for SARA-indicators in early lactating cows should be reconsidered.

Prediction of reticular and ruminal pH progressions with transponder based feed intake

A. Mensching[1], K. Bünemann[2], U. Meyer[2], D. Von Soosten[2], S. Dänicke[2], J. Hummel[1] and A.R. Sharifi[1]
[1]University of Goettingen, Department of Animal Sciences, Albrecht Thaer Weg 3, 37075 Goettingen, Germany, [2]Friedrich-Loeffler-Institut (FLI), Institute of Animal Nutrition, Bundesallee 37, 38116 Braunschweig, Germany; andre.mensching@uni-goettingen.de

In the dairy industry, the prevention and controlling of metabolic and digestive disorders is an enormous challenge. It can be assumed that particularly subclinical forms such as subacute rumen acidosis (SARA) severely impair both animal health and economic efficiency. Due to the lack of specific and definitive clinical symptoms of this disorder, the diagnosis is not well established and many cases will remain undetected. In scientific studies, indicators derived from the ruminal pH fluctuations are currently considered to be the main indicator of SARA. The objective of this study was to examine, if the daily pH progression in the reticulum as well as in the ventral rumen can be predicted with data from a transponder based feeding system. The data were collected at the experimental station of the Institute of Animal Nutrition, Friedrich-Loeffler-Institut (FLI) in Germany and include observations from 13 ruminally cannulated cows. The individual intakes of a partially mixed ration, additional concentrate feed and water were recorded with transponder based feeding stations. In addition, the records of the continuously measured pH in the reticulum and in the ventral rumen of the cows were provided. The pH in the reticulum was measured using an intra-reticular pH measurement bolus and the pH in the ventral rumen was recorded using the LRCpH measurement system. In order to take the cumulative effect of successive meals into account, the animal-specific transponder data were modelled over time using recursive and convolutional time series filters. The final statistical analysis was performed with linear mixed models in a time series context to predict the two different pH progressions. The modelling showed that the daily pH profile in both the reticulum and the ventral rumen depends primarily on the animal's individual feed intake behaviour, given by the meal pattern in the course of the day.

Genomic prediction of serum biomarkers of health in early lactation dairy cows

T.D.W. Luke[1,2], T.T.T. Nguyen[2], S. Rochort[1,2], W.J. Wales[3], C.M. Richardson[1,2], M. Abdelsayed[4] and J.E. Pryce[1,2]
[1]La Trobe University, School of Applied Systems Biology, 5 Ring Rd, Bundoora 3083, Australia, [2]Agriculture Victoria, AgriBio Centre for Agribioscience, 5 Ring Rd, Bundoora 3083, Australia, [3]Department of Jobs, Precincts and Regions, Ellinbank Centre, 1301 Hazeldean Rd, Ellinbank 3821, Australia, [4]Datagene Ltd, AgriBio, Centre for AgriBioscience, 5 Ring Rd, Bundoora 3083, Australia; tim.luke@ecodev.vic.gov.au

Improving animal health and resilience is an increasingly important breeding objective for the dairy industry. In this study we estimated genetic parameters of serum biomarkers of health in early lactation dairy cows. A single serum sample was taken from 1,301 cows, located on 14 farms in south eastern Australia, within 35 days after calving. Sera were analysed for biomarkers of energy balance (BHBA and fatty acids), mineral status (Ca and Mg), protein nutrition (urea and albumin) and inflammation (globulins and haptoglobin). After editing, 47,162 single nucleotide polymorphism marker genotypes were used for estimating genomic heritabilities and breeding values (gEBV) for these traits in ASReml. Heritabilities were low for Ca, BHBA, urea and fatty acids (0.056, 0.102, 0.160 and 0.169, respectively), and moderate for Mg, albumin and globulins (0.231, 0.238 and 0.372, respectively). Standard errors for all heritabilities were less than 0.064. The heritability of haptoglobin was 0. The magnitude of genetic correlations between traits (estimated using bivariate models) varied considerably (0.007 to 0.538). Interestingly, most genetic correlations were favourable suggesting that selecting for normal concentrations of one biomarker may result in improvements in the concentrations of other biomarkers. Accuracies of gEBVs were evaluated using 5-fold cross validation, and by calculating the theoretical accuracy from prediction error variances associated with the gEBVs. Accuracies of gEBVs using 5-fold cross validation were low (0.179 to 0.348), while theoretical accuracies were higher, ranging from 0.314 to 0.447. While increasing the size of the reference population should theoretically improve accuracies, our results suggest that genomic prediction may allow identification of healthier cows that are less susceptible to diseases in early lactation.

Genetic analysis of ruminal acidosis resistance in dairy cows using fat percentage in milk

E.C. Verduijn, G. De Jong, M.L. Van Pelt and J.J. Bouwmeester-Vosman

CRV u.a., Animal Evaluation Unit, P.O. Box 454, 6800 AL Arnhem, the Netherlands; lisette.verduijn@crv4all.com

Ruminal acidosis (RA) is a large metabolic issue in dairy cattle, especially in high producing herds. RA occurs when the production of volatile fatty acids in the rumen is faster than the time it takes for the rumen to neutralize or absorb these acids. When the pH of the rumen is lower than 5.5 a cow has RA. This disease causes a lower milk production, damage to the gastrointestinal tract and related health issues, and possibly death. Not all cows have an equal risk of RA, so altering the feed strategy for the whole herd is not always the best option. We researched the possibility to improve the resistance to RA by breeding. Because it is not yet possible to continuously measure the pH of the rumen, an indicator for RA was used, based on the fat percentage of the milk: a cow has RA when her milk fat percentage is below 4 and lower than the protein percentage between day 60 and 120 in lactation. This definition is used in the Dutch/Flemish milk recording. Because of the difference in genetic predisposition between cows, fat and protein percentages were corrected with the breeding values for fat and protein percentage. For this research three traits were analysed, for parity one, two and three and higher (3+). The prevalence of RA in the Dutch population ranged from 4 to 8%. The estimated heritability of RA ranged from 0.04 to 0.05. Despite a low heritability, the variability between sires was large, with a genetic standard deviation of 2%. The genetic correlation between parity 2 and 3+ was high (0.90). The correlations between parity 1 and the other parities were lower, with 0.65 and 0.59 for parity 2 and 3+ respectively. This shows that RA is a different trait across parities. We calculated correlations between the EBV for RA and EBVs for other traits, where a higher EBV for RA means more acidosis. The correlations with fat and protein percentage were negative and the correlation with feed intake was positive. This shows that breeding for more efficient cows leads to a higher prevalence of RA. Although there is no golden standard for this trait, the definition that was used is the best indicator for RA available for each cow in milk recording at the moment.

Genetic parameters for ketosis and newly developed ketosis risk indicators based on MIR spectra

A. Köck[1], L.M. Dale[2], A. Werner[2], M. Mayerhofer[1], F.J. Auer[3] and C. Egger-Danner[1]

[1]ZuchtData EDV-Dienstleistungen GmbH, Dresdner Str. 89, 1200 Vienna, Austria, [2]LKV-Baden-Wuerttemberg, Heinrich Bauman Str. 1-3, 70190 Stuttgart, Germany, [3]LKV-Austria, Dresdner Str. 89, 1200 Vienna, Austria; koeck@zuchtdata.at

Ketosis is the most frequent metabolic disease in dairy cows and is associated with a wide range of changes in the milk. The basic idea is currently to build spectrometric tools for ketosis risk determination of dairy cows based on veterinary diagnosis and milk mid-infrared (MIR) spectra from routine milk recording. The first approach, KetoMIR 1, was based on milk components predicted from standardised milk MIR spectra and is routinely applied by LKV Baden Württemberg and LKV Austria since 2015 respective 2017. Recently, a new ketosis risk indicator (KetoMIR 2) was developed directly from the milk MIR spectra by using GLMNET modelling approaches. As a result, the KetoMIR models offer three classes of ketosis warning such as not, moderately and severely endangered. The specific objective of this study was to estimate genetic parameters for clinical ketosis and these newly developed ketosis risk indicators in Austrian Fleckvieh. Frequency of clinical ketosis was low (0.60%), whereas frequencies for KetoMIR 1 (moderately endangered = 14.38% and severely endangered = 1.87%) and 2 (moderately endangered = 16.34% and severely endangered = 7.69%) were higher. Heritability of clinical ketosis was 0.003 (0.0017). For KetoMIR 1 and KetoMIR 2 higher heritabilities of 0.14 (0.02) and 0.10 (0.01), respectively, were found. Genetic correlation estimates between clinical ketosis and these newly developed ketosis risk indicator traits were moderate 0.74 (0.22) and 0.73 (0.22), respectively. As genetic correlation estimates were associated with high standard errors, further analyses with larger data sets are needed to confirm these results. As soon as these results are confirmed in larger data sets, our strategy will be to use the ketosis risk indicator traits based on milk MIR spectra in genetic evaluations.

Evaluation of customized dry period management in dairy cows

A. Kok, R.J. Van Hoeij, B. Kemp and A.T.M. Van Knegsel
Wageningen University & Research, Adaptation Physiology group, P.O. Box 338, 6700 AH Wageningen, the Netherlands;
akke.kok@wur.nl

Dairy cows typically have a sharp increase in milk production and a concomitant negative energy balance in early lactation, associated with increased risk of metabolic diseases. Shortening or omitting the dry period before calving improves the energy balance after calving. Trade-offs of this strategy are a reduction in milk production and a loss of opportunity for selective dry cow therapy (DCT). Customised dry period management, i.e. deciding upon DCT and dry period length per cow, could mitigate negative impacts of shortened dry periods on milk production and udder health, and at the same time retain benefits such as improved energy balance, metabolic status and fertility. In this study, we evaluated two decision trees to customize dry period management based on individual cow characteristics. Milk production and composition, SCC, and disease incidence from 8 weeks before to 14 weeks after calving were compared between the two decision trees (T1, n=59;T2, n=63) and standard dry period management (C, n=61). In C, all cows had a 60 d dry period, with DCT if SCC≥150,000 cells/ ml prior to dry-off. In T1, heifers and older cows were assigned DCT if SCC≥150,000 cells/ml and SCC≥50,000 cells/ml, respectively, whereas in T2 the threshold for DCT was SCC≥200,000 cells/ml for all cows. In T1 and T2, cows with DCT were assigned a 60 d dry period, whereas cows without DCT were assigned a 30 d or 0 d dry period if their milk production was >12 kg/ d at the last test-day prior to potential dry-off. Effects of dry period length and decision trees on variables were analysed separately with mixed models; effects on udder health status and disease incidence with logistic regressions and chi-square tests. Compared with C, cows in T1 and T2 on average produced more milk in the 8 weeks before calving (0.1 vs 4.1 vs 7.3 kg/d in C vs T1 vs T2), and less in the 14 weeks after calving (40.1 vs 37.1 vs 35.1 kg/d). There was no difference in udder health status between treatment groups, but disease incidence tended to be lower for T2 (0.54 incidents/ cow) than C (0.90) and T1 (0.85). Overall, shortened dry periods reduced milk revenues, but with customized dry period management this may be financially compensated by improved health.

Comparison of a voluntary calcium drink to a calcium bolus administered to dairy cows after calving

S. Van Kuijk[1], A. Klop[2], R. Goselink[2] and Y. Han[1]
[1]Nutreco Nederland BV, R&D, P.O. Box 36, 5830 AA Boxmeer, the Netherlands, [2]Wageningen Livestock Research, P.O. Box 338, 6700 AH Wageningen, the Netherlands; sandra.van.kuijk@trouwnutrition.com

Recent advances in dry cow feeding strategies have reduced the incidence of clinical hypocalcaemia in modern dairies, but the prevalence of subclinical hypocalcaemia may still be substantial. Stimulating feed and water intake after calving, supported with extra minerals and energy, can be helpful. By supplying specific supportive nutrients in a voluntary cow drink, both cow and farmer benefit an easy, welfare-friendly application. The objective of this study was to compare a Ca drink on voluntary basis to application of a force-fed Ca bolus and a negative control in periparturient dairy cows. In total 72 pregnant cows were allocated to three different treatments. Within 15 min after calving the cows were offered: (1) lukewarm water as negative control; (2) Ca supplement dissolved in lukewarm water (Ca-drink); or (3) Ca-bolus (equal amount of Ca as Ca-drink) and lukewarm water. The lukewarm water was withdrawn after 30 min and voluntary intake was measured. Blood was sampled 2 weeks before the expected calving date to serve as a baseline (T-14). In addition, blood was sampled within 15 minutes after administering the treatments (T0), and 6, 12, 24 and 48 h after calving. Blood was analysed for Ca, non-esterified fatty acids (NEFA) and β-hydroxybutyrate (BHB) (all timepoints) and cortisol (T-14 and T0 only). Milk yield and dry matter intake were measured during the first 3 weeks after calving. Cows that received Ca-drink consumed more water than the other two groups and tended to have a higher dry matter intake (P=0.095). There was no effect on milk yield. In multiparous cows, but not in heifers, higher blood Ca levels were observed at 24 h after calving in the group receiving the Ca-drink compared to the group receiving the Ca-bolus. In general, cows did not show signs of (subclinical) ketosis, with BHB and NEFA values within the normal ranges; treatments did not have an effect on BHB or NEFA. The plasma cortisol level seemed lower in the Ca-drink group compared to the other groups, however due to differences before calving (T-14) this was not significant. The results showed that the Ca-drink improved blood Ca and appetite after calving.

Knockdown of patatin-like phospholipase domain-containing protein 3 increased cellular triglyceride

S.J. Erb and H.M. White
University of Wisconsin-Madison, 1675 Observatory Dr, 53706, USA; heather.white@wisc.edu

Accumulation of lipids during the periparturient period in dairy cows may have negative impacts on liver function. Abundance of a key hepatic lipase, patatin-like phospholipase domain-containing protein 3 (PNPLA3), during this period, and involved with fatty liver onset in humans, may be associated with onset or severity of fatty liver. The objective of this study was to determine if knockdown (KD) of PNPLA3 protein abundance using bovine-specific PNPLA3 siRNA increased accumulation of triglyceride (TG) in primary bovine hepatocytes. Hepatocytes isolated from bull calves (n=3; age<7 d) were isolated and cultured in monolayer. After 24 h, cells were randomly assigned to two parallel blocks and treated in triplicate with nonspecific (siNON) or PNPLA3-targeted (siKD1 or siKD2) siRNA. Media was replenished after 4 h, and cells incubated for an additional 20 h before harvest in dissociation buffer for subsequent TG and DNA isolation, or in TRIzol for subsequent protein isolation. Cellular TG were quantified and normalized to DNA concentration. Protein abundance of PNPLA3 was determined by blot analysis, normalized to total lane protein, and expressed relative to a BSA control. Data were analysed (PROC MIXED, SAS 9.4) with fixed effect of treatment and random effect of calf. Differences were declared when P<0.05 and means compared by preplanned contrast (siNON vs siKD1 and siKD2). Knockdown successfully reduced (P=0.04) PNPLA3 protein abundance by 33% with siKD1 (1.18 vs 0.79±0.32 arbitrary units (AU)) and 39% by siKD2 (1.18 vs 0.72±0.32 AU) compared to siNON. Accumulation of cellular TG increased (P=0.04) with knockdown compared to siNON (siNON vs siKD1 and siKD2: 0.24 vs 0.56 and 1.48±0.30 AU). This is a 6.17-fold increase in cellular TG with the 39% reduction in PNPLA3 abundance with siKD2. This data demonstrates a causative role of PNPLA3 in lipid accumulation in bovine hepatocytes. Given the association between PNPLA3 protein abundance and liver lipids *in vivo* and the clear role of PNPLA3 in TG accumulation in primary bovine hepatocytes, future research should focus on understanding regulation of PNPLA3. Furthermore, the role of PNPLA3 function in TG accumulation and recovery *in vivo* should be examined further.

Effect of corn starch digestibility on milk fat depression in Holstein dairy cattle

A. Rahimi[1], A. Naserian[1], R. Valizadeh[1], A. Tahmasebi[1], H. Dehghani[1], A. Shahdadi[1], K. Park[2], J. Ghasemi Nejad[3] and M. Ataallahi[2]
[1]Ferdowsi university of Mashhad, Mashhad, 9177948974, Iran, [2]kangwon National university, Chuncheon, 24341, Korea, South, [3]Konkuk University, Seoul, 05029, Korea, South; kpark74@kangwon.ac.kr

In this study we aimed to investigate the effects of starch digestibility of processed corn on milk fat depression and milk fat to protein ratio in Holstein dairy cattle. Twenty-four dairy cattle (milk yield 44.8±3 kg and DIM 159.6±10 d) were randomly assigned to one of four experimental diets: (1) grounded corn (G); (2) pre-grinding super-conditioned corn (PSC); (3) extruded corn (E); and (4) steam-flaked corn (SF). Concentrate was adjusted to have 32% starch. Statistical analysis was performed using the GLM procedure based on completely randomized design of SAS (P<0.05). Milk yield was increased (P<0.05) using PSC corn than other processed corn (42.98, 44.40, 37.63 and 42.85, in G, PSC, E and SF corn, respectively). FCM 3.5% and ECM in both of PSC and SF had high tendency (P>0.05) than G and E corn. Percentage of milk fat was significantly decreased (P<0.05) in PSC than others (3.04, 2.51, 3.22 and 3.16 in G, PSC, E and SF corn, respectively), but kg of milk fat, percentage and kg of milk protein and percentage of milk lactose were not affected (P>0.05) using processed corn in diets. Fat to protein ratio in PSC was shown 0.84 that is under 1.00 that is specified for inducing of milk fat depression syndrome, whereas fat to protein ratio in G, E and SF corn was 0.99, 1.06 and 1.14, respectively. Percentage of MUN in both of PSC and SF was lower (P<0.05) and percentage of SNF was higher than G and E diets. It is concluded that PSC corn due to more starch degradation in the rumen may induce milk fat depression in diets containing high levels of starch (more than 32%) and SF corn may recovery from milk fat depression induced with feeding of high starch digestibility corn in dairy cattle diet.

Comparison of three buffers for the simulation of SARA in an *in vitro* rumen incubation

H. Yang, S. Debevere and V. Fievez
Department of Animal Sciences and Aquatic Ecology, Ghent University, Campus Coupure, building F, 1ˢᵗ floor, Coupure Links 653, 9000 Gent, Belgium; hong.yang@ugent.be

Subacute ruminal acidosis (SARA) is characterized by a decreased ruminal pH for a longer period of the day. To find a stable low pH buffer for the *in vitro* simulation of a subacute ruminal acidosis condition, three biological buffers, 3.7 g/l tricarballylate (the citrate analog) buffer, 7.53 g/l Bis-tris buffer and 0.25 N nonmetabolizable Nmorpholinoethane sulfonic acid (MES) buffer were chosen based on their useful pH range. The pH of the three buffers was adjusted to 5.8 and 6.8 before mixing it with the three cow rumen fluids sampled in early lactation for testing the pH stability at 5.8 and 6.8 after 24 and 48 hours *in vitro* incubation. Furthermore, a second *in vitro* experiment was performed for 0, 1.5, 3, 6, 24, and 48 hours incubation at 5.8 and 6.8 pH. Each of the treatments has 3 replicates. Gas and VFA production as well as pH were measured after incubation at each time point. Data was analysed by ANOVA in SAS 9.4 (SAS Inst. Inc., Cary, NC). The low pH simulation (start pH=5.8) in a MES buffer remained at 5.82±0.01 and 5.73±0.02 after 24 hours and 48 hours incubation respectively, which was significantly higher than the pH of the tricarballylate buffer (5.29±0.02 and 5.34±0.32) and the Bis-tris buffer (5.2±0.02 and 5.06±0.03) ($P<0.05$). Also in the high pH simulation (start pH=6.8), the MES buffer allowed to keep a stable pH after 24 and 48 hours of incubation. Accordingly, the MES buffer was chosen for an *in vitro* fermentation kinetics experiment with sampling at 6 time points. Only after 3 hours of incubation, the pH of the (6.01±0.01) deviated from the start pH, but decreased again to 5.91 (±0.02) at 6 hours after which it remained stable at the later time points. At each time point, the methane and VFA production of the high pH was significantly higher than that of the lower pH ($P<0.05$). In conclusion, the MES buffer could allow to keep stable pH conditions both at 5.8 as well as 6.8. Accordingly, this buffer allows simulate the pH range from normal to SARA conditions when incubated for 0, 1.5, 3, 6, 24 and 48 hours.

Hyperketonaemia genome-wide association study in Holstein cows

R.S. Pralle, N.E. Schultz, K.A. Weigel and H.M. White
University of Wisconsin – Madison, 1675 Observatory Drive, 53706, USA; rpralle@wisc.edu

Hyperketonemia (HYK) is a common metabolic disorder in early postpartum dairy cows with an estimated economic loss of $289 US per case. The research objectives were to associate single nucleotide polymorphism (SNP) genotypes with HYK and identify gene pathways enriched with these associations. Holstein cows (n=1,903; 4 herds in Wisconsin, USA) were enrolled onto the experiment after parturition. Blood samples were collected at 4 timepoints between 5 to 18 days postpartum for each cow. Concentration of blood β-hydroxybutyrate (BHB) was quantified cow-side via an electronic, handheld BHB meter. Cows were labelled a HYK case (n=640) when at least one blood sample achieved a BHB≥1.2 mmol/l and all other cows were labelled non-HYK controls (n=1,263). The SNP genotypes (n=60,476) were provided by the Council of Dairy Cattle Breeding (USA). All SNP were annotated to the closest gene (*Bos taurus* assembly, release 106) using bedtools (version 2.25). After quality control procedures, 1,710 cows and 58,699 genotypes were available for the genome-wide association study (GWAS). Association was performed using the forward feature select linear mixed model method, with HYK as a binary response, herd-year-season as a random intercept, and parity as a continuous covariate. Association P-values were corrected for multiple testing (Q-values) by the false discovery rate method. Genome-wide significance and marginal significance was declared at $Q\leq0.05$ and $0.05<Q\leq0.10$, respectively. Genes with coding regions within 15 kb of a SNP with $P<0.05$ (n=1,356) were supplied to DAVID for pathway analysis. Evaluation of genomic inflation factor ($\lambda=1.01$) suggested negligible genomic inflation. Preliminary results indicated ARS-BFGL-NGS-91238 (chromosome 10, base pair 3856662) had marginal evidence for a genome-wide association with HYK, $P=1.54\times10^{-6}$ and $Q=0.09$. The nearest gene (~55 kb) was *LOC787395*, a ras-related nuclear protein pseudogene. Focal adhesion (bta04510) was a significantly enriched pathway ($Q=0.02$) for HYK susceptibility. This work identifies a potentially novel SNP associated with HYK and suggests that genetic variation of focal adhesion pathway could contribute to HYK etiology.

Effect of 1,25-dihydroxycholecalciferol-glycosides given as a rumen bolus on blood pharmacokinetics

M. Meyer[1,2], C. Ollagnier[2], K. Bühler[3], L. Eggerschwiler[2], M. Meylan[1] and P. Schlegel[2]
[1]*University of Bern, Vetsuisse faculty, Clinic for Ruminants, Bremgartenstr. 109a, 3012 Bern, Switzerland, [2]Agroscope, Ruminant Research Unit, Tioleyre 4, 1725 Posieux, Switzerland, [3]Herbonis Animal Health GmbH, Rheinstr. 30, 4302 Augst, Switzerland; martina.meyer@agroscope.admin.ch*

A single oral application of glycosides from the biologically active metabolite of vitamin D3, 1,25-dihydroxycholecalciferol (1,25(OH)2D3), extracted from Solanum glaucophyllum (SG), prior to calving represents a novel approach to the prevention of puerperal hypocalcaemia. The aim of this study was to determine the time response of blood parameters following the application of 1,25(OH)2D3 containing boluses with different concentrations and physical properties. Thirty pregnant dry dairy cows (220-257 days p.i.) were allocated in blocs of 5 animals to 6 treatments: one bolus containing uncoated tablets with 191, 310 or 501 μg 1,25(OH)2D3, one bolus containing coated tablets with 310 or 501 μg 1,25(OH)2D3, and 2 boluses containing uncoated tablets resulting in a total of 1,002 μg 1,25(OH)2D3 as a safety for eventual erroneous application procedure. Nineteen blood samples were collected at regular intervals between 96 hours before and 336 hours after bolus application. Samples were determined for 1,25(OH)2D3 and Ca concentrations. Data were analysed using a mixed model procedure. Serum 1,25(OH)2D3 and Ca were increased ($P<0.001$) from 12 to 120 and 12 to 264 hours, respectively after bolus application with a mean of 484 and 22%, respectively at peak of 30 and 72 hours, respectively. Overall serum 1,25(OH)2D3 was higher ($P<0.001$) with treatments 300 uncoated, 501, 1002 than with 191 μg. Overall serum Ca was higher ($P<0.001$) with 501 coated and 1002 than with 191 and 310 μg. Coating had no effect on serum values ($P>0.10$). No time × treatment interaction ($P>0.10$) was observed for serum 1,25(OH)2D3 and Ca. In conclusion, the optimal time-window for an ante partum oral administration is estimated to be between 9 and 0.5 days. The most efficient dosage was 501 μg 1,25(OH)2D3 per bolus. The potentially erroneous application of two of such boluses was not of any concern regarding the measured parameters. Additional investigations with this approach are currently conducted during the peripartum period.

The transition period as a monitoring window for resilience of high yielding dairy cows

S. Heirbaut[1], X. Jing[1], L. Vandaele[2] and V. Fievez[1]
[1]*Ghent University, Department of Animal Sciences and Aquatic Ecology, Coupure Links 653 – block F, 9000 Gent, Belgium, [2]Research institute for agriculture, fisheries and food – ILVO, Eenheid Dier, Scheldeweg 68, 9090 Melle-Gontrode, Belgium; stijn.heirbaut@ugent.be*

The transition period, the period of 60 to 90 days around calving, is a crucial period for high yielding dairy cattle. Problems during transition can have significant negative effects on production, animal health and welfare. Since there is an important inter-animal variation in susceptibility to these problems, animal-specific measures are required in addition to general herd-based transition management. This project aims to identify risk animals for key transition problems (ketosis, rumen acidosis, mastitis and hypocalcaemia) using sensor data and biomarkers during the transition period and to assess the added value of specific cow sensors and biomarkers. A monitoring program at the ILVO research institute during the transition period has been set up from November 2018 onwards. For about one year milk metabolites (protein, lactose, urea, cell count, fatty acids and ketones) are monitored in freshly calved dairy cows, aiming to monitor 100 cows in total. These data are supplemented with information on milk production, blood metabolites and daily sensor data (activity (IceTag©), body condition and weight). Blood samples are taken one week before expected calving and three, six, nine and 21 days post partum (pp). These samples are analysed for non-esterified fatty acids (NEFA), ketones, glucose, insulin and insulin-like growth factor (IGF-1). Additionally, samples taken on day 21 are analysed for aspartate aminotransferase (AST), urea and lipopolysaccharide (LPS) and electrophoresis is performed. Preliminary results (12 cows) indicated roughage intake during the last 14 days before calving tended to negatively correlate with the NEFA serum concentration three days pp (Rpearson=-0.55; P=0.061). Furthermore the variance in body weight during the first five weeks in lactation is positively correlated with the NEFA serum concentration (Rpearson=0.60; P=0.023). Updated results will be presented at the conference.

Consequences of maternal fatty acid supplementation on muscle development in the neonatal calf
N. Dahl, E. Albrecht, K. Uken, W. Liermann, D. Dannenberger, H.M. Hammon and S. Maak
Leibniz Institute for Farm Animal Biology, W.-Stahl-Allee 2, 18196 Dummerstorf, Germany; dahl@fbn-dummerstorf.de

Essential fatty acids (EFA) and conjugated linoleic acids (CLA) have been recognized as crucial for the development of the foetus and neonatal calf. As they cannot be synthesized endogenously, their supply depends on the maternal transfer. Cows may have a reduced status of these fatty acids in modern production systems with consequences for the offspring. The aim of the current study was to investigate the effect of maternal supplementation with EFA and CLA on the calf skeletal muscle development. Holstein cows (n=40) were abomasaly supplemented with coconut oil (76 g/d, control), or EFA (78 g/d linseed and 4 g/d safflower oil) or CLA (38 g/d Lutalin) or a combination of both during the transition period. After birth, the calves received milk from their respective mothers for 5 days and were then slaughtered. Samples of *M. longissimus* were collected for analyses of the fatty acid composition, muscle structure and myosin heavy chain (MYH) isoform abundance. The data were analysed with the MIXED model of SAS with group and sex as fixed factors. The supplemented EFA, their metabolites and CLA were enriched in the calf muscle of respective supplementation groups (P<0.05), indicating successful transfer from cow to calf. Muscle structure including muscle fibre size and fibre type composition was not different among groups (P>0.05). However, the capillary density was highest in the CLA group (P<0.05) corresponding to a better capillary supply index for CLA calves. Type I fibres were confirmed by immunohistochemistry with a MYH7 antibody. Double staining with MYH2 indicated highly variable number of hybrid fibres within all groups. This suggests progression of fibre type conversion. A significant influence of the treatment could not be confirmed (P>0.05). The relative protein abundance of MYH isoforms within the muscle sample was similar between the groups (P>0.05). The results confirm that maternal supplementation can influence the offspring development via fatty acid transfer. The maternal supplementation led to changes in the nutritional supply network, but not in the muscle fibre characteristics of the calves.

Physiological concentrations of fatty acids impact lipolytic genes in primary bovine hepatocytes
S.J. Erb and H.M. White
University of Wisconsin-Madison, 1675 Observatory Dr, 53706, USA; serb2@wisc.edu

Fatty acids (FA) released from adipose tissue during the periparturient period in dairy cows are known regulators of hepatic gene expression and protein. In rodents and humans, hepatic lipases are responsive to fatty acids; however, this regulation has not been examined in bovine. The objective of this research was to determine the effect of FA on gene expression and protein abundance of lipid-related genes in bovine. Primary bovine hepatocytes were isolated from bull calves (n=4, <7 d of age) and cultured in monolayers. After 24 h, cells were treated in triplicate with either a physiologically relevant concentrations (PRC) or 0.25 mM treatment (trt) of individual FA (C14:0, C16:0, C18:0, C18:1, C18:2, C20:5, and C22:6) observed in serum of peripartum dairy cows. Cells incubated for 24 h and were then harvested in TRIzol for subsequent RNA and protein isolation. Gene expression of carbohydrate regulatory element-binding protein 1 (ChREBP1), sterol (*S*) *REBP1c*, and liver X receptor (LXR) α and β were quantified and made relative to reference genes. Protein abundance of SREBP1c and patatin-like phospholipase domain-containing protein 3 (PNPLA3) were determined by Western blot analysis, normalized to total lane protein, expressed relative to the control, and log transformed for normality. Data were analysed (PROC MIXED, SAS, 9.4) with fixed effect of FA and random effect of calf and preplanned contrasts. Expression of ChREBP1 and LXRβ were not altered by FA or concentration (P>0.1). Expression of LXRα tended to decrease with PRC C18:0 (0.20 vs 0.10±0.04 arbitrary units (AU); P=0.07) but not with other treatments or concentrations (P>0.1). Treatment with PRC C14:0 increased (P=0.05) expression of *SREBP1c* (0.22 vs 0.62±0.13 AU) as did C16:0 (0.22 vs 0.75±0.12 AU); PRC C18:0 tended to increase expression (P=0.09), but was not altered (P>0.1) by other FA. Interestingly, treatment with 0.25 mM FA did not affect (P>0.1) gene expression or protein abundance in primary bovine hepatocytes. Despite some shifts in gene expression with PRC FA treatment, SREBP1c and PNPLA3 protein abundance were not altered by any FA trt (P>0.1). This research indicates that individual FA, even at PRC, are not sufficient to regulate SREBP1c and PNPLA3 protein abundance.

Hyperketonaemia SNP by parity group genome-wide interaction study in Holstein cows

R.S. Pralle, N.E. Schultz, K.A. Weigel and H.M. White
University of Wisconsin – Madison, 1675 Observatory Drive, 53706, USA; rpralle@wisc.edu

Our research objective was to perform a genome-wide interaction study (GWIS) to discover single nucleotide polymorphism (SNP) genotype effects, dependent on parity group (primiparous vs multiparous), for hyperketonaemia (HYK). Holstein cows (n=1,903) were enrolled onto the experiment after parturition on 4 dairy herds in Wisconsin, USA. Blood samples were collected at 4 timepoints between 5 to 18 days postpartum for each cow. Concentration of blood β-hydroxybutyrate (BHB) was quantified cow-side via an electronic, handheld BHB meter. Cows were labelled as a HYK case (n=640) when at least one blood sample achieved a BHB≥1.2 mmol/l and all other cows were labelled non-HYK controls (n=1,263). The SNP genotypes (n=60,476) were provided by the Council of Dairy Cattle Breeding (USA). All SNP were annotated to the closest gene (*Bos taurus* assembly, release 106) using bedtools (version 2.25). After quality control procedures, 1,710 cows and 58,699 genotypes were available for the GWIS. A two-stage procedure was implemented for GWIS. Stage 1 selected SNP for GWIS via a genome-wide association study (GWAS); SNP with a P<0.001 (n=64) proceeded to the next stage. Stage 2 was the GWIS, which evaluated the SNP×parity group effect. The GWAS and GWIS were performed using the forward feature select linear mixed model method, with HYK as a binary response, herd-year-season as a random intercept, and parity group as a binary covariate. Multiple test correction of GWIS P-values was performed by the false discovery rate method (Q-values). Significant and marginally significant interactions were declared $Q\leq0.05$ and $0.05<Q\leq0.10$, respectively. Preliminary results indicated BovineHD0600024247 (chromosome 6, base pair 86882515) had a significant interaction (P=5.33×10^{-4} and Q=0.03) and BovineHD1400023753 (chromosome 14, base pair 86882515) had a marginally significant interaction (P=2.73×10^{-3} and Q=0.09). The nearest genes were *LOC782958* and *COL14A1* for BovineHD0600024247 and BovineHD1400023753, respectively. This work provides novel evidence of SNP associations with HYK dependent on parity status.

Responses of lipid associated proteins to FA treatment in bovine primary hepatocytes

H.T. Holdorf and H.M. White
University of Wisconsin-Madison, 1675 Observatory Dr, Madison, WI 53502, USA; hholdorf@wisc.edu

Lipid associated proteins may allow for dynamic storage or utilization of liver triglyceride (lvTG) and are regulated by fatty acids (FA) in other species. The objective of this study was to determine if the abundance of lipid associated proteins is responsive to incremental changes in concentration of FA in primary hepatocytes. Primary hepatocytes isolated from 4 Holstein calves were maintained as monolayer cultures for 24 hours before treatment. FA treatments (FA trt) were palmitic acid (C16:0, PA), oleic acid (C18:1n6, OA), α-linolenic acid (C18:3n3, ALA), and a FA cocktail (FAC) with a profile of FA reflective of plasma FA at parturition (3% C14:0, 27% C16:0, 23% C18:0, 31% C18:1n6, 8% C18:2 n6, and 8% C18:3n3) applied at 0, 0.25, 0.5, 0.75, and 1.0 mM. Protein abundance of abhydrolase domain containing 5 (ABHD5), hormone sensitive lipase (HSL), perilipin 1 (PLIN), patatin-like phospholipase domain containing 2 and 3 (PNPLA3) was determined by Western blot analysis. Data were transformed as natural log(abundance), to achieve normal residuals. Data were analysed for main effects of FA trt, concentration, and the interaction of FA trt × concentration with random effect of calf using PROC GLIMMIX (SAS 9.4). Linear, quadratic, and cubic contrasts were tested. Data are presented as lsmeans ± standard error and considered significant at P<0.05. Treatment with ALA and FAC increased (P<0.05) HSL but decreased (P<0.05) PPLIN, compared to PA and OA. PLIN was decreased (P<0.05) by FAC while PNPLA3 was increased (P<0.05) by ALA (15.75±0.20), compared to OA (15.39±0.20) and PA (15.29±0.20). Abundance of ABHD5, PPLIN, PHSL, PNPLA3 were all decreased (P<0.05) linearly by increasing FAC. Increasing FA concentration quadratically affected (P<0.05) HSL abundance. No interactions of FA trt × concentration were detected. The responsiveness of lipases to changes in NEFA concentrations may play a role in peripartum regulation. Responses to ALA were similar to FAC responses, whereas responses to other individual FA were of an opposing pattern. Given that the FAC contained ALA, responses to FAC may have been mediated through ALA presence and should be further examined.

Claw health traits and mastitis in breeding of the Czech Holstein cows

Z. Krupová, E. Krupa, M. Wolfová, J. Přibyl and L. Zavadilová

Institute of Animal Science, Přátelství 815, 10400 Prague Uhříněves, Czech Republic; krupova.zuzana@vuzv.cz

Breeding of the Czech Holstein cows is currently aimed to improve the complex of 10 production and functional traits (clinical mastitis as the only direct health trait) and breeding values are estimated for 15 selection criteria (without a direct health candidate). Genomic evaluation of the female dairy population (calves, heifers and cows) has started in the last year to enhance the reference populations. To improve health status of the population the overall claw disease incidence as the new breeding objective and three traits indicating the claw health (overall, infectious and non-infectious claw diseases incidence) and clinical mastitis incidence as the new selection criteria of cows were added. Economic weights were calculated for all traits in the breeding objective for two-year-old cows by applying a bio-economic model. General principles of the selection index theory were applied when calculating the selection response in the new as well as in the current breeding objectives. The annual economic weights were -193 and -168 € per case for clinical mastitis and overall claw disease incidence, respectively. Based on the construction of the actual selection index of cows (where i.a. the somatic cells score and udder exterior are included) the slightly favourable selection response in clinical mastitis (-0.001 cases) was recorded. Using the comprehensive health selection index, the incidence of both udder and claw diseases was reduced by 0.004 cases per cow per year. Simultaneously, a favourable selection response was obtained for other functional traits, e.g. 0.024% for cow conception rate and 0.037 years for cow productive life when using the health index for Czech Holstein cows. A direct selection of cows for claw and udder health is recommended to improve the health status of herds and to increase the safety of production and the overall efficiency of production system. The study was supported by project MZE-RO0719 and QK1910320 of the Czech Republic.

Association analyses for ketosis indicators ketone bodies and fatty acid profiles in Holstein cows

S.-L. Klein[1], C. Scheper[1], K. May[2], H.H. Swalve[3] and S. König[1]

[1]Justus-Liebig-University Gießen, Institut of Animal Breeding and Genetics, Ludwigstr. 21B, 35390 Gießen, Germany, [2]Vereinigte Informationssysteme Tierhaltung w.V., Heinrich-Schröder-Weg 1, 27283 Verden, Germany, [3]Martin Luther University Halle-Wittenberg, Institute of Agricultural and Nutritional Sciences, Karl-Freiherr-von-Fritsch-Str. 4, 06120 Halle, Germany; sarah.klein@agrar.uni-giessen.de

Accurate detection of subclinical ketosis implies to determine concentrations of acetone and β-hydroxybutyrate (BHB) in blood or in milk. Besides acetone and BHB, fatty acid (FA) profiles in milk reflect the energy status in early lactation. Aim of the study was to investigate associations between clinical ketosis (KET) with the FTIR traits (FTIR = Fourier transform infrared spectroscopy in milk) acetone, BHB and FA concentrations (unsaturated FA, monounsaturated FA and saturated FA C18:0) phenotypically, quantitative-genetically and genomically. Considering 4,385 Holstein cows, relationships between KET with acetone and BHB were analysed. 10,649 cows were used to investigate associations between KET with FA concentrations. Cows were phenotyped for KET according to a veterinarian diagnosis key. One entry for KET in the first six weeks after calving implied a score = 1 (diseased); otherwise, a score = 0 (healthy) was assigned. For genomic analyses, cows were genotyped with the Illumina BovineSNP50 v2 BeadChip and EuroGenomics 10k chip. Phenotypically, we detected strong positive associations between first test-day acetone, BHB and FA with KET. Using pedigree-based and genomic relationship matrices, we estimated large genetic correlations between KET with acetone and BHB concentrations (0.88-0.99), and low to moderate genetic correlations between KET and FA concentrations (0.29-0.76). Heritabilities for acetone, BHB and FA were in the range from 0.01 to 0.14. Genome-wide association studies identified potential candidate genes INS, BCL2L11, DGAT1 involved in diabetes, insulin secretion, lipid and energy metabolism. Regarding the large correlations between KET, we suggest utilization of acetone and BHB as indicator traits for improvements of the metabolic health status of dairy cows via genetic selection. However, due to the moderate genetic correlations between KET and FA, FA biomarkers are less valuable indicators in selection indices.

Does BCS at drying-off and calving provide information on prepartum energy balance of dairy cows?

N. Siachos[1], N. Panousis[1], G. Oikonomou[2] and G.E. Valergakis[1]
[1]Faculty of Veterinary Medicine, School of Health Sciences, Aristotle University of Thessaloniki, Box 393, 54124, Thessaloniki, Greece, [2]Institute of Veterinary Science, Faculty of Health and Life Sciences, University of Liverpool, Leahurst, Neston CH64 7TE, United Kingdom; nsiachos@vet.auth.gr

The objective was to evaluate energy balance of dairy cows during dry period. One-hundred and fifty-five Holstein cows from 5 dairy farms (A: n=32; B: n=39; C: n=19, D: n=35 and E: n=30) in different parities (2nd: n=63; 3rd: n=48 and >3rd: n=44) were included in the study. The BCS of each cow was assessed by the same evaluator at 4 time-points relative to calving (0d): at drying-off; -21d; -8d and 0d, resulting in a total of 609 assessments. The mean (±sd) dry period duration was 65.4 (±28.3) days. Mean (±sd) BCS at each time-point was 3.12 (±0.51); 3.23 (±0.48); 3.24 (±0.85) and 3.10 (±0.49), respectively. From drying-off until -21d, 76 (48.5%) cows gained 0.25-0.75 units of BCS (group A); the BCS remained unchanged for 60 (38.8%) cows (group B), whereas 19 (12.3%) cows lost 0.25-0.50 units (group C). During this period, BCS gain was significantly higher for cows with BCS<3.00 (P<0.05) at drying-off. A total of 78 (50.3%) cows lost 0.25-0.75 units of BCS during the last 3 weeks of gestation; 45/76 cows (group A), 27/60 cows (group B) and 6/19 (group C). The BCS loss was significantly higher for cows with BCS>3.50 (P<0.05) at drying-off. Cows losing BCS during the last 3 weeks of gestation represented 37.5, 61.5, 42.1, 62.9 and 40.0% of total cows assessed in herds A, B, C, D and E, respectively. In 50 of these 78 cows (64.1%) the BCS decline was detectable even during the last week of gestation (period from -8 to 0 d). At calving, 91/155 (58.2%) of cows had a BCS within the target range (3.00-3.50); however, almost half of them lost BCS during the last 3 weeks of gestation. The same was the case for 25/48 (52.1%) and 10/16 (62.5%) of cows that calved at lower and higher BCS than the target range, respectively. BCS estimation at drying-off and at calving does not provide useful information on energy balance during the dry period. To improve prepartum management of dairy cows, body condition scoring should be performed at drying-off, at -21 days and at calving.

Selection against clinical mastitis changed the level of somatic cell count throughout lactation

M. Heringstad[1], K.B. Wethal[1], G. Klemetsdal[1], G. Dalen[2] and B. Heringstad[1]
[1]Norwegian University of Life Sciences, Faculty of Biosciences, Department of Animal and Aquacultural Sciences, P.O. Box 5003 NMBU, 1432 Ås, Norway, [2]Faculty of Veterinary Medicine, NMBU, Oslo, Norway; bjorg.heringstad@nmbu.no

Breeding is an efficient tool to reduce the incidence of mastitis in dairy cattle. Results from a selection experiment with Norwegian Red illustrate the potential with large observed genetic differences in susceptibility to clinical mastitis between two groups, one selected for high protein yield (HPY) the other for low incidence of clinical mastitis (LCM). The aim of this study was to examine whether selection against clinical mastitis also has changed the level and trajectory of somatic cell count (SCC) throughout lactation. Data from the automatic milking systems (AMS) with online cell counters (OCC) at the research herd at NMBU was analysed. The OCC provides SCC measurements from every milking. Data were from August 2015 to October 2017, with cows from 5 to 305 days in milk (DIM). The final dataset included OCC data from 99,241 AMS visits made by 173 cows from the 2 selection groups (79 HPY and 94 LCM) with a total of 279 lactations. SCC was log transformed to Somatic Cell Score (SCS) and analysed using a linear model that included fixed effects of selection group (HPY, LCM) by parity (1, 2, ≥3), month-year of calving, DIM, and random effect of cow. Least Squares Mean (LSM) of SCS for group by parity ranged from 3.26 (LCM 1st parity) to 4.18 (HPY parity≥3), equivalent to SCC varying from 26,000 to 65,000. The LSM of SCS increased with parity for both selection groups, but all parities of LCM had lower SCS than the lowest HPY group. The difference between selection groups corresponded to a difference in SCC of around 23,000, 15,000, and 24,000 for parity 1, 2 and ≥3, respectively. Plots of mean SCS by DIM revealed largest differences between HPY and LCM cows in early and mid-lactation of 1st parity. The observed differences in SCC between selection groups suggest indirect selection responses in SCC after selection for high protein yield or low incidence of clinical mastitis.

Changes in colostrum bioactive components depend on cytological quality

K. Puppel[1], T. Sakowski[2], G. Grodkowski[1], P. Solarczyk[1], M. Stachelek[2] and M. Klopčič[3]
[1]Warsaw University of Life Sciences, Ciszewskiego 8, 02-786 Warszawa, Poland, [2]Institute of Genetics and Animal Breeding PAS, Postępu 36A, 05-552 Magdalenka, Poland, [3]University of Ljubljana, Kongresni trg 12, 1000 Ljubljana, Slovenia; m.stachelek@ighz.pl

Although good colostrum quality is highly desired, not many researches show what factors have impact on its components. It is believed that udder diseases, ketosis and chronic acidosis significantly lower the level of immunoglobulines in colostrum simultaneously causing its poor quality. The aim of this study was to indicate which intramammary infections, and as a result cytological quality, impact on the content of immunostymulating components of colostrum. The experiment was conducted on 250 cows all kept in a freestall housing system. Animals were divided into two groups, depending on somatic cell counts in colostrum (SCC) per ml: <400×10^3 (GCC) and >400×10^3 (LCC). Obtained results ensured that variety of colostrum quality is determined by its cytology. In GCC group not only protein content but also fat was higher. Noticeable changes in concentration of immunoglobulines were proved according to Lf and Lz increase, moreover very meaningful varieties in amount of unsaturated fatty acids were noticed. It is crucial to improve colostrum quality in dairy breeding, so all factors that affect it must be recognized in detail. The quality of colostrum varies, with that variability being determined by cytological quality. Colostrum rich in health-promoting ingredients (especially IgG) and with a low level of SCC shortens the period in which calves are at high risk for disease. Bacteria in colostrum may bind free IgG in the gut lumen or block the uptake and transport of IgG molecules into the enterocytes, thus reducing absorption of IgG. 'The authors acknowledge the financial support for this project provided by transnational funding bodies, being partners of the Horizon 2020, ERA-net project, CORE Organic Cofund, and the cofound from the NCBR'

Disposal reasons as indicator traits for health traits in German Holsteins

J. Heise[1], K.F. Stock[1], S. Rensing[1], H. Simianer[2] and F. Reinhardt[1]
[1]IT Solutions for Animal Production (vit), Heinrich-Schroeder-Weg 1, 27283 Verden, Germany, [2]University of Goettingen, Center for Integrated Breeding Research, Animal Breeding and Genetics, Albrecht-Thaer-Weg 3, 37075 Goettingen, Germany; johannes.heise@vit.de

In Germany, routine genetic evaluations for direct health traits started in April 2019. Phenotypic data come from manual documentations of farmers, hoof trimmers and veterinarians. Population-wide data recording has not been established yet. Furthermore, the historic depth of health data is relatively low compared to data used in other genetic evaluation systems in Germany. In contrast, disposal reasons have been recorded for decades for every disposed cow under milk recording. Previous analyses have shown the potential to use this information for improving the reliabilities of EBV for direct health. Therefore, a routine evaluation system was developed for 4 of the 10 routinely recorded disposal reasons: 'udder diseases', 'claw and leg diseases', 'infertility', and 'metabolic diseases', which correspond to the health trait complexes 'mastitis resistance', 'hoof health', 'reproduction', and 'metabolic stability'. The trait definition for the genetic evaluation of disposal reasons is 'survival against a certain disposal reason'. Phenotypic data are binary coded (0: culled for respective reason, 1: period survived or culled for another reason) for each of nine periods (lactations 1 to 3: days 0-49, 50-249 and 250 to subsequent calving). For 'infertility', only the last period of each lactation is considered, and for 'metabolic diseases' only the first period. Genetic evaluations are performed using a multi-trait animal model with herd-year-season as fixed effect. For each disposal reason, period-wise EBV are combined to an index using uniform weights. Heritabilities of the indices range from 0.024 ('metabolic diseases') to 0.053 ('claw and leg diseases'). Genetic correlations between the disposal reason indices and respective health complex indices used for the blending of disposal EBV into health EBV are 0.85 for 'udder diseases' to 'mastitis resistance', 0.6 for 'claw and leg diseases' to 'hoof health', 0.55 for 'infertility' to 'reproduction', and 0.8 for 'metabolic diseases' to 'metabolic stability'. Reliabilities of health EBV improve by ca. 0.05 to 0.10 on average for sires.

New strategies: prevention of diarrhoea in piglets

Y. Yin[1,2]

[1]*College of Animal Science and Technology, Hunan Agriculture University, Changsha, China, P.R.,* [2]*National Engineering Laboratory for Pollution Control and Waste Utilization, Ministry of Agriculture; Institute of Subtropical Agriculture, CAAS, Changsha, Hunan, 410128, China, P.R.; yinyulong@isa.ac.cn*

Post-weaning diarrhoea (PWD) is one of the most serious threats for the swine production worldwide, and is characterized by watery diarrhoea, dehydration, loss of body weight, and high mortality. An in-depth understanding of the pathogenic mechanism is vital to improve the control methods on PWD. PWD can be caused by a serious of chain reactions: the stress of adapting to new environment and less digestible feed may induce low voluntary food intake and poor intestinal digestibility, thereby cause enteral nutrition deficiency and intestinal barrier dysfunction; the pathogenetic antigen and enterotoxin are absorbed through impaired mucosal barrier, and drive the onset of PWD. Based on the pathogenic mechanism of PWD, we mainly implemented the following four strategies: (1) establishing friendly animal welfare and farm environment; (2) improving feed processing technology; (3) optimizing creep feed; (4) utilization of antimicrobial, protective agent, and astringent. Briefly, to reduce the rate of diarrhoea in piglets, we maintained the hygienic farm environment by using new-type house, increased the digestibility and utilization of nutrients by improving the physical, biological, and chemical feed processing methods, and conducted studies on both accurate nutrition formula and alternatives of antibiotic. In our studies, various natural materials such as lysozyme, acidifier, plant-derived active substance, probiotics, yolk antibody, yeast nucleotides, peptides, as well as the functional amino acids (aromatic amino acids, arginine and taurine) have been tested as effective nutrients to prevent diarrhoea, maintain the intestinal barrier function and improve intestinal microbial homeostasis in piglets. Furthermore, we found that high protein diet and vitamin B3 supplementation could increase and aggravate the post-weaning diarrhoea. Based on the above results, we will establish new strategies to prevent and alleviate post-weaning diarrhoea in piglets, and reduce the utilization of antibiotics, which further provide the theoretical and technical basis for improving intestinal health and growth performance in piglets.

Antimicrobial usage evolution between 2010, 2013 and 2016 in a group of French pig farms

A. Hemonic[1]*, A. Chiffre*[2]*, I. Correge*[1]*, C. Belloc*[2] *and M. Leblanc-Maridor*[2]

[1]*IFIP, La Motte au Vicomte, 35650 Le Rheu, France,* [2]*Oniris & INRA, 101 route de Gachet CS 40706, 44307 Nantes, France; catherine.belloc@oniris-nantes.fr*

Monitoring antimicrobial usage in pig farms is a key element of a reduction plan. The objective of this study was to analyse the antimicrobial usage evolution in the same farms between 2010-2013-2016 and to identify the factors of variations. The study monitored antimicrobial usage by weight group in 2016 in 33 farrow-to-finish farms in the West of France. The antimicrobial usage had ever been registered twice for 23 of them in 2010 and 2013 and once for 10 of them in 2013. It was quantified by the number of Course Doses per produced pig per year (nCD/pig). Farmers were asked about the factors that could explain the evolution between 2013-2016. On average, antimicrobial usage significantly decreased over six years (-38%). However, a high variability of individual evolutions was observed: among the 23 farms with three annual data, 43% decreased their use between 2010-2013 (-3 nCD/pig on average) but had a stable use between 2013-2016 (-0,2 nCD/pig). 26% decreased their use between 2010-2013 (-4 nCD/pig on average) and also between 2013-2016 (-2 nCD/pig). 9% increased then decreased their use during the two periods (+4 then -7 nCD/pig). One farm had the opposite trajectory (-9 then +2 nCD/pig) and another always increased its use (+2 then +5 nCD/pig). Among the 33 farms with data in 2013-2016, 36% decreased their use (-2 nCD/pig on average), 39% had a stable use and 24% increased their use (+3 nCD/pig). For sows, suckling piglets and fattening pigs, most of the farms had stable usage between 2013-2016. Only antimicrobial usage for weaned piglets was more frequently reduced. Increases were explained by occurrence of sanitary problems (mainly urogenital, digestive and respiratory problems on sows, piglets and fatteners respectively). Decreases were explained by vaccination, stop of preventive treatments and improvement of herd management. This study highlights the variability of individual trajectories in antimicrobial usage, due to sanitary issues that may be different according to each weight group. It usefully complements the monitoring of average evolution at the country level.

Antimicrobial usage: pig farmers' perceptions, attitudes and management

Y. Piel[1], A. Le Gall[1], C. Belloc[2] and M. Leblanc-Maridor[2]
[1]Sanders, rue Monge, 22600 Loudeac, France. [2]Oniris & INRA, CS 40706, 101 route de Gachet, 44307 Nantes, France;
catherine.belloc@oniris-nantes.fr

European and national initiatives have been developed to mitigate the risk from antimicrobial use in animals. However, they are facing challenges and information gaps, especially related to how to quantify, explain and reduce antimicrobial use. In order to update data concerning the different practices and perception for antimicrobial usage in pig farms, a study has been conducted to compare antimicrobial usage, technical performances, management practices, farmers' perception of their antimicrobial usage and farmers' attitudes toward antimicrobial resistance. he survey was carried out in 20 selected pig farms located in the West region in France among 156 monitored since 2015 for their antimicrobial usage (based on vet prescriptions). A questionnaire has been filled during an interview with farmers. The association between practices, technical performances or perceptions has been analysed with Chi^2 or Kruskal-Wallis tests. No link has been seen between technical performances and antimicrobial usage. Vaccination is considered as a major tool of antimicrobial reduction like biosecurity (17/20) or the use of 'alternatives' (15/20). Nevertheless, during an opened question concerning the ineffective measures, these 'alternatives' have been cited (5/20). In all the interviews, pig farmers underline the strong advisory role of the veterinarian and state that antimicrobial resistance is a main concern. Among the farmers' proposals as key measures for antimicrobial reduction, a better training and more information/knowledge concerning diseases, treatments or alternatives were cited (8/25). This study highlights brakes and levers for antimicrobial reduction in pig farms. Many positive and encouraging points like the strong implication of the veterinarian as the main advisor on animal health. Strengthened advisory role implies veterinarians to have better communication skills as asked by the farmers and to adjust their advices to the perceptions and attitudes of the farmers. Finally, the absence of link between the technical performances and the consumption of antimicrobials is also a major point to help veterinarians to engage farmers to reduce their antimicrobial use and comply with the alternative measures they recommend.

DISARM: community of practice for antibiotic resistance management via multi-actor farm health plans

E. Wauters and F. Leen
Flanders research institute for agriculture, fisheries and food, Social Sciences Unit, Burg. Van Gansberghelaan 115 box 2, 9820 Merelbeke, Belgium; frederik.leen@ilvo.vlaanderen.be

Without urgent, coordinated action of many stakeholders, we are headed for a post-antibiotic era. To reduce the threat of antibiotic resistance, part of the solution will come from a more prudent and responsible use of antibiotics in livestock farming. In the new EU Horizon 2020 thematic network DISARM, we build a Community of Practice (COP) in which farmers, veterinarians, suppliers, processors, policy makers and researchers collaborate to identify, co-create and disseminate best livestock farming practices. The obtained knowledge and insights from the COP will be used in piloting and promoting a multi-actor farm health plan (MAFHP) approach. Previous research has well documented the potential beneficial effect of coaching farmers for improving animal health while reducing the use of antibiotics. Further, previous studies have also highlighted the need for farm-specific approaches to improving agricultural production. Based on this knowledge, it is the fundamental idea of DISARM that, in order to stimulate a more responsible use of antibiotics, we need viable approaches for coaching farmers towards farm-specific improvement plans. In DISARM we pilot and test a coaching approach to facilitate multi-actor teams consisting of the farmer, the herd veterinarian, nutritionist, the equipment supplier and potentially other relevant stakeholders to collaborate in the development of farm-specific action plans. Guidelines for the successful development of such MAFHPs will be published in an on-line toolbox. Forty farms across 8 European countries evenly divided across the poultry, pig, dairy, beef and sheep industry will pilot this MAFHP-approach. The results obtained at these farms will be used as dissemination material to further stimulate the use of the toolbox and the adoption MAFHP's in the European livestock industry. Additionally, these innovator pilot farms will act as ambassadors to promote the approach among their colleagues during events and workshops.

First step to increase swine farmers' trust in their veterinarians: development of a scale

M. Leblanc-Maridor, J. Le Mat, F. Beaugrand, M. De Joybert and C. Belloc

Oniris & INRA, CS 40706, 101 route de Gachet, 44307 Nantes, France; mily.leblanc-maridor@oniris-nantes.fr

Because of the rising threat from antimicrobial resistance, pig farmers are strongly encouraged to reduce their antimicrobial usage. In order to achieve national and European reduction targets, herd level action is needed. An intervention study showed that farms with higher compliance with the intervention plans tended to achieve bigger reduction. Moreover, farmers who followed the vet recommendations often trust the measure. In human medicine, patients' trust in their physician is considered essential for good quality and effective medical care. Several questionnaires/scales exist in human medicine, but none of them have been developed in veterinary medicine. Assessing farmers' trust in their veterinarians is probably an essential step to improving adherence to treatments or recommendations. Thus, a scale is needed, which is capable of identifying the different levels and dimensions of the trust. The aim of this study was to develop and validate the Trust in Veterinarian Scale (TiVS), which aims to measure swine farmers' trust in their veterinarian. The scale construction process used a literature review and involved a panel of voluntary professionals through focus groups and open-ended qualitative interviews. A list of items, based on a multidimensional theoretical framework, explored the different dimensions of trust. Dimensionality, internal consistency, test–retest reliability and construct validity were investigated. The scale comprised at the moment 52 items, divided in seven dimensions (Competence, Availability, Integrity, Fidelity, Honesty, Caring, Global trust) with excellent psychometric properties. TiVS is the first scale adapted to pig farmers' trust in veterinarian. Many factors have an impact: history, psychology, competence, economical context, production organisation, agricultural advisor, etc. So, trust between farmers and veterinarians is a dynamic concept, changing over time and circumstances. Assessing farmers' trust in their veterinarians is probably an essential step to improve adherence to treatments or recommendations. Thus, this scale could be used to identify the different levels and dimensions of trust. Further studies are needed to confirm our result and validate TiVS in other production.

Early disease detection for weaned piglet based on live weight, feeding and drinking behaviour

M. Marcon, Y. Rousseliere and A. Hemonic

IFIP, La motte au Vicomte, 35650 Le Rheu, France; anne.hemonic@ifip.asso.fr

Early disease detection is one of the key to effective disease control in farms and reducing antibiotics usage. A batch of 153 weaned piglets was used to test a first machine learning algorithm in order to predict the individual health state of each animal. In order to build the early disease detection algorithm, nine boxes of 17 piglets has been set up with automata. In real time within this section we knew the number of times each animal went to the drinker or the feeder, the quantity of water and feed it took and its weight. As the golden standard to know either a piglet seems healthy or not, the clinical signs will be observed by trained operators on each pig every workday and recorded on a standardized grid (diarrhoea, cough, lameness…). Then, data collected from this batch of 153 piglets were used to create an algorithm with the software R, based on bagging and random forest machine-learning method. The database was split into learning (70%) and testing (30%). We obtained a global success of 86% of good prediction. In order to validate the accuracy of the model, a second batch of 153 piglets was used. Every day, a list of predicted sick pigs was printed automatically, indicating the individual identification of the animal, and its pen. Then, the results of these predictions were compared with the golden standard (observations of clinical signs by trained operators). Out of 3,437 observations (including predictions that the piglet is not sick), the algorithm correctly predicted the status of the piglets 2,462 times. Artificial intelligence has made 72% of good predictions. Regarding the true positive results, 96 alerts out of 117 were actually associated with observations of animals suffering mainly from diarrhoea within two days (82% of success). Now, the aim is to improve this algorithm in different ways: to test accelerometers to check the activity of each piglet; to be more accurate on recording cough by a microphone (SOMO, Soundtalks); to test if some trajectories of behavioural change are linked to specific diseases (lameness, digestive or respiratory disease) and not only to generic disease. These studies will be part of the Healthylivestock project (EC funded H2020 research project).

Added value of data integration to reduce the use of antimicrobials on dairy farms

C. Firth[1], C. Egger-Danner[2], K. Fuchs[3], A. Kaesbohrer[1], M. Suntinger[2], T. Wittek[1], B. Fuerst-Waltl[4] and W. Obritzhauser[1]
[1]Univ.Vet.Med, Vet.pl.1, 1210 Vienna, Austria, [2]ZuchtData, Dresd.Str.89/19, 1200 Vienna, Austria, [3]AGES, Zinzendorfg. 27/1, 8010 Graz, Austria, [4]BOKU, Greg.-Mendel.-Str.33, 1190 Vienna, Austria; suntinger@zuchtdata.at

With respect to the increasing emergence of antimicrobial resistance, the use of antibiotics in livestock production is an issue of growing concern. In an observational study in 249 dairy herds (6,475 cow-years) in Austria, antimicrobial treatments were analysed by the number of animals treated daily per 365 production days (DDD/cow/year). Treatment data that were provided by 17 different veterinary practices showed very diverse patterns of antimicrobials used for treatment of mastitis and for drying-off. For all diseases indications, each animal was treated 1.20 (median) times per year with a defined daily dose (DDD) of antibiotics (of which 0.65 (median) times per year was for the treatment of mastitis). Adjusted by calving interval and replacement rate, approximately one out of every two cows was dried off using an antibiotic agent (0.47 (median) defined course doses (DCD) per cow per year). After adjusting DCDs of antibiotic drying off products to make them comparable with DDDs of antibiotics for systemic, intrauterine and intramammary use, 60% of all antimicrobials were used for udder health. This study also showed that the pathogens isolated from mastitis milk were predominantly contagious on some farms and mainly environmental on others. These results support the need to develop tools which lead to a more evidence-based prudent use of antimicrobials when treating mastitis and drying off dairy cattle. Analyses across routinely-recorded production data, health data and antimicrobial use provide valuable information on disease-risks as well as the cow groups at risk. Assessing the infection status of the udder, by means of milk culture results, can assist in decision-making processes regarding more precise control and prevention measures to improve udder health. The more information available, the more targeted a treatment can be. Standardization and integration of data, therefore, play a crucial role in the prudent use of antimicrobials on dairy farms.

Elevated platforms as enrichment and to monitor activity and weight in broilers

J. Malchow, H. Schomburg and L. Schrader
Friedrich-Loeffler-Institut, Institute of Animal Welfare and Animal Husbandry, Dörnbergstraße 25/27, 29223 Celle, Germany; julia.malchow@fli.de

Elevated platforms are an enrichment in broiler housings and, in addition, possibly can be used to continuously monitor activity and weight of chickens. To validate this method a total of 81 Ross 308 chickens were kept in one pen for five weeks. The pen provided one elevated platform (h×l×w: 50×300×60 cm) and single-phase feed and water were given *ad libitum*. Within the framework of the platform, a weighing system with four single point load cells were installed and connected to a transmitter. As output, the electronic voltage was continuously logged and transformed into weights. We develop algorithms for automatic analysis of usage of the elevated platforms using the data of weighing system as input data. A model describing the relationship between measured total weight on the platform (P), number of birds on the platform (n), and the average individual bird weight (w), over time (t) is given by $P(t) = n(t)w(t)$. Hence, if the average bird weight is known the number of birds on the platform can be estimated as the quotient of total and individual weight. An approximation of the average bird weight over time can be obtained by resorting to breeder specifications or by analysing subsequent changes in total weight indicating birds entering or leaving the platform. For a first validation, we compared results obtained as described above using breeder information of the average bird weight to results of the manual evaluation of video data where the number of chickens was counted by using hourly scan samplings. Average use of the platform from 7 am to 5 pm was calculated for two days of each week of life during the entire fattening period, respectively. Spearman's rho test performed in R shows a significant correlation between the two samples (R=0.84, P=0.0045). Using the proposed methods to monitor platform usage, the aim of future work is to automatically detect significant differences in the use of elevated platforms compared to normal behaviour as an indicator for health problems.

A smart ear-attached sensor for real-time sows behaviour classification

L.S. Liu[1], M.X. Shen[1], W. Yao[2], R.Q. Zhao[3] and M.Z. Lu[1]

[1]College of Engineering, Nanjing Agricultural University, Nanjing, Jiangsu, 210031, China, P.R., [2]College of Animal Science & Technology, Nanjing Agricultural University, Nanjing, Jiangsu, 210095, China, P.R., [3]College of Veterinary Medicine, Nanjing Agricultural University, Nanjing, Jiangsu, 210095, China, P.R.; liulongshen@njau.edu.cn

Monitoring animal behaviour has always been a subject of great interest since behaviour is one of the most important indexes when we evaluate their health and welfare. There are recent research projects focused on designing monitoring systems of measuring sows' activities based on accelerometer. Traditionally, the acceleration data is transmitted to a server to classify the behaviours. However, wireless network unreliability and high-power consumption have limited their applicability. The aim of this research was to develop an animal behaviour recognition, classification and monitoring system based on a wireless sensor network and ear-attached device, provided with accelerometer and an embedded multi-layer neural network, to classify the different behaviours based on the acceleration data. The main novelty of this study is the full implementation of a reconfigurable neural network embedded into the sows' ear-attached device, which carries out a real-time behaviour classification and sends the results to receiver. Moreover, this approach reduces the amount of acceleration data transmitted to the receiver, achieving a significantly energy saving. The ear-attached device is developed, while the embedded neural network is need to test in a real scenario.

Individual pig tracking across multiple cameras for the extraction of behavioural metrics

J. Cowton[1,2], J. Bacardit[1] and I. Kyriazakis[2]

[1]School of Computing, Newcastle University, Newcastle Upon Tyne, NE4 5TG, United Kingdom, [2]School of Natural and Environmental Sciences, Newcastle University, Newcastle Upon Tyne, NE1 7RU, United Kingdom; ilias.kyriazakis@newcastle.ac.uk

Individual pig tracking from video feeds is key to stepping away from pen-level treatment and towards individual pig care as it enables the estimation of individual-level measures of activity. By doing so we can monitor behavioural changes over time and use these as an indicator of health and well-being, which will assist in the early detection of disease allowing for earlier and more effective intervention. However, it is a much more computationally challenging task than pen-level analysis, as data becomes sparser, and mistakes in identification and tracking can accumulate and, over time, provide noisy measures. Deep learning algorithms are well suited to tackle these challenges in order to create a low-cost, automated system for individual pig behaviour tracking. We combined deep convolutional neural networks (CNN) with multiple deep learning-powered multi-object tracking methods to create a system that could autonomously localise, track, and extract metrics pertaining to individual pig behaviours across multiple cameras without requiring specialised setup or inter-camera synchronisation. We used Deep Simple Online Real-time Tracking for intra-camera tracking. This method tracks pigs between frames using their frame-to-frame distance and how similar a pig looks compared to the previous frame. On top of this, we used Siamese CNNs for inter-camera tracking. This method is trained to be able to identify whether two images of an object are the same object, allowing to re-identify a pig after it leaves one camera and enters another. Our detection method was capable of 0.9 mAP (mean average precision), our tracking implementation achieved 95% MOTA (multi-object tracking accuracy), and our individual behavioural metrics were calculated with less than 0.01 MSE (mean squared error). We show that deep learning is competent at solving large-scale, individual pig tracking, with data coming from a commercial farm setting and recorded with standard, cheap cameras. We have also shown that this tracking information was able to be used to compute reliable behavioural metrics for each of the pigs.

How many pigs within a group need to be sick to cause a diagnostic change in the group's behaviour?

H.A. Dalton[1], A.L. Miller[1], T. Kanellos[2] and I. Kyriazakis[1]
[1]Newcastle University, School of Natural and Environmental Sciences, Agriculture Building, NE1 7RU, Newcastle upon Tyne, United Kingdom, [2]Zoetis International, Cherrywood, Loughlinstown, D18 K7W4, Dublin, Ireland; ilias.kyriazakis@newcastle.ac.uk

In commercial pig operations, the early detection of disease is critical to limit negative impacts to welfare and productivity and prevent further spread throughout the herd. Early disease diagnosis can be achieved by identifying specific behavioural changes that occur at disease onset to allow for faster intervention and improved treatment success. In our study, the objective of Trial 1 was to evaluate the behavioural disturbance in groups of pigs when using total pen vaccination as a model of an acute health challenge. The objective of our second trial was to quantify the minimum proportion of individuals required within a pen to detect these behavioural changes at the group level using three levels of vaccination treatments: a control (Con; 0% pigs), Low ($\pm$20% pigs), or a High ($\pm$50% pigs) number of pigs vaccinated in the groups. In Trial 1, total pen vaccination resulted in group level reductions in rates of standing, ($P<0.001$), drinking ($P<0.001$), feeding ($P<0.001$), enrichment interaction ($P<0.001$), and non-nutritive visits to the feeder ($P<0.01$). These changes coincided with increased rectal temperatures ($P<0.001$) and increased lying behaviour ($P<0.01$), which confirmed vaccination serves as suitable model of acute illness in pigs. In Trial 2, the groups of pigs showed decreased standing ($P=0.064$) and enrichment interaction ($P<0.001$), but increased lying rates ($P<0.001$) from the Con to Low treatment. In contrast, the rates of group level feeding (treatment $\times$ time of day: $P<0.01$) and drinking (treatment: $P=0.07$) only showed a reduction from the Con to High treatments, suggesting these behaviours would be better suited for confirming disease spread within a herd. Our results suggest group level changes in standing, lying, and enrichment interaction are valuable for confirming the presence of disease when only a few individuals are acutely ill within a group. Therefore, automated early warning systems for pig production should focus on variation in these specific behaviours to improve early disease detection.

Veterinary precision medicine and reducing veterinary drugs in animal production in China:case study

S.M. Yang and R. Liu
Institute of Quality Standards and Testing Technology for Agro-Products (IQSTAP) of CAAS, No.12 Zhongguancun South St., Haidian District, Beijing, 100081, China, P.R.; yangshuming@caas.cn

In 2018, Action on Reduction of Antimicrobial Drugs Use was initiated by Ministry of Agriculture and Rural Affairs with the aim of zero growth of antimicrobial drugs. Veterinary Precision Medicine (VPM), defined as an optimized preventive or therapeutic approach (right animals, right drugs, right dose, right time) based on use of technologies of disease and antimicrobial resistance monitoring and targeting drugs, was focused. The paper reviewed reported measures for practical animal production and relevant research papers, included their functions, achievements, profit analysis and limits. Several pilot farms were investigated as case study. It was concluded that the reduction of antimicrobial drugs in animal production with both preventive and therapeutic aims is a general trend in China, and the VPM techniques will play a key role on reduction of antimicrobial drugs in animal production.

Monitoring individual water consumption for optimization of antibiotic treatments in herds

B.B. Roques, M.Z. Lacroix, A.A. Ferran, D. Concordet and A. Bousquet-Mélou

INTHERES, Université de Toulouse, INRA, ENVT, National Veterinary School of Toulouse, 31076 Toulouse, France; a.bousquet-melou@envt.fr

The metaphylactic use of antibiotics consists in a collective treatment of the herd as soon as the disease is detected in some animals. The collective delivery of antibiotics via drinking water is a promising strategy in this context, but the actually ingested doses are dependent on the individual drinking behaviours, which could lead to an under-exposure of some animals to the antibiotics potentially responsible for therapeutic failures. In order to optimize this strategy, we characterized the individual drinking behaviours of fattening lambs and measured their impacts on the interindividual variations of the ingested doses. About 800 lambs were studied in a fattening house over 4 periods. For each period, the individual real-time water consumptions of 200 lambs were recorded using water meters connected to drinking troughs that detected the lamb thanks to RFID chips in ear tags. One hundred of these lambs received antibiotics: sulfadimethoxine/trimethoprim (SDM/TMP, period 1), sulfadimethoxine (SDM, period 2), oxytetracycline or tilmicosin (OTC or TIL, period 3) or flumequine (FLU, period 4), and antibiotics were measured in plasma and in the drinking water. The pharmacokinetics of the antibiotics was also determined in other lambs during classical PK studies performed in the laboratory. The average individual daily water consumption of each group was surprisingly much lower than the expected one (2.5 l/day) with a very high variability between lambs. Concomitantly, plasma concentrations of antibiotics varied between lambs from 4-fold (SDM) to 10-fold (TIL, FLU). A pharmaco-statistical model was built by incorporating the time series of individual water consumptions in a classical pharmacokinetic model which was able to predict plasma concentrations actually observed in the lambs. Such model will allow to predict for any antibiotic the impact of individual drinking behaviours on individual plasma exposures, and to propose improved dosage regimens adapted to the specificity of animal behaviours. The same experimental and modelling strategy will be used for pigs in the context of HealthyLivestock.

An overview of inVALUABLE: insect value chain in a circular bioeconomy

L.H.L. Heckmann

Danish Technological Institute, Life Science, Danish Technological Institute, Kongsvang Alle 29, 8000, Denmark; lhlh@dti.dk

inVALUABLE is, to-date, one the largest publicly funded R&D projects in Europe on insects as feed and food. The project involves 10 partners and runs from January 2017 to December 2019 with a total budget of 3.7M EUR. The vision of inVALUABLE is to create a sustainable resource-efficient industry for animal protein production based on insects. Overall, inVALUABLE addresses three major challenges for the insect industry: (1) upscaling of production to industrial level; (2) regulatory issues; and (3) consumer acceptance. Together with other large European R&D initiatives, trade associations and networks, inVALUABLE is expected to have a large impact on shaping the growing insect industry. The goal is that inVALUABLE will facilitate Danish industrial insect production and be an enabler of new market opportunities for insects as feed, food and other high-value components. inVALUABLE aims to demonstrate the potential of using insects to meet the increasing demand for protein in the food chain by assessing the following specific objectives: (1) developing an insect value chain using low-value by-products, reintroducing valuable resources back into the food chain; (2) document the nutritional potential of insects using state-of-the-art animal models; (3) combine the best technologies to enable market penetration, focusing on large-scale production, automation and processing; and (4) support Danish/EU authorities on feed/food legislation providing data to ensure safe insect products. The presentation will introduce the audience to the session by providing an overview of the activities that have been conducted during the majority of the project; highlighting some of the key external collaborations as well as the strong focus on stakeholder interaction and public dissemination.

Optimizing temperature for adult reproduction and humidity for eggs and young larvae of *T. molitor*

J.L. Andersen[1], S.F. Thormose[2], N. Skytte[2], I.E. Berggreen[1] and L.-H.L. Heckmann[1]
[1]Danish Technological Institute, Insect and Protein Technology, Kongsvang Alle 29, 8000 Aarhus C, Denmark, [2]Aarhus university, Bioscience, C.F. Moellers Alle 3-5, 8000 Aarhus C, Denmark; jlan@dti.dk

The optimal temperature range for growth of the *Tenbrio molitor* larvae has been determined and optimized over the past couple of years. However, the optimal temperature and relative humidity (RH) for the adults and newly hatched larvae is less explored. The objectives of these studies were to: (1) determine the optimal temperature for the adults to maximize egg production; and (2) determine the optimal RH for maximized survival and growth of the eggs and newly hatched larvae. To investigate the optimal temperature for the beetles' reproduction, two experiments were conducted. First, a broad range of temperatures were tested to find the most optimal temperatures and second, a complete experiment was done to investigate the fecundity at the predicted 4 most optimal temperatures (25, 26, 27 and 28 °C, n=3). We found that the fecundity for beetles at the 4 temperatures were not statistically different. However, the survival of the beetles increased with decreasing temperature, suggesting that 25 °C (perhaps lower) is the optimal temperature to maximize egg production while lowering daily maintenance and adult culture turnover. Optimizing the RH for *T. molitor* eggs and newly hatched larvae was done in a pilot experiment spanning from 11 to 92% RH at 30 °C. From the pilot data 5 RHs were chosen for a larger experiment (43, 51, 68, 75 and 84% RH, at 27 °C). One hundred eggs were collected and set up in pairs 4 times for a total of 8 replicates. Survival and mass of the larvae was collected every week until week 3 (week 2 after hatching). We saw no significant difference in the survival of the larvae exposed to the different RHs. However, we did find a significant increase in the fresh mass of the larvae from around 1.1 mg at 43% RH to around 2.1 mg at 84%, after 3 weeks. This data leads to the conclusion that survival might not, as predicted, be strongly correlated with RH, but the initial growth of the larvae is affected and increases with increasing RH, at least up to 84%.

Insights on feed and nutritional requirements of *Tenebrio molitor*

I.E. Berggreen, J.L. Andersen, N.S. Jönsson and L.H. Heckmann
Danish Technological Institute, Life Science, Kongsvang Allé 29, 8000, Denmark; ideb@teknologisk.dk

Feed is important for various aspect of insect production, from growth and health to economic and environmental considerations. During the inVALUABLE project feed and nutrition for *Tenebrio molitor* larvae have been investigated. The feed-mixtures were designed from vegetable-based former foodstuffs to implement a circular insect production. The objectives in these studies was to explore and optimize growth of the larvae in relation to: (1) the feed composition (typically; flour mixtures from several grains, seeds and legumes); and (2) the addition of existing and custom-made premixes and/or salts to the previously designed feed mixes. Also, a brief investigation into the protein requirement of larvae in different life stages has been initiated. All the feed-optimization experiments were conducted in the insect pilot facility lab, at the Danish Technological Institute, at either 30 or 27 °C with 65±5% relative humidity. Some experiments were carried out in our pilot production in 60×40 cm boxes, while other was conducted in smaller experimental boxes (20×30 or 10×10 cm). Most of the experiments with positive results have been implemented directly in our continuously running *Tenebrio* pilot production. The presentation will show results from experiments conducted during inVALUABLE and address some of the challenges in relation to feed-optimization in a mealworm production. These will include experiments on; general feed composition, salt, premixes, proteins and amino acids and particle size of the feed.

Insect diseases in production systems: prevention and management

J. Eilenberg, A. Lecocq and A.B. Jensen

University of Copenhagen, Plant and Environmental Sciences, Thorvaldsensvej 40, 1871 Frederiksberg C, Denmark; jei@plen.ku.dk

Insects in production systems may become infected by insect pathogens like bacteria, fungi and viruses and develop disease. In some cases, such insect pathogens can develop epidemics in the insect stock, causing economic losses for the producer. It is essential to prevent and/or manage insect pathogens. We will outline some major challenges and options. First, we describe three types of insect production systems, namely closed production systems, semi-open production systems and open production systems. Then, we list some important insect diseases to occur in different main production insects. An important aspect is that there seems to be different levels of insect disease pressure on the most commonly grown insect species. Several insect diseases are known from mealworm (*Tenebrio molitor*) a house cricket (*Acheta domesticus*), while black soldier fly (*Hermetia illucens*) seems unaffected by diseases. We perform currently studies to learn, if the use of probiotics can improve mealworm health as one of several management tools.

Probiotic *P. pentosaceus* inhibits bacterial growth and improves *Tenebrio molitor* larvae fitness

A. Lecocq

University of Copenhagen, PLEN, Thorvaldsensvej 40, 1871 Frederiksberg C, Denmark; antoine@plen.ku.dk

Insect production for food and feed around the world is burgeoning. With it, comes new challenges associated with mass rearing of livestock species. As such, there is a growing need for timely diagnosis, management and prevention of emerging and existing diseases. One hypothesized solution is the introduction of beneficial probiotic bacteria to the diet of the insects. In this study, we evaluated the inhibitory potential of a *Pediococcus pentosaceus* strain, extracted from the gut of *Tenebrio molitor*, on selected pathogens. We also carried out an assessment of the fitness benefits of the bacteria to the development of the mealworm larvae. Our bacterial strain successfully inhibited the growth of all tested bacterial pathogens. Higher growth rate, survival to adulthood, egg-laying capacity and adult weight in the presence of the bacteria all point to a significant probiotic effect of *P. pentosaceus* on *T. molitor*. Both *in vitro* and *in vivo* results support the promise of the use of probiotics in improving future industrial mass rearing facilities of the mealworm *T. molitor*.

Salmonella Typhimurium level in mealworm larvae after exposure to contaminated substrate

A.N. Jensen, C. Johnsen and S.H. Hansen
Technical University of Denmark, National Food Institute, Kemitorvet Bygning 202, 2800 Kgs. Lyngby, Denmark;
anyj@food.dtu.dk

Mealworms (*Tenebrio molitor*) for feed and food must be reared on feed-grade materials and findings of *Salmonella* spp. bacteria, which are important foodborne pathogens, have not yet been reported. However, *Salmonella* spp. belongs to the family of Enterobacteriaceae, which is found in levels up to 7.5 log cfu/g in mealworms. This indicates that *Salmonella* spp. may be able to thrive in the mealworms if introduced into the production. This study aimed to assess the level of *Salmonella* to be found in mealworms after exposure to different *Salmonella* contamination levels in the substrate. The initial contamination level was 2, 4 or 6 log cfu/g of a *Salmonella enterica* serotype Typhimurium strain resistant to rifampicin (in-house strain). Samples of mealworm and substrate were collected at start and additional four times until termination at day 7. For each sample, ten-fold dilutions series were prepared from 1 g of homogenized sample in 9 ml saline (0.9% NaCl). Each dilution was plated (0.1) on Nutrient Agar (NA) with rifampicin 50 mg/l to suppress competing bacteria, and incubated overnight at 37 °C before counting of colonies. Presumptive *S.* Typhimurium[rif] colonies were verified by slide-agglutinion. Preliminary results indicated that the *S.* Typhimurium level found in mealworm and substrate depended on the initial contamination level while there was no clear effect over time (7 days). In future studies it would be of interest to see for how long *Salmonella* will persist in the mealworms and the substrate after introduction. Also, the quantitative method is limited by a detection level of minimum 1 log cfu/g and a qualitative approach with enrichment of the samples would be necessary to exclude the presence of *Salmonella* in low numbers. Finally, the entire mealworm was homogenized and analysed, not allowing concluding whether the *Salmonella* had been ingested or only present on the surface. Adequate surface decontamination methods must be established to elucidate if *Salmonella* is ingested or not. However, as the entire mealworm is normally used, this may from a food safety perspective be less important.

Effect of feeding managements on the milk concentrations of short- and medium-chain fatty acids

E. Vargas-Bello-Pérez[1], R. Dhakal[1], M. Kargo[2], A.J. Buitenhuis[2], M.O. Nielsen[1] and N.A. Poulsen[3]
[1]University of Copenhagen, Department of Veterinary and Animal Sciences, Grønnegårdsvej 3, 1870, Denmark, [2]Aarhus University, Department of Molecular Biology and Genetics, Blichers Allé 20, 8830 Tjele, Denmark, [3]Aarhus University, Department of Food Science, Blichers Allé 20, 8830 Tjele, Denmark; evargasb@sund.ku.dk

The objective of this study was to identify the feeding managements that are responsible for production of milk with high concentration of SCFA (6 to 10 carbons) and MCFA (from 12 to 14 carbons). Milk samples were taken from 7 Danish organic dairy farms and 388 samples were analysed for milk fatty acid profile during indoor and outdoor seasons in 2017. The farms (F) had the following conditions for the previous 6 years based on % of grass silage in the ration, % of corn silage in the ration, dry matter intake per day and milk fat %: F1=66%, no corn silage, 21.3 kg/DM/d and 4.26%. F2=53%, 19%, 21.1 kg/DM/d and 4.13%. F3=56%, 10%, 25.1 kg/DM/d and 3.96%. F4=70%, 12%, 20.3 kg/DM/d and 4.37%. F5=51%, 14%, 24 kg/DM/d and 4.15%. F6=57%, 14%, 23.1 kg/DM/d and 3.85%. F7=50%, 24%, 22 kg/DM/d and 4.20%. The model considered interactions of herd (n=7) and season (outdoor=1 and indoor=2); parity (n=5), days in milk, and fat (%) as fixed effects and cow as the random effect. SCFA contents were affected (P<0.001) by farm, fat % and farm × season interactions. SCFA (0.51 g/100 g milk) were higher (P<0.05) in F4 in season 2. Saturated FA and MCFA contents were affected (P<0.001) by parity, farm, season, days in milk, fat %, and farm × season interactions. Saturated FA (2.95 g/100 g milk) and MCFA (0.69 g/100 g milk) were higher (P<0.05) in F7 in season 1. Monounsaturated FA and polyunsaturated FA were affected (P<0.001) by parity, farm, season, days in milk, fat %, and farm × season interactions. Monounsaturated FA (1.10 g/100 g milk) were higher in F7 in season 2 and polyunsaturated FA (0.17 g/100 g milk) were higher (P<0.05) in F4 in season 2. Overall, major FA groups had seasonal changes and the level (% of dry matter) of dietary inclusion of grass (G) and/or corn silages (C) appeared to have an effect on SCFA (70G+12C) and MCFA (50G+24C). Results from this study can be useful for developing nutritional strategies targeting to increase specific milk FA groups.

Effect of wheat soluble concentrate on performances and nitrogen metabolism in lactating dairy cows

A. Palmonari[1], D. Cavallini[1], L. Mammi[1], J. Vettori[2], P. Parazza[1], M. Dall'Olio[1], E. Valle[3] and A. Formigoni[1]
[1]University of Bologna, Dipartimento di Scienze Mediche Veterinarie, Via Tolara di Sopra 50, 40064, Italy, [2]Santa Catarina State University, Veterinary College, Santa Catarina, 2090, Brazil, [3]University of Torino, Dipartimento di Scienze Mediche Veterinarie, Grugliasco, 10095, Italy; alberto.palmonari2@unibo.it

Nowadays the use of by-products in dairy cows nutrition is becoming crucial for sustainability and profitability. The objective of this study was to test the wheat soluble concentrate (WSC) on performances and nitrogen metabolism in lactating dairy cows. Eight homogenous dairy cows were enrolled in a crossover design with 2 wks of diet adaptation and 1 wk of samples and data collection. Diets were isonitrogenous, and the major protein source was WSC in treated group (TRT, 5% inclusion on DM), while soybean meal was used in control group (CTR). DMI, rumination time (RT), rumen and manure samples, and milk production and quality were collected and analysed through the experimental weeks. Statistical analysis was performed using a MIXED procedure in R(2.5.3.1) with diet as fix effect, cow as random effect, and sampling time point as repeated measurement. Results showed no differences in productive parameters (DMI, RT, milk production), however long hay consumption resulted lower in TRT group (-0.3 kg/d, $P<0.10$) showing a better rumen environment. On the other hand, nitrogenous balance was improved in TRT group, in which milk urea content decreased (-3.21 mg/dl in TRT, $P<0.05$) as ruminal ammonia (-1.32 mg/dl in TRT 24 h after feeding, $P<0.10$). Moreover, milk fatty acids profile changed with higher de-novo content (+0.96%/100%Fain TRT group, $P<0.05$) and lower CLA production (-0.05%/100%FA of C18:2 Cis9, Trans11 in TRT group, $P<0.05$) in TRT group. Total tract fibre digestibility showed high values and no differences (78.63 on avg. $P=0.74$) among groups. These results underline that WSC utilization in dairy cows feeding would have no negative effects on production, intakes and physiology, and a positive effect on nitrogenous balance, with a better utilization of by-pass WSC protein source. These positive effects suggest that even a major WSC inclusion could be used in dairy cows nutrition.

Lower *in vitro* rumen metabolization of mycotoxins at conditions of rumen acidosis and dry conditions

S. Debevere[1,2], S. De Baere[2], G. Haesaert[3], S. Croubels[2] and V. Fievez[1]
[1]Ghent University, Faculty of Bioscience Engineering, Department of Animal Sciences and Aquatic Ecology, Coupure links 653, 9000 Ghent, Belgium, [2]Ghent University, Faculty of Veterinary Medicine, Department of Pharmacology, Toxicology and Biochemistry, Salisburylaan 133, 9820 Merelbeke, Belgium, [3]Ghent University, Faculty of Bioscience Engineering, Department of Plants and Crops, Coupure Links 653, 9000 Ghent, Belgium; sandra.debevere@ugent.be

Ruminants are generally considered less susceptible to the effects of mycotoxins than monogastric animals as the rumen microbiota are able to detoxify some of these toxins. Despite this potential degradation, mycotoxin associated subclinical health problems are seen in high productive dairy cows. In this *in vitro* research, two hypotheses were tested: (1) a lower rumen pH leads to a decreased degradation of mycotoxins; and (2) rumen fluid (RF) of lactating cows degrade mycotoxins better than RF of dry cows given their metabolically more active microbial population. Maize silage was spiked with a mixture of deoxynivalenol (DON), nivalenol (NIV), enniatin B (ENN B), mycophenolic acid (MPA), roquefortine C (ROQ-C) and zearalenone (ZEN). Fresh RF of 2 lactating cows (L) and 2 dry cows (D) was added to a buffer of normal pH (6.8) and low pH (5.8), leading to 4 combinations (L6.8, L5.8, D6.8, D5.8) that was combined with the spiked substrate. After 24 h of incubation, a complete detoxification of DON occurred at pH 6.8, in contrast with pH 5.8, where 14 and 91% of DON was still present for L5.8 and D5.8, respectively. NIV could not be detected anymore after 24 h of incubation at pH 6.8, in contrast with pH 5.8, where 38% was still present for L5.8 and 100% for D5.8. For ZEN, after 48 h of incubation, a partial transformation of ZEN to α-ZEL (13-22%) and β-ZEL (4-5%) was only observed at pH 6.8. After 48 h of incubation, disappearance of ENN B could be seen at pH 6.8 (L: 71%, D: 43%), but not at pH 5.8. For MPA and ROQ-C, neither pH nor lactation stage had an effect on the metabolization. In conclusion, low ruminal pH (e.g. SARA conditions) and bacterial inoculum of dry cows can have a lower ruminal degradation of certain mycotoxins such as DON, NIV, ZEN and ENN B. Hence, these mycotoxins could reach the small intestine intact and exert their toxic effects.

Evolution of the rumen fluid enzymatic activity during *in vitro* incubation

M. Simoni[1], F. Righi[1], A. Foskolos[2], E. Tsiplakou[3] and A. Quarantelli[1]
[1]University of Parma, Department of Veterinary Science, via del Taglio, 10, 43126 Parma, Italy, [2]University of Thessaly, Department of Animal Science, Campus Larisa, 41110, Greece, [3]Agricultural University of Athens, Department of Nutritional Physiology, Iera odos 75, 11855, Greece; federico.righi@unipr.it

The aim of this study was to evaluate the evolution of the rumen fluid enzymatic activity (EA) during *in vitro* incubation. Rumen fluids (RFs) were collected from dairy cows in 2 physiological stages fed 4 diets (DT): dry cows administered 100% hay (TH) or 80:20 forage:concentrate ratio (F:C) diets; and lactating cows fed hay *ad libitum* and concentrate separately at about 60:40 F:C or total mixed rations (TMR) with a 60:40 F:C. Three farms per each DT were involved and rumen collection was performed on 3 donor cows/farm. RFs were pooled by farm, divided in 2 flasks, inoculated at a ratio of 1:4 with medium and 5 g of a common substrate – a TMR diet- in an *in vitro* batch fermentation system. Flasks were maintained at 39 °C under anaerobic conditions for 48 hours. During the incubation process, diluted RFs were sampled in duplicate at 0, 1, 2, 4, 8, 24 and 48 h of incubation. Samples were centrifuged and filtered through 0.45 μm porosity filters for the EA determination. EA was evaluated through the radial enzyme diffusion method. Petri dishes with the specific substrate were inoculated with RFs, incubated for 16 h. Halos dimension was measured and expressed as area of the halos surface. Statistical analysis was performed through the repeated measures procedure of the general linear model using DT as a fixed factor, farm as random effect and intervals as repeated measures. With the exception of Xyl, starting from different EA, after an initial irregular peak of activity, a gradual reduction of EA was observed over time. RFs derived from TH showed an opposite trend for both cellulose and xylanase. RFs showed similar A activity at 4 h. Overall, DT showed significant effect for C and A ($P \le 0.001$) while only a trend was observed for Xyl ($P = 0.065$). At 48 h considering A and at 24 h considering C and Xyl, EA were similar between DT and only TH rumen fluid was different compared to the other RFs. Generally, it appears that the incubation of RFs of different origin with a common substrate tend to homogenise their EA.

Effect of mixing time of the total mixed ration on beef cattle performance

G. Marchesini, M. Cortese, N. Ughelini, M. Chinello and I. Andrighetto
University of Padova, Animal Medicine, Productions and Heath, Viale dell'Università, 16, 35020 Legnaro (PD), Italy; giorgio.marchesini@unipd.it

The aim of this study was to assess the effect of mixing time of the total mixed ration (TMR) on its physical and chemical characteristics, consistency over time, dry matter intake (DMI), rumination, activity and performance of beef cattle. The trial was conducted on 98 charolais bulls (476±36 kg) which were randomly assigned to two groups (A: 54 animals in 6 pens; B: 44 animals in 5 pens) which were fed rations with the same composition (DM=53.1; NDF=33.0% DM; Starch=32.6% DM; CP=12.3% DM), but that were mixed for a standard mixing time (SMT) or for a 10 minutes longer mixing time (LMT). The trial lasted 60 days and was divided into 2 periods (P1 and P2). After P1 the diets were swop between groups according to a cross-over design. Animals were assessed for average daily weight gain (ADG), DMI, rumination and activity. Data on TMR, ADG, DMI, rumination and activity were submitted to a mixed model using the pen as random effect and period, mixing time and their interaction as fixed effects. The differences on TMR composition between consecutive days within each mixing time length were tested using a mixed model with the pen as random effect and the day of sampling as fixed effect. LMT ration, as expected, had a lower particle geometric mean length (GML) compared with SMT (4.12 vs 5.08 mm). Bulls raised with the LMT showed a significantly higher ADG (1.96 vs 1.87 kg/day; P<0.05) and a lower conversion ratio (4.97 vs 5.39; P<0.05) compared with animals raised with SMT. Average daily rumination time and activity level did not differ between mixing time and were 381 minutes and 494 binary digits, respectively, but LMT bulls showed a lower (0.083 vs 0.095; P<0.05) index of dishomogeneity in activity (DA) and a lower sorting activity against long particles (99.8 vs 96.3%; P=0.005). SMT led to a more significant difference between consecutive days in NDF (P=0.223 vs P<0.001) and starch content (P=0.077 vs P<0.001) of TMR than LMT. These results indicate that LMT improves consistency of TMR over time, ADG, conversion ratio and reduces feed sorting. Acknowledgements: This research was funded by SCR Engineers Ltd.

Milk production and urination patterns of dairy cows grazing second year chicory-based herbage

M. Mangwe, R. Bryant, M. Beck and P. Gregorini

Lincoln University, Faculty of Agriculture and Life Science, Lincoln University, 7647, New Zealand; mancoba.mangwe@lincolnuni.ac.nz

Diurnal variation patterns of urination behaviour and therefore urinary nitrogen (UN) excretion between forage species present opportunities to manage high risk nitrogen loading periods by grazing ruminants. Chicory (*Cichorus intybus* L.) is an alternative forage that has shown potential to alter urination behaviour of dairy cows compared to traditional ryegrass/white clover (*Lolium perenne* L./*Trifolium repens* L) herbage. This study examined the effects of timing of chicory allocation on milk production and urination behaviour of grazing dairy cows. Replicated groups (4 cows per replication) of dairy cows were allocated to one of three feeding regimes: (1) perennial ryegrass/white clover only (RGWC), (2) ryegrass/white clover + morning allocation of chicory (CHAM) and (3) ryegrass/white clover + afternoon allocation of chicory (CHPM). Both RGWC and chicory treatments were grazed *in situ* using similar herbage dry matter (DM) allocation for all treatments (34 kg DM/cow/d above ground). Automated urine sensors and urine spot samples were collected to determine the diurnal patterns of urine excretion. There were no differences in apparent dry matter intake between treatments (16.3±0.78 kg/cow/day; P=0.11) but cows grazing CHPM had greater milk (22.0 vs 19.6±0.42 kg/cow/day, P<0.001) and milk solid yields (fat + protein; 1.97 vs 1.69±0.013 kg/cow/day; P<0.001) than those grazing RGWC. Cows grazing CHAM produced intermediate milk yield and solids. Cows offered chicory urinated more frequently than those grazing RGWC only (16.47 vs 9.64±1.1 events/cow/day) regardless of time of allocation, resulting in greater urine volumes per day than those grazing RGWC (48.25 vs 26.9±3.1 l/cow/day; P<0.001). The most urination events were recorded during peak gazing hours (i.e. 09.00 h to 13.00 h and 16.00 h to 21.00 h). Compared to RGWC, cows grazing CHAM and CHPM had the lowest UN and urine urea concentration, with greatest reductions observed within 4 hours in their chicory allocation. The study reveals that incorporating chicory into the management regime alters urination and UN excretions, while maintaining milk production when compared to the traditional management practices.

Feeding value of lactating cow diets popularly used by Vietnamese household dairy farms

B.N. Nguyen[1,2], C.V. Nguyen[3], H.T. Nguyen[3], T.X. Nguyen[1], K.D. Nguyen[3], S.H. Nguyen[1], H.T. Nguyen[1], J.B. Gaughan[2], R. Lyons[2], B.J. Hayes[2] and D.M. McNeill[2]

[1]Vietnam National University of Agriculture, Hanoi, 100000, Viet Nam, [2]The University of Queensland, Gatton, 4343, Australia, [3]Nong Lam University, Ho Chi Minh, 700000, Viet Nam; nn.bang@uqconnect.edu.au

A limited literature on household dairy cow diets in Vietnam suggests relatively simple formulations such as *ad libitum* Napier grass and commercial dry concentrate offered at approximately 0.5 kg per litre of milk. This could be limiting domestic milk production as these herds comprise more than 90% of Vietnamese dairy cattle. We evaluated the potential to improve production by characterising typical diets in four contrasting dairying regions of Vietnam relative to milk yield. Eight farms from each region were selected, from high or low altitude in the north and south of Vietnam. Each farm was visited over a 24-hour period in autumn 2017 to collect the data necessary to match nutrient supply to cow demand using the formulation model PC Dairy. Each feed ingredient offered to the herd was weighed to determine amount fed per cow. Samples of popular ingredients were pooled across farms within regions and subsampled to determination % dry matter and sent for nutrient analysis at the DairyOne laboratory, USA, to estimate whether energy or protein was first limiting to milk yield. Cow weight was estimated by girth circumference, days in milk by farmer records, and milk yield by direct measurement per cow into a bucket and subsampled for milk fat and protein composition. ANOVA was used to compare diets between regions. All diets included commercial dry concentrate in about the expected amount, but diets varied dramatically between regions in the types of forage and by-products used. Use of maize silage was predominant in the north but not the south. Use of by-products such as brewers grain and especially cassava pulp and rice straw was prevalent in the low altitude southern region. All diets lacked energy relative to protein. Maximum potential milk yields were higher than actual in all regions which could indicate that husbandry factors other than dietary energy or protein could be more important drivers of milk yield. Increasing dietary net energy in household dairy farms could improve the domestic milk supply in Vietnam.

Effects of diet composition on nitrogen use efficiency of beef cattle
A. Angelidis[1], L.A. Crompton[1], T. Misselbrook[2], T. Yan[3], C.K. Reynolds[1] and S. Stergiadis[1]
[1]*University of Reading, School of Agriculture, Policy and Development, PO Box 237, Earley Gate, RG6 6AR, Reading, United Kingdom, [2]Rothamsted Research, North Wyke, EX20 2SB, Okehampton, United Kingdom, [3]Agri-Food and Biosciences Institute, Large park, BT26 6DR, Hillsborough, United Kingdom; a.angelidis@pgr.reading.ac.uk*

Understanding the influence of diet composition on nitrogen (N) use efficiency (NUE) can inform the development of beef rations that minimise N outputs in faeces (FN), urine (UN) and manure (MN); and the subsequent nitrous oxide and ammonia emissions and nitrate leaching to groundwater. The aim of the present study was to investigate the influence of dietary N, neutral detergent fibre (NDF), acid detergent fibre (ADF), ether extract (EE), starch and metabolizable energy (ME) on FN, UN, MN, retained N (RN), and NUE (expressed by the ratios FN/MN, UN/MN, FN/nitrogen intake (NI), UN/NI, MN/NI). Individual animal data (n=300) from digestibility trials with beef cattle were grouped according to diet N content (low, <21; medium, 21-27; high, >27; g N/kg diet dry matter (DM)). A multivariate redundancy analysis was performed in CANOCO5, using diet energy and nutrient composition as drivers and UN, FN, MN, and the NUE ratios as response variables. Diet ME and starch concentrations were negatively correlated with N outputs, in all diet groups. Diet N concentration was positively correlated with N outputs and negatively correlated with NUE, in all diet groups. Diet NDF and ADF concentrations were: (1) positively correlated with N outputs when N diet content was medium or high, but this effect was not seen in low N diets; and (2) positively correlated with NUE but this effect was not seen in high N diets. EE concentration was positively correlated to RN and NUE when dietary N was medium or high, but this effect was not seen in low N diets. In order to mitigate N outputs and improve NUE in commercial beef production, improved feed quality (e.g. greater ME and starch) and lower N content (but to a level that still supports sufficient growth rates) may be warranted. However, in diets with low N content (e.g. 13.6-21 g/kg DM), higher NDF and ADF concentrations and less ME and N concentrations may shift N outputs from urine to faeces, which is also preferable from an environmental point of view.

New equations to predict OM digestibility of concentrates and by-products used for ruminants
D. Sauvant[1], G. Tran[2], V. Heuzé[2] and P. Chapoutot[1]
[1]*UMR MoSAR, INRA, AgroParisTech, Université Paris-Saclay, 16 rue C.Bernard, 75231 Paris Cedex 5, France, [2]Assoctiation Française Zootechnie, 75005 PARIS, France;, 16 rue C.Bernard, 75231 Paris Cedex 5, France; sauvant@agroparistech.fr*

Concentrates and by-products used for ruminant feeding are obtained from different plant species, organs and technological processes. The aim of this work is to propose a set of specific equations according groups of feeds to predict accurately Organic Matter digestibility (OMd) from chemical composition. Methods: A database of *in vivo* OMd and chemical values[Crude Protein (CP), Crude Fibre (CF), NDF and ADF and Fat, in g/kg dry matter (DM)] was built from data published in the literature or from the major feed tables. Two types of regressions were calculated: (1) to predict NDF or ADF from CF,(2) to predict OMd either from CF (2a) or from NDF or ADF (2b) (with CP an dFat or not as supplementary variables). Each dataset was treated by a model of analysis of variance-covariance with qualitative factors: the data source (literature and tables)nested by type of product (sub-group) within a family (group). When the sub-group(e.g. seed vs meal) appeared as a significant effect, several sub-equations were built, generally differing by the intercept. Fitting outliers were removed when their normalized residue was >3. Results:24 families of products were identified: wheat, maize, barley, oats, rice, pea, horsebean, lupine, vetch, soybean, rapeseed, sunflower, palmkernel and coconut, cotton, peanut, linseed, beet and citrus pulp, apple and pear, potato and tomato, olive and grape. For each one, 0 to 5 sub-groups, were distinguished when statistically different in the analyses. Thus, a set of 42 regressions has been calculated to predict NDF and ADF contents from CF. Root mean square error (RSME)were lower than NDF (10-20 vs 15-50 g/kg DM). For OMd prediction, 82 regressions have been calculated with a RMSE range of 1.5-6.0% point of digestibility. Significant systematic differences were observed between feed tables, with NRC values generally higher for a same product. These results allow to calculate OMd values with more precision, and therefore to have more accurate energy values for major concentrates and by-products in ruminants.

Repeatability of feed efficiency of lactating dairy cows fed high and low starch diets

A. Fischer and K.F. Kalscheur
USDA-ARS, U.S. Dairy Forage Research Center, 1925 Linden Drive, Madison, WI, USA; amelie.fischer@inra.fr

Improving feed efficiency is a way to maintain dairy farms production level while using fewer resources. Feed price volatility and the decrease in arable land availability drive dairy farms to use more diverse diets, which are also less edible for human consumption. The objective of this research was to analyse whether lactating cows maintained their feed efficiency across a control diet (HiStarch) formulated with corn grain (27% starch, 29% NDF) or a low human edible diet (LoStarch) formulated without corn grain (13% starch, 37% NDF). High moisture corn (25% of the diet DM) and part of canola meal in HiStarch diet were replaced with beet pulp, corn distiller grains and more forages (47.1% vs 66.4% forages) to formulate the LoStarch diet. Lactating Holstein cows (DIM=137±3.8 d), including 29 primiparous and 33 multiparous cows, followed a crossover design with 2 experimental periods of 10 weeks each, with the first 11 days being discarded for diet transition and adaptation. Feed efficiency was estimated as a residual feed intake (RFI), defined as the repeatable part of the difference between actual and expected intake. To do so, RFI was the sum of the random effect of cow on both, intercept and diet, of the mixed model predicting metabolizable energy intake within diet, parity and period with net energy in milk, metabolic BW, BW gain and loss and a repeated effect of 2-weeks nested within period. For both cohorts, RFI variability was higher when the cows were fed the HiStarch diet (SD=2.89 for cohort starting with HiStarch; 2.82 MCal/d for the other cohort) than when fed the LoStarch diet (SD=2.54; 2.05 MCal/d). RFI in diet HiStarch and RFI in diet LoStarch were positively correlated (r=0.73; 0.72). The 10% most efficient and 10% least efficient cows within cohort, as potential targets for a genetic selection, do not match across diets: 33% of these cows in both cohorts are no longer efficient or inefficient, without moving to the opposite group after diet change. The results suggest that feed efficiency is repeatable on average, but not necessarily for all cows, across diets high or low in starch. Given this repeatability variability across cows, the identified most or least efficient cows may vary with the diet and has to be considered in a selection approach.

Shredlage or whole-crop maize silage: conservation, digestibility and use in fattening-bulls ration

I. Morel, M. Rothacher, J.-L. Oberson, Y. Arrigo and U. Wyss
Agroscope, Ruminant Research Unit, Tioleyre 4, 1725 Posieux, Switzerland; isabelle.morel@agroscope.admin.ch

A novel technique for harvesting whole-crop maize silage called 'shredlage'(SHR) has been developed in the USA. The crop is chopped more coarsely than in the standard method (STD), the stalk is ripped longitudinally and the corn grains are crushed by a cracker roller. The aim of this process is to improve digestibility and nutrient availability for ruminants. Three experiments comparing SHR to STD were implemented: one with the aim of evaluating conservation parameters, a second studying the forage fibrosity and digestibility, and the third assessing the performances of fattening bulls in a feeding experiment. The maize silage was harvested on the same plot field at a theoretical length of cut of 30 mm (SHR) or 10 mm (STD) and ensiled in bales or in a tower silo (STD only). The proportion of fibres larger than 19 mm in SHR was not as high as announced by the manufacturer (21 vs >25%). The conservation test revealed that the fermentation parameters were similar between STD and SHR. However, the SHR contained more yeasts than the STD, which could be explained by higher temperature increase of SHR silage noticed during a post-fermentation test. The apparent organic-matter digestibility measured *in vivo* in sheep was not significantly different between the SHR and STD silages (76.3 vs 75.4%), resulting in close nutritional value. In the fattening-bull experiment, the maize silage accounted for 72% of total Dry Matter Intake (DMI). The average DMI (6.97 *vs* 7.30 kg/d; P=0.05) and average daily gain (1.50 vs 1.58 kg/d for SHR and STD, respectively; P=0.07) tended to be lower for SHR than for STD diets. Particularly at the beginning and at the very end of fattening period, the SHR ration was ingested at a lower rate than the STD one. This could be due to a reduced ingestibility linked to the coarse structure for the young animals and to the higher fill value of the feed at the end of the fattening period. Carcass quality was not significantly different between the two groups. Such results highlight that the potential advantages of the novel SHR technique are reduced and may not impact positively the economic profitability of farm.

Genetic parameters for the energy balance predicted using milk traits in Holsteins in Japan

A. Nishiura[1], O. Sasaki[1], M. Aihara[2], T. Tanigawa[3] and H. Takeda[1]
[1]Institute of Livestock and Grassland Science NARO, Tsukuba Ibaraki, 3050901, Japan, [2]Livestock Improvement Association of Japan, Tokyo, 1350041, Japan, [3]Hokkaido Research Organization, Nakashibetsu Hokkaido, 0861135, Japan; akinishi@affrc.go.jp

It is important to improve the energy balance (EB) of dairy cows to prevent the deterioration of health and fertility. Our objective was to estimate genetic parameters of EB predicted using milk traits measured in dairy herd performance test. The data to create the equation for prediction of EB consisted of records of 250 lactations of 147 Holstein cows in Hokkaido Research Organization. We measured milk yield and dry matter intake daily and milk components and body weight weekly. We calculated the average EB per 10 DIM, then the number of EB values was 30 per one lactation. Multiple regression model was built to predict EB. Independent variables were DIM, milk yield (MY), fat % (F), fat yield (FY), protein % (P), protein yield (PY), lactose % (L), lactose yield (LY), fat to protein ratio (FPR), fat to lactose ratio (FLR) and protein to lactose ratio (PLR). These variables except DIM were together with 'd' variables, which were the current minus the previous value. Model reduction was carried out by stepwise regression. The data to estimate genetic parameters for predicted EB consisted of 3,457,401 test-day milk records for the first three lactations of 228,731 Holstein cows. Genetic parameters were estimated by using a multi-parity random regression model, in which herd-test-day, region-calving-month and calving age were included as the fixed effects. We adopted two equations for prediction of EB. 7 variables (DIM, F, FY, PY, FPR, L, dPY) were included in the equation 1 (EB1, R^2=0.50) and 4 variables (DIM, L, FPR, dPY) were included in the equation 2 (EB2, R^2=0.47). Heritability estimates for EB1 were 0.17-0.37 and lower than those for EB2 which were 0.25-0.42. Heritability estimates for EB1 in the 1st lactation and EB2 in all lactations were 0.25-0.35 in the early stage, and then increased to 0.35-0.42 in the mid to late stage. However, heritability estimates for EB1 in the 2nd and 3rd lactations were 0.31-0.35 in the early stage, and then decreased to 0.17-0.24. Our results indicated that it would be possible to improve EB predicted by using our equation.

Effect of protein and rumen protected fat supplements on lipid metabolism of beef cattle

Y.H. Kim[1,2], K. Thirugnanasambantham[1], R. Bharanidharan[2], J.S. Woo[3] and S.H. Choi[4]
[1]Seoul National University, Institute of Green Bio Science and Technology, 1447, Pyeongchang-ro, Daehwa, Gangwon, 25354, Korea, South, [2]Seoul National University, Department of International Agricultural Technology, Daehwa, Gangwon, 25354, Korea, South, [3]Cheonan City Agricultural Technology Center, Cheonan-si, Chungcheongnam, 31233, Korea, South, [4]Chungbuk National University, Cheongju, Cheongbuk, 28644, Korea, South; khhkim@snu.ac.kr

The present study was conducted to evaluate the effect of dietary level of protein and rumen protected fat (RPF) supplementation on fatty acid composition and expression of 14 genes involved in lipid metabolism in intramuscular and subcutaneous adipose tissues of Hanwoo steers. Forty steers with an initial live weight of 486±37 kg and age of 18 months were assigned to a completely randomized design with 2×2 factorial arrangements for 150 days. One concentrate contained 14.5% and the other contained 17.0% protein level. Half of the steers that received each of the concentrate were supplemented with or without RPF during the entire experimental period. Each steer was able to consume each own concentrate diet with or without top dressed RPF (200 gram/head/day) twice a day (08.00 and 18.00 h). After 30 min., steers were allowed *ad libitum* access freely to rye grass. Intramuscular expression of *GPAT1* was up regulated by high protein diet (P=0.004) and positively correlated to fatty acid content. This suggested that increased expression of *GPAT1* may further regulate the incorporation of the fatty acids to first step of triacyglycerol synthesis. The intramuscular expression of *SNAP23* was down regulated (P=0.039) in high protein diet over the low protein diet without any significance in subcutaneous adipose tissues. It is worth to note that the high protein feed with RPF could contribute to the fineness in marbling texture of carcass traits via down regulation of *SNAP23*. All of observations revealed that feeding high protein diet with RPF in middle fattening stage (18-22 month ages) could increase fineness marble with high intramuscular fatty acid content in the longissimus dorsi.

Evaluating cicer milkvetch as a feed source for cattle including forage yield and nutritive value

H.A. Lardner, L.T. Pearce and D. Damiran
University of Saskatchewan, Animal and Poultry Science, 2D30 – 51 Campus Drive Saskatoon, Saskatchewan, S7N 5A8, Canada; leah.pearce@usask.ca

Cicer milkvetch (CMV; *Astragalus cicer* L.) is a non-bloating forage legume that can be fed to livestock as an alternative to alfalfa (*Medicago sativa*). A 2-yr study was conducted to compare three CMV varieties (Oxley, Oxley II and Veldt) to a commonly grown alfalfa variety (AC Grazeland). Each year, samples were harvested in late summer and analysed for nutrient profile. Crude protein (CP) did not differ (P>0.05) significantly between CMV and alfalfa varieties. All three CMV varieties had similar levels of total digestible nutrients (TDN) (63.6±0.6%; P>0.05), but had higher TDN content than the alfalfa forage (54.9±1.4%). No anti-quality factors were detected in the CMV cultivars. Nutrient yield per ha was calculated by multiplying crop forage yield (kg/ha) by nutrient concentration to compare nutrient yield potential (NYP) of the four crops. Oxley NYP for CP and TDN was 3% less than that of the alfalfa. Oxley II NYP was 15 and 13% greater than alfalfa for CP and TDN, respectively. Veldt NYP for CP and TDN was 50% greater than the alfalfa. Beef cows were presented freshly harvested CMV and alfalfa forage in order to detect preference and palatability. Study dry matter intake (DMI) data indicated that CMV was more preferred by beef cows compared to alfalfa (Oxley II, Veldt, and Oxley CMV consumption was 4, 18 and 28% higher than alfalfa, respectively). These study results suggest that CMV is a suitable alternative forage to alfalfa for late summer grazing.

Rumen-protected methionine product in lactating dairy cows

V. Sáinz De La Maza[1], B. Rossi[2], R. Paratte[2], A. Piva[2,3] and E. Grilli[3,4]
[1]University of Lleida, Department of Animal Production, Av. Alcalde Rovira Roure, 191, 25198 Lleida, Spain, [2]Vetagro SpA, Via Porro, 2, 42124 Reggio Emilia, Italy, [3]University of Bologna, DIMEVET, Via Tolara di Sopra, 50, 40064 Ozzano dell'Emilia (Bologna), Italy, [4]Vetagro Inc, 230 South Clark Street, # 320, 60604 Chicago (IL), USA; richard.paratte@vetagro.com

Methionine is the first limiting amino acid for milk and protein production in lactating dairy cows fed corn-based diets. A field trial was conducted in a commercial dairy farm in Northeast Spain with the aim to evaluate the effect of supplementing a rumen-protected methionine product (Timet®; VETAGRO S.p.A.; Reggio Emilia, Italy) on lactation performance in high-yielding dairy cows. Ninety-nine multiparous Holstein cows were used in a 3-period switchback design: Control 1 (CTR1, 21-d), Timet (TMT, 21-d) and CTR2 (12-d). All the cows that entered the study were between 0 and 120 DIM. Cows were fed a TMR formulated to provide 16.7% CP, 29% Starch, and 32% NDF with 2,793 g/d of MP and Lys:Met 2.93. During TMT treatment, Timet (55% of DL-Methionine) was supplemented 25 g/d per head. Milk yield was recorded daily and milk was sampled for protein, fat, and urea every 2 days. Data were analysed using MIXED model ANOVA including cow as random effect (JMR pro 13®). Although there was not significant difference in milk yield, milk protein was significantly increased by TMT (3.43%) compared to CTR1 (3.35%) and CTR2 (3.38%) (P<0.05), and milk fat was also significantly increased by TMT (3.75%) compared with CTR1 (3.63%) and CTR2 (3.71%); (P<0.05). During TMT treatment it was possible to observe a trend in lowering milk urea (180.8 mg/l) compared to CTR1 (193.0 mg/l) and CRT2 (189.9 mg/l) (P=0.43). In conclusion, addition of Timet to a corn-based diet increased milk protein and fat concentration and reduced milk urea, thus improving the nitrogen metabolism.

Effect of protein sources in calf starter on growth performance of pre-weaned calves

S.H. Rasmussen[1,2], C. Brøkner[2] and M. Vestergaard[1]
[1]Aarhus University, Department of Animal Science, Blichers Alle 20, 8830 Tjele, Denmark, [2]Hamlet Protein A/S, Saturnvej 51, 8700 Horsens, Denmark; mogens.vestergaard@anis.au.dk

Soybeans are the preferred choice of plant protein in calf starters because of a high CP content, an excellent AA profile and good palatability. Heat treatment only reduces the content of heat labile ANF's, e.g. trypsin inhibitor, which is not sufficient to ensure a high digestibility of soybean meal in the un-weaned calf. Thus, further processing are needed to reduce the content of heat stable ANF's, e.g. β-conglycinin and oligosaccharides. We investigated the effect of an enzyme-treated soybean meal in calf starter for pre-weaned rosé veal calves on feed intake and growth performance. The experiment was divided into two sub-experiments (A and B) each consisting of two blocks of 16 calves. A total of 64 calves were purchased from 7 commercial herds [age(d), LW(kg) (mean ± SE): A=12.8±0.57, 49.6±1.11; B=12.7±0.62, 51.7±0.67]. Within block of a sub-experiment, calves were randomly allocated on two treatment (trt) groups: control (CON) or test (TEST). From 2 wk of age, CON was offered a traditional calf starter with 30% SBM and TEST was offered a calf starter with 23% HP RumenStart for *ad libitum* consumption. Both calf starters had (on a DM basis) the same CP (24%), starch (27-31%), aNDF (14-16%) and Net Energy (8.2-8.3 MJ) content. Additionally, calves were offered skimmed-milk milk replacer up to 8 l/d (a total of 224 l for 6 wk) and artificially-dried grass hay for *ad libitum* consumption. Calves were weaned at 8 wk of age and followed until 10 wk of age. Calves were weighed at 2 (before start), 4, 6, 8 and 10 wk of age and feed consumption was recorded on pen level daily. Initial LW (wk 2 of age) did not differ between CON and TEST. There was an effect of wk of age (P<0.001) on LW and ADG but no effect of trt across the experimental period. For ADG, there was a trt x wk interaction showing a 150 g higher ADG of TEST compared to CON at 10 weeks of age. This, together with a numerically lower calf starter intake throughout the experiment resulted in a numerically improved feed conversion efficiency for TEST as compared to CON. In conclusion, the soy protein source may marginally affect growth performance of rosé veal calves.

Pre-grinding super-conditioned corn affects performance of Holstein dairy cattle

A. Rahimi[1], A. Naserian[1], M. Malekkhahi[2] and K.H. Park[3]
[1]Ferdowsi University of Mashhad, Mashhad, 9177948974, Iran, [2]R & D manager of Dordaneh Razavi, Mashhad, 9177948971, Iran, [3]Kangwon national University, Chounchan, 24341, Korea, South; kpark74@kangwon.ac.kr

The objective of this study was to investigate the effects of pre-grinding super-conditioning in comparison with steam flaking of corn on performance of Holstein dairy cattle in early lactation. Eighteen dairy cattle (milk yield 46.7±3 kg and DIM 131.6±14 d) were randomly assigned to one of three experimental diets: (1) concentrate containing of 50% pre-grinding super-conditioned corn (PSC; moisture 20%, retention time 6 min and conditioning temperature 95 °C); (2) concentrate containing of 50% steam flaked corn (SF; moisture 20%, retention time 60 min, rolling temperature 105 °C); and (3) concentrate containing of 50% grounded corn (G; 3 mm mill screen). Statistical analysis was performed using the GLM procedure based on completely randomized design of SAS (P<0.05). Geometric mean particle size and peNDF were increased (P<0.05) with SF corn than others. Dry matter intake, feed efficiency, DM, OM and ADF digestibility were not affected using experimental diets, but CP, starch and EE digestibility were increased (P<0.05) and NDF digestibility was decreased (P<0.05) in PSC than others. Rumen fluid pH, total VFA and molar proportions of acetate, propionate and butyrate were not different among diets, but ammonia nitrogen was decreased (P<0.05) in PSC than others. The pH and manure score were not affected, but the excretion of starch from faeces decreased (P<0.05) in PSC than others (0.61, 1.35 and 9.83% in PSC, SF and G, respectively). Milk yield was not affected using processed corn, but was 2 kg higher in PSC than SF and G corn. Milk fat content was shown a low tendency in PSC, but milk protein content was increased (P<0.05) in PSC than others. Milk fat to protein ratio was decreased (P<0.05) in PSC than other processed corn. It is concluded that pre-grinding super-conditioned corn in diet of early lactation dairy cattle increased starch digestibility, milk production and efficiency of starch utilization and decreased starch excretion from faeces.

Potential of Brussels sprout stems in dairy cattle diets

L. Vandaele[1], J. Leenknegt[2], B. Van Droogenbroeck[1] and J. De Boever[1]
[1]Flanders Research Institute for Agriculture, Fisheries and Food (ILVO), Scheldeweg 68, 9090 Melle, Belgium, [2]INAGRO, Ieperseweg 87, 8800 Roeselare, Belgium; leen.vandaele@ilvo.vlaanderen.be

After the harvest of Brussels sprouts, the leaves and stems remaining on the field lead to odour nuisance and extensive nitrate leakage into the soil. As these left-overs may yield 18 to 25 tons per ha, their potential as cattle feed was investigated. First, the variation in nutritive value was examined by determining the chemical composition and *in vitro* digestibility of 7 batches fresh sprout stems from different varieties and fields. The dry matter (DM) content amounted (mean ± SD) to 187±20 g/kg and contained per kg DM 122±17 g crude protein, 269±31 g crude fibre, 233±62 g sugars and 75±9 g crude ash. The net energy value was estimated at 4.67±0.39 MJ/kg DM and the digestible protein content at 41±6 g/kg DM, indicating a quite low nutritive value. Because fresh stems have a shelf time of only a few days, the possibility of ensiling was examined. A total of 7,000 kg chopped stems were mixed with 14,000 kg of pressed beet pulp and ensiled in a silo bag. Core samples of the mixed silage taken after 8 weeks had a low pH (4.23), a low ammonia fraction (6.4%) and 41 g lactic acid, 9 g acetic acid and 30 g ethanol per kg DM, indicating a good fermentation. Then, 16 lactating dairy cows fed a control diet for 3 weeks were divided in two groups based on parity, lactation stage, DMI and milk production. One group of cows continued with the control diet (35% maize silage, 50% grass silage and 15% pressed beet pulp on DM base), whereas the 15% beet pulp in the diet of the other group was replaced by 15% of the mixed silage. Apart from the roughage the cows were fed on average 6 kg DM concentrates to meet their energy and protein requirements. The individual roughage and total DM intake and milk production were monitored and were analysed using a linear mixed model with week and treatment as fixed factor and cow as random factor. Inclusion of the mixed silage significantly (P<0.01) decreased daily intake of roughage (16.7 vs 14.8 kg DM) and total DM (23.1 vs 20.8 kg) as well as milk production (31.3 vs 28.2 kg). In conclusion, sprout stems can be well preserved as mixed silage, but are an undesirable feed component in the diet for high producing dairy cattle.

Effect of a *Saccharomyces cerevisiae* fermentation product on milk production of dairy goats

I. Yoon[1], F. Ysunza[2] and A.R. Boerenkamp[3]
[1]Diamond V, 2525 60th Ave SW, 52404 Cedar Rapids IA, USA, [2]Diamond V, P.O. Box 10022, 9400 CA Assen, the Netherlands, [3]InsoGoat, Johan van Wyttenhorstweg 7, 5816 AH Vredepeel, the Netherlands; fysunza@diamondv.com

A trial was conducted in the SE of the Netherlands: 403 mostly Saanen crossbred, non-pregnant goats (>365 DIM) blocked by lactation, DIM, milk production and quality, were randomly assigned to either Control (basal diet) or *Saccharomyces cerevisiae* fermentation product (SCFP, basal diet with daily dose of 3.5 g of XPC_{LS}). Goats were individually fed a concentrate and housed in a ventilated barn with *ad libitum* access to water and straw; grass silage was offered at a daily rate of 350 g (DM) per head. Individual records were collected from all animals for concentrated feed intake (FI) and milk yield (MY). Milk components and SCC were determined on days 7, 36 and 78 from 50 selected goats per treatment. Body condition (BCS) was scored on days -5, 38 and 80 of the trial in all goats; manure was assessed for consistency and colour on same days as BCS from a subset of 10 goats per group. All goats reached a level of FI for the target dosage of XPC_{LS} in SCFP at week 4. Control had an advantage in milk yield over SCFP on starting day; producers above 6 kg/d average where removed and initial milk yield levelled. MY was compared by production level: up to 6, 5 and 4 kg/d. FI averaged 1,471 and 1,493 g/d, BCS 3.4 and 3.3 for Control and SCFP, respectively (P>0.1); faecal colour and consistency did not differ between treatments. SCFP supplemented goats averaged 0.14, 0.12 and 0.21 kg/d of MY above Control goats by the end of week 11 for production levels of up to 6, 5 and 4 kg/d, respectively (P≥0.1). The difference became evident after 6 weeks at all production levels and goats producing up to 4 kg/d and fed SCFP increased MY by 0.19 kg/d (5.7%, P=0.11), 0.24 kg/d (7.2%, P=0.05), and 0.21 kg/d (5.9%, P=0.10) over Control counterparts at weeks 9, 10 and 11, respectively. Supplementation of XPC_{LS} increased goat milk production between 2 and 7% without affecting components and intake. Response to supplementation was greater for goats producing up to 4 kg/d. Results suggest that XPC_{LS} can be effective for maximizing profitability of dairy goat production.

Rumen liquor from slaughtered and fistulated cattle as sources of inoculum

P. Lutakome[1,2], F. Kabi[1] and F. Tibayungwa[1]
[1]Makerere University Agricultural Producion Uganda 7062 Kampala Uganda. [2]Ghent University Animal Sciences and Aquatic Ecology, Ghent, 9000 Ghent, Belgium; plutakome@gmail.com

Use of rumen liquor from fistulated cattle as a reference source of inoculum for feed evaluation using *in vitro* gas production (GP) techniques is currently limited by the high cost associated with surgical preparation and maintenance of fistulated cattle. The current *in vitro* study aimed to investigate the effect of rumen liquor source (i.e. fistulated steers vs slaughtered cattle) on GP kinetics, partitioning factor (PF) and efficiency of microbial biomass production of reference diets. The fistulated steers were fed on 4 reference diets mixed on dry matter basis according to the following ratios [g/kg DM]: (D1) 1000:0; (D2) 900:100; (D3) 800:200 and (D4) 700:300 of hay:concentrate. Rumen liquor was obtained either from these 4 fistulated steers prior to morning feeding or from 4 freshly slaughtered cattle, which were deprived from feed for 18 hours prior to slaughtering. Sampling was done at 4 separate occasions synchronized with the time of rumen liquor collection from the fistulated steers. Gas production was recorded at 2, 4, 8, 10, 18, 24, 36, 48, 72, 96 and 120 h and used to derive GP kinetics, PF, microbial biomass production and its efficiency according to a 2×4 factorial design. Rumen liquor from slaughtered cattle had ideal conditions in terms of pH (7.2) and ammonia (235 mg/l) necessary for microbial activity. Inoculum from slaughtered cattle also produced the highest asymptotic GP ($P<0.001$) with the 4 dietary substrates and the maximum rate of GP followed a similar trend. Gas production profile curves of the 4 diets incubated with both inoculum sources did not differ ($P>0.05$) in their shape and were sigmoidal. The partitioning factor was not influenced by source of inoculum but it varied ($P<0.05$) per incubated substrate. Comparisons made between current results and literature data indicate that rumen liquor from slaughtered cattle is an effective alternative source of inoculum for ruminant feed evaluation.

Correlations between foetal weight, organ weight and placenta weight at day 57 of gestation in sows

A.V. Strathe[1], T.S. Bruun[2] and C. Amdi[1]
[1]University of Copenhagen, Department of Veterinary and Animal Sciences, Grønnegårdsvej 3, 1870 Frederiksberg, Denmark, [2]SEGES Danish Pig Research Centre, Axeltorv 3, 1609 Copenhagen, Denmark; avst@sund.ku.dk

The objective of the study was to characterize correlations between foetal weight and placenta and organ weight during mid-gestation. The sows were slaughtered at day 57 of gestation. The uterus was removed and dissected. Forty –eight individual foetuses with placenta were examined. Organs (brain, heart, liver, intestine, lungs, kidneys and spleen) from the foetuses and their placentas were weighed and the length of the umbilical cord was recorded. Pearson correlations coefficients were calculated between foetal weight and the other measured variables. Organ weight as percentage of foetal weight and Pearson correlations to foetal weight were calculated. Average foetal weight at day 57 of gestation was 83.3±19.6 g. There was a positive correlation between foetal weight and placenta weight (r=0.80; $P<0.001$) and length of umbilical cord (r=0.34; $P<0.05$). Weight of heart (r=0.82; $P<0.001$), brain (r=0.32; $P<0.05$), kidneys (r=0.92; $P<0.001$), lungs (r=0.92; $P<0.001$), intestine(r=0.91; $P<0.001$), spleen (r=0.79; $P<0.001$) and liver (r=0.91; $P<0.001$) were positively correlated to foetal weight. The foetal weight was negatively correlated to the weight of brain (r=-0.85; $P<0.001$) and heart r=-0.49; ($P<0.001$) measured as percentage of body weight, whereas there was a positive correlation to liver (r=0.63; $P<0.001$) and kidney (r=0.68; $P<0.001$) weight measured as percentage of foetal weight. A Large placenta and length of the umbilical cord seems to be of great importance for foetal growth and development. Larger foetuses had both heavier placentas and longer umbilical cord compared to smaller foetuses, which implies that the larger foetuses had a larger supply of nutrients. The smaller foetuses had larger brains and heart and smaller liver and kidneys relative to body weight compared to the larger foetuses, indicating that the growth and development of these organs were either prioritized or compromised because of insufficient nutrient supply. In conclusion, there was already at day 57 of gestation clear differences between small and large foetuses.

Variability in gestating sows' nutrient requirements

C. Gaillard, R. Gauthier and J.Y. Dourmad

PEGASE, INRA, Agrocampus Ouest, 35590, Saint-Gilles, France; charlotte.gaillard@inra.fr

In practice, sows often receive the same diet during the gestation even though their nutrient requirements vary over gestation and among individuals. The objective of this study, realized within the Horizon 2020 EU Feed-a-Gene program (grant agreement n°633531), was to report the variability in nutrient requirements among sows and over time, in order to develop a precision feeding approach. A dataset of 2,511 gestations reporting sows' characteristics at insemination and farrowing performance was used as an input for a Python model based in InraPorc® predicting nutrient requirements over gestation. The influence of parity (P1, P2, P3+), litter size, and week of gestation on metabolizable energy (ME), SID lysine (Lys) and threonine (Thr), digestible phosphorus (dP) and total calcium (Ca) requirements was analysed using a mixed model with R software. Total ME requirement increased with increasing litter size, with week of gestation, and with parity (30.6 vs 35.5 MJ/d for P1 and P3+). Lys and Thr requirements per kg diet increased from week 1 to week 6, remained stable from week 7 to 10, and increased again from week 11 until the end of gestation. Lys and Thr requirements increased with increasing litter size and decreased when parity increased (Lys: 4.04 vs 3.09 g/kg for P1 and P3+; Thr: 3.06 vs 2.49 g/kg for P1 and P3+). dP and Ca requirements increased markedly after week 9, increased with litter size, and decreased when parity increased (dP: 1.36 vs 1.31 g/kg for P1 and P3+; Ca: 4.28 vs 4.10 g/kg for P1 and P3+). Based on empiricalcumulative distribution functions, a strategy with 4 diets, varying in Lys and dP content with parity and gestation week (before B or after A week 11), can be suggested to achieve the requirements of 90% of the sows (two diets for multiparous sows: B: 3.2 g Lys/kg and 1.1 g dP/kg; A: 4.9 g Lys/kg and 2.3 g dP/kg; and two diets for primiparous sows: B: 3.8 g Lys/kg and 1.1 g dP/kg; diet A: 5.5 g Lys/kg, 2.2 g dP/kg). Better considering the high variability of sows' requirements in practice should thus allow optimizing their performance whilst reducing feeding cost. A first step can be achieved by grouping and feeding sows by week of gestation and parity, and a second step by feeding the sows individually.

Nutritional measures to improve viability, birth weight and uniformity of the litter

X. Benthem De Grave and F. Molist

Schothorst Feed Research, Meerkoetenweg 26, 8218 NA Lelystad, the Netherlands; xbenthemdegrave@schothorst.nl

The objective of the study was to determine the effect of insulin stimulating diets during the last week of lactation and weaning to oestrus interval (WOI) on birth weight and uniformity. The set-up of the experiment was according to a randomized block design with four dietary treatments and 15 sows per treatment (parity 2-5). The treatments were: (1) control diet; (2) diet with 4.5% dextrose; (3) diet with 20% sugar beet pulp (SBP); and (4) diet with 5.8% expanded maize. The diets were fed from day 21 of lactation until insemination. During WOI, plasma insulin growth factor-1 (IGF-1) was measured daily. Glucose and insulin levels were measured on day 2 at a 12 min. interval after feeding. Follicle diameters were measured at weaning and daily from day 3 of WOI until ovulation. The effect of diet on the subsequent litter was analysed. A higher insulin increase after feeding was found for expanded maize compared to dextrose and SBP (P=0.04). This resulted in a higher insulin AUC360 for expanded maize compared to SBP (P=0.05). Despite the effect on insulin, no effect on IGF-1 levels or follicle development was found. However, regardless of treatment young sows had higher IGF-1 levels than older sows at days 4 and 5 (P=0.02 and P=0.03). A significant higher total born (TB) was observed for the control (16.6 TB) and dextrose (17.3 TB) compared to expanded maize (13.2 TB; P=0.02), resulting in higher number of live born (LB) for dextrose (15.6 LB) compared to expanded maize (12.7 LB; P=0.08). Dietary treatment had no effect on piglet weight or uniformity at birth on day 7 of lactation. However, dietary treatment did have an effect on piglet vitality where expanded maize had a higher piglet growth (P=0.02) and weaning weight (P=0.02) and a lower mortality the first 7 days of lactation (P=0.08). In conclusion, regardless of parity, insulin stimulating diets did increase insulin level after feeding. However, no effect on IGF-1 secretion and therefore on follicle development and subsequent birth weight and uniformity was found. Insulin stimulating diets may only be beneficial in sows with suboptimal follicle development, for example young sows and sows with a low feed intake with high body weight loss during lactation.

Effect of positive handling of sows on litter performance and pre-weaning mortality

D. De Meyer[1], I. Chantziaras[2], A. Amalraj[2], L. Vrielinck[1], T. Van Limbergen[2], M. Fockedey[1], I. Kyriazakis[3] and D. Maes[2]
[1]Vedanko BVBA, Keukelstraat 66, 8750 Wingene, Belgium, [2]Ghent University, Faculty of Veterinary Medicine, Salisburylaan 133, 9820 Merelbeke, Belgium, [3]Newcastle University, School of Agriculture, Food and Rural Development, King's Road, NE1 7RU Newcastle upon Tyne, United Kingdom; ilias.chantziaras@ilvo.vlaanderen.be

Fear over humans may affect maternal behaviour in pigs, increasing farrowing duration and piglet mortality. This study investigated the effect of positive handling of sows (scratching, music) in the farrowing room on litter performance and pre-weaning piglet mortality. The study was conducted in a sow herd (n=1,014 PIC sows) that practiced a 2-week farrowing system. The sows were moved to the farrowing unit one week before farrowing. Lactating sows were housed in conventional farrowing crates. Twenty successive farrowing batches were included and randomly allocated to either treatment (T) (n=6,299 sows; parity 3.25) or control (C) (n=14,715 sows; parity 3.27). In the T batches, backscratching of the sows was done daily for 15 seconds per sow from entry into the farrowing unit until farrowing, and music (commercial radio station) was played from 6 am until 6 pm from entry into the farrowing unit until weaning (21 days). Litter performance and piglet mortality data were recorded. Regarding sow parity, two groups were formed (group 1: parities 1-2, group 2: parities 3-8). Two-sided t-tests and linear mixed models were used to analyse the data. Comparing the T and C groups, differences (P<0.05) were seen in total born piglets (13.88 vs 14.58), piglets born alive (13.26 vs 13.85), piglets born dead (0.62 vs 0.75) and preweaning mortality (9.96% vs 13.37%). Mummified piglets (0.38 vs 0.38) and piglets weaned per sow (11.94 vs 12.04) did not significantly differ (P>0.05). In the mixed models, mortality had negative associations (P<0.05) with parity group 1 and treated sows. Total born and live born piglets respectively, each had negative associations with parity group 1 and treated sows. For born dead piglets, only a negative association with parity group 1 was seen. To conclude, treatment was associated with lower piglet mortality and a smaller litter size at birth. Further research is warranted to confirm the present results and to assess the separate effects of back scratching and music.

Effect of a beneficial flora colonization of pen surfaces on health and performance of pig weaners

E. Royer[1], F. Bravo De Laguna[2], J. Plateau[2] and E. Chevaux[2]
[1]Ifip-institut du porc, 34 bd de la gare, 31500 Toulouse, France, [2]Lallemand SAS, 19 rue des Briquetiers, 31700 Blagnac, France; eric.royer@idele.fr

The objective was to test the effects of a positive biofilm formation on the surfaces of post-weaning piglet facilities. In total, 494 piglets were used in two experiments using a sanitary challenge. 48 h (d-2) before introduction of piglets, 2 identical rooms of 14 pens were sprayed either with water (Control) or a mix (LP) of selected bacteria strains. Rooms were exchanged between Exp.1 and Exp.2. In Exp.1 rooms were sprayed again at d 15 and in Exp.2 at d 5, 12, 19, 26 and 33. Environmental challenge for piglets was stronger in Exp.1 than in Exp.2. Wiping samples indicated significantly (P<0.05) higher loads of aerobic bacteria (Lactobacillus spp., Bacillus spp.) in LP pen surfaces in Exp.1 at d 0, 5, and 14 and at d 0, 5, 7 and 35 in Exp.2, suggesting the development of the positive biofilm. Percentage of piglets with regular consistency of faeces was continuously higher in LP rooms in Exp.1 (from d 8 to 21) and Exp.2 (from d 5 to 28). Furthermore, mean scores were significantly improved at d 8 in Exp.1 (3.13 vs 4.50; P<0.01) and in Exp.2 at d 9 (2.19 vs 3.19; P=0.01) and 28 (2.03 vs 2.50; P<0.01). Disease outbreaks occurred two days later in Exp.1 (d 9 vs 7) and five days later in Exp.2 (d 12 vs 7) in LP rooms. However, total numbers of deaths from diarrhoea were similar in both treatments in Exp.1 and 2. In Exp.1, LP piglets had numerically better overall ADFI (794 vs 781 g/d; P>0.10) and ADG (510 vs 499 g/d; P>0.10), and had slightly higher weight at d 42 (29.8 vs 29.4 kg; P>0.10). In Exp.2, ADFI (259 vs 219 g/d; P<0.001) and ADG (211 vs 154 g/d; P<0.001) were significantly increased in the LP treatment in phase 1 (d 0 to 15). Weight was significantly higher for LP piglets at d 15 (11.9 vs 11.0 kg; P<0.001), although it was similar at d 41 (P>0.10). In conclusion, the spraying of a beneficial flora on surfaces may result in a protective positive biofilm that would help the piglets to deal better with the weaning challenges.

Effect of L-Selenomethionine on feed intake and selenium deposition in milk from high-yielding sows

S. Van Beirendonck[1,2], B. Driessen[1], G. Du Laing[3], B. Bruneel[4] and L. Segers[4]
[1]*Dier&Welzijn, Kleinhoefstraat 4, 2440 Geel, Belgium,* [2]*KU Leuven, Faculty of Engineering Technology, Kleinhoefstraat 4, 2440 Geel, Belgium,* [3]*Ghent university, Faculty of Bioscience Engineering, Coupure Links 653, 9000 Gent, Belgium,* [4]*Orffa, Vierlinghstraat 51, 4251 LC Werkendam, the Netherlands; bruneel@orffa.com*

A field study in periparturient sows fed different dietary concentrations of either sodium selenite (SS) or L-selenomethionine (L-SeMet) was conducted to evaluate feed intake and total selenium (Se) increase in blood and milk. Eleven sows were allotted to two groups from 14 days (d) prepartum throughout on average a 26 d lactation period. SS and L-SeMet supplemented diet contained 0.2 mg added Se/kg feed. Blood samples were collected at weaning for both sows and piglets. Milk samples were collected 14 d after farrowing and at weaning. Total Se was determined using inductively coupled plasma mass spectrometry. Data was analysed with a MIXED model, with treatment as a fixed effect and parity as a covariate in the statistical package SAS 9.4. Results show that there was a dietary effect on average daily feed intake (ADFI) from d 14 post partum until the end of the study (P<0.05) and on the total milk Se content. The ADFI was significantly higher for this period in the L-SeMet supplemented group (6.366±0.212 kg) compared to the SS supplemented group (5.605±0.193 kg). Total milk Se was significantly higher, 47.63 and 27.25%, in the L-SeMet supplemented group and this at d 14 after farrowing (P<0.001) and at weaning (P<0.002), respectively. Total serum Se at weaning showed a numerical increase in the L-SeMet supplemented group compared to the SS supplemented group and this for both sows and piglets. In conclusion, this study shows that L-SeMet supplementation of sow diets before farrowing and during lactation has a significant effect on feed intake during lactation and total Se content in the milk. Similar results have been published in literature.

The effect of dietary protein levels on reproduction performance and *E. coli* shedding in sows

K.C. Kroeske[1,2], M. Heyndrickx[2], N. Everaert[1], M. Schroyen[1] and S. Millet[2]
[1]*Gembloux Agro-Bio Tech, TERRA Teaching and Research Centre, Precision Livestock and Nutrition Unit, Passage des Déportés 2, 5030 Gembloux, Belgium,* [2]*ILVO, Dier 68, scheldeweg 68, 9090, Belgium; kikianne.kroeske@ilvo.vlaanderen.be*

Low protein levels in sows' gestation diets have been associated with foetal development. Low dietary protein levels have also demonstrated to increase protein absorption efficiency when maintaining the levels of digestible essential amino acids and to decrease the available amino acids for microbial fermentation in the large intestine. The microbiota of the sow affects the intestinal colonization of the piglets. Moreover, an important production disease for piglets is weaning diarrhoea, caused by enterotoxigenic *Escherichia coli*. Therefore, we investigated the effect of two levels of crude protein in gestation diet and the effects on reproduction parameters, the abundance of *E. coli* in faeces of the sow and piglets performance until weaning. The hypothesis was tested with high and low protein levels in sow diets (12 (LP)vs 17% (HP) crude protein (CP)) during the last five weeks of gestation until farrowing. After farrowing, all sows received the same lactation diet (16% CP). From each sow, four faecal samples were taken over time: at 5 weeks before (before treatment), 1 week before (gestation stable), 4 days before (farrowing stable) and 2 weeks after farrowing. Faecal samples were diluted and cultivated on BIO-RAD RAPID'E.coli 2 Medium. Microbial colony-forming units (cfus) were calculated and transformed with LOG. Statistical analysis was performed using RStudio with linear mixed-effect models (lme4 package) and pairwise comparison (emmeans package). No treatment effect was found in weight gain/-loss before farrowing (LP 260.76±13.04 kg, HP 249.86±10.96 kg), back fat thickness (LP 16.76±1.09 mm, HP 15.36±1.13 mm) or litter size (17±0.8, HP 16.28±1.1). No difference was found for piglet weight at farrowing (LP 1.42±0.06 kg, HP 1.35±0.42 kg) or weaning (LP 7.2±0.26 kg, HP 7.45±0.23 kg). For *E. coli* counts (log), an average was found for LP 6.99±0.14 (se) and HP 7.08±0.14 (se), but no treatment differences were found. In conclusion, no effect on sow performance, or *E. coli* shedding were found. Future research will focus on nitrogen digestion in sows and microbiota in piglets.

Body minerals content of reproductive sows

J.Y. Dourmad, M. Etienne, J. Noblet, S. Dubois and A. Boudon
INRA Agrocampus Ouest, Pegase, 35590 Saint-Gilles, France; jean-yves.dourmad@inra.fr

Minerals are essential for the development and maintenance of the skeletal system as well as for many essential physiological functions. The factorial determination of minerals requirement is generally based on potential total body minerals retention as driving force. However, only very limited information is available in reproductive sows, data from fattening pigs being often used, which makes the predictions unprecise. Complete dissection (lean, fat, bones, skin, organs) was performed on 189 sows, among which 23 were also chemically analysed. Body minerals and protein contents were determined in all sows using the double-regression technique. Sows were from different parities and different physiological status: 42 and 66 primiparous sows at farrowing and at weaning, and 25 and 56 multiparous sows at mating and at farrowing, respectively. Body minerals weight (BM) was greater (P<0.001) in multiparous (6.99 kg) than in primiparous sows (4.87 kg) and differed according to physiological status in primiparous (lower at weaning than at farrowing, 4.68 vs 5.07 kg), but not in multiparous sows. When expressed per kg empty body weight (EBW), body minerals content was higher (P<0.01) in multiparous sows at mating (33.8 g/kg EBW) than in the others groups which did not differed among each other (29.2 g/kg EBW on average). When expressed per kg body protein, differences among groups almost disappeared (193 to 198 g/kg BP for extreme values). The prediction equation of BM content (kg) was more precise with body protein (BP, kg) as predictor [BM = 0.258 (±0.065) + 0.187 (±0.002) BP; R^2=0.97] than with EBW (kg) as predictor [BM = 0.330 (±0.194) + 0.028 (±0.001) EBW; R2=0.81]. This was partly related to body fat tissues content which varied according to the physiological status of sows (from 17 to 23% of carcass weight). Indeed, when backfat thickness (P2, mm) was included in the equation, the precision of the prediction equation based on EBW increased: [BM = 0.978 (±0.165) + 0.0318 (±0.0007) EBW – 0.105 (±0.008) P2; R^2=0.90]. It is concluded that body protein mass is a better predictor of body minerals mass than EBW. However, although less precise, the use of EBW and P2 as predictors could be an interesting alternative and easier to apply in practice.

Maternal undernutrition determines postnatal growth and meat traits in Iberian pigs

M. Vázquez-Gómez[1], C. García-Contreras[2], L. Torres-Rovira[1], J.L. Pesantez-Pacheco[3], S. Astiz[4], C. Ovilo[2], A. Gónzalez-Bulnes[4] and B. Isabel[1]
[1]Faculty of Veterinary Medicine, UCM, Av.Puerta de Hierro, 28040 Madrid, Spain, [2]INIA, Animal Breeding, Carr.de La Coruña, 28040 Madrid, Spain, [3]School of Veterinary Medicine, UCuenca, Cuenca, Ecuador, [4]INIA, Animal Reproduction, Av. Puerta de Hierro, 28040 Madrid, Spain; martavazgomez@gmail.com

A severe reduction in maternal feed during pregnancy, especially during its last third, leads to low birth-weight (BW) piglets. However, there is scarce data about the effects of light maternal undernutrition during other less demanding pregnancy periods on the BW and postnatal growth of offspring under farm conditions. In this study, we studied the developmental traits of offspring (Iberian×Duroc) from Iberian sows fed with either a diet fulfilling total requirements (control, n=47) or only 70% requirements between days 38 and 90 of gestation (underfed, n=33). At birth, piglets were classified as low BW (LBW; <1 kg) or normal BW (NBW, ≥1 kg). There were no effects of maternal nutrition on sow productivity or piglet phenotype at birth, but the effects of maternal undernutrition were found from weaning onwards. At 110 days-old, LBW pigs showed a catch-up growth, although NBW pigs were heavier than LBW pigs (P<0.05). At the fattening phase, underfed pigs showed slower growth and higher feed conversion ratio than control pigs (P<0.01). In addition, the total average daily gain was higher in NBW, male and control pigs than in LBW (P<0.01), female (P<0.01) and underfed pigs (P<0.05) and the slaughter age was lower in NBW pigs than in LBW pigs (P<0.01). After slaughter, underfed pigs also had lower carcass yield and back-fat depth than control pigs (P<0.01). The maternal undernutrition caused main changes in the membrane fatty acids (FA) of liver and loin and underfed pigs also had higher C18:2n−6 levels than control pigs (P<0.01) in the back-fat, which is related to impaired water migration and rancidity problems. In conclusion, our results showed that maternal undernutrition during mid-gestation did not have significant effects on the phenotype of newborn piglets, but has adverse effects on growth patterns and meat characteristics from weaning to slaughter.

The role of pasture type and fattening in the lifecycle impacts of Portuguese meat production

T.G. Morais, R.F.M. Teixeira, M.P. Santos, T. Valada and T. Domingos
MARETEC – Marine, Environment and Technology Centre, Universidade de Lisboa, Av. Rovisco Pais, 1, 1049-001, Portugal; tiago.g.morais@tecnico.ulisboa.pt

In Alentejo, Portugal, beef cattle rearing is substantially pasture-based. Cows are mostly fed with a mixture of grazing in semi-natural pastures and forages/concentrate feed. Calves are kept on farm up until weaning or slightly after (6-9 months) and then sold for intensive, concentrate-based fattening. Many farmers have been installing sown biodiverse permanent pastures rich in legumes (SBP) to provide quality animal feed and offset concentrate consumption. SBP also sequester large amounts of carbon in soils during the first 10 years after installation. This system can also sometimes support steer fattening on the farm. Here, we used a comparative life cycle assessment (LCA) approach to assess the effects of replacing concentrate through installation of high-yield SBP, controlling for the effect of the fattening system (on or off farm) and age at slaughter. The goal was to identify the main environmental and economic trade-offs involved in the implementation of these systems. We used field data from farms in the region, covering different production systems in terms of pasture system, approach for fattening of steers and age at slaughter. We used global warming potential (GWP) and land use indicators, using recent and innovative highly-regionalized models for the role of land use on biotic production potential of soils and biodiversity loss, as well as an economic assessment of costs and revenues. The analysis was carried out using software OpenLCA. Preliminary results show that the most beneficial scenario, with successful installation of SBP and steer fattening done on-farm, avoids the emission of about 3 t CO_2eq/ha even after SBP stop sequestering carbon. This scenario can avoid 25% emissions from beef production per kg of live animal weight. This effect is mostly due to avoiding concentrate consumption. We also found that each hectare of SBP installed avoids the occupation of 0.5 hectares per year that would have been used to produce feed ingredients. It also avoids soil organic carbon depletion and loss of species.

Environmental study of dairy sheep production system in the region of Basilicata (Italy)

E. Sabia[1], M. Gauly[1], G.F. Cifuni[2] and S. Claps[2]
[1]Free University of Bolzano, Faculty of Science and Technology, Piazza Università 5, Bolzano, 39100, Italy, [2]CREA, Research Centre for Animal Production and Acquaculture, Via Appia, Muro Lucano (PZ), 85054, Italy; emilio.sabia@unibz.it

Sheep milk production is an important livestock sector for the European Mediterranean countries including the regions of southern Italy. The main objective of this study was to identify the resource use categories that have the major environmental impact on dairy sheep farming systems in Basilicata region using a Life Cycle Assessment (LCA) approach. The environmental impact was evaluated on four dairy sheep farms with an average farm size of 66 ha. The functional unit was 1 kg of Fat and Protein Corrected Milk (FPCM) with a reference milk fat and protein content of 8.0 and 5.3%, respectively. Average herd size was 400 sheep and the average FPCM per lactating sheep was 76.90 kg/year. The evaluated farm activities were: energy consumption, manure management, transports, enteric fermentation, on-farm feed productions and management of the sheds. Enteric methane emissions used the national emission factor (8 kg CO_2-eq/head/year). The investigated impact categories were: Climate Change (CC), Terrestrial Acidification (TA), Marine eutrophication (ME) and non-renewable energy use (NRE). The LCA was determined with the use of the commercial software package: SimaPro 8.01. The average environmental impacts associated with 1 kg of FPCM were: CC 3.38 kg CO_2-eq, TA 37.74 g SO_2-eq, ME 8.67 g PO_4^{-3}-eq and NRE 6.31 MJ-eq. The CC showed the highest impact generated mainly by enteric fermentation (72.9%) with biogenic CH_4 as the prevalent chemical compound, followed by feeding management (22.9%). The environmental impact due to TA and ME was allocated between feeding management and emissions from manure inside the shed and direct emissions on the pasture. For the TA and ME the main polluting compounds were atmospheric NH_3 followed by N_2O and PO_4 in water. The component with the highest environmental impact in terms of NRE was crude oil followed by energy produced from natural gas. In conclusion, the sustainability of dairy sheep farming can be achieved with its continuous monitoring in order to minimize the impact without impairing its competitiveness.

Environmental impact assessment of Danish pork – focusing on mitigation options

T. Dorca-Preda, L. Mogensen, T. Kristensen and M. Trydeman Knudsen
Aarhus University, Department of Agroecology, Blichers Alle 20, P.O. Box 50, 8830 Tjele, Denmark;
lisbeth.mogensen@agro.au.dk

Pig production systems represent a sector where environmental mitigation options have to be identified and implemented because of concerns related to the increasing world population, pork increased consumption at global level or feed-food competition issues. Given the inter-relationships in terms of productivity and environmental performance of livestock production systems, and in order to avoid pollution swapping, these mitigation options have to be assessed from a whole-system perspective. For this reason, an environmental analysis was conducted by using a life cycle assessment approach from cradle to slaughterhouse gate to estimate the environmental impact of Danish pork. National average primary production data (sows, weaners and slaughter pigs) and data from the largest slaughterhouse in Denmark were used in the study. Six impact categories were assessed: climate change, eutrophication potential, acidification potential, abiotic depletion, land occupation and biodiversity damage. Additionally, by analysing the development between 2005 and 2016 of Danish pork, successful mitigation options were identified. The environmental impact of Danish pig 'meat' (= the functional unit = the total amount of pig products used for human consumption) in 2016, was estimated to be 2.8 kg CO_2 eq., 19 g PO_4 eq., 33 g SO_2 eq., 15 MJ, 5.5 m^2yr and 3.7 PDF (Potentially Disappeared Fraction) for climate change, eutrophication potential, acidification potential, abiotic depletion, land occupation and biodiversity damage, respectively. It was also found that the environmental impact of Danish pig 'meat' has decreased considerably since 2005 due to the changes that have been introduced in the systems during the last decade. Consequently, the most successful mitigation options were determined: increased share of pig live weight used for human consumption, increased feed and animal production efficiency and use of renewable energy sources.

Comparing greenhouse gas emissions and nutrition between non vegan and vegan

G.W. Park
Kangwon National University, Dept of Animal Life Sciences, Gangwondo, Chuncheonsi, Gangwondaehak gil 1, College of Animal Life Science, 24341, Korea, South; pgweu@naver.com

Producing animal products from farm to table has emitted significant greenhouse gas (GHG). Some have recommended meal plans are mainly including vegetables and grains to reduce GHG emissions. However, the meal plans based on the energy requirements with grains and vegetables, which could show the nutritional imbalance. For this reason, by using Korean suggested meal plans, we compared GHG emissions by using life cycle inventory database as well as nutritional pros and cons. To calculate GHG emissions, food's carbon footprint data was from the foundation of agriculture technology commercialization and transfer suggested meal plans (SMP) including meat referred to the Korean Nutrition Society for adult men age 19 to 29, and changed it to vegan meal plans (VMP) which substituted for meat and other animal-based protein sources to meat alternatives such as beans, tofu, etc., which had food's carbon footprint data. In addition, for comparing nutritional differences, 9th revision Korean Food Composition Table I and II were used. According to per daily intake, VMP GHG emission was 12.3% lower than SMP. Korean daily meat intake reported in Korea Agricultural Statistics Service was 37.1% lower than SMP, whose conversion to protein intake was 17% lower. In SMP, essential amino acid and non-essential amino acid were 2.25 and 2.03 times higher than VMP, respectively. Therefore, this study could suggest that SMP had more GHG emissions but more nutritional advantage. Following this reason, additional researches related to GHG emission's change with meal plans by using daily meat intake per person should be conducted.

Exploring the sustainability of livestock systems using yeast as a next-generation protein source
H.F. Olsen and M. Øverland
Norwegian University of Life Sciences, Department of Animal and Aquacultural Sciences, P.O. Box 5003, 1432 Ås, Norway; hanne.fjerdingby@nmbu.no

In Norway, sustainable intensification is an overall political goal to meet the expected future population growth of about 14% in the coming 20 years. The goal is to increase food production to maintain self-sufficiency while minimizing environmental impact. Today, only about 3% of Norway is farmed land, and it is decreasing due to the expansion of national and regional infrastructures. Still, 12.5 million decares of arable land are unused, but more than 98% of this is suitable for grass production, which is a resource not suitable for direct human consumption. Further, the use of compound feed is increasing during the last 20 years, mainly due to the increase in the consumption of pork and poultry meat. The annual Norwegian production of compound feed amount to 2 million tons, but Norway is overreliant on imported feed resources, especially protein sources such as soybean meal and rapeseed (>95%). Due to the limited possibility to cultivate feed crops, a promising opportunity to increase the share of local protein sources is to develop novel feed ingredients. Foods of Norway is a centre for research-based innovation at the Norwegian University of Life Sciences. The centre develops sustainable feed resources from local novel renewable natural resources such as forest by-products or seaweed. By using new technology, the biomass is converted into sugar and other nutrients, which can be used as a growth medium for the production of yeast. The yeast is a high-quality protein source and can potentially replace imported protein sources. However, we need to document the sustainability of yeast production compared to the existing feed resources used in our livestock systems. Through the project LIVESTOCK, we will develop LCA-models for domestic livestock systems (both ruminants and monogastric), with a focus on novel feed ingredients and assess potential trade-offs between climate gas mitigation and other sustainability aspects. In addition, the economic effects of the different options for improving sustainability will be documented, and the possibilities for adapting the livestock production towards the low-emission society in 2050 within resource limits both nationally, regionally and on a local scale will be explored.

Mixed crop-suckler cattle farming systems: between economies of scale and economies of scope
P. Veysset, M. Charleuf and M. Lherm
INRA, UMR Herbivores, 63122 Saint-Genès-Champanelle, France; patrick.veysset@inra.fr

Over the last few decades, we observed a constant expansion of the French beef cattle farms' labour productivity. While the volume of agricultural production per hectare (ha) of agricultural area (UAA) has stagnated, the volumes of equipment use per ha have increased by more than 1% per year. According to the concept of economy of scope, the shared use of production factors for complementary productions could be a source of economic gain. Mixed crop-livestock farms combining fodder and cash crops could limit the equipment costs by sharing it between these two productions. The 66 farms in the Charolais suckler-cattle farms' network monitored between 2012 and 2016 (307 observations) show a great variability in the proportion of their agricultural area allocated to non-fodder crops (from 0 to 70%, 19% on average). The 11 farms with the largest area under non-fodder crops (>75 ha, 43% of their UAA) have the highest mechanisation costs per ha of UAA, while the 25 farms with less than 20 ha of non-fodder crops (or less than 8% of their UAA) have the lowest mechanisation costs per ha. There is a significant negative correlation between mechanisation costs per ha and the percentage of forage area in the UAA. There is therefore no sharing of the equipment factor between fodder and non-fodder crop, no economy of scope on this item. Farms with the largest crop area are also those with the largest UAA, and, contrary to the concept of economy of scale, the correlation between the total area of the farm and mechanisation costs per ha is significantly positive. Surveys were carried out to understand farmers' choices in terms of mechanization. It appears that there is a certain lack of knowledge of this expense item. The expansion of farms and the substitution labour/capital leads to a sharp increase in equipment needs and related costs. These charges are not diluted with size, nor are they shared between animal and plant production. A real awareness of this phenomenon should lead farmers to more rational choices of equipment and especially to the sharing of equipment. Or is it better to stay 'small' and/or 'specialized'?

Elucidating producer behaviour of providing positive life opportunities to farm animals

J.E. Stokes[1], S. Mullan[2], T. Takahashi[2,3], F. Monte[4] and D.C.J. Main[1]
[1]Royal Agricultural University, Cirencester, Gloucestershire, GL7 6JS, United Kingdom, [2]University of Bristol, Langford, Somerset, BS40 5DU, United Kingdom, [3]Rothamsted Research, Okehampton, Devon, EX20 2SB, United Kingdom, [4]ADAS, Avonmouth Way, Bristol, BS11 8DE, United Kingdom; taro.takahashi@rothamsted.ac.uk

While welfare certification schemes for animal-originated food products are designed to reduce negative behavioural, health and physical outcomes on the farm, it has now been widely accepted that animal welfare should not only be defined by the absence of negative subjective states but also by the presence of positive life experiences. There is little consensus, however, as to how best to quantify positive welfare of animals from observed outcomes. The present study investigated the feasibility of utilising input-based measures of positive welfare as part of evaluation criteria for food certification protocols. Assessments of welfare-enhancing resource inputs and welfare outcomes were carried out on 49 free-range laying hen farms in the UK. For input-based measures of positive welfare, the resource tier framework was applied to each study farm. The framework consists of 13 resource needs, with each graded on a scale of 0 to 3 (no score, Welfare +, ++ and +++) based on physical resources, farm environment and proactive management activities above what is stipulated by UK law and codes of practice. For outcome-based measures of negative welfare, six indicators commonly used by assurance schemes were collected on each farm. In addition, the financial cost associated with resource provision on each farm was estimated. The results of data analysis showed that improved on-farm resources are generally associated with reduction of negative outcomes, with the resource tier score and the estimated total cost also strong correlated (ρ=0.822, P<0.001). A closer investigation revealed, however, that farms are spending considerably more resources than theoretically required to achieve the same score and a large proportion of overspending occurs on resources more visible to human eyes. On average across 49 farms, the cost saving potential was 81% of current total expenses, suggesting that significant room exists to reduce costs without compromising welfare or to improve welfare without incurring additional costs.

Typologies of smallholder pig farmers in Hoima and Kamuli districts of Uganda

B.M. Babigumira[1], J. Sölkner[1], E. Ouma[2] and K. Marshall[3]
[1]University of Natural Resources and Life Sciences, Vienna-BOKU, Department of Sustainable Agricultural Systems, Gregor-Mendel-Straße 33, 1180, Vienna, Austria, [2]International Livestock Research Institute, Kampala, P.O. Box 24384, Kampala, Uganda, [3]International Livestock Research Institute, Nairobi, P.O. Box 30709, Nairobi 00100, Kenya; bbabigumira@gmail.com

Designing client-focused interventions can be challenging particularly for a heterogeneous group like smallholder farmers. Farm typologies overcome such limitations and attempt to cluster farmers into more realistic and representative groups. The aim of this study was to delineate and describe types of latent subgroups among 200 smallholder pig farmers in two districts of Uganda- Kamuli and Hoima. Principal Component Analysis and K-means Cluster Analysis identified three farm types based on six variables-household head's gender, age and education; household size, total livestock units and land owned. Analysis of variance showed gender, age and education of the household head and household size significantly (P<0.05) differentiated the three farm types. Type 1 was male-headed with an average household head age of 41.6+11.5 years, average household membership of 6.14+2.6 and an average of 2.32+1.6 acres. Type 2 was male-headed with an average household age of 57.9+11.1 years, average household membership of 6.6+2.6 and an average of 6.9+4.6 acres. Type 3 were female-headed households with an average household age of 56.2+14.8 years, household membership of 4.8+1.9 and an average of 1.9+2.0. For type 1 and 2, the main decision makers for the farmers' livelihoods were adult male or adult male and adult female jointly. The adult female or adult male and female jointly provided labour. The main decision maker for type 3 farmers' livelihoods and labour provider was the adult female. The results indicate that gender plays a significant role in smallholder pig production particularly the participation of the household adult female in decision-making and provision of labour.

Peer community in animal science: a free & open science recommendation process of scientific papers

R. Muñoz-Tamayo

UMR Modélisation Systémique Appliquée aux Ruminants, INRA, AgroParisTech, Université Paris-Saclay, 75005, Paris, France; rafael.munoz-tamayo@inra.fr

Publishing is at the backbone of the scientific work. Scientific publications contribute to the progress of science and its dissemination. However, the current publication system is particularly very expensive. The high fees for publishing and accessing to scientific papers in journals contrast with the fact that the manuscript evaluation process (the hub of the publication system) is performed at nearly no cost for the journals. To this controversy, it is added the use of public funding that research institutes must allocate to cover the costs of publishing and accessing scientific papers. Although open access journals remove the barrier of the access, their high publication fees create a barrier for publishing that depends on the financial capacities of the researchers. In this context, our goal in this contribution is to introduce the initiative Peer Community In (PCI) Animal Science; an alternative to the current publication system under the umbrella of the Peer Community In project (https://peercommunityin.org/). PCI Animal Science is a community of researchers working in animal science that promotes open science and research transparency. PCI Animal Science is not a journal but it operates similarly with editors and reviewers. PCI Animal Science is managed by scientists without the intermediation of any commercial publisher. The PCI Animal Science community performs, at no cost for readers and authors, rigorous reviews of unpublished preprints (deposited in repositories such as bioRxiv and HAL) from a wide range of research areas applied to animal science. On the basis of independent reviews, a recommender/editor decides whether a paper is recommended or not. The preprints recommended by PCI Animal Science are complete, reliable and citable articles of high scientific value that do not need to be published in traditional journals (although the authors can further submit their recommended preprints afterwards). On February 2019, 35 international scientists have joined the initiative: https://goo.gl/Qfp9px. In the long term, we expect PCI Animal Science to have an important position in the publication landscape of scientific outputs. Join us to contribute to the change!

Development of a bronchoalveolar lavage sampling method in goat kids

S. Verberckmoes[1], M. Willockx[2] and B. Pardon[2]

[1]DAP Verberckmoes, Beerveldestraat, 9160 Lokeren, Belgium, [2]Ghent University, Department of Large Animal Internal Medicine, Salisburylaan, 9820 Merelbeke, Belgium; steven.verberckmoes@skynet.be

Respiratory infections are a major health issue and the main indication for antimicrobial use in goat kids. Pressure to reduce antimicrobial use is high, and despite its organic image the goat sector needs to be proactive. Availability of an economic and practical sampling method for lower respiratory tract infections is essential to rationalize antimicrobial use. The objective was to develop a bronchoalveolar lavage (BAL) technique for sampling lower airways in goat kids, with minimal impact on animal welfare. Two techniques were attempted: (1) a non-endoscopic BAL technique in unsedated animals, as currently frequently used in calves; (2) a laryngoscope-guided bronchoalveolar lavage after sedation. The flushing volume was 15 ml of sterile saline in every case. Technique 1 used a central venous catheter, which was introduced blindly through the nasal cavity into the trachea, after rinsing the nares. It was essential that the animal's head was fixated in an (over)stretched position, which was not well tolerated. The technique resulted in a proper BAL sample in 20% (3/15) of the animals sampled. The laryngoscopic BAL was done after sedation with xylazine (0,2 mg/kg IM). A modified universal AI sheath (was moved over the epiglottis and introduced into the trachea. Subsequently, a flexible polypropylene catheter (Buster 1.3 mm × 50 cm) was introduced through the rigid tube into the lungs, until the wedge position was reached. On 9 farms, a total of 134 goat kids were sampled (19, 1-2 weeks old; 115, 2-3 months old). The technique was successful in 93% (125/134) of the goat kids sampled and the sampling rate was 10 animals an hour. Complications were no or insufficient amount of BAL fluid recovered in 9 animals, and mortality in 1 animal, which was evidenced septicemic. In conclusion, the laryngoscopic BAL technique in sedated animals is preferred, as it resulted in a high success rate, a realistic speed of sampling and was better tolerated by the animals. This technique can aid in better diagnosis, control, prevention and targeted therapy for respiratory diseases in goat kids in the future.

The effect of two-week access to willow foliage on the immune status of goats in late lactation

H. Muklada[1,2], H. Voet[1], T. Deutch[2], M. Zachut[2], S. Blum[3], O. Krifuks[3], T.A. Glasser[4], J.D. Klein[2] and S.Y. Landau[2]
[1]The Faculty of Agriculture, Food and Environment, P.O. Box 12, Rehovot 76100, Israel, [2]ARO, 64 HaMakabim Road, 7505101 Rishon LeZion, Israel, [3]Kimron Veterinary Institute, 62 HaMakabim Road, 7505101 Rishon LeZion, Israel, [4]Ramat Hanadiv Nature Park, Zikhron Yaakov, P.O. Box 325, Israel; vclandau@agri.gov.il

In a previous study, we showed that access to willow fodder decreased somatic cell counts (SCC) in the milk of local Mamber goats grazing in brushland at the end of lactation. In the present study 48 Alpine crossbred grazing in the same environment were offered free access to freshly cut willow fodder (W, n=24) or not (C, n=24) for two weeks. Udder health status was determined before the experiment and each the two groups included 6 goats defined as infected by cfu in milk and 18 non-infected goats. Goats ingested, on average, 600 g of DM from willow, resulting in minor change in dietary quality. The willow contained 13 g/kg DM of salicin. Goats in W and C did not differ in milking performance and in milk attributes. Initial SCC and milk neutrophils (CD18+ and PG68) cells were initially higher (P<0.01) in infected than in non-infected goats and decreased significantly in W (P<0.05 and P<0.01, in the same order) but not in C uninfected goats, throughout experiment. Throughout experiment, the percentage of CD8+ T-cells increased strongly (P<0.001) in all C goats. In contrast, a significant increase was found in the W group (P<0.01) only for the infected goats. Overall, eating willow mitigated the increase of CD8+ (P<0.05). When treatments were spliced, eating willow mitigated (P<0.01) the increase in percentage of CD8+ in non-infected goats, but not in infected goats. Another interesting, but puzzling finding is related with reticulocytes, whose counts in blood increased in W but not in C goats. Data suggests anti-inflammatory and anti-stress effects for willow fodder in late-lactating goats. However, no effect was found in SCC in milk, which raises the question of the validity of this estimate in the late lactation of goats. This is the first report of a direct nutraceutical effect on the immune status of goats.

Differences in growth between Corriedale sheep divergent lines for resistance to nematodes

G.F. Ferreira[1], I. De Barbieri[1], D. Castells[2], E.A. Navajas[1], D. Giorello[1], J.T.C. Costa[1], G. Banchero[1] and G. Ciappesoni[1]
[1]INIA, Meat & Wool, Ruta 48 km 10, 90200, Uruguay, [2]SUL, Uruguayan Wool Secretariat, S. Gómez 2408, 12100, Uruguay; gciappesoni@inia.org.uy

The association between genetic resistance of lambs to gastrointestinal parasites (GIP) with dry matter intake (DMI) average daily gain (ADG) and residual feed intake (RFI) was studied. Sixty-four Corriedale lambs (357±14 days old), from divergent lines for resistance to GIP (27 resistant-R and 37 susceptible-S) developed by the Uruguayan Wool Secretariat were used. The animals were allotted to one of five outdoor pens, they were stratified by sex, body weight, and sire. Each pen was equipped with five automated feeding systems and two automatic weighing platforms allowing individual records of feed intake and body weight. After 14 days of acclimatization to diet (*ad libitum* Lucerne silage: DM 36.5%, CP 21.7%, ME 2.51%) and feeding system, two tests were run over two periods of 44 (P1) and 42 days (P2), respectively. Firstly, the animals were maintained worm-free (P1) followed by an artificial infestation of *Haemonchus contortus* (P2). The infestation occurred in three consecutive days with 2,000 L3 larvae per day. The second period was split into two subperiods from 0-23 and 24-42 days post infestation (P2a and P2b, respectively). Records for faecal egg count (FEC) were taken in days 9, 23, 27, 30, 42 post infestation. The DMI (kg/day) was computed as the average of the individual daily intake, ADG (kg/day) was calculated by regression using all weights for each period, RFI is the residuals resulting from the model DMI = ADG + metabolic weight (defined as mid-weight for each period ⊥0.75) + pen (1-5) + type of birth (1 or 2). There were no differences between lines in DMI, ADG, RFI for both periods. However, statically significative differences (P<0.05) were found in P2b for ADG and FEC (at day 23). The R line showed higher ADG than S line (0.132±0.017 vs 0.091±0.014 kg/day) and lower parasite infestation (1,049 vs 2,479 back-transformed FEC mean). Probably, the high CP diet content and the age of the animals contribute to decreasing the differences in FEC between lines. These preliminary results suggest a difference in growth pattern between R&S lines during the infestation period without effects on DMI.

Exploring the success or failure of passive antibody transfer in goat kids in Flanders

M. Willockx[1], S. Verberckmoes[2], J. Vicca[3], E. Van Mael[4] and B. Pardon[1]
[1]Ghent University, Faculty of Veterinary Medicine, Large Animal Internal Medicine, Salisburylaan 133, 9820 Merelbeke, Belgium, [2]Private practive, Beerveldestraat 118, 9160 Lokeren, Belgium, [3]Odisee UC, Agro- and biotechnology, Hospitaalstraat 23, 9100 St.-Niklaas, Belgium, [4]DGZ Vlaanderen vzw, Hagenbroeksesteenweg 167, 2500 Lier, Belgium; jo.vicca@odisee.be

In industrialized countries, dairy goat farmers are more reluctant to give mother-own colostrum to their offspring, because of the possible transfer of *Mycobacterium avium* subspecies *paratuberculosis* and caprine arthritis and encephalitis virus to the kids. Both micro-organisms are widely spread in the dairy goat sector in Flanders. Various colostrum management strategies are currently practiced, such as the use of commercially available lyophilized cow colostrum, fresh cow colostrum from a neighbouring cattle herd, on-farm pasteurized goat colostrum, fresh goat colostrum or a combination of different kinds of colostrum. To date, it is unknown whether this situation results in a higher prevalence of insufficient uptake of maternal antibodies (failure of passive transfer (FPT)). Therefore, the objective of this study was to determine the prevalence of FPT in a convenience sample of Belgian professional dairy goat farms. Between February and May 2018, 14 of the 46 (30%) registered Belgian dairy goat farms with at least 100 animals, were sampled based on willingness to cooperate. Blood samples were collected from 12-15 lambs on each herd between 2 to 7 days after birth. The immunoglobulin (Ig) level of 196 sera was determined using capillary electrophoresis. Ig levels ranged between 1.3 and 27.6 g/l, with an average of 10 ± 5.4 g/l. Using the cut-off value of 12 g Ig/l to define FPT in goat kids, as suggested by O-Brein and Sherman, 69% of the sampled goat kids would have FPT. At this cut-off, mean within herd prevalence was $65.4\pm25.4\%$), ranging from 13.3 to 100%. Under the conditions of this study, FPT appears to be a highly prevalent issue in the Flemish dairy goat sector. Given the important health and growth consequences of FPT, alternative colostrum management practices, ensuring sufficient antibody intake and avoiding infection, should be urgently explored.

Sustainable prevention and eradication of footrot in sheep

A. Wirth, D. Vasiliadis and O. Distl
University of Veterinary Medicine Hannover, Institute for Animal Breeding and Genetics, Bünteweg 17p, 30559 Hannover, Germany; ottmar.distl@tiho-hannover.de

Ovine footrot is a widespread problem in German sheep farming. The objectives of our research project are to determine the dynamics of virulent strains of *Dichelobacter nodosus* within and between flocks and for individual sheep and to implement sustainable prevention and eradication programs based on herd management and breeding. For this purpose, competitive RT PCR is used to differentiate benign and virulent *D. nodosus* strains for classification of the infectious status of flocks. Stable schools will be introduced to demonstrate in living examples how prevention, veterinary health and breeding programs can be successfully operated. On farm data recording is done using a mobile electronic hand-held system for individual animal ear tags and data input on animals. Project coordination is done by the University of Veterinary Medicine Hannover, Germany, and project partners are the Union of all German sheep breeding associations, sheep health organizations and veterinary sheep practitioners. The project is supported by the Federal Ministry of Food and Agriculture. A standardized questionnaire is used to collect data on flock management and practises, livestock movements among flocks, vaccinations, claw trimming and treatments with antibiotics and/ or footbaths as well as housing and pasture conditions. Breeding program will be developed using genotypings of ovine SNP50 genotyping beadchip, ovine infinium HD SNP beadchip and next generation sequencing data of sheep. Phenotyping is only done in flocks infected with virulent *D. nodosus* strains in order to distinguish among resistant and susceptible individuals. Further discrimination of animals can be done on the severity and fast course of clinical signs of footrot. The project is open for sheep farms under footrot risk including herdbook and non-herdbook flocks. At least two breeds are included for genotypings and genomics to develop footrot resistant flocks. Sample sizes envisaged per breed are at 1,500-2,000.

Genetic parameter estimates for gastrointestinal parasites resistance traits in Santa Inês breed

M.P.B. Berton[1], R.P.S. Da Silva[2], F.E.C. De Carvalho[2], P.S.O. Oliveira[2], J.P.E. Eler[2], G.B. Banchero[3], J.B.S.F. Ferraz[2] and F.B. Baldi[1]
[1]Faculdade de Ciências Agrárias e Veterinárias – FCAV / UNESP, Via de Acesso Prof.Paulo Donato Castellane s/n, 14884-900 Jaboticabal, SP, Brazil, [2]Faculdade de Zootecnia e Engenharia de Alimentos – FZEA / USP, Av. Duque de Caxias Norte, 225, 13635-900 Pirassununga/SP, Brazil, [3]Instituto Nacional de Investigación Agropecuária -INIA, Ruta 50, Km. 11, Colonia, Uruguay, Uruguay; mapberton@gmail.com

The study aimed to estimate genetic parameters for gastrointestinal parasites resistance indicator traits in Santa Inês breed in tropical conditions. The phenotypic records were collected from 700 naturally infected animals of Santa Inês breed, belonging to four flocks located in Southeast, Northeast states of Brazil. The following traits were evaluated: degree of anaemia assessed by the famacha card (FMC), haematocrit (HCT), white blood cell (WBC), red blood cell (RBC), the egg counts per gram of faeces (EPGlog). A total of 576 animals were genotyped with the Ovine SNP12k BeadChip (Illumina, Inc.). The variance components were estimated using a single animal trait model by the ssGBLUP procedure. The fixed effects considered were contemporary groups (farm and management group), sample collection month, sex, covariable body condition (linear effect), and age at the collection (linear and quadratic effect). The analyses were performed using the REMLF90. Moderate heritability estimates were obtained for FMC (0.31), HCT (0.29), RBC (0.27) and WBC (0.21), and low for EPGlog (0.10). The genetic correlation estimates were low between FMC with EPGlog (0.23) and WBC (-0.097); and EPGlog with RBC (-0.17), moderated between EPGlog with HCT (-0.43) and high between FMC with HCT (-0.92) and RBC (-0.95); and EPGlog with WBC (-0.85). The high FMC heritability and its high genetic correlation with the resistance traits indicators enable the selection response. Also, lower EPGlog values should favour the FMC and conduct the blood cells concentrations to an increase in animal resistance. Thus, the genetic parameters obtained shows the feasible genetic progress for the traits studied allowing to improve the animal resistance. Acknowledgments: São Paulo Research Foundation (FAPESP) grant #2014/07566-2 and #2017/10493-5.

Distribution of ewes in SCC groups during first and second lactation

K. Tvarožková[1], V. Tančin[1,2], M. Uhrinčat[2], L. Mačuhová[2] and I. Holko[3]
[1]Slovak University of Agriculture, FAFR, Tr. A. Hlinku 2, Nitra, 94976, Slovak Republic, [2]NPPC, Research Institute for Animal Production Nitra, Hlohovecká 2, Lužianky, 95141, Slovak Republic, [3]Vetservis, s.r.o., Kalvária 715/3, Nitra, 949 01, Slovak Republic; kristina.tvarozkova@gmail.com

Somatic cell count (SCC) in milk is used for detection of subclinical mastitis. Among scientists it is still a big discussion about the physiological level of SCC in milk of ewes. The aim of this study was to describe the frequency of distribution of ewes in SCC groups on the basis SCS (somatic cells score) per lactation and estimate changes of SCC from1[st] lactation on 2nd lactation. The experiment was carried at seven farms in1[st] observed period (2016 and 2017) and at eight farms in2[nd] observed one(2017 and 2018). Within each of periods the same animals were sampled on their first and following second lactation in next year of study, only. Totally 1,199 milk samples from 159 ewes and 1,653 milk samples from 219 ewes were collected during 1st period and 2nd period, respectively. Milk sampling were taken monthly from April to August in both periods. For evaluation only ewes with minimum three sampling per year (min 6 samples per animal) were included in the study within both periods. The ewes were divided into the five SCC groups on basis of their SCS per lactation: G1= $SCC < 200 \times 10^3$ cells/ml, G2= $SCC \geq 200 < 400 \times 10^3$ cells/ml, G3= $SCC \geq 400 < 600 \times 10^3$ cells/ml, G4= $SCC \geq 600 < 1000 \times 10^3$ cells/ml and G5= $SCC \geq 1000 \times 10^3$ cells/ml. In total statistically significant positive impact of parity on SCC in 2nd period was detected (P<0.0001) only. From the farm point of view in 1st period only in two farms and in second one in 4 farms significant effect of parity was found out. Thus in some farms no increase of SCC from first to second lactation was observed. When comparing the changes in SCC from the first to the second lactation in both first and second periods, 6.92 and 10.96%, respectively ewes moved from SCC group G1 to G5. In conclusion the significant effect of farm management and parity on SCC was demonstrated. This study was funded by the projects: APVV-15-0072.

Possibilities of mastitis detection by measuring the electrical conductivity of ewe's milk

M. Uhrinčat[1], V. Tančin[1,2], K. Tvarožková[2], L. Mačuhová[1], M. Oravcová[1] and M. Vršková[1]
[1]NPPC-Research Institute for Animal Production Nitra, Hlohovecká 2, 951 41 Lužianky, Slovak Republic, [2]Slovak University of Agriculture, Tr. A. Hlinku 2, 949 76 Nitra, Slovak Republic; uhrincat@vuzv.sk

Measurement of electrical conductivity (EC; mS/cm) is a method often used in dairy cows during milking in milking parlours, but especially in robotic milking as a low-cost mastitis detection method. The aim of this study was to evaluate the relationship between somatic cell count (SCC; cells/ml) and EC of milk in sheep reared in Slovakia as factors for monitoring subclinical mastitis. The bacteriological examination of milk samples were performed too. Samples were collected individually from both halves of the udder from 295 ewes of different breeds from eight farms during evening milking. Based on SCC, the samples (590) were divided into three categories with classes (SCC<200,000, 200,000≤SCC<400,000, 400,000≤SCC<600,000, and SCC≥600,000), (SCC<700,000 and SCC≥700,000) and (SCC<100,000 and SCC≥100,000) respectively. Based on the presence of pathogens in the udder halve, they were classified as 'without pathogens (W)' (415), 'minor (MI)' (161) and 'major (MA)' (14) pathogens. The presence of a pathogen had a significant effect on the increase in EC (W; MI; MA (Mean ± SD)) 4.63±0.76; 5.29±1.24; 5.88±1.68, LogSCC 4.90±0.76; 5.85±0.76; 6.50±0.91 log cells/ml and protein content 5.66±0.68; 5.82±0.97; 6.55±1.43% and also decrease in content of lactose 4.97±0.48; 4.59±0.80; 4.10±0.80%. Significant correlation between ES and SCC was found when all data were analysed together (r=0.531). In the first category, a significant difference in EC by SCC was only between SCC≥600,000 (5.81±1.38) and SCC<200,000 (4.50±0.54) classes. In the second and third category, we found significant differences in both cases, the SCC<700,000 (4.54±0.56) and the SCC≥700,000 (5.86±1.40) and SCC<100,000 (4.46±0.54) and the SCC≥100,000 (5.24±1.19). We can assert that for the detection of mastitis in ewes we can use also the method of measuring the EC of ewe's milk. EC can be useful in detecting animals with level of SSC greater than 600,000, but we cannot estimate a threshold for healthy animals yet. This study was supported by APVV 15-0072.

Somatic cell count and presence of mastitis pathogens in milk of ewes

V. Tančin[1,2], K. Tvarožková[1], M. Uhrinčat[2], L. Mačuhová[2], M. Oravcová[2] and I. Holko[3]
[1]Slovak University of Agriculture, FAFR, Tr. A. Hlinku 2, Nitra, 94976, Slovak Republic, [2]NPPC Research Institute for Animal Production Nitra, Hlohovecká 2, Lužianky, 951 41, Slovak Republic, [3]Vetservis, s.r.o., Kalvária 715/3, Nitra, 949 01, Slovak Republic; tancin@vuzv.sk

The aim of this study was to find out the frequency of pathogens occurrence and a possible relationship to somatic cells count (SCC) in milk samples from the individual halves of the udder of ewes. Sampling was carried out during the month of May and June in 2017 and 2018 during the evening milking under practical conditions in Slovakia. In 2017 the 116 samples of milk were collected for SCC analysis and bacteriological examination for the presence of mastitis pathogens. Based on SCC, ewes were divided into five categories: up to 0.2×10^6/ml; 0.2-0.4×10^6/ml; 0.4-0.6×10^6/ml; 0.6×10^6/ml; over 10^6/ml. In 2018 bacteriological examination of milk samples (310 samples) were performed only. In the first category of SCC there were up to 84.03% of milk samples, in second 6.90%, in third 0.89%, in fourth 5.17% and in fifth one 6.03%. Higher SCC values in samples with pathogens presence (log 5.28±0.09/ml) compared to free of pathogens samples (log 4.73±0.06/ml, P<0.001) were found. Frequency of pathogens occurrence in 2017 and 2018 were as followed: two important infectious mastitis pathogens were isolated: *Streptococcus agalactiae* (23.33%, 10.7% resp.) and *Staphylococcus aureus* (3.33%, 6,9% resp.). Most often, there were coagulase negative staphylococci (CNS) (60%, 76.6% resp.), than followed by *Streptococcus dysagalactiae* (5%, 4.6% resp.), *Candida* sp. (1.67%), *Enterocuccus faecalis* (1.67%, 1.5% resp.), *Streptococcus parauberis* (1.67%), *Streptococcus uberis* (1.67%) and moulds (1.67%). In 2018 *S. uberis, S. parauberis, Candida* sp., *Klebsiella* sp., and moulds were below 1%. In conclusion, the most frequent pathogens cultivated in ewes milk were CNS, and SCC was significantly affected by the presence of the pathogen. This study was funded by the projects: APVV-15-0072.

Effect of weaning system on milk yield and the frequency of individual milk flows

L. Mačuhová[1], V. Tančin[1,2], J. Mačuhová[3], M. Uhrinčať[1] and M. Margetín[1,2]
[1]NPPC Research Institute for Animal Production Nitra, Hlohovecká 2, 951 41 Lužianky, Slovak Republic, [2]Slovak Agricultural University Nitra, FAFR, Tr. A. Hlinku 2, 949 76 Nitra, Slovak Republic, [3]Institute for Agricultural Engineering and Animal Husbandry, Vöttinger Str. 36, 85354 Freising, Germany; macuhova@yuzv.sk

The aim of this study was to evaluate the effects of three weaning systems on milk yield and the frequency of occurrence of different milk flow types of crossbred ewes of Improved Valachian (IV×LC; n=38) and Tsigai (TS×LC; n=41) with Lacaune. Prior to parturition, ewes were assigned to one of following three treatments for the first 53 day of lactation: (1) ewes weaned from their lambs at 24 h postpartum and afterwards machine milked twice daily (MT); (2) ewes, beginning 24 h postpartum, kept during the daytime with their lambs and allowed them to suckle for 12 h, nights separated from their lambs for 12 h and machine milked once daily in the morning (MIX); and (3) ewes exclusively suckled by their lambs (ES). After the treatment period, lambs were weaned from MIX and ES ewes, and all three groups were machine milked twice daily. The recording of milk yield and milk flow was performed on 110±5 day of lactation. The milk flow curves were classified into four types: one peak (1P), two peaks (2P), plateau I (maximal milk flow over 0.4 l/min (PLI)), plateau II (maximal milk flow less 0.4 l/min (PLII)). In ewes with 2P and PLI milk flows, the release of oxytocin can be supposed in response to machine milking. Milk yield did not differ significantly (P=0.4421) between weaning systems (0.17±0.02 l in ES, 0.20±0.01 l in MIX, 0.21±0.02 l in MT). The frequency of occurrence of different milk flow types (1P:2P:PLI:PLII) was 17:48:31:4%, 19:42:25:14%, and 50:20:25:5% in ES, in MIX, and MT system, resp. The weaning system and also crossbreed had not significant effect on occurrence milk flow type (P=0.1594 and P=0.6279, resp.). In conclusion, no effect of weaning system on milk yield and milk flow type could be observed in mid lactation. This publication was written during carrying out of the project APVV-15-0072.

Breeding value estimates of fertility traits in five small Swiss sheep populations

A. Burren[1], C. Hagger[1], C. Aeschlimann[2], B. Brunold[2] and H. Joerg[1]
[1]Bern University of Applied Sciences, School of Agricultural, Forest and Food Sciences, Länggasse 85, 3052 Zollikofen, Switzerland, [2]Swiss Sheep Breeding Association, Industriestrasse 9, 3362 Niederönz, Switzerland; hannes.joerg@bfh.ch

In 2018 breeding value estimates of fertility traits were introduced for the Swiss minor sheep breeds Charolais (CHS, litters=2,540), Dorper (DOP, litters=1,763), Shropshire (SHR, litters=1,130), Suffolk (SU, litters=2,030) and Texel (TEX, litters=2,355). There was used the same principle as for the four major Swiss sheep breeds namely White Alpine (WAS, litters=63,445), Black-Brown Mountain (SBS, litters=21,353), Valais Blacknose (SN, litters=31,668), and Brown Headed Meat sheep (BFS, litters=25,300). We selected the following four traits to assess the breeding value (BV) using REML and BLUP methods: age at first lambing (BV1), lambing interval (BV2), litter size one (BV3) and litter size two (BV4). The last two traits are the number of lambs at the first and second lambings. The pedigree data comprised 30,963, 12,322, 9,191, 18,593 and 18,743 for CHS, DOP, SHR, SU and TEX, respectively. The statistical model accounted for fixed effects of herd×year of the first lambing (BV1, 2, 3 and 4), season of the first lambing (BV1 and 3) and random effects of animals and residuals (BV1, 2, 3 and 4). The heritability estimates vary from 0.13-0.51 (age at first lambing), 0.05-0.13 (lambing interval), 0.08-0.11 (litter size one) and 0.01-0.05 (litter size two) for the four traits and the five breeds, respectively. This heritabilities correspond to the values from the major sheep breeds and are comparable between the breeds. The phenotypic correlations between all traits were positive and range from 0.02 to 0.34 for all breeds. On the other hand, 12 of the 30 genetic correlations were negative and range between -0.03 to -0.4. Most of the negative correlation coefficients were found for BV2-BV3 (DOP, TEX=-0.07, SU=-0.17, SHR=-0.4) and BV1-BV3 (TEX=-0.11, SU=-0.18, DOP=-0.36). The genetic correlations between BV3 and BV4 were 1.0 for the breeds DOP and SHR. The reasons for these high values could be the low number of litters. The other results seem to be plausible and show, that genetic improvement trough selection would be possible in sheep fertility traits.

Response of Dorper females exposed to the 'male effect' with males of different social hierarchy

A. Gonzalez-Tavizon[1], J.M. Guillen-Muñoz[1], F. Arellano-Rodriguez[1], S.O. Yong-Wong[1], R. Delgado-Gonzalez[1], C. Meza-Herrera[2], F.G. Veliz-Deras[1] and V. Contreras-Villarreal[1]
[1]*Universidad Autonoma Agraria Antonio Narro, Periferico Raul Lopez Sanchez, 27054, Torreon, Coahuila, Mexico,* [2]*Universidad Autonoma de Chapingo-URUZA, Carretera Gómez Palacio, Ciudad Juárez, 35230, Bermejillo, Durango, Mexico; dra.viridianac@gmail.com*

The aim of this study was to evaluate the reproductive response of anoestrous females exposed to rams with either high or low social hierarchy. In order to determine the social status of each male, a competition behavioural test was performed in April (26°N) on Dorper rams to select 2 high and 2 low hierarchy males. Besides, anovulatory females (n=76) were treated with an intravaginal sponge impregnated with 20 mg of progesterone (cronolone) and separated into two homogeneous groups regarding body condition score and body weight. While in Group 1 (LHG; n=38) ewes were exposed to low hierarchy male, ewes from Group 2 (HHG; n=38) were exposed to high hierarchy males. Ewe´s oestrus response was evaluated every 12 h (08:00 and 18:00 h) to determine the percentage of oestrus; the percentage of ovulation was evaluated by ultrasonography at day 10; percentages were analysed with a Chi-square test (MYSTAT 12). No differences (P>0.05) occurred regarding oestrus (81.5% vs 89.5%, LHG vs HHG, respectively), yet, ovulation percentage was higher in the HHG (73.7% vs 89.5%, LHG vs HHG, respectively). Results of this research suggest that males with a high hierarchy are able to induce a better reproductive response than males with a low hierarchy in Dorper ewes exposed to the 'male effect' in northern Mexico.

Oestrous induction with the ram effect in postpartum ewes during the non-breeding season

R. Pérez-Clariget[1], M.G.K. Rodriguez[2], M. Olivera[3] and R. Ungerfeld[3]
[1]*Facultad de Agronomía, Universidad de la República, Departamento de Producción Animal y Pasturas, Garzón 780, 11900, Uruguay,* [2]*Faculdade de Ciencias Agrárias e Veterinárias, Universidade Estadual Paulista, Departamento Medicina Veterinaria Preventivas e Reprodução Animal, Jaboticabal, 14884-900, Brazil,* [3]*Facultad de Veterinaria, Universidad de la República, Departamento de Fisiología, Lasplaces 1620, 11600, Uruguay; raquelperezclariget@gmail.com*

The aim of this study was to induce out-of-season oestrus, ovulations and pregnancy with the ram effect in postpartum spring-lambing ewes. The study was performed during October (spring, seasonal anoestrus) with 163 multiparous Corriedale ewes. At the beginning of the study, single lambing nursing ewes had 47 (group PP47; n=54) or 68 (group PP68; n=23) days postpartum. Other 86 ewes that were not pregnant that year were included as a control group (group CON). Ewes remained isolated from rams during 5 months (minimum distance=2,000 m). An intravaginal sponge impregnated with 30 mg of acetate of medroxyprogesterone was inserted to all ewes, and withdrawn 6 days later (Day 0). Immediately after, 19 adult Corriedale marking rams were joined with the ewes. The rams were removed from the flock 28 days later. The experiment had two replicates. Oestrus was detected once daily until Day 6, and the presence of corpus luteum (CL) was determined by ultrasound on Day 8. Pregnancy was confirmed by ultrasound on Day 73. Reproductive variables were analysed using general lineal models (GENMOD procedure of SAS). There were no differences between replicates, so data were analysed all together. More CON than PP47 and PP68 ewes came into oestrus [66.3% (57/86) vs 25.9% (14/54) and 26.1% (6/23); P=0.0001], but there was no difference in the proportion of ewes with a CL on Day 8 [66.3% (45/86) vs 33.3% (18/54) and 52.2% (12/23)]. However, at the end of the study more CON than PP47 and PP68 ewes were pregnant [66.3% (57/86) vs 18.5% (10/54) and 30.4% (7/23); P=0.0001]. Although less postpartum ewes responded to the ram effect and resulted pregnant, the percentage that ovulated was no different than the CON ewes, opening interesting possibilities to study more in depth how to improve the results of that ovulation considering that the ram effect is an inexpensive and clean tool.

Variability in some morphological traits of indigenous Nigerian and Sudanese goat breeds

D.M. Ogah[1], E.S. Shuiep[2], S.I. Daikwo[3], H. Abdullahi[4] and Z. Dankoli[4]
[1]Nasarawa State University Keffi, Animal Science, Faculty of Agriculture, Shabu-lafia campus, +234, Nigeria, [2]University of Nyala, Department of Molecular Genetics, Institute of molecular Biology, Nyala South Darfur State, +253, Sudan, [3]Federal University Wukari, Taraba State., Department of Animal Production and Health, Faculty of Agriculture Wukari, +234, Nigeria, [4]Bayero University kano, Animal Science, Faculty of Agric. BUK, +234, Nigeria; mosesdogah@gmail.com

A total of 360 male and female goats were sampled for morphological traits assessment to distinguish the populations from Nigeria and Sudan. Three breeds from Nigerian goat population (West African Dwarf (WAD), Red sokoto and sahelian breeds) and two Sudanese goat population (Local and Desert breeds) were considered for the study. Nine quantitative traits were used, the include, body weight, height at wither, body length, face length, heart girth, neck length, ear length, rump length and tail length. The Sudanese desert goat are heavier 28.57±1.38, longer body 58.75±0.61, better heart girth 69.36±1.26, longer neck, ears, rump and tail 23.23±1.26, 22.65±0.38, 14.76±0.26, 14.38±0.20 respectively while the Nigerian red sokoto had longer face 20.13±0.39. The Mahalanobis distance of the morphological traits between the Sudanese and Nigerian indigenous goats showed that the Sudanese local and the desert breeds are the closest 1.27. while the highest distance was among the Nigerian goat population red sokoto and the sahelian 68.79. The result of classification into appropriate group clearly show Nigerian goats all classified as 100% with the Sudanese breed sharing their classification across breed. The information generated from this study on the morphology of the goats will provide an insight to the genetic details of the goat populations in the two countries, thus will be valuable information for the genetic characterization of the goat populations from the two countries.

State of the art in equine reproduction techniques

K. Smits, J. Govaere and A. Van Soom
Faculty of Veterinary Medicine, Ghent University, Department of Reproduction, Obstetrics and Herd Health, University, Salisburylaan 133, 9820 Merelbeke, Belgium; katrien.smits@ugent.be

Artificial insemination, uterine flushing of *in vivo* embryos and embryo transfer are routinely performed in equine practice. In recent years, horse embryos are also increasingly being produced *in vitro* (IVP) by ovum pick up (OPU) and intracytoplasmic sperm injection (ICSI). Also the number of equine practitioners performing OPU and sending the oocytes to a specialized ICSI laboratory has grown substantially. While results are improving and an average of one transferable embryo per OPU can be expected, the outcome in clinical settings is influenced by several biological and technical aspects. When equine IVP blastocysts are obtained, they can be cryopreserved successfully by slow freezing or by vitrification. Cryopreservation of *in vivo* derived horse embryos is also possible, provided that the embryos are small or that large embryos are collapsed by removal of the blastocoel fluid. Cryopreservation of equine oocytes is far less efficient and further optimization is needed to allow clinical application. Culture and cryopreservation of (stem) cell lines to preserve equine genetics is feasible and can be applied for cloning. Cloning of valuable horses is performed successfully in clinical practice even though the technique is less efficient than ICSI and has been associated with perinatal problems. Long term effects of OPU-ICSI on the foals' health on the other hand have not been documented in the horse, even though morphological, developmental and genetic differences between IVP and *in vivo* derived embryos have been observed. Transfer to a recipient mare can induce epigenetic alterations, but these are dependent on the mare and not on the *in vivo* or *in vitro* origin of the embryo. Application of all these assisted reproductive technologies makes the horse an excellent model for optimization of genetic resource banking in endangered equids and even in other species, like the rhinoceros.

Developmental origins of health and diseases in horses: proof of concept and significance
P. Chavatte-Palmer, E. Derisoud and M. Robles
UMR BDR, INRA, ENVA, Université Paris Saclay, 78350 Jouy en Josas, France; pascale.chavatte-palmer@inra.fr

In mammals, environmental conditions in the periconceptional and gestational periods are known to contribute to offspring phenotype, through the apposition of epigenetic marks modulating gene expression without affecting genetics. Epigenetic modifications have not yet been thoroughly studied in horses. Nevertheless, developmental programming has been firstly demonstrated using embryo transfer experiments between horses of different breeds and size, with long term consequences of being born to a smaller or larger breed. In field conditions, maternal body condition, age and/or parity, metabolic status and nutrition during pregnancy have been shown either to affect foal growth, osteo-articular health and/or metabolism. For example, primiparous mares produce lighter foals than multiparous mares. These foals remain lighter and smaller until after weaning and their glucose clearance is faster after a glucose tolerance test at 6 months of age. Ultimately, their performances are slightly reduced compared to foals born to multiparous mares. In terms of maternal nutrition, feeding mares with carbohydrate-rich diets as well as maternal obesity during pregnancy also affect foal carbohydrate metabolism and increase the risk of developing osteochondrosis. The placenta is considered as a major programming agent as its function will adapt depending on maternal environmental conditions, thus affecting feto-maternal exchanges. In the horse, maternal insulin resistance has been associated to placental inflammation, possibly participating to the chronic inflammation observed in offspring. Finally, a 'predictive adaptive response' is induced in the offspring, favouring adaptation to post-natal nutritional conditions matching the prenatal experience. The role of the periconceptional environment as well as preventive or correcting strategies, either by supplementing mares with amino-acids such as arginine, or by adapting post-natal nutrition to reduce the effects of programming, are currently being explored in our laboratory.

Importance of strong data base systems in the study of genetic characteristics in horses
M. Wobbe[1,2], K.F. Stock[1,2], F. Reinhardt[1] and R. Reents[1]
[1]IT Solutions for Animal Productions (vit), Heinrich-Schroeder-Weg 1, 27283 Verden (Aller), Germany, [2]University of Veterinary Medicine Hannover (Foundation), ABG, Buenteweg 17p, 30559 Hanover, Germany; mirell.wobbe@vit.de

The monitoring of genetic defects plays an important role in many different animal species. It is the responsibility of breeding organizations to consider defects and other genetic characteristics in their breeding programs. In horse breeding some hereditary defects are also known and routinely screened, but not systematically yet in Warmblood breeds. After a case of Warmblood Fragile Foal Syndrome (WFFS) occurred in the USA and was widely discussed in 2018, awareness of breeders increased. The autosomal recessively inherited connective tissue disorder was first described by a US research group in 2012. Since 2013, a commercial genetic test for the point mutation in the PLOD1 gene is available. Homozygous foals are not viable, but little is known about frequencies of WFFS related losses and their timing during foal development. WFFS was therefore chosen to demonstrate how comprehensive data base systems can help elucidating genetic characteristics in horses. Covering data of the last 10 years were provided by 10 German studbooks, so more than 420,000 coverings could be used for analyses of variance using procedures GLM and MIXED of SAS software (version 9.2). The dependent variable reflected the risk of no living foal: It was coded 0 if the foal was born and 1 if it was not born or died within the first two days. In the model, we included WFFS status (carrier or free), individual sire considering his WFFS status, breeding station as fixed effects and a random residual. Assuming a carrier frequency of around 10% in Warmblood horse populations, as previously reported, one would (following Hardy-Weinberg) expect a difference of about 2.8% in the foaling rates of carriers and free sires. Depending on the model and data restrictions (e.g. minimum of 5 potential foals per sire) we found differences of 2.7-3.0% in the foaling rates. These figures fit to the expectations rather well. However, data completeness and recording quality have major impact on the power of statistical analyses, implying the need of strong data base systems especially for more complex genetic characteristics.

Validation of breeding applications for sport horses based on linear profiling across age groups

K.F. Stock[1], I. Workel[2], A. Hahn[2] and W. Schulze-Schleppinghoff[2]
[1]IT Solutions for Animal Production (vit), Heinrich-Schroeder-Weg 1, 27283 Verden (Aller), Germany, [2]Oldenburger Pferdezuchtverband e.V., Grafenhorststrasse 5, 49377 Vechta, Germany; friederike.katharina.stock@vit.de

Linear profiling has been introduced by studbooks for sport horses to increase objectivity and refine phenotypic information available through their routine assessments. With increasing amounts of linear data and of breeding applications for linear traits becoming available, guidance is needed by the breeders regarding how to use the new tools. In a system where not only mares and stallions, but also foals are linearly described and contributing to genetic evaluation, validation of the multiple-trait approach with joint analyses of corresponding linear traits in young and adult horses is of particular interest. For this study, we used linear data of the Oldenburg studbooks collected in 2012-2018 for analyses focusing on the predictive value of linear data of foals. Breeding values (BV) were estimated for in total 46 linear traits relating to conformation and performance and defined within age group, using either the full dataset (20,655 linear profiles; GE2018) or truncated data (9,656 linear profiles; GE2015). In each case, foals contributed about 62% of the data, and pedigree information including four ancestral generations was considered. BV were estimated in uni- and bivariate linear trait animal models using PEST software. Of the 2,396 sires in GE2018, 1,493 sires had also BV from GE2015. Comparisons between evaluation results revealed BV correlations of mostly ≥0.85 in all stallions and sub-groups (at least 10 progeny in GE2015; young stallions with only foals in GE2015). Correlation patterns reflected genetic parameters with highest correlations for linear gait traits, indicating high predictive value of foal data especially for trot aspects. According to the validation results, genetic linear profiles of stallions provide targeted support of decision making and benefit from integrated use of linear data of young and adult horses.

Genetic parameters between ages for jumping ponies using a structured antedependence model

H. Crichan[1], L. Arrialh[1], M. Sabbagh[1] and A. Ricard[1,2]
[1]Institut Français du Cheval et de l'Equitation, Pôle développement, Innovation et Recherche, La jumenterie, 61310 Exmes, France, [2]Institut National de la Recherche Agronomique, Génétique Animale et Biologie Intégrative, Domaine de Vilvert, 78350 Jouy-En-Josas, France; harmony.crichan@ifce.fr

To provide breeding values for ponies in jumping competition, we analysed data from ponies' official competitions reserved for children. All results in competition from 1996 to 2018 were taken into account involving 73,915 ponies aged 4 to 20 with a sum of 301,453 performances. Annual performances were measured by two traits: (1) logarithm of annual sum of points exponentially distributed according to ranking and technical difficulty of the event (points); and (2) annual summary of ranking based on an underlying liability responsible for ranks (ranking). Pedigree data contained ancestors over four generations with 583,963 ponies and horses. Each ages was considered as a different trait. An animal model with fixed effects of height (2nd order Legendre polynomial), year, and sex was applied. To reduce the number of covariance parameters to be estimated, we used a structured antedependent model. Heritability (h^2) for points was 0.24 at 4 years old, 0.34 at 5 and then decreases slowly, at 12, h^2=0.25. For ranking, heritability is 0.18 at 4, 0.20 at 5 and go through the same decline as points, at 12, h^2=0.16. After 8 years old, genetic correlations between ages were higher than 0.80. They were even higher than 0.90 if the interval between two ages was equal or less than 4 years. The age of 4 years old is a specific trait as heritability and genetic correlations were lower than the others, we obtained correlations from 0.70 to 0.60 between 4 years old and adult ages (8 and more). At adult ages, residual or environmental correlations are influenced mainly by the interval between ages. For points and ranking, at consecutives ages correlation were higher than 0.65, within an interval of 3 years correlation were around 0.3 and decreased with the interval increasing. Here again, young ages (4 and 5) display low correlation with adults ages, less than 0.1. Breeding values are summarized using the first two principal components of the genetic variance covariance matrix. They are now officially published.

Potential of inbreeding depression in morphological traits of the Pura Raza Español horse

D.I. Perdomo-González[1], J. Poyato-Bonilla[1], M.J. Sánchez-Guerrero[1], J. Casellas[2], A. Molina[3] and M. Valera[1]
[1]*Universidad de Sevilla, Dpto. de Ciencias Agroforestales, ETSIA, Ctra. Utreta Km 1, 41013 Sevilla, Spain,* [2]*Universitat Autònoma de Barcelona, Dpt. de Ciència Animal i dels Aliments, Travessera dels Turons, s/n, 08193 Bellaterra, Spain,* [3]*Universidad de Córdoba, Dpto. de Genética, CN IV Km 396, 14071, Spain; daviniapergon@gmail.com*

One of the main objectives of the Pura Raza Español horse (PRE) Breeding Program is to improve the breed´s morphology while maintaining the genetic variability. However, related matings for the selection of type traits can lead to an increase in inbreeding levels, which, in turn, has been usually associated to reductions in fitness and loss of performance, mainly in the reproductive and physiological efficiency, but also in any trait under selection. Thus, the objective of this study was to perform a genetic analysis of the transmission potential in 7 type traits (height at withers, height of withers, scapular-ischial length, length of shoulder, cresty neck, ewe neck and frontal angle of knee) of this phenomenon, known as 'inbreeding depression', of certain breeding animals responsible for this increase of homozygosis in PRE offspring. The partial inbreeding (F_p) was estimated for the active PRE animals (alive or with preserved semen) as reference population (283,948 animals). Horses with at least 4 or more offspring, a $F_p \geq 6.25$ coming from an ancestor and type traits measured were selected (639 horses) and a pedigree with the last 5 generations was generated (5,026 animals). The genetic model used included sex, age and stud as fixed effects. Direct additive genetic (DAG) and potential of inbreeding depression transmission (PIDT) variances for the 7 traits were estimated using an own software. The DAG variance ranged between 0.03±0.010 (frontal angle of knee) to 16.1±2.97 (height at withers) meanwhile, for PIDT variance ranged from 3.1±1.66 (frontal angle of knee) to 434.7±161.55 (scapular-ischial length). Most of the Pearson correlations between PIDT and DAG estimated breeding value were positive, ranging from -0.36 (length of shoulder) to 0.60 (ewe neck). According to our results, high levels of inbreeding do not always imply a reduction of the phenotypic mean value or an increase of morphological defects, but in some cases can have a positive effect on the trait.

Genomics to identify genetic variants for semen quality traits of fresh and frozen-thawed semen

T. Greiser[1], G. Martinsson[2], M. Gottschalk[1], H. Sieme[3] and O. Distl[1]
[1]*University of Veterinary Medicine Hannover, Institute for Animal Breeding and Genetics, Bünteweg 17p, 30559 Hannover, Germany,* [2]*Lower Saxon National Stud Celle, Spörckenstrasse 10, 29221 Celle, Germany,* [3]*University of Veterinary Medicine Hannover, Unit of Reproductive Medicine, Clinic for Horses, Bünteweg 15, 30559 Hannover, Germany; ottmar.distl@tiho-hannover.de*

A consistently high semen quality plays an important role for the employment of stallions in artificial insemination (AI). Management, environment and individual genomic effects of the stallion lead to significant variation in the quality of fresh and frozen-thawed semen. The evaluation of stallion semen includes evaluation of semen volume, sperm concentration, sperm motility patterns, total number of sperm for fresh semen and post-thaw motility, sperm membrane integrity (non-viable sperm) and sperm chromatin structure assays (DNA fragmentation index). We collected 63,972 reports from fresh semen and 4,681 reports from frozen-thawed semen of 241 and 121 warmblood stallions, respectively. Genetic parameters were estimated using multivariate animal models for all semen traits simultaneously. Heritabilities ranged from 0.13 to 0.28 (fresh semen) and from 0.11 to 0.45 (frozen-thawed semen). Sperm concentration and motility in fresh semen showed high genetic correlations with all frozen-thawed semen traits. Motility in fresh and frozen-thawed semen were moderately correlated with estimated breeding values for stallion fertility (0.31-0.35, P<0.001) indicating a positive selection response for per cycle conception rate. A genome-wide association study was performed for 153 warmblood stallions. In fresh semen, gel-free volume showed a genome-wide significant association on ECA 5 and 28, sperm concentration on ECA 5 and 21, total number of sperm on ECA 3, 9, 19 and 22, progressive motility on ECA 1 and 3, and total number of motile sperm on ECA 1, 8, 9, 15 and 22. In frozen-thawed semen, post-thaw motility was genome-wide associated on ECA 5 and 9, sperm membrane integrity on ECA 1, 8, 20 and 21 and DNA fragmentation index on ECA 1 and 18. These associated regions are being searched and filtered for genetic variants using next generation sequencing data of more than 130 horses to delimit causal variants. The aim is to provide a stallion fertility set of genetic variants.

Mitochondrial DNA haplotypes in the population of Bosnian mountain horse

M. Mesaric[1], A. Dolinsek[2], M. Cotman[1], T. Pokorn[3], M. Zorc[3], J. Ogorevc[3] and P. Dovc[3]
[1]University of Ljubljana, Veterinary Faculty, Gerbiceva 60, 1000 Ljubljana, Slovenia, [2]International association of bosnian moutain horse breeders, Rtice 1, 1414 Podkum, Slovenia, [3]University of Ljubljana, Biotechnical Faculty, Department of Animal Science, Groblje 3, 1230 Domzale, Slovenia; peter.dovc@bf.uni-lj.si

Genotyping of the mtDNA allows us to track maternal inheritance and to reconstruct the history of the breed. The analysis of mtDNA enabled the identification of the genetic relationship of horses within and between breeds, which reflects the presence and representation of different maternal lines in the breed. In the majority of horse breeds, a number of mtDNA haplotypes were found. Some of them are common for a larger number of breeds, which may be either a consequence of the vivid exchange of breeding material among breeds, or to multiple sampling of genetic material from a heterogeneous population during domestication and breed formation. Among more than 50 different mtDNA haplotypes found in modern horse breeds only a few are characteristic of certain populations. In the population of Bosnian mountain horse, 34 maternal lines were established during the history of the breed. Some of them were preserved and according to the pedigree data, nine old maternal lines still exist; however, in addition to that six new maternal lines were established in the second half of the 20th century. We sequenced the 450 bp fragment at the 5'-end of the mtDNA D-loop region of 42 Bosnian mountain horses. Data analysis revealed 15 different mtDNA haplotypes in the population of the Bosnian mountain horse, which belong to the eight main mt DNA haplogroups (A, B, G, H, J, L, M and Q), indicating the intra breed diversity of mtDNA, comparable to other horse breeds. The pedigree data match relatively well with the genetic diversity of mtDNA haplotypes. The majority of mtDNA haplotypes is identical to haplotypes found in other breeds, but two of them could be clearly distinguished from all known haplotypes so far. Our research considerably supports the international efforts for the preservation of genetic base of the Bosnian mountain horse population, which underwent a drastic reduction in the population size in the last three decades and is considered endangered.

A proposal to optimize the linear score system in the Arab Horse in Spain

M. Arcos[1], J.P. Gutiérrez[2], A. Molina[3], M. Valera[4] and I. Cervantes[2]
[1]Asociación Española de Criadores de Caballos Árabes, 28006, Madrid, Spain, [2]Dpto. Producción Animal, Universidad Complutense de Madrid, 28040, Spain, [3]Dpto. Genética, Universidad de Córdoba, 14071, Spain, [4]Dpto. Ciencias Agroforestales, Universidad de Sevilla, 41013, Spain; icervantes@vet.ucm.es

Linear score system is applied in the Spanish Arab Horse Breeding Program since 2010. The main selection objective is the performance in endurance competitions. This methodology allows recording conformation information aiming at attaining a functional morphology. The evaluation system consists of 58 variables (scale 1-7): 29 variables related with a body measurement, 19 subjective variables, 9 traits of movements and 1 of temperament. The number of variables included makes the recording process slow, and then a decrease in the number of traits could increase the collection of information. The objective of this study was to reduce the number of variables to optimize collecting time and the genetic response in the future. The selection of variables was made using genetic parameters between them and the genetic correlations with endurance performance using three traits: recovery time, time in race and placing. Data contained 602 linear records belonging to 199 Arab horses and 1,476 endurance records from to 765 horses. Several multivariate models were applied grouping the linear traits by anatomical regions. The model included the gender, age, event and appraiser as fixed effects with the animal and residual as random effects for linear traits. For the functional traits the model includes the gender, age, event, as fixed effect and the rider, the animal and the residual as random effects. The time in race included the number of kilometres as covariate. The pedigree matrix contained 6,819 animals. Genetic correlations between linear and functional traits were moderate-low magnitude (-0.39-0.82) being the placing trait the most determined by morphology traits. Linear traits heritabilities ranged between 0.06 and 0.45. The selected variables were: neck shaper upper line, thoracic perimeter, forelegs front view, leg length, fore-cannon perimeter, temperament and stride length of the walk. More analysis should be done to test the relationships between these traits and endurance performance before implementing a selection index.

Heritability of coat colour parameters in Old Kladruber horses

B. Hofmanova[1], L. Vostry[2], H. Vostra-Vydrova[3] and I. Majzlik[1]
[1]Czech University of Life Sciences Prague, Department of Ethology and Companion Animal Science, Kamycka 129, 165 00 Praha 6, Czech Republic, [2]Czech University of Life Sciences Prague, Department of Genetics and Breeding, Kamycka 129, 165 00 Praha 6, Czech Republic, [3]Institute of Animal Science, Pratelstvi 815, 104 00 Praha 10, Czech Republic; hofmanova@af.czu.cz

The Old Kladruber Horse is the only Czech autochthonous warmblood baroque breed, originated at the end of the 16th century, nowadays kept in two colour varieties – grey and black. Since this breed is used especially for ceremonial purposes, coat colour is considered to be an important exterior trait. The aim of this study was to estimate heritability of coat colour parameters. A total of 628 horses (376 greys and 252 blacks) aged between 1 and 25 years, were included in the analysis. Coat colour of each individual was recorded by Konica Minolta Spectrophotometer CM-2500d. The parameters measured under the CIE system consisted of: L* – lightness (0=black, 100=white), a* – redness (-128=green, +127=red), b* – yellowness (128=blue, +127=yellow). Taking into account selected fixed effects (age, sex, season, housing), the variances of genetic and non-genetic effects were estimated by the REML method using the DMU software package. The estimated heritability values between 0.14 and 0.52 (according to colour variety and spectrophotometric parameter) suggest possible multifactorial inheritance, especially with regard to the level of greying in greys and the reddish tinge in black horses. This study was supported by the project QK1910156.

Genetic variability within the Murgese horse inferred from genealogical data: an update study

G. Bramante[1], G. Carchedi[2], E. Pieragostini[3] and E. Ciani[4]
[1]Freelance, Via F. S. Sergio 19, 74016 Massafra, Italy, [2]A.I.A. Ufficio Equidi, Via G. Tomassetti 9, 00161 Roma, Italy, [3]UNIBA, DETO, SP per Casamassima km 3, 70010 Valenzano, Italy, [4]UNIBA, DBBB, Via Amendola 165/A, 70126 Bari, Italy; elena.ciani@uniba.it

The Murgese is an historical horse breed from Southern Italy, assumed to trace back its origin to the steeds of Frederick II (XIII cent. CE). This breed was suitable for both military and civilian use but nowadays its use has decreased. The official registration started in 1926 with 46 founder brood mares. In the subsequent decades the breed first experienced a demographic expansion, followed by a gradual contraction of the population size. Today, the breed is appreciated in equestrian tourism for its rusticity and good temperament, with increasing interest also in equestrian sports, mainly dressage. The aim of this work is to follow-up on a previous study monitoring the population genetic variability by using genealogical information. The dataset included 5,542 horses, out of which 1,494 males and 4,048 females, registered in the studbook from its foundation to 2017, out of which 5,179 with both father and mother known (reference population). The analyses were conducted, on the reference population, by the software package ENDOG48. N_e of founders was 40.38. The maximum number of generations was 14 (mean 9.43), the maximum number of complete generations was 6 (mean 3.53). Number of ancestors contributing to the reference population was 315. N_e of founders and ancestors for the reference population was, respectively, 36 and 19. Seven ancestors were able to explain 50% of the population diversity. Matings occurred between full sibs (0.04%), half sibs (1.80%) and parent/offspring (0.45%). Mean generation interval (±SE) was 11.21±9.93 years. Average F and AR were 1.22 and 2.45% in 2007, and 4.46 and 8.51% in 2016, respectively. In 2017, average F was 4.82% and average AR was 9.14%. Increase in inbreeding was 1.07% by equivalent generation, and 1.44% by complete generation. Based on the above, it may be inferred that the inbreeding rate in Murgese is growing faster than generally recommended to avoid possible reduction in population fitness, thus pointing to the need for a more balanced genetic contribution of the registered animals.

Phenotypic differences of the Sokólski mares in distinct regions of Poland

G.M. Polak

National Research Institute of Animal Production, Department of Horse Breeding, Sarego, 2, 31-047 Cracow, Poland; grazyna.polak@izoo.krakow.pl

In 2018, the population of Sokolski horses in the program of genetic resources conservation reached 1,650 animals, including 1,229 mares and 421 stallions recognized for their mating. The herds were located in 10 regions. The largest number was found in the east of Poland, in the Podlasie and Lubelskie regions – historical regions of their region origin. The aim of the study was to analyse three basic body measurements in mares population. The measurements were collected in 6 regions of the country with at least 50 mares. The average size for the height at the withers ranged from 155.8 to 157.1 cm, for the thorax circumference the values were between 24 and 24.8 cm, and the chest circumference ranged from 208.6 to 216.1 cm. There were statistically significant differences between groups of mares in distinct regions of Poland at the $P \leq 0.05$ level. Mares with the highest height at the withers they were located in the region of Warmia-Mazury and Podlasie (north-east of Poland). The smallest mares were in the Małopolska and Lubelskie regions (south-east of Poland).

Adaptation of dromedary camels to harsh environmental conditions in arid and semi-arid zones

P. Nagy

Emirates Industries for Camel MIlk and Products, Farm and Veterinary, Umm Nahad, Dubai, 294236, United Arab Emirates; peter@camelicious.ae

Bactrian and dromedary camels show a remarkable adaptation to harsh environmental condition and are very important livestock species in arid and semi-arid regions of the world. With growing desertification and increasing ambient temperature, the number of animals has been constantly increasing and it exceeded 28 million in 2016. In parallel, the production of camel meat and milk have also been growing at a rate of over 3% yearly. Despite low reproductive efficiency, the expansion of the population is due to (1) the ability to utilize marginal feed sources and high salt-content plants (halophytes), (2) high feed conversion efficiency, and (3) physiological mechanisms to tolerate high heat loads associated with dehydration. The role of these adaptation mechanisms is to conserve water and energy, and to reduce heat production. It includes: (1) high diurnal variation in body temperature (heterothermy); (2) thick fur with controlled sweating; (3) unique behaviour and plant selection in desert environment; (4) low water turnover rates; (5) reduced urine and saliva production with increased concentration; (6) reduced metabolic rate; (7) selective brain cooling; (8) fluid shift between compartments to maintain plasma volume resulting in stable blood viscosity; (9) high drinking capacity and the ability for fast recovery from rehydration, etc. Camels can tolerate water losses of up to 30% of their body weight and survive water deprivation and dehydration in hot climate (41 °C) for 10 to 15 days. If there is free access to water, thirsty camels are able to drink 60 to 110 litres within 10 minutes and may drink 50% of their dehydrated body weight within 24 hours In addition, dromedaries do not only survive in arid, semi-arid climate, but are also able to maintain milk production even if they have access to water only once a week. Therefore, nomadic tribes have constant supply of fresh milk in the desert and can travel large distances with their livestock between watering points. The importance of this species in sustainable livestock production systems is likely to increase in the future.

The role of camels in food security in the arid zone: meat and milk production potential
B. Faye
CIRAD, Environnement et sociétés, 1479 avenue du Père Soulas, 34090 Montpellier, France; bjfaye50@gmail.com

The estimated camel population in the world (35 million heads) is low compared to the other domestic farm animals but represents a significant part of the domestic herbivorous biomass in arid countries. As multipurpose animal, the large camelids (dromedary and Bactrian camels) are not only used for packing, pulling or riding, but also for dairy, meat and wool production. Although their estimated milk production (2.85 million tons) and meat production (630,000 tons) worldwide is 0.35 and 0.75% of the total milk and red meat respectively consumed in the world, their contribution could reach more than 10% in Africa. Characterized by hypo-allergic properties, exceptional richness in vitamin C and D, potential health benefit for diabetic patients and lactose-intolerants, effect against diseases affecting liver, dietetic interest (richness in iron, in long-chain fatty acids and essential amino-acids), camel milk is also contributing to the health welfare of people living in remote areas. Camel meat is appreciated for its dietetic virtues low-fat and cholesterol, high essential amino-acid index, beneficial effect on hypertension). Contrary to cow milk and meat produced in almost all the ecosystems of the earth, camel products are produced in desert areas only, from Mauritania to Mongolia, contributing to the pastoralist's diet. Even in rich Gulf countries, more than 70% of the camel milk, the 'white gold of the desert', is self-consumed by the Bedouins providing animal proteins in places where the access to other proteins is difficult. In pastoralist households, camel milk can reach 70% of the dietary calories. Moreover, camel meat is involved in a regional market with important live camel export from Sahelian countries to North-Africa and Arabian Peninsula. Thus, camel milk and meat are contributing to the food security in many remote areas of the old-world, sometimes as exclusive source of animal proteins. In addition, the remarkable increase of urban demand in camel products occurred both in southern and western countries. Indeed, climatic changes, globalization of economy and interest for health benefit of camel products contribute to boost camel rearing in the world and to its spatial expansion.

Phenotypic and genotypic evaluation of camelids, why these species lag behind other livestock
P.A. Burger[1] and E. Ciani[2]
[1]Research Institute of Wildlife Ecology, Vetmeduni Vienna, Departement of Integrative Biology and Evolution, Savoyenstrasse 1, 1160 Vienna, Austria, [2]Università degli Studi di Bari Aldo Moro, Dipartimento di Bioscienze, Biotecnologie e Scienze Farmacologiche, Via Orabona, 4, 70125, Bari, Italy; pamela.burger@vetmeduni.ac.at

Increasing desertification and constant human population growth pose a challenge on the field of animal science and production. Old World camels, the one-humped dromedary and two-humped Bactrian camel, seem to be a perfect answer to the demand for sustainable food production in arid environments. As the last livestock species to be domesticated, camels also rank last on the list of available genomic resources. While there are genome assemblies on chromosome-level for *Camelus dromedarius* and on scaffold-level for *Camelus bactrianus* and *Camelus ferus*, tools are still missing for large-scale genomic studies. International cooperations (e.g. International Camel Consortium for Genomic Improvement and Conservation; ICC-GIC) aim to develop necessary high-quality and high-density genomic tools to investigate genotypes underlying traits of interest. With the recently awarded 2019 Illumina® Agricultural Greater Good Initiative grant, this goal is getting within reach. The collection of economical and physiological important phenotypes is currently confined to local farms or regional associations; except from a number of large camel farms (e.g. UAE, CH, KZ) where regular performance recording is practiced on different quantitative and qualitative levels. The lack of governmental support to build infrastructure for animal identification and recording under national breeders' associations hampers the development of a successful camel production sector. There is an urgent need for international promotion and cooperation, like the ARIMnet2 project CARAVAN (Toward a CAmel tRAnsnational VAlue chain) linking scientists and breeders in Mediterranean countries. A recent ICAR initiative aims to establish the *status quo* of animal identification and performance recording practices in Old World camels on an international scale. Using the results of this baseline survey, we will identify the immediate next steps to start closing the gap between camels and other livestock in terms of phenotypic and genotypic evaluation.

Toward a CamelHD BeadChip: the Illumina Greater Good Initiative 2019

E. Ciani[1], S. Brooks[2], F. Almathen[3], A. Eggen[4] and P. Burger[5]
[1]University of Bari, DBBB, Via Amendola 165/a, 70126, Italy, [2]University of Florida, Dept. of Animal Sciences, UF Genetics Institute, 2250 Shealy Drive, 32608 Gainesville FL, USA, [3]King Faisal University, Faculty of Veterinary Medicine and Animal Resources, P.O. Box 1757, Al-Ihssa 31982, Saudi Arabia, [4]Illumina, Agrigenomics, 5 Rue Desbrures, 91030 Evry Cedex, France, [5]Vetmeduni, Research Institute of Wildlife Ecology, Savoyenstrasse 1, 1160 Vienna, Austria; elena.ciani@uniba.it

Camels represent a key livestock resource in several low-income countries. Due to their unique assortment of biological and physiological traits specifically adapted to extreme harsh desert conditions, in these areas, camels better than other livestock species can thrive and produce high-quality protein for food consumption. Currently rapid changes are ongoing in the camel sector, with camel farming systems in peri-urban areas commonly evolving toward more intensive management practices. Implementation of a selection of Single Nucleotide Polymorphisms (SNPs) in a genotyping platform may allow rapid and cost-effective genome-wide genotyping in large numbers of animals, thus boosting downstream applications such as genome-wide association studies for production traits and genome-based selective breeding. To reach this aim, the risk of 'SNP ascertainment' bias, i.e. polymorphisms that are specific or highly frequent in one population but not present in another, must be carefully considered. Here, we present the first world-wide camel diversity study where whole genomes from more than 400 samples, representative of the entire geographic range of the camel distribution, are going to be sequenced, for a total of 20 tera-bases of Illumina NovaSeq sequencing data, as a first step toward the development of an Illumina® CamelHD BeadChip. This unprecedented opportunity, offered by the 2019 Illumina Agricultural Greater Good Initiative grant, will also contribute to deepening the understanding of evolutionary processes that shaped the camel genomes and to deciphering the molecular basis of the peculiar physiological adaptation traits of camels. Parallel progresses in large-scale recording of camel phenotypes following standardized procedures are advisable and necessary for an ecologically and economically sensible exploitation of the Illumina® CamelHD BeadChip genotyping tool.

CAMELMILK project

M. Garrón Gómez
IRTA, Food Technology, Finca Camps i Armet, s/n, Building A, 17121 Monells (Girona), Spain; marta.garron@irta.cat

CAMELMILK is focused to promote the production, processing and consumption of camel milk and camel dairy products in the Mediterranean area. The objective is to empower smallholder and small SMEs of the camel milk sector with the required tools to ensure an increase of competitiveness, company growth and job creation at both shores of the Mediterranean. Camel play a significant role in the lifestyle of many communities in different countries of the Mediterranean basin being an important source of milk, meat and transport in areas living under harsh conditions. Camel milk has traditionally been used for its medicinal values in rural regions, however, its special nutritional properties (rich source of bioactive, antimicrobial and antioxidant substances, no allergic respond, support diabetic treatment, etc.) has led to a recent growing demand of urban population. CAMELMILK will: (1) improve and bring existing camel milk production systems in Algeria and Turkey closer to EU standards to increase efficiency, quality and safety and setting the bases for a short-future exportation to EU; and (2) adapt processing technologies to camel milk technological properties to produce pasteurized camel milk, fermented camel products and different types of cheese. To achieve a successful market implementation and to support market uptake of the developed products CAMELMILK will take into account different aspects: (1) regulatory issues and legislation; (2) improve the camel milk value-chain to allow the smallholders and small SMEs involved in the project to occupy a new market niche with a high potential in the coming years; (3) define the most convenient business models and market strategies for each industrial partner involved in the project; and (4) determine consumers acceptance, knowledge and willingness to consume/pay for camel milk products. CAMELMILK will contribute to the development, interaction and collaboration of the areas of the Mediterranean basin.

Review of the genetic variants of milk protein in Old and New World Camelids

A. Pauciullo
University of Torino, Agricultural, Forest and Food Sciences, Largo Paolo Braccini 2, 10095, Italy;
alfredo.pauciullo@unito.it

Milk proteins in camelids consist of caseins (CN) and whey proteins (WPs) in the ratio 75 *vs* 25%. CN are coded by single autosomal genes: *CSN1S1* (αs1-), *CSN2* (β-), *CSN1S2* (αs2-) and *CSN3* (κ-), clustered in a DNA stretch of ~190 kb mapped on chromosome 2. Conversely, several WPs exist including α-lactalbumin (α-LA), serum albumin (SA), whey acid protein (WAP), etc., each encoded by the respective genes. In dromedary, genetic polymorphism was found in 3 out of 4 CN genes. Two genetic variants (A and B) were identified at *CSN1S1*. They differ for 8 amino acids (aa) (EQAYFHLE), skipped in A variant as consequence of the alternative splicing of the exon 18. The C variant was identified at protein level by isoelectrofocusing (IEF) and confirmed at DNA level as SNP at the exon 5 (c.150G>T) responsible for the aa change p.30Glu>Asp. Recently, another variant (D) was identified by IEF. Apparently, the sequence coding for this variant does not differ from that of the A allele. Genetic variants were also described for the *CSN2* and *CSN3*. The SNP g.2126A>G at *CSN2* and g.1029T>C at *CSN3* are particularly relevant for affecting consensus sequences for transcription factors. Conversely, no genetic variants were reported at *CSN1S2*. To date, only one polymorphism affecting the *CSN2* was reported in Bactrian, that is the SNP c.666G>A, responsible for the aa change p.201Met>Ile. The CN genes cluster was investigated also in llama at transcript and protein level, whereas only partial information is known for alpaca. Four αs1-CN variants corresponding to four haplotypes were reported for the *CSN1S1* in llamas. The molecular bases of these differences were identified in two SNPs (c.366G>A, exon 12 and c.690C>T, exon 19) responsible for the aa changes p.86Val>Ile and p.194His>Tyr, respectively. Three αs1-CN variants were observed in alpaca; however, it was not possible to establish a link with IEF despite the SNP c.366G[Val]>A[Ile] was also found so in this species. Genetic polymorphism of WPs was less investigated in camelids. Two variants of WAP were found in Bactrian (transition c.119G[Val]>A[Met]). No genetic variants were found in the other WPs.

Cheese-making ability of dromedary camel milk: comparison with cattle, buffalo, goat and sheep milk

N. Amalfitano[1], M. Bergamaschi[1], N. Patel[1], M.L. Haddi[2], H. Benabid[2], F. Tagliapietra[1], S. Schiavon[1] and G. Bittante[1]
[1]University of Padova, Department of Agronomy, Food, Natural resources, Animals and Environment, viale dell'Università 16, Padova, 35020, Italy, [2]Université des Frères Mentouri, Laboratoire de Mycologie, biotechnologie et activité Microbienne and INATAA, Route Ain el Bey, Constantine, 25000, Algeria; nicolo.amalfitano@studenti.unipd.it

This preliminary study was carried out to compare dromedary camel milk with the milk from the other major dairy species in terms of coagulation traits and cheese-making ability. Ten bulk milk samples (2.0 l each) per species were collected from different Italian farms for cattle, buffalo, goat and sheep species, while 10 milk samples were collected from dromedaries reared in two Algerian areas. We processed the milk to assess, and analyze using mixed models: (1) milk composition; (2) milk coagulation properties, through lactodynamography with two replicates per sample; (3) curd firming and syneresis equation parameters, through modelling of 240 point measures from each replicate; (4)% cheese-yield traits, through individual model-cheese procedure on each milk sample; (5) milk nutrients recovery in curd, from cheese-making. Dromedary milk presented (mixed model): a fat and protein content not different from bovine, and caprine milk, but smaller than bubaline and ovine milk; a coagulation time interval after rennet addition (LSM: 13.9 min) about half than bovine milk and similar to other dairy species; a potential curd firmness (41.4 mm) and a curd firming rate (9.5%/min) similar to bovine milk, but a greater syneresis rate (1.9%/min) and then a much lower maximum curd firmness (17.8 mm) measured 36.2 min after rennet addition. The cheese yield of dromedary milk (13.8%) was lower than that of buffaloes (25.7%) and ewes (23.2%), and intermediate to that of cows (15.2%) and goats (11.9%). The recovery of milk fat (58.2%), total solids (39.7%) and energy (49.4%) in the fresh cheese was lower than for other dairy species, whereas recovery of protein (78.0%) was similar. In conclusion, the peculiar pattern of coagulation, curd firming and syneresis should be considered to optimize milk processing and increase the efficiency of dromedary camel cheese production.

Kisspeptin and RFRP neurons control breeding season but not induction of ovulation in the camel

H. Ainani[1,2], N. El Bousmaki[2], M.R. Achaâban[2], M. Ouassat[2], M. Piro[2], V. Simonneaux[1] and K. El Allali[2]
[1]Institute of Cellular and Integrative Neurosciences, University of Strasbourg, Neurobiology of rhythms, B.P.6202 Rabat-Instituts, 10101, France, [2]Hassan II Institute of Agronomy and Veterinary Medicine, Comparative Anatomy Unit, B.P.6202 Rabat-Instituts, 10101, Morocco; hassanainani@gmail.com

The dromedary camel (*Camelus dromedarius*), a desert mammal, is sexually active during short photoperiod and display provoked ovulation induced by the nerve growth factor beta (β-NGF) present in the seminal plasma of males. However, the mechanisms involved in the control of seasonality and induction of ovulation are still unknown in this species. Recently, two neuropeptides, kisspeptin (Kp) and RFRP, were involved in the control of breeding season in many species and several arguments point Kp as a target by which β-NGF induces ovulation. The aim of this study is, first to map hypothalamic Kp and RFRP neurons and assess their seasonal variations and secondly identify the targets of β-NGF when injected to camel females during breeding season. Results showed that Kp neurons are present in the preoptic area (POA) and arcuate nucleus (Arc); while those of RFRP are localized in the dorsomedian (DMH) and paraventricular (PVN) nucleus. Neuronal quantification showed that Kp neurons were more abundant during breeding season; while, RFRP neurons were more numerous during non-breeding season. In the second part of the study females were injected intramuscularly with 0 (0.9% NaCl/control) or 1 mg of β-NGF. Two hours later, brains were sampled and prepared for immunohistochemistry analysis of neuronal activation using c-Fos marker. Only the 1 mg β-NGF injection induced c-Fos expression, observed in POA, Arc, DMH and PVN. Unexpectedly, activated neurons didn't include Kp, GnRH and RFRP neurons but instead the oxytocin and vasopressin neurons within the PVN and DMH. In conclusion, in camel, as in other species, Kp seems to stimulate the onset of breeding season; while, RFRP inhibits it. These two RF-amides are now well identified to act in the seasonal synchronization of camel breeding activity but, apparently, not involved in the mechanisms of β-NGF central induction of ovulation, which likely require hypothalamic (DMH/PVN) vasopressin and oxytocin neurons.

Production potential of Llama and Alpaca (Domestic South American Camelids) in the Andean Region

C. Renieri
University of Camerino, Via Gentile III da Varano s.n., 62032 Camerino, Italy; carlo.renieri@unicam.it

Llamas (*Lama glama*) and alpacas (*Vicugna pacos*) are domestic mammals classed in the Tilopods suborder together with guanacos (Lama guanicoe) and vicuñas (*Vicugna vicugna*). Domesticated by the Preconquest Andean cultures, the distribution is a result of the onset of the European colonisation and the introduction of animals of European origin. Llamas and alpacas were progressively forced away from the coasts and the Interandean Valles and confined to the higher altitudes of the Puna. Two varieties of llama are described, 'Q'ara', characterised by a sparsely distributed coat of hair and coarse fibre quality, and 'Ch'aku', which displays an increased coat cover and a superior fibre quality. In alpacas, a differentiation is made between two types of fleece, Huacaya and Suri. Huacaya is more common and is characterised by compact, soft and highly crimped fibres. Suri has straight, less crimped fibres and 'cork-screw' shaped locks. Only primary populations ('primitive breeds') have been identified in both llamas and alpacas. These populations are defined by high genetic variability. Camelid farming systems are considered pastoral activities, carried out in extreme environmental conditions, closely connected to environmental variation, in particular water availability. In many cases, camelid breeding is the only source of income for human Andean populations. Llama, especially the 'Q'ara' type, is more efficient than Alpaca in meat production. In alpacas, meat production is secondary to fibre production. Charqui is a meat that is dried or dehydrated through salt-curing processes and the climate, in particular cold nights. It is the system through which the Andean 'campesinos' preserve meat. This know-how is ancient because it was already present in the Inca culture. Llama and Alpaca populations should be managed according to different strategies: Alpacas as single purpose animals (fibre production). Meat should be not considered in the selection plan but utilised as secondary product obtained in the animals discarded from selection plan. Llamas should be managed as dual-purpose animals, primarily for meat and secondarily for fibre.

Pharmacokinetics of a long-acting progesterone formulation in female camels

H. Chhaibi[1], A. Tibary[2] and A.J. Campbell[2]
[1]Institut Agronomique et Veterinaire Hassan II, Comparative Anatomy Unit, Madinat alirfane BP 6202 rabat instituts, 10112, Morocco, [2]College of Veterinary Medicine, Washington State University, Comparative Theriogenology, Department of Veterinary Clinical Sciences, Washington state university, Pullman, Pullman, WA, 99164-6610, USA; hamid.chhaibi@gmail.com

Progesterone administration is used extensively in camel embryo transfer programs for synchronization of recipients and donors. Daily intramuscular administration (IM) of 100 to 150 mg of progesterone in oil for 14 days is recommended in order to achieve the desired effect. The present experiment was designed to evaluate progesterone pharmacodynamics following a single standard dose administration of compounded proprietary long-acting progesterone that was formulated for mares. Fourteen (n=14) nulliparous female camels or 3.5 years of age and of similar weights were included in the study. Each female was given an intramuscular injection of 5 ml of a proprietary progesterone formulation (BioRelease P4 LA300, 300 mg of progesterone per ml). All females were examined by transrectal ultrasonography and only females with no corpora lutea present on the ovaries were included in the study. Blood samples were collected daily starting one day prior (Day 0) and continuing for 14 days after injection. Serum was harvested and stored at -20 °C until assayed for progesterone using radio-immunoassay. Change in daily progesterone level following treatment was examined using a repeated measurement ANOVA. As expected progesterone level was low (Mean ± SEM=0.2 ± 0.07 ng/ml) prior to injection and increased significantly (36.76 ± 3.8 ng/ml, $P<0.05$) within 24 hours of treatment. Serum progesterone level remained above 2 ng/ml in all animals for 10 days. By 12 days after injection only 50% of the females had progesterone levels below 2 ng/ml. By 14 days after treatment, five females (36%) had serum progesterone between 1 and 2 ng/ml while all the other has less than 1 ng/ml. In conclusion, this study demonstrated that administration of 5 ml of BioRelease P4 LA300 to female camels provides elevated serum progesterone levels that comparable to those expected during the luteal phase luteal function for at least 10 days. This treatment may be useful to eliminate the need for repeated daily administration for at least that period of time.

Some hormonal values in relation to foetal age in dromedary camel

S.G. Hassan
National Research Centre, Animal Reproduction and AI, 33 Al Bohooth St., Dokki, 12622, Giza, Egypt; Hassan9372003@yahoo.com

Few studies were carried out concerning the hormonal profile during gestation in one humped camel. The foetuses were collected from Cairo abattoir at different gestational ages and taken to the laboratory where the foetal age were determined by crown vertebral rump length (CVR), the developmental standard morphometrically methods were used. Biometrically the weight of the foetus, CVR increased with the advancement of gestational ages (1st trimester to the 3rd trimester). The hormonal profile of progesterone was estimated by RIA. The progesterone values showed increasing pattern of 1.8 ± 0.79 ng/ml, 2.38 ± 1.00 ng/ml and 2.95 ± 1.3 ng/ml for the first (up to 4 months), second (4 to 8 months) and third stage of pregnancy (9 months up to full term) respectively. Moreover, this pregnancy may be confirmed by the ultrasound. While in the second stage of pregnancy (4-8 month) it is quite difficult to detect the age of the foetus by ultrasonography; since the foetus is large in size. So, the hormonal level may be of value to determine (approximately) the foetal age.

Transfer of persistent organic pollutants in camel milk

F. Amutova[1,2,3], M. Delannoy[3], M. Nurseitova[3], G. Konuspayeva[1] and S. Jurjanz[1]
[1]Antigen LLP, Scientific and Production Enterprise, Abay vil., Azerbaeva str.4, 040905, Almaty, Kazakhstan, [2]Al-Farabi Kazakh National University, Faculty of Geography and Environmental Sciences, 71 al-Farabi Ave, 050040, Almaty, Kazakhstan, [3]University of Lorraine-INRA, URAFPA, 2 avenue de la Forêt de Haye, 54505 Vandoeuvre, France; stefan.jurjanz@univ-lorraine.fr

Camels as well as other mammalians can transfer environmental pollutants to milk when they live in contaminated areas. Nevertheless, these animals run through large areas and can so ingest feed or water contaminated at very different levels by environmental contaminants as persistent organic pollutants (POPs). As camel milk consumption corresponds to approximately 7% of total milk, this pathway can become a real risk for human exposure. The aim of present work was to summarize existing knowledge about POP concentrations in milk of camels. The work is focused on Kazakhstan as all available data came from this country. Monitoring field studies on pooled individual milks showed concentrations of PCDD/Fs up to 1.36 and 1.33 pg TEQ/g fat respectively from 4 regions in 2011 and Mangistau region in 2015-2016. These studies reported maximal concentrations of DL-PCB of respectively 4,7 and 47 and concentrations of NDL-PCBs were reported at 6,3 and 44,6 ng/g fat. These maximal concentrations respected the European regulation (1259/2011/UE) for PCDD/Fs but can overpass thresholds for PCBs (DL and NDL). Huge variations between sampling points, even within the same region, were reported but no differences between Dromedaries and Bactrians. Analyses revealed also presence of pesticides (up to: 3,6 HCB, 20,4 HCH and 2,4 DDT ng/g fat) as well as PAHs (718 ng/g fat). By the way, a study in controlled conditions using a daily exposure of 1.3 μg of Aroclor 1254 PCBs/kg BW during 56 days reported that a subsequent 60-day depuration which decreased PCBs differently: congeners 101 (-10%), 138 (-47%), 153(-57%) and 180 (-68%). Existing data showed the POP transfer to camel milk resulting in quantifiable but tolerable levels of PCDD/Fs or OCPs. Nevertheless, PCB concentrations are sometimes elevated. Thereby consumption of polluted camel milk or products may pose health risks. Further research is needed to investigate the transfer rates and depuration of POPs in camels.

Proprieties of young Sahraoui dromedary's meat

H. Smili[1,2], S. Becila[2], A. Ayad[1], B. Babelhadj[3], A. Della Malva[4], M. Albenzio[4], M. Caroprese[4], A. Santillo[4], A. Sevi[4], A. Adamou[3], A. Boudjellal[2] and R. Marino[4]
[1]Laboratoire des Bioressources Sahariennes: préservation et valorisation, Université Kasdi Merbah, Ouargla, 30000, Algeria, [2]Laboratoire Bioqual, Equipe Maquav, INATAA, Université Freres Mentouri Constantine, route Ain El Bey, 25000, Algeria, [3]Laboratoire de protection des écosystèmes en zones arides et semi-arides, University Kasdi Merbah Ouargla, 30000, Algeria, [4]Dipartimento di Scienze Agrarie, degli Alimenti e dell'Ambiente (SAFE), University of Foggia, Via Napoli, 25, Foggia, 71121, Italy; samirabecila@gmail.com

During the last years, a growing production and demand for dromedary meat is observed. In Algeria, the slaughter volumes have triplet over the last 20 years going from 13,000 head in 1997 to 40,000 head in 2017. In the Southeastern region, Sahraoui dromedary is an important population with morphological characteristics that makes it an interesting meat producer. However, in order to understand the potentialities of Sahraoui dromedary in terms of meat quality proprieties, the present study has been conducted. The *Longissimus lumborum* muscle of six young Sahraoui dromedaries (1-2 years old) were used to estimate the quality of meat during the ageing time (6, 8, 10, 12, 24, 48 and 72 hours). pH, drip loss, myofibrillar fragmentation index (MFI) and total collagen content were estimated at each ageing time. The measurements revealed a slow acidification process of the muscle with a pH value of 6.43 at 6 h *post-mortem* which decreased to reach 5.97 at 24 h *post-mortem*. Drip loss increased from 1.65% at 24 h to 5.01% at 72 h *post-mortem*. MFI showed the highest value at 6 hours *post-mortem* (97.89) after decrease with a minimum value recorded at 8 h *post-mortem* and starting from 12 h increase. While, collagen content showed a lowest value of 1.95 mg/g at 6 hour that increase and tends to stabilize after 24 h *post-mortem* around 2.7 mg/g. Beside the high meat quality potential of this category of age, other important factors need to be mastered in the management of dromedary meat production in order to ensure a constant high-quality product.

Changes in genomic predictions by ssGBLUP when using different core animals in the APY algorithm

I. Misztal, S. Tsuruta, I. Pocrnic and D. Lourenco

University of Georgia, Animal and Dairy Science, 425 River Road, GA 30602, USA; ignacy@uga.edu

For populations with a small effective population size, a genomic relationship matrix constructed for a large number of animals is singular. The APY algorithm exploits the reduced dimensionality of the matrix for lower computing cost, by splitting animals into core and noncore, and using recursion to predict noncore animals from core animals. Typically, the core animals are randomly selected, and their number is approximately equal to the number of eigenvalues explaining 98% variance in the matrix. While correlations in GEBV obtained when using two random cores is >0.99, some animals rerank. The purpose of this study was find the extent and origins of reranking using simulated and field data sets across species, and propose methods to reduce the reranking. In general, changes in GEBV obtained with two random cores are small but for some animals can be as large as 0.5 additive SD but predictivity and genetic merit of top 20 or 100 animals are nearly identical. Large changes occur nearly always for lower accuracy animals. Animals with low residual in the recursion do not show large changes. The changes are smaller for core than noncore animals. Some changes originate from blending. The changes can be minimized by increasing the number of core animals and by treating important animals as core. Keeping same core animals reduces the changes for existing animals. GEBV generated with the APY algorithm exhibit some variations with the choice of core animals without affecting accuracy of selection.

Genomic prediction with missing pedigrees in single-step GBLUP for production traits in US Holstein

Y. Masuda[1], S. Tsuruta[1], E. Nicolazzi[2] and I. Misztal[1]

[1]University of Georgia, Animal and Dairy Science, 425 River Road, Athens, GA, 30602, USA, [2]Council of Dairy Cattle Breeding, 4201 Northview Drive, Suite 302, Bowie, MD, 20716, USA; yutaka@uga.edu

Single-step GBLUP (ssGBLUP) is a method for genomic prediction combining phenotypes, pedigree information, and genotypes in mixed model equations. It needs the inverse of unified relationships (H^{-1}) as a function of pedigree relationship matrix (A^{-1}), genomic relationship matrix (G^{-1}), and a subset of pedigree relationships for genotyped animals (A_{22}^{-1}). These matrices should be compatible in scale, but it is rarely true in dairy cattle. The pedigree is long but partially missing and the genotypes are mainly available for the last few generations. In such a situation, A and A_{22} lack the relationships among some animals and G has a different genetic base compared with A_{22}. Unknown parent-groups (UPG) and metafounders (MF) can be used to fill the missing relationships in these matrices and may make them compatible in scale. The objectives of this study were to derive reasonable relationship-matrix with UPG or MF in ssGBLUP, to implement this matrix in genetic-prediction software supporting a large-scale genotyped population, and to apply the relationship matrices to production traits in US Holstein. We have derived H^{-1} in which UPG are considered only for pedigree relationships. We also derived an alternative H^{-1} from a joint density function including both real animals and UPG. We have shown that this alternative form is equivalent to one derived from the MF theory. Although UPG and MF are equivalent under some assumptions, MF seems to be more flexible to keep the compatibility among the relationship matrices because the genomic relationships among MF will be properly considered. The alternative H^{-1} showed that the computation of A_{22}^{-1} with MF is simplified with sparse-matrix techniques. The newly-derived H^{-1} was implemented in the BLUPF90 programs. The genotypes, pedigrees, and phenotypes of 305-d milk, fat, and protein yield were provided by the Council of Dairy Cattle Breeding (CDCB). The data included more than 72 million phenotypes for each trait, 80 million pedigree animals, and 2.3 million genotyped animals. The validation predictability and inflation of genomic predictions with UPG or MF will be presented.

Genomic prediction of a binary trait using a threshold model

J. Ten Napel[1], I. Stranden[2], M. Taskinen[2], J. Vandenplas[1], R.F. Veerkamp[1] and K. Matilainen[2]
[1]Wageningen University & Research Animal Breeding & Genomics, P.O. Box 338, 6700 AH Wageningen, the Netherlands, [2]Natural Resources Institute Finland (Luke), Production Systems, Animal Genetics, 31600 Jokioinen, Finland; jan.tennapel@wur.nl

Breeding value estimation increasingly uses both pedigree and genomic information in the covariance structure for specifying genetic similarity between animals. For binary traits, an animal threshold model is more appropriate than a linear animal model, but the latter model is often preferred in practice due to its computational ease. In this study, we compare the benefits of using genomic information for an animal threshold model with a linear animal model or using pedigree information only. We simulated a population of 10 discrete generations having 1,080 individuals each from a base population of 90 females and 15 males, in total 10,905 individuals. A number of 760 randomly-spaced QTL determined the true breeding values for a liability, based on a heritability of 0.20. The liability determined a binary trait with a prevalence of 15%. The heritability for the binary scale was 0.09 All animals born in generations 7-10 were genotyped for 19,000 randomly-spaced SNP markers. Training and validation populations consisted of generations 1-9 and 10, respectively. The animal threshold model was solved with both a Newton Raphson algorithm (NR) in MiXBLUP and Gibbs sampling. The linear animal model was solved with MiXBLUP. Posterior means of animal effects from the Gibbs sampler were virtually identical to solutions of NR for all analyses (r>0.999). Realised reliability, calculated as the squared correlation between estimated and true breeding values, was higher for the threshold than the linear model (0.218 vs 0.201) and higher for use of genomic information than pedigree information only (0.218 vs 0.190). Genetic evaluation of binary traits benefits from using both genomic and pedigree information and an animal threshold model.

Exploring different strategies to implement a weighted single-step GBLUP in Belgian Blue beef cattle

J.L. Gualdron Duarte[1], A.S. Gori[2], X. Hubin[2], C. Charlier[1] and T. Druet[1]
[1]GIGA-R University of Liège, Unit of Animal Genomics, 11 Avenue de l'Hôpital (B34), 4000-Liège, Belgium, [2]Association Wallonne de l'Elevage, Ciney, 5590, Belgium; jlgualdron@uliege.be

In order to implement a genomic evaluation in the Belgian Blue cattle (BBC) breed, the Walloon breeders association has started to genotype individuals in 2015. As the number of artificial insemination sires with accurate estimated breeding values is limited in BBC, the reference population consists mainly in phenotyped individuals that are genotyped with low-density genotyping arrays. The evaluation will rely on a single-step GBLUP (ssGBLUP) approach that allows to easily combine data from all individuals (genotyped AI sires and females and non-genotyped animals). Several variants with large effects on size or muscular development have previously been reported in BBC and ongoing research projects are expected to identify more. The objective of the present work is to develop a strategy to optimally weight the SNP effects in the ssGBLUP approach by using a weighted genomic relationship matrix. To that end, we first performed genome-wide association studies to determine whether the traits are highly polygenic or not and to identify variants with large effects. Noteworthy, three previously identified recessive deleterious variants presented heterozygote advantage and were among the most significant SNPs for many traits. Using the MultiBLUP approach we determined that a few significant regions (2 to 11) explained from 8 to 30% of the genetic variance. Next, we assessed predictive abilities (PA) by cross-validation and compared them for different genomic prediction methods including GBLUP, BayesR, Bayesian Sparse Linear Mixed Models or Adaptive MultiBLUP. The average PA using GBLUP was equal to 0,400. Bayesian methods resulted in a moderate gain of PA (+0.009 on average) but using the posterior variances of SNP effects as weights in the GBLUP reduced its reliability. The Adaptive MultiBLUP approach presented modest benefits (+0.003 on average) and was sometimes less reliable than the GBLUP, probably because selected genomic regions were not identified precisely (10 Mb on average). We are now exploring other solutions to exploit the results of the Bayesian approaches in order to efficiently weight the GBLUP.

A diagonal preconditioner for solving single-step SNPBLUP efficiently

J. Vandenplas[1], M.P.L. Calus[1], H. Eding[2] and C. Vuik[3]
[1]Wageningen University and Research, P.O. Box 338, 6700 AH Wageningen, the Netherlands, [2]CRV BV, Wassenaarweg 20, 6843 NW Arnhem, the Netherlands, [3]TU Delft, Van Mourik Broekmanweg 6, 2628 XE Delft, the Netherlands; jeremie.vandenplas@wur.nl

The preconditioned conjugate gradient (PCG) method is an iterative solver of systems of linear equations commonly used in animal breeding. However, applied to single step Single Nucleotide Polymorphism BLUP (ssSNPBLUP) models traditional PCG methods have shown convergence issues. Efficiency of the PCG method depends on the quality of the preconditioner. Therefore, the poor convergence patterns suggest that the preconditioners commonly used in animal breeding are not suitable for solving ssSNPBLUP efficiently. Recently we showed that applying a deflated PCG (DPCG) method to ssSNPBLUP solved the convergence issues. However, the DPCG method requires substantial additional computations in comparison to the PCG method. Hence, we developed a second-level diagonal preconditioner for the PCG method, that leads to better convergence patterns than the traditional PCG method, but avoids the additional costs of the DPCG method. The proposed preconditioner is easy to implement, and does not lead to additional computing costs, as it can be combined with the commonly used preconditioner. The performance of the proposed second-level diagonal preconditioner was compared against the (D)PCG methods, using two different ssSNPBLUP models. Implemented in a shared-memory Fortran program, it was tested on a four-trait ssSNPBLUP model with >6 million animals in the pedigree and >90,000 genotyped animals in the data. The traditional PCG method did not reach convergence within 10,000 iterations. With the proposed second-level preconditioner, convergence was achieved after between 2,600 and 3,900 iterations, depending on the ssSNPBLUP model. These numbers were higher than the numbers of iterations to reach convergence using the DPCG method (between 740 and 1,600 iterations). However, because additional costs of the DPCG method were avoided, the PCG method with the proposed preconditioner was as fast as, or faster than, the DPCG method. The results suggest that the PCG method combined with the proposed second-level diagonal preconditioner is a promising approach for solving large multivariate ssSNPBLUP within manageable amounts of memory and time.

Effect of fitting a genotypic mean μg on bias and accuracy of single-step genomic prediction

T.K. Belay[1], S. Eikje[2], A. Gjuvsland[2] and T. Meuwissen[1]
[1]Norwegian University of Life Sciences, Animal and Aquacultural Sciences, Arboretveien 6, 1432, Norway, [2]GENO Breeding and A.I Association, Arboretveien 6, 1432, Norway; tesfaye.kebede.belay@nmbu.no

Single-step genomic best linear unbiased prediction (SS-GBLUP) model combines all available data from genotyped and non-genotyped animals. In such model, missing genotypes are implicitly imputed for animals that are not genotyped, and observed genotypes are centred using means of unselected founders, which are typically unavailable for populations under selection. For marker effects models, a simulation study proposed an alternative analysis that did not necessarily require centring genotypes but fitting of an additional fixed covariate (J) that estimates the mean μ_g of the linear component of the genotypic value. Here, we evaluated the effect of fitting the additional fixed covariate on bias and accuracy in the SS-GBLUP model using Norwegian Red data. For non-genotyped animals, the covariate vector J was calculated using the pedigree relationship matrix, where the relationship among non-genotyped animals and between non-genotyped and genotyped individuals were used. A vector of J=-1 was used for genotyped individuals. The SS-GBLUP analyses with or without fitting the additional fixed covariate were undertaken on first lactation milk yield. Inflation, level bias (mean difference in breeding values (BV) scaled by their standard deviation) and accuracy of BV were evaluated using cross-validation, where phenotypes (P), or genotypes (G) or both phenotypes and genotypes (PG) of young animals were masked. The BV from such reduced analyses were compared to the BV or corrected-phenotypes from complete analysis. Fitting J reduced inflation and level bias, but had no effect on accuracy of BV. Regression coefficients as measures of inflation-bias were closer to one when fitting J than not fitting it (1.0008 vs 1.0062, 1.8770 vs 1.9419, and 1.0218 vs 1.0484 when P, G and PG were masked, respectively). Level biases were substantially reduced with the J model compared with the model without J when G or PG were masked (-0.0831 vs -0.1457 or -0.0937 vs -0.1836 (almost halved), respectively). Results from the current study confirmed that fitting J in the SS-GBLUP model reduces biases, but hardly affects prediction accuracies.

Computation of many relationships between metafounders replacing phantom parents

M.P.L. Calus and J. Vandenplas
Wageningen University & Research, Animal Breeding and Genomics, P.O. Box 338, 6700 AH Wageningen, the Netherlands; mario.calus@wur.nl

Phantom parents are used in genetic evaluations to obtain unbiased estimated breeding values and genetic trends when missing parents appear throughout the pedigree. In routine genetic evaluations several hundred phantom parents may be present, while accurately estimating their individual effects may be a challenge. Phantom parents are assumed to be mutually unrelated and non-inbred. For generations after the base of the pedigree this assumption can be relaxed by assigning phantom parents the mean relationships and inbreeding of other ancestors in the same period. In genomic evaluations, this assumption can be relaxed by using metafounders, which are pseudo-individuals with relationships between them estimated based on genotyped descendants. Amongst others, these relationships may help to estimate effects for phantom parents with few data. In our experience, replacing hundreds of phantom parents one-to-one by metafounders, induces problems in estimating metafounder relationships, due to too few or too weak connections between the genotyped animals and some metafounders. Often this can be solved by drastically reducing the number of metafounders to as few as only one per breed or line, however, at the expense of losing the ability to model the change in genetic level of phantom parents across time. We propose the following approach to enable computation of relationships among many metafounders within a single breed or line: (1) Assign one or more metafounders at the 'true' base of the pedigree, e.g. the oldest generations of the pedigree, and use a conventional method to estimate their allele frequencies; (2) Estimate allele frequencies for a limited number of metafounders in between the true base and genotyped population, selecting those with the highest average relationships with the genotyped population; (3) Compute average allele frequencies by birth year for the genotyped population; (4) Compute allele frequencies for birth years between those for the base, the metafounders considered in step 2, and the genotyped population, using interpolation; and (5) Compute metafounder relationships as 8 times the covariance between their allele frequencies. Results will be presented at the conference.

The impact of non-additive effects on the genetic correlation between populations

P. Duenk[1] and J.H.J. Van Der Werf[2]
[1]Wageningen University and Research, Animal Breeding and Genomics, Droevendaalsesteeg 1, 6708 PB Wageningen, the Netherlands, [2]University of New England, School of Environmental and Rural Science, W074, University of New England, 2351 Armidale, Australia; pascal.duenk@wur.nl

The genetic correlation between populations (r_g) is in part affected by non-additive effects (i.e. dominance and epistasis) in combination with differences in allele frequencies between populations. For populations in the same environment, r_g is equal to 1 in the absence of non-additive effects, or in the absence of allele frequency differences. Although this phenomenon is widely recognized, the relationships between the nature of non-additive effects, differences in allele frequencies, and the value of r_g, remain unclear. Our objective was therefore to study the effects of dominance and epistasis on r_g. For this purpose, we simulated genotypes of quantitative trait loci (QTL) for two populations that have diverged for a number of generations under either drift or selection, and we simulated traits where the genetic model and size of non-additive effects were varied. Our results showed that r_g depended mostly on the difference in allele frequency between populations, and that it is unlikely that values of r_g below 0.8 are due to dominance effects alone. Overall, we observed that the value of r_g was much lower with epistasis than with dominance. Furthermore, the value of r_g depended on the nature of the epistatic interaction between QTL. As expected, an increase in size of dominance effects resulted in a smaller r_g. Surprisingly, an increase in size of epistatic effects did not consistently result in a smaller r_g. The results in this study may provide valuable insight into the relationship between non-additive effects, allele frequency differences and r_g.

GWAS for genotype by lactation stage interaction for milk production

H. Lu and H. Bovenhuis
Wageningen University and Research, Animal Breeding and Genomics, P.O. Box 338, 6700 AH Wageningen, the
Netherlands; haibo.lu@wur.nl

The genetic background of milk production traits changes during lactation. However, most genome-wide association study (GWAS) for milk production traits assume that genetic effects are constant during lactation. The objective of this study was to identify those QTL whose effects on milk production traits change during lactation. Test-day records were available for eight milk production traits on 1,829 first-parity Holstein cows. On average, each cow had 10.7 test-day records. The lactation was divided in 26 lactation stages of 15 days each. After SNP filtering, 30,348 SNP remained for the GWAS. A GWAS specifically for SNP by lactation stage interaction was performed. Significance for the interaction term was determined based on permutation of the genotypes. Eleven genomic regions showing significant SNP by lactation stage interaction were identified indicating that the effects of those regions change during lactation. The region on BTA 14 containing the diacylglycerol O-acyltransferase 1 (*DGAT1*) K232A polymorphism showed highly significant SNP by lactation stage interaction on milk yield, lactose yield, protein content and fat content. Another strong interaction was detected for a region on BTA 19 on lactose content. Less pronounced but significant interaction signals were detected for chromosomal regions on BTA 3, 9, 10b, and 27 on protein content, for regions on BTA 4 and 16 on fat yield and for regions on BTA 10a, 11, and 23 on fat content. The estimated effects for the lead SNP in the identified regions showed that the interaction was due to changes in SNP effects either in early or late lactation. We hypothesize that different SNP effects in early lactation are due to the negative energy balance which affects fat metabolism. Deviating SNP effects in late lactation might be due to pregnancy. Our study demonstrated that a GWAS for genotype by lactation stage interaction can identify new genomic regions involved in milk production. This will help to elucidate the genetic and biological background of milk production. The performed methods can also be used for other longitudinal traits like body weight or egg production.

Genome-wide association study for leg disorders in Austrian Braunvieh and Fleckvieh

B. Kosińska-Selbi[1,2], T. Suchocki[1,2], M. Frąszczak[1], C.H. Egger-Danner[3], H. Schwarzenbacher[3] and J. Szyda[1,2]
[1]Wroclaw University of Life and Environmental Sciece, Animal Genetics / THETA (Statistical Genetics Group), Kożuchowska 7, 51-631, Poland, [2]National Research Institute of Animal Production, Krakowska 1, 32-083, Poland, [3]ZuchtData EDV-Dienstleistungen GmbH, Dresdner Straße 89/19, 1200, Austria; barbara.kosinska@upwr.edu.pl

The aim of the study was to associate regions significant for the total number of leg disorders scored until 300th day of lactation in 985 Braunvieh and 1999 Fleckvieh cows to the genomic structure expressed by SNP genotypes. The cows were genotyped with the GeneSeek Genomic Profiler HD BeadChip, resulting in 76,932 SNPs. The trait was represented by a pseudophenotype expressed by estimated breeding values. In the first step, a genome-wide association study (GWAS) was performed using a series of single-SNP mixed models, as implemented in the GCTA software. Based on the false discovery rate of 10% 27 SNPs (BTA1, BTA4, BTA6, BTA12, BTA13, BTA14, BTA15, BTA16, BTA24, BTA26, BTA29) were associated with the trait in Braunvieh and only one SNP, located on BTA18, was associated in Fleckvieh. Further on, the principal component analysis of SNP genotypes was performed, also using GCTA software. Differences in a genomic structure between the breeds were assessed for a 2-dimensional space defined by the first two principal components, separately for each of 50-SNP windows defined along the genome. In general, both breeds showed a similar structure of genetic variability, but some differences appeared on BTA19, BTA17, BTA16, BTA15, BTA14, BTA13 and BTA12. Moreover, we assessed individual variation along the genome by measuring how each cows' position in the 2-dimensional spaces changes along the genome. On average, for Fleckvieh no changes were observed, with the average position change amounting to 2.97×10^{-7}. For Braunvieh, changes were larger and varied between 0.0 and 8.32×10^{-7}, with the average 9.88×10^{-9}. Although some differences in the pattern of genomic variability between Fleckvieh and Braunvieh exist, they do not explain the observed differences in significant SNP positions for total number of leg disorders.

Sequence-based genome-wide association study of milk mid-infrared wavenumbers in dairy cattle

K. Tiplady[1,2], T. Lopdell[2], E. Reynolds[1], R. Sherlock[2], M. Keehan[2], T. Johnson[2], J. Pryce[3,4], H. Blair[1], S. Davis[2], M. Littlejohn[1,2], D. Garrick[1], R. Spelman[2] and B. Harris[2]

[1]Massey University, AL Rae Centre, Ruakura, Hamilton 3240, New Zealand, [2]Livestock Improvement Corporation, R & D, Newstead, Hamilton 3240, New Zealand, [3]La Trobe University, School of Applied Systems Biology, Bundoora, VIC 3083, Australia, [4]Agriculture Victoria, AgriBio, Bundoora, VIC 3083, Australia; ksanders@lic.co.nz

Studies of Fourier-transform infrared (FTIR) spectra from milk samples have shown that the transmittance of individual FTIR spectra wavenumbers has moderate to high heritability across most of the mid-infrared region. This indicates that genetic gain may be achieved by directly selecting on linear combinations of wavenumber estimated breeding values (EBV), rather than selection on EBV of composite traits, which are a function of the FTIR spectra wavenumbers. The purpose of this study was to identify associations between individual FTIR wavenumbers and regions of the genome. In total, 100,671 FTIR spectra records for 38,085 mixed-breed New Zealand dairy cattle with imputed whole genome sequence were used in the study. Adjusted phenotypes for each wavenumber were generated using mixed linear models with parity, days in milk, breed, heterosis and herd by test date as fixed effects, and animal as a random effect. Separate GWAS analyses were run for each of 895 adjusted FTIR wavenumber phenotypes using Bolt-LMM, fitting 16,825,207 imputed sequence variants, with a subset of 43,851 SNP used to represent genomic relationships and account for population structure. A conservative Bonferroni significance threshold was set at 6.64×10^{-13}. Significant associations were identified for 830 wavenumbers, distributed across all 29 autosomes. Many of the genomic regions with significant associations coincide with regions previously reported for major milk composition traits, but a number of new regions were also identified. Significant associations on chromosomes 5, 6, 11, 14, 19, 20 and 27 were conserved across multiple wavenumbers. Wavenumbers in bands of the infrared spectrum associated with water absorption also had significant associations, particularly on chromosomes 5, 6, 14 and 27. Further work is required to determine how this information can be used to improve genomic prediction.

Genome-wide association for metabolic adaptation in early lactation dairy cows

M. Salamone[1], H. Atashi[1,2], M. Salavati[3], J. De Koster[1], M.A. Crowe[4], G. Opsomer[1] and M. Hostens[1]

[1]Ghent Univeristy, Department of Reproduction, Obstetrics and Herd Health, Salisburrylaan 133, 9820 Merelbeke, Belgium, [2]Shiraz University, College of agrigculture, departement of animal Science, Shiraz, Iran, [3]University of Edinburgh, The Roslin Institute and Royal (Dick) School of Veterinary Studies, Easter Bush, 76065 Midlothian, United Kingdom, [4]University College Dublin, Dublin, Dublin, Ireland; matthieu.salamone@ugent.be

A genome-wide association study (GWAS) was performed to identify genetic markers associated with metabolic adaptation in early lactation Holstein dairy cows. The k-means clustering method based on log-transformed and standardized concentrations of blood glucose, insulin like growth factor I (IGF-I), free fatty acid (FFA) and β-hydroxybutyrate (BHB) at 14 days in milk (DIM) and 35 DIM was used to divide 105 cows in metabolic clusters. Cows with high glucose and IGF-I and low FFA and BHB at 14 and 35 DIM were grouped in a BALANCED metabolic cluster (n=42). Cows with low glucose and IGF-I and high FFA and BHB at 14 and 35 DIM were grouped in an IMBALANCED cluster (n=19). The cows were genotyped using Illumina BovineHD BeadChip for a total of 777,962 single nucleotide polymorphisms (SNP). The GWAS was done using a case–control approach through the GEMMA software accounting for the population structure. There were 2 and 12 SNP associated with either BALANCED or IMBALANCED metabolic clusters, respectively. Most SNP associated with IMBALANCED metabolic clusters are located within quantitative trait loci (QTL) for body weight, dry matter intake, and milk composition. K-means clustering based on the genome-wide ancestry proportions estimated by ADMIXTURE, grouped animals into three sub-populations. The logistic regression analysis showed a significant association between derived sub-populations and the metabolic clusters (P<0.05). Distribution of sires by country of origin showed that the cows were mainly sired by American, German, Scandinavian and Irish bulls. Most of the BALANCED cows originated from Scandinavian followed by American bulls. This study provides new insights into the genetic basis of metabolic adaptation in early lactation Holstein dairy cows.

Genome-wide association for metabolic adaptations early in lactation Holstein cows

H. Atashi[1,2], M. Salavati[3], J. De Koster[2], M.A. Crowe[4], G. Opsomer[2] and M. Hostens[2]
[1]Shiraz University, College of Agriculture, Department of Animal Science, Shiraz, Iran, [2]Ghent University, Department of Reproduction, Obstetrics and Herd Health, Salisburylaan 133, 9820 Merelbeke, Belgium, [3]University of Edinburgh, The Roslin Institute and Royal (Dick) School of Veterinary Studies, Easter Bush, Midlothian, United Kingdom, [4] University College Dublin, Dublin, Ireland; hadi.atashi@ugent.be

Genomic-wide association studies (GWAS) are a powerful tool for detecting genomic regions explaining variation in phenotype. The objective of this study was to detect genomic regions associated with metabolic adaptations early in 3,437 Holstein cows. Milk samples collected starting from the first week in milk until 50 days in milk (DIM) were used to determine Fourier-transform mid-infrared spectra and subsequently used to predict metabolic cluster of animals using an earlier created prediction model. Breeding values of animals for the defined metabolic cluster were estimated using a Bayesian approach. Deregressed estimated breeding values of the animals were used as the response variable. The genome-wide association study was performed using a single SNP (single nucleotide polymorphism) regression mixed linear model and imputed high-density panel (777k) genotypes. We identified 10 significant SNPs on BTA14 at position of 1.58-2.05 and the 19 significant SNPs on BTA27 at position of 36.1-36.3 Mb. The SNPs identified on BTA14 were mapped inside or close to genes including PPP1R16A, NAPRT, ARHGAP39, CPSF1, VPS28, SCRIB, SLC52A2, SPATC1, CYHR1, DGAT1, EEF1D, FAM83H, FOXH1, GRINA, LRRC14, MAF1, MAPK15, TONSL, ZC3H3, ZNF623. The SNPs identified on BTA27 were mapped inside or close to functional genes including ANK1, GINS4, AGPAT6, AP3M2, GOLGA7, KAT6A, MIR486, NKX6-3, PLAT, SFRP1. Most SNPs associated with metabolic adaptation in early lactation cows, were located inside quantitative trait loci (QTL) for body weight, dry matter intake, milk yield, and milk composition. The regions detected in this study support previous associations reported for blood glucose level, blood free fatty acid level and dry matter intake in cattle. This study provides new insights into the genetic basis of metabolic adaptation in early lactation Holstein dairy cows.

TheSNPpitGroup – managing large scale genotyping data in Open Source

E. Groeneveld[1], H. Schwarzenbacher[2], J. McEwan[3], F. Seefried[4], M. Jafarikia[5], P. Von Rohr[4] and J. Jakobsen[6]
[1]Institute of Farm Animal Genetics (FLI), Höltystrasse 10, 31535 Neustadt, Germany, [2]ZuchtData GmbH, Dresdner Straße 89/19, 1200 Vienna, Austria, [3]AgResearch, Invermay, Mosgiel 9053, New Zealand, [4]Qualitas AG, Chamerstr 56, 6300 Zug, Switzerland, [5]CCSI, 960 Carling Ave, Ottawa, Canada, [6]NSG, Box 104, 1431 Ås, Norway; eildert.groeneveld@gmx.de

TheSNPpit is an ultrafast database for large scale genotyping (SNP) data. It functions as a backend data storage to be integrated in a SNP pipeline, for instance in gBLUP. As such it is able to handle millions of genotyped individuals with any panel size with mio of SNP each. Pipeline integration is facilitated through the command line interface. Released under the Open Source license, TheSNPpit is freely available without licensing cost. To address maintenance, further development and support TheSNPpitGroup was established which runs the project website https://thesnppit.net. A Memorandum of Understanding (MoU) was developed to provide an organizational structure to achieve 3 main objectives: (1) It serves as a platform for the maintenance of the Open Source TheSNPpit software. (2) It coordinates bug fixes and further development of the core software as well as contributed modules. (3) It creates and maintains an infrastructure for mutual support and fosters discussion of ideas around TheSNPpit. The creation of contributed libraries built on the core SNPpit software facilitates sharing software across genomic programs and species. As users contribute such modules, TheSNPpitGroup will ensure its proper integration and dissemination. Parentage verification based on ICAR rules is one example. According to personal interests/capabilities the member structure consists of users, developers and contributors. A further substantial development of the core TheSNPpit is the inclusion of genotyping by sequencing (GBS) data. While its data structure is different from SNP data it can nonetheless be incorporated in the existing data framework by creating a variable entity size in terms of bits. As a result, the total disk space requirement per individual will increase substantially. However, given the excellent scaling of genotype access in TheSNPpit, this will only increase storage requirements but not processing time for data export.

Weighting genomic relationship matrix by simulated annealing

M. Martín De Hijas-Villalba[1], L. Varona[2], J.L. Noguera[3], N. Ibañez-Escriche[4], J.P. Rosas[5] and J. Casellas[1]
[1]Universitat Autonoma de Barcelona, Departament de Ciència Animal i dels Aliments, Plaza Cívica s/n, 08193 Bellaterra, Spain, [2]Universidad de Zaragoza, Departamento de Anatomia, Embriología y Genética Animal, c/ Miguel Servet 177, 50013. Zaragoza., Spain, [3]IRTA, Genética i Millora Animal, C/ Rovira Roure 191, 25198. Lleida, Spain, [4]Universitat Politécnica de Valencia, Departament de Ciencia Animal, c/ Camino de Vera, 46071. Valencia, Spain, [5]INGA FOOD S.A., c/ Ronda Poniente 9, 28760. Tres Cantos, Spain; lvarona@unizar.es

In standard genomic Best Linear Unbiased Prediction (gBLUP) models (model G) information from all single nucleotide polymorphism (SNP) has equal weight in the construction of the genomic relationship matrix (G). This study proposed a new model (model W) where all SNPs were weighted by a value ranging between 0 and 1. Both models were compared according to their accuracy when predicting breeding values (BV) in simulated datasets. Mixed model equations were solved by Gauss-Seidel. Model W was solved by simulated annealing based on the mean squared error (MSE) between simulated and predicted BV. Two scenarios were tested. The first one (P1) involved 100 populations with 1000 individuals and one autosomal chromosome per individual (h2 ranged from 0.1 to 0.5). The second scenario relied on three populations (P2) of 1000 individuals with 30 chromosomes per individual (h^2=0.1, 0.25 and 0.4, respectively). All chromosomes were 100 cM long, with 6,000 SNPs and 250 quantitative trait loci (QTL). All populations evolved during 1000 generations (Ne=100) and five final genotyped generations (Ne=200); all generations were non-overlapped and originated under random mating accounting for recombination, mutation and linkage disequilibrium. Populations P1 showed a reduction in MSE between 7.7 and 26.7% this value being positively correlated with heritability (r=0.827). Accuracy for predicted BV increased between 4 and 11% and tended to reduce this increase with heritability (r=-0.625). Regarding populations P2, and after 6,000 iterations of simulated annealing, all populations showed a reduction in MSE between 10 and 15%, and an increase of accuracy for predicted breeding values between 1.5 and 3%.

EMABG 2.0 – European master in animal breeding and genetics

G. Mészáros[1], J. Sölkner[1], T. Heams[2], I. Laissy[2], H. Simianer[3], E. Strandberg[4], A.M. Johansson[4], H. Komen[5], D. Lont[5], G. Klementsdal[6] and P. Berg[6]
[1]University of Natural Resources and Life Sciences, Vienna, Gregor Mendel Str. 33, 1180 Vienna, Austria, [2]AgroParisTech, 16 Rue Claude Bernard, 75231 Paris Cedex 5, France, [3]University of Goettingen, Albrecht-Thaer-Weg 3, 37075 Goettingen, Germany, [4]Swedish University of Agricultural Sciences, Box 7023, 75007 Uppsala, Sweden, [5]Wageningen University, P.O. Box 338, 6700 AH Wageningen, the Netherlands, [6]Norwegian University of Life Sciences, Arbotetsveien 6, 1433 Ås, Norway; gabor.meszaros@boku.ac.at

The previous European Master in Animal Breeding and Genetics (EMABG) programme has proven to be very successful in recruiting highly skilled students and educate graduates with a very high uptake in research and industry. Funding from the EU Erasmus Mundus programme has allowed the launch of a re-designed EMABG. The domain of animal breeding and genetics has undergone a revolution with the availability of high throughput genomic markers as well as phenotypic data from automated recording systems, which resulted in a dramatic change in the understanding of genetic mechanisms and in the way livestock and aquaculture species are bred. The new EMABG, a 2-year Master programme, addresses demands of industry and academia for highly trained people with skills to use new concepts and tools to face both increasingly complex problems in science and to address evolving societal issues. The EMABG, coordinated by BOKU, allows students to specialize in one of four areas: (1) Breeding programmes in low-income countries; (2) Biological and societal context of breeding; (3) Genomic selection and precise genotypes; (4) Integrative biology. Studies combine a first year at either University of Göttingen (DE) or Wageningen University (NL) with a second year at either AgroParisTech (FR), SLU (SE), NMBU (NO) or BOKU (AT). In addition, the programme provides integrated joint activities, internships, collaborative student projects on sustainable development goals, and a Breeding Lab course promoting collaborative work on assignments developed in collaboration with associated industry partners. EMABG offers a limited number of highly competitive Erasmus Mundus scholarships for the programme, covering participation, living and travel costs. See www.emabg.eu for more information.

Prediction of genomic breeding values in Montana Tropical Composite cattle using single-step GBLUP

L. Grigoletto[1], L.F. Brito[2], H.R. Oliveira[2], J.P. Eler[1], F. Baldi[3] and J.B. Ferraz[1]
[1]FZEA/University of Sao Paulo, Basic Science, 225 Duque de Caxias Norte, 13635-900, Brazil, [2]Purdue University, Animal Science, 270 S. Russel St, 47907, USA, [3]FCAV/Sao Paulo State University, Animal Science, Prof.Paulo Donato Castellane Rd, 14884-900, Brazil; britol@purdue.edu

The Montana Tropical® Composite population was developed in Brazil based on the crossing among four different biological types (i.e. breed groups) derived from various *Bos taurus* and *Bos indicus* breeds. As the majority of genomic predictions methods have been performed and implemented on purebred beef cattle, the complex genetic composition of this population is a challenge for the implementation of genomic selection. Therefore, the objective of this study was to evaluate the accuracy and bias of genomic breeding values (GEBVs) predicted using the single-step GBLUP (ssGBLUP) and weighted single-step GBLUP (WssGBLUP2; second and WssGBLUP3; third iteration) methods. The full data set included around 300,000 records for the main traits included in the selection index: weaning weight (WW), yearling weight (YW), muscling (MUSC) and scrotal circumference (SC). The pedigree file included 682,101 animals, from which 1,900 animals were genotyped using 28,195 SNPs. Thereafter, genotyped animals were divided into training (n=1,452) and validation (n=448) populations based on their year of birth. Accuracy was calculated as the Pearson correlation coefficient between current pedigree-based EBVs and GEBVs predicted using a reduced data set. Prediction bias was assessed using the regression coefficient estimated using a linear regression of current EBVs on GEBVs. The ssGBLUP model showed higher accuracy (0.69 average accuracy between traits) compared to the second (0.61) and third iteration (0.57) on WssGBLUP. In addition, GEBVs were less biased (i.e. regression coefficient was closer to 1.0) for ssGBLUP (1.03 average regression coefficient between traits) compared to WssGBLUP (0.79 and 0.67 for the 2nd and 3rd iteration, respectively). These results demonstrated the application of ssGBLUP to perform genomic predictions in a composite population. Alternative models and evaluation approaches will be investigated next in order to improve the accuracy and reduce bias of genomic breeding values estimates.

The use of novel phenotyping technologies in animal breeding

H. Bovenhuis
Wageningen University & Research, Animal Breeding and Genomics, P.O. Box 338, 6700 AH Wageningen, the Netherlands; henk.bovenhuis@wur.nl

In the coming decades livestock production faces tremendous challenges to produce high quality animal products in a sustainable and animal friendly way. In order to meet future market needs breeding goals need to be adjusted and extended with new traits. Automation and technological innovations can contribute to recording new phenotypes that provide information on both existing and new breeding goal traits. In addition, this information can be used by farmers to better manage animals according to their individual needs, i.e. precision livestock farming. However, criteria for management indicators and breeding indicators (selection index traits) are not the same. Whereas for management indicators false-negative (e.g. missing a disease occurrence) as well as false-positive alerts (e.g. an alert but no disease) are highly undesirable, this might be less relevant for selective breeding purposes. Even if an indicator is strongly affected by random measurement errors, collecting repeated observations on multiple animals might be used to reduce effects of random errors and result in accurate estimates of breeding values. However, if indicator phenotypes are subjected to systematic errors, the issue is more serious as these cannot be accounted for in statistical analyses and they might be relevant for selective breeding in case these systematic errors are partly genetic. Even in this case the indicator phenotype might provide valuable information about breeding goal traits, depending upon the genetic correlation between the indicator phenotypes and the breeding goal trait.

Session 66

Theatre 2

Precision phenotyping using Fourier-transform infrared spectroscopy for expensive-to-measure traits

F. Tiezzi[1], H. Toledo-Alvarado[2], G. Bittante[3] and A. Cecchinato[3]
[1]North Carolina State University, Department of Animal Science, 120 W Broughton dr, 27695, Raleigh NC, USA, [2]Universidad Nacional Autónoma de México, Departamento de Genética y Bioestadística, Ciudad Universitaria, Ciudad de México, Mexico, [3]University of Padova, Department of Agronomy, Food, Natural resources, Animals and Environment (DAFNAE), Viale dell'Università 16, 35020 Legnaro, Padova, Italy; f_tiezzi@ncsu.edu

Large-scale phenotyping is rapidly becoming the major operational bottleneck in precision farming and animal breeding. The Fourier-transformed infrared spectroscopy (FTIR) technique has been proven to be a powerful tool for high-throughput milk quality assessment. The objective of this study was to evaluate the ability of FTIR-derived phenotyping in the prediction of protein composition for Brown Swiss cows, as opposed to different genotyping strategies. Data were from 1,011 cows, phenotyped using the High Performance Liquid Chromatography method. These records will be considered as LAB in the present study, and include fractions of αS1-casein (CN), αS2-CN, β-CN, k-CN, β-lactoglobulin (LG), and α-lactalbumin (LA). FTIR calibration equations were developed. At the end, it was possible to obtain ~600,000 FTIR predictions from the test-day records for ~30,000 cows. This dataset will be considered as FIELD in this study. Genotyped individuals consisted of: 1,011 LAB cows, 1,493 of the FIELD cows, the 181 sires with both LAB and FIELD daughters and 540 sires with FIELD daughters only. Individuals were genotyped with different SNP panels and imputed to 50k. A 4-fold cross-validation was used to assess the predictive ability of models, meant as the ability to predict masked LAB records from daughters of progeny testing bulls, the correlation between observed and predicted LAB measures in validation was averaged over the 4 training-validation sets. Predictions were obtained using the Single-Step GBLUP method. Different sets of phenotypic information were defined, including sequentially LAB cows from the training set; FIELD cows from the training set; FIELD cows from the validation set. Results show that the use of genomic information does not provide any advantage in predictive ability, but prediction models that included FIELD records showed an advantage for most of the traits.

Session 66

Theatre 3

Towards a genomic evaluation of cheese-making traits including candidate SNP in Montbéliarde cows

M.P. Sanchez[1], T. Tribout[1], V. Wolf[2], M. El Jabri[3], N. Gaudillière[2], S. Fritz[1,4], C. Laithier[3], A. Delacroix-Buchet[1], M. Brochard[5] and D. Boichard[1]
[1]INRA, GABI, AgroParisTech, Université Paris Saclay, 78350 Jouy-en-Josas, France, [2]ECEL25-90, 6 rue des épicéas, 25640 Roulans, France, [3]Idele, MNE, 75012 Paris, France, [4]Allice, MNE, 75012 Paris, France, [5]Umotest, CS 10002, 01250 Ceyzériat, France; marie-pierre.sanchez@inra.fr

In the FROM'MIR project, milk cheese-making and composition traits were predicted from 6.6 million mid-infrared (MIR) spectra in 410,622 Montbéliarde cows (19,862 with genotypes). In this study, we assess the reliability of single step GBLUP breeding values (ssEBV) for 11 traits (cheese yields, coagulation, casein and calcium contents). The dataset was split in two independent training and validation (VAL) sets that respectively contained cows with the oldest and the most recent lactations. The training set, including 155,961 cows (12,850 with genotypes), was used to predict ssEBV of 2,125 VAL genotyped cows. For each trait, the reliability of ssEBV (ratio of squared correlation between ssEBV and adjusted phenotypes of VAL cows to the heritability) was estimated. We first tested 4 models including lactation (LACT) or test-day (TD) records from the first (1) or the first three (3) lactations, giving the 50k SNP effects equal variances. Mean reliabilities were 56, 57, 64 and 65% for the LACT1, LACT3, TD1 and TD3 models, respectively. Keeping the most accurate model (TD3), we then added to the 50k SNP, the research SNP of the EuroG10k chip that included candidate SNP for cheese-making traits detected from GWAS on imputed whole genome sequences. SNP effect variances were either equal (50k+) or increased for 5 to 14 candidate SNP, depending on the trait (CAND). The 50k+ and CAND scenarios led to similar mean reliabilities (67%) and both outperformed the 50k scenario (65%). The CAND scenario gave the less biased ssEBV. These results were implemented in a genomic evaluation for cheese-making traits predicted from MIR spectra in Montbéliarde cows. This study was funded by the French Ministry of Agriculture, Agro-food and Forest (CASDAR), the French Dairy Interbranch Organization (CNIEL), the Regional Union of Protected Designation cheeses of Franche-Comté (URFAC) and the Regional Council of Bourgogne-Franche-Comté, under the project FROM'MIR.

Feed efficiency: the next step in animal breeding

J.J. Bouwmeester-Vosman[1], P. Van Goor[2], E.C. Verduijn[1], C. Van Der Linde[2] and G. De Jong[1]
[1]CRV U.A., AEU, Wassenaarweg 20, 6800 AL Arnhem, the Netherlands, [2]CRV B.V., Wassenaarweg 20, 6800 AL Arnhem, the Netherlands; jorien.vosman@crv4all.com

After years of breeding for production and health, we are now able to breed for improved feed efficiency. Although the CO_2 emission per animal has increased over the past years, the CO_2 emission per kg of milk produced has decreased. Breeding for an animal that converts feed in an efficient way into milk (or meat) is key to enlarge environmental sustainability. Besides the long-term advantage of a smaller carbon footprint the direct benefit is a reduction of feed costs for the farmer. The heritability of dry matter intake (DMI) is around 28% (overall trait), with a genetic standard deviation of 1.4 kg/day. The breeding value for DMI is available in the Netherlands since 2014. More recently a breeding value saved feed costs for body maintenance is added. EBVs are estimated based on more than 130,000 weekly DMI records, measured on more than 5,500 Holstein-Friesian cows. In July 2017 CRV started with data collection on feed intake in a commercial dairy farm with 260 cows. The coming years more commercial dairy farms will be equipped with roughage intake bins and concentrate feeders to increase the number of animals with DMI measurements. Based on these data we found large differences in feed efficiency between farms. Farm averages vary between 1.2 and 1.6 kg milk/kg DMI. Differences between animals are even larger (between 1.1 and 1.8 kg milk/kg DMI). These results show that there are considerable differences in efficiency of dairy production and in profit per cow. The use of breeding value is a way to improve the feed efficiency of the herd. The goal is not to reduce the amount of DMI, but to increase kg milk per kg dry matter. The 25% best cows for feed efficiency produce 9 kg milk per day compared to the 25% worst animals, on the same amount of feed. Animals with a higher feed efficiency provide more profit for the farmer and have a lower carbon footprint per kg milk.

Genetic parameters for environmental traits in Australian dairy cattle

C.M. Richardson[1,2], T.T.T. Nguyen[2], M. Abdelsayed[3], B.G. Cocks[1,2], L. Marett[4], B. Wales[4] and J.E. Pryce[1,2]
[1]La Trobe University, School of Applied Systems Biology, 5 Ring Road, Bundoora 3083, Australia, [2]Agriculture Victoria, AgriBio, Centre for AgriBioscience, 5 Ring Road, Bundoora 3083, Australia, [3]DataGene Ltd., AgriBio, 5 Ring Road, Bundoora 3083, Australia, [4]Department of Jobs, Precincts and Regions, Ellinbank Centre, 1301 Hazeldean Rd, Ellinbank 3820, Australia; caeli.richardson@ecodev.vic.gov.au

As enteric methane accounts for 57% of the Australian dairy industry's environmental impact, mitigating emissions is key to improve efficiency and sustainability. Genetic selection is a cumulative, permanent reduction option with minimal labour or cost. However, measuring methane is difficult and expensive, resulting in a small sample size to use as a reference population for genomic prediction equations. Therefore, alternative options for obtaining methane phenotypes should be explored. Methane predicted using mid-infrared spectroscopy (MIR predicted methane) offers and inexpensive and relatively simple method of obtaining may methane phenotypes. In this study, we estimated the genetic parameters for methane, dry matter intake (DMI), and MIR predicted methane. Methane and DMI records were obtained from 327 cows over a 5-day period at the Ellinbank Dairy Research Institute using the SF_6 method and an electronic feed recording system, respectively. MIR predicted phenotypes were from a mid-infrared prediction equation applied to 3,425 commercial cows from 18 herds across Australia. All animals were genotyped and imputed to 6,333,374 SNP genotypes with Beagle3 software. Heritabilities were estimated using univariate models, correcting for fixed effects and fitting a genomic relationship matrix, for DMI (0.27), methane (0.11), and MIR predicted methane (0.34). The genetic correlation estimated using bivariate models between methane and MIR predicted methane was 0.93 (0.34) while methane and DMI was 0.53 (0.04). Accuracies of gEBV were validated using a 6-fold cross validation, as r(gEBV, methane) where SF_6 methane phenotypes were corrected for fixed effect. Obtained accuracies were low (0.01-0.16). Our results indicate that the improvement in accuracy gained by using MIR predicted methane is low and more data is required to determine the effect of MIR predicted methane on gEBV accuracies.

Development of a NIRS method to assess the digestive ability in growing pigs

E. Labussière, P. Ganier, A. Condé, E. Janvier and J. Van Milgen
Pôgôʊe. INRA Agrocampus Ouest, 16 Le Clos, 35590 Saint-Gilles, France; etienne.labussiere@inra.fr

In growing animals, feed efficiency is determined by the digestive and metabolic efficiency. The digestive efficiency is difficult to assess in a large number of animals. In pigs, its measurement requires to house animals in a digestibility cage during a prolonged period to measure feed intake and totally collect the faeces that should be analysed for the nutrient contents. The objective of this study is to identify an alternative methodology to characterize the digestive ability in pigs. For this purpose, 63 pigs from Large White, Pietrain, and Duroc breeds were fed alternatively two diets differing in crude fibre content (3.1 and 8.5% of dry matter) during four periods of three weeks each. The diets were supplemented with three molecules that may be used as indigestible marker: silicone oil, the plastic resin Kynar®, and polyethylene glycol. At each period, digestibility was measured using the gold standard as previously described. At the end of each period, a sample of faeces was also collected directly from the rectum of the pigs. All samples were also analysed by Near InfraRed Spectroscopy (NIRS). The digestive ability was variable among pigs during the growing period, especially when the diet contained a high level of fibre. The digestibility of this type of diet increased when body weight increased. The utilization of indigestible markers did not give satisfying results, because of low recovery in the faeces (plastic resin or polyethylene glycol) or because of large variability in the recovery (silicone oil). The NIRS prediction of digestibility coefficients from a sample collected directly in the faeces was adequate (bias of digestibility coefficients for dry matter, organic matter, energy, and N did not differ from 0%) when pigs were heavier than 60 kg and when they are fed a diet with a high fibre content. The method of collecting a single sample of faeces directly from the rectum of the pig and analysed by NIRS can be used in selection farms of breeders to give information on the digestive ability of individual pigs in a population. The Feed-a-Gene Project has received funding from the European Union's H2020 Programme under grant agreement no 633531.

Keeping up with a healthy milk fatty acid profile require selection

M. Kargo[1,2], L. Hein[1], N.A. Poulsen[3] and A.J. Buitenhuis[2]
[1]SEGES, Agro Food Park 15, 8200 Aarhus N, Denmark, [2]Aarhus University, Department of Molecular Biology and Genetics, Blickers alle 20, 8830, Denmark, [3]Aarhus University, Department of Food Science, Blichers Allé 20, 8830, Denmark; morten.kargo@mbg.au.dk

The effect of various fatty acids and fatty acid groups on human health is under discussion, and results are not unambiguously. However, current knowledge shows that a health-promoting milk fat profile is likely to be obtained by increasing the content of unsaturated fatty acids and reducing the proportion of palmitic acid. From a milk product point of view this would contribute to butter with a softer texture which is easier to spread. The objective of this study was to investigate the genetic impact on the milk fatty acid profile measured as g/100 g total fat for 7 FA fractions and 4 individual FA. Milk samples from all cows in the Danish herd testing scheme were collected from May 2015 to December 2017 and analysed at a certified laboratory using MilkScanTM FT+/FT600 equipped with special software for predicting 7 FA fractions: saturated FA (SFA), Mono-unsaturated FA (MUFA), Poly-unsaturated FA (PUFA), short chain FA (SCFA), medium chain FA (MCFA), long chain FA (LCFA), and transFA; and 4 individual FA: C14:0, C16:0, C18:0, and C18:1. Heritabilities for test day measurement of the different fractions and individual FA were at the same level (0.08 to 0.16) as heritability for overall fat yield for both Danish Holstein and Danish Jersey. Genetic correlations between the fraction of unsaturated FA and total fat yield were however unfavourable indicating an unfavourable trend for the healthy FA contents unless selected for. Correlations between breeding values for FA (groups and individuals) and the different traits in the Nordic Total merit index were in general below 0.15 for both Holstein and Jersey. This suggest that selection for FA (groups and individual FA) can be carried out without harming other traits in the breeding goal. It is therefore proposed to substitute the present Fat index in the total merit index with and index composed of the FA's weighed together based on their health promoting values and their values in relation to product quality.

New phenotypes from milk MIR spectra: challenges to obtain reliable predictions

C. Grelet[1], P. Dardenne[1], H. Soyeurt[2], A. Vanlierde[1], J.A. Fernandez[1], N. Gengler[2] and F. Dehareng[1]
[1]Walloon Agricultural Research Centre, 5030, Gembloux, Belgium, [2]ULiège-GxABT, Passage des Déportés 2, 5030 Gembloux, Belgium; c.grelet@cra.wallonie.be

In the recent years, the research aiming to predict new phenotypes from the FT-MIR analysis of milk was very active. Models were developed to predict phenotypes such as fine milk composition, cow health and environmental impact or technological properties of milk. Those models could be of great interest in order to perform genetic studies as they could allow generating large amount of data at large scale and with reasonable cost. To achieve this, it is nonetheless necessary to insure that the models provide reliable predictions when applied on the large diversity of spectral data met on real field conditions. The robustness of models -its capacity to be 'all terrain' and provide good results in various conditions- is therefore essential to ensure reliability of predictions. Robustness could be estimated by evaluating the error in external validation (RMSEP), the reproducibility of predictions between instruments and the ability of the calibration dataset to cover the variability of routine field data. However, in current literature, the model robustness is often omitted. Models are frequently developed on reduced dataset, with limited number of herds, breeds and diets. Additionally, models are evaluated by looking to the statistical performances, through the R^2 and the standard error (RMSE or SEC), while the robustness is rarely assessed. Finally, only a limited number of models is used in routine and faces the large variability of real field conditions to provide phenotypes for management of cows or genetic studies. The objective of this work is consequently to evaluate the impact of different factors influencing robustness on prediction quality. The impact of sampling scheme (oriented vs random), and model development are investigated. Effect of inclusion of variability in the model by adding countries, breeds, MIR instruments and days in milk are also investigated. The obtained results encourage for international collaborations in order to constitute large and robust datasets and enable the use of models in routine conditions.

Zootechnical parameters added to the milk MIR spectra as predictive value to estimate CH$_4$ emissions

A. Vanlierde[1], F. Dehareng[1], N. Gengler[2], E. Froidmont[1], S. McParland[3], M. Kreuzer[4], M. Bell[5], P. Lund[6], C. Martin[7], B. Kuhla[8] and H. Soyeurt[2]
[1]CRA-W, 5030, Gembloux, Belgium, [2]ULiège-GxABT, 5030, Gembloux, Belgium, [3]Teagasc Moorepark, Fermoy, Cork, Ireland, [4]ETH Zürich, 8092, Zürich, Switzerland, [5]AFBI, Large Park, Hillsborough, United Kingdom, [6]Aarhus University, 8830, Tjele, Denmark, [7]INRA, 63122, Saint-Genès-Champanelle, France, [8]FBN, 18196, Dummerstorf, Germany; a.vanlierde@cra.wallonie.be

Breeding needs tools to quantify at large scale greenhouse gases emissions in order to develop levers to reduce them. Milk mid-infrared (MIR) spectrum is a promising proxy to estimate daily methane (CH$_4$) emissions of individual dairy cows. A model has been developed based on 1,089 reference measurements combining milk MIR spectra and CH$_4$ records, collected using the SF$_6$ tracer technique (n=513) and respiration chambers (n=576). These data came from 7 countries (BE, IRL, CH, UK, FR, DK and D) and from cows of 5 major breeds: Holstein (74%), Brown Swiss (13%), Jersey (3%), Red Holstein (3%) and Swedish Red Crossed (2.5%). Using a 5 groups cross-validation (CV), statistics reached an R^2 of 0.64 with a SE of 61 g of CH$_4$/day. A cow and country dependent external validation (CCDEV) has been performed: all the data from 20% of the cows per country are removed, calibration was then done on the 80% remaining cows from all countries and validated on the 20% removed. After 500 repetitions, the R^2 and the RMSEP of CCDEV were 0.55±0.07 and 70±4.5 g of CH$_4$/day respectively. To asses improving these results, parity, milk yield and breed information have been added individually and by combination to the MIR spectra. This led to 7 new calibration models. Statistics were improved with the parameters included, with an optimum found for the version combining milk MIR spectra, milk yield, parity and breed as predictors. The statistics reached a R^2cv of 0.68, a SECV of 57 g of CH$_4$/day, and R^2 and RMSEP of CCDEV of 0.6±0.06 and 65±4.1 respectively. Including these routinely available cow parameters improves the model prediction performance. Practical applications are still required to observe and confirm the relevance of these new predictions.

Potential use of MIR spectra in the prediction of hoof disorders in Holstein Friesians

A. Mineur[1], E.C. Verduijn[2], H.M. Knijn[2], G. De Jong[2], H. Soyeurt[1], C. Grelet[3] and N. Gengler[1]
[1] ULiège-GxABT, Passage des Déportés 2, 5030 Gembloux, Belgium, [2] CRV, Wassenaarweg 20, 6843 NW Arnhem, the Netherlands, [3] CRA-W, Rue de Liroux 9, 5030 Gembloux, Belgium; axelle.mineur@uliege.be

The prediction of hoof disorders using MIR (mid-infrared) could be a promising approach for genomic selection of locomotion in dairy cattle and management of hoof problems in herds. Previously, we studied the temporal relationship between locomotion scores and MIR based biomarkers. This time we focussed on hoof disorder scores and MIR spectra directly instead of the biomarker concentrations derived from the spectra. The data provided by CRV through the ClawMIR project, consisted of 638,904 hoof disorder records and 5,708,128 MIR records, coming from 261,647 Holstein Friesian cows in 1,983 herds between 2013 and 2018. Each lactation was subdivided into 30-day month classes. Pre-processing of the spectral data consisted of the first derivative, applied with a window size of 5 wavenumbers. The MIR data and hoof disorder scores were corrected for animal-lactation, herd-testday and lactation group (1, 2 or 3+) with a fixed effect model. The spectral data and hoof disorder severity scores were then averaged over animals and month classes. Only the first five months of lactation were investigated as that is the period when the cow is most at risk of developing a metabolism related locomotion disorder. The first step of this research was to establish correlations between the hoof disorders at a specific month and each of the 212 wavenumbers during the 1,2, 3 or 4 months before the hoof disorders. Looking at the results, certain patterns appeared in these correlations. White Line disorders have the highest correlations with absorbance values taken the last month before their occurrence (r between 0.06 and 0.08). Sole Haemorrhage and Sole Ulcer scores did not make a distinction between the months preceding their occurrence, but inside each month, they have the highest correlations with the same groups of wavenumbers (r between 0.03 and 0.05 for certain wavenumbers for Sole Haemorrhage and r between 0.06 and 0.09 for similar wavenumbers for Sole Ulcer). These patterns showed time-dependent but also intra-waveband patterns that are potentially very useful in the establishment of early warning equations for diseases.

Validation of the prediction of body weight from dairy cow characteristics and milk MIR spectra

H. Soyeurt[1], E. Froidmont[2], I. Dufrasne[3], Z. Wang[4], N. Gengler[1], F. Dehareng[2], GplusE Consortium[5] and C. Grelet[2]
[1] ULiège, GxABT, 5030 Gembloux, Belgium, [2] Walloon Research Centre, 9, rue de Liroux, 5030 Gembloux, Belgium, [3] ULiège, FMV, 4000 Liège, Belgium, [4] University of Alberta, 116 St. and 85 Ave., Edmonton, Canada, [5] Lead partner UCD, Belfield, 4 Dublin, Ireland; hsoyeurt@uliege.be

Body weight (BW) recording is of interest to optimize herd management and environmental fingerprint. Due to its cost, a weighing system is not installed in many farms. Linear type traits are mostly available only once in the lifetime of the cow. So, an interest exists to develop a method to predict routinely BW with traits easily recorded and cheap. First investigation was conducted one year ago to build a prediction equation of BW from parity, test month, milk yield, days in milk and milk MIR spectrum. This study based on 717 records obtained a herd validation root mean square error (RMSEv) ranged from 37 to 64 kg. Cross-validation (cv) R^2 was equal to 0.51 with RMSE of 50 kg. This equation was applied on 1,161 milk MIR spectra collected in the GplusE project from Holstein cows. Validation R^2 was 0.51 with RMSEv of 65 kg. A total of 109 spectra had a global H distance higher than 3 suggesting spectral outliers. After their removal, R^2v increased slightly (0.54). Difference of RMSEv can be explained by a lack of spectral variability in the calibration set. So, the 2 datasets were merged to build a PLS regression including the same predictive traits as the first study. After a spectral cleaning based on GH and residual analysis, the best equation used 1,837 records and gave a 10 fold R^2cv of 0.64±0.02 with a RMSEv of 46±2 kg. The ability to predict BW was improved by adding this new data. Lowest errors were observed for BW ranged from 500 to 750 kg which represents 94% of the set. So, low and high BW lack in the calibration set. This study confirms the preliminary results and the potentiality to predict an indicator of body weight. This alternative BW prediction could explain a part of the BW variability not already covered by other BW predictors as those based on linear scores. Moreover, this method allows to consider the past information if the spectral data is available. This approach should help research on BW changes and selection for this trait in the future.

Chromosome regions affecting MIR spectra indirect predictors of cheese-making properties in cattle

R.R. Mota, H. Hammami, S. Vanderick, S. Naderi, N. Gengler and GplusE Consortium
Gembloux Agro-Bio Tech, University of Liège, Terra teaching and Research Centre, Passage des Deportes 2, 5030,
Gembloux, Belgium; rrmota@uliege.be

The current study aimed to identify single nucleotide polymorphism (SNP) markers and candidate genes associated with titratable acidity (TA), total caseins (TC) and dry cheese yield (DC) predicted from MIR spectra as indirect predictors of milk cheese-making properties in early lactation cows. These predicted MIR novel phenotypes were obtained by using equations developed through GplusE and former EU projects. Only autosomes SNPs (n=29) with call rates >0.95, minor allele frequencies (MAF) >0.05 and significant deviations from Hardy-Weinberg equilibrium ($P>10^{-7}$) were used for the genome-wide association study. After quality control edits, 32,687 SNPs remained for further analysis. The detection of SNPs affecting TA, TC and DC were obtained based on the proportion of variation explained by each SNP. The SNPs were considered as outliers if the proportion of variance explained was higher than 5×IQR+Q3, where IQR is the interquartile range and Q3 is the third quartile of the distribution. The putative genes located within or close to an outlier SNP were further identified. A total of 1,109 SNPs were considered as outliers with 8, 13, and 4% in common to all, between TC and DC, and between TC and TA traits, respectively. The remaining 833 were trait specific SNPs, with 18, 28, and 29% affecting TC, TA and DC, respectively. The 84 common SNPs to all traits were identified on BTA14 (75%), 1 (19%), 19 (5%) and 6 (1%). The trait specific markers were located on 26 out of 29 chromosomes. For TA, 86% of SNPs were located on chromosomes 1, 20, 7, and 14. On the other hand, TC showed the majority of SNPs on BTA6, BTA20 and BTA10 whereas for DC, SNPs were mainly on BTA5, BTA14 and BTA27. The top 5% outlier SNPs were located on BTA14, BTA1 and BTA20 with 20 (RPL8; FOXH1; CYHR1; OPLAH; HSF1; CPSF1; DGAT1; CYC1; GRINA; PARP10; NRBP2; PUF60; RNF19A; RRM2B; C8orf33; FBXO43; LY6H; ANKRD46; NAPRT; LY6E), 11 (FAM3B; MX1; HSF2BP; RRP1B; CSTB; C21orf2; LRRC3; DSCAM; ADARB1; ITGB2; MX2) and 3 (ANKH; TRIO; OTULIN) potential candidate genes affecting MIR indirect predictors (TA, TC and DC) of cheese-making properties in early lactation, respectively.

Using milk MIR spectra to identify candidate genes associated with climate-smart traits in cattle

H. Hammami[1], R.R. Mota[1], S. Vanderick[1], S. Naderi[1], C. Bastin[2], N. Gengler[1] and GplusE Consortium[3]
[1]ULiège-GxABT, Passage des Déportés, 2, 5030 Gembloux, Belgium, [2]Walloon Breeding Association, R&D, 5590, Ciney,
Belgium, [3]GplusE Consortium, http://www.gpluse.eu, Dublin, Ireland; hedi.hammami@uliege.be

Enhancing the resilience to extreme climate events and reducing greenhouse gases (mitigation), while improving productivity are of large importance in future breeding objectives. However, gaps to get large-scale phenotyping as well as limited knowledge about the biological basis are still challenges. This study tried to exploit the milk MIR spectra jointly with genomics to identify potential candidate genes affecting mitigation and resilience abilities. For mitigation, methane emissions (CH_4) and phosphorus (P) could be viewed as major sources of environmental pollution. Biomarkers reflecting the equilibrium between mobilisation and intake and body ketones associated with energy status are potential indicator traits for resilience to thermal stress. Predicted phenotypes form MIR data, CH_4 and P (mitigation), and C18:1cis-9, Acetone, BHB in addition to milk yield (resilience) were used in this study. The proportion of variance explained by SNPs was evaluated for all studied traits. The putative genes located within or close to outliers SNPs were further identified. Concerning CH_4 and P, 31; 28 and 7% of the what we called outliers SNPs were located at BTA14; BTA1 and BTA15, respectively. The markers in the most informative windows on BTA1 were close or within several genes such as RIPK4, PRDM15, ABCG1, SLC37A1, UBASH3A and FAM3B. On BTA14, the top outlier SNPs were located in LD block containing 3 genes (DGAT1, CYHR1, and PLEC). For resilience traits, the two outlier markers associated with acetone slope were located on BTA7 at approximately 45.5-45.7 Mb. Possible candidate genes located within this interval are UQC411 and ATP5F1D. The SNPs ARS-BFGL-NGS-4939 and ARS-BFGL-NGS-34135, located on BTA14 were top SNPs and were also detected to affect milk, C18:1cis9, acetone, and BHB slopes. Interesting candidate genes such as DGAT1, HSP1 and SLC52A2 were identified surrounding those two markers. In conclusion, markers identified for resilience and mitigation traits could prove useful in genomic selection for climate-smart breeding programs.

Using of spectral global H distance improves the accuracy of milk MIR based predictions

L. Zhang[1], C.F. Li[2], F. Dehareng[3], C. Grelet[3], F. Colinet[1], N. Gengler[1] and H. Soyeurt[1]
[1]ULiège-GxABT Passage des Déportés 2, 5030 Gembloux, Belgium, [2]Hebei Livestock Breeding Station, XueFu Road 7, Shijiazhuang City, China, P.R., [3]Walloon Research Centre, Chaussée de Charleroi 24, 5030 Gembloux, Belgium; lei.zhang@doct.uliege.be

Milk MIR spectrometry predicts traits related to quality, health and environment. Standardizing MIR data enables to share prediction equations. The prediction accuracy depends partly on the calibration spectral variability. So, the calculation of spectral global H (GH) distance is of interest. The effect of GH on the accuracy of MIR predictions were studied using 198,394 milk samples collected from Chinese cows and analysed on 3 Bentley FTS machines. The content of fat, protein, monounsaturated fatty acid (MFA), unsaturated FA (UFA), saturated FA (SFA), polyunsaturated FA (PFA) predicted by manufacturer's models were the reference values. Fat, protein, MFA, PFA, SFA and UFA averages were 3.97, 3.43, 0.86, 0.07, 2.62 and 0.93 g/dl of milk. Bentley MIR spectra were standardized to master MIR spectra according to the method developed by the European Milk Recording network. Then the studied traits were predicted on those spectra using published MIR equations. The averaged predicted content of fat, protein, MFA, PFA, SFA, and UFA were 3.99, 3.53, 1.15, 0.15, 2.64, 1.29 g/dl of milk. GH ranged from 0 to 475. Correlation values between predicted and reference contents ranged from 0.92 to 0.98, except for PFA (0.59). Root mean square errors (RMSE) were 0.19, 0.18, 0.32, 0.09, 0.21, and 0.39 g/dl for fat, protein, MFA, PFA, SFA, and UFA. Correlation values between the squared residuals and GH were positive (0.17-0.42). This suggests a positive effect of GH on the prediction accuracy. When a threshold of GH$\leq$5 was applied, the data loss ranged from 3.83 to 9.27%. Correlation values between predicted and reference contents increased (0.94 to 0.98, 0.64 for PFA). RMSE decreased from 1.26 to 7.37%. Considering GH limits spectral extrapolation. To improve the accuracy, samples with GH>5 must be included in the calibration set to cover the spectral variability. Moreover, the spectral standardization must be performed on a regular basis in order to check the potential spectral deviation of an instrument. In this study, the standardization was performed once.

Including the scapula width from CT-scans improves selection accuracy for sow longevity

Ø. Nordbø[1,2]
[1]Geno SA, Storhamargata 44, 2317 Hamar, Norway, [2]Norsvin SA, Storhamargata 44, 2317 Hamar, Norway; oyvind.nordbo@norsvin.no

The longevity of sows is a key economical trait for pig breeders. In the breeding nucleus, sows are mainly selected based on their breeding values. Longevity traits are therefore rather collected in multipliers or in commercial crossbreed herds, on animals with long genetic distance from selection candidates. To improve selection accuracy for longevity traits, it is useful to find indicator traits which could be measured closer to the nucleus. In this study we utilize the width of scapula measured in CT-scanned purebred Norwegian Landrace (NL) boars as such an indicator trait. We estimate the genetic correlations against two sow longevity traits in addition to shoulder lesions (SL) and body condition score (BCS) of sows at weaning of the first litter. The longevity traits, LGY12, which is a binary trait indicating whether the sow was inseminated after the first weaning, and LGY15, which is the number of litters achieved within 570 days after first farrowing, were measured on purebred NL sows in multipliers outside of Norway. BCS and SL were measured on purebred NL sows both in the nucleus and in multipliers. We found that the scapula width had high heritability (0.72±0.06) and genetic correlations with BCS (-0.27±0.08), SL (0.27±0.08), LGY12 (-0.18±0.15) and LGY15 (-0.37±0.15). Through a cross validation study, we also measure how including the scapula width on a selection candidate improves the accuracy of sow traits. This was done by removing data for all daughters of 151 genotyped boars with the scapula width phenotype. Then we predicted genomic breeding values for the sow traits from one model where scapula width was included as a correlated trait and one where it was excluded. The breeding values was then compared with daughter yield deviations for the 151 boars. By including scapula width as a correlated trait in a single step genomic prediction, the accuracy of sow traits was increased from 0.32 to 0.55 for LGY15, from 0.63 to 0.64 for SL, and 0.57 to 0.59 for BCS, while for LGY12, the accuracy was not affected. As all NL nucleus boars are being CT-scanned, this study quantifies the potential for improvement of sow longevity traits by including novel precision phenotypes from CT-scan before selection.

How to inform dairy farmers and breeding companies about breeding values in practice

M.L. Van Pelt, J.J. Vosman-Bouwmeester and G. De Jong

CRV u.a., Animal Evaluation Unit, P.O. Box 454, 6800 AL Arnhem, the Netherlands; mathijs.van.pelt@crv4all.com

Over the past decades the number of estimated breeding values (EBV) has increased dramatically with the development of genetic evaluation (GE) for new traits. For a long time, only EBV for milk production and conformation were available. These traits have a moderate to high heritability, and the relation between EBV and phenotype is perceived as good by dairy farmers as well as employees of breeding companies. New traits that followed, have in general a lower heritability. Difficulty with low heritable traits is that the relation between EBV and phenotype at farm level or cow level is more often perceived as doubtful. On top of the abundance of available EBV, the evolution of estimation techniques for GE and the adaptation of genomic information in GE and EBV resulted in reduced acceptance by certain farmers that EBV work. EBV are under constant scrutiny by farmers, and the acceptance of GE can be affected if one or few bulls change drastically. GE is driven by population genetics and the aim is to shift the whole population in a certain direction. A remark often made by farmers is that a farmer still focuses on the individual cow and every animal should perform good and not just the average cow. As the complexity of GE increases, it is important to explain results in an understandable manner and as simple and clear as possible. To assist in this need to show that EBV work in practice, information material is continuously being developed. Short papers, articles in farmer magazines, and fact sheets were and are being created on single topics. Main focus is to show the relation between EBV and phenotype. This is shown in multiple ways: (1) relation between sire EBV and mean offspring performance where a birthyear of bulls is grouped in four quartiles for EBV; (2) calculation tool to show what the impact of a bull will be on expected herd performance; and (3) relation between direct genomic values and realized performance. Other topics that are explained, are for example how reliability is working at cow and bull level, or how much a EBV can change with a certain reliability. By creating information material that EBV are working, based on facts, we aim to increase the acceptance and understanding of GE and EBV.

Derivation of economic values for German dairy breeds

C. Schmidtmann[1], M. Kargo[2], J. Ettema[2], D. Hinrichs[3] and G. Thaller[1]

[1]Kiel University, Institute of Animal Breeding and Husbandry, Hermann-Rodewald-Str. 6, 24118 Kiel, Germany, [2]Aarhus University, Center for Quantitative Genetics and Genomics, Department of Molecular Biology and Genetics, Blichers Allé 20, 8830 Tjele, Denmark, [3]University of Kassel, Department of Animal Breeding, Nordbahnhofstr. 1a, 37213 Witzenhausen, Germany; cschmidtmann@tierzucht.uni-kiel.de

A precondition for successful dairy cattle breeding is the specification of an appropriate and balanced breeding goal. Economic values (EV) are indicating the marginal utility of breeding goal traits and are used to weight traits according to their economic importance. The objective of the present study was to derive EV of breeding traits for the German dairy breeds Holstein Friesian (HOL), Angler (ANG) and Red-and-White Dual Purpose (RDN) using the stochastic herd simulation model SimHerd. To obtain the EV of each trait, two scenarios (basic vs extended) were simulated and compared in terms of the average annual net return (NR) caused by a marginal change in the respective trait. In order to avoid double counting of effects, the NR was corrected by multiple regression with mediator variables. EV were derived for traits regarding production, reproduction, health as well as calving. They are indicated as the change in NR from a 1% increase in the mean of the respective trait. The results have shown the largest differences for fertility traits among the considered dairy breeds (EV$_{conception rate}$: HOL=220€, ANG=249€, RDN=141€; EV$_{insemination rate}$: HOL=170€, ANG=211€, RDN=115€). Moreover, these EV were more pronounced for lactating cows than for heifers. Sensitivity analyses confirmed the robustness of calculated EV. Currently, different breeding goal scenarios for different production systems are simulated for the regarded breeds to study the implications on the genetic gain as well as genetic diversity.

Impact of multiple ovulation and embryo transfer on genetic gain and diversity in dairy cattle

A.-C. Doublet[1,2], G. Restoux[2], S. Fritz[1,2], A. Michenet[1,2], C. Hozé[1,2], D. Laloë[2] and P. Croiseau[2]
[1]ALLICE, 149 rue de Bercy, 75012 Paris, France, [2]INRA, AgroParisTech, Université Paris-Saclay, GABI, Allée de Vilvert, 78350 Jouy-en-Josas, France; anna-charlotte.doublet@inra.fr

With the beginning of genomic selection, annual genetic gain increased in major French dairy cattle breeds. In some breeds, it was linked with an increase of annual loss of genetic diversity. Yet, a loss of genetic diversity can have detrimental economic effects due to inbreeding depression and loss of adaptive potential. The recent spread of new technologies such as multiple ovulation and embryo transfer might increase selection intensity on dams of bulls. Consequently, the use of these technologies presents a risk for genetic diversity and should be managed carefully to ensure the sustainability of breeding schemes in the future while maintaining acceptable levels of genetic gains. The impact of the use of embryo transfer and multiple ovulation on genetic gain and genetic diversity was addressed by simulating different breeding schemes scenarios, based on different French dairy cattle breeds. We analysed the impact of changes in: (1) the proportion of embryo-donating heifers relative to the total number of dams of bulls; (2) the number of flushes per donor heifer; (3) the proportion of pre-selected calves entering breeding station born from donor females compared to those born from dams of bulls under conventional insemination; and (4) the proportion of pre-selected calves becoming sires of bulls. We also analysed the impact of maximum allowable thresholds of kinship within groups of donor or conventionally inseminated dams, as well as within groups of calves entering the breeding station. The outcomes of these different scenarios in terms of genetic diversity were evaluated thanks to genomic measures of inbreeding, based on runs of homozygosity. The results of these simulations will help to adapt the use of multiple ovulation and embryo transfer within selection schemes in order to make them more sustainable in terms of genetic diversity while maintaining genetic gain.

How does Holstein cattle in Switzerland react to heat stress?

B. Bapst[1], M. Bohlouli[2], S. König[2] and K. Brügemann[2]
[1]Qualitas AG, Genetic/Genomic evaluation, Chamerstrasse 56, 6300 Zug, Switzerland, [2]Justus-Liebig-University Gießen, Institute of Animal Breeding and Genetics, Ludwigstraße 21B, 35390 Gießen, Germany; beat.bapst@qualitasag.ch

Switzerland is very diverse in terms of topography and climate. This raises the question of how Holstein cattle, one of the most important dairy breeds in Switzerland, cope with these different circumstances and whether the consideration of environmental factors in the genetic evaluation would lead to a reranking of the most important sires which is an indicator for the existence of genotype × environment interactions (G×E). In order to investigate these questions we developed a random regression test day model (RRTDM), which was derived from the official Swiss genetic evaluation system. Meteorological data was included in this model as environmental descriptor. We got daily temperature and humidity measurements over a period of 10 years from 60 official federal weather stations. These values were combined to temperature-humidity indexes (THI), which we assigned to each relevant test day (TD). The covariates days in milk (DIM) and THI, both with Legendre polynomials of order 3, were added to the model for estimating the additive genetic and the permanent environment effects. After the data editing process a dataset of 7,340,498 TD records of 363,472 Holstein cows in 9,231 herds were available for the analysis. THI varied from 5 to 77 with a mean of 48.8 and data records between the 5th and 305th DIM were considered. Breeding values were estimated for daily protein yield, once using an RRTDM including THI and once with a RRTDM without THI as covariate. The rankings of the most important sires, derived from the number of daughters, were compared. Preliminary results show rerankings of these bulls, which is indicating the existence of G×E. Further investigations to quantify the extent of the reranking and to detect patterns of the largest displacements will be performed.

Heat stress and productivity in lactating Vietnamese household dairy cows

B.N. Nguyen[1,2], J.B. Gaughan[2], C.V. Nguyen[3], H.T. Nguyen[3], T.X. Nguyen[1], K.D. Nguyen[3], B.J. Hayes[2], R. Lyons[2] and D.M. McNeill[2]
[1]Vietnam National University of Agriculture, Hanoi, 100000, Viet Nam, [2]The University of Queensland, Gatton, 4343, Australia, [3]Nong Lam University, Ho Chi Minh, 700000, Viet Nam; nn.bang@uqconnect.edu.au

Despite it being well accepted that heat stress limits milk production, the demand to develop household dairying in the lower altitude/hotter regions of Vietnam continues. Our aim was to scope the extent to which heat stress could limit milk production in Vietnam by associating cow productivity to heat stress, dietary and housing parameters across four key dairying regions that cover a wide range of heat stress risk. Infrared thermal temperature of the cow's vulva (VIRT) was included as a novel measure of heat stress as compared to the more traditional measure of the cow's panting score. A total of 345 lactating cows from 32 farms, 8 each from the Moc Chau (high altitude, north: cool), Ha Nam (low altitude, north: hot), Da Lat (high altitude, south: cool), and Cu Chi (low altitude, south: hot) regions were monitored over a 24 hour period per farm, in the autumn of 2017. VIRT was determining by placing a children's infra-red ear thermometer against the moist pink inner surface of the vulval lips. Across regions, VIRT was positively correlated with panting score (r=0.23, P<0.001). Multiple regression analysis (R^2=0.709) indicated that yield of milk solids (average kg/100 kg BW, fat + protein per cow, pm + am milking) was related (coefficient, P-value) to VIRT (-0.014, 0.046) but not to panting score; was related to dietary dry matter intake (0.034, 0.049), dry matter % (0.004, <0.001), fat % (0.025, 0.005), and lignin % (0.014, 0.001), but not to dietary crude protein, neutral detergent fibre, non-fibre carbohydrate, and starch; and not related to parameters of airflow in the dairy shed. Since VIRT ranked with dietary parameters as being associated with propductivity but panting score did not, and since VIRT was correlated with panting score, we conclude that VIRT deserves further evaluation as a measure of heat stress for dairy cows in tropical environments.

Impact of heat stress on weights, weight gain and fertility in the local breed 'Rotes Hoehenvieh'

K. Halli, K. Brügemann, M. Bohlouli and S. König
Justus-Liebig-University, Institute Animal Breeding and Genetics, Ludwigstrasse 21B, 35390 Giessen, Germany; kathrin.halli@agrar.uni-giessen.de

Climate change causes rising temperatures and extreme weather events worldwide, with detrimental impact on performance, health, and fertility traits of cattle, kept in outdoor production systems. The aim of the present study was to analyse the influence of mean daily temperature-humidity index (mTHI) and number of heat stress days (nHS) during three observation periods (7d, 42d and 56d) before (a.p.) and after calving (p.p.) on birth weight, 200d- and 365d-weight gain of calves, and calving interval of their dams in the local dual-purpose cattle breed 'Rotes Hoehenvieh' (RHV). Climate data from public weather stations with minimal distance to the farms were used to calculate the mTHI. Number of days representing heat stress (HS) (mTHI≥60) were counted within the three observation periods. Mixed models were applied to infer associations between climatic effects and RHV traits. Data included 3,611 birth weights, 2,634 200d- and 2,078 365d-weight gains of calves, and 2,298 calving intervals of dams. Across all observation periods, HS during pregnancy contributed to low birth weights. Birth weight was significantly lowest when nHS comprised the maximal number of six to seven days in the 7d period (P≤0.05). 200d-weight gain and 365d-weight gain were considerably lower when calves suffered HS in the 56d p.p. period (P≤0.05 and P≤0.001 respectively). In the same period, significantly lowest 200d-weight gain (P≤0.01) and 365d-weight gain (P≤0.001) were found, when nHS comprised 41 to 56 days. Calving interval was significantly prolonged for HS conditions in the 56d p.p. period (P≤0.001). Furthermore, the highest nHS classes across all p.p. periods contributed to the longest CI. Generally, the 56 days p.p. were identified as the most critical period regarding HS effects on weight development traits in RHV calves and on calving interval in their dams. Results underline the importance of climatic impact on primary and functional traits also for 'robust' cattle breeds kept in outdoor production systems. We suggest improving the grazing management based on the identified thresholds, or a further enhancement of robustness via optimized breeding strategies.

Influence of ambient temperature-humidity index on sperm motility in Holstein bulls

A.M. Livernois[1,2], A. Canovas[2], S. Miller[3], F.S. Schenkel[2] and B.A. Mallard[1]

[1]University of Guelph, Department of Pathobiology, Ontario Veterinary College, Guelph, ON, Canada, [2]University of Guelph, Centre for Genetic Improvement of Livestock, Department of Animal Biosciences, Guelph, ON, Canada, [3]The Semex Alliance, 5653 Hwy 6 N., Guelph, ON, Canada; livernoa@uoguelph.ca

Heat stress in bulls has been shown to increase the incidence of sperm abnormalities and reduce fertilization rates. Examination of the effect of heat stress on semen quality could allow for its quantification and indicate the optimum conditions for semen collection based on animal and environmental characteristics. This type of investigation is increasingly important as future predictions for the patterns of heat waves point to increases in frequency, severity, and duration. As such, the objective of this study was to examine the linear and quadratic effect of temperature and humidity (combined into an index called the temperature-humidity index: THI), as well as bull age (7 classes) and ejaculate number on semen quality (post-thaw total motility) in a group of 2,365 Holstein bulls using a repeatability animal model, which also included year-barn-season and age class-barn-season of birth effects, in the ASReml software. Semen data from an artificial insemination centre collected over the years 2010-2018 were used. In total, 134,650 ejaculates from 2,365 bulls aged 7.5 months to 12 years old were analysed for post-thaw total motility. Daily average THI was calculated using hourly temperature and humidity data from Environment Canada weather stations. Significant quadratic regressions ($P<0.001$) of motility on the collection day average THI within age classes were found. Both the intercept and shape of the curves varied among age classes – older bulls were markedly negatively affected by high THI, while young bulls were, in contrast, mildly positively affected by high THI. The age class least affected by THI included bulls between 20 and 25 months old. Additionally, significant linear regressions ($P<0.001$) of motility on average THI within barn were observed, being negative for one barn and positive for the other two barns, suggesting that perhaps good climate management could reduce the impact of heat stress on post-thaw total motility in Holstein bulls.

Emission and mitigation of greenhouse gases from dairy farms: the cow, the manure and the field

M.A. Wattiaux, M.E. Uddin, P. Letelier, R.D. Jackson and R.A. Larson
University of Wisconsin-Madison, Madison, WI, USA; wattiaux@wisc.edu

In this review, we identify the primary drivers of the three main on-farm sources of greenhouse gases (GHG): the cow [enteric methane (CH_4)], the manure [CH_4 and nitrous oxide (N_2O)], and the field [N_2O and carbon dioxide (CO_2)]. Then, we explore the contribution of distinct feeding, manure handling, and cropping practices to mitigating emissions from the mixed livestock-crop dairy systems typical of the Midwest of the United States. Our last step is to summarize lessons learned from cradle-to-farmgate life cycle assessments (LCA) that compared specific management practices. Dry matter intake drives enteric CH_4 production (g/d) of cows, but reduction in CH_4 intensity (kg/kg of milk) is driven by feed conversion efficiency. Dietary carbohydrates and fats (and additives) influence enteric CH_4 whereas dietary crude protein influences N_2O. Better herd health, increased productive life, reduction in the size of replacement herd, and genetic selection for greater feed efficiency are effective herd-level mitigation practices. In the manure management chain, long-term storage of unprocessed liquid manure leads to greater GHG emissions from mid-size farms than from small farms with no storage. On large farms, the adoption of manure solid-liquid separation and anaerobic digestion reduce substantially the carbon balance. Nitrogen fertilization practices are key to reducing N_2O emissions from fields, which are highly variable depending on weather, type of soil, and soil conditions (temperature, moisture, nutrient content). Reduced or no till, introduction of perennial crops or winter crops in rotations that include *leguminosae*, and organic fertilization are management practices that reduce CO_2 losses from fields. Cradle-to-farm gate GHG emissions of 10 studies reviewed here averaged 1.1 kg CO_2-eq per kg of milk. Differences among farms within a system (e.g. conventional, grazing, organic) are large and often greater than among systems. Most current (attributional) LCA tend to favour intensification as a means to reduce milk carbon footprint. However new research tool will need to be developed if the goal is to reduce emissions of the dairy sector per unit of economic activity generated resulting from socially acceptable methods of production.

Comparative study on production traits between dual-purpose and dairy breeds in South Tyrol

I. Poulopoulou, C. Lambertz and M. Gauly
Free University of Bolzano, Faculty of Science and Technology, Piazza Università 5, 39100 Bolzano, Italy;
ioanna.poulopoulou@unibz.it

Mountain areas can for economic reasons not compete with other milk producing areas in Europe. Therefore, they need to improve their production systems on certain levels. The aim of the present study is to compare different systems on all levels (from performance to environmental impact) and propose recommendations for a sustainable dairy system. In a first step a breed comparison, including Holstein (H), Braunvieh (BV), Jersey (J), Fleckvieh (FV), Alpine Grey (AG), and Pinzgauer (PG), was performed during the years 2012-2017 on a dataset consisted of 1,372,393 observations. Milk yield was recorded and individual milk samples analysed monthly for fat, protein, lactose, casein, urea and somatic cell count. The concentrations of saturated fatty acids (SFA), mono- and poly-unsaturated fatty acids (MUFA, PUFA) were also determined. Cows categorized according to their parity (1, 2, 3, 4, $\geq$5). The year was allocated in four quarters: Q1 (December to February), Q2 (March to May), Q3 (June to August) Q4 (September to November). The highest milk amount was produced by H cows (25 ± 0.02 kg/d) and the lowest by AG (14 ± 0.03 kg/d). J cows had a significant higher fat ($4.9\pm0.01\%$) and protein ($3.9\pm0.01\%$) concentration while the lowest concentrations determined in AG and H, respectively. The effect of parity showed that milk yield at fourth parity was significantly higher (20 ± 0.04 kg/d) compared to all other parity groups while animals at first parity had the lowest milk yield (17 ± 0.03 kg/d). Season significantly affected milk traits, with the higher milk yield determined during Q2 and Q3 and lower during Q4, while fat and protein were significantly lower during Q3. The milk yield decreased as days in milk increased. SFA determined significantly higher in J cows (71%). Significantly, lower concentrations were estimated in PG (66%) and GV (67%). The concentrations of MUFA and PUFA estimated higher (26 and 3.3%) for PG while the lowest concentration determined in J cows (23 and 2.8%). In the next steps, additional parameters of dairy farms will be evaluated in order to propose strategies for their economic sustainability.

Consumer acceptance and willingness to pay for cow housing systems in eight EU countries

M. Waldrop and J. Roosen
Technical University of Munich, Chair of Marketing and Consumer Research, TUM School of Management, Alte Akademie 16, 85354 Freising, Germany; megan.waldrop@tum.de

Consumers are increasingly concerned about the welfare of livestock animals. Free walk housing systems for cows aim at creating higher welfare; however, little work has evaluated consumer preferences for cow housing systems in multiple countries. The purpose of the study is to estimate consumer acceptance and willingness to pay (WTP) of products from tie-stall, cubicle, compost bedded, and artificial floor housing systems. Moreover, consumer attitudes towards animal welfare, grazing, and re-using compost from the compost bedded system are assessed. Focus groups were held in October 2018 in Germany, Austria, and Slovenia to identify concepts for a quantitative survey. Countries were chosen to represent different consumer WTP and animal welfare attitudes based on previous survey results. Two 90-minute sessions of six to 10 people were held in each country. A quantitative survey is being conducted online in Austria, Germany, Italy, the Netherlands, Norway, Slovakia, Slovenia, and Sweden in early March 2019. There will be nationally representative samples of 400 to 600 participants per country. A choice experiment with milk as a representative product varying in price, grazing, housing type, and production type (organic and conventional) attributes is included for estimating WTP. Results from the choice experiment will be analysed using a mixed logit model. Participants also answer demographic, food purchase behaviour, and other attitude related questions. The focus groups generally agreed the tied-stall barn is the worst system, the cubicle system was slightly better, and the compost bedded or artificial floor were the best systems. Grazing was thought to be very important to the quality of milk and meat. Austria and Germany had some concerns with re-using the compost for other food products, whereas in Slovenia everyone approved. The quantitative survey will be used to gain deeper insights into different consumer attitudes and explanations for any varying WTP and animal welfare preferences. This research helps gain a better understanding of how European consumers value animal welfare and different cow housing systems, which can help farmers make more informed production decisions.

Explored and unexplored topics in mountain livestock farming scientific literature

A. Zuliani, G. Arsenos, A. Bernués, G. Cozzi, P. Dovc, M. Gauly, Ø. Holand, B. Martin, C. Morgan-Davies, M.K. Schneider and W. Zollitsch
EAAP Across Commissions Working Group on Mountain Livestock Farming, E4AP Via G. Tomassetti 3/A1, 00161 Roma, Italy; zuliani.anna.2@spes.uniud.it

Mountain areas represent more than 40% of European land. In the mountains, the primary sector has a key role in terms of employment, livelihood structures in rural communities, landscape management and conservation and provision of high quality food products. In addition, mountain farming is tightly bound to High Value Nature areas since farming activities significantly contribute to the maintenance of biodiversity and the provision of ecosystem services, due to their low-input/extensive character. However, these features limit productivity which impairs the economic viability of mountain agriculture, despite its pivotal role in providing key services for society. The new EAAP Mountain Livestock Farming Working Group performed a review of the existing scientific literature to investigate explored and unexplored topics in Mountain Livestock Farming (MLF) research. An electronic search on literature published from 1960 to 2017 was carried out using Scopus database. A string of key words Livestock and Farm and Mountain was chosen to retrieve the most relevant scientific papers. A total of 545 papers were identified. The year with more publications was 2014 with 47 papers. During this timeframe, 20 peer reviewed scientific journals have published at least 5 papers related to MLF. Text mining analysis on the 547 available abstracts aimed at identifying the most frequent words used in these papers. Most of them were dealing with livestock management (e.g. sheep, cattle, transhumance, stocking density) and with landscape management (e.g. land, forest, grazing, water). Fewer papers were targeting socio-economic (e.g. livelihood quality and income) and environmental sustainability issues (e.g. climate change, biodiversity). The results of this exercise highlight a growing interest towards mountain farming related topics by the scientific community and reveal the need to build on current knowledge by engaging with new disciplines and expertise in order to strengthen the key role of MLF for the survival of agroecosystems and local communities.

Livestock farming systems and the French society: key controversies

E.D. Delanoue
Institut de l'Elevage, ifip Institut du Porc, Itavi, 149 Rue de Bercy, 75595 Paris, France; elsa.delanoue@idele.fr

For several years, French livestock farming has been frequently questioned by society. Those questions concern its environmental impact, sanitary risks or animal treatment and ask, more generally, livestock farming's place among a society that is more and more concerned about its alimentation. To understand this phenomenon, analyse representations on livestock farming that coexist within the society and enlighten agricultural actors on those social evolutions at work, the project ACCEPT, funded by the CASDAR and led by the French Pork and Pig Institute (IFIP), has been carried out between 2014 and 2018. With the aim of identifying the subjects of controversy about breeding in France, all animal productions considered, and to describe the diversity of actors and arguments, both qualitative and quantitative studies were conducted: around seventy interviews were conducted, in France and five other UE countries, with the main stakeholders (livestock farming professionals, farmers, companies, journalists, NGOs, consumers) and more than 2007 French citizens were questioned on a online survey. The analysis of discourses led to classify the debates on the French livestock farming in four major areas: environmental impact, animal welfare, risks to human health and socioeconomic model of livestock farming. It reflects expectations for different types of system: some want a gradual disappearance of intensive farming systems for the benefit of under official quality signs or implementing alternative practices; others want the development of intensive farming to produce more and become more competitive; and between them many want a gradual improvement in the intensive system, with stronger environmental and animal welfare requirements. We consider that we face a global controversy on livestock farming: indeed, beyond specific controversies on practices, debates regarding the very legitimacy of livestock farming or the consumption of animal products gather all the different topics, and are linked with larger society problematics (consumption habits, growth models, globalization, etc.). This feature may complicate the resolution of the controversy and make it hazardous and longer.

Artificial grass increases dairy cows' walking speed

S. Buijs and D. McConnell
Agri-Food and Biosciences Institute, Large Park, BT26 6DR Hillsborough, United Kingdom; stephanie.buijs@afbini.gov.uk

Dairy cows often walk long distances between parlour and pasture. A softer, less abrasive walking surface may improve cow comfort, resulting in an increased walking speed. Knowledge on how laneway surface types affect dairy cow locomotion is currently lacking. This cannot be extrapolated from studies on barn floor types, as both the way the cow uses the surface and the available top-surface materials differ. To evaluate how a softer top-surface affects walking speed, a standard laneway (with a stonedust-over-gravel surface) was split into two sides and one side was covered with artificial grass. Dairy cows were each observed twice as they returned from parlour to pasture after their morning milking, once on each surface type. Video recordings were used to determine walking speed on a predefined 20 m long section of the laneway. In addition, the number of times one of the forelimbs was placed down on the 20 m stretch was counted as a rough index of step length. Speed and step length were determined for 31 cows that were observed on each surface type without stopping, trotting, cantering, or being too close to the preceding cow (and thus, likely influenced by her walking speed). Wilcoxon signed-rank tests were used to analyse the data, using each cow as her own control. Cows walked 5% faster on the artificial grass than on the standard surface (grass: 5.3 km/h (IQR: 5.1-5.5), standard: 5.1 km/h (IQR: 4.5-5.3), P=0.024). Although there was a numerical indication of longer steps on the artificial grass (8 cows used less front limb placements on the artificial grass than on the standard surface, whilst only 3 cows used more), this difference did not reach significance (P=0.12). This may be due to the rough index of step length we used. Alternatively, the greater walking speed may be partially due to quicker (instead of larger) steps. In conclusion, a softer, less abrasive laneway surface can increase dairy cows' walking speed between parlour and pasture. As reduced walking speed is a symptom of lameness, this may mean that the softer surface alleviated lameness. This will be analysed further by studying the cows' mobility score on the different surface types and its association with differences in walking speed.

The new EAAP across commissions working group on mountain livestock farming

A. Zuliani, G. Arsenos, A. Bernués, G. Cozzi, P. Dovc, M. Gauly, Ø. Holand, B. Martin, C. Morgan-Davies, M.K. Schneider and W. Zollitsch
EAAP Across Commissions Working Group on Mountain Livestock Farming, EAAP, Via G. Tomassetti 3/A1, 00161 Roma, Italy; zuliani.anna.2@spes.uniud.it

The new EAAP Mountain Livestock Farming (MLF) across-commissions Working Group was established in late 2018 following the inputs received during the 1st Mountain Livestock Conference held in Bolzano, Italy, in June 2018 and the 69th Annual Meeting of EAAP held in Dubrovnik, Croatia, in August 2018. This initiative aims at joining research efforts in the field of mountain livestock farming, fostering transdisciplinary collaborations across disciplines, the development of new research ideas and the dissemination of relevant projects and results related but not limited to: animal nutrition, farming systems, land use/management, biodiversity conservation, organic/low input production, food quality, economy/policies and environmental sustainability/climate change in mountain areas. The working group is now in a process of engaging experts in the many fields that are relevant for MLF, identifying emerging areas of research, and encouraging collaboration among researchers from different countries and disciplines. It wants to be a welcoming platform for discussion on livestock production in mountain areas in a broad and transdisciplinary perspective. A further mandate is to disseminate relevant outcomes by organizing conferences (e.g. Mountain Livestock Conference), conference sessions (e.g. within EAAP) and joined events with other related initiatives that will enable researchers and stakeholders to contribute and exchange their new knowledge, experiences and innovations.

Understanding maternal behaviour of cattle to optimize calving management and ensure animal welfare

M.V. Rørvang[1], B.L. Nielsen[2], M.S. Herskin[3] and M.B. Jensen[3]
[1]Swedish University of Agricultural Sciences, Biosystems and Technology, Sundsvägen 16, 23053 Alnarp, Sweden, [2]INRA, UMR Modélisation Systémique Appliquée aux Ruminants, AgroParisTech, Université Paris-Saclay, 75005 Paris, France, [3]Aarhus University, Animal Science, Blichers Allé 20, 8830 Tjele, Denmark; mariav.rorvang@slu.se

Successful reproduction is central for a sustainable dairy production. Housing peri-parturient cows in individual pens have been suggested as a means to ensure a successful calving and safeguard animal welfare. This practice is recommended in several countries worldwide. However, the underlying pre-partum maternal motivations of dairy cows are currently not fully understood and management of calving cows in individual calving pens can be a challenge. These studies reviewed the behaviour of parturient cows and their use of maternity pens to suggest future improvement of calving management. Pre-parturient cows in a commercial calving environment are affected by a number of factors modulating their pre-partum maternal motivation and behaviour and their use of a calving facility. Examples of these are the physical environment, disturbances from group-mates, as well as olfactory cues from for example birth fluids. The behaviour observed prior to calving is likely influenced by a motivation to locate an appropriate calving site, as cows appear to seek, by means of a combination of distance from the herd and physical cover, to isolate themselves from disturbances or potential threats (e.g. predators, other cows). Irrespective of the chosen types of calving pen, this motivation may increase with increasing levels of disturbance. Future research into the decision-making of cows prior to calving is recommended to answer questions such as: Do parturient cows asses their environment in terms of degree of physical cover, distance from the herd, risk of disturbances, or perhaps ability to escape? If so, how do they obtain this information and how do they weigh up these factors? This knowledge may be used to design calving facilities, for example if provision of hides allows cows to fulfil the motivation to locate an appropriate calving site. It may be possible to facilitate entry into individual secluded areas by exploiting other motivations of the cow, such as attraction to birth fluids.

Different techniques to determine the sperm concentration in an artificial insemination center

E. Henrotte, C. Latour, C. Bertozzi and C. Boccart
The Walloon breeding association, Research and development, Chemin du tersoit 32, 5590, Belgium; ehenrotte@awenet.be

Number of straws produced in an AI (artificial insemination) centre are dependent on sperm concentration that is usually measured by a photometer on the fresh semen. The aim of the present work was to evaluate how precise is the concentration obtained by the photometer, in comparison to other existing techniques. A dilution curve of undiluted and diluted fresh semen was established using NaCl on 3 highly concentrated semen. The concentration was determined in triplicates using the photometer (Minitüb), the CASA system IVOS II (computer assisted sperm analyser, IMV Technologies), the nucleocounter (Chemometec) and the flow cytometer (Easycyte, IMV Technologies). None of the alternative methods were significantly different from the photometer, and the coefficients of correlation for the dilution curve were all very high (>0.98). Nevertheless, the concentrations given by the CASA system were significantly lower than the ones predicted by the nucleocounter. In terms of repeatability (coefficient variation = CV), the nucleocounter and the photometer were very efficient, with 5.02 and 3.02% CV respectively, by comparison with the CASA system and the flow cytometer that were more variable (9.80 and 9.19% CV, respectively). Further assays were done on the concentration of the commercial straws, for the quality control. In that case, only the CASA system, the nucleocounter and the flow cytometer were compared. The results obtained with the nucleocounter were closer to the theoretically values predicted by the photometer, and were not significantly different from the ones given by the flow cytometer. The concentrations measured with the CASA system were significantly lower (P<0.005). The repeatability was better with the flow cytometer and nucleocounter measurements (6.18 and 6.14% CV respectively), than with the CASA system (CV: 10.22%). In conclusion, from a price value and repeatability point of view, the photometer is recommended to measure the concentration of the fresh semen. Concerning the post thawing control quality, flow cytometry or nucleocounter are more reliable. Thanks are due to Alycia Marteau for her technical assistance during the experiment.

The good practices in artificial insemination: straw handling and its consequences on semen quality

E. Henrotte, C. Latour, C. Bertozzi and C. Boccart

The Walloon breeding association, Research and Development, 32 Chemin du tersoit, 5590, Belgium; ehenrotte@awenet.be

Responsible artificial insemination (AI) organizations strive to ensure that semen sold has the potential to achieve acceptable levels of fertility when used in herds of fertile cows and heifers. Accordingly, straw handling by the inseminator should be limited, in order to reduce the damages to the semen quality. This study tested different conditions of handling before AI: (1) straws are kept out of the liquid nitrogen during 0-5-10-15 sec before thawing (Time condition); (2) straws are kept out of the liquid nitrogen 0-3-6-9 times during 5 sec before thawing (Frequency condition); (3) or straws are kept in vapour nitrogen (instead of liquid) during 24-48 h before previous described handling conditions and thawing. The semen tested were provided by 5 bulls (Belgian blue and Holstein bulls), 3 straws per bull for each condition. The motility (using CASA system- computer assisted sperm analyser), the viability, the acrosomal integrity and the mitopotential (using flow cytometer) were measured. Data were analysed by analysis of variance. Semen quality was not significantly affected by straw handling (Time and Frequency conditions) under liquid nitrogen storage. After 24 h in vapour nitrogen storage condition, the acrosomal integrity and the mitopotential level significantly decreased even without additional straw handling, while keeping the straws during 48 h in vapour nitrogen with additional handling (the straws were kept out of the tank 9 times during 5 sec) decreased the semen quality for all the measured parameters. The results of this study suggest that caution should be taken before AI by the users, especially if the straws are maintained under atmospheric vapour of nitrogen instead of liquid nitrogen immersion, to preserve the fertility potential of the bulls. Thanks are due to Mathilde Macors for her technical assistance during the experiment.

Performance and oxidative status of beef cows facing short nutritional challenges during lactation

I. Casasús, K. Orquera, J.R. Bertolín, J. Ferrer and M. Blanco

Centro de Investigación y Tecnología Agroalimentaria de Aragón-IA2, Montañana 930, 50059 Zaragoza, Spain; icasasus@cita-aragon.es

The performance and plasma oxidative status of 16 Parda de Montaña suckler cows were analysed in response to a 4-day restriction in months 2, 3 and 4 post-calving, in order to evaluate their strategies to cope with undernutrition throughout lactation. Prior to restriction and after the challenge, the cows received a diet meeting 100% of their energy requirements (7.0 kg DM hay, 2.7 kg DM concentrate). In the 4-day challenge, the diet met 55% of cow requirements (6.2 kg DM hay). Dam and calf live weight (LW), dam milk yield (MY, weigh-suckle-weigh technique) and plasma malondialdehyde (MDA) were measured twice the week before the restriction (basal), daily during the 4-d restriction and on the first 2 days of realimentation. Basal dam LW, milk yield and MDA concentrations decreased from month 2 to 4 (P<0.001), implying that decreasing performance after peak production reduced the oxidative status. Dam LW dropped immediately on the first day of restriction and did not recover by day 2 of realimentation in any month. Milk yield also dropped during restriction, and thereafter recovered the basal values by day 2 of realimentation in months 2 and 3 but not in month 4. The percent reduction of MY during restriction increased through lactation (-11.4, -16.0 and -21.0% in months 2, 3 and 4, P<0.001) and recovery was lower in month 4 than the rest (+3.5, +1.6 and -9.8%, P<0.001). Calf gains decreased during restriction in months 2 and 4 and did not fully recover during realimentation, but remained unchanged in month 3. Plasma MDA, indicator of lipid peroxidation, fell to a minimum at the start of restriction, increased sharply to a maximum in days 3 or 4 of restriction and returned to basal values in realimentation in all months. The difference between maximum and minimum MDA concentrations increased as lactation advanced. Although other indicators of nutritional status and inter-individual variations remain to be analysed, these preliminary results indicate that the patterns with which beef cows cope with short but severe nutritional challenges change throughout lactation.

Effects of isoquinolene alkaloids on antioxidative status and inflammation of dairy cows

H. Scholz[1], A. Ahrens[2] and D. Weber[3]
[1]Anhalt University of Applied Sciences, faculty LOEL, Strenzfelder Allee 28, 06406 Bernburg, Germany, [2]Tierseuchenkasse Thüringen, Victor-Goerttler-Straße 4, 07745 Jena, Germany, [3]Universität Leipzig, Veterinärmedizinische Fakultät, An den Tierkliniken 1-43, 04103 Leipzig, Germany; heiko.scholz@hs-anhalt.de

It has been demonstrated that isoquinoline alkaloids (IQ), secondary plant components extracted from *Macleaya cordata* (*Papaveraceae*), have anti-inflammatory and immunomodulatory effects in various species. Due to their possible positive effects in ruminants the main objective of this study was to determine the effect of supplementation with an IQ-containing plant extract (IQPE) on antioxidative status and inflammation of dairy cows during the transit period. The experiment was conducted in a commercial farm in Thuringia with more than 600 milking cows. A pool of 93 cows were randomly divided into two groups: control group (CG, without additive) and experimental group (EG) supplemented with 800 g IQPE/cow/d (Sangrovit®) from 3 weeks ante partum to 15 weeks post-partum. Both groups were fed the same total mixed ration (TMR) according to the recommendations of GfE (2009). Data were statistically analysed by use of SPSS (21. 0) and R (3. 2. 2). Parameters of metabolism, with the only exception of CRP, were influenced by time of sampling ($P<0.05$). Several of those parameters (BHBA, retinol, TAG, ASAT, γ-tocopherol, TEAC, TBARS) were also influenced by cow parity ($P<0.05$). Supplementation of Sangrovit (EG) has less effects on the plasma variables determined. The only significant change due to Sangrovit (EG) supplementation over all the sampling times was an increase of the albumin concentration in plasma (28.8 vs 27.6 g/l, in average of the four sampling times). The present study indicates that use of alkaloids in dairy cows during the transition period show no effects on metabolic status.

Effect of breed type in grass-fed production systems

G. Scaglia
LSU AgCenter, Iberia Research Station, 70544, Jeanerette, USA; gscaglia@agcenter.lsu.edu

The Gulf Coast region has a humid subtropical climate with temperature humidity index (THI) that can cause extreme heat load for most of the summer. Breed adaptation may help improve beef production for grass-fed beef operations. Eighteen steers from each Angus, Holstein, Brangus and Pineywoods were evaluated during three consecutive summer grazing seasons while rotationally stocked on 'Jiggs' bermudagrass pastures (3 replicates per breed type) at the same grazing pressure (3,500 kg bodyweight (BW)/hectare). Dry matter intake (DMI) was determined in three 7-day periods during the grazing season by pre and post grazing sampling. At the same time, bite rate (visually) and grazing behaviour was recorded using IceTag monitors (v.2.004) on two steers per replicate. A weather station located near the site provide data that served to determine THI. Data indicated that cattle were in mild heat stress (THI>72) between 6 am and midnight and on severe heat load (THI>78) between 6 am and 6 pm in July and August. Brangus and Pineywoods are breeds adapted to heat and humidity. There were no differences ($P=0.07$) in DMI (% BW) between breed types (1.97, 1.80, 2.01, 1.75% for Brangus, Angus, Pineywoods and Holstein, respectively). Bite rate (bites/minute) was not different ($P=0.09$) between breed types; however, it was smaller ($P=0.03$) at 35.7 bites/minute in period 2 (late July), when compared to period 1 (mid-June) and 3 (late August-early September) at 39.8 and 42.1 bites/min, respectively. Throughout the grazing season, grazing time (minutes/day) was longer ($P=0.003$) for Pineywoods (542) and Brangus (502) than Angus (473) and Holsteins (407), while Holstein and Angus steers spent more time ($P<0.05$) standing and less time lying down. Average daily gains were greater for Angus (0.61 kg) than for Brangus and Holstein (0.57 and 0.50 kg, respectively) and smaller for Pineywoods (0.40 kg); however, production per hectare was greater ($P=0.04$) for Pineywoods (442 kg/ha) and Angus (421 kg/ha) than for Brangus and Holstein (391 and 380 kg/ha, respectively). Pineywoods (a heritage breed of small mature size and lean beef) might be a great alternative for small farms with limited resources while providing new genetics with great potential for grass-fed beef production in the southeastern US.

Peri-implantational subnutrition decreases dam IGF-1, but does not impair fertility in suckled cows

A. Sanz, A. Noya, J.A. Rodríguez-Sánchez and I. Casasús
CITA de Aragón – IA2, Avenida Montanana 930, 50059 Zaragoza, Spain; asanz@aragon.es

The effect of a negative energy balance during the first term of pregnancy on metabolic and endocrine parameters and pregnancy was studied in two beef cattle breeds, Parda de Montaña (PA, n=75) and Pirenaica (PI, n=40). Seventy-six days after calving, lactating cows were synchronised with a Cosynch protocol and artificially inseminated (AI). Then, cows were allocated to two diets (CONTROL (100% of their requirements) or SUBNUT (65%)) until day 82 of gestation. Cow and calf live-weights, and dam body condition score (BCS) were recorded fortnightly. Dam blood samples were assayed for glucose, cholesterol, NEFA, β-hydroxybutyrate and urea (fortnightly); insulin-like growth factor 1 (IGF-1) (monthly); and progesterone (days 14, 18, 21, 28, 42, 56, 69, 82 post-AI). Pregnancy was recorded 37 days post-AI by ultrasounds. Pirenaica cows showed higher BCS at AI than PA (2.9 vs 2.7, P<0.001). CONTROL cows maintained BCS and live-weights, whereas SUBNUT had negative daily gain (0.12 vs -0.36 kg, P<0.001). CONTROL lactating calves had higher daily gains than SUBNUT (0.63, 0.56, 0.62 and 0.44 kg for PA-CONTROL, PA-SUBNUT, PI-CONTROL and PI-SUBNUT, P<0.05). An interaction occurred in metabolites and IGF-1 profiles among breed, nutrition and time. SUBNUT cows showed higher NEFA values than CONTROL ones. NEFA values on day 56 were related to BCS at AI (r=0.39, P<0.001). From day 69 onwards, PI-SUBNUT had lower cholesterol than PI-CONTROL (P<0.05). At day 82 post-AI, IGF-1 values were higher in CONTROL groups (74.6, 57.2, 90.4 and 73.6 ng/ml, for PA-CONTROL, PA-SUBNUT, PI-CONTROL and PI-SUBNUT, P<0.001). Surprisingly, pregnant dams showed lower IGF-1 values than non-pregnant ones (70.4 vs 92.5 ng/ml, P<0.01). Progesterone level was greater in pregnant cows from d 21 on (P<0.001), providing the earliest and accurate day to pregnancy diagnose. A high fertility rate was obtained (77%) with no breed, nutrition or sire effect. Peri-implantational subnutrition affected cow performance and physiological profiles, and impaired lactating calf growth, but did not damage progesterone level from day 21 post-AI or fertility rate, confirming that undernourished pregnant dams prioritize the partition of dietary energy to reproductive functions.

Effects of pasture grazing on growth performance and physicochemical traits in Hanwoo

J.S. Jung, K.C. Choi, B.R. Choi and W.H. Kim
National Institute of Animal Science, Grassland and Forage Division, 114, Sinbang 1-gil, Seongwhan-eup, Seobuk-gu, Cheonan-si, Chungcheongnam-do, 31000, Korea, South; jjs3873@gmail.com

This study investigated the effect of pasture grazing on growth performance, quality grade factors and physicochemical traits of longissimus dorsi in 10 Hanwoo (native Korean cattle) steers. Steers were randomly assigned to two groups (grazing and housing). In the grazing group, steers had access to pasture and were supplemented with concentrates (1.2% of body weight (BW) per day) during the rearing period. Pasture grazing (PG) was performed at a 7 ha grassland from April to November by rotational grazing method. The housed group (Barn feeding system, BFS) was fed with rice straw and concentrate (1.5 and 1.2% of BW per day, respectively) during the rearing period. In the fattening period (for 10 months after growing stage), the grazing group was supplied with fermented TMR feed while housed group received rice straw and concentrates. Initial BWs of BFS and PG were 231±20.33 kg and 245±22 kg, respectively. Average daily gains of both groups did not differ significantly, at 0.85 kg and 0.87 kg, respectively. Final BWs of BFS and PG (27 months) were 728±36.79 and 732+26.97, respectively. The carcass weight and longissimus muscle back-fat thickness were higher in PG. Waner-Bratzler shear force was higher in PG than that of BFS but other physicochemical traits of m. longissimus dorsi were not different among all treatments. The ratio of w-6/w-3 fatty acids was lower in PG, but amino acids were not different between BFS and PG. These results indicate that grazing from April to November in the rearing period had no effect on physicochemical traits of Hanwoo steers.

Development and structural change of dairy farms sector in Province of Vojvodina

D. Kučević, S. Trivunović, M. Plavšić, D. Janković, L.J. Štrbac and M. Radinović
University of Novi Sad, Faculty of Agriculture, Department of Animal Science, Trg D.Obradovića 8, 21000 Novi Sad, Serbia; denis.kucevic@stocarstvo.edu.rs

There are many methods that could be used to measure structural change in the milk producer sector. Traditionally, one of the most acceptable has been placed on tracking changes in farm numbers, herd sizes (the number of cows on farms), asset values, and employment in the sector. In addition, activities of the Main breeding organization and thereby applied breeding programs in recorded cattle's population has made a huge impact on the development of the dairy sector in Province of Vojvodina. Data provided by the main breeding organization (recorded population) at the Faculty of Agriculture, University of Novi Sad and Statistical Office of the Republic of Serbia (total population) was used as a material for this research. Collected data were studied and analysed by methods of descriptive statistics within the Statistica software package. The dairy farms sector ranged from the existence of large state farms – several thousand cows per farm to the end of the 1980s, to the breakdown of sector into small farmers with one to five heads, in the nineties, and the re-establishment of the system of organized production and larger privates farms after 2000. In this regard, during the mentioned period has decreased the total number of dairy cows and heifers by 54%. In the meanwhile, the number of recorded and total dairy cows in Vojvodina has trended upward by around 25%. Dairy farms located in the north of the country are typically diversified crop/livestock agricultural holdings. A trend in development of milk production in Vojvodina goes in the direction of decreasing the number of agricultural holdings which raise breeding stocks of dairy cows and simultaneous increase in the number of cows per farms engaged in milk production. While the number of dairy farms with small herds (up to 5 cows, from 5 to 10 and 10 to 15 dairy cows) has been declining, the number of farms with large herds (from 15 to 50, from 50 to 100 and over 100 dairy cows) has been increasing. With increasing size of farms increases the milk yield per cow.

Influence of temperature-humidity relations during year on milk production and quality

M. Klopčič
University of Ljubljana, Biotechnical Faculty, Dept. of Animal Science, Groblje 3, 1230 Domžale, Slovenia; marija.klopcic@bf.uni-lj.si

The objective of this study was to assess the relationship of temperature-humidity index (THI) with milk production and cow behaviour in different seasons of the year. Two herds of Holstein-Friesian cows in two different housing systems (loose housing system vs compost barn system) were monitored for more than one year. The heat stress reduced daily milk yield and milk content during the summer when THI was higher than 68. In conclusion, summer heat stress negatively affects milk yield and milk content of dairy cows. Therefore, management strategies are needed to minimize the heat stress and attain optimal dairy cow productivity.

Cooling of fattening pigs during warm thermal conditions improves behaviour and environment

A.-C. Olsson[1], K.-H. Jeppsson[1] and A. Nasirahmadi[2]
[1]*Swedish University of Agricultural Sciences, Department of Biosystems and Technology, P.O. Box 103, 230 53 Alnarp, Sweden,* [2]*University of Kassel, Department of Agricultural and Biosystems Engineering, Nordbahnhofstr. 1a, 37213 Witzenhausen, Germany; anne-charlotte.olsson@slu.se*

In Sweden fattening pigs are kept in pens with partly slatted floors. These pens permit usage of straw for bedding and allow the pigs to lie down on solid floor surfaces. Straw to pigs in barren environments stimulates the rooting and exploration behaviour, which is positive for the welfare of the pigs. Pens with partly slatted floor show lower levels of ammonia compared to pens with fully slatted floor. This is partly explained by a reduced slurry pit area. However, during warm thermal conditions for the pigs, problems with fouled pens and high ammonia emissions arise. Heat stressed pigs try to cool themselves by lying on the slatted floor instead of the solid floor. They also start to urinate and defecate on the solid floor since the slatted floor is occupied and/or since the pigs want to wet their skin to cool down. Within the ERA-NET SusAn project PigSys, the overall aim is to develop a decision support system for optimal climate control in the house to raise fattening pigs in a more animal-friendly and resource-efficient way. One part of the project is to test technical solutions for improving pen hygiene in partly slatted pens for growing-finishing pigs. The pigs are either cooled by sprinkling of low-pressure water above the slatted floor or by higher air velocity on the lying area (convective cooling), or both. The effect on pig behaviour (e.g. standing, lying posture and position in the pen) is evaluated by means of an automated machine vision technique and correlated with pen hygiene and ammonia emission. The studies are performed in a commercial fattening pig house with 10 identical compartments with 16 pens and 160 pigs per compartment. Two compartments are filled simultaneously (control versus treatment). Preliminary results from the first summer, demonstrate that the automated monitoring technique, could be a reliable method for monitoring of pigs under commercial farm conditions. It is shown that water sprinkling on the slatted area changes the pig behaviour, reduces the pen fouling and decreases the ammonia emission.

Is immunocastration a reliable and sustainable alternative for pig production?

K. Kress[1], U. Weiler[1], M. Čandek-Potokar[2], N. Batorek Lukač[2], M. Škrlep[2] and V. Stefanski[1]
[1]*University of Hohenheim, Garbenstr. 17, 70599 Stuttgart, Germany,* [2]*Agricultural Institute of Slovenia, Hacquetova 17, 1000 Ljubljana, Slovenia; kress.kevin@uni-hohenheim.de*

The study was designed to test the reliability of immunocastration in pigs under enriched, standard and stressful (mixing of groups) housing conditions evaluating endocrine and growth parameter (ADG) and welfare. The study was carried out with immunocastrates (IC, n=48) and entire males (EM, n=48) in 2 consecutive trials, 6 animals were assigned to each housing by treatment group/trial. For mixing, 2 groups/trial were required and the animals were mixed around the 1st (V1) and 2nd vaccination (V2). ADG was evaluated during 3 growing/fattening stages. In addition, 4 blood samples per pig were collected by venipuncture during fattening (B1-B3) and at slaughter (B4) to measure GnRH-antibody levels (^{125}I-GnRH binding), testosterone and cortisol levels (^{3}H-RIA). At slaughter, fat samples were collected for androstenone and skatole determinations (HPLC) and penile injuries were evaluated. Statistical analysis was performed by using U-Test, H-Test and Spearmans Rho (SPSS). Testosterone levels in IC (0.34 ± 0.03 ng/ml plasma; mean+SEM) and GnRH-binding ($48.44\pm0.88\%$) at slaughter revealed, that immunocastration was successful under all housing conditions. Thus, all IC had low androstenone and skatole levels (A:<0.24μg/g; S:<0.03μg/g liquid fat). Compared to EM, IC had lower plasma testosterone concentrations 2 weeks after V2 and at slaughter (EM vs IC: B3: 3.65 ± 0.60 vs 0.13 ± 0.01 ng/ml; B4: 11.50 ± 1.15 vs 0.34 ± 0.03 ng/ml; P$\leq$0.001). EM had more penile injuries than IC (EM vs IC: 5.19 ± 0.98 vs 1.60 ± 0.39 injuries/animal; P$\leq$0.001). After V2, IC had higher ADG than EM (990 ± 10 vs 830 ± 30 g/d; P$\leq$0.001) and were heavier before slaughter than EM (131.2 ± 1.43 vs 125.4 ± 1.74 kg live weight (LW); P$\leq$0.05). During fattening stage (90 kg LW – slaughter), EM under stressful housing conditions (n=24) had lower ADG than EM under enriched (n=12) housing conditions (770 ± 40 vs 960 ± 40 g/d; P$\leq$0.01; Bonferroni-corrected). We therefore conclude that immunocastration is a reliable and sustainable alternative to meet the objectives of pork markets. These results were accomplished within SuSi project, financed by SusAn ERA-NET.

Protein requirements in finishing immunocastrated male pigs

E. Labussière[1], L. Brossard[1] and V. Stefanski[2]
[1]Pegase, INRA, AGROCAMPUS OUEST, 16 Le Clos, 35590 Saint-Gilles, France, [2]Institute of Animal Science, Behavioral Physiology of Livestock, University of Hohenheim, Garbenstraße 17, 70599 Stuttgart, Germany; etienne.labussiere@inra.fr

Immunocastration is a method for raising male pigs without surgical castration and with reduced boar taint. It consists in two vaccinations against GnRH. After the second vaccination (V2) performed a few weeks before slaughter, pigs increase their feed intake whereas protein deposition remains constant. The dietary protein content should then be reduced to limit N excretion but the kinetics of protein reduction may differ between animals. The objective of the trial was to determine individually the optimal dietary amino acid to energy ratio. For that purpose, 75 male pigs were housed in a stable equipped with one weighing scale and eight automatic feeders filled with two diets varying in their amino-acid content (0.96 and 0.40 g DIS lys/MJ NE). The automatic feeders allowed mixing the two diets in proportions depending on the body weight of the pig and the expected daily gain and feed intake. From 30 kg body weight to V2, individual proportions of the mixture were determined. The V2 was performed at 127 days of age (74.3 kg body weight) and animals were slaughtered at 164 days of age. From V2 onwards, the pigs were allocated to 3 groups, that either 1/received until slaughter a mixture of the two diets whose proportions equalled the individual proportions measured the week before V2; 2/received a mixture of the two diets, whose proportions were daily and individually calculated as before V2 or 3/were proposed the high and the low protein diets so that each pig can adapt itself the composition of its intake. Feed intake from V2 to slaughter did not significantly differ between groups (3.42 kg/d). The BW at slaughter (115.8 kg) and carcass weight (91.3 kg) did not also differ between groups. The lysine intake was the highest for group 1 (28.4 g/d) and the lowest for group 2 (21.5 g/d). Finally, the efficiency of lysine deposition was higher for groups 2 and 3 than for group 1 (59, 55 and 47% of lysine intake, respectively), which suggests that dietary protein level can be decreased up to 25% after V2 in immunocastrated pigs. This study is part of the ERA-NET project SuSI (ANR-16-SUSN-0004).

Effect of diet and castration method on sensory meat quality

M. Aluwé, S. Millet and A. Van Den Broeke
Flanders Research Institute for Agriculture, Fisheries and Food (ILVO), Animal sciences unit, Scheldeweg 68, 9090, Belgium; marijke.aluwe@ilvo.vlaanderen.be

In most countries, surgical castration is still performed to eliminate boar taint, an off-odour present in pork from non-castrated or entire male pigs. However, alternatives are sought in the light of animal welfare as this method is painful. One alternative is immunocastration, which enables the production of male pigs without boar taint and without surgical castration by applying a 2-shot anti-GnRH vaccine. Immunocastrated male pigs show more efficient growth, because of increased muscle and lower fat deposition compared to surgical castrated pigs, resulting in lower intramuscular fat content and consequently may result in slightly tougher pork. In this study, we evaluated the effect of a low and a high energy diet both in barrows and immunocastrated male pigs (n=30 per treatment group). All samples were evaluated by 6 experts for frying odour and flavour, pig odour and flavour, boar taint odour and flavour, acidness, tenderness, juiciness and overall tastiness. Consumers evaluated the effect of diet (low vs high energy content) in a preference test organized per sex (barrows vs immunocastrates). In each of 30 sessions, 3 consumers evaluated two pairs – a pair of barrows and a pair of immunocastrates – and indicated their preference based on their visual expectations and after tasting, followed by scoring their liking of each sample. Based on the preliminary results of the experts with 10 animals per treatment group, the overall tastiness score was higher for barrows compared to immunocastrated male pigs (P=0.034) and tended to be higher for pigs fed a low compared to a high energy diet (P=0.052). Consumers did not have a strong preference for one of the diets, with 53.3% of the consumers preferring high energy diet for the barrows and 51.1% for the immunocastrated male pigs. Further evaluation of the results will further clarify the potential of this feeding strategy on pork quality. The research was done within ERA-Net Cofund SusAn (696231) Sustainability in pork production with immunocastration (SuSI).

Characterisation of adipose tissue in immunocastrated pigs

K. Poklukar[1], M. Čandek-Potokar[1], N. Batorek Lukač[1], M. Vrecl Fazarinc[2], G. Fazarinc[2], K. Kress[3], U. Weiler[3], V. Stefanski[3] and M. Škrlep[1]

[1]Kmetijski inštitut Slovenije, Hacquetova 17, 1000, Slovenia, [2]Univerza v Ljubljani, Veterinarska Fakulteta, Gerbičeva 60, 1000 Ljubljana, Slovenia, [3]Universität Hohenheim, Garbenstr. 17, 70599 Stuttgart, Germany; klavdija.poklukar@kis.si

The objective of the study was to evaluate the effect of immunocastration on fat accretion by comparing immunocastrated pigs (IC; n=12) to entire males (EM; n=12) and surgical castrates (SC; n=12), all of Landrace × Pietrain crossbreed. IC pigs were vaccinated at the age of 12 and 21 weeks and all sex categories were slaughtered at the age of 26 weeks reaching 121.7±1.6 kg. Adipose tissue (AT) thickness was measured on the carcass split line. AT on withers (both layers) was sampled for determination of lipogenic enzyme (LE) activity, fatty acid (FA) composition and histomorphological analysis. The data were analysed by one-way ANOVA using R statistical software. IC exhibited lower AT thicknesses at the level of the last rib ($P<0.05$) and Gluteus medius muscle ($P<0.05$) than SC. Additionally, IC exhibited higher PUFA ($P<0.05$) and lower SFA ($P<0.05$) than SC. At the histological level, adipocyte and fascicle surface area in the outer backfat layer, as well as fascicle surface area in the inner backfat layer were higher in IC compared to EM ($P<0.05$). There was a trend of higher number of adipocytes per individual fascicle in the outer backfat layer ($P<0.10$) of IC than EM. These results are consistent with 1.4 to 2.7-fold higher LE activity (fatty acid synthase, malic enzyme, glucose 6 phosphate dehydrogenase and citrate synthase) in IC than EM but also SC. Altogether, in comparison to EM, immunocastration influenced AT morphology and lipid metabolism, as exhibited by larger adipocytes/fascicles and higher LE activities. The differences between the IC and SC were noted for AT thicknesses, FA composition and LE activities. These results were accomplished within project SuSI, co-financed by Susan EraNet and Slovenian Ministry of Agriculture, Forestry and Food. Collaboration within COST action CA12215 IPEMA, and core financing of Slovenian Research Agency (grant P4-0133 and P4-0053, PhD scholarship for KP) are also acknowledged.

Towards the valorisation of pork with boar taint

L.Y. Hemeryck[1], J. Wauters[1], L. Dewulf[2], A.I. Decloedt[1], M. Aluwé[3], S. De Smet[4], I. Fraeye[2] and L. Vanhaecke[1,5]

[1]Ghent University, Lab. for Chemical Analysis, Salisburylaan 133, 9820 Merelbeke, Belgium, [2]KU Leuven Technology Campus Ghent, Research Group for Technology and Quality of Animal Products, Gebroeders De Smetstraat 1, 9000 Gent, Belgium, [3]Institute for Agricultural, Food and Fisheries Research, Animal Husbandry, Scheldeweg 68, 9090 Melle, Belgium, [4]Ghent University, Lab. for Animal Nutrition and Animal Product Quality, Coupure Links 653, 9000 Gent, Belgium, [5]Queen's University, Institute for Global Food Security, University Road, BT7 1NN Belfast, United Kingdom; lieseloty.hemeryck@ugent.be

Boar taint is an off-odour in pork (products) from male pigs, which can be avoided by castration. However, due to welfare concerns, surgical castration of piglets without anaesthesia/analgesia is to be abolished, whilst immunocastration is not accepted throughout the entire European Union. Hence, the pork supply chain is in search of valid alternatives, taking into account social, economic as well as product-technological factors. This work aims at the valorisation of pork with boar taint by exploring alternative processing practices that reduce boar taint perception by consumers. Carcasses with moderate to severe boar taint were selected through sensory (soldering iron method) and chemical (UHPLC-HRMS for indole, skatole and androstenone) analysis. Several pork products including Frankfurter and dry fermented sausage, marinated bacon and pork tenderloin, cooked ham and hamburger were produced, after which the smell, taste and several other sensory characteristics of these products were evaluated by a boar taint expert panel. The results demonstrate several opportunities for the valorisation of pork with boar taint: (1) mixing with meat from gilts, depending on the severity of the taint (e.g. by applying an androstenone threshold of 300 ppb); (2) processing into Frankfurter sausages, as it appears that the typical smoky flavour of the sausage successfully masks boar taint smell and taste; (3) the addition of specific smoke condensates to the brine during production of cooked ham (model system); and (4) the use of specific marinades for the preparation of fresh pork cuts. This work contributes to the establishment of a workflow for the allocation and further processing of tainted carcasses.

Sensory meat quality differences for crossbred offspring of different terminal sire lines

E. Kowalski[1,2], E. Vossen[1], M. Aluwé[2], S. Millet[2] and S. De Smet[1]
[1]UGent, Department of Animal Sciences and Aquatic Ecology, Coupure Links 653, 9000 Ghent, Belgium, [2]Flanders research institute for agriculture, fisheries and food (ILVO), Scheldeweg 68, 9090 Melle, Belgium; eline.kowalski@ilvo.vlaanderen.be

In Flanders, pig production is characterized by a low feed conversion ratio and carcasses with a high lean meat content. However, nowadays there is an increasing concern about inferior sensory and technological meat quality. One problem is the high incidence of PSE meat and in addition, as a result of the selection for a high carcass lean meat content, the meat is nowadays characterized by a low intramuscular fat content. The objective of this study was to evaluate the eating quality of the loin of crossbred offspring of three terminal sire lines. A homozygous stress positive sire line with high lean meat percentage (Belgian Piétrain (BP)) was compared with homozygous stress negative sire lines with a potentially better sensory meat quality (French Piétrain (FP) and Canadian Duroc (CD)). A total of 120 pigs (40 pigs/sire line) were sampled for assessment of sensory quality. This was evaluated on the one hand by a trained expert panel (EP) and on the other hand by a consumer panel (CP). The trained EP consisted of 6 experts and 10 sessions. The CP was executed by 120 households (i.e. a cook and a taster). Each household received three packages with each one sample to prepare and evaluate at home. A linear mixed model was used for statistical analysis of both panels. No significant difference was observed for tenderness (EP: P=0.060, CP: P=0.117), but the EP results showed a trend towards lower tenderness in BP compared to the FP and CD. The EP evaluated the juiciness of FP and CD samples higher compared to BP (P=0.049), but this was not observed by the CP (P=0.181). The overall liking of FP and CD samples was higher compared to BP by the EP (P=0.005). This was partly confirmed by the CP, which evaluated the overall liking of the CD higher compared to FP and BP (P=0.030). In conclusion, it was not possible to differentiate meat of crossbred progeny from three sire lines regarding tenderness, whereas BP had lower scores for juiciness compared to the other two sire lines. The overall liking of BP was lower compared to CD.

Regulatory genes and pathways influencing FE traits in Norwegian Landrace pigs fed fibre-rich diet

A. Skugor[1], L.T. Mydland[1], W.M. Rauw[2] and M. Øverland[1]
[1]Norwegian University of Life Sciences (NMBU), Department of Animal and Aquacultural Sciences, P.O. Box 5003, 1432, Ås, Norway, [2]Instituto Nacional de Investigación y Tecnología Agraria y Alimentaria (INIA), Dept Mejora Genética Animal, Crta de la Coruña km 7.5, 28040, Spain; liv.mydland@nmbu.no

Because of systematic selection for reduced feed conversion ratio, the Norwegian Landrace pig is a fast growing and highly feed efficient breed. However, animals have been selected based on growth performance when fed high-quality cereal, soybean meal (SBM)-based diets with moderate fibre content. In the SusPig project we aim to evaluate the performance of efficient pigs when fed local rapeseed meal (RSM)-based diets that are rich in fibre and contain bioactive phytochemicals such as glucosinolates. In a three-month growth performance experiment with growing-finishing pigs, animals were fed a diet with 20% RSM vs a commercial SBM-based diet. In general, pigs fed RSM had reduced growth performance and nutrient and energy digestibility compared to pigs receiving the SBM diet, but no differences in meat quality traits were observed. A large observed within-group variation in feed intake and daily weight gain in pigs fed RSM, suggests the existence of individual variation in the ability to perform on fibre-rich RSM diets. Global transcriptional profiling of skeletal muscle, gut and liver tissues revealed significant differences in gene expression patterns between pigs fed SBM or RSM. Results showed differential regulation of several genes and associated biological processes that were common for all tissues, including those regulating carbohydrate, lipid and energy metabolism, immunity, mitochondrial function and response to reactive oxygen species. Reduced growth performance in RSM pigs was accompanied by the activation of numerous genes orchestrating adaptive gene expression responses, including changes in energy metabolism and tissue remodelling. Some of these genes may be key contributors to the observed variation in feed efficiency traits. Our results provide valuable insights into the biological background of feed efficiency in growing pigs, and generate knowledge that can help facilitate strategies to breed pig genotypes that are better adapted to novel fibre-rich diets.

Molecular drivers of the phosphorus homeostasis in pigs

H. Reyer, M. Oster, D. Wittenburg, E. Murání, S. Ponsuksili and K. Wimmers
Leibniz Institute for Farm Animal Biology (FBN), Institute of Genome Biology, Wilhelm-Stahl-Allee 2, 18196 Dummerstorf, Germany; reyer@fbn-dummerstorf.de

For all organisms, phosphorus (P) is a vital element and its usage and application across agricultural production systems requires great attention. Pigs depend on a sufficient supply of mineral inorganic P in the diet to develop and maintain a healthy musculoskeletal system via appropriate bone mineralisation and physiologic immune cell function. However, pigs are also major excretors of P, therefore animal husbandry at regional level is an important source of P discharges into the environment. The sustainable use of P ensuring sufficient but not excessive P supply makes a significant contribution to global food security and environmental protection. In animal sciences, this implies the need of an improved understanding of genetics and physiology with regard to absorption (intestinal transport), retention (storage in bones) and excretion (renal function) processes of P. In fact, P homeostasis is maintained via a number of known and yet to be elucidated regulators, transporters, hormones, and paracrine signals. Indeed, pigs show a wide variation in blood P levels, which is partly attributed to genetics. A targeted genetic study of single actors involved in mineral and bone metabolism revealed only minor contributions to the phenotypic variation in pigs. However, a genome-wide analysis of markers adds a number of genomic regions to the list of quantitative trait loci, some of which overlap with previous results in humans and farm animals. Despite the gaps in knowledge of genes involved in calcium and P metabolism, candidate genes like *THBS2, SHH, PTPRT, PTGS1* and *FRAS1* with reported connections to bone metabolism were derived from the significantly associated genomic regions. Additionally, genomic regions harboured *TRAFD1* and genes coding for P transporters (*SLC17A1-A4*), which are linked to P homeostasis. The study calls for improved functional annotation of the proposed candidate genes to derive features involved in maintaining calcium and P balance. The biodiversity of pigs regarding genetics and physiology provides a huge potential for breeding and management towards improved P efficiency.

Feed efficiency × diet interaction on acorns vs a commercial diet in Iberian pigs

W.M. Rauw[1], L.A. García Cortés[1], F. Gómez Carballar[2], J.M. García Casco[3], E. De La Serna Fito[2] and L. Gomez-Raya[1]
[1]INIA, Dept Genética Animal, Ctra de La Coruña km 7.5, 28040, Spain, [2]Sánchez Romero Carvajal, Jabugo S.A., Ctra San Juan del Puerto s/n, 21290 Jabugo, Spain, [3]INIA, INIA-Centro de I+D en Cerdo Ibérico, Ctra EX101 km 4,7, 06300 Zafra, Spain; rauw.wendy@inia.es

Improved growth of pigs must be supported by increased quality of nutrients that allows for the expression of production potential. Therefore, modern livestock diets have high nutritional value that is sourced from international markets making Europe heavily reliant on feed ingredient imports. The European parliament calls for growing on-farm animal feed, and using by-products of oilseed and agrofuels production. However, this may require an animal with different performance characteristics than those currently selected in high quality input – high output production systems. The present study aims to investigate feed efficiency × diet interaction on a commercial vs a local diet in Iberian pigs. Body weight gain was measured individually in 30 Iberian pigs, who were provided consecutively, individually, with 4 kg/d of a regular concentrate diet for 30 days (Concentrate; 21.5% corn, 15% wheat, and 38.2% barley), and with 8 kg/d of acorns for 21 days (Acorn). After the Acorn period, all animals fed extensively on acorns, roots, and herbs for 19 days (Montanera) and were group-supplemented with 4 kg/d/animal of concentrate diet. Body weight was recorded at the beginning and end of each period. Animals were slaughtered two days after Montanera and slaughter yield was estimated as carcass weight/live weight × 100%. Pigs that grew faster (with higher feed conversion efficiency) on Concentrate grew slower on Acorns (r=-0.55, P<0.01) but faster in Montanera (r=0.43, P<0.05). Pigs that grew faster on Concentrate and in Montanera had lower slaughter yields (r=-0.32, P=0.087 and r=-0.65, P<0.0001, respectively). The results support the hypothesis that selection for feed efficiency would favour a different type of animal on different diets. This work was funded by ERA-Net SusAn project 'SusPig' (www.suspig-era.net).

Intensively manipulated silage to growing pigs – influence on nutrient utilization and behaviour

M. Presto Åkerfeldt
SLU Swedish University of Agricultural Sciences, Dept. of Animal Nutrition and Management, Box 7024, 75007 Uppsala, Sweden; magdalena.akerfeldt@slu.se

Locally produced silage in diets to pigs is interesting as a potential nutrient source and behaviour enrichment for sustainable pig production. A major challenge is to find applicable ways of feeding the silage without pigs' sorting out parts of the silage. The aim of this study was to investigate how grass/clover silage with a smaller particle size than in chopped silage, fed as a total mixed ration (TMR), influenced the nutrient utilization and behaviour of growing pigs. The study included two experiments: (1) nutrient utilization; and (2) behaviour. Exp1 included 10 YH pigs with an average start weight of 28.1 (±1.7) kg housed in individual pens. Exp2 included 64 YH pigs from two batches with an average start weight of 40.2 kg (±12.1 kg) in batch 1 and 79.3 kg (±13.6 kg) in batch 2, housed in 8 pens (4 pens/batch, 8 pigs/pen). The dietary treatments were cereal-based feed mixed with either intensively manipulated silage (SI) or chopped silage (SC) in an 80:20 (DM basis) ratio for cereal and silage respectively, and a control diet (C) without silage in Exp1. All diets in Exp1 contained an indigestible TiO_2 marker. The same silage was used for SI and SC in both Exp1 and 2, a mixture of grass/red clover. Apparent total tract nutrient digestibility was calculated using the indicator technique and behaviour recordings were performed with direct observations, instantaneous and continuous sampling. Statistics was performed with GLM in Minitab, and PROC MIXED in SAS. SI pigs had greater silage consumption than SC pigs ($P=0.001$) and spent more time eating from the feed through ($P<0.01$), while SC pigs rooted more ($P<0.01$). No difference was found for nutrient digestibility in diets ($P>0.05$) or in silage ($P>0.05$). SC pigs performed more social interactions compared with SI pigs ($P=0.029$) prior to feeding. After and separate from feeding SC pigs were more active ($P=0.003$ and $P=0.002$, respectively), however not more social interactive ($P=0.834$ and $P=0.466$, respectively) than SI pigs. We conclude that feeding pigs TMR with intensively treated silage might benefit the consumption and nutrient utilization of silage and give opportunity longer eating times and abilities for pigs to perform feed related behaviours.

Rotational grazing for organic pregnant sows: a mechanism to reduce feed consumption

A. Roinsard[1], F. Maupertuis[2], C. Gain[1] and P. Pierre[3]
[1]ITAB, 9, rue André Brouard, 49105 Angers, France, [2]Chambre d'agriculture Pays de la loire, La Géraudière, 44939 Nantes Cedex 9, France, [3]IDELE, 9, rue André Brouard, 49105 Angers, France; antoine.roinsard@itab.asso.fr

In organic farming, 44% of pregnant sows are raised outdoors. To reduce concentrated feed intake, it would be useful to develop an innovative feeding management system that would increase the value of the free-range area. Rotational grazing on grassland rich in legumes was established for pregnant sows at the Trinottières station (Maine-et-Loire, France) for 2 consecutive years. To enhance the value of grazing, the experimental concentrated feed distributed was restricted to 80% of that of the control groups (those outside the grazing period) and was less rich in protein (10.2% crude protein vs 13.6% for the control). The objectives were to: (1) assess the impact on sow performance; (2) assess the contribution of grazing to fulfilling feed requirements; (3) quantify grass intake; and (4) describe sow preferences for specific species. Grazing sows had the same back fat thickness gain but a slightly lower live weight gain, related to reduced motivation to graze at the end of gestation. The nutritional needs of 'grazing' sows were calculated using INRAPorc software. As a result, grazing contributed on average to about 22% of the metabolizable energy requirement and 33% of the digestible lysine requirement. The biomass ingested was 1.75 kg of dry matter (DM)/day/sow and showed high animal variability (0,2 to 4 kg DM/day). Finally, sows expressed a strong preference for consuming legumes. Implementing efficient pasture management for pregnant sows reduced annual feed costs by 16% and concentrated feed consumption by 9%. However, because of reduced grass intake, special attention should be paid at the end of gestation (increase the quantity of concentrated feed) to ensure safe weight gain for sows.

Evaluating pen-allocation strategies for uniform weights in finishing pigs estimating age at 120 kg

J.A.N. Filipe[1], E.F. Knol[2], R.H. Vogelzang[2] and I. Kyriazakis[1]
[1]*Newcastle University, Newcastle upon Tyne, NE1 7RU, United Kingdom,* [2]*Topigs Norsvin Research Center, 6640 AA Beuningen, the Netherlands; joao.filipe@newcastle.ac.uk*

Variable performance of finishing pigs within pens causes losses to producers through delayed availability of pen space and deviation from target market weight. We explored alternative pen allocation strategies to assess their potential to reduce losses. We estimated retrospectively the maximum benefit gained if individual growth to target weight could be predicted at allocation. If a strategy's retrospective benefit is significant there is a case to invest resources to implement it. We used a dataset with three routine weight measurements per pig during finishing (n=240, range (kg): 14.8-31.8, 43.0-86.5, 100.0-138.0, CV: 0.12, 0.12, 0.06) at points where age differed considerably and increasingly (range (d): 53-66, 103-128, 144-194 (0.03, 0.04, 0.07)), and comprising six cohorts of contemporary pigs. Three pen-allocation strategies were compared: the actual strategy used by producers (PS) against an 'optimal' (OS) and a 'random' 'strategy (RS), both applied to contemporaries. Bayesian modelling was used to predict individual growth and age at 120 kg (A120) due to the deviations of end weights from 120 kg. In the OS, cohort pens were filled with pigs of similar predicted A120. In the RS, sorting was random. The economic loss of each strategy quantified revenue lost in unused pen space and in deviation from target weight in each pen until empty, and for each cohort. We estimated, based on current market values, that OS can generate up to +28% in profit per pig in relation to the PS (cohort average +15%); i.e. a gain of up to €29,000 per year in a 4,000-pig farm. The RS lead more often to considerably greater loss than profit compared to the PS. Therefore, the producer's strategy is clearly better than the RS but can be improved through direct or proxy estimation of A120. The level of improvement suggests it is worth using or collecting further data to allow this estimation at the start of finishing. Precision management will become more relevant in future systems where optimised pen sorting is applied and animals are given designed feed and treatment on a pen (or individual) basis. This study was supported by the EU's H2020 program (agreement no. 633531) and UK's BBRSC.

Comparing meat quality of entire male, immunocastrated and surgically castrated pigs

K. Poklukar[1], M. Škrlep[1], N. Batorek Lukač[1], M. Vrecl Fazarinc[2], G. Fazarinc[2], K. Kress[3], U. Weiler[3], V. Stefanski[3] and M. Čandek-Potokar[1]
[1]*Kmetijski inštitut Slovenije, Hacquetova 17, 1000, Slovenia,* [2]*Univerza v Ljubljani, Veterinarska fakulteta, Gerbičeva 60, 1000 Ljubljana, Slovenia,* [3]*Universität Hohenheim, Gerbenstr. 17, 70599 Stuttgart, Germany; klavdija.poklukar@kis.si*

The objective of the study was to determine technological and physiochemical properties of meat in entire male (EM; n=48), imunocastrated (IC; n=48) and surgically castrated pigs (SC; n–48). Pigs were of a commercial crossbreed (Landrace × Pietrain). IC were vaccinated at the age of 12 and 21 weeks and all sex categories were slaughtered at the age of 27 weeks reaching 124.7±0.9 kg. After the slaughter, lean meat percentage was determined according to SEUROP system. Cooled carcasses were cut at the level of the last rib, and longissimus dorsi (LD) muscle was sampled for determination of physiochemical parameters and meat quality traits. The data were analysed by 2-way ANOVA using R software. EM exhibited higher and SC lower (P<0.05) leanness than IC. Protein oxidation (carbonyl groups concentration) in cooked meat was higher (P<0.01) in EM compared to IC (13%) and SC (14%). Moreover, peroxidation in cooked LD was higher (P<0.05) in EM compared to SC (23%), as well as total collagen content (8 and 17% compared to IC and SC, respectively; P<0.01). Several differences in meat quality were observed. In comparison to EM, the meat of IC exhibited higher Minolta L (P<0.10) and Minolta b value (P<0.05), while meat of SC had lower Minolta a value (P<0.01). Lower (P<0.01) water holding capacity (WHC) assessed as cooking losses was higher in EM (8%) and IC (6%), than in SC. EM had also tougher meat (P<0.05) than IC (9%) or SC (12%). We can conclude that compared with IC and SC, EM exhibited inferior meat quality (tougher meat, lower WHC) differentiating also in physiochemical parameters (higher oxidation and collagen content). These results were accomplished within project SuSI, co-financed by Susan EraNet and Slovenian Ministry of Agriculture, Forestry and Food. Collaboration within COST action CA12215 IPEMA and core financing of Slovenian Research Agency (grant P4-0133, PhD scholarship for KP) are also acknowledged.

Specifications and return of investment for an automated production of insects

J. Fynbo[1], J.L. Andersen[2] and M. Ebbesen Nielsen[3]
[1]Hannemann Engineering ApS, Sivlandvænget 3, 5260 Odense S, Denmark, [2]DTI, Kongsvang Allé 29, 8000 Aarhus, Denmark, [3]Ausumgaard, Strandbjerggårdvej 8, 7000 Struer, Denmark; jlan@teknologisk.dk

In this article we will address aspect of automizing insect production that today is mostly done using manual labour. Four main questions will be answered: (1) which insect should be chosen for an automized production; (2) how and which production processes should be automized; (3) will it be possible to make a fully automized production unit for insects; and (4) return of investment (ROI) for an automized production unit. The interest in large scale production has exploded during the last decade. The main reason for this is the increasing population – we need to feed 9 billion people in 2050. To meet the food and nutrition challenges in the near future, we need to develop a more sustainable way to 'harvest' protein. Production of insects at industrial scale are just beginning. Even though insects have been on the 'menu carte' for decades in non-western countries, we are just starting to scale this type of production. To industrialise this type of production we need to understand the insects. A lot of knowledge is shared in academia. Lots of papers can be found on optimal temperature, humidity, etc. that have been studied at lab scale. But looking into large scale production, we see very little knowledge sharing. This sharing is critical due to the fact, that we are only at the beginning of several decades of development. We need to share our knowledge and experience in order to gain momentum and scale – at the moment we are looking at a very efficient way of producing protein but the barriers of shifting from producing traditional livestock to insect production are very large. The 'few' that have invested in a scaled production facility are, generally, not interested in sharing their knowledge. But if that attitude grows, we will lose valuable time in the overall race, not against the potential competitors but against time and the planetary boundaries of mother nature. Therefor we will try to present a model, that may be used when ROI is calculated for an insect production facility; including some thoughts and ideas of a cooperative way of solving some of the challenges that might be a barrier for 'SME' producers of insects.

Processing: lessons with mealworms

S. Hvid, A.L. Dannesboe Nielsen, S. Mikkelsen, L.H. Lau Heckmann and J. Hinge
Danish Technological Institute, Kongsvang Alle 29, 8000 Aarhus C, Denmark; simh@dti.dk

Different approaches in processing of a new product is a major focus, as different processing techniques may have an impact on the final product. Different processing technologies have been tested on both the lesser and common mealworm (*Alphitobius diaperinus* and *Tenebrio molitor*). Freeze drying, defattening, enzymatic treatment, extrusion, hydrolysis, industrial drying (heat oven) and vacuum drying was performed on the insects, initially to investigate the effect on the amino acids. Next, a feed trial on rats including these different processed insect meals was performed to evaluate protein digestibility and available essential amino acids. The defattening of the common mealworm is believed to be crucial, if the mealworm meal is going to be part of a regular diet for Europeans; particularly in processed food. The grinding of non-defatted common mealworms has proven to cause problems, due to too high fat content. The mill will simply clock up. To prevent this, several methods have been tested. The defattening process has been investigated, with regard to the choice of method. Originally di-ethyl ether was used, but several other lipid extraction methods have also been tested and compared with regard to total output and fatty acid composition. The results and experience learned from working with Tenebrio will be shared in the presentation.

Nutritional evaluation by PDCAAS and DIAAS of insect protein in rats and pigs

J.V. Nørgaard[1], L.D. Jensen[1], R. Miklos[2], T.K. Dalsgaard[1], N. Malla[3], N. Roos[3] and L.H. Heckmann[2]
[1]Aarhus University, Foulum, Blichers Alle 20, 8830, Denmark, [2]Danish Technological Institute, Kongsvang Allé 29, 8000 Aarhus C, Denmark, [3]University of Copenhagen, Department of Nutrition, Exercise and Sports, Rolighedsvej 26, 1958 Frederiksberg, Denmark; janvnoergaard@anis.au.dk

The objective of this study was to evaluate the nutritional value of crude protein and amino acids from different insect species. Exp. 1: protein digestibility-corrected amino acid score (PDCAAS) assessing two mealworm species *Alphitobius diaperinus* (AD; lesser/buffalo mealworm), *Tenebrio molitor* (TM; common/yellow mealworm). Exp. 2: digestible indispensable amino acid score (DIAAS) assessing AD, TM, *Hermetia illucens* (HI; black soldier fly larvae), *Gryllodes sigillatus* (GS; banded cricket), and Acheta domesticus (ACD; house cricket). The Exp. 1 used rats (n=6) where total collection of faeces and urine was obtained during 4 days. Data was analysed by ANOVA including treatment and period as fixed effects. Exp. 2 used ileum cannulated pigs in a Latin square design with 6 pigs fed the 5 insects and a N-free diet. Ileal digesta samples are in process of analysis. The analysed content of crude protein in Exp. 1 was 62.3 and 59.3 g /100 g dry matter for AD and TM, respectively. The analysed content in Exp. 2 of crude protein in dry matter (g/100 g) was 58.4 for AD, 45.3 for TM, 39.7 for HI, 53.2 for GS, and 63.7 for ACD. Crude fat in dry matter (g/100 g) was 26.6 for AD, 31.1 for TM, 27.3 for HI, 32.0 for GS, and 20.3 for ACD. True protein digestibility in rats was 0.94 and 0.92 for AD and TM. The PDCAAS was higher for AD (0.82) compared to TM (0.76), both with Met+Cys as the limiting amino acids. The biological value of the crude protein was 0.63 for AD and 0.55 for TM. In conclusion, the 5 insects may be promising as food and feed due to their high protein content and digestibility. The PDCAAS results were modest. However, DIAAS is recognized as a superior method for assessment of protein quality, and the completion of Exp. 2 data analysis will provide a better picture of protein quality.

Characteristics and functionality of different *Tenebrio* nutrients

S. Hvid[1], A.L. Dannesboe Nielsen[1], L.H. Lau Heckmann[1] and N. Roos[2]
[1]Danish Technological Institute, Kongsvang Alle 29, 8000 Aarhus C, Denmark, [2]University of Copenhagen, Nutrition Exercise and Sports (NEXS), Rolighedsvej 26, 1958 Frederiksberg, Denmark; simh@dti.dk

Tenebrio molitor (common mealworm) is one of the most promising insects, when it comes to food in the future. The rearing of *Tenebrio* is well studied and is already put to use in several production facilities today. However, the characteristics and functionality of *Tenebrio* as an ingredient in food is not as well studied. We have tested the physical properties of grinded *Tenebrio* meal, both with mealworms reared on a diet, resulting in a high fat content, as well as *Tenebrio* meal reared under different parameters, resulting in a lower fat content. A fat reduced *Tenebrio* meal was also tested. The physical properties tested include water binding, lipid binding, foam stability, emulsion stability, solubility with more. The focus of the study was to compare the three different *Tenebrio* meals to estimate how significant the fat content in the meal is, when it comes to physical properties. A complete fatty acid analysis was also performed, with focus on the potential fractionation of specific fatty acids and the possibilities of using them in other applications. Tests are still ongoing, but one of the potential markets are cosmetics for e.g. palmitic and oleic acid. Different micronutrients have also been screened to identify possible branding-opportunities regarding e.g. vitamin content, when using the *Tenebrio* meal in food. Chitin from insects are assumed to decrease the digestibility, solubility and several other things. Different enzymes that are able to make chitosan from chitin have been tested on *Tenebrio* meal. The enzymatically treated *Tenebrio* meal is compared in physical properties with the no-treatment *Tenebrio* meal to see, if the degradation from chitin to chitosan is able to change the functionality of the *Tenebrio* meal.

The role of edible insects in sustainable healthy human diets

N. Roos[1], N. Malla[1] and L.H.L. Heckmann[2]
[1]University of Copenhagen, Department of Nutrition, Exercise and Sports, Rolighedsvej 26, 1958, Denmark, [2]The Danish Technological Institute, Bioengineering and Environmental Technology, Kongsvang Allé 29, 8000 Århus Denmark; nro@nexs.ku.dk

Edible insects as food components in diverse human diets are to be recognized as animal-source foods, contributing intakes of high-quality protein, important micronutrients, and fat of variable nutritional importance. Insects may also contain bioactive compounds with health impacts beyond meeting nutritional requirements. The exact nutritional contributions of insects to human diets are depending on the insect species, the metamorphological life-stages, the growth and feeding history of the insect, as well as the post-harvest processing. While the knowledge of the nutritional composition of the commonly farmed insect species is rapidly growing, also with important contributions from the Invaluable project, the direct testing of the impact of insect consumption on human nutrition and health is still in an early phase and much research is needed. This presentation will contribute an updated overview of the known nutritional properties of the most commonly farmed insect species for food, including impacts of management practices of insect farming on nutritional quality. Also, the limited current knowledge of the nutritional and health impacts directly assessed in humans will be presented. Due to the biological complexity of edible insects, they may be a source of bioactive compounds with additional beneficial health impacts. *In vitro* and animal model studies have indicated possible impacts on e.g. hypertension, immunity and diabetes; however, well-designed studies in humans' remains to confirm the indicated benefits. In January 2019 the 'Lancet EAT Commission' released a framework for global standards for sustainable and healthy diets, suggested as guidelines for a transition of food systems to environmentally sustainable diets for a future global population of 10 billion. Insects are mentioned for their potentials, however, absent in the recommendations due to lack of sufficient scientific evidence to qualify environmental and health impacts. These knowledge gaps need to be addressed for the recognition of edible insects in future policy recommendations for sustainable healthy diets.

Tenebrio applications and sensory qualities in processed foods

S. Hvid[1], A.L. Dannesboe Nielsen[1], L.H. Heckmann[1], M. Holse[2], N. Roos[3] and M. Bom Frøst[3]
[1]Danish Technological Institute, Kongsvang alle 29, 8000 Aarhus C, Denmark, [2]Novozymes A/S, Krogshoejvej 36, 2880 Bagsvaerd, Denmark, [3]University of Copenhagen, Rolighedsvej 26, 1958 Frederiksberg, Denmark; aln@dti.dk

Incorporating insect flour into processed foods can enhance the protein content and quality and possibly also contribute with vitamins and minerals. However, the physical and sensory qualities of the Tenebrio flour, alone or flour as part of a product, need to be described to understand its optimal applications in processed products. The *Tenebrio* flour itself was tested by a trained sensory panel, aiming to create a 'flavour wheel', describing the *Tenebrio* flour and a defatted version of the same flour. The flavour wheel will contribute as one of several flavour wheels, useful to describing the quality differences between insect species and different processing methods, like defatting. This tool will be useful to decide which kind of insect is optimal for a specific type of product, or vice versa. *Tenebrio* flour was tested as an ingredient in several cereal-based products, ranging from pita bread, crackers and Finnish flatbread to crisps (snacks) and regular whole flour white bread. The products were developed with varying concentration of *Tenebrio* flour and included in different consumer tests, to identify the maximum content of *Tenebrio* flour, regarding flavour and visual appearance of the product, as well as estimating the potential consumer profile for products containing insects. When developing products and substituting flour with *Tenebrio* flour, the gluten content and amount of carbohydrates in the product decreases. For bread, this had negative impact on the bread quality, such as leavening abilities. In collaboration with Novozymes A/S, different combinations of enzymes were tested, aiming to reconstitute the raising quality of the bread. Enzymes can contribute to overcoming this barrier, and the concentration of *Tenebrio* flour can be even further increased. The levels of insect flour which then can be applied to foods can allow the food producer to obtain more than 20% protein, and thus claim the product being 'high in protein', in accordance to EFSA regulations for health clams. Results from these tests will be shared in this presentation.

European consumer acceptance of insects as food and feed: key findings and considerations

J.A. House

Wageningen University, Sociology of Consumption and Households, De Leeuwenborch (Building 201), Hollandseweg 1, 6706 KN Wageningen, the Netherlands; jonas.house@wur.nl

The potential of insects as an environmentally-friendly and resource-efficient protein source is now clearly established, and considerable advances have been made across the insect value chain in Europe. Nevertheless, consumer acceptance remains a key challenge for the broader success of the sector. This presentation draws on research with European consumers, outlining a number of key findings and considerations regarding consumer acceptance of insects as food and feed. The presentation highlights a 'demand-side' tendency in current research, in which public attitudes are identified as the main barrier to consumer acceptance. Some drawbacks of this perspective are discussed, and an alternative 'supply-side' approach is proposed. In this view, attainment of consumer acceptance is the responsibility of actors in the insect value chain, who must work together to create insect-based products which are tasty, affordable, easily available, and which integrate into existing diets in a coherent manner. Key recommendations include a focus on potential 'early adopter' consumers rather than whole populations, and consideration of appropriate dishes and cuisines for the introduction of insect-based products.

Feed efficiency and methane emissions in dairy cattle: animal, herd and population considerations

P.C. Garnsworthy

University of Nottingham, School of Biosciences, Sutton Bonington Campus, Loughborough LE12 5RD, United Kingdom; phil.garnsworthy@nottingham.ac.uk

Strategies to improve feed conversion efficiency (FCE) and reduce methane (CH_4) emissions in dairy cattle can target the individual animal, the dairy herd, or a population of dairy cattle. Strategies are different for CH_4 emissions per day (MPD), per unit feed intake (MPI), and per unit product (MPP). Higher milk yields, for example, are associated with greater FCE and MPD, but lower MPI and MPP for individuals. A herd with high milk yield, however, might have poorer fertility and increased replacement rate. High replacement rates lower herd FCE and increase MPD and MPP because greater numbers of non-productive replacement animals have to be maintained in the herd. Energy lost as CH_4 represents 2 to 12% of gross energy intake, so lowering MPD or MPI should increase energetic efficiency, especially if energy spared is diverted to milk synthesis. This may be true when MPD is lowered by altering diet components to reduce the ratio of rumen acetate to propionate concentrations. For a fixed diet, however, cows that digest forage more efficiently have greater MPI, and can have greater MPD. Conversely, if these efficient cows produce more milk, fewer will be required for a given level of milk production, which will lower MPP for the herd. Some CH_4 inhibitors act through reducing forage digestibility and feed intake, which can be detrimental for FCE. Dietary strategies to lower CH_4 emissions should also avoid trade-offs with feed carbon footprint, which accounts for nearly half the greenhouse gas emissions of milk production. Genetic selection for low CH_4 emissions must be approached with caution because low emitters might have low milk yield and FCE; low CH_4 emissions per animal could be offset by increased animal numbers needed to maintain milk supply. Any selection index must include milk production, efficiency and resilience traits alongside CH_4 emissions. In conclusion, interactions between FCE and CH_4 emissions are not straightforward; understanding responses at animal, herd and population level is essential for designing optimum strategies to achieve goals in different scenarios.

Restricting feed intake of inefficient lactating cows affects feed efficiency and methane emissions

A. Fischer, N. Edouard and P. Faverdin
PEGASE, INRA, Agrocampus-Ouest, 16 Le Clos, 35590 Saint-Gilles, France; amelie.fischer@inra.fr

An emerging hypothesis claims that feed inefficiency can be due to overconsumption. The objective was therefore to test feed restriction as a lever to improve feed efficiency. Feed efficiency was estimated with the residual feed intake (RFI), as the difference between observed and expected dry matter intake (DMI). A cohort of 32 lactating Holstein cows was identified among 70 lactating cows as the 25% least efficient cows and the 25% most efficient cows. This identification was done using the last 2.5 months of *ad libitum* feeding, before feed restriction started. For a given expected DMI during *ad libitum* period, the offered DMI during restriction was set to the observed DMI of the 10% most efficient cows during the last 2 months of *ad libitum* feeding. Feed restriction was applied during 2.5 months, starting on average at 194±16 days in lactation. A single diet was fed during the whole experimentation based on 65.8% of corn silage and 34.2% of concentrates on a DM basis. Individual daily DMI, morning BW, daily milk yield, weekly milk composition, monthly BCS were recorded individually, as well as methane (CH_4) emissions with two Greenfeed units during the whole study. RFI was the residual of the linear regression predicting each period's average DMI with period's averages of net energy in milk, metabolic BW, BCS, BCS gain and BW loss, the fixed effects of parity and period. RFI was averaged per period. The least efficient group ate 2.7 kg DM/d ($P<0.01$) and emitted 23 g CH_4/d ($P=0.23$) more than the most efficient group during *ad libitum* feeding. Feed restriction decreased DMI ($P<0.01$) and daily CH_4 emissions ($P<0.01$), with a 1.8 kg DMI/d and a 20 g CH_4/d higher decrease for the least efficient group, but had no effect on CH_4 yield per DMI ($P=0.81$). Feed restriction reduced RFI SD from 1.14 kg DM/d to 0.72 kg DM/d. The 25% least efficient cows remained the least efficient cows in both periods (RFI *ad libitum* vs RFI restriction, r=0.59). Feed efficiency was not correlated to CH_4 emissions, but was negatively correlated to CH_4 yield per DMI (*ad libitum* r=-0.43, restriction r=-0.25). Overall, feed restriction applied to inefficient lactating dairy cows seems to reduce feed efficiency variability and their CH_4 emissions.

Comparing feed efficiency in Italian Holstein Friesian heifers, lactating cows and bulls

F. Omodei Zorini[1], R. Finocchiaro[2], M. Cassandro[3], G. Savoini[1], E. Olzi[2] and G. Invernizzi[1]
[1]University of Milan, Department of Health, Animal Science and Food Safety, via Celoria 10, 20133 Milan, Italy, [2]National Association of Holstein and Jersey Breeders, Via Bergamo 292, 26100 Cremona, Italy, [3]University of Padova, Department of Agronomy, Food, Natural Resources, Animals and Environment, Viale dell'Università 16, 35020 Legnaro, Italy; raffaellafinocchiaro@anafi.it

The aim of this work was to study the associations between different indices of FE and performance parameters in Italian Holstein-Friesian heifers, bulls and dairy cattle. Trials were conducted at the Experimental farm of Animal Production Research and Teaching Centre of Lodi (CZDS) and at Genetic Centre of the National Association of Holstein Breeders. Sixteen heifers were investigated for dry matter intake (DMI), body weight (BW), body condition score (BCS) and wither height. Similarly, 33 young bulls were controlled for DMI and BW. Data obtained were used to calculate the residual feed intake (RFI) and feed conversion ratio (FCR) of each animal. After the ex-post division of each batch into two groups with the higher (H-RFI) and the lower (L-RFI) RFIs, data were analysed by a MIXED procedure of SAS. For heifers, significant differences were highlighted between DMI values of the two groups ($P=0.01$) and also for the wither height gain character ($P<0.05$). There was a tendency for FCR data of the two groups of bulls ($P=0.07$). Furthermore, 30 lactating cows were investigated for DMI, BW, BCS, milk yield and composition. Data obtained were used to calculate the RFI and milk to feed ratio (M:F) of each animal. The ex-post division of the batch into two groups with the higher and the lower RFIs highlighted a statistically significant difference between DMI values of the groups but not between milk production-related parameters. There was also a positive correlation between RFI and DMI and a negative correlation between M:F values and milk energy output and energy corrected milk. Our results confirm the goodness of RFI as an index of FE, even if differences between functional groups of animals could be an important variable to account for.

Reliability of genomic prediction for feed and residual feed intake in Holstein cows

D. Gordo[1], C. Manzanillap-Pech[1], G. Su[1], M. Lund[1] and J. Lassen[1,2]
[1]Center for Quantitative Genetics and Genomics, Aarhus University, Department of Molecular Biology and Genetics, Blichers Alle 20, 8830 Tjele, Denmark, [2]Viking Genetics, Ebeltoftvej 16, 8960 Randers, Denmark; danielmansangordo@mbg.au.dk

The objective of the present study was to investigate the reliability of genomic breeding values (GEBV) for dry matter intake (DMI) and residual feed intake (RFI) in Holstein cows using a joint international reference population. The data used in this investigation belongs to The Efficient Dairy Genome Project database (EDGP), which contains herds from six countries. We included 4,453 cows with 296,066 weekly records of DMI, milk yield (MY), energy corrected milk (ECM), metabolic body weight (MBW), and body weight change (BWC). In the data, 3,313 cows were genotyped with 50k SNP. Three models for RFI were applied due to the heterogeneity of management and recording protocols across herds. The difference in the models were the partial regression of DMI on distinct sources of energy sink: (1) MY and MBW (RFI_1); (2) ECM and MBW (RFI_2); (3) MY, MBW, and BWC (RFI_3). The prediction of GEBV were carried out by using ssGBLUP. We assessed the reliabilities of GEBV for DMI and RFI_i by applying two scenarios: (1) A five-fold cross-validation procedure, where half-sib families in the EDGP database were randomly divided (Scenario 1); (2) A straightforward validation procedure, where cows that were born within the last two years in Denmark were the test set. GEBVs were validated by using either the EDGP data or the Danish data as reference population (Scenario 2). The prediction reliabilities were calculated as the squared correlation between GEBV and the adjusted phenotype, divided by the approximated reliability of the adjusted phenotypes. In Scenario 1, the predictive reliabilities were 0.38 (DMI), 0.13 (RFI_1), 0.08 (RFI_2), and 0.14 (RFI_3). In Scenario 2, the prediction reliability increased, in average, by 35% when using the EDGP data, compared with using the Danish data only. In conclusion, our study has shown a low and moderated reliability for RFI_i and DMI, and that an international reference population improved the prediction reliabilities for DMI and RFI in Danish Holstein cows.

Sequence-based GWAS on feed efficiency and carcass traits in French Charolais bulls

S. Taussat[1,2], R. Saintilan[2], A. Michenet[2], S. Fritz[2], C. Hoze[2], E. Venot[1], P. Martin[1], D. Krauss[3], D. Maupetit[3] and G. Renand[1]
[1]GABI, INRA, AgroParisTech, Université Paris-Saclay, 78350, Jouy-en-Josas, France, [2]Allice, 75012, Paris, France, [3]UE0332 Domaine expérimental Bourges-La Sapinière, 18390, Osmoy, France; sebastien.taussat@inra.fr

Improvement of feed efficiency is of a major interest for beef breeders to increase their profitability. However, this trait is difficult and expensive to measure on large scale. Using causal mutations in genomic evaluation could be useful to facilitate selection. A sequence-based GWAS was performed on 789 Charolais terminal bulls for feed intake (FI), final weight (FW) and two feed efficiency criteria: residual feed intake (RFI) and ratio of feed efficiency (FE). Carcass yield (CY), percentage of muscle (MU) and fat (FAT) in carcass were also analysed. Animals were fed *ad libitum* with pelleted diet and slaughtered at seventeen months of age on average. Bulls were genotyped with the BovineSNP50 (50k) chip and their genotypes were successively imputed to HD density using Fimpute and up to the sequence level with Minimac software. GWAS analyses were performed using the GCTA software and a corrected Bonferroni threshold was used (-logP>5.75). In total, 1,404 SNPs were highly associated with one or several traits and 42 genes were identified. DIAPH3 and MMP13, already detected for feed efficiency in literature, were associated with RFI. Some SNPs associated with FE were presents in TYR genes which were already detected for body and carcass weight. Both FE and RFI were significantly associated with USH2A on BTA 16. Feed intake was associated with ZNF248 on BTA 28. Final weight was associated with SNPs in intergenic region only. Feed efficiency traits seemed to have different genomic control than FI and FW. Carcass yield, MU and FAT were all found highly associated with the same region of the BTA2. This result showed the major effect of the MSTN on the carcass composition, however, no association was found with RFI and FE. Carcass muscle content was also specifically associated with SLC38A1 in BTA 5, already found for carcass traits. These results showed that several genes were specifically associated with feed efficiency traits. Further investigation will be focus on genomic selection on these traits in French Charolais breed.

Transcriptomics analysis of key metabolic tissues to identify regulatory patterns of feed efficiency
H.Z. Sun, K. Zhao, M. Zhou, Y. Chen and L.L. Guan
University of Alberta, Department of Agricultural, Food and Nutritional Science, T6G2P5, Canada; lguan@ualberta.ca

To date the fundamental understanding of regulatory mechanisms behind the biological processes of feed efficiency traits is lacking. In this study, the global gene expression patterns across the rumen, liver, muscle and backfat tissues, which are the four key tissues involved in digestion and metabolism in beef cattle, were assessed to determine their roles in contributing to feed efficiency traits. The transcriptome profiling of 189 samples across four tissues from 48 beef steers with varied feed efficiency were generated using Illumina HiSeq4000 platform. The analyses of global gene expression profiles of four tissues were performed using mfuzz soft clustering, principle component analysis and hierarchical clustering. In addition, gene ontology terms functional analysis of tissue-shared and -unique genes, co-expressed network construction of tissue-shared genes using *Spearman* correlation and Cytoscape were performed. The weighted gene correlation network analysis (WGCNA) was further conducted to relate modules of highly correlated genes to the dry matter intake (DMI), average daily gain (ADG), residual feed intake (RFI) and feed conversion ratio (FCR) data using WGCNA package in R software. Among the four tissues, the transcriptome of muscle tissue was distinctive from others, while those of rumen and backfat tissues were similar. The associations between co-expressed genes and feed efficiency traits at single or all tissues level exhibited that the gene expression in the rumen, liver, muscle and backfat were the most correlated with FCR, DMI, ADG and RFI, respectively. We identified 19 genes from each feed efficiency traits-related gene module in four tissues, which may serve as potential generic gene markers for feed efficiency. Our data could provide new insights and perspectives on the functional genomic basis for the potential genetic selection of feed efficiency in beef cattle and create shareable reference data for the future research in cattle.

Metagenomic analysis indicates an association of rumen microbial genes with appetite in beef cattle
J. Lima[1], M.D. Auffret[1], R.D. Stewart[2], R.J. Dewhurst[1], C.-A. Duthie[1], T.J. Snelling[3], A.W. Walker[3], T.C. Freeman[2], M. Watson[2] and R. Roehe[1]
[1]SRUC (Scotland's Rural College), Roslin Institute Building, EH25 9RG Midlothian, United Kingdom, [2]The Roslin Institute and R(D)SVS, University of Edinburgh, Division of Genetics and Genomics, Roslin Institute Building, EH25 9RG Midlothian, United Kingdom, [3]The Rowett Institute – University of Aberdeen, Aberdeen, AB25 2ZD Aberdeen, United Kingdom; joana.lima@sruc.ac.uk

The rumen microbiome has been shown to influence the feed conversion efficiency of the host animal most likely by fermenting ingested feed and releasing nutrients, which are subsequently absorbed by the host. However, there is a lack of knowledge as to how the rumen microbiome might affect appetite in beef cattle. We used metagenomic analyses to characterize the microbiome based on relative abundances of rumen microbial genes. The main aim of this research was to generate novel insights into the biological processes and microbial pathways underlying the influence of the microbiome on host' appetite, through: (1) investigation of the association of rumen microbial genes with the appetite of the host animal (assessed by daily feed intake); and (2) identification of rumen microbial genes that could be used as biomarkers for appetite in beef cattle. Forty-two animals from 4 different breeds (crossbred Charolais, Aberdeen Angus, Limousin and purebred Luing, fed 2 basal diets differing in forage:concentrate ratio) were allocated to two groups with extreme differences in feed conversion ratio. Microbial DNA was extracted from rumen digesta collected at the abattoir and the relative abundance of genes was determined by alignment of metagenome sequence reads. Using partial least squares, we identified 18 microbial genes whose relative abundances explained 73% of the variation of daily feed intake. Microbial genes identified as important for the prediction of daily feed intake such as *rfbG*, *tolC* and *glnB* were associated with functions of cell wall biosynthesis, resistance to and synthesis of antibiotics, and environmental sensing. These results indicate that the relative abundances of the identified microbial genes may be suitable biomarkers of the animals' appetite, potentially associated with the microbiome-host interaction.

Searching for low methane dairy cows

P.J. Moate, S.R. Williams, L.C. Marett, J.B. Garner, J.L. Jacobs, M.C. Hannah, J.E. Pryce, W.J. Wales and C.M. Richardson
Agriculture Victoria Research, Dairy Production, 1301 Hazeldean Road, 3821 Ellinbank, Australia; caeli.richardson@ecodev.vic.gov.au

Enteric methane production from ruminants contributes a substantial proportion of global greenhouse gas emissions. Globally, research groups are screening large numbers of dairy cows for their enteric methane emissions in the hope that a gene marker will be found to identify low methane-emitting cows. Given the expensive and challenge to employ respiration chambers to screen large numbers of animals, we modified the SF_6 technique so that it can now be used to accurately measure enteric methane production from dairy cows. Over five years (2013-2017), we screened 480 Holstein dairy cows for methane emissions. Each year during spring (October – December), cows of 3 to 6 years of age, 40 to 100 days in milk and producing 25.5±4.1 kg/d (mean ± standard deviation) of energy corrected milk (ECM) were used in this research. Cows had individual access to automatic feed bins and were offered feed cubes of 75% lucerne hay and 25% crushed barley *ad libitum*. The cubes contained 191 g/kg DM of crude protein and 10.6 MJ ME/kg DM. After 4 weeks adaptation to the diet, individual cow daily feed intakes (DMI) and methane emissions were measured over 5 consecutive days. Dry matter intake was 25.4±3.4 kg DM/day, methane emissions 468±83.1 g/d, methane yield 18.6±3.2 g/kg DMI and methane intensity 18.7±3.9 g/kg ECM. As far as we can ascertain, this data set is unique because it is the only dairy cow database where methane emissions have been measured using a single method in the same season across multiple years, using animals of similar stage of lactation and offered a similar diet. This was deliberately done with the expectation that this would result in genetic variation constituting a considerable proportion of the observed variation in measured parameters. The ranges in values for methane yield (9.9 to 32.6 g/kg DMI) and methane intensity (8.6 to 31.5 g/kg ECM) indicate considerable opportunity for selecting low methane-emitting animals. We are currently analysing these data to try to find a gene marker for low methane intensity animals.

Enteric methane emission of dairy cows on practical farms

L. Koning and L.B. Šebek
Wageningen Livestock Research, Animal Nutrition, De Elst 1, 6708 WD Wageningen, the Netherlands; lisanne.koning@wur.nl

According to the Paris Climate Agreement of December 2015, the Netherlands aim to reduce greenhouse gas (GHG) emissions by 49% from 221 Tg CO_2-equivalents (eq.) in 1990 to 113 Tg CO_2-eq. in 2030. The dairy sector needs to reduce methane (CH_4) emissions by 1.0 Tg CO_2-eq. Around 80% of these CH_4 emissions originates from enteric CH_4. Observations on CH_4 emission are mainly based on data of respiration chambers with relatively low numbers of cows. It is unknown if the average emission and variation measured is representative for conditions in practice. Besides, not all variation is explained by feed intake and composition, other factors that cause variation are taken as random cow effects. This variation potentially offers opportunities to reduce CH_4 emission as well. The first objective of this study was inventory on the level of CH_4 emission from the Dutch dairy herd. The second objective was to investigate whether variation in CH_4 emission was related to the factors cow and farm. Enteric CH_4 emission was measured using the Greenfeed (C-lock Inc. Rapid City, SD, US), a feeding station that measures both CH_4 concentration and quantitative airflow. In total measurements were obtained for more than 500 dairy cows of 18 farms for a period of 3 weeks from September 2018 to May 2019. In addition, individual milk samples were taken to analyse milk composition and on herd level feed intake and composition data were collected. Linear Mixed Models fitted with Residual Maximum Likelihood were used to analyse the data. The preliminary results show an average CH_4 emission of Dutch dairy cattle of 429±63 g CH_4/cow/day, expressed in fat protein corrected milk 13.9±2.1 g CH_4/kg FPCM. These results also show differences between cows and farms. Within-farm variation is shown to be greater than between-farm variation, because at every farm a group representative for the composition of the herd was selected with considerable variation in parity, stage of lactation and milk yield. Comparable patterns and less variation is shown across farms. These preliminary results on CH_4 emission correspond reasonably with expected CH_4 emission based on model calculations (IPCC Tier 3), which is for a period between 2009 and 2012 13,1 g CH_4/kg FPCM.

Association between exhaled methane concentration and concentrate composition and rumination in cows

J. Rey[1], R. Atxaerandio[1], E. Ugarte[1], O. González-Recio[2], A. Garcia-Rodriguez[1] and I. Goiri[1]
[1]Neiker tecnalia, animal production, NEIKER-Granja Modelo de Arkaute, Apdo. 46. 01080 Vitoria-Gasteiz, Spain, [2]INIA, Animal production, Instituto Nacional de Investigación y Tecnología Agraria y Alimentaria (Madrid), 28040 Madrid, Spain; jrey@neiker.eus

Enteric methane emissions imply a loss of feed energy efficiency, which supposes an economic loss for the farms. In addition, methane (CH_4) emissions have a great impact on global warming. To address this problem, it is necessary to collect CH_4 measurements from a large number of animals individually, together with information on their potentially associated factors. The objective of this study was to study the association between methane concentration in exhaled air and concentrate composition and rumination activity in Holstein dairy cows under commercial conditions. For this reason, an infrared CH_4 detector (SNIFFER) was installed in the feeding trough of the automatic milking system (AMS) along 5 commercial farms, and methane concentration was recorded in each visit to the AMS. Methane concentration data was recorded for 14 days in each farm, registering the methane emissions of a total of 429 lactating dairy cows. The concentration of CH_4 was calculated from the average of the maximum concentration peaks during the stay of the cow in the AMS. Emissions were calculated with an ad hoc program, and then a 14-day average for each cow was calculated. Information about concentrate composition and rumination was also collected. Spearman correlation between CH_4 concentration and the associated factors was calculated. Average CH_4 concentration was 819.2 ppm (SD= ±406.40 ppm). Negative correlation was found between CH_4 concentration and rumination time (r=-0.306, P<0.001), average daily intake of protein (r=-0.246, P<0.001), calcium (r=-0.139, P=0.005), sodium (r=-0.200, P<0.001) and phosphorus (r=-0.205, P<0.001) in the concentrate. Finally, a positive correlation was observed with the average daily cellulose intake (r=0.242, P<0.001) in the concentrate. In conclusion, rumination time and concentrate composition supplied in the AMS, especially the crude protein and cellulose content, were related to methane concentration.

Predictive capabilities of a dynamic mechanistic model of *in vitro* fermentation by rumen microbiota

R. Muñoz-Tamayo[1], A.S. Atzori[2], A. Cannas[2], F. Masoero[3], S. Giger-Reverdin[1], D. Sauvant[1] and A. Gallo[3]
[1]UMR Modélisation Systémique Appliquée aux Ruminants, INRA, AgroParisTech, Université Paris-saclay, 75005, Paris, France, [2]Dipartimento di Agraria, Università di Sassari, 07100, Sassari, Italy, [3]Department of Animal Science, Food and Nutrition (DIANA), Facoltà di Scienze Agrarie, Alimentari e Ambientali, Università Cattolica del Sacro Cuore, 29100, Piacenza, Italy; rafael.munoz-tamayo@inra.fr

We have previously developed a mechanistic dynamic model to describe *in vitro* rumen fermentation. The model includes stoichiometry of fermentation, microbial metabolism, acid–base reactions and liquid–gas transfer. The rumen microbiota is represented by three functional microbial groups namely sugars utilisers, amino acids utilisers and hydrogen utilisers (methanogens). From the first calibration procedure, the model was efficient to represent data from experiments using rumen inocula from goats. The objective of this work was to evaluate the model predictive capabilities with an independent data set using rumen inocula from two cows. Five total mixed rations (TMR) were collected from different dairy farms. All TMR were characterized for chemical parameters, along with NDF, crude protein and starch degradability rates determined by previously developed enzymatic based methods. Each TMR was incubated in diluted rumen fluid. The dynamics of fermentation were tracked at six sampling times (2, 5, 9, 24, 36, and 48 h) by measuring gas production and composition, as well as the concentrations of acetate, butyrate, propionate, and ammonia. The model was calibrated using volatile fatty acids (VFA) and CH_4 data. The model represented satisfactorily VFA concentrations and CH_4 production with Lin's concordance correlation coefficient (CCC) of 0.93 in average. We further assessed whether the hydrolysis rate constants of dietary carbohydrates and proteins determined enzymatically could be used as known parameters in the model while keeping satisfactory predictions. When using the enzymatically determined hydrolysis rate constants, the CCC was in average of 0.89, indicating the need of further improvements. These results strengthen the capabilities of our model structure to predict the *in vitro* fermentation pattern and CH_4 production from rumen microbiota.

Rumen microbiome to predict methane emissions from cattle

L. Zetouni[1], M.K. Hess[1], R. Brauning[1], H. Henry[1], T. Van Stijn[1], A. McCulloch[1], J. Budel[1,2], J.C. McEwan[1], H. Flay[3], M. Camara[3] and S.J. Rowe[1]
[1]*AgResearch, Animal Genomics, 159-183 Puddle Alley, 9092, Mosgiel, New Zealand,* [2]*Para Federal University, Animal Science, Augusto Correa St., 01, Guama, 66075-110, Belem, Para, Brazil,* [3]*DairyNZ, Cnr Ruakura & Morrinsville Rds (SH26), RD6 Newstead, 3240, Hamilton, New Zealand; lzetouni@gmail.com*

Cattle are responsible for 62% of livestock methane (CH_4) emissions worldwide. The mix of microbial species in the rumen affects the amount of CH_4 released and the animal's efficiency in extracting nutrients and energy from feed. Current methods for obtaining microbial DNA are too expensive to implement in commercial selection programs. A methodology that offers fast, low-cost, high throughput profiling of rumen microbiomes using Genotyping-by-sequencing (GBS) has been developed and successfully used to generate microbial profiles from freeze-dried New Zealand sheep rumen samples. After digestion by restriction enzymes and fragment selection based on size, the method captures 0.3% of each microbial genome on average. Two approaches were used to determine individual microbial profiles: a reference-based, which BLASTs the sequences obtained against the Hungate1000 collection; and a reference-free, which uses non-redundant DNA sequences of at least 65 bp to obtain a count matrix with rows and tags for downstream analysis. We used 440 rumen samples from New Zealand dairy cattle to validate both approaches in a different ruminant species. Samples were freeze-dried, mechanically ground and used for DNA extraction, generating high quality DNA and ~ 800k reads per sample. The reference-based approach assigned 9.3±1.6% of reads while the reference-free approach assigned 64.3±6.8%, consistent with results in sheep. Phenotypic data from the 440 dairy cows in this study are being obtained to estimate the genetic correlations between the microbial profile and CH_4 yield and feed efficiency. This study is part of a major project funded by the Global Research Alliance on Agricultural Greenhouse Gases that aims to obtain samples from different production systems and cattle species worldwide to deepen global understating of how the rumen adapts to different production systems, the effect of this adaptation on rumen outputs and how this affects livestock performance.

Multi-trait genomic prediction of methane emission in Danish Holstein

C.I.V. Manzanillap-Pech[1], P. Løvendahl[1], D. Gordo[1] and J. Lassen[2]
[1]*Aarhus University, Faculty of Science and Technology, Department of Molecular Biology and Genetics, P.O. Box 50, 8830, Tjele, Denmark,* [2]*Vikings Genetics, Ebeltofvej 16, Assenstoft, 8960, Randers, Denmark; cmanzanillap@gmail.com*

Reducing methane (CH_4) emissions in livestock production is a challenge for this century. In dairy cattle, selecting for lower methane emitting animals is one of the best approaches to reduce CH_4 as genetic progress is cumulative and permanent. However, genetic selection requires a large amount of animals with records in order to get accurate estimated breeding values (EBV). Given that, CH_4 records are scarce, the use of information on routinely recorded traits highly correlated with CH_4 has been suggested to increase the accuracy of genomic EBV (GEBV) through multi-trait genomic prediction. Therefore, the objective of this study was to estimate and compare the accuracies of prediction of GEBV for CH_4 including or not, energy corrected milk (ECM) and body weight (BW) information in a multi-trait analysis. A total of 14,013 weekly records on CH_4 from 2,241 Danish Holstein multiparous cows with ~60,000 weekly records (from ~2,700 cows) on ECM and BW were included in the analyses. The reference population contained 2,714 cows with phenotypes (in CH_4, ECM or BW), 1,920 with genotypes with 50k markers. Four cross validation groups were created randomly grouping daughters per sire. Several scenarios were analysed: the base scenario was single trait of CH_4 not including additional information, and posteriorly including ECM, BW or both traits as information in: (1) reference and validation populations; or (2) only in reference population. Accuracies were obtained using single step GBLUP (SS-GBLUP). Fixed effects included herd-year×season-year, lactation week, and parity number as covariate. Random effects included additive genetic effect, permanent environmental effect and residual. The average accuracy of GEBV for CH_4 in the base scenario was 0.28 (SE=0.03), whereas in the other scenarios ranged from 0.21 to 0.62. Based on our results, we conclude that including information on ECM can increase significantly the accuracy of GEBV for CH_4, whereas, BW information only improved the accuracy slightly.

Animal breeding as a mitigation tool for enteric methane emissions of dairy cattle

Y. De Haas[1], G. De Jong[2], R. Van Der Linde[2] and E.P.C. Koenen[2]
[1]*Wageningen University and Research, Animal Breeding and Genomics, P.O. Box 338, 6700 AH Wageningen, the Netherlands,* [2]*CRV, P.O. Box 454, 6800 AL Arnhem, the Netherlands; yvette.dehaas@wur.nl*

Climate change is a growing international concern and the emission of greenhouse gases (GHG) is a contributing factor. The European Union has committed itself to reduce its GHG emissions by 50% by the year 2050 relative to 1990 levels. The global livestock sector, particularly ruminants, contribute substantially to the total anthropogenic GHG. Management and diet solutions to reduce enteric methane (CH_4) emissions have been, and continue to be, extensively researched. Animal breeding that exploits natural animal variation in CH_4 emissions is an additional mitigation solution that is cost-effective, permanent, and cumulative. We quantified the effect of including CH_4 production in the Dutch breeding goal using selection index theory. The current Dutch national index contains 15 traits, related to milk yield, longevity, udder health, fertility, conformation of udder and feet & legs, calving ease, claw health and feed efficiency. So far, very few studies have published genetic parameters for enteric CH_4 production. From literature we obtained a heritability of 0.21 for enteric CH_4 production, and genetic correlations with milk lactose, protein, fat and dry matter intake of 0.43, 0.37, 0.77 and 0.42, respectively. Correlations between enteric CH_4 production and other traits in the breeding goal were set to zero. When including CH_4 production in the current breeding goal with a zero economic value, CH_4 production increases each year with 1.5 g/d as a correlated response. When extrapolating this, the average daily CH_4 production of 392 g/d in 2018 will increase to 440 g/d in 2050 (+12%). However, in the same period the CH_4 intensity (CH_4 production/kg milk) will reduce by 13% due to the continuous genetic selection for higher milk production. The reduction of CH_4 intensity could go down even faster when specifically selecting for lower enteric CH_4 emissions of dairy cattle, showing that breeding is a valuable contribution to the whole set of solutions that could be applied in order to achieve the goals for 2050 set by the EU.

Digestibility of OM during the grazing period and the winter feeding conditions in suckler cows

H. Scholz[1], P. Kühne[1] and G. Heckenberger[2]
[1]*Anhalt University of Applied Sciences, faculty LOEL, Strenzfelder Allee 28, 06406 Bernburg, Germany,* [2]*State Institute for Agriculture and Horticulture Saxony-Anhalt, Lindenstraße 18, 39606 Iden, Germany; heiko.scholz@hs-anhalt.de*

To achieve a good reproductive performance of suckler cows and a high daily gain of their calves particularly a performance based supply is important. In the EU often there are two opposite interests: Extensive use of grassland versus achieving high animal performance to ensure an economical production. An increased extensivation can lead to a decrease in digestibility of grass during the grazing period. During the grazing season was sampling faeces from 5 suckler cows in each 6 beef cattle farm´s every month (May-September). The farms can be divided into: (1) intensive pasture management with more than 100 kg nitrogen per hectare; or (2) semi-intensive pasture management with less than 100 kg N/ha; and (3) ecologically farms without nitrogen fertilizer on grassland. Faces were analysed for digestibility of organic matter after the method from Lukas *et al*. During the winter period was sampling faeces two times and analysed for digestibility. Statistical analysis took place with ANOVA with fixed effects of farms (1-6), month (May until September) and number of lactations (1-7 lactations) using SPSS Version 25.0 for windows. An alpha of 0.05 was used for all statistical tests. Digestibility of organic matter during the grazing period 2018 was 71% with variation from average 65 to 74% in the individual farms. The cows on (1) shows an average of 74% digestibility, (2) of 72% and (3) was estimated with 66% digestibility in the faces. From Mai (76%) to September (70%) was a tendency decreased digestibility or organic matter observed in all farms. In the winter feeding period was an average digestibility of 66% observed. An effect of crude fibre in the ration on digestibility wasn´t found. The investigation shows an effect of the degree of extensivation of pasture on digestibility of organic matter during the grazing period of suckler cows. However, this reduction in digestibility of organic matter can be controlled within certain limits through good grazing management and sufficient grazing.

The effect of grass and maize silage ratio on milk and methane production of lactating dairy cows

M. Lamminen, C.F. Børsting, A.L. Frydendahl Hellwing and P. Lund
Aarhus University, Department of Animal Science, P.O. Box 50, 8830 Tjele, Denmark; marjukka.lamminen@anis.au.dk

The aim of this study was to examine the effect of linear substitution of grass silage with maize silage on methane production of dairy cows measured in respiration chambers. Five multiparous Holstein cows were used in an extended 4×4 Latin square study. Treatments consisted of total mixed rations with varying proportions of first cut grass-clover silage and maize silage (100:0, 66:33, 33:66, 0:100 on dry matter (DM) basis, respectively), and concentrate:forage ratio of 30:70 in all diets. Diets were formulated in NorFor to meet the energy and metabolizable protein requirements. The concentration of crude protein and starch were balanced between treatments by adjusting concentrate composition. The fifth cow was added to the trial in the last two periods, due to a declining yield of one of the original cows. The milk yield data of the low yielding cow was omitted, resulting in n=14 for milk parameters and *n*=18 for others. Data were analysed by analysis of variance using the Mixed-procedure of SAS 9.4. Pure maize silage (0:100) resulted in higher DM intake (DMI) than pure grass silage (100:0) (20.4 vs 18.1 kg/d, P<0.05), and DMI increased linearly (P<0.05) with increasing proportion of maize. Forage source had no effect on milk or energy corrected milk yield (on average 33.0 and 30.8 kg/d, respectively). The daily CH_4 and CO_2 production did not differ between the pure grass and maize silage diets (on average 549 and 6,714 l/d, respectively), but CH_4 tended to be higher for mixtures of grass and maize silage (66:33 and 33:66) than for pure grass or maize (P=0.06 for quadratic effect). Ratio of CH_4:DMI was lower for pure maize than for grass silage (32.0 vs 26.0 l/kg P<0.05), and it decreased linearly (P<0.05) with increasing proportion of maize silage. Pure maize silage resulted in lower ratios of CH_4:CO_2 than pure grass silage (0.080 vs 0.084, P<0.05). In conclusion, grass and maize silage were equally suitable for dairy cows in terms of daily milk and CH_4 production, but CH_4 intensity per DMI and per CO_2 were lower for maize than for grass silage.

Feed efficiency and blood immune system indicators in Nellore beef cows

S.F.M. Bonilha, A.C.A.R. Paz, M.E.Z. Mercadante, J.N.S.G. Cyrillo, R.C. Canesin and R.H. Branco
Instituto de Zootecnia, Rodovia Carlos Tonani, km 94, 14.174-000 Sertãozinho, SP, Brazil; sbonilha@iz.sp.gov.br

Blood analysis as a practical measure to identify efficient animals has relevance, once there is scientific evidence of associations between immune system indicators and feed efficiency in beef cattle. This study aimed to investigate relationships between feed efficiency and immune system blood parameters in Nellore beef cows. Fifty-eight Nellore cows (27 with and 31 without calves), having 491±37.6 kg and 1,157±20 d of initial body weight and age, respectively, were fed over a 73 d feeding period to evaluate differences in dry matter intake (DMI), residual feed intake (RFI), white blood cells (WBC), and lymphocytes (LYM), monocytes (MON), and segmented neutrophils (SNE) counts. RFI was calculated by regression of DMI in relation to average daily gain and mid-test metabolic body weight. Blood samples were obtained on days 0 and 73 and were analysed for WBC, LYM, MON and SNE. Cows were placed into a negative (NEG; RFI<0) or positive (POS; RFI>0) RFI group. Data were analysed using the SAS MIXED procedure, including as fixed effects RFI group and calf presence, and as random effect blood collection day. Means were compared by t test at 5% of probability. RFI mean detected for NEG RFI cows was -0.727±0.091 kg/d and for POS RFI cows was 0.946±0.104 kg/d. DMI was 18% (P<0.001) greater for POS (11.6±0.170 kg/d) compared to NEG (9.82±0.149 kg/d) RFI group. No significant differences were detected between NEG and POS RFI group, respectively, for WBC (12.2±1.42 vs 11.6±1.51 10^3 cells/µl; P=0.6489), LYM (47.5±2.82 vs 44.8±2.93 WBC%; P=0.1931), and MON (4.76±2.84 vs 4.78±2.84 WBC%; P=0.9613). NEG RFI cows had lower SNE count than POS RFI cows (39.8±7.39 vs 43.8±7.43 WBC%; P=0.0395). During a SNE inflammatory response, increased oxygen consumption results in greater energy expenditure. Thus, the higher abundance of SNE may be related to increased basal energy requirements. More studies are needed to determine if WBC, LYM, MON and SNE counts may be useful traits to identify efficient Nellore animals, however it is possible to infer that cellular defence system is linked with feed efficiency. FAPESP Process 2017/06709-2.

Organic rose veal meat: an environmentally friendly alternative for beef meat production?
A. Mertens, V. Decruyenaere, M. Mathot, A.-M. Faux and D. Stilmant
Walloon Agricultural Research Center, rue du serpont, 100, Libramont, Belgium; a.mertens@cra.wallonie.be

While ruminant production systems are criticized for their contribution to climate change, grass-based systems offer promising opportunities to reduce the environmental impact of beef production. We explored the environmental, zootechnical and ecoconomic performances of producing organic grass-based rose veal meat from calves slaughtered before the age of eight months. Twelve dual-purpose Belgian blue suckler cows (genotype 'mh/mh') and their calves, eight males and four females, were followed in 2018 at the experimental farm of the CRA-W located in Libramont (Belgium). Calves were born between February 16 and April 2. Grazing season started on May 4 and lasted until December 6. A rotational grazing including five paddocks (0.6 ha each) with an instantaneous stocking rate of 27 LU per hectare was adopted. During the grazing period, cows had only access to grass while calves were supplemented with spelt and an organic concentrate (16% protein content). Calves were weighed monthly and their enteric methane emissions measured in July and August using a Greenfeed system. Male calves were valorised through the organic value chain. Carcasses were weighed at the slaughter house and the extra-muscular fat content of a rib was measured. Economic performances of this production scheme were calculated. Calves achieved an average daily growth of 1.2 ± 0.2 kg/day during their whole life span. The total amount of concentrate-spelt mixture supply was 250 kg per calf. Males were slaughtered from 7.35 to 8.05 months and sold at 6.6 ± 0.1 € per kg of carcass. The carcass weight was 209 ± 29 kg (64.6±2% carcass yield), and the extra-muscular fat content (w/w) was $7.0 \pm 1.2\%$. Six males visited the Greenfeed system resulting in 114 methane emissions measurements. The mean methane emission value was 82 ± 14 g/day. Grass-based rose veal meat might be a relevant option for organic beef meat production in the Belgian Ardenne. Satisfactory animal performances were obtained with a relatively low concentrate supply. Perspectives include a second year experiment to quantify the environmental performances of the whole production system.

Changes of nutritive value by mixing ratio of Korean native forages in paddy field
B.R. Choi, T.Y. Hwang, J.S. Jung, G.J. Choi and W.H. Kim
National Institute of Animal Science, 114, Sinbang 1-gil, Seonghwan-eup, Seobuk-gu, Cheonan-si, Chungcheongnam-do, 31000, Korea, South; qaz8715@gmail.com

Rice consumption is a very important issue because rice is a major food in my country. Rice consumption has been decreasing since the mid-2000s. Accordingly, it is necessary to expand the cultivation of alternative forage crops using rice paddy field in Korea. This study was conducted to examine the yields, nutritive values and *in vitro* dry matter digestibility (IVDMD) of forage using domestic native forage resources such as barnyard millet, foxtail and crabgrass in no-till paddy field. This experiments were consists of 4 groups of mixing ratio of domestic native forage resources (barnyard millet, foxtail, crabgrass), such as group 1 (10:10:4), group 2(15:7.5:3), group 3(7.5:15:3) and group 4(7.5:7.5:6). IVDMD was analysed using Daisy[II] incubator. All data were analysed by the GLM procedure of SAS package 9.4. Yields of dry matter and total digestible nutrient (TDN) in group 2 were the highest (P<0.05). The contents of acid detergent fibre and TDN in group 1 and group 4 had the highest (P<0.05). However there is no significantly difference in the nutritive values and IVDMD of all groups. Therefore, we suggest that group 2 is the best mixing ratio of forages in paddy field during the summer season.

Effect of garlic-citrus based additive on *in vitro* rumen methanogenesis and abundance of microbes

J. Jeyanathan[1], B. Vlaeminck[2], O. Riede[3], C.J. Newbold[4] and V. Fievez[1]
[1]Ghent University, Faculty of Bioscience Engineering, 9000 Ghent, Belgium, [2]Ghent University, Faculty of Sciences, 9000 Ghent, Belgium, [3]Mootral GmbH, Mainzer Landstr. 251, 60326 Frankfurt, Germany, [4]Scotland's Rural College, West Mains Road, Edinburgh, EH9 3JG, United Kingdom; jeyamalar.jeyanathan@ugent.be

Use of plant-based feed additives is one preferred strategy to reduce rumen methanogenesis, which can be easily accepted by producers as well as customers alike. In this study we investigated the effect of a garlic-citrus based additive, Mootral (TM), on production of volatile fatty acids (VFA), methane (CH_4) and, abundance of bacteria and methanogens *in vitro* by using batch culture system and the rumen simulation technique (RUSITEC). Mootral, a combination of garlic (Allium sativum) powder and a flavonoid-rich extract from bitter orange (*Citrus aurantium*), was incubated at different concentrations (0.5, 2.5 and 5 mg/ml incubation media for low, medium and high dose, respectively) with substrate (mixture of grass silage, maize silage and dairy concentrates) and rumen inoculum collected from dairy cows. A control treatment without additive was included in all incubations. The 'medium' and 'high' dose of additive decreased ($P \leq 0.05$) the CH_4 production in batch culture system, whereas in RUSITEC system, the decrease in CH_4 production ($P \leq 0.05$) was observed with the 'high' dose. The VFA production was not affected by the additive irrespective of the doses used, however changes in VFA profile were observed. Molar proportions of propionate increased ($P \leq 0.05$) with the 'medium' and 'high' dose in the batch culture system, whereas molar proportions of butyrate increased with the 'high' dose in the RUSITEC system. Quantitative PCR, targeting total bacteria and total methanogens, indicated that the additive is specifically targeting methanogens. The garlic-citrus based additive showed strong anti-methanogenic ability without impairing the rumen fermentation activity *in vitro*.

Effects of nutritionally improved straw in dairy cow diets at two starch concentrations

M.E. Hanlon[1], J.M. Moorby[1], H. McConochie[2] and A. Foskolos[1,3]
[1]Aberystwyth University, Institute of Biological, Environmental and Rural Sciences, Gogerddan, SY23 3EE, Aberystwyth, United Kingdom, [2]Wynnstay Group Plc, Dairy Technical Services, Eagle House, Llansantffraid, SY22 6AQ Powys, United Kingdom, [3]University of Thessaly, Department of Animal Science, Campus Larisa, 413 34 Larisa, Greece; andreasfosk@hotmail.com

The objective of this experiment was to explore effects of different dietary neutral detergent fibre (NDF) sources within diets of lactating dairy cattle with low or high starch concentrations. Holstein cows in early-to mid-lactation (n=12; 666±67 kg of body weight) and dry cannulated cows (n=4; 878±67 kg of body weight) were used in multiple 4×4 Latin square design and fed four different diets. The treatments were 50:50 forage to concentrate diets within a total mixed ration consisting on a dry matter (DM) basis of 42% grass silage, as the main forage, 7.8% chopped untreated straw or nutritionally improved straw, and 50% of two different concentrates with low or high starch level (16.0 vs 24.0, respectively). There were four experimental periods, each consisting of a 21-day (d) adaptation period and 7-d sampling period. During the sampling periods for each experimental period, DMI and rumen fluid pH, ammonia-N and volatile fatty acid concentrations were recorded for each trial. Sampling for trial 1 also included milk yield and composition, body weight (BW), body condition score, digestibility of nutrients from the experimental diets and N balance. All statistical analyses were conducted using JMP. Dry matter intake and milk yield was affected by the type of straw included in the diet. A 1.45 kg/d higher DMI was seen when NIS was fed compared to untreated straw. The higher DMI was a result of a 4.3% higher NDF digestibility when fed NIS. The higher DMI resulted in a 1.05 kg/d higher milk yield and greater milk protein yield. Milk protein concentration was only affected by straw type and was 3.87% higher when NIS was fed due to a 30 g/d greater N intake. The dietary starch level of the diets also affected DMI, which was 1.55 kg/d greater when fed diets with low starch levels, resulting in a 1.05 kg/d higher milk yield. In conclusion, addition of NIS and low starch levels in the diet were beneficial for milk production.

Effect of cicer milkvetch on invitro digestion, methane production and microbial protein synthesis

B.M. Kelln[1], J.J. McKinnon[1], G.B. Penner[1], T.A. McAllister[2], S.N. Acharya[2], A.M. Saleem[2,3], B. Biligetu[1] and H.A. Lardner[1]
[1]University of Saskatchewan, 52 Campus Dr., Saskatoon, Saskatchewan, S7N 5B4, Canada, [2]Agriculture and Agri-Food Canada, Lethbridge Research Center, Lethbridge, Alberta, T1J 4B1, Canada, [3]South Valley University, Faculty of Agriculture, Qena, 83523, Egypt; bmk371@mail.usask.ca

The objective of this study was to investigate the effects of cicer milkvetch (*Astragalus cicer* L.; CMV) and alfalfa (*Medicago sativa* L.; ALF) on dry matter disappearance (DMD), methane (CH_4) production, short chain fatty acid (SCFA) concentration, and microbial protein synthesis in an artificial rumen system (RUSITEC). The experiment was a completely randomized block design with 4 treatments assigned to sixteen fermentation vessels (4/treatment) in two RUSITEC systems. The study period was 15-d, with an 8-d adaptation and 7-d sample collection period and DMD was measured after 48-h of incubation. Cicer milkvetch and ALF were incubated in ratios of 25:75, 50:50, 75:25 and 100:0 (as DM). Dry matter disappearance was significantly increased (P<0.01) (75.42, 78.48, 81.20, 83.07%) with increasing level of CMV inclusion (25, 50, 75, 100% CMV). Total gas production (1.04 vs 1.21 l/24 h) and CO_2 concentration (12.40 vs 14.12%) were increased numerically, however CH_4 production (mg/g DMD) and CH_4 concentration (%) were not changed for 25 to 100% CMV inclusion. Numerically, total dissolved hydrogen decreased (0.35 to 0.15 µmol/l) for 25 vs 100% CMV, and effluent production and rumen fluid pH were not changed by treatment. Total ruminal SCFA concentration did not differ (P=0.17) (166.59, 175.38, 181.66, 170.81 mM) among treatments. The acetate to propionate ratio was significantly lower (P<0.05) (3.14 vs 2.85 mM) and there was an increased trend (P<0.07) in propionate concentration (from 34.26 to 38.51 mM) for the 25 vs 100% CMV treatment. Total microbial N production was increased (P<0.05) for 25% CMV (35.8 mg/d) than 100% CMV treatment (52.4 mg/d). Results indicate that inclusion of CMV into forage diets can be beneficial by improving DM digestibility, lowering acetate to propionate ratio and improving microbial protein synthesis.

Role of TGFB1 as key regulator of feed efficiency in beef cattle

J.B.S. Ferraz[1], P.A. Alexandre[1,2], L.R. Porto-Neto[2], A. Reverter[2] and H. Fukumasu[1]
[1]University of Sao Paulo, FZEA, Veterinary Medicine, Rua Duque de Caxias Norte, 225, Campus da USP, 13635-970 Pirassununga, SP, Brazil, [2]CSIRO, Agriculture & Food, Level 5, Queensland Bioscience Precinct 306 Carmody Rd, St Lucia, Brisbane, QLD 4067, Australia; jbferraz@usp.br

To Identify central regulatory genes of complex phenotypes as feed efficiency (FE) is important to achieve some of the main animal selection goals and increase productivity of livestock. We aimed to do so by analysing RNAseq data (Illumina HiSeq2500, 100 pb, pair-ended) of adrenal gland, hypothalamus, liver, skeletal muscle and pituitary samples collected from 18 young bulls classified as high and low FE by residual feed intake. Samples were aligned to bovine reference genome (UMD3.1) using STAR and read counts was performed using HTseq. Expression values were estimated by FPKM, keeping only genes presenting an average value of 0.2 across all samples and tissues. From the 17,354 expressed genes considering all tissues, 1,335 were prioritized by five selection categories (differentially expressed, harbouring SNPs associated with FE, tissue-specific, secreted in plasma and key regulators) for network construction using PCIT algorithm. TGFB1, the sixth most connected gene in the co-expression network, was validated as a key regulator of FE in cattle muscle by motif discovery analysis of its target genes and using publicly available ChIPseq data of skeletal muscle cells to determine its target's accessibility. The analysis of the 217 genes co-expressed with TGFB1 showed enriched motifs for master regulators of muscle differentiation, MEF2 (NES=10.42) and MYOD1 (NES=5.09), with 136 direct target genes. Indeed, TGFB1 signalling triggers the phosphorylation of SMAD2/3 transcription factors, which in skeletal muscle cells co-bind with MYOD1 leading to specific transcriptional changes. Using ChIPseq data we found an overlap between MYOD1 and SMAD3 target genes demonstrating the significant association between both genes in skeletal muscle (hypergeometric test 1.98×10^{-6}). TGFB1 has been previously pointed as a master regulator of FE in beef cattle using genomics and metabolomics data and, in pigs, there is evidence that increased FE is associated with stimulation of muscle growth by TGFB1 signalling pathway. That was corroborated by these results.

Carcass composition is linked to residual feed intake level in crossbreed growing bulls

S. Lerch, J.L. Oberson, P. Silacci and I. Morel

Agroscope, Ruminant Research Unit; Animal Biology, Tioleyre 4, 1725 Posieux, Switzerland; isabelle.morel@agroscope.admin.ch

Feed efficiency is often determined by the residual feed intake (RFI) that is the difference between measured and predicted feed intake [the latter is obtained by a linear regression relating feed intake on metabolic body weight ($BW^{0.75}$) and average daily gain (ADG)] within a group of cattle receiving the same diet. This study aimed to assess the effect of RFI level on carcass composition of crossbreed bulls. Bulls (n=84, 155±9 kg BW, 124±13 days old; mean ± SD) cross from dairy mother and dairy (n=22), mixed (n=24) or beef (n=38) breed father were allocated into three growing diets. Diets composed of maize (whole crop or ear-enriched) silages, grass silage and concentrates (76:24 forage/concentrate) were iso-proteinic and had low, mid or high levels of net energy for meat production (7.0, 7.4 and 7.5 MJ/kg DM, respectively). Bulls were slaughtered at a BW of 534 kg (n=41, 340 days old) or 603 kg (n=43, 390 days old). Carcass weight, proportions of adipose tissue, muscle and bone of the 11th rib were determined in order to predict carcass composition using the linear models developed by Geay and Béranger. Within each treatment, RFI was computed over a 152 days period (164 to 315 days old) before bulls were classified as high, mid or low feed efficient (relative to the RFI mean, HiE>0.5 SD below, MiE0.5 from and LoE>0.5 SD above, respectively). Carcass traits were analysed by ANOVA using the MIXED procedure of SAS with diet, breed type, BW at slaughter, RFI class and their interactions as fixed effects, and bull as random effect. On average, ADG was 1.46 kg/d and was unaffected (P>0.10) by diet, breed type or RFI class. Among breeds, beef crossbreed bulls had higher (P<0.01) carcass weight and yield, whereas dairy crossbred tended to have lower (P<0.10) muscle and had higher (P<0.05) bone proportions in carcass. For LoE bulls, amount and proportion of adipose tissue were higher (13.3%; P<0.05) and that of muscle tended to be lower (68.8%; P<0.10), than for MiE and HiE bulls (11.8% adipose, 70.4% muscle). Such results highlight that animal selection based only on RFI could affect carcass composition and thus carcass sale incomes.

Breed effect on the rumen microbiome and its association with feed efficiency in beef cattle

F. Li[1], T. Hitch[2], Y. Chen[1], C. Creevey[3] and L.L. Guan[1]

[1]University of Alberta, Department of Agricultural, Food and Nutritional Science, Edmonton, T6G2P9, Canada, [2]Aberystwyth University, Institute of Biological, Environmental and Rural Sciences, Aberystwyth, United Kingdom, [3]Queen's University Belfast, School of Biological Sciences, Institute for Global Food Security, Belfast, United Kingdom; lguan@ualberta.ca

Rumen microorganisms are responsible for the rumen fermentation that produces volatile fatty acids for cattle growth and therefore, the rumen microbiome has been suggested to play a key role in affecting cattle feed efficiency. However, to what extent the breed affects the rumen microbiome and its association with host feed efficiency is unknown. In this study, we analysed metagenomes and metatranscriptomes of rumen microbiomes from beef cattle (n=48) with varied breeds (Angus, Charolais, Kinsella composite hybrid) and feed efficiencies. Rumen metagenomes were more conserved among individuals than metatranscriptomes, suggesting that inter-individual functional variations at the RNA level were higher than those at the DNA level. Our results indicate that rumen microorganisms possess a diverse, genetically encoded, functional potential to enable survival and proliferation, but with a large emphasis on activities for feed degradation and fermentation. While distinguishable active rumen microbial composition (metatranscriptome) and functional genetic potential (metagenome) were observed among three breeds, the active functional variations at the metatranscriptomic level were less pronounced. Our results suggest that metatranscriptomics is a more powerful approach to better associate rumen microorganisms with host performances. The differentially abundant rumen microbial features detected between cattle with high and low feed efficiency were mostly breed-specific, suggesting interactions between host breed and the rumen microbiome contribute to the variation of feed efficiency. The observed breed-associated differences represent potential superiorities of each breed, which could be further applied to manipulate the rumen microbiome through genetic selection and breeding.

Relationships between feed efficiency, live weight and faecal composition of Italian Holstein cows

G. Carlino[1], G. Niero[1], F. Cendron[1], F. Omodei Zorini[2], R. Finocchiaro[3], G. Invernizzi[2], G. Savoini[2] and M. Cassandro[1]
[1]University of Padova, Department of Agronomy, Food, Natural resources, Animals and Environment, Via dell' Università, 16, 35020, Legnaro, Italy, [2]University of Milano, Department of Health, Animal Science and Food Safety, Via Celoria, 10, 20133, Milano, Italy, [3]National Association of Holstein and Jersey Breeders (ANAFIJ), Via Bergamo, 292, 26100, Cremona, Italy; gabriele.carlino@phd.unipd.it

Thirty Italian Holstein lactating cows were investigated for the relationships between live body weight (LBW), faecal composition and feed efficiency (FE). Cows were from the University of Milano experimental farm and were sampled in June and July 2018. Data were recorded daily for individual dry matter intake (DMI), and weekly for milk production, LBW and faecal sample. Days in milk, DMI and LBW averaged 177.6 60.5, 19.9±3.2 kg/d and 597.4±65.5 kg, respectively. Means of milk yield (MY), fat percentage and protein percentage were 26.3±5.8 kg/d, 3.8±0.8% and 3.2±0.2%, respectively. Feed efficiency, calculated as the ratio of energy corrected milk to DMI, averaged 1.3±0.4. Individual faecal samples were analysed for neutral detergent fibre (NDF), acid detergent fibre (ADF), acid detergent lignin (ADL), undigested neutral detergent fibre on the % of total NDF (uNDF), acid detergent insoluble ashes (ADIA) and ashes (ASH); averages of these compounds (on dry matter basis) were 64.6±3.5, 37.1±3.0, 15.7±0.7, 76.4±5.0, 12.8±0.5 and 5.8±0.8%, respectively. Pearson's correlations of DMI and MY with FE were -0.33 and 0.46 ($P<0.05$), respectively, and a not significant association was estimated between LW and FE (0.10). Overall, DMI and FE did not correlate with faecal composition, except for DMI with NDF (-0.30; $P<0.05$) and FE with ASH (0.31; $P<0.05$). Moderate associations were assessed between MY and NDF (-0.40; $P<0.01$) and between MY and ADF (-0.33; $P<0.05$). Cows were divided in three groups, according to mean and standard deviation of FE: low FE, medium FE and high FE. Faecal composition was analysed through ANOVA using a model that included FE group, days in milk classes and parity as fixed effects. Results showed that faecal composition of cows in the high FE group had significantly higher ADL, ASH and uNDF than the other two groups.

Effects of enzymes feed additives on beef cattle performance

C.J. Härter[1], M.C. Tamiozzo[1], L.S. Monteiro[2] and J.M. Santos Neto[2]
[1]Universidade Federal de Pelotas, Animal Science, Eliseu Maciel, s/n, Capão do Leão, RS, 96050-500, Brazil, [2]Universidade de São Paulo, Animal Scince, Pádua Dias,11, Piracicaba, SP, 13418-900, Brazil; harter.carla@gmail.com

Several studies using enzyme additives in ruminant diets have reported controversial results on beef cattle performance and compiling the literature data may help find more conclusive results. Therefore, we conducted a meta-analysis to investigate the impact of fibrolytic enzyme supplements on performance of finishing beef cattle. Data were compiled from 18 peer-reviewed publications and the enzymes supplements were classified as endoglucanase and xylanase (EX), cellulase and xylanase (CX) and compared to a non-enzyme supplemented diet used as control. The variables evaluated were: average daily gain (ADG), gain per fed (G:F), dry matter intake (DMI), dressing, ribeye area, and fat thickness. The meta-analysis was performed as mixed models regressing the variables against the fixed effect of enzyme type. Average body weight and crude protein intake were used as covariate in the model. Study was considered a random effect and standard error of means or number of replicates by treatment was used as weighing factor. Dry matter intake and fat thickness were similar among the enzymes supplements, with averages of 9.81±0.46 kg/d and 9.50±1.3 cm respectively. The CX supplement promoted the greatest ADG (1.47±0.108 kg, $P<0.05$), which led G:F of cattle fed CX to be the greatest (0.156±0.012 kg/kg, $P<0.01$) as no change between treatments was observed on DMI. Following these results, CX combination was the greatest as well on dressing (58.6±1.44%, $P<0.01$), and ribeye area (77.4±3.19 cm, $P<0.01$). The ADG and G:F observed for the control and EX enzymes were 1.41±0.108 kg and 0.146±0.012 kg and 1.43±0.112 kg and 0.150±0.013 kg, respectively. The dressing and ribeye area observed for the control and EX enzymes were 58.1±1.43% and 75.4±3.19 cm^2 and 57.4±1.44% and 75.4±3.20 cm^2, respectively. Cellulases are composed by endoglucanases, exoglucanases and β-glucosidases which may help to better digest the diet cellulose than endoglucanase only. Therefore to improve finishing beef cattle performance, cellulase and xylanase enzymes are indicated to be added in the diet.

Feed efficiency of dairy cattle in an urbanizing environment – insights from Bangalore, India

M. Reichenbach[1], A. Pinto[2], S. König[2], P.K. Malik[3] and E. Schlecht[1]
[1]Animal Husbandry in the (Sub-)Tropics, University of Kassel, 37213 Witzenhausen, Germany, [2]Animal Breeding and Genetics, Justus Liebig University Gießen, 35390 Gießen, Germany, [3]National Institute of Animal Nutrition and Physiology, Adugodi, 560030 Bangalore, India; marion.reichenbach@gmail.com

The urbanization of our environment is an indisputable feature of the 21st century, pushing agroecosystems towards intensification and/or more efficient resources use. As an emerging megacity in India, Bangalore combines rapid urbanisation with a high demand for dairy products and, interestingly, presence of dairy cattle in its urban landscape. The aim of this study was to quantify on-farm resources use with a focus on feed efficiency of dairy cattle, and thus, to assess the efficiency of dairy units in relation to the urbanization of its surroundings. Twenty-eight dairy units were selected across the rural-urban interface of Bangalore: 4 urban, 8 peri-urban and 16 rural ones. Over one year, each dairy unit was visited at a 6-weeks interval to collected data on offered forages and milk offtake at animal level. Forages were analysed for nutrient content, *in vitro* dry matter digestibility and methane production. Most commonly offered forages were self-cultivated maize and elephant grass, and during the dry season, finger millet straw. Dairy producers also offered grass collected on field margins or public land, and in urban areas, market waste. 77% of them sent their cows to pasture. Across dairy units and seasons, the range of fresh green forage offered was between 0 and 44.4 kg/cow/day. In addition, each cow was fed twice a day between 0 and 4.3 kg of concentrate. Methane production of offered forage or concentrate was between 8.7 and 45.7 litres CH_4/kg DM. Daily milk offtake in the early lactation varied from 5.8 to 18.1 kg/cow. These preliminary data show the high variability in feeding strategies and subsequent feed (in-) efficiency. Urban dairy units combined a higher milk production with the (environmentally beneficial) conversion of market waste into a highly demanded animal product. However, manure management in the inner city was poor, suggesting that peri-urban dairy units holds the highest potential for efficient resources use as they combine opportunities of rural and urban localities.

Opportunities to improve the productive potential of dairy cattle from conception through calving

M.E. Van Amburgh and R.A. Molano
Cornell University, Department of Animal Science, 272 Morrison Hall, Ithaca, New York 14850, USA; mev1@cornell.edu

The productive potential of dairy cattle is initially determined by genetics, and further shaped by the environment and the interaction of both factors. From conception forward, environmental signals have been identified that play a key role in the phenotypic expression of the innate capacity of the animal. During gestation, factors such as nutrient status of the dam and heat stress impact maternal blood flow and nutrient transfer, conceptus development, fetal organ development, muscle development and ultimately growth and BW at maturity. Similarly other environmental signals after birth have been recognized to alter the calf's productive capacity. There are significant data characterizing the role of non-nutrient, non-immunoglobulin components in colostrum on immunity, gastrointestinal development, nutrient absorption and anabolic status of the calf. Longitudinal studies demonstrate greater growth and anabolic status in calves receiving more colostrum over 1 to 4 days of life. Further, the heifer's nutrient intake above maintenance during development is also a predominant stimuli resulting in improved productivity as a lactating cow. In large a study, milk replacer intake, and consequently growth rate pre-weaning, were significant and positively correlated with milk production during the first and subsequent lactations. In this data, pre-weaning growth accounted for 22% of the variation in first lactation milk production, which was approximately 4 times the variation explained by genetic merit. These investigators also detected the same relationship between rate of gain from birth to first breeding and first lactation milk production, and between average daily gain from weaning to breeding and the cumulative milk production until third lactation. Additional work has associated this enhanced nutrient availability with increased mammary gland development, and improved body weight and composition at calving, as part of the mechanisms facilitating improved performance. Likewise, several studies demonstrate these effects are interdependent and could be cumulative. It is important to recognize that with this new and emerging information on epigenetics, basic growth outcomes need to be met in order to benefit from this emerging capacity to improve productivity.

Why dry or why not? – pros and cons of shortening or omitting the dry period for cow and calf health

A. Van Knegsel[1], B. Kemp[1] and A. Kok[1,2]
[1]*Wageningen University, Adaptation Physiology group, De Elst 1, 6708 WD Wageningen, the Netherlands,* [2]*Wageningen University, Animal Production Systems group, De Elst 1, 6708 WD Wageningen, the Netherlands; ariette.vanknegsel@wur.nl*

It is well known that shortening or omitting the dry period in dairy cows results in a more positive energy balance and improves metabolic status in early lactation, due to lower milk yields and sometimes also improved energy intake. Moreover, the improved energy balance due to shortening or omitting of the dry period is related with improved reproductive performance. Besides these beneficial health effects, shortening or omitting of the dry period has potential detrimental consequences for animal health and performance. Shortening and omitting of the dry period results in a reduction in milk yield in the subsequent lactation. Furthermore, specifically omitting of the dry period results in: (1) lower antibody levels both in colostrum and in plasma of calves during the early weeks of life; and (2) loss of opportunity to treat cows with an elevated somatic cell count with dry cow antibiotics during the prepartum period. Dry period length in relation to animal health, performance and management was the focus of 2 large research projects and several PhD theses at our group in the last decade. This contribution gives an overview of benefits and drawbacks of shortening or omitting of the dry period for both cow and calf health and discusses how customising dry period management based on individual cow characteristics could make use of the benefits and limit the drawbacks of shortening or omitting of the dry period. Specific cow characteristics like somatic cell count, milk yield level, parity and disease history were input variables for a decision support model to optimize dry cow management (dry period length and use of dry cow antibiotics) at cow level. Customising dry period management using this decision support model resulted in a reduction in milk yield losses at herd level, compared with shortening or omitting of the dry period for all cows. Furthermore, customising dry period management resulted in a reduction in use of dry cow antibiotics, did not affect udder health, but tended to reduce the incidence of other diseases in the peripartum period.

Effect of dry period length on culling and pregnancy in the subsequent lactation of dairy cattle

P. Pattamanont[1], M.I. Marcondes[1,2] and A. De Vries[1]
[1]*University of Florida, Animal Sciences, Gainesville, 32608 FL, USA,* [2]*Federal University of Vicosa, Animal Science, Viçosa, 36570-900 MG, Brazil; ppattamanont@ufl.edu*

Cows could be individually managed to optimize their DD. However, it is unclear how dry period length (days dry: DD) is associated with culling and pregnancy risk in the subsequent lactation. The objective of this study was to quantify the effect of DD on culling and pregnancy risk in the subsequent lactation. For this retrospective study, we obtained DHI milk test records of 413,537 Holstein cows from 4,544 herds in 42 states with the last dry date in 2014 or 2015 from DRMS, NC, USA. Three groups of adjacent lactations were constructed: 1 and 2 (lac1-2), 2 and 3 (lac2-3), and 3 and greater (lac3+). Included cows had DD from 5 to 120 d which were classified into 14 categories. Risks were investigated for two survival times after calving: 0 to 250 and 251 to 500 d for culling, 51 to 150 and 151 to 300 d for pregnancy. Cox regressions were fit to determine the effects of DD categories on hazard ratios (HR). Control covariates of 305-d energy corrected milk in the current lactation (ECM), year and season of dry off, and the random effect of herd were added to the models. Pregnancy status in the subsequent lactations was added to the culling model. The HR at each DD category were compared with the baseline HR for 55 to 60 DD. The culling HR were 20 to 40% greater when DD was >60 d for all lactations. However, cows in lac1-2 had greater culling HR when DD was longer than cows in later lactations. The lowest HR, 8 to 13% lower than baseline, were for cows with 40 to 45 DD. The culling HR after 250 d tended to increase with DD. High ECM reduced culling HR by 2 to 9% per 1000 kg. The culling HR of pregnant cows were <14% of non-pregnant cows. The pregnancy HR by DD showed less clear trends. Pregnancy HR were 10 to 30% greater when DD was shorter only in the first 51 to 150 d after calving. Cows with higher ECM had 3 to 4% lower pregnancy HR per 1000 kg. In conclusion, both HR of culling and pregnancy were affected by DD and need to be considered when optimizing DD for individual cows.

Calf and young stock management practices on 189 Flemish dairy farms

S.A. Curial, L. Vandaele and K. Goossens
Flanders Institute for Agricultural, Fisheries and Food Research (ILVO), Animal sciences, scheldeweg 68, 9090 Melle, Belgium; sabrina.curial@ilvo.vlaanderen.be

The development from new-born calf to heifer plays an important role in the age at first calving and the performance, resilience and longevity of the dairy herd. To enlarge the implementation of best practices during the rearing period it is important to know the current rearing practices in Flanders. The objective of this online survey was to obtain more insight into the young stock rearing management strategies in Flanders. A total of 184 Flemish dairy farmers and 5 specialized young stock rearing farmers correctly completed the online questionnaire. The average age at first calving (24,9 months) and milk production (9,851 kg/year) indicate that these farms perform better than the average Flemish farm (age at first calving of 26 months). On 67% of the farms cows are milked within 4 h after calving, only 48% test colostrum quality. Colostrum was stored (frozen) on 71% of the farms. Milk replacer was used on a majority of the farms (61%). The vast majority of the farmers (93%) weaned their calves gradually. The moment of weaning was determined mostly by age of the calves (67%), a small amount of farmers took concentrate intake into account (20%). Only 24% of respondents measures the growth of their young stock. Age was the main factor determining timing of first insemination. In 78% of the farms all employees have access to the calf housing facilities and 40% of the participants are not taking any biosecurity measures. Housing the bull calves separately is the mostly implemented biosecurity measure (53%). In conclusion, Flanders calf rearing can be improved by optimizing colostrum and weaning management, implementing biosecurity measures and frequently monitoring the growth of the young stock at critical control points.

Dairy farms and cows 50 years in the future

J.H. Britt
North Carolina State University, Animal Science, 212 Eagle Chase Lane, 28729-8712, USA; jackhbritt@gmail.com

Milk provides several essential dietary nutrients more efficiently than other foods, and global milk intake is estimated to increase from 652 billion to 1.3 trillion kg over the next 50 years. To meet this demand sustainably, milk yield per cow will need to increase 1.25% annually over this period. The greatest opportunity for increasing average yield per cow is among 20 countries that produce 75% of today's global supply of milk. Among these countries, average annual yield per cow is 3,300 kg and ranges from 1,200 to 10,500 kg. Yield is limited primarily by amount and quality of feed among smallest herds, but genetic potential may also limit output. Lowest yielding cows and herds produce up to 7-fold more greenhouse gas equivalents per kg of milk at the farm gate than highest producing cows and herds; therefore, it is important to increase yield per cow to improve sustainability of the global dairy industry. Improved genetics, driven by application of genomic technologies, will lead to greater output per cow in all settings, and emerging technologies for improving crop yields and soil quality will boost quality and quantity of feed. Among the most technologically developed dairy sectors, yield per cow will continue to climb, but there will be greater emphasis on increasing protein and fat yield and less emphasis on volume. Dairy cows in these sectors will produce 17,000 to 22,000 kg per cow per year 50 years in the future. Changes in climate will cause dairying to move northward in the northern hemisphere, especially in regions where there is more arable land per person. Average herd size will increase globally, but size will be limited by land use policies and government regulations. Automation, robotics, sensors and artificial intelligence systems will be incorporated into herd management to improve precision and consistency of practices on the dairy farm. Water, nutrients and energy will be recovered and recycled on larger dairy farms. There will be an accelerated increase in understanding how various microbiomes affect cow health and performance and the ecological sustainability of the dairy farm. Epigenetics will receive much greater attention in the future and practices will be adopted to manage the epigenome. Natural products will replace many synthetic products for care of cow health and welfare.

Relation between the prevalence of claw lesions in Flemish dairies and the perception of the farmers

B. Nivelle[1,2], S. Van Beirendonck[1,2], J. Van Thielen[1,2,3], J. Buyse[2] and B. Driessen[1]
[1]Research Group Animal Welfare, Wilbroek 25, 3583 Paal, Belgium, [2]KU Leuven, Kasteelpark Arenberg 30, 3001 Leuven, Belgium, [3]Thomas More Kempen, Kleinhoefstraat 4, 2440 Geel, Belgium; beke.nivelle@dierenwelzijn.eu

Claw lesions are an important reason for compromised dairy cow welfare, reduced production, financial losses and culling. Farmers are not always aware of the prevalence and impact of claw problems on their dairy. The aim of this study was to address the prevalence of claw lesions in lactating cows in Flanders. Therefore farmers were asked to complete a survey involving the management and housing practices they apply. Next they were asked to estimate the prevalence of the following claw problems among lactating cows on their farm: interdigital dermatitis (ID), digital dermatitis (DD), sole haemorrhages (SoH), white line disease (WLD), sole ulcer (SU), interdigital hyperplasia (IH), footrot (FR) and swollen heels (SH). Even though SH is not a claw disease, it was included because it can be seen without detailed inspection of the foot. Forty dairies (16.4%) were selected to be visited to assess the claw health. Selection was based on access to pasture, number of lactating cows, estimate of ID and DD and herd production level. These 40 dairies are representative for all respondents of the survey. At least 13 cows per farm (fixed number) were examined. Selection was based on parity, lactation stage and milk production corrected for age, calving season and lactation stage. ID was scored according to Berry *et al.*, while other lesions were scored on a 4-point scale. Mobility, body condition, leg condition, hygiene, claw hook and length, etc. were determined. About 31.7 (+ 14.2)% of the examined cows per farm were clinically lame, but percentages varied between 0 and 66.7%. ID (heel horn erosion), DD, and SoH were the most common, with respectively 55.0, 27.9 and 31.8% of the cows affected. Farmers could best estimate the presence of FR, WLD and SU, which often cause severe lameness, but underestimated ID, SoH and DD strongly. Cows only affected with ID, SoH and DD often display less severe lameness. Although SH is a condition for which thorough inspection of the animal is not necessary, SH was strongly underestimated by 40.0% of the farmers.

Physiological limits of milk production in dairy cows

J.J. Gross and R.M. Bruckmaier
University of Berne, Veterinary Physiology, Bremgartenstrasse 109a, 3001 Berne, Switzerland;
josef.gross@vetsuisse.unibe.ch

Milk production of dairy cows continues to increase through animal breeding which is required to ensure a sufficient income for the farmer. However, production systems vary tremendously between countries in terms of environment, feeding and husbandry, intensity, and genetics of cows. Milk production does not appear to be limited by the productive capacity of the mammary gland. Limits are much more determined by the provision of energy and nutrients to the animal challenging feed intake as well as the activity of intermediary metabolism. Increasing the energy density in the diet by including energy-rich ingredients such as fat and starch has limitations as the breakdown of dietary fibre in rumen ecosystem decreases under acidotic conditions and excessive lipid supply. Rumen acidosis and constraints in rumen fermentation process can be avoided by feeding rumen-protected or by-pass starch and fat. However, rumen microbes may not be limited in their supply with crude protein, essential amino acids and energy. On the other hand duodenal absorption capacity for starch and fat is limited. By-pass components in dairy cow rations (e.g. soybean meal as protein source) are controversially discussed in terms of their ecologic footprint. Limited feeding of concentrates (e.g. due to economic considerations or exclusive herbage-based production systems) gives the impression of dairying close to nature, but lowers milk production in high-producing animals and further affects animal health. Low-genetic merit cows could be an alternative for these production systems. A reduced metabolic load during peak lactation can be achieved by cows with a more moderate start into lactation at simultaneously high lactational persistency. This may reduce the propensity for various production diseases. Management tools such as extended lactation through short dry periods or late service can support the genetic adaptations. Furthermore, functional feed additives improving the metabolic status result in an elevated feed intake and save resources for milk production and the immune system. Despite efforts in management physiological limitations remain unsolved in high-yielding dairy cows.

Initial and residual sward height on high stocking rate dairy grazing system on a farmlet desing

G. Ortega[1], P. Chilibroste[2] and Y. Lopez[3]

[1]Agronomy Faculty, Animal Science Department, Av Gral Eugenio Garzon 809, 12900, Uruguay, [2]Agronomy Faculty, Animal Science Department, Ruta 3 km 363, 60000, Uruguay, [3]Agronomy Faculty, Centro Regional Sur, Camino folle km 35500, 90300, Uruguay; gortegaconforte@gmail.com

A farmlet study was conducted to determine the effect of stocking rate and pasture management on milk and pasture production and feeding management of autumn calving grazing dairy cows. Four farmlets under a rotation of annual and perennial pastures were grazed by one of the four 2×2 factorial arrangements of treatments: two stocking rates 1.5 (MSR) or 2.0 (HSR) milking cows per hectare and two pasture managements: A (4 cm residual sward height all year round), and B (6 cm residual sward height for autumn and winter, and 9 cm for spring). The four treatments grazed during 2017 and 2018 from March till December. Nighty six cows were randomized to the farmlets based on parity (2.1 + 1.6; 2.5 + 1.1), BW (520 + 87; 548 + 80) and BCS (2.9 + 0.5; 3.7 + 0.6) for 2017 and 2018, respectively. Sward mass and mean growth rate (GR, kg DM/ha/day) of each individual plot in each farmlet was weekly assessed through the double sample technique. Additionally, mean sward height and phenologic stage before and after grazing were assessed on a daily basis. The data was analysed with a mixed model which included treatment and season as fixed effects and paddock as a random effect. Differences were declared significant when P<0.05 (Tukey). There was an effect of treatments (30.1 + 0.6 vs 28.6 + 0.6 cm for 2.0A and 2.0B) and season (31.1 + 0.8 vs 26.7 + 0.8 for spring and autumn respectively) on initial sward height. The treatments on the B management had grater residual sward height (8.7 + 0.3; 8.8 + 0.3 for 1.5B and 2.0B, respectively) than treatments on A management (7.6 + 0.3; 7.2 + 0.3). Mean sward mass between farmlets was higher for B (1,674 + 40.3 P<0.05; 1,608 + 40.3 P>0.05 kg DM/ha for 1.5B and 2.0B) than for A management (1,585 + 40.8 P>0.05; 1,581 + 40.8 P>0.05 DM/ha for 1.5A and 2.0A). However, neither phenologic stage before grazing (3.5 + 0.05 P=0.26) nor GR (27.2+5.8; P=0.67) were different between treatments. Productivity (GR) and phenologic stage before grazing were independent of the stocking rate and grazing management applied.

Utilizing dairy farmers' stated and realized preferences in AI bull selection

E.M. Paakala[1,2], D. Martín-Collado[3], A. Mäki-Tanila[2] and J. Juga[2]

[1]Faba co-op, Urheilutie 6D, 01370 Vantaa, Finland, [2]University of Helsinki, Agricultural Sciences, Koetilantie 5, 00014 Helsinki, Finland, [3]Centro de Investigación y Tecnología Agroalimentaria de Aragón, Avda. Montañana 930, 50059 Zaragoza, Spain; elina.paakala@helsinki.fi

In the Nordic Total Merit index (NTM) the traits are weighted according to their economic importance in the Nordic production environment. Farms may deviate from this in their production system and trait genetic levels. Hence customizing the total merit index to better reflect herd's characteristics and farmers' preferences could result in a faster economic response at the herd level. Dairy farmers' stated selection preferences in AI bull selection have been analysed increasingly in recent years. Taking the stated preferences into consideration in bull selection could improve customer satisfaction and increase the adoption of a coordinated breeding program at herd level; however, it is poorly known how farmers' stated preferences relate to their real choices. We compared the Finnish Ayrshire (AY) and Holstein (HOL) farmers' stated and realized preferences for bull selection. An Analytic Hierarchy Process (AHP) based online survey was conducted to find out farmers' stated preferences. The realized preferences were derived from the herds' insemination data. We found substantial differences between stated and realized preferences. Yield was the most important (AY) and the second most important (HOL) trait in realized selection but showed low relevance in the stated selection preferences analysis. Also conformation was more favoured in realized selection. Health and fertility were quite poorly favoured in realized selection considering that health had the highest stated preference and fertility had also high stated preference. There was, however, much consistency for longevity, it was the second most important trait in stated and most (HOL) or second most important (AY) trait in realized selection. We propose a recommendation system which is jointly using the information on both farmers' past realized and recent stated preferences in bull selection. This allows the modification of trait assessment to be compared to past selection preferences. The genetic merit of recommended AI bulls is compared to the expected NTM value.

Ingestive behaviour of Holstein cows grazing fescue managed at different defoliation intensities
G. Menegazzi[1], P.Y. Giles[2], M. Oborsky[1], T.C. Genro[3], D.A. Mattiauda[1] and P. Chilibroste[1]
[1]Facultad de Agronomía. Universidad de la República Uruguay, Paysandú, 60000, Uruguay, [2]Facultad de Agronomía de Azul. Universidad Nacional del Centro de la Provincia de Buenos Aires, Azul, 7300, Argentina, [3]Empresa Brasileira de Pesquisa Agropecuária, Bagé, 96401-970, Brazil; menegazzi95@gmail.com

An experiment was carried out at the University of Republic, Uruguay, in spring, to evaluate the effect of three defoliation intensities on a Fescue (*Festuca arundinacea*) based pasture on the animal's displacement at feeding station (FS) scale. Three treatments were compared: Lax (L), Medium (M) and Control (C), with residual sward height of 15, 12 and 9 cm, respectively. The criteria to start grazing was when the pasture reached the three leaf physiologic state and/or 18-20 cm height. Twenty seven Holstein mid lactation dairy cows were blocked by parity (2.6±0.8), body weight (618±48) kg and body condition score (2.8±0.2) and randomized to the treatments. Grazing method was rotational (6 d occupation period (OP)) and the pasture accessed from 8 am to 2 pm and 5 pm to 3 am. Cows were milked twice a day (at 4 am and 3 pm). The number of FS visited and the number of steps between FS were recorded sequentially during 15 minutes for each individual cow during the am and pm grazing session at the beginning and end of the OP. Response variables were analysed with Proc Glimmix of SAS and means values were declared different when Tukey test <0.05. The number of FS visited was higher for L and M than C cows (4.7 and 4.5 vs 4.0 FS/min) and at the beginning than at the end of the OP (4.6 vs 4.3). In contrast, the time at each FS was higher for C than M and L (15.6 vs 14.5 and 13.5 seconds/FS) and higher at the end than at the beginning of the OP (15.2 vs.13.9 seconds/FS). The number of searching steps between FS was lesser for L than M and C (0.26 vs 0.38 and 0.37). The cows visited a greater number of FS per minute in the evening than in the morning grazing session (4.7 vs 4.2) and remain less time at each FS in the evening (13.4 vs 14.7 seconds/FS). The number of FS visited, the time at each FS and the steps between FS changed according to residual sward height, parcel occupation period (initial vs final) and grazing session (am vs pm).

Effect of early lactation milking frequency on peak milk yield and persistency of milk production
J.O. Lehmann, L. Mogensen and T. Kristensen
Aarhus University, Foulum, Department of Agroecology, Blichers Alle 20, 8830 Tjele, Denmark; jespero.lehmann@agro.au.dk

Many high-yielding dairy herds in Denmark practice milking cows 3 times per day, which research has shown to increase milk production per lactation when compared with milking 2 times per day. Often, milking 3 times per day is done from the very beginning of the lactation with the aim of maximising peak yield, but this may also worsen the negative energy balance and associated health problems due a higher energy expenditure of the cow when compared with milking 2 times per day. Our hypothesis was that milking cows 2 times per day in early lactation before switching to milking 3 times per day would reduce peak milk yield, and that the later switch to milking 3 times per day would increase milk yield persistency. We tested this by randomly allocating cows in second lactation or higher to one of 3 treatments in 2 commercial dairy herds with an average milk production of 11,500 and 12,000 kg ECM per cow per year, respectively. The three treatments were milking cows 2 times per day for less than one week after calving, 4 weeks or 7 weeks before switching to milking 3 times per day for the rest of the lactation. Cows milked 2 times per day were milked first in the morning, last in the afternoon and skipped during the evening milking, which effectively gave them milking intervals of 10 and 14 hours. A total of 183 cows have completed minimum 125 days of lactation. Individual lactation curves were fitted with a standard Wilmink-function and used to estimate peak milk yield and milk yield persistency. Afterwards both variables were analysed with a linear model that included herd, treatment and the interaction between them. Preliminary analyses show that average peak milk yield ranged from 30.1 to 63.4 kg per day, and it was not affected by the treatment (P=0.16), but it was affected by herd (P<0.001). There was no effect of herd (P–0.30) or treatment (P–0.27) on persistency of kg milk production per day. However, these analyses only explain little of the variation in data as R2 was in the range of 0.16-0.20. Hence, further analyses may reveal potential interactions with other recorded factors including parity, dry period, milk yield in previous lactation, welfare and calving season.

Does prostaglandin F2α induce ovulation in the absence of corpora lutea in cattle?
L. Chaufard, Z. Dehghani Madiseh, F. Fatahnia and D.M.W. Barrett
Dalhousie University, 58 Sipu Awti, B2N 5E3, Truro, NS, Canada; david.barrett@dal.ca

Often controlled breeding protocols using prostaglandin F2α (PGF2$_\alpha$) need to be used to maximize reproductive herd performance. It is unclear if a luteolytic dose of PGF2$_\alpha$ induces ovulation in the absence of corpora lutea (CL) in cattle and no one has examined this in pre-pubertal dairy heifers or the effects of a sub-luteolytic dose. The objective of this study was to determine if a luteolytic and sub-luteolytic dose of PGF2$_\alpha$ will induce ovulation in the absence of CL in pre-pubertal dairy heifers. Ovarian ultrasonography (U/S) was conducted 11 days apart on pre-pubertal heifers to determine if they had CL or not. If they had no CL, the largest follicle diameter was noted and this diameter determined when they would undergo U/S again. When a growing follicle was first observed to be 10-12 mm in diameter (Day 0), calves were randomly treated with an injection of saline (CTL; n=4) or PGF2$_\alpha$ (250 µg (P250; n=4) or 500 µg (P500; n=5)). Calves were balanced for dam parity, sire, and calf age, BCS, height, and weight. Calves underwent U/S daily until ovulation or Day 5. From Days 1 to 5, calves were observed for oestrus. Data was analysed using SAS PROC MIXED and then by Tukey Test. There were no signs of oestrus and only one calf ovulated- a P500 calf on Day 5. From Days 1 to 5, the largest diameter follicle observed was bigger in the CTL (15.8±2.2 mm) than P500 calves (14.5±2.2 mm; P<0.0001); P250 calves were intermediate (14.9±2.3 mm; P>0.05). From Days 0 to 5, daily maximum follicle diameter was not affected by PGF2$_\alpha$ (P>0.05), but was larger (P<0.0001) in calves from dams with a parity of 1 (14.5±0.4 mm) than a parity of ≥2 (11.5±0.4 mm). It is unclear if P500 induces ovulation in the absence of CL; it appears that P250 does not. But, follicle diameter in pre-pubertal dairy heifers was affected by P500.

Productive performance and profiles of metabolic parameters of Holstein Friesian and Normand cows
A. Oxiley[1], L. Astigarraga[1], C. Loza[1], D. Rico[2] and E. Jorge-Smeding[1]
[1]Universidad de la Republica, Faculty of Agronomy, Garzon 780, 12800, Uruguay, [2]University of Laval, Animal Production, Pavillon Paul-Comtois 2425, rue de l'Agriculture, G1V 0A6, Canada; aalvarezoxiley@gmail.com

The goal of this experiment was to study the effect of dairy breed (DB; Holstein, H; Normand, N) and week of lactation (WL) on milk production, body condition score (BCS), metabolic parameters and insulin profile of multiparous cows (H, n=14; N, n=13) with similar body weight at calving (H: 569.2±10.1 kg, N: 589.9±10.5 kg; P=0.1689) during the whole lactation cycle in a low input pastoral system. The experiment was carried out at the Universidad de la Republica (Uruguay). All data was analysed as repeated measures on time, where DB, WL and its interaction was considered fixed effects using the PROC MIXED. H cows produced more milk with less solid content than N cows (17.4±1.0 vs 12.1±1.0 kg/day P<0.05). N cows reached their lowest BCS earlier than H cows (6th vs 10th WL), and it was also higher in N compared to H cows (2.39±0.08 vs 1.84±0.08, P=0.001). Plasma glucose was affected by the interaction between the DB and WL (P=0.0087) and no differences between DB were observed on average across lactation cycle (N−66.2±0.6 mg/dl; H=66.1±0.5 mg/dl; P=0.866). Neither DB (P>0.05) nor interaction between the DB and WL (P>0.05) affected urea concentration (18.3±0.5 vs.18.6±0.5 mg/dl, H and N, respectively) and BHB (H, 0.38±0.02 vs N, 0.42±0.02 mM). Higher milk yields in H cows were associated with higher body reserve mobilization than N, which was concordant with a longer time (P<0.001) of increased NEFA concentration (average across lactation, 0.19±0.01 vs 0.21±0.01 mM, P>0.05, H and N). N cows seems to have a better balance between milk production and body reserve mobilization. Despite no significant effect of DB was observed on insulin concentration, higher insulin levels on N cows (7.6±0.5 vs 8.9±0.5 mU/ml, H and N, respectively; P=0.065) during early and middle lactation stages (winter and spring) could explain the apparent different energy partition between DB regarding to milk yield and BCS profile during the whole lactation cycle.

Effect of a blend of B vitamins on performance of milking cows

J. Leloup, A. Budan, M. Devine and J. Eouzan
Cargill Animal Nutrition, Rue des Moulins, 22950 Trégueux, France; jacques_eouzan@cargill.com

The B vitamins required for ruminant metabolism come from rumen fermentation and diet. These vitamins are hydrolysed, used by certain microorganisms and absorbed by the animal in varying proportions according to the ration profile and the needs of the animals. Evans and Philippe have shown that intake of B vitamins has a positive effect on milk production in US farms. The aim of this work was to evaluate the effect of a combination of B vitamins on milk performance with French rations. Trials were conducted in 5 commercial farms equipped with milking robots in 2016 and 2017. 270 Holstein dairy cows were divided into four batches: 'Early lactation control', 'Early lactation B vitamins', 'Mid lactation control' and 'Mid lactation B vitamins'. Lactation stages ranged from 0 to 90 days in early lactation and from 90 to 200 days in mid lactation. The only difference between 'control' and 'B vitamins' batches was the delivery of biotin and two rumino-protected vitamins to cows from the 'B vitamins' batches. A significant improvement of milk production was observed in mid lactation multiparous cows (+1.2 kg/cow/day) without dilution of protein and fat contents and in early lactation cows (+1.8 kg/cow/day). The response of the combination of B vitamins on milk production is therefore higher in early lactation, when the need in energy is the greatest.

Genetic and non-genetic characteristics of milk yield and milk composition of Holsteins in Tunisia

M. Khalifa, M. Khalifa, A. Hamrouni and M. Djemali
National Agronomic Institute of Tunis, Animal Genetic and Feed Resources Research Lab (INAT), 43, Avenue Charles Nicole, 1080 Mahragène, Tunis, 1080, Tunisia; mdjemali@webmails.com

Today, it is important to treat the production characteristics of Holsteins and to make non-genetic effects adjustments in order to develop reliable recommendations to dairy managers and applicable to livestock conditions in Tunisia. The objectives of this study were to determine the most important environmental factors limiting milk yield and milk contents, to derive adjustments to standardize milk yield records for DIM, month and age at calving, to estimate genetic parameters of milk yield and percent of fat and protein using 50,376 complete lactation records recorded in 147 herds during 2000-2016. A linear model was developed, including herd-year of calving, calving month, lactation number, age of cow at calving, birth year, country of origin sires, country of origin of maternal grand sires and days in milk for first lactations and second and later lactations separately. The main results showed that all of these factors included in the model are all significant sources of variation for milk yield and percent protein in first and later lactations of Holsteins raised in Northern Africa. Least square solutions of month of calving allowed to identify three seasons of calving with season 1 (September to January), season 2 (February to May) and season 3(June to August). Holstein cows in Tunisia reached their maximum of milk yield between 78 and 90 mo of age. Unadjusted means milk yield, percent fat and percent protein in first lactations were, respectively, 6,343 kg, 3.19, and 2.84. In second lactations and greater, milk yield, percent fat and protein were 6,277 kg, 3.26, and 2.88, respectively. Heritability of milk estimated by MIVEQUE (0) was 21% and both ML and REML gave 16 and 17%, respectively. Percent fat heritability estimates were higher 32, 36 and 37% by MIVEQUE(0), ML and REML, respectively. Phenotypically, the decrease in milk yield was on average 49 kg per year of calving during the period 2000 till 2015, but genetically, the decrease of milk yield was an average 13 kg per year of birth since 1995.

Deficiencies and potentials of dairy bulls testing in Tunisia

A. Arjoune, A. Hamrouni, M. Khalifa and M. Djemali
National Agronomic Institute of Tunisia, Animal Sciences, 43 Avenue Charles Nicolle, Tunisia 1082, Tunisia;
arjouneasma@gmail.com

All criteria to set up a national program for testing dairy bulls are met in Tunisia. The objectives of this project were to: (1) describe this program; (2) identify its current constraints; and (3) make proposals and adjustments to upgrade it into a national progeny testing program. A total of 15 young bulls from three breeds (Holstein (4), Tarentaise (3) and Brown Swiss (8) were hosted at the National Genetic Improvement Center (OEP) in Sidi Thabet. These bulls were selected from herds enrolled in the dairy recorded herds. Bulls' pedigree and seminal characteristics were used and analysed. The main results showed that the genetic breeding values and average seminal characteristics were low. For the Holstein breed, the index of fat percent varied between -0.46 and + 0.04 and the index of protein percent was between -0.03 and + 0.14. For the Brown Swiss breed, the index of milk of bull sires varied between +10 kg to +660 kg, the % fat index was between -0.03 and +0.23 and the index of protein % was between +0.01% and +0.16. For the Tarentaise breed, the index of milk of bull sires varied between +300 kg and +557 kg of milk, however no information was provided for % of fat or % of protein. Seminal characteristics (volume of ejaculation, semen concentration, motility and percentage of live sperm cells and the number of total doses produced by ejaculation) were relatively low. They were on average 3.34±0.97 ml; 1,251±126 spz/ml; 3.7±0.43; 66.8±3.5% and 200±62. Young Holstein bulls were higher than the average in all identified seminal traits except for the number of total doses produced by ejaculate (194). Young Brown Swiss bulls were less to average and Tarentaise young bulls were higher than average. The current method used to select young bulls is far from dairy bulls testing standards. It should be upgraded based on a standard method and scheme of dairy bull's testing.

Understanding the role of livestock farming systems in agroecological transitions

R. Ripoll-Bosch[1] and B. Dumont[2]
[1]Wageningen university & Research, P.O. Box 338, 6700 AH Wageningen, the Netherlands, [2]Université Clermont Auvergne, INRA, VetAgro Sup, UMR Herbivores, Theix, 63122 Saint Genès Champanelle, France; raimon.ripollbosch@wur.nl

Agroecology is gaining momentum as a principle to transition towards more sustainable food systems. Agroecology promotes system redesign, enhances interactions between system components, and benefits from functional diversity for the production of goods and services. It has been defined as a science, a set of practices and a social movement, and overlaps with other frameworks such as ecological intensification, organic farming or crop-livestock integration. The role of livestock, however, is usually overlooked. Recently, a review has addressed the prospects for agroecology across a large diversity of livestock systems (i.e. ruminants, integrated crop-livestock, pigs and aquaculture). Based on ecological processes, five agroecological principles were identified, including management options reducing inputs use and promoting biological diversity. Each of the five principles can help in the design of a large range of livestock systems, but the options may vary considerably between agro-ecological zones and according to the social, economic and human dimensions of livestock farming. Management practices derived from agroecological principles also relate to farmers' worldviews about animal status, modernity and the role of technology, and their relation with nature. Though the adoption of some agroecological principles may impair productivity, farm revenues might increases due to input reduction. Diversification reduces profit variability and is thus likely to increase system robustness, but little is known on how farm performances are related to the strength of interactions among system components. Combining ecological network analysis to bioeconomic farm models could provide such knowledge. Finally, it is highly relevant to consider agroecological transition at the territory scale. This calls for analysing interconnection of agricultural activities embedded in local territories that lead to valorise local resources (feed, manure and equipment). Agroecology thus questions the mainstream developments in livestock production and challenges researchers, extension services and policy makers by accounting for the diversity of local and regional situations.

A new technical referencing system for organic rabbit farming in France: first results

T. Gidenne[1], L. Fortun-Lamothe[1], A. Roinsard[2], J.P. Goby[3], M. Cormouls[1] and D. Savietto[1]
[1]INRA, GenPhySE, Animal Physiology and farming sysems, BP 52627, 31326 Castanet-Tolosan, France, [2]ITAB, Institut Technique de l'Agriculture Biologique, 49100 Angers cedex 2, France, [3]University of Perpignan, IUT agronomy, ch. Passo Vella, 66962 Perpignan, France; thierry.gidenne@inra.fr

Free-ranged organic rabbit correspond to the consumer's demand since animals are raised on pastures all the year around, in either movable cages or paddocks. Presently, the offer in organic rabbit meat did not cover the market demand. But the development of organic rabbit farming is greatly impaired by the lack of validated technical references. Thus, we aimed to develop a new system to obtained technical references from 6 organic rabbit farms, and with data collected over 3 years of production (2015 to 2017) for the maternity unit (movable cages on pasture or individual pens, average herd = 33 females). The productive lifespan of a female was, on average over one year, and can reach 2 years (variability of 75%). Female mortality averaged 21% over the period. With 5 mating, 3 litters per female/year (60% fertility rate) were obtained, for a total of 25 live kits and 19 weanlings (26% mortality before weaning). Thus, from these French average technical data, the income of a full-time organic rabbit farmer could potentially be around 26.3k €, with a flock of 80 reproducing does and about 9 ha for pastures (5 ha) and cereal production (4 ha). This first referencing system will be extended to a larger number of breeders, with the deployment of a smartphone application (GAELA) managing the rabbit farm, and synchronized to a secured database.

Agroecological practice on Flemish beef farms: a closer look at integrated animal health management

L. Tessier[1,2], J. Bijttebier[2], F. Marchand[2,3] and P.V. Baret[1]
[1]UCL, ELIA, Croix du Sud 2, L7.05.14. – 1, 1348 Louvain-la-Neuve, Belgium, [2]ILVO, L&M, Burg. Van Gansberghelaan 115, 9820 Merelbeke, Belgium, [3]UA, ECOBE, Campus Drie Eiken C1.15, 2610 Wilrijk, Belgium; louis.tessier@ilvo.vlaanderen.be

As it remains largely unexplored what agroecology may practically entail for the Flemish beef sector, we've interviewed 37 beef farmers in this region to reveal what actions farmers may take to put agroecology into practice. Agroecology cannot be reduced to a set of concrete practices but requires contextualized solutions, and we therefore take principles as the methodological starting point for the design and study of agricultural systems. We confronted farmers in a semi-structured interview format with a set of principles, formulated to fit the requirements of this study, yet consistent with the agroecological literature. Our hypothesis is that inductive coding of mentioned practices and comparative analysis of these coded practices may be an interesting method to explore the agroecological nature of farmers' practices and to draw more general insights for agroecosystem management and design. We discuss our preliminary findings on how farmers pursue one of the thirteen principles discussed with farmers, namely the topic of integrated animal health management. We were struck by opposing practices and viewpoints farmers had about this principle. We saw parallels with the Control management and Adaptive management models conceptualized by ten Napel *et al.*, as we noticed that these distinct practices co-appeared. We sorted therefore coded practices under these two approaches and grouped farmers based on whether they mention primarily practices that fit within one of these models. It resulted in four groups: an adaptive management group, a control management group, a group with mixed practices and a group of farmers from whom we could not elicit enough practices to group them adequately. Preliminary analysis suggests that the adaptive management model is more inspired by agroecology, yet this will become clear when we study in similar fashion how the farmers pursue other principles, such as commercial and financial autonomy, biodiversity preservation and use of external input.

Nematode control on French sheep farms: farmer's practices and knowledge

M. Sautier[1], A. Somera[2], J.M. Astruc[3] and P. Jacquiet[4]
[1]Université de Toulouse, INRA, INPT, ENVT, GenPhySE, 24 chemin de Borde-Rouge, 31326 Castanet Tolosan, France, [2]CDEO, Quartier Ahetzia, 64130 Ordiarp, France, [3]IDELE, 24 chemin de Borde-Rouge, 31326 Castanet Tolosan, France, [4]UMT Santé des Petits Ruminants, Ecole Nationale Vétérinaire de Toulouse, UMR INRA/ENVT 1225 IHAP, 23 chemin des capelles, 31076 Toulouse, France; marion.sautier@inra.fr

Agroecology applied to livestock systems calls for interdisciplinary research to unlock barriers to sustainable livestock farming practices. In ruminant systems one barrier is the trade-offs between grazing, animal health, and anthelmintic resistance. Grazing allows for reducing feeding inputs but exposes animals to gastrointestinal nematodes. Such parasites affect animal performances, and can result in clinical symptoms in severe infestations. Lastly, the heavy reliance on anthelmintics to manage nematode populations generated resistance, i.e. non-effective easy options against infections. Here and now, integrated management of parasitism is required. This study aims at identifying and characterising farmers' practices and knowledge in managing nematode risk on dairy sheep farms. It relies on a questionnaire-based survey among 700 dairy-sheep farmers in Pyrénnées-Atlantiques, France. The sample includes a diversity of pedoclimatic contexts and of livestock farming systems (herd size and dynamics, breed, animal performance, work organisation, grazing management…). First, both widespread and unusual nematode control practices were identified. Second, types of parasitism management were created through hierarchical cluster analysis on grazing and veterinary practices. Second, the specificities of each type were highlighted in terms of characteristics of livestock farming systems and farmer's knowledge and preferences using a multiple correspondence analysis. We conclude that nematode control strategies relate to farmers' worldviews regarding animal status on a farm, as well as structural constraints. We assessed opportunities for upcoming research and development in this region in regards with farmers' preferences for strategies using tannins, genetic selection or phytotherapy. A such grasp on actual practices, knowledge and preferences of farmers is crucial for the relevance of future research and development programs.

A method combining sociological and biotechnological approaches to study agroecological transitions

G. Brunschwig[1], A. Mondiere[1], A. Jarousse[1], P. Cayre[2] and J.P. Goron[3]
[1]VetAgro Sup, Inra, UMR Herbivores, 89 avenue de l'Europe, 63370 Lempdes, France, [2]AgroParisTech, UMR Territoires, 9 boulevard Blaise Pascal, 63170 Aubière, France, [3]PEP bovins lait chambre agriculture, 40, avenue Marcelin Berthelot, 38100 Grenoble, France; gilles.brunschwig@vetagro-sup.fr

Mixed farming-livestock (MFL) farms are emerging as a sustainable model and a pathway to agroecological transition. We studied the organization and sustainability as seen by farmers, of MFL systems vs specialised systems (S) which recently ended MFL. By a semi-directive survey in the plains of Isère and Ain, we collect data from 16 MFL varied farms (cattle/sheep, dairy/suckling, conventional/organic) with large components vs 14 S farms (crops/dairy/crops with suckling cattle). From farmers' speeches, we made in parallel 2 qualitative analysis. One of the sustainability pillars they expressed to create a typology, and one of the sociological notion of 'modernity' according to the motivations expressed to define archetypes. A graphical analysis was utilised to associate types and archetypes. Complementary interviews in 11 farms (7 MFL and 4 S) enabled to validate results and shown farms trajectories. We obtained 6 types related with sustainability (from – expressed to +) and 5 archetypes related with modernity (from + expressed to –). The analysis of the link between the sustainability pillars and social reference values highlights two trends. Some farmers remain 'modern' with economic pillar as a priority and tend, however, to better take into account the social and environmental pillars. In contrast, some farmers speak together about the three pillars of sustainability, change their view about 'nature' and question the modern system. The analysis of trajectories shows for MFL and S, a global move towards a greater consideration of the 3 pillars of sustainability, but 2 major trends about modernity: either the trajectory remains anchored in modernity, or it shows an extraction of modernity, with a strong challenge to modern values, and often conversions to Organic. The S farms fit mainly with the first trend while the MFL farms are more part of the second trend. The originality of this study is in the combined use of social and biotechnical approaches.

Use of ecological network to analyse how interactions drive performance in mixed livestock farms

L. Steinmetz, P. Veysset, M. Benoit, C. Troquier and B. Dumont
Université Clermont Auvergne, INRA, VetAgro Sup, UMR Herbivores, Theix, 63122 Saint-Genès-Champanelle, France;
lucille.steinmetz@inra.fr

Agricultural production systems have become specialized in recent decades, making intensive use of various production factors (labour, inputs, capital). In organic farming (OF), characterized by a strong link to the soil and a ban on chemicals, farms have developed various strategies to be viable, among which several forms of diversification: integration between crops and livestock, mixed grazing, on-farm transformation and sales, etc. Diversification can be seen as a risk-spreading strategy against market price fluctuations. In mixed grazing systems, the use of animal species with different ecological niches is likely to make a better use of pastures and to dilute parasite burden, both limiting dependence on external inputs. Combining cattle and monogastric on-farm, which have a long and a short production cycle, respectively, is assumed to generate more regular cash inflows over the year. In mixed crop-livestock systems, closing nutrient cycles and promoting feed autonomy reduce dependence on external inputs. There are many references and methods for analysing specialized systems (either conventional or in OF), but only few references on processes operating in mixed livestock farms, and even less that enable to link interactions among system components and farm multiperformance. Here, we used ecological network analysis to quantify nutrient fluxes and to simulate the strength and direction of interactions among system components (i.e. among categories of animals, grasslands and crops). Our aim is to analyse whether farms with the more complex interaction network are also those with the higher performance. Fifteen surveys were carried out in northern Massif central in mixed farms associating beef cattle with either sheep, pig or poultry. First results suggest that mixed livestock systems have good economic performances per unit of workload, but that performances are only weakly related to the strength of interactions among system components.

What have been the advantages of mixed livestock farming systems under past prices and policies?

C. Mosnier, C. Verdier and Z. Diakité
Université Clermont Auvergne, Inra, VetAgro Sup, UMR Herbivores, Theix, 63122 Saint-Genès-Champanelle, France;
claire.mosnier@inra.fr

Farmers have to face numerous risks and to cope with a changing agricultural policy. Diversification can be an effective way of reducing farmers' risk exposure. Risk management generally comes at a cost since diversification requires producing products besides the most profitable one. Complementarity and synergy between products can decrease the cost of diversification owing to a more efficient use of farm resources. The objective of this presentation, arising from the NewDeal project, is to estimate how the economic advantages of diversification have varied according to past prices and policies, taking into account some complementarities between productions. We simulate with the bioeconomic farm model Orfee, three farming systems from the French Massif Central area, that combine dairy and beef cattle, beef cattle and sheep for meat or beef cattle and cash crops. For each farming system, we consider several scenarios characterized by different levels of diversification: from totally specialized in dairy, beef, sheep or cash crop production to half-half situations. The sources of complementarity considered in the model are the capacity of beef cattle to graze remote pasture contrary to dairy cattle, the capacity of sheep to graze during winter and to use cattle barn for spring lambing, the precedent value of forage crops for cash crops and the exchange of litter and organic fertilizer between cash crops and livestock. The variations of economic results over time result from the market price fluctuations and policy changes that have occurred between 1990 and 2017. Main results show that the profitability of each activity in relation to others have varied over time on the short run because of market fluctuations and on the mid to long run because of policy changes. Diversification reduces profit variability but the lowest level of variability doesn't correspond to the half-half situation. In many cases, the complementarity between the different enterprises enables to increase average profit due to reduced input consumption or due to increased production per hectare. Nonetheless, this economic added value depend on the difference of profitability between enterprises.

Agronomic and economic interest of straw-manure exchanges between crop and livestock farms

E. Thiery[1,2], G. Brunschwig[1], P. Veysset[1] and C. Mosnier[1]
[1]Université Clermont Auvergne, VetAgro Sup, Inra, UMR Herbivores, INRA Saint-Genès-Champanelle, 63122 Saint-Genès-Champanelle, France, [2]AgroSupDijon, 26 boulevard Docteur Petitjean CS 87999, 21079 Dijon Cedex, France; eglantine.thiery@agrosupdijon.fr

Since World War II, mechanization and intensive use of synthetic fertilizers have driven increasing specialization in crop and livestock production. Today, stocks of organic matter in specialized crop-farm soils are in decline. Organic matter is a key element of soil fertility. Reintroducing organic fertilizers may be a way to increase or maintain organic matter stock in soils. Straw–manure transactions between specialized crop and livestock farms could improve the overall fertility of the land and hence the long-term income of farmers. However, the costs and benefits of these transactions are still poorly quantified. This study simulates the economic value of straw-manure transactions between specialized farms as part of a soil fertility improvement strategy. The simulations take into account the effects of manure on fertilizer supply via short- and long-term mineralization of organic matter, on crop yields, and on costs related to transport and spreading of organic fertilizers. The added value of this study is the consideration of the value of manure beyond its fertilizing value, which is most often used in other studies. The results show that it makes smart economic sense for a conventional-system crop farm to purchase manure at a price between 9 and 14 €/t, to enjoy expected benefits of short-term and long-term increase in soil organic matter. On the other hand, it also seems economically sound practice for the livestock farm to sell some of its manure, even at a price of 3 €/t.

Relationships between cattle grazing, soil microbiology and nutrients cycle in highland pastures

S. Raniolo, A. Squartini, M. Ramanzin and E. Sturaro
University of Padova, DAFNAE, viale dell'Università 16, 35020 Legnaro (PD), Italy; enrico.sturaro@unipd.it

Highland pastures are particular agro-ecosystems with a high level of plant and animal biodiversity, and for this reason semi-natural grassland are classified as High Nature Value Farmland. This study aims at characterizing the agroecological relationships among grazing animals-plants-microorganisms in alpine summer pastures. The study area included two summer farms in the Dolomites, Eastern Italian Alps. Both summer farms were located at an elevation around 2,000 m asl, and were grazed by dairy cows from mid-June to mid-September. In order to assess the impact of dairy cattle grazing on the functional microbial biodiversity, surveys were carried out using topsoil cores before ('point zero') and during the grazing period. The distribution of cattle stocking rate at temporal and spatial scale was analysed by using GPS tracking of grazing animals and farmer's interviews. The samples of soils were collected in patches representative of different vegetation types and stocking rates, with three repetitions for each sampling points. A molecular ecological approach was applied, analysing microbial communities in terms of abundances relative to copies of target genes through the application of real-time PCR. The studied genes were *nosZ* for denitrification, *amoA* archea and bacteria for nitrification. The analysis highlighted different outcomes between the considered areas, showing a likely impact of grazing on microbial communities but in different terms depending on the local pedoclimatic and vegetational conditions and on the cattle stocking rates. This kind of approach will be integrated with indicators on carbon sink, in order to have an in-depth view of the ecosystem support services, such as nutrient cycle and soil formation regulation, carbon storage and supply in soil, and forage production. These preliminary results could be used to develop specific and minimally-invasive biophysical indicators of supporting (and regulating) ecosystem services. These tools will assist the definition of strategies aiming at enhancing the sustainability of grazing livestock systems in compliance to agroecology principles.

Using phenotypic distribution models to predict livestock performance

M. Lozano-Jaramillo[1], J. Bastiaansen[1], T. Dessie[2], S. Alemu[2] and H. Komen[1]
[1]Wageningen University & Research, Animal Breeding and Genomics, Droevendaalsesteeg 1, 6708 PB Wageningen, the Netherlands, [2]International Livestock Research Institute, Addis Ababa, P. O. Box 5689, Ethiopia; m.lozanojaramillo@gmail.com

Some commercial breeds have been developed for production in tropical conditions. However, there is a lack of understanding of how the environment is affecting the productivity of these breeds under the diverse conditions encountered in Africa. Here we present a novel approach to animal breeding, phenotypic distribution models, to make predictions on suitable areas for the different breeds based on phenotypes and local environmental conditions. We tested the method by predicting the response of body weight in 5 different breeds of chicken as a function of the diverse environmental conditions in Ethiopia. Male body weight was collected for five introduced breeds in five regions in Ethiopia. Environmental data and a cultivated land variable were used to model the relationship between environmental variables and body weight. Gradient boosting, a machine learning technique that improves model accuracy, was applied to a generalized additive model (GAM). The predicted body weight as a function of environment is represented as a heatmap. The model also predicts which environmental variable has the highest contribution in building the models.The phenotypic distribution models predicted for the Sasso breed to have the highest body weight. The Horro and Koekoek breeds had the lightest body weight predicted. The environmental variables of strongest influence for male body weight distribution for the Horro, Koekoek, Sasso and Sasso-RIR, were associated with temperature. For the Kuroiler, precipitation was the variable that had the strongest influence on the predicted distribution. These environmental parameters correspond to the conditions encountered in the area of origin of each breed. Our results highlight the importance of taking environmental variables into account in breeding programs, as it shows that different breeds can respond differently to the same environmental conditions. We recommend the use of these models in animal breeding to capture the breeds' phenotypic variation as a response to different agro-ecologies.

Several pathways of agroecological transitions for agropastoral systems

C.H. Moulin, M. Jouven, J. Lasseur, M. Napoléone, E. Vall and A. Vidal
UMR SELMET, Montpellier Université, INRA, CIRAD, Monptellier SupAgro, 2 place Viala, 34060 Montpellier, France; charles-henri.moulin@supagro.fr

Agropastoral farming systems are under pressure of various drivers in territories and exhibit dynamics of change. We propose to examine those dynamics in two areas (French Mediterranean and South-West Savanna of Burkina-Faso), in order to examine how the dynamics observed could be considered as an agroecological transition. In French Mediterranean, the stake is to reinforce the adaptive capacities of the farms, as preserving their spatial inscription in the territory, in order to maintain production system based on the use of spontaneous vegetation. The analysis of on-going dynamics shows that there is no design of new farming systems, but new combination of relationships with others stakeholders. In Burkina-Faso, the stake is to produce more milk, peculiarly in dry season, in order to provide local urban markets with fresh local milk, in substitution to imported milk powder. Conventional intensification is a possible pathway to reach this target. However, other pathways with agroecological practice changes that enable increasing milk yield and producing all the year-through, maintaining the agroecological state of pastoral systems. Those pathways are thus a form of transition, with a higher self-sufficiency of the local food system, valorising local by-products (crop residues, cotton-grain from industry. Those two case studies show that: (1) is more relevant to consider the agroecological transition at the territory scale in agropastoral situations; (2) agroecological transition is not only the ecologization of intensive systems, but also an intensification of the production, maintaining the agroecological functioning of farming systems.

Addressing territorial resources to foster agroecological transitions in livestock farming systems

V. Thenard[1], G. Martel[2], M. Moraine[3] and J.P. Choisis[4]
[1]INRA, UMR AGIR, 31326 Castanet-Tolosan, France, [2]INRA, UMR BAGAP, 49007 Angers, France, [3]INRA, UMR Innovation, 34060 Montpellier, France, [4]INRA, UMR SELMET, 34398 Montpellier, France; vincent.thenard@inra.fr

Animal sector is strongly impacted by climate change, economic instability, environmental pressure and new social expectations. Despite these issues, animal farming have a place in global health through its potential role on soils, ecosystems, agriculture and food systems. For this, major transformations are required. Still, some research works have enlighten promising pathways for such agroecological transitions, around principles such as diversification and interconnection of activities, and embeddedness in local territory, to valorise the diverse local resources. The territorial resources supporting agroecological transitions combine material (soils, water, ecosystems), technical (equipment, factories), cognitive (training, collective organization), and socioeconomic (supply chains, public policies, local community support, work organization) resources. Analysing combinations of resources and describing how agroecological systems mobilize them can help supporting agroecological transitions at local scale. In this communication, we propose to identify the territorial resources mobilized in four case studies for agroecological livestock systems. The case studies are chosen to represent a wide diversity of livestock systems (different levels of self-sufficiency, degree of crop-livestock integration, etc.) and a diversity of natural and socio-economical contexts (PDO label, mountainous or intensive region, etc.). The analysis is based on the diversity of uses of territorial resources and understanding of how the farming systems are embedded in their territories. In the different case studies, we explain how farmers could shift the management the territorial resources: on one hand with determinist innovations and on the other hand toward an open-ended transition. Finally, we propose a transversal framework analysis to discuss to what extent the sets of territorial resources should determine or not the agroecological transitions of livestock systems. Based on this we suggest possible developments for supporting such transitions in different territories.

Reducing competition in crop-livestock-forest integrated system by thinning eucalyptus trees

J.R.M. Pezzopane[1], W.L. Bonani[2], C. Bosi[3], E.L.F. Rocha[4], A.C.C. Bernardi[1], P.P.A. Oliveira[1] and A.F. Pedroso[1]
[1]Embrapa Pecuaria Sudeste, Rod. Washington Luis, Km 334, 1356-3970, São Carlos, Brazil, [2]UNIARA, Rua Carlos Gomes, 1338, 14801-340, Araraquara, Brazil, [3]USP, ESALQ, Av. Padua Dias, 9, 13418-900, Piracicaba, Brazil, [4]UNICEP, Rua Miguel Petroni, 5111, 13563-470, São Carlos, Brazil; jose.pezzopane@embrapa.br

Crop-livestock-forest integrated systems are important alternatives for pasture intensification. However, depending on the arrangement and population of trees, strong competition may occur between the plant species in these systems, reducing yield of at least one of them. Trees thinning is considered a good management practice to reduce competition. This study aimed to evaluate the effects of eucalyptus trees thinning on yield and nutritive value of corn for silage and palisadegrass (Urochloa brizantha cv. Piatã) in a crop-livestock-forest integrated system. Assessments of plant variables, as well as of photosynthetically active radiation (PAR) incidence and soil moisture were performed between October 2016 and March 2018 in São Carlos, SP, Brazil, in a crop-livestock-forest and, for comparison, in a crop-livestock system. In the crop-livestock-forest system, eucalyptus trees (Eucalyptus urograndis clone GG100) were planted in April 2011, in single rows, with a 15×2 m spacing. In 2016, trees were thinned and the spacing changed to 15×4 m. In this system, the assessments were performed in four equidistant positions in relation to the eucalyptus rows. Corn was sown first, and palisadegrass was sown after corn harvest. PAR transmission during the corn cycle and the pasture cycles was 65.3 and 60%, respectively. Soil moisture was lower at the 50 to 100 cm depth layer near the tree line than at the other positions, during the dry period. Corn yields were similar between systems with an average of 13.6 Mg of dry matter/ha. Corn silage produced in the crop-livestock-forest system presented higher percentage of grain (41.4 and 42.1%) than that produced in the crop-livestock system (35.6%). No differences in forage yield were observed. Crude protein of corn for silage and palisadegrass was higher in the crop-livestock-forest than in the crop-livestock system. Such results indicate that tree thinning was favourable for production in the crop-livestock-forest system.

Importance of Vlashko Vitoroga sheep in agroecological farming regarding food quality and landscape

M. Savic[1], S. Vuckovic[2], M. Baltic[1], Z. Becskei[1], R. Trailovic[1] and V. Dimitrijevic[1]
[1]Faculty of Veterinary Medicine, University of Belgrade, Department for Animal Breeding and Genetics, Bul. Oslobodjenja 18, 11000 Belgrade, Serbia, [2]Faculty of Agriculture, University of Belgrade, Nemanjina 6, 11000 Belgrade, Serbia; mij@beotel.net

Considering the importance of biodiversity for agriculture, global food and health security, the Republic Project was focused on supporting the biodiversity and the role of autochthonous Vlashko Vitoroga sheep in agroecological farming system. Vlashko Vitoroga type of Zackel sheep is an important element of regional agro-biodiversity, relevant to aesthetic value of landscape, the tradition and the cultural heritage of Serbia. Vlashko Vitoroga sheep is an endangered type, registered at the Endangered–Maintained breeds list. This sheep is unique for its adaptation to extreme climatic conditions in the South Banat region, at the edge of Deliblato Sands area. Vlashko Vitoroga sheep is traditionally reared in the pasture based farming systems. The objective of this study is to evaluate the integration of natural resources of the sensitive habitat on robustness and meat characteristics of Vlashko Vitoroga sheep. The results of the study could be important for decision-making strategy for future management of the agroecological livestock system and landscape conservation. The analysis of the botanical composition revealed a high degree of floristic biodiversity, the dominantly present families were *Fabaceae* family, *Poaceae* family and *Lamiaceae* family. Health status and robustness of Vlashko Vitoroga sheep, important for pasture based farming system, were examined by clinical and laboratory investigations. In scope of food security and meat quality the analyses of lamb meat (m. longissimus dorsi) has been performed. The results has shown that fatty acid content and the values of n-6:n-3 fatty acid ratio (1.92±0.44) are in the accordance with results of pasture fed lamb meat. Sensory meat characteristics make consumers prefer Vlashko Vitoroga pasture fed lamb meat as a local product. The obtained results of added value of Vlashko Vitoroga sheep should increase the interest for this endangered breed, sustainable production and regional development.

Grazing frequency using goats to control *Salsola ibérica* scrub in rangelands of northern Mexico

E. Ruiz-Martinez[1], O. Angel-García[1], V. Contreras-Villarreal[1], C.A. Meza Herrera[2], F.G. Véliz-Deras[1], J.L. Morales-Cruz[1] and L.R. Gaytán-Alemán[1]
[1]Universidad Autónoma Agraria Antonio Narro, Periferico Raul López Sánchez, 27256, Mexico, [2]universidad autonoma de Chapingo-URUZA, Carretera Gomez Palacio Cd Juárez, 35230, Bermejillo, Mexico; zukygay_7@hotmail.com

The aim of this study was to develop a grazing frequency-based model to control *Salsola ibérica* weed by goats in rangeland from northern Mexico. The planting of *S. ibérica* plants was carried out on a plot of 135 m^2 divided in 15 plots (3×3 m, each). Once the plant´s growth achieved 35 cm height, the production of total green biomass (TGB; kg/ha), total dry biomass (TDB; kg/ha), number of plants (NP; n) and foliage area (FA; cm), were determined. Two weekly grazing frequencies were used: GF1 (once) and GF2 (twice), introducing 2 goats/plot during 3 h/d × 5 weeks. The third group (CG) had no animals in the assigned plots. Analyses considered an ANOVA and a Student-T to compare NP, FA, TGB and TDB means and a difference of $P<0.05$ was considered as significant. At the beginning of the study, no differences occurred regarding NP, FA, TGB and TDB among groups. While the CG had a higher NP, FA, TGB and TDB ($P<0.05$), the NP and FA favoured to the GF1 group ($P<0.05$). To conclude, the use of goats demonstrated to be an effective biological control regarding the control of *S. ibérica*; a higher grazing frequency decreased the number of plants and foliage area.

Agroecological evaluation of four production systems in very dry tropical forest in Colombia
A. Conde-Pulgarin, C.P. Alvarez-Ochoa, R. Frias-Navarro and D.E. Rodriguez-Robayo
Universidad de la Salle, Facultad de Ciencias Agropecuarias, Carrera 7 número 179 – 03, 110141665 Bogotá. D.C.,
Colombia; aconde@unisalle.edu.co

The objective of this study was to identify the constraints and potentialities of four agricultural production systems: plantain (*Mussa* spp.), maize (*Zea mais* L.), cassava (*Manihot esculenta*) and a dual purpose livestock, in order to contribute to a program of sustainable development in the Baja Guajira region, a very dry tropical forest in Colombia, South America. Characterization of the social and economic components of the region was carried out and the quick agroecological evaluation methodology proposed by Altieri was applied, with the participation of the producers. Ten indicators of soil quality and ten indicators for the health of crops and pastures were recorded, both on a scale of one to ten. The characterization allowed identifying the importance of the model of community work, as an alternative for the consolidation of the peace in the region, since it constitutes one of the main alternatives of economic and social reinstatement of the demobilized people from the FARC guerrilla in Colombia. In the four zones and production systems established, very important differences were found in terms of soil quality and health of crops and pastures, of which it is emphasized that, in the case of the plantain, an average of seven were obtained which is considered as a level of sustainability close to strong. Maize and cassava reported an average of 3.4 indicating a weak level of sustainability, especially for the low biological activity. The paddocks showed weak levels of sustainability in particular by the criteria of structure of soil, with the presence of compaction to a depth of 15 cm and low Biological activity with an average of 3.7. The information obtained will be part of a multicriterio decision model for regional development.

Dietary zinc:copper ratios and net postprandial portal appearance of these minerals in pigs
D. Bueno Dalto, I. Audet and J.J. Matte
Agriculture and Agri-Food Canada, Sherbrooke Research and Development Centre, 2000 College street, Sherbrooke,
QC, J1M 0C8, Canada; jacques.matte2@canada.ca

Zinc (Zn) and copper (Cu) are considered as essential micronutrients in feed and hazardous heavy metals when excreted in manure. Knowledge on their gastrointestinal fate is limited as well as strategies to optimize their intestinal absorption. The present study compared the net intestinal absorption of Zn and Cu after meals with different dietary ratios among them. Ten 46-kg pigs were used in a cross-over design to assess the 10-hours net portal-drained viscera (PDV) flux of serum Cu and Zn after ingestion of boluses of ZnSO4 and CuSO4 in different Zn:Cu ratios (mg:mg): 120:20; 200:20; 120:8; and 200:8. Arterial Zn concentrations peaked within the first hour post-meal and responses were greater with 200 (0.9 to 1.8 mg/l) than with 120 mg (0.9 to 1.6 mg/l) of dietary Zn (dietary Zn × time, P=0.05). Net PDV flux of Zn was greater (P=0.02) with 200 than with 120 mg of dietary Zn, and tended to be greater (P=0.10) with 20 than with 8 mg of dietary Cu. The cumulative PDV appearance of Zn (in % of dietary intake) was greater with 120 than 200 mg of dietary Zn from 8 hours post-meal (P≤0.04) and with 20 than 8 mg of dietary Cu from 7 hours post-meal (P≤0.05). At the end of the postprandial period (10 hours), estimated PDV appearance of Zn was 16.0, 18.4, 12.0 ad 15.3% of Zn intake for 120:8, 120:20, 200:8 and 200:20 ratios, respectively. For Cu, whatever treatment, arterial values varied (P<0.01) by less than 5% across post-meal times. Net PDV flux was not affected by treatments (P≥0.12) but the value for ratio 120:20 was different from zero (P=0.03). There was an interaction dietary Zn × dietary Cu on cumulative PDV appearance of Cu (in % of dietary intake) at 30 minutes post-meal (P=0.04) and thereafter at 3 hours post-meal (P=0.04). For the whole postprandial period (10 hours), estimated PDV appearance of Cu was 61.9, 42.1, -17.1 and 23.6% of Cu intake for 20:8, 120:20, 200:8 and 200:20 ratios, respectively. In conclusion, the present dietary amounts and ratios of dietary Zn and Cu can regulate the amount of both trace minerals that are metabolically available for pigs. Ratios with 120 mg of dietary Zn maximized the post-intestinal availability of both Zn and Cu.

Nutritional evaluation of feeding lambs different supplementary dietary levels of *Morenga* dry leaves

A.Y.M.A.N. Hassan
Animal Production Research Institute, Agriculture Research Center, Dokki, Giza, Egypt., Animal Nutrition, Agriculture Research Center, Dokki, Giza, Egypt., 12618, Egypt; aymanan19@hotmail.com

The effect of feeding different levels of dried *Moringa oleifera* leaves (ML) on body weight gain (BWG), digestibility, nitrogen utilization, and enteric methane production was studied on male Barki lambs aged 4 months and weighed 24.4±1.25 kg. Five lambs were randomly allocated to each of the four treatment diets: concentrate mixture + corn silage + rice straw (control, R1), control ration + 2.5 g, 5.0 g and 10 g ML for R2, R3, and R4, respectively. The feeding period was for 70 days, where animals within groups were fed individually and bi-weekly weighed. Dry matter intake (DMI) was decreased (P<0.05) with the highest supplemented ML level (10 g). Lambs fed R2 and R3 diets gained 164.3 and 171.4 g/h/day respectively, vs 158.6 for control, while R4 had the lowest daily gain (85.7 g). Nitrogen balance and N utilization were highly (P<0.01) recorded for R3. Rumen fermentation parameters showed that R3 had the highest (P<0.05) TVFA's concentration than other diets. Ruminal enteric methane production calculated relative to DMI (g/kg) was the lowest and the microbial protein production showed the highest yield compared to animals in other groups. It is concluded that *M. oleifrra* dry leaves could be safely supplemented at 5.0 g/h/day or 170 mg/kg BW to enhance growth performance of lambs.

A general method to relate feed intake and body mass across individuals and species

J.A.N. Filipe[1], M. Piles[2], W.M. Rauw[3] and I. Kyriazakis[1]
[1]Newcastle University, Newcastle upon Tyne, NE1 7RU, United Kingdom, [2]Institute for Food and Agriculture Research and Technology, Barcelona, 08140, Spain, [3]INIA, Dept Mejora Genética Animal, Madrid, 28040, Spain; joao.filipe@newcastle.ac.uk

We propose a simple two-parameter allometric relationship between body weight gain (Y) and cumulative voluntary feed intake (X) during a given growth period: $Y = A\ X^b$. The relationship was supported by fitting datasets on individuals and groups across livestock species, including pig, rabbit, chicken, turkey, sheep and cattle. The datasets span over ~100 years of selection, different breeds, diets and sexes. Comparison was made with existing alternative relationships. Differences among species, breeds or diets were reflected in the parameters 'b' and 'A' and associated with efficiency of utilisation of feed for growth (and feed conversion ratio) and with history of selective breeding. (1) Variation in 'b' related to efficiency but not to animal size; b=0.8-0.95 and 0.75-0.8 in the more- and less-intensively-selected animals, respectively; and 'b' was lower in ruminant species than in monogastric species. (2) Wider variation in 'A', A=0.08-2.2, related both to efficiency and to animal size; e.g. 'A' was largest in cattle breeds and smallest in rabbits. The model is easily fitted to empirical data offering a flexible way of predicting feed intake from weight gain, and vice-versa, during any period prior to the slowdown of growth towards maturity. Changes in daily feed intake and feed efficiency during growth and their variation among individuals and groups are estimated more accurately via this model than using the raw data. The approach to maturity is accounted for through a third parameter in the relationship, but is not required in many livestock applications. We suggest this approach has a range of potential applications in selective breeding and livestock management. This study received funding from the European Union's H2020 program through grant agreement no. 633531.

A meta-analysis assessing the bacterial inoculation of corn silage over 37-years period
C.H.S. Rabelo[1], A. Bernardi[2], C.J. Härter[1], A.W.L. Silva[2] and R.A. Reis[3]
[1]*Federal University of Pelotas, Capão do Leão, 96050-500, RS, Brazil,* [2]*Santa Catarina State University, Chapecó, 89815-630, SC, Brazil,* [3]*São Paulo State University, Jaboticabal, 14884-900, SP, Brazil; carlos.zoo@hotmail.com*

Inoculation of corn silage with different lactic acid bacteria (LAB) on fermentation, aerobic stability and productive response of dairy cows, beef cattle and sheep was examined through a meta-analysis approach. A database containing 141 articles published in journals (731 treatment means evaluated) was used to examine dry matter (DM) loss and aerobic stability of corn silage. Moreover, the effect of silage inoculation on livestock production was examined through a second database comprising 37 articles (99 treatment means evaluated) with 775 individual cattle and 183 sheep. Data were analysed as mixed models using the MIXED procedure of SAS. Studies were considered random and the inverse of the squared standard error (SSE) of each treatment mean or the inverse of the number of observations of each study (when SSE was lacking) were used as a factor in the WEIGHT statement of the model. Differences between means were determined using the P-DIFF option of the LSMEANS statement at $P \leq 0.05$, and trends discussed at $0.05 > P \leq 0.10$. As results, DM loss increased from 8 to 50% ($P<0.01$) by inoculating corn silage either with homolactic LAB ([ho]LAB) or heterolactic LAB ([he]LAB), respectively. Inoculation of corn silage with [he]LAB increased ($P<0.01$) the aerobic stability by 71.3 h. Inoculation of corn silage did not affect DM intake (DMI) in dairy cows ($P=0.13$), but using both [ho]LAB and [he]LAB simultaneously ([mix]LAB) lowered milk yield (–2.5 kg/d; $P<0.01$). Inoculation of corn silage with [ho]LAB increased DMI in sheep (+0.15 kg/d; $P=0.02$), but decreased it in beef cattle (–0.26 kg/d; $P=0.01$). The ADG of beef cattle ($P=0.06$) and sheep ($P=0.14$) was not affected by silage inoculation. In conclusion, this meta-analysis provided a quantitative summarization of 37-yrs period and indicated, in general, that inoculation of corn silage is needless to improve silage quality and animal performance.

The effect of hemp and Camelina cake feeding on the fatty acid composition of duck muscles
R. Juodka[1], V. Juskiene[1], R. Juska[1], R. Leikus[2], R. Nainiene[3] and D. Stankeviciene[1]
[1]*LUHS Animal Science Institute, Department of Ecology, R. Zebenkos 12, 82317, Lithuania,* [2]*LUHS Animal Science Institute, Department of Animal Feeding and Feedstuffs, R. Zebenkos 12, 82317, Lithuania,* [3]*LUHS Animal Science Institute, Department of Animal Breeding and Reproduction, R. Zebenkos 12, 82317, Lithuania; violeta.juskiene@lsmuni.lt*

Camelina (*Camelina sativa* L. Crantz) and hemp (*Cannabis sativa* L.) seed cakes are rich sources of n-3 polyunsaturated fatty acids (PUFAs). This study was carried out to investigate the effects of Camelina and hempseed cakes in the diet of ducks on the intramuscular fatty acid profile. Male ducks (n=99) were randomly allocated to 3 dietary treatments: Control or C group (wheat-soybean-meal-barley-based diet with 15-20% rapeseed cake), Experimental 1 or HEM group (with hempseed cake added at 15-20% instead of rapeseed cake), and Experimental 2 or CAM group (with Camelina cake added at 15-20% instead of rapeseed cake). All groups received the diets *ad libitum*. At 49 days of age, six ducks from each group were slaughtered for analysis of the fatty acid composition in the breast and leg muscles. Feeding ducks with the diet enriched with Camelina cake resulted in significantly higher amounts of n-3 a-linolenic fatty acid (ALA) ($P<0:01$) and total n-3 PUFA ($P<0.01$) in breast and leg muscles, while eicosatrienoic fatty acid (ETE) ($P<0.01$) was higher in the leg muscle. The ratios of n-6/n-3 and linoleic / a-linolenic fatty acids ($P<0.01$) decreased significantly compared to the Control and HEM groups of ducks. Feeding ducks with the diet enriched with hempseed cake resulted in significantly higher amounts of linoleic (LA) ($P<0.01$), total n-6 PUFA ($P<0.05$-$P<0.01$) and n-6 g-linolenic (GLA) ($P<0.01$) fatty acid. Our study showed that using Camelina cake as supplementation in duck diets opens a possibility to develop functional food, i.e. meat with a significantly higher content of ALA, total n-3 PUFA and the lowest ratios of n-6 / n-3 fatty acids in ducks' muscles. A duck diet with hempseed cake produces exceptional-quality meat with an enriched content of n-6 GLA.

Defining a management goal that maximizes the intake rate in palisadegrass

G.F. Silva Neto[1,2], J. Bindelle[1,2], J. Rossetto[1], P.A. Nunes[1], Í.M. Monteiro[1], R. Becker[1], J.V. Savian[3], G.D. Farias[1], R.T. Schons[1], E.A. Laca[4] and P.C.F. Carvalho[1]
[1]*Federal University of Rio Grande do Sul, D. Forage Plants, and Agrometeorology, Brazil,* [2]*Liège University, Gembloux Agro-Bio Tech, Precision Livestock and Nutrition Unit, Belgium,* [3]*National Institute of Agricultural Research, INIA, Treinta y Tres, Uruguay,* [4]*University of California, D. Plant and, Environmental Sciences, USA; neto.gentilfelix@gmail.com*

'Rotatinuous' stocking (RS) is an innovative sward management concept based on the ingestive behaviour of grazing animals. It considers that grazing time is the major limitation to allow animals to take the best from the vegetation on offer and limit supplementation to cover the daily requirements. It hypothesizes that there is an ideal sward structure that allows the animals to maximize herbage short-term intake rate (STIR, g dry matter/min). In this work, we tested which sward structure should be used as a management goal under RS in palisadegrass pastures (*Urochloa brizantha*). Treatments consisted of four sward heights (20, 30, 40 and 60 cm) in palisadegrass cv. Marandu pasture. The experimental design was a randomized complete block with four replicates, which consisted of the interaction between the evaluation time (morning or afternoon) and the period (two experimental cycles). The sward height was measured at 200 points per paddock pre- and post-grazing using a sward stick. 16 grazing tests of 45 minutes were performed using the double-weighing technique for determining the STIR of three adult Brangus cows. The STIR presented a quadratic response ($P<0.01$) to the sward heights, being maximized at approximately 42 cm (57 g DM/min). Previous studies suggested that the sward height at which the intake rate was maximized should be adopted as the pre-grazing management target and that the grazing down should not exceed 40% of the initial sward height. In our study, this represents the use of 42 cm and 25 cm as pre- and post-grazing sward heights in rotational stocking. We conclude that palisadegrass pastures must be managed at 42 cm of sward height in pre-grazing according to the RS concept to maximize intake rate.

Introducing *Lupinus* spp. seeds in Churra da Terra Quente lambs diets and its effect on growth

M. Almeida[1,2], A. Nunes[1], S. Garcia-Santos[1,3], C. Guedes[1,2], S. Silva[1,2] and L. Ferreira[1,3]
[1]*UTAD, Quinta de Prados, 5000-801 Vila Real, Portugal,* [2]*Veterinary and Animal Research Centre (CECAV), Quinta de Prados, 5000-801 Vila Real, Portugal,* [3]*Centre for the Research and Technology Agro-Environmental and Biological Sciences (CITAB), Quinta de Prados, 5000-801 Vila Real, Portugal; mdantas@utad.pt*

The legumes of the genus *Lupinus* spp. are well-adapted species to the soil and climatic conditions of the Mediterranean region. Most of them are good sources of protein (~20-45%) and energy. Therefore, these could be alternatives to imported protein sources for animal feed, such as soybean meal. In this context, the objective of this work was to study the replacement of soybean meal by *Lupinus albus* grain (26%) or *Lupinus luteus* grain (33%) on Churra da Terra Quente lambs feeding and its effect in animal growth. Twelve male lambs (ages between 92 and 110 days and initial mean live weight of 18±2.8 kg) were divided into 3 groups of 4 animals and fed with 3 isoproteic and isoenergetic diets: the control diet (CTR) consisting of meadow hay, wheat grain and soybean meal, the diet with *L. luteus* replacing soybean meal (33%, LL) and the diet with soybean meal being replaced by *L. albus* (26%, LA). During the growth period, the animals were weighed weekly and the amount of feed supplied and ingested by the animals was recorded. After the growth period, an evaluation of the digestibility of the diets was performed in metabolic cages. The results indicate that the partial substitution of soybean meal by *L. luteus* and *L. albus*, in diets of growing lambs, had no effect on mean daily gain, feed efficiency, and feed conversion. The dry matter intake of lambs fed the CTR diet was significantly lower ($P<0.05$) than the groups fed with the LL and LA diets. However, the amount of digestible organic matter ingested did not differ among groups (547.7, 538.0 and 543.6 g OM digestible/day for the CTR, TC and TR diet, respectively). The diet did not affect the digestibility of any of its fractions, although a trend was observed for higher digestibility values of the NDF fraction of the CTR diet, compared to the others. The results indicate the possibility of using these protein sources as feed materials for growing lambs, although higher incorporations need to be studied.

The effect of introduction of *Lupinus* spp. seeds in diets on lamb eating behaviour

M. Almeida[1,2], A. Nunes[1], S. Garcia-Santos[1,3], C. Guedes[1,2], L. Ferreira[1,3], G. Stilwell[4] and S. Silva[1,2]
[1]*UTAD, Quinta de Prados, 5000-801, Vila Real, Portugal,* [2]*Veterinary and Animal Research Centre (CECAV), Quinta de Prados, 5000-801, Vila Real, Portugal,* [3]*Centre for the Research and Technology Agro-Environmental and Biological Sciences (CITAB), Quinta de Prados, 5000-801, Portugal,* [4]*Animal Behaviour and Welfare Laboratory, Centre of Interdisciplinary Investigation in Animal Health, Faculty of Veterinary Medicine, Lisbon University, 1300-477, Lisboa, Portugal; mdantas@utad.pt*

The objective of this study was to evaluate the partial replacement of soybean meal by *Lupinus albus* or *Lupinus luteus* grains on the eating behaviour of Churra da Terra Quente lambs. Twelve male lambs were divided into three groups of four animals and fed with three isoproteic and isoenergetic diets during 35 days. Control diet (CTR) consisted of meadow hay, wheat grain and soybean meal and, in the other two diets, soybean meal was replaced by *L. luteus* (LL, 33%) and *L. albus* (LA, 26%). Eating behaviour was evaluated by registering the frequency of lambs eating hay and the time spent eating the concentrate feeds in three different moments, before the introduction of *Lupinus* seeds (BI), on the day (DI) and 30 days after of introduction (AI). Eating hay (H) behaviour was recorded every 10 minutes, between 8 am and 8 pm, using a scan sampling. For each diet, differences between moments on the frequency of eating hay behaviour was compared using a Pearson's Chi-square contingency table analysis. The effect of diet and moment on the time spent eating the concentrate feeds was examined using a two-way ANOVA procedure. The moment did not have an effect on hay eating for CTR and LA diets (CTR: χ^2=8.58, df=8, P=0.379 and LA: χ^2=7.97, df=8, P=0.437). In contrast, for LL diet the hay eating behaviour changed across the period (χ^2=30.65, df=8, P=0.002), with eating hay making up the greatest percentage at AI (25%), and the smallest percentage on DI (19%). The moment had an effect (P<0.001) on the time spent eating the concentrate feeds with the less time reported at the BI (3.5 vs 7.2 and 8.6 min., P<0.05, for BI, DI and AI, respectively). Overall, data indicates that the replacement of soybean meal by *Lupinus* grains does not affects lamb eating behaviour.

Productivity and feeding behaviour of finishing pigs fed with a calming herbal extract

M. Fornós[1], E. Jiménez-Moreno[1], J. Gasa[2], V. Rodríguez-Estévez[3] and D. Carrión[1]
[1]*Cargill Animal Nutrition, Mequinenza, 50170 Zaragoza, Spain,* [2]*Department of Animal and Food Sciences, Universitat Autònoma de Barcelona, 08193 Barcelona, Spain,* [3]*Department of Animal Production, Universidad de Córdoba, 14071 Córdoba, Spain; marta_fornos@cargill.com*

Finishing pigs reared under intensive conditions are exposed to different stressors which might jeopardize growth performance by a change in feeding behaviour. The aim of this study was to evaluate the effect of the inclusion of ConverMax® in a diet on growth performance, feeding behaviour and carcass characteristics of pigs during the period from day 84 to 130 (75 to 110 kg of BW). Seventy-two [Pietrain × (Landrace × Large White)] pigs, were sexed and allotted randomly to two treatments with three pens of 12 pigs each (24 males and 12 females per treatment). ConverMax is a mineral feed containing natural compounds and plant extracts with the target to improve animal welfare by reducing aggressive behaviour in pigs. ConverMax was added to the control diet at level of 0.2%. Pigs were fed *ad libitum* with one automatic system per pen. Fasting, as a stressor, started on day 110 (18:30 h) and finished on day 112 (12:00 h). Pigs fed with ConverMax diet tended to have a greater average daily gain (P=0.07) than pigs fed with the control diet, however, no significant differences between treatments were observed for final BW and feed conversion ratio (P>0.1). Before fasting, no significant differences between treatments were observed for average daily feed intake (FI), number of feeding visits, time spent eating and feeding rate (P>0.05). However, the next four days after fasting, pigs fed with ConverMax had a greater daily FI (P<0.05) and longer time spent eating per day (P<0.05) than pigs fed with control diet. Loin depth but not backfat thickness, was increased by 3 mm in pigs fed with ConverMax diet (P<0.05). In conclusion, the supplementation of 0.2% of ConverMax in a diet for finishing pigs can increase both, daily FI and time spent in visits per day, especially in the days post-fasting, and carcass loin depth.

Reliability of genomic predictions for feed efficiency traits based on different pig lines

A. Aliakbari, E. Delpuech, Y. Labrune, J. Riquet and H. Gilbert
INRA, GenPhySE, 24, chemin de Borde-Rouge, Auzeville Tolosane, 31320 Castanet-Tolosan, France; amir.aliakbari@inra.fr

The majority of genomic predictions use a unique population split between a reference and a validation set. However, a genomic evaluation using genetically different reference and validation sets could provide more flexibility for the choice of reference sets in small populations. The aim of our study was to investigate the potential of genomic evaluation for feed efficiency related traits using a reference set that combines two different lines. Data came from two lines divergently selected for residual feed intake during 9 generations. Genomic breeding values (GBVs) of animals for five production traits were predicted using the single-step genomic BLUP method with six scenarios. All scenarios aimed to predict GBVs of pigs of the three last generations (~ 400 pigs, G7 to G9) in one or in the other line (validation line). To compare the scenarios prediction accuracy, a first scenario (control) had a reference set with animals from G1 to G6 (~ 400 pigs) of the validation line. In scenario 2, in addition to those of the control scenario, the reference set included about 600 pigs from G4 to G9 of the alternate line. Scenario 3 had ~ 800 pigs in the reference set, by excluding animals from G4 to G6 of the alternate line from the reference set compared to scenario 2. For the last three scenarios, fewer animals from the validation line were included in the reference set (~200 pigs from G4 to G6). In scenario 4, G4 to G9 animals from the alternate line (~600 pigs, as in scenario 2) were included in the reference set. In scenario 5, only ~400 pigs from G7 to G9, and in scenario 6 ~200 pigs from G9, were used. In scenarios 2, 3 and 4, genotyping 400 to 600 additional individuals from the alternate line provided on average limited improvement the prediction accuracies for the five traits (<14%, except in 3 cases), and sometimes led to reduced accuracies. Scenarios 5 and 6 had similar accuracies as the control scenario, with less genotyping in scenario 6. It indicates that if samples from earlier generations are missing in a line, part of them can be replaced by recent samples from a related different line, giving more flexibility to design training populations in small lines.

The effect of protein and salt level on performance and faecal consistency in weanling piglets

S. Millet[1], B. Ampe[1] and M.D. Tokach[2]
[1]ILVO (Flanders Research Institute for Agriculture, Fisheries and Food), Scheldeweg 68, 9090 Melle, Belgium, [2]College of Agriculture, Kansas State University, Department of Animal Sciences and Industry, 246 Weber Hall, Manhattan KS 66506, USA; sam.millet@ilvo.vlaanderen.be

Recent research at Kansas State University (KSU) showed that salt levels of minimal 3.5 g/kg diet give best performance in young piglets. This is higher than in common European diets. However, in the US, CP levels are higher than in most European diets. The higher CP levels may lead to higher protein breakdown. Higher salt levels may lead to increased water consumption, which might facilitate nitrogen clearance through the kidneys. Therefore, it was hypothesized that high salt levels for pigs are more important when the pigs receive a high CP diet. For this experiment, 96 pens of 5 piglets were divided over 8 treatments in a two by four factorial design with 2 protein levels (189 vs 223 g/kg) and 4 salt levels (1.9, 2.5, 3.1, and 3.8 g/kg in the low protein diet and 1.9, 2.6, 3.2, and 3.9 g/kg in the high protein diet). Different salt levels were obtained by exchanging SiO_2 for NaCl. The low protein diet was formulated to contain 11.8 g SID lysine and 9.85 MJ net energy/kg diet, while in the high protein diet, this was 12.5 g SID lysine and 10.40 MJ net energy/kg diet. The initial hypothesis could not be confirmed as an interaction between salt and protein level on performance was not detected. The higher nutrient density in the high protein diet resulted in lower feed intake with similar growth, leading to improved feed efficiency. Salt level did not affect performance. This is in contrast with the findings at KSU. At the moment, the reason is unclear. Faecal consistency scores deteriorated with increasing salt levels the first week after weaning. The latter might be a result of increased water consumption or increased osmotic pressure in combination with limited absorption capacity of piglets shortly after weaning, especially when receiving diets without zinc oxide or antibiotics, which were included in the KSU study.

Semen and sperm quality parameters of Potchefstroom koekoek cockerels fed dietary Moringa oleifera l

A. Sebola
University of South Africa, Agriculture and Animal Health, Cnr Christiaan de Wet and Pioneer Ave, Florida, 1710, South Africa; sebolana@outlook.com

The study was designed to evaluate semen and sperm quality of Potchefstroom Koekoek roosters fed Moringa oleifera leaf meal (MOLM) aged 40 weeks. At four weeks of age, the chickens were randomly allotted to two dietary treatments diets consisting of MOLM0 (control) and MOLM70 g/kg. A completely randomised design (CRD) experiment, with 5 birds, replicated 4 times was used for this experiment. Semen was collected six times a week by the dorso-abdominal massage method. Semen was used to evaluate ejaculate volume, sperm concentration, live in total sperm, live normal sperm, sperm quality factor and abnormal sperm using a computer-aided sperm analysis (CASA) system. Rosters fed MOLM70 g/kg resulted in higher (57.73%) progressive motility (PM) than the control diet (39.38%). Control diet had higher non progressive motility (NPM) and static of 48.59 and 12.57%, respectively. Moringa oleifera leaf meal diet resulted in higher ($P>0.05$) sperm velocity/rapid (89.06%) than the control diet. Feeding MOLM70 also resulted in higher VCL (106.64 µm/s), VAP (66.19 µm/s) and VSL (45.54 µm/s). In conclusion, MOLM has proven to be a good supplement that can be used to improve fertility in poultry.

The effect of lactic acid bacteria inoculation or molasses on the fermentation quality and nutritive

M. Abbasi[1], Y. Rouzbehan[1], J. Rezaei[1] and S.E. Jacobsen[2]
[1]Tarbiat Modares University, Animal Science Department, P.O. Box 14115-336, 1411713116 Tehran, Iran, [2]Faculty of Science, University of Copenhagen, Department of Plant and Environmental Sciences, Højbakkegård Allé 13, 2630 Taastrup, Denmark; rozbeh_y@modares.ac.ir

This study assessed the influence of wilting, lactic acid bacteria (LAB), and/or molasses on the chemical composition, phenolic compounds, CP *in situ* degradability, and *in vitro* ruminal fermentation parameters of amaranth (var. Maria) silage using a randomized complete block design with 6 replicates. Treatments were fresh amaranth forage (FAF), ensiled amaranth without additive (EA), EA inoculated with LAB (EAB), EA + 5% of molasses (EAM), WA inoculated with LAB + 5% of molasses (EABM), and 24 h wilted EA (WEA). The ensiled materials were stored anaerobically for a period of 45 d. Chemical composition, oxalic acid and nitrate levels, silage fermentation characteristics, DM disappearance (DMD), OM disappearance (OMD), *in vitro* ruminal total volatile fatty acids (TVFA), cellulolytic bacteria and total protozoa counts, and *in situ* CP degradability were determined. Compared with FAF, EA had significantly lesser values (based on % DM) of water-soluble carbohydrates (2.0 vs 6.3), nitrate (0.02 vs 0.3), total phenolics (0.8 vs 1.1), total tannins (0.5 vs 0.8), DMD (67.0 vs 70.3), metabolizable protein (9.0 vs 9.8), *in vitro* ruminal cellulolytic bacteria (7.57 vs 8.41, log 10/g digesta), with non-significant effect on TVFA, total protozoa counts (6.62 vs 6.38, log10/g digesta), and oxalic acid (0.75 vs 0.76), respectively. Inoculation EA with LAB had no effect on all the measured parameters. Adding molasses to EA had significantly increase ash (19.0 vs 17.0% DM), and lactate concentrations (70.0 vs 57.0% DM), and decreased ammonia-N (51.0 v. 55.0) mg/dl), but had no effect on DMD (66.8 vs 65.7) and OMD (61.7 vs 60.3), respectively. The WEA had significantly lesser ammonia-N (45.9 vs 54.3, kg of total N) and greater (5.9 vs 5.7% DM) compared with EA. Overall, FAF var. Maria can be preserved as a valuable silage to feed ruminants. Ensiling amaranth var. Maria decreased the antiquality compounds, and molasses addition improved the quality of the silage.

Impact of varying levels of liquid fermented potato hash diets on growth performance of weaned pigs

R.S. Thomas[1], R.T. Netshirovha[1], B.D. Nkosi[1,2], P. Sebothoma[1], F.V. Ramukhithi[1], Z.C. Raphalalani[1] and O.G. Makgothi[1]
[1]Agricultural Research Council, Animal Production, Private Bag x2, Irene, 0062, South Africa, [2]Centre for Sustainable Agriculture, University of the Free State, P.O. Box 339, Bloemfontein, 9300, South Africa; ronaldt@arc.agric.za

The study evaluated the impact of varying levels of liquid fermented potato hash (LFPH) diets on growth performance of weaned pigs. Diets were formulated to provide 14 MJ/kg digestible energy (DE), 180 g crude protein (CP)/kg and 11.6 g lysine /kg and fermented for 8 hours before being fed to weaner pigs. A back-slopping fermentation approach was used to prepare liquid fermented diets. Dietary treatments were control (commercial diet with no LFPH), low (100 g/kg) liquid fermented potato hash (LLFPH), medium (150 g/kg) liquid fermented potato hash (MLFPH), and high (200 g/kg) liquid fermented potato hash (HLFPH). The dietary treatments were fed *ad libitum* to 80 Large White × Landrace crossbred weaner pigs (10±0.9 kg body mass) that were individually housed in 1.54×0.8 m pens. Pigs were allocated to treatments in a complete randomized design. The experiment lasted 50 days during which growth and feed intake were recorded. The control and LLFPH had higher (P<0.001) final weight (45.7 kg and 49.2 kg respectively) and average daily gains (ADG) (670 and 659 g/d respectively) compared to pigs fed diets containing MLFPH and HLFPH. However, pigs on HLFPH had a lower (P<0.001) average daily feed intake (ADFI) and a better feed conversion rate (<2.10) compared to those on the other diets. Feeding LFPH to weaner pigs improved the health and growth performance of pigs. This study implies that FLPH might be an alternative feed source for weaned pigs.

The determination of antioxidant activity and content of polyphenols in grape pomace

R. Kolláthová[1], B. Gálik[1], M. Juráček[1], D. Bíro[1], M. Šimko[1], M. Rolinec[1], E. Ivanišová[2], O. Hanušovský[1] and P. Vašeková[1]
[1]Slovak University of Agriculture, Faculty of Agrobiology and Food Resources, Department of Animal Nutrition, Trieda Andreja Hlinku 2, 94976 Nitra, Slovak Republic, [2]Slovak University of Agriculture, Faculty of Biotechnology and Food Sciences, Department of Plant Storage and Processing, Trieda Andreja Hlinku 2, 94976 Nitra, Slovak Republic; xkollathova@uniag.sk

The aim of this study was to determine antioxidant activity, total polyphenol, flavonoid and phenolic acid content of dried grape pomace (*Vitis vinifera* L., variety Pinot Gris). The antioxidant activity was detected by DPPH method, total polyphenol content with Folin – Ciocalteu reagent, flavonoids content by aluminium chloride method and total phenolic acid content by Arnova reagent. Absorbance of the reaction mixtures was determined using the spectrophotometer Jenway (6405 UV/Vis). The analysed grape pomace was characterized by high antioxidant activity (9.06 mg Trolox equivalent antioxidant capacity per g). The total polyphenol content (42.69 mg gallic acid equivalent per g), as well as the total phenolic acid content (16.69 mg caffeic acid equivalent per g) has also proven to be high. The analysed samples of grape pomace had a lower content of total flavonoids (1.08 mg quercetin equivalent per g). Results have shown that grape pomace is rich for polyphenol compounds with antioxidant activity and can be used as potential source of bioactive compounds in animal nutrition. This work was supported by the Slovak Research and Development Agency under the contract no. APVV-16-0170.

Effect of lysine level at late gestation on body condition and reproductive performances in sows

S. Seoane[1], J.M. Lorenzo[2], P. González[2], L. Pérez-Ciria[3] and M.A. Latorre[3]
[1]COREN, Av. da Habana, 32003 Ourense, Spain, [2]Cent. Tecnol. de Carne, Parq. Tecnol. de Galicia, 32900 Ourense, Spain, [3]IA2-Univ. de Zaragoza, C/ Miguel Servet 177, 50013 Zaragoza, Spain; leticiapcgm@gmail.com

The aim was to evaluate the effect of the dietary standardized ileal digestible lysine (SID Lys) level provided during the last month in the gestation facilities on backfat and loin muscle depths (BFD and LMD, respectively) and on reproductive performances. For that, a total of 62 hyperprolific DanBred sows, between second and fourth parity, was used. During pregnancy, the feeding program for all the animals was the following: 3 kg/d from day 1 to 30, 2.5 kg/d from day 30 to 77 and 3 kg/d from day 77 to 107. At day 107, sows were moved to the parturition facilities and a common lactation diet was provided. During the last phase of gestation (day 77-107), dams received a diet with 0.6% SID Lys (control) (n=31) or with 1% SID Lys (n=31). Data were analysed using the GLM procedure of SAS. At farrowing, BFD was thicker in sows fed with 1% SID Lys than in those fed with 0.6% SID Lys (16.8 vs 15.3 mm, for 1 and 0.6% SID Lys, respectively; P=0.032), but these differences disappeared at weaning (P>0.05). Also, sows eating 1% SID Lys showed higher LMD than those receiving 0.6% SID Lys at farrowing and at weaning, although it was not significant (P>0.05). When the period studied was longer (a full cycle; from the previous weaning to the next one), it was found that BFD was not affected by diet (P>0.05) but the LMD losses were lower with the diet including 1% SID Lys than with that including 0.6% SID Lys (1.25 vs 2.83 mm; P=0.0002). The highest dietary SID Lys level improved some parameters related to reproductive performances: at farrowing, it enhanced the number of live-born piglets (17.1 vs 15.7; P=0.043) and the litter weight (25.9 vs 23.1 kg; P=0.017); at weaning, it increased the weight of the litter (73.0 vs 67.1 kg; P=0.024) and of the piglet (5.79 vs 5.40 kg; P=0.034); and at next parity, it improved the total number of born (18.6 vs 16.4; P=0.005) and live-born piglets (16.9 vs 15.3; P=0.018). We concluded that the increase in the SID Lys level from 0.6 to 1% during the last month of gestation improved body condition of the sows and their performance results even at next cycle.

Effect of low-fibre sunflower meal and phytase on broiler chicks' performance and meat quality

D. Grigore[1,2], G. Ciurescu[2], M. Hăbeanu[2], S. Mironeasa[3], M. Iuga[3] and N. Băbeanu[1]
[1]University of Agronomic Science and Veterinary Medicine, 59 Marasti Street, Bucharest, 011464, Romania, [2]National Research-Development Institute for Biology and Animal Nutrition, Calea Bucuresti no.1, Baloteşti, Ilfov, 077015, Romania, [3]Stefan cel Mare University Suceava, Str. Universitatii, no. 13, Suceava, 720229, Romania; daniella_31190@yahoo.com

High costs of conventional protein feed sources including soybean meal (SBM) generated the need for finding other alternatives. Thus, the present study was designed to evaluate the impact of graded replacements of SBM by low-fibre sunflower meal (SFM) with or without phytase supplementation on growth performance, carcass traits, and plasma profile of broilers. A total of 800 mixed sexes 1-d-old broiler chicks (Cobb 500) were randomly assigned to eight dietary treatment groups (five replicates each) in a 4×2 factorial arrangement, including four levels of SFM (0, 25, 50, and 75% replacing SBM) and two levels of enzyme added (0 or 0.2 g/kg diet). Feed conversion ratio (FCR), and body weight gain (BWG) were significantly (P<0.001) decreased with increasing SFM more than 25% substitution for SBM. There was no significant effect (P>0.05) between the main factors interaction (SFM × phytase) on growth performance during the starter, grower and finisher as well as overall period. Enzyme addition had no beneficial effects on performance traits. The evaluated carcass yield, breast yield and digestive organ weights (i.e. gizzard, liver, pancreas and spleen) were not (P>0.05) influenced by feeding SFM or enzyme addition. However, increasing the level of SFM (50 and 75% respectively) in diets reduced the abdominal fat (P<0.001). The selected physicochemical properties of breast meat, including pH, colour, texture hardness, springiness, resilience, cohesiveness, guminess and chewiness were not influenced (P>0.05) by dietary treatments. It is concluded that the diets with low-fibre SFM levels up to 25% led to growth performance and carcass traits comparable with diet contained SBM.

Linseed oil supplementation affects IGF1 expression in turkey (Meleagris gallopavo)

K. Szalai, K. Tempfli, E. Zsédely and Á. Bali Papp
Széchenyi István University, Department of Animal Science, 2 Vár Square, 9200 Mosonmagyaróvár, Hungary;
tempfli.karoly@sze.hu

Linseed oil supplementation is a potential way to improve polyunsaturated fatty-acid profile of animals in order to enhance dietary value of meat products. In an experiment, male turkeys were divided into two groups (n=12 each) according to diet; control (without linseed) and linseed oil-supplemented animals were sampled for muscle (breast, thigh), abdominal fat, and liver tissue at a local abattoir. Total RNA was isolated, DNase treated, reverse-transcribed, and expression of insulin-like growth factor 1 gene (*IGF1*) was analysed by means of quantitative PCR. *GAPDH* and *beta-actin* were used as reference genes. Higher ($P<0.05$) *IGF1* expression was revealed in muscle samples of linseed oil-fed animals (breast: 1.66 ± 0.41; thigh: 2.29 ± 0.33) compared to control (breast: 1.00 ± 0.19; thigh: 1.00 ± 0.30), whereas *IGF1* expression in fat was lower ($P<0.05$) in the linseed oil group (0.32 ± 0.10 vs 1.00 ± 0.27). Expression of *IGF1* in the liver was not affected by the dietary treatment (control: 1.00 ± 0.38; linseed oil: 1.05 ± 0.44; $P>0.05$). The work was supported by the EFOP-3.6.3-VEKOP-16-2017-00008 project. The project is co-financed by the European Union and the European Social Fund.

Antioxidant effects of phytogenic herbal mixtures additives used in chicken feed on meat quality

Ż. Zdanowska-Sąsiadek[1], K. Damaziak[2], M. Michalczuk[2], A. Jóźwik[1], J. Horbańczuk[1], N. Strzałkowska[1], D. Róg[3], W. Grzybek[1], K. Jasińska[1], K. Kordos[1], K. Kosińska[1], M. Łagoda[1] and J. Marchewka[1]
[1]Institute of Genetics and Animal Breeding, Postepu 36A, 05-552 Magdalenka, Poland, [2]Warsaw University of Life Sciences, Ciszewskiego 8, 02-786 Warsaw, Poland, [3]General Industrial, Trading and Service Company Bellako Limited, Gocławska 11, 03-810 Warsaw, Poland; aa.jozwik@ighz.pl

Plants including herbs, vegetables and fruits are important source of antioxidants. When fed to production animals they may protect the characteristics of the final product. The aim of the current study was to demonstrate the influence of herbal and vegetable blends fed to broiler chickens on the oxidative status and physiochemical properties of their breast muscles. Four hundred male broiler chickens (ROSS 308) were randomly divided into four feeding groups (100 birds each) fed *ad libitum* and reared on litter until 42 days of age. The control group (C) received the standard diet, while groups E1, E2 and E3 diet with 2% herbal-vegetables mixtures as an additive. Mixtures were composed of 70% onion, 25% thyme, 5% mint or 90% ginger, 7% rosemary, 3% chili or 30% onion, 20% garlic, 25% oregano, 10% fennel, 5% mint, 5% turmeric, 5% ginger in group E1, E2 and E3 respectively. A generalised linear mixed model was applied on all measured parameters using PROC GLIMMIX of SAS. Application of the vegetable and herbs mixtures affected meat physiochemical characteristics, increasing pH24 to 6.00 ± 0.04 in E3 and 5.98 ± 0.03 in E1, WHC to 13.0 ± 0.94 cm2/g in E1, reducing shear force to 40.0 ± 1.98 N, 39.4 ± 2.02 N, 45.7 ± 1.93 N in E1, E2 and E3 respectively) ($P<0.05$). Chickens receiving herbal and vegetable additives showed highest content of polyphenols and anthocyanins in E3: 1.20 ± 0.06 mg GAE/g and 0.592 ± 0.03 mg/100 g respectively, and retinol and α-tocopherol in E1: 0.242 ± 0.02 mg/100 g and 1.53 ± 0.09 mg/100 g respectively in the breast muscle ($P<0.05$). Herbal-vegetable mixtures applied into broiler diet improved the physiochemical properties of the meat, enriched it with health-promoting substances and protected against the unfavourable oxidation process.

Influence of overfeeding duration and intensity on health and behaviour indicators measured in ducks

J. Litt[1], L. Fortun-Lamothe[2], D. Savietto[2] and C. Leterrier[3]
[1]ITAVI, 1076 route de Haut-Mauco, 40280 Benquet, France, [2]GenPhySE, Université de Toulouse, INRA, ENVT, 31326 Castanet Tolosan, France, [3]UMR Physiologie de la Reproduction et des Comportements, INRA, CNRS, IFCE, Université de Tours, Centre Val de Loire, 37380 Nouzilly, France; litt@itavi.asso.fr

During overfeeding, ducks are housed in a space of reduced size and forced-fed. In order to improve the condition of birds during this period, it is necessary to have a simple and objective assessment of their status. Since overfeeding intensity and duration influence the duck status, the purpose of the study was to measure the impact of both factors on various health and behavioural indicators measured on the animal. 320 ducks were randomly distributed into 80 pens from 70 to 97 days and forced-fed twice a day by a feed dispenser. Overfeeding was carried out from day 87 to day 97 with a moderate intensity (MI, 376 g of maize flour per meal as a mean during the period, n=160) or a high intensity (HI, 414 g, n=160) procedure. Various indicators were used to measure 'Good feeding' (n=3), 'Good housing' (n=5), 'Good health' (n=10) and 'Appropriate behaviours' (n=11). Each indicator was noted as present (1) or absent (0) for each animal in the pen, without handling or restraint, by an observer circulating among the pens. Statistical analyses were done using logistic regression on the 24 indicators having a prevalence higher than 2.5%. 13 indicators were significantly impacted by at least one of the tested factors and preselected. All of 13 were significantly impacted by the days on overfeeding, whereas the intensity of overfeeding degraded all preselected 'Good feeding' parameters (2/2) but had no effect on 'Good health' parameters (4/4). Most of 'Good housing' (2/3) and 'Appropriate behaviours' (2/4) parameters were affected by the intensity of overfeeding only at some days due to an interaction between both factors. This experiment evidenced 11 indicators that could be used for on-farm evaluation of the duck status during overfeeding (feed stain, drinking, resting, panting, dirty eyes, leg lesion, blood stain, preening, active, motionless and ruffled). However, questions are still raised to know when this evaluation has to be carried out. Reference values are also still needed for some of them to allow a diagnosis.

Evaluation of four commercial fish feed in the growth of pikeperch (*Sander lucioperca*) reared in RAS

E. Panana and S. Teerlinck
Inagro, Aquaculture, Rumbeke-Beitem, 8800, Belgium; edson.pananavillalobos@inagro.be

Pikeperch (*Sander lucioperca*) is considered to have the highest potential for inland aquaculture diversification in Europe due to its good market acceptance and rapid growth rate. During the last decades the culture of pikeperch in recirculating aquaculture systems (RAS) has increased. However, until now, commercial feeds for pikeperch are not available in the market and feed for other fish species (trout, turbot, salmon) are used during rearing. The aim of this research was to compare the growth performance of pikeperch fed with commercial feeds in a pilot-scale RAS culture. Four different commercial feeds (Skretting, Alltech Coppens, Biomar, Le Gouessant) were tested in triplicate using a completely random design. Pikeperch (80±8 g initial body weight) were cultured for four months at 22 °C water temperature at a density of 30 kg/m^3 in twelve (1.8 m^3) RAS tanks. To maintain production conditions, fish were fed until apparent satiation with automatic feeder, once per month grading was performed for reduction of size heterogeneity and to keep constant density. Production parameters such as feed conversion ratio (FCR), specific growth rate (SGR) amongst others were statistically compared. No significant differences (P>0.05) were found in the FCR and SGR amongst treatments, however, fish fed with Skretting and Coppens showed a slightly better FCR performance in compared with AquaBio and Le Gouessant (1.14±0.4, 1.18±0.4, 1.23±0.4 and 1.28±0.5 respectively). The growth performance of pikeperch was similar amongst the commercial feed tested. This indicates that the commercial feeds for other carnivorous fish species apparently provide enough nutritional requirements during the ongrowing phase of pikeperch in RAS. These results can help producers to take decisions concerning the selection of the feed, where other aspects such as price and availability of the feed may play a major role.

Ammonia-nitrogen release urea samples with different protection to *in vitro* ruminal degradation

M. Simoni[1], M. Rodriguez[2] and S. Calsamiglia[2]
[1]University of Parma, Department of Veterinary Science, via del Taglio, 10, 43126 Parma, Italy, [2]Università autonoma de Barcelona, Departement of Animal and Food Science, Campus de la UAB, Bellaterra, Spain; marica.simoni@studenti.unipr.it

The aim of the study was to compare the protection degree against rumen degradation of 2 slow release urea (SR) prototypes. The study was conducted in a closed batch fermentation system including a negative control (C- no urea) and 3 treatments (T): Urea (U), SR1 (Protigen®- Karizoo, Caldes, Spain) and SR2. Incubations were conducted in triplicate for U, SR1 and SR2 and duplicate for C within period, and in two replicated periods. Isonitrogenous samples (155 mg N) were weighed in glass tubes. Ruminal fluid (RF) was collected from a ruminally fistulated fasted adult cow, filtered through 4 layers of cheesecloth and kept at 39 °C. The incubation medium was prepared anaerobically by mixing a N-free buffer solution with RF in a 4:1 ratio. The medium (200 ml) was added to each tube and placed in a 39.0 °C water bath with agitation for 24 h. Each tube was sampled in duplicate at 0, 1, 2, 4, 6, 8 and 24 h of incubation, acidified, and frozen. The day of analysis, samples were centrifuged at 15,000×g for 15 min and the supernatant analysed for ammonia concentration by spectrophotometry. The NH_3-N concentrations at each incubation time (IT) was expressed as the difference between mg of NH_3-N/100 ml corrected by their C. Results were analysed as repeated measures using the mixed procedure of SAS, were T, IT and their interaction were used as fixed factors, and period and the GT as random effects. In case of significant interactions, the slice output option was used to determine the IT at which T were different. The concentrations of NH_3-N were different among T (P=0.029). The highest mean value was observed with U. SR had slower release of NH_3-N compared to U, and SR1 was numerically slower than SR2. IT was significant (P<0.001). Starting from similar N-NH_3 among T at time 0, differences became wider over time (26.22, 20.59, 23.43 mg/100 ml at 24 h for U, SR1 and SR2 respectively). The type of coating used to protect U can be a tool to modulate the rate of urea release. Trial was supported by 'CowficieNcy' project-H2020-777974. We thank Mr. Ramon Vila (Laboratorios Karizoo s.a.) for providing samples.

The effect of cannabinoids in turkey fattening

E. Zsédely[1], Z.S. Turcsán[2], L. Turcsán[3] and J. Schmidt[1]
[1]Széchenyi István University, Faculty of Agricultural and Food Sciences, Vár tér 2, 9200, Hungary, [2]Y Takarmányipari Ltd., Törökmalom street 2, 7370 Sásd, Hungary, [3]Mezőpergamen Ltd., Hársfa street 11, 5820 Mezőhegyes, Hungary; zsedely.eszter@sze.hu

The aim of trial was to assay the effect of cannabinoid (medical marijuana) supplementation on the performance of turkey fattening. Some articles reported about the beneficial effects of them for disorders of the cardiovascular system. Rupture of aorta is common in 12-16 week old heavy and fast growing male turkeys. The tested cannabinoid was extracted by supercritical extraction. The fattening period was 22 weeks long. The turkeys of control treatment (C) fed commercial diets *ad libitum*. The experimental birds fed the same commercial diets except the fact that cannabinoid extraction was sprayed on the surface of the feed in the dose of 0.5 or 1%/body mass kg (K05; K1) between 3-16 weeks of age. The amount of treated feed was calculated according to the hybrid standard parameters (body weight, feed consumption). 210 Converter hybrid turkeys were divided into cages, where the stocking density was 2 birds/m2. All treatments were in 2 replicates. The values were evaluated by the method of one-way ANOVA using SPSS 13. for windows. The average BW was significantly higher (at least P<0.05) in K05 compared to C during the growing I, growing II. and finishing I. feeding phase (5.83±0.45, 10.15±0.66 and 13.33±0.76 kg vs.5.64±0.62, 9.72±1.04 and 12.62±1.38 kg). That of the higher dose (K1) was similar to C except growing II., because it was significantly higher (10.05±0.78 kg). Experimental birds consumed more feed than that of C in the whole trial, however the weight gain values were varied. Therefore the most favourable FCR was observed in C (2.99 kg/kg vs 3.07 and 3.22 kg/kg). Rupture of aorta or did not occurred during the experiment. The breast meat quality was analysed (6 birds/treatment C and K1) at the end. As we expected, nutrient content including the fatty acid profile was not influenced by cannabinoid supplement. Its effect on the other quality parameters, like color, chewiness, drip loss, cooking loss, sensory properties, was different. According to our results cannabinoids in turkey fattening could be an interesting research topic in the future.

Relationship between feed processing with expander, particle size and mill type on broiler growth

M.A. Ebbing[1], V.D.N. Haro[2], W. Sitzmann[3], K. Schedle[1] and M. Gierus[1]
[1]*Institute of Animal Nutrition, Livestock Products and Nutrition Physiology (TTE), BOKU, Muthgasse 11, 1190 Vienna, Austria,* [2]*Evonik Nutrition & Care GmbH, Rodenbacher Chaussee 4, 63457 Hanau, Germany,* [3]*Amandus Kahl GmbH & Co. KG, Dieselstraße 5, 21465 Reinbek, Germany; martin.gierus@boku.ac.at*

Grinding is the first processing step applied to the raw ingredients in a feed mill processing line. In a diet based on corn-soybean for broilers, corn has quantitatively the biggest effect on the particle size regulation of the compound diet. In broilers, coarse particle size is associated to higher digestibility by promoting a higher development and activities of the gizzard. This results in longer retention time, lower pH, more refluxes cycles, and consequently improved digestive enzymatic reactions, resulting in higher absorption rates of nutrients. Indifferent to diet composition, different feed processing steps affect broiler performance. In the present study, three factors in feed processing were applied before pelleting to investigate their influence on broiler performance. The factors were two mill types, i.e. hammer (HM) and roller (RM) mill; three particle sizes, i.e. 800, 1,200 and 1,600 μm; and two conditionings, i.e. with and without expander, which were applied to produce grower and finisher diets. In total 864 unsexed Ross-308 chickens were randomly distributed in a 2×3×2 factorial arrangement based on a block design. Feed intake (FI), body weight gain (BWG) and feed conversion ratio (FCR) were measured on day 22 and 38. The results showed that FI was higher in broilers fed pellets without expander processing in the finishing (2,803 vs 2,736 g; P=0.0330) and in the overall period (3,917 vs 3,839 g; P=0.0496). In the grower phase, mill type and conditioner interacted (P=0.0336), i.e. FI of broilers was higher when diets were milled with RM and pelleted without expander. BWG was not affected by treatments. For the grower phase (P=0.0315) and the overall period (P=0.0130) the best FCR was obtained when 1,600 μm of particle size was obtained after RM grinding and expander processing before pelleting, i.e. 1.593 vs 1.745 obtained with HM, with expander and 1,600 μm particle size. In conclusion, feed processing improved broilers FCR from 8 to 38 d of age when using RM, ground coarsely and expanding before pelleting.

Nutritional value of DL-methionine and hydroxy analogue-free acid in common Greek broiler diets

G.K. Symeon[1], M. Anastasiadou[1], V. Dotas[2], A. Athanasiou[3] and M. Müller[4]
[1]*HAO-Demeter, Research Institute of Animal Science, Paralimni, 58100 Giannitsa, Greece,* [2]*School of Agriculture, AUTh, Thessaloniki, 54124, Greece,* [3]*Dr. D.A. DELIS SA, P. Benizelou 5, 10556 Athens, Greece,* [4]*Evonik Nutrition & Care GmbH, Rodenbacher Chaussee 4, 63457 Hanau, Germany; gsymewn@yahoo.gr*

Methionine is the first limiting amino acid in broiler diets, which are commonly supplemented with DL-methionine (DLM) or methionine hydroxy analogue (MHA-FA). MHA-FA has a relative methionine equivalence of about 65% compared to DLM. The objective of this experiment was to determine the biological efficacy of liquid MHA-FA relative to DLM in common Greek broiler diets. 420 Cobb 500 day-old broiler chicks were randomly allocated to five treatments, each consisting of seven replicates of 12 birds each: T1: Control (C) – a commercial diet deficient in Met+Cys; T2: MHA50 – as C, but calculated to partly fulfil (50%) the Met+Cys recommendations; T3: DLM50 – as treatment 2, but replacing 65% of supplemented MHA-FA with DLM; T4: MHA100 – as C, but calculated to fulfil the Met+Cys recommendations; and T5: DLM100 – as treatment 4, but replacing 65% of supplemented MHA-FA with DLM. Live weight and feed intake were measured pen-wise at the end of each feeding period (days 10, 22 and 42). 24 hours after slaughter (42 days) the weights of cold carcass, carcass yield, breast meat, bone, fat, thigh, drumstick, wing, and abdominal fat were recorded. Average daily gain and feed conversion ratio were calculated respectively. Data were statistically processed by means of analysis of variance and Tukey's test. Significance level was set at P<0.05. At 42 days of age, weights of groups DLM100 and MHA100 were the highest (2,892 g and 2,921 g, respectively), followed by DLM50 and MHA50 (2,661 g and 2,703 g, respectively), while C showed the lowest body weight (1,841 g). C consumed significantly less feed than the other four groups (3,753 g vs 4,503 g, 4,381 g, 4,404 g and 4,350 g for DLM50, MHA50, DLM100 and MHA100, respectively). With respect to the parameters of carcass yield and quality, C had lower values than the other groups being comparable in the majority of parameters investigated. The results of the present experiment verified the assumed bioavailability of 65% of MHA-FA compared to DLM on product base.

Energy deficiency in dairy farms, a problem – farmers and stakeholders perceptions

Y. Pénasse[1], N. Devriendt[1], V. Gotti[1], Y. Le Cozler[2], M. Gelé[3], J. Jurquet[3] and J. Guinard-Flament[2]
[1]Agrocampus Ouest, 65 rue de Saint-Brieuc, 35042 Rennes Cedex, France, [2]UMR PEGASE, INRA, Agrocampus Ouest, Domaine de la Prise, 35590 Saint-Gilles, France, [3]Institut de l'élevage, 42 rue Georges Morel, 49071 Beaucouzé, France; jocelyne.flament@agrocampus-ouest.fr

Energy deficiency (ED) is known to have many origins on farms (beginning of lactation, climatic conditions…) and to impact dairy cows' performances. If some indicators regarding ED have been highlighted in the past, their on-farm use remains unclear and possible use of other indicators or demands on new expectations from both farmers and stakeholders are unknown. Students involved in animal sciences at Agrocampus Ouest (France) led two semi-quantitative surveys in order to collect such information. Stakeholders (vets, specialists in nutrition…; n=14) were exclusively selected from existing network, while farmers (n=67) were selected either from existing networks (n=32) or randomly (n=35). Farmers' surveys were completed by phone (n=42) or on-farm (n=25). The surveyed holdings were 33 to 320-ha wide, organic (n=13) or not and mainly from western part of France (n=52), the most significant area of dairy farms in the country. 8/67 of farmers were unable to define ED but all knew how to handle this problem when occurring. For both stakeholders and farmers, the sensitive periods were early lactation (33/81) and summer season (32/81). Farmers also included dietary transitions in the sensitive periods (22/67). Almost half of them regularly experienced energy deficiency on their farm (38/67). The most-cited indicator was the change in milk composition (64% of both stakeholders and farmers); the second was body condition score (BCS) (31% of them). Farmers seem to rely more on visual indicators, such as BCS (44/67), coat appearance (7/67) and lameness (4/67), whereas stakeholders prefer to use technical indicators, such as protein (11/14) and fat (4/14) content of milk. Most of them don't need any additional indicator or tool to follow the energy deficiency (33/81) but they would appreciate to have a more sensitive daily indicator (18/81).

Using olive cake in dairy sheep diets: effects on ruminal bacterial communities in rusitec fermenter

I. Mateos[1,2], C. Saro[1], M.D. Carro[3] and M.J. Ranilla[1,2]
[1]Instituto de Ganadería de Montaña (CSIC-ULE), Finca Marzanas, 24346 León, Spain, [2]Universidad de León, Dpto. Producción Animal, Campus de Vegazana, s/n, 24071 León, Spain, [3]Universidad Politécnica de Madrid, Dpto. de Producción Agraria, Ciudad Universitaria, 28040 Madrid, Spain; mjrang@unileon.es

Spain is the highest olive oil producer in the world. Oil extraction generates several by-products that can be used in animal feeding, such as olive cake (OC) which has bioactive compounds with antimicrobial and antioxidant activities that could be beneficial for the animals, but might also affect the ruminal microbial communities. The aim of this study was to assess the effect of partially replacing straw and corn silage by olive cake in a dairy sheep diet on bacterial populations in Rusitec fermenters. Two diets were incubated in 4 Rusitec fermenters in a cross-over design with two 14-day incubation periods. Fermenters were given daily 30 g of diet (50:50 forage:concentrate), and in each period half of them received the diet with no olive cake (CON) and the other half received the OC diet (16.7% OC). Samples of solid and liquid digesta phase were collected from the fermenters, DNA was extracted, and bacterial profiling was performed using 16S rRNA gene amplicon sequencing. The reads generated were processed using the FROGS pipeline. The statistical analysis of the sequences and the principal coordinate analysis were done with lme4 and phyloseq packages in R program. Alpha diversity was not affected (P>0.05) by the diet. Regarding to beta-diversity, samples were segregated according to the type of sample (solid or liquid) but there was not a clear division according to the diet. Linear discriminant analysis (LDA) effect size (LEfSe) was used to identify differentially abundant bacterial taxa between diets. From the top 50 genera detected, 12 were enriched in the CON diet and 6 in the OC diet. Some Streptococcaceae, Christensenellaceae a *Ruminococcus* and a *Eubacterium* were associated to the OC diet and some Actinobacteria, Spirochaetes, Acidaminococcaceae and Saccharofermentans to the CON diet. In summary, the inclusion of OC in a dairy sheep diet did not affect the alpha or beta diversity in Rusitec fermenters, but influenced the distribution of the OTUs.

Comparison of three markers to determine nutrient digestibility and net energy content of pig feeds
L. Paternostre, S. Millet and J. De Boever
Flanders Research Institute for Agriculture, Fisheries and Food (ILVO), Animal Sciences Unit, Scheldeweg 68, 9090 Melle, Belgium; louis.paternostre@ilvo.vlaanderen.be

Pig performances are strongly correlated with the net energy (NE) content of the feed. The gold standard to determine NE is the use of respiration chambers, but this method requires a lot of work. As an alternative, faecal digestibility of the nutrients is used to estimate NE more easily. Digestibility can be determined either by total faeces collection or by spot sampling of faeces using an indigestible internal or external marker. The goal of the present study was to compare three markers to measure the digestibility of organic matter (DC_{OM}) and the NE of pig compound feeds. The markers were acid insoluble ash (AIA) as internal marker and AIA and titanium dioxide (TiO_2) as external markers. The study was carried out with 12 compound feeds to which 0.5% SiO_2 and 0.5% TiO_2 was added and among them 8 feeds were composed without addition of a marker to enable comparison of internal and external AIA. The 20 feeds were fed *ad libitum* to 120 pens (6 pens/feed) with 3 growing pigs per pen. Faeces were spot sampled from 2 pigs per pen twice daily during 5 consecutive days, were pooled per pen and freeze dried. Feed and faeces were analysed by accredited ISO-methods. The mean and sd of DC_{OM} and NE were compared with a student t-test. The external markers resulted in similar (P>0.05) results; the DC_{OM} of the 12 feeds amounted (mean ± sd) to 82.8±1.8 and 83.0±1.6% and the NE to 9.64±0.33 and 9.66±0.32 MJ/kg, for AIA and TiO_2 respectively. The 8 feeds without external marker contained 0.31±0.11% AIA, compared with 1.17±0.14% AIA in the supplemented feeds. The DC_{OM} of the 8 feeds without and with marker amounted to 81.7±2.4 and 82.8±2.0% and the NE to 9.74±0.32 and 9.69±0.35 MJ/kg, respectively and were not significantly (P>0.05) different. On the other hand, the average variation between the 6 pens was higher for the feeds without marker than for those with marker, being 1.4% vs 1.0% for DC_{OM} and 0.17 vs 0.11 MJ/kg for NE, respectively, but these differences were not significant (P>0.05). One may conclude that the AIA content naturally present in a compound feed is high enough to determine DC_{OM} and NE and there is no difference between TiO_2 and SiO_2 used as external marker.

Nutritive value of different pastures and soil intake in free-ranged organic growing rabbit
S. Jurjanz[1], H. Legendre[2], J.P. Goby[3] and T. Gidenne[2]
[1]Université de Lorraine – INRA, UR AFPA, BP 20163, 54505 Vandoeuvre, France, [2]Université de Toulouse – INRA, GenPhySE, 24, chemin de Borde-Rouge, 31320 Castanet Tolosan, France, [3]Université de Perpignan, IUT, 77 Chemin de la Passio Vella, 66962 Perpignan, France; thierry.gidenne@inra.fr

Free-ranged organic rabbit correspond to the consumer's demand of improved animal welfare conditions but lacks in technical reference values. Moreover, the risk of exposure to environmental contaminants of rabbits in these conditions has to be evaluated by an estimation of soil intake. Therefore, 30 growing rabbits have been enrolled (from weaning at the age of 43 days to slaughter at day 100) in two groups of 5 movable cages on two pasture types: sainfoin or fescue. Each cage contained 3 animals and had a shelter of 0.4 m² and an pasture surface of 1.2 m². The cages are moved daily on the pasture. A commercial pelleted organic feed was daily given at a level of 64 g/animal. Grass intake was evaluated by the difference between offered and residual grass (cut over 3 cm). Soil intake was estimated by an internal marker method and dry matter digestibility measure. Therefore, concentrations of acid insoluble ash were analysed in soil, all components of diet (i.e. vegetation and feed) and faeces. Moreover, digestibilities on pasture were compared with digestibilities of the same pelleted feed, both vegetation types and their mixtures were determined in indoor 'controlled' conditions. Pelleted feed was totally consumed and the pasture dry matter (DM) intake averaged 55 g (from age of 43 to 100 d) without difference between sainfoin or fescue plot. The rabbit growth was half-higher on sainfoin compared to fescue plot (29.3 vs 20.0 g/d, P<0.05). The digestibility of fescue varied from 37 to 43% for pasture and indoor measurement respectively, while that of sainfoin was stable (65.5 to 66.5 respectively). The fescue-feed mixture had a DM digestibility of 37 and 47% in indoor and pasture trial, while for sainfoin-feed mixture digestibility were respectively 65.3 and 61.7. Soil intake was lower (P<0.01) on fescue (1.3% of total DM intake or 1.9 g/d) compared to sainfoin pasture (3.0% or 4.2 g/d). The lower soil intake on fescue pasture could be due to the better soil cover by grass forming a buffer between animals and soil in comparison to sparser sainfoin cover.

Mushroom polysaccharides in the control of mycotoxins: an update

J. Loncar[1], M. Reverberi[2], A. Parroni[2], P. Cescutti[3], R. Rizzo[3] and S. Zjalic[1]
[1]University of Zadar, Department of ecology, agronomy and aquaculture, trg kneza Viseslava 9, 23000 Zadar, Croatia, [2]Sapienza University of Rome, Department of environmental biology, P.le Aldo Moro 5, 00185 Roma, Italy, [3]University of Trieste, Department of life sciences, via Weiss 2, 34128 Trieste, Italy; szjalic@unizd.hr

Mushroom metabolites are useful tools in limiting of the presence of mycotoxins in food and feed stuff. Different mushroom polysaccharides demonstrated able in controlling the synthesis of several mycotoxins simultaneously, while mushroom enzymes showed to be active in toxin degradation. Among the polysaccharides most active in the control of the mycotoxin biosynthesis are those produced by *Trametes versicolor*, mushroom commonly known as Turkey tail. The active polysaccharide, in previous years, was isolated and characterized and its name, tramesan, has been registered. Tramesan has proven able to provide a long-lasting control of biosynthesis of the most important mycotoxins like aflatoxins, ochratoxin A and fumonisins both in field and during the storage. Moreover it has demonstrated some positive influence on animal health and protection against toxic effects of mycotoxins. These results indicates that tramesan could actually represent an environmental-friendly tool in mycotoxin control. But the production costs and variability in its yield could negatively influence the application of this polysaccharide. To solve out this problem and to evaluate the possibility of its industrial synthesis the characterization of the smallest polysaccharide's active fraction has been undertaken. The research has focused on the valuation of the tramesan's fragments length and their ability to inhibit toxin biosynthesis. The results showed that oligosaccharides longer than 7 units were active in toxin inhibition, while the smaller ones showed no biological activity. Even if the inhibition of aflatoxin and ochratoxin A biosynthesis performed by the hepta oligosaccharide was lower than the one observed with the entire polysaccharide (around 50% against over 90%), the results indicate that shorter portions of tramesan could be effective in mycotoxin control. The research on optimal length and composition of active oligosaccharide is ongoing.

The effect of feed crude protein content and feed form on broiler performance and litter quality

M. Brink[1], E. Delezie[1], P. Demeyer[2], Ö. Bagci[2] and J. Buyse[3]
[1]Institute for Agricultural and Fisheries Research, Animal Sciences Unit, Scheldeweg 68, 9090, Melle, Belgium, [2]Institute for Agricultural and Fisheries Research, Agricultural Engineering Unit, Burg. Van Gansberghelaan 115, 9820 Merelbeke, Belgium, [3]KU Leuven Laboratory of Livestock Physiology, Department of Biosystems, Kasteelpark Arenberg 30, 2456 Heverlee, Belgium; madri.brink@ilvo.vlaanderen.be

A reduction in dietary crude protein (CP) is an effective way to reduce nitrogen excretion in broilers provided that the amino acid (AA) requirements are met. Feed form (mash vs pellets) influences the development of the gastrointestinal tract as well as feed and water intake and therefore the nitrogen and moisture level of the excreta. This study investigated the effect of a reduction in dietary CP and feed form on performance as well as meat and litter quality of 2,232 Ross 308 male broilers. All birds received a starter diet with a CP content of 22.0%. Dietary treatments with a CP content of 20.5, 18.8 and 17.5% during the grower phase and 19.5, 18.0 and 16.6% during the finisher phase were fed. Diets were supplemented with individual AA to maintain digestible AA:Lysine ratios. Each dietary treatment was fed in mash and pellet form, which resulted in a 3×2 (CP level × feed form) experimental design with six treatments, each with six replications. Pellets affected (P<0.01) feed intake (FI), body weight gain (BWG) and feed conversion ratio (FCR) during the starter phase with birds receiving pellets performing better than birds receiving mash. Pelleted diets led to an increase in FI (g/bird/day; P<0.01) during the grower (90.37) and finisher (176.03) phase and a CP content of 16.6% led to a lower (P<0.01) FI (167.60) than a CP content of 18.0% (173.11) in the finisher phase. Birds receiving pellets had the lowest FCR (P<0.01; 1.61) in the finisher phase. Pelleted treatments with the lowest CP resulted in the lowest BWG (g/bird/day; P<0.01) during the grower (61.50) and finisher (104.73) phase. From grower phase onwards, litter quality appeared better when birds were fed mash diets. In addition, a lower incidence of footpad dermatitis was observed with mash diets but also when CP content was reduced in both mash and pelleted diets. To conclude, CP content during the grower and finisher phase, respectively, can be decreased to 18.8 and 18.0%.

Effect of *Cistus ladanifer* L. tannins as silage additives

M.T. Dentinho[1,2], K. Paulos[2], A.T. Belo[2], P.V. Portugal[2], S. Alves[1,3], O.C. Moreira[2], J. Santos-Silva[1,2] and R.J.B. Bessa[1,3]
[1]Centro Investigação Interdisciplinar em Sanidade Animal (CIISA), Avenida Universidade Técnica, 1300-477 Lisboa, Portugal, [2]Instituto Nacional de Investigação Agrária e Veterinária (INIAV), Fonte Boa, 2005-048 Vale de Santarém, Portugal, [3]Faculdade de Medicina Veterinária, Universidade Lisboa, Avenida da Universidade Técnica, 1300-477 Lisboa, Portugal; teresa.dentinho@iniav.pt

The effects of ensiling lucerne with *Cistus ladanifer* condensed tannins (CT) on in silo fermentative parameters and on *in vitro* organic matter digestibility (OMD) were studied. Lucerne forage was sprayed with different solutions of *C. ladanifer* CT extract in 60 ml of water to dose 0 (control), 40 (L40), 80 (L80) and 120 (L120) g of CT per kg of lucerne DM and was ensiled. The inclusion of CT in the silages caused a proteolysis reduction. Also the rumen undegradable protein increased linearly with CT inclusion (P<0.01). However, a linear decrease of 5, 13 and 22% of OMD was observed for the silages L40, L80 and L120. A level of 40 g/kg DM of CT applied to ensiling lucerne seems adequate to reduce proteolysis in silo and improve protein utilization with a slight depression in OMD. So, a metabolic trial was conducted with six crossbred Romani rams averaging 69±5 kg live weight, in three simultaneous 2×2 Latin square design. The rams were fed with Lucerne silage containing 9 l of molasses per ton, diluted in water (50:50 v/v) and treated with 0 (Control) and 40 g of *C. ladanifer* CT/kg of Lucerne DM (L40). The effect of *C. ladanifer* CT as lucerne silage additives on digestion, nitrogen balance, and nitrogen excretion was evaluated. No significant differences were observed between *in vivo* DMD and OMD between silages, that averaged 52 and 53%, respectively. The apparent digestibility of CP was lower in L40 than in control silage (64 vs 69%), but no differences in nitrogen retained by the animals were observed. The pattern of N excretion was affected by treatments, with lower urinary N and higher faecal N with L40 than in Control. The results suggest CT added to silage protect protein from ruminal degradation. The shift in N excretion from urine to faeces is environmentally beneficial since urine N can rapidly volatilized into the environment and faecal N is incorporated into soil increasing nutrient availability.

Supplementation of sheep fed low-quality *Eraggrostis curvula*: a meta-analytical study

H. Mynhardt, W.A. Van Niekerk, L.J. Erasmus, R.J. Coertse and C.J.L. Du Toit
University of Pretoria, Department of Animal Science, Hatfield, 0028, South Africa; willem.vanniekerk@up.ac.za

The objective of this study was to evaluate relative effects of supplemental starch (1.407 and 5.210 g starch/kg BW) and urea (0.186 and 0.701 g urea/kg BW) on the efficiency of N utilisation in sheep fed low-quality *Eragrostis curvula* hay (ranging between 0.4% N and 0.7% N, >65% NDF) using meta-analytical methods. The dataset used was compiled from four supplementation studies and six trials, totalling 123 data points, using linear mixed model meta-analysis (REML analysis). Prior to meta-analysis, the discriminating variables were evaluated using canonical variate analysis (GenStat®). The relative importance of starch and/or urea supplementation were determined using Akaike's information criterion, as well as the R-square of the adjusted models. Neither starch nor urea supplementation had an effect on NDF intake or digestibility (P>0.05). However, starch supplementation influenced the microbial N supply (MNS) model the most, with MNS increasing linearly as starch supplementation was increased (P<0.05, R^2=0.602). In contrast, rumen ammonia N concentration (RAN) and the ratio between MNS and N intake (NI) models were best described by urea supplementation. As such the MNS: NI ratio decreased curvilinearly as urea supplementation was increased (P<0.05, R^2=0.402), while RAN increased linearly as urea supplementation was increased (P<0.05, R^2=0.585). In addition, a relationship was established between RAN and the starch supplemented to available crude protein (CP) intake ratio (P<0.05, R^2=0.761), where available CP intake was calculated from CP intake from urea and forage intake – acid digestible insoluble CP intake. Sheep fed low-quality *E. curvula* hay needed to be supplemented with at least 2.2 g starch/kg BW/day to fulfil maintenance N requirements through MNS. Based on MNS: NI ratios, the minimum quantity of urea needed to be supplemented to sheep in this meta-analysis was 0.5 g urea/kg BW. In addition, the relationship between RAN and starch: available CP ratio suggested that the optimal ratio of starch supplemented and available CP needed to be at least 2:1.

Author index

Bardou, P.	125, 295	Ben Zaabza, H.	121
Bareille, N.	297, 387	Benzertiha, A.	399
Baret, P.V.	654	Benzie, J.A.H.	517
Baric, R.	516	Berard, J.	156
Barreau, S.	277	Berben, G.	325
Barrett, D.M.W.	651	Beretta, V.	260
Bartley, D.	271, 273	Berg, J.	161
Bartolomé, E.	279	Berg, P.	110, 122, 210, 600
Bassols, A.	182, 269	Bergamaschi, M.	589
Bastiaansen, J.	658	Berge, A.A.	512
Bastiaansen, J.W.M.	122, 248, 486	Berger, B.	362, 369
Bastiaens, L.	159, 163, 238, 243, 396, 400	Berggreen, I.E.	552
Bastin, C.	533, 607	Berghof, T.V.L.	500, 507
Bateman, K.S.	516	Bergsma, R.	248, 257, 287, 288, 306, 486
Batonon Alavo, D.I.	170	Bergsten, C.	465
Batorek Lukač, N.	621, 623, 627	Beriain, M.J.	440
Battacone, G.	234	Bernal, B.	364
Battheu-Noirfalise, C.	144, 200	Bernal-Juarez, B.	364
Bauer, E.A.	141	Bernardi, A.	663
Baumgard, L.H.	259	Bernardi, A.C.C.	659
Baumont, R.	380, 442	Bernard, L.	320
Baumung, R.	284	Bernués, A.	425, 614, 615
Bawa, M.	245	Berodier, M.	210
Baxter, M.	499	Berry, D.P.	210, 379
Beaugrand, F.	547	Bersabé, D.	108
Becila, S.	592	Berthelot, V.	308, 309
Beck, M.	557	Bertocchi, M.	307
Becker, R.	664	Bertolini, F.	108
Becker Scalez, D.C.	128	Bertolín, J.R.	617
Beckers, Y.	145, 178, 265	Berton, M.	427
Becskei, Z.	660	Berton, M.P.	454
Bedhiaf-Romdhani, S.	231, 232	Berton, M.P.B.	576
Bedhiaf, S.	131	Bertozzi, C.	616, 617
Bednarczyk, M.	496	Besbes, B.	284
Bee, G.	234, 336	Besche, G.	138
Belay, T.K.	595	Bessa, R.J.B.	230, 328, 332, 677
Belghit, I.	403	Bevilacqua, C.	307
Bell, A.W.	261	Bezen, R.	473
Bell, D.	350	Bezerra, H.V.A.	174
Bell, M.	605	Bharanidharan, R.	560
Bellenot, D.	180	Biancarosa, I.	403
Belloc, C.	386, 387, 393, 545, 546, 547	Biasato, I.	395
Belo, A.T.	224, 677	Biau, S.	281
Belo, C.C.	224	Bieber, A.	390
Beltrán De Heredia, I.	388	Bignon, L.	506
Beltran-Legaspi, J.A.	504	Bijma, P.	142, 160, 289, 326, 368, 513
Ben Abdelkrim, A.	471	Bijttebier, J.	654
Benabid, H.	589	Bikker, P.	250
Benavent, M.	440	Biligetu, B.	166, 642
Benavides-Reyes, C.	268	Billon, Y.	305, 443
Benedet, A.	532	Bilton, T.P.	129
Bengtsson, C.	465	Bindelle, J.	189, 664
Benjelloun, B.	444, 445	Bink, M.	410
Bennewitz, J.	525	Birch, K.	161
Benoit, M.	348, 656	Bíro, D.	253, 668
Benthem De Grave, X.	565	Bisutti, V.	217

De Boer, I.J.M.   194, 316, 345, 346
De Boever, J.   183, 416, 563, 675
Debournoux, P.   268
Debus, N.   138
De Campeneere, S.   144, 376
De Carvalho, F.E.   233
De Carvalho, F.E.C.   576
Decloedt, A.I.   190, 623
De Cocq, P.   160
De Craene, N.   325
Decruyenaere, V.   144, 178, 200, 377, 640
De Cuyper, C.   183, 480, 481
Dedieu, B.   193
De Feijter, H.F.   249, 252
Degeest, P.   303
Degrande, R.   135
De Greeff, A.   389
De Greef, K.H.   289
Degroote, J.   336, 341
De Haas, Y.   136, 289, 317, 638
Dehareng, F.   443, 605, 606, 608
Dehghani, H.   537
Dehghani Madiseh, Z.   651
De Jong, G.   129, 142, 535, 603, 606, 609, 638
De Jong, I.C.   389
De Joybert, M.   547
De Ketelaere, B.   529
Dekkers, J.C.M.   446
De Klerk, B.   136, 289
De Koning, D.J.   268, 448
De Koster, J.   458, 464, 598, 599
Delacroix-Buchet, A.   602
De La Fuente, G.   166, 312, 416
De La Fuente, M.A.   333
Delannoy, M.   592
Delanoue, E.D.   614
De La Serna Fito, E.   625
De La Torre, A.   475
Delattre, L.   304
De Lauwere, C.C.   386
Delbare, D.   510, 516
Del Corvo, M.   444
Delezie, E.   507, 508, 676
Delgado-Gonzalez, R.   579
Delhez, P.   145
Della Malva, A.   592
Dellar, M.   352
Delouard, J.M.   304
Del Pino, L.   357
Del Prado, A.   352
Delpuech, E.   666
Delval, E.   236
De Marchi, M.   217, 532
Demarquet, F.   135
Demartini, B.L.   180
De Mercado De La Peña, E.   482

De Meyer, D.   566
De Meyer, F.   314
Demeyer, P.   676
Deneux, V.   518
Dentinho, M.T.   677
De Olde, E.M.   197, 345, 346, 349
De Oliveira Silva, R.   363
De Palo, P.   438
Depuydt, J.   409
Derisoud, E.   581
Derk, M.F.L.   106
Derks, M.   426
Derks, M.F.L.   255
De Roest, K.   482
Derrien, C.   158
De Ruiter, B.   160
Déru, V.   415
Deruytter, D.   161, 162, 240, 242
Dervishi, E.   152, 153
Deschandelliers, S.   406
De Smet, J.   394
De Smet, S.   336, 341, 487, 623, 624
Desmet, T.   341
De Spiegeleer, M.   190
Dessie, T.   501, 658
Detilleux, J.   396
De Tourdonnet, S.   426
Deutch, T.   574
Devaney, E.   271
De Vega, A.   166
De Vera, M.   512
Devine, M.   652
De Visscher, A.   374
De Vos, M.   494
De Vos, S.   184
Devriendt, N.   674
De Vries, A.   646
De Vries, S.   316
De Vuyst, M.L.   298
Dewanckele, L.   309
Dewhurst, R.   442
Dewhurst, R.J.   149, 334, 634
De Wit, A.   364
De Wit-Joosten, K.   372
Dewulf, J.   314, 492, 494
Dewulf, L.   623
Dhakal, R.   554
Diakité, Z.   656
Dias, E.F.   248
Dias, E.F.F.   180
Díaz, C.   264, 353
Díaz De Otalora, X.   185
Diaz-Olivares, J.   530, 531
Dicksved, J.   313
Diercles Francisco Cardoso, D.F.C.   128
Dierks, C.   366

Fonseca, M.	422	Gálik, B.	253, 668
Fonseca, M.F.	421	Gallardo, B.	333
Fontanesi, L.	105, 108	Gallego, R.	353
Forier, R.	184	Galliano, D.	145, 195
Formigoni, A.	555	Galli, M.C.	344
Formoso-Rafferty, N.	335, 459, 521	Gallo, A.	636
Forneris, N.S.	121	Gallo, D.	452
Fornós, M.	343, 665	Gallo, L.	286, 427
Forsberg, S.	161	Gallo, M.	105
Forte, C.	307	Gallo, S.B.	174
Fortun-Lamothe, L.	654, 671	Gama, L.T.	445
Føske Johnsen, J.	306	Gambacorta, G.	437
Foskolos, A.	199, 377, 381, 556, 641	Gambade, P.	387, 393
Fossaert, C.	380	Gameiro, A.H.	321, 376, 485
Foucras, G.	269, 292	Gandini, G.	117
Fougère, H.	320	Ganesan, A.	366
Foulquié, D.	235, 236	Ganier, P.	415, 604
Fourdin, S.	347	Gao, H.	411, 467
Fourichon, C.	386	Garcez Neto, A.F.	228, 472
Fourie, P.J.	487	Garcia, A.	130
Fradinho, M.J.	436	García, A.I.	231
Fraeye, I.	623	Garcia, A.L.S.	463
Francione, G.L.	283	Garcia, N.P.	382
Francis, F.	163, 325	García Casco, J.M.	482, 625
Francisco, A.	230, 328, 332	García-Contreras, C.	337, 568
Franco, V.	419	García Cortés, L.A.	482, 625
François, D.	519	García-Dorado, A.	117
Franzoi, M.	532	García-García, J.J.	358
Frąszczak, M.	296, 597	Garcia-Launay, F.	409, 424
Freeman, T.C.	634	García-Rodríguez, A.	185, 388, 636
Freire, F.	353	García-Rodríguez, J.	361
Frias, A.	421, 422	Garcia-Santos, S.	664, 665
Frias-Navarro, R.	661	Gardner, G.E.	380
Friedrich, L.	322	Garibay, S.V.	116
Friggens, N.C.	529	Garner, J.B.	635
Fritz, S.	293, 602, 610, 633	Garnsworthy, P.C.	284, 631
Froidmont, E.	178, 475, 605, 606	Garrick, D.	353, 598
Frooninckx, L.	241	Garrido, C.	231
Früh, B.	304	Garrido, J.M.	139
Frydendahl Hellwing, A.L.	639	Garrido-Varo, A.	371
Fuchs, K.	548	Garrón Gómez, M.	588
Fuchs, S.	179	Gasa, J.	665
Fuerst, C.	143, 235, 525	Gasco, L.	241, 327, 395
Fuerst-Waltl, B.	143, 235, 334, 470, 548	Gaspa, G.	452
Fukumasu, H.	642	Gaudillière, N.	602
Fumière, O.	325	Gaughan, J.B.	557, 611
Furlan, E.	382	Gauly, M.	204, 244, 509, 569, 613, 614, 615
Fusi, E.	254	Gauthier, R.	188, 565
Fynbo, J.	628	Gautier, D.	135
		Gautier, J.M.	132, 135
		Gave, M.	231
G		Gavojdian, D.	369, 461
Gabler, N.K.	259	Gay, M.	272
Gado, H.M.	172, 173	Gaytán-Alemán, L.R.	660
Gai, F.	241	Gebreyesus, G.	124
Gaillard, C.	188, 565	Gebska, M.	478, 482
Gain, C.	626		

Geerinckx, K.	529	Gómez, M.A.	383
Geffen, O.	473	Gómez-Maqueda, I.	371
Geibel, J.	132	Gomez-Raya, L.	446, 482, 625
Gelé, M.	533, 674	Gonçalves, V.	436
Gengler, N.	115, 443, 605, 606, 607, 608	Gondret, F.	170
Genro, T.C.	030	Gonen, S.	514
Gentz, M.	204	Gonzalez-Añover, P.	337
Gérard, A.	325	Gónzalez-Bulnes, A.	337, 568
Gerberich, J.	161	González-Diéguez, D.	287
Gerhardy, H.	374, 479	González-García, E.	132, 227, 235
Gerken, M.	520	González, L.	388
Germain, K.	432	González, P.	669
Germon, P.	263, 269	González-Recio, O.	636
Gerritsen, A.M.	440	Gonzalez-Tavizon, A.	579
Gerrits, W.J.J.	316	Goossens, K.	141, 376, 647
Gertz, M.	301	Gordo, D.	633, 637
Ghaffari, M.H.	148	Göres, N.	422
Ghasemi Nejad, J.	537	Gori, A.S.	594
Ghassemi Nejad, J.	266	Gorjanc, G.	124, 233
Gheorghe, A.	184	Goron, J.P.	655
Gianola, D.	120	Görs, S.	340
Gianotten, N.	159, 243	Gorssen, W.	409, 524
Gibbons, J.	377	Goselink, R.	536
Gidenne, T.	248, 654, 675	Goselink, R.M.A.	329
Giersberg, M.F.	201, 389	Gottardo, F.	217, 344
Gierus, M.	673	Gotti, V.	674
Giger-Reverdin, S.	531, 636	Gottschalk, M.	583
Gil, M.G.	368	Goudet, G.	489
Gilbert, H.	291, 316, 415, 443, 483, 666	Govaere, J.	580
Giles, P.Y.	650	Goyache, F.	113, 335
Giller, K.	332	Granado-Tajada, I.	236, 388
Gilmour, A.R.	462	Granato, A.	224
Giorello, D.	574	Grandl, F.	470
Giovannini, S.G.	114, 116	Grando, C.	220
Giráldez, F.J.	361	Grant, J.R.	209
Girard, C.L.	218	Grasseau, I.	364
Girard, M.	336	Graulet, B.	218
Gjerde, B.	512	Gregorini, P.	167, 557
Gjøen, H.M.	511	Greiser, T.	583
Gjuvsland, A.	595	Grelet, C.	443, 533, 605, 606, 608
Gjuvsland, A.B.	460	Gress, L.	443
Glasser, T.	134	Griffith, B.A.	223
Glasser, T.A.	574	Grigoletto, L.	290, 454, 601
Glauser, D.	533	Grigore, D.	669
Goby, J.P.	654, 675	Grilli, E.	561
Godinho, R.M.	486	Grisot, P.G.	135
Godo, A.	134, 473	Grivault, D.	489
Goers, S.	337	Grodkowski, G.	302, 544
Gohin, A.	424	Groenen, M.A.M.	106, 160, 209, 255
Goiri, I.	636	Groeneveld, E.	442, 446, 599
Gombeer, L.	277	Groeneveld, L.F.	110
Gomes, P.	308, 309	Gromboni, C.F.	152
Gomes, S.	224	Gross, J.J.	648
Gómez Carballar, F.	625	Große-Brinkhaus, C.	255, 335, 410
Gómez-Cortés, P.	333	Grosse-Kleimann, J.	479
Gómez Izquierdo, E.	482	Grøva, L.	110, 132

Jacquiet, P.	655	Jonker, A.	225, 405
Jacquot, A.-L.	429	Jönsson, I.	161
Jafarikia, M.	259, 599	Jonsson, N.	371
Jaffrezic, F.	284	Jönsson, N.S.	552
Jagusiak, W.	141, 455	Joost, S.	442, 444
Jakob, M.	393	Jordan, H.R.	324
Jakobsen, J.	290, 599	Jorge-Smeding, E.	651
Jakobsen, J.H.	451	Joshi, R.	511, 512
Jalava, T.	244	Jost, J.	218
Jammes, H.	292, 293	Jouneau, L.	292, 293
Janíček, M.	230	Jousset, T.J.	139
Janković, D.	620	Jouven, M.	658
Jansen, H.	533	Józefiak, A.	399
Jansman, A.J.M.	257, 306	Józefiak, A.J.	394
Janssen, P.H.	225	Józefiak, D.	399
Janssens, G.G.P.	507	Jóźwik, A.	670
Janssens, S.	409, 524	Jozwik, A.A.	176
Janvier, E.	495, 604	Juga, J.	649
Jara, A.	375	Juncker, E.	249
Jarousse, A.	655	Junge, W.	301
Jasińska, K.	176, 670	Jung, J.S.	619, 640
Jauhiainen, L.	244	Jung, S.O.	343
Jawasreh, K.I.	356	Juodka, R.	663
Jean, H.	272	Juráček, M.	253, 668
Jenko, J.	124	Jurgens, G.	492
Jensen, A.B.	242, 553	Jurjanz, S.	354, 432, 435, 592, 675
Jensen, A.N.	554	Jurquet, J.	674
Jensen, D.	474	Juska, R.	663
Jensen, J.	448, 457, 467, 498, 502	Juskiene, V.	663
Jensen, L.D.	629		
Jensen, M.B.	199, 616	**K**	
Jensen, S.K.	170, 215	Kaart, T.	527
Jeppsson, K.H.	414, 621	Kabi, F.	564
Jérôme, B.	311	Kadarmideen, H.N.	150, 152, 294
Jerónimo, E.	230	Kadi, S.A.	248
Jeyanathan, J.	309, 641	Kadlečík, O.	112, 528
Ji, S.	398	Kadri, N.K.	105
Jibrila, I.	449	Kaesbohrer, A.	548
Jiménez, M.A.	353	Kuić, A.	434
Jiménez-Moreno, E.	343, 665	Kakanis, M.	203
Jing, L.	309	Kalscheur, K.F.	559
Jing, X.	539	Kaltenegger, A.	247
Jing, X.P.	230	Kamionka, E.M.	417
Jo, J.H.	264	Kanellidou, F.	361
Jo, Y.H.	264	Kanellos, T.	550
Joerg, H.	462, 578	Kanengoni, A.T.	487
Johansen, L.-H.	513	Kang, J.P.	230
Johansen, M.	215	Karapandža, N.	111, 417
Johansson, A.M.	208, 600	Karger, V.	351
Johnsen, C.	554	Kargo, M.	115, 215, 285, 526, 554, 604, 609
Johnson, P.L.	225	Karhapää, M.	244
Johnson, T.	598	Karimi, Z.	259
Johnson, V.	327	Karlskov-Mortensen, P.	457
Johnstone, P.	225	Kasarda, R.	106, 112, 462, 475, 528
Joly, F.	348	Kaseja, K.	226
Jones, A.G.	223	Kato, K.	299

Katoele, J.	398	Klaffenböck, M.	369
Katoh, K.	218	Klambeck, L.	498
Kaufmann, F.	498, 500, 509	Kleefisch, M.-T.	151
Kaufmann, L.	346	Klein, J.D.	574
Kauppinen, T.	216	Klein, S.-L.	542
Kaur, K.	272	Klementsdal, G.	600
Kavlak, A.T.	423	Klemetsdal, G.	543
Kawecka, E.	298, 466, 467	Klevenhusen, F.	151
Kawecka, K.	324	Kliemann, R.D.	472
Kazemi, H.	108	Kling-Eveillard, F.	429
Keating, A.F.	259	Klišanić, V.	417
Kecman, J.	422	Klont, R.	372, 474
Keehan, M.	598	Klop, A.	536
Kefalas, G.	386	Klopčič, M.	544, 620
Kelln, B.M.	166, 642	Klosa, J.	120
Kemp, B.	200, 384, 388, 389, 536, 646	Klotz, D.	498
Kemper, N.	201, 338, 422, 498	Kluivers-Poodt, M.	386, 497
Kenez, A.	147	Knapp, P.	334
Kennedy, E.	380	Knap, P.W.	334
Kenny, D.A.	292, 379	Knijn, H.M.	606
Kenyon, F.	132, 134	Knol, E.F.	186, 256, 288, 486, 501, 627
Ketavong, S.	272	Knowler, K.	225
Khalifa, M.	652, 653	Ko, H.L.	192
Khamis, F.M.	240	Ko, N.	358
Khanal, P.	307	Koch, C.	147
Khattab, H.M.	172	Köck, A.	535
Khaw, H.L.	513, 517	Koenen, E.P.C.	638
Khayatzadeh, N.	362, 369	Koerkamp, P.W.G.G.	426
Khiaosa-Ard, R.	151, 177	Kofler, J.	470
Kholif, M.A.	173	Kohl, C.	325
Kiefer, H.	292, 293	Koivula, M.	127, 468
Kierończyk, B.	399	Kok, A.	349, 536, 646
Kim, D.W.	488	Kokkonen, T.	328
Kim, H.B.	312, 313	Kolega, M.	516
Kim, H.R.	312, 313	Kolláthová, R.	253, 668
Kim, I.H.	175, 178, 504	Kołodziejski, P.	399
Kim, J.Y.	343	Komatsu, T.	218
Kim, K.	398	Komen, H.	122, 486, 517, 600, 658
Kim, K.A.	504	Kominakis, A.	457
Kim, M.	483, 484	Kommadath, A.	209
Kim, W.H.	619, 640	Kompan, R.	174
Kim, W.S.	264	König, S.	222, 542, 610, 611, 645
Kim, Y.H.	560	Koning, L.	635
Kim, Y.J.	162	Konta, A.	339
Kim, Y.M.	175	Konuspayeva, G.	592
Kimata, M.	339	Koopmans, A.	309
Kindmark, A.	268	Kopke, G.	464
King, J.	223	Kordos, K.	176, 670
King, M.	223	Kościuczuk, E.	298, 466
Kinyuru, J.	245	Kosińska, K.	670
Kirchhofer, M.	270	Kosińska-Selbi, B.	597
Kirchhofer, M.K.	220	Kostyra, E.	490
Kirk, T.W.D.	198, 350	Kowalski, E.	487, 624
Kisielewska, J.	402	Krauss, D.	453, 633
Kitazawa, H.	299	Krauss, M.	156
Klaaborg, J.	342	Krawczyk, W.	320

Kreienbrock, L.	479	Landau, S.Y.	574
Kress, K.	480, 489, 621, 623, 627	Langton, M.	161
Kreuzer, M.	156, 239, 332, 605	Lannoo, K.	494
Krieter, J.	191, 201, 205, 279, 300, 301, 322, 413	Laper, Y.	134
Krifuks, O.	574	Lardner, H.A.	166, 561, 642
Kristensen, T.	570, 650	Larroque, H.	263
Krivushin, K.	209	Larsgard, A.G.	460
Kroeske, K.C.	189, 567	Larson, K.	166
Krogenaes, A.	352	Larson, R.A.	612
Krogh, K.	170	Larzul, C.	291, 443
Krooneman, J.	330	Lasek, J.	246
Krudewig, K.H.	301	Lashkari, S.	170, 215
Krugmann, K.L.	191	Lassen, J.	633, 637
Krupa, E.	486, 542	Lasseur, J.	658
Krupiński, J.	107, 219	Latorre, M.A.	414, 669
Krupová, Z.	486, 542	Latour, C.	616, 617
Kučević, D.	620	Latvala, T.	216
Kudinov, A.A.	126	Laubier, J.	295
Kugler, W.	362	Lau Heckmann, L.H.	628, 629
Kuhla, B.	442, 605	Launay, C.	495
Kuhlmann, A.	478	Launay, F.	295
Kühne, P.	638	Laurent, C.	218
Kulig, B.	302	Laurino, D.	326
Kunene, N.W.	171	Lauvie, A.	113
Kunze, T.	202	Lavado, N.	422
Kwakkel, R.P.	331	Lavín, P.	333
Kyle, J.	352	Lebas, N.	415
Kyriazakis, I.	191, 257, 258, 420, 480, 549, 550,	Leblanc-Maridor, M.	387, 393, 545, 546, 547
	566, 627, 662	Lebret, B.	170
		Lebret, E.	194
		Lecaro, L.	218
L		Lecchi, C.	314
Labatut, J.	195	Lecocq, A.	242, 553
Laborde, D.	286	Le Cozler, Y.	674
Labrune, Y.	305, 666	Lecrenier, M.C.	323
Labussière, E.	415, 604, 622	Le Danvic, C.	293
Laca, E.A.	664	Ledda, A.	358
Laclef, E.	354	Leder, E.H.	514
Lacroix, M.Z.	551	Leduc, A.	295
Lagadec, S.	319, 484	Lee, A.	406
Łagoda, M.	176, 670	Lee, H.G.	264
Lagriffoul, G.	132	Lee, J.H.	312, 313
Lahoz, B.	453	Lee, J.S.	264
Lai, F.	358	Lee, M.R.F.	197, 214, 223, 261, 322, 425
Laing, R.	271	Lee, S.	398, 483, 484
Laissy, I.	600	Lee, S.H.	123, 312, 313
Laithier, C.	602	Lee, S.I.	175, 178
Lakhssassi, K.	453	Lee, Y.L.	209
Lall, S.P.	517	Leeb, C.	334, 478, 482
Laloë, D.	208, 295, 610	Leen, F.	187, 546
Lambe, N.	352	Leenknegt, J.	563
Lambe, N.R.	221	Lefaucheur, L.	263
Lamberton, P.	268	Lefebvre, R.	295
Lambertz, C.	204, 407, 510, 613	Lefebvre, T.	238
Lammers, A.	200	Le Floc'h, N.	170
Lamminen, M.	328, 639	Lefter, N.A.	184
Lanari, M.R.	445		

Le Gall, A.	546	Liu, L.S.	473, 549
Legarra, A.	123, 126, 222, 236, 522	Liu, R.	550
Legendre, H.	675	Liu, X.	178
Le Guillou, S.	295	Liu, X.L.	392
Lehmann, J.O.	199, 650	Livernois, A.	450
Lenocka, K.	112	Livernois, A.M.	612
Leiber, F.	156, 239, 407, 444	Lizardo, R.	188, 421, 488
Leikus, R.	663	Llonch, P.	192
Leite, J.N.P.	382	Lobo, R.B.	454
Leite, N.G.	463	Lobón, S.	383
Leloup, J.	652	Lock, E.-J.	403
Lemaire, B.	180	Loh, G.	510
Leman, A.	246	Loiseau, P.	387
Le Mat, J.	547	Loke, B.J.	469
Leme, P.R.	174	Lollivier, V.	268
Leni, G.	159, 163, 400	Loncar, J.	676
Leopold, J.	407	Lonergan, P.	292
Lepar, J.	473	Long, C.	225
Le Provost, F.	295	Long, R.J.	230
Lerch, S.	643	Lont, D.	600
Leroy, G.	207, 284	Loor, J.J.	140
Letelier, P.	612	Loos, C.M.M.	440
Leterrier, C.	671	Lopdell, T.	598
Leupi, S.	239	Lopes, M.S.	248, 410
Leury, B.J.	261, 354	Lopes, P.S.	458, 463
Levart, A.	174	Lopes Pinto, F.	268
Lévêque, P.A.	263	López, D.	231
Levit, H.	134, 473	Lopez, M.	243
Lherm, M.	571	López, M.C.	493, 496
Li, C.F.	608	Lopez, V.	469
Li, F.	643	Lopez, Y.	139, 649
Li, J.X.	391	López-Bote, C.	343
Li, L.	152	López-Cortegano, E.	107
Li, Y.J.	175	López-Lujan, M.C.	493
Li, Z.	337, 340	Lopez-Villalobos, N.	223, 286, 353
Liang, Z.J.	391	Lopreiato, V.	140
Lidauer, M.H.	143	Lorenzo, J.M.	438, 440, 669
Liebhart, D.	270	Lourenco, D.	123, 130, 593
Liebscher, V.	120	Lourenco, D.A.L.	288, 463
Liere, P.	489	Lourenco, J.M.	310
Liermann, W.	540	Lourenço, M.	507
Ligda, C.	361	Louveau, I.	263, 419
Likar, C.P.	434	Love, S.	148
Lima, J.	634	Løvendahl, P.	457, 637
Lindqvist, B.	244	Lovito, C.	307
Linke, C.	470	Loyon, L.	319
Link, Y.C.	168	Loza, C.	651
Linton, M.	402	Lozano-Jaramillo, M.	658
Lipinska-Palka, P.	176	Lu, D.	308
Lippens, E.	370	Lu, H.	597
Liptoi, K.	364	Lu, J.	517
Litt, J.	506, 671	Lu, M.Z.	473, 549
Littlejohn, M.	598	Lucero, C.	445
Liu, A.T.	385	Luciano, A.	239, 254
Liu, F.	261	Lucinao, A.	247
Liu, H.	526	Łuczycka, D.	520

Martel, G.	659	McCulloch, A.	637
Martell, J.E.	260	McDermott, K.	225
Martemucci, G.	437	McEwan, J.C.	129, 131, 225, 290, 451, 599, 637
Marthey, S.	295	McGilchrist, P.	380
Martikainen, K.	523	McGoldrick, J.	271
Martin, B.	614, 615	McGrew, M.	364
Martin, B.M.	146	McGuire, M.A.	313
Martin, C.	442, 475, 605	McHugh, N.	352
Martin, G.	424	McIntyre, J.	271
Martin, G.B.	261	McKinnon, J.J.	166, 642
Martin, L.	421, 422	McLaren, A.	352
Martin, O.	471, 529	McLean, A.	436
Martin, P.	453, 633	McLeod, K.R.	440
Martín-Collado, D.	198, 425, 649	McNeill, D.M.	557, 611
Martín De Hijas-Villalba, M.	600	McNulty, N.P.	308
Martínez-Álvaro, M.	310, 311	McParland, S.	605
Martínez-Paredes, E.M.	493, 496	Meade, K.G.	352
Martini Gnezda, M.	434	Meciano, G.P.	382
Martins, A.	411	Medrano, J.F.	149
Martins, A.P.	224	Mees, K.	413
Martinsen, K.H.	203	Megens, H.-J.	106
Martinsson, G.	583	Mehaba, N.	148, 171
Martyniuk, E.	362	Mehrgardi, A.	291
Maselyne, J.	303, 304, 370, 372, 474, 476	Mehtiö, T.	143
Masoero, F.	636	Meignan, T.	297
Massacci, F.R.	307	Meijer, N.P.	157, 164
Massart, X.	533	Mejdandzic, D.	516
Masuda, Y.	130, 593	Mele, M.	165
Mata, L.	231	Melo, A.D.B.	421
Matej, S.	346, 347	Melo-Durán, D.	169
Mateos, I.	361, 674	Melucci, L.M.	445
Mathot, M.	200, 377, 640	Melville, L.	134
Mathys, A.	155	Menassol, J.-B.	135, 138
Matilainen, K.	450, 594	Menegazzi, G.	650
Matte, J.J.	661	Meneguz, M.	241, 395
Matthews, K.	350	Meneses, C.	264
Mattiauda, D.A.	650	Mengistu, S.B.	517
Mattos, E.	290	Mensching, A.	533, 534
Mauclert, J.	354	Menz, C.	260
Maupertuis, F.	481, 489, 626	Menzer, K.	344
Maupetit, D.	453, 633	Mercadante, M.E.Z.	376, 639
Maxa, J.	133	Mercat, M.J.	291
Maxfield, M.	280	Mercier, Y.	170
Mayer, A.	346	Merlot, E.	263
Mayeres, P.	115	Mertens, A.	377, 640
Mayerhofer, M.	535	Mertens, K.	372
Mayer, M.	296	Mesaric, M.	584
May, K.	542	Messelink, G.J.	246
Mazzoni, G.	152	Messori, S.	390
McAllister, T.A.	166, 642	Mészáros, G.	143, 362, 369, 600
McAuliffe, G.A.	197, 425	Metges, C.C.	274, 337, 340
McBean, D.	134	Méthot, S.	490
McConnell, D.	615	Metuki, E.	473
McConochie, H.	641	Metwally, H.M.	172, 173
McCormack, H.A.	268	Meuwissen, M.	427
McCracken, D.	132, 428	Meuwissen, T.H.E.	110, 210, 290, 451, 511, 595

Mevissen, M.	390	Molist, F.	168, 250, 257, 306, 565
Meyer, A.M.	242	Molle, G.	132
Meyer, H.	479	Mollenhorst, H.	317
Meyer, J.	192	Møller, S.H.	412
Meyer, M.	539	Momen, M.	120, 291
Meyer, U.	534	Mondiere, A.	655
Meyermans, R.	409, 524	Montaldo, H.H.	237
Meylan, M.	539	Montalvo, G.	179
Meza-Herrera, C.A.	504, 579, 660	Montanaro, T.	372
Miana-Mena, F.J.	414	Montañés, M.	358
Miar, Y.	452	Montanholi, Y.R.	260
Michalczuk, M.	176, 670	Monte, F.	572
Michel, G.	336	Monteils, V.	378
Michenet, A.	610, 633	Monteiro, A.	179
Michie, C.	371, 375, 530	Monteiro, A.N.T.R.	316
Michiels, J.	336, 341	Monteiro, Í.M.	664
Mielczarek, M.	296	Monteiro, L.S.	644
Migdałek, G.	461	Moorby, J.M.	381, 641
Migliorati, L.	378	Moraine, M.	659
Miglior, F.	149, 209, 450	Morais, T.G.	569
Mihali, C.V.	474	Morales-Cruz, J.L.	660
Mikawa, S.	125	Morales, J.	169, 179
Mikkelsen, S.	628	Moran, D.	362, 363
Miklos, R.	629	Morante, R.	521
Millen, D.D.	180	Mora-Ortiz, M.	149
Miller, A.L.	550	Moravčíková, N.	106, 112, 462, 475, 528
Miller, S.	123, 130, 612	Moreira, O.	397, 400
Miller, S.P.	260	Moreira, O.C.	677
Millet, S.	183, 187, 189, 303, 416, 480, 481, 487,	Moreira, V.E.	248
	567, 622, 624, 666, 675	Morek-Kopec, M.	454
Miltiadou, D.	221	Morel, I.	559, 643
Mindus, C.	506	Moreno, C.	294, 522
Minero, M.	278	Moreno, E.	113
Minervini, S.	344	More, S.J.	385
Mineur, A.	606	Morgan-Davies, C.	132, 352, 428, 614, 615
Minozzi, G.	117	Morgan, E.	134, 271, 273
Mintegui, L.	353	Morgan, L.	192, 305
Minuti, A.	140	Morisset, A.	236
Mironeasa, S.	669	Mormede, P.	443
Mirzaei, F.	429	Morota, G.	120
Misiura, M.	480	Morra, P.	359
Misselbrook, T.	558	Morrison, A.	134, 271
Misztal, I.	123, 130, 288, 593	Morris, R.S.	223
Mizeranschi, A.E.	474	Morris, V.	322
Moakes, S.	444	Mortari, I.	382
Moate, P.J.	265, 635	Moskała, P.	219
Mock, B.	224	Mosnier, C.	346, 381, 428, 656, 657
Möddel, A.	179	Mosoni, P.	308, 309
Mogensen, L.	199, 570, 650	Mostafa, T.	356
Mohammadpanah, M.	291	Mota, R.R.	607
Mohnl, M.	182	Moula, N.	396
Mohr, J.	302	Moulin, C.H.	426, 658
Mol, W.V.	273	Mourão, G.B.	526, 527
Molano, R.A.	645	Moya, J.	493
Molenaar, R.	389	Mrode, R.	501
Molina, A.	231, 232, 583, 584	Mucha, A.	246, 418, 420

Rekik, M.R.	274, 276	Rocha, B.	251
Remus, A.	490	Rocha, E.L.F.	659
Renand, G.	453, 633	Rochetti, R.C.	382
Renard, J.	217	Rochort, S.	534
Renaudeau, D.	249, 263	Rodenas, L.	402
Renieri, C.	590	Ródenas, L.	493, 496
Rensing, S.	544	Rodenburg, T.B.	136, 200, 389, 469, 508
Renzi, F.	372	Rodrigues, S.	216, 220
Resino-Talaván, P.	368	Rodríguez, M.G.K.	579
Resmond, R.	269	Rodríguez, M.	169, 179, 672
Ressi, W.	427	Rodríguez-Arias, L.	357
Restoux, G.	208, 610	Rodríguez-Estévez, V.	665
Reverberi, M.	676	Rodriguez-Navarro, A.B.	268
Revermann, R.	334	Rodríguez-Ramilo, S.T.	222, 236, 237
Reverter, A.	222, 642	Rodríguez-Robayo, D.E.	661
Revilla, I.	401	Rodríguez-Sánchez, J.A.	619
Rey, J.	636	Roehe, R.	149, 334, 634
Reyer, H.	625	Roels, J.	159, 243
Reynolds, C.	442	Róg, D.	670
Reynolds, C.K.	558	Rogel-Gaillard, C.	307
Reynolds, E.	598	Roh, S.	218
Rezaei, J.	667	Roininen, H.	244
Rezar, V.	174	Roinsard, A.	180, 489, 626, 654
Rhoads, R.P.	259	Rolfes, A.	140
Ribani, A.	105, 108	Rolinec, M.	253, 668
Ribeiro, E.	450	Rollet, P.	433, 434
Ribeiro, J.R.	224	Romé, H.	448, 498
Ricard, A.	137, 582	Roméo, A.	184
Ricard, E.	305	Rommers, J.M.	497
Richard, C.	292	Romulo Carvalho, M.	450
Richards, C.S.	395	Roos, N.	629, 630
Richards, J.	350	Röös, E.	485
Richardson, C.M.	534, 603, 635	Roosen, J.	613
Rico, D.	651	Ropka-Molik, K.	153, 418, 420
Riede, O.	641	Roques, B.B.	551
Riek, A.	520	Roques, G.	415
Righi, F.	199, 556	Rørvang, M.V.	616
Rigueiro, A.L.N.	180	Rosa, A.F.	174
Rijk, T.C.	164	Rosa, A.P.	234
Rikkola, K.	492	Rosa, G.J.M.	120
Rinaldi, L.	271, 273	Rosa, H.J.D.	251
Ringenier, M.	314	Rosas, J.P.	600
Ringuet, A.	354	Rosenfeld, L.	134
Riosa, R.	148	Rosner, F.	202, 422, 449, 464
Ripoll-Bosch, R.	349, 653	Ross, J.W.	259
Ripoll, G.	414	Rossetto, J.	664
Riquet, J.	105, 666	Rossi, B.	561
Riuzzi, G.	217, 344	Rossignol, M.-N.	295
Rivero, M.J.	261	Rothacher, M.	559
Rizzi, R.	326	Roth, K.	255, 410
Rizzo, R.	676	Rothrock Jr., M.J.	310
Robert-Granié, C.	125	Rouchez, L.	470
Roberts, A.J.	524	Rousseliere, Y.	547
Robles, J.A.	269	Rouzbehan, Y.	667
Robles, L.	431	Rowe, S.J.	129, 225, 637
Robles, M.	581	Royer, E.	415, 421, 566

Ruckli, A.K.	477, 478, 482	Santiago Moreno, J.	364
Ruesche, J.	443	Santillo, A.	592
Ruf, T.	520	Santos, A.S.	280
Rufener, L.	272	Santos, M.P.	569
Ruis, M.A.W.	398	Santos Neto, J.M.	644
Ruiz, M.	440	Santos-Silva, J.	230, 328, 332, 677
Ruiz, R.	185, 388	Santos-Silva, M.F.	234
Ruiz-López, F.J.	237	Sanz, A.	383, 619
Ruiz-Martinez, E.	660	Sargison, N.	271
Rupp, R.	125	Sargolzaei, M.	260
Rustas, B.O.	313	Saric, T.	516
Ruy, I.M.	382	Saridaki, A.	457
Rydhmer, L.	285, 485	Sarmiento, A.	401
Rymer, C.	406	Saro, C.	361, 674
Rzewuska, M.	298, 324, 466	Sarrazin, S.	492
		Sarriés, M.V.	440
S		Sarri, L.	416
Sabady, M.	321	Sarti, F.M.S.	114, 116
Sabbagh, M.	582	Sarto, P.	453
Sabia, E.	569	Sartorello, G.L.	376
Sadri, H.	147	Sartori, C.	525
Sae-Lim, P.	513	Sasaki, O.	560
Sæther, N.	110	Sasaki, O.S.	128
Saeys, W.	529, 530, 531	Sato, S.	502
Saha, S.	286	Satoh, M.	339, 456
Sahana, G.	124	Sauerwein, H.	147, 148
Saintilan, R.	633	Saukh, O.	470
Sáinz De La Maza, V.	561	Sautier, M.	655
Sakowski, T.	302, 544	Sauvant, D.	472, 558, 636
Sala, G.	314	Savian, J.V.	664
Salama, A.	171	Savic, M.	660
Salama, A.A.K.	148, 355	Savietto, D.	654, 671
Salamone, M.	458, 598	Savoie, J.	489
Salas-Barboza, E.	431, 432	Savoini, G.	239, 632, 644
Salau, J.	279, 300, 322	Scaglia, G.	618
Salavati, M.	464, 598, 599	Scampicchio, M.M.	244
Salazar, L.C.	198	Schader, C.	316
Salazar-Diaz, E.	357	Schafberg, R.	449
Saleem, A.M.	642	Schalch Junior, F.J.	382
Salimei, E.	249, 252, 323, 441	Schäler, J.	109
Saliminen, J.P.	272	Schedle, K.	673
Sallam, M.T.	355	Schellander, K.	255, 335
Salmanzadeh, M.H.	266	Schenkel, F.	509
Salobir, J.	174	Schenkel, F.S.	209, 383, 612
Samson, A.	495	Scheper, C.	542
Sanad, Y.M.	310	Schetelat, E.	495
Sánchez-Guerrero, M.J.	279, 583	Scheu, T.	147
Sánchez-Mayor, M.	459	Schiavo, G.	105, 108
Sanchez, M.P.	602	Schiavone, A.	241, 395
Sanchez-Rodriguez, E.	268	Schiavon, S.	589
Sandercock, D.A.	205	Schibler, L.	292, 293
Sandrock, C.	156, 239	Schiefler, I.	410
Sanna, R.	323	Schild, S.-L.A.	338
Santana, B.	290	Schillebeeckx, C.	308
Santana, M.H.A.	382	Schillewaert, S.	237
Santana, T.E.Z.	463	Schlather, M.	366

Schlecht, E.	645	Seok, W.J.	504
Schlegel, P.	539	Seppala, A.	492
Schmid, M.	525	Seradj, A.R.	416
Schmid, S.M.	418	Serradilla, J.M.	231
Schmidt, J.	672	Serrano, M.	353, 453
Schmidtmann, C.	609	Serviento, A.M.	263
Schmitt, A.O.	511	Serviere, G.	193
Schmitt, E.	155, 158	Servin, B.	367
Schmitz-Hsu, F.	270, 462	Seshoka, M.L.	487
Schmoelzl, S.	132	Sevane, N.	367
Schneider, M.K.	614, 615	Sevi, A.	592
Schodl, K.	334	Sevillano, C.A.	288
Schoen, H.	509	Sforza, S.	159, 163, 243, 400
Schokker, D.	257, 306, 389	Shafey, H.	293
Scholtens, M.	353	Shahdadi, A.	537
Scholtz, M.M.	291	Shajahan, F.	299
Scholz, H.	618, 638	Shanmugasundaram, R.	182
Schomburg, H.	548	Sharifi, A.R.	132, 511, 533, 534
Schönherz, A.A.	111	Sheng, H.	473
Schön, H.-G.	498	Shen, M.X.	473, 549
Schons, R.T.	664	Sherlock, R.	598
Schrader, L.	497, 548	Shinder, D.	297
Schregel, J.	337, 340	Shin, D.J.	162
Schrooten, C.	211	Shirali, M.	412, 457, 467
Schroyen, M.	189, 567	Shourrap, M.	262
Schuh, K.	147	Shuiep, E.S.	580
Schukat, S.	478	Shull, C.	308
Schuler, U.	533	Shumo, M.	240
Schultz, N.E.	538, 541	Siachos, N.	357, 532, 543
Schulze-Schleppinghoff, W.	582	Sieme, H.	583
Schurink, A.	118, 367	Silacci, P.	643
Schütz, A.	258	Siljander-Rasi, H.	244
Schwab, C.	307, 308	Silva, A.A.	458, 463
Schwarzenbacher, H.	143, 597, 599	Silva, A.W.L.	663
Schwarz, S.	318	Silva, D.A.	458, 463
Schwarz, T.	186, 246, 251	Silva, F.F.	458, 463, 486
Sciascia, Q.	340	Silva, H.T.	458, 463
Sciascia, Q.L.	337, 340	Silva, M.F.	321, 485
Scollo, A.	344	Silva, R.P.	454
Šebek, L.B.	635	Silva, S.	664, 665
Sebola, A.	667	Silva, S.L.	174
Sebothoma, P.	668	Silva Neto, G.F.	664
Sedefoglu, H.	325	Silvestre, A.M.	180
Seefried, F.	599	Simčič, M.	233
Seefried, F.R.	127, 523	Simeone, A.	260
Segato, S.	217	Simianer, H.	119, 132, 366, 511, 544, 600
Segers, L.	503, 567	Šimko, M.	253, 668
Seifert, J.	404	Simm, G.	334
Sele, V.	403	Simoni, M.	556, 672
Sellem, E.	292, 293	Simonneaux, V.	590
Sell-Kubiak, E.	186, 256	Simon, N.R.	120
Selmoni, O.	442, 444	Simonová, D.	462
Selsby, J.T.	259	Simons, Q.	163
Selvaraj, R.	182	Singh, Y.	401
Sendor, P.	320, 321	Sinković, T.	111
Seoane, S.	669	Siqueira, T.T.S.	145, 195

Sitzmann, W.	673	Spiekermeier, I.	479
Siwek, M.	495, 499	Spigarelli, C.	427
Skaarud, A.	512	Spindler, B.	201, 498
Skarwecka, M.	460	Spoolder, H.	482
Skrede, A.	330	Spoolder, H.A.M.	384, 388, 506
Škrlep, M.	490, 621, 623, 627	Spranghers, T.	237
Skugor, A.	624	Sprock, C.	338
Skytte, N.	552	Squartini, A.	657
Slagboom, M.	215	Stachelek, M.	544
Slawinska, A.	495, 521	Stadnicka, K.	496, 501
Sleator, R.D.	379	Stålhammar, H.	208, 465
Słoniewska, D.	298, 324, 466, 467	Stankeviciene, D.	663
Smagghe, G.	510, 516	Star, L.	398
Smart, A.	359	Stauder, A.	247
Smeets, N.	329	Stead, S.	190
Smetana, S.	155	Stefanska, B.	309
Smets, R.	327	Stefanski, V.	480, 489, 621, 622, 623, 627
Smili, H.	592	Stegmann, V.	351
Smirnov, K.	527	Stehr, M.	274
Smith, E.M.	221, 226	Steinhoff-Wagner, J.	418
Smits, C.	168	Steininger, F.	470
Smits, K.	580	Steinmetz, L.	656
Snelling, T.J.	634	Stella, A.	365, 442
Sobry, L.	275	Stentiford, G.D.	516
Sodre, E.	426	Štepec, M.	233
Soetemans, L.	163, 400	Stergiadis, S.	149, 214, 272, 558
Sokolov, A.	442	Stewart, R.D.	634
Solà-Oriol, D.	169, 187	Stewart, S.M.	380
Solarczyk, P.	544	Stilmant, D.	144, 200, 377, 640
Soleimani Jevinani, F.	483	Stilwell, G.	665
Sölkner, J.	143, 362, 369, 572, 600	Stockhofe, N.	497
Somenzi, E.	114	Stock, K.F.	449, 544, 581, 582
Somera, A.	655	Stokes, J.E.	572
Sonck, B.	304, 372, 379, 476, 479	Stokvis, L.	331
Song, M.	312	Stolze, M.	444
Song, M.H.	313	Stölzl, A.	520
Song, W.	386	Stoop, W.M.	129
Song, W.P.	392	Storlien, H.	460
Songsermpong, S.	245	Stothard, P.	209
Sørensen, A.C.	124, 215, 526	Strabel, T.	142
Soriano, B.	427	Stracke, J.	498
Sorjonen, J.	244	Strandberg, E.	208, 285, 465, 600
Sosa Bruno, J.	505	Strandén, I.	121, 126, 127, 212, 468, 594
Sossidou, E.	203, 506	Stratakos, A.	402
Soteriades, A.D.	377	Strathe, A.V.	564
Sotiraki, S.	271, 273, 276	Štrbac, L.J.	620
Soufleri, A.	532	Strzałkowska, N.	176, 670
Sousa, F.	216	Stuffe, S.	161
Souza Da Silva, C.	389	Stupart, C.	267
Soyeurt, H.	145, 408, 443, 605, 606, 608	Sturaro, E.	427, 657
Spaepen, R.	329	Sturaro, E.S.	146
Sparaggis, D.	360	Styles, D.	377
Speakman, J.R.	520	Su, G.	124, 411, 467, 633
Špehar, M.	111, 417	Subramanian, S.	240
Spelman, R.	598	Succi, M.	496
Spengeler, M.	127, 523	Such, X.	148, 171, 355

Toro, M.A.	108	Usai, G.	105
Toro-Mujica, P.M.	431	Usala, M.	264
Torres, O.	368	Utnik-Banaś, K.	503
Torres, R.	521	Utzeri, V.J.	105, 108
Torres-Rovira, L.	337, 568		
Tosser-Klopp, G.	125, 295	**V**	
Tourneix, L.	281	Vaarst, M.	199
Tozawa, A.	502	Vaarten, J.	384
Trailovic, R.	660	Vaes, R.	379
Trakovická, A.	112, 528	Våge, D.I.	110
Tran, G.	558	Vajana, E.	442, 444
Traulsen, I.	204, 338	Valada, T.	569
Travel, A.	180	Valares, J.A.	231
Tråvén, M.	448	Valdemarsson, S.	512
Tretola, M.	247	Valera, M.	279, 583, 584
Trevisan, G.	243	Valergakis, G.	357
Trevisi, E.	140	Valergakis, G.E.	532, 543
Trevisi, P.	307, 492	Valizadeh, R.	537
Tribout, T.	211, 602	Vall, E.	658
Trincanato, S.	224	Valle, E.	436, 555
Trivunović, S.	620	Valleix, S.	217
Troquier, C.	656	Valloto, A.A.	526, 527
Trunk, W.	491	Valros, A.	482
Trydeman Knudsen, M.	570	Van Amburgh, M.E.	645
Tsiokos, D.	361	Van Arendonk, J.	510
Tsiplakou, E.	381, 556	Van Beirendonck, S.	503, 567, 648
Tsuruta, S.	123, 463, 593	Van Breukelen, A.E.	365
Tuchscherer, A.	337	Van Campenhout, L.	394
Tudor, M.	427	Van Cruchten, S.	262, 339, 341, 342
Tulli, F.	159	Vandaele, L.	141, 144, 376, 539, 563, 647
Turcsán, L.	672	Van Damme, D.	374
Turcsán, Z.S.	672	Vande Ginste, J.	494
Turkaspa, A.	470	Van De Gucht, T.	370
Turkmen, S.	299	Vande Maele, L.A.	181
Turner, S.P.	256	Vandenbossche, C.	176
Turri, F.	326	Van Den Brand, H.	389
Tusell, L.	287, 291	Vanden Broeck, J.	245
Tuyttens, F.	300, 379, 479	Van Den Broeke, A.	480, 481, 490, 622
Tuyttens, F.A.M.	303, 508	Van Den Brulle, I.	529
Tuz, R.	186, 246, 251, 418	Vandenbussche, C.	372, 476
Tvarožková, K.	576, 577	Vandenheede, M.	280
Tyra, M.	153, 417	Vanden Hole, C.	262, 339
Tzamaloukas, O.	360	Vandenplas, J.	211, 212, 288, 447, 449, 594, 595, 596
		Vandepoel, G.	379, 479
U		Van Der Aa, A.	181
Uddin, M.E.	612	Van Der Borght, M.	241, 327, 394
Uemoto, Y.	339, 456	Van Der Burg, S.	373
Uerlings, J.	189	Van Der Eijk, J.A.J.	200
Ugalde, E.	440	Van Der Fels-Klerx, H.J.	157, 158, 164
Ugarte, E.	185, 232, 236, 353, 636	Van Der Gaag, M.A.	303
Ughelini, N.	556	Van Der Heide, M.E.	397
Uhrinčať, M.	506, 576, 577, 578	Vanderick, S.	115, 607
Uimari, P.	126, 423, 523	Van Der Linde, C.	603
Uken, K.	540	Van Der Linde, R.	638
Ungerfeld, R.	579	Van Der Linden, A.	345, 346
Urschel, K.L.	440		

Villalba, D.	312	Wallgren, T.	206
Villanueva, B.	108	Wall, S.-C.	168
Villca, B.	188, 488	Walsh, S.W.	210
Villumsen, T.M.	412	Walter, C.	156
Vineer, H.	271	Wang, G.B.	392
Vinet, A.	453	Wang, H.	175
Viola, I.	359	Wang, J.Q.	408
Visker, M.H.P.W.	507	Wang, L.	391
Visscher, J.	500	Wang, W.J.	230
Visscher, P.M.	211	Wang, X.	150, 294
Visser, B.	302	Wang, X.Z.	391
Visser, E.K.	469	Wang, Z.	606
Vitezica, Z.G.	287, 522	Warin, L.	506
Vlaeminck, B.	309, 641	Warnken, F.J.	191
Vlček, M.	475	Waterhouse, A.	132, 428
Voet, H.	574	Watson, M.	149, 634
Vogelzang, R.H.	627	Watteyn, A.	379, 479
Vogt-Kaute, W.	407	Wattiaux, M.A.	612
Von Borell, E.	344	Wauters, E.	427, 546
Von Meyer-Höfer, M.	258	Wauters, J.	190, 623
Von Rohr, P.	599	Weary, D.	306
Von Samson-Himmelstjerna, G.	271, 273	Webb, E.C.	138
Von Soosten, D.	534	Weber, D.	618
Von Tavel, L.	270	Weerman, E.	160
Von Tavel, L.V.T.	220	Wegge, B.	507
Vosman-Bouwmeester, J.J.	609	Wegner, B.	479
Vosough Ahmadi, B.	363	Wehrend, A.	222
Voss, B.	422	Weigel, K.A.	538, 541
Voß, B.	413	Weigend, S.	110, 366
Vossen, E.	487, 624	Weiler, U.	489, 621, 623, 627
Vostra-Vydrova, H.	528, 585	Weisbjerg, M.R.	215
Vostry, L.	528, 585	Weller, B.	160
Vouzela, C.F.M.	251	Wendin, K.	161
Vrecl Fazarinc, M.	623, 627	Wensch-Dorendorf, M.	344, 464
Vreysen, S.	156	Wensman, J.J.	448
Vrielinck, L.	566	Werner, A.	533, 535
Vršková, M.	577	Wester, J.	200
Vuckovic, S.	660	Wethal, K.B.	543
Vu Dinh, T.	311	White, H.M.	154, 537, 538, 540, 541
Vuik, C.	595	Whitelaw, C.B.A.	198
Vujasinović, N.	233	Wientjes, Y.C.J.	289
Vuylsteke, D.	510	Wierzbicki, M.	521
		Wiesner, L.	489
W		Wilcock, P.	188, 488
Wagner, H.	222	Wilder, T.	201
Wahner, M.	202	Wilhelm, M.	179
Walczak, J.	320, 321	Willam, A.	334
Waldmann, P.	119	Willems, E.	189
Waldrop, M.	613	Willems, O.	509
Wales, B.	603	Williams, A.R.	190
Wales, W.J.	265, 534, 635	Williams, S.R.	265, 635
Walkenhorst, M.	390, 393	Willockx, M.	573, 575
Walker, A.W.	634	Wilmot, H.	115
Wall, E.	225, 352	Wimel, L.	137, 441
Wall, H.	268	Wimmers, K.	625
Wallace, M.	257, 258, 420	Winckler, C.	334